Geometry

Ray C. Jurgensen
Richard G. Brown
John W. Jurgensen

HOUGHTON MIFFLIN COMPANY • BOSTON
Atlanta Dallas Geneva, Ill. Lawrenceville, N.J. Palo Alto Toronto

THE AUTHORS

Ray C. Jurgensen, formerly Chairman of the Mathematics Department and holder of the Eppley Chair of Mathematics, Culver Academies, Culver, Indiana.

Richard G. Brown, Mathematics Teacher, Phillips Exeter Academy, Exeter, New Hampshire.

John W. Jurgensen, teaches mathematics at the University of Houston–Downtown and is a mathematician for the National Aeronautics and Space Administration (NASA) at the Johnson Space Center.

TEACHER CONSULTANTS

Camille Bishop, Mathematics Teacher, New Hanover High School, Wilmington, North Carolina.

Joseph L. Gates, Mathematics Teacher and Chairman, Mathematics Department, Polytechnic High School, Long Beach, California.

Jack B. Skelton, Instructional Assistant for Secondary Mathematics, Tulsa Public Schools, Tulsa, Oklahoma.

John A. Terhune, Mathematics Teacher, Marion Adams High School, Sheridan, Indiana.

ABOUT THE COVER

The photograph on the cover shows the geometric sculpture "Constellation, 1971" by Morton C. Bradley, Jr. This sculpture is an assembly of 12 Great Stellated Dodecahedra. The Great Stellated Dodecahedron is a regular polyhedron that was discovered by Johannes Kepler more than 300 years ago. Each facet is an isosceles triangle in which the two long edges are 1.618 times longer than the short edge. The ratio of the length of a long edge to a short one is often referred to as the *golden ratio;* it was known to ancient Greek mathematicians and artists. The sculpture is painted with a set of 20 colors that are equally spaced on the surface of a sphere. Light colors are near the top, dark colors are near the bottom, and the various hues—red, orange, yellow, and so forth—of the spectrum follow an orderly progression around the vertical neutral axis.

Printed in U.S.A.

ISBN: 0-395-43060-7

LMNOP-RM-96543

Contents

6 Right Triangles

7 Circles

11 Coordinate Geometry

12 Transformations

METRIC UNITS OF MEASURE

Prefixes	kilo	hecto	deka	deci	centi	milli
	1000	100	10	0.1	0.01	0.001

Length
1 centimeter (cm) = 10 millimeters (mm)
1 meter (m) = 100 cm
1 kilometer (km) = 1000 m

Area
1 square centimeter (cm^2) = 100 square millimeters (mm^2)
1 square meter (m^2) = 10,000 cm^2

Volume
1 cubic centimeter (cm^3) = 1000 cubic millimeters (mm^3)
1 cubic meter (m^3) = 1,000,000 cm^3
1 liter (L) = 1000 cm^3
1 liter = 1000 milliliters (mL)

Mass
1 gram (g) = 1000 milligrams (mg)
1 kilogram (kg) = 1000 g
1 metric ton (t) = 1000 kg

Symbols

$\lvert x \rvert$	absolute value of x (p. 6)
adj. $\angle$s	adjacent angles (p. 12)
alt. int. $\angle$s	alternate interior angles (p. 56)
$\angle$, $\angle$s	angle(s) (pp. 10, 12)
a	apothem (p. 397)
$\approx$	is approximately equal to (p. 267)
$\overset{\frown}{BC}$	arc with endpoints B and C (p. 301)
A	area (p. 370)
B	area of base (p. 424)
b	length of base; y-intercept (p. 380; p. 492)
$\odot O$	circle with center O (p. 293)
C	circumference (p. 327)
comp. $\angle$s	complementary angles (p. 30)
$\cong$	congruent, is congruent to (p. 7)
$\leftrightarrow$	corresponds to (p. 105)
corr. $\angle$s	corresponding angles (p. 56)
cos	cosine (p. 273)
$^\circ$	degrees (p. 12)
diag.	diagonal (p. 171)
d	diameter; distance; length of diagonal (p. 327; p. 470; p. 386)
$D_{O,k}$	dilation with center O and scale factor k (p. 530)
e	edge length (p. 426)
$=$	equal(s); equality (pp. 7, 18)
ext. $\angle$	exterior angle (p. 83)
$>$, $\geq$	greater than; greater than or equal to (p. 10)
H_O	half turn about point O (p. 526)
h	height, length of altitude (p. 248; p. 380)
hyp.	hypotenuse (p. 130)
T^{-1}	inverse of transformation T (p. 540)
I	identity transformation (p. 540)
int. $\angle$	interior angle (p. 83)
JL	length of $\overline{JL}$, distance between points J and L (p. 8)
$<$, $\leq$	less than; less than or equal to (p. 10)
$\overleftrightarrow{AB}$	line containing points A and B (p. 1)
$S: A \to A'$	S maps point A to point A'. (p. 512)
$m\angle A$	measure of $\angle A$ (p. 12)
$\not\cong$	not congruent (p. 180)
$\neq$	not equal (p. 17)
$\not>$	not greater than (p. 183)
opp. $\angle$s	opposite angles (p. 171)
(x, y)	ordered pair (p. 107)
$\parallel$	parallel, is parallel to (p. 55)
$\square$	parallelogram (p. 159)
p	perimeter (p. 382)
$\perp$	perpendicular, is perpendicular to (p. 36)
π	pi (p. 327)
n-gon	polygon with n sides (p. 80)
$S \circ T$	product of S and T (p. 534)
quad.	quadrilateral (p. 160)
r	radius (p. 308)
$\frac{a}{b}$, $a : b$	ratio of a to b (pp. 205, 206)
$\overrightarrow{AB}$	ray with endpoint A, passing through point B (p. 6)
R_j	reflection in line j (p. 515)
rt. $\angle$	right angle (p. 12)
rt. $\triangle$	right triangle (p. 252)
$\mathcal{R}_{O,90}$	rotation about point O through 90° (p. 525)
s-s. int. $\angle$s	same-side interior angles (p. 56)
$\overline{AB}$	segment with endpoints A and B (p. 6)
s	length of a side of a regular polygon (p. 380)
$\sim$	similar, is similar to (p. 213)
sin	sine (p. 273)
l	slant height (p. 430)
m	slope (p. 483)
$\sqrt{x}$	positive square root of x (p. 247)
supp. $\angle$s	supplementary angles (p. 30)
T.A.	total area (p. 424)
tan	tangent (p. 266)
trap.	trapezoid (p. 176)
$\triangle$, $\triangle$s	triangle(s) (pp. 72, 117)
vert. $\angle$s	vertical angles (p. 31)
V	volume (p. 424)

READING GEOMETRY

You probably realize that there is more than one kind of reading. You don't need to give the same kind of concentrated attention to a newspaper article or a story that you give to a textbook. The following paragraphs will tell you why reading geometry, in particular, calls for much care and concentration. Don't try to hurry through the explanations in this book. You will find that you are well repaid if you read slowly and think about what you are reading.

Vocabulary

You will come upon many new words in this book. Some of them are unique to mathematics—for example, *hypotenuse, isosceles,* and *secant*. Others are words that occur in everyday speech but have quite different meanings in geometry—for example, *plane, line,* and *construction*. When you begin a new chapter, you may want to skim through it, looking for important words whose meanings you will learn as you study the chapter. These words, printed in heavy type, are defined, and often illustrated by diagrams, when they first occur. For example, notice the word **intersection,** defined and illustrated on page 2. If you can't recall the meaning of *intersection* when you come upon the word later, you can look it up in the Glossary or in the Index at the back of the book. The Index gives you the number of the page where the word is defined, and the other page references may lead you to additional information.

Symbols

In order to understand geometry, you must know how to read symbols. Symbols are very useful because they present a great deal of information in a small space. For example, look at what is to be proved in Theorem 4-1, page 159. The statement "$\overline{EF} \cong \overline{HG}$; $\overline{FG} \cong \overline{EH}$" means "The segment with endpoints E and F is congruent to the segment with endpoints H and G, and the segment with endpoints F and G is congruent to the segment with endpoints E and H." Be sure to look for *all* the information given by such a statement in condensed form. If you have trouble reading a statement expressed in symbols, first make sure you understand all the symbols. Use the list of symbols on page ix as a handy checklist if you need to refresh your memory. Reread the statement slowly. Sometimes you may find it helpful to write the statement out in words or just to say the words aloud. Many exercises, such as Classroom Exercises 1–7 and Written Exercises 5–16 on page 8, will help you check your understanding of symbols.

Diagrams

Throughout this book you will find many diagrams. Like symbols, diagrams present a lot of information in condensed form. The diagrams are essential parts of the explanations and exercises, and you need to read them along with the text. Look carefully for all the information given by a diagram, but be sure that you don't read more into a diagram than is actually there. For instance, don't assume that all the sides of a triangle are the same length just because they *look* that way; if the sides are the same length, they will be marked to show it.

You may want to make sketches of your own from time to time to help you understand what you are reading. Sometimes you may need to redraw a diagram in a different position or to draw the parts of a complicated diagram separately. Be sure your drawings are marked to illustrate the relationships correctly. On page 40 you will find a list of many of the marks that are used, together with some suggestions for reading and drawing diagrams.

Many of the exercises throughout will help you test your ability to read diagrams correctly. Exercises 1–8 on page 3 and Exercises 1–10 on page 4 are just a few of these.

Organization of the Textbook

As the preceding discussion suggests, your textbook has been planned to make reading and learning geometry as easy as possible. New words and phrases are printed in heavy type so that you can find them easily and are carefully defined for you. The Glossary and the Index are useful time-savers if you want to refresh your memory about some item previously studied. Theorems, postulates, and summaries of methods are highlighted to emphasize their importance and to help you find them quickly. Theorems are also listed at the back of the book in the order in which they occur in the text.

Within every chapter you will find Self-Tests that provide a means of checking your progress on your own. Every chapter also includes a summary, a review, and a test on the whole chapter. There are cumulative reviews as well. You will have ample opportunity to check your understanding of geometry.

Your appreciation of your surroundings will be increased if you take time to observe the geometric forms in both natural and manufactured objects. Points, lines, angles, and planes—the subject of this first chapter—can be seen in the building reflections below.

Points, Lines, Planes, and Angles

1

Undefined Terms and Basic Definitions

Objectives

1. Understand *point, line,* and *plane;* draw representations of them.
2. Use undefined terms to define some basic terms in geometry.
3. Use symbols for lines, segments, rays, and distances; find distances.
4. Name angles and find their measures.

1-1 Points, Lines, and Planes

When you look at a color television picture, how many different colors do you see? Actually, the picture is made up of just three colors—red, green, and blue. Most color television screens are covered with more than 300,000 colored dots, as shown in the enlarged diagram on the right below. Each dot glows when it is struck by an electron beam. Since the dots are so small, and so close together, your eye sees three superimposed images rather than individual dots.

Each small dot on a television screen suggests the simplest figure studied in geometry—a *point*. Although a point has no size, it is often represented by a dot and named by a capital letter. Two points, A and B, are pictured at the right.

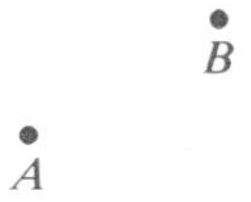

All geometric figures consist of points. One familiar geometric figure is a *line,* which extends in two directions without ending. Although a picture of a line has some thickness, the line itself has no thickness.

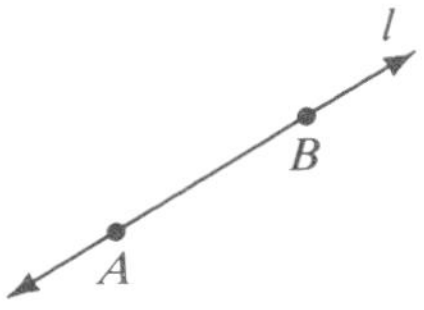

Often a line is referred to by a single lower-case letter, such as *line* l. If you know that a line contains the points A and B, you can also call it *line AB* (*denoted* $\overleftrightarrow{AB}$) or *line BA* ($\overleftrightarrow{BA}$).

A geometric *plane* is suggested by a floor, wall, or table top. Unlike a table top, a plane extends without ending and has no thickness. Although a plane has no edges, we usually picture a plane by drawing a four-sided figure as shown below. We often label a plane with a capital letter.

Plane M

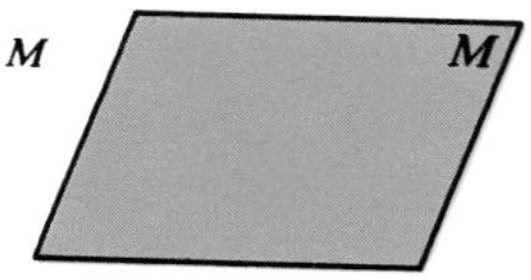

Plane N

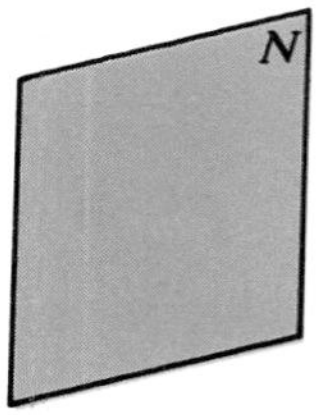

In geometry, the terms *point, line,* and *plane* are accepted as intuitive ideas and are not defined. These *undefined terms* are then used in the definitions of other terms, such as those below.

Space is the set of all points.

Collinear points are points all in one line.

Collinear points

Noncollinear points

Coplanar points are points all in one plane.

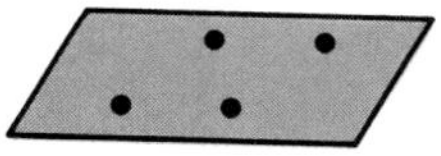

Coplanar points

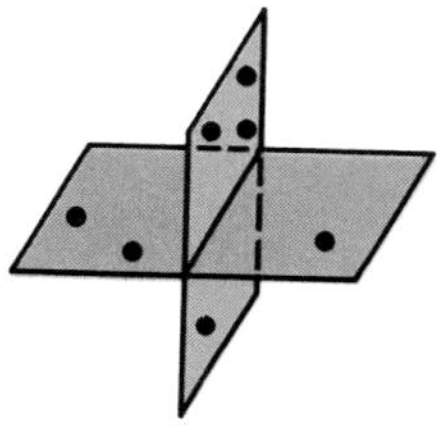

Noncoplanar points

Some expressions commonly used to describe relationships between points, lines, and planes follow. In these expressions, *intersects* means "meets" or "cuts." The **intersection** of two figures is the set of points that are in both figures. Dashes in the diagrams indicate parts hidden from view in figures in space.

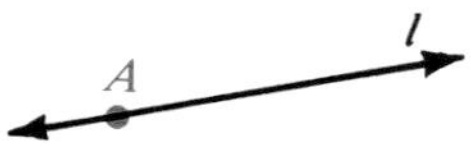

A is in l.
A is on l.
l contains A.
l passes through A.

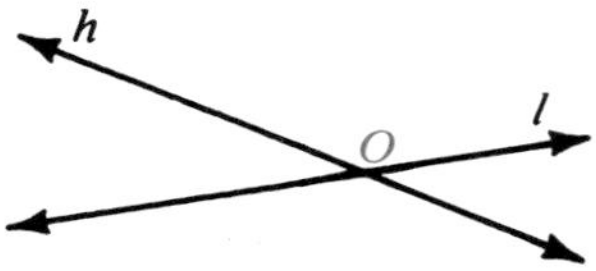

l and h intersect in O.
O is the intersection of l and h.

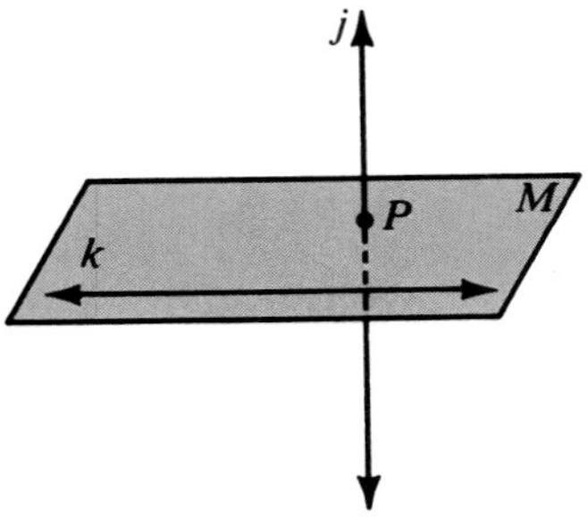

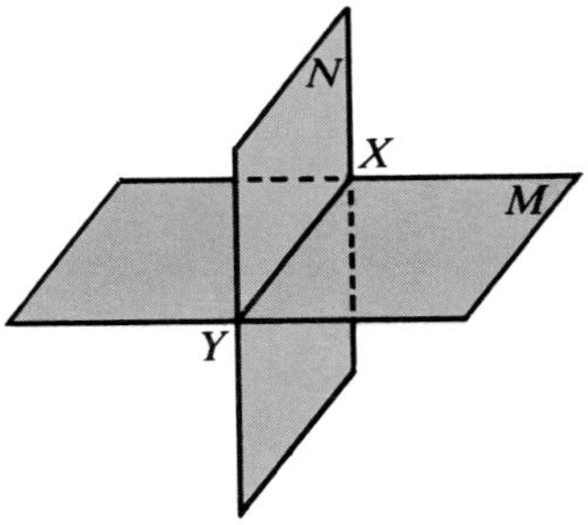

k and *P* are in *M*.
M contains *k* and *P*.
j intersects *M* in *P*.
P is the intersection of *j* and *M*.

M and *N* intersect in $\overleftrightarrow{XY}$.
$\overleftrightarrow{XY}$ is the intersection of *M* and *N*.
$\overleftrightarrow{XY}$ is in *M* and *N*.
M and *N* contain $\overleftrightarrow{XY}$.

In this book, whenever we refer, for example, to "two points" or "three lines," we will mean *different* points or lines (or other geometric figures).

Classroom Exercises

Classify each statement as true or false.

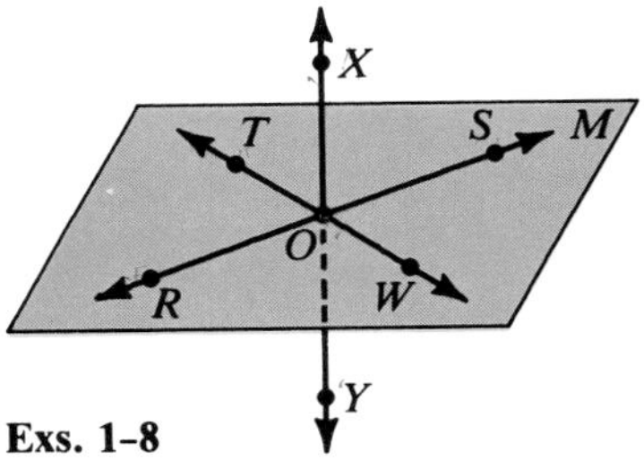

Exs. 1–8

1. $\overleftrightarrow{XY}$ intersects plane *M* at point *O*.
2. *R*, *O*, and *S* are collinear.
3. *T*, *O*, and *R* are collinear.
4. *X*, *O*, and *Y* are collinear.
5. *R*, *O*, *S*, and *W* are coplanar.
6. *R*, *S*, *T*, and *X* are coplanar.
7. *R*, *X*, and *O* are coplanar.
8. *R*, *X*, *O*, and *Y* are coplanar.
9. Does a plane have edges?
10. Can a given point be in two lines? in ten lines?
11. Can a given line be in two planes? in ten planes?

Name a fourth point that is in the same plane as the given points.

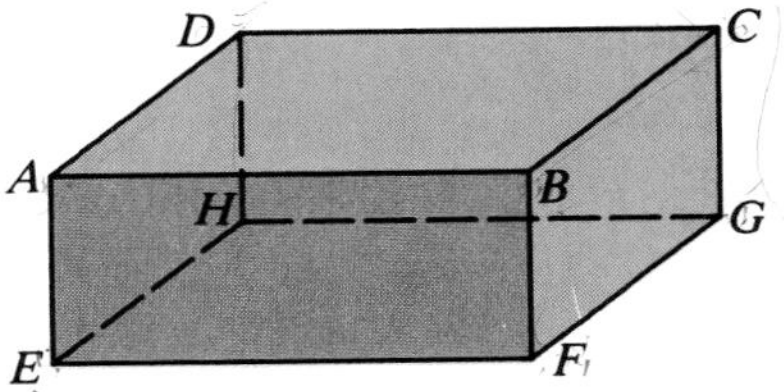

Exs. 12–21

12. *A*, *B*, *C*
13. *E*, *F*, *H*
14. *D*, *C*, *H*
15. *A*, *D*, *E*
16. *B*, *E*, *F*
17. *B*, *G*, *C*

The plane that contains the top of the box can be called plane *ABCD*.

18. Are there more than four points in plane *ABCD*?
19. Are there any other points in $\overleftrightarrow{CG}$ besides *C* and *G*?
20. The intersection of planes *ABFE* and *BCGF* is __?__.
21. Name two planes that do not intersect.

Written Exercises

Classify each statement as true or false.

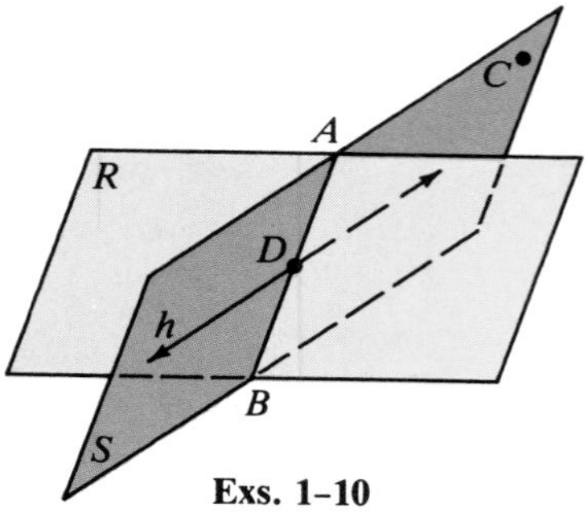

Exs. 1–10

A 1. $\overleftrightarrow{AB}$ is in plane R.
2. S contains $\overleftrightarrow{AB}$.
3. Plane R intersects plane S in $\overleftrightarrow{AB}$.
4. Point C is in R and S.
5. R and S contain D.
6. D is on line h.
7. h is in S.
8. h is in R.
9. A, B, and C are collinear.
10. A, B, C, and D are coplanar.

11. a. Make a sketch showing four coplanar points, no three of which are collinear.
b. Make a sketch showing four points that are not coplanar.

A plane can be named by three noncollinear points it contains. In Chapter 10 you will study *pyramids* like the one shown at the right below.

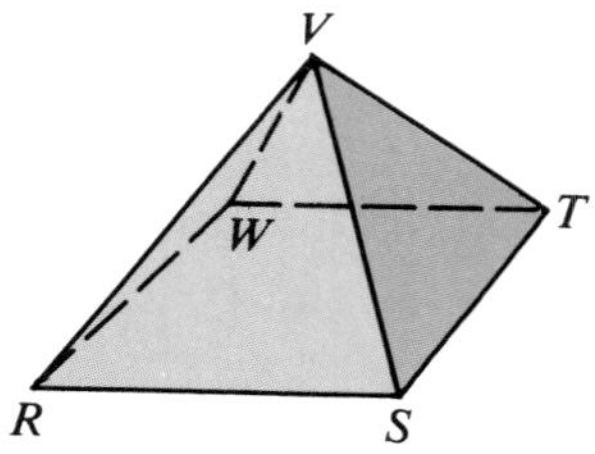

Exs. 12–18

12. Name five planes that contain sides of the pyramid shown.
13. Of the five planes containing sides of the pyramid, are there any that do not intersect?
14. Name three lines that intersect at point R.
15. Name two planes that intersect in $\overleftrightarrow{ST}$.
16. Name three planes that intersect at S.
17. Name a line and a plane that intersect in a point.
18. Name a line and a plane whose intersection is the line.

19. To practice drawing figures in space, follow the three steps below to draw a diagram of a barn. (As you gain more practice in drawing figures in space, you will probably be able to go directly from Step 1 to Step 3.)

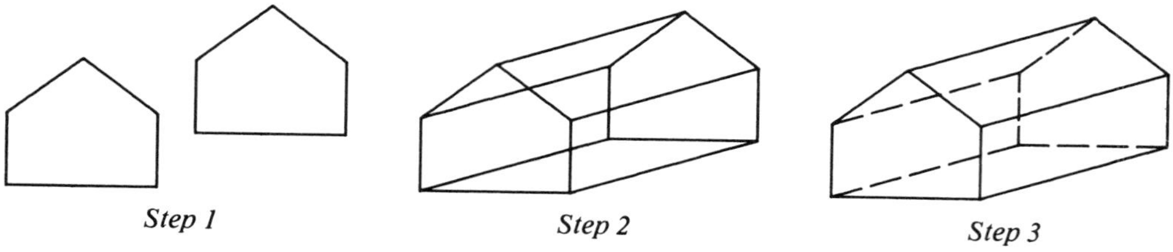

20. Name two planes that intersect in $\overleftrightarrow{FG}$.

21. Name three lines that intersect at E.

22. Name three planes that intersect at B.

23. a. Are points A, D, and C collinear?
    b. Are A, D, and C coplanar?

24. a. Are points R, S, G, and F coplanar?
    b. Are points R, S, G, and C coplanar?

25. a. Name two planes that do not intersect.
    b. Name two other planes that do not intersect.

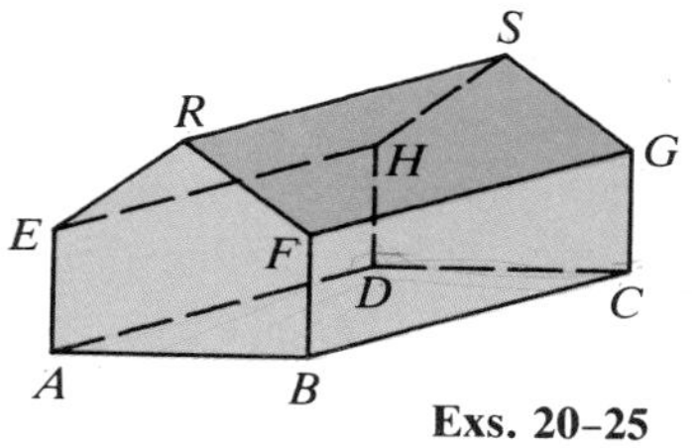

Exs. 20–25

You can think of the ceiling and floor of a room as parts of *horizontal planes*. The walls are parts of *vertical planes*. Vertical planes are represented by figures like those shown in which two sides are vertical. A horizontal plane is represented by a figure having two sides horizontal and no sides vertical.

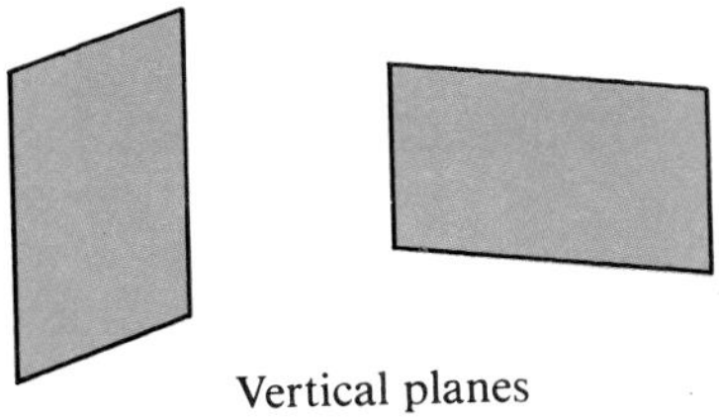

Vertical planes

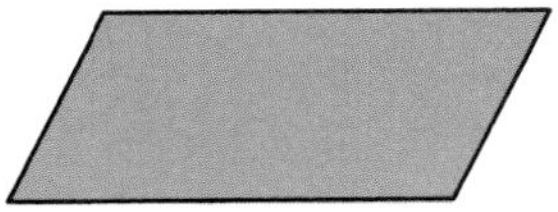

Horizontal plane

B 26. a. Can two horizontal planes intersect? b. Can two vertical planes intersect?

Sketch and label the figures described. Use dashes for parts hidden from view.

27. Vertical line l intersects a horizontal plane M at point O.

28. Horizontal plane P contains two lines k and n that intersect at point A.

29. Horizontal plane Q and vertical plane N intersect.

30. Vertical planes X and Y intersect in $\overleftrightarrow{AB}$.

C 31. Three vertical planes intersect in a line.

32. Point P is not in plane N. Three lines through P intersect N in points A, B, and C.

33. Two horizontal planes intersect a vertical plane in lines l and n.

34. Three planes intersect in a point.

1-2 Segments, Rays, and Distance

In the diagram, point B is *between* points A and C. The term is undefined, but note that B lies on $\overleftrightarrow{AC}$.

Segment AC, denoted $\overline{AC}$, consists of points A and C and all points that are between A and C. Points A and C are called the *endpoints* of $\overline{AC}$.

Ray AC, denoted $\overrightarrow{AC}$, consists of $\overline{AC}$ and all other points P such that C is between A and P. The *endpoint* of $\overrightarrow{AC}$ is A, the point named first.

$\overrightarrow{SR}$ and $\overrightarrow{ST}$ are called **opposite rays** if S is between R and T.

The hands of the clock shown suggest opposite rays.

On a *number line* every point is paired with a number and every number is paired with a point. Below, point J is paired with -3, the *coordinate* of J.

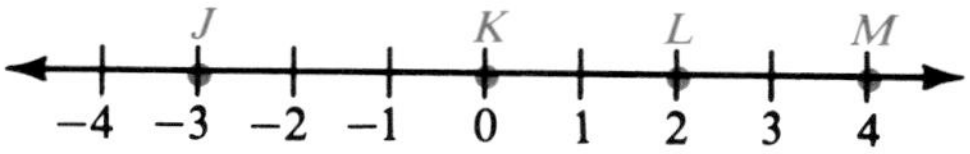

The **length** of $\overline{JL}$, denoted by JL, is the distance between point J and point L. You can find the length of a segment on a number line by subtracting the coordinates of its endpoints:

$$MJ = 4 - (-3) = 7$$

Notice that since a length must be a positive number, you subtract the lesser coordinate from the greater one. Actually, the distance between two points is the absolute value of the difference of their coordinates. When you use absolute value, the order in which you subtract coordinates doesn't matter.

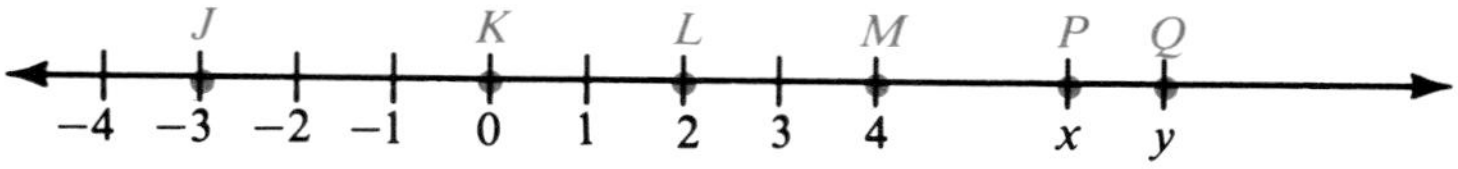

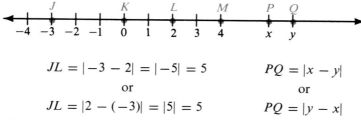

$$JL = |-3 - 2| = |-5| = 5 \qquad PQ = |x - y|$$

or

$$JL = |2 - (-3)| = |5| = 5 \qquad PQ = |y - x|$$

Using a number line involves the following basic assumptions. Statements such as these that are accepted without proof are called **postulates** or **axioms.**

Postulate 1 *Ruler Postulate*

The points on a line can be paired with the real numbers in such a way that:

1. **Any two desired points can have coordinates 0 and 1.**
2. **The distance between any two points equals the absolute value of the difference of their coordinates.**

Postulate 2 *Segment Addition Postulate*

If B is between A and C, then

$$AB + BC = AC.$$

Example B is between A and C, with $AB = x$, $BC = x + 6$, and $AC = 12$. Find:

a. The value of x **b.** BC

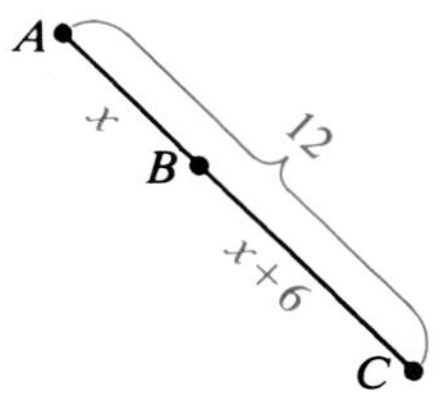

Solution **a.** $AB + BC = AC$

$$x + (x + 6) = 12$$
$$2x + 6 = 12$$
$$2x = 6$$
$$x = 3$$

b. $BC = x + 6$
$= 3 + 6$
$= 9$

Congruent segments are segments that have equal lengths. To indicate that $\overline{DE}$ and $\overline{FG}$ are congruent, you can write either

$DE = FG$ or $\overline{DE} \cong \overline{FG}$ (read "$\overline{DE}$ is congruent to $\overline{FG}$").

Note that these statements are equivalent and may be used interchangeably.

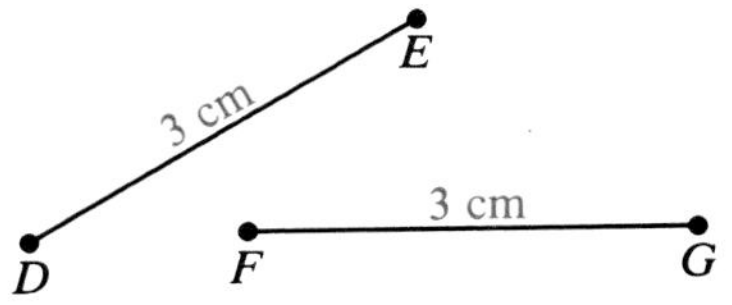

The **midpoint of a segment** is the point that divides the segment into two congruent segments. In the figure, $\overline{AP} \cong \overline{PB}$. P is the midpoint of $\overline{AB}$.

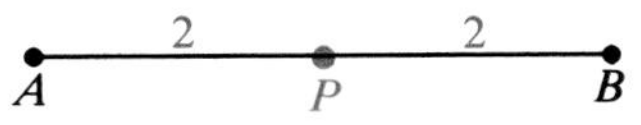

A **bisector of a segment** is a line, segment, ray, or plane that intersects the segment at its midpoint. Line l is a bisector of $\overline{AB}$. $\overline{PQ}$ and plane X also bisect $\overline{AB}$.

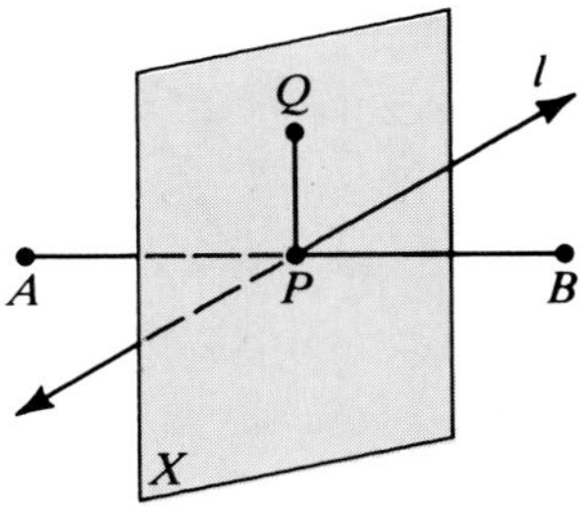

Classroom Exercises

1. What does each symbol represent?
   a. $\overline{PQ}$ **b.** $\overrightarrow{PQ}$ **c.** $\overleftrightarrow{PQ}$ **d.** PQ
2. How many endpoints does a segment have? a ray? a line?
3. Is $\overline{AB}$ the same as $\overline{BA}$?
4. Is $\overrightarrow{AB}$ the same as $\overrightarrow{BA}$?
5. Is $\overleftrightarrow{AB}$ the same as $\overleftrightarrow{BA}$?

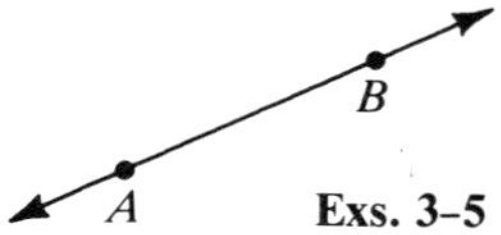

Exs. 3–5

6. What is the coordinate of P? of R?
7. Name the point with coordinate 2.
8. Find each distance: **a.** RS **b.** RQ **c.** PT
9. Name three segments congruent to $\overline{PQ}$.
10. Name the ray opposite to $\overrightarrow{RQ}$.
11. Name the midpoint of $\overline{RT}$.
12. What number is halfway between 1 and 2?
13. What is the coordinate of the midpoint of $\overline{ST}$?
14. Could you list all the numbers between 1 and 2?
15. Is there a point on the number line for every number between 1 and 2?
16. Is there any limit to the number of points on the number line between S and T?
17. Simplify: **a.** $|-4|$ **b.** $|5-(-2)|$ **c.** $|-7-12|$ **d.** $|-9-(-6)|$

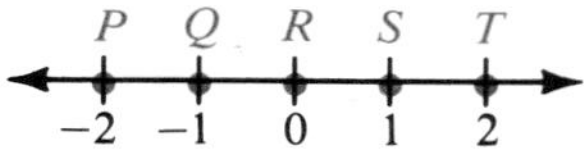

Exs. 6–16

The numbers given are the coordinates of two points on a number line. State the distance between the points.

18. -5 and 12
19. -5 and -15
20. $3\frac{1}{3}$ and $6\frac{2}{3}$

Written Exercises

The numbers given are the coordinates of two points on a number line. State the distance between the points.

A
1. -6 and 9
2. -3 and -17
3. -1.2 and -5.7
4. -2.5 and 4.6

In the diagram, $\overline{HL}$ and $\overleftrightarrow{KT}$ intersect at the midpoint of $\overline{LH}$. Classify each statement as true or false.

5. $\overline{LM} \cong \overline{MH}$
6. KM must equal MT.
7. $\overline{MT}$ bisects $\overline{LH}$.
8. $\overleftrightarrow{KT}$ is a bisector of $\overline{LH}$.
9. $\overrightarrow{MT}$ and $\overrightarrow{TM}$ are opposite rays.
10. $\overrightarrow{MT}$ and $\overrightarrow{MK}$ are opposite rays.
11. $\overleftrightarrow{LH}$ contains only three points.
12. $\overrightarrow{KT}$ is the same as $\overrightarrow{KM}$.
13. $\overleftrightarrow{KT}$ is the same as $\overleftrightarrow{KM}$.
14. $\overline{KT}$ is the same as $\overline{KM}$.
15. $HM + ML = HL$
16. $TM + MH = TH$

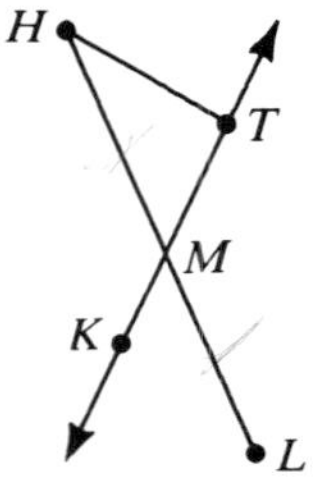

Exs. 5–16

Name each of the following.

Exs. 17–24

17. The point on $\overrightarrow{DA}$ whose distance from D is 2
18. The point on $\overrightarrow{DG}$ whose distance from D is 2
19. Two points whose distance from E is 2
20. The ray opposite to $\overrightarrow{BE}$
21. The midpoint of $\overline{BF}$
22. The coordinate of the midpoint of $\overline{BD}$
23. The coordinate of the midpoint of $\overline{AE}$
24. A segment congruent to $\overline{AF}$

In Exercises 25–28, draw $\overline{CD}$ and $\overline{RS}$ so that the conditions are satisfied.

25. $\overline{CD}$ and $\overline{RS}$ intersect, but neither segment bisects the other.
26. $\overline{CD}$ and $\overline{RS}$ bisect each other.
27. $\overline{CD}$ bisects $\overline{RS}$, but $\overline{RS}$ does not bisect $\overline{CD}$.
28. $\overline{CD}$ and $\overline{RS}$ do not intersect, but $\overrightarrow{CD}$ and $\overrightarrow{RS}$ do intersect.

B 29. In the diagram, $\overline{PR} \cong \overline{RT}$, S is the midpoint of $\overline{RT}$, $QR = 4$, and $ST = 5$. Complete.

a. $RS = \underline{\ ?\ }$ b. $RT = \underline{\ ?\ }$
c. $PR = \underline{\ ?\ }$ d. $PQ = \underline{\ ?\ }$

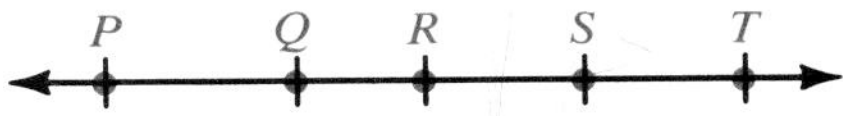

30. In the diagram, X is the midpoint of $\overline{VZ}$, $VW = 5$, and $VY = 20$. Find the coordinates of W, X, and Y.

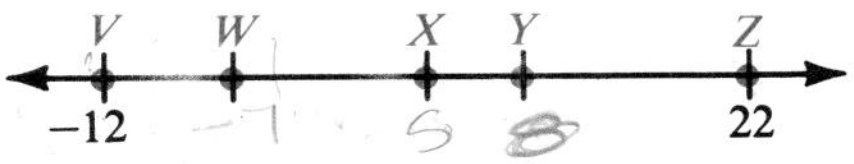

***E* is the midpoint of $\overline{DF}$. Find the value of *x*.**

31. $DE = 5x + 3$, $EF = 33$
32. $DE = 45$, $EF = 5x - 10$
33. $DE = 3x$, $EF = x + 6$
34. $DE = 2x - 3$, $EF = 5x - 24$

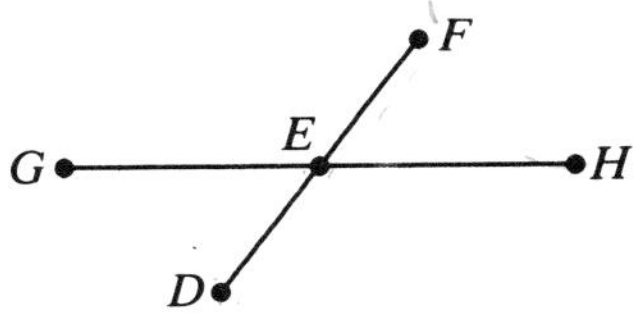

Exs. 31–38

Find the value of *y*.

35. $GE = y$, $EH = y - 1$, $GH = 11$
36. $GE = 3y$, $GH = 7y - 4$, $EH = 24$

Find the value of *z*. Then find *GE* and *EH* and state whether *E* is the midpoint of $\overline{GH}$.

37. $GE = z + 2$, $GH = 20$, $EH = 2z - 6$
38. $GH = z + 6$, $EH = 2z - 4$, $GE = z$

39. a. On $\overrightarrow{AB}$, how many points are there whose distance from point A is 10 cm?
    b. On $\overleftrightarrow{AB}$, how many points are there whose distance from point A is 10 cm?

C 40. On $\overrightarrow{AB}$, how many points are there whose distance from point B is 10 cm?

Name the graph of the given equation or inequality.

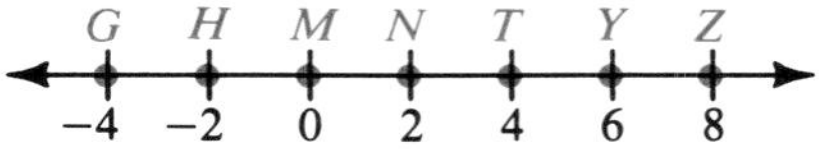

Exs. 41–48

Example **a.** $x \geq 2$ **b.** $4 \leq x \leq 6$

Solution **a.** $\overrightarrow{NT}$ **b.** $\overline{TY}$

41. $-2 \leq x \leq 2$ **42.** $x \leq 0$ **43.** $|x| = 0$ **44.** $|x| \geq 0$ **45.** $|x| \leq 4$

Write an inequality whose graph is described.

Example **a.** $\overrightarrow{HG}$ **b.** All points in $\overline{MT}$ that are not in $\overline{NT}$

Solution **a.** $x \leq -2$ **b.** $0 \leq x < 2$

46. $\overrightarrow{MZ}$ **47.** $\overline{HZ}$ **48.** All points in $\overrightarrow{MY}$ that are not in $\overrightarrow{NG}$

49. The Ruler Postulate suggests that there are many ways to assign coordinates to a line. The Fahrenheit and Celsius temperature scales on a thermometer indicate two such ways of assigning coordinates. A Fahrenheit temperature of 32° corresponds to a Celsius temperature of 0°. The formula, or rule, for converting a Fahrenheit temperature F into a Celsius temperature C is

$$C = \frac{5}{9}(F - 32).$$

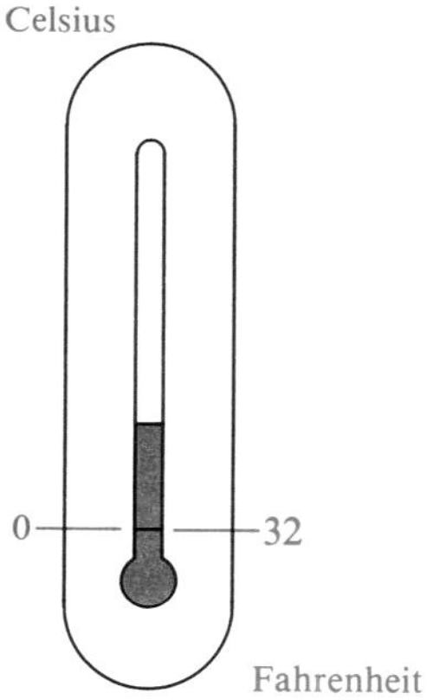

a. What Celsius temperatures correspond to Fahrenheit temperatures of 98.6° and −40°?
b. Solve the equation above for F to obtain a rule for converting Celsius temperatures to Fahrenheit temperatures.
c. What Fahrenheit temperatures correspond to Celsius temperatures of 100° and −10°?

1–3 Angles

An **angle** (∠) is a figure formed by two rays that have the same endpoint. The two rays are called the **sides** of the angle, and their common endpoint is the **vertex** of the angle.

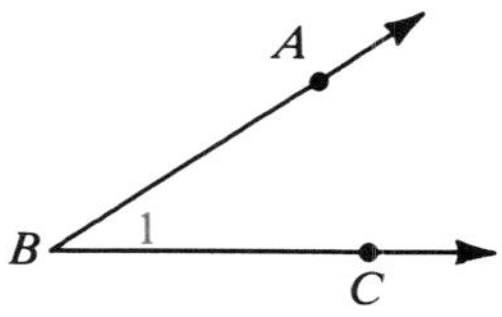

The sides of the angle shown are $\overrightarrow{BA}$ and $\overrightarrow{BC}$. The vertex is point B. The angle can be called $\angle B$, $\angle ABC$, $\angle CBA$, or $\angle 1$. If three letters are used to name an angle, the middle letter names the vertex.

When there is no possibility of confusion, an angle can be named by just its vertex, as $\angle B$ above. On the other hand, it would be incorrect to refer to $\angle Q$ in the diagram at the right, since point Q is the vertex of three angles: $\angle RQT$, $\angle RQS$, and $\angle SQT$.

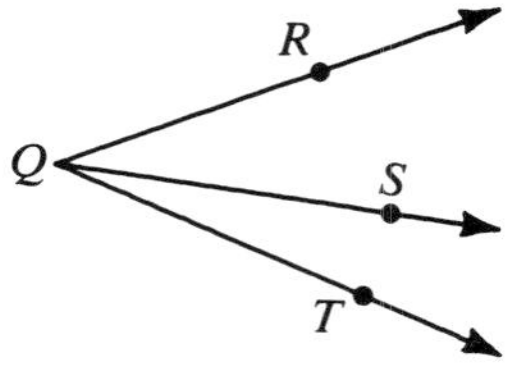

You can use a protractor like the one shown below to find the *measure in degrees* of an angle. Although angles are sometimes measured in other units, you may assume that any angle measure in this book is in degrees.

Using the outer (red) scale of the protractor, you can see that the (degree) measure of $\angle XOY$ is 40. You can indicate this by writing

$$m\angle XOY = 40.$$

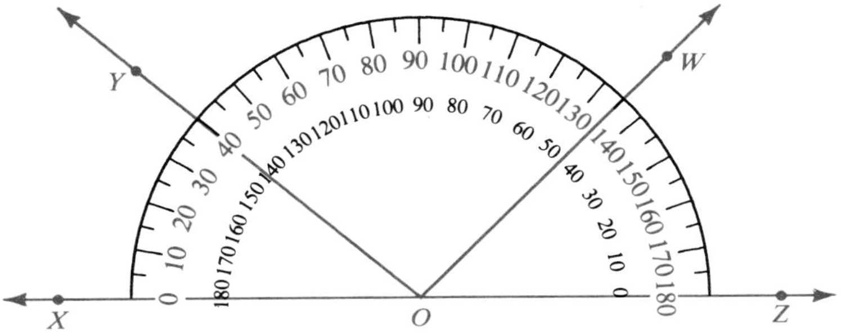

With just one placement of a protractor, you can find the measures of several angles that have a common vertex. Using the inner scale of the protractor, you find that:

$m\angle YOZ = 140 \qquad m\angle WOZ = 45 \qquad m\angle YOW = 140 - 45 = 95$

Using a protractor involves the following basic assumption.

Postulate 3 Protractor Postulate

On $\overleftrightarrow{AB}$ in a given plane, choose any point O between A and B. Consider $\overrightarrow{OA}$ and $\overrightarrow{OB}$ and all the rays that can be drawn from O on one side of $\overleftrightarrow{AB}$. These rays can be paired with the real numbers from 0 to 180 in such a way that:

a. $\overrightarrow{OA}$ is paired with 0, and $\overrightarrow{OB}$ with 180.

b. If $\overrightarrow{OP}$ is paired with x, and $\overrightarrow{OQ}$ with y, then $m\angle POQ = |x - y|$.

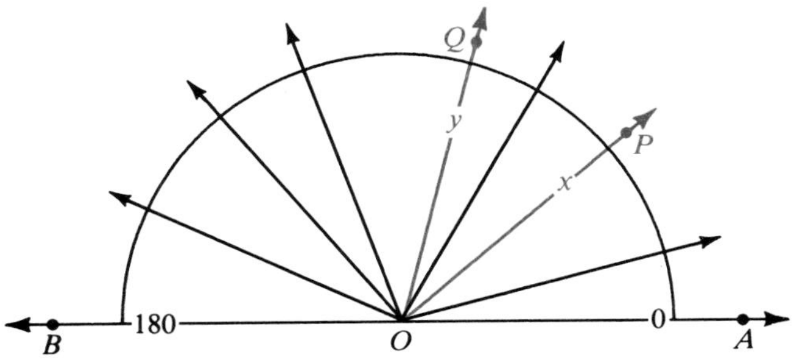

Angles are classified according to their measures.

Acute angle:	Measure between 0 and 90
Right angle:	Measure 90
Obtuse angle:	Measure between 90 and 180
Straight angle:	Measure 180

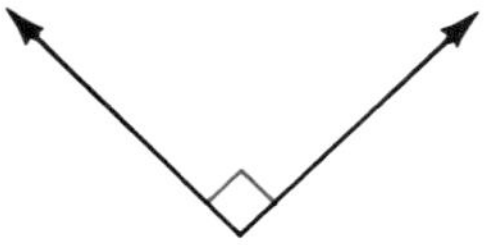

The small square indicates a right angle (rt. ∠).

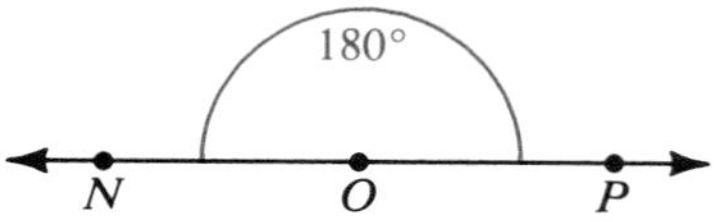

∠ *NOP* is a straight angle.

Another basic assumption we make about angles is the following.

Postulate 4 Angle Addition Postulate

If point *B* lies in the interior of ∠*AOC*, then

$$m\angle AOB + m\angle BOC = m\angle AOC.$$

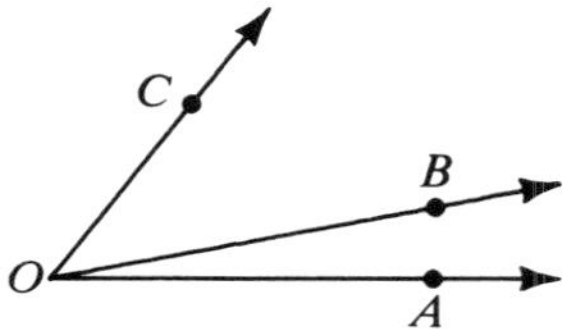

If ∠*AOC* is a straight angle and *B* is any point not on $\overleftrightarrow{AC}$, then

$$m\angle AOB + m\angle BOC = 180.$$

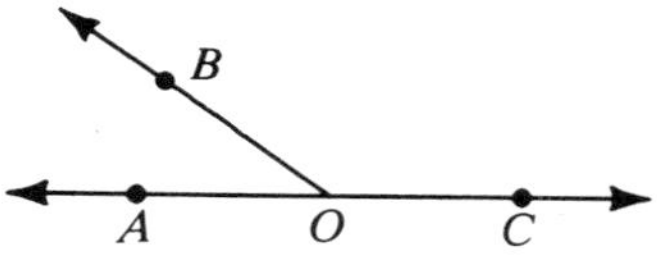

Congruent angles are angles that have equal measures. Since ∠*DEF* and ∠*GEH* each have measure 45, you can write either

$$m\angle DEF = m\angle GEH \quad \text{or} \quad \angle DEF \cong \angle GEH.$$

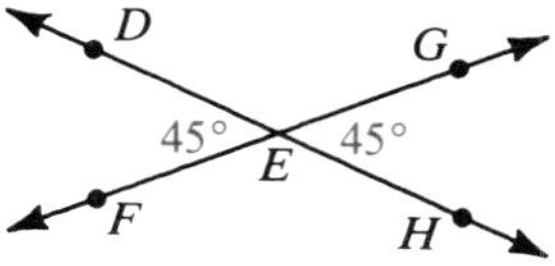

Adjacent angles (adj. ∠s) are two angles in a plane that have a common vertex and a common side but no common interior points.

∠1 and ∠2 are adjacent angles.

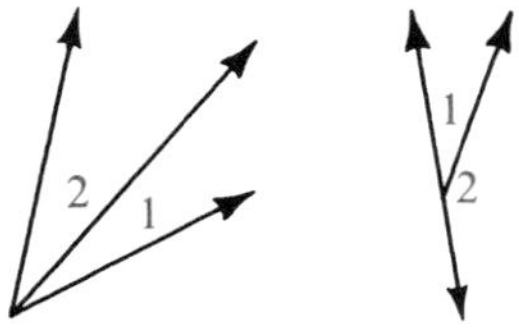

∠3 and ∠4 are not adjacent angles.

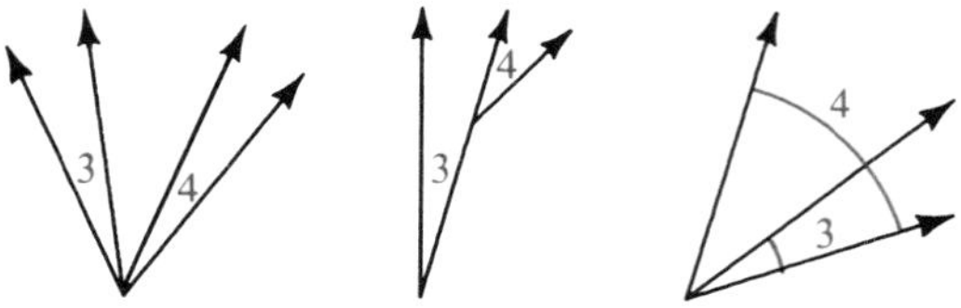

The **bisector of an angle** is a ray that divides the angle into two congruent adjacent angles. In the diagram, $m\angle XYW = m\angle WYZ$ and thus $\overrightarrow{YW}$ bisects $\angle XYZ$.

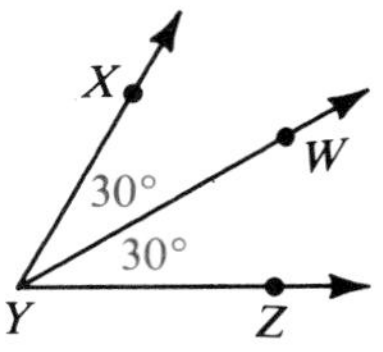

Classroom Exercises

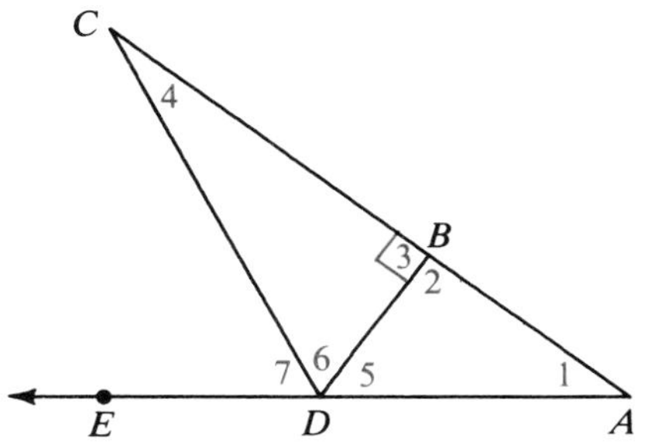

Exs. 1–21

1. What is the vertex of $\angle 4$?
2. Name the sides of $\angle 4$.
3. Name all angles adjacent to $\angle 6$.

State another name for the angle.

4. $\angle ACD$
5. $\angle ABD$
6. $\angle EDC$
7. $\angle 6$
8. $\angle 3$
9. $\angle 5$
10. Why is it confusing to refer to $\angle B$?
11. Name three angles that have B as the vertex.
12. How many angles shown in the diagram have D as the vertex?

State whether the angle appears to be acute, right, obtuse, or straight.

13. $\angle 1$
14. $\angle 3$
15. $\angle EDB$
16. $\angle CDB$
17. $\angle ADC$
18. $\angle ADE$

Complete.

19. $m\angle 7 + m\angle 6 = m\angle$ _?_
20. $m\angle 6 + m\angle 5 = m\angle$ _?_
21. If $\overrightarrow{DB}$ bisects $\angle CDA$, then $\angle$ _?_ $\cong \angle$ _?_.

State the measure of each angle.

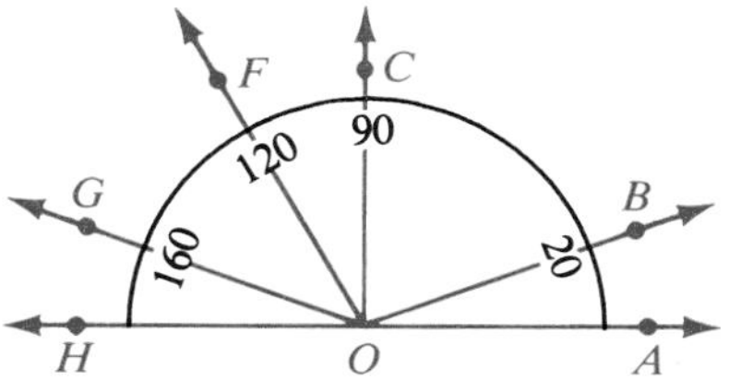

Exs. 22–31

22. $\angle BOC$
23. $\angle HOG$
24. $\angle FOG$
25. $\angle FOC$
26. $\angle BOG$
27. $\angle HOA$
28. Name four angles that are adjacent to $\angle FOG$.
29. Name some angles that are not adjacent to $\angle FOG$.
30. What ray bisects an angle? Which angle(s)?
31. Name a pair of:
    a. congruent acute angles
    b. congruent right angles
    c. congruent obtuse angles

Estimate the measure of each angle.

32.

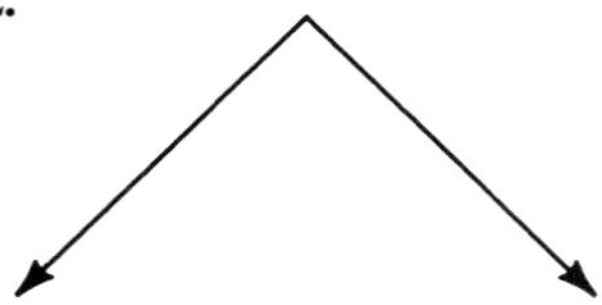

33.

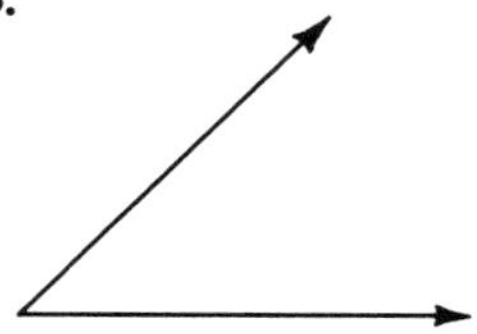

34.

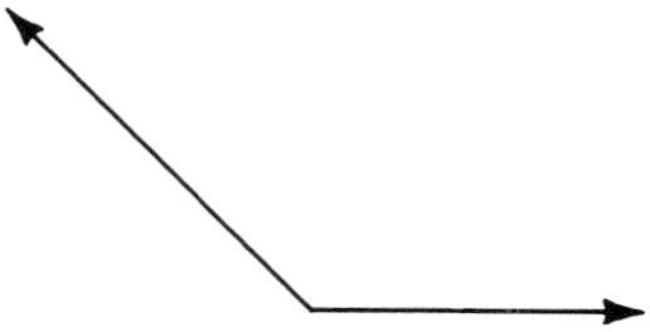

State an equation or inequality that describes $m\angle A$ if $\angle A$ is the type of angle named.

35. right **36.** straight **37.** acute **38.** obtuse

Written Exercises

A **1.** Name the vertex of $\angle 5$. **2.** Name the sides of $\angle 4$.

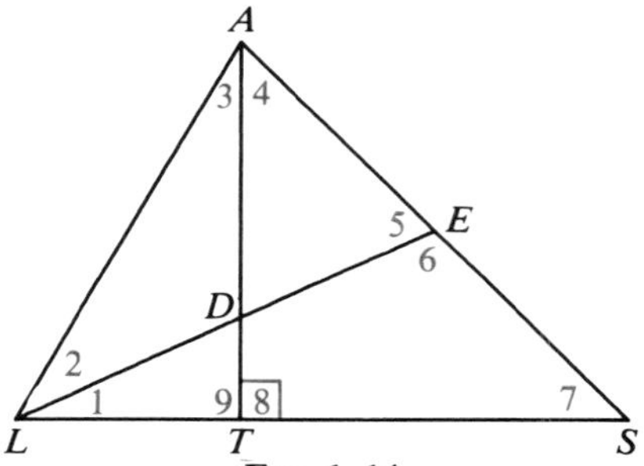

Exs. 1–14

State another name for the angle.

3. $\angle 1$ **4.** $\angle 3$ **5.** $\angle 5$

6. $\angle ALD$ **7.** $\angle AST$ **8.** $\angle LES$

State whether the angle appears to be acute, right, obtuse, or straight.

9. $\angle 2$ **10.** $\angle LAS$ **11.** $\angle ATL$

12. $\angle S$ **13.** $\angle LTS$ **14.** $\angle EDT$

Complete.

15. $m\angle 1 + m\angle 2 = m\angle$ _?_

16. $m\angle MKN - m\angle 2 = m\angle$ _?_

17. If $m\angle 1 = m\angle 2$, then _?_ bisects _?_.

18. If $\overrightarrow{KP}$ bisects $\angle MKN$, then _?_.

19. $m\angle LNK + m\angle KNP =$ _?_

20. $m\angle LKM = m\angle LKP + m\angle$ _?_
$= m\angle$ _?_ $+ m\angle$ _?_ $+ m\angle$ _?_

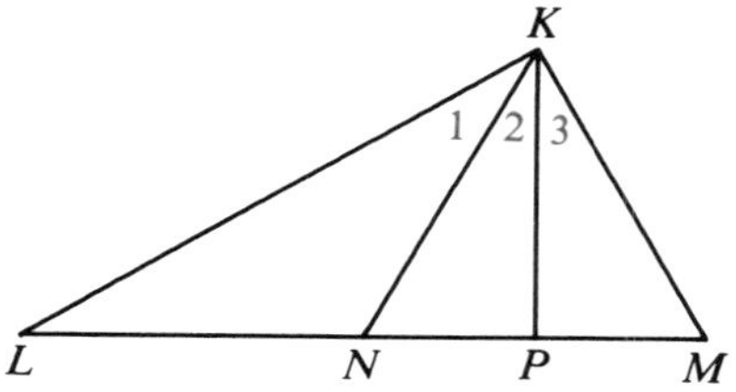

Exs. 15–20

Without measuring, sketch each angle. Then use a protractor to check your accuracy.

21. $90°$ angle **22.** $45°$ angle **23.** $150°$ angle **24.** $10°$ angle

B **25.** Using a ruler, draw a large triangle. Then use a protractor to find the approximate measure of each angle and compute the sum of the three measures. Repeat this exercise for a triangle with different shape. Did you get the same result?

26. Find $m\angle 2$, $m\angle 3$, and $m\angle 4$ if the measure of $\angle 1$ is:
a. 90 **b.** 93 **c.** x

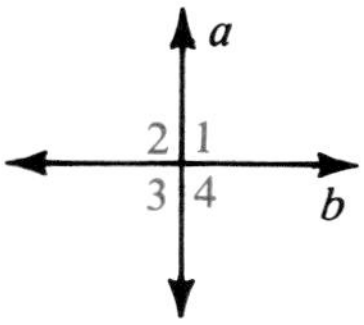

27. Name four right angles.

28. Name nine acute angles.

29. Name three obtuse angles and give their measures.

30. Name a pair of congruent obtuse angles.

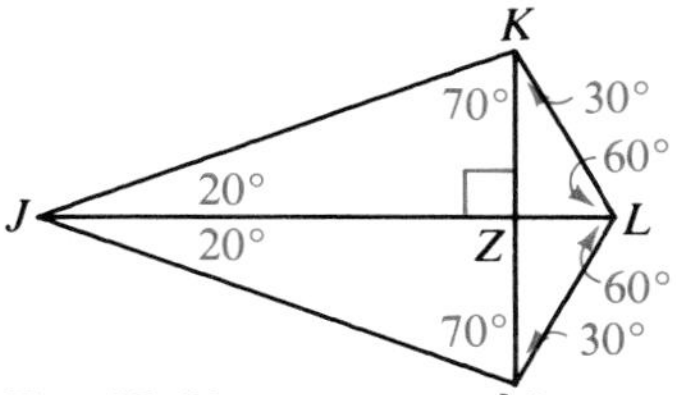

Exs. 27–30

31. Draw three angles, $\angle AOB$, $\angle BOC$, and $\angle AOC$, for which it is *not* true that $m\angle AOB + m\angle BOC = m\angle AOC$.

32. $\overrightarrow{OC}$ bisects $\angle BOD$, $\overrightarrow{OD}$ bisects $\angle COE$, and $m\angle BOC = 33$. Find $m\angle BOE$.

33. $\overrightarrow{OD}$ bisects $\angle BOF$, $\angle 1 \cong \angle 2$, $m\angle 2 = 34$, and $m\angle 3 = 37$. Find $m\angle 4$.

34. $m\angle 1 = x$, $m\angle COE = 2x - 5$, and $m\angle BOE = 100$. Find the value of x.

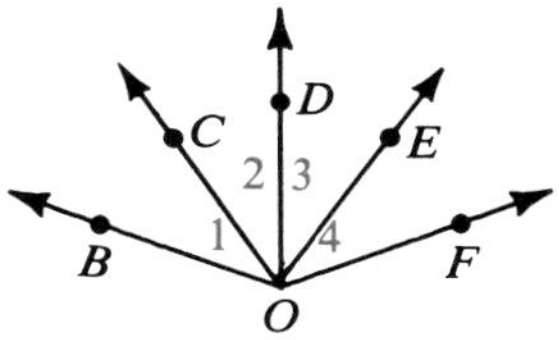

Exs. 32–34

$\overrightarrow{AL}$ bisects $\angle KAT$. Find the value of x.

35. $m\angle 1 = x$, $m\angle 3 = 4x$

36. $m\angle 1 = 8x + 3$, $m\angle 2 = 6x + 9$

37. $m\angle 1 = 5x - 5$, $m\angle 2 = x + 19$

38. $m\angle 1 = 3x - 7$, $m\angle 3 = 128$

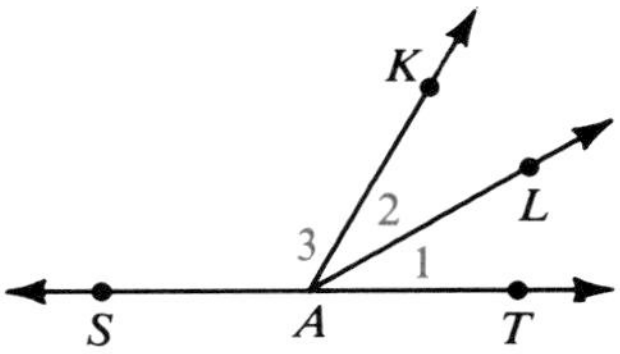

Exs. 35–38

C **39. a.** Complete.

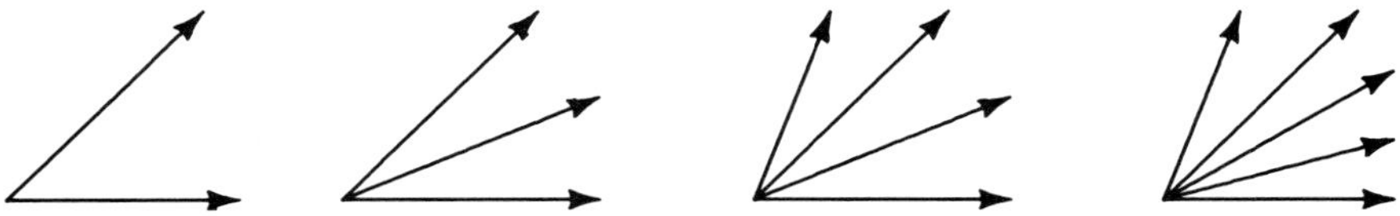

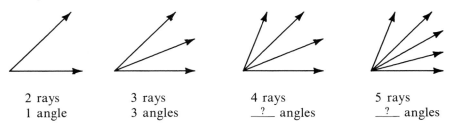

2 rays	3 rays	4 rays	5 rays
1 angle	3 angles	_?_ angles	_?_ angles

b. Without making a drawing, predict the number of angles formed by six noncollinear rays that have the same endpoint.

c. Which of the expressions below gives the number of angles formed by n noncollinear rays that have the same endpoint?

$n - 1$ $2n - 3$ $n^2 - 3$ $\dfrac{n(n-1)}{2}$

Self-Test 1

Classify each statement as true or false.

1. $\overleftrightarrow{FG}$ intersects plane M in point E.
2. $\overleftrightarrow{FG}$ and $\overleftrightarrow{AC}$ intersect.
3. A, B, and D are collinear.
4. A, B, and D are coplanar.
5. $m\angle DBA + m\angle ABE = 180$

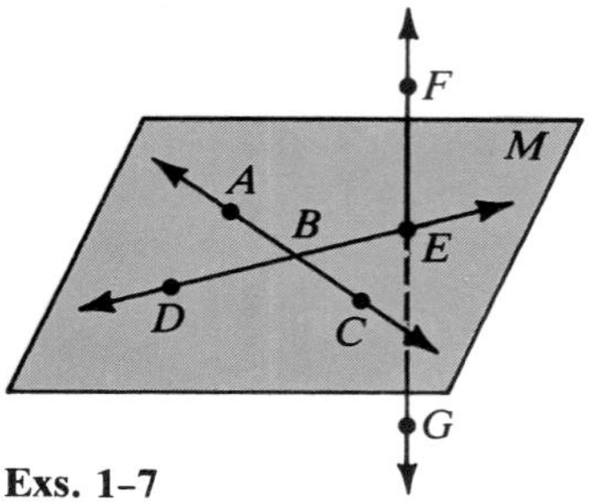

Exs. 1–7

Complete.

6. The ray opposite to $\overrightarrow{EF}$ is _?_.
7. B is the midpoint of $\overline{AC}$, $AB = 17$, and $BC = 2x - 5$. The value of x is _?_.

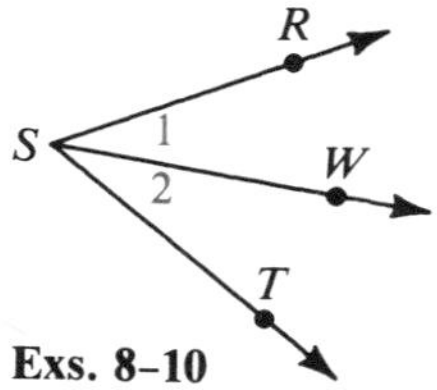

Exs. 8–10

8. Point S is the _?_ of $\angle 1$.
9. If $m\angle 1 = m\angle 2$, then _?_ bisects _?_.
10. If $m\angle RST = 63$ and $m\angle 1 = 29$, then $m\angle 2 =$ _?_.

Introduction to Proof

Objectives

1. Use properties from algebra and the first geometric postulates in two-column proofs.
2. Know four kinds of reasons used in proofs.
3. Apply the definitions of complementary, supplementary, and vertical angles.
4. Apply the Midpoint Theorem, the Angle Bisector Theorem, and the theorem about vertical angles.

1-4 Properties from Algebra

Since the length of a segment and the measure of an angle are given in terms of real numbers, the facts about real numbers and equality that you learned in algebra will be used in your study of geometry. The properties of equality that will be used most often are listed on the following page.

Properties of Equality

Addition Property	If $a = b$ and $c = d$, then $a + c = b + d$.
Subtraction Property	If $a = b$ and $c = d$, then $a - c = b - d$.
Multiplication Property	If $a = b$, then $ca = cb$.
Division Property	If $a = b$ and $c \neq 0$, then $\frac{a}{c} = \frac{b}{c}$.
Substitution Property	If $a = b$, then either a or b may be substituted for the other in any equation (or inequality).
Reflexive Property	$a = a$
Symmetric Property	If $a = b$, then $b = a$.
Transitive Property	If $a = b$ and $b = c$, then $a = c$.

Recall that $DE = FG$ and $\overline{DE} \cong \overline{FG}$ can be used interchangeably, as can $m\angle D = m\angle E$ and $\angle D \cong \angle E$. Thus the following properties of congruence follow directly from the related properties of equality.

Properties of Congruence

Reflexive Property	$\overline{DE} \cong \overline{DE}$ $\qquad$ $\angle D \cong \angle D$
Symmetric Property	If $\overline{DE} \cong \overline{FG}$, then $\overline{FG} \cong \overline{DE}$. If $\angle D \cong \angle E$, then $\angle E \cong \angle D$.
Transitive Property	If $\overline{DE} \cong \overline{FG}$ and $\overline{FG} \cong \overline{JK}$, then $\overline{DE} \cong \overline{JK}$. If $\angle D \cong \angle E$ and $\angle E \cong \angle F$, then $\angle D \cong \angle F$.

The properties of equality and other properties from algebra, such as the **Distributive Property,**

$$a(b + c) = ab + ac,$$

can be used to justify your steps when you solve an equation.

Example 1 Solve $3x = 6 - \frac{1}{2}x$ and justify each step.

Solution

Steps	*Reasons*
1. $3x = 6 - \frac{1}{2}x$	1. Given equation
2. $6x = 12 - x$	2. Multiplication Property of Equality
3. $7x = 12$	3. Addition Property of Equality
4. $x = \frac{12}{7}$	4. Division Property of Equality

Example 1 shows a proof of the statement "If $6 - \frac{1}{2}x = 3x$, then x *must* equal $\frac{12}{7}$." In other words, when given the information that $6 - \frac{1}{2}x = 3x$, we then can use the properties of algebra to conclude, or *deduce,* that $x = \frac{12}{7}$.

Our first proofs in geometry will follow this same pattern. We will use certain given information along with the properties of algebra and accepted statements, such as the Segment Addition Postulate and Angle Addition Postulate, to show that other statements *must* be true. Often a geometric proof is given in two-column form, with statements on the left and a reason for each statement on the right.

In the following examples, congruent segments are marked alike and congruent angles are marked alike. For example, in the diagram below, the marks show that $\overline{RS} \cong \overline{PS}$ and $\overline{ST} \cong \overline{SQ}$.

Example 2

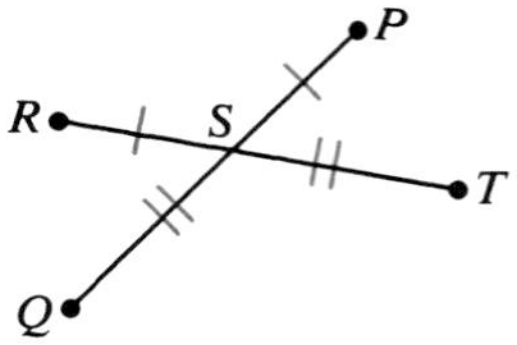

Given: $\overline{RT}$ and $\overline{PQ}$ intersecting at S so that $RS = PS$ and $ST = SQ$.

Prove: $RT = PQ$

Proof:

Statements	Reasons
1. $RS = PS$; $ST = SQ$	1. Given
2. $RS + ST = PS + SQ$	2. Addition Prop. of =
3. $RS + ST = RT$; $PS + SQ = PQ$	3. Segment Addition Postulate
4. $RT = PQ$	4. Substitution Prop.

In Steps 1 and 3 of Example 2, notice how statements can be written in pairs when justified by the same reason.

Example 3

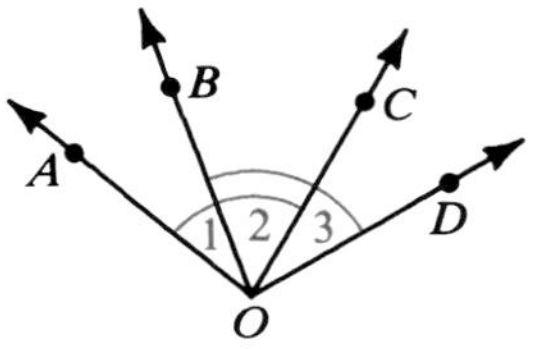

Given: $m\angle AOC = m\angle BOD$

Prove: $m\angle 1 = m\angle 3$

Proof:

Statements	Reasons
1. $m\angle AOC = m\angle BOD$	1. Given
2. $m\angle AOC = m\angle 1 + m\angle 2$; $m\angle BOD = m\angle 2 + m\angle 3$	2. Angle Addition Postulate
3. $m\angle 1 + m\angle 2 = m\angle 2 + m\angle 3$	3. Substitution Prop.
4. $m\angle 2 = m\angle 2$	4. Reflexive Prop.
5. $m\angle 1 = m\angle 3$	5. Subtraction Prop. of =

Notice that the reason given for Step 4 is "Reflexive Property" rather than "Reflexive Property of Equality." Since the reflexive, symmetric, and transitive properties of equality are so closely related to the corresponding properties of congruence, we will simply use "Reflexive Property" to justify either

$$m\angle BOC = m\angle BOC \quad \text{or} \quad \angle BOC \cong \angle BOC.$$

Sometimes you can use either "Transitive Property" or "Substitution Property" to justify a statement. For example, either reason would be a justification for

If $m\angle R = m\angle S$ and $m\angle S = m\angle T$, then $m\angle R = m\angle T$.

Similarly, if you know that

$$
\begin{aligned}
&(1)\ m\angle R = m\angle S,\\
&(2)\ m\angle S = m\angle T,\\
\text{and}\quad &(3)\ m\angle T = m\angle W,
\end{aligned}
$$

you can use the Transitive Property twice to get

$$m\angle R = m\angle W,$$

or you can get this equation by substituting from both (1) and (3) into (2).

Classroom Exercises

Justify each statement with a property from algebra or a property of congruence.

1. $m\angle B = m\angle B$

2. If $\overline{AB} \cong \overline{CD}$ and $\overline{CD} \cong \overline{EF}$, then $\overline{AB} \cong \overline{EF}$.

3. If $RS = TW$, then $TW = RS$.

4. If $x + 5 = 16$, then $x = 11$.

5. If $5y = -20$, then $y = -4$.

6. If $\frac{z}{5} = 10$, then $z = 50$.

7. $2(a + b) = 2a + 2b$

8. If $2z - 5 = -3$, then $2z = 2$.

9. If $2x + y = 70$ and $y = 3x$, then $2x + 3x = 70$.

10. If $m\angle ABC = m\angle 1 + m\angle 2$, and $m\angle 1 + m\angle 2 = 90$, and $90 = m\angle 3 + m\angle 4$, then $m\angle ABC = m\angle 3 + m\angle 4$.

Complete each proof by supplying missing reasons and statements.

11. Given: $m\angle 1 = m\angle 3$;
$m\angle 2 = m\angle 4$

Prove: $m\angle ABC = m\angle DEF$

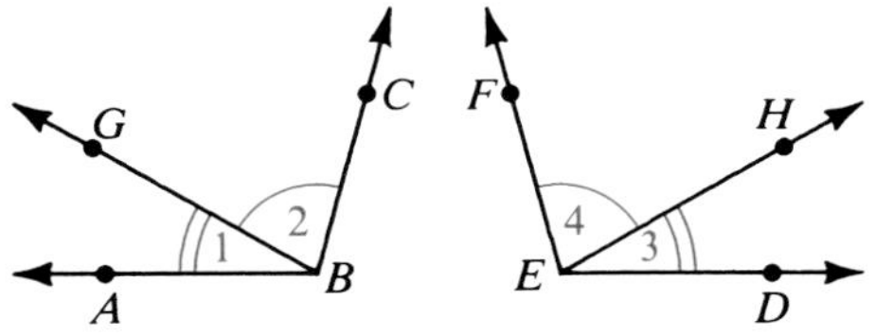

Proof:

Statements	Reasons
1. $m\angle 1 = m\angle 3$; $m\angle 2 = m\angle 4$	1. ?
2. $m\angle 1 + m\angle 2 = m\angle 3 + m\angle 4$	2. ?
3. $m\angle 1 + m\angle 2 = m\angle ABC$; $m\angle 3 + m\angle 4 = m\angle DEF$	3. ?
4. $m\angle ABC = m\angle DEF$	4. ?

12. Given: $ST = RN$; $IT = RU$

Prove: $SI = UN$

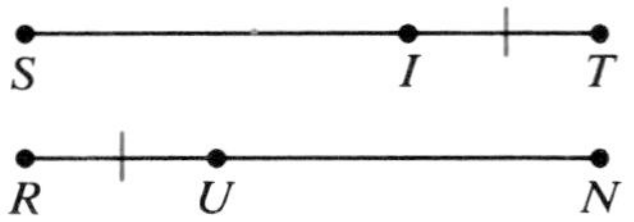

Proof:

Statements	Reasons
1. $ST = RN$	1. ?
2. ? $= SI + IT$; ? $= RU + UN$	2. ?
3. $SI + IT = RU + UN$	3. ?
4. $IT = RU$	4. ?
5. ?	5. ?

Written Exercises

Justify each step.

A **1.** $4x - 5 = -2$
$4x = 3$
$x = \frac{3}{4}$

2. $\frac{3a}{2} = \frac{6}{5}$
$3a = \frac{12}{5}$
$a = \frac{4}{5}$

3. $\frac{z+7}{3} = -11$
$z + 7 = -33$
$z = -40$

4. $15y + 7 = 12 - 20y$
$35y + 7 = 12$
$35y = 5$
$y = \frac{1}{7}$

5. $\frac{2}{3}b = 8 - 2b$
$2b = 3(8 - 2b)$
$2b = 24 - 6b$
$8b = 24$
$b = 3$

6. $\frac{x-2}{2} = \frac{4+x}{5}$
$5(x - 2) = 2(4 + x)$
$5x - 10 = 8 + 2x$
$3x - 10 = 8$
$3x = 18$
$x = 6$

Copy everything shown and supply any missing statements and reasons.

7. Given: $\angle AOD$ as shown

Prove: $m\angle AOD = m\angle 1 + m\angle 2 + m\angle 3$

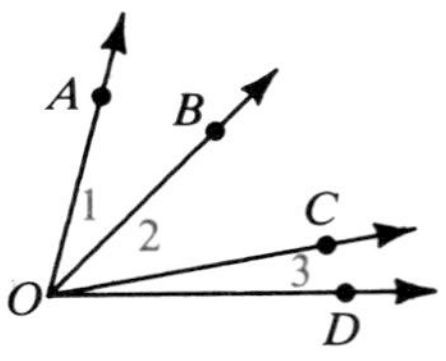

Proof:

Statements	Reasons
1. $m\angle AOD = m\angle AOC + m\angle 3$	1. ?
2. $m\angle AOC = m\angle 1 + m\angle 2$	2. ?
3. ?	3. ?

8. Given: $FL = AT$
Prove: $FA = LT$

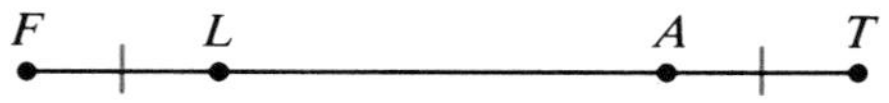

Proof:

Statements	Reasons
1. ?	1. Given
2. $LA = LA$	2. ?
3. $FL + LA = AT + LA$	3. ?
4. $FL + LA = FA$; $LA + AT = LT$	4. ?
5. ?	5. Substitution Prop.

9. Given: $DW = ON$
Prove: $DO = WN$

Proof:

Statements	Reasons
1. $DW = ON$	1. ?
2. $DW = DO + OW$; $ON =$? + ?	2. ?
3. ?	3. Substitution Prop.
4. $OW = OW$	4. ?
5. ?	5. ?

B **10.** Given: $m\angle 4 + m\angle 6 = 180$
Prove: $m\angle 5 = m\angle 6$

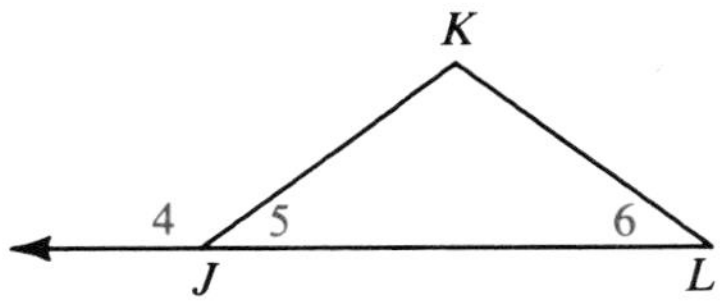

Proof:

Statements	Reasons
1. $m\angle 4 + m\angle 6 = 180$	1. ?
2. $m\angle 4 + m\angle 5 = 180$	2. ?
3. $m\angle 4 + m\angle 5 = m\angle 4 + m\angle 6$	3. ?
4. $m\angle 4 \quad = m\angle 4$	4. ?
5. ?	5. ?

Copy everything shown and write a two-column proof.

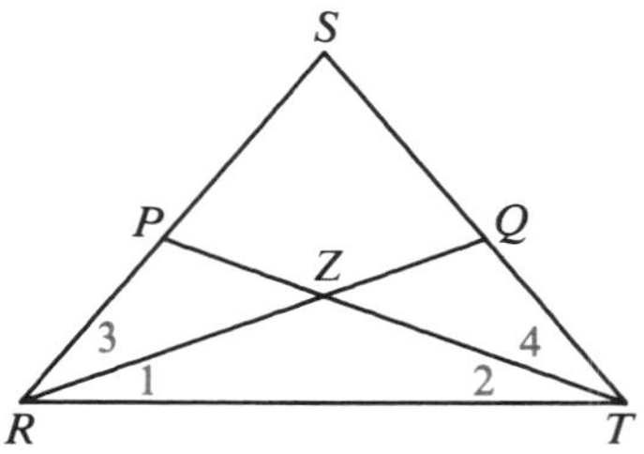

Exs. 11–14

11. Given: $m\angle 1 = m\angle 2$; $m\angle 3 = m\angle 4$
Prove: $m\angle SRT = m\angle STR$

12. Given: $RP = TQ$; $PS = QS$
Prove: $RS = TS$

13. Given: $RQ = TP$; $ZQ = ZP$
Prove: $RZ = TZ$

14. Given: $m\angle SRT = m\angle STR$; $m\angle 3 = m\angle 4$
Prove: $m\angle 1 = m\angle 2$

C **15.** Consider the following statements:

Reflexive Property: Robot A is as rusty as itself.
Symmetric Property: If Robot A is as rusty as Robot B, then Robot B is as rusty as Robot A.
Transitive Property: If Robot A is as rusty as Robot B and Robot B is as rusty as Robot C, then Robot A is as rusty as Robot C.

A *relation* that is reflexive, symmetric, and transitive is an *equivalence relation.* The relation "is as rusty as" is an equivalence relation. Which of the following are equivalence relations?

a. is rustier than
b. has the same length as
c. is opposite (for rays)
d. is coplanar with (for lines)

B I O G R A P H I C A L N O T E

Julia Morgan

Julia Morgan (1872–1959), the first successful woman architect in the United States, was born in San Francisco. Though best known for her design of San Simeon, the castle-like home of William Randolph Hearst that is now the property of the State of California, she designed numerous public buildings and private homes. Even today, to own "a Julia Morgan house" carries considerable prestige.

To become an architect, Morgan needed great determination as well as a brilliant mind. Since the University of California did not have an architecture curriculum at that time, she prepared for graduate work in Paris by studying civil engineering. In Paris the École des Beaux-Arts, which had just begun to admit foreigners, was particularly reluctant to admit a foreign woman. She persisted, however, and became the school's first woman graduate.

1-5 Proving Theorems

Recall that statements that are accepted without proof are called *postulates.* You have already seen statements of the following postulates:

The Ruler Postulate	The Segment Addition Postulate
The Protractor Postulate	The Angle Addition Postulate

We will also accept properties from algebra as postulates in our study of geometry.

Statements that are proved are called *theorems.* Our first theorem uses the definition of a midpoint to prove additional properties of a midpoint that are not explicitly given in the definition. Although this theorem states something very obvious, later theorems in this book will not be so obvious. In fact you may find many of them surprising or amazing.

Theorem 1-1 *Midpoint Theorem*

If M is the midpoint of $\overline{AB}$, then:

$$2\,AM = AB \text{ and } AM = \tfrac{1}{2}AB$$
$$2\,MB = AB \text{ and } MB = \tfrac{1}{2}AB$$

Given: M is the midpoint of $\overline{AB}$.

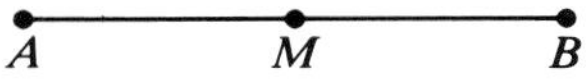

Prove: $2\,AM = AB$; $AM = \frac{1}{2}AB$;
$2\,MB = AB$; $MB = \frac{1}{2}AB$

Proof:

Statements	Reasons
1. M is the midpoint of $\overline{AB}$.	1. Given
2. $AM = MB$	2. Definition of midpoint
3. $AM + MB = AB$	3. Segment Addition Postulate
4. $AM + AM = AB$, or $2\,AM = AB$	4. Substitution Prop.
5. $AM = \frac{1}{2}AB$	5. Division Prop. of =
6. $MB = \frac{1}{2}AB$; $2\,MB = AB$	6. Substitution Prop. (Steps, 2, 4, and 5)

Example 1 Given: M is the midpoint of $\overline{AB}$;
N is the midpoint of $\overline{CD}$;
$AB = CD$

What can you conclude?

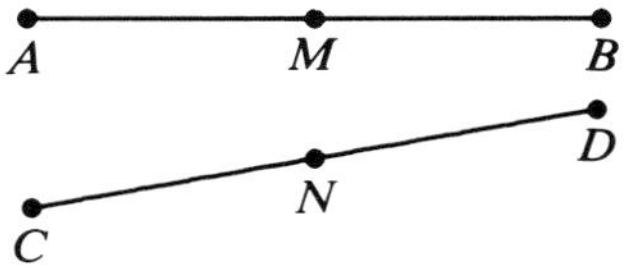

Solution From the definition of midpoint, you know that $AM = MB$ and $CN = ND$. From the Midpoint Theorem, you know that $AM = \frac{1}{2}AB$ and $CN = \frac{1}{2}CD$. Since $AB = CD$, you can deduce that $\frac{1}{2}AB = \frac{1}{2}CD$. Thus, $AM = MB = CN = ND$.

The next theorem is similar to the Midpoint Theorem. It proves properties of the angle bisector that are not given in the definition. The proof is left as Classroom Exercise 12.

Theorem 1-2 Angle Bisector Theorem

If $\overrightarrow{BX}$ is the bisector of $\angle ABC$, then:

$$2m\angle ABX = m\angle ABC \text{ and } m\angle ABX = \tfrac{1}{2}m\angle ABC$$
$$2m\angle XBC = m\angle ABC \text{ and } m\angle XBC = \tfrac{1}{2}m\angle ABC$$

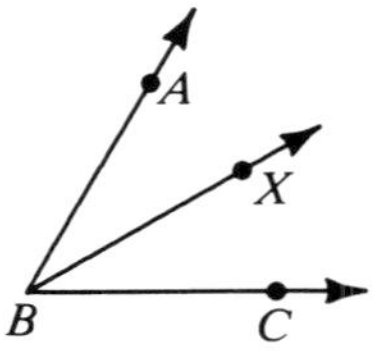

Given: $\overrightarrow{BX}$ is the bisector of $\angle ABC$.

Prove: $2m\angle ABX = m\angle ABC$; $m\angle ABX = \frac{1}{2}m\angle ABC$;
$2m\angle XBC = m\angle ABC$; $m\angle XBC = \frac{1}{2}m\angle ABC$

In addition to postulates and definitions, theorems may be used to justify steps in a proof. Notice the use of the Angle Bisector Theorem in Example 2.

Example 2

Given: $\overrightarrow{EG}$ is the bisector of $\angle DEF$;
$\overrightarrow{SW}$ is the bisector of $\angle RST$;
$m\angle DEG = m\angle RSW$

Prove: $m\angle DEF = m\angle RST$

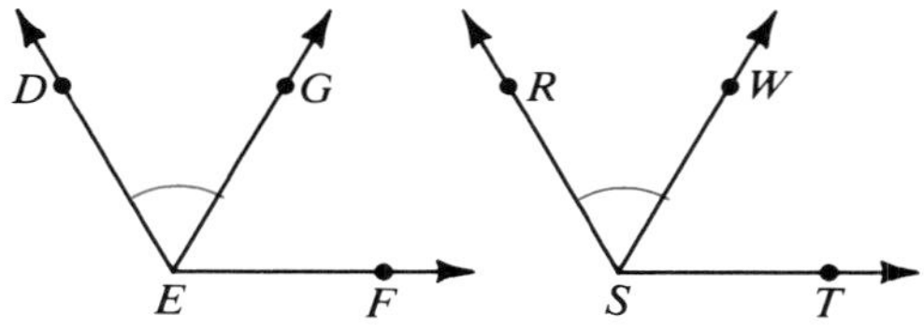

Proof:

Statements	Reasons
1. $m\angle DEG = m\angle RSW$	1. Given
2. $2m\angle DEG = 2m\angle RSW$	2. Multiplication Prop. of =
3. $\overrightarrow{EG}$ is the bisector of $\angle DEF$; $\overrightarrow{SW}$ is the bisector of $\angle RST$.	3. Given
4. $2m\angle DEG = m\angle DEF$; $2m\angle RSW = m\angle RST$	4. Angle Bisector Theorem
5. $m\angle DEF = m\angle RST$	5. Substitution Prop. (Steps 4 and 2)

The two-column proofs you have seen in this section and the previous one are examples of *deductive reasoning*. We have proved statements by reasoning from accepted statements such as postulates, definitions, theorems, and given information. The kinds of reasons that may be used to justify statements in a proof are listed on the following page.

Reasons Used in Proofs

Given information
Definitions
Postulates (These include properties from algebra.)
Theorems that have already been proved

Classroom Exercises

What postulate, definition, or theorem justifies the statement about the diagram?

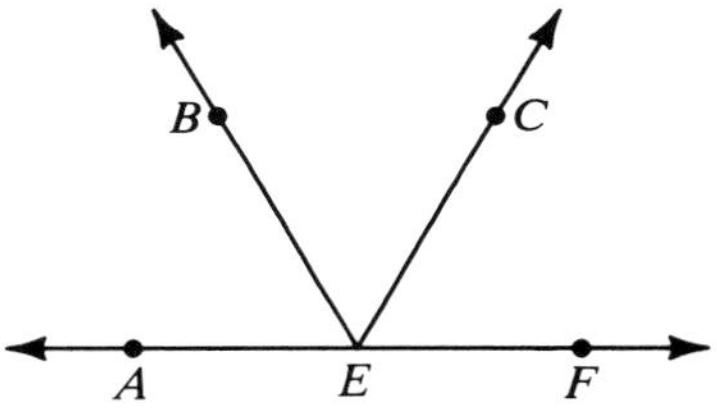

Exs. 1–11

1. If E is the midpoint of $\overline{AF}$, then $AE = EF$.
2. If E is the midpoint of $\overline{AF}$, then $2AE = AF$.
3. If $\overline{AE} \cong \overline{EF}$, then E is the midpoint of $\overline{AF}$.
4. If $\overrightarrow{EB}$ is the bisector of $\angle AEC$, then $m\angle AEB = \frac{1}{2}m\angle AEC$.
5. If $m\angle BEC = m\angle CEF$, then $\overrightarrow{EC}$ bisects $\angle BEF$.
6. If $\overrightarrow{EC}$ is the bisector of $\angle BEF$, then $\angle BEC \cong \angle CEF$.
7. If E is the midpoint of $\overline{AF}$, then $\overrightarrow{EC}$ bisects $\overline{AF}$.
8. If $\overrightarrow{EB}$ bisects $\overline{AF}$, then E is the midpoint of $\overline{AF}$.
9. $m\angle AEB + m\angle BEC = m\angle AEC$
10. $AE + EF = AF$
11. $m\angle AEB + m\angle BEF = 180$
12. Complete the proof of Theorem 1–2.
    Given: $\overrightarrow{BX}$ is the bisector of $\angle ABC$.
    Prove: $2m\angle ABX = m\angle ABC$; $m\angle ABX = \frac{1}{2}m\angle ABC$; $2m\angle XBC = m\angle ABC$; $m\angle XBC = \frac{1}{2}m\angle ABC$

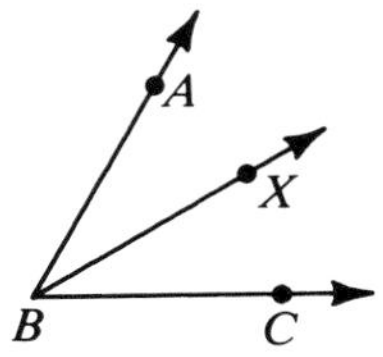

Proof:

Statements	Reasons
1. $\overrightarrow{BX}$ is the bisector of $\angle ABC$.	1. ?
2. $m\angle ABX = $?	2. ?
3. $m\angle ABX + m\angle XBC = m\angle ABC$	3. ?
4. $m\angle ABX + m\angle ABX = m\angle ABC$, or $2m\angle ABX = m\angle ABC$	4. ?
5. $m\angle ABX = \frac{1}{2}m\angle ABC$	5. ?
6. $2m\angle XBC = m\angle ABC$; $m\angle XBC = \frac{1}{2}m\angle ABC$	6. ?

13. M is the midpoint of $\overline{PQ}$. Complete the table.

Coordinate of *P*	1	19	-2	a	1	b
Coordinate of *Q*	25	7	24	$3a$	?	?
Coordinate of *M*	?	?	?	?	-2	$4b$

Written Exercises

Name the definition, postulate, or theorem that justifies the statement about the diagram.

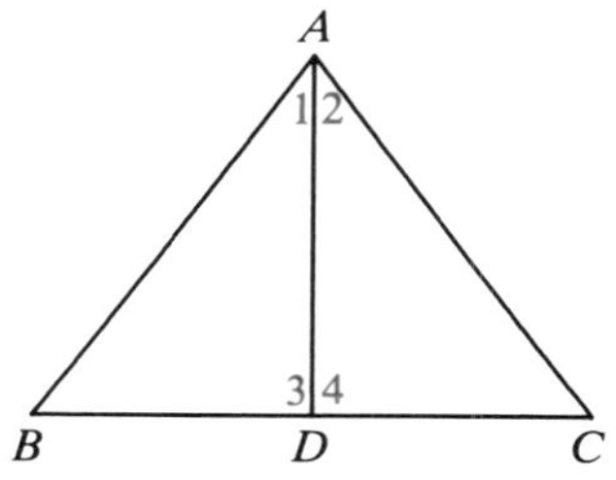

Exs. 1–9

A 1. If D is the midpoint of $\overline{BC}$, then $\overline{BD} \cong \overline{DC}$.

2. If $\angle 1 \cong \angle 2$, then $\overrightarrow{AD}$ is the bisector of $\angle BAC$.

3. If $\overrightarrow{AD}$ bisects $\angle BAC$, then $m\angle 1 = \frac{1}{2}m\angle BAC$.

4. $m\angle 3 + m\angle 4 = 180$

5. If $BD = DC$, then D is the midpoint of $\overline{BC}$.

6. If D is the midpoint of $\overline{BC}$, then $2BD = BC$.

7. $m\angle 1 + m\angle 2 = m\angle BAC$

8. $BD + DC = BC$

9. If $\overrightarrow{AD}$ is the bisector of $\angle BAC$, then $m\angle 1 = m\angle 2$.

In the diagrams below, what number is paired with the bisector of $\angle CDE$?

10.

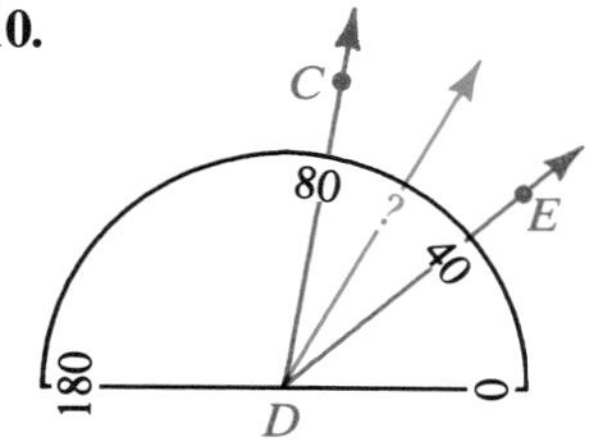

11.

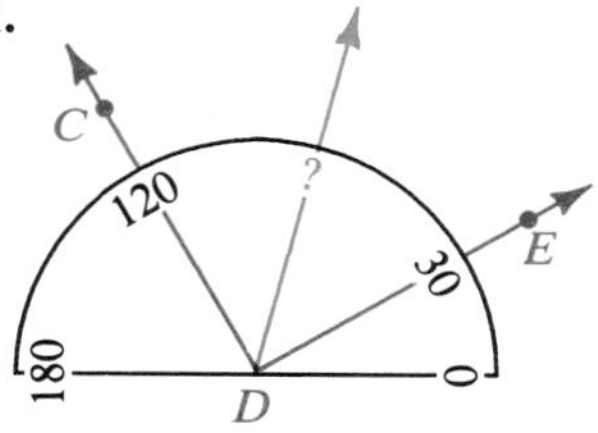

12.

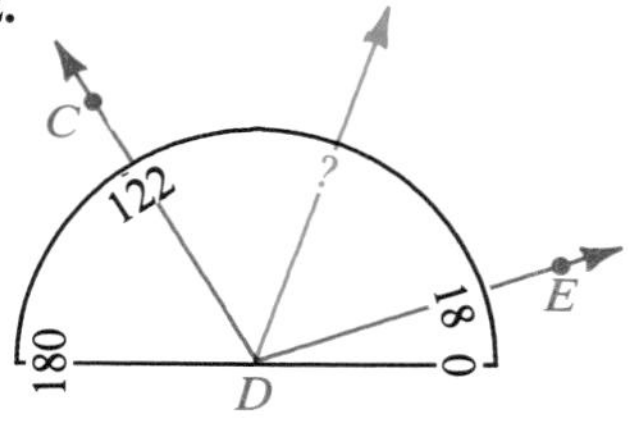

***M* is the midpoint of $\overline{AB}$. Complete.**

	13.	**14.**	**15.**	**16.**	**17.**	**18.**
Coordinate of *A*	0	24	-12	-3	2	?
Coordinate of *B*	17	6	-28	13	?	4
Coordinate of *M*	?	?	?	?	20	-6

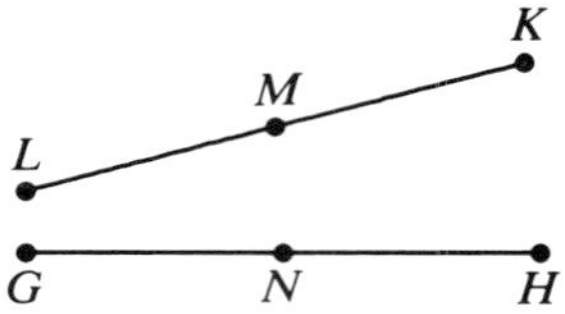

B 19. a. Suppose M and N are the midpoints of $\overline{LK}$ and $\overline{GH}$, respectively. What segments are congruent?

b. What additional information would enable you to deduce that $LM = NH$?

20. a. Suppose $\overrightarrow{SV}$ bisects $\angle RST$ and $\overrightarrow{RU}$ bisects $\angle SRT$. What angles are congruent?

b. What additional information would enable you to deduce that $m\angle VSU = m\angle URV$?

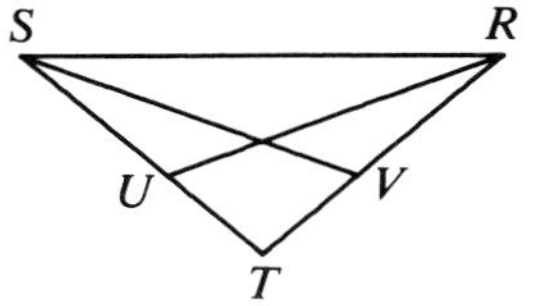

What can you conclude from the given information?

21. Given: $AE = DE$;
$CE = BE$

22. Given: $\overline{AC}$ bisects $\overline{DB}$;
$\overline{DB}$ bisects $\overline{AC}$;
$CE = BE$

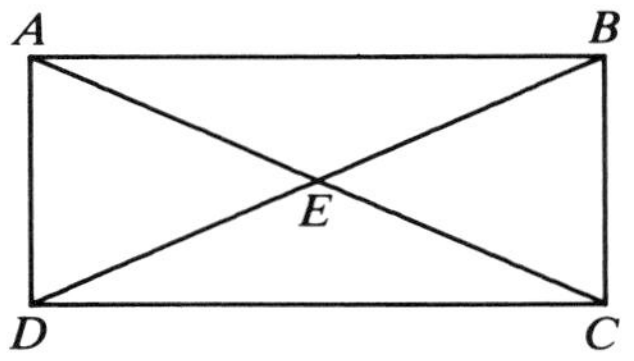

Exs. 21, 22

23. Point N is the midpoint of $\overline{LX}$, and point Y is the midpoint of $\overline{LN}$. The coordinates of L and X are 16 and 40, respectively. Sketch a diagram and find:

a. LN **b.** LY
c. The coordinate of Y **d.** The coordinate of N

24. $\overrightarrow{SW}$ is the bisector of $\angle RST$, $\overrightarrow{SZ}$ is the bisector of $\angle RSW$, and $\overrightarrow{SR}$ is the bisector of $\angle NSW$. If $m\angle RST = 72$, find:

a. $m\angle RSZ$ **b.** $m\angle NSZ$

25. Copy and complete the following proof of the statement: If points A and B have coordinates a and b, with $b > a$, and midpoint M of $\overline{AB}$ has coordinate x, then $x = \dfrac{a+b}{2}$.

Given: Points A and B have coordinates a and b;
$b > a$; midpoint M of $\overline{AB}$ has coordinate x.

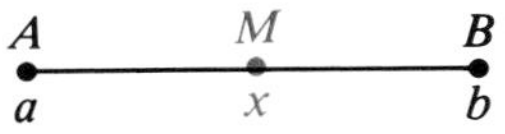

Prove: $x = \dfrac{a+b}{2}$

Proof:

Statements	Reasons
1. A, M, and B have coordinates a, x, and b, respectively; $b > a$	1. ?
2. $AM = x - a$; $MB = b - x$	2. ?
3. M is the midpoint of $\overline{AB}$.	3. ?
4. $AM = MB$	4. ?
5. $x - a = b - x$	5. ?
6. $2x =$?	6. ?
7. $x = \dfrac{a+b}{2}$	7. ?

26. Suppose $\overrightarrow{OP}$ and $\overrightarrow{OQ}$ are paired with the real numbers p and q and that $p > q$. Let $\overrightarrow{OX}$, the bisector of $\angle POQ$, be paired with the real number x. Derive an expression for x in terms of p and q.

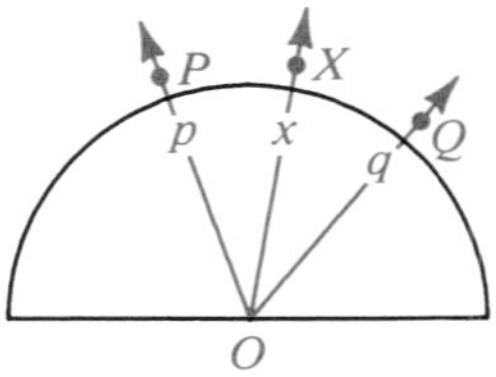

27. $\overrightarrow{BD}$ bisects $\angle ABC$ and $m\angle ABD = x$. Write an equation or inequality that describes x if $\angle ABC$ is:
a. a right angle **b.** an acute angle **c.** an obtuse angle

Copy everything shown and write a two-column proof.

28. Given: $\overrightarrow{AC}$ bisects $\angle DAB$;
$\overrightarrow{CA}$ bisects $\angle BCD$;
$m\angle 1 = m\angle 3$
Prove: $m\angle DAB = m\angle DCB$

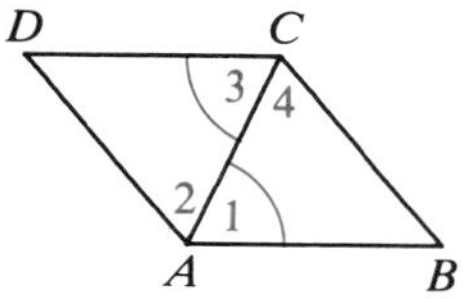

29. Given: M is the midpoint of $\overline{LK}$;
N is the midpoint of $\overline{GH}$;
$LK = GH$
Prove: $MK = NH$

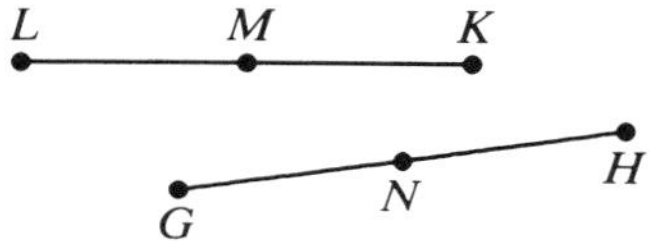

C 30. Fold down a corner of a rectangular sheet of paper as in Figure 1. Then fold the next corner so that the edges touch as in Figure 2.

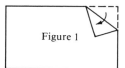

Open the paper and measure the angle between the fold lines. Repeat with another sheet of paper, folding the corner at a different angle. Explain why the measured angle is the same in both cases.

31. Point T is the midpoint of $\overline{RS}$, W is the midpoint of $\overline{RT}$, and Z is the midpoint of $\overline{WS}$. If the length of $\overline{TZ}$ is x, find the following lengths in terms of x.
a. RW **b.** ZS **c.** RS **d.** WZ
(*Hint:* Sketch a diagram and let $y = WT$.)

COMPUTER KEY-IN

A bee starts at point P_0, flies to point P_1, and lands there. The bee then returns half of the way to P_0, landing at P_2. From P_2, the bee returns half of the way to P_1, landing at P_3, and so forth. Can you predict the bee's location after 10 trips?

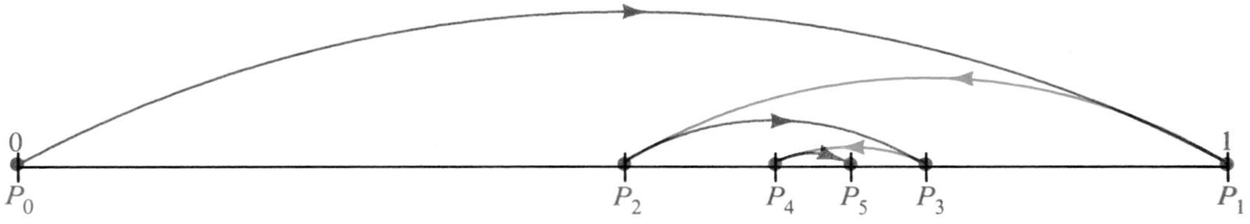

Assuming that P_0 and P_1 have coordinates 0 and 1, respectively, the BASIC program below will compute and print the bee's location at the end of trips 2 through 10. Notice that P_n represents the position of the bee after n trips. Since P_n is the midpoint of the bee's previous two positions, P_{n-1} and P_{n-2}, line 50 calculates $P(N)$ by using the statement proved in Exercise 25, page 27.

```
10  DIM P(50)
20  LET P(0) = 0
30  LET P(1) = 1
40  FOR N = 2 TO 10 STEP 1
50  LET P(N) = (1/2) * (P(N - 2) + P(N - 1))
60  PRINT N, P(N)
70  NEXT N
80  END
```

Exercises

1. Enter the program on your computer and RUN it. Do you notice any patterns or trends in the coordinates? Change line 40 so that the computer will print the coordinates up to P_{40}. What simple fraction is approximated by P_{40}?

2. In line 50, $P(n)$ could instead be computed from the *series*

$$1 - \tfrac{1}{2} + \tfrac{1}{4} - \tfrac{1}{8} + \cdots + (-\tfrac{1}{2})^{n-1}$$

where each term of the series reflects the bee's return half of the way from P_{n-1} to P_{n-2}. Thus, line 50 could be replaced by

```
50  P(N) = P(N - 1) + (-1/2) ↑ (N - 1).
```

Change your line 50 and RUN the new program. Check to make sure that both programs produce the same results. (Some slight variations will be expected, due to rounding off.)

3. (Optional) What series, similar to the one in Exercise 2, do you think would reflect the bee's movements if on each trip, it returned one-third of the way to the previous point instead of half of the way? Use this series to modify your program to produce coordinates for the bee's new flight pattern. RUN the program for 30 trips. Determine what point the bee is approaching. What simple fraction is the coordinate of this point?

4. (Optional) Repeat Exercise 3 for a bee that returns one-fourth of the way to the previous point.

1-6 Special Pairs of Angles

Complementary angles (comp. ∠s) are two angles whose measures have the sum 90. Each angle is called a *complement* of the other.

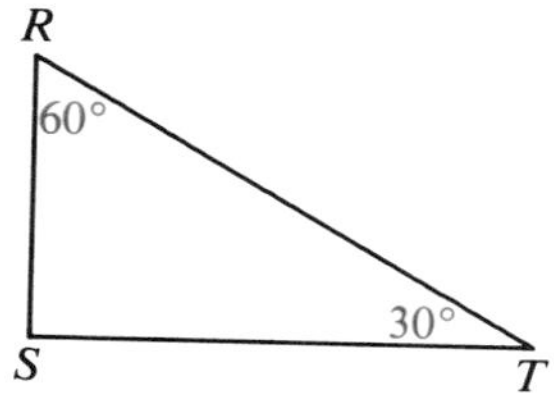

$\angle R$ and $\angle T$ are complementary.

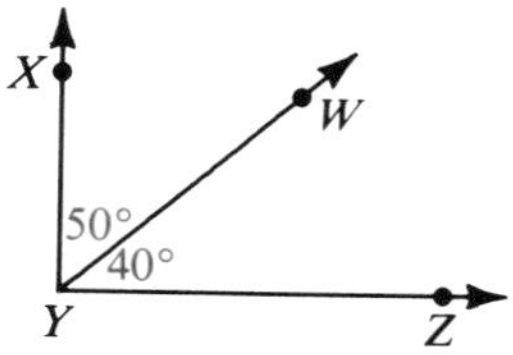

$\angle XYW$ is a complement of $\angle WYZ$.

Supplementary angles (supp. ∠s) are two angles whose measures have the sum 180. Each angle is called a *supplement* of the other.

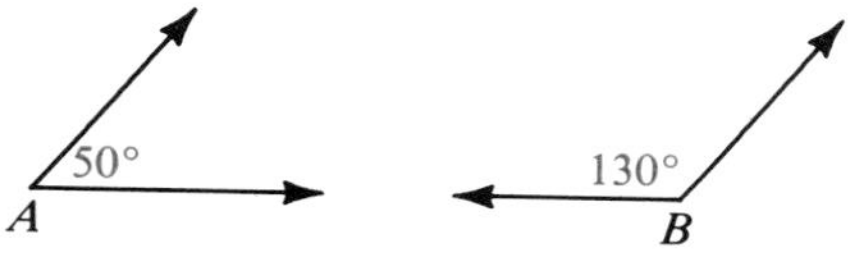

$\angle A$ and $\angle B$ are supplementary.

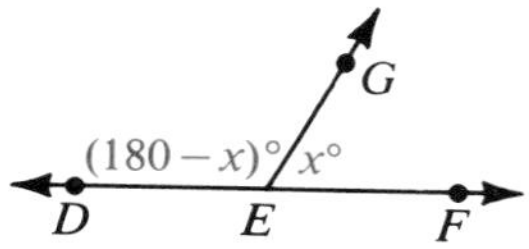

$\angle DEG$ is a supplement of $\angle GEF$.

Example 1 The measure of a supplement of an angle is 24 less than twice the measure of the angle. Find the measures of the angles.

Solution Let x = the measure of one angle.
Then $2x - 24$ = the measure of its supplement.

$$\begin{aligned} x + (2x - 24) &= 180 \\ 3x - 24 &= 180 \\ 3x &= 204 \\ x &= 68 \end{aligned} \qquad 2x - 24 = 112$$

The measures of the angles are 68 and 112.

Example 2 A supplement of an angle is three times as large as a complement of the angle. Find the measures of all three angles.

Solution Let x = the measure of the angle.
Then $180 - x$ = the measure of its supplement,
and $90 - x$ = the measure of its complement.

$$\begin{aligned} 180 - x &= 3(90 - x) \\ 180 - x &= 270 - 3x \\ 2x &= 90 \\ x &= 45 \end{aligned} \qquad \begin{aligned} 90 - x &= 90 - 45 \\ &= 45 \end{aligned} \qquad \begin{aligned} 180 - x &= 180 - 45 \\ &= 135 \end{aligned}$$

The measures of the angle, its complement, and its supplement are 45, 45, and 135, respectively.

Vertical angles (vert. $\angle$s) are two angles whose sides form two pairs of opposite rays. When two lines intersect, they form two pairs of vertical angles.

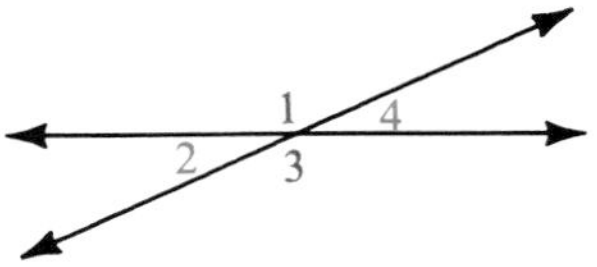

$\angle 1$ and $\angle 3$ are vertical angles.

$\angle 2$ and $\angle 4$ are vertical angles.

Theorem 1-3

Vertical angles are congruent.

Given: $\angle 1$ and $\angle 2$ are vertical angles.

Prove: $\angle 1 \cong \angle 2$ (or $m\angle 1 = m\angle 2$)

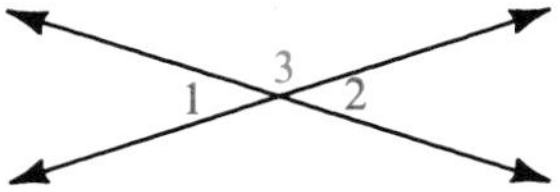

Proof:

Statements	Reasons
1. $m\angle 1 + m\angle 3 = 180$; $m\angle 2 + m\angle 3 = 180$	1. Angle Addition Postulate
2. $m\angle 1 + m\angle 3 = m\angle 2 + m\angle 3$	2. Substitution Prop.
3. $m\angle 3 = m\angle 3$	3. Reflexive Prop.
4. $m\angle 1 = m\angle 2$	4. Subtraction Prop. of =

Example 3 In the diagram, $\angle 4 \cong \angle 5$.
Name two other angles congruent to $\angle 5$.

Solution $\angle 8 \cong \angle 5$
Since $\angle 7 \cong \angle 4$ and $\angle 4 \cong \angle 5$, $\angle 7 \cong \angle 5$.

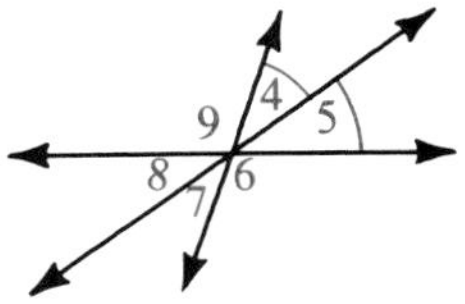

Classroom Exercises

Find the measures of a complement and a supplement of $\angle A$.

1. $m\angle A = 10$ **2.** $m\angle A = 75$ **3.** $m\angle A = 89$ **4.** $m\angle A = y$

5. Name two right angles.

6. Name two adjacent complementary angles.

7. Name two complementary angles that are not adjacent.

8. a. Name a supplement of $\angle MLQ$.
b. Name another pair of supplementary angles.

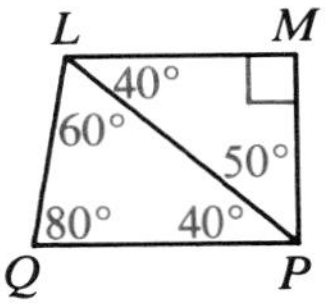

Exs. 5–8

9. In the diagram, assume that $m\angle CDB = 90$. Name:
a. Two congruent supplementary angles
b. Two supplementary angles that are not congruent
c. Two complementary angles **d.** A straight angle

10. In the diagram, assume that $m\angle DFB = 90$ and $\overrightarrow{FE}$ bisects $\angle AFD$. Find each measure.
a. $m\angle AFD$ **b.** $m\angle AFE$ **c.** $m\angle BFE$

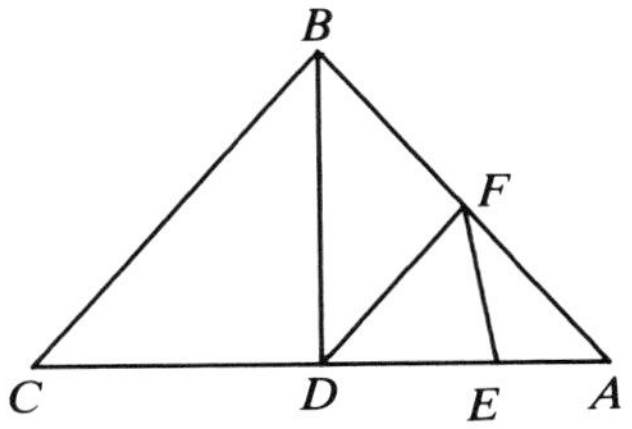

Exs. 9, 10

Complete.

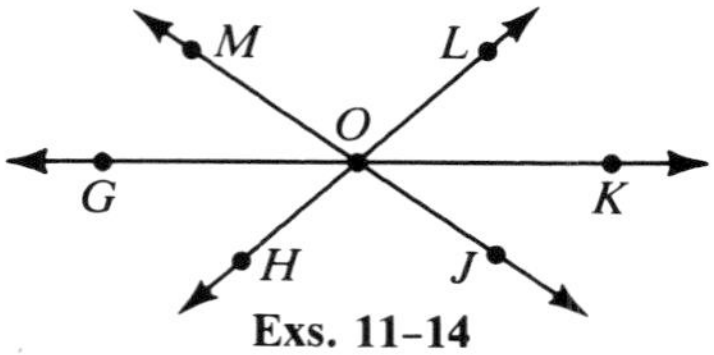

Exs. 11–14

11. $\angle GOH \cong$ ___?___
12. $\angle GOM \cong$ ___?___
13. $\angle MOK \cong$ ___?___
14. $\angle LOG \cong$ ___?___

15. 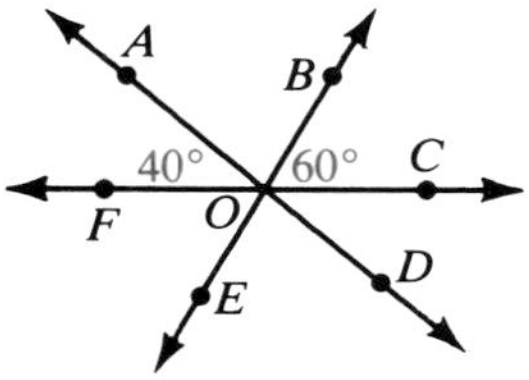

a. $m\angle FOE =$ ___?___
b. $m\angle COD =$ ___?___
c. $m\angle BOD =$ ___?___
d. $m\angle AOB =$ ___?___
e. $m\angle DOE =$ ___?___

16. 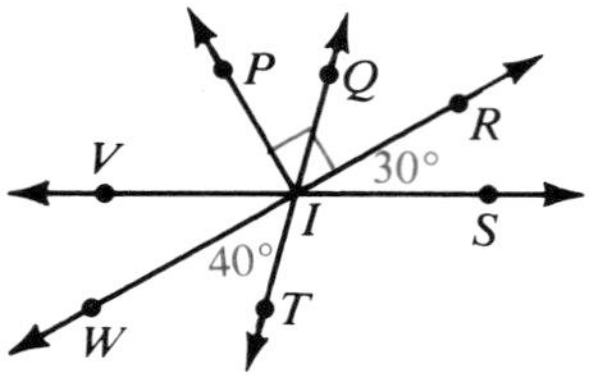

a. $m\angle QIR =$ ___?___
b. $m\angle PIQ =$ ___?___
c. $m\angle VIT =$ ___?___
d. $m\angle VIQ =$ ___?___
e. $m\angle SIT =$ ___?___

17. A supplement of
 a. an acute angle is ___?___. b. an obtuse angle is ___?___. c. a right angle is ___?___.
18. a. A complement of an acute angle is ___?___.
 b. Can a right or an obtuse angle have a complement?

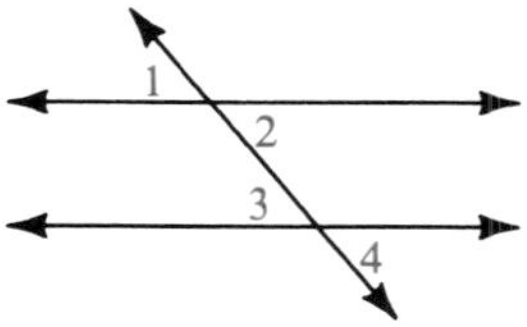

19. Given: $\angle 2 \cong \angle 3$
 a. What can you conclude?
 b. Explain how you would prove your conclusion.

Written Exercises

Find the measures of a complement and a supplement of $\angle B$.

A 1. $m\angle B = 55$ 2. $m\angle B = 1$ 3. $m\angle B = 72.5$ 4. $m\angle B = 3x$

5. Two angles are both congruent and complementary. Find their measures.
6. Two angles are both congruent and supplementary. Find their measures.

In the diagram, $m\angle KAE = 90$.

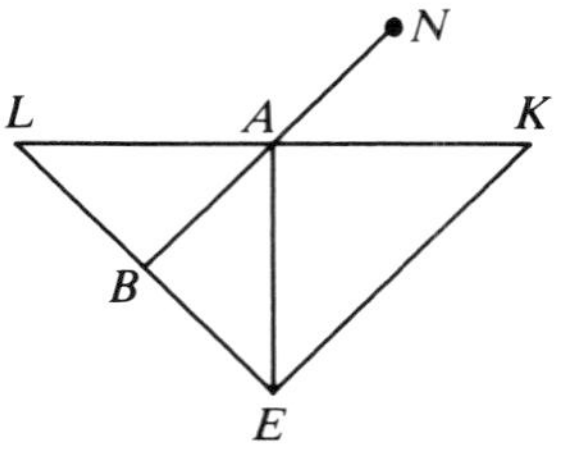

7. Name another right angle.
8. Name two congruent supplementary angles.
9. Name two noncongruent supplementary angles.
10. Name two supplementary angles that may or may not be congruent.
11. Name two complementary angles.
12. Name a pair of vertical angles.

In the diagram, $\overrightarrow{OT}$ bisects $\angle SOU$, $m\angle UOV = 35$, and $m\angle YOW = 120$. Find the measure of each angle.

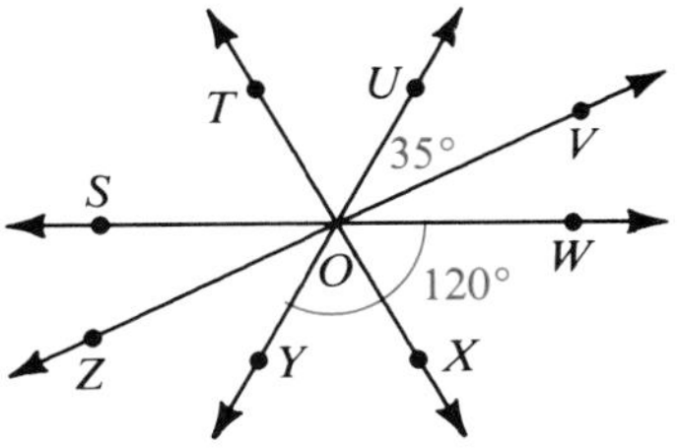

13. $m\angle ZOY$
14. $m\angle ZOW$
15. $m\angle VOW$
16. $m\angle SOU$
17. $m\angle TOU$
18. $m\angle ZOT$

Find the value of x.

19.

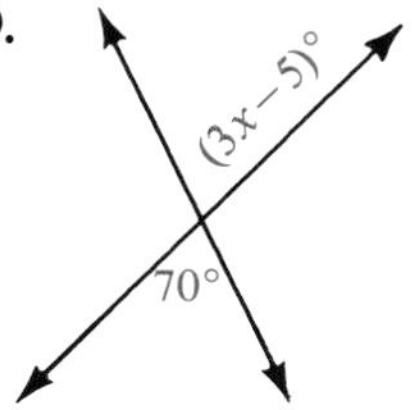

20.

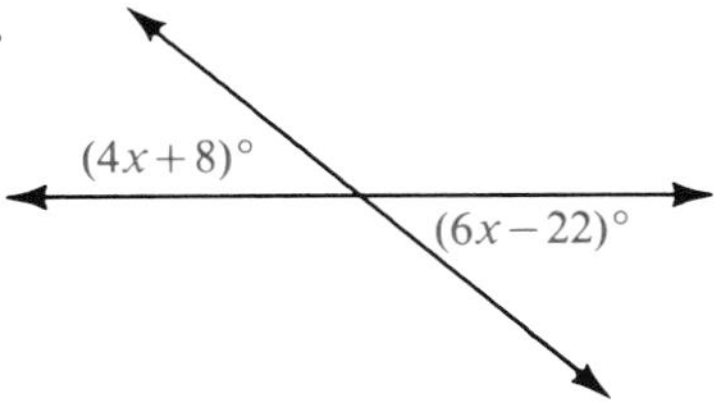

21.

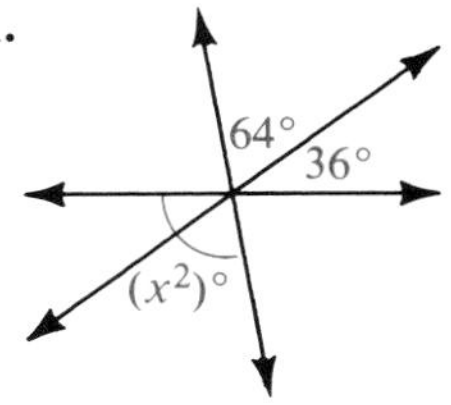

22. $\angle 1$ and $\angle 2$ are supplements.
$\angle 3$ and $\angle 4$ are supplements.

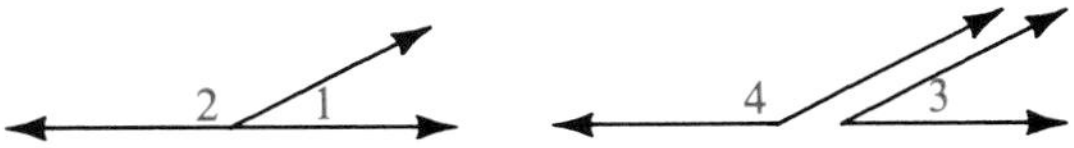

   a. If $m\angle 1 = m\angle 3 = 27$, find the measures of $\angle 2$ and $\angle 4$.
   b. If $m\angle 1 = m\angle 3 = x$, find the measures of $\angle 2$ and $\angle 4$ in terms of x.
   c. If two angles are congruent, must their supplements be congruent?

If $\angle A$ and $\angle B$ are supplementary, find the value of x and the measures of the angles.

B 23. $m\angle A = 2x$, $m\angle B = x - 15$

24. $m\angle A = x + 16$, $m\angle B = 2x - 16$

If $\angle C$ and $\angle D$ are complementary, find the value of y and the measures of the angles.

25. $m\angle C = 3y + 5$, $m\angle D = 2y$

26. $m\angle C = y - 8$, $m\angle D = 3y + 2$

Use the given information to write an equation. Solve the equation to find the measures of the two angles described.

27. A supplement of an angle is twice as large as the angle.
28. A complement of an angle is five times as large as the angle.
29. The measure of one of two complementary angles is six less than twice the measure of the other.
30. The difference between the measures of two supplementary angles is 42.

Find the measures of the angle, its complement, and its supplement.

31. A supplement of an angle is six times as large as a complement of the angle.
32. Three times the measure of a supplement of an angle is eight times the measure of a complement of the angle.

33. Copy everything shown. Complete the proof.
Given: $\angle 2 \cong \angle 3$
Prove: $\angle 1 \cong \angle 4$

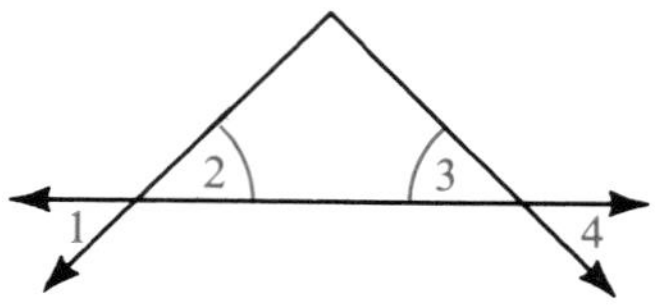

Proof:

Statements	Reasons
1. $\angle 1 \cong \angle 2$	1. ?
2. $\angle 2 \cong \angle 3$	2. ?
3. $\angle 3 \cong \angle 4$	3. ?
4. ?	4. Transitive Property (used twice)

34. Copy the figure and the statement of what is given and what is to be proved. Then write a two-column proof.
Given: $\angle 2 \cong \angle 3$
Prove: $\angle 1 \cong \angle 4$

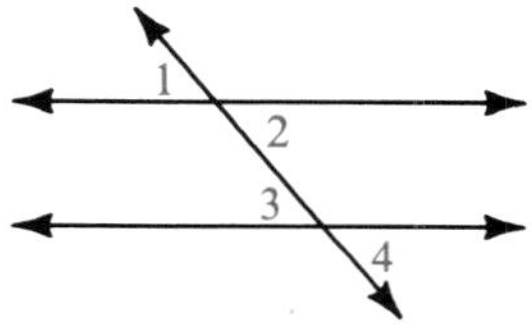

Find the values of x and y for each diagram.

35.

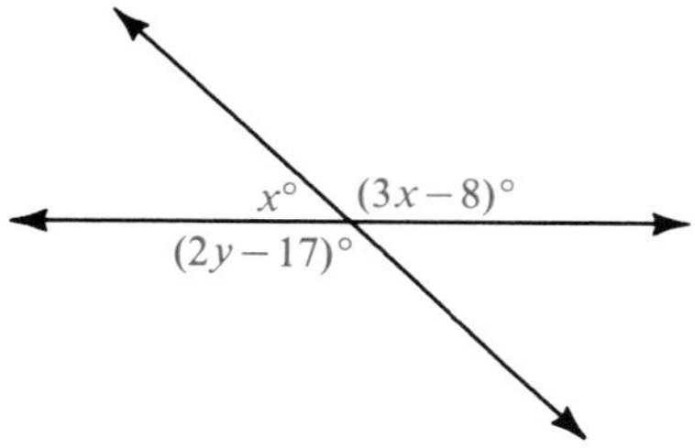

36.

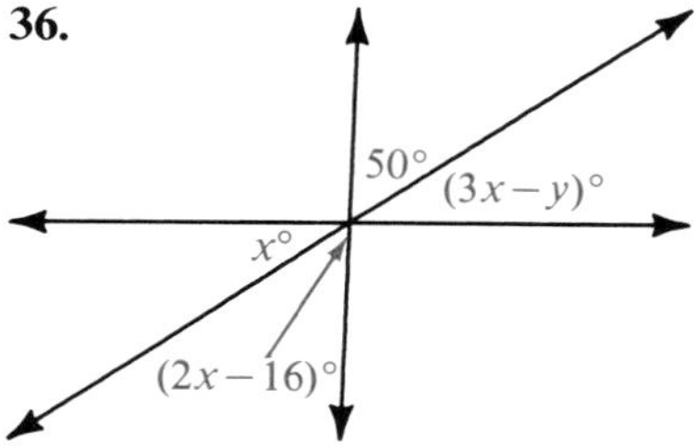

C 37.

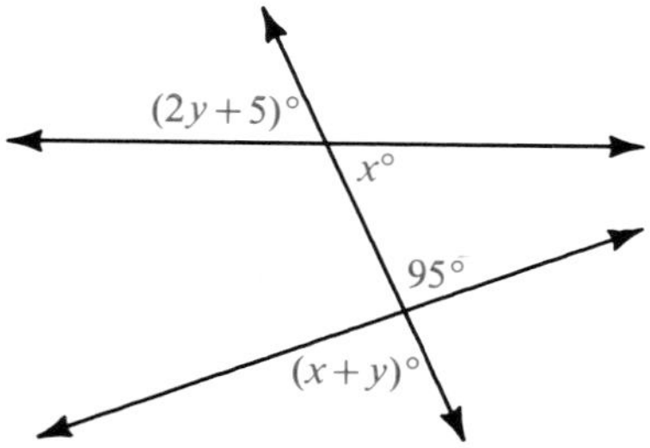

38.

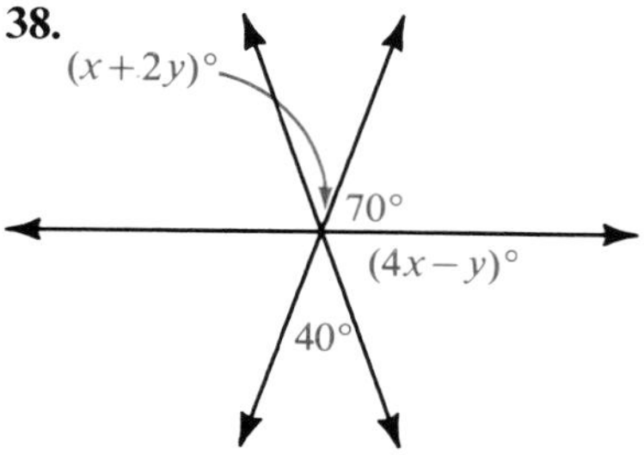

39. Explain why the measure of a complement of an angle can never be exactly half the measure of a supplement of the angle.

40. Describe all angles whose measure is equal to the difference between the measure of a supplement of the angle and twice the measure of a complement of the angle.

Self-Test 2

1. Name the four kinds of reasons that may be used to justify the statements in a proof.

Write the name or the statement of the property or theorem that justifies the given statement.

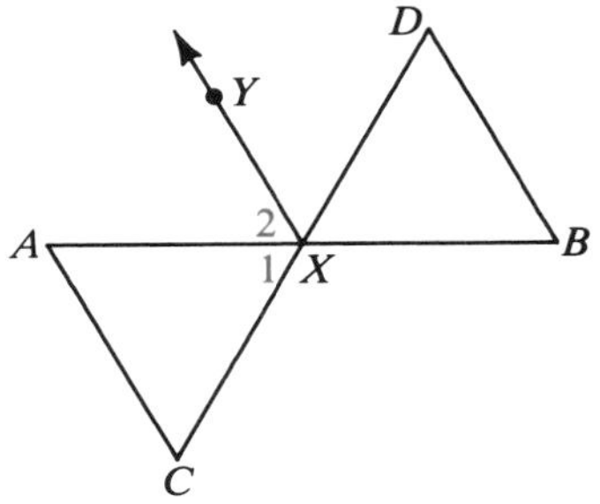

Exs. 2–7

2. If $AB = CD$ and $AX = CX$, then $AB - AX = CD - CX$.
3. If $\overrightarrow{XY}$ bisects $\angle AXD$, then $m\angle 2 = \frac{1}{2}m\angle AXD$.
4. If $\angle A \cong \angle C$ and $\angle C \cong \angle D$, then $\angle A \cong \angle D$.
5. If $AX = CX$ and $AX + XB = AB$, then $CX + XB = AB$.
6. $m\angle 1 = m\angle DXB$
7. Name two angles that are supplements of $\angle CXB$.
8. The measure of a supplement of an angle is four times as large as the measure of a complement of the angle. Find the measures of all three angles.

More about Proof

Objectives

1. State and apply the theorems about perpendicular lines, supplementary angles, and complementary angles.
2. Recognize the information conveyed by a diagram.
3. Plan and write two-column proofs.
4. Understand the relationships described in the postulates and theorems of Section 1–9.

1-7 Perpendicular Lines

Since a supplement of a right angle is a right angle, you know that if two lines intersect to form one right angle, they actually form four right angles. The photo shows the reflection of a building in a grid of window panes. Notice that where any two of the grid lines intersect, four right angles are formed.

Perpendicular lines ($\perp$ lines) are two lines that form right angles. This definition can be used in the following situations.

1. If $\overleftrightarrow{AB}$ is perpendicular to $\overleftrightarrow{CD}$ ($\overleftrightarrow{AB} \perp \overleftrightarrow{CD}$), then each numbered angle is a right angle.
2. If any of the numbered angles is a right angle, then $\overleftrightarrow{AB} \perp \overleftrightarrow{CD}$.

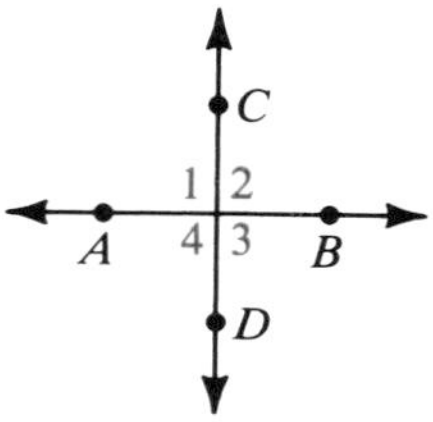

The word "perpendicular" is also used for intersecting rays and segments that are parts of perpendicular lines. For example, if $\overleftrightarrow{CD} \perp \overleftrightarrow{AB}$ in the diagram, then we can also say that $\overline{CD} \perp \overline{AB}$.

The following theorems are easily deduced by using the definition of perpendicular lines. You will complete the proofs of the theorems in Classroom Exercise 10 and Written Exercises 9 and 10.

Theorem 1-4

Adjacent angles formed by perpendicular lines are congruent.

Theorem 1-5

If two lines form congruent adjacent angles, then the lines are perpendicular.

Theorem 1-6

If the exterior sides of two adjacent acute angles are perpendicular, then the angles are complementary.

Given: $\overrightarrow{BA} \perp \overrightarrow{BD}$

Prove: $\angle ABC$ and $\angle CBD$ are comp. $\angle$s.

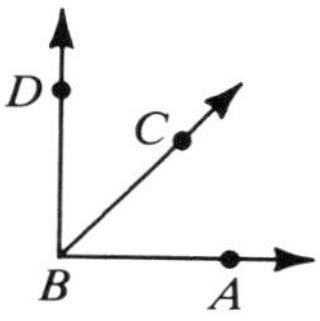

The following example shows how these theorems can be used in a proof.

Example

Given: $\angle 1 \cong \angle 2$

Prove: $\angle 3$ and $\angle 4$ are comp. $\angle$s.

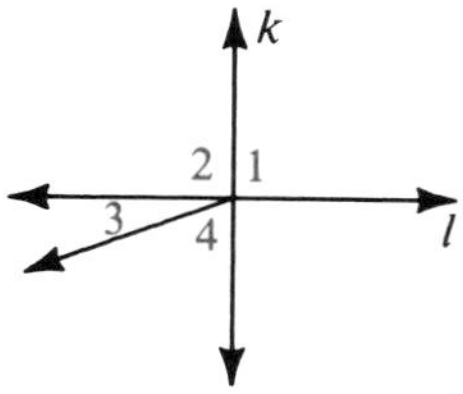

Proof:

Statements	Reasons
1. $\angle 1 \cong \angle 2$	1. Given
2. $k \perp l$	2. If 2 lines form $\cong$ adj. $\angle$s, then the lines are $\perp$.
3. $\angle 3$ and $\angle 4$ are comp. $\angle$s.	3. If the ext. sides of 2 adj. $\angle$s are $\perp$, then the $\angle$s are comp.

Classroom Exercises

In the diagram, $\overrightarrow{BE} \perp \overleftrightarrow{AC}$ and $\overrightarrow{BD} \perp \overrightarrow{BF}$. Find the measures of the following angles.

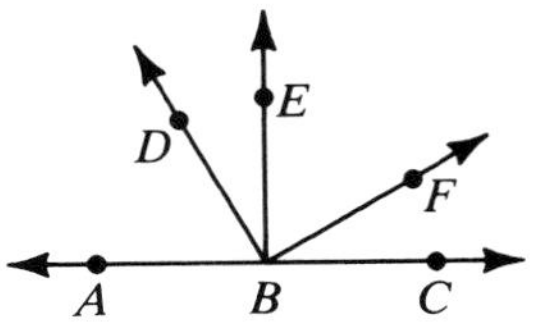

Exs. 1–8

1. **a.** $\angle ABE$ **b.** $\angle DBF$

	$m\angle CBF$	$m\angle EBF$	$m\angle DBE$	$m\angle DBA$	$m\angle DBC$
2.	40	?	?	?	?
3.	x	?	?	?	?

State the definition or theorem that justifies the statement about the diagram.

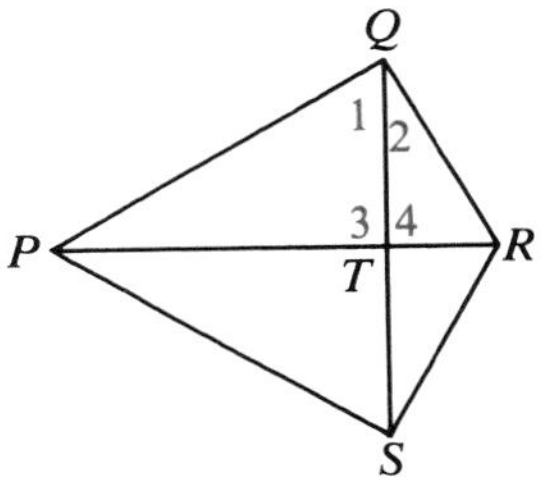

4. If $\overrightarrow{QP} \perp \overrightarrow{QR}$, then $\angle 1$ and $\angle 2$ are complementary.
5. If $\overleftrightarrow{PR} \perp \overleftrightarrow{QS}$, then $\angle 4$ is a right angle.
6. If $\angle PSR$ is a right angle, then $\overleftrightarrow{PS} \perp \overleftrightarrow{SR}$.
7. If $\angle 3 \cong \angle 4$, then $\overline{PR} \perp \overline{QS}$.
8. If $\angle 3$ is a right angle, then $m\angle 3 = 90$.
9. If $m\angle PSR = 90$, then $\angle PSR$ is a right angle.
10. Complete the proof of Theorem 1–4: Adjacent angles formed by perpendicular lines are congruent.

    Given: $a \perp b$

    Prove: $m\angle 1 = m\angle 2$

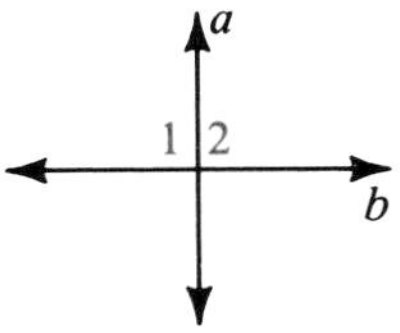

Proof:

Statements	Reasons
1. $a \perp b$	1. ___?___
2. $\angle 1$ and $\angle 2$ are right angles.	2. ___?___
3. $m\angle 1 =$ ___?___ and $m\angle 2 =$ ___?___	3. ___?___
4. ___?___	4. ___?___

Written Exercises

Write the definition or theorem that justifies the statement about the diagram.

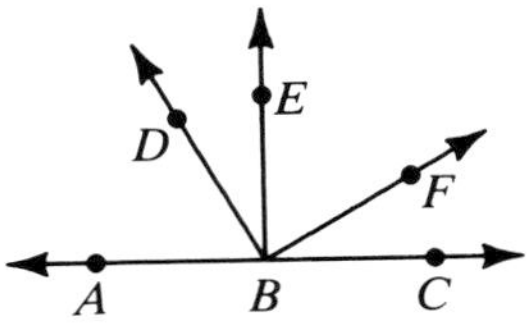

A 1. If $\angle EBC$ is a right angle, then $\overrightarrow{BE} \perp \overleftrightarrow{AC}$.

2. If $\angle ABE$ is a right angle, then $m\angle ABE = 90$.
3. If $\overrightarrow{BE} \perp \overleftrightarrow{AC}$, then $\angle ABD$ and $\angle DBE$ are complementary.
4. If $\angle ABE \cong \angle CBE$, then $\overleftrightarrow{BE} \perp \overleftrightarrow{AC}$.

Write the definition or theorem that justifies the statement about the diagram.

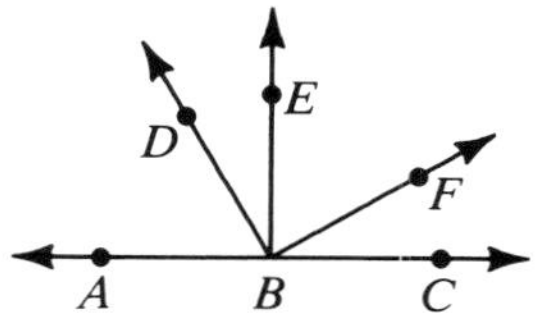

Exs. 1–8

5. If $m\angle DBF = 90$, then $\angle DBF$ is a right angle.

6. If $\overleftrightarrow{AC} \perp \overleftrightarrow{BE}$, then $m\angle ABE = m\angle CBE$.

7. If $\overleftrightarrow{AC} \perp \overleftrightarrow{BE}$, then $\angle ABE$ is a right angle.

8. If $\angle ABD$ and $\angle DBE$ are complements, then $m\angle ABD + m\angle DBE = 90$.

Copy and complete the proofs of Theorems 1–5 and 1–6.

9. If two lines form congruent adjacent angles, then the lines are perpendicular.

Given: $m\angle 1 = m\angle 2$

Prove: $a \perp b$

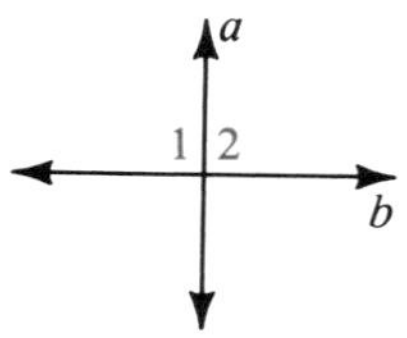

Proof:

Statements	Reasons
1. $m\angle 1 + m\angle 2 = 180$	1. ?
2. $m\angle 1 = m\angle 2$	2. ?
3. $m\angle 1 + m\angle 1 = 180$, or $2m\angle 1 = 180$	3. ?
4. ?	4. Division Prop. of =
5. $\angle 1$ is a rt. $\angle$.	5. ?
6. ?	6. ?

10. If the exterior sides of two adjacent acute angles are perpendicular, then the angles are complementary.

Given: $\overrightarrow{BA} \perp \overrightarrow{BD}$

Prove: $\angle ABC$ and $\angle CBD$ are comp. $\angle$s.

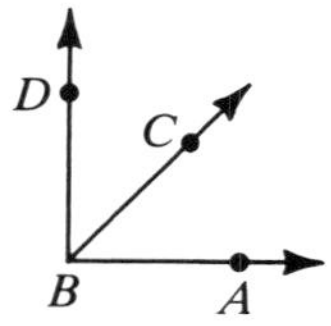

Proof:

Statements	Reasons
1. $\overrightarrow{BA} \perp \overrightarrow{BD}$	1. ?
2. ?	2. Def. of $\perp$ lines
3. $m\angle ABD = 90$	3. ?
4. $m\angle ABD = m\angle ABC + m\angle CBD$	4. ?
5. ?	5. Substitution Prop.
6. ?	6. Def. of comp. $\angle$s

Copy everything shown and write a *short* two-column proof, using a theorem stated in this section.

B **11.** Given: $\overline{SW} \perp \overline{RT}$
Prove: $m\angle 1 = m\angle 2$

12. Given: $m\angle 1 = m\angle 2$
Prove: $\overline{SW} \perp \overline{RT}$

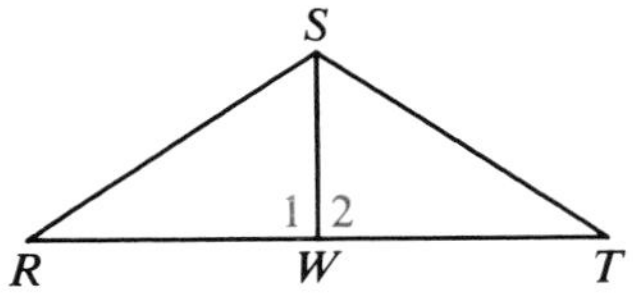

In the figure, $\overleftrightarrow{BF} \perp \overleftrightarrow{AE}$, $m\angle BOC = x$, and $m\angle HOG = y$. Express the measure of the angle in terms of x, y, or both.

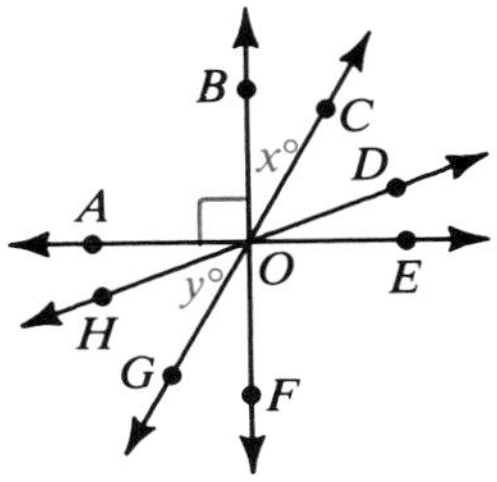

13. $\angle COA$

14. $\angle COH$

15. $\angle HOF$

16. $\angle DOE$

Can you conclude from the given information that $\overrightarrow{BA} \perp \overrightarrow{BC}$?

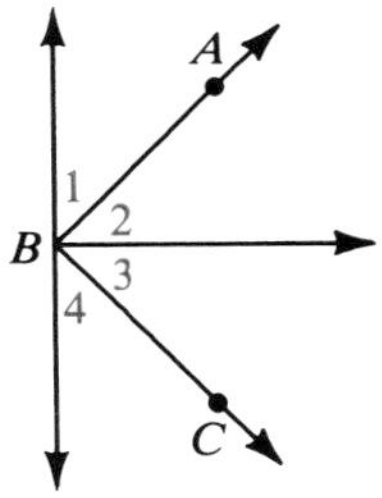

17. $m\angle 1 = 46$ and $m\angle 4 = 44$

18. $\angle 1$ and $\angle 3$ are complementary.

19. $\angle 2 \cong \angle 3$

20. $m\angle 1 = m\angle 4$

21. $\angle 1$ and $\angle 3$ are congruent and complementary.

22. $m\angle 1 = m\angle 2$ and $m\angle 3 = m\angle 4$

23. $\angle 1 \cong \angle 3$ and $\angle 2 \cong \angle 4$

24. $\angle 1 \cong \angle 4$ and $\angle 2 \cong \angle 3$

What can you conclude from the given information?

25. Given: $\overrightarrow{SX} \perp \overrightarrow{SY}$

26. Given: $\angle RSX$ and $\angle YST$ are comp. $\angle$s.

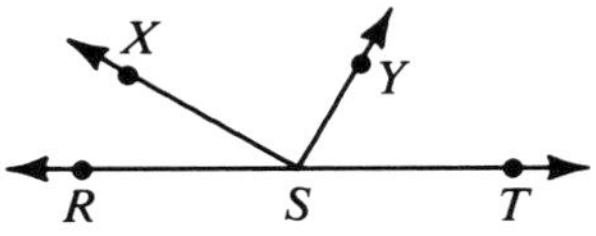

27. Given: $\overrightarrow{AB}$ bisects $\angle DAC$;
$\overrightarrow{CB}$ bisects $\angle ECA$;
$m\angle 2 = 45$;
$m\angle 3 = 45$

28. Given: $\overrightarrow{AD} \perp \overleftrightarrow{AC}$; $\overrightarrow{CE} \perp \overleftrightarrow{AC}$; $m\angle 1 = m\angle 4$

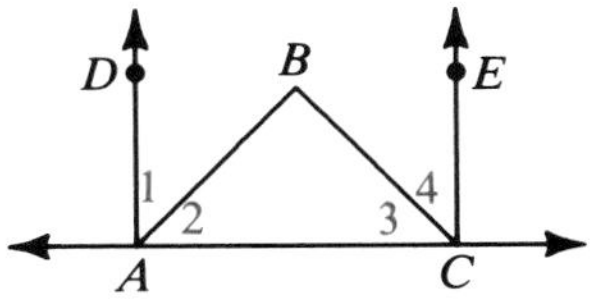

Copy everything shown and write a two-column proof.

C **29.** Given: $\angle 1$ and $\angle 2$ are comp. $\angle$s.
Prove: $\overleftrightarrow{AO} \perp \overleftrightarrow{CO}$

30. Given: $\overleftrightarrow{AO} \perp \overleftrightarrow{CO}$
Prove: $\angle 1$ and $\angle 3$ are comp. $\angle$s.

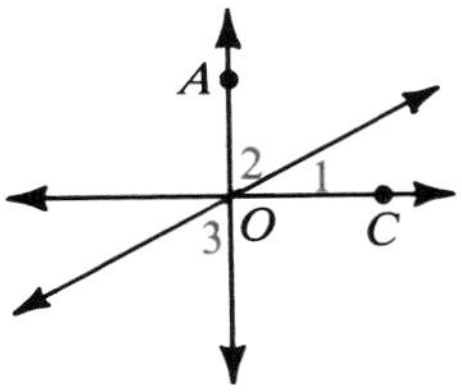

1-8 Planning a Proof

As you have seen in the last few sections, a proof of a theorem consists of five parts:

1. *Statement* of the theorem
2. A *diagram* that illustrates the given information
3. A list, in terms of the figure, of what is *given*
4. A list, in terms of the figure, of what you are to *prove*
5. A series of *statements and reasons* that lead from the given information to the statement that is to be proved

In many of the proofs in this book, the diagram and the statements of what is given and what is to be proved will be supplied for you. Sometimes you will be asked to provide them.

When you draw a diagram, try to make it reasonably accurate, avoiding special cases that might mislead. For example, when a theorem refers to *an* angle, don't draw a *right* angle.

You may use information provided by the diagrams of your text. For example, the diagram at the right tells you that:

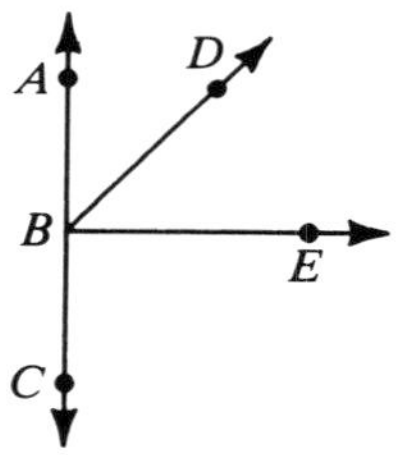

All points shown are coplanar.
$\overleftrightarrow{AB}$, $\overrightarrow{BD}$, and $\overrightarrow{BE}$ intersect at B.
A, B, and C are collinear.
B is between A and C.
$\angle ABC$ is a straight angle.
D is in the interior of $\angle ABE$.
$\angle ABD$ and $\angle DBE$ are adjacent angles.

The diagram above does *not* tell you that $\overline{AB} \cong \overline{BC}$, that $\angle ABD \cong \angle DBE$, or that $\overrightarrow{BE} \perp \overleftrightarrow{AC}$. This additional information is indicated by the marks in the diagram at the right.

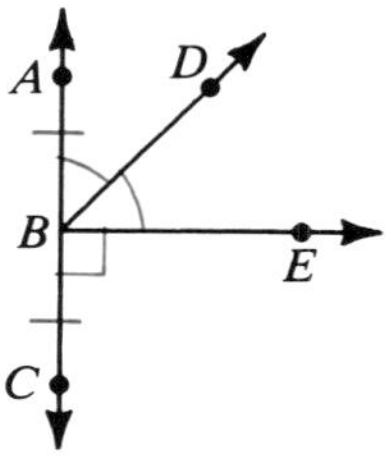

Before you write the steps in a two-column proof you will need to plan your proof. Some theorems are such that as soon as you see the theorem, you know how to prove it. Other times you may need to work out your plan for proof on scratch paper.

Unless you discover a method of proof at once, try reasoning back from what you would like to prove. Think: "This conclusion will be true if __?__ is true. This, in turn, will be true if __?__ is true. . . ." Sometimes this procedure leads back to a given statement. If so, you have found a method of proof.

Studying the proofs of previous theorems may suggest methods to try. For example, the proof of the theorem that vertical angles are congruent suggests the proof of the following theorem.

Theorem 1–7

If two angles are supplements of congruent angles (or of the same angle), then the two angles are congruent.

Given: $\angle 1$ and $\angle 2$ are supplementary;
$\angle 3$ and $\angle 4$ are supplementary;
$\angle 2 \cong \angle 4$ (or $m\angle 2 = m\angle 4$)

Prove: $\angle 1 \cong \angle 3$ (or $m\angle 1 = m\angle 3$)

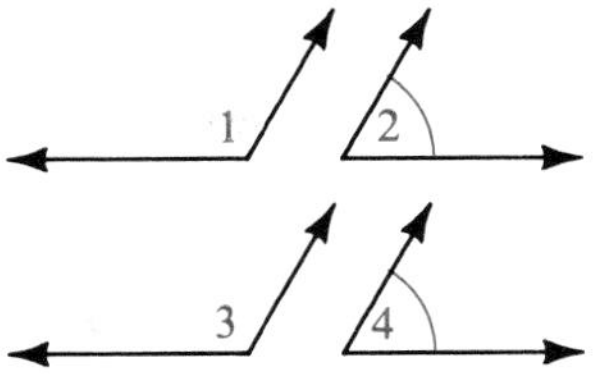

Proof:

Statements	Reasons
1. $\angle 1$ and $\angle 2$ are supplementary; $\angle 3$ and $\angle 4$ are supplementary.	1. Given
2. $m\angle 1 + m\angle 2 = 180$; $m\angle 3 + m\angle 4 = 180$	2. Def. of supp. $\angle$s
3. $m\angle 1 + m\angle 2 = m\angle 3 + m\angle 4$	3. Substitution Prop.
4. $m\angle 2 = m\angle 4$	4. Given
5. $m\angle 1 = m\angle 3$	5. Subtraction Prop. of =

The proof of the following theorem is left as Exercise 23.

Theorem 1–8

If two angles are complements of congruent angles (or of the same angle), then the two angles are congruent.

There is often more than one way to prove a particular statement, and the amount of detail one includes in a proof may differ from person to person. You should show enough steps so the reader can follow your argument and see why the theorem you are proving is true. As you gain more experience in writing proofs, you and your teacher may agree on what steps may be combined or omitted. For example, suppose we are given $\overrightarrow{BA} \perp \overrightarrow{BD}$. Then instead of writing the three steps shown at the left below, we will write the two steps shown at the right.

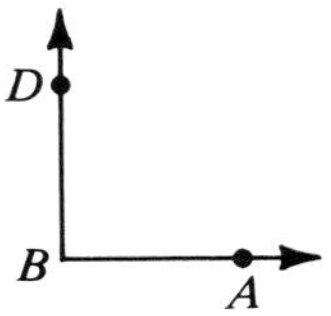

1. $\overrightarrow{BA} \perp \overrightarrow{BD}$	1. Given
2. $\angle ABD$ is a rt. $\angle$.	2. Def. of $\perp$ lines
3. $m\angle ABD = 90$	3. Def. of rt. $\angle$

1. $\overrightarrow{BA} \perp \overrightarrow{BD}$	1. Given
2. $m\angle ABD = 90$	2. Defs. of $\perp$ lines and rt. $\angle$

In proofs applying the Addition Property of Equality or the Subtraction Property of Equality, a step involving the Reflexive Property is often omitted. For example, Step 3 in the proof of Theorem 1–3 on page 31 could be omitted.

Classroom Exercises

Given the figure, state whether you can reach the conclusion shown.

1. $m\angle FOB = 50$ **2.** $m\angle AOC = 90$ **3.** $m\angle DOC = 180$

4. $AO = OB$ **5.** $\angle AOC \cong \angle BOC$ **6.** $\overleftrightarrow{AB} \perp \overleftrightarrow{CD}$

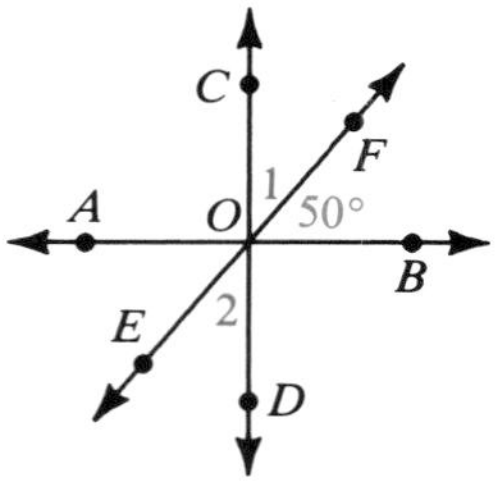

Exs. 1–12

7. Points E, O, and F are collinear.

8. Point C is in the interior of $\angle AOF$.

9. $\angle AOE$ and $\angle AOD$ are adjacent angles.

10. $\overrightarrow{OA}$ and $\overrightarrow{OB}$ are opposite rays.

11. O is between A and B.

12. $\angle 1$ and $\angle 2$ are vertical angles.

13. Given: $\angle 1$ is a supplement of $\angle 2$;
$\angle 2$ is a supplement of $\angle 3$.
State the theorem that allows you to conclude that $\angle 1 \cong \angle 3$.

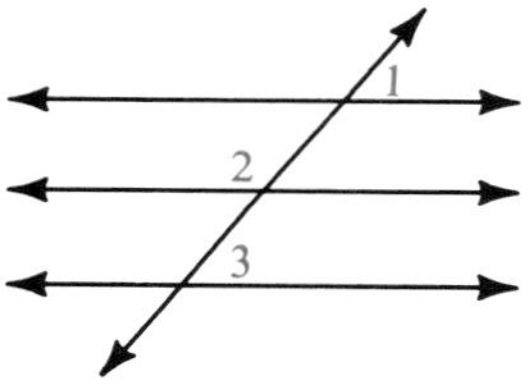

14. Given: $\angle 4$ is a supplement of $\angle 5$;
$\angle 6$ is a supplement of $\angle 7$;
$\angle 5 \cong \angle 7$
State the theorem that allows you to conclude that $\angle 4 \cong \angle 6$.

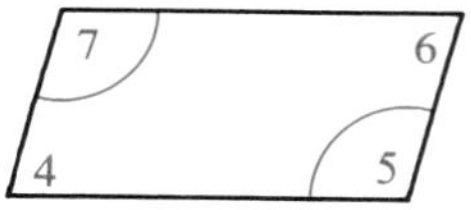

What can you deduce from the given information? State the definitions, postulates, and theorems that justify your deduction.

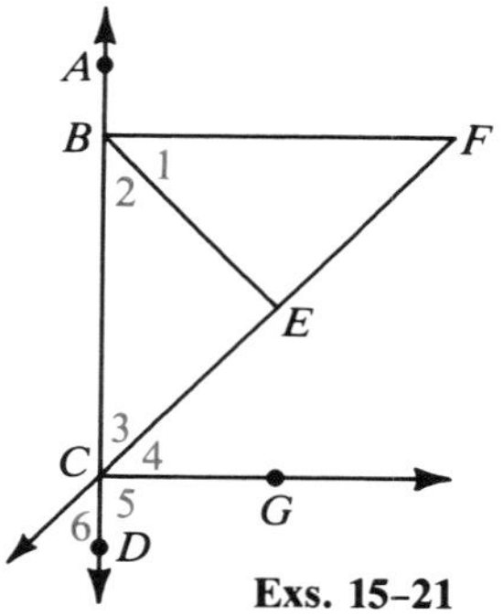

Exs. 15–21

15. Given: $m\angle 1 = m\angle 4$; $m\angle 2 = m\angle 3$

16. Given: $AB = CD$

17. Given: $m\angle 6 = m\angle 4$

18. Given: $\overline{FB} \perp \overline{AD}$; $\overrightarrow{BE}$ bisects $\angle FBC$.

19. Given: $BE = EF$; E is the midpoint of $\overline{FC}$.

20. Given: $\angle 1$ and $\angle 2$ are complements.

21. Given: $\angle 1$ and $\angle 3$ are complements; $\overline{GC} \perp \overline{AD}$

Describe your plan for proving the following.

22. Given: $\overline{AC} \perp \overline{BC}$;
$\angle 3$ is comp. to $\angle 1$,
Prove: $\angle 3 \cong \angle 2$

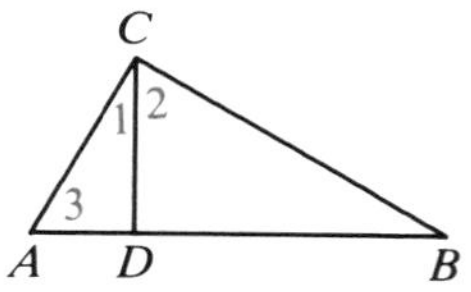

23. Given: $\angle 2 \cong \angle 3$;
$\angle 4 \cong \angle 5$
Prove: $\angle 1$ is supp. to $\angle 6$.

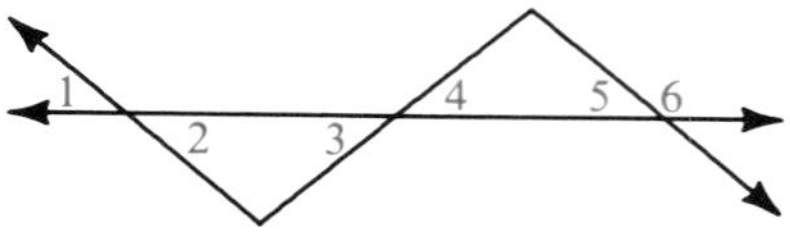

24. Refer to the proof of Theorem 1–7 on page 41 and consider the case when the two angles are supplementary to the same angle.
a. Draw a figure, and state what is given and what is to be proved.
b. Describe how you would change the proof on page 41 to prove the conclusion in part (a).

Written Exercises

A

1. **a.** Name a supplement of $\angle 2$.
b. Name a supplement of $\angle 3$.
c. What postulate or theorem, along with the definition of supplementary angles, justifies your answers to parts (a) and (b)?

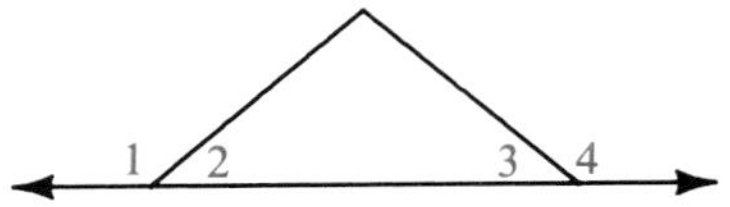

d. If $\angle 2 \cong \angle 3$, write the theorem that allows you to conclude that $\angle 1 \cong \angle 4$.

2. In the diagram, $\overline{LM} \perp \overline{MN}$ and $\overline{KN} \perp \overline{MN}$.
a. Name a complement of $\angle 2$.
b. Name a complement of $\angle 3$.
c. Write the theorem that justifies your answers to parts (a) and (b).
d. If $\angle 2 \cong \angle 3$, write the theorem that allows you to conclude that $\angle 1 \cong \angle 4$.

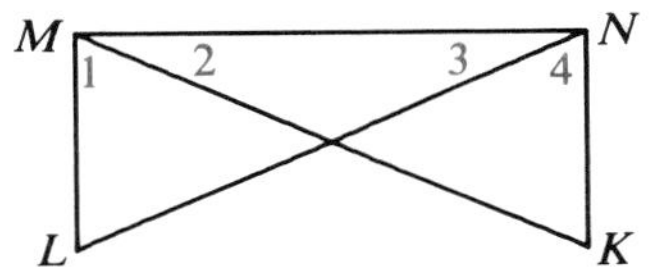

Write the name or statement of the definition, postulate, property, or theorem that justifies the statement about the diagram.

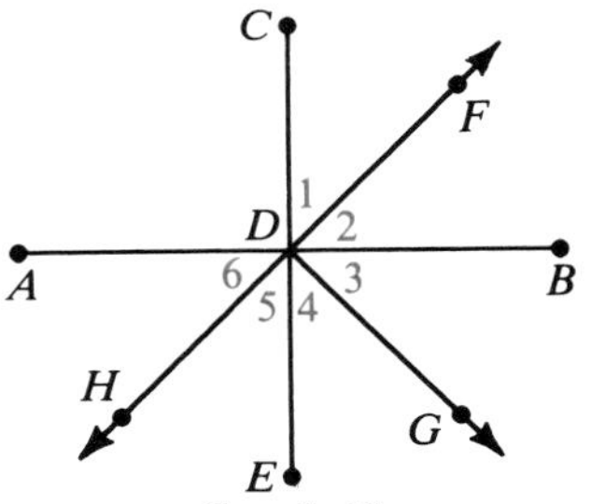

Exs. 3–18

3. $AD + DB = AB$

4. $m\angle 1 + m\angle 2 = m\angle CDB$

5. If $AD = DB$ and $CD = DE$, then $AD + CD = DB + DE$.

6. $\angle 2 \cong \angle 6$

7. If $\overrightarrow{DF}$ bisects $\angle CDB$, then $m\angle 1 = m\angle 2$.

8. If D is the midpoint of $\overline{AB}$, then $AD = DB$.

9. If $\overline{CD} \perp \overline{AB}$, then $m\angle CDB = 90$.

10. $m\angle ADF + m\angle FDB = 180$

11. If $m\angle 3 + m\angle 4 = 90$, then $\angle 3$ and $\angle 4$ are complements.

12. If $\angle ADF$ and $\angle 4$ are supplements, then $m\angle ADF + m\angle 4 = 180$.

13. If D is the midpoint of $\overline{CE}$, then $CE = 2 \cdot DE$.

14. If $m\angle 4 = m\angle 3$, then $\overrightarrow{DG}$ bisects $\angle BDE$.

15. If $\overline{AB} \perp \overline{CE}$, then $\angle ADC \cong \angle ADE$.

16. If $\angle 4$ is complementary to $\angle 5$ and $\angle 6$ is complementary to $\angle 5$, then $\angle 4 \cong \angle 6$.

17. If $m\angle FDG = 90$, then $\overrightarrow{DF} \perp \overrightarrow{DG}$.

18. If $m\angle FDG = m\angle GDH$, then $\overrightarrow{DG} \perp \overleftrightarrow{HF}$.

19. a. Complete the proof.

Given: $\overline{PQ} \perp \overline{QR}$;
$\overline{PS} \perp \overline{SR}$;
$\angle 1 \cong \angle 4$

Prove: $\angle 2 \cong \angle 5$

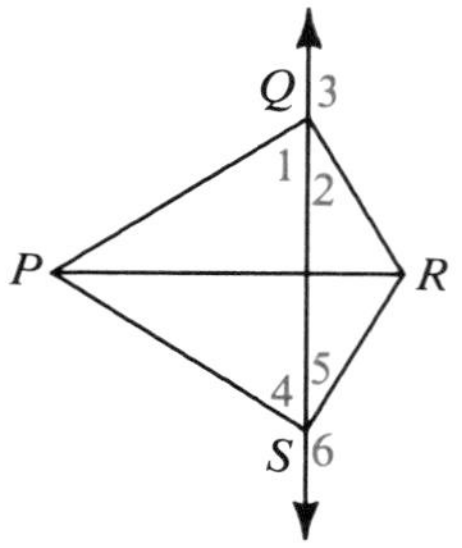

Proof:

Statements	Reasons
1. $\overline{PQ} \perp \overline{QR}$; $\overline{PS} \perp \overline{SR}$	1. ?
2. $\angle 2$ is comp. to $\angle 1$; $\angle 5$ is comp. to $\angle 4$.	2. ?
3. $\angle 1 \cong \angle 4$	3. ?
4. $\angle 2 \cong \angle 5$	4. ?

b. Now that you have proved that $\angle 2 \cong \angle 5$, describe a plan for proving that $\angle 3 \cong \angle 6$.

20. a. Are there any angles in the diagram that must be congruent to $\angle 4$? Explain.

b. If $\angle 4$ and $\angle 5$ are supplementary, name all angles shown that must be congruent to $\angle 4$.

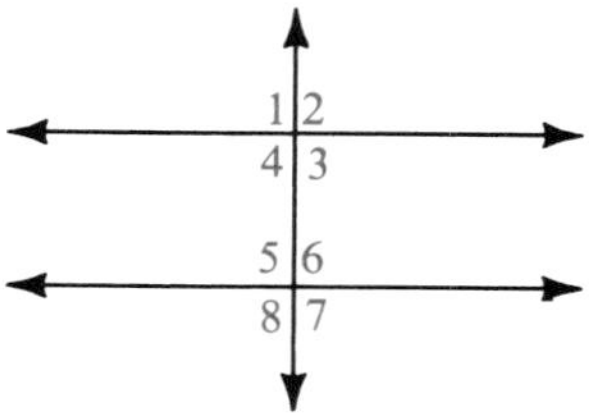

Copy everything shown and write a two-column proof.

B **21.** Given: $\angle 2 \cong \angle 3$
Prove: $\angle 1 \cong \angle 4$

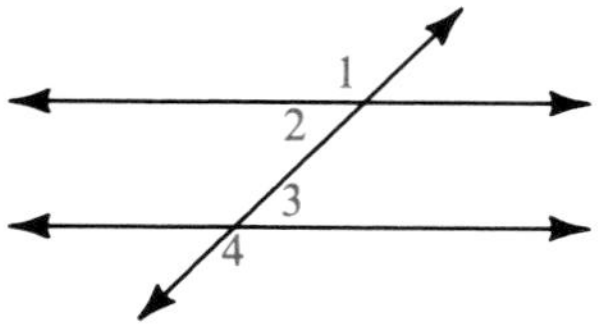

22. Given: $\overline{AC} \perp \overline{BC}$
$\angle 3$ is comp. to $\angle 1$.
Prove: $\angle 3 \cong \angle 2$

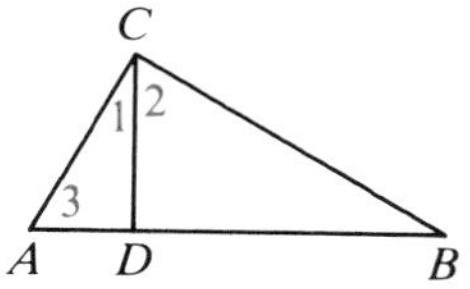

23. Prove Theorem 1–8: If two angles are complements of congruent angles, then the two angles are congruent. (*Hint:* See the proof of Theorem 1–7 on page 41.)

24. Write a two-column proof of Theorem 1–3 that is different from the one on page 31.

25. Given: $\overline{RS} \perp \overline{ST}$;
$\angle 1$ and $\angle 4$ are comp. $\angle$s;
$\angle 2$ and $\angle 3$ are comp. $\angle$s.

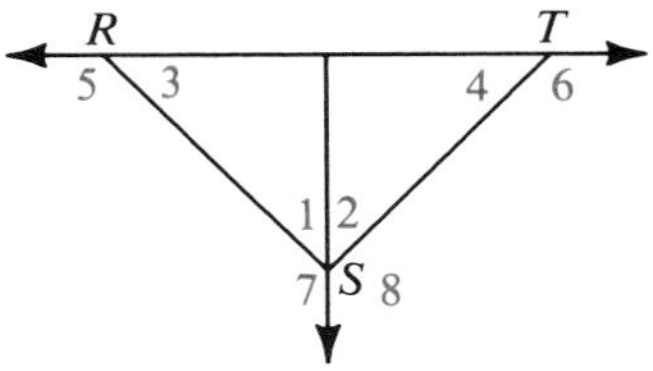

Complete:

a. $\angle 1 \cong \angle$ _?_ **b.** $\angle 2 \cong \angle$ _?_

c. $\angle 5 \cong \angle$ _?_ **d.** $\angle 6 \cong \angle$ _?_

Copy everything shown and write a two-column proof.

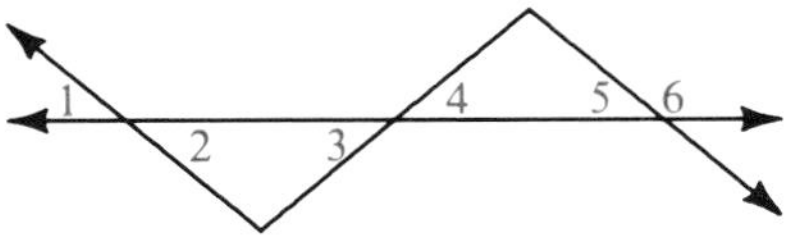

26. Given: $\angle 4$ is supp. to $\angle 6$.
Prove: $\angle 3 \cong \angle 5$

27. Given: $\angle 2 \cong \angle 3$;
$\angle 4 \cong \angle 5$
Prove: $\angle 1$ is supp. to $\angle 6$.

28. Given: $m\angle 1 = m\angle 2$; $m\angle 3 = m\angle 4$
Prove: $\overrightarrow{OC} \perp \overleftrightarrow{AE}$

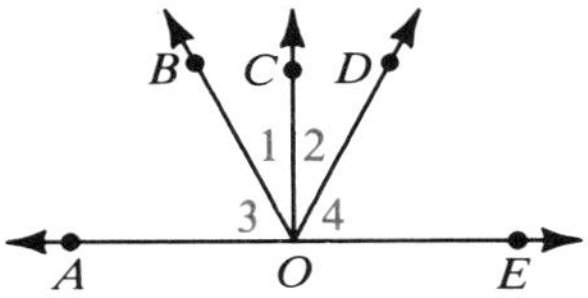

29. Given: $\overrightarrow{OC} \perp \overleftrightarrow{AE}$; $\overrightarrow{OC}$ bisects $\angle BOD$.
Prove: $m\angle 3 = m\angle 4$

C 30. Draw any $\angle AOB$ and its bisector $\overrightarrow{OX}$. Now draw the rays opposite to $\overrightarrow{OA}$, $\overrightarrow{OB}$, and $\overrightarrow{OX}$. What can you conclude? Prove it.

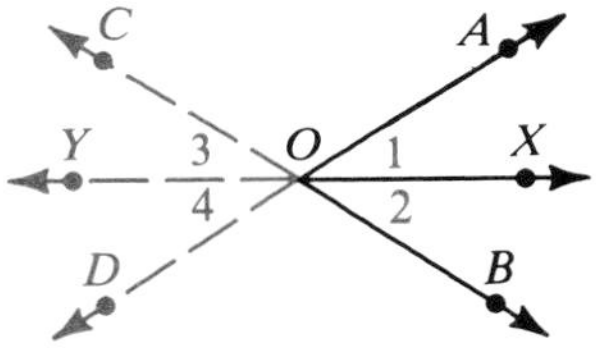

31. Make a diagram showing $\angle PQR$ bisected by $\overrightarrow{QX}$. Choose a point Y on the ray opposite $\overrightarrow{QX}$.
Prove: $\angle PQY \cong \angle RQY$

32. Given: $m\angle DBA = 45$;
$m\angle DEB = 45$
Prove: $\angle DBC \cong \angle FEB$

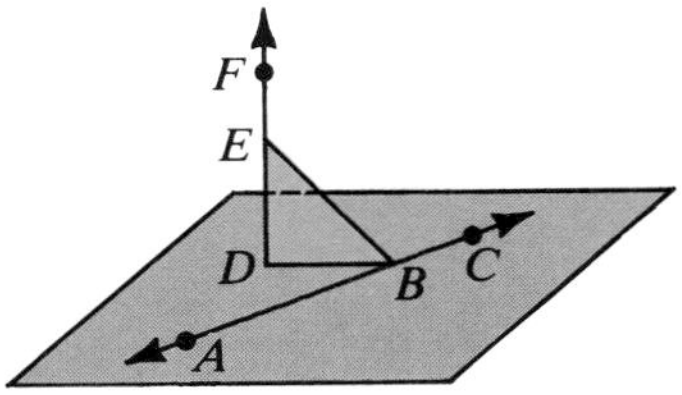

1-9 Postulates Relating Points, Lines, and Planes

Our first postulates and theorems have dealt primarily with segments and their lengths and with angles and their measures. We will be able to prove many more things about geometric figures after we make the following basic assumptions about relationships between points, lines, and planes.

Postulate 5

A line contains at least two points; a plane contains at least three points not all in one line; space contains at least four points not all in one plane.

Postulate 6

Through any two points there is exactly one line.

Postulate 7

Through any three points there is at least one plane, and through any three noncollinear points there is exactly one plane.

Postulate 8

If two points are in a plane, then the line that contains the points is in that plane.

Postulate 9

If two planes intersect, then their intersection is a line.

These postulates can be used to prove the following theorems. For example, Theorem 1-9 follows from Postulate 6.

Theorem 1-9

If two lines intersect, then they intersect in exactly one point.

Theorem 1-10

If there is a line and a point not in the line, then exactly one plane contains them.

Theorem 1-11

If two lines intersect, then exactly one plane contains them.

The phrase "exactly one" appears several times in the postulates and theorems of this section. The phrase "one and only one" has the same meaning. For example, here are two correct forms of Theorem 1-11:

If two lines intersect, then *exactly one* plane contains them.
If two lines intersect, then *one and only one* plane contains them.

Either statement is called an *existence and uniqueness statement*. The theorem states that (1) the plane described exists and (2) it is unique (no more than one such plane exists).

Classroom Exercises

1. The diagram suggests what would happen if we tried to show two "lines" drawn through two points. State the postulate that makes this situation impossible.

2. Plane *M* and plane *N* both contain point *P*.
   a. Do the planes have any other points in common?
   b. State the postulates that answer this question.
3. A carpenter checks to see if a board is warped by laying a straightedge across the board in several directions. State the postulate that is related to this procedure.
4. State how many planes contain the given figure.
   a. a triangle
   b. an obtuse angle
   c. a straight angle
   d. a ball
5. Reword the following statement as two statements, one describing existence and the other describing uniqueness:

   A segment has exactly one midpoint.
6. State the postulate that allows you to name a plane by three noncollinear points it contains.

In Exercises 7–9, you will have to visualize certain planes and lines that are not shown in the diagram. Name each plane by four points it contains, no three of which are collinear.

7. Where does plane *DCFE* intersect plane *BCGF*?
8. Name three planes that intersect in $\overleftrightarrow{BF}$.
9. Name five planes that contain point *A*.
10. When a three-dimensional object is represented by a two-dimensional drawing, it is usually drawn in perspective. Notice that ∠*BFG* appears to be an acute angle in the diagram, but it represents a right angle in a box. Using this as an example, complete the table.

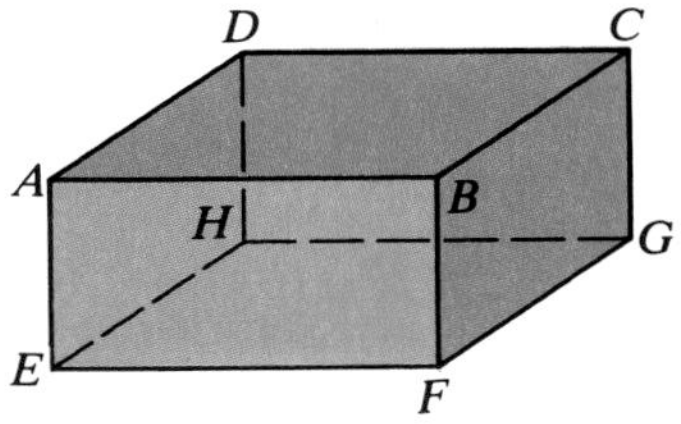

Exs. 7–11

	∠*BFG*	∠*FBC*	∠*ABC*	∠*EFB*	∠*EFC*
In the diagram	acute	?	?	?	?
In the box	right	?	?	?	?

11. a. Name three lines that do not intersect $\overleftrightarrow{AE}$.
    b. Name a line that does not intersect $\overleftrightarrow{DB}$.
    c. Name three segments perpendicular to $\overline{CG}$ at point *C*.
    d. How many lines in plane *BCGF* are perpendicular to $\overline{CG}$ at point *C*?

Written Exercises

Complete with *always, sometimes,* or *never.*

A 1. Two points __?__ lie in exactly one line.

2. Three points __?__ lie in exactly one line.

3. Three points __?__ lie in exactly one plane.

4. Three collinear points __?__ lie in exactly one plane.

5. Two planes __?__ intersect.

6. Two intersecting planes __?__ intersect in exactly one point.

7. Two intersecting lines __?__ intersect in exactly one point.

8. Two lines __?__ intersect in exactly one point.

9. Two intersecting lines __?__ lie in exactly one plane.

10. A line and a point not on that line __?__ lie in more than one plane.

11. A line __?__ contains exactly one point.

12. When A and B are in a plane, $\overleftrightarrow{AB}$ is __?__ in that plane.

B 13. Rewrite the following statement as two statements, one describing existence and the other describing uniqueness:

An angle has exactly one bisector.

14. Draw a figure for Theorem 1-9 and state in terms of the figure what is given and what is to be proved. Do not complete the proof.

15. Suppose R, S, and T are three noncollinear points.

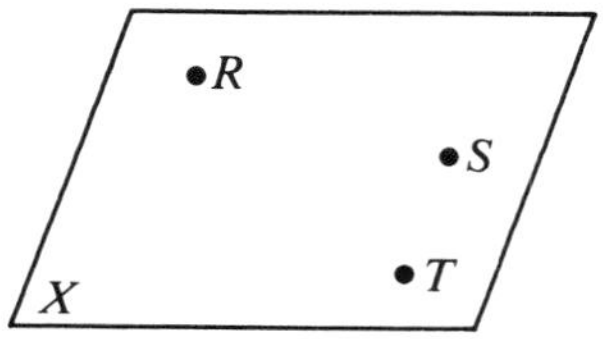

a. State the postulate that guarantees the existence of a plane X containing R, S, and T.

b. State the postulate that guarantees that any point P on $\overline{RS}$ is in plane X.

c. State the postulate that guarantees that $\overleftrightarrow{TP}$ exists.

d. State the postulate that guarantees that $\overleftrightarrow{TP}$ is in plane X.

16. Suppose A, B, C, and D are four noncoplanar points.

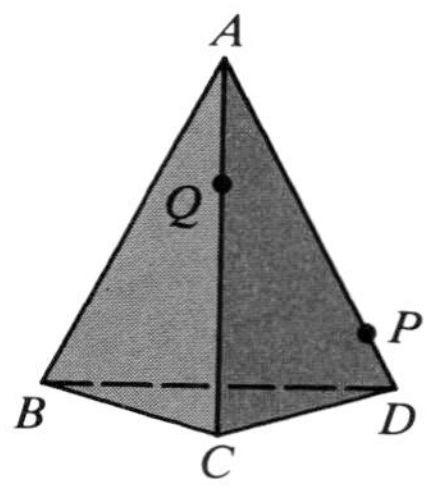

a. State the postulate that guarantees the existence of planes ABC, ABD, ACD, and BCD.

b. Explain how the Ruler Postulate guarantees the existence of a point P between A and D and a point Q between A and C.

c. State the postulate that guarantees the existence of plane BPQ.

d. Explain why there are an infinite number of planes through point P.

C 17. State how many segments can be drawn between the points in each figure. No three points are collinear.

a. 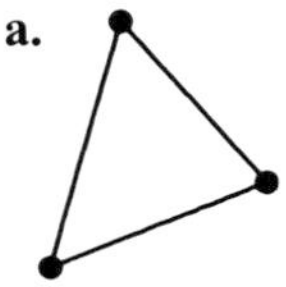

3 points
_?_ segments

b. 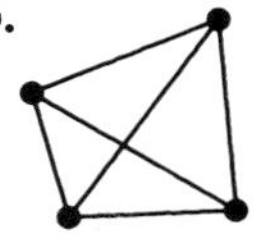

4 points
_?_ segments

c. 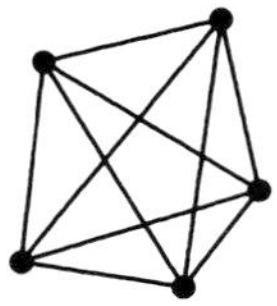

5 points
_?_ segments

d. 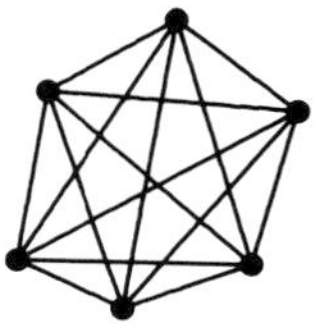

6 points
_?_ segments

e. Without making a drawing, predict how many segments can be drawn between seven points, no three of which are collinear.

f. How many segments can be drawn between n points, no three of which are collinear?

Self-Test 3

In the diagram, $\overline{BD} \perp \overline{AC}$ and $\angle 1 \cong \angle 4$. Write the definition or theorem that justifies the conclusion.

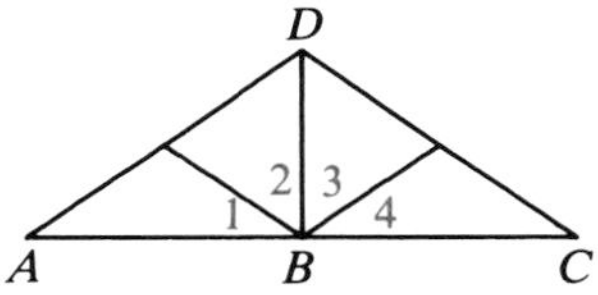

1. $\angle ABD \cong \angle CBD$
2. $\angle DBC$ is a right angle.
3. $\angle 3$ and $\angle 4$ are complements.
4. Complete: If two lines form congruent adjacent angles, then _?_.
5. Name the five parts of a proof.
6. If planes P and Q intersect, what is their intersection?
7. If points A and B lie in plane M, what do you know about $\overleftrightarrow{AB}$?
8. Complete: Through points X, Y, and Z, there is _?_ one plane.
9. Complete in three different ways: _?_ are contained in exactly one plane.
10. Explain a plan for the following proof.

Given: $\overrightarrow{OC} \perp \overrightarrow{AE}$;
$\overrightarrow{OB}$ bisects $\angle AOC$;
$\overrightarrow{OD}$ bisects $\angle COE$

Prove: $\angle AOB \cong \angle DOE$

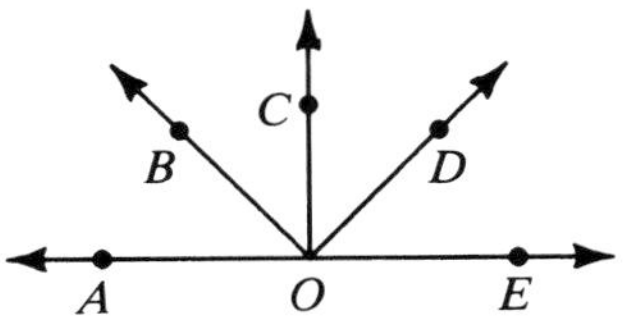

Chapter Summary

1. The concepts of *point, line,* and *plane* are basic to geometry. These undefined terms are used in the definitions of other terms.
2. $\overleftrightarrow{AB}$ represents a line, $\overline{AB}$ a segment, and $\overrightarrow{AB}$ a ray. AB represents the length of $\overline{AB}$; AB is a positive number.
3. Two rays with the same endpoint form an angle.

4. Congruent segments have equal lengths. Congruent angles have equal measures.
5. Angles are classified as acute, right, obtuse, or straight, according to their measures.
6. Properties of equality and congruence (p. 17) are used to reach conclusions about segments and their lengths and about angles and their measures.
7. Statements that are accepted without proof are called postulates. You should understand the Ruler Postulate, the Segment Addition Postulate, the Protractor Postulate, and the Angle Addition Postulate.
8. Statements that are proved are called theorems.
9. Deductive reasoning is a process of logical reasoning from accepted statements (given information, definitions, postulates, and previously proved theorems) to a conclusion.
10. Perpendicular lines are two lines that form right angles.
11. Two angles are congruent if they are:
    a. vertical angles
    b. adjacent angles formed by perpendicular lines
    c. supplements of congruent angles or of the same angle
    d. complements of congruent angles or of the same angle

Chapter Review

Sketch and label the figures described.

1. Points A, B, C, and D are coplanar, but A, B, and C are the only three of those points that are collinear. 1-1
2. Line l intersects plane X in point P.
3. Plane M contains intersecting lines j and k.

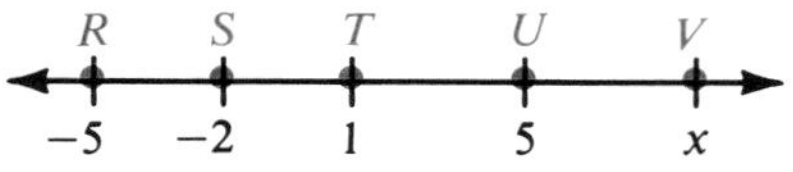

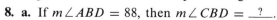

4. Name a point on $\overrightarrow{ST}$ that is not on $\overline{ST}$. 1-2
5. a. Find RS and ST.
   b. Complete: $\overline{RS}$ and $\overline{ST}$ are _?_.
6. If U is the midpoint of $\overline{TV}$, find the value of x.

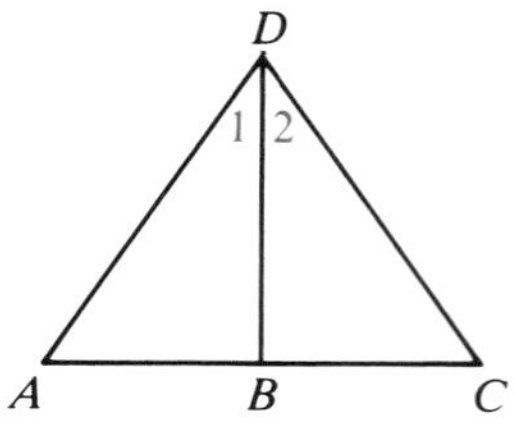

7. Name three angles that have vertex D. Which angles with vertex D are adjacent angles? 1-3
8. a. If $m\angle ABD = 88$, then $m\angle CBD =$ _?_.
   b. Name the postulate that justifies your answer in part (a).
   c. What kind of angle is $\angle CBD$?
9. $\overrightarrow{DB}$ bisects $\angle ADC$, $m\angle 1 = 5x - 3$, and $m\angle 2 = x + 25$. Find the value of x.

Justify each statement with a property from algebra or a property of congruence.

10. If $\angle A \cong \angle B$ and $\angle B \cong \angle C$, then $\angle A \cong \angle C$. 1–4

11. If $RS = XY$ and $ST = YZ$, then $RS + ST = XY + YZ$.

12. If $m\angle 1 + m\angle 2 = m\angle 3$ and $m\angle 2 = m\angle 4$, then $m\angle 1 + m\angle 4 = m\angle 3$.

Refer to the diagram for Exercises 7–9. Name the postulate, definition, or theorem that justifies the statement.

13. If B is the midpoint of $\overline{AC}$, then $AB = \frac{1}{2}AC$. 1–5

14. If $m\angle 1 = m\angle 2$, then $\overrightarrow{DB}$ bisects $\angle ADC$.

15. If $\overrightarrow{DB}$ is the bisector of $\angle ADC$, then $2m\angle 2 = m\angle ADC$.

16. Two vertical angles are complementary. Find the measure of each angle. 1–6

Exercises 17 and 18 refer to the diagram.

17. Find the measure of $\angle RXT$. State the theorem that justifies your answer.

18. Name two pairs of supplementary angles.

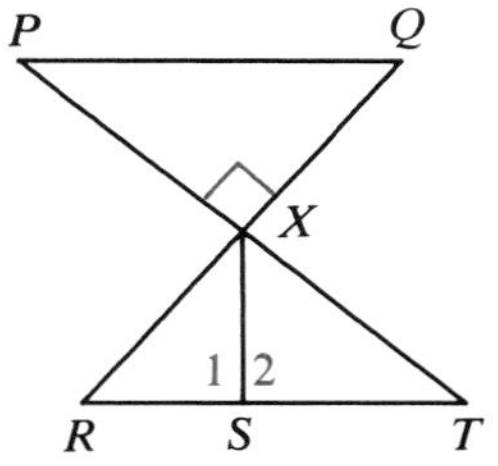

Exs. 17–20

State the definition or theorem that justifies the statement about the diagram at the right.

19. If $\overline{XS} \perp \overline{RT}$, then $\angle 1 \cong \angle 2$. 1–7

20. If $\angle 1 \cong \angle 2$, then $\overline{XS} \perp \overline{RT}$.

21. Given: $\overrightarrow{BA} \perp \overrightarrow{BC}$;
$m\angle 3 = 4t - 13$;
$m\angle 4 = 2t + 19$
Find the value of t.

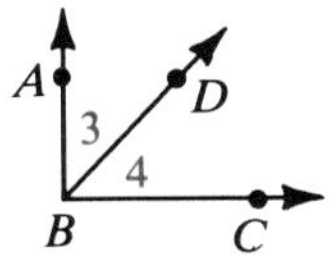

Ex. 21

What can you deduce from the given information?

22. Given: $\angle 2$ and $\angle 5$ are complements. 1–8

23. Given: $\overline{FG} \perp \overline{GH}$;
$\angle 4$ and $\angle 5$ are comp. $\angle$s.

24. Given: $\angle 1 \cong \angle 3$

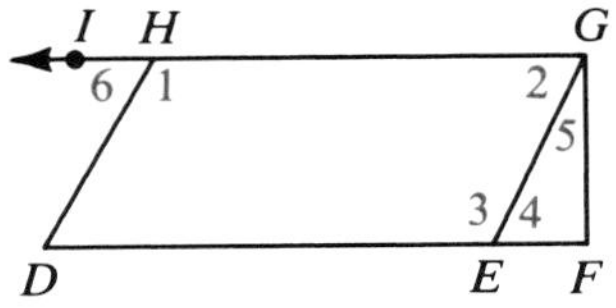

Classify each statement as true or false.

25. Through any three points there is exactly one plane. 1–9

26. Two intersecting lines may be noncoplanar.

27. Through any three points there is at least one line.

28. If points A and B lie in plane P, then so does the midpoint, M, of $\overline{AB}$.

Chapter Test

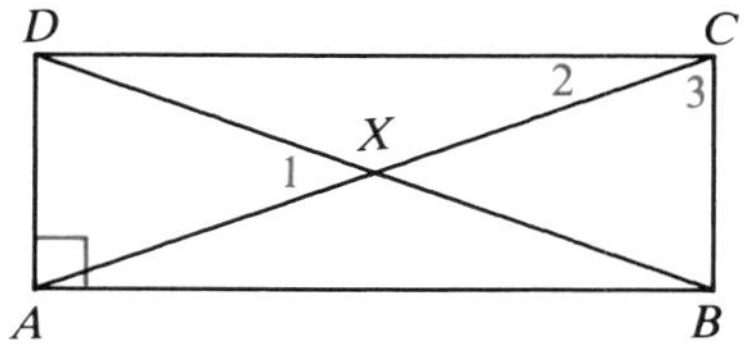

Exs. 1–12

1. Name three collinear points.
2. Name the intersection of $\overrightarrow{CX}$ and $\overleftrightarrow{AB}$.
3. Which postulate justifies the statement $AX + XC = AC$?
4. If $\overline{AC}$ bisects $\overline{BD}$, name two congruent segments.
5. Name the vertex and sides of $\angle 1$.
6. Name a right angle.
7. Name two adjacent supplementary angles.
8. Name two complementary angles.
9. Name the property that justifies the following statement:
   If $\overline{AX} \cong \overline{XC}$ and $\overline{XC} \cong \overline{XD}$, then $\overline{AX} \cong \overline{XD}$.
10. Complete: If X is the midpoint of $\overline{AC}$, then $AX = \frac{1}{2}$ _?_.
11. If $m\angle 1 = 46$, find $m\angle DXC$ and $m\angle CXB$.
12. If $m\angle DAX = 70$, find the measure of $\angle XAB$.
13. On a number line, J has coordinate -7 and K has coordinate -2. K is the midpoint of $\overline{JL}$. Find the coordinate of L.

Complete with *always, sometimes,* or *never.*

14. If two planes intersect, then their intersection is _?_ a segment.
15. A line and a point not on the line are _?_ coplanar.

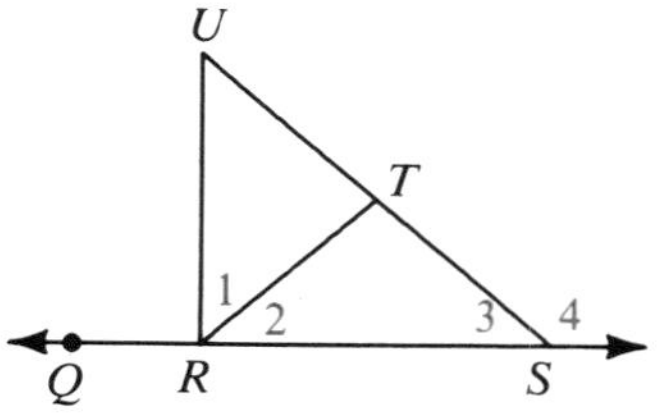

Exs. 16–18

16. If $\overrightarrow{RT}$ bisects $\angle URS$, $m\angle 1 = 6x - 30$, and $m\angle URS = 5x + 24$, find the value of x.
17. Justify each statement.
    a. $m\angle 1 + m\angle 2 = m\angle URS$
    b. If $RT = ST$, then $\frac{1}{3}RT = \frac{1}{3}ST$.
    c. If $\overleftrightarrow{RU} \perp \overleftrightarrow{QS}$, then $\angle URQ \cong \angle URS$.
18. Write a two-column proof.
    Given: $\angle 2$ and $\angle 4$ are supp. $\angle$s.
    Prove: $\angle 2 \cong \angle 3$

Algebra Review

Solve each equation.

1. $5a - 22 = 8$
2. $2z = 5z + 12$
3. $3x - 4 = 2x + 4$
4. $3(9 - t) = 5 + t$
5. $4(90 - x) = 180 - x$
6. $90 - y = 3 + 2y$
7. $(n - 2)180 = 160n$
8. $b(b - 3) = b^2 + 12$
9. $2k(k - 1) = k(2k + 1)$
10. $5(2d + 1) = 3(5d - 5)$
11. $5 + 1.6m = 1$
12. $0.3q - 8 = 6 + q$

Solve each system of equations.

Example 1 (1) $x - 3y = 6$
(2) $4x + 5y = 7$

Solution 1 Substitution method

From (1): $x = 3y + 6$
Substituting $3y + 6$ for x in (2):
$4(3y + 6) + 5y = 7$
$17y + 24 = 7;\ y = -1$
Substituting -1 for y in (1):
$x - 3(-1) = 6;\ x = 3$

Solution 2 Addition-or-subtraction method

$4 \times$ (1): $4(x - 3y) = 4(6)$
$4x - 12y = 24$
(2) $4x + 5y = 7$
Subtract: $-17y = 17;\ y = -1$
Substituting -1 for y in (1):
$x - 3(-1) = 6;\ x = 3$

Both methods lead to $x = 3,\ y = -1$.

13. $2x - 3y = -2$; $5x + 3y = 37$
14. $-4x + 7y = 2$; $4x - 5y = 10$
15. $2x - 5y = 0$; $x - 5y = 10$
16. $y = 5x - 3$; $8x - y = 9$
17. $y = x - 8$; $x - 4y = 5$
18. $4x + 3y = -9$; $2x - y = 3$
19. $5x + y = 29$; $2x - 3y = 32$
20. $x + 4y = 7$; $2x - y = -1$
21. $2x - 3y = 21$; $8x + 5y = -1$
22. $8x - 9y = 14$; $5x + 3y = 26$
23. $7x + 4y = 2$; $3x - 8y = 13$
24. $12x - 7y = -6$; $4x - 9y = -2$

Example 2 $7x - 6y = 4$ $\times 2 \Rightarrow$ $14x - 12y = 8$
$3x + 4y = -18$ $\times 3 \Rightarrow$ $9x + 12y = -54$
Add: $23x = -46;\ x = -2$

Substitute: $7(-2) - 6y = 4;\ -6y = 18;\ y = -3$
The solution is $x = -2,\ y = -3$.

25. $4x - 5y = 0$; $3x + 2y = -46$
26. $3x + 7y = 1$; $4x + 11y = 8$
27. $13x + 11y = -1$; $2x - 3y = 28$
28. $4x + 5y = -9$; $5x - 2y = 8$
29. $2x = 7(1 - y)$; $3x + 8y = -2$
30. $3(2x - 5) = y$; $5x - y = 11$

The rows of crops shown in the photograph, and the strips of earth between them, suggest parallel lines lying in the plane of the field. The roads cutting across the rows suggest transversals.

Parallel Lines and Planes

2

When Lines and Planes Are Parallel

Objectives

1. Distinguish between intersecting lines, parallel lines, and skew lines.
2. State and apply the theorem about the intersection of two parallel planes by a third plane.
3. Identify the angles formed when two lines are cut by a transversal.
4. State and apply the postulates and theorems about parallel lines.
5. State and apply the theorems about a parallel and a perpendicular to a given line through a point outside the line.

2-1 Definitions

Two lines that do not intersect are either *parallel* or *skew*.

Parallel lines (|| lines) do not intersect and are coplanar.

Skew lines do not intersect and are not coplanar.

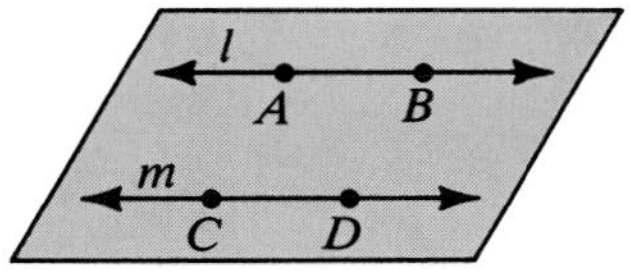

l and *m* are parallel lines.
l is parallel to *m* ($l \parallel m$).

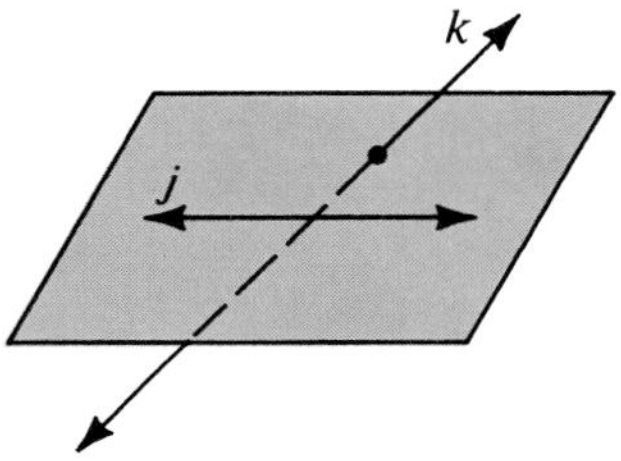

j and *k* are skew lines.

Segments and rays contained in parallel lines are also called parallel. For example, in the figure at the left above, $\overline{AB} \parallel \overline{CD}$ and $\overrightarrow{AB} \parallel \overrightarrow{CD}$.

Thinking of the top of the box pictured below as part of plane *X*, and the bottom as part of plane *Y*, may help you understand the following definitions.

Parallel planes (|| planes) do not intersect.
Plane *X* is parallel to plane *Y* ($X \parallel Y$).

A line and a plane are parallel if they do not intersect.
For example, $\overleftrightarrow{EF} \parallel Y$ and $\overleftrightarrow{FG} \parallel Y$.
Also, $\overleftrightarrow{AB} \parallel X$ and $\overleftrightarrow{BC} \parallel X$.

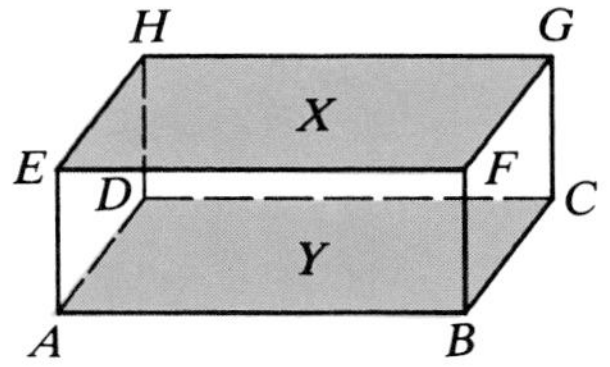

Our first theorem about parallel lines and planes is given on the next page. Notice the importance of definitions in the proof.

Theorem 2-1

If two parallel planes are cut by a third plane, then the lines of intersection are parallel.

Given: Plane $X \parallel$ plane Y;
plane Z intersects X in line l;
plane Z intersects Y in line m.

Prove: $l \parallel m$

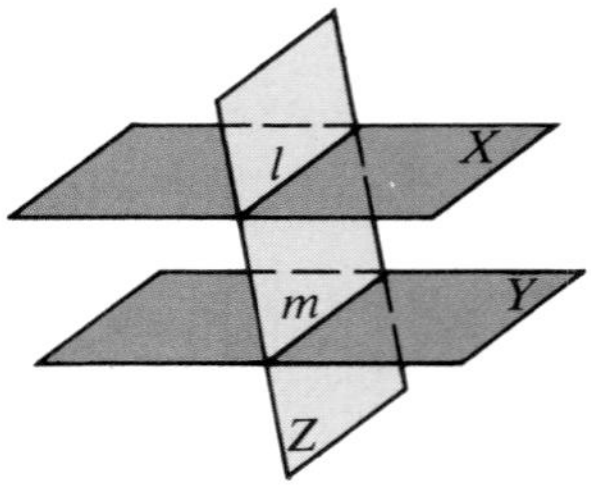

Proof:

Statements	Reasons
1. l is in X; m is in Y; $X \parallel Y$	1. Given
2. l and m do not intersect.	2. Parallel planes do not intersect. (Def. of $\parallel$ planes)
3. l is in Z; m is in Z.	3. Given
4. l and m are coplanar.	4. Def. of coplanar
5. $l \parallel m$	5. Def. of $\parallel$ lines (See steps 2 and 4.)

The following terms, which are needed for future theorems about parallel lines, apply only to coplanar lines.

A **transversal** is a line that intersects two or more coplanar lines in different points. In the next diagram, t is a transversal of h and k. The angles formed have special names.

Interior angles: angles 3, 4, 5, 6 *Exterior angles:* angles 1, 2, 7, 8

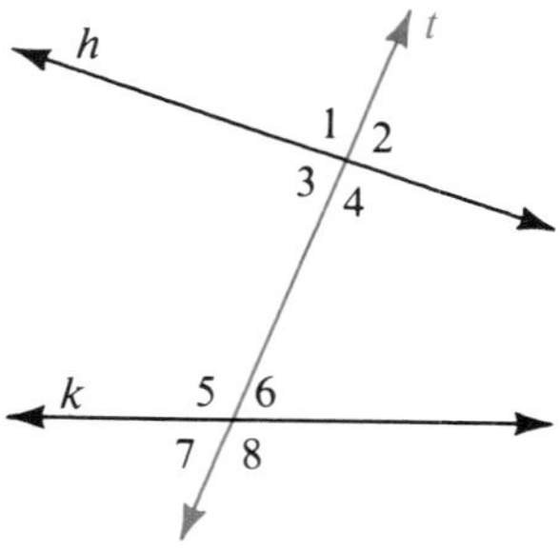

Alternate interior angles (alt. int. $\angle$s) are two nonadjacent interior angles on opposite sides of the transversal.

$\angle 3$ and $\angle 6$ $\angle 4$ and $\angle 5$

Same-side interior angles (s-s. int. $\angle$s) are two interior angles on the same side of the transversal.

$\angle 3$ and $\angle 5$ $\angle 4$ and $\angle 6$

Corresponding angles (corr. $\angle$s) are two angles in corresponding positions relative to the two lines.

$\angle 1$ and $\angle 5$ $\angle 2$ and $\angle 6$ $\angle 3$ and $\angle 7$ $\angle 4$ and $\angle 8$

Classroom Exercises

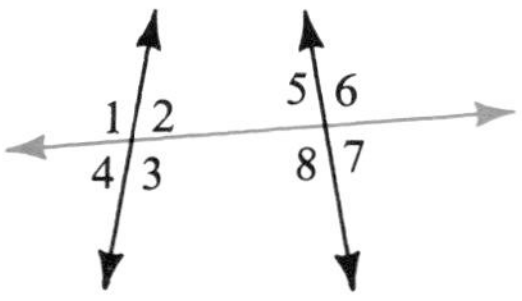

1. The blue line is a transversal.
a. Name four pairs of corresponding angles.
b. Name two pairs of alternate interior angles.
c. Name two pairs of same-side interior angles.

Classify each pair of angles as alternate interior angles, same-side interior angles, corresponding angles, or none of these.

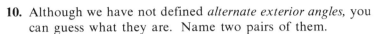

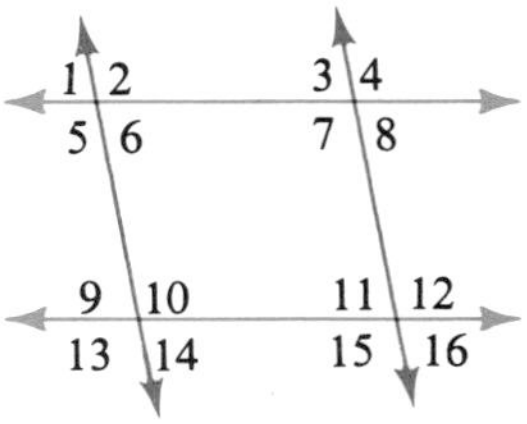

2. $\angle 2$ and $\angle 4$
3. $\angle 6$ and $\angle 10$
4. $\angle 7$ and $\angle 15$
5. $\angle 7$ and $\angle 12$
6. $\angle 7$ and $\angle 10$
7. $\angle 14$ and $\angle 15$
8. $\angle 11$ and $\angle 14$
9. $\angle 1$ and $\angle 11$

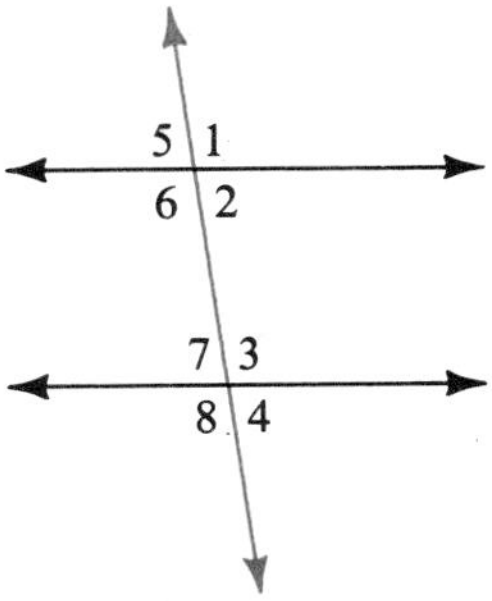

10. Although we have not defined *alternate exterior angles,* you can guess what they are. Name two pairs of them.
11. Name two pairs of angles we would call *same-side exterior angles.*
12. Suppose one pair of alternate interior angles are congruent (say, $\angle 2 \cong \angle 7$). Explain why the other pair of alternate interior angles must also be congruent.
13. Suppose a pair of same-side interior angles (say, $\angle 2$ and $\angle 3$) are supplementary. What must be true of any pair of corresponding angles?
14. Classify each pair of lines as intersecting, parallel, or skew
 a. $\overleftrightarrow{AB}$ and $\overleftrightarrow{EJ}$ **b.** $\overleftrightarrow{AB}$ and $\overleftrightarrow{FK}$ **c.** $\overleftrightarrow{AB}$ and $\overleftrightarrow{ID}$
 d. $\overleftrightarrow{EF}$ and $\overleftrightarrow{IH}$ **e.** $\overleftrightarrow{EF}$ and $\overleftrightarrow{NM}$ **f.** $\overleftrightarrow{CN}$ and $\overleftrightarrow{FG}$

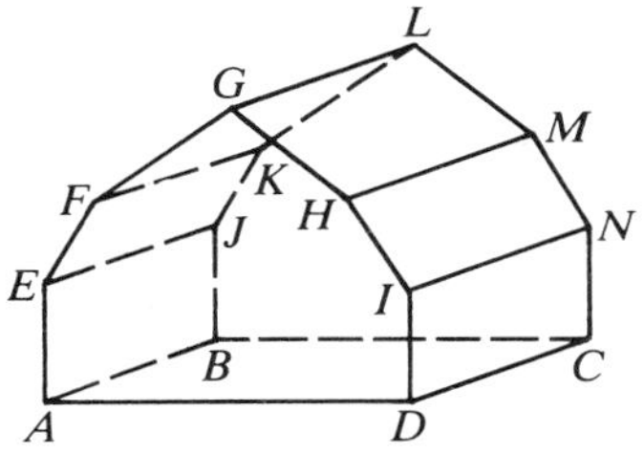

15. Name six lines parallel to $\overleftrightarrow{GL}$.
16. Name several lines skew to $\overleftrightarrow{GL}$.
17. Name five lines parallel to plane *ABCD*.
18. Name a plane parallel to plane *ADIHGFE*.
19. Are there any other parallel planes in the diagram?

Complete each statement with the word *always, sometimes,* or *never.*

20. Two skew lines are _?_ parallel.
21. Two parallel lines are _?_ coplanar.
22. A line in the plane of the ceiling and a line in the plane of the floor are _?_ parallel.
23. Two lines in the plane of the floor are _?_ skew.

Written Exercises

Classify each pair of angles as alternate interior angles, same-side interior angles, or corresponding angles.

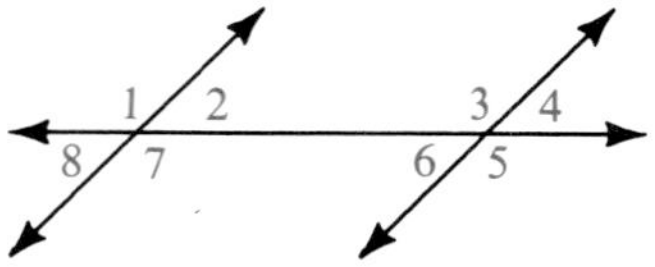

A 1. ∠2 and ∠6
2. ∠8 and ∠6
3. ∠2 and ∠3
4. ∠3 and ∠7
5. ∠5 and ∠7
6. ∠3 and ∠1

Name the two lines and the transversal that form each pair of angles.

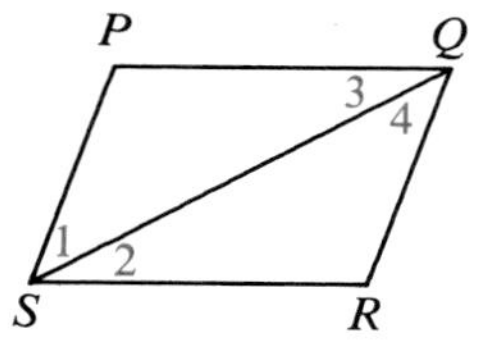

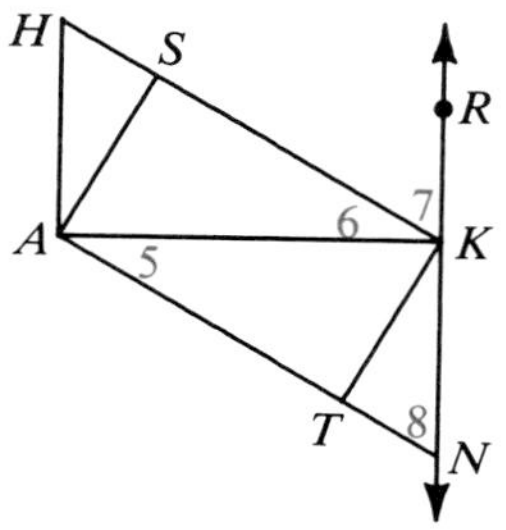

7. ∠2 and ∠3
8. ∠1 and ∠4
9. ∠*P* and ∠*PSR*
10. ∠5 and ∠6
11. ∠7 and ∠8
12. ∠8 and ∠*HAN*

Classify each pair of angles as alternate interior, same-side interior, or corresponding angles.

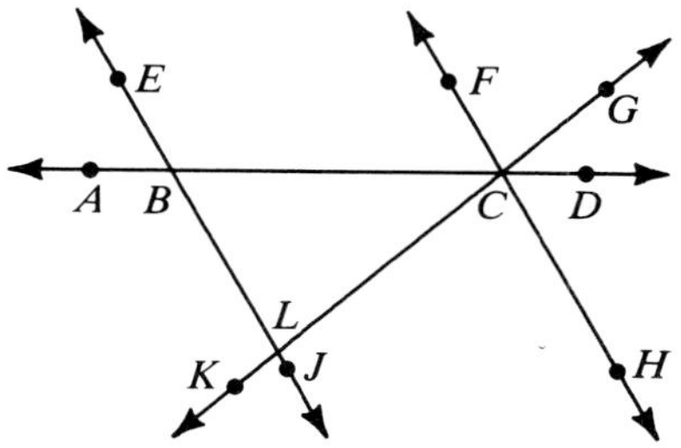

13. ∠*EBA* and ∠*FCB*
14. ∠*DCH* and ∠*CBJ*
15. ∠*FCB* and ∠*CBL*
16. ∠*FCL* and ∠*BLC*
17. ∠*HCB* and ∠*CBJ*
18. ∠*GCH* and ∠*GLJ*
19. Make a drawing that shows two coplanar segments that do not intersect and yet are not parallel.

In Exercises 20–22, use two lines of notebook paper for parallel lines and draw any transversal. Use a protractor to measure.

20. Measure one pair of corresponding angles. Repeat the experiment with another transversal. What appears to be true?
21. Measure one pair of alternate interior angles. Repeat the experiment with another transversal. What appears to be true?
22. Measure one pair of same-side interior angles. Repeat the experiment with another transversal. What appears to be true?

B **23.** Draw a diagram of a six-sided box by following the steps below.

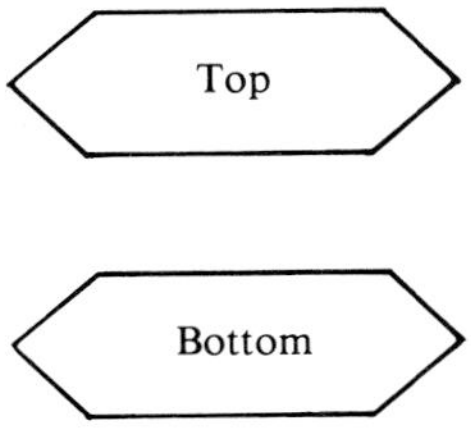

Step 1

Draw a six-sided top. Then draw an exact copy of the top directly below it.

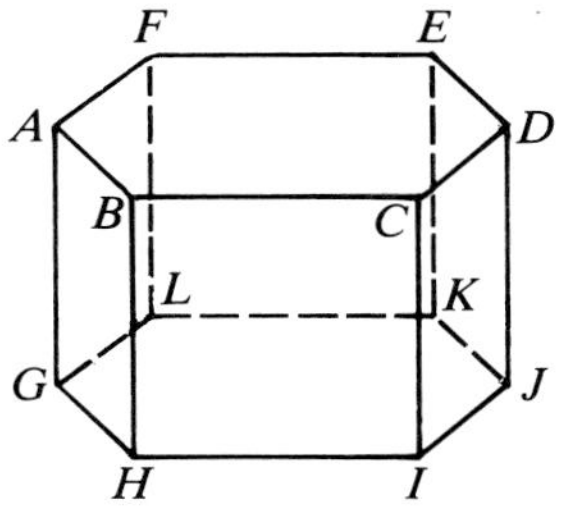

Step 2

Draw vertical edges. Make invisible edges dashed.

Exercises 24–30 refer to the diagram in Step 2 of Exercise 23.

24. Name five lines that appear to be parallel to $\overline{AG}$.

25. Name three lines that appear to be parallel to $\overline{AB}$.

26. Name four lines that appear to be skew to $\overline{AB}$.

27. Name two planes parallel to $\overleftrightarrow{AF}$.

28. Name four planes parallel to $\overleftrightarrow{FL}$.

29. How many pairs of parallel planes are shown?

30. Suppose the top and bottom of the box lie in parallel planes. Explain how Theorem 2-1 can be used to prove $\overline{CD} \parallel \overline{IJ}$.

Complete each statement with the word *always, sometimes,* or *never.*

31. When there is a transversal of two lines, the three lines are __?__ coplanar.

32. Two lines that are not coplanar __?__ intersect.

33. Two lines skew to a third line are __?__ skew to each other.

34. Two lines perpendicular to a third line are __?__ perpendicular to each other.

35. Two planes parallel to the same line are __?__ parallel to each other.

36. Two planes parallel to the same plane are __?__ parallel to each other.

37. If a line is parallel to a plane, a plane containing that line is __?__ parallel to the given plane.

38. Two lines parallel to the same plane are __?__ parallel to each other.

Draw the figures described.

C **39.** Lines a and b are skew, lines b and c are skew, and $a \parallel c$.

40. Lines d and e are skew, lines e and f are skew, and $d \perp f$.

41. Line $l \parallel$ plane X, plane $X \parallel$ plane Y, and l is not parallel to Y.

2-2 Properties of Parallel Lines

By experimenting with parallel lines, transversals, and a protractor in the last set of exercises, you probably discovered that corresponding angles are congruent. There is not enough information in our previous postulates and theorems to deduce this property as a theorem. We will accept it as a postulate.

Postulate 10

If two parallel lines are cut by a transversal, then corresponding angles are congruent.

From this postulate we can easily prove the following theorems.

Theorem 2-2

If two parallel lines are cut by a transversal, then alternate interior angles are congruent.

Given: $k \parallel m$; transversal t cuts k and m.

Prove: $\angle 1 \cong \angle 2$

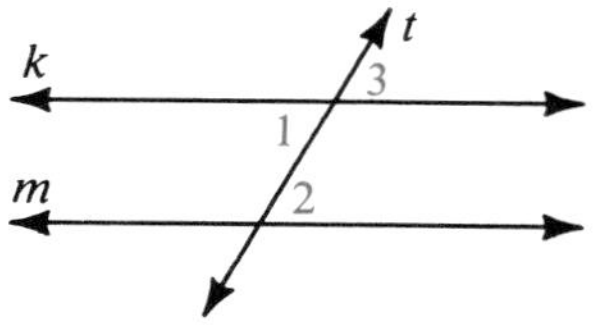

Proof:

Statements	Reasons
1. $k \parallel m$; t is a transversal.	1. Given
2. $\angle 1 \cong \angle 3$	2. Vert. ∠s are ≅.
3. $\angle 3 \cong \angle 2$	3. If two parallel lines are cut by a transversal, then corr. ∠s are ≅.
4. $\angle 1 \cong \angle 2$	4. Transitive Property

Theorem 2-3

If two parallel lines are cut by a transversal, then same-side interior angles are supplementary.

Given: $k \parallel m$; transversal t cuts k and m.

Prove: $\angle 1$ is supplementary to $\angle 4$.

The proof is left as Exercise 19.

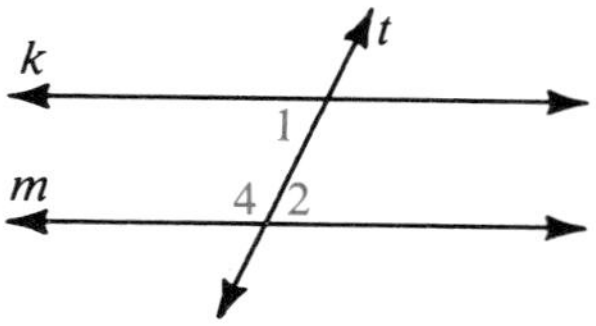

Theorem 2-4

If a transversal is perpendicular to one of two parallel lines, then it is perpendicular to the other one also.

Given: Transversal t cuts l and n;
$t \perp l$; $l \parallel n$

Prove: $t \perp n$

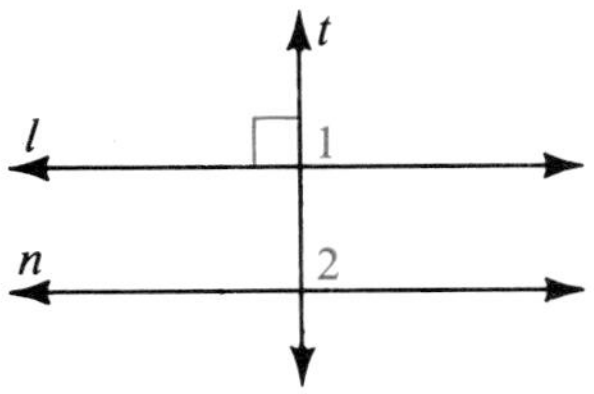

Proof:

Statements	Reasons
1. $t \perp l$	1. Given
2. $m\angle 1 = 90$	2. Def. of $\perp$ lines and def. of a rt. $\angle$
3. $l \parallel n$	3. Given
4. $m\angle 2 = m\angle 1$	4. If two parallel lines are cut by a transversal, then corr. $\angle$s are $\cong$.
5. $m\angle 2 = 90$	5. Substitution Property
6. $t \perp n$	6. Def. of a rt. $\angle$ and def. of $\perp$ lines

At this point in your study of geometry, we believe that it is no longer necessary to use arrowheads in diagrams to suggest that a line extends in both directions without ending. Instead, pairs of arrowheads (and double arrowheads when necessary) will be used to indicate parallel lines, as shown in the following examples.

Example 1 Find the measure of $\angle PQR$.

Solution The diagram shows that

$$\overleftrightarrow{QR} \perp \overleftrightarrow{RS} \text{ and } \overleftrightarrow{QP} \parallel \overleftrightarrow{RS}.$$

Thus, by Theorem 2-4, $\overleftrightarrow{QR} \perp \overleftrightarrow{QP}$ and therefore

$$m\angle PQR = 90.$$

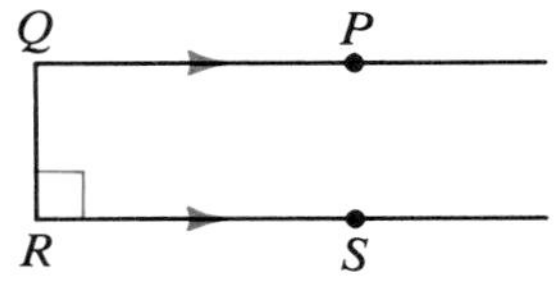

Example 2 Find the values of x and y.

Solution Since $a \parallel b$, $2x = 60$. (Why?)
Thus, $x = 30$.
Since $c \parallel d$, $60 + y = 180$. (Why?)
Thus, $y = 120$.

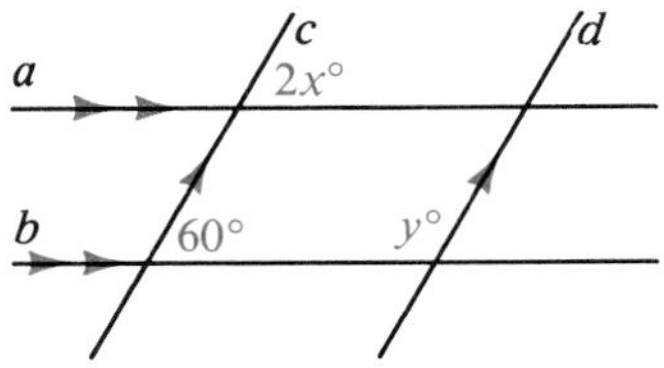

Classroom Exercises

1. What do the arrowheads in the diagram tell you?

2. a. How are lines k and l related?
b. How are lines k and p related? Why?

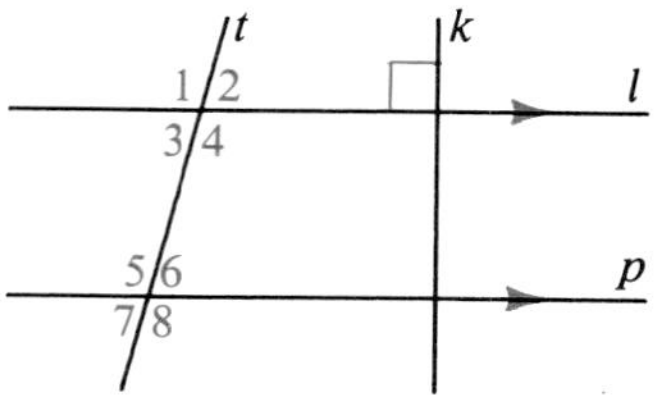

Exs. 1–11

State the postulate or theorem that justifies each statement.

3. $\angle 1 \cong \angle 5$
4. $\angle 3 \cong \angle 6$
5. $m\angle 4 + m\angle 6 = 180$
6. $m\angle 4 = m\angle 8$
7. $m\angle 4 = m\angle 5$
8. $\angle 6 \cong \angle 7$
9. $k \perp p$
10. $\angle 3$ is supplementary to $\angle 5$.

11. If $m\angle 1 = 130$, what are the measures of the other numbered angles?

12. Alan tried to prove Postulate 10 as shown below. However, he did *not* have a valid proof. Explain why not.

If two parallel lines are cut by a transversal, then corresponding angles are congruent.

Given: $k \parallel l$; transversal t cuts k and l.

Prove: $\angle 1 \cong \angle 2$

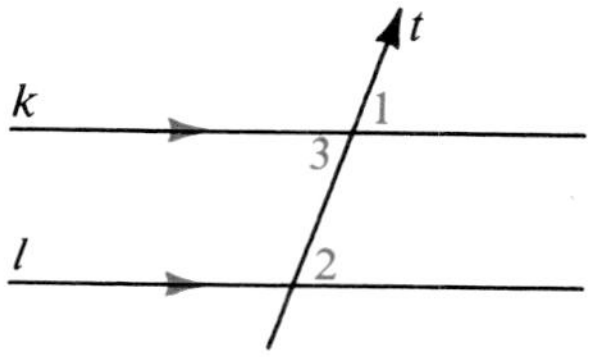

Proof:

Statements	Reasons
1. $k \parallel l$; t is a transversal.	1. Given
2. $\angle 3 \cong \angle 2$	2. If 2 parallel lines are cut by a transversal, then alt. int. ∠s are ≅.
3. $\angle 1 \cong \angle 3$	3. Vert. ∠s are ≅.
4. $\angle 1 \cong \angle 2$	4. Transitive Prop. of ≅

Written Exercises

A

1. If $a \parallel b$, name all angles that must be congruent to $\angle 1$.

2. If $c \parallel d$, name all angles that must be congruent to $\angle 1$.

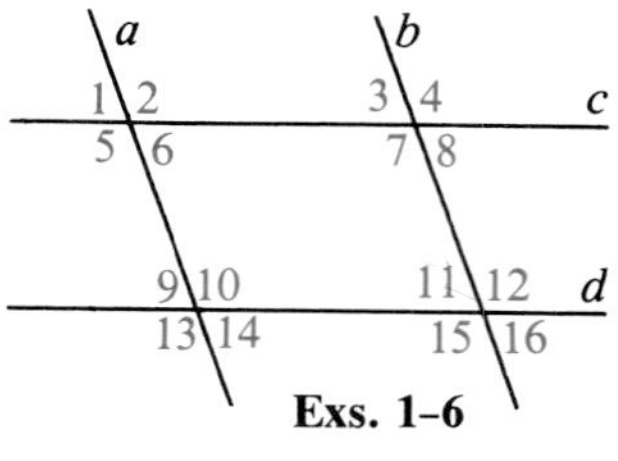

Exs. 1–6

Assume that $a \parallel b$ and $c \parallel d$.

3. Name all angles congruent to $\angle 4$.

4. Name all angles supplementary to $\angle 4$.

5. If $m\angle 16 = 50$, then $m\angle 14 =$ __?__ and $m\angle 2 =$ __?__.

6. If $m\angle 9 = x$, then $m\angle 12 =$ __?__ and $m\angle 7 =$ __?__.

Find the values of x and y.

7.

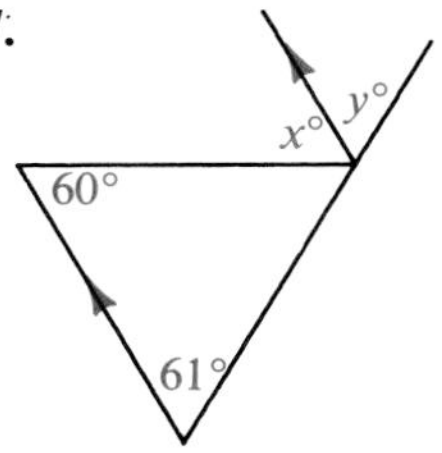

8.

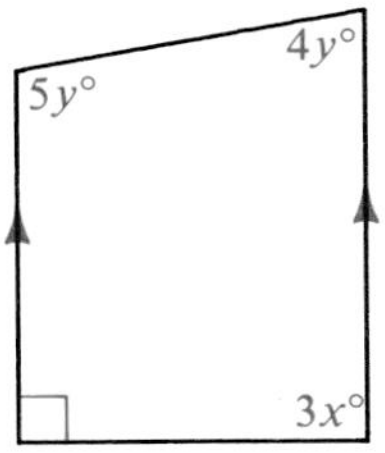

9.

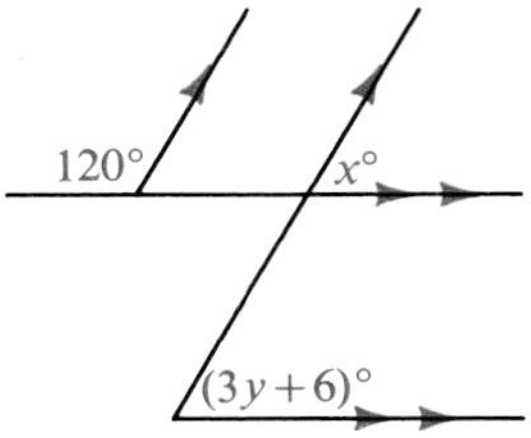

10.

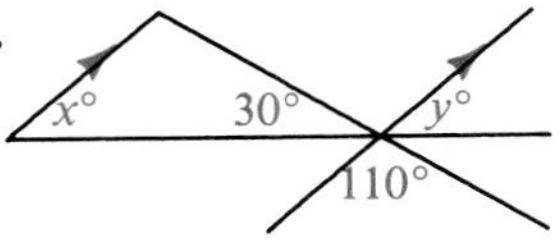

11.

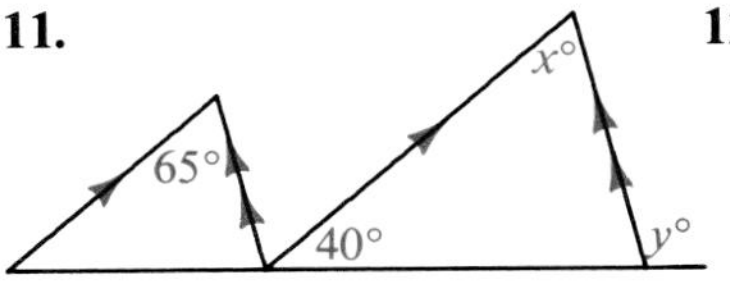

12.

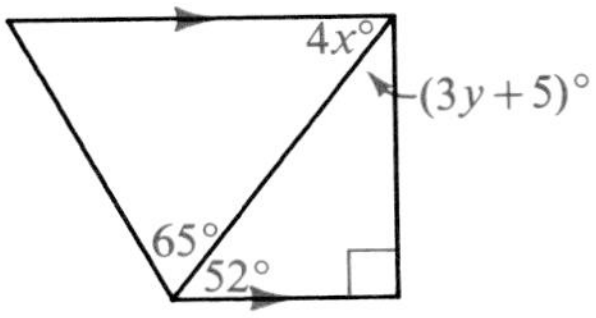

13. Write the reasons.

Given: $k \parallel l$

Prove: $\angle 6$ is supp. to $\angle 7$.

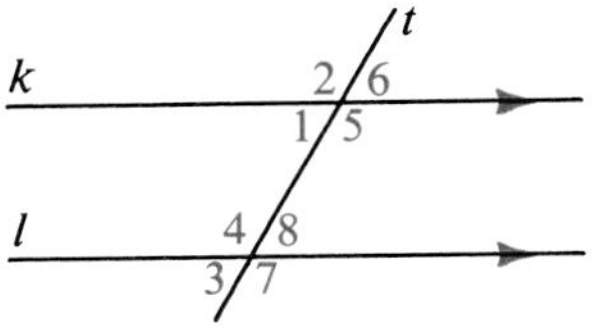

Proof:

Statements	Reasons
1. $k \parallel l$	1. ?
2. $m\angle 6 = m\angle 8$	2. ?
3. $m\angle 8 + m\angle 7 = 180$	3. ?
4. $m\angle 6 + m\angle 7 = 180$	4. ?
5. $\angle 6$ is supp. to $\angle 7$.	5. ?

B **14.** Given: $\overline{PQ} \perp \overline{QR}$; $\overline{ST} \parallel \overline{QR}$; $\overrightarrow{QT}$ bisects $\angle PQR$.

a. Find the measures of $\angle QST$, $\angle SQT$, and $\angle STQ$.

b. If you are also given that $m\angle R = 60$, find the measure of $\angle QTR$.

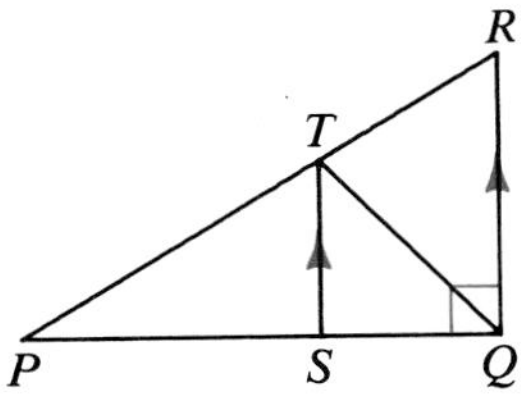

15. Given: $\overline{AB} \parallel \overline{CD}$; $m\angle D = 116$; $\overrightarrow{AK}$ bisects $\angle DAB$.

a. Find the measures of $\angle DAB$, $\angle KAB$, and $\angle DKA$.

b. Is there enough given information for you to conclude that $\angle D$ and $\angle C$ are supplementary, or is more information needed?

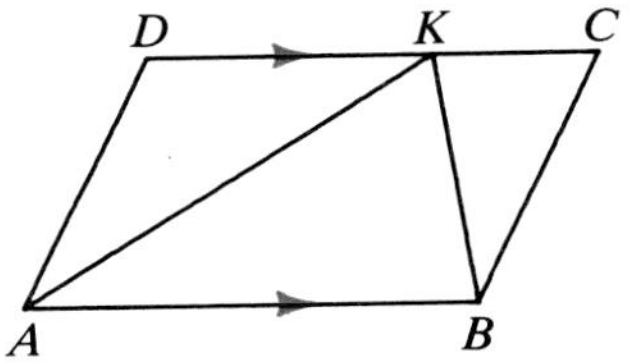

16. Find the values of x and y.

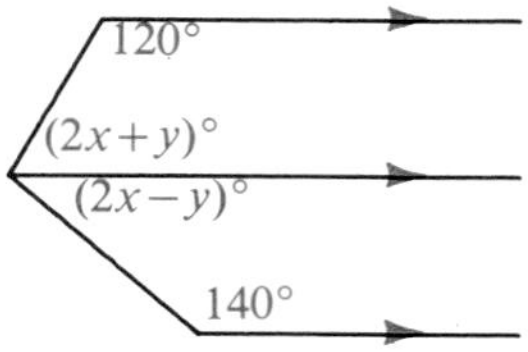

Use the diagram in Exercise 13. Write proofs in two-column form.

17. Given: $k \parallel l$
Prove: $\angle 2 \cong \angle 7$

18. Given: $k \parallel l$
Prove: $\angle 1$ is supplementary to $\angle 7$.

19. Copy what is shown for Theorem 2–3 on page 60. Then write a proof in two-column form.

20. Draw a four-sided figure $ABCD$ with $\overline{AB} \parallel \overline{DC}$ and $\overline{AD} \parallel \overline{BC}$.
a. Prove that $\angle A \cong \angle C$.
b. Is $\angle B \cong \angle D$?

C **21.** Given: $\overline{AS} \parallel \overline{BT}$;
$m\angle 4 = m\angle 5$
Prove: $\overrightarrow{SA}$ bisects $\angle BSR$.

22. Given: $\overline{AS} \parallel \overline{BT}$;
$m\angle 4 = m\angle 5$;
$\overrightarrow{SB}$ bisects $\angle AST$.
Find the measure of $\angle 1$.

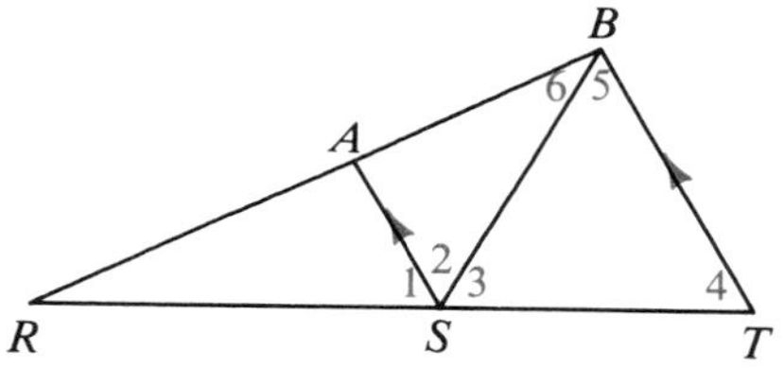

Exs. 21, 22

2-3 Proving Lines Parallel

In the preceding section, you saw situations in which two lines were given as parallel. You then concluded that certain angles were congruent or supplementary. In this section, the situation is reversed. From two angles being congruent or supplementary you will conclude that certain lines forming the angles are parallel. The key to doing this is Postulate 11 below. Postulate 10 is repeated so you can compare the wording of the postulates.

Postulate 10

If two parallel lines are cut by a transversal, then corresponding angles are congruent.

Postulate 11

If two lines are cut by a transversal and corresponding angles are congruent, then the lines are parallel.

The next three theorems can be deduced from Postulate 11.

Theorem 2-5

If two lines are cut by a transversal and alternate interior angles are congruent, then the lines are parallel.

Given: Transversal t cuts lines k and n;
$\angle 1 \cong \angle 2$

Prove: $k \parallel n$

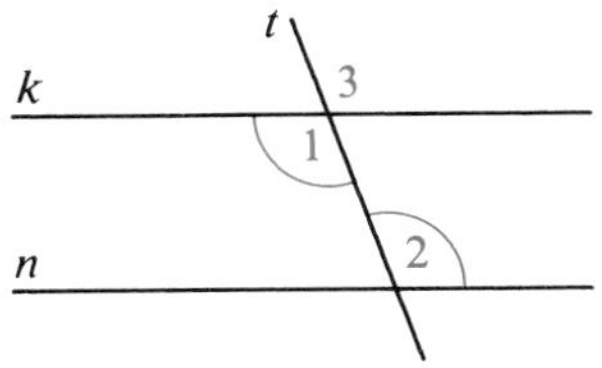

Proof:

Statements	Reasons
1. Transversal t cuts k and n.	1. Given
2. $\angle 3 \cong \angle 1$	2. Vert. $\measuredangle$s are $\cong$.
3. $\angle 1 \cong \angle 2$	3. Given
4. $\angle 3 \cong \angle 2$	4. Transitive Property
5. $k \parallel n$	5. If two lines are cut by a transversal and corr. $\measuredangle$s are $\cong$, the lines are $\parallel$.

Theorem 2-6

If two lines are cut by a transversal and same-side interior angles are supplementary, then the lines are parallel.

Given: Transversal t cuts lines k and n;
$\angle 1$ is supplementary to $\angle 2$.

Prove: $k \parallel n$

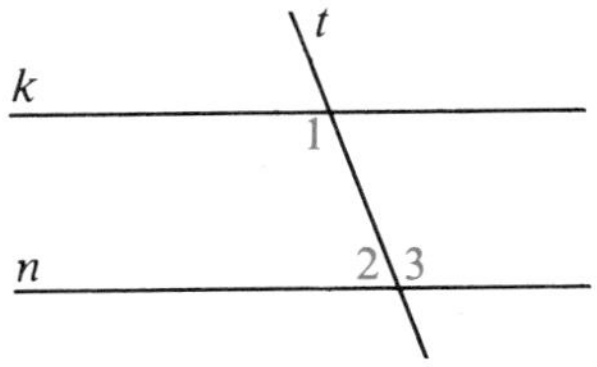

The proof is left as Exercise 19.

Theorem 2-7

In a plane, two lines perpendicular to the same line are parallel.

Given: $k \perp t$; $n \perp t$

Prove: $k \parallel n$

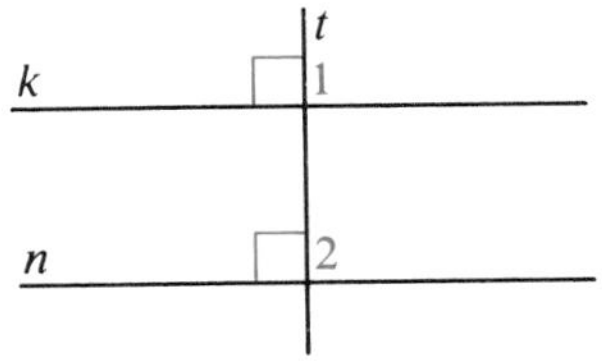

The proof is left as Exercise 20.

Example 1 Which segments are parallel?

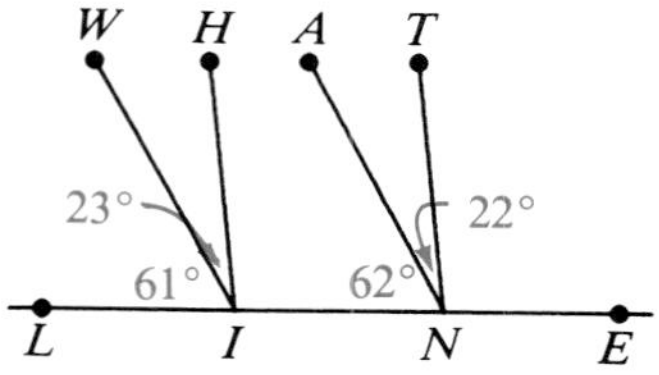

Solution (1) $\overline{HI}$ and $\overline{TN}$ are parallel since corresponding angles have the same measure:
$m\angle HIL = 23 + 61 = 84$
$m\angle TNI = 22 + 62 = 84$

(2) $\overline{WI}$ and $\overline{AN}$ are *not* parallel since $61 \neq 62$.

Example 2 Find the values of x and y that make $\overline{AC} \parallel \overline{DF}$ and $\overline{AE} \parallel \overline{BF}$.

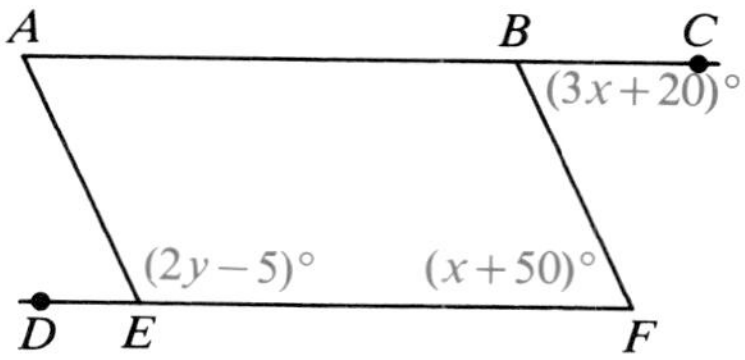

Solution $\overline{AC} \parallel \overline{DF}$ if $m\angle CBF = m\angle BFE$. (Why?)

$$3x + 20 = x + 50$$
$$2x = 30$$
$$x = 15$$

$\overline{AE} \parallel \overline{BF}$ if $\angle AEF$ and $\angle F$ are supplementary. (Why?)

$$(2y - 5) + (x + 50) = 180$$
$$(2y - 5) + (15 + 50) = 180$$
$$2y = 120$$
$$y = 60$$

The following theorems can be proved using previous postulates and theorems. We state the theorems without proof, however, for you to use in future work.

Theorem 2-8

Through a point outside a line, there is exactly one line parallel to the given line.

Theorem 2-9

Through a point outside a line, there is exactly one line perpendicular to the given line.

Given this:

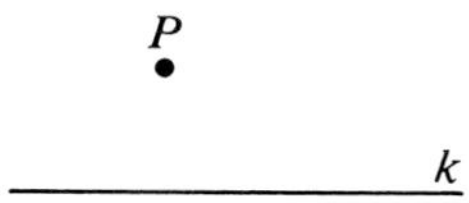

Then line m exists and is unique.

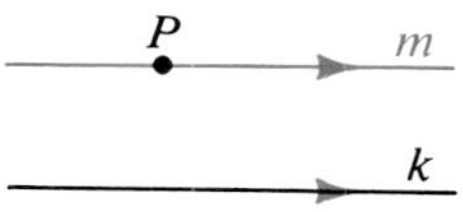

Given this:

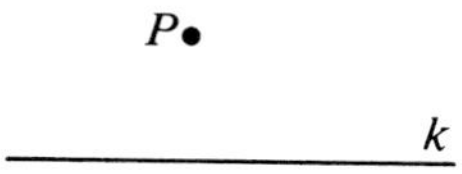

Then line h exists and is unique.

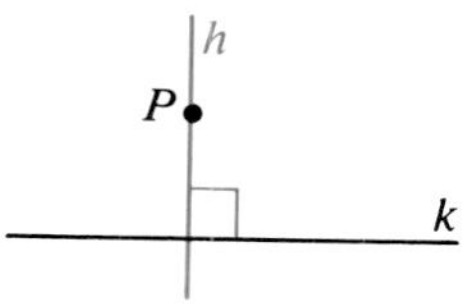

You may use the next theorem whether or not the three lines are coplanar. The case for coplanar lines will be justified as a classroom exercise.

Theorem 2-10

Two lines parallel to a third line are parallel to each other.

Ways to Prove Two Lines Parallel

1. Show that a pair of corresponding angles are congruent.
2. Show that a pair of alternate interior angles are congruent.
3. Show that a pair of same-side interior angles are supplementary.
4. In a plane, show that both lines are perpendicular to a third line.
5. Show that both lines are parallel to a third line.

Classroom Exercises

State which segments (if any) are parallel. State the postulate or theorem that justifies your answer.

1.

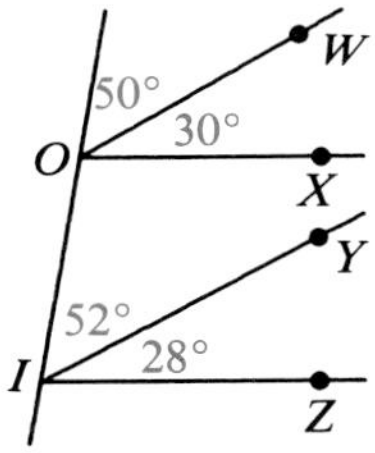

2.

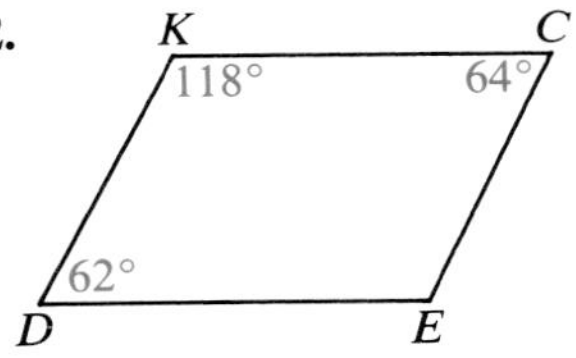

3.

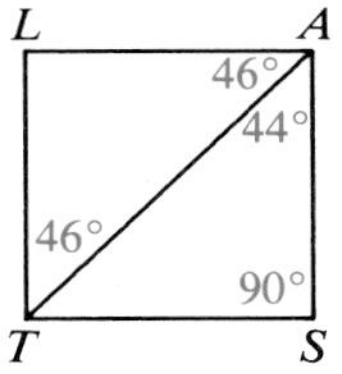

Use the given information to name the segments that must be parallel. If there are no such segments, say so.

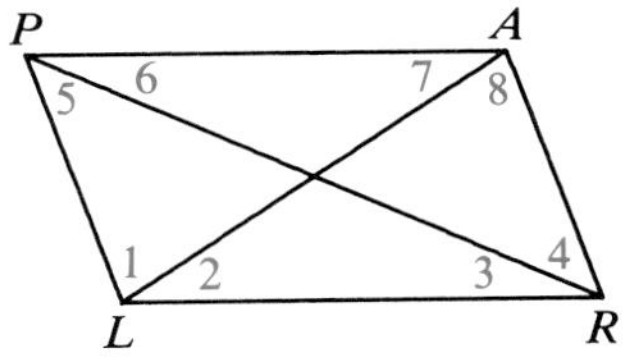

4. $m\angle 1 = m\angle 8$

5. $\angle 2 \cong \angle 7$

6. $\angle 5 \cong \angle 3$

7. $m\angle 5 = m\angle 4$

8. $m\angle 5 + m\angle 6 = m\angle 3 + m\angle 4$

9. $m\angle APL + m\angle PAR = 180$

10. Reword Theorem 2–8 as two statements, one describing existence and the other describing uniqueness.

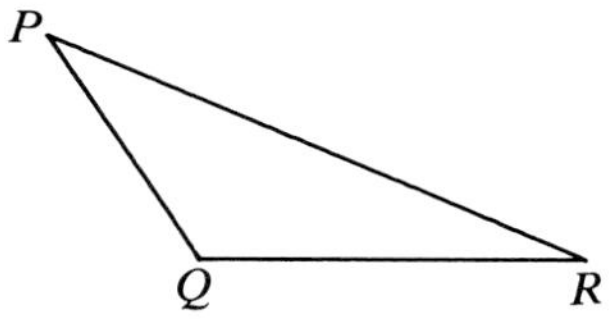

11. How many lines can be drawn through P parallel to $\overleftrightarrow{QR}$?

12. How many lines can be drawn through Q parallel to $\overleftrightarrow{PR}$?

13. How many lines can be drawn through P perpendicular to

a. $\overleftrightarrow{QR}$? **b.** $\overline{QR}$?

14. In the plane containing P, Q, and R, how many lines can be drawn through R perpendicular to $\overleftrightarrow{QR}$? What postulate or theorem enables you to answer the question?

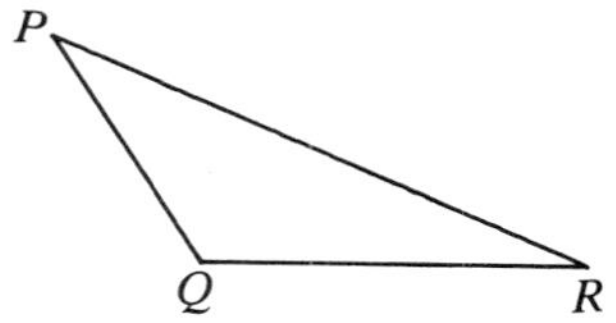

15. If you are not confined to the plane containing P, Q, and R, how many lines can be drawn through R perpendicular to $\overleftrightarrow{QR}$?

16. True or False?
   a. Two lines perpendicular to a third line must be parallel.
   b. In a plane two lines perpendicular to a third line must be parallel.

17. In a plane, $k \parallel l$ and $k \parallel n$. Use the diagram to explain why $l \parallel n$.

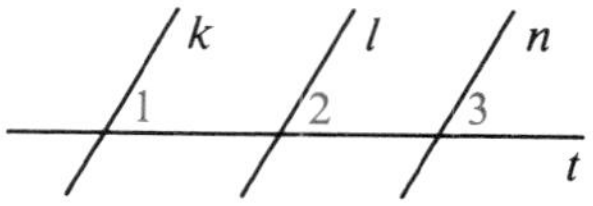

Written Exercises

Use the given information to name the segments that must be parallel. If there are no such segments, write *none*.

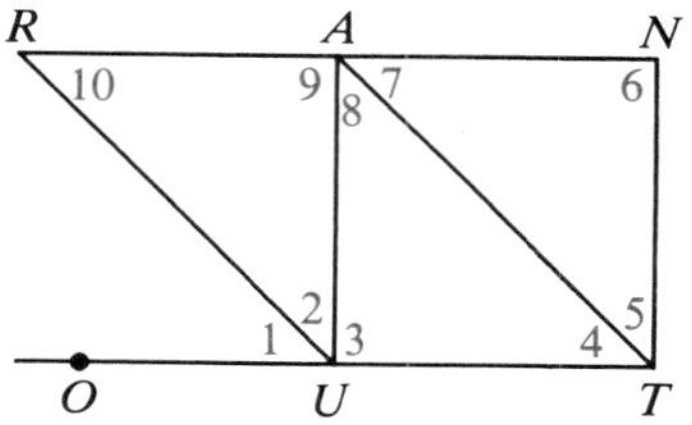

A 1. $\angle 1 \cong \angle 4$
2. $m\angle 2 = m\angle 10$
3. $m\angle 5 = m\angle 7$
4. $\angle 5 \cong \angle 8$
5. $m\angle 6 = m\angle 9 = 90$
6. $m\angle 6 = m\angle 3 = 90$
7. $m\angle 7 = m\angle 10 = m\angle 1$
8. $\overline{AU} \perp \overline{OT}$, $\overline{NT} \perp \overline{OT}$
9. $\angle 2 \cong \angle 5$
10. $m\angle 2 = m\angle 5 = m\angle 8$

11. Write the reasons.
   Given: Transversal t cuts lines l and n;
   $\angle 1 \cong \angle 2$
   Prove: $l \parallel n$

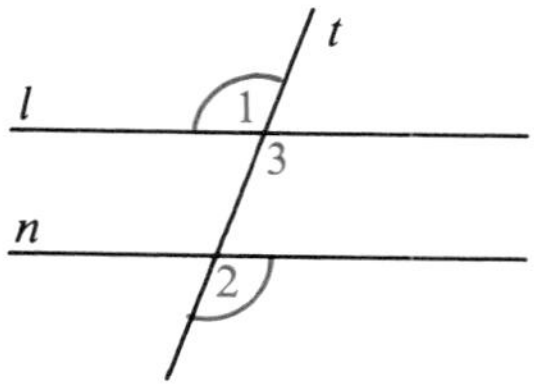

Proof:

Statements	Reasons
1. Transversal t cuts l and n.	1. ?
2. $\angle 1 \cong \angle 3$	2. ?
3. $\angle 2 \cong \angle 1$	3. ?
4. $\angle 2 \cong \angle 3$	4. ?
5. $l \parallel n$	5. ?

12. Restate Theorem 2-9 as two statements, one describing existence and the other describing uniqueness.

Name *two* pairs of parallel lines in each figure. Which congruent or supplementary angles did you use to determine the parallel lines?

13. 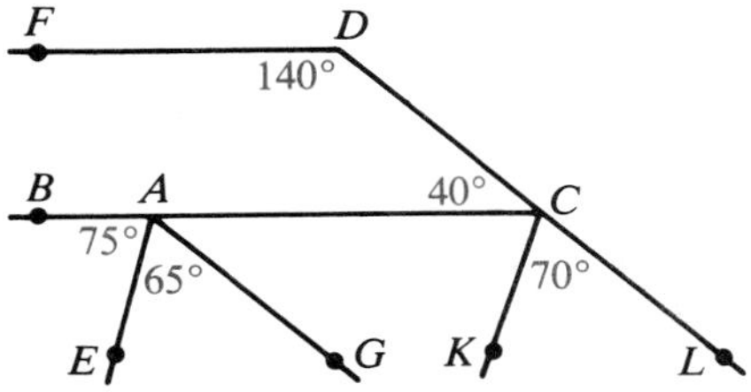

14. 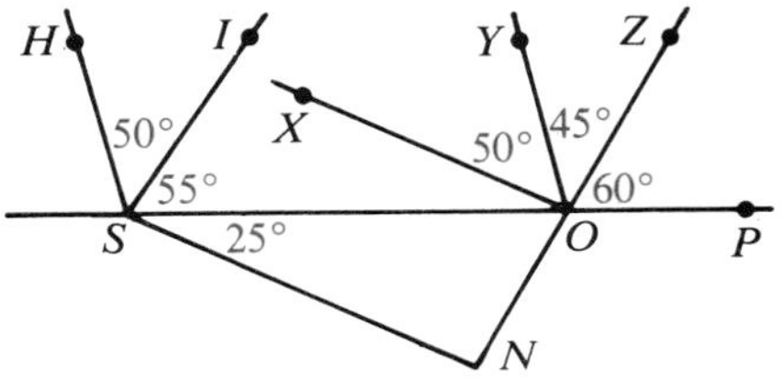

Find the values of x and y that make the red lines parallel *and* the blue lines parallel.

B **15.** 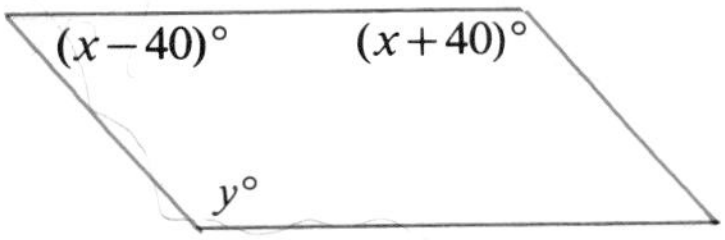

16. 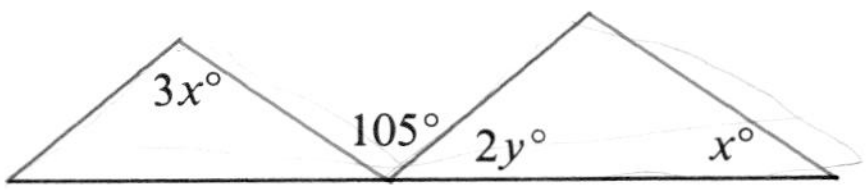

17. Given: $\angle 1 \cong \angle 2$; $\angle 4 \cong \angle 5$
What can you prove about $\overline{PQ}$ and $\overline{RS}$? Be prepared to give your reasons in class, if asked.

18. Given: $\angle 3 \cong \angle 6$
What can you prove about other angles? Be prepared to give your reasons in class, if asked.

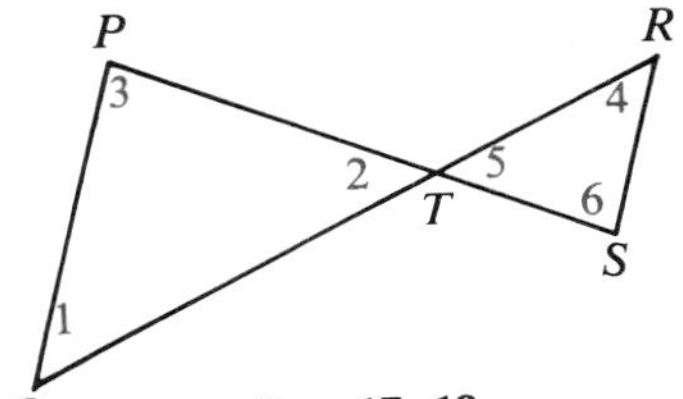

Exs. 17, 18

19. Copy what is shown for Theorem 2-6 on page 65. Then write a proof in two-column form.

20. Copy what is shown for Theorem 2-7 on page 65. Then write a proof in two-column form.

21. Given: $\overline{BE} \perp \overline{DA}$; $\overline{CD} \perp \overline{DA}$
Prove: $\angle 1 \cong \angle 2$

22. Given: $\angle C \cong \angle 3$; $\overline{BE} \perp \overline{DA}$
Prove: $\overline{CD} \perp \overline{DA}$

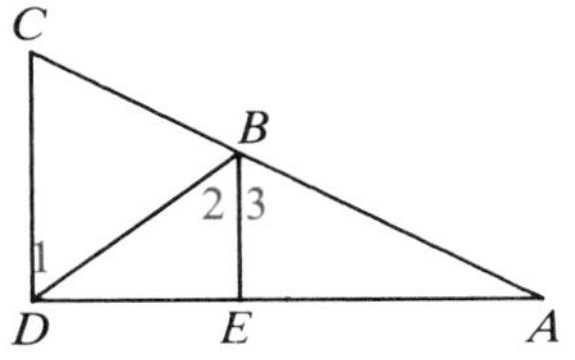

Exs. 21, 22

Find the measure of $\angle RST$. (*Hint:* Draw a line through S parallel to $\overline{RX}$ and $\overline{TY}$.)

23. 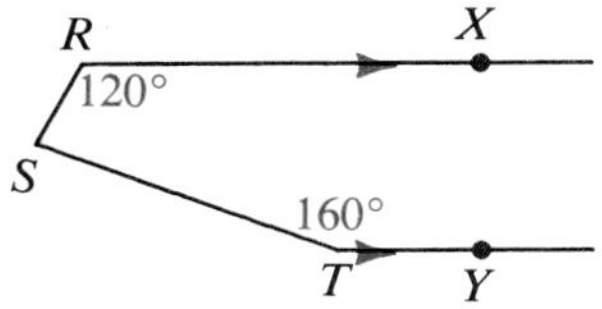

24. 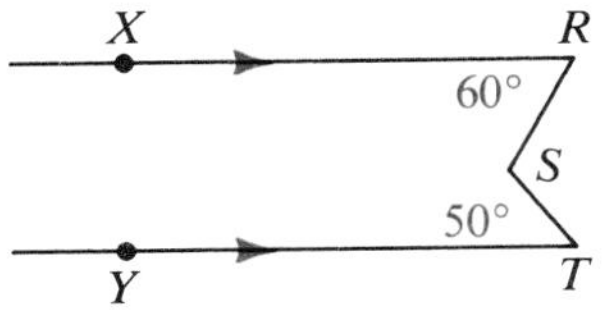

25. Given: $m\angle 1 = m\angle 4$; $\overline{BC} \parallel \overline{ED}$
Prove: $\overline{AB} \parallel \overline{DF}$

26. Given: $m\angle ABD = m\angle FDB$; $m\angle 1 = m\angle 4$
Prove: $\overline{BC} \parallel \overline{ED}$

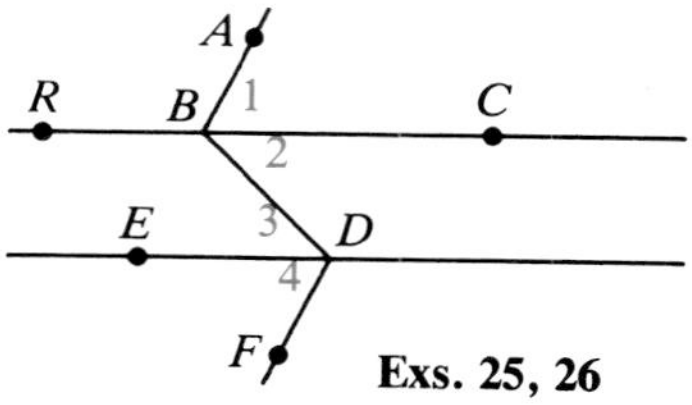

Exs. 25, 26

27. Find the values of x and y that make the lines shown in red parallel.

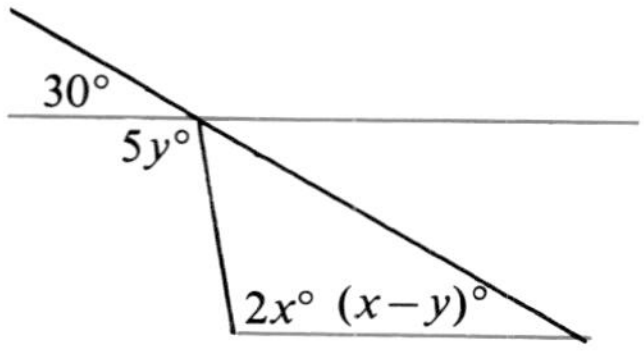

Ex. 27

C 28. Draw two parallel lines cut by a transversal. Then draw the bisectors of two corresponding angles. What appears to be true about the bisectors? Prove that your conclusion is true.

29. Find the value of x that makes the lines shown in red parallel.

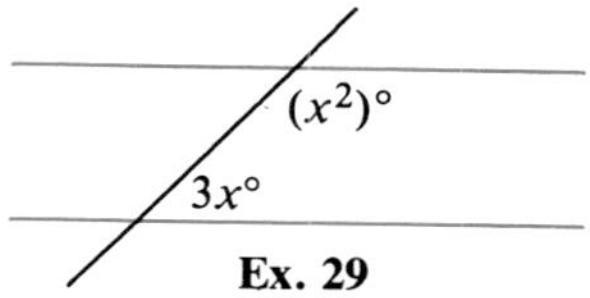

Ex. 29

Self-Test 1

Complete each statement with the word *always, sometimes,* or *never*.

1. Two lines that do not intersect are __?__ parallel.
2. Two skew lines __?__ intersect.
3. If two parallel lines are cut by a transversal, then the same-side interior angles are __?__ supplementary.
4. Two lines perpendicular to a third line are __?__ parallel.
5. If a line is parallel to plane X and also to plane Y, then plane X and plane Y are __?__ parallel.
6. Complete: If $\overline{AE} \parallel \overline{BD}$, then $\angle 1 \cong$ __?__ and $\angle 9 \cong$ __?__.
7. If $\overline{ED} \parallel \overline{AC}$, name all pairs of angles that must be congruent.
8. If $\overline{ED} \parallel \overline{AC}$ and $\overline{EB} \parallel \overline{DC}$, name all angles that must be congruent to $\angle 5$.

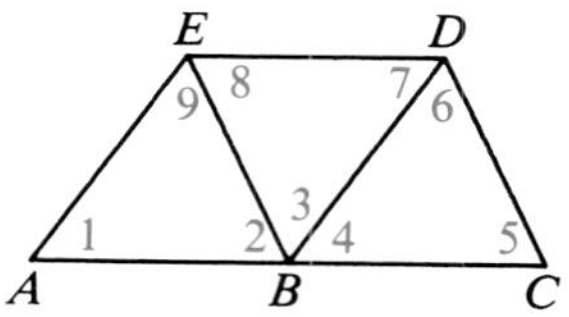

Exs. 6–11

Use the given information to name the segments (if any) that must be parallel.

9. $\angle 3 \cong \angle 6$ 10. $\angle 9 \cong \angle 6$ 11. $m\angle 7 + m\angle AED = 180$

CAREERS

Geologist

Geologists study rock formations such as those at Checkerboard Mountain in Zion National Park. Rock formations often occur in *strata,* or layers, beneath the surface of the Earth. Earthquakes occur at *faults,* breaks in the strata. In search of a fault, how would you determine the position of a stratum of rock buried deep beneath the surface of the Earth?

A geologist might start by picking three noncollinear points, A, B, and C, on the surface and drilling holes to find the depths of points A', B', and C' on the stratum. These three points determine the plane of the surface of the stratum.

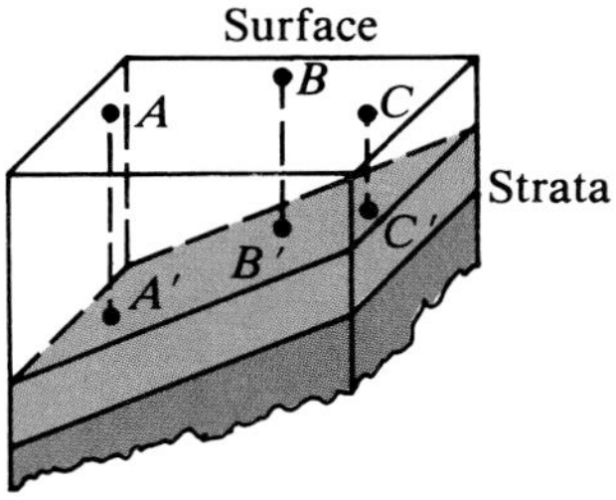

Geologists may work for industry, searching for oil or minerals. They may work in research centers, developing ways to predict earthquakes.

Today, geologists are trying to locate sources of geothermal energy, generated by the Earth's internal heat. A career in geology usually requires knowledge of mathematics, physics, and chemistry, as well as a degree in geology.

Applying Parallel Lines to Polygons

Objectives

1. Classify triangles according to sides and to angles.
2. State and apply the theorem and the corollaries about the sum of the measures of the angles of a triangle.
3. State and apply the theorem about the measure of an exterior angle of a triangle.
4. Recognize and name convex polygons and regular polygons.
5. Find the measures of interior angles and exterior angles of convex polygons.

2-4 Angles of a Triangle

A **triangle** is the figure formed by three segments joining three noncollinear points. Each of the three points is a **vertex** of the triangle. (The plural of *vertex* is *vertices*.) The segments are the **sides** of the triangle.

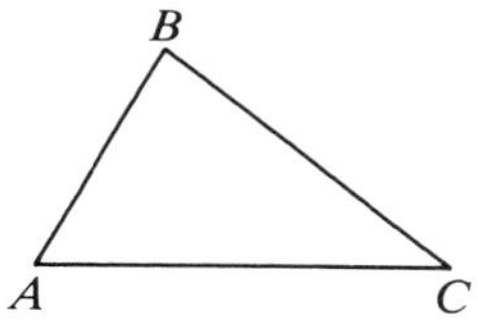

Triangle ABC ($\triangle ABC$) is shown.
Vertices of $\triangle ABC$: points A, B, C
Sides of $\triangle ABC$: $\overline{AB}$, $\overline{BC}$, $\overline{CA}$
Angles of $\triangle ABC$: $\angle A$, $\angle B$, $\angle C$

A triangle is sometimes classified by the number of congruent sides it has.

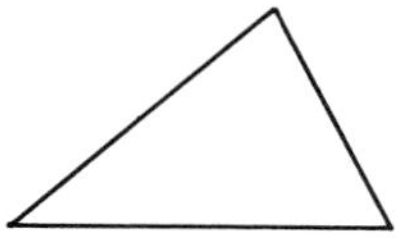

Scalene triangle
No sides congruent

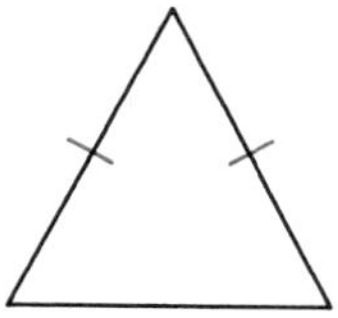

Isosceles triangle
At least two sides congruent

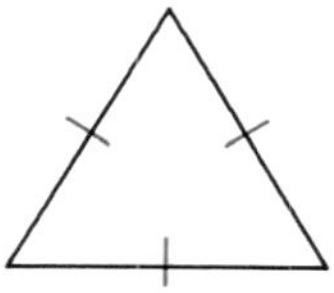

Equilateral triangle
All sides congruent

Triangles can also be classified by their angles.

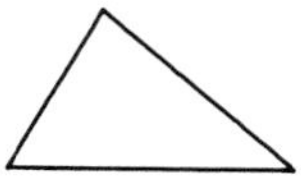

Acute △
Three acute ∠s

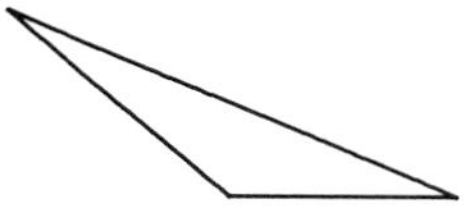

Obtuse △
One obtuse ∠

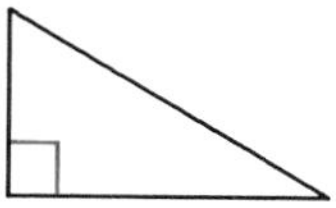

Right △
One right ∠

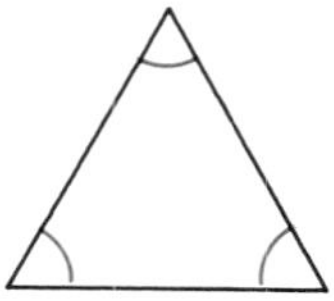

Equiangular △
All ∠s congruent

An **auxiliary line** is a line (or ray or segment) added to a diagram to help in a proof. An auxiliary line is used in the proof of the next theorem, one of the best-known theorems of geometry.

Theorem 2-11

The sum of the measures of the angles of a triangle is 180.

Given: $\triangle ABC$

Prove: $m\angle 1 + m\angle 2 + m\angle 3 = 180$

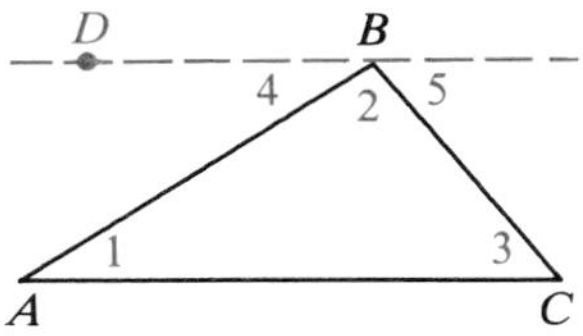

Proof:

Statements	Reasons
1. Through B draw $\overleftrightarrow{BD}$, the line parallel to $\overleftrightarrow{AC}$.	1. Through a point outside a line, there is exactly one line $\parallel$ to the given line.
2. $m\angle DBC + m\angle 5 = 180$; $m\angle DBC = m\angle 4 + m\angle 2$	2. Angle Addition Postulate
3. $m\angle 4 + m\angle 2 + m\angle 5 = 180$	3. Substitution Property
4. $m\angle 4 = m\angle 1$; $m\angle 5 = m\angle 3$	4. _?_
5. $m\angle 1 + m\angle 2 + m\angle 3 = 180$	5. Substitution Property

A statement that can be proved easily by applying a theorem is often called a **corollary** of the theorem. Corollaries, like theorems, can be used as reasons in proofs. Each of the four statements that are shown below is a corollary of Theorem 2-11.

Corollary 1

If two angles of one triangle are congruent to two angles of another triangle, then the third angles are congruent.

Corollary 2

Each angle of an equiangular triangle has measure 60.

Corollary 3

In a triangle, there can be at most one right angle or obtuse angle.

Corollary 4

The acute angles of a right triangle are complementary.

You will justify these corollaries as classroom exercises.

Example 1 Is $\angle P \cong \angle V$?

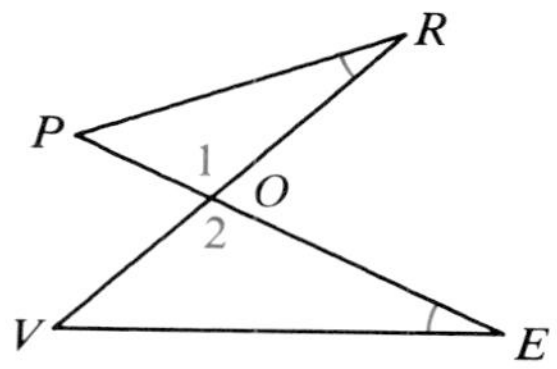

Solution $\angle R \cong \angle E$ (Given in diagram)
$\angle 1 \cong \angle 2$ (Vertical angles)
Thus, two angles of $\triangle PRO$ are congruent to two angles of $\triangle VEO$, and $\angle P \cong \angle V$ by Corollary 1.

When one side of a triangle is extended, an *exterior angle* is formed. Because an exterior angle of a triangle is always a supplement of the adjacent interior angle of the triangle, its measure is related in a special way to the measure of the other two angles of the triangle, called the *remote interior angles*.

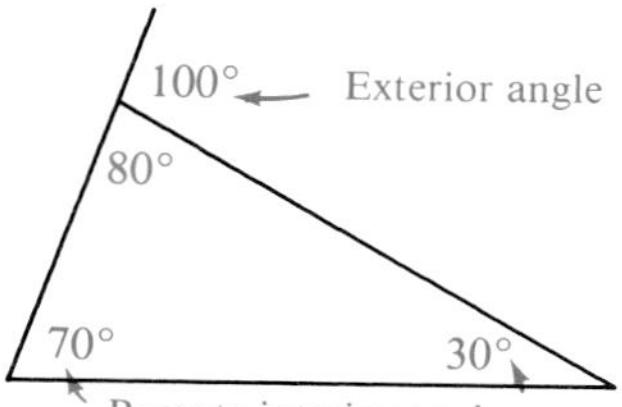

$100 = 70 + 30$

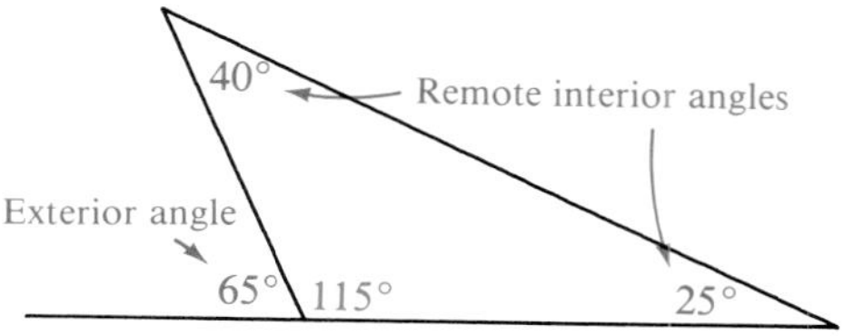

$65 = 40 + 25$

Theorem 2-12

The measure of an exterior angle of a triangle equals the sum of the measures of the two remote interior angles.

The proof of Theorem 2-12 is left as a classroom exercise.

Example 2 In $\triangle RST$, $m\angle R = 80$ and an exterior angle at T is three times as large as $\angle S$. Find the measure of $\angle S$.

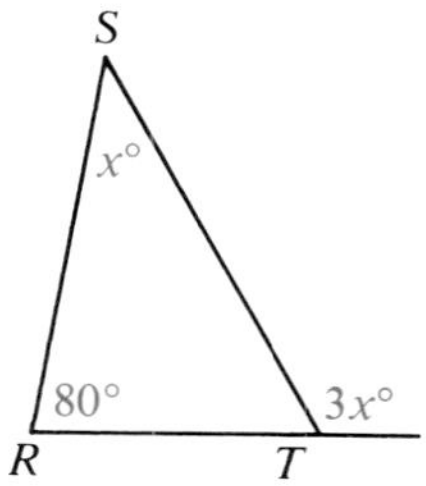

Solution Draw a diagram that shows the given information. Then apply Theorem 2-12.

$$3x = 80 + x$$
$$2x = 80$$
$$x = 40$$
$$m\angle S = 40$$

Classroom Exercises

Complete each statement with the word *always, sometimes,* or *never*.

1. If a triangle is isosceles, then it is __?__ equilateral.

2. If a triangle is equilateral, then it is __?__ isosceles.

3. If a triangle is scalene, then it is __?__ isosceles.

4. If a triangle is obtuse, then it is _?_ isosceles.

5. If a triangle has two complementary angles, then it is _?_ a right triangle.

6. A corollary of a theorem is _?_ an acceptable reason in a proof.

Explain how each corollary of Theorem 2–11 follows from the theorem.

7. Corollary 1
8. Corollary 2
9. Corollary 3
10. Corollary 4

Find the value of x.

11.

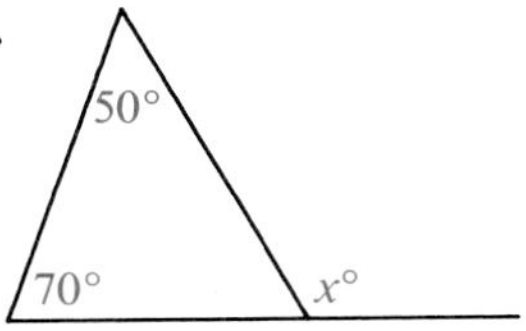

12.

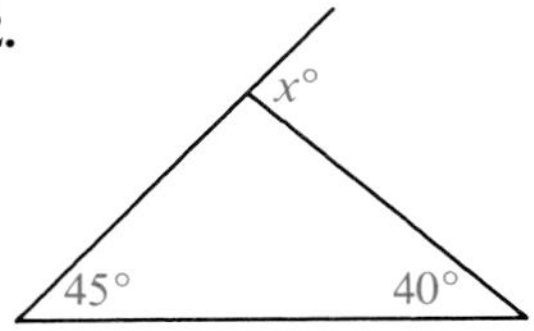

13.

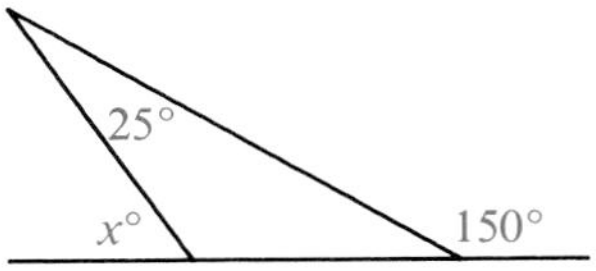

Find the measures of the angles of each triangle described.

14. An equiangular triangle

15. A right triangle with one angle whose measure is 40

16. A right triangle with two congruent angles

17. A triangle whose angles have measures x, $x + 10$, and $x - 10$ (Find the numerical measures.)

What is wrong with each of the following instructions?

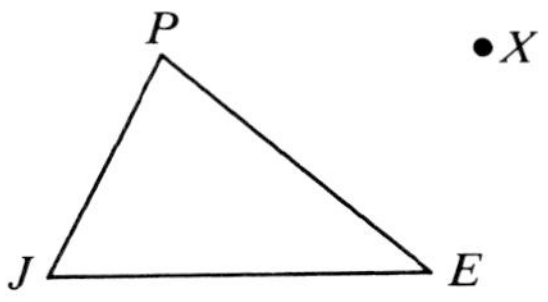

18. Draw the bisector of $\angle J$ to the midpoint of $\overline{PE}$.

19. Draw the line from P perpendicular to $\overline{JE}$ at its midpoint.

20. Draw the line through P and X parallel to $\overleftrightarrow{JE}$.

21. Complete the proof of Theorem 2–12:
The measure of an exterior angle of a triangle equals the sum of the measures of the two remote interior angles.

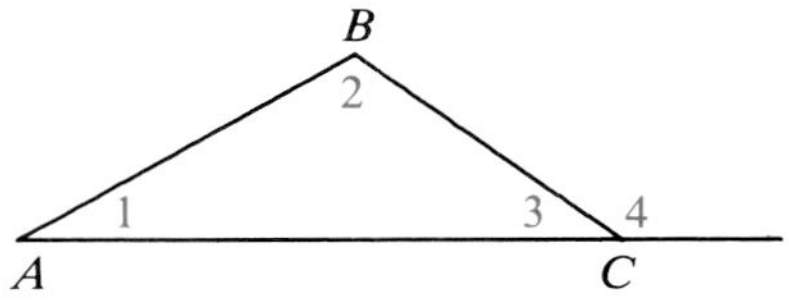

Given: $\triangle ABC$ with exterior $\angle 4$

Prove: $m\angle 4 = m\angle 1 + m\angle 2$

Proof:

Statements	Reasons
1. $m\angle 1 + m\angle 2 + m\angle 3 = 180$	1. _?_
2. $m\angle 3 + m\angle 4 = 180$	2. _?_
3. $m\angle 3 + m\angle 4 = m\angle 1 + m\angle 2 + m\angle 3$	3. _?_
4. $m\angle 4 = m\angle 1 + m\angle 2$	4. _?_

Written Exercises

Draw a triangle that satisfies the conditions stated. If no triangle can satisfy the conditions, write *not possible*.

A

1. **a.** An acute isosceles triangle
   b. A right isosceles triangle
   c. An obtuse isosceles triangle

2. **a.** A scalene right triangle
   b. A scalene isosceles triangle
   c. A scalene obtuse triangle

3. A triangle with two acute exterior angles

4. An obtuse equilateral triangle

Complete.

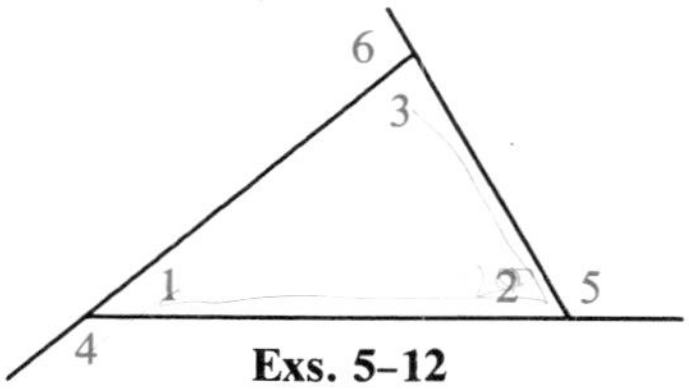

Exs. 5–12

5. If $m\angle 1 = 40$ and $m\angle 2 = 60$, then $m\angle 6 = \underline{\ ?\ }$.
6. If $m\angle 1 = 45$ and $m\angle 3 = 70$, then $m\angle 5 = \underline{\ ?\ }$.
7. If $m\angle 2 = 50$ and $m\angle 3 = 65$, then $m\angle 4 = \underline{\ ?\ }$.
8. If $m\angle 4 = 135$ and $m\angle 2 = 60$, then $m\angle 3 = \underline{\ ?\ }$.
9. If $m\angle 5 = 120$ and $m\angle 1 = 40$, then $m\angle 3 = \underline{\ ?\ }$.
10. If $m\angle 1 = x$, $m\angle 2 = x + 10$, and $m\angle 6 = 120$, then $x = \underline{\ ?\ }$.
11. If $m\angle 2 = 2x - 5$, $m\angle 3 = 3x + 10$, and $m\angle 4 = 140$, then $x = \underline{\ ?\ }$.
12. $m\angle 4 + m\angle 5 + m\angle 6 = \underline{\ ?\ }$.

Find the values of *x* and *y*.

13.

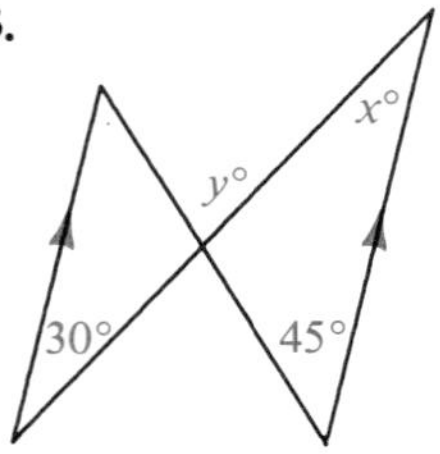

14.

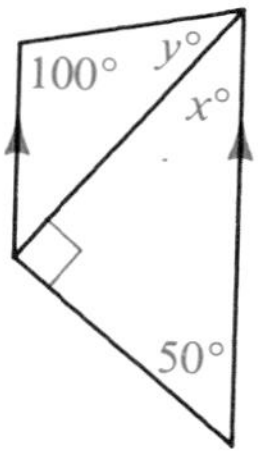

15.

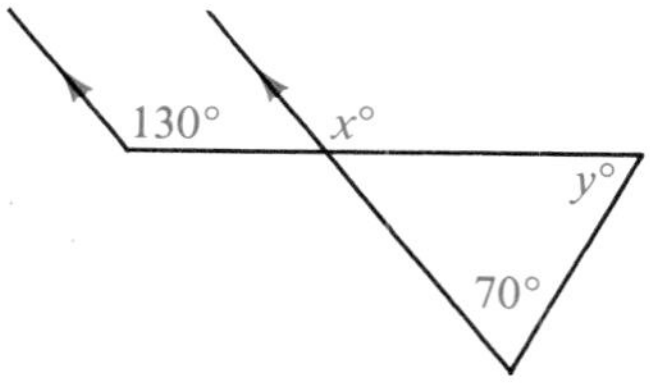

16.

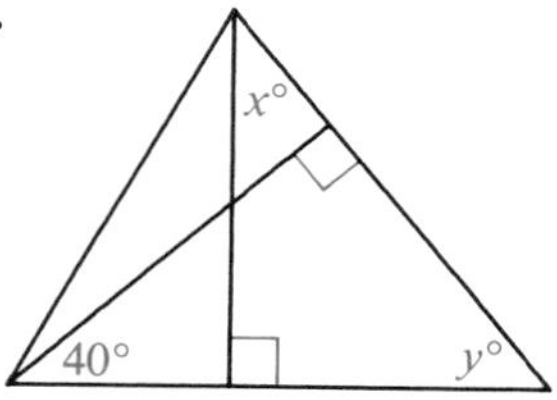

17.

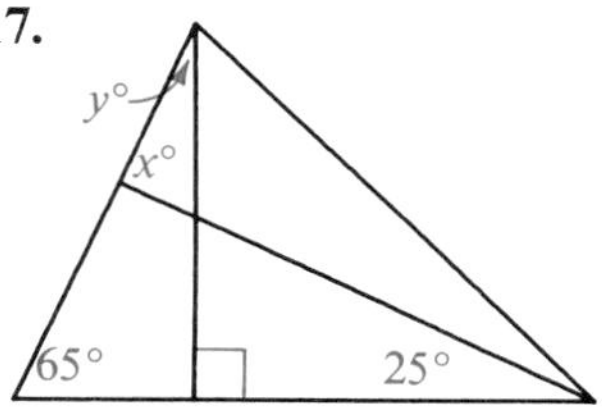

18.

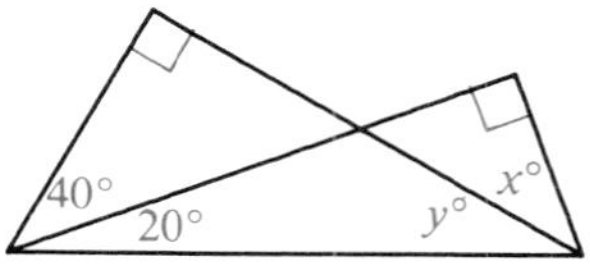

B

19. The lengths of the sides of a triangle are $4n$, $2n + 10$, and $7n - 15$. Is there a value of n that makes the triangle equilateral?
20. The largest two angles of a triangle are two and three times as large as the smallest angle. Find all three measures.
21. In $\triangle ABC$, $m\angle A = 60$ and $m\angle B < 60$. What can you say about $m\angle C$?
22. In $\triangle RST$, $m\angle R = 90$ and $m\angle S > 20$. What can you say about $m\angle T$?

23. Given: $\overline{AB} \perp \overline{BC}$; $\overline{BD} \perp \overline{AC}$
  a. If $m\angle C = 22$, find $m\angle ABD$.
  b. If $m\angle C = 23$, find $m\angle ABD$.
  c. Explain why $m\angle ABD$ always equals $m\angle C$.

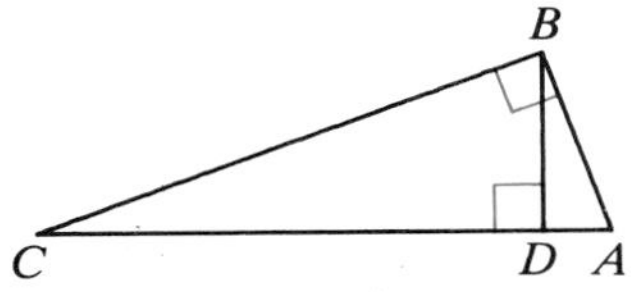

24. The bisectors of $\angle EFG$ and $\angle EGF$ meet at I.
  a. If $m\angle EFG = 40$, find $m\angle FIG$.
  b. If $m\angle EFG = 50$, find $m\angle FIG$.
  c. Explain your results in (a) and (b).

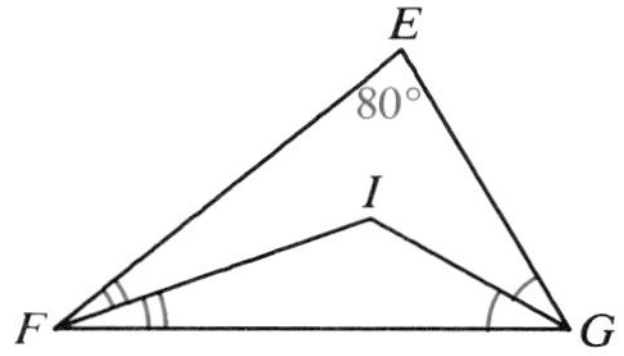

25. Given: $\angle ABD \cong \angle AED$
Prove: $\angle C \cong \angle F$

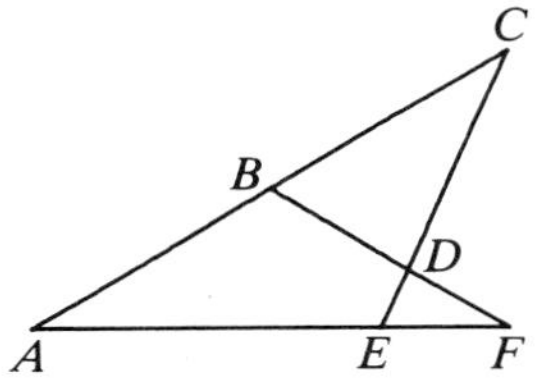

26. Prove Theorem 2-11 by using the figure below.

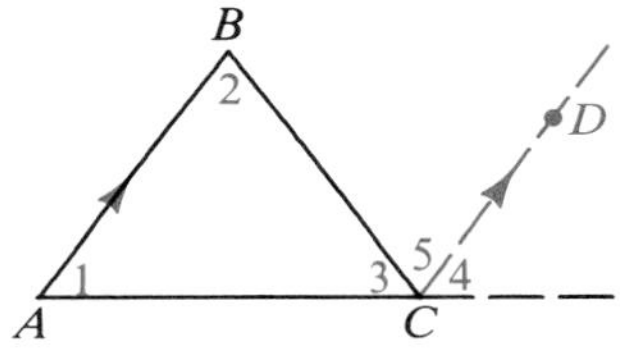

Find the sum of the measures of the angles of each figure. (*Hint:* Divide each figure into triangles.)

27.

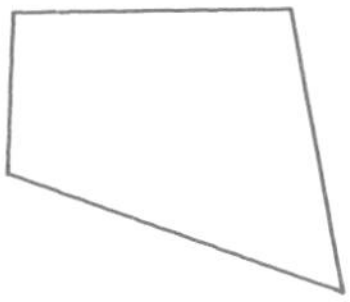

28.

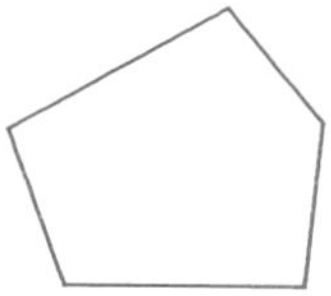

29.

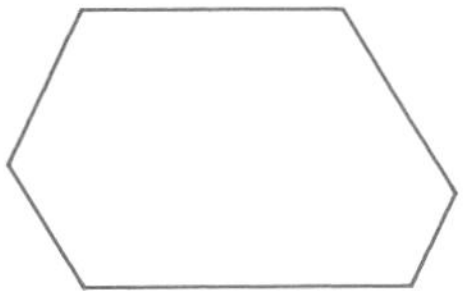

30. Find the measures of $\angle 1$ and $\angle 2$.

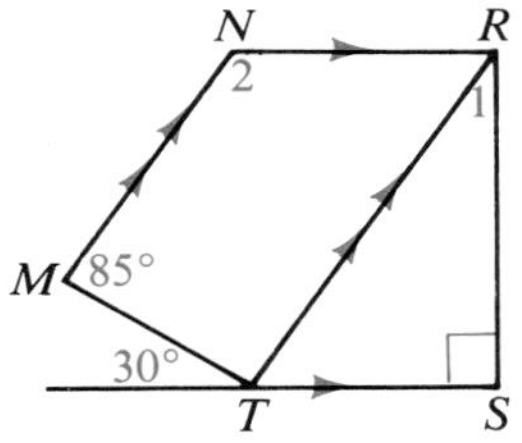

31. Given: $\overrightarrow{GK}$ bisects $\angle JGI$;
$m\angle H = m\angle I$
Prove: $\overline{GK} \parallel \overline{HI}$

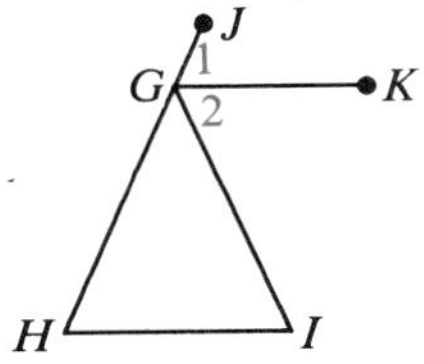

Find the values of x and y.

32.

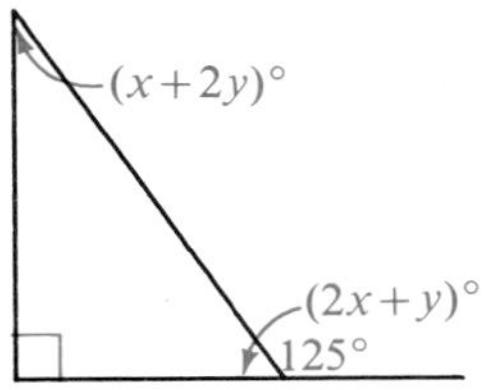

33.

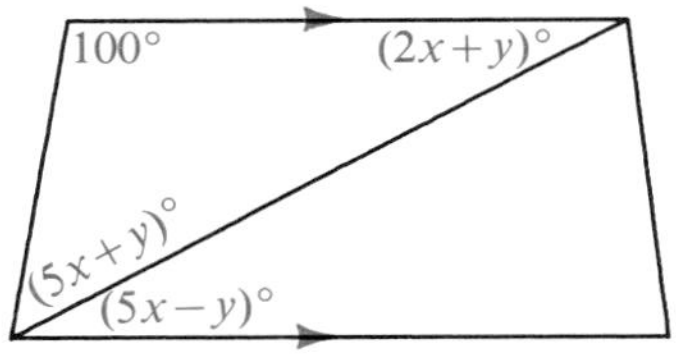

34. Given: $\overline{AB} \perp \overline{BF}$; $\overline{HD} \perp \overline{BF}$;
$\overline{GF} \perp \overline{BF}$; $\angle A \cong \angle G$

Which numbered angles must be congruent?

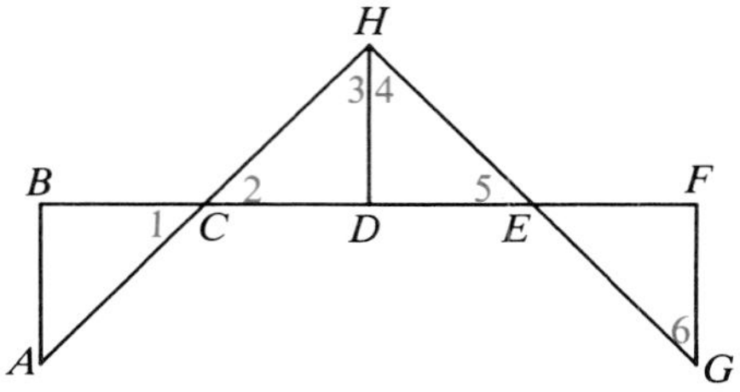

C 35. Given: $\overrightarrow{PR}$ bisects $\angle SPQ$;
$\overline{PS} \perp \overline{SQ}$; $\overline{RQ} \perp \overline{PQ}$

Which numbered angles must be congruent?

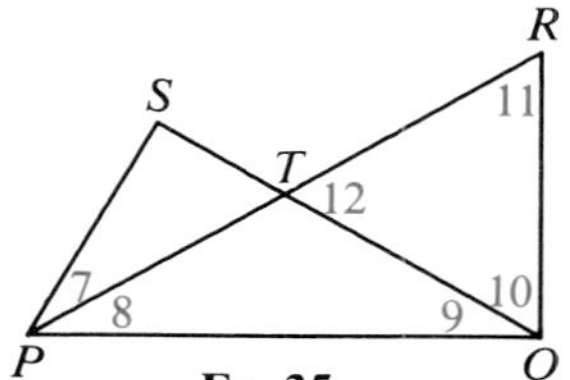

Ex. 35

36. a. Draw two parallel lines and a transversal.
b. Use a protractor to draw bisectors of two same-side interior angles.
c. Measure the angle formed by the bisectors. What do you notice?
d. Prove your answer to (c).

37. A pair of same-side interior angles are trisected (divided into three congruent angles) by the red lines in the diagram. Find out what you can about the angles of $ABCD$.

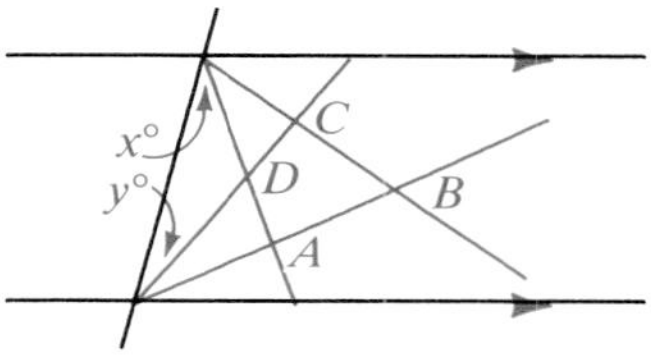

Application

ORIENTEERING

The sport of orienteering involves finding your way from point to point in a wilderness area, using a map and magnetic compass for guidance. You can often locate your position by sighting on specific objects shown on your map. For example, suppose you can see a mountain peak and a lookout tower, both of which are marked on your map (Lily Bay Mountain at ▲ and the lookout tower on Number Four Mountain at ●).

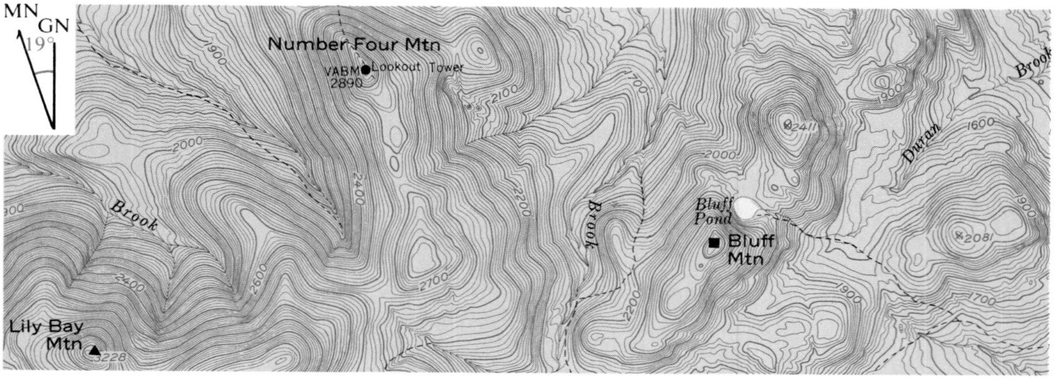

You sight across your compass and find that the lookout tower is 33° east of magnetic north (MN). On your map you draw a line through the tower that makes this same angle with the direction of magnetic north shown on your map. (Be sure to use magnetic north rather than true north. Hiking maps and nautical charts usually give both, but they may differ by as much as 20°. All compass readings used here are given in terms of magnetic north.)

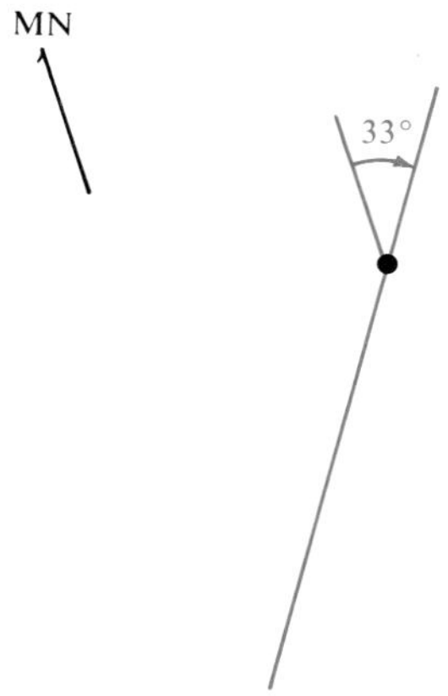

You are somewhere on this line. To find out where, take a sighting on the peak of Lily Bay Mountain. It is 50° west of north. Draw a line through the peak making a 50° angle with magnetic north. You are close to the point where the two lines cross.

If a third landmark is visible, say the summit of Bluff Mountain (■ on the map), you may want to check your position with a third sighting. The three lines should cross at a single point. Usually there is some error in sighting and drawing angles, and instead of meeting exactly at a point the three lines form a triangle. If the triangle is small, you know that your true position is very close to it.

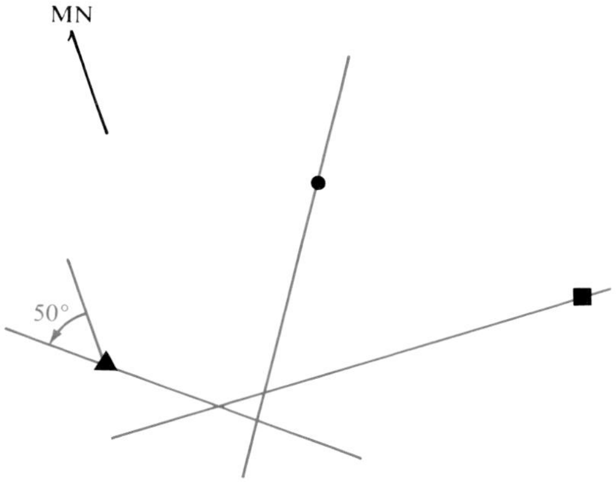

Exercises

1. Sailors use this method of finding their position when navigating near shore, sighting on lighthouses, smokestacks, and other landmarks shown on their charts. They call the small triangle formed by three sighting lines a "cocked hat," and usually mark their position at the corner closest to the nearest hazard. Why is this a sensible rule?
2. Another orienteering party sights on Lily Bay Mountain and the lookout tower and finds the following angles: mountain, 58° west of north; tower, 40° east of north. Are they north of you or south of you?
3. Lillian and Ray both sight Lily Bay Mountain at 70° west of north, but Lillian sees the lookout tower at 40° east of north, while Ray sees it at 20° east of north. Which person is closer to Bluff Mountain?
4. If you head due east from Lily Bay Mountain (90° east of magnetic north), will you pass Bluff Mountain on your right or on your left?

2-5 Angles of a Polygon

The word **polygon** means "many angles." Look at the figures at the left below and note that each polygon is formed by coplanar segments (called *sides*) such that:

(1) Each segment intersects exactly two other segments, one at each endpoint.

(2) No two segments with a common endpoint are collinear.

Polygons

Not Polygons

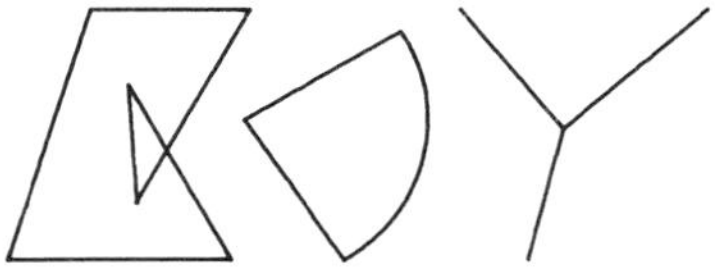

A **convex polygon** is a polygon such that no line containing a side of the polygon contains a point in the interior of the polygon. If you picture the kite shown as lying in a plane, then the outer edge of the kite is not convex, but the outer edge of the yellow part is convex.

When we refer to a polygon in this book we will mean a convex polygon.

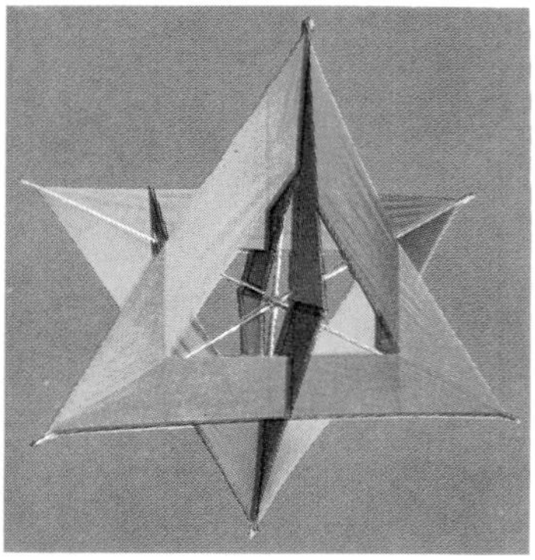

As shown below, polygons are named according to the number of sides. A triangle is the simplest polygon. The terms that we applied to triangles (such as *vertex* and *exterior angle*) also apply to other polygons.

3 sides:	triangle	8 sides:	octagon
4 sides:	quadrilateral	10 sides:	decagon
5 sides:	pentagon	*n* sides:	*n*-gon
6 sides:	hexagon		

When referring to a polygon, we list its consecutive vertices in order. Pentagon *ABCDE* and pentagon *BAEDC* are two of the many correct names for the polygon shown.

A segment joining two nonconsecutive vertices is a **diagonal** of the polygon. The diagonals of the pentagon at the right are indicated by dashes.

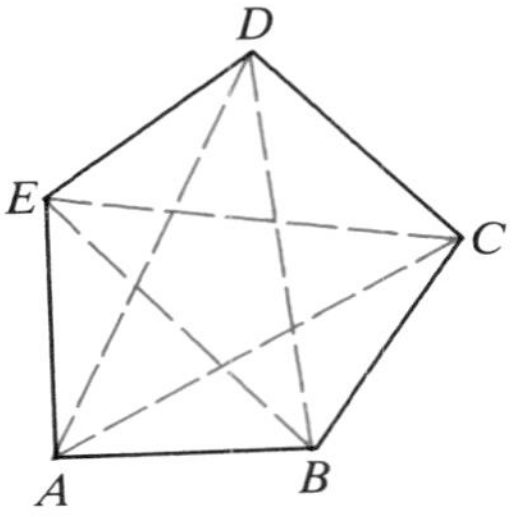

If you draw all the diagonals from just *one* vertex of a polygon, you can determine the sum of the measures of the angles of the polygon.

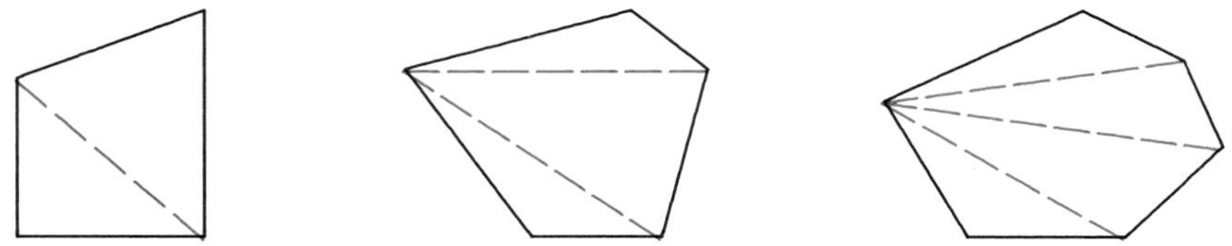

4 sides	5 sides	6 sides
2 triangles	3 triangles	4 triangles
Angle sum = 2(180)	Angle sum = 3(180)	Angle sum = 4(180)

Note that the number of triangles formed in each polygon is two less than the number of sides. This result suggests the following theorem.

Theorem 2-13

The sum of the measures of the angles of a convex polygon with n sides is $(n - 2)180$.

Theorem 2-13 deals with the sum of the measures of the *interior* angles of a polygon and states that the sum depends on the number of sides of the polygon. On the other hand, the sum of the exterior angles does *not* depend on the number of sides of the polygon. It is always 360.

Theorem 2-14

The sum of the measures of the exterior angles of any convex polygon, one angle at each vertex, is 360.

To prove Theorem 2-14, we reason as follows. At each vertex of a polygon, the interior and exterior angles are supplementary:

$$m\angle 1 + m\angle 2 = 180,\ m\angle 3 + m\angle 4 = 180, \text{ and so on.}$$

If the polygon has n vertices, then there are n pairs of supplementary angles and the sum of the measures of all these angles is $180n$.

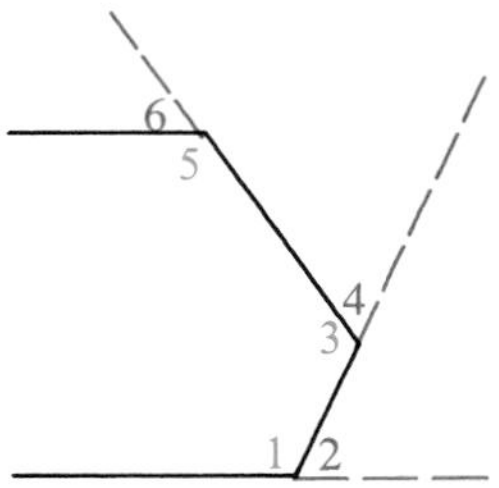

Thus:
$$\begin{aligned} 180n &= (\text{interior angle sum}) + (\text{exterior angle sum}) \\ 180n &= 180(n - 2) + (\text{exterior angle sum}) \\ 180n &= 180n - 360 + (\text{exterior angle sum}) \\ 360 &= (\text{exterior angle sum}) \end{aligned}$$

Example 1 A polygon has 22 sides. Find (a) the sum of the measures of the interior angles and (b) the sum of the measures of the exterior angles, one angle at each vertex.

Solution (a) Interior angle sum = (22 − 2)180 = 3600 (Theorem 2–13)
(b) Exterior angle sum = 360 (Theorem 2–14)

Polygons can be equiangular or equilateral. If a polygon is both equiangular and equilateral, it is called a **regular polygon.**

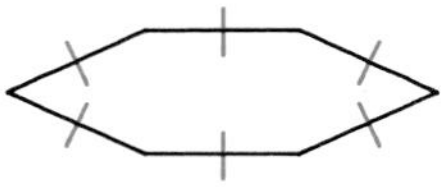

Equilateral hexagon

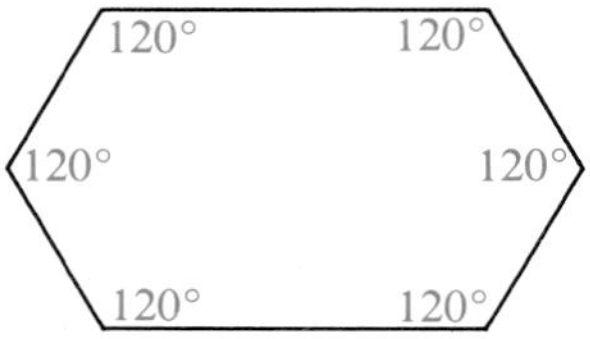

Equiangular hexagon

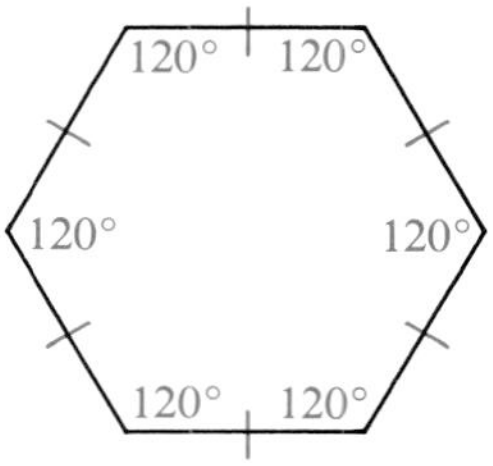

Regular hexagon

Example 2 A regular polygon has 12 sides. Find the measure of each interior angle.

Solution 1 Interior angle sum = (12 − 2)180 = 1800
Each of the 12 congruent angles has measure 1800 ÷ 12, or 150.

Solution 2 Exterior angle sum = 360
Each of the 12 congruent exterior angles has measure 360 ÷ 12, or 30. Since each interior angle is a supplement of an exterior angle, each interior angle has measure 180 − 30, or 150.

Classroom Exercises

Is the red outline of the figure a convex polygon? If not, is it a nonconvex polygon?

1. 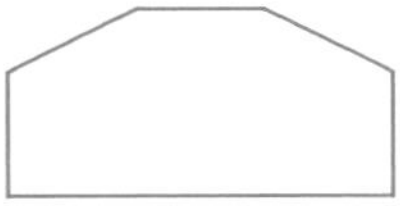

2. 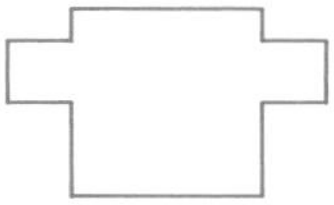

3. 

4. 

5. 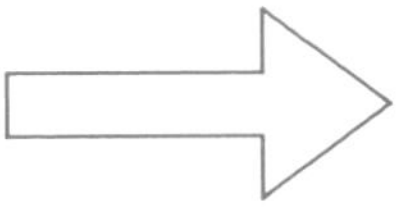

6. 

7. Imagine stretching a rubber band around each of the figures in Exercises 1–6. What is the relationship between the rubber band and the figure when the figure is a convex polygon?

8. A polygon has 102 sides.
a. What is its interior angle sum?
b. What is its exterior angle sum?

9. Complete the table for regular polygons.

Number of sides	6	10	20	?	?	?	?
Measure of each ext. ∠	?	?	?	10	20	?	?
Measure of each int. ∠	?	?	?	?	?	179	90

Written Exercises

For each polygon, what is (a) the interior angle sum? (b) the exterior angle sum?

A

1. Pentagon
2. Hexagon
3. Quadrilateral
4. Octagon
5. Decagon
6. n-gon
7. Complete the table for regular polygons.

Number of sides	9	15	30	?	?	?	?
Measure of each ext. ∠	?	?	?	6	8	?	?
Measure of each int. ∠	?	?	?	?	?	165	178

8. A baseball diamond's home plate has three right angles. The other two angles are congruent. Find their measure.

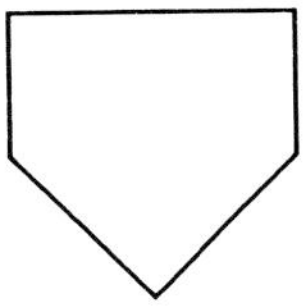

Sketch the polygon described. If no such polygon exists, write *not possible*.

9. A quadrilateral that is equiangular but not equilateral.
10. A quadrilateral that is equilateral but not equiangular.
11. A pentagon that is equilateral but not equiangular.
12. A triangle that is equilateral but not equiangular.
13. A regular polygon, one of whose angles has measure 110.

14. The face of a honeycomb consists of interlocking regular hexagons. What is the measure of each angle of these hexagons?

B **15.** The sum of the measures of the interior angles of a polygon is four times the sum of the measures of its exterior angles, one angle at each vertex. How many sides does the polygon have?

16. The measure of each interior angle of a regular polygon is eight times that of an exterior angle. How many sides does the polygon have?

17. Make a sketch showing how to tile a floor using both squares and regular octagons.

18. **a.** What is the measure of each interior angle of a regular pentagon?
b. Can you tile a floor with tiles shaped like regular pentagons?

19. **a.** Is it possible to tile a floor with tiles shaped like equilateral triangles? (Ignore the difficulty in tiling along the edges of the room.)
b. Make a sketch showing how such tiles could be placed together to form a regular hexagon.

20. The cover of a soccer ball consists of interlocking regular pentagons and regular hexagons, as shown at the right. The second diagram shows that pentagons and hexagons cannot be interlocked in the same pattern to tile a floor. Why not?

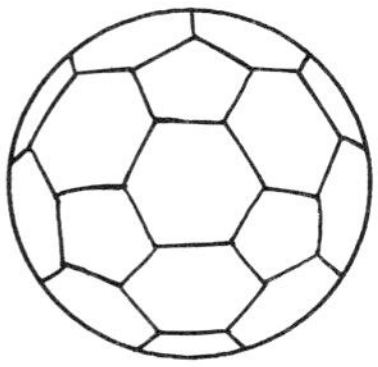
Possible

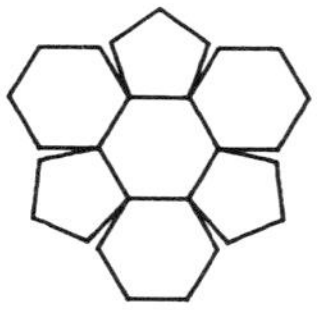
Impossible

21. *ABCDEFGHIJ* is a regular decagon. If sides $\overline{AB}$ and $\overline{CD}$ are extended to meet at *K*, find the measure of $\angle K$.

22. In quadrilateral *ABCD*, $m\angle A = x$, $m\angle B = 2x$, $m\angle C = 3x$, and $m\angle D = 4x$. Find the value of *x* and then state which pair of sides of *ABCD* must be parallel.

23. In pentagon *PQRST*, $m\angle P = 90$ and $m\angle Q = 150$. $\angle S$ and $\angle T$ are each twice as large as $\angle R$.
a. Find the measures of $\angle R$, $\angle S$, and $\angle T$.
b. Which pair of sides of *PQRST* must be parallel?

24. The sum of the measures of the interior angles in a polygon is known to be between 2500 and 2600. How many sides does the polygon have?

25. Find $m\angle A + m\angle B + m\angle C + m\angle D + m\angle E$.

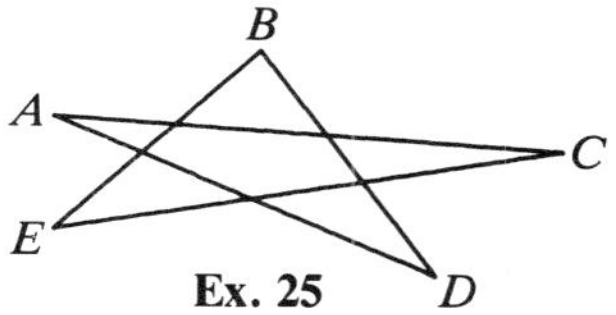

Ex. 25

C **26.** The sum of the measures of the interior angles of a polygon with *n* sides is *S*. Not using *n* in your answer, express in terms of *S* the sum of the measures of the angles of a polygon with:
a. $n + 1$ sides **b.** $2n$ sides

27. The formula $S = (n - 2)180$ can apply to nonconvex polygons if you allow the measure of an interior angle to be more than 180.
a. Illustrate this with a diagram that shows interior angles with measures greater than 180.
b. Does the reasoning leading up to Theorem 2–13 apply to your figure?

COMPUTER KEY-IN

Theorem 2–13 gives a formula for the sum of the measures of the angles of a convex polygon. Using this formula you can compute the measure of an angle of a convex polygon if you know the measures of its other angles.

Computers are often used to generate random numbers, usually by means of a built-in operation called RND. Since usage of RND varies with different computers, check with the manual for your computer. In line 20 of the following BASIC program, the computer randomly selects a polygon of 3, 4, or 5 sides. After printing the measures of all but one of the angles, the computer then asks the user to find the measure of the missing angle.

```
10   DIM A(5)
15   REM  N REPRESENTS THE NUMBER OF SIDES OF POLYGON
20   LET N = INT(RND(1) * 3) + 3
30   LET S = (N - 2) * 180
35   REM  T REPRESENTS SUM OF THE CHOSEN INTERIOR ANGLES
40   LET T = 0
50   FOR I = 1 TO N - 1
60   LET R = S - T
70   IF R > 180 THEN R = 180
75   REM  A(I) IS A RANDOM ANGLE BETWEEN 0 AND 180
80   LET A(I) = INT(RND(1) * R)
90   LET T = T + A(I)
100  NEXT I
105  REM  M REPRESENTS THE MISSING INTERIOR ANGLES
110  LET M = S - T
120  IF M >= 180 THEN 40
130  PRINT "HERE ARE THE MEASURES OF ALL BUT ONE OF THE INTERIOR
     ANGLES OF A POLYGON."
140  FOR J = 1 TO N - 1
150  PRINT A(J)
160  NEXT J
170  PRINT "-----"
180  PRINT "WHAT IS THE MEASURE OF THE MISSING ANGLE";
190  INPUT X
200  IF X = M THEN 230
210  PRINT "SORRY, THE MISSING ANGLE HAS MEASURE"; M
220  END
230  PRINT "YOU ARE CORRECT"
240  END
```

Notice that lines 70 and 80 make it certain that A(I) will be between 0 and 180. The purpose of line 120 is to check whether the computer has generated a convex polygon. If not, the computer is asked to generate a new set of angles before printing the measures.

Type the program into your computer and RUN it several times. Give both correct and incorrect responses, checking the computer results.

Self-Test 2

Complete.

1. If the measure of each angle of a triangle is less than 90, the triangle is called _?_.
2. If a triangle has no congruent sides, it is called _?_.
3. In the diagram, $m\angle 1 =$ _?_ and $m\angle 2 =$ _?_.
4. If the measures of the acute angles of a right triangle are $2x + 4$ and $3x - 9$, then $x =$ _?_.

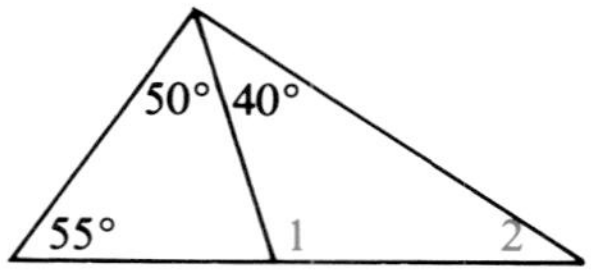

Ex. 3

5. A regular polygon is both _?_ and _?_.
6. In a regular decagon, the sum of the measures of the exterior angles is _?_ and the measure of each interior angle is _?_.
7. If the measure of each angle of a polygon is 174, then the polygon has _?_ sides.

Conditional Statements

Objectives

1. Recognize the hypothesis and the conclusion of a conditional.
2. State conditionals in if-then form.
3. State the converse of an if-then statement.
4. Understand the meaning of *if and only if.*
5. Use circle diagrams to represent if-then statements.
6. State the contrapositive and the inverse of an if-then statement.
7. Understand the relationship between logically equivalent statements.
8. Draw correct conclusions from given statements.

2-6 If-Then Statements and Their Converses

Each of the following statements is an **if-then statement.**

> *If* Jane is on our team, *then* we are sure to win.
> *If* two lines intersect, *then* exactly one plane contains them.
> *If* $x = 3$, *then* $x^2 = 9$.

We can use the letter p to represent the words in red, and q for those in blue. Then we have the basic form for an if-then statement that is shown at the top of the next page.

If p, then q.

↑ ↑

***p*: hypothesis** ***q*: conclusion**

An if-then statement is sometimes called a **conditional statement,** or simply a **conditional.** The following conditionals are equivalent.

General Form	*Example*
If p, then q.	If $x = 3$, then $x^2 = 9$.
p implies q.	$x = 3$ implies $x^2 = 9$.
p only if q.	$x = 3$ only if $x^2 = 9$.
q if p.	$x^2 = 9$ if $x = 3$.

Although conditionals are sometimes expressed without the words "if" and "then," they can always be restated in if-then form. For example:

Conditional: All panthers are cats.
If-then form: If an animal is a panther, then it is a cat.

Conditional: The diagonals of a square are congruent.
If-then form: If a figure is a square, then its diagonals are congruent.

Conditional: The sum of two odd integers is even.
If-then form: If two numbers are odd integers, then their sum is even.

An if-then statement is false if an example can be found for which the hypothesis is true and the conclusion is false. Such an example is called a **counterexample.** It takes only one counterexample to disprove a statement.

Example Do you think the statement is true or false?
If it is false, draw a diagram that shows a counterexample.
If it appears to be true, draw a diagram. List, in terms of the diagram, what is given and what is to be proved.

a. All isosceles triangles are acute.
b. The diagonals of an equiangular quadrilateral are congruent.

Solution **a.** False (The diagram shows an isosceles triangle that is not acute.)

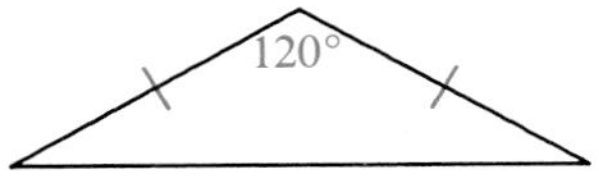

b. True

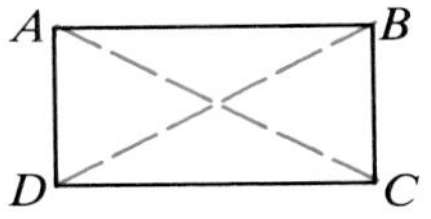

Given: $ABCD$ is an equiangular quadrilateral.
Prove: $\overline{AC} \cong \overline{BD}$

When the hypothesis and conclusion of an if-then statement are interchanged, the statement formed is called the **converse** of the original statement.

Statement: If p, then q.
Converse: If q, then p.

Some true statements have false converses.

True statement: If a figure is a triangle, then it is a polygon.
False converse: If a figure is a polygon, then it is a triangle.

Some true statements have true converses.

True statement: If a triangle is equilateral, then it is equiangular.
True converse: If a triangle is equiangular, then it is equilateral.

When a statement and its converse are both true, we can combine them into one statement by using the words "if and only if" as follows:

A triangle is equilateral *if and only if* it is equiangular.

This means that:

1. A triangle is equilateral if it is equiangular.
2. A triangle is equilateral only if it is equiangular.

Similarly the statement

p if and only if q

means:

1. p if q. (Equivalent to "If q, then p.")
2. p only if q. (Equivalent to "If p, then q.")

All the definitions you have learned could have been stated as if-and-only-if statements. For example, the definition *Perpendicular lines are two lines that form right angles* tells us two things:

1. If two lines are perpendicular, then the lines form right angles.
2. If two lines form right angles, then the lines are perpendicular.

Thus, the definition could have been stated in this form:

Two lines are perpendicular if and only if the lines form right angles.

Classroom Exercises

State the hypothesis and conclusion of each statement.

1. If $2x - 1 = 5$, then $x = 3$.

2. If I'm smart, then she's a genius.

3. $x \div (-2) = 9$ implies that $x = -18$.

4. I'll go if you go.

5. There is smoke only if there is fire.

6. You can if you try.

7. You can only if you try.

Give the converse of the statement in the exercise listed.

8. Exercise 1 **9.** Exercise 4 **10.** Exercise 5

Express each statement in if-then form.

11. All equilateral triangles are isosceles. (Begin "If a figure is . . .")
12. The diagonals of a rectangle are congruent.
13. All students love vacations. (Begin "If a person is . . .")
14. Every rectangle is equiangular.
15. He will drive provided it doesn't snow.
16. $x > -2$ whenever $-4x < 8$.

17. Give an example of a true statement that has a false converse.
18. Give an example of a true statement that has a true converse.

Replace each statement by two if-then statements that are converses of each other.

19. A triangle is obtuse if and only if it contains an obtuse angle.
20. Two angles are complementary if and only if the sum of their measures is 90.

Written Exercises

For each statement, state (a) the hypothesis, (b) the conclusion, and (c) the converse.

A **1.** If $3x - 7 = 32$, then $x = 13$.
2. If $\overline{AB} \perp \overline{BC}$, then $m\angle ABC = 90$.
3. I'll try if you will.
4. I can't sleep if I'm not tired.
5. $|x| = 0$ only if $x = 0$.
6. $a + b = a$ implies $b = 0$.

For each statement, (a) rewrite the statement in if-then form and state whether it is true or false. Then (b) write the converse and state whether it is true or false.

7. All Olympic competitors are athletes.
8. Every positive number has two square roots.
9. $x^2 = 0$ only if $x = 0$.
10. $x^2 = 16$ when $x = 4$.
11. The product of two odd integers is odd.
12. The sum of two even integers is even.
13. $-2x < 2$ implies $x > -1$.
14. $\triangle XYZ$ is obtuse only if $\angle XYZ$ is obtuse.
15. Every regular polygon is equiangular.
16. All even integers are divisible by 4.

Rewrite each statement as two if-then statements that are converses of each other.

17. Two angles are congruent angles if and only if their measures are equal.
18. Two angles are supplementary if and only if the sum of their measures is 180.
19. $ab > 0$ if and only if a and b are both positive or both negative.
20. $(x - 4)(x + 6) = 0$ if and only if $x = 4$ or $x = -6$.
21. Postulate 10 may also be worded in this way: "If two lines are parallel, then corresponding angles formed by the two lines and a transversal are congruent." Write the converse of this statement. Is the converse true?
22. Theorem 2–5 may also be worded in this way: "If alternate interior angles formed by two lines and a transversal are congruent, then the lines are parallel." Write the converse of this statement. Is the converse true?
23. **a.** Write the converse of the Subtraction Property of Equality: If $a = b$ and $c = d$, then $a - c = b - d$.
    b. Choose values of a, b, c, and d to show that the converse is false.

Tell whether you think each statement is true or false.
If false, draw a diagram that shows a counterexample.
If true, draw a diagram. List, in terms of the diagram, what is given and what is to be proved. Do *not* write a proof.

B 24. If a triangle has two congruent sides, then the angles opposite those sides are congruent.
25. If a triangle has two congruent angles, then the sides opposite those angles are congruent.
26. Two triangles have equal perimeters only if they have congruent sides.
27. All diagonals of a regular pentagon are congruent.
28. If both pairs of opposite sides of a quadrilateral are parallel, then each side is congruent to the side opposite it.
29. If the diagonals of a quadrilateral are congruent and also perpendicular, then the quadrilateral is a regular quadrilateral.
30. The diagonals of an equilateral quadrilateral are congruent.
31. The diagonals of an equilateral quadrilateral are perpendicular.

In the conditional statement "if p, then q," p is said to be a *sufficient* condition for q to occur. Similarly, q is said to be a *necessary* condition for p. In Exercises 32–37 tell whether the first statement is necessary for the second statement, sufficient for it, or both necessary and sufficient for it.

Example	*First statement*	*Second statement*
	a. An integer is divisible by 2.	An integer is even.
	b. Lines l and m are coplanar.	Lines l and m are parallel.

Solution **a.** necessary and sufficient (An integer is divisible by 2 if and only if the integer is even.)

b. necessary (If lines l and m are parallel, then they are coplanar. Note that the first statement is not sufficient for the second because two lines may be coplanar without being parallel.)

	First statement	*Second statement*
C **32.**	$x > 4$	x is positive.
33.	An integer is odd.	The square of an integer is odd.
34.	Lines l and m do not intersect.	Lines l and m are parallel.
35.	$\angle A$ is a right angle.	$\triangle ABC$ is a right triangle.
36.	A polygon is equilateral.	A polygon is regular.
37.	Alternate interior angles formed by lines l and m and transversal t are congruent.	Lines l and m are parallel.

38. a. Given: $\overline{AB} \parallel \overline{DC}$; $\overline{AD} \parallel \overline{BC}$
Prove: $\angle A \cong \angle C$; $\angle B \cong \angle D$

b. Tell what is given and what is to be proved in the converse of part (a). Then write a proof of the converse.

c. Combine what you have proved in parts (a) and (b) into an if-and-only-if statement.

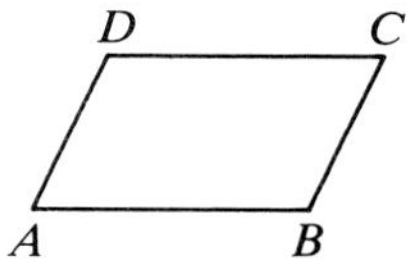

2-7 Converse, Contrapositive, Inverse

To show the relationship between an if-then statement and its converse, it is helpful to use circle diagrams (also called Venn diagrams or Euler diagrams).

To represent a statement p, we draw a circle named p. If p is true, we think of a point inside circle p. If p is false, we think of a point outside circle p.

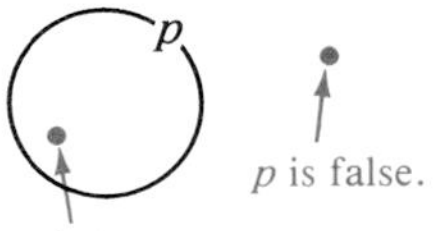

In the diagram at the left below, a point that lies inside circle p must also lie inside circle q. In other words: *If p, then q.* Check to see that the middle diagram represents the converse: *If q, then p.* Check the diagram at the right also.

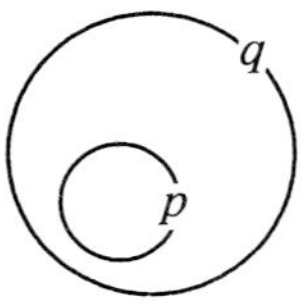

If p, then q.

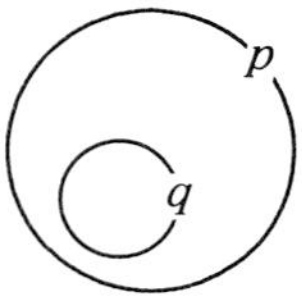

If q, then p.

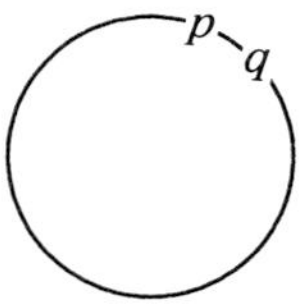

p if and only if q.

Compare the following if-then statements.

Statement: If p, then q.
Contrapositive: If not q, then not p.

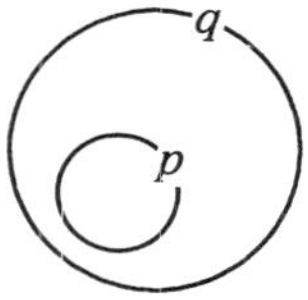

You already know that the diagram at the right represents "If p, then q." The diagram also represents "If not q, then not p," because a point that isn't inside circle q can't be inside circle p either. Since the statement and its contrapositive are both true or else both false, they are called **logically equivalent.** The following statements are logically equivalent.

True statement: If a figure is a triangle, then it is a polygon.
True contrapositive: If a figure is not a polygon, then it is not a triangle.

Since a statement and its contrapositive are logically equivalent, we may prove a statement by proving its contrapositive. Sometimes that is easier.

There is one more conditional related to "If p, then q" that we will consider. A statement and its *inverse* are *not* logically equivalent.

Statement: If p, then q.
Inverse: If not p, then not q.

True statement: If a figure is a triangle, then it is a polygon.
False inverse: If a figure is not a triangle, then it is not a polygon.

Summary of Related If-Then Statements

Given statement: If p, then q.
Contrapositive: If not q, then not p.
Converse: If q, then p.
Inverse: If not p, then not q.

A statement and its contrapositive are logically equivalent.
A statement is *not* logically equivalent to its converse or to its inverse.

The relationships just summarized permit us to base conclusions on the contrapositive of a true if-then statement but *not* on the converse or inverse. For example, suppose we accept this statement as true:

All Olympic competitors are athletes.
(If a person is an Olympic competitor, then that person is an athlete.)

This statement is paired with four different statements below.

1. *Given:* If p, then q. — All Olympic competitors are athletes.
   p — Ozzie is an Olympian.
   Conclude: q — Ozzie is an athlete.

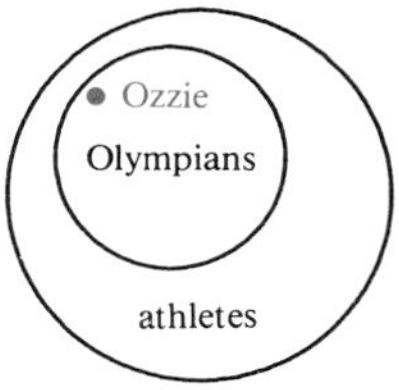

2. *Given:* If p, then q. — All Olympic competitors are athletes.
   not q — Ned is not an athlete.
   Conclude: not p — Ned is not an Olympic competitor.

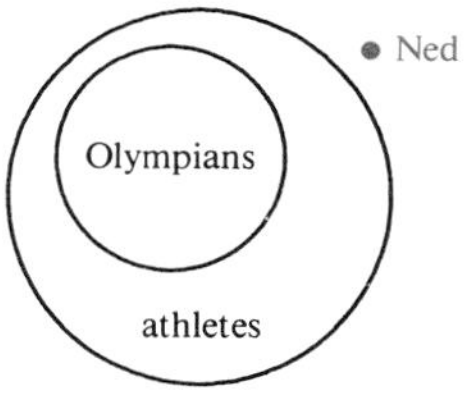

3. *Given:* If p, then q. — All Olympic competitors are athletes.
   q — Anne is an athlete.
   No conclusion follows. — Anne might be an Olympic competitor or she might not be.

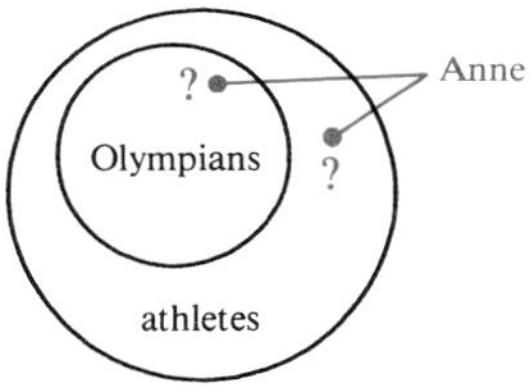

4. *Given:* If p, then q. — All Olympic competitors are athletes.
   not p — Nancy is not an Olympic competitor.
   No conclusion follows. — Nancy might be an athlete or she might not be.

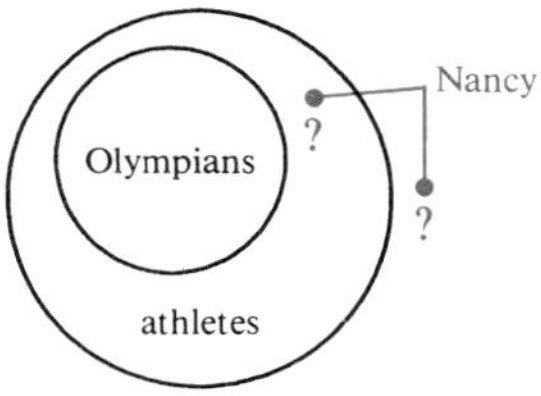

Classroom Exercises

1. State the contrapositive of each statement.
   a. If $x = 3$, then $x^2 + 1 = 10$.
   b. If $y < 5$, then $y \neq 6$.
   c. If a polygon is a triangle, then the sum of the measures of its angles is 180.
   d. If you can't do it, then I can't do it.

2. State the converse of each statement in Exercise 1.

3. State the inverse of each statement in Exercise 1.

4. A certain conditional is true. Must its converse be true? Must its inverse be true? Must its contrapositive be true?

5. A certain conditional is false. Must its converse be false? Must its inverse be false? Must its contrapositive be false?

Classify each statement as true or false. Then classify its contrapositive, converse, and inverse as true or false.

6. If $x = -5$, then $x^2 = 25$.
7. If $\angle 1 \cong \angle 2$, then $l \parallel n$.
8. If $l \parallel n$, then t is perpendicular to both l and n.
9. If l is not parallel to n, then $m\angle 1 + m\angle 3 \neq 180$.

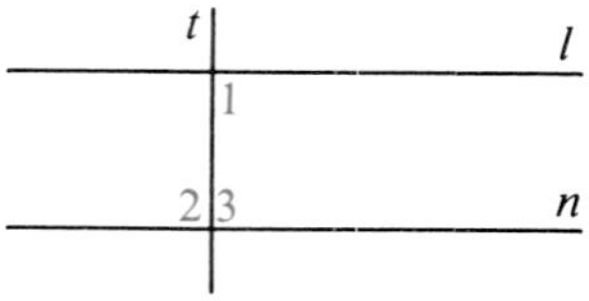

Exs. 7–9

Given statement: All whales are mammals.

10. Reword the statement in if-then form.
11. Explain how the diagram represents both the statement and its contrapositive.
12. Accept the statement below as true and pair it with the given statement. What can you conclude? If no conclusion is possible, say so.
    a. Moby is a whale.
    b. Mabel is not a whale.
    c. Manfred is a mammal.
    d. Myrtle is not a mammal.

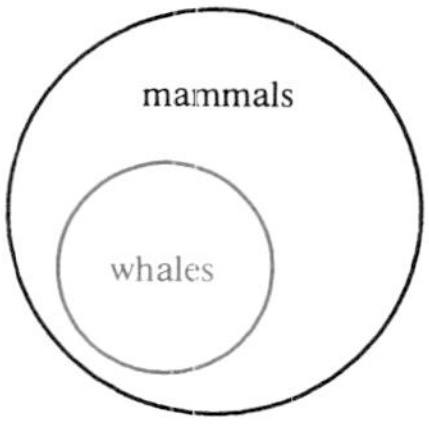

Exs. 11–13

13. Copy the diagram on the chalkboard. Locate points to represent statements (a)-(d) in Exercise 12. Use the points to check your conclusions in Exercise 12.

Accept the first two statements as true. Does the third statement follow as a conclusion?

14. An elephant never forgets.
    Ed is not an elephant.
    Thus, Ed sometimes forgets.
15. An elephant never forgets.
    Ellie never forgets.
    Thus, Ellie is an elephant.
16. An elephant never forgets.
    Edith sometimes forgets.
    Thus, Edith is not an elephant.

Written Exercises

State (a) the contrapositive, (b) the converse, and (c) the inverse of each statement.

A
1. If $n = 21$, then $5n - 5 = 100$.
2. If that is red, then this is white.
3. If x is not even, then $x + 1$ is not odd.
4. If Gregory is not here, then he is not well.

Make a circle diagram to illustrate each statement.

5. If the car has airbags, then it must be new.
6. A triangle is equilateral if and only if it is equiangular.
7. All mice like cheese.
8. No musician likes noise.

For each statement in Exercises 9–15, copy and complete a table like the one shown below.

		True/False?
Statement	?	?
Contrapositive	?	?
Converse	?	?
Inverse	?	?

9. If $\angle 1 \cong \angle 2$, then $\angle 1$ and $\angle 2$ are vertical angles.
10. If $x > 6$, then $x = 5$.

B 11. If $AM = MB$, then M is the midpoint of $\overline{AB}$.

12. If $x^2 - 1 = 99$, then $x = 10$.
13. If a is not negative, then $|a| = a$.
14. If $-3a > -6$, then $a < 2$.
15. If $x^2 > y^2$, then $x > y$.

16. Given: All senators are at least 30 years old.
    a. Reword this statement in if-then form.
    b. Make a circle diagram to illustrate the statement.
    c. If the given statement is true, what can you conclude from each of the following additional statements? If no conclusion is possible, say so.
    (1) Jose Avila is 48 years old.
    (2) Rebecca Castelloe is a senator.
    (3) Constance Brown is not a senator.
    (4) Ling Chen is 29 years old.

In Exercises 17–21, assume that the given statement is true. Then tell what you can conclude if each statement in (a)-(d) is also true. If no conclusion is possible, say so.

17. Given: If a quadrilateral is equiangular, then its diagonals are congruent.
    a. $ABCD$ is equiangular.
    b. $PQRS$ has congruent diagonals.
    c. In $WXYZ$, $m\angle X = 80$.
    d. In $EFGH$, $EG > FH$.
18. Given: If it is not raining, then I am happy.
    a. I am not happy.
    b. It is not raining.
    c. I am overjoyed.
    d. It is raining.
19. Given: All my students love geometry.
    a. Stu is my student.
    b. Luis loves geometry.
    c. Stella is not my student.
    d. George does not love geometry.

For Exercises 20 and 21, see the directions on page 95.

20. Given: If a triangle has two congruent sides, then the angles opposite those sides are congruent.
   a. In $\triangle ABC$, $AB = AC$.
   b. In $\triangle DEF$, $m\angle D \neq m\angle F$.
   c. In $\triangle LMN$, $m\angle L = m\angle M$.
   d. $\triangle RST$ is scalene.

C 21. Given: If both p and q are true, then r or s is false.
   a. r is false.
   b. r is true.
   c. r is true and s is true.
   d. Neither p nor q is true.

22. a. Draw a circle diagram for the converse of the statement "If p, then q."
   b. Does the diagram you have drawn also represent the statement "If not p, then not q"? Explain.
   c. What have you shown about the converse and the inverse of a statement?

Prove each of the following statements by proving its contrapositive. Begin by writing what is given and what is to be proved.

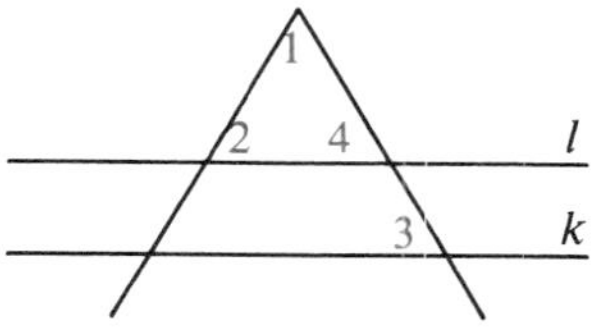

23. If l is not parallel to k, then

$$m\angle 1 + m\angle 2 + m\angle 3 \neq 180.$$

24. If the square of an integer n is odd, then n is odd.

Self-Test 3

Given: If $\triangle ABC$ is a right triangle, then $\angle C$ is a right angle.

1. What is the hypothesis of the statement?
2. What is the conclusion of the statement?
3. What is the converse of the statement? Is it true or false?
4. What is the inverse of the statement? Is it true or false?
5. What is the contrapositive of the statement? Is it true or false?
6. Use the statement: If Dan can't go, then Valerie can go.
   Pair that statement with each statement below. Accept both statements as true. What can you conclude? If no conclusion is possible, say so.
   a. Dan can't go.
   b. Dan can go.
   c. Valerie can go.
   d. Valerie can't go.
7. Use the statement: All squares are rectangles.
   a. Rewrite the statement in if-then form.
   b. Make a circle diagram to illustrate the statement.
8. Rewrite the following statement as two if-then statements: A triangle is equilateral if and only if all its sides are congruent.

EXTRA

Finite Geometry

Recall that this text treats *point, line,* and *plane* as undefined terms. Since the terms aren't defined, we are free to interpret them in any way that we wish. Our interpretation has been the ordinary one, the one in which a line is filled with points. Let's try to find some other interpretations.

Geometry A

We use a diagram to indicate the way to interpret the terms *point, line,* and *plane.* You should read the diagram in such a way that you identify the following:

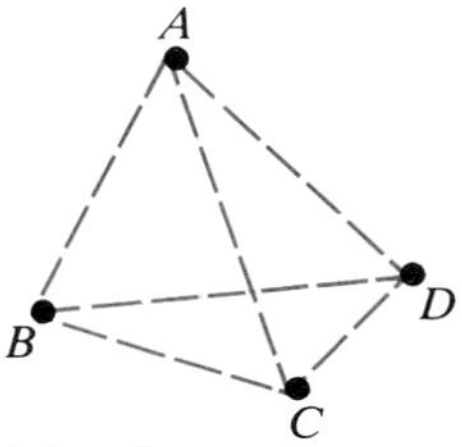

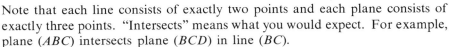

Exactly four points: *A, B, C, D*
Exactly six lines: (*AB*), (*AC*), (*AD*), (*BC*), (*BD*), (*CD*)
Exactly four planes: (*ABC*), (*ABD*), (*ACD*), (*BCD*)

Note that each line consists of exactly two points and each plane consists of exactly three points. "Intersects" means what you would expect. For example, plane (*ABC*) intersects plane (*BCD*) in line (*BC*).

Check the diagram against Postulates 5–9 on page 46. (In this Extra we will set Postulates 1–4 aside and work just with the five postulates introduced on page 46.) Since all five postulates are satisfied, the diagram shows that Postulates 5–9 can be interpreted in such a way that a *finite geometry* is possible. All of space could consist of just four points! (But the postulates don't limit us to that kind of space.)

In Geometry A, the terms *point, line,* and *plane* had interpretations not entirely unrelated to the usual ones. The next geometry will be far different. This time we won't use a diagram at all, but we'll explain in words how to interpret the basic undefined terms.

Geometry B

Term:	point	line	plane
Interpretation:	person	committee	club

Let's make some minor adjustments in vocabulary. For example, Postulate 6 states that there is exactly one line through any two points. We can interpret this to mean that there is exactly one committee containing any two persons. Also, "two intersecting planes" can be interpreted to mean "two clubs having at least one person in common." Any statement already established, under some other interpretation, is automatically established under our new interpretation. For instance, take four people: Ann, Bea, Carlos, and Dick. Refer to the finite geometry shown in Geometry A. You can, in a mechanical way, list six committees and four clubs so that Postulates 5–9 are satisfied. Furthermore, theorems based on Postulates 5–9 are automatically available. For example, you know: If two committees have a person in common, then they have exactly one person in common (Theorem 1-9, page 46).

Postulate 5 requires space to contain *at least* four points not all in one plane and any line to contain *at least* two points. Geometry A works with exactly four points, and we have developed Geometry B using just four points (persons). But a finite geometry satisfying Postulates 5–9 can have more than four points, and more than two points on a line. Some of the exercises below explore finite geometries with more than four points.

However, no finite geometry can satisfy all the postulates of this book. Consider Postulate 1, page 7. That postulate calls for infinitely many points, so you see that the finite geometry shown in Geometry A does not satisfy all the postulates of this book. How about Geometry B? Since the number of living persons is finite, you know that we cannot use the interpretation discussed in Geometry B when all the postulates of the book are to be satisfied.

Exercises

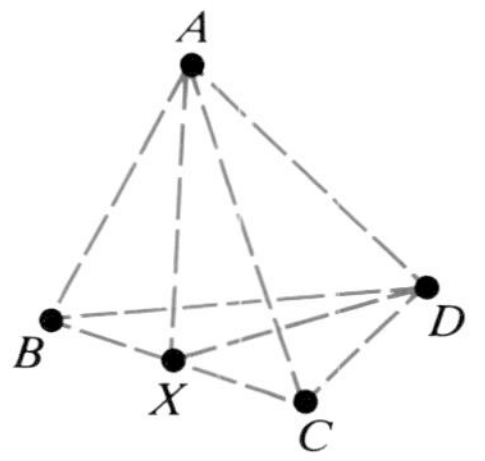

1. Refer to Geometry B and restate Postulates 5–9 in terms of *person, committee,* and *club.* Also restate Theorems 1–9, 1–10, and 1–11.

2. a. Check the five-point geometry shown to decide whether Postulates 5–9 are satisfied.

b. Apply the Geometry B interpretation to the diagram. For the fifth point X, use a fifth person, Xavera. List the members of a three-person committee and the members of two four-member clubs.

3. Limiting yourself to Postulates 5–9, write the key steps of a proof of the theorem: If there are five points not all in one plane, then there are at least eight lines and five planes. (*Hint:* Let A, B, C, and D be four points not all in one plane. Consider three cases for a fifth point X.

(1) X lies on one of the lines—use (BC)—determined by A, B, C, D.

(2) X lies on one of the planes, but not on any line, determined by A, B, C, D.

(3) X does not lie on any plane determined by A, B, C, D.

4. Refer to Exercise 3. State the theorem in terms of persons. Note that no further proof is necessary.

5. Suppose a finite geometry satisfying Postulates 5–9 contains exactly six points.

a. What is the largest possible number of different lines?

b. What is the smallest possible number of different lines?

c. What is the largest possible number of different planes?

d. What is the smallest possible number of different planes?

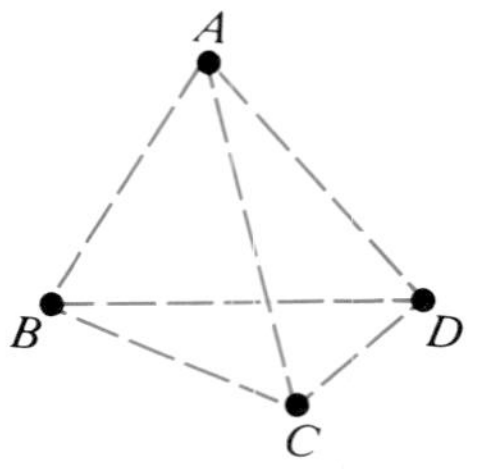

6. For this exercise, we define two lines to be parallel when they do not have any point in common. Consider the five-point geometry pictured above and Geometry A, which is pictured again at the right. Decide whether each of these geometries satisfies the following statement: Through a point outside a line, there is exactly one line parallel to the given line.

Chapter Summary

1. Lines that do not intersect are either parallel or skew.
2. When two parallel lines are cut by a transversal:
   a. corresponding angles are congruent;
   b. alternate interior angles are congruent;
   c. same-side interior angles are supplementary;
   d. if the transversal is perpendicular to one of the two parallel lines, it is also perpendicular to the other one.
3. The chart on page 67 lists five ways to prove lines parallel.
4. Through a point outside a line, there is exactly one line parallel to, and exactly one line perpendicular to, the given line.
5. Two lines parallel to a third line are parallel to each other.
6. Triangles are classified (page 72) by the lengths of their sides and by the measures of their angles. In any $\triangle ABC$, $m\angle A + m\angle B + m\angle C = 180$.
7. The measure of an exterior angle of a triangle equals the sum of the measures of the two remote interior angles.
8. The sum of the measures of the angles of a convex polygon with n sides is $(n - 2)180$. The sum of the measures of the exterior angles, one angle at each vertex, is 360.
9. The summary on page 92 gives the relationships between an if-then statement and its converse, its contrapositive, and its inverse. An if-then statement and its contrapositive are logically equivalent.

Chapter Review

2-1

1. $\angle 5$ and $\angle$ _?_ are same-side interior angles.
2. $\angle 5$ and $\angle 1$ are _?_ angles.
3. $\angle 5$ and $\angle 3$ are _?_ angles.
4. Line j, not shown, does not intersect line r. Must lines r and j be parallel?

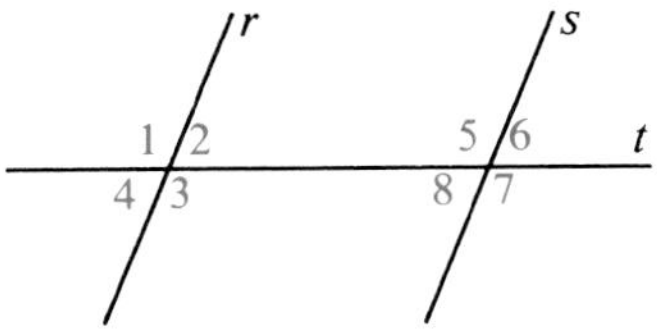

Exs. 1–7

In the diagram above, $r \parallel s$.

5. If $m\angle 1 = 105$, then $m\angle 5 =$ _?_ and $m\angle 7 =$ _?_. 2-2
6. Solve for x: $m\angle 2 = 70$ and $m\angle 8 = 6x - 2$
7. Solve for y: $m\angle 3 = 8y - 40$ and $m\angle 8 = 2y + 20$
8. Lines a, b, and c are coplanar, $a \parallel b$, and $a \perp c$. What can you conclude? Explain.

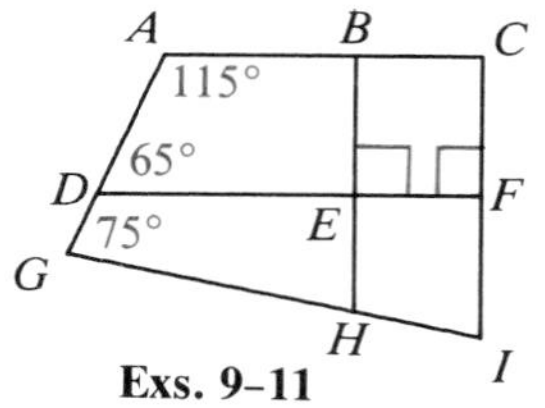

Exs. 9–11

9. Which line is parallel to $\overleftrightarrow{AB}$? Why? 2–3

10. Name a pair of parallel lines other than the pair in Exercise 9. Why must they be parallel?

11. Find the measure of $\angle I$.

12. Name five ways to prove two lines parallel.

13. If x and $2x - 15$ represent the measures of the acute angles of a right triangle, find the value of x. 2–4

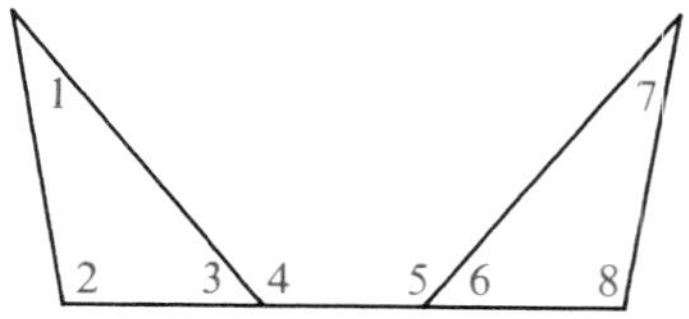

Exs. 14–16

14. $m\angle 6 + m\angle 7 + m\angle 8 =$ __?__

15. If $m\angle 1 = 30$ and $m\angle 4 = 130$, then $m\angle 2 =$ __?__.

16. If $\angle 4 \cong \angle 5$ and $\angle 1 \cong \angle 7$, name two other pairs of congruent angles and give a reason for each answer.

17. **a.** Sketch a hexagon that is equiangular but not equilateral. 2–5
b. What is its interior angle sum?
c. What is its exterior angle sum?

18. A regular polygon has 18 sides. Find the measure of each interior angle.

19. A regular polygon has 24 sides. Find the measure of each exterior angle.

20. Each interior angle of a regular polygon has measure 156. How many sides does the polygon have?

Consider the statement "A quadrilateral is equilateral if it is a square."

21. Write the statement in if-then form. 2–6

22. Name the hypothesis and the conclusion.

23. Write the converse and state whether it is true or false.

24. Rewrite the following statement as two if-then statements that are converses of each other: Two segments are congruent if and only if their lengths are equal.

You are given the true statement "Toads are amphibians." In each exercise, accept the additional information as also true. What can you conclude, if anything?

25. Toddie is a toad.

26. A frog is an amphibian. 2–7

27. A dog isn't an amphibian.

28. A tortoise isn't a toad.

Chapter Test

Complete each statement with the word *always, sometimes,* or *never.*

1. Two lines that have no points in common are __?__ parallel.
2. If a line is perpendicular to one of two parallel lines, then it is __?__ also perpendicular to the other one.
3. If two lines are cut by a transversal and same-side interior angles are complementary, then the lines are __?__ parallel.
4. An obtuse triangle is __?__ a right triangle.
5. In $\triangle ABC$, if $\overline{AB} \perp \overline{BC}$, then $\overline{AC}$ is __?__ perpendicular to $\overline{BC}$.
6. As the number of sides of a regular polygon increases, the measure of each exterior angle __?__ decreases.
7. The converse of a true if-then statement is __?__ true.

Find the value of x.

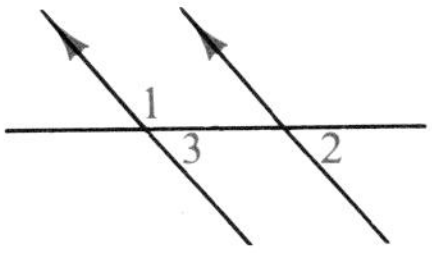

8. $m\angle 1 = 3x - 20$, $m\angle 2 = x$
9. $m\angle 2 = 2x + 12$, $m\angle 3 = 4(x - 7)$

Find the measures of the numbered angles.

10. XYZ is regular.

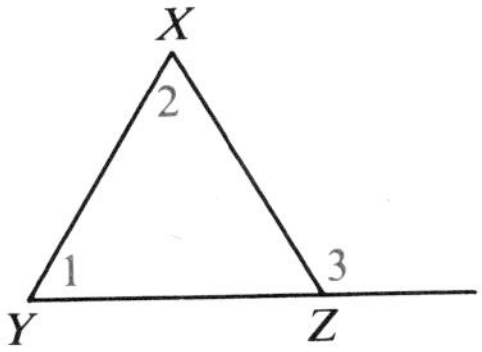

11.

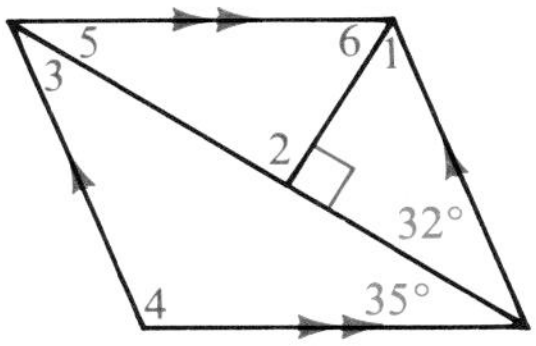

12. $ABCDE$ is regular.

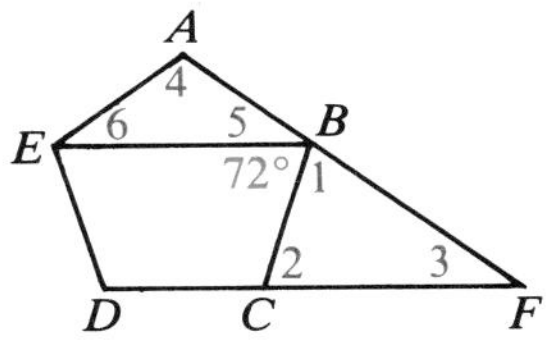

13. In the diagram for Exercise 12, explain why $\overline{EB}$ and $\overline{DF}$ must be parallel.

Use the statement "If $x = y$, then $x^2 = y^2$."

14. Write the: **a.** hypothesis **b.** conclusion **c.** converse **d.** contrapositive
15. Pair each statement below with the given statement above and tell what conclusion, *if any,* must follow.
    a. $x^2 = y^2$ **b.** $x^2 \neq y^2$ **c.** $x = y$ **d.** $x \neq y$

16. Given: $\overleftrightarrow{AB} \parallel \overleftrightarrow{CD}$; $\overrightarrow{BF}$ bisects $\angle ABE$;
    $\overrightarrow{DG}$ bisects $\angle CDB$.
    Prove: $\overleftrightarrow{BF} \parallel \overleftrightarrow{DG}$

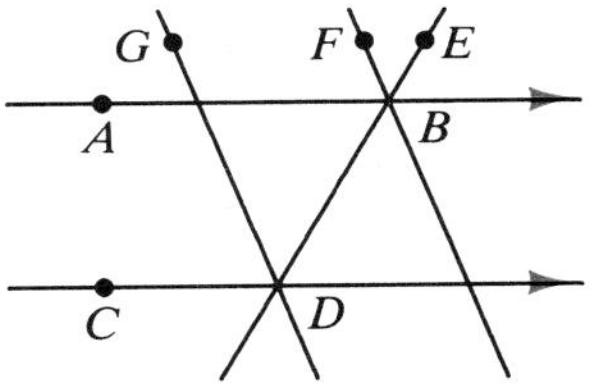

Preparing for College Entrance Exams

Strategy for Success

When you are taking a college entrance exam, be sure to read the directions, the questions, and the answer choices very carefully. You may want to underline important words such as *not, exactly, false, never,* and *except,* and to cross out answer choices that are clearly incorrect.

Indicate the best answer by writing the appropriate letter.

1. On a number line, point M has coordinate -3 and point R has coordinate 6. Point Z is on $\overrightarrow{RM}$ and $RZ = 4$. Find the coordinate of Z.
(A) -7 **(B)** 1 **(C)** 2 **(D)** 10 **(E)** cannot be determined

2. The measures of the angles of a triangle are $2x + 10$, $3x$, and $8x - 25$. The triangle is:
(A) acute **(B)** right **(C)** obtuse **(D)** equilateral **(E)** isosceles

3. Which statement would not guarantee that $\angle 1$ and $\angle 2$ are congruent?
(A) $\angle 1$ and $\angle 2$ are adjacent angles formed by perpendicular lines.
(B) $\angle 1$ and $\angle 2$ are vertical angles.
(C) $\angle 1$ and $\angle 2$ are both complements of $\angle 3$.
(D) $\angle 1$ and $\angle 2$ are same-side interior angles formed by two lines and a transversal.
(E) $\angle 1$ and $\angle 2$ are consecutive angles of a regular decagon.

4. A regular polygon has an interior angle of measure 120. How many vertices does the polygon have?
(A) 3 **(B)** 5 **(C)** 6 **(D)** 9 **(E)** 12

5. Given: (1) If $x = -2$, then $|x| = 2$.
(2) $|x| \neq 2$
Use the given statements to determine the statement that must be true.
(A) $|x| = -2$ **(B)** $x = -2$ **(C)** $x = 2$ **(D)** $x \neq -2$ **(E)** $x \neq 2$

6. Which of the following would allow you to conclude that $\overline{PQ} \parallel \overline{SR}$?
(A) $\angle PTQ \cong \angle TQR$ **(B)** $m\angle R = m\angle PSZ$
(C) $\angle R \cong \angle SPQ$ **(D)** $m\angle SPT + m\angle PTS = 90$
(E) $\angle STP \cong \angle QPT$

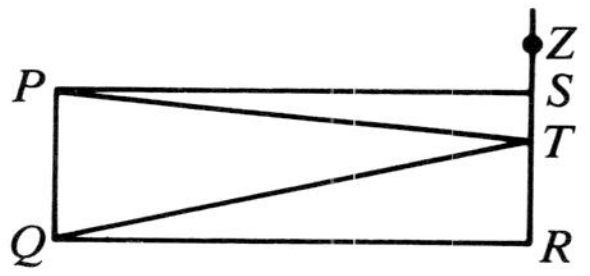

7. In $\triangle JKL$, $m\angle J = 40$. An exterior angle at L is a supplement of $\angle K$. The measure of $\angle K$ is:
(A) 25 **(B)** 70 **(C)** 80 **(D)** 100 **(E)** impossible to determine

8. If $\overrightarrow{AX}$ and $\overrightarrow{BX}$ are angle bisectors, which angles are not necessarily complementary?
(A) $\angle 1$ and $\angle 3$ **(B)** $\angle 3$ and $\angle 5$
(C) $\angle 1$ and $\angle 5$ **(D)** $\angle 2$ and $\angle 3$
(E) $\angle 2$ and $\angle 4$

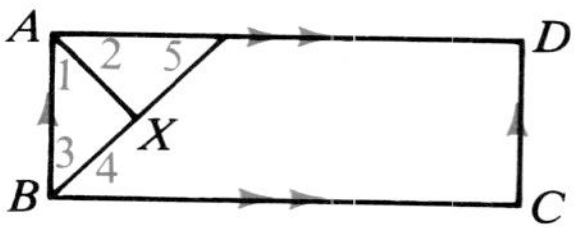

Cumulative Review: Chapters 1 and 2

Complete each sentence with the most appropriate word, phrase, or value.

A 1. If $AB = CD$, then $\overline{AB}$ and $\overline{CD}$ are __?__.

2. If M is the midpoint of $\overline{RS}$, $RM = 3x - 7$, and $RS = 4x + 2$, then $x =$ __?__.

3. If $\angle XYZ$ is a right angle and $\overrightarrow{YW}$ bisects $\angle XYZ$, then $m\angle XYW =$ __?__.

4. The statement "If $m\angle 1 = m\angle 2$, then $2m\angle 1 = 2m\angle 2$" can be justified by the reason __?__.

5. If two intersecting lines form congruent adjacent angles, then the lines are __?__.

6. Two adjacent angles formed by two intersecting lines are __?__.

7. If two planes intersect, then their intersection is a(n) __?__.

8. If $\angle 1$ and $\angle 2$ are complements and $m\angle 1 = 74$, then $m\angle 2 =$ __?__.

9. Every triangle has at least two __?__ angles.

10. If the measure of each angle of a regular polygon is 144, then the polygon has __?__ sides.

11. A line and a point not on the line are contained in __?__.

12. A conditional and its __?__ are logically equivalent.

Find the measure of each numbered angle.

13.

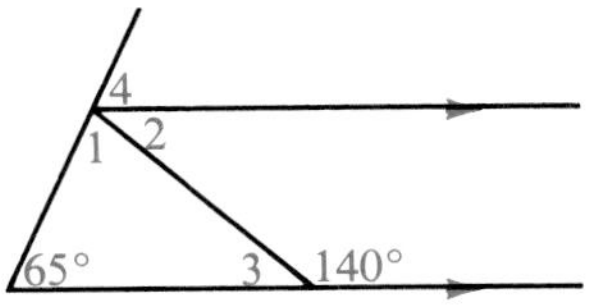

14.

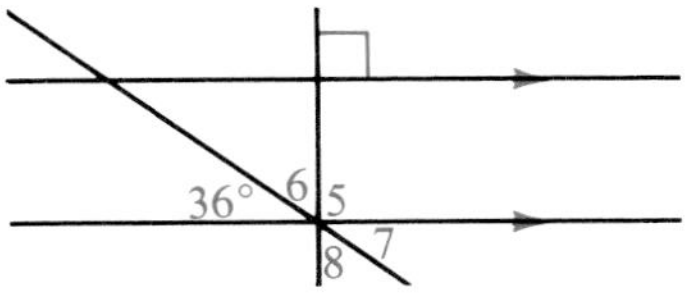

Could the given information be used to prove that two lines are parallel? If so, which lines?

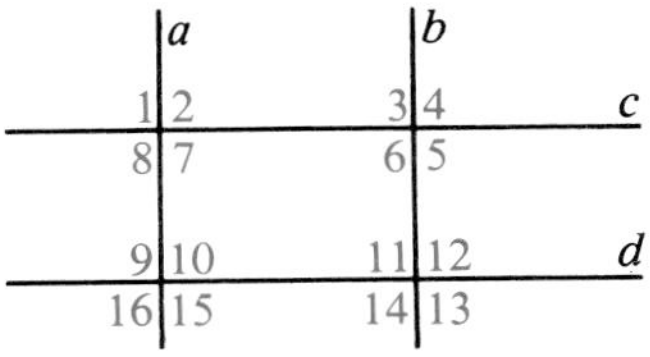

15. $m\angle 8 + m\angle 9 = 180$

16. $\angle 1 \cong \angle 4$

17. $m\angle 2 = m\angle 6$

18. $\angle 8$ and $\angle 5$ are rt. $\angle$s.

Complete each statement about the diagram. Then state the definition, postulate, or theorem that justifies your answer.

19. $m\angle 1 + m\angle 2 + m\angle 3 =$ __?__

20. $m\angle 1 + m\angle 4 =$ __?__

21. $m\angle 4 + m\angle 5 + m\angle 6 =$ __?__

22. $m\angle 1 + m\angle 2 = m\angle$ __?__

23. If $\overleftrightarrow{EC} \parallel \overleftrightarrow{BD}$, then $\angle 7 \cong$ __?__.

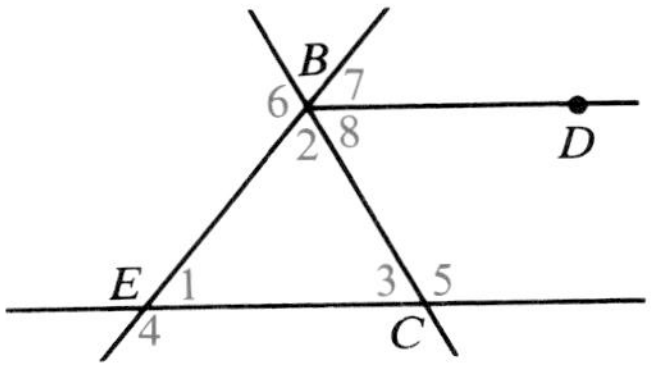

Handicrafts such as the Navajo rug shown often feature repeated, or *congruent*, designs. Notice that the figures in the top row have the same size and shape as those in the bottom row.

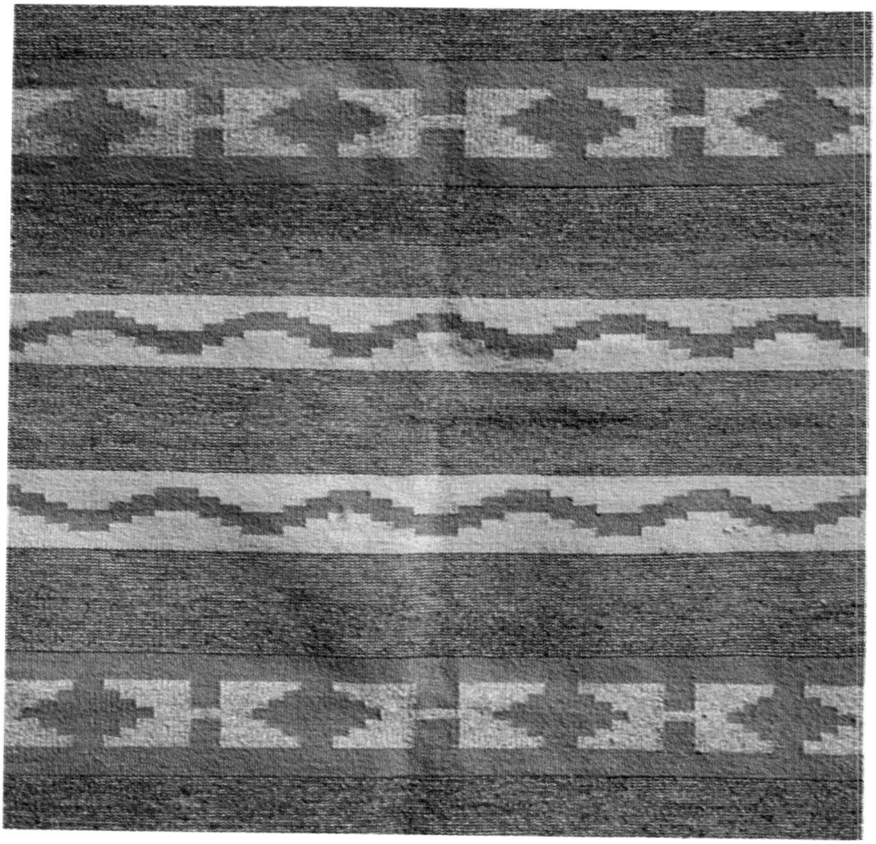

Congruent Triangles

3

Corresponding Parts in a Congruence

Objectives

1. Identify the corresponding parts of congruent figures.
2. Use the SSS Postulate, the SAS Postulate, and the ASA Postulate to prove two triangles congruent.
3. Deduce information about segments or angles by first proving that two triangles are congruent.

3-1 Congruent Figures

The Navajo rug on the facing page features the design shown below.

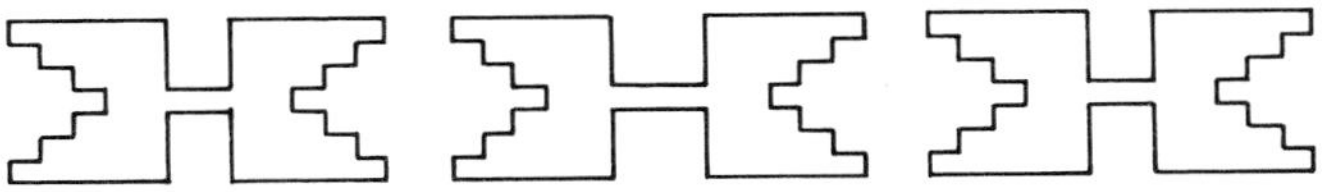

Do you think the three figures have the same size and shape? If you were to trace them, you would find that the first and third figures have the same size and shape, but the one in the middle is slightly wider.

Whenever two figures have the same size and shape, they are called **congruent.** You are already familiar with congruent segments (segments that have equal lengths) and congruent angles (angles that have equal measures). In this chapter you will learn about congruent triangles.

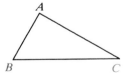

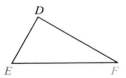

Triangles ABC and DEF are congruent. If you mentally slide $\triangle ABC$ to the right, you can fit it exactly over $\triangle DEF$ by matching up the vertices like this:

$A \longleftrightarrow D$ $\quad$ $B \longleftrightarrow E$ $\quad$ $C \longleftrightarrow F$

The sides and angles will then match up like this:

Corresponding angles	*Corresponding sides*
$\angle A \longleftrightarrow \angle D$	$\overline{AB} \longleftrightarrow \overline{DE}$
$\angle B \longleftrightarrow \angle E$	$\overline{BC} \longleftrightarrow \overline{EF}$
$\angle C \longleftrightarrow \angle F$	$\overline{AC} \longleftrightarrow \overline{DF}$

Do you see that

(1) since congruent triangles have the same shape, their corresponding angles are congruent, and

(2) since congruent triangles have the same size, their corresponding sides are congruent?

We have the following definition:

Two triangles are **congruent** if and only if their vertices can be matched up so that the corresponding parts (angles and sides) of the triangles are congruent.

The congruent parts of the triangles shown are marked alike. Imagine sliding $\triangle SUN$ to the right and then flipping it over so that its vertices are matched with the vertices of $\triangle RAY$ like this:

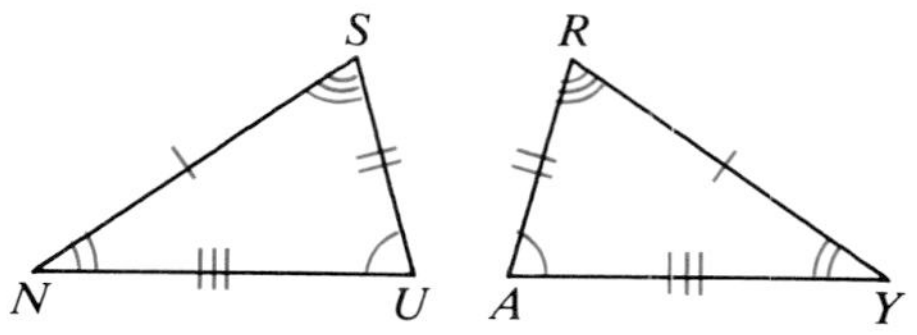

$S \longleftrightarrow R$ $\qquad$ $U \longleftrightarrow A$ $\qquad$ $N \longleftrightarrow Y$

The corresponding parts are then congruent and we see that the triangles are congruent.

When referring to congruent triangles, we name their corresponding vertices in the same order. For the triangles above,

$\triangle SUN$ is congruent to $\triangle RAY$.
$\triangle SUN \cong \triangle RAY$

The following statements about these triangles are also correct, since corresponding vertices of the triangles are named in the same order.

$\triangle NUS \cong \triangle YAR$ $\qquad$ $\triangle SNU \cong \triangle RYA$

What has been said about congruent triangles applies to all congruent polygons. For example, the two pentagons shown are congruent.

The statement $BRAKE \cong CHOKE$ is just one way to list this congruence with corresponding vertices written in the same order. Notice that side $\overline{KE}$ of pentagon $BRAKE$ corresponds to side $\overline{KE}$ of pentagon $CHOKE$. $\overline{KE}$ is called a *common side* of the two pentagons.

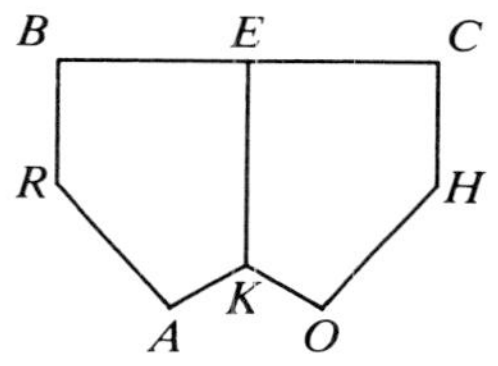

$BRAKE \cong CHOKE$

Classroom Exercises

Suppose you know that $\triangle FIN \cong \triangle WEB$.

1. Name the three pairs of corresponding vertices.
2. Name the three pairs of corresponding sides.
3. Name the three pairs of corresponding angles.
4. Is it correct to say $\triangle NIF \cong \triangle BEW$?
5. Is it correct to say $\triangle INF \cong \triangle EWB$?

The two triangles shown are congruent. Complete.

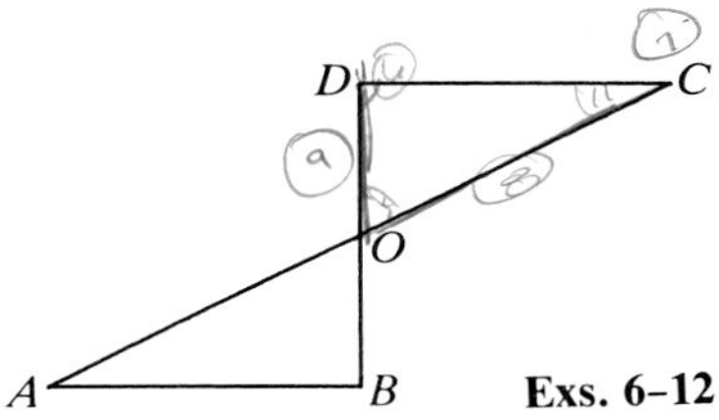

Exs. 6–12

6. $\triangle ABO \cong$ __?__
7. $\angle A \cong$ __?__
8. $\overline{AO} \cong$ __?__
9. $BO =$ __?__
10. Can you deduce that O is the midpoint of any segment? Explain.
11. If $\overline{AB} \perp \overline{BD}$, can you deduce that $\overline{CD} \perp \overline{BD}$? Explain.
12. Suppose you don't know whether $\angle B$ and $\angle D$ are right angles. Can you deduce that $\overline{AB} \parallel \overline{DC}$? Explain.

The pentagons shown are congruent. Complete.

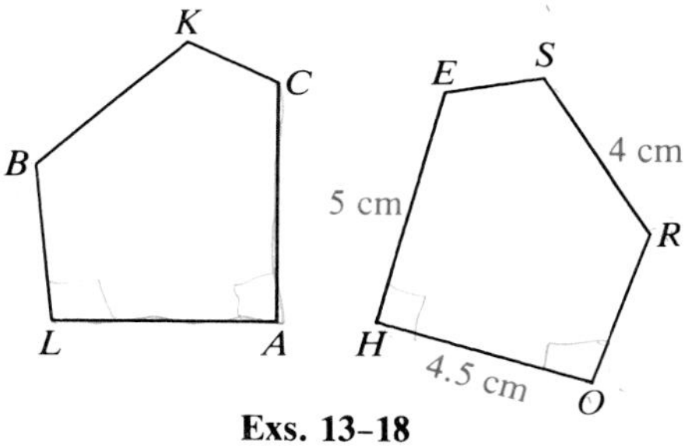

Exs. 13–18

13. B corresponds to __?__.
14. $BLACK \cong$ __?__
15. $KB =$ __?__ cm
16. __?__ $= m\angle E$
17. If $\overline{CA} \perp \overline{LA}$, name two right angles in the figures.
18. **a.** Find the sum of the measures of the interior angles of each pentagon.
    b. Would the sums of the measures of the interior angles be different if the pentagons were not congruent?

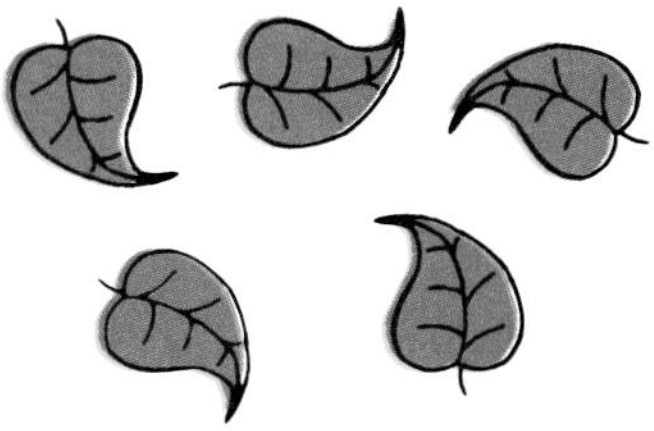

19. The five leaves shown are all congruent, but one differs from the others. Which one is different and how?

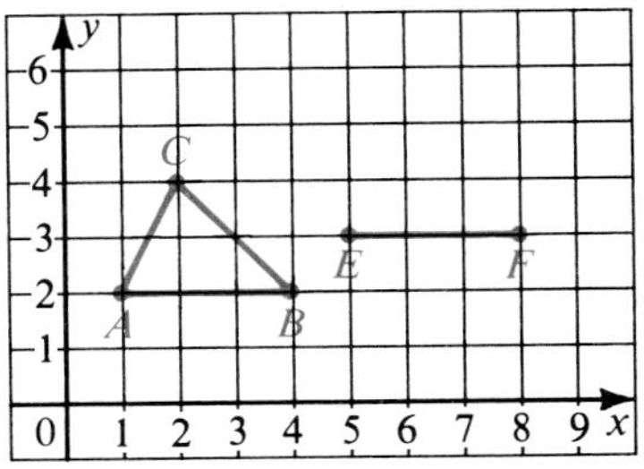

20. To get to point A on the grid shown, you start at zero, go 1 unit right, and then 2 units up. We say that (1, 2) are the *coordinates* of A.
    a. Name the coordinates of B and C.
    b. Name the coordinates of point D such that $\triangle ABC \cong \triangle ABD$.
21. **a.** Name the coordinates of a point G such that $\triangle ABC \cong \triangle EFG$. Is there another location for G such that $\triangle ABC \cong \triangle EFG$?
    b. Name two other locations for a third vertex for a triangle congruent to $\triangle ABC$ that has $\overline{EF}$ as one side.

Written Exercises

Suppose $\triangle BIG \cong \triangle CAT$. Complete.

A

1. $\angle G \cong \underline{\ ?\ }$
2. $\underline{\ ?\ } = m\angle A$
3. $BI = \underline{\ ?\ }$
4. $\underline{\ ?\ } \cong \overline{AT}$
5. $\triangle IGB \cong \underline{\ ?\ }$
6. $\underline{\ ?\ } \cong \triangle CTA$
7. If $\triangle BIG \cong \triangle CAT$, $m\angle B = 100$, and $m\angle G = 40$, name four congruent angles.
8. Is the statement "Corresponding parts of congruent triangles are congruent" based on a definition, postulate, or theorem?
9. Suppose $\triangle LXR \cong \triangle FNE$. Write six congruences that follow from this statement and the definition of congruent triangles.

The two triangles shown are congruent. Complete.

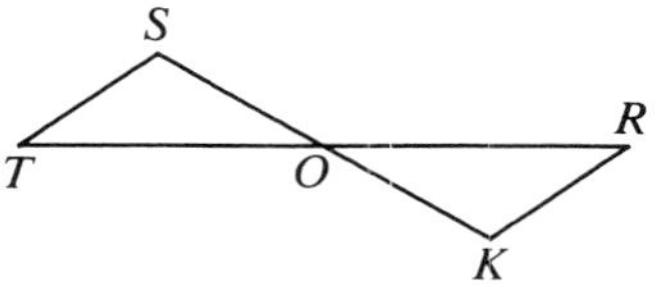

10. $\triangle STO \cong \underline{\ ?\ }$.
11. $\angle S \cong \underline{\ ?\ }$ because $\underline{\ ?\ }$.
12. $\overline{SO} \cong \underline{\ ?\ }$ because $\underline{\ ?\ }$. Thus point O is the midpoint of $\underline{\ ?\ }$.
13. Since $\underline{\ ?\ } = RO$, O is the midpoint of $\underline{\ ?\ }$.
14. **a.** $\angle T \cong \underline{\ ?\ }$ because $\underline{\ ?\ }$.
    b. $\overline{ST} \parallel \overline{RK}$ because $\underline{\ ?\ }$.

The two triangles shown are congruent. Complete.

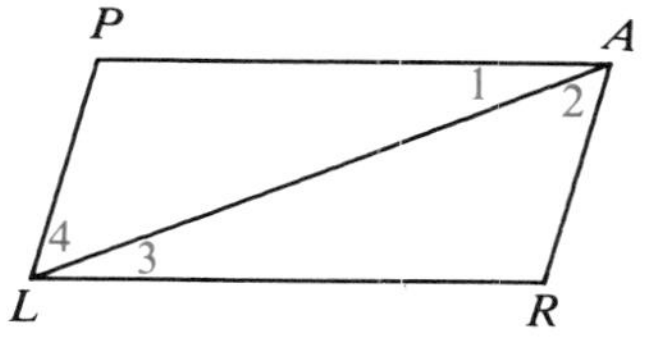

15. $\triangle PAL \cong \underline{\ ?\ }$.
16. **a.** $\angle 1 \cong \underline{\ ?\ }$ because $\underline{\ ?\ }$.
    b. $\overline{PA} \parallel \underline{\ ?\ }$ because $\underline{\ ?\ }$.
17. **a.** $\angle 2 \cong \underline{\ ?\ }$ because $\underline{\ ?\ }$.
    b. $\underline{\ ?\ } \parallel \underline{\ ?\ }$ because $\underline{\ ?\ }$.
18. **a.** $\overline{PA} \cong \underline{\ ?\ }$ **b.** $\overline{PL} \cong \underline{\ ?\ }$.

Plot the given points on graph paper. Draw $\triangle FAT$. Locate point C so that $\triangle FAT \cong \triangle CAT$.

19. $F(1, 2)$ $A(4, 7)$ $T(4, 2)$
20. $F(7, 5)$ $A(-2, 2)$ $T(5, 2)$

Plot the given points on graph paper. Draw $\triangle ABC$ and $\triangle DEF$. Copy and complete the statement $\triangle ABC \cong \underline{\ ?\ }$.

B

21. $A(-1, 2)$ $B(4, 2)$ $C(2, 4)$
    $D(5, -1)$ $E(7, 1)$ $F(10, -1)$
22. $A(-7, -3)$ $B(-2, -3)$ $C(-2, 0)$
    $D(0, 1)$ $E(5, 1)$ $F(0, -2)$
23. $A(-3, 1)$ $B(2, 1)$ $C(2, 3)$
    $D(4, 3)$ $E(6, 3)$ $F(6, 8)$
24. $A(1, 1)$ $B(8, 1)$ $C(4, 3)$
    $D(3, -7)$ $E(5, -3)$ $F(3, 0)$

Plot the given points on graph paper. Draw $\triangle ABC$ and $\overline{DE}$. Find two locations of point F such that $\triangle ABC \cong \triangle DEF$.

25. $A(1, 2)$ $B(4, 2)$ $C(2, 4)$ $D(6, 4)$ $E(6, 7)$

26. $A(-1, 0)$ $B(-5, 4)$ $C(-6, 1)$ $D(1, 0)$ $E(5, 4)$

$\overline{OR}$ is a common side of two congruent quadrilaterals.

27. Complete: quad. $NERO \cong$ quad. __?__

28. In your own words explain why each of the following statements must be true.

a. O is the midpoint of $\overline{NM}$.

b. $\angle NOR \cong \angle MOR$

c. $\overline{RO} \perp \overline{NM}$ **d.** $\overline{OR} \cong \overline{OR}$

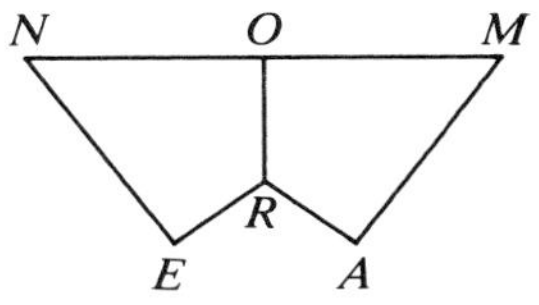

Exs. 27, 28

29. a. Use a protractor to draw a triangle whose angles have measure 50, 60, and 70.

b. Can you draw another triangle whose angles have measure 50, 60, and 70 but which is not congruent to the first triangle?

Accurately draw the pairs of triangles described. Measure all the corresponding parts of each pair. Are the two triangles congruent?

30. In $\triangle ABC$, $AB = 4$ cm, $m\angle B = 45$, and $BC = 6$ cm.
In $\triangle DEF$, $DE = 4$ cm, $m\angle E = 45$, and $EF = 6$ cm.

31. In $\triangle RST$, $m\angle R = 30$, $RS = 5$ cm, and $m\angle S = 100$.
In $\triangle XYZ$, $m\angle X = 30$, $XY = 5$ cm, and $m\angle Y = 100$.

C 32. Does congruence of triangles have the reflexive property? the symmetric property? the transitive property?

33. Suppose twelve toothpicks are arranged as shown to form a regular hexagon.

a. Copy the figure. Sketch six more toothpicks of the same size inside the hexagon so that it is divided into three congruent regions.

b. Now "move" only four toothpicks so that the hexagon is divided into two congruent regions.

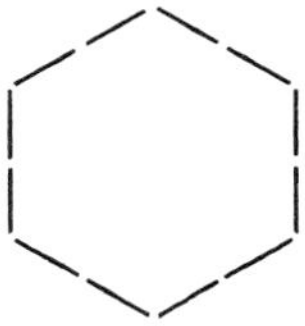

Challenge

The two blocks of wood are congruent. It is possible to cut a hole in one block in such a way that you can pass the other block completely through the hole. How?

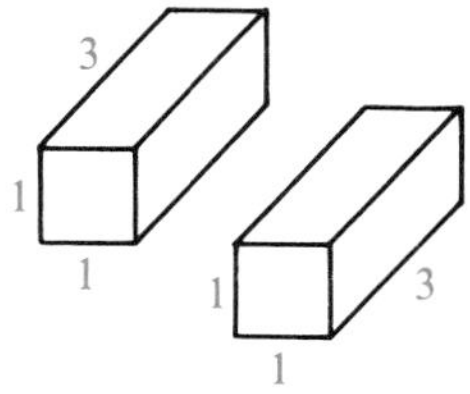

3-2 Some Ways to Prove Triangles Congruent

If you know that two triangles are congruent, you can conclude that the six parts of one triangle are congruent to the six parts of the other triangle. If you are not sure whether two triangles are congruent, however, it is not necessary to compare all six pairs of parts. As you saw in Exercises 30 and 31 of the preceding section, sometimes three pairs of congruent corresponding parts will guarantee that two triangles are congruent. The following postulates give you three different ways to show that two triangles are congruent by comparing three pairs of corresponding parts.

Postulate 12 SSS Postulate

If three sides of one triangle are congruent to three sides of another triangle, then the triangles are congruent.

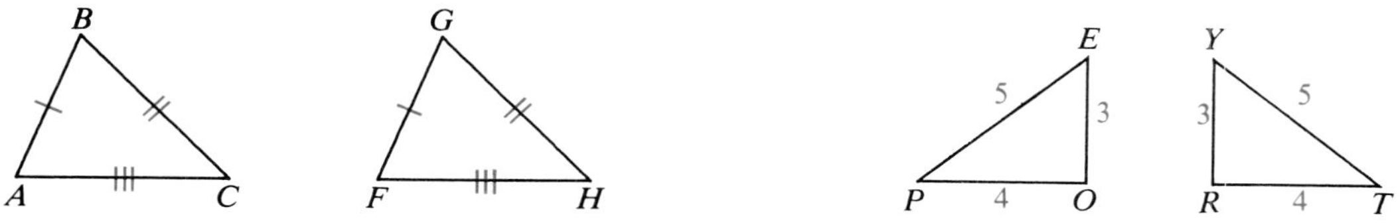

By the SSS Postulate, $\triangle ABC \cong \triangle FGH$ and $\triangle POE \cong \triangle TRY$.

Sometimes it is helpful to describe the parts of a triangle in terms of their relative positions.

$\overline{AB}$ is *opposite* $\angle C$.
$\overline{AB}$ is *included* between $\angle A$ and $\angle B$.
$\angle A$ is *opposite* $\overline{BC}$.
$\angle A$ is *included* between $\overline{AB}$ and $\overline{AC}$.

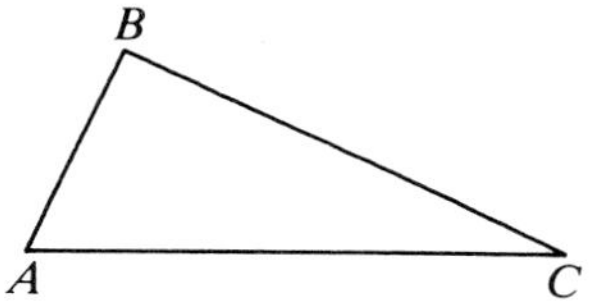

Postulate 13 SAS Postulate

If two sides and the included angle of one triangle are congruent to two sides and the included angle of another triangle, then the triangles are congruent.

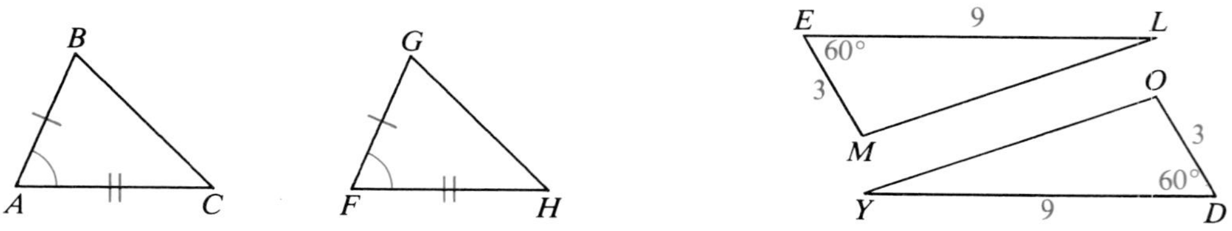

By the SAS Postulate, $\triangle ABC \cong \triangle FGH$ and $\triangle MEL \cong \triangle ODY$.

Postulate 14 ASA Postulate

If two angles and the included side of one triangle are congruent to two angles and the included side of another triangle, then the triangles are congruent.

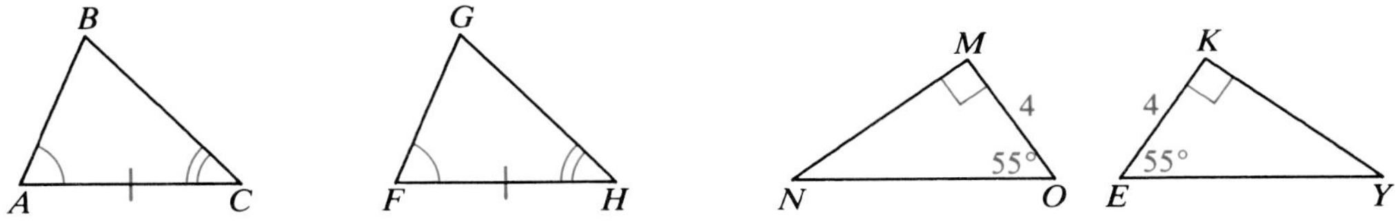

By the ASA Postulate, $\triangle ABC \cong \triangle FGH$ and $\triangle MON \cong \triangle KEY$.

Example Supply the missing statements and reasons in the following proof.

Given: E is the midpoint of $\overline{MJ}$;
$\overline{TE} \perp \overline{MJ}$

Prove: $\triangle MET \cong \triangle JET$

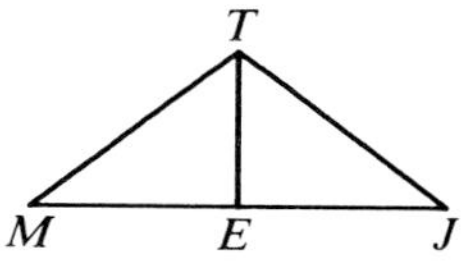

Proof:

Statements	Reasons
1. E is the midpoint of $\overline{MJ}$.	1. Given
2. _?_ $\cong$ _?_	2. Def. of midpoint
3. $\overline{TE} \perp \overline{MJ}$	3. _?_
4. $\angle MET \cong \angle JET$	4. _?_
5. $\overline{TE} \cong$ _?_	5. _?_
6. $\triangle MET \cong \triangle JET$	6. _?_

Solution 2. $\overline{ME} \cong \overline{JE}$
3. Given
4. Adj. $\angle$s formed by $\perp$ lines are $\cong$.
5. $\overline{TE}$; Reflexive Prop.
6. SAS Postulate

Classroom Exercises

1. In $\triangle PQR$, what angle is included between $\overline{PQ}$ and $\overline{QR}$?
2. In $\triangle PQR$, what angle is included between $\overline{PR}$ and $\overline{QR}$?
3. In $\triangle PQR$, what side is included between $\angle P$ and $\angle Q$?
4. In $\triangle PQR$, what side is included between $\angle P$ and $\angle R$?

Tell whether each pair of triangles *must* be congruent by the SAS postulate.

5.

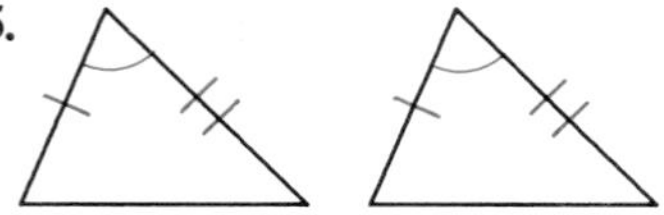

6.

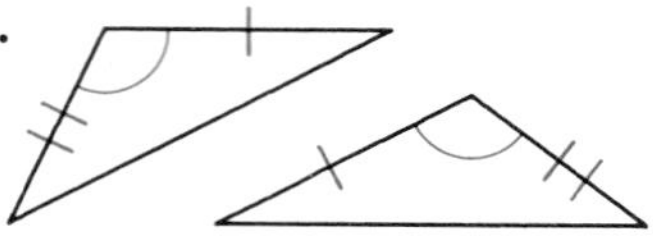

7.

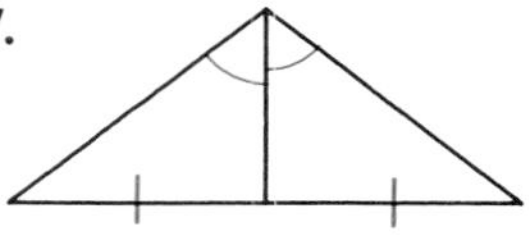

Can the two triangles be proved congruent? If so, what postulate can be used?

8.

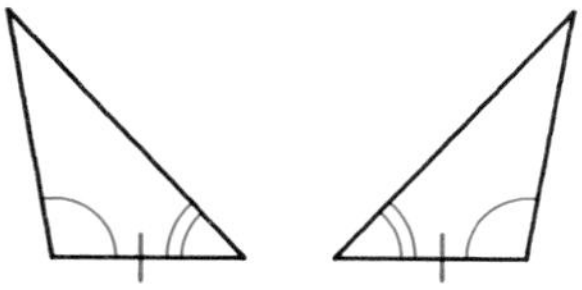

9.

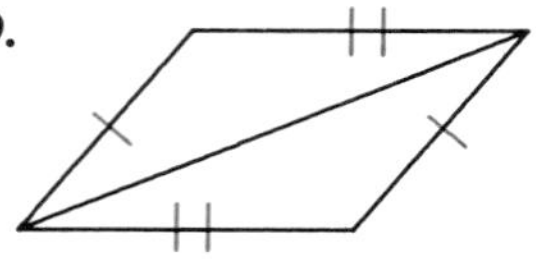

10.

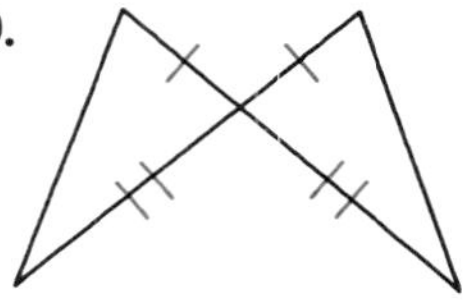

11.

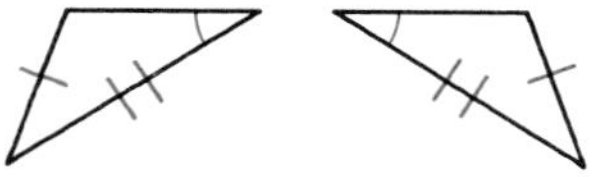

12.

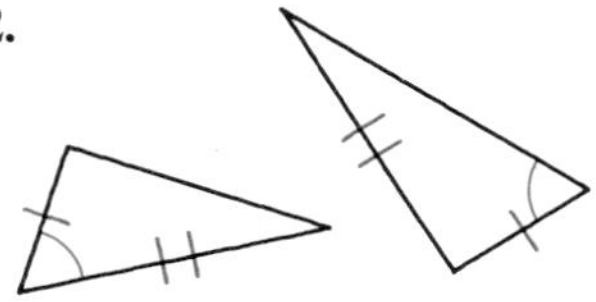

13.

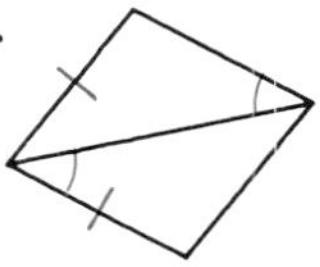

14. Supply the missing statements and reasons in the following proof.

Given: $\overline{HY} \cong \overline{LY}$;

$\overline{WH} \parallel \overline{LF}$

Prove: $\triangle WHY \cong \triangle FLY$

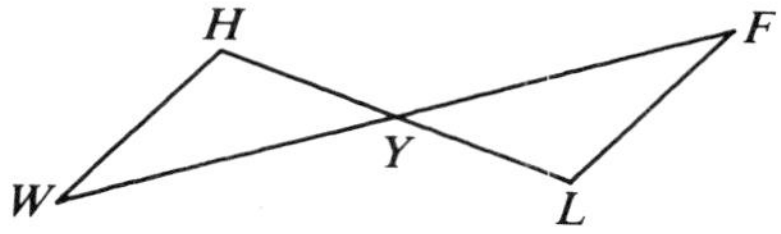

Proof:

Statements	Reasons
1. $\overline{WH} \parallel \overline{LF}$	1. ___?___
2. $\angle H \cong \angle L$	2. ___?___
3. $\overline{HY} \cong \overline{LY}$	3. ___?___
4. $\angle$ ___?___ $\cong \angle$ ___?___	4. ___?___
5. $\triangle WHY \cong \triangle FLY$	5. ___?___

15. Explain how you would prove the following.

Given: $\overline{WH} \cong \overline{FL}$; $\overline{WH} \parallel \overline{LF}$

Prove: $\triangle WHY \cong \triangle FLY$

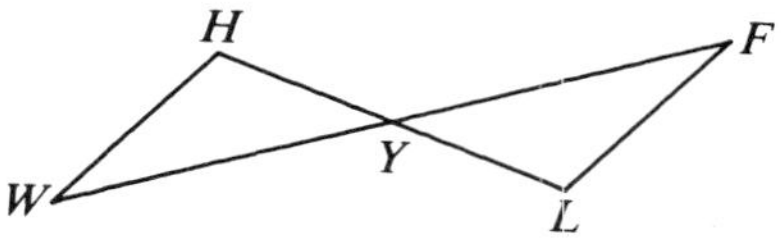

Written Exercises

Decide whether there is a triangle congruent to $\triangle ABC$. If so, write the congruence and name the postulate used. If not, write *no congruence can be deduced.*

A

1. 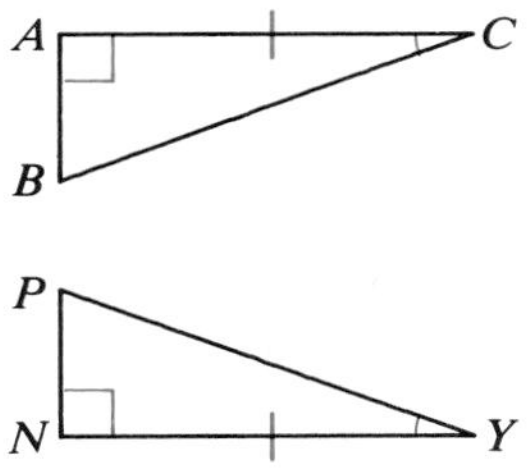

2. 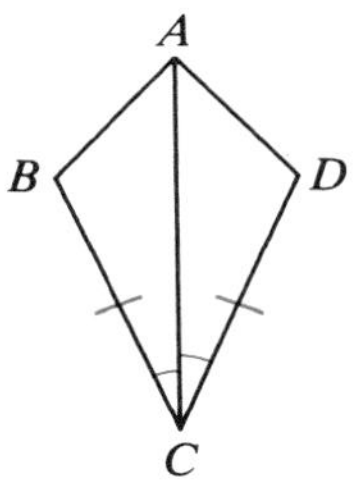

3. 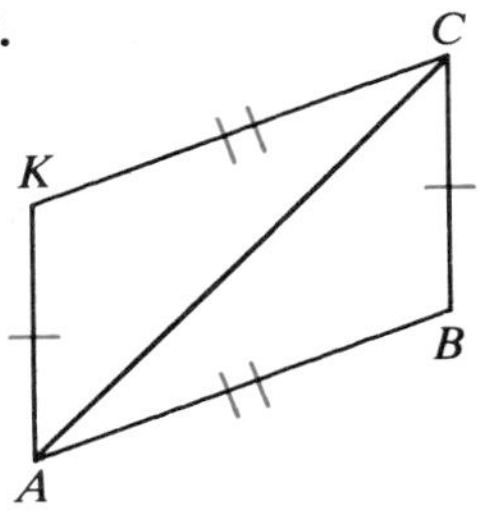

4. 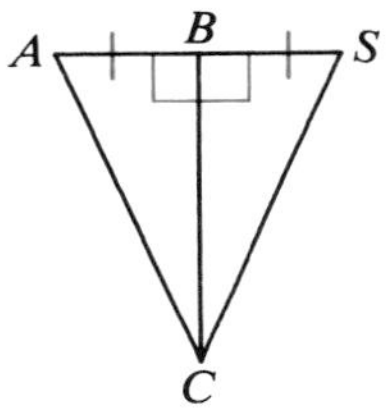

5. 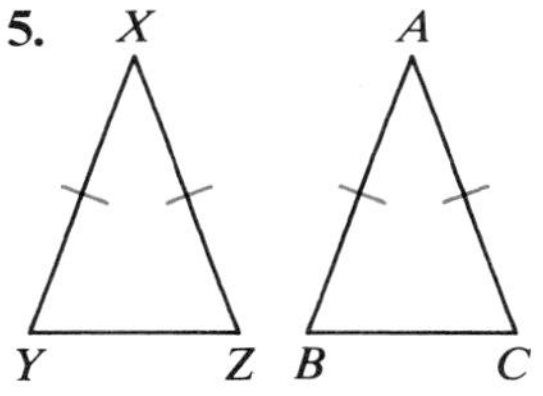

6. 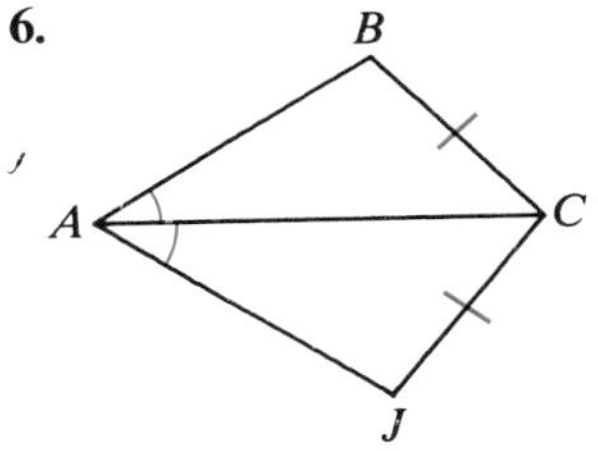

7. 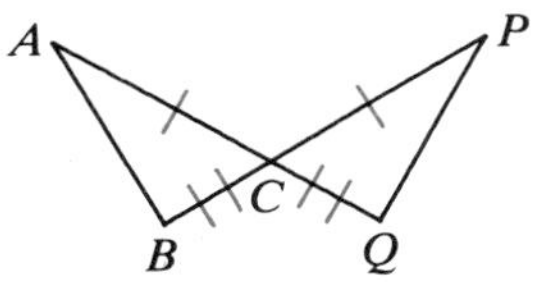

8. 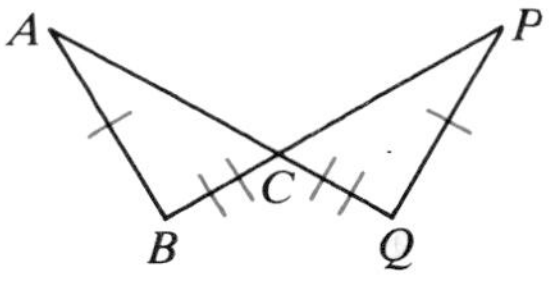

9. 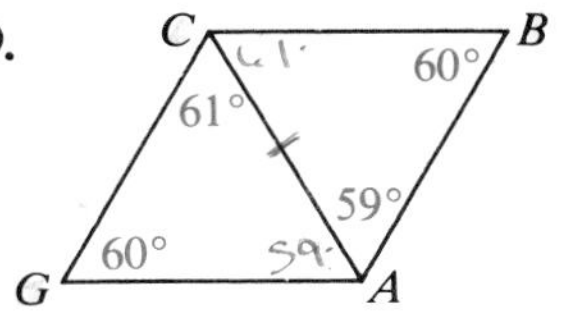

10. 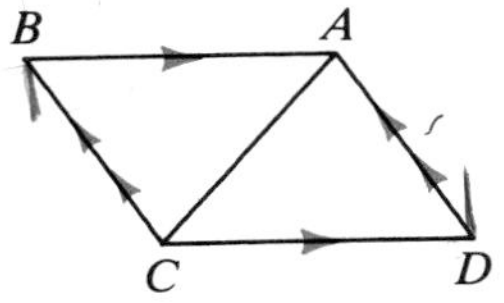

11. 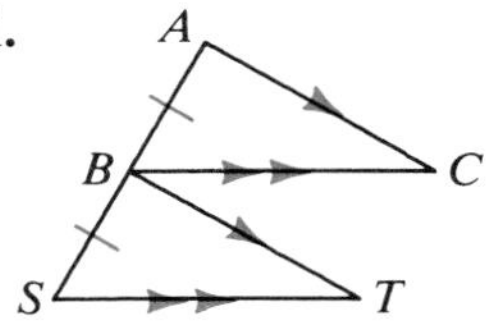

12. 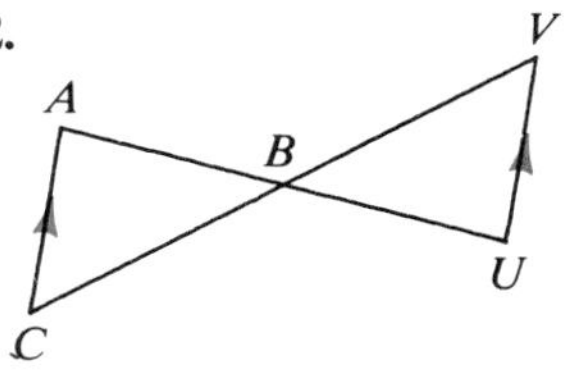

13. 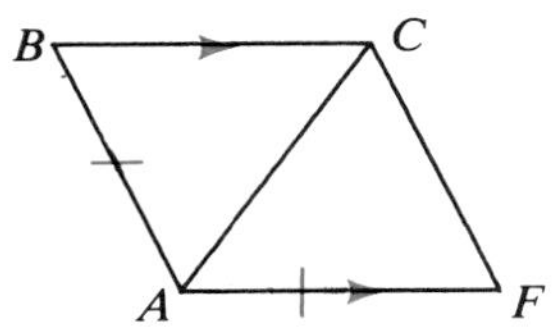

14. 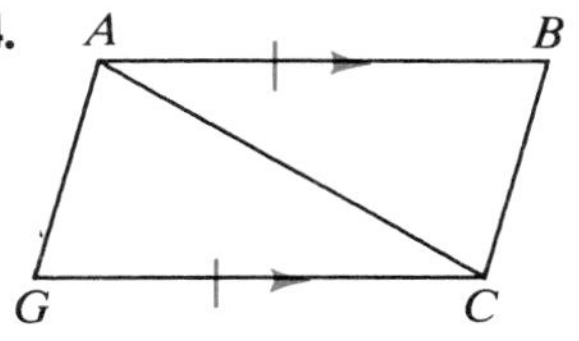

15. 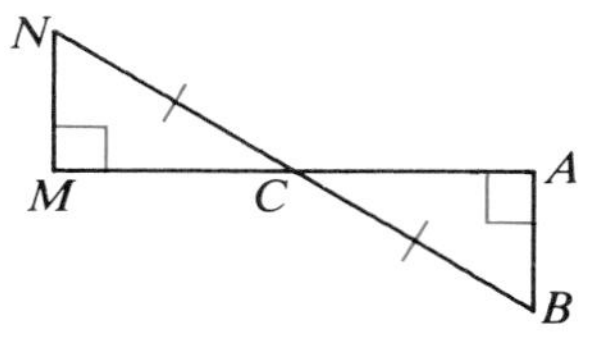

Supply the missing statements and reasons.

16. Given: $\overline{AB} \parallel \overline{DC}$; $\overline{AB} \cong \overline{DC}$

Prove: $\triangle ABC \cong \triangle CDA$

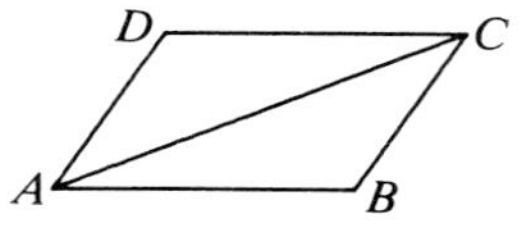

Proof:

Statements	Reasons
1. $\overline{AB} \cong \overline{DC}$	1. __?__ Given
2. $\overline{AC} \cong \overline{AC}$	2. __?__
3. $\overline{AB} \parallel \overline{DC}$	3. __?__
4. $\angle BAC \cong \angle DCA$	4. __?__
5. $\triangle ABC \cong \triangle CDA$	5. __?__

17. Given: $\overline{RS} \perp \overline{ST}$; $\overline{TU} \perp \overline{ST}$; V is the midpoint of $\overline{ST}$.

Prove: $\triangle RSV \cong \triangle UTV$

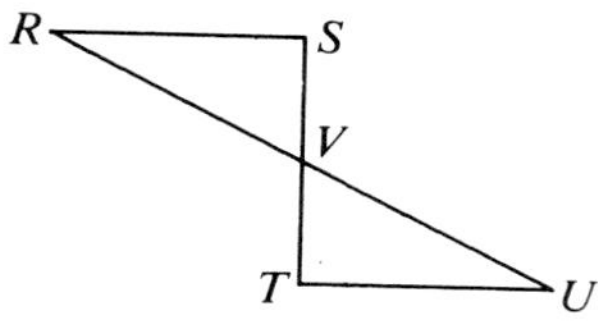

Proof:

Statements	Reasons
1. $\overline{RS} \perp \overline{ST}$; $\overline{TU} \perp \overline{ST}$	1. __?__
2. $m\angle S = 90 = m\angle$ __?__	2. Def. of $\perp$ lines and def. of rt. $\angle$
3. V is the midpoint of $\overline{ST}$.	3. __?__
4. $\overline{SV} \cong$ __?__	4. __?__
5. $\angle RVS \cong \angle$ __?__	5. __?__
6. $\triangle$ __?__ $\cong \triangle$ __?__	6. __?__

18. Given: $\overleftrightarrow{SA}$ bisects $\angle DAR$ and $\angle DTR$.

Prove: $\triangle DAT \cong \triangle RAT$

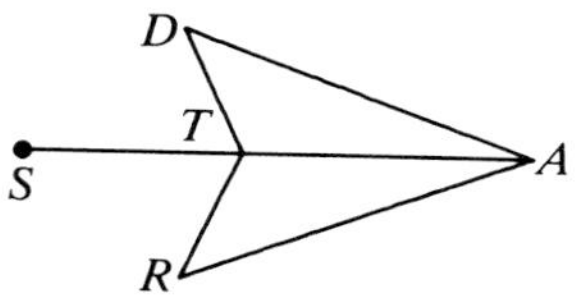

Proof:

Statements	Reasons
1. $\overleftrightarrow{SA}$ bisects $\angle DAR$ and $\angle DTR$.	1. __?__
2. $\angle DAT \cong$ __?__; __?__ $\cong$ __?__	2. Def. of $\angle$ bisector
3. $\angle DTA$ and $\angle DTS$ are suppl. $\angle$s; $\angle$ __?__ and $\angle$ __?__ are suppl. $\angle$s.	3. Angle Addition Postulate and def. of suppl. $\angle$s.
4. $\angle$ __?__ $\cong$ __?__	4. Supplements of $\cong$ $\angle$s are $\cong$.
5. __?__ $\cong$ __?__	5. Reflexive Prop.
6. __?__	6. __?__

Write proofs in two-column form.

B **19.** Given: M is the midpoint of $\overline{AB}$;
M is the midpoint of $\overline{CD}$.
Prove: $\triangle MAD \cong \triangle MBC$

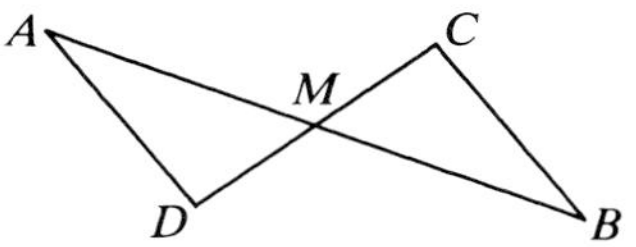

20. Given: Plane M bisects $\overline{AB}$; $\overline{PA} \cong \overline{PB}$
Prove: $\triangle POA \cong \triangle POB$

21. Given: Plane M bisects $\overline{AB}$; $\overline{PO} \perp \overline{AB}$
Prove: $\triangle POA \cong \triangle POB$

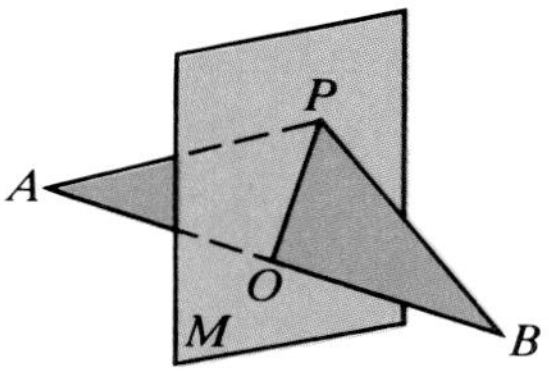

22. a. Draw an isosceles $\triangle ABC$ with $\overline{AC} \cong \overline{BC}$. Let D be any point on $\overrightarrow{AB}$ such that B is between A and D. Draw $\overline{CD}$.
b. Name the pairs of congruent parts of $\triangle ACD$ and $\triangle BCD$.
c. Do you think that SSA is enough to guarantee that two triangles are congruent?

The following diagrams represent three-dimensional figures. Copy each figure and with colored pencils outline the pair of triangles listed. Tell which postulate guarantees that these triangles are congruent.

C **23.**

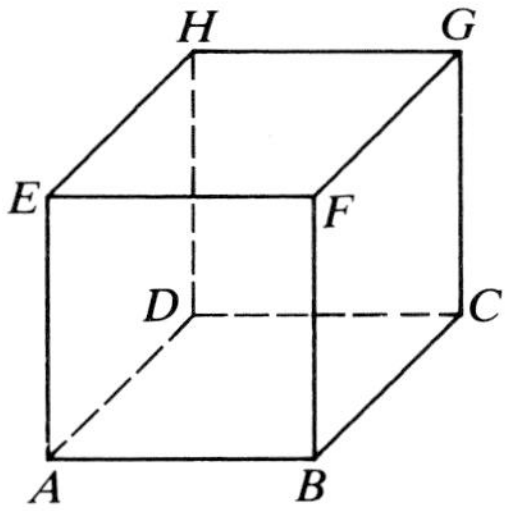

Given: Cube whose faces are congruent squares
Show: $\triangle ABF$, $\triangle BCG$

24.

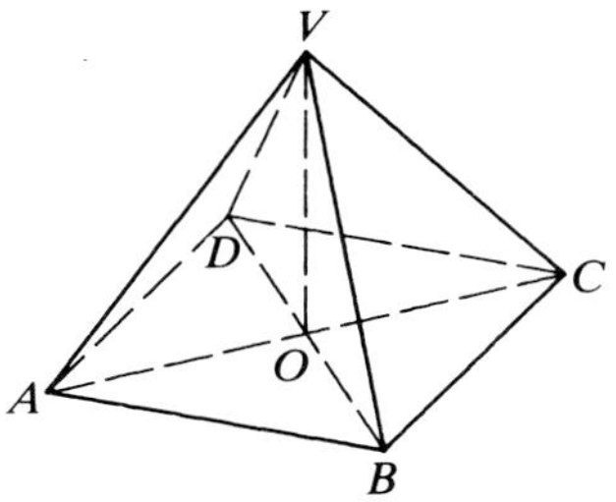

Given: Pyramid with square base;
$VA = VB = VC = VD$
Show: $\triangle VAB$, $\triangle VBC$

Challenge

You can cut a cubical block into 8 cubes by using a slicer 3 times, as shown. To cut a cube into 27 cubes, you need 6 slicings even though you are permitted to rearrange parts in any way you choose after each cutting. To cut a cube into 216 cubes requires only 9 slicings.

How many slicings does it take to divide a cube into 64 cubes? into 125 cubes?

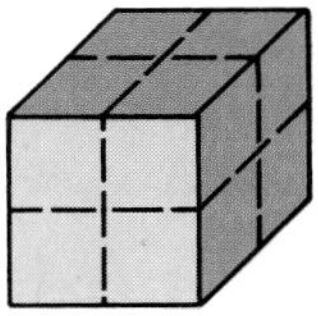

CAREERS

Computer Software Developer

Large or small, mainframe or microcomputer, computers are well on their way to becoming an ordinary part of everyday life. When computers were first becoming available on a widespread basis, emphasis was primarily on *hardware,* or the machines themselves. Soon, however, those involved with computers realized that effective *software* was important if computer use was to become practical for people who were not themselves programmers.

Software is the technical name for *programs,* which are logical, step-by-step sequences of instructions that tell the computer how to accomplish a particular task. Software may be developed "in-house" for use by a single company or organization or may be produced by a company that specializes in developing programs that are expected to have a wide application.

When software is developed in-house, a programmer works for the particular company or organization, writing a program that will solve a specific problem. The programmer must work closely with the person requesting the program, to be sure that the problem is clearly understood and that the program will really do the job.

Commercial software companies try to identify the fields in which software is needed. They employ entire staffs of programmers who develop appropriate software. This type of software is intended for a mass market and is usually distributed nationally or internationally. The products are as diverse as home recipe files, school attendance registers, word processing, and corporate accounting systems. Software is usually marketed on a disc, at prices ranging from $20 to $20,000 and more.

People who develop software generally are proficient in more than one computer language. Employment often requires a college degree in computer science, although there are other ways to attain the necessary level of skill.

3-3 Using Congruent Triangles

Our goal in the preceding section was to learn when to conclude that two triangles are congruent. Our goal in this section is to deduce information about two segments or angles once we have shown that they are corresponding parts of congruent triangles. The following example will illustrate this technique.

Example 1

Given: $\overline{AB}$ and $\overline{CD}$ bisect each other at M.

Prove: $\overline{AD} \parallel \overline{BC}$

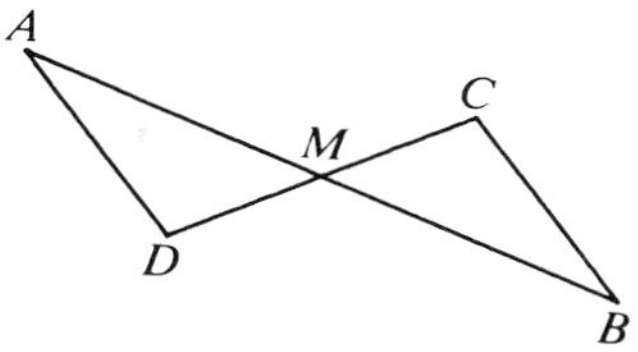

■ ***Plan for Proof:*** We can prove $\overline{AD} \parallel \overline{BC}$ if we can show that alternate interior angles $\angle A$ and $\angle B$ are congruent. To show that they are congruent, we can show that they are corresponding parts of congruent triangles. Thus our first goal is to prove $\triangle AMD \cong \triangle BMC$.

Proof:

Statements	Reasons
1. $\overline{AB}$ and $\overline{CD}$ bisect each other at M.	1. Given
2. $\overline{AM} \cong \overline{MB}$; $\overline{DM} \cong \overline{MC}$	2. Def. of bisect and def. of midpoint
3. $\angle AMD \cong \angle BMC$	3. Vert. $\angle$s are $\cong$.
4. $\triangle AMD \cong \triangle BMC$	4. SAS Postulate
5. $\angle A \cong \angle B$	5. Corr. parts of $\cong$ $\triangle$s are $\cong$. (Def. of $\cong$ $\triangle$s)
6. $\overline{AD} \parallel \overline{BC}$	6. If two lines are cut by a transversal and alt. int. $\angle$s are $\cong$, then the lines are $\parallel$.

Some proofs require the idea of a line perpendicular to a plane. **A line and a plane are perpendicular** if and only if they intersect and the line is perpendicular to all lines in the plane that pass through the point of intersection. Suppose you are given $\overleftrightarrow{PO} \perp$ plane X. Then you know that $\overleftrightarrow{PO} \perp \overleftrightarrow{OA}$, $\overleftrightarrow{PO} \perp \overleftrightarrow{OB}$, $\overleftrightarrow{PO} \perp \overleftrightarrow{OC}$, and so on. The ice-fishing equipment shown below suggests a line perpendicular to a plane.

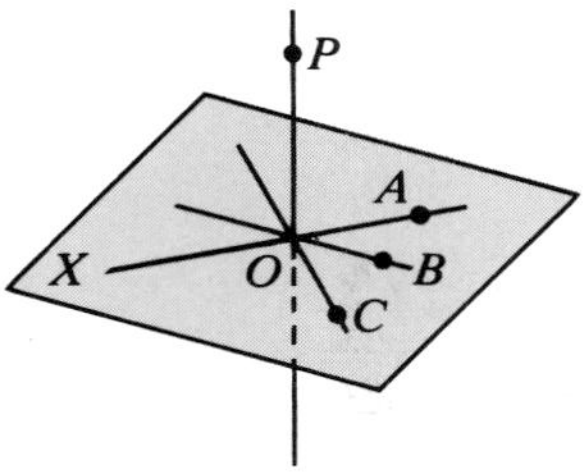

Example 2

Given: $\overline{PO} \perp$ plane X;
$\overline{AO} \cong \overline{BO}$

Prove: $\overline{PA} \cong \overline{PB}$

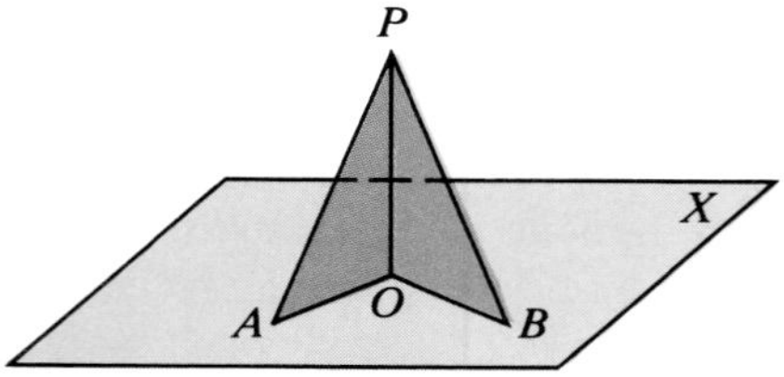

Plan for Proof: We can prove $\overline{PA} \cong \overline{PB}$ by showing that $\overline{PA}$ and $\overline{PB}$ are corresponding sides of the congruent triangles POA and POB.

Proof:

Statements	Reasons
1. $\overline{PO} \perp$ plane X	1. Given
2. $\overline{PO} \perp \overline{OA}$; $\overline{PO} \perp \overline{OB}$	2. Def. of a line perpendicular to a plane
3. $m\angle POA = m\angle POB$	3. Def. of $\perp$ lines and def. of rt. $\angle$
4. $\overline{AO} \cong \overline{BO}$	4. Given
5. $\overline{PO} \cong \overline{PO}$	5. Reflexive Prop.
6. $\triangle POA \cong \triangle POB$	6. SAS Postulate
7. $\overline{PA} \cong \overline{PB}$	7. Corr. parts of $\cong$ ≜ are $\cong$.

A Way to Prove Two Segments or Two Angles Congruent

1. Identify two triangles in which the two segments or angles are corresponding parts.
2. Prove that the triangles are congruent.
3. State that the two parts are congruent, supporting the statement with the reason: Corr. parts of $\cong$ ≜ are $\cong$.

Classroom Exercises

1. Explain your plan for proof.
Given: E is the midpoint of $\overline{DF}$;
$\angle 1 \cong \angle 2$; $\overline{CE} \cong \overline{GE}$
Prove: $\overline{CD} \cong \overline{GF}$

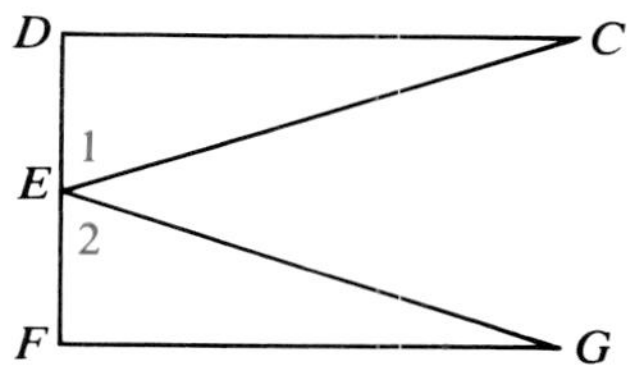

2. In Exercise 1, suppose you are also given that $m\angle D = 90$. What else can you prove about $\overline{CD}$ and $\overline{GF}$?

3. Supply the reasons.

Given: $\overleftrightarrow{RP}$ bisects $\angle QRS$ and $\angle QPS$.

Prove: $\overline{RQ} \cong \overline{RS}$

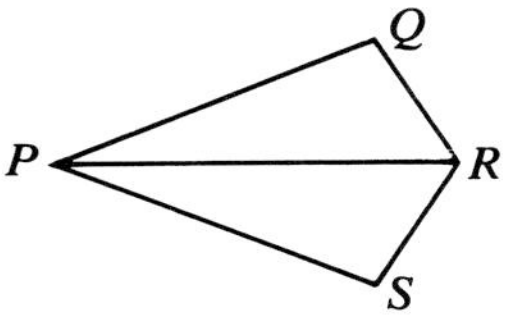

Proof:

Statements	Reasons
1. $\overleftrightarrow{RP}$ bisects $\angle QRS$ and $\angle QPS$.	1. ?
2. $\angle QRP \cong \angle SRP$; $\angle QPR \cong \angle SPR$	2. ?
3. $\overline{RP} \cong \overline{RP}$	3. ?
4. $\triangle QRP \cong \triangle SRP$	4. ?
5. $\overline{RQ} \cong \overline{RS}$	5. ?

4. Given: M is the midpoint of $\overline{AB}$; plane $X \perp \overline{AB}$ at M. What can you deduce about the figure? Explain your plan for proving that your conclusion is correct.

5. In Exercise 4, suppose Q is some point in plane X but not on $\overleftrightarrow{MP}$. Locate such a point in a diagram on the chalkboard. Then join Q to points A, B, M, and P. Now what more can you prove about the figure?

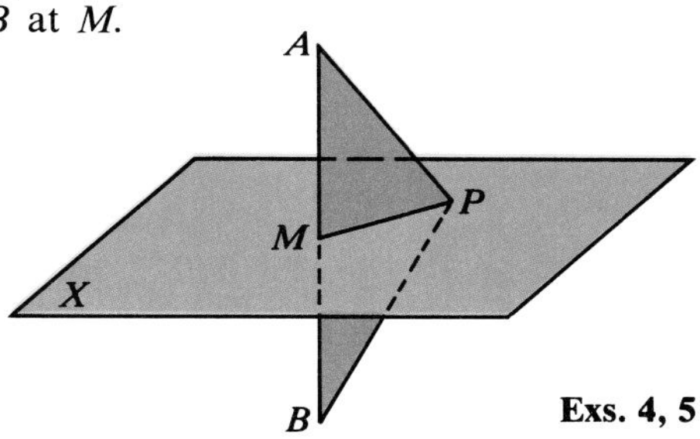

Exs. 4, 5

Written Exercises

A **1.** Supply the reasons.

Given: $\angle P \cong \angle S$;
O is the midpoint of $\overline{PS}$.

Prove: O is the midpoint of $\overline{RQ}$.

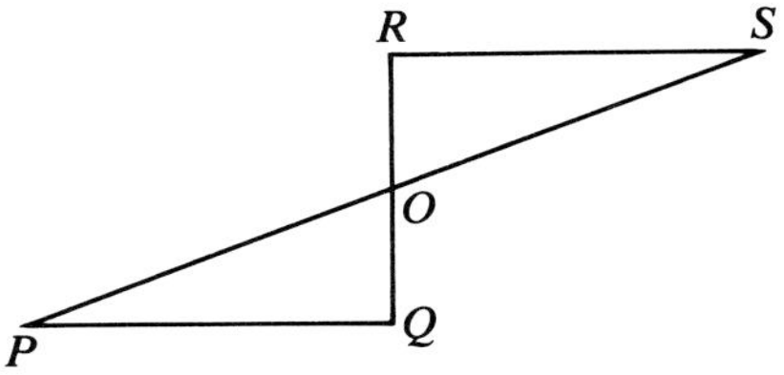

Proof:

Statements	Reasons
1. $\angle P \cong \angle S$	1. ?
2. O is the midpoint of $\overline{PS}$.	2. ?
3. $\overline{PO} \cong \overline{SO}$	3. ?
4. $\angle POQ \cong \angle SOR$	4. ?
5. $\triangle POQ \cong \triangle SOR$	5. ?
6. $\overline{QO} \cong \overline{RO}$	6. ?
7. O is the midpoint of $\overline{RQ}$.	7. ?

2. Supply the reasons.

Given: $\overline{CD} \perp \overline{AB}$;
D is the midpoint of $\overline{AB}$.

Prove: $\overline{CA} \cong \overline{CB}$

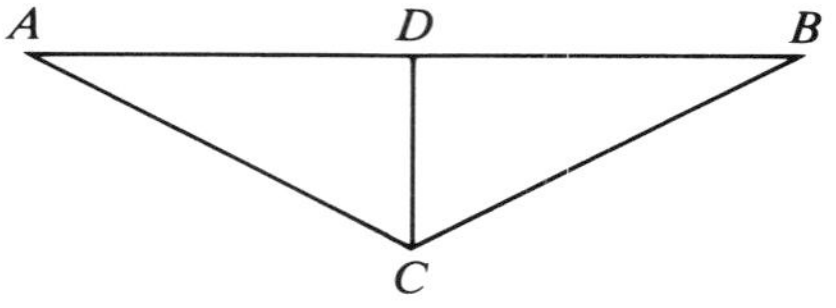

Proof:

Statements	Reasons
1. $\overline{CD} \perp \overline{AB}$	1. ___?___
2. $\angle CDA \cong \angle CDB$	2. ___?___
3. D is the midpoint of $\overline{AB}$.	3. ___?___
4. $\overline{AD} \cong \overline{DB}$	4. ___?___
5. $\overline{CD} \cong \overline{CD}$	5. ___?___
6. $\triangle CDA \cong \triangle CDB$	6. ___?___
7. $\overline{CA} \cong \overline{CB}$	7. ___?___

Write proofs in two-column form.

3. Given: $\overline{WO} \cong \overline{ZO}$; $\overline{XO} \cong \overline{YO}$
   Prove: $\angle W \cong \angle Z$

4. Given: $\angle X \cong \angle Y$; $\overline{XO} \cong \overline{YO}$
   Prove: $\overline{WO} \cong \overline{ZO}$

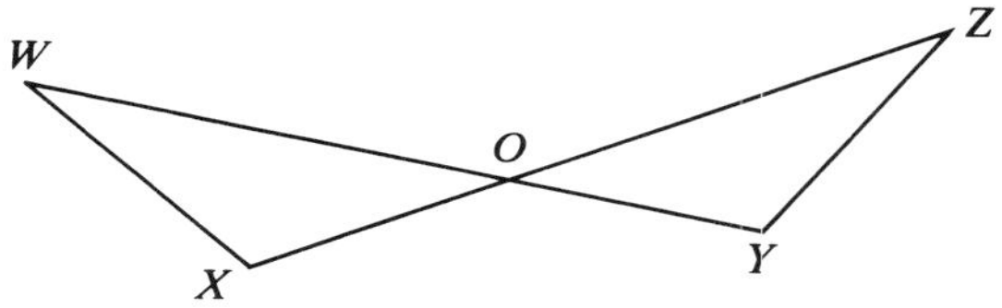

5. Given: $\overline{SK} \parallel \overline{NR}$; $\overline{SN} \parallel \overline{KR}$
   Prove: $\overline{SK} \cong \overline{NR}$; $\overline{SN} \cong \overline{KR}$

6. Given: $\overline{SK} \cong \overline{NR}$; $\overline{SN} \cong \overline{KR}$
   Prove: $\overline{SK} \parallel \overline{NR}$; $\overline{SN} \parallel \overline{KR}$

7. Exercise 5 proves that "If the opposite sides of a quadrilateral are parallel, then they are also congruent."
   a. Write the converse of this statement.
   b. Does Exercise 6 prove the converse?

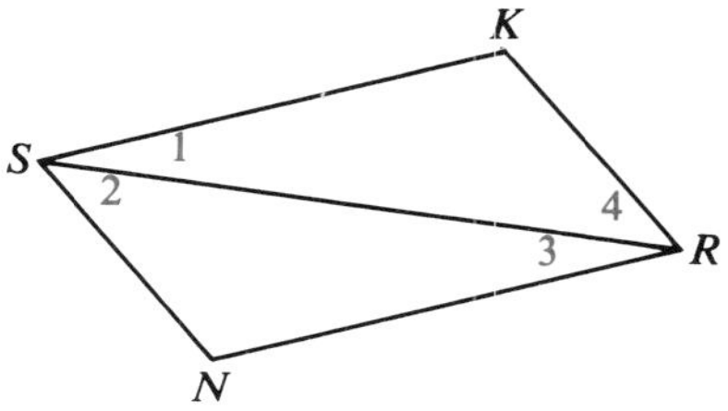

Write proofs in two-column form.

B 8. Given: $\overline{AD} \parallel \overline{ME}$; $\overline{MD} \parallel \overline{BE}$;
M is the midpoint of $\overline{AB}$.
   Prove: $\angle D \cong \angle E$

9. Given: M is the midpoint of $\overline{AB}$;
$\overline{AD} \cong \overline{ME}$; $\overline{AD} \parallel \overline{ME}$
   Prove: $\overline{DM} \parallel \overline{BE}$

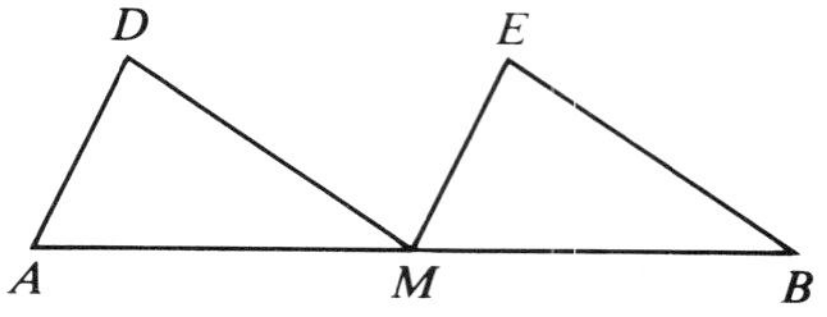

10. A young tree on level ground is supported at P by three wires of equal length. The wires are staked to the ground at points A, B, and C, which are equally distant from the base of the tree, T. Explain in a paragraph how you can prove that the angles the wires make with the ground are all congruent.

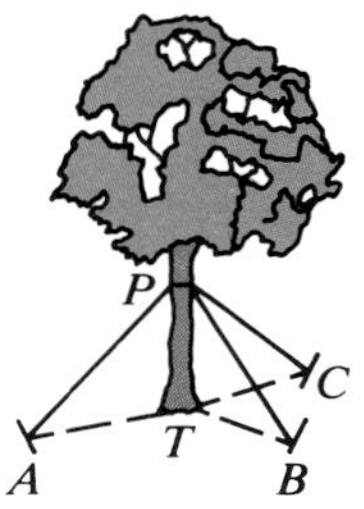

In Exercises 11 and 12, you are given more information than you need. In each exercise state one of the pieces of given information that you do not need for the proof. Then give a two-column proof that does not use that piece of information.

11. Given: $\overline{PQ} \cong \overline{PS}$; $\angle Q \cong \angle S$; $\overline{QR} \cong \overline{SR}$

Prove: $\angle QPR \cong \angle SPR$

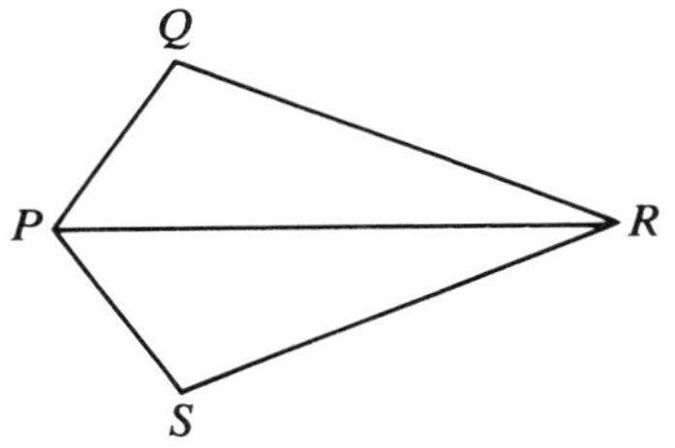

12. Given: $\overline{LM} \cong \overline{LN}$; $\overline{KM} \cong \overline{KN}$; $\overrightarrow{KO}$ bisects $\angle MKN$.

Prove: $\overrightarrow{LO}$ bisects $\angle MLN$.

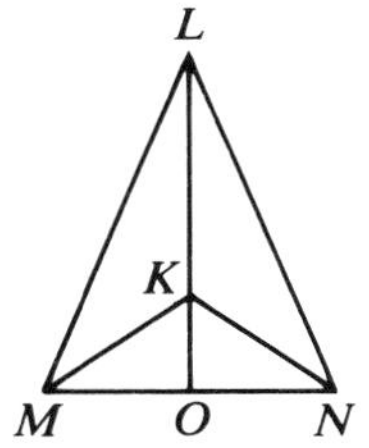

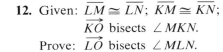

13. Given: $\overline{WS} \cong \overline{RQ}$; $\overline{ST} \cong \overline{QP}$; $WP = RT$

Prove: $\overline{WS} \parallel \overline{RQ}$

14. Given: $\overline{WS} \parallel \overline{RQ}$; $\overline{ST} \parallel \overline{PQ}$

Prove: $\angle S \cong \angle Q$

(*Hint:* If you can't find congruent triangles, try another method.)

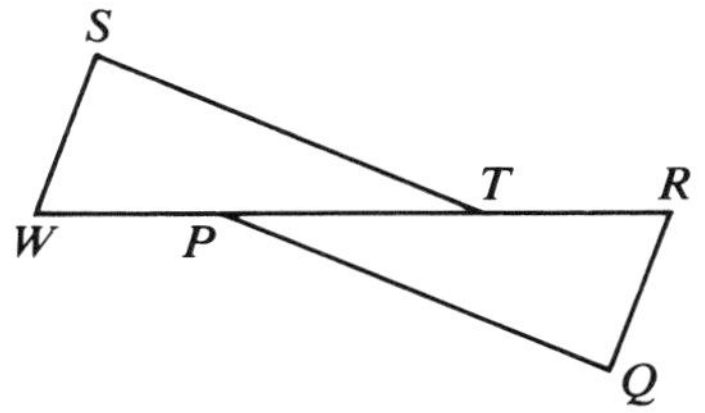

15. Given: $\overline{WX} \perp \overline{YZ}$; $\angle 1 \cong \angle 2$; $\overline{UX} \cong \overline{VX}$

Which one(s) of the following statements *must* be true?

(1) $\overline{XW} \perp \overline{UV}$ (2) $\overline{UV} \parallel \overline{YZ}$ (3) $\overline{VX} \perp \overline{UX}$

16. Given: $\overline{WX} \perp \overline{UV}$; $\overline{WX} \perp \overline{YZ}$; $\overline{WU} \cong \overline{WV}$

Prove whatever you can about angles 1, 2, 3, and 4.

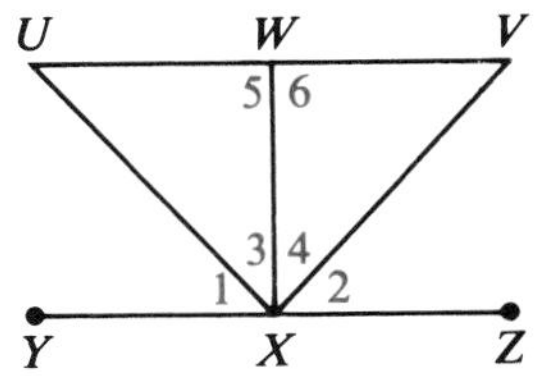

17. Given: $\overline{PA}$ and $\overline{QB}$ are perpendicular to plane X; O is the midpoint of $\overline{AB}$.

Prove: O is the midpoint of $\overline{PQ}$.

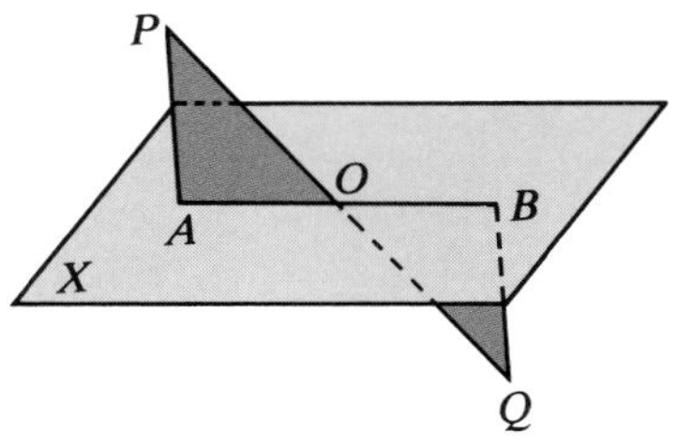

In each of Exercises 18–20, first draw a figure and state what is given and what is to be proved. Then write a proof in two-column form.

18. $\overline{PQ}$ and $\overline{PR}$ are the congruent sides of isosceles triangle PQR. The bisector of $\angle P$ meets $\overline{QR}$ at K. Prove $\overline{PK} \perp \overline{QR}$.

19. Pentagon $ABCDE$ is equilateral and has right angles at B and E. Prove that diagonals $\overline{AC}$ and $\overline{AD}$ are congruent.

20. One pair of opposite sides of a quadrilateral are both congruent and parallel. Prove that the other pair of opposite sides are also congruent and parallel. (*Hint:* Label the vertices of the quadrilateral with any letters you wish. Draw one diagonal.)

C 21. Copy the cube shown. Then draw $\overline{BE}$, $\overline{BG}$, and $\overline{EG}$. What kind of triangle is $\triangle BEG$? Write a paragraph explaining your reasoning. (You may use the fact that the faces of a cube are congruent squares.)

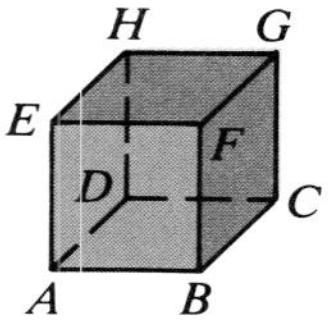

22. Napoleon, on a river bank, wanted to know the width of the stream. A young soldier faced directly across the stream and adjusted the visor of his cap until the tip of the visor was in line with his eye and the opposite bank. Next he did an about-face and noted the spot on the ground now in line with his eye and visor-tip. He paced off the distance to this spot, made his report, and earned a promotion. What postulate is this method based on? Draw a diagram to explain.

23. Given: X is the midpoint of $\overline{US}$ and $\overline{RV}$;
Y is the midpoint of $\overline{VT}$ and $\overline{SW}$.
Prove whatever you can about $\overline{UR}$ and $\overline{WT}$.

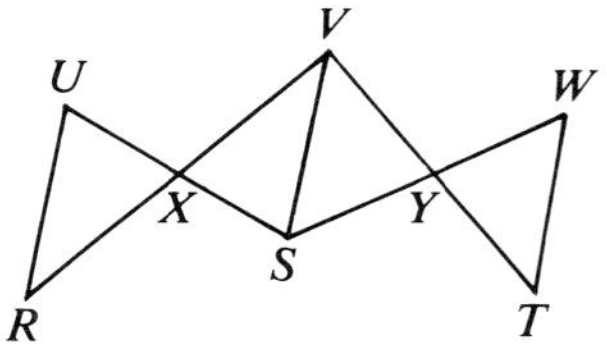

Self-Test 1

You are given that $\triangle RED \cong \triangle CAB$.

1. What can you conclude about $\angle R$? Why?
2. Name three pairs of corresponding sides.

Decide whether two triangles must be congruent. If so, write the congruence and name the postulate used. If not, write *no congruence can be deduced.*

3.

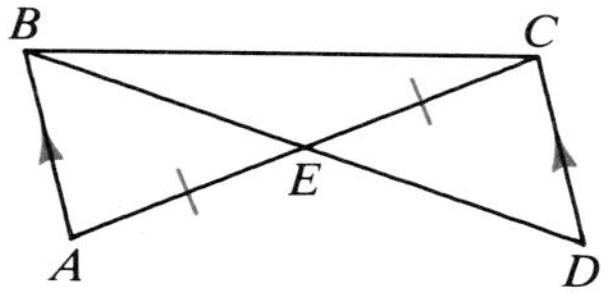

4.

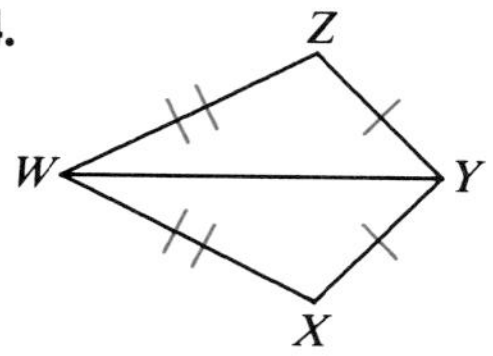

5.

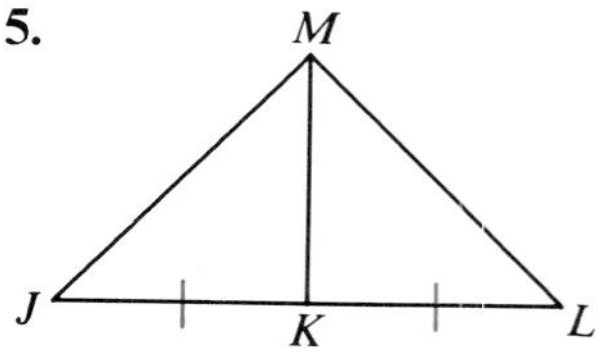

Write proofs in two-column form.

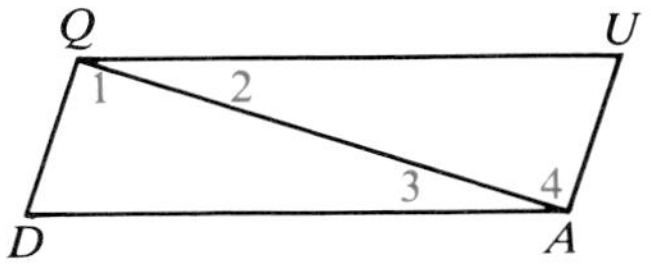

6. Given: $\overline{QU} \cong \overline{AD}$; $\overline{QU} \parallel \overline{DA}$
   Prove: $\overline{UA} \cong \overline{DQ}$
7. Given: $\overline{UA} \perp \overline{QA}$; $\overline{DQ} \perp \overline{QA}$; $\angle 2 \cong \angle 3$
   Prove: $\angle U \cong \angle D$

Application

BRACING WITH TRIANGLES

The two famous landmarks pictured above have much in common. They were completed within a few years of each other, the Eiffel Tower in 1889 and the Statue of Liberty in 1886. The French engineer Gustave Eiffel designed both the tower's sweeping form and the complex structure that supports Liberty's copper skin. And both designs gain strength from the rigidity of the triangular shape.

The strength of triangular bracing is related to the SSS Postulate, which tells us that a triangle with given sides can have only one shape. A rectangle formed by four bars joined at their ends can flatten into a parallelogram, but the structural triangle cannot be deformed except by bending or stretching the bars.

The Eiffel Tower's frame is tied together by a web of triangles. A portion of the statue's armature is shown in the photograph on the right. The inner tower of wide members is strengthened by double diagonal bracing. A framework of lighter members, also joined in triangular patterns, surrounds this core.

Structural engineers use geometry in designing bridges, towers, and large-span roofs. See what you can find out about Eiffel's bridges and about the work of some of the other great modern builders.

Some Theorems Based on Congruent Triangles

Objectives

1. Apply the theorems and corollaries about isosceles triangles.
2. Use the AAS Theorem to prove two triangles congruent.
3. Use the HL Theorem to prove two right triangles congruent.
4. Prove that two overlapping triangles are congruent.

3-4 The Isosceles Triangle Theorems

Recall (page 72) that an isosceles triangle, such as each face of the Transamerica Pyramid in San Francisco, has two congruent sides. These congruent sides are called **legs** and the third side is called the **base.** The angles at the base are called *base angles* and the angle opposite the base is called the *vertex angle* of the isosceles triangle.

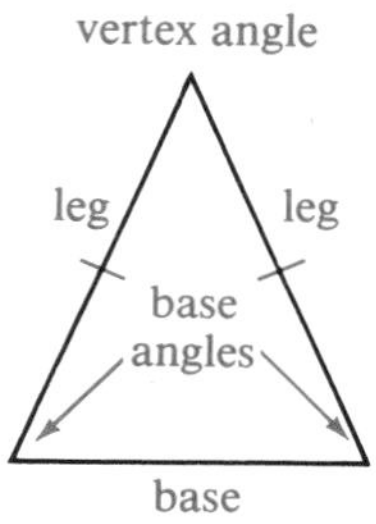

We can prove that the base angles of an isosceles triangle are congruent.

Theorem 3-1 The Isosceles Triangle Theorem

If two sides of a triangle are congruent, then the angles opposite those sides are congruent.

Given: $\overline{AB} \cong \overline{AC}$

Prove: $\angle B \cong \angle C$

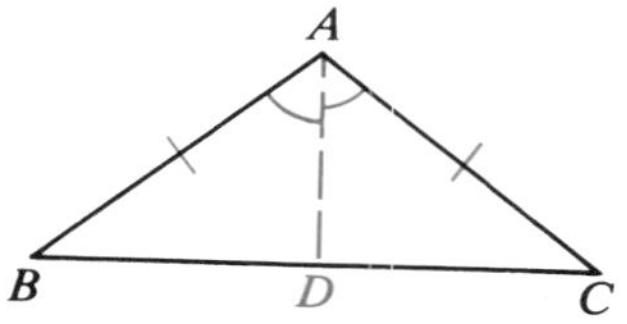

■ ***Plan for Proof:*** Draw the bisector of $\angle A$. Prove that two triangles are congruent and use corresponding parts.

The proof of Theorem 3-1 will be discussed as a classroom exercise, as will the proofs of the following corollaries of Theorem 3-1.

Corollary 1

An equilateral triangle is also equiangular.

Corollary 2

An equilateral triangle has three 60° angles.

Corollary 3

The bisector of the vertex angle of an isosceles triangle is perpendicular to the base at its midpoint.

Remember that not every theorem has a true converse. However, the converse of Theorem 3-1 *is* true and is proved below as Theorem 3-2.

Theorem 3-2

If two angles of a triangle are congruent, then the sides opposite those angles are congruent.

Given: $\angle B \cong \angle C$

Prove: $\overline{AB} \cong \overline{AC}$

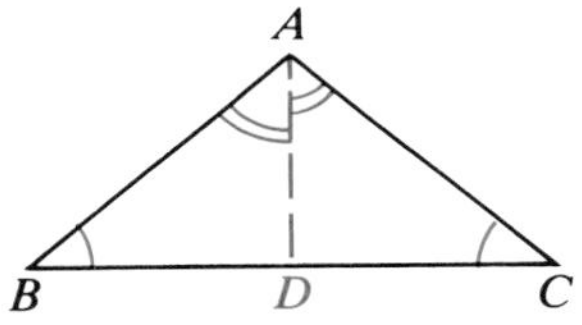

Proof:

Statements	Reasons
1. Draw the bisector of $\angle A$, intersecting $\overline{BC}$ at D.	1. By the Protractor Postulate, an angle has exactly one bisector.
2. $\angle BAD \cong \angle CAD$	2. Def. of angle bisector
3. $\angle B \cong \angle C$	3. Given
4. $\angle BDA \cong \angle CDA$	4. If two ∠s of one △ are ≅ to two ∠s of another △, then the third ∠s are ≅.
5. $\overline{AD} \cong \overline{AD}$	5. Reflexive Property
6. $\triangle BAD \cong \triangle CAD$	6. ASA Postulate
7. $\overline{AB} \cong \overline{AC}$	7. Corresponding parts of ≅ △s are ≅.

Corollary

An equiangular triangle is also equilateral.

Classroom Exercises

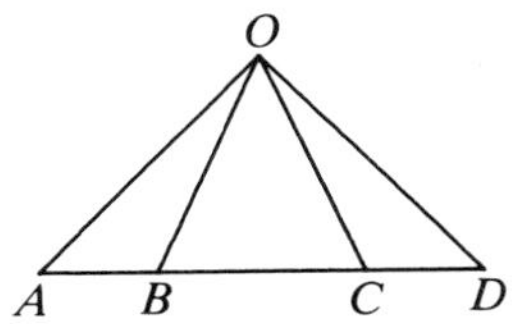

1. If $\overline{OA} \cong \overline{OD}$, then $\angle$ __?__ $\cong \angle$ __?__.
2. If $\overline{OB} \cong \overline{OC}$, then $\angle$ __?__ $\cong \angle$ __?__ and $\angle$ __?__ $\cong \angle$ __?__.
3. If $\triangle AOD$ is an isosceles right triangle, then the measure of $\angle A$ is __?__.

4. $\triangle DEF$ and $\triangle RST$ are isosceles, with legs 8 cm long. Name all pairs of angles that are congruent.

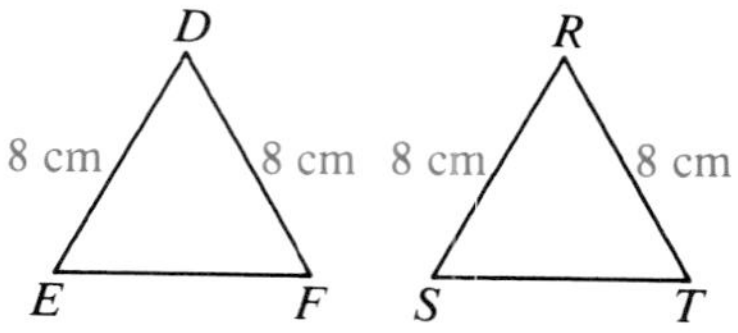

5. If $\angle 1 \cong \angle 2$, then $\underline{\ ?\ } \cong \underline{\ ?\ }$.
6. If $m\angle 3 = m\angle 4$, then $\underline{\ ?\ } = \underline{\ ?\ }$.
7. If $\angle 5 \cong \angle 6$, then $\underline{\ ?\ } \cong \underline{\ ?\ }$.
8. True or false? $MK = NK$ if and only if $m\angle 3 = m\angle 4$.

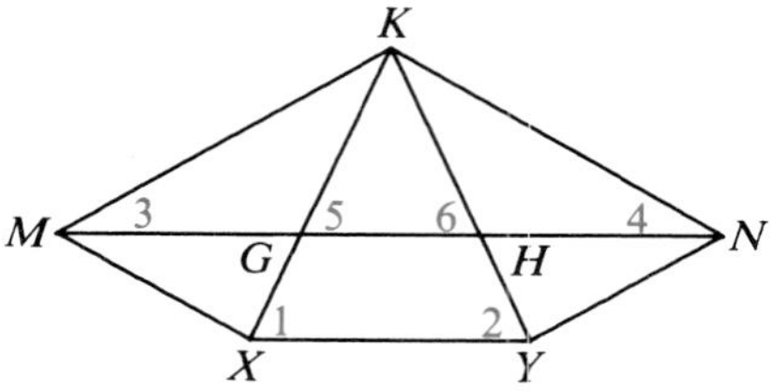

9. Prove Theorem 3–1.
10. Explain how Corollary 1 follows from Theorem 3–1.
11. Explain how Corollary 2 follows from Corollary 1.
12. What is the converse of Corollary 1? Is this converse true?
13. Use the diagram for Theorem 3–1 to explain why Corollary 3 is true.

Written Exercises

Find the value of *x*.

A **1.**

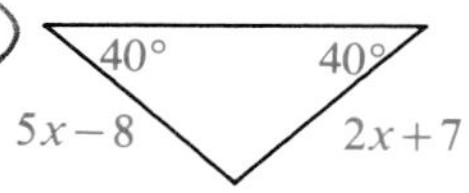

2.

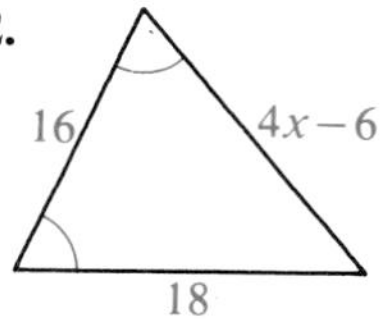

3.

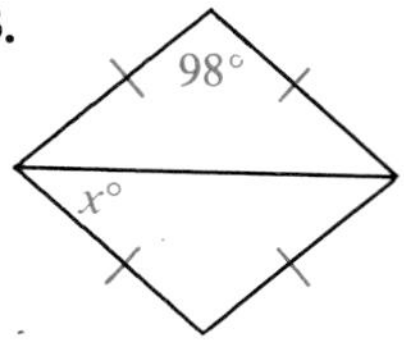

4.

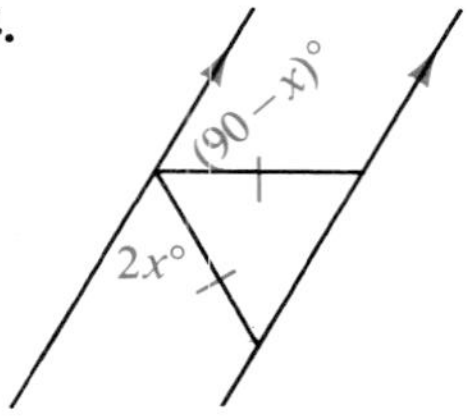

5. Supply the missing statements and reasons.

Given: $\overline{KA} \cong \overline{KB}$; $\overline{PQ} \parallel \overline{AB}$

Prove: $\overline{KP} \cong \overline{KQ}$

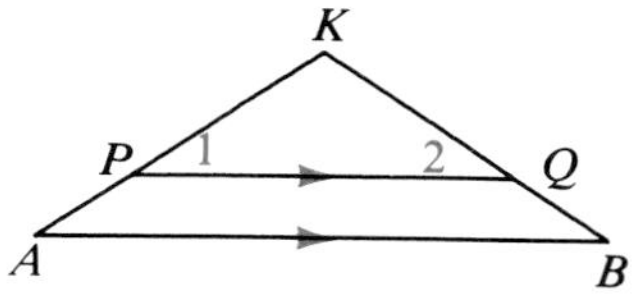

Proof:

Statements	Reasons
1. $\overline{KA} \cong \overline{KB}$	1. ?
2. $\angle A \cong \underline{\ ?\ }$	2. ?
3. $\overline{PQ} \parallel \overline{AB}$	3. ?
4. $\angle KPQ \cong \underline{\ ?\ }$; $\angle B \cong \underline{\ ?\ }$	4. ?
5. ?	5. Transitive Prop. (Steps 2 and 4)
6. $\overline{KP} \cong \overline{KQ}$	6. ?

Write proofs in two-column form.

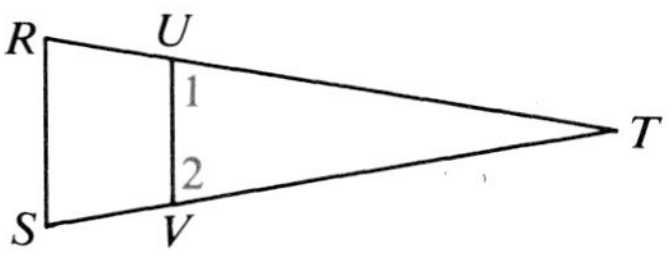

6. Given: $\angle R \cong \angle S$; $\angle TUV \cong \angle TVU$
   Prove: $RU = SV$

7. Given: $\overline{TU} \cong \overline{TV}$; $\overline{UV} \parallel \overline{RS}$
   Prove: $\angle R \cong \angle S$

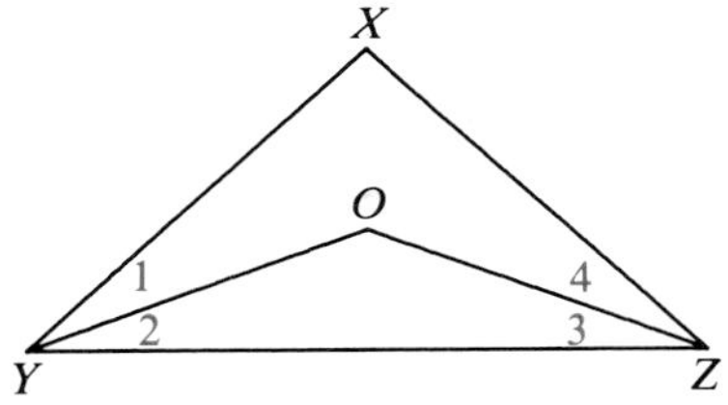

8. Given: $\overline{XY} \cong \overline{XZ}$; $\overline{OY} \cong \overline{OZ}$
   Prove: $m\angle 1 = m\angle 4$

9. Given: $\overline{XY} \cong \overline{XZ}$;
   $\overrightarrow{YO}$ bisects $\angle XYZ$;
   $\overrightarrow{ZO}$ bisects $\angle XZY$.
   Prove: $\overline{YO} \cong \overline{ZO}$

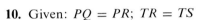

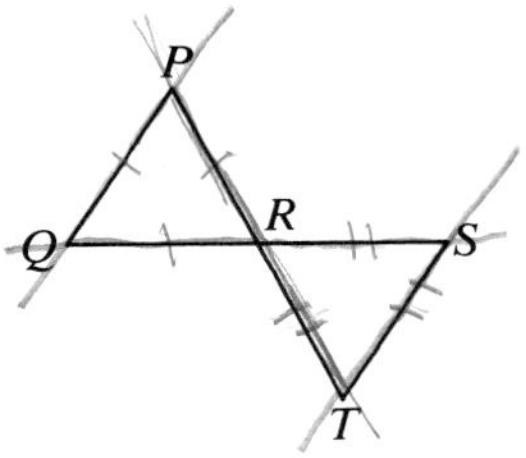

10. Given: $PQ = PR$; $TR = TS$
    Which one(s) of the following *must* be true?
    (1) $\angle T \cong \angle P$ (2) $\overline{ST} \cong \overline{QP}$ (3) $\overline{ST} \parallel \overline{QP}$

11. Given: $\angle S \cong \angle T$; $\overline{ST} \parallel \overline{QP}$
    Which one(s) of the following *must* be true?
    (1) $\angle P \cong \angle Q$ (2) $PR = QR$
    (3) R is the midpoint of $\overline{PT}$.

Write proofs in two-column form.

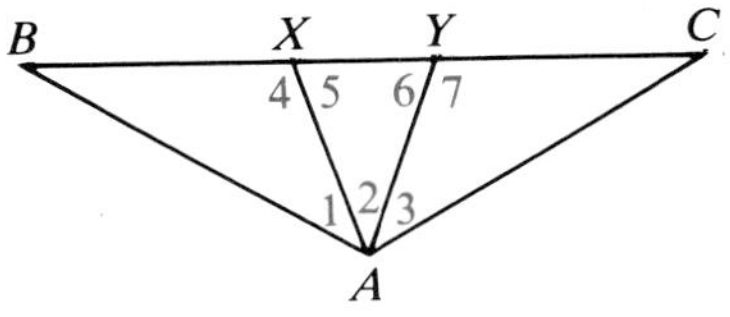

12. Given: $\overline{AB} \cong \overline{AC}$; $\overline{AX}$ and $\overline{AY}$ trisect $\angle BAC$.
    (This means $\angle 1 \cong \angle 2 \cong \angle 3$.)
    Prove: $\overline{AX} \cong \overline{AY}$

B 13. Given: $\angle 4 \cong \angle 7$; $\angle 1 \cong \angle 3$
    Prove: $\triangle ABC$ is isosceles.

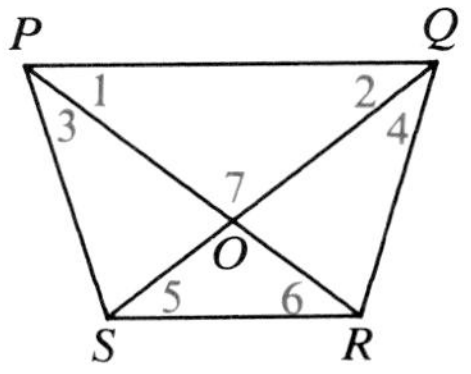

14. Given: $\angle 1 \cong \angle 2$; $\angle 3 \cong \angle 4$
    Prove: $\angle 5 \cong \angle 6$

15. Given: $PO = QO$; $RO = SO$
    a. If you are also given that $m\angle 1 = 40$, find the measures of $\angle 2$, $\angle 7$, $\angle 5$, and $\angle 6$. Then decide whether $\overline{PQ}$ must be parallel to $\overline{SR}$.
    b. If $m\angle 1 = x$, find the measures of $\angle 2$, $\angle 7$, $\angle 5$, and $\angle 6$. Is $\overline{PQ} \parallel \overline{SR}$?

16. Draw an isosceles $\triangle ABC$ whose vertex angle, $\angle A$, has measure 80.
    a. Draw $\overrightarrow{AX}$, the bisector of an exterior angle at A. Is $\overrightarrow{AX} \parallel \overline{BC}$? Explain.
    b. Would your answer change if the measure of $\angle A$ changed?

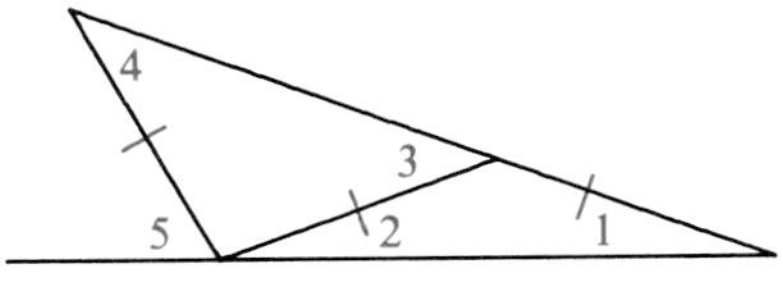

17. a. If $m\angle 1 = 20$, then $m\angle 3 =$ _?_, $m\angle 4 =$ _?_, and $m\angle 5 =$ _?_.
    b. If $m\angle 1 = x$, then $m\angle 3 =$ _?_, $m\angle 4 =$ _?_, and $m\angle 5 =$ _?_.

18. **a.** If $m\angle 1 = 35$, then $m\angle ABC =$ __?__.
b. If $m\angle 1 = x$, then $m\angle ABC =$ __?__.

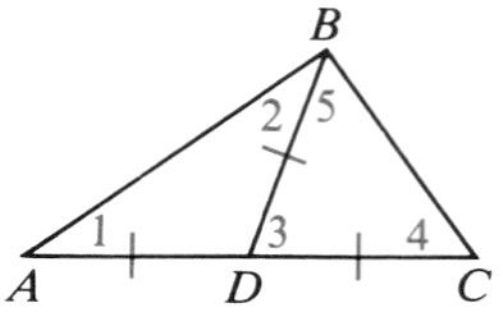

19. **a.** If $m\angle 1 = 23$, then $m\angle 7 =$ __?__.
b. If $m\angle 1 = x$, then $m\angle 7 =$ __?__.

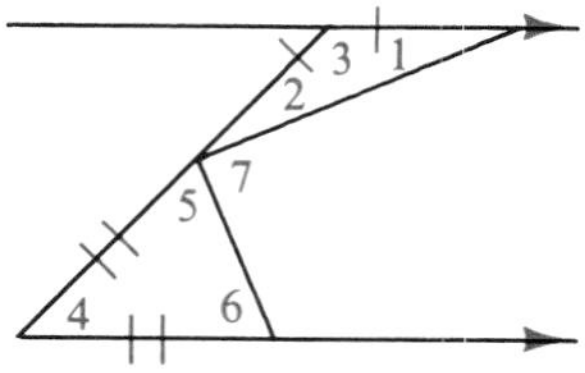

Solve for *x* and *y*.

20. $\triangle ABC$ is equiangular, $AB = 4x - y$, $BC = 2x + 3y$, and $AC = 7$.

21. $\triangle DEF$ is equilateral, $m\angle D = x + y$, and $m\angle E = 2x - y$.

22. In $\triangle JKL$, $\overline{JK} \cong \overline{KL}$, $m\angle J = 2x - y$, $m\angle K = 2x + 2y$, and $m\angle L = x + 2y$.

23. Given: $\triangle ABC$ in plane M; D not in plane M;
$\angle ACB \cong \angle ABC$; $\angle DCB \cong \angle DBC$
Name a pair of congruent triangles.
Prove your answer correct.

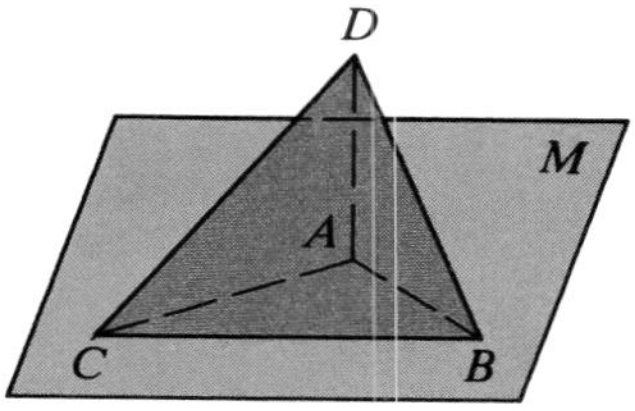

24. $ABCDE$ is a regular pentagon and $DEFG$ is a square. Find the measures of $\angle EAF$ and $\angle AFD$.

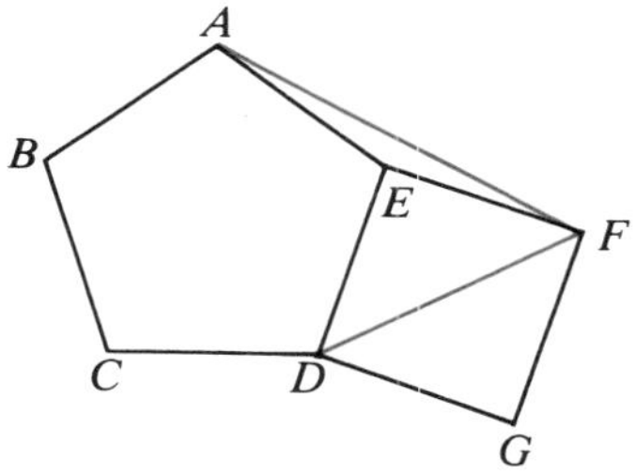

C 25. Draw an isosceles triangle and then join the midpoints of its sides to form another triangle. What can you deduce about this second triangle? Explain.

26. Given: $\triangle ABC$ is equilateral;
$\angle CAD \cong \angle ABE \cong \angle BCF$
Prove something interesting about $\triangle DEF$.

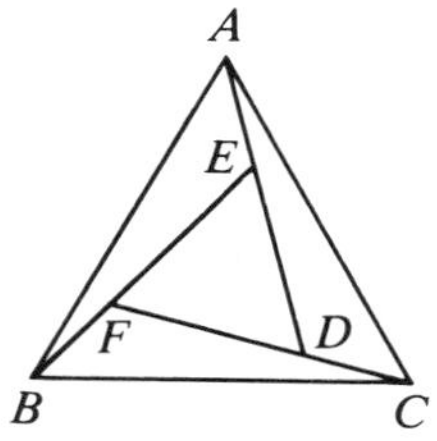

27. Given: $\overline{RX} \cong \overline{SX} \cong \overline{SY}$;
$\overline{XY} \cong \overline{YT}$
Prove: $\overline{ST} \cong \overline{RY}$

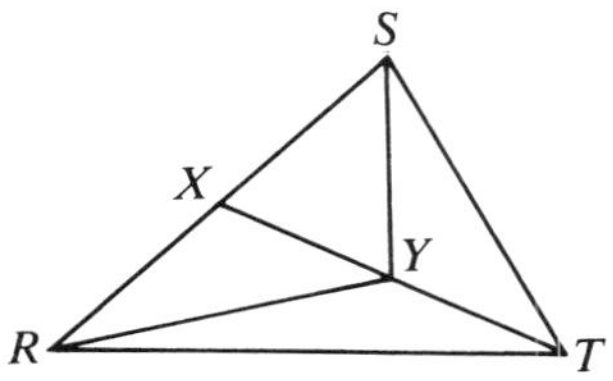

28. **a.** The figure on the left is a regular pentagon. Find the measures of $\angle 1$, $\angle 2$, and $\angle 3$.
b. The figure on the right is a regular hexagon. Find the measures of $\angle 1$, $\angle 2$, $\angle 3$, and $\angle 4$.

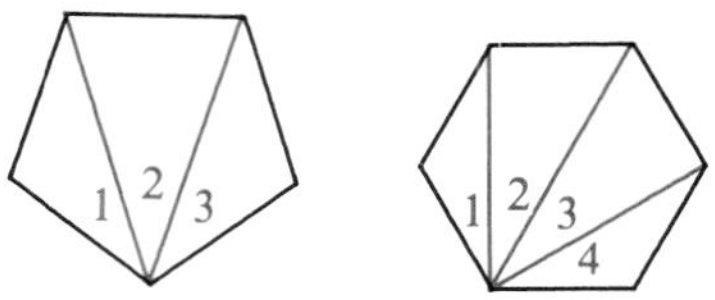

3-5 Other Methods of Proving Triangles Congruent

The SSS, SAS, and ASA Postulates give us three methods of proving triangles congruent. In this section we will discuss two other methods.

Theorem 3-3 AAS Theorem

If two angles and a non-included side of one triangle are congruent to the corresponding parts of another triangle, then the triangles are congruent.

Given: $\triangle ABC$ and $\triangle DEF$; $\angle B \cong \angle E$;
$\angle C \cong \angle F$; $\overline{AC} \cong \overline{DF}$

Prove: $\triangle ABC \cong \triangle DEF$

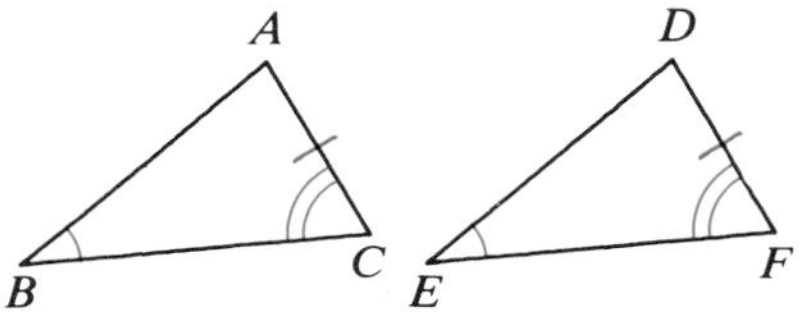

■ ***Plan for Proof:*** From the fact that two angles of one triangle are congruent to two angles of the other, conclude that $\angle A \cong \angle D$. Apply the ASA Postulate to reach the desired conclusion.

Do you see overlapping triangles in the photograph? Sometimes you want to prove that certain overlapping triangles are congruent. For example, suppose you have the following problem:

Given: $\overline{AD} \cong \overline{AE}$;
$\angle B \cong \angle C$

Prove: $\triangle ABD \cong \triangle ACE$

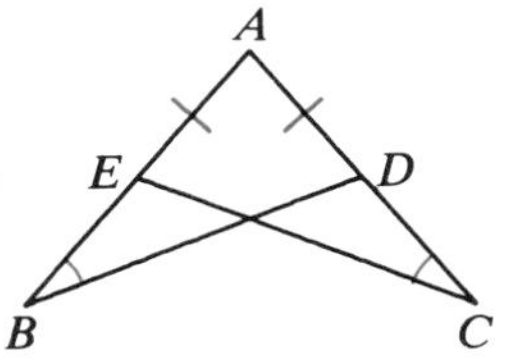

You may find it helps you visualize the congruence if you redraw the two triangles, as shown below. Now you can see that since $\angle A$ is common to both triangles, the triangles must be congruent by the AAS Theorem.

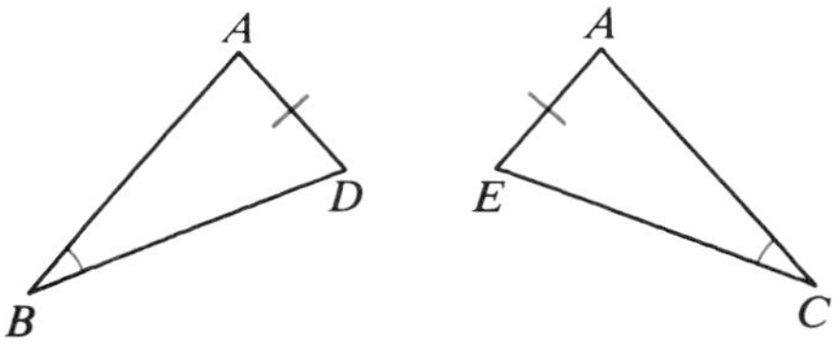

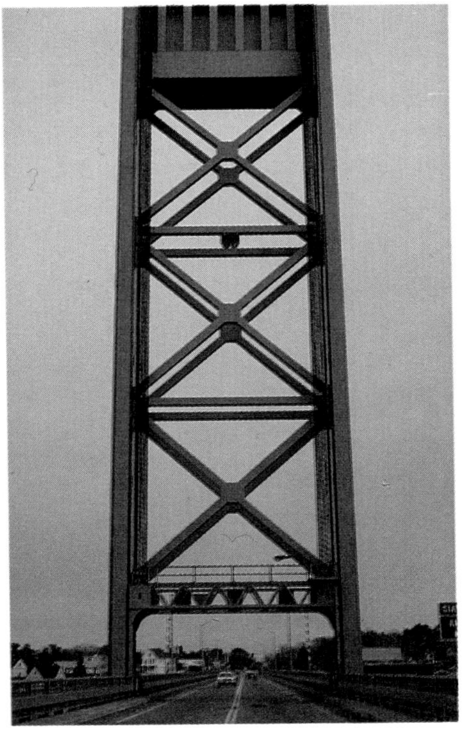

Our final method of proving triangles congruent applies only to right triangles. In a right triangle the side opposite the right angle is called the **hypotenuse** (hyp.). The other two sides are called **legs.**

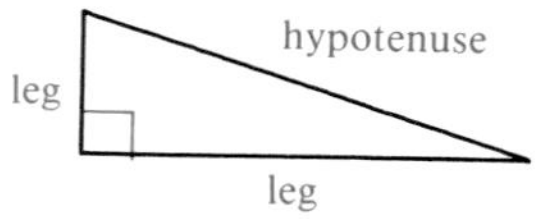

Theorem 3-4 HL Theorem

If the hypotenuse and a leg of one right triangle are congruent to the corresponding parts of another right triangle, then the triangles are congruent.

Given: $\triangle ABC$ and $\triangle DEF$;
$m\angle C = 90$; $m\angle F = 90$;
$\overline{AB} \cong \overline{DE}$; $\overline{BC} \cong \overline{EF}$

Prove: $\triangle ABC \cong \triangle DEF$

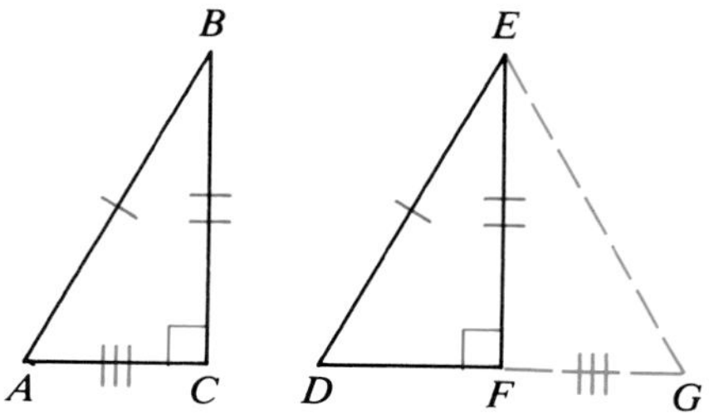

Plan for Proof: Construct $\triangle GEF$ with G on $\overrightarrow{DF}$ such that $\overline{FG} \cong \overline{AC}$. By the SAS Postulate, $\triangle ABC \cong \triangle GEF$ and hence $\angle A \cong \angle G$ and $\overline{AB} \cong \overline{GE}$. Thus $\overline{DE} \cong \overline{GE}$. The angles opposite those sides, $\angle D$ and $\angle G$, are congruent. This leads to $\angle A \cong \angle D$. Finally, use the AAS Theorem to reach the desired conclusion.

Proof:

Statements	Reasons
1. Let G be the point on the ray opposite $\overrightarrow{FD}$ such that $\overline{FG} \cong \overline{CA}$.	1. The Ruler Postulate guarantees exactly one such point.
2. Draw $\overline{EG}$.	2. Through any two points there is _?_.
3. $m\angle C = 90$; $m\angle DFE = 90$	3. Given
4. $m\angle EFG = 180 - m\angle DFE = 90$	4. Angle Addition Postulate
5. $m\angle C = m\angle EFG$, or $\angle C \cong \angle EFG$	5. Substitution Prop.
6. $\overline{BC} \cong \overline{EF}$	6. Given
7. $\triangle ABC \cong \triangle GEF$	7. SAS Postulate
8. $\overline{AB} \cong \overline{GE}$	8. Corr. parts of $\cong$ ⧍ are $\cong$.
9. $\overline{AB} \cong \overline{DE}$, or $\overline{DE} \cong \overline{AB}$	9. Given and Symmetric Prop.
10. $\overline{DE} \cong \overline{GE}$	10. Transitive Prop.
11. $\angle G \cong \angle D$	11. If 2 sides of a $\triangle$ are $\cong$, then _?_.
12. $\angle A \cong \angle G$	12. Corr. parts of $\cong$ ⧍ are $\cong$. (Step 7)
13. $\angle A \cong \angle D$	13. Transitive Prop.
14. $\triangle ABC \cong \triangle DEF$	14. AAS Theorem

Now that we have an AAS method for proving two triangles congruent, it is natural to ask whether there is an SSA method (for two sides and a non-included angle). As you saw in Exercise 22, page 115, the answer is *no*. For example, both triangles below have a side 3 cm long and a side 2 cm long and a non-included angle of 30°, but the triangles are not congruent.

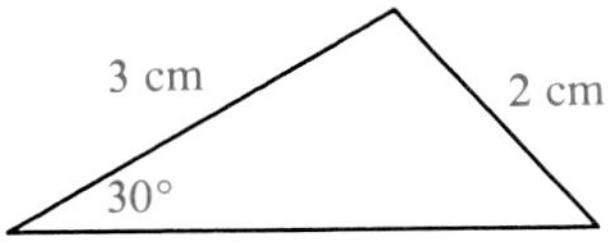

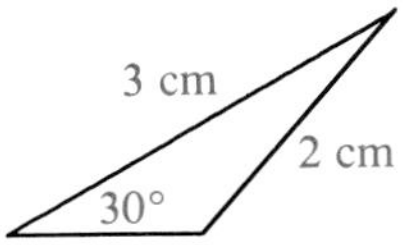

> **Summary of Ways to Prove Two Triangles Congruent**
>
> All triangles: SSS SAS ASA AAS
> Right triangles: HL

Classroom Exercises

1. Which of the following congruence methods are postulates and which are theorems?

ASA AAS SAS SSS HL

State which congruence method(s) can be used to prove the triangles congruent. If no method applies, say *none*.

2.

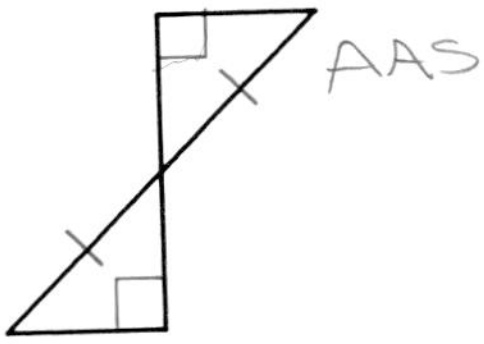

3.

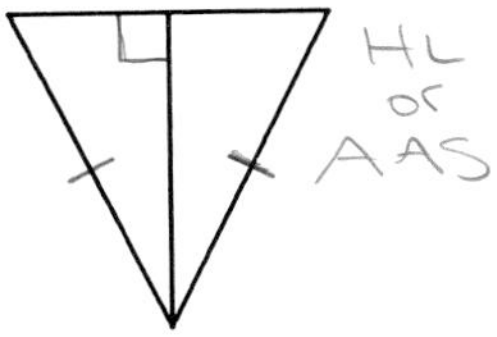

4.

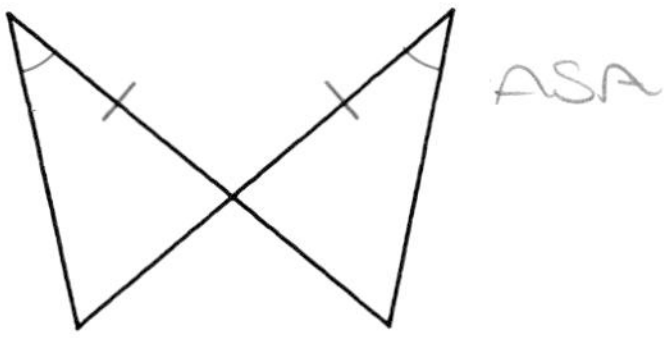

5.

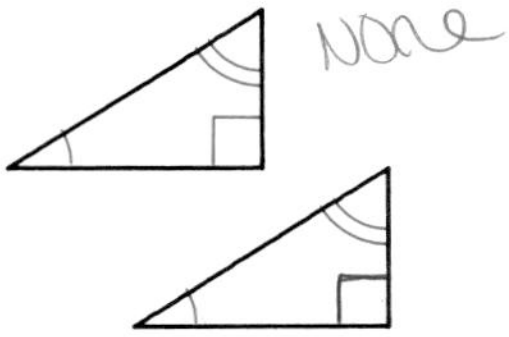

6.

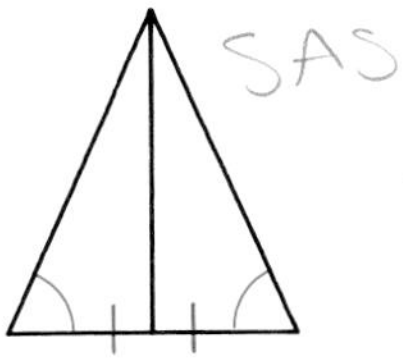

7.

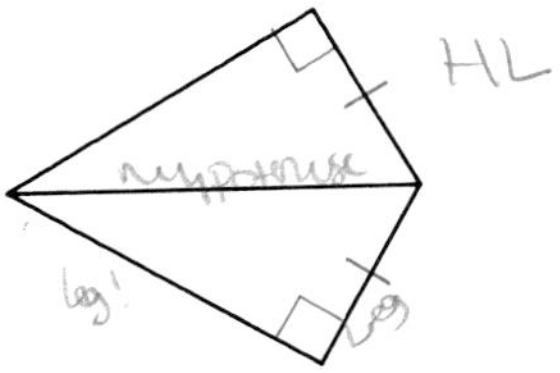

8.

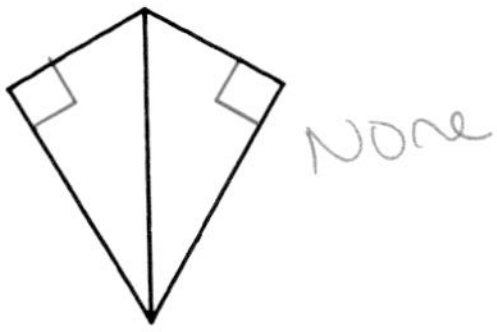

9.

10.

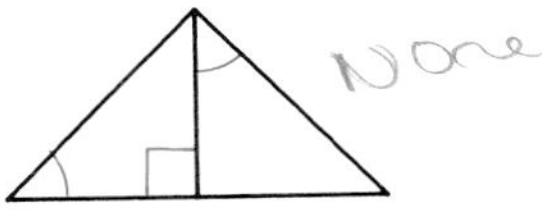

For each diagram, name a pair of overlapping congruent triangles. Tell whether the triangles are congruent by the SSS, SAS, ASA, AAS, or HL method.

11. Given: $AB = DC$; $AC = DB$

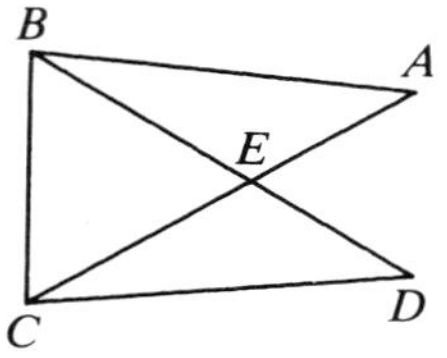

12. Given: $\angle ABC \cong \angle ACB$

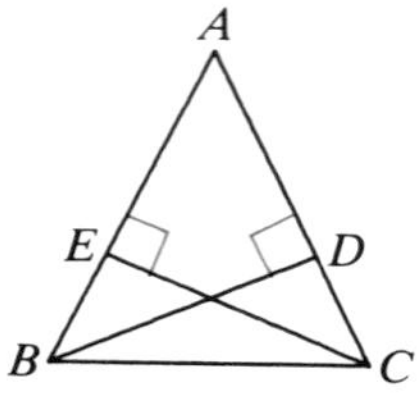

13. Given: $\angle 1 \cong \angle 2$; $\angle 3 \cong \angle 4$

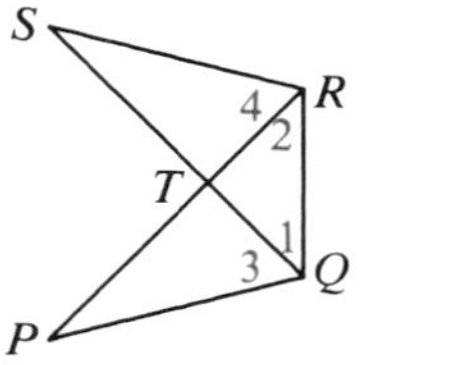

14. Given: $\angle 1 \cong \angle 2$; $\angle 5 \cong \angle 6$
Prove as much as you can about the diagram.

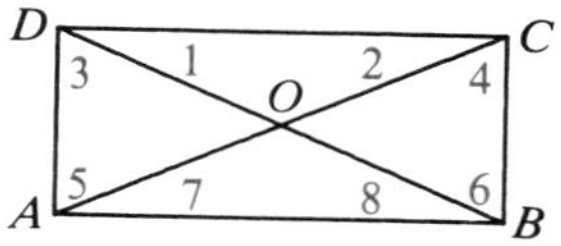

15. To prove that right triangles are congruent, some geometry books also use the methods stated below. Tell which of our methods (SSS, SAS, ASA, AAS, or HL) could be used instead of each method listed.

a. Leg–Leg Method (LL) If two legs of one right triangle are congruent to the two legs of another right triangle, then the triangles are congruent.

b. Hypotenuse–Acute Angle Method (HA) If the hypotenuse and an acute angle of one right triangle are congruent to the hypotenuse and an acute angle of another right triangle, then the triangles are congruent.

c. Leg–Acute Angle Method (LA) If a leg and an acute angle of one right triangle are congruent to the corresponding parts in another right triangle, then the triangles are congruent.

Written Exercises

Supply the missing statements and reasons.

A **1.** Given: $\angle W$ and $\angle Y$ are rt. $\angle$s; $\overline{WX} \cong \overline{YX}$

Prove: $\overline{WZ} \cong \overline{YZ}$

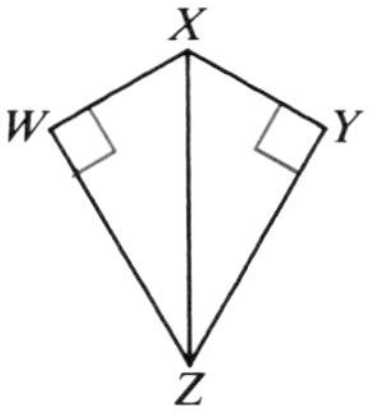

Proof:

Statements	Reasons
1. $\angle W$ and $\angle Y$ are rt. $\angle$s.	1. ?
2. $\triangle XWZ$ and $\triangle XYZ$ are rt. $\triangle$s.	2. ?
3. $\overline{WX} \cong \overline{YX}$	3. ?
4. ?	4. Reflexive Prop.
5. $\triangle WXZ \cong$?	5. ?
6. ?	6. ?

2. Given: $\overline{KL} \perp \overline{LA}$; $\overline{KJ} \perp \overline{JA}$;
$\overrightarrow{AK}$ bisects $\angle LAJ$.

Prove: $\overline{LK} \cong \overline{JK}$

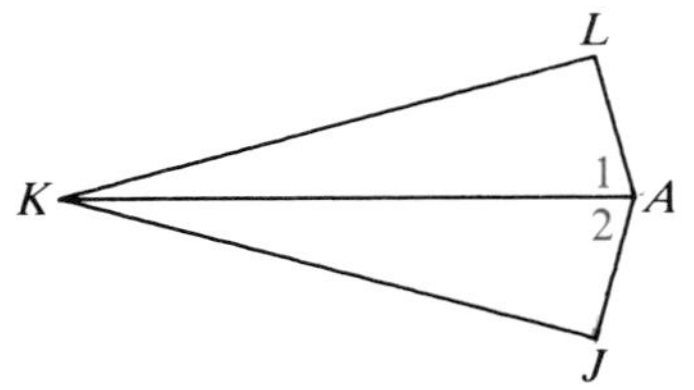

Proof:

Statements	Reasons
1. $\overline{KL} \perp \overline{LA}$; $\overline{KJ} \perp \overline{JA}$	1. ?
2. $m\angle L = 90 = m\angle$?	2. Def. of $\perp$ lines and def. of rt. $\angle$
3. $\overrightarrow{AK}$ bisects $\angle LAJ$.	3. ?
4. ? $\cong$?	4. Def. of $\angle$ bisector
5. $\overline{KA} \cong$?	5. ?
6. $\triangle LKA \cong$?	6. ?
7. ?	7. ?

In Exercises 3 and 4 write proofs in two-column form.

3. Given: $\overline{EF} \perp \overline{EG}$; $\overline{HG} \perp \overline{EG}$;
$\overline{EH} \cong \overline{GF}$

Prove: $\angle H \cong \angle F$

4. Given: $\overline{EF} \parallel \overline{HG}$;
$\angle H \cong \angle F$

Prove: $\overline{HE} \cong \overline{FG}$

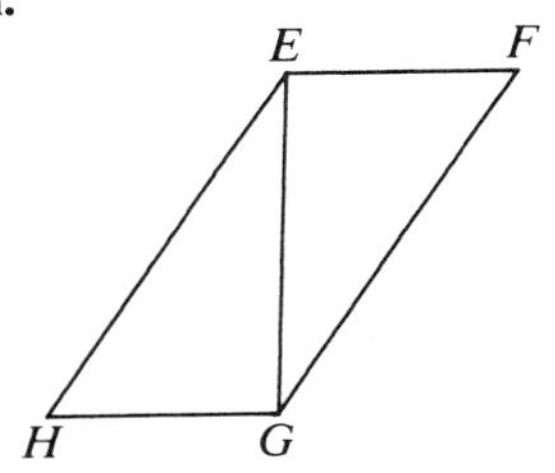

5. Given: $\overline{PR} \cong \overline{PQ}$;
$\overline{SR} \cong \overline{TQ}$

Prove: $\overline{QS} \cong \overline{RT}$

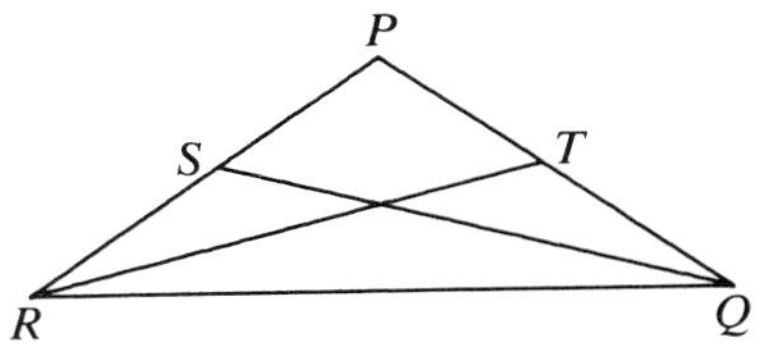

■ ***Plan for Proof:*** $\overline{QS}$ and $\overline{RT}$ are corresponding parts of $\triangle PQS$ and $\triangle PRT$ and also of $\triangle RQS$ and $\triangle QRT$. The second set of triangles is easier to prove congruent than the first set.

Proof:

Statements	Reasons
1. $\overline{PR} \cong \overline{PQ}$	1. ?
2. $\angle PQR \cong$?	2. If 2 sides of a $\triangle$ are $\cong$, then ? .
3. $\overline{SR} \cong \overline{TQ}$	3. ?
4. $\overline{RQ} \cong \overline{RQ}$	4. ?
5. $\triangle RQS \cong$?	5. ?
6. $\overline{QS} \cong \overline{RT}$	6. ?

6. Given: $m\angle 1 = m\angle 2$;
$m\angle 3 = m\angle 4$

Prove: $\overline{MJ} \cong \overline{NL}$

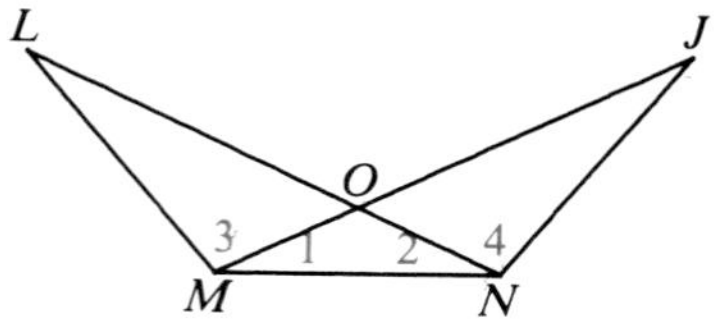

■ ***Plan for Proof:*** $\overline{MJ}$ and $\overline{NL}$ are corresponding parts of $\triangle MJN$ and $\triangle NLM$. Try to prove these triangles congruent.

Proof:

Statements	Reasons
1. $m\angle 1 = m\angle 2$; $m\angle 3 = m\angle 4$	1. ?
2. $m\angle 1 + m\angle 3 = m\angle 2 + m\angle 4$	2. ?
3. $m\angle 1 + m\angle 3 = m\angle LMN$; $m\angle 2 +$? $=$?	3. ?
4. $m\angle LMN = m\angle JNM$	4. ?
5. ?	5. Reflexive Prop.
6. $\triangle MJN \cong$?	6. ?
7. ?	7. ?

7. Given: $\overline{RT} \cong \overline{AS}$;
$\overline{RS} \cong \overline{AT}$

Prove: $\angle TSA \cong \angle STR$

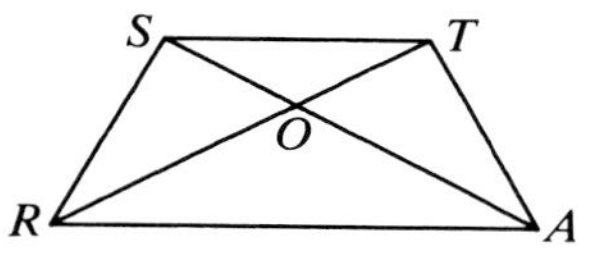

8. Given: $\overline{DH} \perp \overline{DJ}$; $\overline{JK} \perp \overline{DJ}$;
$\overline{JH} \cong \overline{DK}$

Prove: $\angle H \cong \angle K$

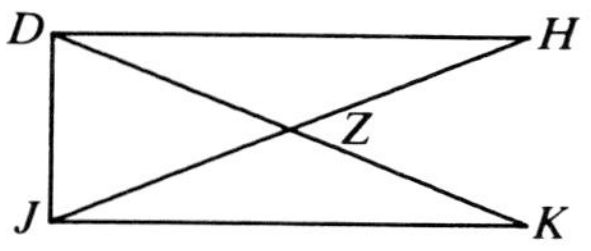

9. Given: $\overline{AO} \perp$ plane M
State the definition that allows you to conclude that $\overline{AO} \perp \overline{BO}$ and $\overline{AO} \perp \overline{CO}$.

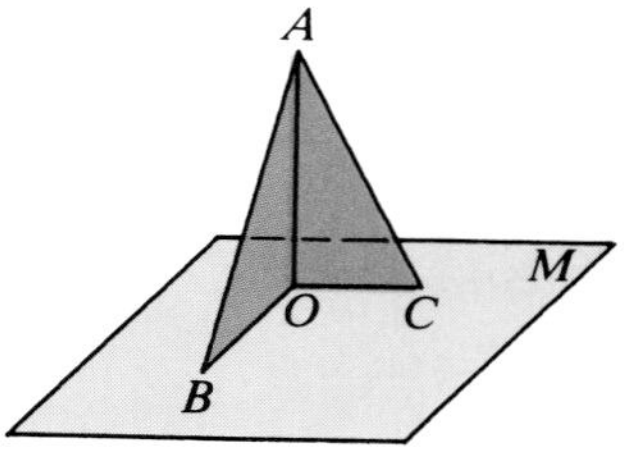

Use the figure and the given information below to tell what method (SSS, SAS, ASA, AAS, or HL) can be used to prove $\triangle ABO \cong \triangle ACO$. You need not write the proofs.

10. Given: $\overline{AO} \perp$ plane M; $\overline{BO} \cong \overline{CO}$

11. Given: $\overline{AO} \perp$ plane M; $\angle B \cong \angle C$

12. Given: $\overline{AO} \perp$ plane M; $\overline{AB} \cong \overline{AC}$

B 13. Given: The figure above with no information except that $\overline{AB} \cong \overline{AC}$ and $\overline{OB} \cong \overline{OC}$.

a. Is it possible to prove that $\angle AOB \cong \angle AOC$?

b. Is it possible to prove that $\angle AOB$ and $\angle AOC$ are right angles?

14. Copy the figure for Exercises 9–14 and draw $\overline{BC}$. Suppose $\overline{AO} \perp \overline{OB}$ and $\overline{AO} \perp \overline{OC}$. Classify the following as true or false.
   a. If $\overline{AB} \cong \overline{AC}$, then $\angle OBC \cong \angle OCB$.
   b. The converse of the statement in (a).

Tell which pairs of congruent parts and what method (SSS, SAS, ASA, AAS, or HL) you would use to prove the triangles are congruent.

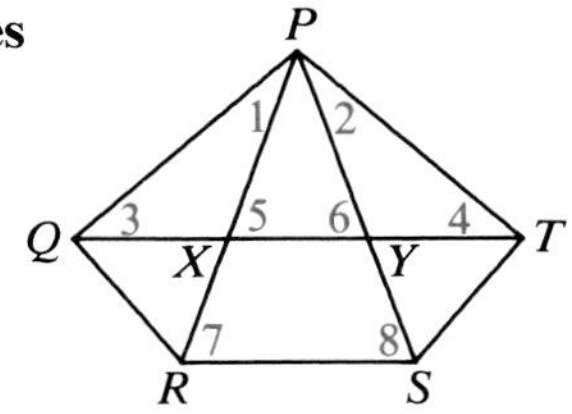

15. Given: $\angle 1 \cong \angle 2$; $\angle 3 \cong \angle 4$; $\angle 5 \cong \angle 6$
$\triangle PQX \cong \triangle PTY$ by what method?

16. Given: $\angle 7 \cong \angle 8$; $\angle 3 \cong \angle 4$; $\overline{QR} \cong \overline{ST}$
$\triangle QPR \cong \triangle TPS$ by what method?

17. a. Draw an isosceles triangle RST with $RS = RT$. Let M be the midpoint of $\overline{RT}$ and N be the midpoint of $\overline{RS}$. Draw $\overline{SM}$ and $\overline{TN}$ and label their common point O. Now draw $\overline{MN}$.
   b. Name four *pairs* of congruent triangles.

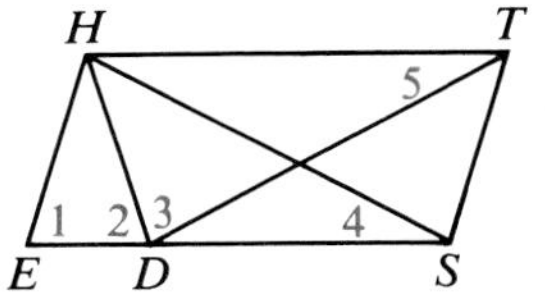

18. Write a two-column proof.
Given: $\angle 1 \cong \angle 2 \cong \angle 3$;
$\overline{ES} \cong \overline{DT}$
Prove: $\angle 4 \cong \angle 5$

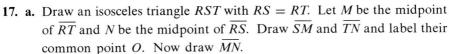

19. Draw an isosceles triangle ABC with $AB = AC$. Also draw the bisector of $\angle B$, intersecting $\overline{AC}$ at X, and the bisector of $\angle C$, intersecting $\overline{AB}$ at Y. Prove that $BX = CY$.

For Exercises 20 and 21, write a detailed plan for proof instead of a two-column proof.

20. Draw an isosceles triangle. From the midpoint of each leg draw a perpendicular segment to the base. Prove that these segments are congruent. (First label your figure and state what is given and what is to be proved.)

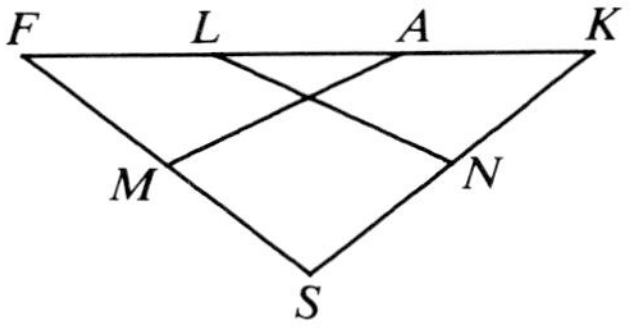

21. Given: $\overline{FL} \cong \overline{AK}$;
$\overline{SF} \cong \overline{SK}$;
M is the midpoint of $\overline{SF}$;
N is the midpoint of $\overline{SK}$.
Prove: $\overline{AM} \cong \overline{LN}$

Write proofs in two-column form.

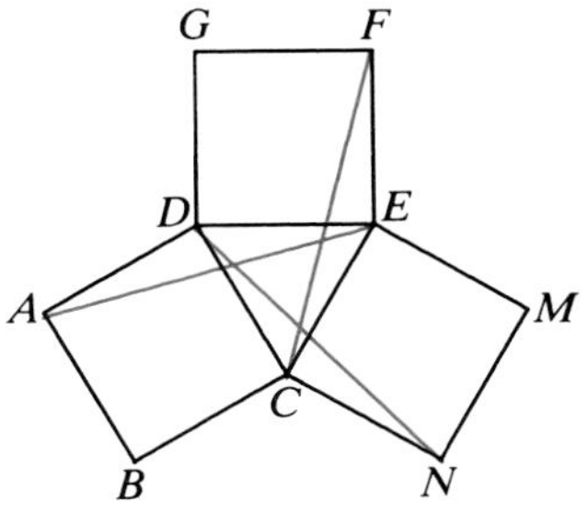

C 22. The diagram shows three squares and an equilateral triangle.
Prove: $AE = FC = ND$

23. Use the results of Exercise 22 to prove that $\triangle FAN$ is equilateral.

Self-Test 2

Find the value of x.

1.

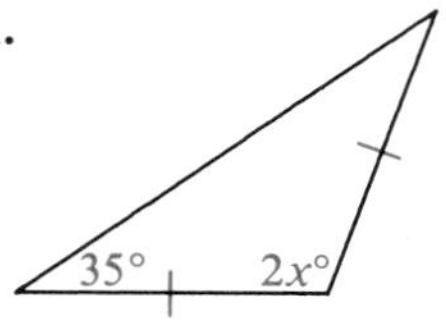

2.

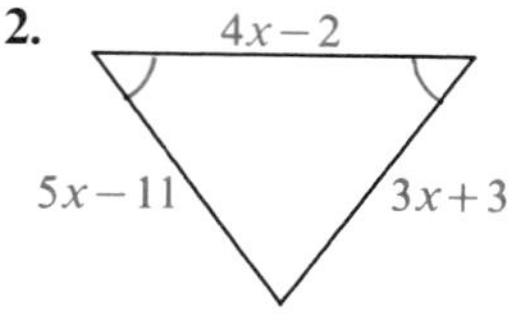

3. If you know that $\overline{WX} \cong \overline{WZ}$, can you conclude that $\angle X \cong \angle Z$? Explain.

4. Given: $\angle X \cong \angle Z$; $\overline{WY} \perp$ plane P
Prove: $\triangle WXY \cong \triangle WZY$

5. Given: $\overline{WX} \cong \overline{WZ}$; $\overline{WY} \perp$ plane P
Prove: $\overline{XY} \cong \overline{ZY}$

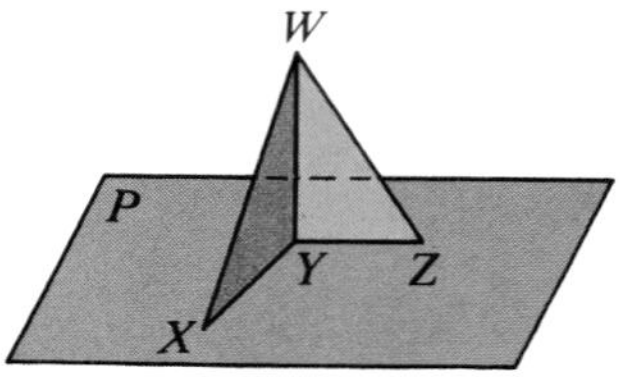

Exs. 3–5

6. Given: $\overline{AB} \cong \overline{AC}$; $\overline{BN} \perp \overline{AC}$; $\overline{CM} \perp \overline{AB}$
Explain how you could prove that $\triangle ABN \cong \triangle ACM$.

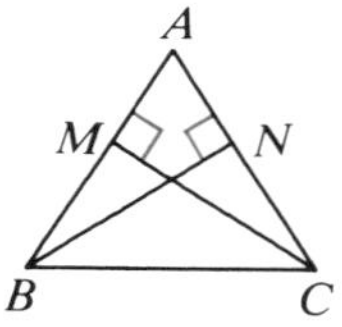

READING GEOMETRY

Diagrams

Do you realize how long this book would be if we weren't allowed to use diagrams? A simple-looking diagram may contain information that would take many words to describe. To test this statement, look at the diagram on page 111 that shows congruent triangles *MON* and *KEY*. Try writing down all the information you can get from this diagram.

Many of the explanations in this book are accompanied by diagrams that illustrate the relationships discussed. For a clear understanding, you need to read the text and the diagrams *together*. Reading diagrams calls for concentration so that you won't miss any of the given information.

You can improve your skill in *reading* diagrams by *drawing* diagrams. If you find an explanation or a theorem hard to understand, make a sketch to illustrate it. Be sure to show all the given information. You may want to look back at page 40 for additional suggestions about reading and drawing diagrams.

As you saw on page 129, when a diagram shows overlapping triangles, you will often find it helpful to draw the triangles separately. Another technique is to mask part of a complicated diagram with your hand or a slip of paper. Try this with quadrilateral *ABCD* in Exercise 14, page 132. Can you pick out eight triangles? Now look at the diagram for Exercise 21, page 147, which shows a pentagon with all its diagonals. See how many triangles you can find.

More about Proof in Geometry

Objectives

1. Apply the definitions of the median and the altitude of a triangle and the perpendicular bisector of a segment.
2. State and apply the theorem about a point on the perpendicular bisector of a segment, and the converse.
3. State and apply the theorem about a point on the bisector of an angle, and the converse.
4. Prove two triangles congruent by first proving two other triangles congruent.
5. Understand and use inductive reasoning.

3-6 Medians, Altitudes, and Perpendicular Bisectors

A **median** of a triangle is a segment from a vertex to the midpoint of the opposite side. The three medians of $\triangle ABC$ are shown below in red.

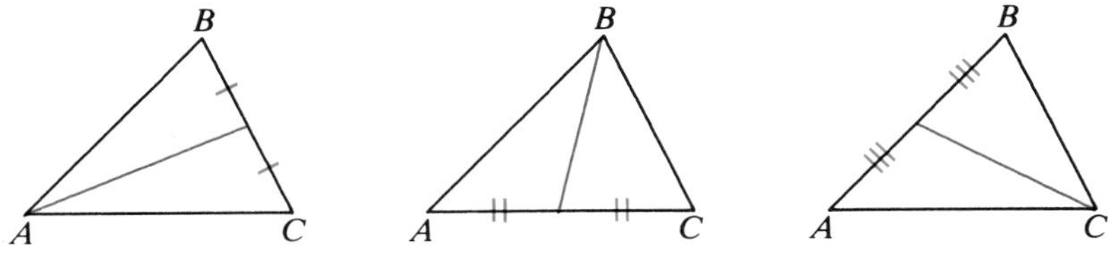

An **altitude** of a triangle is the perpendicular segment from a vertex to the line that contains the opposite side. The three altitudes of acute $\triangle ABC$ are shown below.

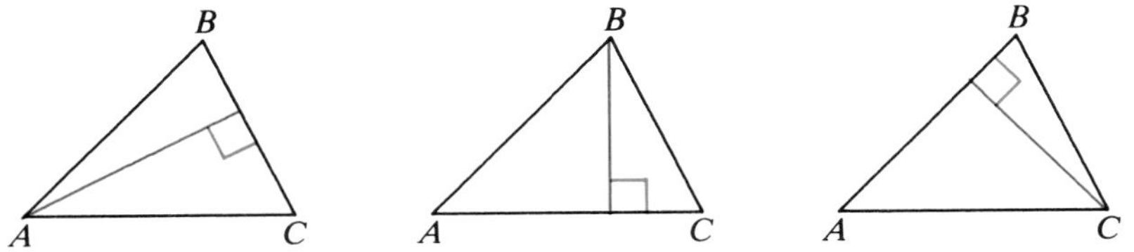

The three medians of a triangle are always inside the triangle. In $\triangle ABC$ above, all three altitudes are also inside the triangle. Do you think that the altitudes of a triangle are *always* inside the triangle?

Two of the altitudes of a right triangle are also legs of the triangle. Two of the altitudes of an obtuse triangle are outside the triangle. For obtuse $\triangle KLN$, $\overline{LH}$ is the altitude from L, and $\overline{NI}$ is the altitude from N.

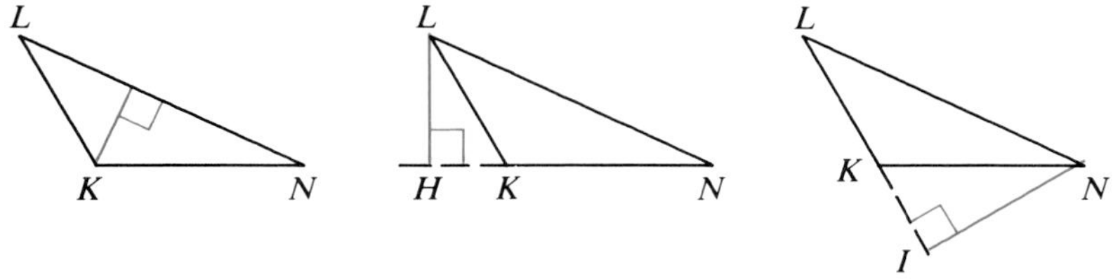

A **perpendicular bisector** of a segment is a line (or ray or segment) that is perpendicular to the segment at its midpoint. In the figure at the right, line l is a perpendicular bisector of $\overline{JK}$.

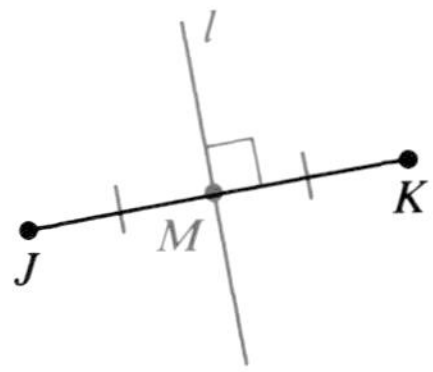

In a given plane, there is exactly one line perpendicular to a segment at its midpoint. We speak of *the* perpendicular bisector of a segment in such a case.

Proofs of the following theorems are left as Exercises 13 and 16.

Theorem 3-5

If a point lies on the perpendicular bisector of a segment, then the point is equidistant from the endpoints of the segment.

Given: A is on l, the perpendicular bisector of $\overline{BC}$.

Prove: $AB = AC$

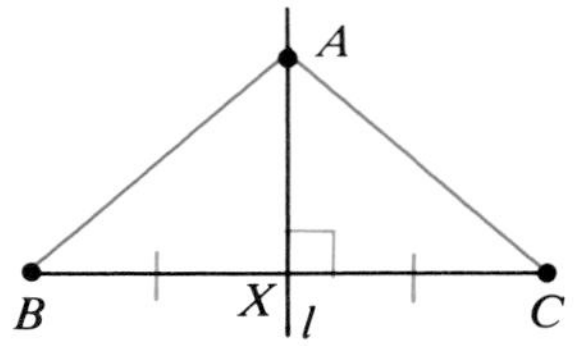

Theorem 3-6

If a point is equidistant from the endpoints of a segment, then the point lies on the perpendicular bisector of the segment.

Given: $AB = AC$

Prove: A is on the perpendicular bisector of $\overline{BC}$.

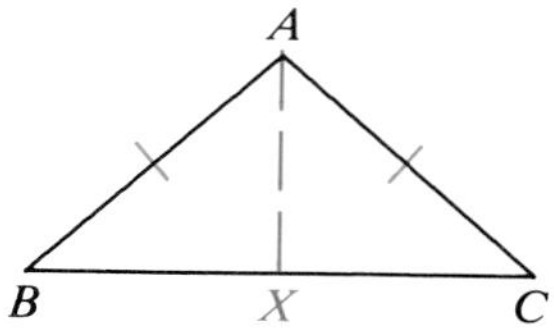

Plan for Proof: We can draw median $\overline{AX}$ and use corresponding parts of congruent triangles to show that $\overline{AX} \perp \overline{BC}$.

Example What can you deduce from the given information?

a. $RS = RT$
b. $ZS = ZT$
c. $RS = RT$ and $WS = WT$
d. $\overleftrightarrow{RW}$ is the perpendicular bisector of $\overline{ST}$.

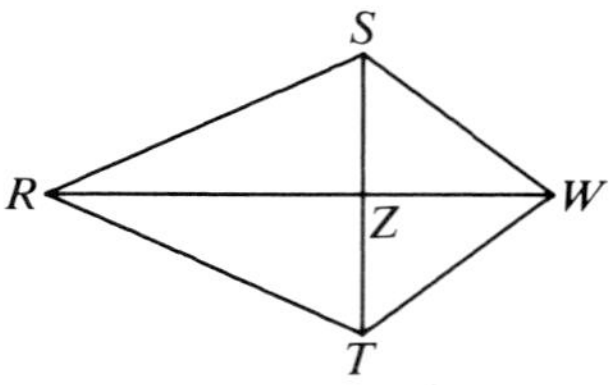

Solution **a.** R lies on the perpendicular bisector of $\overline{ST}$.

b. Z is the midpoint of $\overline{ST}$.

c. Both R and W lie on the perpendicular bisector of $\overline{ST}$. Thus $\overleftrightarrow{RW}$ is the perpendicular bisector of $\overline{ST}$, and Z is the midpoint of $\overline{ST}$. Also, $\overline{RZ}$ is a median and an altitude of $\triangle RST$. $\overline{WZ}$ is a median and an altitude of $\triangle WST$.

d. $RS = RT$, $ZS = ZT$, $WS = WT$; $\angle RZS$, $\angle RZT$, $\angle WZS$, and $\angle WZT$ are right angles.

The **distance from a point to a line** (or plane) is defined to be the length of the perpendicular segment from the point to the line (or plane). Since $\overline{RS} \perp t$, RS is the distance from R to line t.

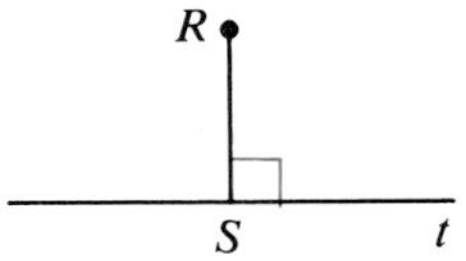

In Exercise 17 you will prove the following theorems, which are very similar to Theorems 3-5 and 3-6.

Theorem 3-7

If a point lies on the bisector of an angle, then the point is equidistant from the sides of the angle.

Given: $\overrightarrow{BZ}$ bisects $\angle ABC$; P lies on $\overrightarrow{BZ}$;
$\overline{PX} \perp \overrightarrow{BA}$; $\overline{PY} \perp \overrightarrow{BC}$

Prove: $PX = PY$

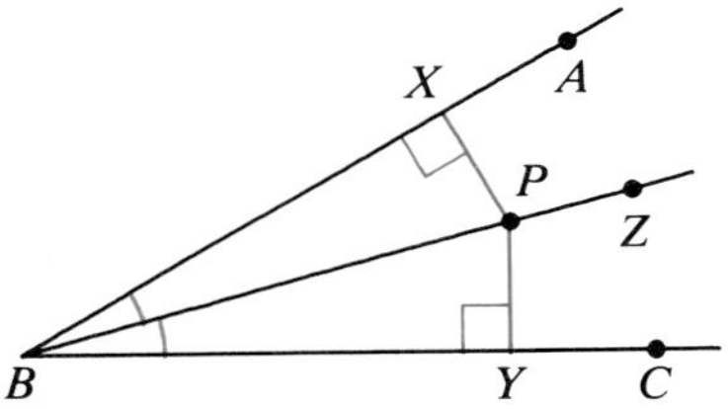

Theorem 3-8

If a point is equidistant from the sides of an angle, then the point lies on the bisector of the angle.

Given: $\overline{PX} \perp \overrightarrow{BA}$; $\overline{PY} \perp \overrightarrow{BC}$;
$PX = PY$

Prove: $\overrightarrow{BP}$ bisects $\angle ABC$.

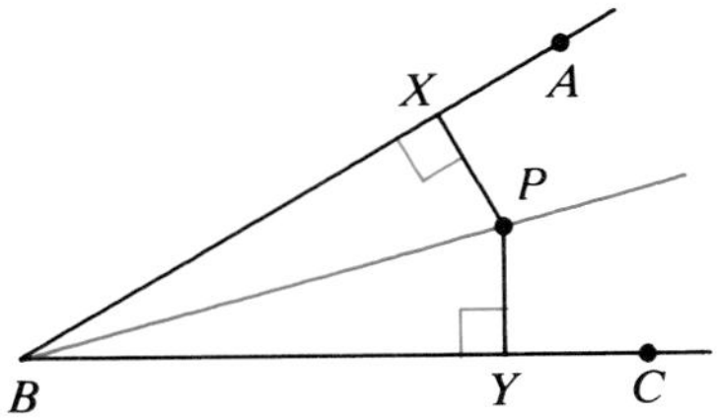

Theorem 3-5 and its converse, Theorem 3-6, can be combined into a single *if and only if* statement. The same is true for Theorems 3-7 and 3-8.

A point is on the perpendicular bisector of a segment if and only if it is equidistant from the endpoints of the segment.

A point is on the bisector of an angle if and only if it is equidistant from the sides of the angle.

Classroom Exercises

Complete.

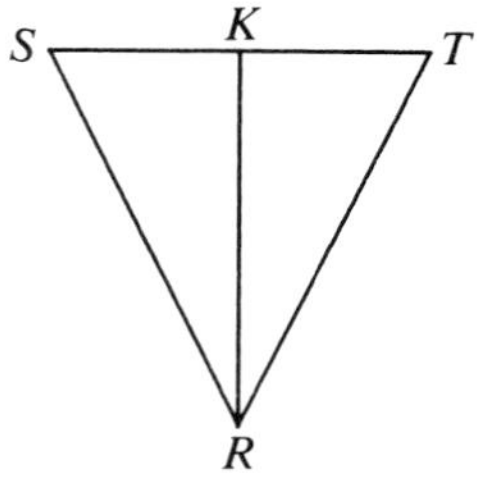

Exs. 1–6

1. If K is the midpoint of $\overline{ST}$, then $\overline{RK}$ is called a(n) __?__ of $\triangle RST$.
2. If $\overline{RK} \perp \overline{ST}$, then $\overline{RK}$ is called a(n) __?__ of $\triangle RST$.
3. If K is the midpoint of $\overline{ST}$ and $\overline{RK} \perp \overline{ST}$, then $\overline{RK}$ is called a(n) __?__ of $\overline{ST}$.
4. If $\overline{RK}$ is both an altitude and a median of $\triangle RST$, then:
   a. $\triangle RSK \cong \triangle RTK$ by __?__
   b. $\triangle RST$ is a(n) __?__ triangle.
5. If R is on the perpendicular bisector of $\overline{ST}$, then R is equidistant from __?__ and __?__. Thus __?__ = __?__.
6. If K is on the angle bisector of $\angle SRT$, then K is equidistant from __?__ and __?__.

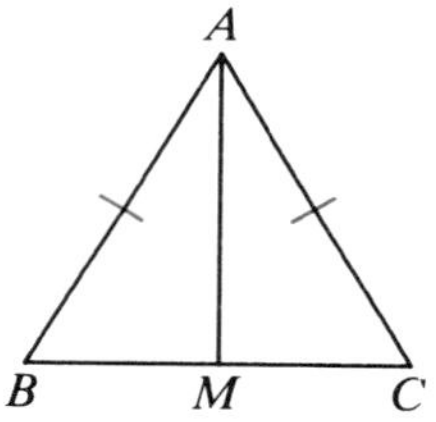

Exs. 7–9

7. Given: Isosceles $\triangle ABC$; $\overrightarrow{AM}$ bisects $\angle BAC$.
   What postulate, theorem, or corollary leads to the conclusion that $\overrightarrow{AM}$ is the perpendicular bisector of $\overline{BC}$?
8. Given: Isosceles $\triangle ABC$; $\overline{AM}$ is the median to base $\overline{BC}$.
   Explain why:
   a. $\triangle AMB \cong \triangle AMC$
   b. $\overline{AM}$ is an altitude.
   c. $\overline{AM}$ is a perpendicular bisector of $\overline{BC}$.
   d. $\overrightarrow{AM}$ bisects vertex angle A.
9. Given: Isosceles $\triangle ABC$; $\overline{AM}$ is the altitude to base $\overline{BC}$.
   Explain why:
   a. $\triangle AMB \cong \triangle AMC$ **b.** $\overline{AM}$ is a median.
10. Do you think it is ever possible for a triangle to have
    a. two congruent medians? **b.** three congruent medians?
    c. two congruent altitudes? **d.** three congruent altitudes?

Written Exercises

A

1. **a.** Draw a large scalene triangle ABC. Carefully draw the bisector of $\angle A$, the altitude from A, and the median from A. These three should all be different.
   b. Draw a large isosceles triangle ABC with vertex angle A. Carefully draw the bisector of $\angle A$, the altitude from A, and the median from A. Are these three different?
2. Draw a large obtuse triangle. Then draw its three altitudes in color.
3. Draw a right triangle. Then draw its three altitudes in color.
4. Draw a large acute scalene triangle. Then draw the perpendicular bisectors of its three sides.

5. Draw a large scalene right triangle. Then draw the perpendicular bisectors of its three sides and tell whether they appear to meet in a point. If so, where is this point?

Complete each statement.

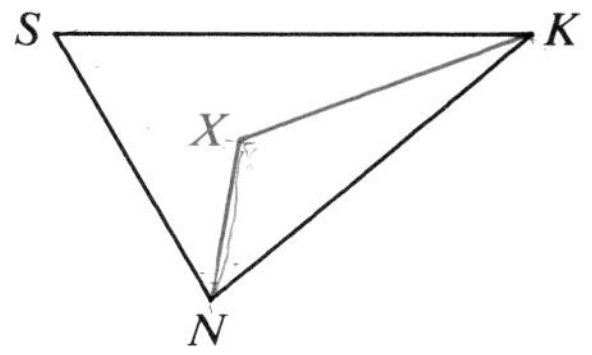

6. If X is on the bisector of $\angle K$, then X is equidistant from _?_ and _?_.

7. If X is on the bisector of $\angle N$, then X is equidistant from _?_ and _?_.

8. If X is equidistant from $\overline{SK}$ and $\overline{SN}$, then X lies on the _?_.

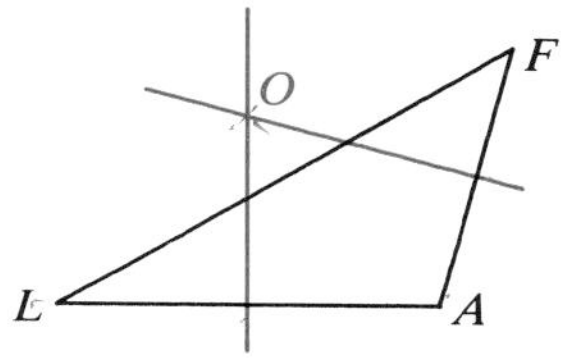

9. If O is on the perpendicular bisector of $\overline{LA}$, then O is equidistant from _?_ and _?_.

10. If O is on the perpendicular bisector of $\overline{AF}$, then O is equidistant from _?_ and _?_.

11. If O is equidistant from L and F, then O lies on the _?_.

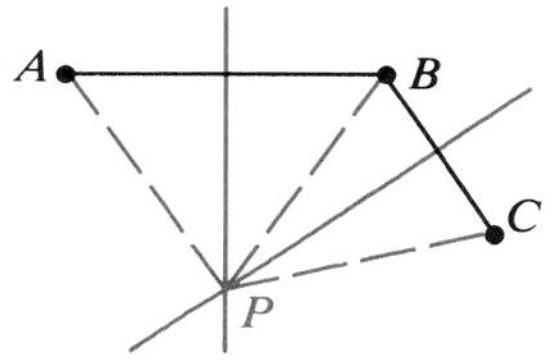

12. Given: P is on the perpendicular bisector of $\overline{AB}$;
P is on the perpendicular bisector of $\overline{BC}$.
Prove: $PA = PC$

B 13. Prove Theorem 3–5. Use the diagram on page 138.

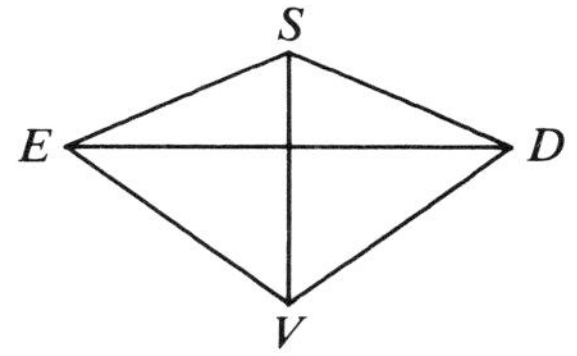

14. Given: S is equidistant from E and D;
V is equidistant from E and D.
Prove: $\overleftrightarrow{SV}$ is the perpendicular bisector of $\overline{ED}$.

15. a. A town wants to build a beach house on the lake front equidistant from the recreation center and the school. Copy the diagram and show the point B where the beach house should be located.

b. The town also wants to build a boat-launching site that is equidistant from Elm Road and Main Street. Find the point L where it should be built.

c. On your diagram, locate the spot F for a flagpole that is to be the same distance from the recreation center, the school, and the courthouse.

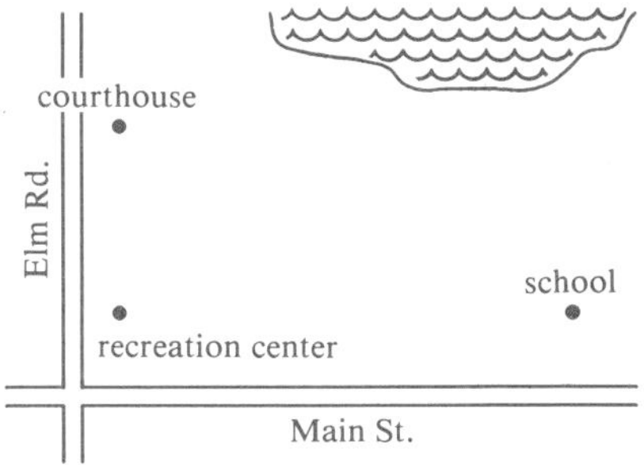

16. Prove Theorem 3–6. Use the diagram on page 138.

17. Prove **(a)** Theorem 3–7 and **(b)** Theorem 3–8. Use the diagrams on page 139.

18. Given: $\overline{BE} \cong \overline{CD}$; $\overline{BD} \cong \overline{CE}$
Prove: $\triangle ABC$ is isosceles.

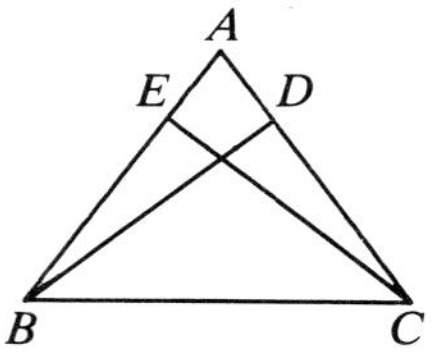

19. **a.** Given: $\overline{AB} \cong \overline{AC}$; $\overline{BD} \perp \overline{AC}$; $\overline{CE} \perp \overline{AB}$
Prove: $\overline{BD} \cong \overline{CE}$
b. You have just proved a theorem about altitudes. State this theorem in your own words.

20. Prove that the medians drawn to the legs of an isosceles triangle are congruent.

21. Given: $\overleftrightarrow{SR}$ is the perpendicular bisector of $\overline{QT}$;
$\overleftrightarrow{QR}$ is the perpendicular bisector of $\overline{SP}$.
Prove: $PQ = ST$
(*Hint:* One theorem will make your proof short.)

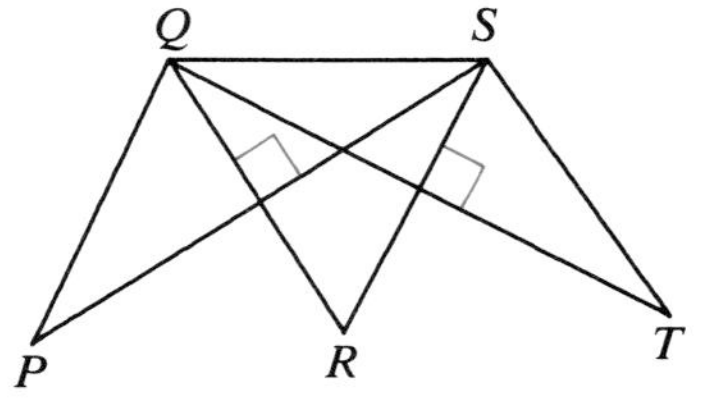

C 22. Given: $\overrightarrow{DP}$ bisects $\angle ADE$;
$\overrightarrow{EP}$ bisects $\angle DEC$.
Prove: $\overrightarrow{BP}$ bisects $\angle ABC$.
(*Hint:* There are two theorems that will make your proof short.)

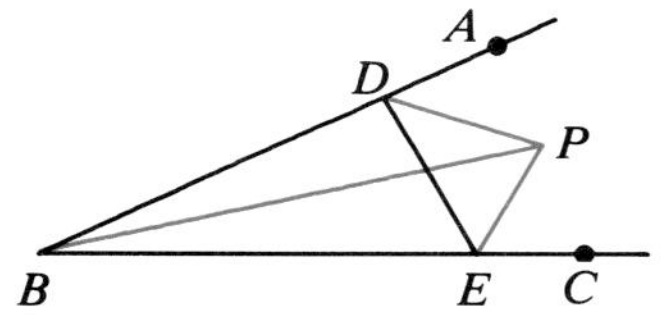

23. Given: $m\angle RTS = 90$;
$\overleftrightarrow{MN}$ is the $\perp$ bisector of $\overline{TS}$.
Prove: $\overline{TM}$ is a median.

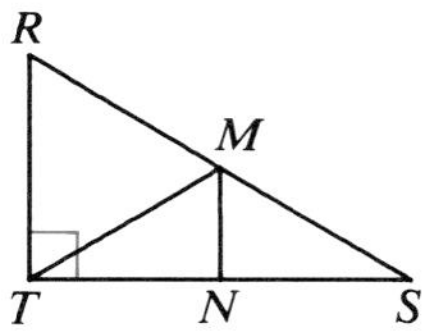

24. Given: $\overline{AB} \perp$ plane M;
O is the midpoint of $\overline{AB}$.
Prove: **a.** $\overline{AD} \cong \overline{BD}$ (*Hint:* In the plane determined by $\overleftrightarrow{AB}$ and D, $\overleftrightarrow{OD}$ is the $\perp$ bisector of $\overline{AB}$.)
b. $\overline{AC} \cong \overline{BC}$
c. $\angle CAD \cong \angle CBD$

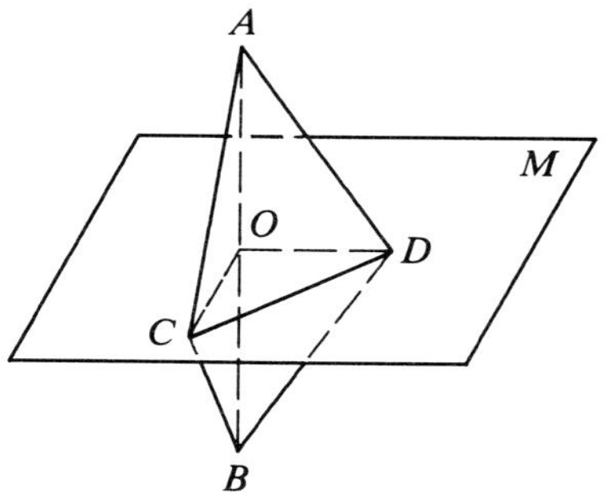

For Exercises 25 and 26, give a detailed plan for proof instead of a two-column proof.

25. Given: $\overline{QM}$ and $\overline{RN}$ are altitudes to the legs of isosceles $\triangle PQR$; $\overline{QM}$ and $\overline{RN}$ intersect at O.
Prove: $\triangle MNO$ is isosceles.

26. Given: $\overleftrightarrow{OZ}$ is the $\perp$ bisector of $\overline{AB}$;
$\angle 1 \cong \angle 3$; $\angle 2 \cong \angle 4$
Prove: The distance from A to $\overrightarrow{OX}$ equals the distance from B to $\overrightarrow{OY}$.

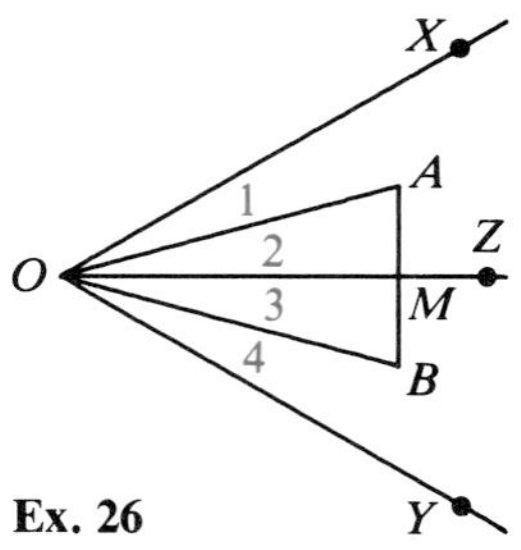

Ex. 26

3-7 Using More than One Pair of Congruent Triangles

To prove one pair of triangles congruent, sometimes you must first prove another pair congruent.

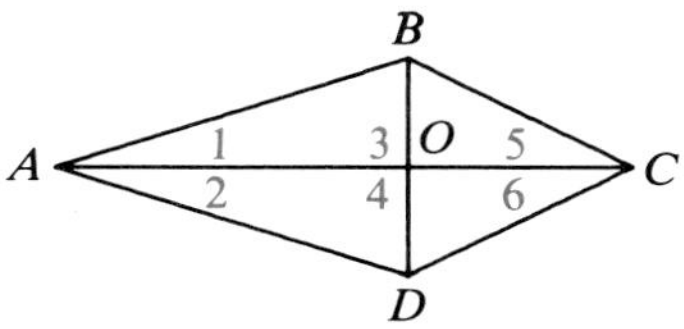

Example

Given: $\angle 1 \cong \angle 2$; $\angle 5 \cong \angle 6$

Prove: $\overline{AC} \perp \overline{BD}$

Plan for Proof: It is helpful here to *reason backward* from what we want to prove. We can show $\overline{AC} \perp \overline{BD}$ if we can show $\angle 3 \cong \angle 4$. $\angle 3$ and $\angle 4$ are corresponding parts of $\triangle ABO$ and $\triangle ADO$. We can prove $\triangle ABO \cong \triangle ADO$ by SAS if we first prove $\overline{AB} \cong \overline{AD}$. $\overline{AB}$ and $\overline{AD}$ are corresponding parts of $\triangle ABC$ and $\triangle ADC$. We can prove $\triangle ABC \cong \triangle ADC$ by ASA. We will do this first.

Proof:

Statements	Reasons
1. $\angle 1 \cong \angle 2$; $\angle 5 \cong \angle 6$	1. Given
2. $\overline{AC} \cong \overline{AC}$	2. Reflexive Property
3. $\triangle ABC \cong \triangle ADC$	3. ASA Postulate
4. $\overline{AB} \cong \overline{AD}$	4. Corr. parts of $\cong$ ⚠ are $\cong$.
5. $\overline{AO} \cong \overline{AO}$	5. Reflexive Property
6. $\triangle ABO \cong \triangle ADO$	6. SAS Postulate (Steps 1, 4, and 5)
7. $\angle 3 \cong \angle 4$	7. Corr. parts of $\cong$ ⚠ are $\cong$.
8. $\overline{AC} \perp \overline{BD}$	8. If 2 lines form $\cong$ adj. $\measuredangle$, then the lines are $\perp$.

A proof in mathematics can take many different forms. Arranging a proof in two columns helps to clarify the justifications for the steps in a proof. But a proof can also be written in a paragraph form. In a *paragraph proof,* justifications that are expected to be clear to the reader are often omitted. The following paragraph proof might be given for the example above.

Proof:

Since $\angle 1 \cong \angle 2$, $\angle 5 \cong \angle 6$, and $\overline{AC} \cong \overline{AC}$, $\triangle ABC \cong \triangle ADC$ by ASA. Thus the corresponding parts $\overline{AB}$ and $\overline{AD}$ are congruent. Together with $\angle 1 \cong \angle 2$ and $\overline{AO} \cong \overline{AO}$, this fact implies that $\triangle ABO \cong \triangle ADO$ by SAS. Thus, $\angle 3 \cong \angle 4$ (corr. parts of $\cong$ ⚠ are $\cong$), and $\overline{AC} \perp \overline{BD}$.

Classroom Exercises

In Exercises 1–3 you are given a diagram that is marked with given information. Give the reason for each key step of the proof.

1. Prove: $\overline{AS} \cong \overline{DT}$

Key steps of proof:

a. $\triangle ABC \cong \triangle DEF$

b. $\angle C \cong \angle F$

c. $\triangle ACS \cong \triangle DFT$

d. $\overline{AS} \cong \overline{DT}$

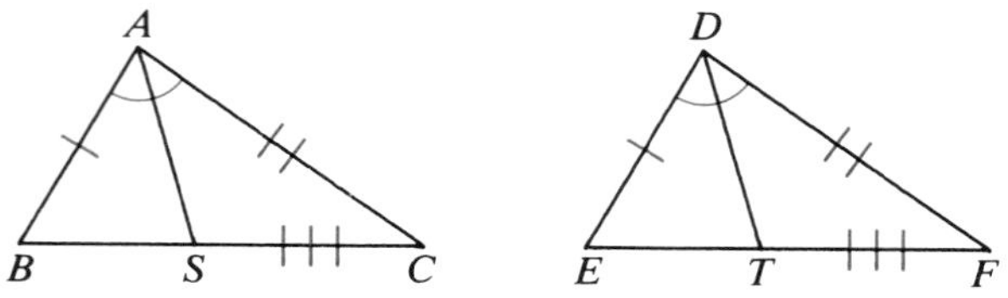

2. Prove: $\overline{AX} \cong \overline{AY}$

Key steps of proof:

a. $\triangle PAL \cong \triangle KAN$

b. $\angle L \cong \angle N$

c. $\triangle LAX \cong \triangle NAY$

d. $\overline{AX} \cong \overline{AY}$

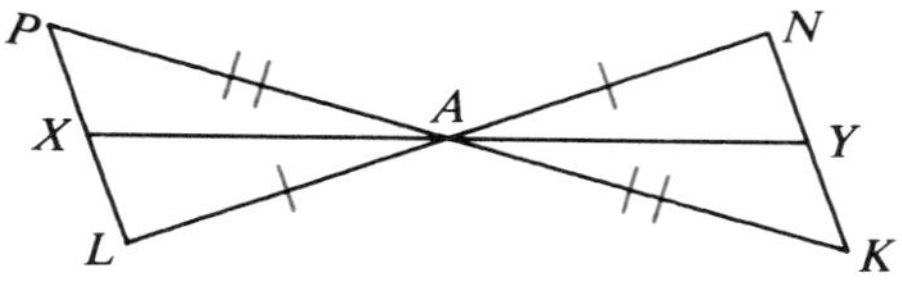

3. Prove: $\angle 3 \cong \angle 4$

Key steps of proof:

a. $\triangle LOB \cong \triangle JOB$

b. $\angle 1 \cong \angle 2$

c. $\triangle LBA \cong \triangle JBA$

d. $\angle 3 \cong \angle 4$

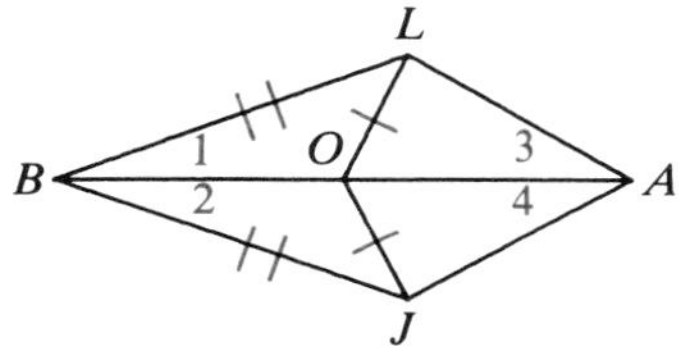

4. State a plan for proving that $\angle D \cong \angle F$.

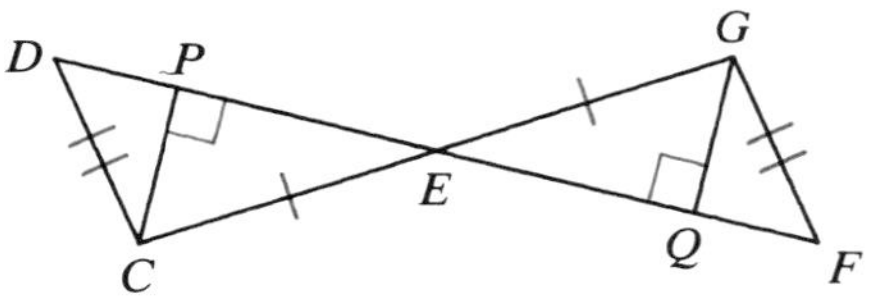

Written Exercises

In Exercises 1–5 you are given a diagram that is marked with given information. Give the reason for each key step of the proof.

A **1.** Prove: $\overline{NE} \cong \overline{OS}$

Key steps of proof:

a. $\triangle RNX \cong \triangle LOY$

b. $\angle X \cong \angle Y$

c. $\triangle NEX \cong \triangle OSY$

d. $\overline{NE} \cong \overline{OS}$

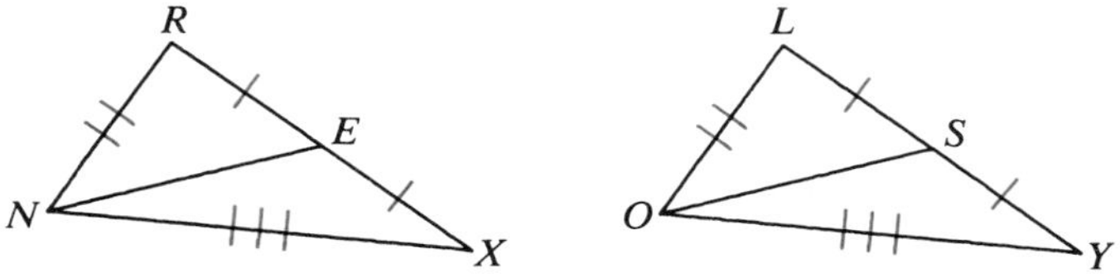

2. Prove: $\overline{BE} \cong \overline{DF}$

Key steps of proof:

a. $\triangle ABC \cong \triangle CDA$

b. $\angle 1 \cong \angle 2$

c. $\triangle ABE \cong \triangle CDF$

d. $\overline{BE} \cong \overline{DF}$

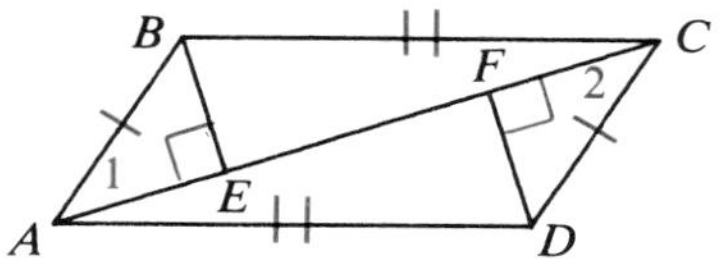

3. Prove: $\angle G \cong \angle T$
Key steps of proof:
a. $\triangle RAJ \cong \triangle NAK$
b. $\overline{RJ} \cong \overline{NK}$
c. $\triangle GRJ \cong \triangle TNK$
d. $\angle G \cong \angle T$

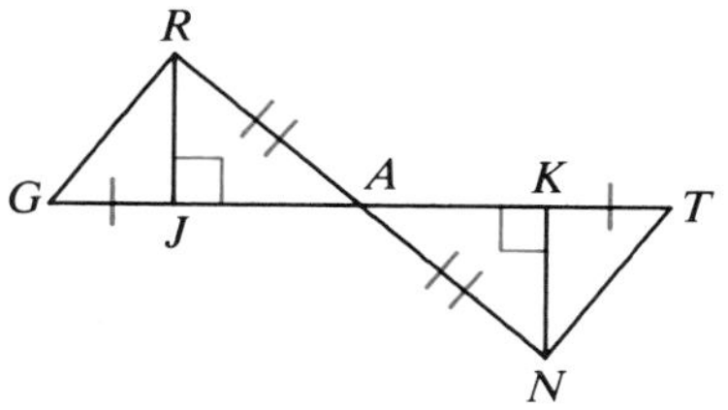

4. Prove: $\overline{AL} \cong \overline{CM}$
Key steps of proof:
a. $\triangle ABD \cong \triangle CDB$
b. $\overline{AD} \cong \overline{CB}$; $\angle 1 \cong \angle 2$
c. $\triangle ADL \cong \triangle CBM$
d. $\overline{AL} \cong \overline{CM}$

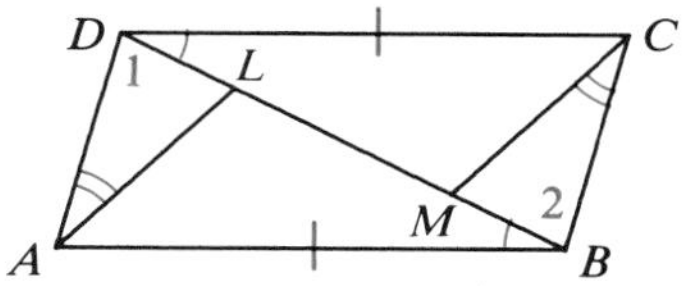

5. Prove: $\overline{DX} \cong \overline{EX}$
Key steps of proof:
a. $\triangle POD \cong \triangle POE$
b. $\overline{PD} \cong \overline{PE}$
c. $\triangle PDX \cong \triangle PEX$
d. $\overline{DX} \cong \overline{EX}$

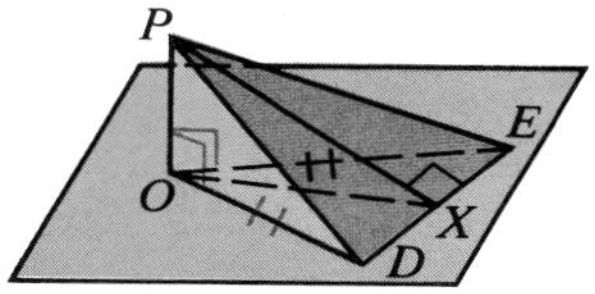

6. Plane M is the *perpendicular bisecting plane* of $\overline{AB}$ at O (that is, the plane that is perpendicular to $\overline{AB}$ at its midpoint, O). Points C and D also lie in plane M. List three pairs of congruent triangles and tell which congruence method can be used to prove each pair congruent.

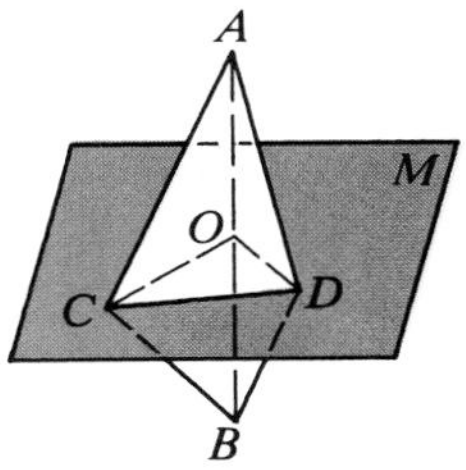

Write proofs in two-column form or paragraph form, as directed by your teacher.

B 7. Given: $\overline{FL} \cong \overline{FK}$; $\overline{LA} \cong \overline{KA}$
Prove: $\overline{LJ} \cong \overline{KJ}$

8. Given: $\overleftrightarrow{FA}$ bisects $\angle LFK$ and $\angle LAK$.
Prove: $\overleftrightarrow{FA}$ bisects $\angle LJK$.

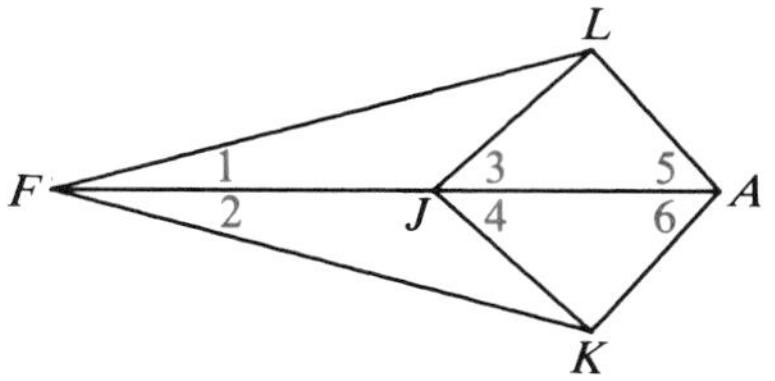

9. Given: $\triangle RST \cong \triangle XYZ$;
$\overrightarrow{SK}$ bisects $\angle RST$;
$\overrightarrow{YL}$ bisects $\angle XYZ$.
Prove: $\overline{SK} \cong \overline{YL}$

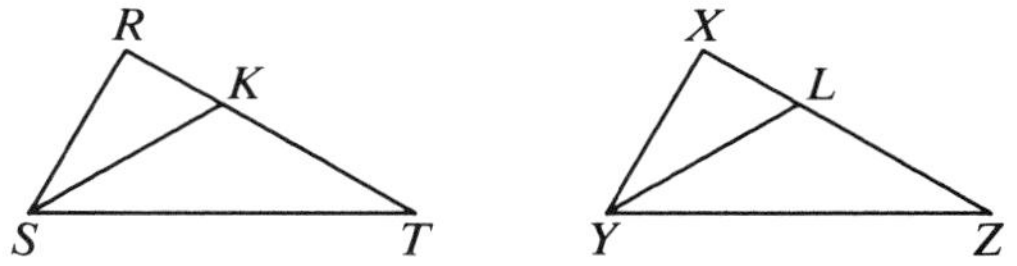

10. Given: Congruent parts as marked in the diagram.
Prove: $\angle B \cong \angle F$
(*Hint:* First draw two auxiliary lines.)

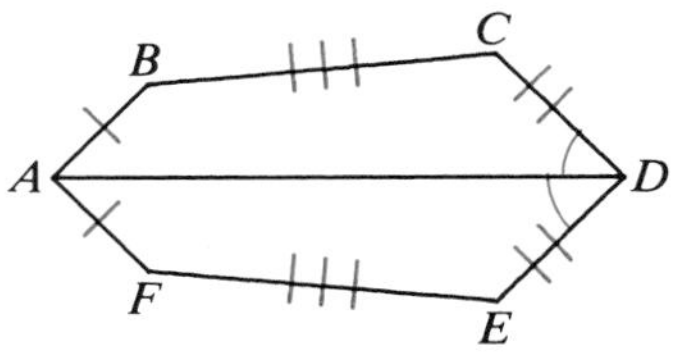

11. Draw two congruent acute triangles, $\triangle PAY$ and $\triangle NOW$. Draw the altitudes $\overline{PE}$ and $\overline{NF}$ and prove that they are congruent. (First state what is given and what is to be proved.)

12. Given: $\triangle LMN \cong \triangle RST$;
$\overline{LX}$ and $\overline{RY}$ are altitudes.
Prove: $\overline{LX} \cong \overline{RY}$

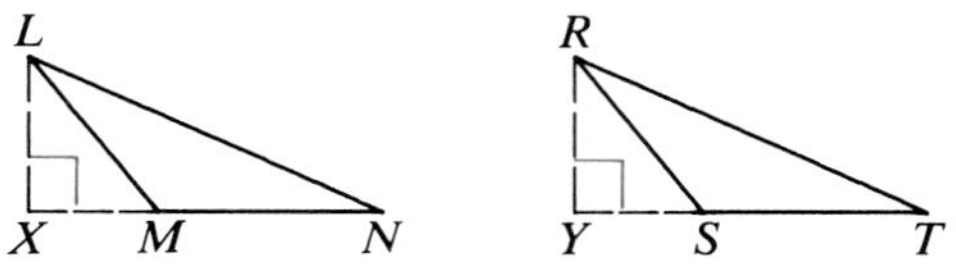

13. Given: $\overline{DE} \cong \overline{FG}$; $\overline{GD} \cong \overline{EF}$;
$\angle HDE$ and $\angle KFG$ are right angles.
Prove: $\overline{DH} \cong \overline{FK}$

14. Given: $\overline{GD} \parallel \overline{EF}$; $\angle GDE \cong \angle GFE$;
$\overline{GH} \cong \overline{EK}$
Prove: $\overline{DH} \parallel \overline{FK}$

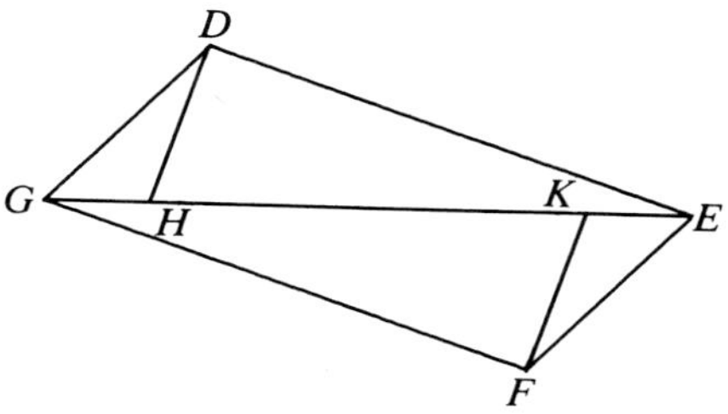

15. Draw two line segments, $\overline{SX}$ and $\overline{TY}$, that bisect each other at O. Choose any point P on $\overline{ST}$ and let Q be the point where $\overrightarrow{PO}$ intersects $\overline{XY}$. Prove that O is the midpoint of $\overline{PQ}$. (First state what is given and what is to be proved.)

16. This figure is like the one that Euclid used to prove that the base angles of an isosceles triangle are congruent (our Theorem 3-1). Write a paragraph proof following his key steps shown below.
Given: $\overline{AB} \cong \overline{AC}$;
$\overline{AB}$ and $\overline{AC}$ are extended so $\overline{BD} \cong \overline{CE}$.
Prove: $\angle ABC \cong \angle ACB$

Euclid's Key Steps
1. Prove $\triangle DAC \cong \triangle EAB$.
2. Prove $\triangle DBC \cong \triangle ECB$.
3. Prove $\angle DBC \cong \angle ECB$ and then $\angle ABC \cong \angle ACB$.

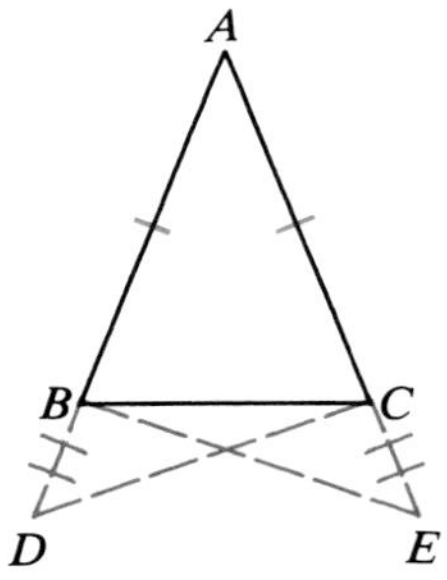

C 17. Given: $\overline{AM} \cong \overline{MB}$; $\overline{AD} \cong \overline{BC}$;
$\angle MDC \cong \angle MCD$
Prove: $\overline{AC} \cong \overline{BD}$

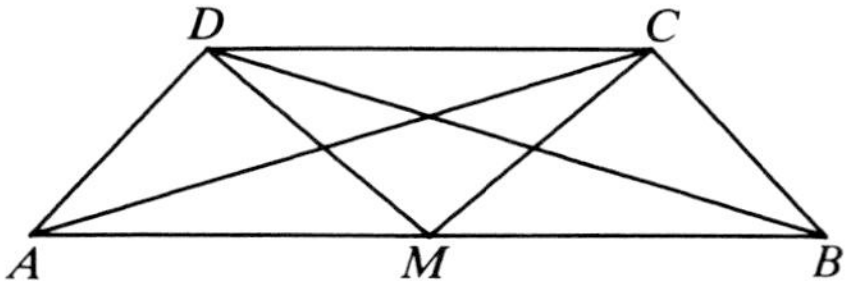

18. Given: $\angle 1 \cong \angle 2$;
$\angle 3 \cong \angle 4$;
$\angle 5 \cong \angle 6$
Prove: $\overline{BC} \cong \overline{ED}$

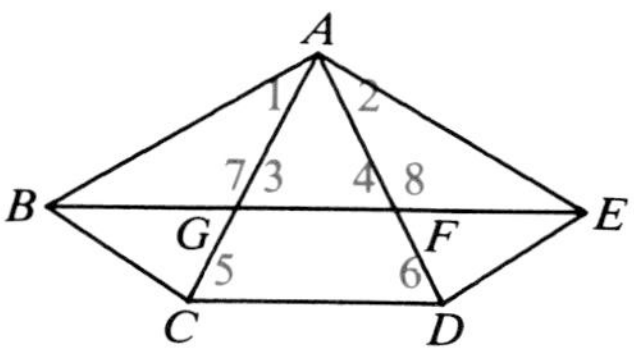

19. A, B, C, and D are noncoplanar. $\triangle ABC$, $\triangle ACD$, and $\triangle ABD$ are equilateral. X and Y are midpoints of $\overline{AC}$ and $\overline{AD}$. Z is a point on $\overline{AB}$. What kind of triangle is $\triangle XYZ$? Explain.

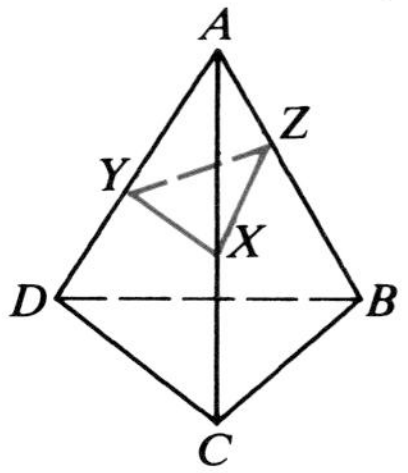

20. Given: $\overline{SN}$ and $\overline{TM}$ are medians of scalene $\triangle RST$;
P is on $\overrightarrow{SN}$ such that $\overline{SN} \cong \overline{NP}$;
Q is on $\overrightarrow{TM}$ such that $\overline{TM} \cong \overline{MQ}$.
Prove: **a.** $\overline{RQ} \cong \overline{RP}$
b. $\overline{RQ}$ and $\overline{RP}$ are both parallel to $\overline{ST}$.
c. P, R, and Q are collinear.

For Exercises 21–23, write paragraph proofs. (In this book a star designates an exercise that is unusually difficult.)

★ **21.** Given: $\overline{AE} \parallel \overline{BD}$; $\overline{BC} \parallel \overline{AD}$;
$\overline{AE} \cong \overline{BC}$; $\overline{AD} \cong \overline{BD}$
Prove: **a.** $\overline{AC} \cong \overline{BE}$
b. $\overline{EC} \parallel \overline{AB}$

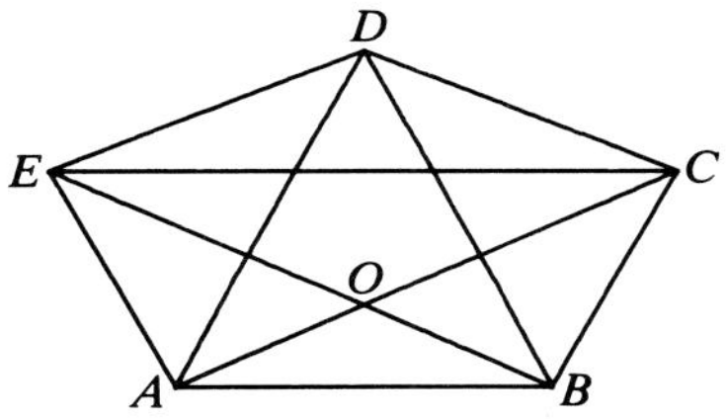

★ **22.** Given: $\overleftrightarrow{AM}$ is the $\perp$ bis. of $\overline{BC}$;
$\overline{AE} \perp \overline{BD}$; $\overline{AF} \perp \overline{DF}$;
$\angle 1 \cong \angle 2$
Prove: $BE = CF$

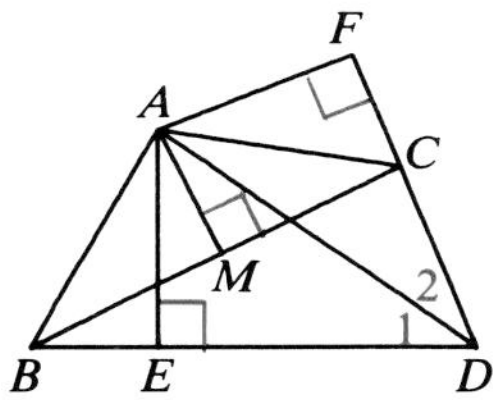

★ **23.** Given: X, R, S, and T are coplanar;
X is the midpoint of $\overline{AB}$;
$\overline{AB} \perp \overline{RX}$; $\overline{AB} \perp \overline{TX}$
Prove: $\overline{AB} \perp \overline{SX}$
Note: This exercise proves that if a line ($\overleftrightarrow{AB}$) is perpendicular to each of two intersecting lines at their point of intersection, then the line is perpendicular to every line of the plane that passes through that point. Thus the line is perpendicular to the plane.

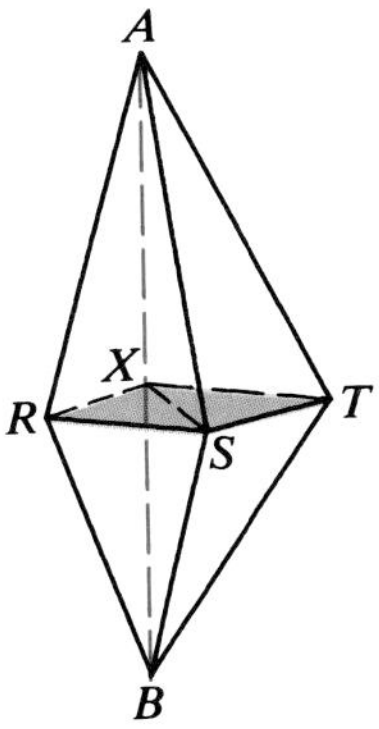

3-8 Inductive Reasoning

Throughout these first three chapters, we have been using *deductive reasoning*. Now we'll consider **inductive reasoning,** a kind of reasoning that is widely used in science and in everyday life.

Example 1 After picking marigolds for the first time, Connie began to sneeze. She also began sneezing the next four times she was near marigolds. Based on this past experience, Connie reasons inductively that she is allergic to marigolds.

Example 2 Every time Pitch has thrown a high curve ball to Slugger, Slugger has gotten a hit. Pitch concludes from this experience that it is not a good idea to pitch high curve balls to Slugger.

In coming to this conclusion, Pitch has used inductive reasoning. It may be that Slugger just happened to be lucky those times, but Pitch is too bright to feed another high curve to Slugger.

From these examples you can see how inductive reasoning differs from deductive reasoning.

Deductive Reasoning	**Inductive Reasoning**
Conclusion based on accepted statements (definitions, postulates, previous theorems, corollaries, and given information)	Conclusion based on several past observations
Conclusion *must* be true if hypotheses are true.	Conclusion is *probably* true, but not necessarily true.

Often in mathematics we reason inductively by observing a pattern. For example, can you see the pattern and predict the next number in each sequence below?

a. 3, 6, 12, 24, _?_ **b.** 11, 15, 19, 23, _?_ **c.** 5, 6, 8, 11, 15, _?_

In (a), we see that each number is twice the preceding number. We expect that the next number will be 2×24, or 48.

In (b), we see that each number is 4 more than the preceding number. We expect that the next number will be $23 + 4$, or 27.

To see a pattern in (c), we look at the differences between the numbers. Can you now predict the next number?

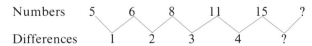

Numbers	5	6	8	11	15	?
Differences		1	2	3	4	?

Classroom Exercises

Tell whether the reasoning process is deductive or inductive.

1. Ramon noticed that spaghetti had been on the school menu for the past five Wednesdays. Ramon decides that the school always serves spaghetti on Wednesday.
2. Andrew did his assignment, adding the lengths of the sides of triangles to find the perimeters. Noticing the results for several equilateral triangles, he guesses that the perimeter of every equilateral triangle is three times the length of a side.

3. By using the definitions of equilateral triangle (a triangle with three congruent sides) and of perimeter (the sum of the lengths of the sides of a figure), Katie concludes that the perimeter of every equilateral triangle is three times the length of a side.
4. Linda observes that $(-1)^2 = +1$, $(-1)^4 = +1$, and $(-1)^6 = +1$. She concludes that every even power of (-1) is equal to $+1$.
5. John knows that multiplying a number by -1 merely changes the sign of the number. He reasons that multiplying a number by an even power of -1 will change the sign of the number an even number of times. He concludes that this is equivalent to multiplying a number by $+1$, so that every even power of -1 is equal to $+1$.
6. Look at the discussion leading up to the statement of Theorem 2–13 on page 81. Is the thinking inductive or deductive?

Written Exercises

Guess the next two numbers of each sequence.

A 1. 1, 4, 16, 64, . . .

2. 18, 15, 12, 9, . . .

3. $1, \frac{1}{3}, \frac{1}{9}, \frac{1}{27}, \ldots$

4. 1, 4, 9, 16, . . .

5. 2, 3, 5, 8, 12, . . .

6. 10, 12, 16, 22, 30, . . .

7. 40, 39, 36, 31, 24, . . .

8. $8, -4, 2, -1, \frac{1}{2}, \ldots$

Accept the two statements as given information. State a conclusion based on *deductive* reasoning. If no conclusion can be reached, write *none*.

9. Chan is older than Pedro.
Pedro is older than Sarah.

10. Valerie is older than Greg.
Dan is older than Greg.

11. Polygon G has more than 6 sides.
Polygon G has fewer than 8 sides.

12. Polygon G has more than 6 sides.
Polygon K has more than 6 sides.

13. All horses like oats.
Muffin likes oats.

14. All horses like oats.
Prince is a horse.

15. There are three sisters. Two of them are athletes and two of them like tacos. Can you be sure that both of the athletes like tacos? Do you reason deductively or inductively to conclude the following? *At least one of the athletic sisters likes tacos.*

For each exercise, write the equation that you think should come next. Then check that the left side of your equation really does equal the right side.

16.
$$1 \times 9 + 2 = 11$$
$$12 \times 9 + 3 = 111$$
$$123 \times 9 + 4 = 1111$$

17.
$$9 \times 9 + 7 = 88$$
$$98 \times 9 + 6 = 888$$
$$987 \times 9 + 5 = 8888$$

18.
$$9^2 = 81$$
$$99^2 = 9801$$
$$999^2 = 998001$$

B **19. a.** Study the diagrams below. Then guess the number of regions for the fourth diagram. Check your answer by counting.

2 points	3 points	4 points	5 points
2 regions	4 regions	8 regions	___?___ regions

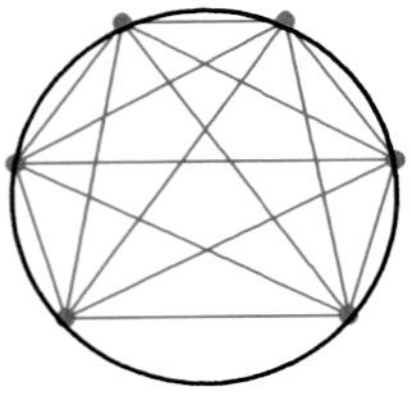

b. Using 6 points on a circle as shown, guess the number of regions within the circle. Carefully check your answer by counting.

Important note: This exercise shows that a pattern predicted on the basis of a few cases may be incorrect. To be sure of a conclusion, use a deductive proof.

20. Draw several quadrilaterals whose opposite sides are parallel. With ruler and protractor measure the opposite sides and the opposite angles of each figure. On the basis of the diagrams and measurements, what do you guess is true for all such quadrilaterals?

C **21. a.** Substitute each of the integers from 1 to 9 for n in the expression $n^2 + n + 11$.

b. Using inductive reasoning, guess what kind of number you will get when you substitute any positive integer for n in the expression $n^2 + n + 11$.

c. Test your guess by substituting 10 and 11 for n.

22. Complete the table for convex polygons.

Number of sides	3	4	5	6	7	8	n
Number of diagonals	0	2	?	?	?	?	?

23. Complete each statement. Using inductive reasoning, suggest a formula for the sum of the angle measures at the tips of an n-pointed star. Can you deductively prove your formula?

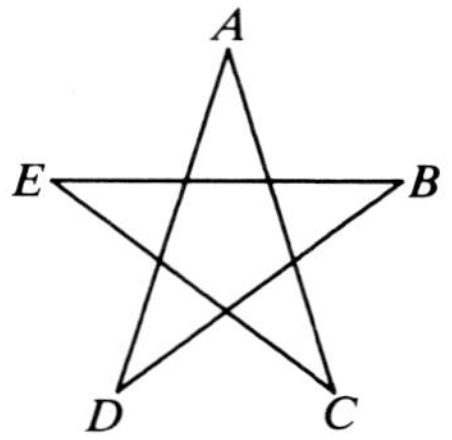

5-pointed star

$m\angle A + m\angle B + m\angle C + m\angle D + m\angle E =$ ___?___

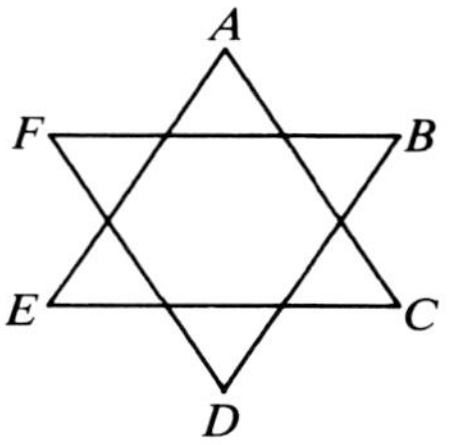

6-pointed star

$m\angle A + m\angle B + m\angle C + m\angle D + m\angle E + m\angle F =$ ___?___

EXTRA

Möbius Bands

Take a long, narrow strip of paper. (Think of the sides of the strip as being different colors.)

Give the strip a half-twist.

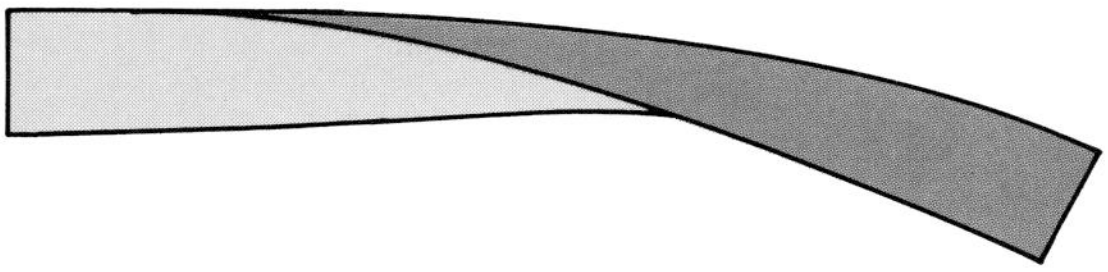

Tape the ends together. The result is a *Möbius band.*

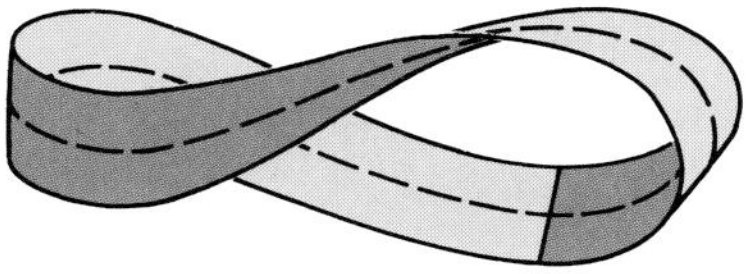

Exercises

1. Make a Möbius band.

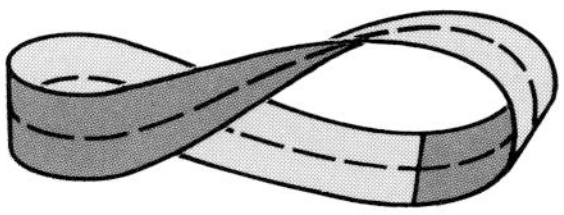

2. Color one side of the Möbius band. How much of the band is left uncolored?

3. The original strip of paper had two sides. How many sides does a Möbius band have?

4. Slit the Möbius band lengthwise down the middle. (First start at a point midway between the edges and trace around the band.) What is the result?

5. Slit the band made in Exercise 4 a second time down the middle. Write a sentence or two describing what happens.

6. Give a *full* twist to a rectangular strip of paper. Tape the ends together. How many sides does the band have?

7. Slit the strip made in Exercise 6 down the middle. Write a brief description of what is formed.

8. Make a Möbius band. Let the band be 3 cm wide. Make a lengthwise cut, staying 1 cm from the right-hand edge. Describe the result.

CALCULATOR KEY-IN

Complete the right side of the first three equations in each exercise. Then use inductive reasoning to guess what the fourth equation would be if the pattern were continued. Check your guess with your calculator.

1. $6 \times 7 = \underline{?}$
$66 \times 67 = \underline{?}$
$666 \times 667 = \underline{?}$
$\underline{?} \times \underline{?} = \underline{?}$

2. $8 \times 8 = \underline{?}$
$98 \times 98 = \underline{?}$
$998 \times 998 = \underline{?}$
$\underline{?} \times \underline{?} = \underline{?}$

3. $7 \times 9 = \underline{?}$
$77 \times 99 = \underline{?}$
$777 \times 999 = \underline{?}$
$\underline{?} \times \underline{?} = \underline{?}$

Self-Test 3

Complete.

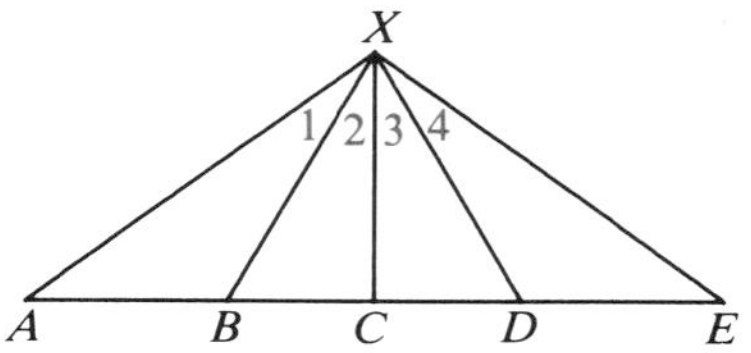

Exs. 1–6

1. If $\overline{XC} \perp \overline{AE}$, then $\overline{XC}$ is a(n) $\underline{?}$ of $\triangle XAE$.
2. If $BC = CD$, then $\overline{XC}$ is a(n) $\underline{?}$ of $\triangle XBD$.
3. If $XB = XD$, then X lies on $\underline{?}$.
4. If C is on the bisector of $\angle AXE$, then C is equidistant from $\underline{?}$ and $\underline{?}$.
5. If $\overline{XC} \perp \overline{AE}$, then XC is the $\underline{?}$ from $\underline{?}$ to $\underline{?}$.
6. Given: $\angle 1 \cong \angle 4$; $\angle 2 \cong \angle 3$; $\angle A \cong \angle E$
   Explain how you could prove that $\overline{XC}$ is a median of $\triangle BXD$.

Use inductive reasoning to guess the next number of each sequence.

7. 1, 3, 7, 13, 21, . . .
8. 1, 4, 2, 8, 4, 16, 8, 32, . . .

Chapter Summary

1. Congruent figures have the same size and shape. Two triangles are congruent if their corresponding sides and angles are congruent.
2. We have five ways to prove two triangles congruent:
   SSS SAS ASA AAS HL (rt. $\triangle$s)
3. A common way to prove that two segments or two angles are congruent is to show that they are corresponding parts of congruent triangles.
4. A line and plane are perpendicular if and only if they intersect and the line is perpendicular to all lines in the plane that pass through the point of intersection.
5. If two sides of a triangle are congruent, then the angles opposite those sides are congruent. An equilateral triangle has three $60°$ angles.
6. If two angles of a triangle are congruent, then the sides opposite those angles are congruent. An equiangular triangle is also equilateral.

7. Every triangle has three medians and three altitudes.
8. The perpendicular bisector of a segment is the line that is perpendicular to the segment at its midpoint.
9. A point lies on the perpendicular bisector of a segment if and only if the point is equidistant from the endpoints of the segment.
10. A point lies on the bisector of an angle if and only if the point is equidistant from the sides of the angle.
11. Inductive reasoning is the process of observing individual cases and then reaching a general conclusion suggested by them. The conclusion is probably, but not necessarily, true.

Chapter Review

The two triangles shown are congruent. Complete. 3–1

1. $\triangle RSP \cong$ __?__
2. $\angle PRS \cong$ __?__
3. Since $PS =$ __?__, S is the __?__.
4. If $m\angle P = 42$, then $m\angle SRQ =$ __?__ (numerical answer).

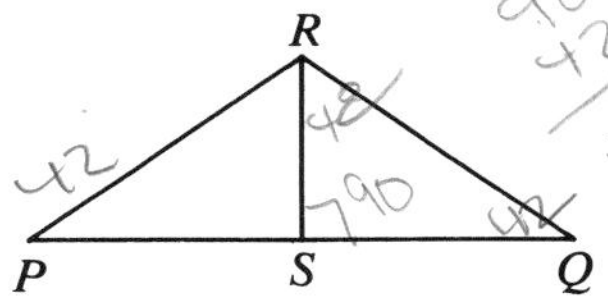

Can two triangles be proved congruent? If so, write the congruence and name the postulate that can be used. 3–2

5. Given: $\overline{BE}$ bisects $\overline{AD}$; $\overline{AD}$ bisects $\overline{BE}$.
6. Given: $\overline{AB} \cong \overline{ED}$; $\overline{BC} \cong \overline{CD}$
7. Given: $\angle A \cong \angle D$; $\overline{AC} \cong \overline{DC}$
8. Given: $\angle A \cong \angle E$; $\angle B \cong \angle D$

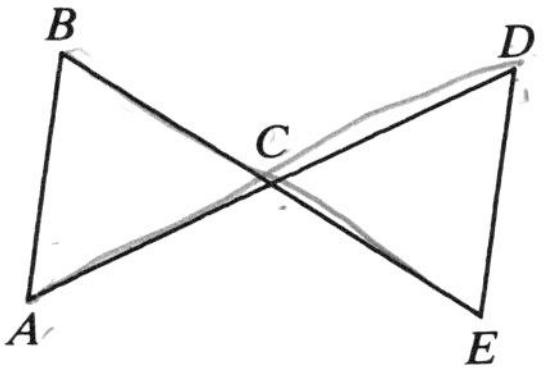

Write two-column proofs. 3–3

9. Given: $\overline{JM} \cong \overline{LM}$; $\overline{JK} \cong \overline{LK}$
   Prove: $\angle MJK \cong \angle MLK$
10. Given: $\angle JMK \cong \angle LMK$; $\overline{MK} \perp$ plane P
    Prove: $\overline{JK} \cong \overline{LK}$

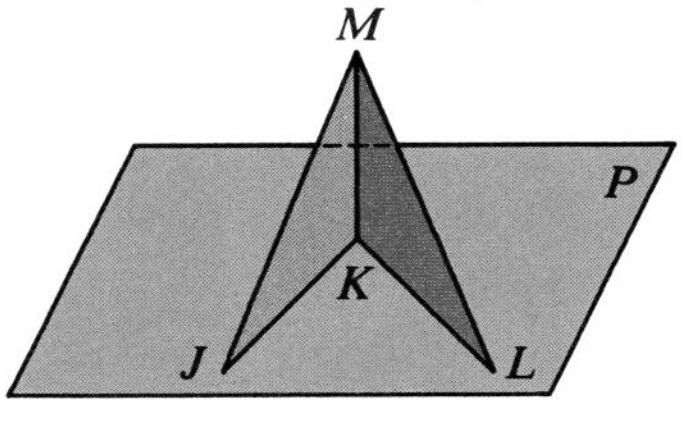

11. If $\overline{DE} \cong \overline{EF}$, which angles must be congruent? State the theorem that justifies your answer. 3–4
12. If $\angle D \cong \angle E$, $DF = 7t - 12$, and $FE = 12 - t$, find the value of t and the length of $\overline{DF}$.
13. If $\triangle DEF$ is equiangular, $DE = x + y$, $EF = 12$, and $DF = 5x - y$, find the values of x and y.

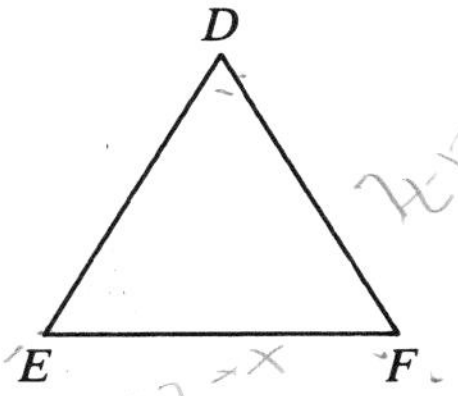

Write two-column proofs.

14. Given: $\overline{GH} \perp \overline{HJ}$; $\overline{KJ} \perp \overline{HJ}$; $\angle G \cong \angle K$ 3-5
Prove: $\triangle GHJ \cong \triangle KJH$

15. Given: $\overline{GH} \perp \overline{HJ}$; $\overline{KJ} \perp \overline{HJ}$;
$\overline{GJ} \cong \overline{KH}$
Prove: $\overline{GH} \cong \overline{KJ}$

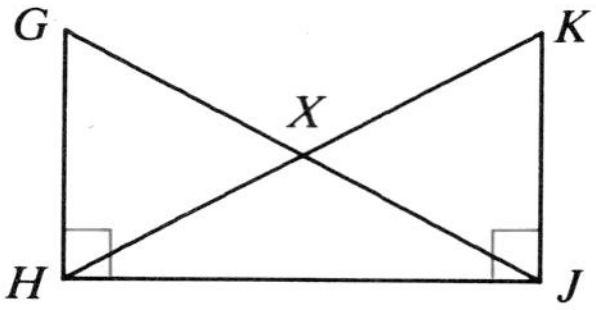

16. Draw a scalene right triangle ABC with hypotenuse $\overline{AC}$. Sketch the altitude from B and the median from A. 3-6

17. If $\overleftrightarrow{PQ}$ is the perpendicular bisector of $\overline{AB}$, name four things you can conclude about the diagram.

18. If P is equidistant from $\overrightarrow{QA}$ and $\overrightarrow{QB}$, then P lies on _?_.

19. If $\overline{QP} \perp \overline{AB}$, then $\overline{QP}$ is a(n) _?_ of $\triangle ABQ$.

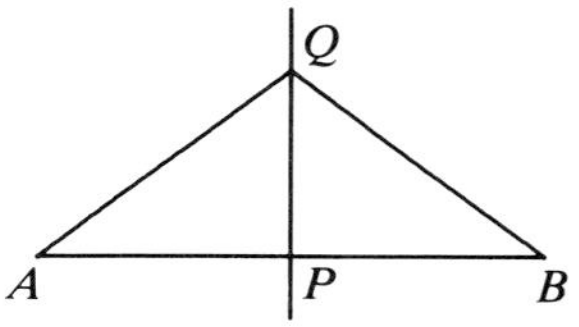

Exs. 17–19

20. Supply the reason for each key step. 3-7
Given: $\angle C \cong \angle E$; $\overline{CG} \cong \overline{EH}$;
$\overline{FG} \perp \overline{CD}$; $\overline{DH} \perp \overline{EF}$
Prove: $\angle 3 \cong \angle 4$
1. $\triangle CGF \cong \triangle EHD$
2. $\overline{GF} \cong \overline{HD}$
3. $\overline{DF} \cong \overline{DF}$
4. $\triangle GFD \cong \triangle HDF$
5. $\angle 3 \cong \angle 4$

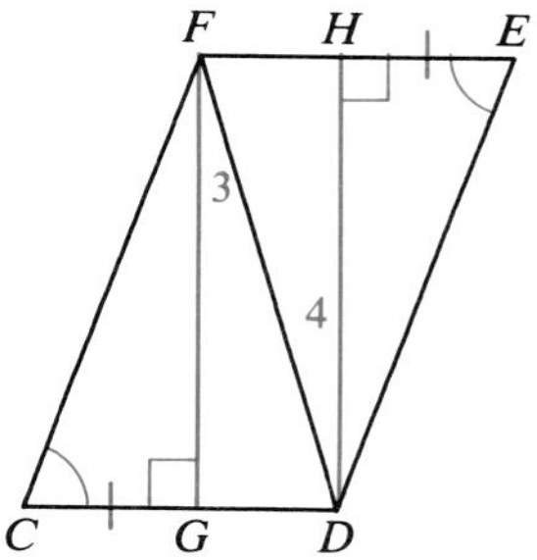

21. Marla tosses a penny six times and gets "tails" on each toss. 3-8
a. What might she conclude about the seventh toss?
b. Is it possible that her conclusion isn't valid? Explain.

Use inductive reasoning to guess the next two numbers of each sequence.

22. 1, 8, 27, 64, . . .

23. 100, -10, 1, $-\frac{1}{10}$, . . .

Chapter Test

Complete.

1. If $\triangle LEG \cong \triangle ARM$, then $\overline{GL} \cong$ _?_ and $\triangle RMA \cong$ _?_.

2. In isosceles $\triangle ABC$, $m\angle A = 130$. The legs are sides _?_ and _?_. $m\angle B =$ _?_ (numerical answer).

3. You want to prove $\triangle RST \cong \triangle XYZ$ by SAS. If you have $\overline{ST} \cong \overline{YZ}$ and $\angle T \cong \angle Z$, you must show that _?_ $\cong$ _?_.

4. The congruence method that applies only to right triangles is the __?__ method.

5. If $\triangle JKL$ is equilateral, then $m\angle K =$ __?__ (numerical answer).

6. A perpendicular segment from a vertex of a triangle to the line that contains the opposite side is called a(n) __?__ of the triangle.

7. A point lies on the perpendicular bisector of a segment if and only if it is equidistant from __?__.

8. Use inductive reasoning to complete: 100, 99, 97, 94, 90, 85, __?__, __?__.

Can the triangles be proved congruent? If so, by which method, SSS, SAS, ASA, AAS, or HL?

9.

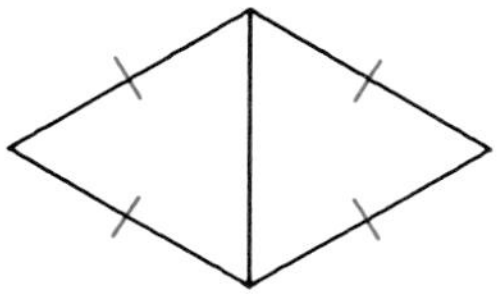

10.

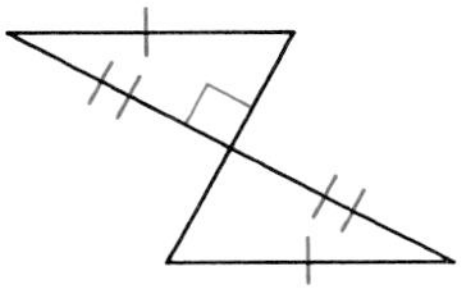

11.

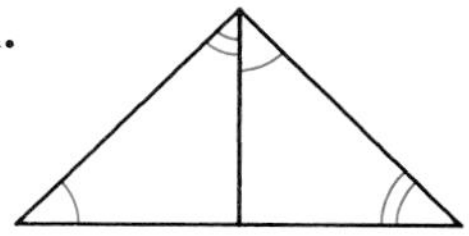

12.

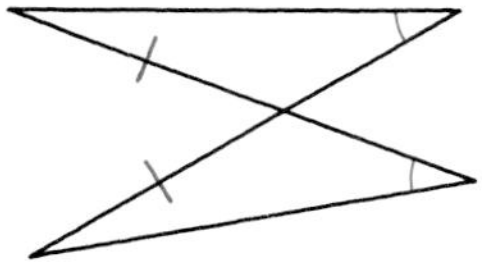

13.

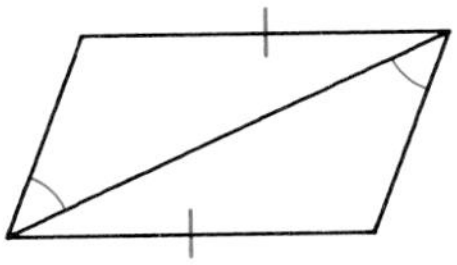

14.

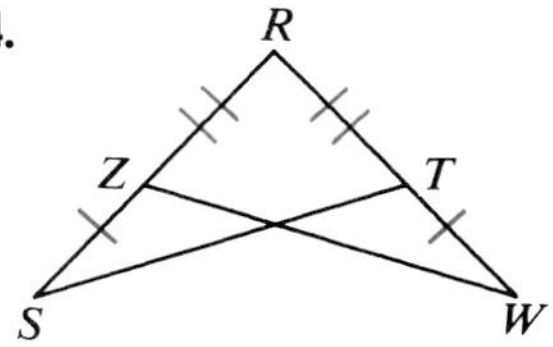

$\overline{AC}$ is the perpendicular bisector of $\overline{BD}$.

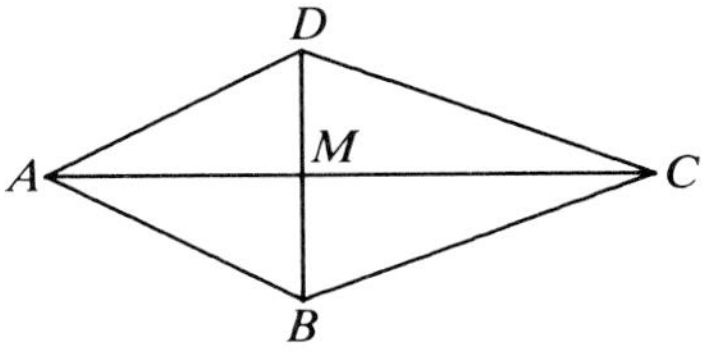

15. a. A is equidistant from __?__ and __?__.
    b. C is equidistant from __?__ and __?__.

16. Name two isosceles triangles.

17. Name three pairs of congruent triangles.

18. In $\triangle JKL$, $\overline{JL} \cong \overline{KL}$. If $m\angle J = x + y$, $m\angle K = 2x + 10$, and $m\angle L = x + 2y$, find the values of x and y.

19. Given: $\angle 1 \cong \angle 2$; $\angle PQR \cong \angle SRQ$
    Prove: $\overline{PR} \cong \overline{SQ}$

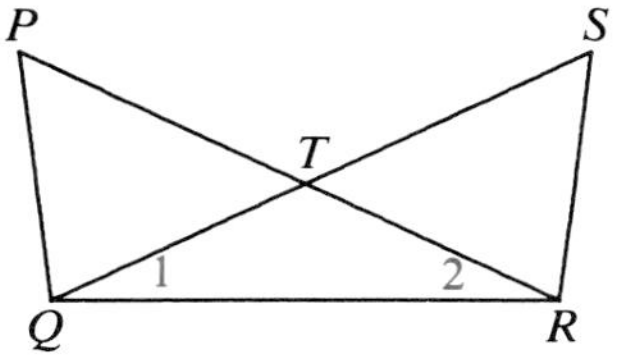

20. Given: $\overline{WZ} \perp$ plane M; $\angle ZXY \cong \angle ZYX$
    Prove: $\overline{WX} \cong \overline{WY}$

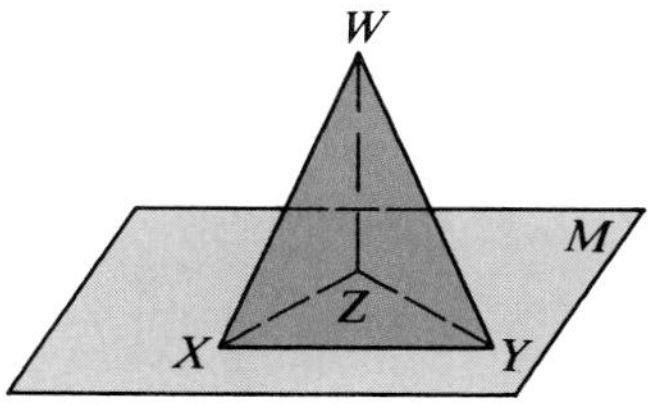

Mixed Review

1. A regular polygon has 40 sides. Find the measure of each interior angle.
2. Given: A triangle is equiangular only if it is isosceles.
   a. Write an if-then statement that is logically equivalent to the given conditional.
   b. State the converse. Sketch a diagram to disprove the converse.
3. Use inductive reasoning to guess the next two numbers in the sequence:

   1, 2, 6, 15, 31, 56, . . .

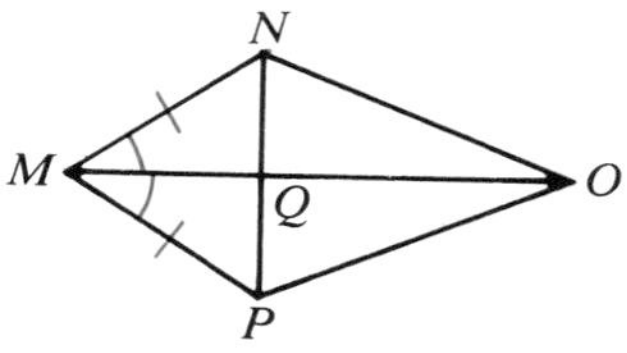

4. Given the information in the diagram, write a paragraph proof showing that $\overleftrightarrow{QO}$ is the perpendicular bisector of $\overline{NP}$.
5. If a diagonal of an equilateral quadrilateral is drawn, what method could be used to show that the two triangles formed are congruent?
6. Name five ways to prove that two lines are parallel.
7. When two parallel lines are cut by a transversal, two corresponding angles have measures x^2 and $6x$. Find the measure of each angle.
8. Prove: If a triangle is isosceles, then exterior angles adjacent to the base angles are congruent. (Begin by drawing a diagram and stating what is given and what is to be proved.)
9. Given: $\overline{RU} \parallel \overline{ST}$; $\angle R \cong \angle T$
   Prove: $\overline{RS} \parallel \overline{UT}$

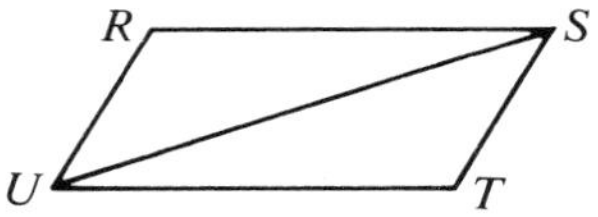

10. How many planes contain two given intersecting lines?
11. $\overrightarrow{BD}$ bisects $\angle ABC$, $m\angle ABC = 5x - 4$, and $m\angle CBD = \frac{3}{2}x + 21$. Is $\angle ABC$ acute, obtuse, right, or straight?
12. In quadrilateral $EFGH$, $\overline{EF} \parallel \overline{HG}$, $m\angle E = y + 10$, $m\angle F = 2y - 40$, and $m\angle H = 2y - 31$. Find the numerical measure of $\angle G$.
13. A statement that is accepted without proof is called a _?_.
14. A statement that can be proved easily by using a theorem is called a _?_.
15. If two parallel planes are cut by a third plane, then the lines of intersection are _?_.
16. A conditional and its _?_ are always logically equivalent.
17. If S is between R and T, then $RS + ST = RT$ by the _?_.
18. Write a paragraph proof: If $\overline{AX}$ is a median and an altitude of $\triangle ABC$, then $\triangle ABC$ is isosceles.

Algebra Review

Solve each equation by factoring or by using the quadratic formula.

Example 1 $3x^2 + 14x = -8$

Solution 1 *By factoring*

Express in the form $ax^2 + bx + c = 0$.

$$3x^2 + 14x + 8 = 0$$
$$(3x + 2)(x + 4) = 0$$
$$3x + 2 = 0 \text{ or } x + 4 = 0$$
$$x = -\tfrac{2}{3} \text{ or } x = -4$$

Solution 2 *By quadratic formula*

For $ax^2 + bx + c = 0$, $a \neq 0$,

$$x = \frac{-b \pm \sqrt{b^2 - 4ac}}{2a}$$

Express in the form $ax^2 + bx + c = 0$.

$$3x^2 + 14x + 8 = 0$$
$$x = \frac{-14 \pm \sqrt{196 - 4(3)(8)}}{2(3)} = \frac{-14 \pm 10}{6}$$

Thus $x = -\tfrac{2}{3}$ or $x = -4$.

Example 2 $x^2 + 3x - 2 = 0$

Solution For $1x^2 + 3x + (-2) = 0$, $x = \dfrac{-3 \pm \sqrt{9 - 4(1)(-2)}}{2(1)} = \dfrac{-3 \pm \sqrt{17}}{2}$

Thus $x = \dfrac{-3 + \sqrt{17}}{2}$ or $x = \dfrac{-3 - \sqrt{17}}{2}$.

1. $x^2 + 5x - 6 = 0$
2. $a^2 - 6a + 8 = 0$
3. $c^2 - 7c - 18 = 0$
4. $x^2 + 8x = 0$
5. $3y^2 = 15y$
6. $2z^2 + 7z = 0$
7. $n^2 - 144 = 20$
8. $50y^2 = 2$
9. $\frac{1}{2}q^2 - 18 = 0$
10. $x^2 - 4x + 4 = 0$
11. $b^2 + 3b - 10 = 0$
12. $3x^2 + 2x - 1 = 0$
13. $9j^2 - 5j - 4 = 0$
14. $2x^2 = x + 10$
15. $x^2 + 7x + 3 = 0$
16. $d^2 + 8d + 12 = 0$
17. $e^2 + 5e = 24$
18. $v^2 + 25 = 10v$
19. $x^2 = 3x + 4$
20. $t^2 - t = 20$
21. $x^2 = 20x - 36$
22. $x^2 - 5x + 3 = 0$
23. $5p^2 - 2p = 7$
24. $4m^2 + 5m = 12m$
25. $\dfrac{x^2}{2} - \dfrac{x}{2} = 1$
26. $\dfrac{a^2}{4} = a - 1$
27. $c + 3 = \dfrac{10}{c}$
28. $(y - 5)^2 = 16$
29. $z^2 = 4(2z - 3)$
30. $x(x + 5) = 14$
31. $b(10 - b) = 24$
32. $r(10 + r) = 24$
33. $(s - 3)^2 = s - 1$
34. $3x^2 + 3x = 4$
35. $15 + 4y^2 = 17y$
36. $6x^2 + 11x + 2 = 0$

Solve each problem by writing and solving a quadratic equation.

37. Refer to the diagram. Find the value of x.

38. In $\triangle RST$, $\angle R \cong \angle S$, $RS = 3x + 2$, $ST = 4x - 2$, and $RT = x^2 + 1$. Find the lengths of the sides of $\triangle RST$.

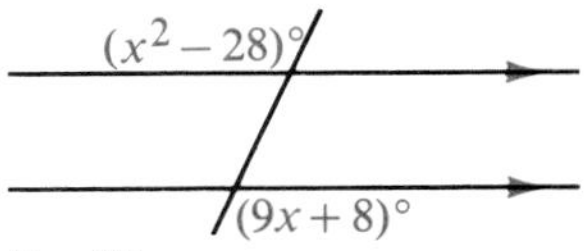

Ex. 37

Minerals classified as carbonates are found in a number of interesting forms. The rose-colored mineral pictured is a rhodochrosite. Its faces are parallelograms, often the special one known as a rhombus.

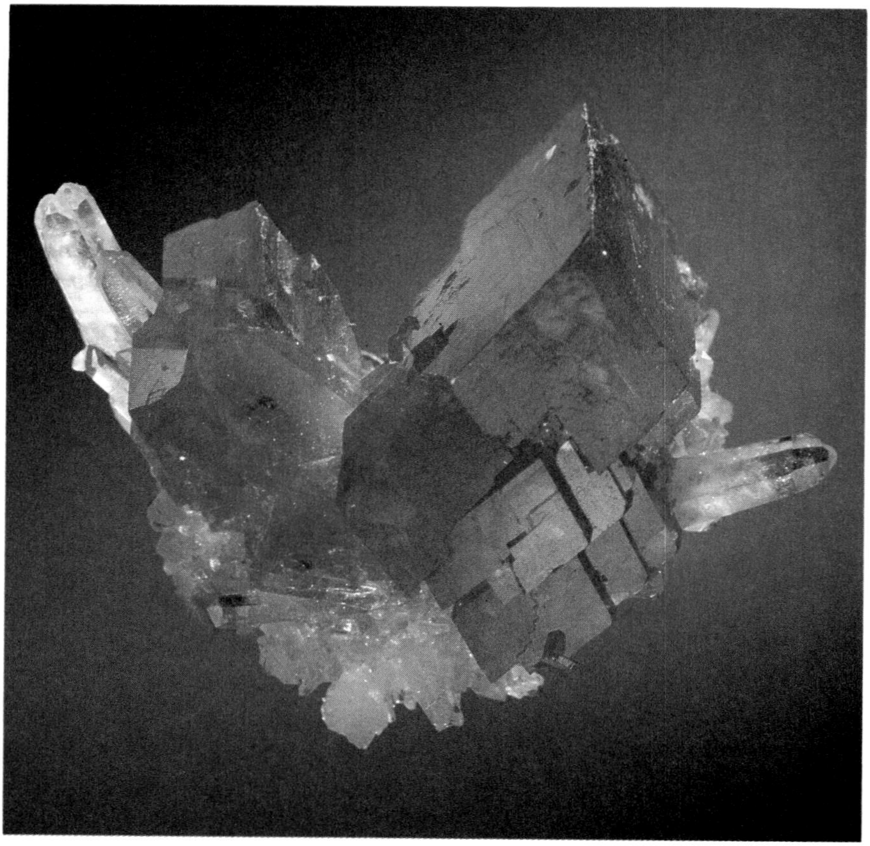

Using Congruent Triangles

4

Parallelograms and Trapezoids

Objectives

1. Apply the definitions of a parallelogram and a trapezoid.
2. State and apply the theorems about properties of a parallelogram.
3. Prove that certain quadrilaterals are parallelograms.
4. Identify the special properties of a rectangle, a rhombus, and a square.
5. State and apply the theorems about the median of a trapezoid and the segment that joins the midpoints of two sides of a triangle.

4-1 Properties of Parallelograms

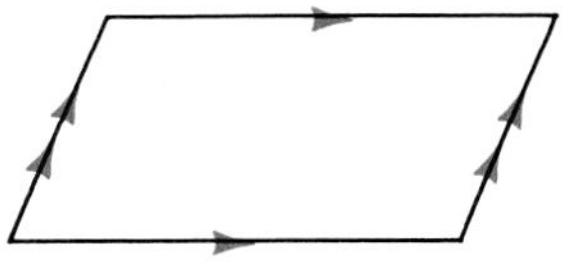

A **parallelogram** (▱) is a quadrilateral with both pairs of opposite sides parallel. The following theorems state some properties common to all parallelograms. Your proofs of these theorems (Exercises 17–19) will be based on what you have learned about parallel lines and congruent triangles.

Theorem 4-1

Opposite sides of a parallelogram are congruent.

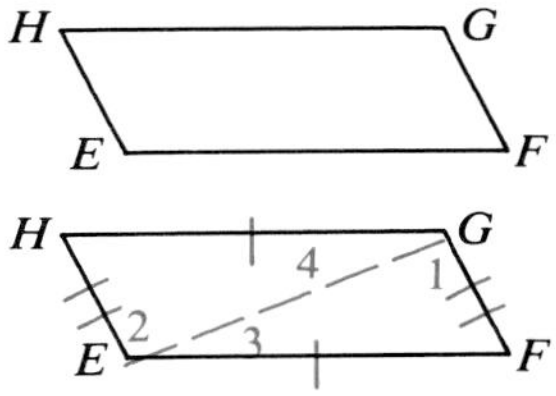

Given: ▱$EFGH$

Prove: $\overline{EF} \cong \overline{HG}$; $\overline{FG} \cong \overline{EH}$

Plan for Proof: Draw $\overline{EG}$ to form triangles that can be proved congruent by ASA. Note that the pairs of alternate interior angles formed are $\angle 1$ and $\angle 2$, $\angle 3$ and $\angle 4$. After showing that the triangles are congruent, you can use corresponding parts to finish the proof.

Corollary

If two lines are parallel, then all points on one line are equidistant from the other line.

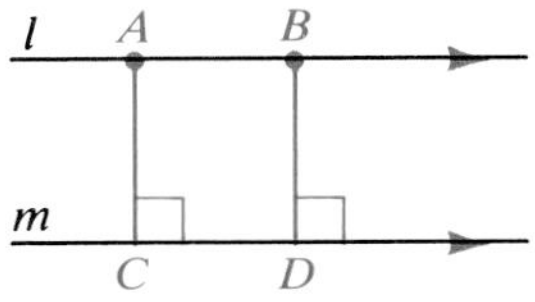

Given: $l \parallel m$; A and B are any points on l; $\overline{AC} \perp m$; $\overline{BD} \perp m$

Prove: $AC = BD$

Proof:

Since $\overline{AB}$ and $\overline{CD}$ are contained in parallel lines, $\overline{AB} \parallel \overline{CD}$. Since $\overline{AC}$ and $\overline{BD}$ are both perpendicular to m, they are parallel. Thus $ABDC$ is a parallelogram, and opposite sides $\overline{AC}$ and $\overline{BD}$ are congruent.

Theorem 4-2

Opposite angles of a parallelogram are congruent.

Theorem 4-3

The diagonals of a parallelogram bisect each other.

Given: $\square QRST$ with diagonals $\overline{QS}$ and $\overline{TR}$

Prove: $\overline{QS}$ and $\overline{TR}$ bisect each other.

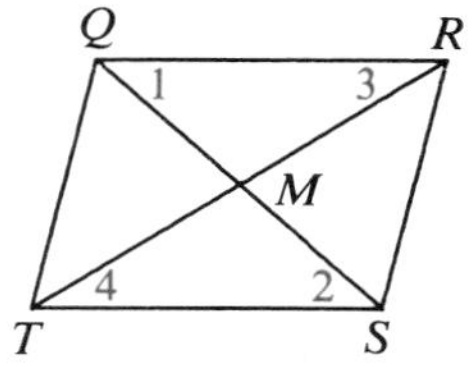

■ ***Plan for Proof:*** You can prove that $\overline{QM} \cong \overline{MS}$ and $\overline{MR} \cong \overline{TM}$ by showing that they are corresponding parts of congruent triangles. Since you have $\overline{QR} \cong \overline{TS}$ by Theorem 4-1, you can use ASA to show that $\triangle QMR \cong \triangle SMT$.

Classroom Exercises

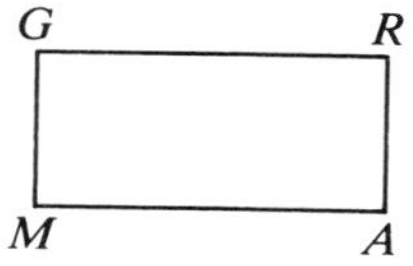

1. Quad. $GRAM$ is a parallelogram.
   a. Why is $\angle G$ supplementary to $\angle M$?
   b. Why is $\angle M$ supplementary to $\angle A$?
   c. Complete: Consecutive angles of a parallelogram are __?__, while opposite angles are __?__.

2. Suppose $\angle M$, in $\square GRAM$, is a right angle. What can you deduce about angles G, R, and A?

Find the measures of $\angle Q$, $\angle S$, $\angle 1$, and $\angle 2$. Quad. $PQST$ is a parallelogram.

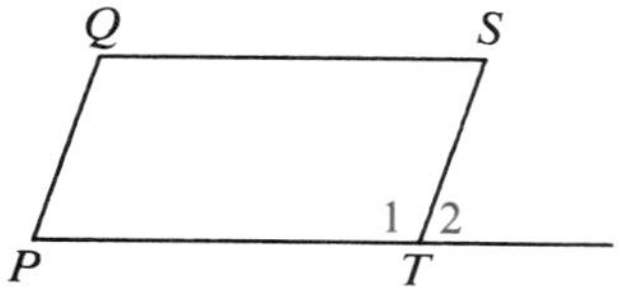

3. $m\angle P = 70$

4. $m\angle P = c$

In Exercises 5–7, quad. $WXYZ$ is a parallelogram.

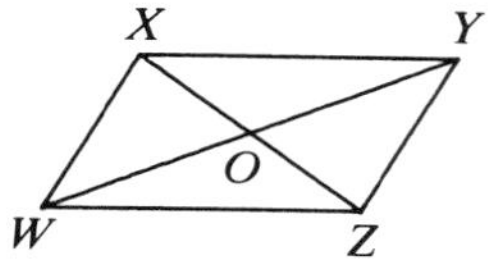

5. Name all pairs of parallel lines in the figure.

6. Name all pairs of congruent segments.

7. Name all pairs of congruent angles.

Must quad. $EFGH$ be a parallelogram? Can it be a parallelogram? Explain.

8.

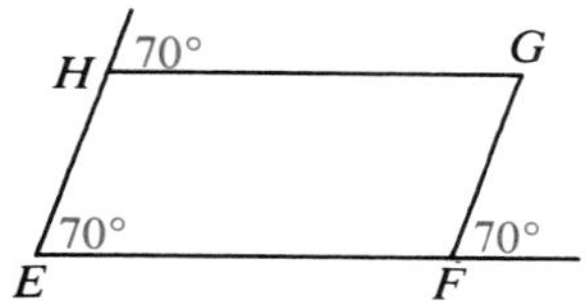

9.

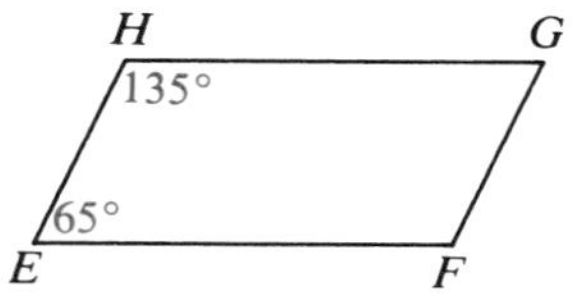

10.

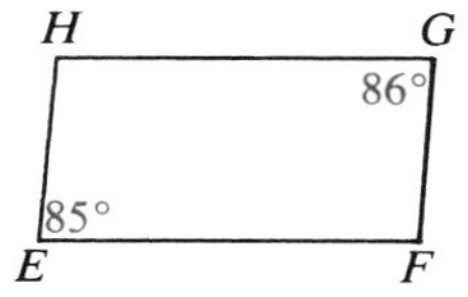

Quad. $ABCD$ is a parallelogram. Name or state the principal theorem or definition that justifies the statement.

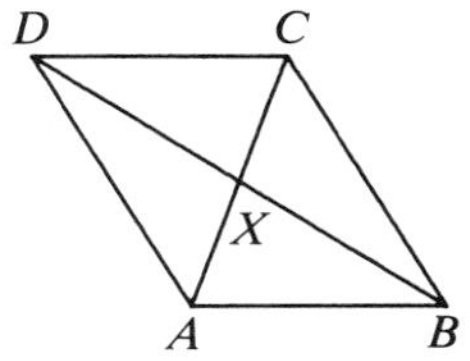

11. $\overline{AD} \parallel \overline{BC}$

12. $\angle ADX \cong \angle CBX$

13. $m\angle ABC = m\angle CDA$

14. $\overline{AD} \cong \overline{BC}$

15. $AX = \frac{1}{2}AC$

16. $DX = BX$

17. Draw a quadrilateral that isn't a parallelogram but does have two 60° angles.

18. What result of this section does each ladder suggest?

Written Exercises

Exercises 1–16 refer to $\square DECK$. Complete each statement in Exercises 1–8.

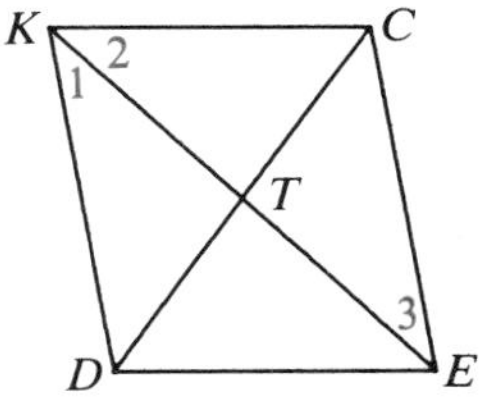

Exs. 1–16

A

1. If $DE = 10$, $KC = \underline{\ ?\ }$.
2. If $DC = 18$, $DT = \underline{\ ?\ }$.
3. If $m\angle EDK = 100$, $m\angle ECK = \underline{\ ?\ }$.
4. If $m\angle DEC = 75$, $m\angle KDE = \underline{\ ?\ }$.
5. If $m\angle 1 = 30$ and $m\angle 2 = 40$, $m\angle KCE = \underline{\ ?\ }$.
6. If $m\angle 1 = 30$ and $m\angle 2 = 40$, $m\angle 3 = \underline{\ ?\ }$.
7. If $m\angle 3 = 36$ and $m\angle 2 = 44$, $m\angle KDE = \underline{\ ?\ }$.
8. If $DT = 7$ and $KT = 9$, $CD = \underline{\ ?\ }$.

Find the value of x or y.

9. $DE = 5x$ and $KC = 3x + 12$
10. $DK = 2x + 5$ and $EC = 47 - 4x$
11. $ET = x + 3$ and $EK = 22$
12. $DT = \frac{1}{2}x$ and $TC = 10$
13. $m\angle KCE = 6y - 20$ and $m\angle EDK = 2y + 80$
14. $m\angle DEC = 80 - y$ and $m\angle DKC = y + 40$
15. $m\angle 1 = y + 10$, $m\angle 2 = 3y$, and $m\angle 3 = \frac{1}{2}y + 15$
16. $m\angle DEC = \dfrac{y}{4}$ and $m\angle ECK = \dfrac{y + 60}{2}$

17. Prove Theorem 4–1.

18. Prove Theorem 4–2. (Draw and label a figure. List what is given and what is to be proved.)

19. Prove Theorem 4–3.

20. Given: Quad. $ABCX$ is a ▱;
quad. $DXFE$ is a ▱.
Prove: $\angle B \cong \angle E$

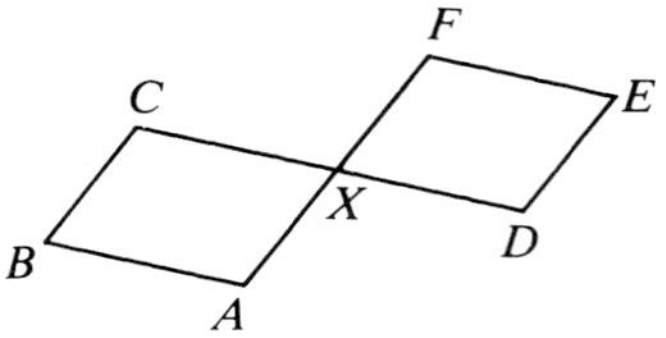

Quad. *DECK* is a parallelogram. Complete.

B 21. If $KT = 2x + y$, $DT = x + 2y$, $TE = 12$, and $TC = 9$, then $x =$ _?_ and $y =$ _?_.

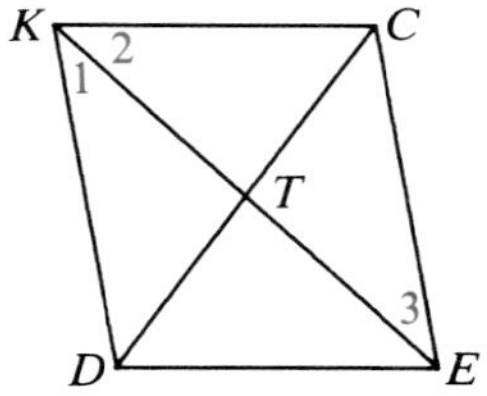

22. If $DE = x + y$, $EC = 12$, $CK = 2x - y$, and $KD = 3x - 2y$, then $x =$ _?_, $y =$ _?_, and the perimeter of ▱$DECK$ = _?_.

23. If $m\angle 1 = 4x$, $m\angle 2 = 3x$, and $m\angle 3 = x^2 - 60$, then $x =$ _?_ and $m\angle CED =$ _?_ (numerical answers).

24. If $m\angle 1 = 20$, $m\angle 2 = x^2$, and $m\angle CED = 9x$, then $m\angle 2 =$ _?_ or $m\angle 2 =$ _?_ (numerical answers).

25. Given: ▱$PQRS$; $\overline{PJ} \cong \overline{RK}$
Prove: $\overline{SJ} \cong \overline{QK}$

26. Given: ▱$JQKS$; $\overline{PJ} \cong \overline{RK}$
Prove: $\angle P \cong \angle R$

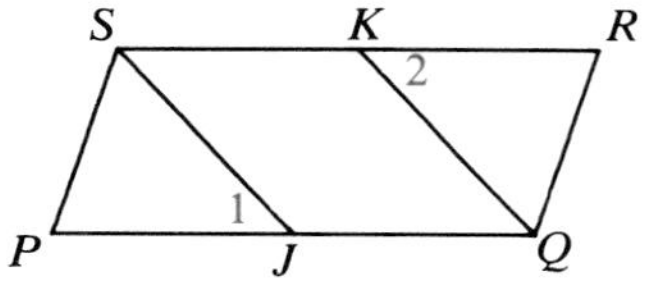

27. Given: $ABCD$ is a ▱; $\overline{CD} \cong \overline{CE}$
Prove: $\angle A \cong \angle E$

28. Given: $RSTQ$ is a ▱.
Prove: $\overline{OX} \cong \overline{OY}$

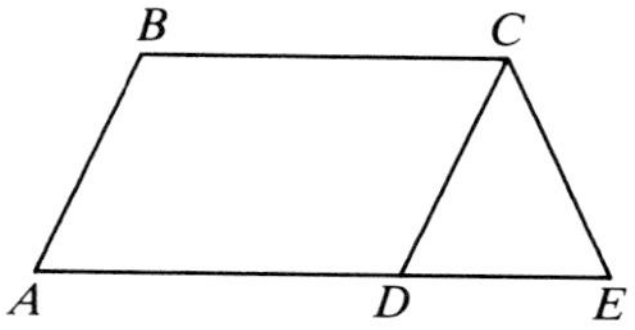

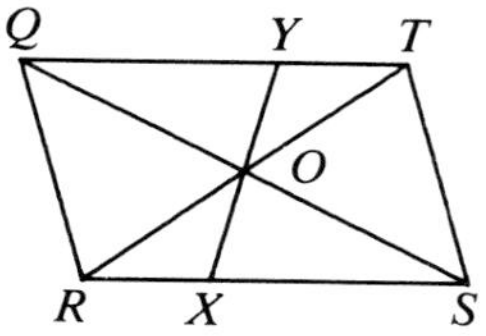

Find something interesting to prove. Then prove it.

29. Given: ▱$ABCD$; $\angle 1 \cong \angle 2$
Prove: _?_

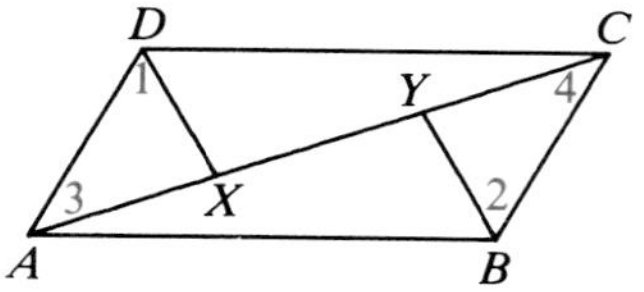

30. Given: ▱$EFIH$; ▱$EGJH$; $\angle 1 \cong \angle 2$
Prove: _?_

31. Given: $GF \neq JI$ and $GE \neq JH$
   a. Can quadrilaterals $GFIJ$ and $EGJH$ be parallelograms? Explain.
   b. Draw a diagram similar to that shown, but such that $EFIH$ is a parallelogram and it is clear that $GF \neq JI$ and $GE \neq JH$.

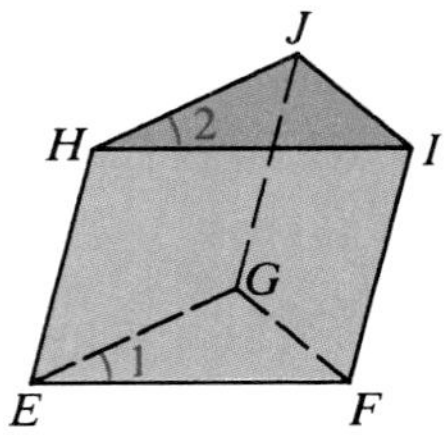

Exs. 30, 31

C **32.** **a.** Given: Plane $P \parallel$ plane Q; $j \parallel k$
Prove: $AX = BY$
b. State, in words, a theorem proved in part (a).

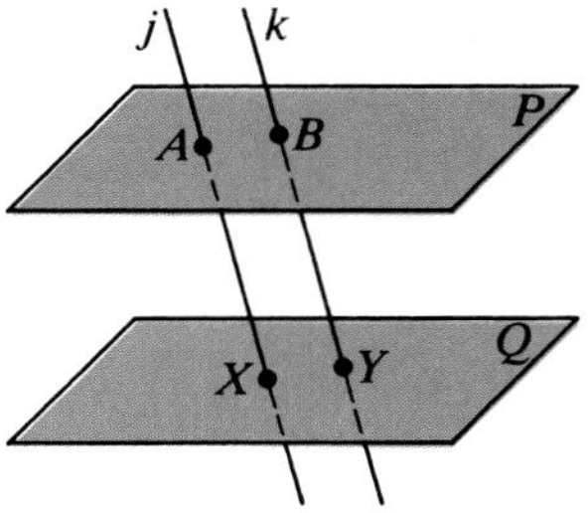

33. Prove: If a segment whose endpoints lie on opposite sides of a parallelogram passes through the midpoint of a diagonal, that segment is bisected by the diagonal.

★ **34.** Write a paragraph proof: The sum of the lengths of the segments drawn from any point in the base of an isosceles triangle perpendicular to the legs is equal to the length of the altitude drawn to one leg from the vertex opposite that leg.

4-2 Ways to Prove that Quadrilaterals Are Parallelograms

If both pairs of opposite sides of a quadrilateral are parallel, then by definition the quadrilateral is a parallelogram. The following theorems, whose proofs are left as exercises, will give you additional ways to prove that a quadrilateral is a parallelogram.

Theorem 4-4

If both pairs of opposite sides of a quadrilateral are congruent, then the quadrilateral is a parallelogram.

Given: $\overline{TS} \cong \overline{QR}$; $\overline{TQ} \cong \overline{SR}$

Prove: Quad. $QRST$ is a ▱.

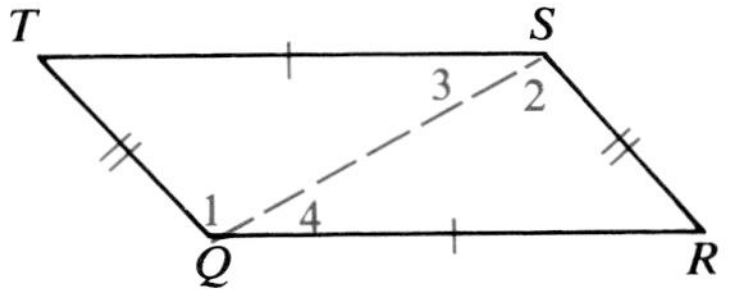

■ ***Plan for Proof:*** Draw $\overline{SQ}$ and prove $\triangle TSQ \cong \triangle RQS$. Then $\angle 1 \cong \angle 2$ and $\angle 3 \cong \angle 4$, and opposite sides are parallel.

Theorem 4-5

If one pair of opposite sides of a quadrilateral are both congruent and parallel, then the quadrilateral is a parallelogram.

Given: $\overline{AB} \cong \overline{CD}$; $\overline{AB} \parallel \overline{CD}$

Prove: Quad. $ABCD$ is a ▱.

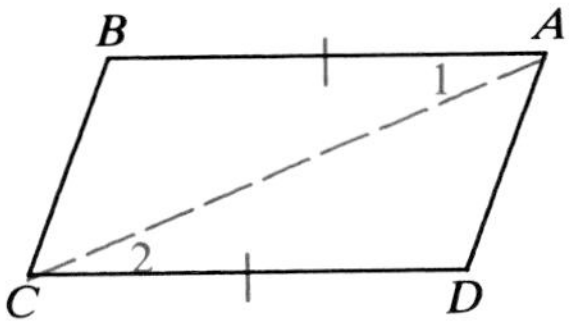

■ ***Plan for Proof:*** Draw $\overline{AC}$ and prove $\triangle ABC \cong \triangle CDA$. Then $\overline{BC} \cong \overline{DA}$, and you can apply Theorem 4-4.

Theorem 4-6

If both pairs of opposite angles of a quadrilateral are congruent, then the quadrilateral is a parallelogram.

Theorem 4-7

If the diagonals of a quadrilateral bisect each other, then the quadrilateral is a parallelogram.

Five Ways to Prove that a Quadrilateral Is a Parallelogram

1. Show that *both* pairs of opposite sides are parallel.
2. Show that *both* pairs of opposite sides are congruent.
3. Show that *one* pair of opposite sides are both congruent and parallel.
4. Show that both pairs of opposite angles are congruent.
5. Show that the diagonals bisect each other.

The proof of the following theorem is based on what you have learned about parallelograms.

Theorem 4-8

If three parallel lines cut off congruent segments on one transversal, then they cut off congruent segments on every transversal.

Given: $\overleftrightarrow{AX} \parallel \overleftrightarrow{BY} \parallel \overleftrightarrow{CZ}$;
$\overline{AB} \cong \overline{BC}$

Prove: $\overline{XY} \cong \overline{YZ}$

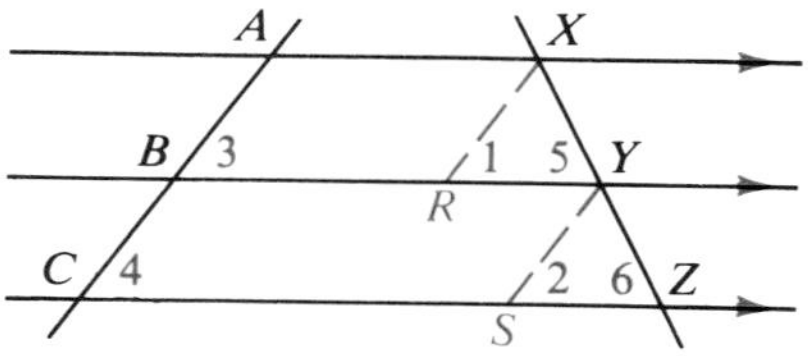

Proof:

Through X and Y draw lines parallel to $\overleftrightarrow{AC}$. Then $AXRB$ and $BYSC$ are parallelograms, by the definition of a parallelogram. Since the opposite sides of a parallelogram are congruent, $\overline{XR} \cong \overline{AB}$ and $\overline{BC} \cong \overline{YS}$. It is given that $\overline{AB} \cong \overline{BC}$, so using the Transitive Property twice gives $\overline{XR} \cong \overline{YS}$. Parallel lines are cut by transversals to form the following pairs of congruent corresponding angles:

$$\angle 1 \cong \angle 3 \qquad \angle 3 \cong \angle 4 \qquad \angle 4 \cong \angle 2 \qquad \angle 5 \cong \angle 6$$

Then $\angle 1 \cong \angle 2$ (Transitive Property), and $\triangle XYR \cong \triangle YZS$ by AAS. Since $\overline{XY}$ and $\overline{YZ}$ are corresponding parts of these triangles, $\overline{XY} \cong \overline{YZ}$.

Corollary

A line that contains the midpoint of one side of a triangle and is parallel to another side bisects the third side.

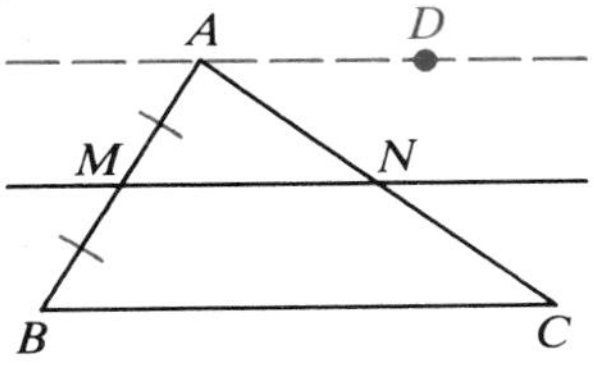

Given: M is the midpoint of $\overline{AB}$;
$\overleftrightarrow{MN} \parallel \overline{BC}$

Prove: $\overleftrightarrow{MN}$ bisects $\overline{AC}$.

Proof:

Let $\overleftrightarrow{AD}$ be the line through A parallel to $\overleftrightarrow{MN}$. Then $\overleftrightarrow{AD}$, $\overleftrightarrow{MN}$, and $\overleftrightarrow{BC}$ are three parallel lines that cut off congruent segments on transversal $\overleftrightarrow{AB}$. By Theorem 4-8 they also cut off congruent segments on $\overleftrightarrow{AC}$. Thus $\overline{AN} \cong \overline{NC}$ and $\overleftrightarrow{MN}$ bisects $\overline{AC}$.

Classroom Exercises

In each exercise decide whether the given information permits you to deduce that quad. *REST* is a parallelogram. If your answer is *yes*, state the definition or theorem that applies.

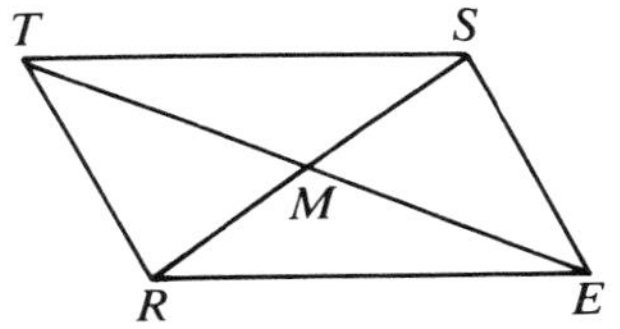

1. $\overline{TM} \cong \overline{EM}$
2. $TM = EM$; $RM = SM$
3. $\overline{TS} \parallel \overline{RE}$; $\overline{TS} \cong \overline{RE}$
4. $TS = RE$; $TR = SE$
5. $\overline{TS} \parallel \overline{RE}$; $\overline{TR} \parallel \overline{SE}$
6. $\overline{TS} \cong \overline{RE}$; $\overline{TS} \cong \overline{TR}$

7. Given: Quad. $ABCD$; $m\angle A = m\angle C = r$; $m\angle B = m\angle D = s$
   a. Tell why $2r + 2s = 360$.
   b. $r + s = \underline{\ ?\ }$
   c. Tell why $\overline{AB} \parallel \overline{CD}$ and why $\overline{AD} \parallel \overline{BC}$.
   d. Tell why quad. $ABCD$ must be a parallelogram.
   e. What theorem have you just proved?

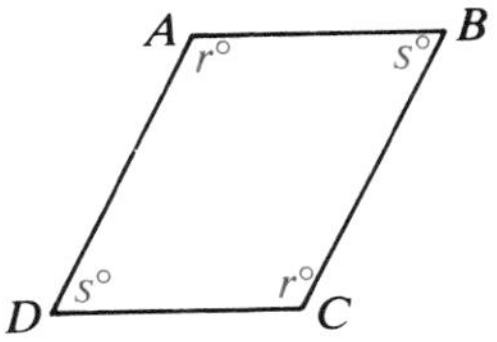

8. *Parallel rulers*, used to draw parallel lines, are constructed so that $EF = HG$ and $HE = GF$. Since there are hinges at points E, F, G, and H, you can vary the distance between $\overline{HG}$ and $\overline{EF}$. Explain why $\overline{HG} \parallel \overline{EF}$.

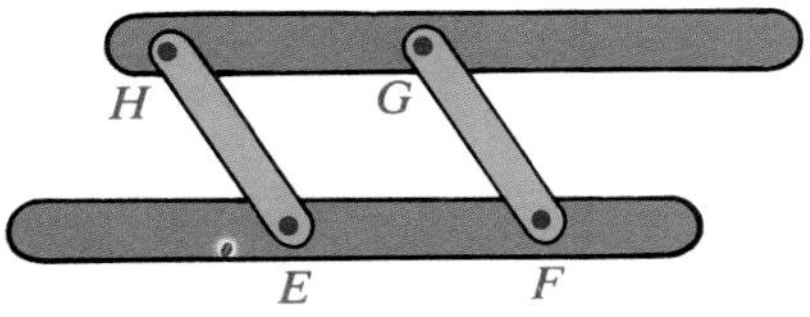

9. The pliers shown are made in such a way that the jaws are always parallel. Explain.

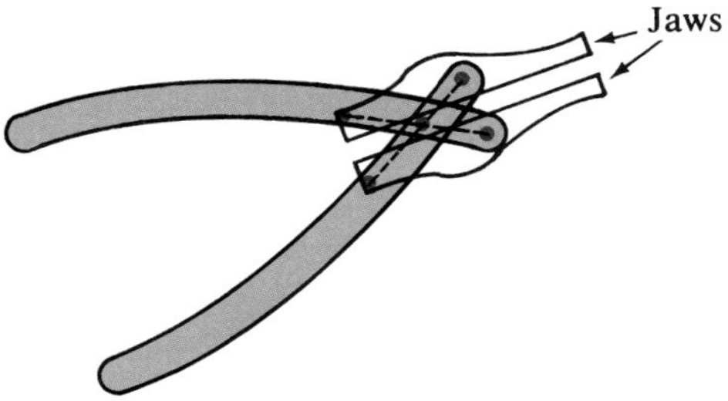

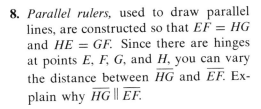

10. Imagine quad. $WXYZ$, with sides $\overline{WX}$ and $\overline{ZY}$ congruent and sides $\overline{WZ}$ and $\overline{XY}$ parallel. Must $WXYZ$ be a parallelogram? Explain.

11. Imagine a quadrilateral with two pairs of sides congruent. Must the quadrilateral be a parallelogram? Explain.

Written Exercises

State the principal definition or theorem that enables you to deduce, from the information given, that quadrilateral $SACK$ is a parallelogram.

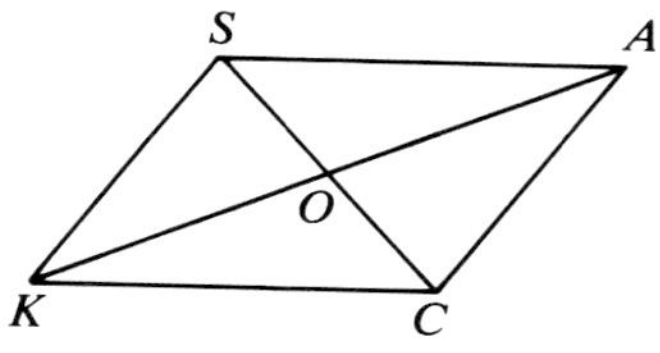

A 1. $\overline{SA} \parallel \overline{KC}$; $\overline{SK} \parallel \overline{AC}$

2. $\overline{SA} \cong \overline{KC}$; $\overline{SK} \cong \overline{AC}$

3. $\overline{SA} \cong \overline{KC}$; $\overline{SA} \parallel \overline{KC}$

4. $SO = \frac{1}{2}SC$; $KO = \frac{1}{2}KA$

5. $\angle SKC \cong \angle CAS$; $\angle KCA \cong \angle ASK$

6. Suppose you know that $\triangle SOK \cong \triangle COA$. Explain how you could prove that quad. $SACK$ is a parallelogram.

$\overleftrightarrow{AE}$, $\overleftrightarrow{BF}$, $\overleftrightarrow{CG}$ and $\overleftrightarrow{DH}$ are parallel, with $EF = FG = GH$. Complete.

7. If $AB = 5$, $AD =$ _?_.

8. If $AC = 12$, $CD =$ _?_.

9. If $AB = 5x$ and $BC = 2x + 12$, $x =$ _?_.

10. If $AC = 22 - x$ and $BD = 3x - 22$, $x =$ _?_.

B 11. If $AB = 15$, $BC = 2x - y$, and $CD = x + y$, $x =$ _?_ and $y =$ _?_.

12. If $AB = 12$, $BC = 2x + 3y$, and $BD = 8x$, $x =$ _?_ and $y =$ _?_.

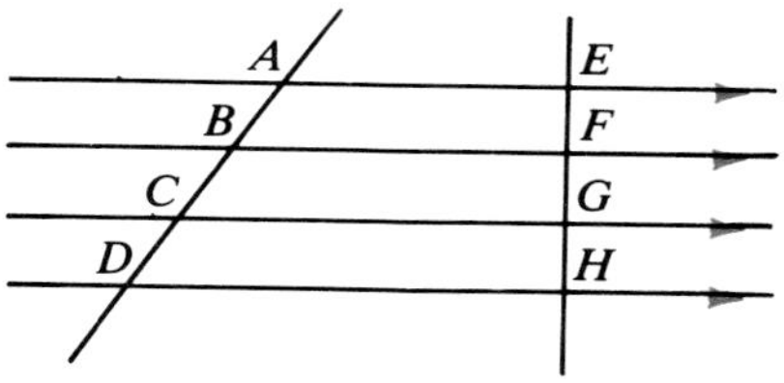

Exs. 7–12

In Exercises 13–15 explain briefly how you would prove that the quadrilateral is a parallelogram.

13. Given: ▱$ABCD$; M and N are the midpoints of $\overline{AB}$ and $\overline{DC}$.
Prove: $AMCN$ is a ▱.

14. Given: ▱$ABCD$; $\overline{AN}$ and $\overline{CM}$ bisect $\angle A$ and $\angle C$.
Prove: $AMCN$ is a ▱.

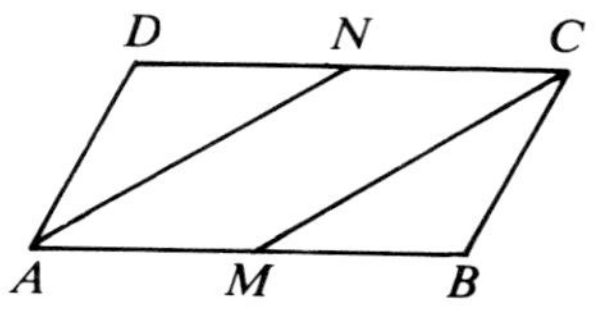

15. Given: ▱$ABCD$; W, X, Y, Z are midpoints of $\overline{AO}$, $\overline{BO}$, $\overline{CO}$, and $\overline{DO}$.
Prove: $WXYZ$ is a ▱.

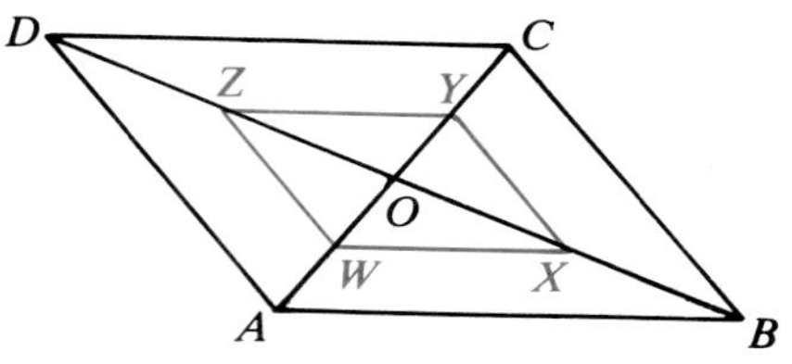

Explain briefly how you would prove that the quadrilateral is a parallelogram.

16. Given: ▱$ABCD$; $DE = BF$
Prove: $AFCE$ is a ▱.

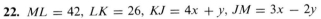

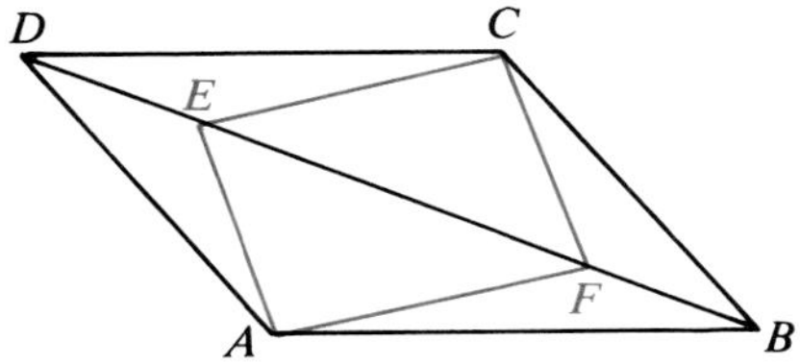

17. Given: ▱$ABCD$ and ▱$CDFE$
Prove: $ABEF$ is a ▱.

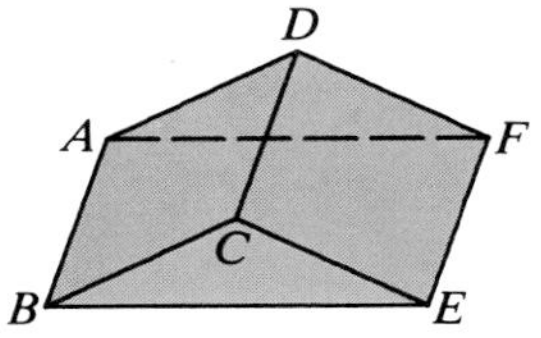

18. **a.** State Theorem 4–1 in if-then form.
b. Which theorem in this section is the converse of Theorem 4–1?

19. Prove Theorem 4–4.

20. Prove Theorem 4–5.

21. **a.** Prove Theorem 4–7.
b. Describe another way to prove Theorem 4–7.

What values must x and y have to make quad. *JKLM* a parallelogram?

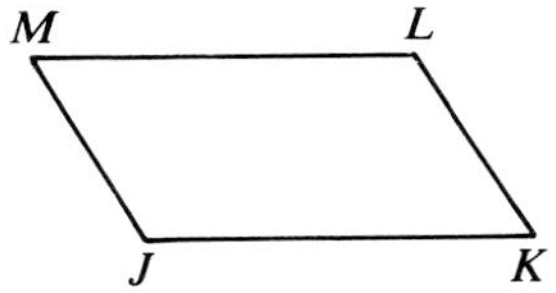

22. $ML = 42$, $LK = 26$, $KJ = 4x + y$, $JM = 3x - 2y$

23. $ML = 5x - 3y$, $LK = x + y$, $KJ = 3x + y$, $JM = 33$

24. $ML = 2y - x$, $LK = 5$, $KJ = 2x - y$, $JM = x - \frac{y}{2}$

25. Given: $\overline{AE} \cong \overline{CD}$;
$\angle DBC \cong \angle C$;
$\angle A \cong \angle DBC$
Prove: Quad. $ABDE$ is a ▱.

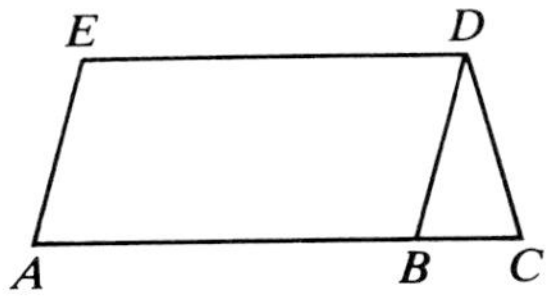

26. Given: ▱$KGLJ$;
$FK = LH$
Prove: Quad. $FGHJ$ is a ▱.

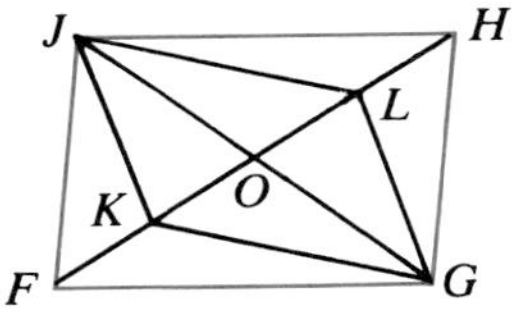

27. Given: Plane $X \parallel$ plane Y;
$\overline{LM} \cong \overline{ON}$
Prove: Quad. $LMNO$ is a ▱.

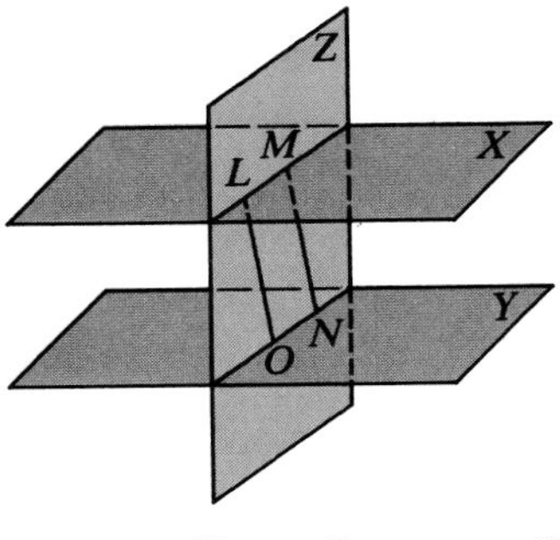

C **28.** Given: Parallel planes P, Q, and R cutting transversals $\overleftrightarrow{AC}$ and $\overleftrightarrow{DF}$; $AB = BC$

Prove: $DE = EF$

(*Hint:* You can't assume that $\overleftrightarrow{AC}$ and $\overleftrightarrow{DF}$ are coplanar. Draw $\overline{AF}$, cutting plane Q at X. Using the plane of $\overline{AC}$ and $\overline{AF}$, apply Theorems 2-1 and 4-8. Then use the plane of $\overline{AF}$ and $\overline{FD}$.)

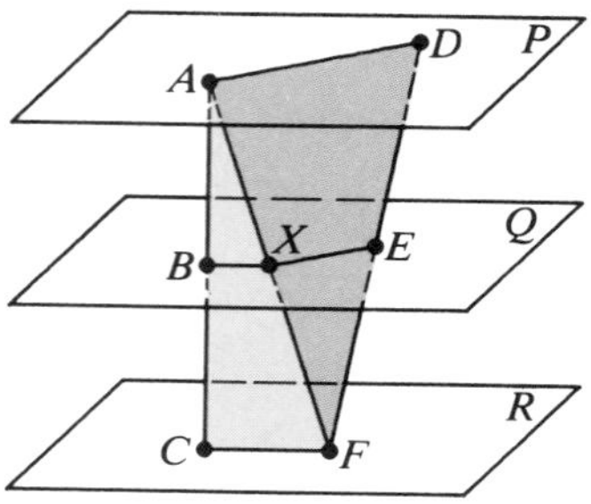

29. Write a paragraph proof.
Given: ▱$ABCD$; ▱$BEDF$
Prove: $AECF$ is a ▱.
(*Hint:* A short proof is possible.)

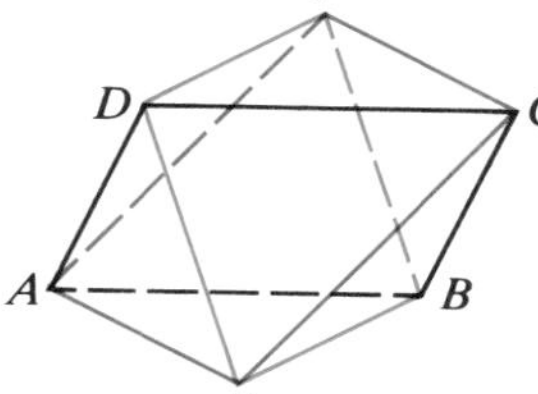

4-3 Special Parallelograms

A quadrilateral with four right angles is a **rectangle.** Since both pairs of opposite angles are congruent, every rectangle is a parallelogram.

A quadrilateral with four congruent sides is a **rhombus.** Since both pairs of opposite sides are congruent, every rhombus is a parallelogram.

A quadrilateral with four right angles and four congruent sides is a **square.** Notice that every square is a special kind of rectangle, as well as a special kind of rhombus.

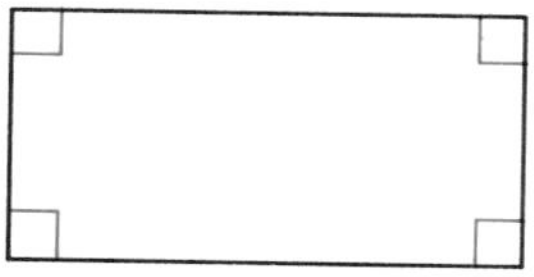
Rectangle

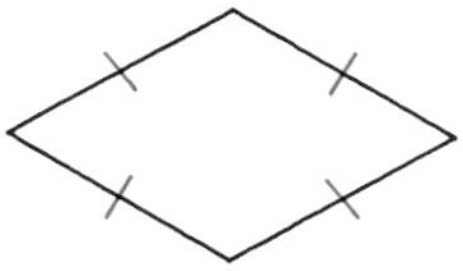
Rhombus

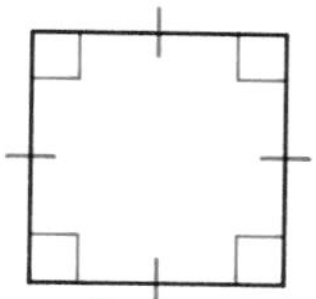
Square

As the photograph suggests, these shapes can be found in many everyday objects.

Since rectangles, rhombuses, and squares are parallelograms, they have all the properties of parallelograms. They also have the following special properties. Proofs of the theorems are left as exercises.

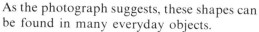

Theorem 4-9

The diagonals of a rectangle are congruent.

Theorem 4-10

The diagonals of a rhombus are perpendicular.

Theorem 4-11

Each diagonal of a rhombus bisects two angles of the rhombus.

Example Given: $ABCD$ is a rhombus.
What can you conclude?

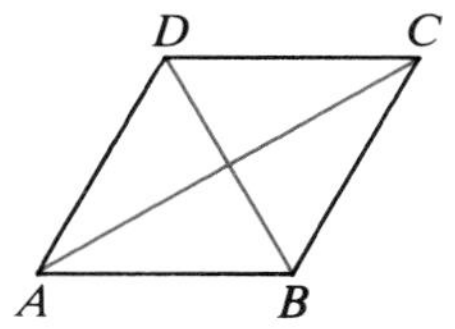

Solution $ABCD$ is a parallelogram, with all the properties of a parallelogram. Also:
By Theorem 4-10, $\overline{AC} \perp \overline{BD}$.
By Theorem 4-11, $\overline{AC}$ bisects $\angle DAB$ and $\angle BCD$;
$\overline{BD}$ bisects $\angle ABC$ and $\angle ADC$.

The properties of rectangles lead to an interesting conclusion about any right triangle.

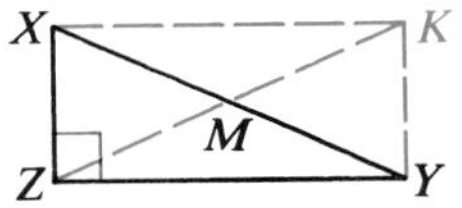

Begin with rt. $\triangle XZY$.

1. Draw lines to form rectangle $XZYK$. (How?)
2. Draw $\overline{ZK}$. $ZK = XY$ (Why?)
3. $\overline{ZK}$ and $\overline{XY}$ bisect each other. (Why?)
4. $MX = MY = MZ = MK$, by (2) and (3).

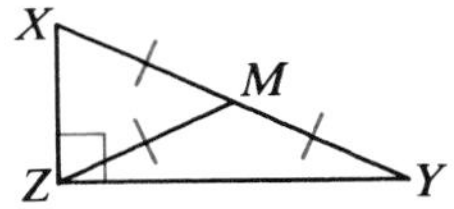

Since $MX = MY = MZ$, we have shown the following.

Theorem 4-12

The midpoint of the hypotenuse of a right triangle is equidistant from the three vertices.

Proofs of the next two theorems will be discussed in the Classroom Exercises.

Theorem 4-13

If an angle of a parallelogram is a right angle, then the parallelogram is a rectangle.

Theorem 4-14

If two consecutive sides of a parallelogram are congruent, then the parallelogram is a rhombus.

Classroom Exercises

1. When you know that one angle of a parallelogram is a right angle, you can prove that the parallelogram is a rectangle. Explain, using the figure below.

Ex. 1

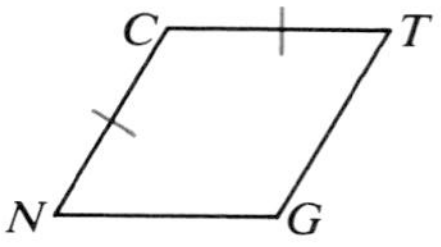

Ex. 2

2. When you know that two consecutive sides of a parallelogram are congruent, you can prove that the parallelogram is a rhombus. Explain, using the figure above.
3. Theorem 4–9 can be stated: If a quadrilateral is a rectangle, then the diagonals are congruent. State the converse. Draw a figure to show that the converse is not true.
4. State the converse of Theorem 4–10. Draw a figure to show that the converse is not true.
5. Can you assert that the figure shown is a polygon? a quadrilateral? a parallelogram? a rectangle? a rhombus? a square? Which term describes the figure most effectively?

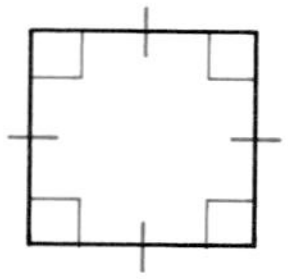

$\angle KAP$ is a right angle, and $\overline{AM}$ is a median.

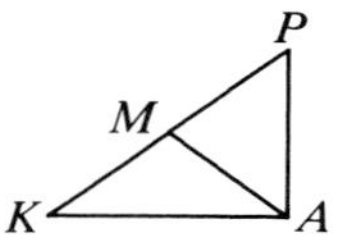

6. If $MP = 6\frac{1}{2}$, $MA =$ __?__.
7. If $MA = t$, $KP =$ __?__.
8. Given: Rhombus $EFGH$
   a. F, being equidistant from E and G, must lie on the __?__ of $\overline{EG}$.
   b. H, being equidistant from E and G, must lie on the __?__ of $\overline{EG}$.
   c. From (a) and (b) you can deduce that $\overline{FH}$ is the __?__ of $\overline{EG}$.
   d. State the theorem of this section that you have just proved.

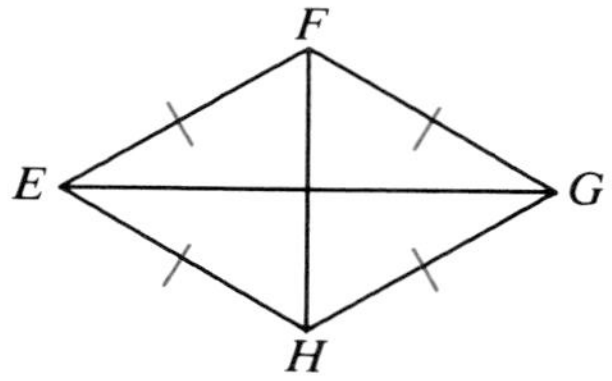

9. In the figure, quad. $ABCD$ is a ▱. $\overrightarrow{AX}$, $\overrightarrow{BX}$, $\overrightarrow{CZ}$, and $\overrightarrow{DZ}$ bisect the angles of the ▱. Let $m\angle 1 = n$. Then
   $m\angle ADC =$ __?__, $m\angle DAB =$ __?__, $m\angle 2 =$ __?__,
   $m\angle 3 =$ __?__, $m\angle 4 =$ __?__
   You can show, similarly, that the measure of each of the other three angles of quad. $WXYZ$ is __?__. Complete the statement of a theorem we have proved: When the bisectors of the angles of a parallelogram are drawn, __?__.

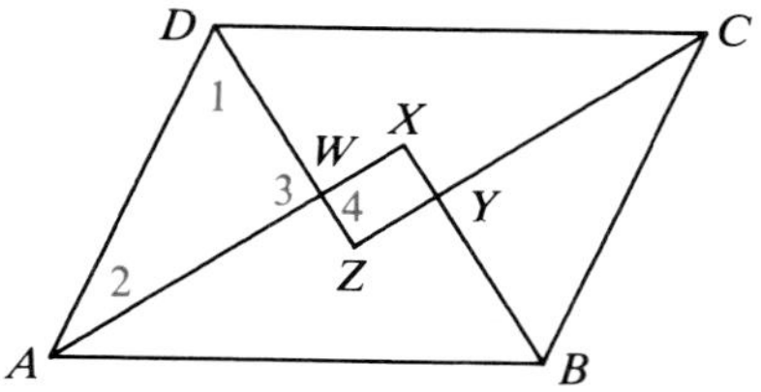

10. Draw a rectangle and bisect its angles. What name best describes the quadrilateral formed?

Written Exercises

Copy the chart. Then place check marks in the appropriate spaces.

	Property	Parallelogram	Rectangle	Rhombus	Square
A **1.**	Opp. sides are $\parallel$.				
2.	Opp. sides are $\cong$.				
3.	Opp. $\angle$s are $\cong$.				
4.	A diag. forms two $\cong$ $\triangle$s.				
5.	Diags. bisect each other.				
6.	Diags. are $\cong$.				
7.	Diags. are $\perp$.				
8.	A diag. bisects two $\angle$s.				
9.	All $\angle$s are rt. $\angle$s.				
10.	All sides are $\cong$.				

11. Explain why an equiangular quadrilateral must be a rectangle.

12. Explain why a quadrilateral that is a regular polygon must be a square.

$\overline{HT}$ is an altitude of $\triangle HOW$. M and N are the midpoints of $\overline{WH}$ and $\overline{OH}$.

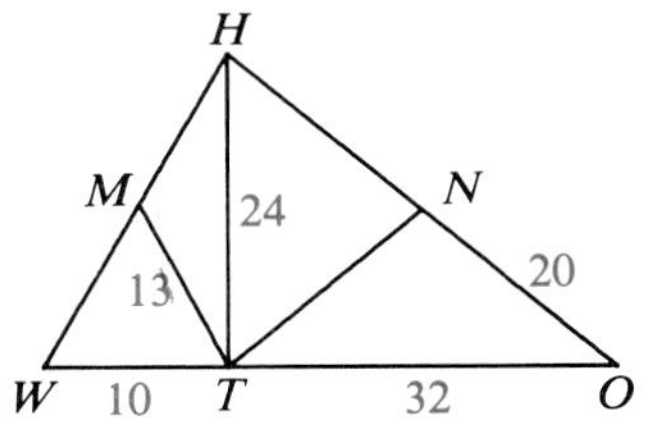

13. $MW =$ _?_ **14.** $MH =$ _?_

15. $NT =$ _?_ **16.** $HO =$ _?_

17. Given: $\square WXYZ$;
$m\angle 1 = 90$
Prove: $WXYZ$ is a rectangle.

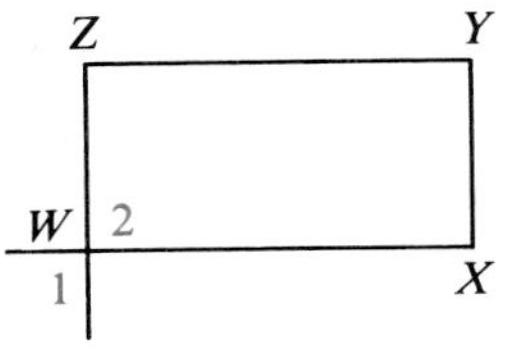

18. Given: $\square ABCD$; $DC = BN$;
$\angle 3 \cong \angle 4$
Prove: $ABCD$ is a rhombus.

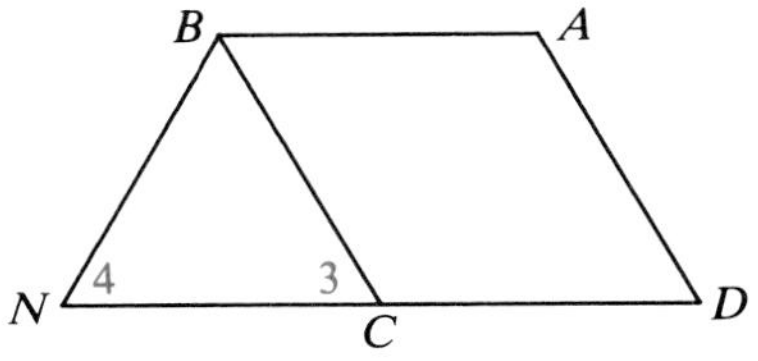

19. Given: Rhombus $ABCD$
Prove: $\angle 1 \cong \angle 2$

20. Given: Rhombus $ABCD$; $\overline{EF} \parallel \overline{AC}$
Prove: $\overline{EF} \perp \overline{DB}$

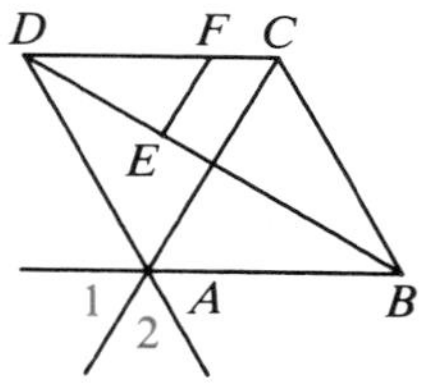

B 21. Given: Rectangle $QRST$; $\square RKST$
Prove: $\triangle QSK$ is isos.

22. Given: Rectangle $QRST$; $\square RKST$; $\square JQST$
Prove: $\overline{JT} \cong \overline{KS}$

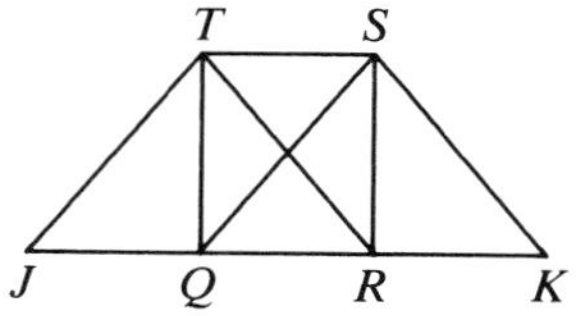

23. Using the figure below, write a complete proof of Theorem 4–9. (*Hint:* Prove $\triangle TQR \cong \triangle SRQ$.)

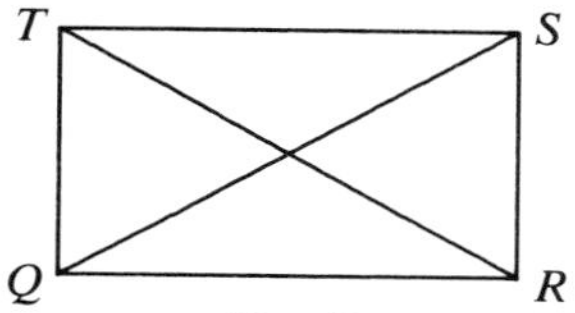

Ex. 23

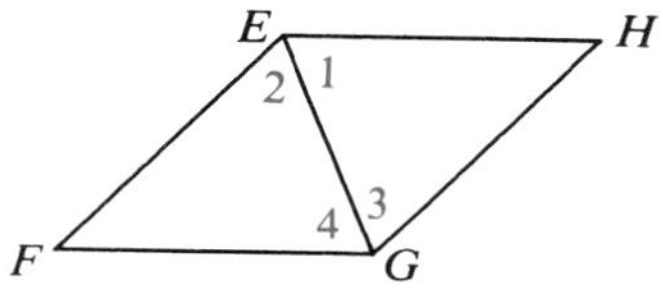

Ex. 24

24. Using the figure above, write a complete proof of Theorem 4–11 for one diagonal of the rhombus. (Note that a proof for the other diagonal would be similar, step-by-step.)

25. Prove: If the diagonals of a parallelogram are perpendicular, then the parallelogram is a rhombus.

26. Prove: If the diagonals of a parallelogram are congruent, then the parallelogram is a rectangle.

In the figure, $m\angle VOZ = 90$.
$\overline{OW}$ is an altitude of $\triangle VOZ$.
$\overline{OX}$ bisects $\angle VOZ$.
$\overline{OY}$ is a median of $\triangle VOZ$.
Find the measures of the four numbered angles.

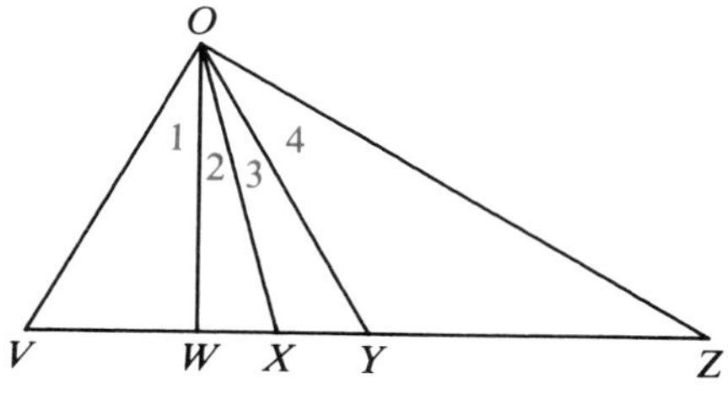

27. $m\angle Z = 30$

28. $m\angle Z = k$

C 29. **a.** It is known that two sides of a quadrilateral are parallel and that one diagonal bisects an angle. Does that quadrilateral have to be special in other ways? If so, write a proof. If not, draw a convincing figure.

b. Repeat part (a) with stronger conditions: It is known that two sides are parallel and that one diagonal bisects two angles of the quadrilateral.

30. Draw a regular pentagon $ABCDE$. Let X be the intersection of $\overline{AC}$ and $\overline{BD}$. What special kind of quadrilateral is $AXDE$? Write a paragraph proof.

31. Given: Rectangle $RSTW$;
equilateral ⚠s YWT and STZ
What is true of $\triangle RYZ$?
Write a paragraph proof.

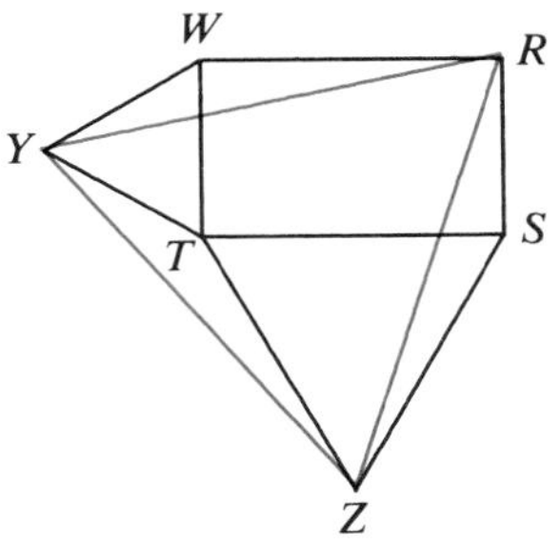

4-4 Trapezoids

A quadrilateral with exactly one pair of parallel sides is called a **trapezoid.** The parallel sides are called **bases;** the other sides are **legs.**

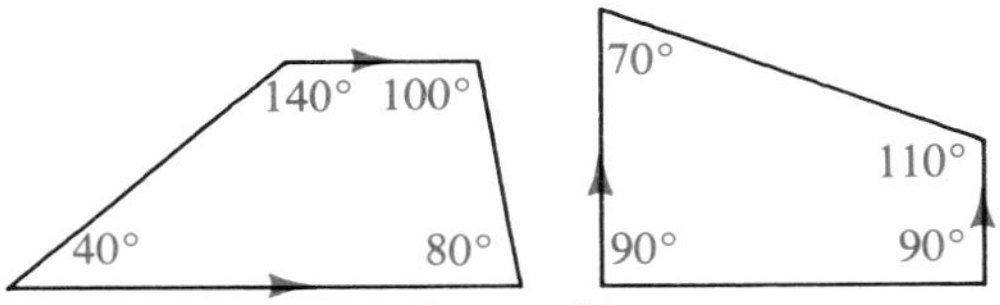

Trapezoids

A trapezoid with congruent legs is called **isosceles.** As the figure below suggests, both pairs of *base angles* of an isosceles trapezoid are congruent.

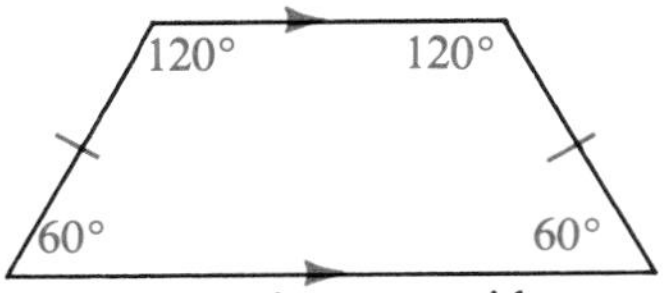

Isosceles trapezoid

A trapezoidal shape can be seen in the photograph. Is it isosceles?

Theorem 4-15

Base angles of an isosceles trapezoid are congruent.

Given: Trapezoid $ABCD$; $\overline{AD} \cong \overline{BC}$

Prove: $\angle A \cong \angle B$; $\angle ADC \cong \angle BCD$

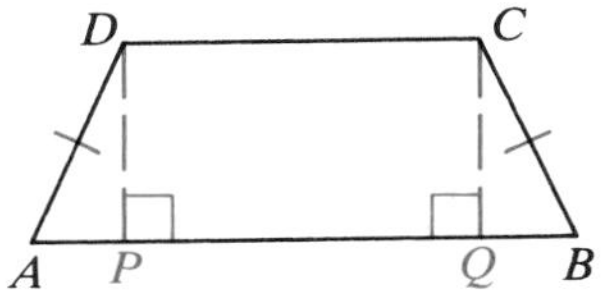

Plan for Proof: Draw segments from D and C perpendicular to $\overleftrightarrow{AB}$. Show that $\overline{DP} \cong \overline{CQ}$, rt. $\triangle APD \cong$ rt. $\triangle BQC$, and $\angle A \cong \angle B$. Since $\angle ADC$ is supp. to $\angle A$, and $\angle BCD$ is supp. to $\angle B$, it follows that $\angle ADC \cong \angle BCD$.

The **median** of a trapezoid is the segment that joins the midpoints of the legs. $\overline{MN}$ is the median of trapezoid *TRAP*.

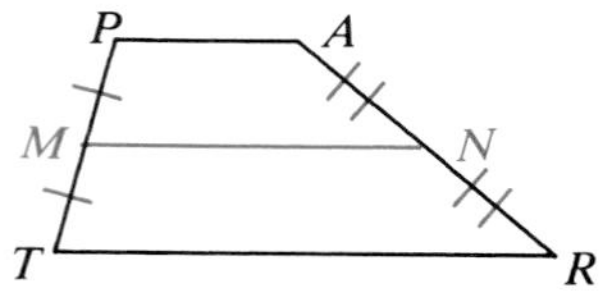

The next theorem could be proved now. Instead you will see a shorter, more appealing proof in Chapter 11.

Theorem 4-16

The median of a trapezoid
(1) is parallel to the bases;
(2) has a length equal to half the sum of the lengths of the bases.

Example A trapezoid and its median are shown. Find the value of x.

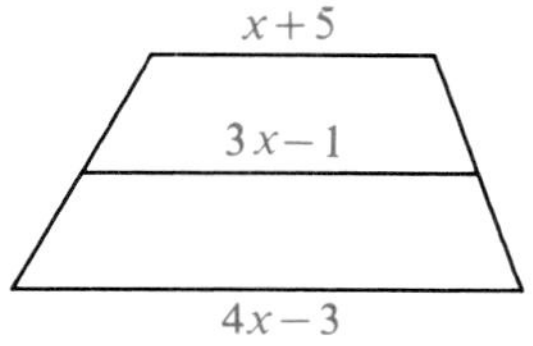

Solution

$$3x - 1 = \frac{(x + 5) + (4x - 3)}{2}$$
$$6x - 2 = x + 5 + 4x - 3$$
$$6x - 2 = 5x + 2$$
$$x = 4$$

In the diagrams below, $\overline{MN}$ joins the midpoints of two segments. As you study the figures from left to right, notice that the upper base of trapezoid *TRAP* becomes shorter and shorter. Finally, $\overline{PA}$ shrinks to a single point and trapezoid *TRAP* becomes $\triangle TRA$.

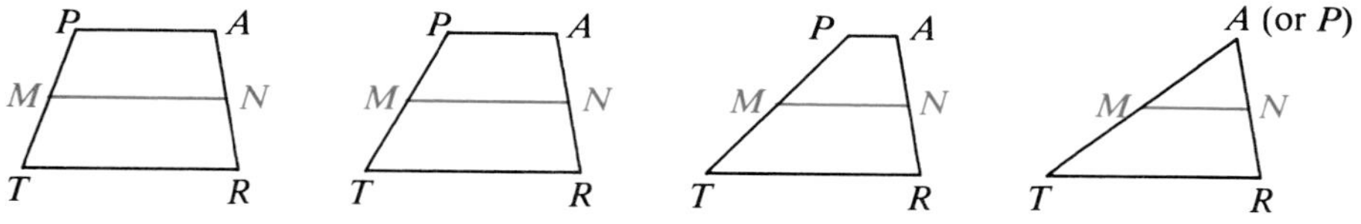

If you think of the last figure as a trapezoid with $PA = 0$, then Theorem 4-16 suggests the following:

(1) $\overline{MN} \parallel \overline{TR}$
(2) $MN = \frac{1}{2}(TR + 0) = \frac{1}{2}TR$

Theorem 4-17, which states these properties for $\triangle TRA$, will be proved in Chapter 11.

Theorem 4-17

The segment that joins the midpoints of two sides of a triangle
(1) is parallel to the third side;
(2) has a length equal to half the length of the third side.

Classroom Exercises

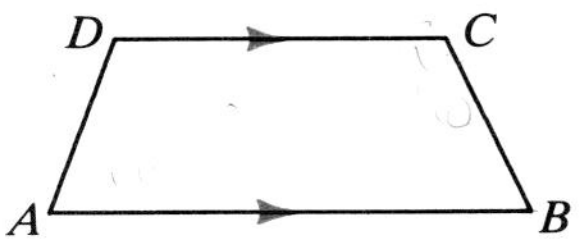

1. In trapezoid $ABCD$, $m\angle A = 70$ and $m\angle C = 120$. Then $m\angle B = \underline{\ ?\ }$ and $m\angle D = \underline{\ ?\ }$.
2. Suppose trapezoid $ABCD$ is isosceles and that $m\angle A = 3j$. Find the measures of $\angle B$, $\angle C$, and $\angle D$ in terms of j.

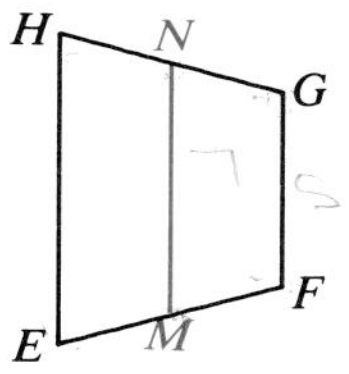

$\overline{MN}$ is the median of trapezoid *EFGH*.

3. If $HN = 4$ and $EM = 6$, $NG = \underline{\ ?\ }$ and $EF = \underline{\ ?\ }$.
4. If $HE = 16$ and $GF = 10$, $MN = \underline{\ ?\ }$.
5. If $GF = 5$ and $NM = 7$, $HE = \underline{\ ?\ }$.
6. If $HE = 12k$ and $NM = 9k$, $GF = \underline{\ ?\ }$.

M, N, and T are the midpoints of the sides of $\triangle XYZ$.

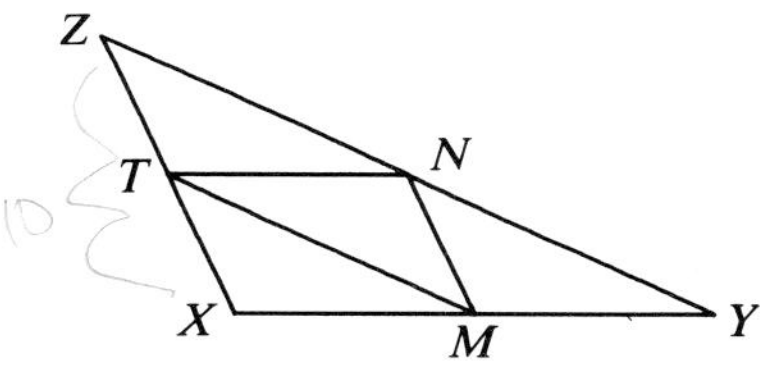

Exs. 7–12

7. If $XZ = 10$, $MN = \underline{\ ?\ }$.
8. If $TN = 7$, $XY = \underline{\ ?\ }$.
9. If $ZN = 8$, $TM = \underline{\ ?\ }$.
10. If $XY = k$, $TN = \underline{\ ?\ }$.
11. Suppose $XY = 10$, $YZ = 14$, and $XZ = 8$. What are the lengths of the three sides of
    a. $\triangle TNZ$? **b.** $\triangle MYN$? **c.** $\triangle XMT$? **d.** $\triangle NTM$?
12. State a theorem suggested by Exercise 11.

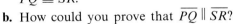

Draw the trapezoid described. If a trapezoid can't be drawn, explain why not.

13. With two right angles
14. With bases shorter than the legs
15. With congruent bases
16. With three acute angles

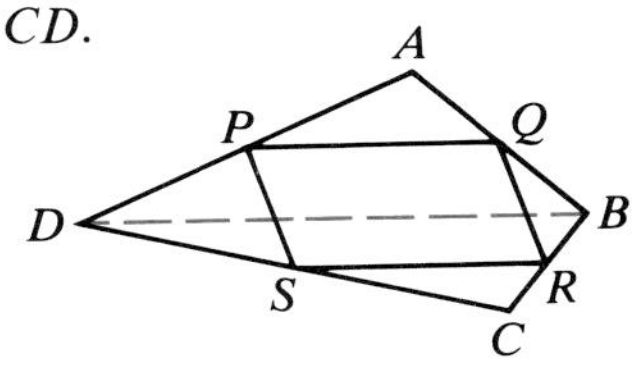

17. P, Q, R, and S are the midpoints of the sides of quad. $ABCD$.
    a. Explain how diagonal $\overline{BD}$ helps you prove that $\overline{PQ} \cong \overline{SR}$.
    b. How could you prove that $\overline{PQ} \parallel \overline{SR}$?
    c. State the theorem that tells you quad. $PQRS$ is a parallelogram.

Written Exercises

Points A, B, E, and F are the midpoints of $\overline{XC}$, $\overline{XD}$, $\overline{YC}$, and $\overline{YD}$. Complete.

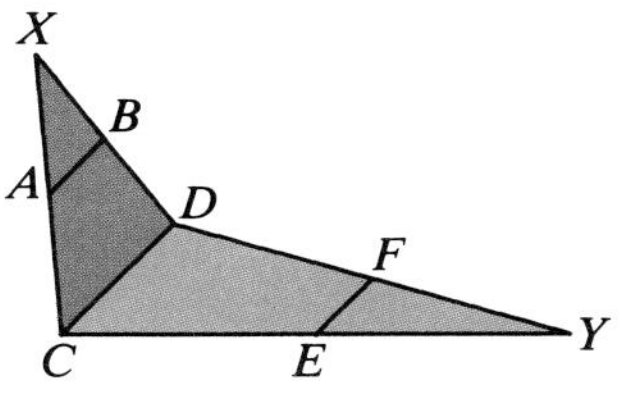

A

1. If $CD = 24$, $AB = \underline{\ ?\ }$ and $EF = \underline{\ ?\ }$.
2. If $AB = k$, $CD = \underline{\ ?\ }$ and $EF = \underline{\ ?\ }$.
3. If $AB = 5x - 8$ and $EF = 3x$, $x = \underline{\ ?\ }$.
4. If $CD = 8x$ and $AB = 3x + 2$, $x = \underline{\ ?\ }$.

$\overline{MN}$ **is the median of trap.** ***PQRS.*** **Complete the table.**

	SP	*MN*	*RQ*
5.	9	?	13
6.	12	14	?
7.	3.4	?	5.2
8.	?	$4\frac{1}{2}$	$5\frac{3}{4}$

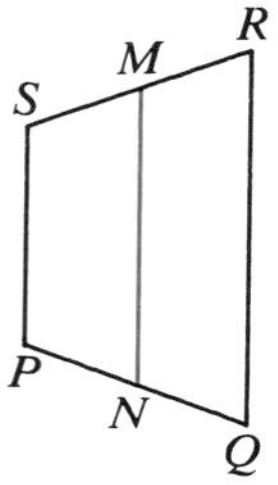

Each figure shows a trapezoid and its median. Find the value of ***x*****.**

9.

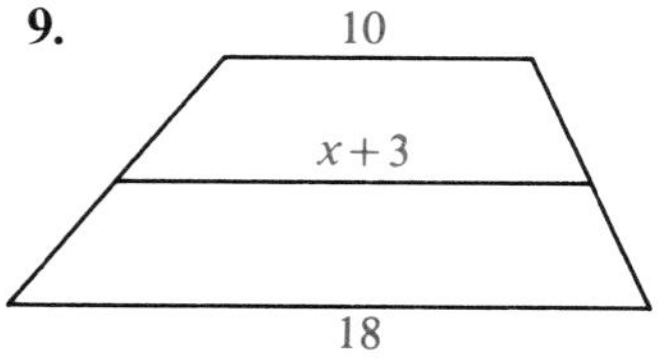

10.

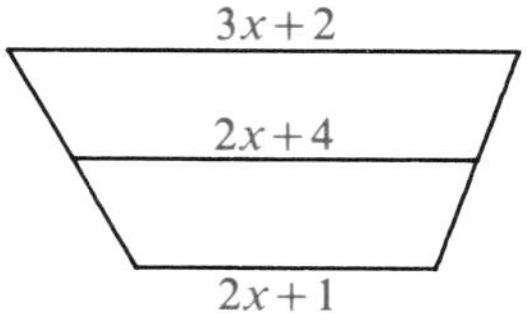

11.

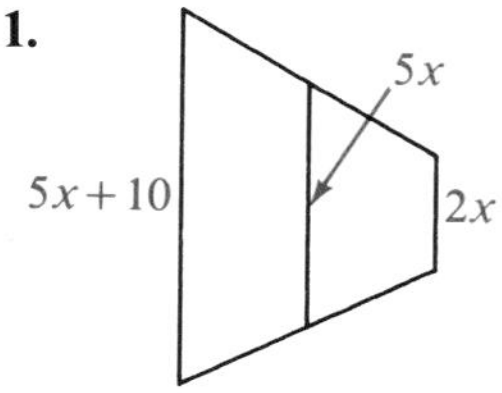

In Exercises 12–16: ***TA = AB = BC*** **and** ***TD = DE = EF.***

12. Compare lengths *AD* and *BE*. (*Hint:* Think of $\triangle TBE$.)

13. Compare the lengths *AD*, *BE*, and *CF*. (*Hint:* Think of trap. *CFDA*.)

14. If $AD = 7$, then $BE =$ _?_ and $CF =$ _?_.

15. If $BE = 26$, then $AD =$ _?_ and $CF =$ _?_.

16. If $AD = x$ and $BE = x + 6$, then $x =$ _?_ and $CF =$ _?_ (numerical answers).

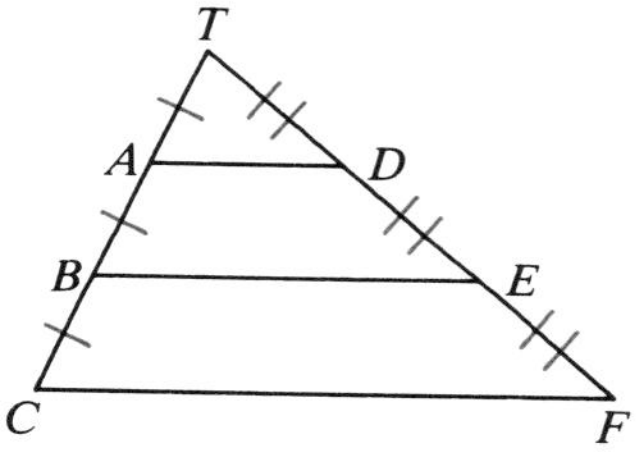

Exs. 12–20

B **17.** If $AD = x + 3$, $BE = x + y$, and $CF = 36$, then $x =$ _?_ and $y =$ _?_.

18. If $AD = x + y$, $BE = 20$, and $CF = 4x - y$, then $CF =$ _?_ (numerical answer).

19. Tony makes up a problem for the figure, setting $AD = 5$ and $CF = 17$. Katie says, "You can't do that." Explain.

20. Mike makes up a problem for the figure, setting $AD = 2x + 1$, $BE = 4x + 2$, and $CF = 6x + 3$ and asking for the value of x. This time Katie says, "Anybody can do that problem." Explain.

Draw a quadrilateral of the type named. Join, in order, the midpoints of the sides. What special kind of quadrilateral do you get?

21. Rhombus

22. Rectangle

23. Trapezoid

24. Isosceles trapezoid

25. Given: ▱$PQRS$;
M is the midpoint of $\overline{QR}$.
Prove: $MX = \frac{1}{2}PQ$

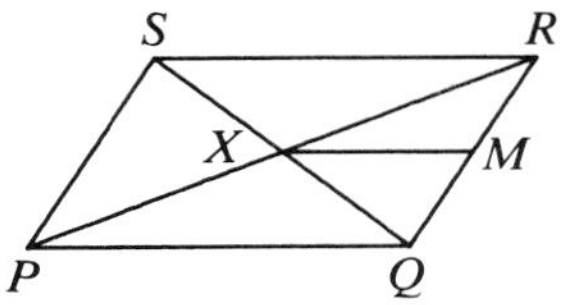

26. Given: ▱$WXYZ$;
$\angle 1 \cong \angle 2$
Prove: $WXEZ$ is an isos. trap.

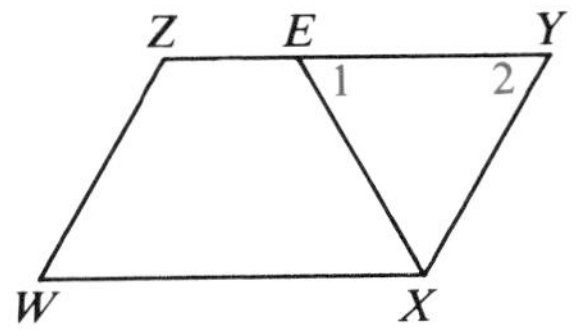

27. Write a proof of Theorem 4–15, following the plan on page 173.

28. Write a proof of Theorem 4–15, using the method suggested by the diagram shown below.

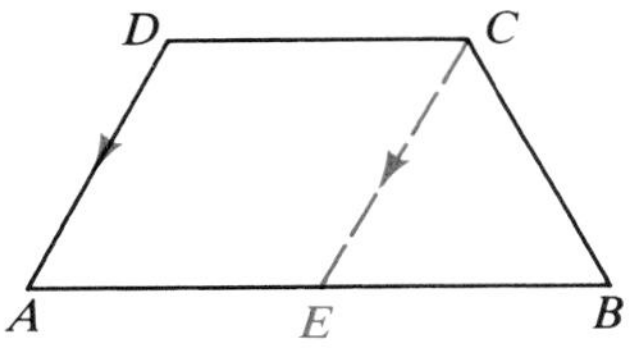

Ex. 28

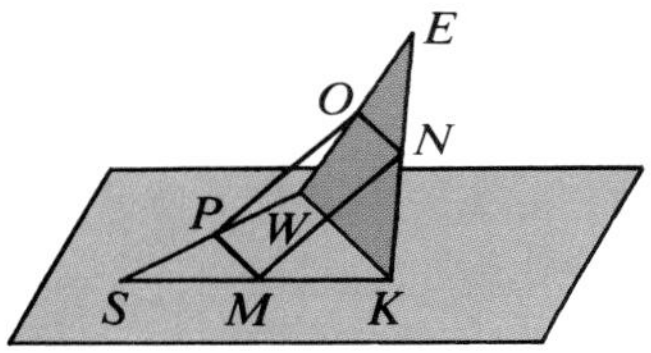

Ex. 29

29. A *skew quadrilateral* $SKEW$ is shown. M, N, O, and P are the midpoints of $\overline{SK}$, $\overline{KE}$, $\overline{WE}$, and $\overline{SW}$. Explain why $PMNO$ is a parallelogram.

30. Discover, state, and prove a theorem about the diagonals of an isosceles trapezoid.

31. Prove that a line drawn through the midpoint of one leg of a trapezoid and parallel to the bases bisects the other leg.

32. $DC = 6$ and $AB = 16$.
Find ME, FN, and EF.

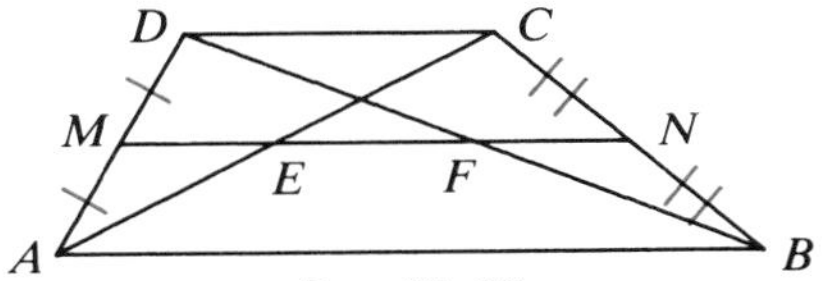

Exs. 32, 33

C 33. $DC = 3x$, $AB = 2x^2$, and $EF = 7$.
Find the value of x.

34. Prove that the perpendicular bisector of one base of an isosceles trapezoid is also the perpendicular bisector of the other base of the trapezoid.

35. State and prove the converse of the theorem you discovered in Exercise 30. (*Hint:* Draw auxiliary lines as in the Plan for Proof for Theorem 4–15 on page 173.)

★36. $\overline{VE}$, $\overline{VF}$, $\overline{VG}$, $\overline{EF}$, $\overline{FG}$, and $\overline{GE}$ are congruent. J, K, L, and M are the midpoints of $\overline{EF}$, $\overline{VF}$, $\overline{VG}$, and $\overline{EG}$. What name best describes $JKLM$? Explain.

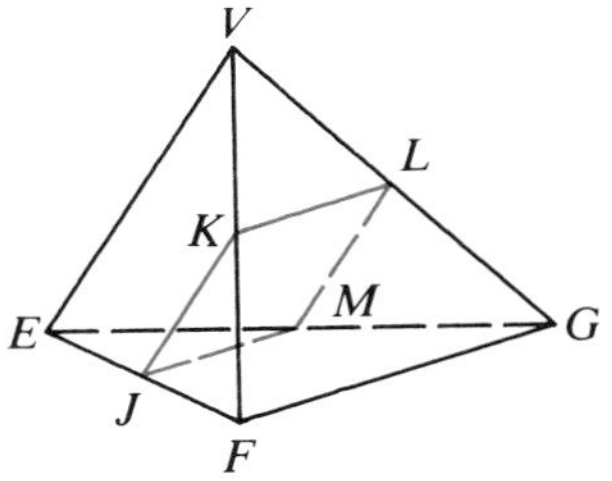

Self-Test 1

The diagonals of $\square ABCD$ intersect at Z. Tell whether each statement *must be*, *may be*, or *cannot be* true.

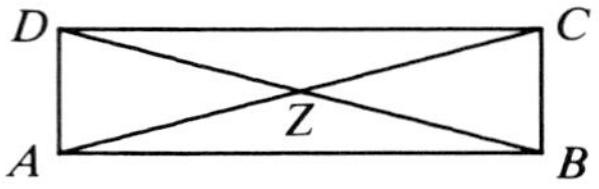

1. $\overline{AC} \cong \overline{BD}$
2. $\overline{DZ} \cong \overline{BZ}$
3. $\overline{AD} \parallel \overline{BC}$
4. $m\angle DAB = 85$ and $m\angle BCD = 95$

Quad. *WXYZ* must be a special figure to meet the conditions stated. Write the best name for that special quadrilateral.

5. $\overline{WX} \cong \overline{YZ}$ and $\overline{WX} \parallel \overline{YZ}$
6. $\overline{WX} \parallel \overline{YZ}$ and $\overline{WX} \not\cong \overline{YZ}$
7. $\overline{WX} \cong \overline{YZ}$, $\overline{XY} \cong \overline{ZW}$, and diag. $\overline{WY} \cong$ diag. $\overline{XZ}$
8. Diagonals $\overline{WY}$ and $\overline{XZ}$ are congruent and are perpendicular bisectors of each other.

In the figure, $\overline{TS} \parallel \overline{JL} \parallel \overline{QR}$.

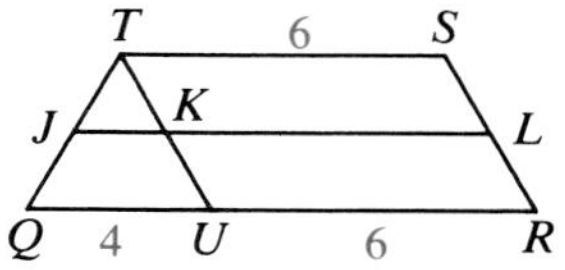

9. Name four trapezoids in the figure.
10. If $\overline{JL}$ is the median of trap. $QRST$, then $JL =$ __?__ and $JK =$ __?__.

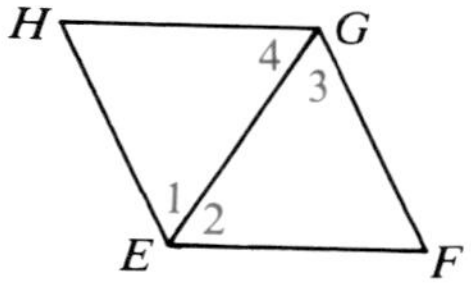

11. Given: $\overline{EH} \cong \overline{FG}$; $\overline{EH} \parallel \overline{FG}$
    Prove: $\overline{EF} \cong \overline{HG}$; $\overline{EF} \parallel \overline{HG}$
12. Given: $\angle 1 \cong \angle 2 \cong \angle 3 \cong \angle 4$
    Prove: Quad. $EFGH$ is a rhombus.

B I O G R A P H I C A L   N O T E

Benjamin Banneker

Benjamin Banneker (1731–1806) was a noted American scholar, largely self-taught, who became both a surveyor and an astronomer.

As a surveyor, Banneker was a member of the commission that defined the boundary line and laid out the streets of the District of Columbia.

As an astronomer, he accurately predicted a solar eclipse in 1789. From 1791 until his death he published almanacs containing not only information on astronomy and tide tables, but also such diverse subjects as insect life and medicinal products. Banneker's almanacs included ideas that were far ahead of their time, for example, the formation of a Department of the Interior and even an organization like the United Nations.

Geometric Inequalities

Objectives

1. Write indirect proofs.
2. State and apply the inequality relations for one triangle.
3. State and apply the inequality relations for two triangles.

4-5 Indirect Proofs

Until now the proofs you have written have been *direct* proofs. Sometimes it is difficult or even impossible to find a direct proof, but easy to reason indirectly. In an **indirect proof** you begin by assuming temporarily that the conclusion is not true. Then you reason logically until you reach a contradiction of the hypothesis or another known fact.

Example 1

Given: n is an integer and n^2 is even.

Prove: n is even.

Proof:

Assume temporarily that n is not even. Then n is odd, and

$$\begin{aligned} n^2 &= n \times n \\ &= \text{odd} \times \text{odd} = \text{odd}. \end{aligned}$$

But this contradicts the given information that n^2 is even. Therefore the temporary assumption that n is not even must be false. It follows that n is even.

Example 2

Prove: The bases of a trapezoid have unequal lengths.

Given: Trap. $PQRS$ with bases $\overline{PQ}$ and $\overline{SR}$

Prove: $PQ \neq SR$

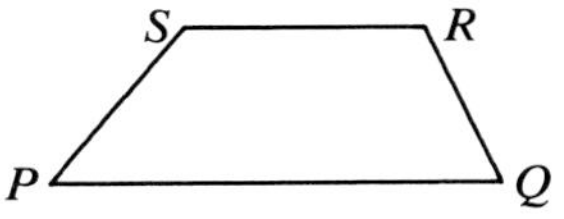

Proof:

Assume temporarily that $PQ = SR$. Then $\overline{PQ} \cong \overline{SR}$. Since $\overline{PQ} \parallel \overline{SR}$, by the definition of a trapezoid, quadrilateral $PQRS$ has two sides that are both congruent and parallel. Therefore quadrilateral $PQRS$ must be a parallelogram, and $\overline{PS}$ must be parallel to $\overline{QR}$. But this contradicts the fact that trapezoid $PQRS$ can have only one pair of parallel sides. Our temporary assumption that $PQ = SR$ must be false. It follows that $PQ \neq SR$.

How to Write an Indirect Proof

1. Assume temporarily that the conclusion is not true.
2. Reason logically until you reach a contradiction of a known fact.
3. Point out that your temporary assumption must be false, and that the conclusion must then be true.

Classroom Exercises

1. An indirect proof is to be used to prove: If $AB = AC$, then $\triangle ABD \cong \triangle ACD$. Which one of the following is the correct way to begin?
Assume temporarily that

$$AB \neq AC.$$

Assume temporarily that

$$\triangle ABD \ncong \triangle ACD.$$

2. Planning to write an indirect proof that $x > 7$, Becky begins by saying, "Assume temporarily that $x < 7$." Using this assumption, she reaches a contradiction. Then she concludes that $x > 7$. Criticize her conclusion.

3. Wishing to prove that l and m are skew lines, John begins an indirect proof by assuming temporarily that l and m are intersecting lines. Using this assumption, he reaches a contradiction and concludes that l and m are skew. Criticize the proof.

What is the first sentence of an indirect proof of the statement shown?

4. $\triangle ABC$ is equilateral.

5. Doug is a Canadian.

6. $a \geq b$

7. Kim isn't a violinist.

8. $m\angle X > m\angle Y$

9. $\overline{CX}$ isn't a median of $\triangle ABC$.

10. Arrange sentences (a)–(e) in an order that completes the indirect proof of the statement: In a plane, two lines perpendicular to a third line are parallel to each other.

Given: Lines a, b, and t lie in a plane;
$t \perp a$; $t \perp b$
Prove: $a \parallel b$

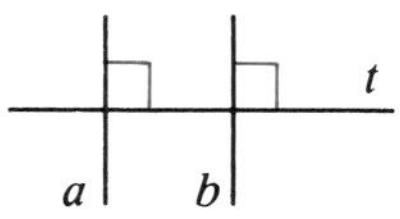

Proof:

(a) Then a intersects b in some point Z.
(b) But this contradicts Theorem 2-9.
(c) It is false that a is not parallel to b, and it follows that $a \parallel b$.
(d) Assume temporarily that a is not parallel to b.
(e) Then there are two lines through Z and perpendicular to t.

Written Exercises

Write indirect proofs in paragraph form.

A

1. Given: $\triangle XYZ$; $m\angle X = 100$
Prove: $\angle Y$ is an acute $\angle$.

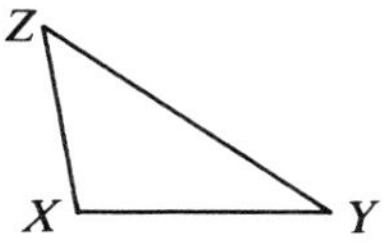

2. Given: Transversal t cuts lines a and b; $m\angle 1 \neq m\angle 2$
Prove: $a \nparallel b$

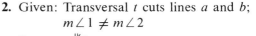

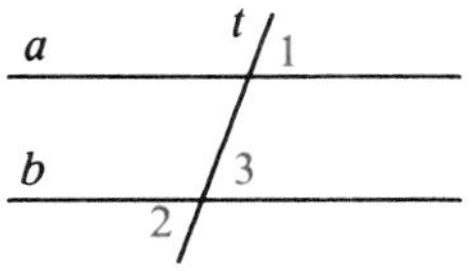

3. Given: $\overline{OJ} \cong \overline{OK}$; $\overline{JE} \ncong \overline{KE}$
Prove: $\overrightarrow{OE}$ doesn't bisect $\angle JOK$.

4. Given: $\angle 1 \cong \angle 2$; $\overline{OJ} \ncong \overline{OK}$
Prove: $\angle J$ and $\angle K$ are not both right angles.

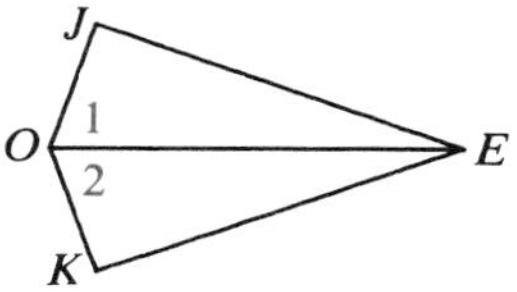

B

5. Given: $\overline{AB} \nparallel \overline{CD}$
Prove: Planes P and Q intersect.

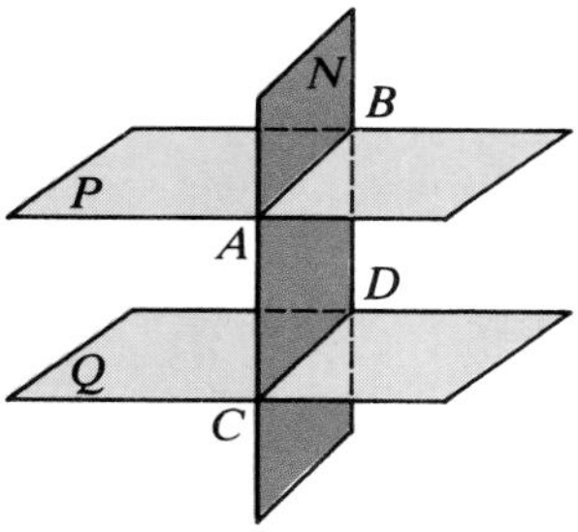

6. Given: $\triangle$s RVT and SVT are equilateral; $\triangle RVS$ is not equilateral.
Prove: $\triangle RST$ is not equilateral.

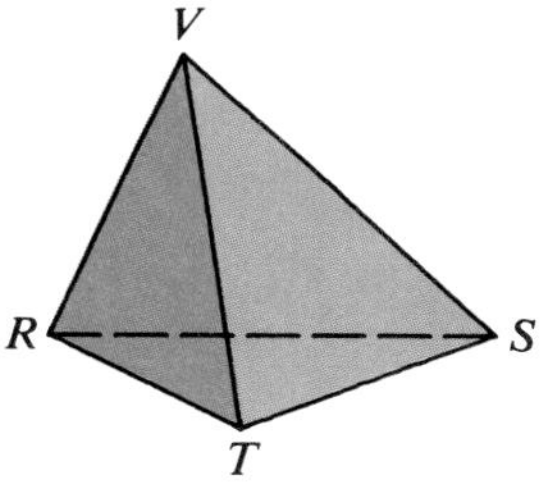

7. Given: Points E, F, G, H; segments $\overline{EF}$, $\overline{FG}$, $\overline{GH}$, $\overline{HE}$; $m\angle EFG = 93$; $m\angle FGH = 70$; $m\angle GHE = 127$; $m\angle HEF = 60$
Prove: E, F, G, and H are not coplanar.

8. Given: $AT = BT = 5$; $CT = 4$
Prove: $\angle ACB$ is not a rt. $\angle$.

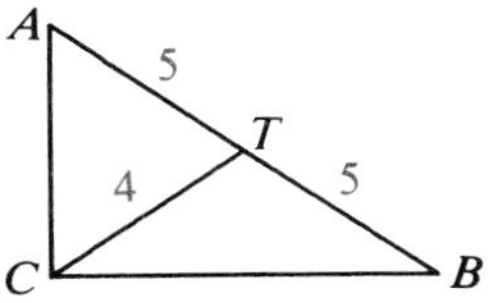

9. Given: Coplanar lines l, m, n; n intersects l in P; $l \parallel m$
Prove: n intersects m.

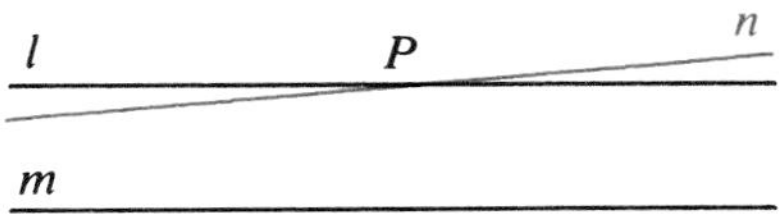

10. Prove that there is no smallest positive number.

11. Prove that a collection of quarters and dimes worth 95¢ must have an odd number of quarters.

12. Prove that no regular polygon has a 155° angle.

13. Prove that the diagonals of a trapezoid do not bisect each other.

C 14. Prove: If r and s are positive integers, then $r + s \geq \sqrt{r^2 + s^2}$.

15. Prove that if two lines are perpendicular to the same plane, then the lines do not intersect.

16. Given: $\overleftrightarrow{RS}$ and $\overleftrightarrow{TW}$ are skew.
    Prove: $\overleftrightarrow{RT}$ and $\overleftrightarrow{SW}$ are skew.

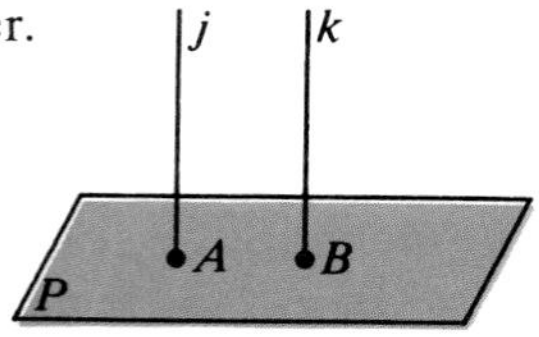

Ex. 15

4-6 Inequalities for One Triangle

From the information given in the figure at the left below you can deduce that $\angle P \cong \angle Q$. From the figure at the right you can deduce that $m\angle S \neq m\angle T$. (If the measures were equal, WS and WT would have to be equal.) The figure suggests a result we will prove in this section:

$$\text{If } TW > SW, \text{ then } m\angle S > m\angle T.$$

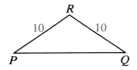

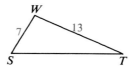

Statements such as $m\angle S \neq m\angle T$ and $m\angle S > m\angle T$ are called *inequalities*. Some of the algebraic properties used in dealing with inequalities are listed below.

Properties of Inequalities

If $a > b$ and $c \geq d$, then $a + c > b + d$.

If $a > b$ and $c > 0$, then $ac > bc$.

If $a > b$ and $b > c$, then $a > c$.

If $a = b + c$, and $c > 0$, then $a > b$.

Here are examples that show how the last property above can be used.

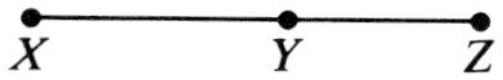

Y is between X and Z.

$XZ = XY + YZ$
Then $XZ > XY$
and $XZ > YZ$.

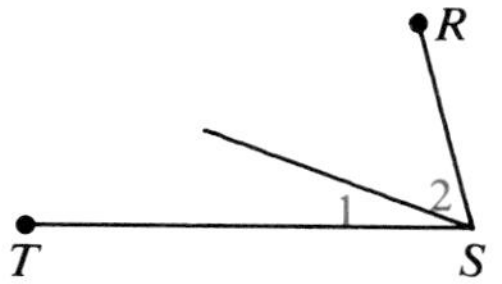

∠1 and ∠2 are adj. ∠s.

$m\angle RST = m\angle 1 + m\angle 2$
Then $m\angle RST > m\angle 1$
and $m\angle RST > m\angle 2$.

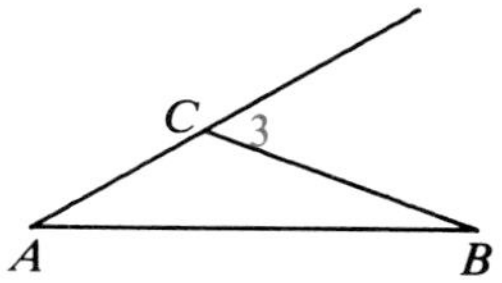

∠3 is an exterior ∠.

$m\angle 3 = m\angle A + m\angle B$
Then $m\angle 3 > m\angle A$
and $m\angle 3 > m\angle B$.

In a proof you could give as a reason for the final sentence in each example above either "a property of inequalities" or "by algebra."

In the proof of the following theorem, look for two of the situations discussed on page 182.

Theorem 4–18

If one side of a triangle is longer than a second side, then the angle opposite the first side is larger than the angle opposite the second side.

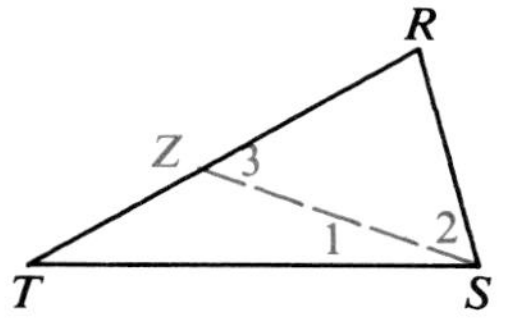

Given: $\triangle RST$; $RT > RS$

Prove: $m\angle RST > m\angle T$

Proof:

By the Ruler Postulate, there is a point Z on $\overline{RT}$ such that $RZ = RS$. Draw $\overline{SZ}$. In isosceles $\triangle RZS$, $m\angle 3 = m\angle 2$.

Since $m\angle RST > m\angle 2$,
$m\angle RST > m\angle 3$ by substitution.

Since $m\angle 3 > m\angle T$,
$m\angle RST > m\angle T$.

Theorem 4–19

If one angle of a triangle is larger than a second angle, then the side opposite the first angle is longer than the side opposite the second angle.

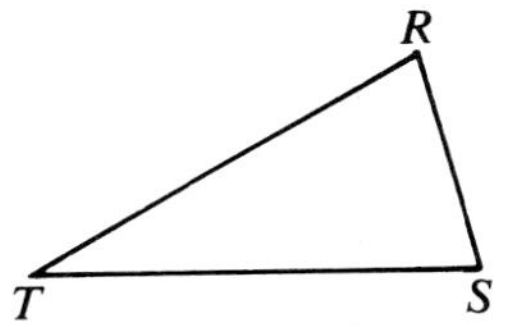

Given: $\triangle RST$; $m\angle S > m\angle T$

Prove: $RT > RS$

Proof:

Assume temporarily that $RT \not> RS$. Then either $RT = RS$ or $RT < RS$.
If $RT = RS$, $m\angle S = m\angle T$.
If $RT < RS$, $m\angle S < m\angle T$ by Theorem 4–18.
In either case we have a contradiction of the fact that $m\angle S > m\angle T$. The property $RT \not> RS$ that we assumed to be true must be false. It follows that $RT > RS$.

Corollary 1

The perpendicular segment from a point to a line is the shortest segment from the point to the line.

Corollary 2

The perpendicular segment from a point to a plane is the shortest segment from the point to the plane.

See Classroom Exercises 17 and 18 for proofs of the corollaries.

Theorem 4-20 The Triangle Inequality

The sum of the lengths of any two sides of a triangle is greater than the length of the third side.

Given: $\triangle ABC$

Prove: (1) $AB + BC > AC$
(2) $AB + AC > BC$
(3) $AC + BC > AB$

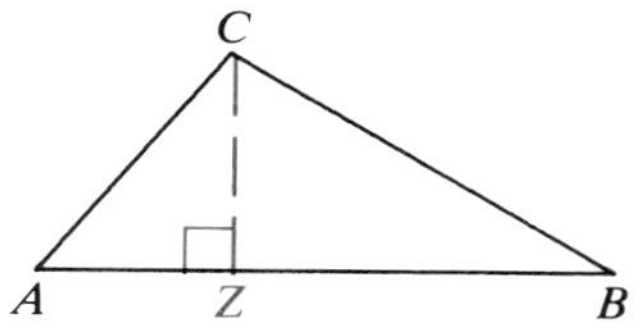

Proof:

One of the sides, say $\overline{AB}$, is the longest side. (Or $\overline{AB}$ is at least as long as each of the other sides.) Then the first two statements to be proved are true. To prove (3), draw a line, $\overleftrightarrow{CZ}$, through C and perpendicular to $\overleftrightarrow{AB}$. (Through a point outside a line, there is exactly one line perpendicular to the given line.) By Corollary 1 of Theorem 4-19, $\overline{AZ}$ is the shortest segment from A to $\overline{CZ}$. Also, $\overline{BZ}$ is the shortest segment from B to $\overline{CZ}$. Thus:

$$AC > AZ \quad \text{and} \quad BC > ZB$$
$$AC + BC > AZ + ZB$$
$$AC + BC > AB$$

Example The lengths of two sides of a triangle are 3 and 5. The length of the third side must be greater than __?__, but less than __?__.

Solution Let x be the length of the third side.

$$x + 3 > 5 \qquad 3 + 5 > x \qquad x + 5 > 3$$
$$x > 2 \qquad 8 > x \qquad x > -2$$

The length of the third side must be greater than 2 but less than 8.

Note that the inequality $x + 5 > 3$ did not give us any useful information. We could have omitted it completely from the solution. (The only inequality that can be omitted is the one in which the length of the third side is added to the longer of the sides with known lengths.)

Classroom Exercises

Name the largest angle and the smallest angle of the triangle.

1.

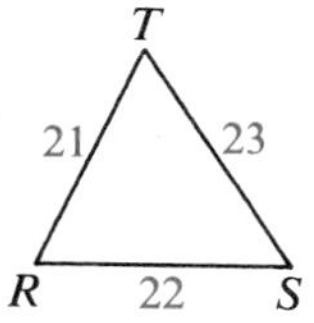

2.

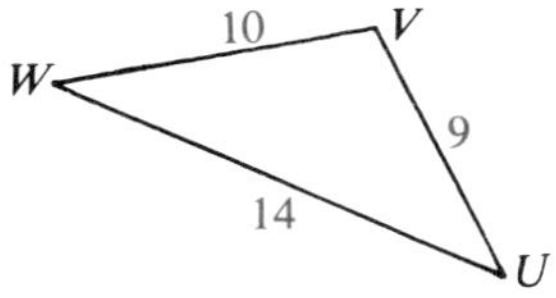

3.

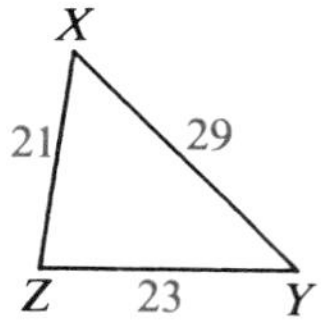

Name the longest side and the shortest side of the triangle.

4. 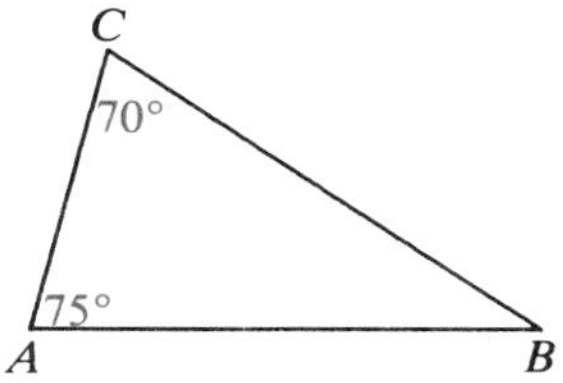

5. 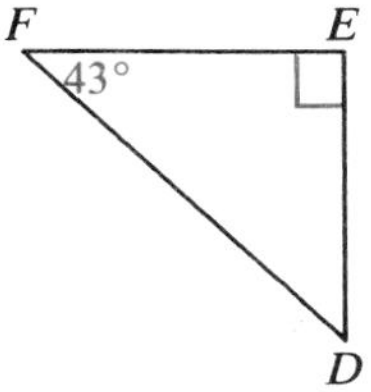

6. 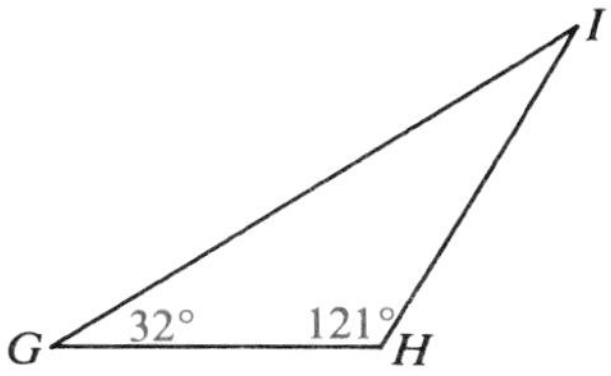

Is it possible for a triangle to have sides with the lengths indicated?

7. 10, 9, 8   **8.** 6, 6, 20   **9.** 7, 7, 14.1

10. 16, 11, 5   **11.** 0.6, 0.5, 1   **12.** 18, 18, 0.06

13. The base of an isosceles triangle has length 10. What can you say about the length of a leg?

14. Two sides of a parallelogram have lengths 10 and 12. What can you say about the lengths of the diagonals?

15. Two sides of a triangle have lengths 15 and 20. The length of the third side can be any number between __?__ and __?__.

16. Suppose you know only that the length of one side of a rectangle is 100. What can you say about the length of a diagonal?

17. Use the figure below to explain how Corollary 1 follows from Theorem 4–19.

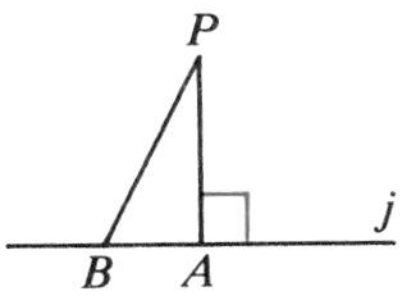

Ex. 17

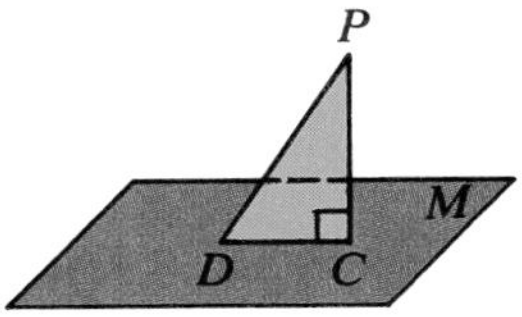

Ex. 18

18. Use the figure, in which $\overline{PC} \perp$ plane M, to explain how Corollary 2 follows from Theorem 4–19 or from Corollary 1.

19. Which is the largest angle of a right triangle? Which is the longest side of a right triangle? Explain.

Written Exercises

The lengths of two sides of a triangle are given. Write the numbers that best complete the statement: The length of the third side must be greater than __?__, but less than __?__.

A **1.** 6, 9   **2.** 15, 13   **3.** 100, 100

4. 2.3, 2.3   **5.** k, $k + 5$ **6.** a, b (where $a > b$)

In Exercises 7–9 the diagrams are not drawn to scale. If each diagram were drawn to scale, which of the numbered angles shown would be the largest?

7.

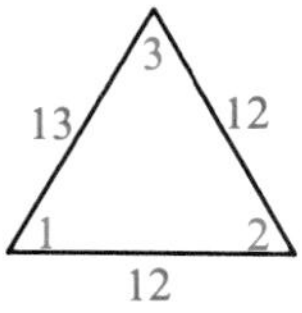

8.

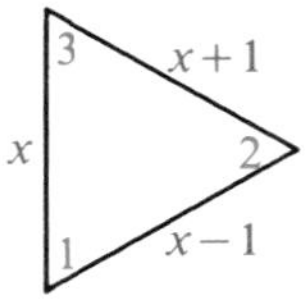

9.

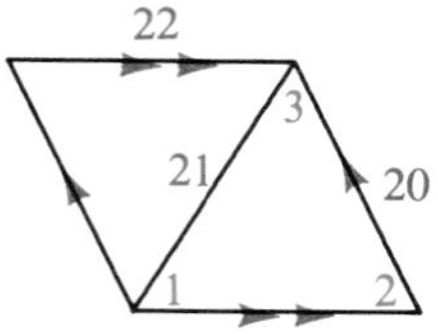

In Exercises 10–14 the diagrams are not drawn to scale. If each diagram were drawn accurately, which segment would be the longest of those shown?

10.

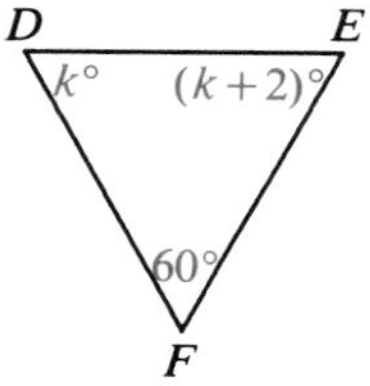

11.

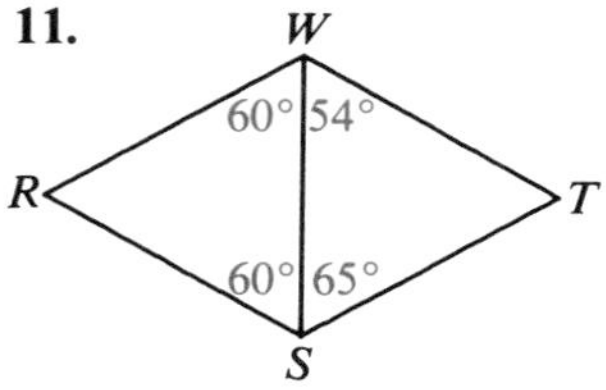

12.

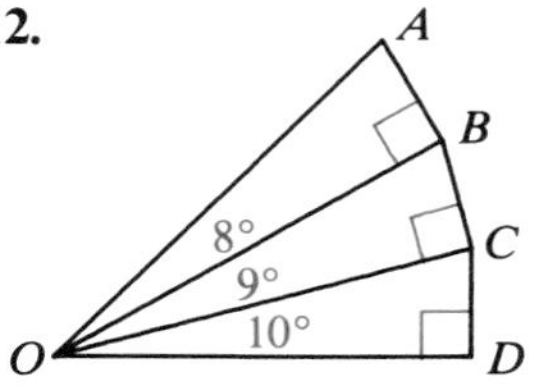

B **13.**

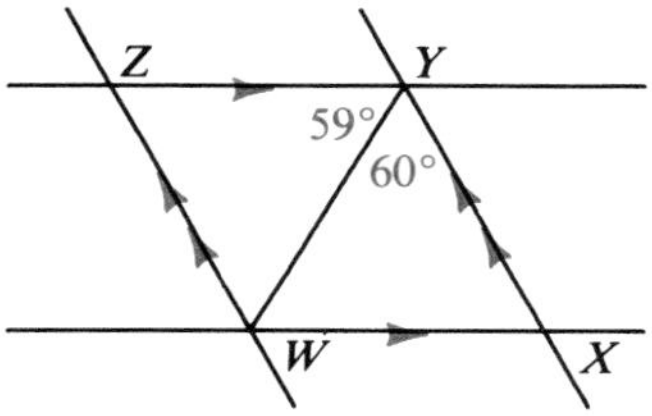

14.

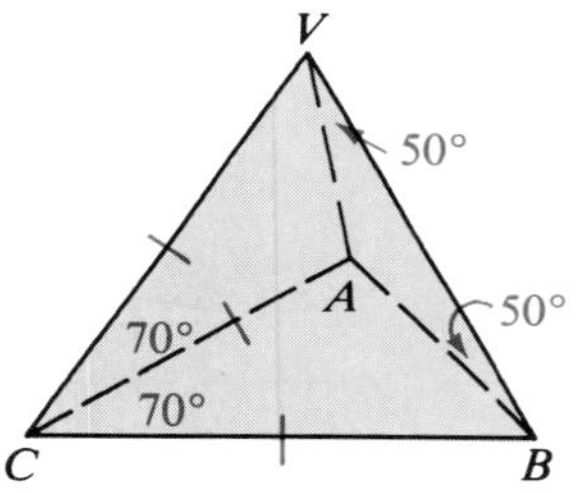

15. Use the lengths a, b, c, d, and e to complete:

$\underline{\ ?\ } > \underline{\ ?\ } > \underline{\ ?\ } > \underline{\ ?\ } > \underline{\ ?\ }$

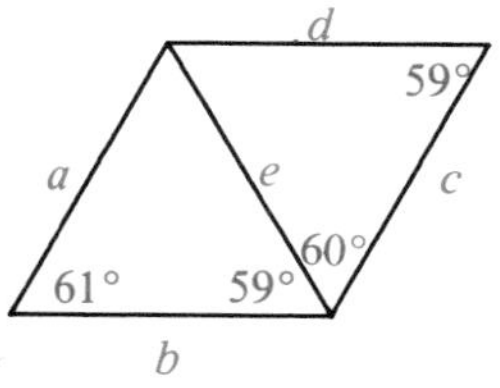

16. Use $m\angle 1$, $m\angle 2$, and $m\angle 3$ to complete:

$\underline{\ ?\ } > \underline{\ ?\ } > \underline{\ ?\ }$

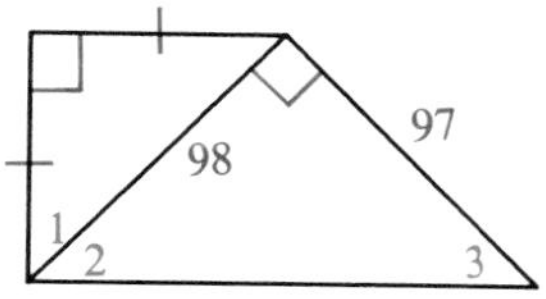

17. The diagram is not drawn to scale. Use $m\angle 1$, $m\angle 2$, $m\angle X$, $m\angle Y$, and $m\angle XZY$ to complete:

$\underline{\ ?\ } > \underline{\ ?\ } > \underline{\ ?\ } > \underline{\ ?\ } > \underline{\ ?\ }$

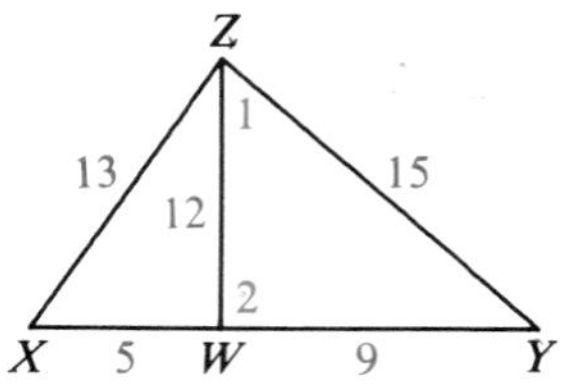

18. Given: Quad. $ABCD$
Prove: $AB + BC + CD + DA > 2(AC)$

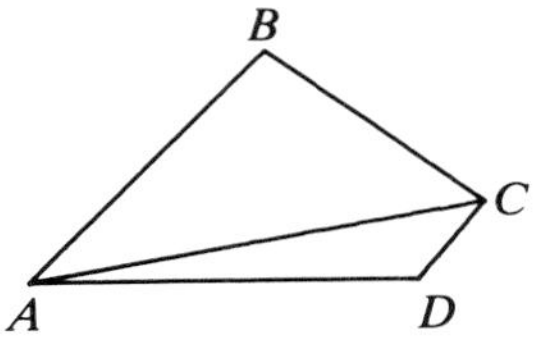

19. Given: $\square EFGH$; $EF > FG$
Prove: $m\angle 1 > m\angle 2$

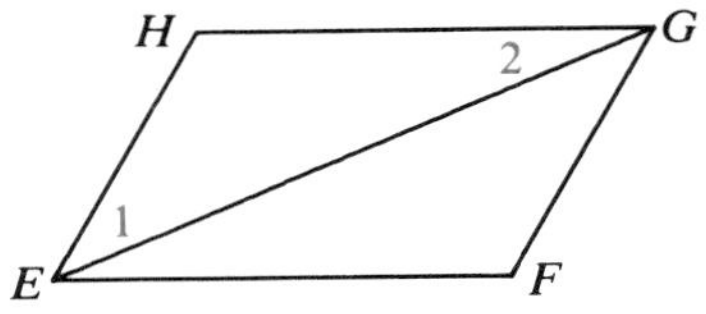

C 20. Discover, state, and prove in paragraph form a theorem that compares the perimeter of a quadrilateral with the sum of the lengths of the diagonals.

21. Prove that the sum of the lengths of the medians of a triangle is greater than half the perimeter.

22. If you replace "medians" with "altitudes" in Exercise 21, can you prove the resulting statement? Explain.

In Exercises 23 and 24, begin your proofs by drawing auxiliary lines.

23. Discover, state, and prove a theorem about how the length of the longest side of a quadrilateral compares with the lengths of the other three sides.

24. Prove: If P is any point inside $\triangle XYZ$, then $ZX + ZY > PX + PY$.

Application

FINDING THE SHORTEST PATH

The owners of pipeline l plan to construct a pumping station at a point S on line l in order to pipe oil to two major customers, located at A and B. To minimize the cost of constructing lines from S to A and B, they wish to locate S along l so that the distance $SA + SB$ is as small as possible.

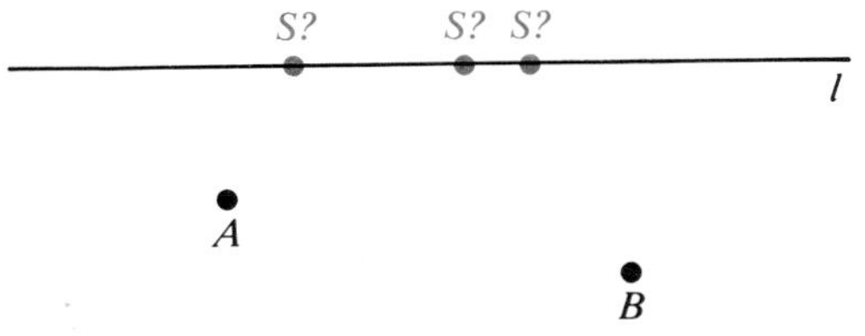

The construction engineer uses the following method to locate S:

1. Draw a line through B perpendicular to l, intersecting l at point P.
2. On this perpendicular, locate point C so that $PC = PB$.
3. Draw $\overline{AC}$.
4. Locate S at the intersection of $\overline{AC}$ and l.

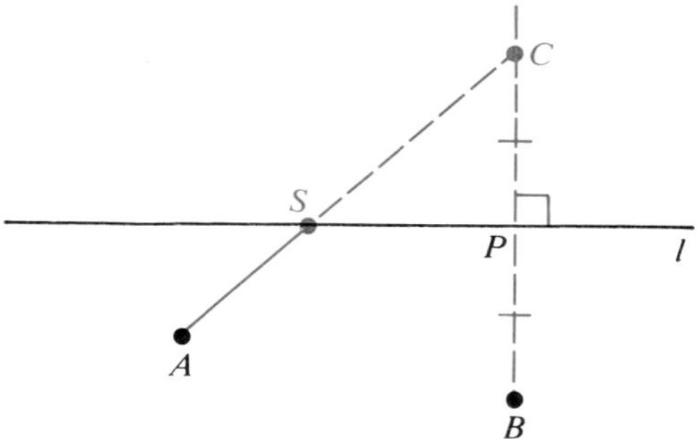

The diagram shows the path of the new pipelines through the pumping station located at S, and an alternative path going through a different point, X, on l. You can use Theorem 4–20 (the Triangle Inequality) to show that if X is any point on l other than S, then

$$AX + XB > AS + SB.$$

So any alternative path is longer than the path through S.

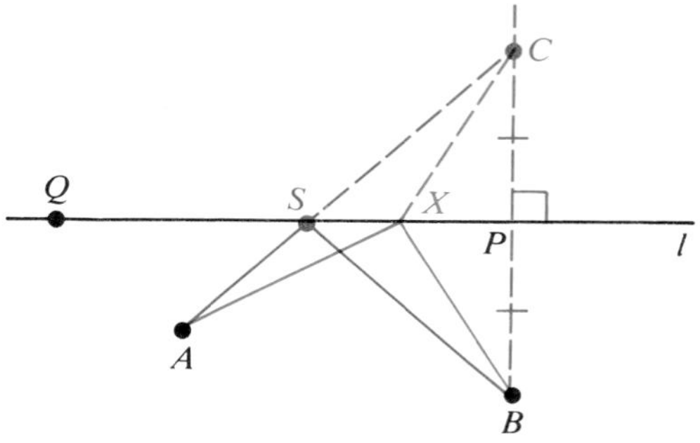

Exercises

Exercises 1–7 outline the proof that the construction given for *S* yields the shortest total length for the pipelines serving *A* and *B*. Supply a reason for each statement.

1. l is the perpendicular bisector of $\overline{BC}$.

2. $SC = SB$

3. $AS + SC = AC$

4. $AS + SB = AC$

5. $XC = XB$

6. $AX + XC > AC$

7. $AX + XB > AS + SB$

8. Point X is shown between points S and P. However, point X could also be to the right of P or to the left of S. Draw a diagram illustrating each of these cases. Will the same proof still work?

9. The construction for S is sometimes called a *solution by reflection,* since it involves *reflecting* point B in line l. (See Chapter 12 for more about reflections.) Show that $\overline{AS}$ and $\overline{SB}$, like reflected paths of light, make congruent angles with l. That is, prove that $\angle QSA \cong \angle PSB$. (*Hint:* Draw your own diagram, omitting the part of the diagram shown in blue.)

4-7 Inequalities for Two Triangles

Begin with two matched pairs of sticks joined loosely at B and E. Open them so that $m\angle B > m\angle E$ and you will find that $AC > DF$. Conversely, if you open them so that $AC > DF$, it will follow that $m\angle B > m\angle E$. Two theorems, one of which is surprisingly difficult to prove, are suggested by these examples.

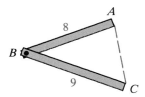

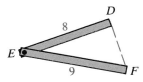

Theorem 4-21 SAS Inequality Theorem

If two sides of one triangle are congruent to two sides of another triangle, but the included angle of the first triangle is greater than the included angle of the second, then the third side of the first triangle is longer than the third side of the second triangle.

Given: $\overline{BA} \cong \overline{ED}$; $\overline{BC} \cong \overline{EF}$;
$m\angle B > m\angle E$

Prove: $AC > DF$

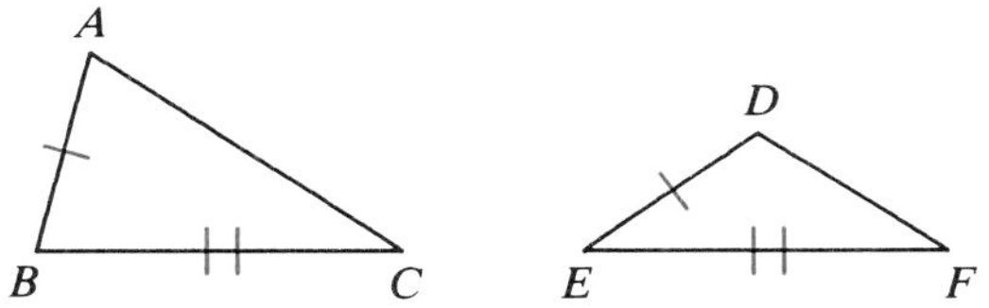

Outline of Proof:

Draw $\overrightarrow{BZ}$ so that $m\angle ZBC = m\angle E$. On $\overrightarrow{BZ}$ take point X so that $BX = ED$.
Then either X is on $\overline{AC}$ or X is not on $\overline{AC}$.
In either case, $\triangle XBC \cong \triangle DEF$ by SAS, and $XC = DF$.

Case 1: X is on $\overline{AC}$.
$AC > XC$ (Seg. Add. Post. and algebra)
$AC > DF$ (Substitution Property, using the equation in red above)

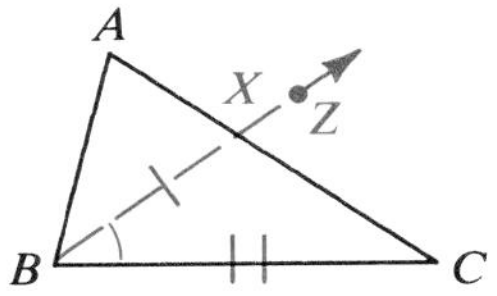

Case 1

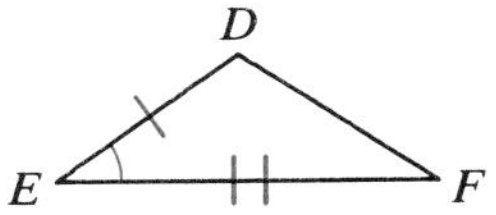

Case 2: X is not on $\overline{AC}$.
Draw the bisector of $\angle ABX$, intersecting $\overline{AC}$ at Y.
Draw $\overline{XY}$ and $\overline{XC}$.
$BA = ED = BX$ (Why?)
Since $\triangle ABY \cong \triangle XBY$ (SAS), $AY = XY$.
$XY + YC > XC$ (Why?)
$AY + YC > XC$ (Why?), or $AC > XC$
$AC > DF$ (Substitution Property)

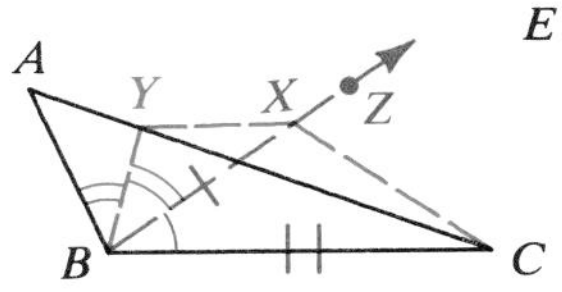

Case 2

Theorem 4-22 SSS Inequality Theorem

If two sides of one triangle are congruent to two sides of another triangle, but the third side of the first triangle is longer than the third side of the second, then the included angle of the first triangle is larger than the included angle of the second.

Given: $\overline{BA} \cong \overline{ED}$; $\overline{BC} \cong \overline{EF}$;
$AC > DF$

Prove: $m\angle B > m\angle E$

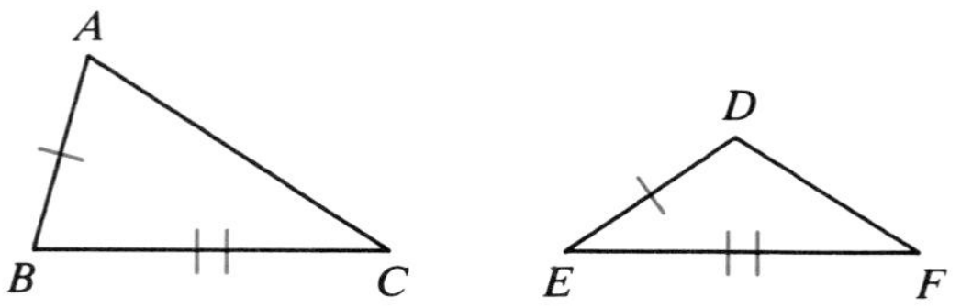

Proof:

Assume temporarily that $m\angle B \not> m\angle E$.
Then either $m\angle B = m\angle E$ or $m\angle B < m\angle E$.

Case 1: If $m\angle B = m\angle E$, then $\triangle ABC \cong \triangle DEF$ by the SAS Postulate, and $AC = DF$.

Case 2: If $m\angle B < m\angle E$, then $AC < DF$ by the SAS Inequality Theorem.

In either case we have a contradiction of the fact that $AC > DF$. What we temporarily assumed to be true, that $m\angle B \not> m\angle E$, must be false. It follows that $m\angle B > m\angle E$.

Example What can you deduce from the given information?

a. $VX = VZ$ and $XY > ZY$

b. $VZ = XZ$ and $m\angle XZY > m\angle VZY$

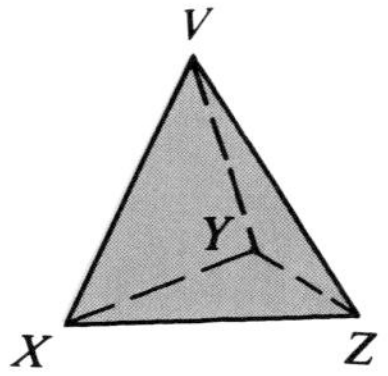

Solution

a. Apply the SSS Inequality Theorem to $\triangle XVY$ and $\triangle ZVY$ to get $m\angle XVY > m\angle ZVY$.
Use the fact that $XY > ZY$ in $\triangle XYZ$ to get $m\angle YZX > m\angle YXZ$.

b. Apply the SAS Inequality Theorem to $\triangle XZY$ and $\triangle VZY$ to get $XY > VY$.
Apply the fact that $XY > VY$ in $\triangle XYV$ to get $m\angle XVY > m\angle VXY$.

Classroom Exercises

In Exercises 1–10, some facts are given. What can you deduce?

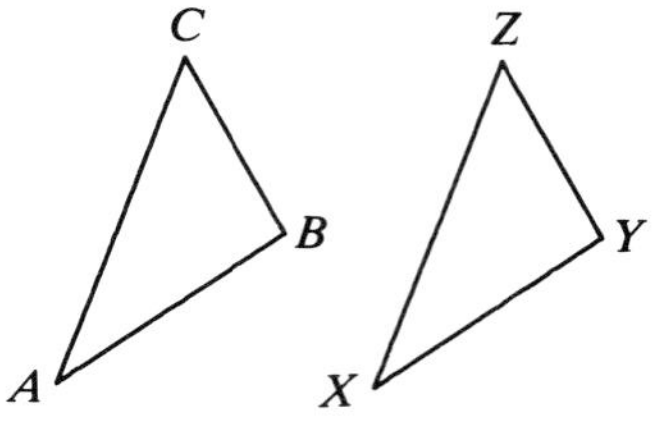

1. $AB = XY$; $AC = XZ$; $m\angle A > m\angle X$
2. $CA = ZX$; $CB = ZY$; $m\angle C < m\angle Z$
3. $BA = YX$; $BC = YZ$; $AC > XZ$
4. $AB = XY$; $AC = XZ$; $\angle B \cong \angle Y$, $m\angle C < m\angle Z$
5. $AB < AC$
6. $m\angle Y > m\angle Z$

7.

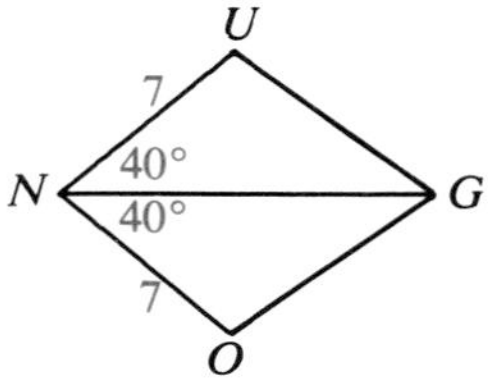

8.

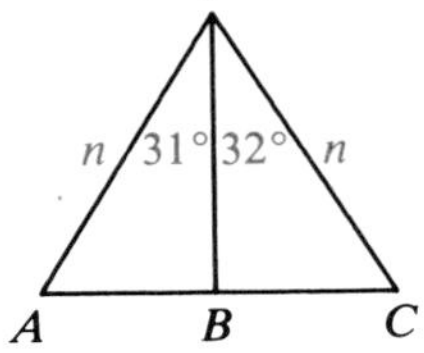

9.

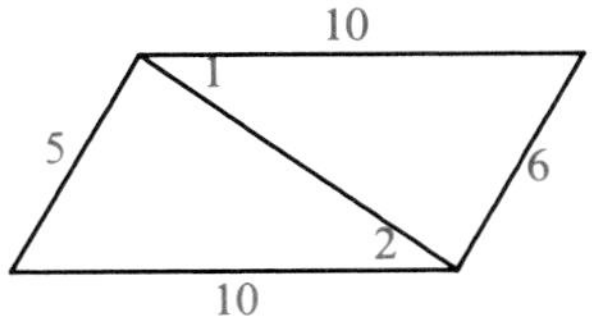

10.

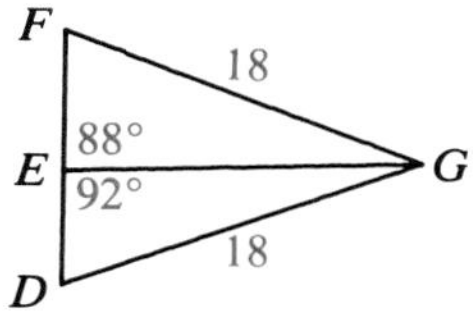

Written Exercises

Choose from the words *always, sometimes,* and *never* to complete the sentence in the best way.

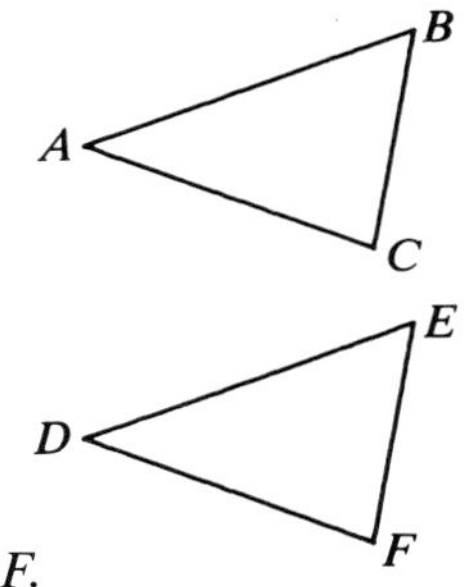

A

1. If $AB = DE$, $BC = EF$, and $AC = DF$, then $m\angle B$ is __?__ equal to $m\angle E$.
2. If $m\angle B > m\angle A$, then AB is __?__ greater than AC.
3. If $m\angle A > m\angle D$ and $m\angle C > m\angle F$, then $m\angle B$ is __?__ greater than $m\angle E$.
4. If $AB > DE$ and $AC > DF$, then BC is __?__ greater than EF.
5. If $AC = DF$, $AB = DE$, and $CB > FE$, then $m\angle A$ is __?__ greater than $m\angle D$.
6. If $BA = ED$, $BC = EF$, and $m\angle B < m\angle E$, then AC is __?__ less than DF.
7. If $m\angle A > m\angle D$ and $m\angle C > m\angle F$, then AC is __?__ less than DF.
8. If $m\angle A = m\angle D$, $m\angle C < m\angle F$, $BA = ED$, and $BC = EF$, then AC is __?__ less than DF.
9. If $AB > DE$, $CA = FD$, and $CB = FE$, then $m\angle C$ is __?__ greater than $m\angle F$.
10. If $AC = AB = DF = DE$ and $m\angle C > m\angle F$, then CB is __?__ greater than FE.
11. In the figure below, which is longer, $\overline{XY}$ or $\overline{XZ}$?
12. Which is larger, $\angle 1$ or $\angle 2$?

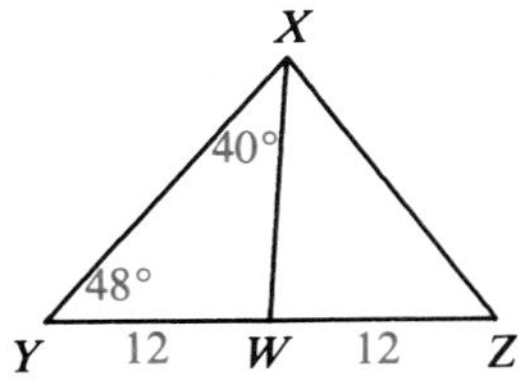

Ex. 11

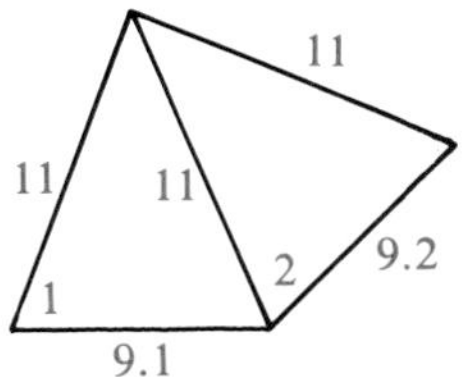

Ex. 12

B **13.** If $RT = RS$ and $VT > VS$, which angle is larger, $\angle VRT$ or $\angle VRS$?

14. If $VR = VS = VT$ and $m\angle RVS > m\angle RVT > m\angle SVT$, which angle of $\triangle RST$ is the largest one?

15. If $TS = TR$ and $m\angle VTS > m\angle VTR$, which angle is larger, $\angle VSR$ or $\angle VRS$?

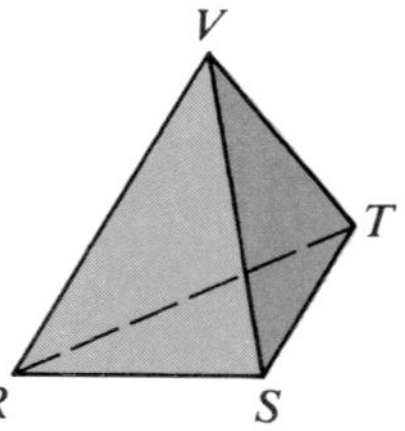

Exs. 13–15

16. Given: $TU = US = SV$
Prove: $ST > SV$

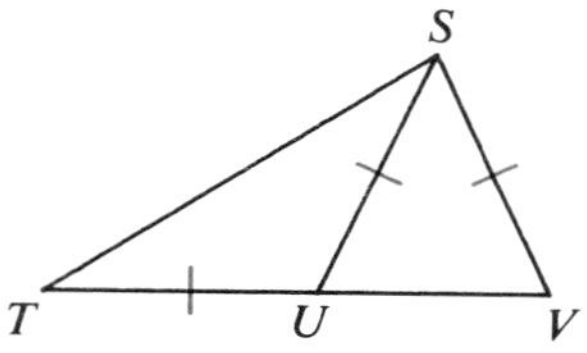

17. Given: $AB > AC$; $BD = EC$
Prove: $BE > CD$

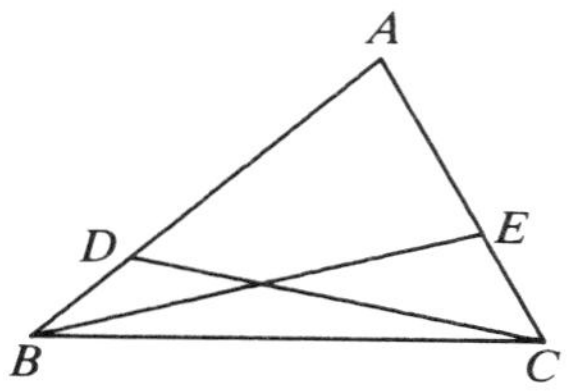

18. Prove: In a parallelogram, the diagonal that joins the vertices of the smaller angles is the longer diagonal.

In Exercises 19–22 write paragraph proofs.

19. Given: M is the midpoint of $\overline{BC}$.
Discover and prove something about BA and CA.

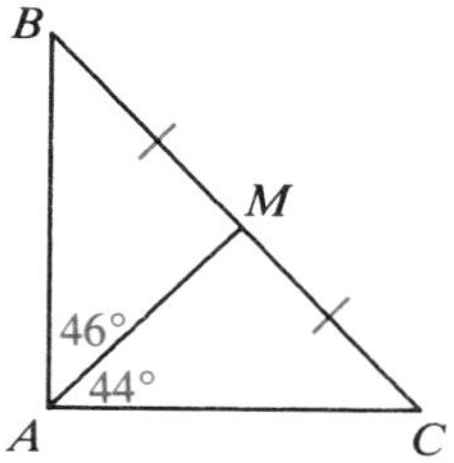

20. Given: Plane P bisects $\overline{XZ}$ at Y; $WZ > WX$.
Discover and prove something about the figure.

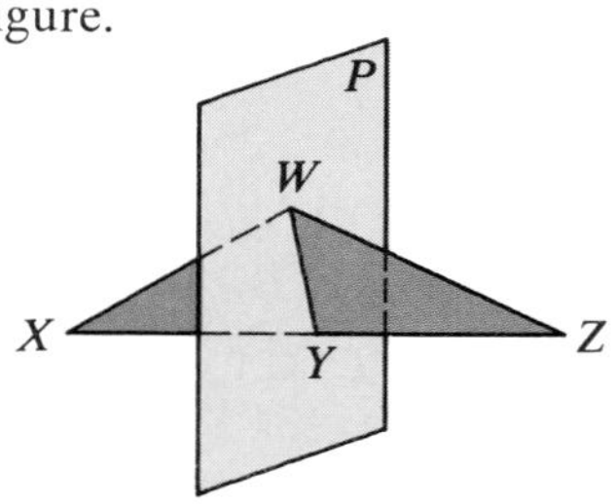

C **21.** Given: $PA = PC = QC = QB$
Prove: $m\angle PCA < m\angle QCB$

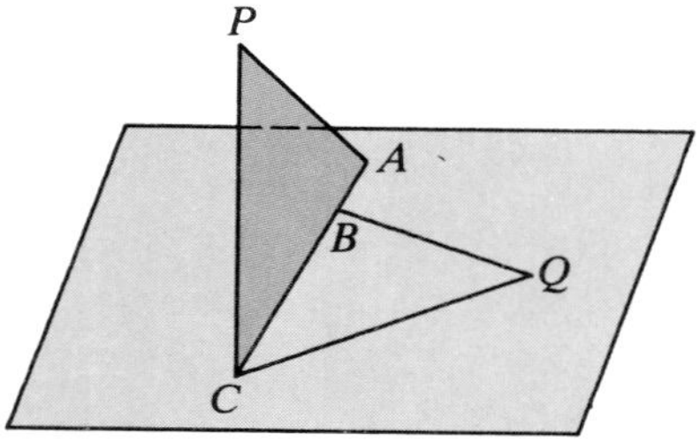

22. Given: $\overline{DE} \perp$ Plane M; $EK > EJ$
Prove: $DK > DJ$
(*Hint:* On $\overline{EK}$ take Z so that $EZ = EJ$.)

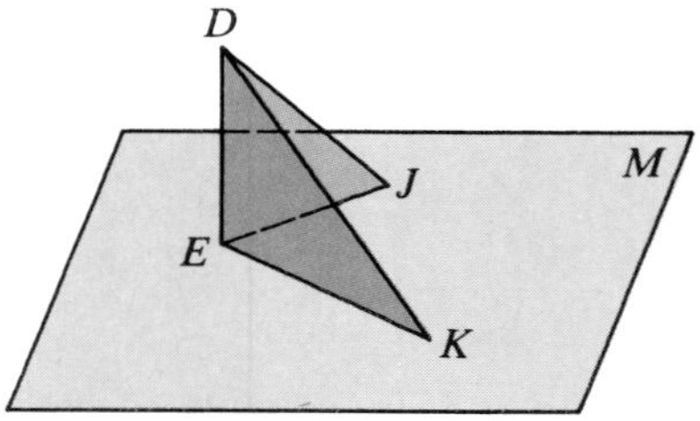

23. In the solid shown, five edges have equal lengths, but $\overline{VC}$ has a different length. What can you say about the largest angles of the twelve angles shown if

a. $\overline{VC}$ is the longest edge?

b. $\overline{VC}$ is the shortest edge?

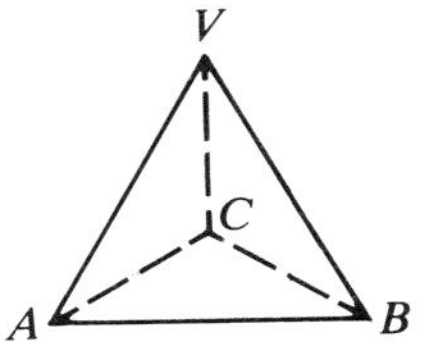

COMPUTER KEY-IN

If you break a stick into three pieces, do you think it is always possible to join the pieces end-to-end to form a triangle?

It's easy to see that if the sum of the lengths of any two of the pieces is less than or equal to that of the third, a triangle can't be formed.

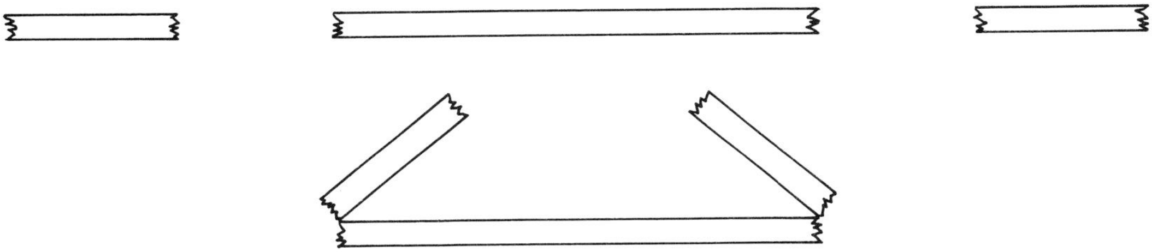

By an experiment, your class can estimate the probability that three pieces of broken stick will form a triangle. Suppose everyone in your class has a stick 1 unit long and breaks it into three pieces. If there are thirty people in your class and eight people are able to form a triangle with their pieces, we estimate that the probability of forming a triangle is about $\frac{8}{30}$.

Of course, this experiment is not very practical. You can get much better results by having a computer simulate the breaking of many, many sticks, as in the program in BASIC on the next page.

In lines 30 and 40 of the following program, you tell the computer how many sticks you want to break. Each stick is 1 unit long, and the computer breaks each stick by choosing two random numbers x and y between 0 and 1. These numbers divide the stick into three lengths r, s, and t.

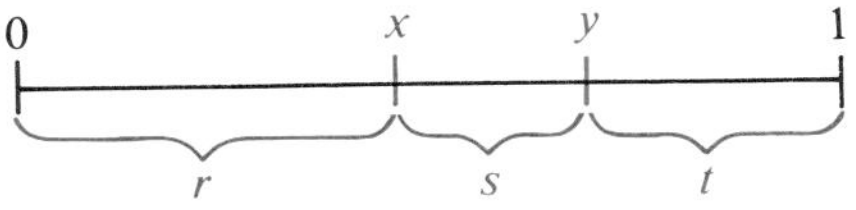

The computer then keeps count of the number of sticks (N) which form a triangle when broken.

Notice that RND is used in lines 70 and 80. Since usage of RND varies, check this with the manual for your computer and make any necessary changes. The computer print-outs shown in this text use capital letters. The x, y, r, s, and t used in the discussion above appear as X, Y, R, S, T.

```
10 PRINT "SIMULATION--BREAKING STICKS TO MAKE TRIANGLES"
20 PRINT
30 PRINT "HOW MANY STICKS DO YOU WANT TO BREAK";
40 INPUT D
50 LET N = 0
60 FOR I = 1 TO D
70 LET X = RND(1)
80 LET Y = RND(1)
90 IF X > Y THEN 120
100 LET R = X
110 GOTO 130
120 LET R = Y
130 LET S = ABS(X - Y)
140 LET T = 1 - R - S
150 IF R + S <= T THEN 210
160 IF S + T <= R THEN 210
170 IF T + R <= S THEN 210
180 PRINT
190 PRINT R,S,T
200 LET N = N + 1
210 NEXT I
220 LET P = N/D
230 PRINT
240 PRINT "THE EXPERIMENTAL PROBABILITY THAT"
250 PRINT "A BROKEN STICK CAN MAKE A TRIANGLE IS";P
260 END
```

Line Number	Explanation
60–140	These lines simulate the breaking of each stick. When I = 10, for example, the computer is "breaking" the tenth stick.
150–170	Here the computer tests to see whether the pieces of the broken stick can form a triangle. If not, the computer goes on to the next stick (line 210) and the value of N is not affected.
200	If the broken stick has survived the tests of steps 150–170, then the pieces can form a triangle and the value of N is increased by 1 here.
210	Lines 60–210 form a loop that is repeated D times. When I = D, the probability P is calculated and printed (lines 220–250).

Exercises

1. Pick any two numbers x and y between 0 and 1 with $x < y$. With paper and pencil, carry out the instructions in lines 90 through 170 of the program to see how the computer finds r, s, and t and tests to see whether the values can be the lengths of the sides of a triangle. Do the same for a pair x and y with $x > y$.

2. If your computer uses a language other than BASIC, write a similar program for your computer.
3. Run the program several times for $D = 40$.
4. Delete the print statements in lines 180 and 190 and then run the program for large values of D, say 100, 400, 800, and compare your results with those of some classmates. Does the probability that the pieces of a broken stick form a triangle appear to be less than or greater than $\frac{1}{2}$?

Self-Test 2

1. Write the letters (a) to (d) in such an order that the sentences provide an indirect proof of the statement: Through a point outside a line, there is at most one line perpendicular to the given line.

   Given: Point P not on line k

   Prove: There is at most one line through P perpendicular to k.

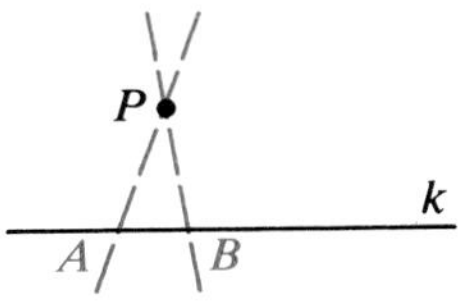

   (a) But this contradicts Corollary 3 of Theorem 2–11: In a triangle, there can be at most one right angle or obtuse angle.
   (b) Then $\angle PAB$ and $\angle PBA$ are right angles, and $\triangle PAB$ has two right angles.
   (c) Thus our temporary assumption is false, and there is at most one line through P perpendicular to k.
   (d) Assume temporarily that there are two lines through P and perpendicular to k at A and B.

Write the correct symbol ($<$, $=$, $>$) to complete the statement.

2. If $ER > EN$, then $m\angle R$ _?_ $m\angle N$.
3. If $\overline{AG} \cong \overline{ER}$, $\overline{AP} \cong \overline{EN}$, and $\angle A \cong \angle E$, then GP _?_ RN.
4. If $\overline{GA} \cong \overline{RE}$, $\overline{GP} \cong \overline{RN}$, and $AP > EN$, then $m\angle G$ _?_ $m\angle R$.
5. If $PG = NR$, $PA = NE$, and $m\angle P < m\angle N$, then GA _?_ RE.

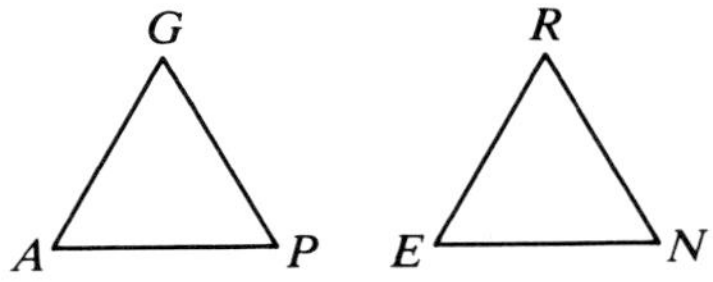

Exs. 2–5

6. The lengths of the sides of a triangle are 5, 6, and x. Then _?_ $< x <$ _?_.
7. In $\triangle DOM$, $\angle O$ is a right angle and $m\angle D > m\angle M$. Which side of $\triangle DOM$ is the shortest side?

The longer diagonal of $\square QRST$ is $\overline{QS}$. Tell whether each statement *must be*, *may be*, or *cannot be* true.

8. $\angle R$ is an acute angle
9. $QS > RS$
10. $RS > RT$

EXTRA

Non-Euclidean Geometries

When you develop a geometry, you have some choice as to which statements you are going to postulate and which you are going to prove. For example, consider these two statements:

(A) If two parallel lines are cut by a transversal, then corresponding angles are congruent.

(B) Through a point outside a line, there is exactly one line parallel to the given line.

In this book, statement (A) is Postulate 10 and statement (B) is Theorem 2-8. In some books, statement (B) is a postulate and statement (A) is a theorem. In still other developments, both of these statements are proved on the basis of some third statement chosen as a postulate.

A natural question to raise is, Do we have to assume either statement, or an equivalent, at all? Couldn't each of them be proved on the basis of the other postulates? The answer, which wasn't known until the nineteenth century, is that some assumption about parallels *must* be made.

A geometry that provides for a unique parallel to a line through a point not on the line is called *Euclidean,* so this book is a book on Euclidean geometry. Geometries that do not provide for a unique parallel are called *non-Euclidean.* Such geometries, developed in the nineteenth century, aren't mere curiosities; for example, Einstein's Theory of Relativity is based on non-Euclidean geometry. The statements below show the key differences between Euclidean geometry and two types of non-Euclidean geometry.

Euclidean geometry	Through a point outside a line, there is *exactly one* line parallel to the given line.
Hyperbolic geometry	Through a point outside a line, there is *more than one* line parallel to the given line. (Bolyai, Lobachevsky, Gauss)
Elliptic geometry	Through a point outside a line, there is *no* line parallel to the given line. (Riemann)

To see a model of a no-parallel geometry, visualize the surface of a sphere. Think of a *line* as being a great circle of the sphere, that is, the intersection of the sphere and a plane that passes through the center of the sphere. On the sphere, through a point outside a line, there isn't any line parallel to the line. All lines, as defined, intersect.

Any development of Euclidean geometry includes an assumption about parallels equivalent to statement (B). For historical reasons statement (B) is often called *the Parallel Postulate,* but it is only one of many possible choices. In this book we state our assumptions about parallels in Postulates 10 and 11 and prove statement (B) as a theorem. Some advantages of this choice are pointed out below.

First we show how statement (B) follows from our postulates. Notice that Postulates 10 and 11 play a crucial role in the proof below. In fact, without such assumptions about parallels there couldn't be a proof. Before the discovery of non-Euclidean geometries people didn't know that this was the case and tried, without success, to find a proof that was independent of any assumption about parallels.

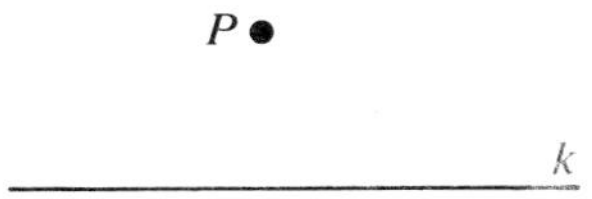

Given: Point P outside line k

Prove: (1) There is a line through P parallel to k.
(2) There is only one line through P parallel to k.

Outline of proof of (1):

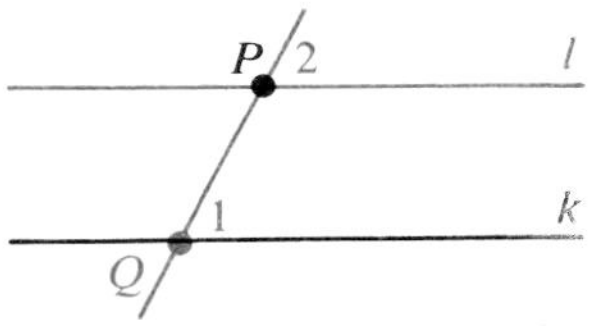

1. Draw a line through P and some point Q on k. (Postulates 5 and 6)
2. Draw line l so that $\angle 2$ and $\angle 1$ are corresponding angles and $m\angle 2 = m\angle 1$. (Protractor Postulate)
3. $l \parallel k$, so there is a line through P parallel to k. (Postulate 11)

Indirect proof of (2):

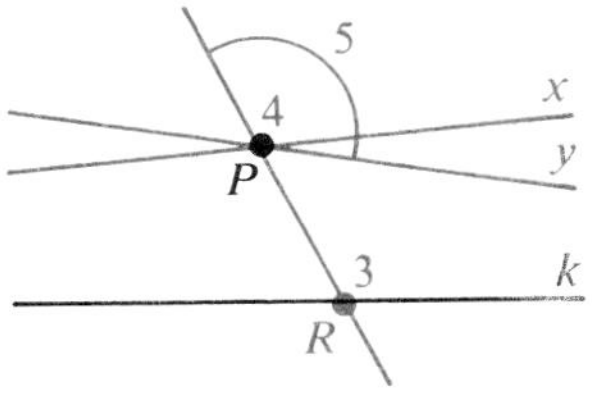

Assume temporarily that there are at least two lines, x and y, through P parallel to k. Draw a line through P and some point R on k. $\angle 4 \cong \angle 3$ and $\angle 5 \cong \angle 3$ by Postulate 10, so $\angle 5 \cong \angle 4$. But since x and y are different lines we also have $m\angle 5 > m\angle 4$. This is impossible, so our assumption must be false, and it follows that there is only one line through P parallel to k.

This theorem, which is our Theorem 2-8, can be used immediately to prove that the sum of the measures of the angles of a triangle is 180 (Theorem 2-11). Thus our choice of postulates allows us to establish basic facts about angles of polygons near the beginning of our study. When statement (B) is taken as a postulate, the most straightforward proof of statement (A) depends on the SAS Congruence Postulate, so that these facts must wait until triangle congruence has been developed.

The original parallel postulate of Euclidean geometry is Postulate V in Euclid's *Elements*. A look at the wording of this postulate, which differs from both statement (A) and statement (B), suggests another advantage of our approach to parallels. It reads as follows: "If a straight line meets two other straight lines so as to make the two interior angles on one side of it together less than two right angles, the other straight lines, if extended indefinitely, will meet on that side on which the angles are less than two right angles." Our Postulate 10 is easier to state and makes a more convenient tool for building proofs.

Chapter Summary

1. A parallelogram has these properties:
   a. Opposite sides are parallel.
   b. Opposite sides are congruent.
   c. Opposite angles are congruent.
   d. Diagonals bisect each other.
2. The chart on page 164 lists five ways to prove that a quadrilateral is a parallelogram.
3. If three parallel lines cut off congruent segments on one transversal, then they cut off congruent segments on every transversal.
4. A line that contains the midpoint of one side of a triangle and is parallel to another side bisects the third side.
5. Rectangles, rhombuses, and squares are parallelograms with additional properties.
6. The median of a trapezoid is parallel to the bases and has a length equal to half the sum of the lengths of the bases.
7. The segment that joins the midpoints of two sides of a triangle is parallel to the third side and has a length equal to half the length of the third side.
8. You begin an indirect proof by assuming temporarily that what you wish to prove true is *not* true. If this temporary assumption leads to a contradiction of a known fact, then your temporary assumption must be false and what you wish to prove true must be true.
9. In $\triangle RST$, if $RT > RS$, then $m\angle S > m\angle T$. If $m\angle S > m\angle T$, then $RT > RS$.
10. The sum of the lengths of any two sides of a triangle is greater than the length of the third side.
11. You can use the SAS and the SSS Inequality Theorems to compare lengths of sides and measures of angles in two triangles.

Chapter Review

In parallelogram *EFGH*, $m\angle EFG = 70$.

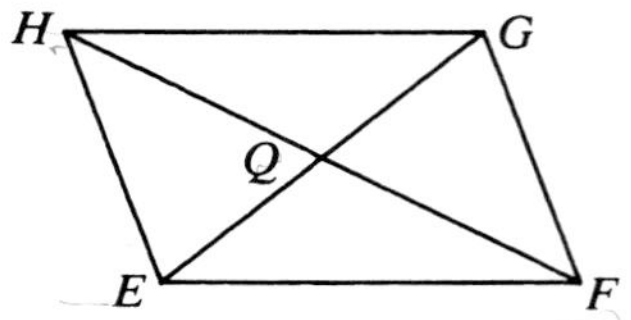

4-1

1. $m\angle HEF =$ _?_
2. If $HQ = 14$, then $HF =$ _?_.
3. If $m\angle EFH = 32$, then $m\angle EHF =$ _?_.
4. If $EH = 8x - 7$ and $FG = 5x + 11$, then $x =$ _?_.

In each exercise you could prove that quad. *SANG* is a parallelogram if one more fact, in addition to those stated, were given. State that fact.

5. $GN = 9$; $NA = 5$; $SA = 9$ 4–2
6. $\angle ASG \cong \angle GNA$
7. $\overline{SZ} \cong \overline{NZ}$
8. $\overline{SA} \parallel \overline{GN}$; $SA = 17$

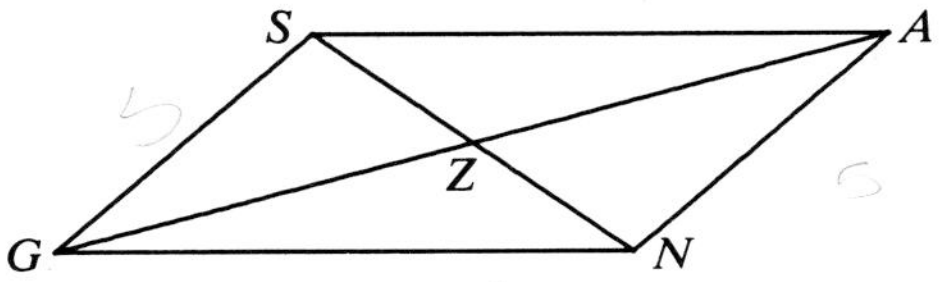

Write the best name for the figure described.

9. A quadrilateral with diagonals that bisect each other. 4–3
10. A rhombus with a right angle.
11. A quadrilateral in which four sides are congruent but not all four angles are congruent.
12. A parallelogram in which two consecutive angles are congruent.

$\overline{MN}$ is the median of trapezoid *ZOID*.

13. The bases of trap. $ZOID$ are _?_ and _?_. 4–4
14. If $ZO = 8$ and $MN = 11$, then $DI =$ _?_.
15. If $ZO = 8$, then $TN =$ _?_.
16. If trap. $ZOID$ is isosceles and $m\angle D = 80$, then $m\angle O =$ _?_.

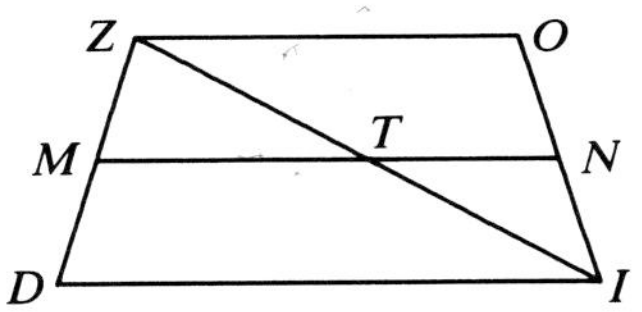

17. Write the letters (a) to (d) in such an order that the sentences provide an indirect proof of the statement: If $n^2 + 6 = 32$, then $n \neq 5$. 4–5
 (a) But this contradicts the fact that $n^2 + 6 = 32$.
 (b) Our temporary assumption must be false, and it follows that $n \neq 5$.
 (c) Assume temporarily that $n = 5$.
 (d) Then $n^2 + 6 = 31$.

18. In $\triangle TEX$, if $TE > XE$, then $m\angle T < m\angle$ _?_. 4–6
19. In $\triangle BAN$, if $m\angle A > m\angle N$, then _?_ $>$ _?_.
20. Two sides of a triangle have lengths 9 and 12. The length of the third side must be greater than _?_ and less than _?_.

Use one of the symbols $<$, $=$, or $>$ to complete the statement.

21. If $\overline{AB} \cong \overline{AC}$ and $m\angle 1 > m\angle 2$, then BT _?_ CT. 4–7
22. If $\overline{TB} \cong \overline{TC}$ and $AB < CA$, then $m\angle 3$ _?_ $m\angle 4$.
23. If $\angle 1 \cong \angle 2$ and $\angle 3 \cong \angle 4$, then AB _?_ AC.
24. If $\overline{TB} \cong \overline{TC}$ and $m\angle 3 > m\angle 4$, then AB _?_ AC.

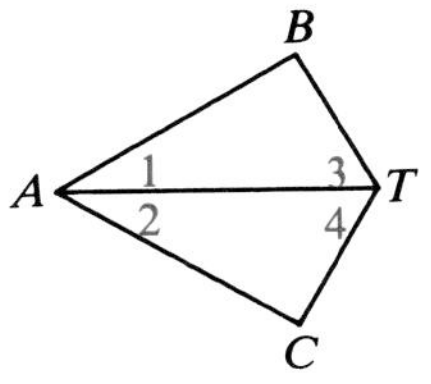

Chapter Test

Tell whether the statement is *always, sometimes,* or *never* true.

1. A square is _?_ a rectangle.
2. A rectangle is _?_ a rhombus.
3. A rhombus is _?_ a square.
4. A rhombus is _?_ a parallelogram.
5. A trapezoid _?_ has three congruent sides.
6. The diagonals of a trapezoid _?_ bisect each other.
7. The sides of a triangle are _?_ 13 cm, 19 cm, and 33 cm long.
8. In $\square ABCD$, if $m\angle A > m\angle B$, then $\angle D$ is _?_ an acute angle.

Trapezoid *ABCD* has median $\overline{MN}$.

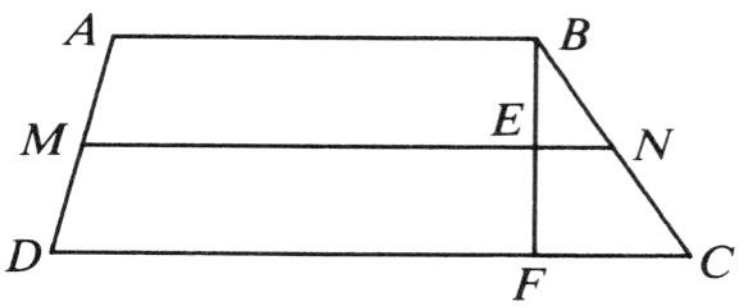

9. If $DC = 17$ and $MN = 12$, then $AB =$ _?_.
10. If $FC = 9$, then $EN =$ _?_.
11. If $AB = 5j + 7k$ and $DC = 9j - 3k$, then $MN =$ _?_.
12. If the sides of a triangle have lengths x, 8, and 12, then _?_ $< x <$ _?_.
13. To write an indirect proof of "If $RS = 10$, then quad. $RSTU$ is a parallelogram," you begin by writing: "Assume temporarily that _?_."

State the theorem that enables you to deduce, from the information given, that quad. *ABCD* is a parallelogram.

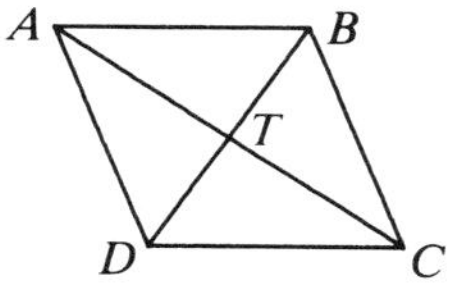

14. $\angle ADC \cong \angle CBA$ and $\angle BAD \cong \angle DCB$
15. $\overline{AD} \parallel \overline{BC}$ and $\overline{AD} \cong \overline{BC}$
16. $AT = CT$ and $DT = \frac{1}{2}DB$
17. $\overline{AB}$, $\overline{BC}$, $\overline{CD}$, and $\overline{DA}$ are all congruent.

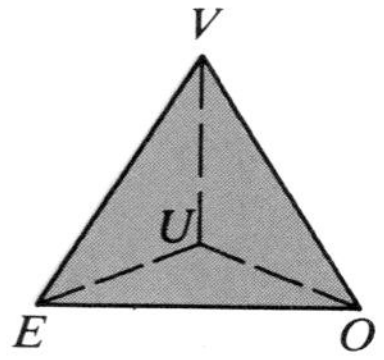

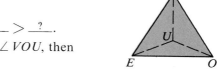

18. If $VE > VU$, then $m\angle$ _?_ $> m\angle$ _?_.
19. If $m\angle EOU > m\angle EUO$, then _?_ $>$ _?_.
20. If $\overline{VE} \cong \overline{VO}$ and $m\angle UVE > m\angle UVO$, then _?_ $>$ _?_.
21. If $m\angle EVU = 60$, $\overline{OE} \cong \overline{OU}$, and $m\angle VOE > m\angle VOU$, then the largest angle of $\triangle UVE$ is $\angle$ _?_.
22. Given: $\square ABCD$; $\angle D \cong \angle 1$
    Prove: Quad. $ASTD$ is a parallelogram.

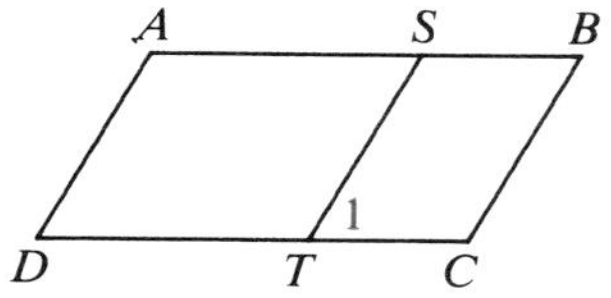

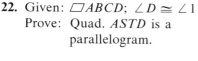

23. Given: $\overline{EM}$ is a median of $\triangle EFG$; $m\angle 2 > m\angle 3$
    Prove: $m\angle G > m\angle F$

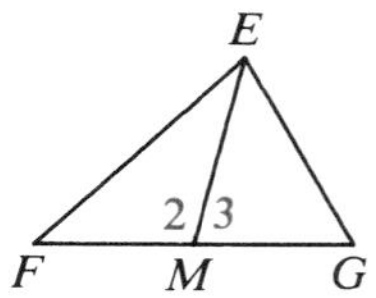

Preparing for College Entrance Exams

Strategy for Success

You may find it helpful to sketch figures or do calculations in your test booklet. Be careful not to make extra marks on your answer sheet.

Indicate the best answer by writing the appropriate letter.

1. Given: $\triangle RGA$ and $\triangle PMC$ with $\overline{RG} \cong \overline{PM}$, $\overline{RA} \cong \overline{PC}$, and $\angle R \cong \angle P$. Which method could be used to prove that $\triangle RGA \cong \triangle PMC$?
(A) SSS **(B)** SAS **(C)** HL **(D)** ASA
(E) There is not enough information for a proof.

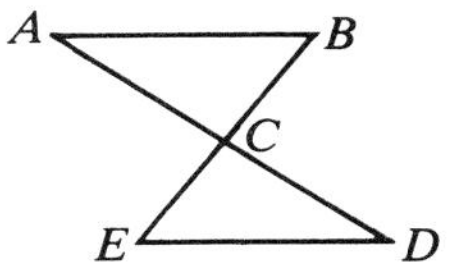

2. Given: $\overline{BE}$ bisects $\overline{AD}$. To prove that the triangles are congruent by the AAS method, you must show that:
(A) $\angle A \cong \angle E$ **(B)** $\angle A \cong \angle D$ **(C)** $\angle B \cong \angle E$
(D) $\angle B \cong \angle D$ **(E)** $\angle \overline{AD}$ bisects $\overline{BE}$.

3. Which statement does *not* guarantee that quadrilateral $WXYZ$ is a parallelogram?
(A) $\overline{WX} \cong \overline{YZ}$; $\overline{XY} \parallel \overline{WZ}$ **(B)** $\angle W \cong \angle Y$; $\angle X \cong \angle Z$
(C) $\overline{WX} \cong \overline{YZ}$; $\overline{XY} \cong \overline{WZ}$ **(D)** $\overline{XY} \parallel \overline{WZ}$; $\overline{WX} \parallel \overline{ZY}$
(E) $\overline{XY} \cong \overline{WZ}$; $\overline{XY} \parallel \overline{WZ}$

4. In $\triangle ABC$, $AB = 7$ and $BC = 10$. AC *cannot* equal:
(A) 7 **(B)** 10 **(C)** 3.14 **(D)** 17 **(E)** $\frac{34}{3}$

5. The next number in the sequence 2, 6, 12, 20, 30, 42, _?_ is:
(A) 56 **(B)** 52 **(C)** 58 **(D)** 54 **(E)** 60

6. Which statement is not always true for every rhombus $ABCD$?
(A) $AB = BC$ **(B)** $AC = BD$ **(C)** $\angle B \cong \angle D$ **(D)** $\overline{AC} \perp \overline{BD}$ **(E)** $\angle ABD \cong \angle CBD$

7. The diagonals of quadrilateral $MNOP$ intersect at X. Which statement guarantees that $MNOP$ is a rectangle?
(A) $MO = NP$ **(B)** $\angle PMN \cong \angle MNO \cong \angle NOP$
(C) $MX = NX = OX = PX$ **(D)** $\overline{MO} \perp \overline{NP}$
(E) Each pair of consecutive angles is supplementary.

8. In $\triangle JKL$, $\overline{KL} \cong \overline{LJ}$, $m\angle K = 2x - 36$, and $m\angle L = x + 2$. Find $m\angle J$.
(A) 50 **(B)** 52 **(C)** 53 **(D)** 55 **(E)** 64

9. In $\triangle RST$, $\overleftrightarrow{SU}$ is the perpendicular bisector of $\overline{RT}$ and U lies on $\overline{RT}$. Which statement(s) must be true?
(I) $\triangle RST$ is equilateral (II) $\triangle RSU \cong \triangle TSU$
(III) $\overrightarrow{SU}$ is the bisector of $\angle RST$
(A) I only **(B)** II only **(C)** III only
(D) II and III only **(E)** I, II, and III

10. In $\triangle ABC$, if $AB = BC$ and $AC > BC$, then:
(A) $AB < AC - BC$ **(B)** $m\angle B > m\angle C$ **(C)** $m\angle B < m\angle A$
(D) $m\angle B = 60$ **(E)** $m\angle B = m\angle A$

Cumulative Review: Chapters 1–4

A

1. On a number line, point A has coordinate -5 and point B has coordinate 3. Find the coordinate of the midpoint of $\overline{AB}$.

2. Name the property that justifies the statement:
   a. If $\angle 1 \cong \angle 2$ and $\angle 2 \cong \angle 3$, then $\angle 1 \cong \angle 3$.
   b. If $x = 3t$ and $t = 4$, then $x = 3 \cdot 4$.

3. Complete: The median to the base of an isosceles triangle _?_ the vertex angle and is _?_ to the base.

4. Given two parallel lines m and n, how many planes contain m and n?

5. **a.** Is it possible for two lines to be neither intersecting nor parallel? If so, what are the lines called?
   b. Repeat part (a), replacing *lines* with *planes.*

6. If two parallel lines are cut by a transversal and an interior angle formed has measure 50, find the measure of the other interior angle on the same side of the transversal.

7. In plane P, line $j \perp$ line l and line $k \perp$ line l. Can you conclude anything about lines j and k?

8. Is it possible for a triangle to be equiangular and scalene? Write a theorem or corollary that supports your answer.

9. Find the sum of the measures of the angles of a pentagon.

10. Find the measure of each exterior angle of a regular polygon with 12 sides.

11. A certain if-then statement is known to be false. Is the contrapositive of the statement true, false, or sometimes true and sometimes false?

12. Write the converse of the statement "If you are a member of the skiing club, then you enjoy winter weather."

13. It is known that $\triangle ART \cong \triangle DEB$.
    a. $\triangle EBD \cong$ _?_ **b.** $m\angle R =$ _?_ **c.** $\overline{DE} \cong$ _?_

14. Can the given information be used to prove the triangles congruent? If so, which congruence postulate or theorem would you use?
    a. Given: $\overline{PC}$ and $\overline{AL}$ bisect each other.
    b. Given: $\angle P \cong \angle C$; U is the midpoint of $\overline{PC}$.
    c. Given: $\overline{PA} \parallel \overline{LC}$
    d. Given: $\overline{PA} \perp \overline{AL}$; $\overline{LC} \perp \overline{AL}$; $\overline{PU} \cong \overline{UC}$

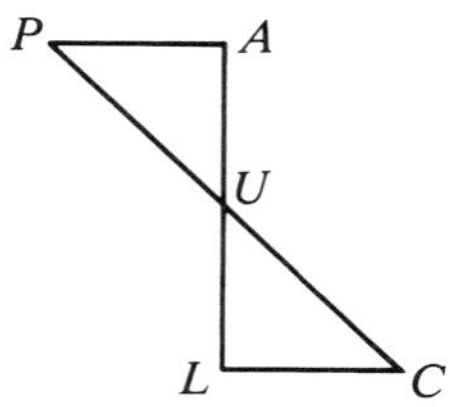

15. **a.** If a point lies on the perpendicular bisector of $\overline{AB}$, then the point is equidistant from _?_.
    b. If a point lies on the bisector of $\angle RST$, then the point is equidistant from _?_.

16. Write the reason for each key step.
Given: $\overline{WX} \cong \overline{WZ}$; $\overline{WY} \perp \overline{XZ}$
Prove: $XY = ZY$

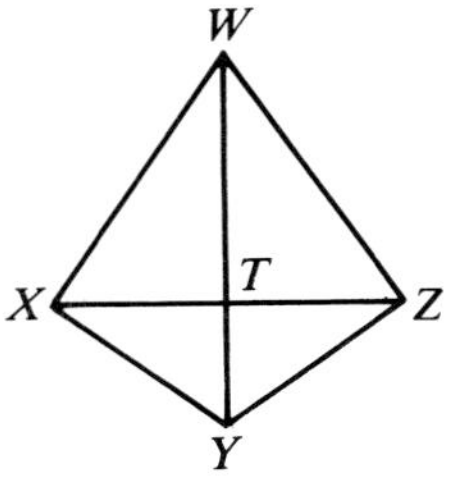

a. $\triangle WXT \cong \triangle WZT$
b. $\overline{XT} \cong \overline{TZ}$
c. $\overline{WY}$ is the perpendicular bisector of $\overline{XZ}$.
d. $XY = ZY$

17. Tell whether the statement is *always, sometimes,* or *never* true for a parallelogram $ABCD$ with diagonals that intersect at P.

a. $AB = BC$ **b.** $\overline{AC} \perp \overline{BD}$ **c.** $\angle A$ and $\angle B$ are comp. $\angle$s.
d. $\angle ADB \cong \angle CBD$ **e.** $\overline{AP} \cong \overline{PC}$ **f.** $\triangle ABC \cong \triangle CDA$

18. In $\triangle RST$, $m\angle R = 64$ and $m\angle S = 54$. Name (a) the longest and (b) the shortest side of $\triangle RST$.

19. **a.** Which segment is longer: $\overline{RS}$ or $\overline{JK}$?
b. Name the theorem that supports your answer.

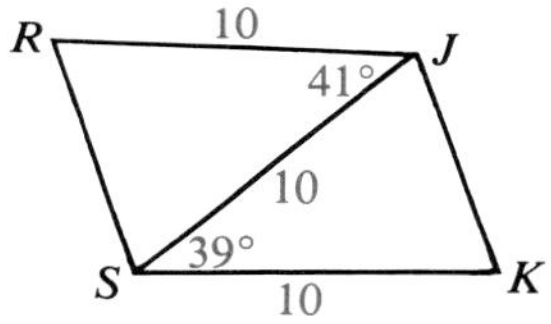

B 20. Give the best name for quadrilateral $MNOP$.

a. $\overline{MN} \cong \overline{PO}$; $\overline{MN} \parallel \overline{PO}$
b. $\overline{MN} \parallel \overline{PO}$; $\overline{NO} \parallel \overline{MP}$; $\overline{MO} \perp \overline{NP}$
c. $\angle M \cong \angle N \cong \angle O \cong \angle P$
d. $\angle M \cong \angle N$; $\angle O \cong \angle P$; $\angle M \not\cong \angle O$
e. $MNOP$ is a rectangle with $MN = NO$.

21. In $\triangle SUN$, $\angle S \cong \angle N$. Given that $SU = 2x + 7$, $UN = 4x - 1$, and $SN = 3x + 4$, find the value of x.

22. The difference between the measures of two supplementary angles is 38. Find the measure of each angle.

23. The lengths of the sides of a triangle are z, $z + 3$, and $z + 6$. What can you conclude about the value of z?

24. M and N are the midpoints of the legs of trapezoid $EFGH$. If bases $\overline{EF}$ and $\overline{HG}$ have lengths $2r + s$ and $4r - 3s$, express the length of $\overline{MN}$ in terms of r and s.

25. B lies between A and C, $AB = 3.2y$, $BC = 2y + 1$, and $AC = 6y - 1$. Is B the midpoint of $\overline{AC}$? Explain.

26. Given: $WP = ZP$; $PY = PX$
Prove: $\angle WXY \cong \angle ZYX$

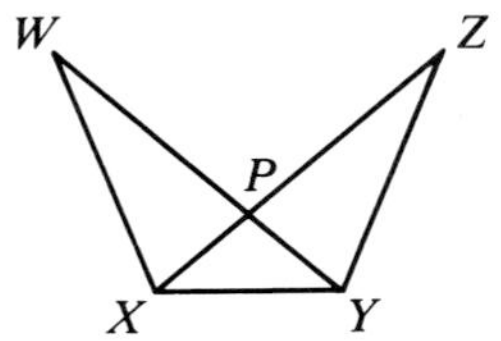

27. Given: $\overline{AD} \cong \overline{BC}$; $\overline{AD} \parallel \overline{BC}$
Prove: $\overline{EF} \cong \overline{FG}$

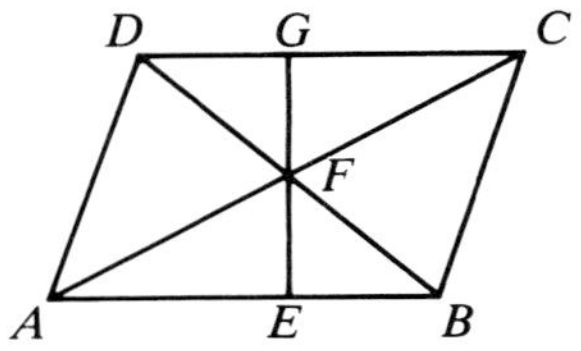

An interesting example of similarity is found in this unusual aquarium that is a scale model of a kitchen. Each object placed in the tank was made to the same scale. Wallpaper and window frames, in an appropriate scale, were applied to the outside of the tank.

Similar Polygons

5

Ratio, Proportion, and Similarity

Objectives

1. Express a ratio in simplest form.
2. Solve for an unknown term in a given proportion.
3. Express a given proportion in an equivalent form.
4. State and apply the properties of similar polygons.

5-1 Ratio and Proportion

The **ratio** of one number to another is the quotient when the first number is divided by the second. This quotient is usually expressed in *simplest form.*

$$\text{The ratio of 8 to 12 is } \frac{8}{12}, \text{ or } \frac{2}{3}.$$

$$\text{If } y \neq 0, \text{ the ratio of } x \text{ to } y \text{ is } \frac{x}{y}.$$

Since we cannot divide by zero, a ratio $\frac{r}{s}$ is defined only if $s \neq 0$. When an expression such as $\frac{r}{s}$ appears in this book, you may assume that $s \neq 0$.

Ratios can be used to compare two numbers. To find the ratio of the lengths of two segments, the segments must be measured in terms of the same unit.

Example 1 **a.** Find the ratio of TP to RA.
b. Find the ratio of the measure of the largest angle of the trapezoid to that of the smallest angle.

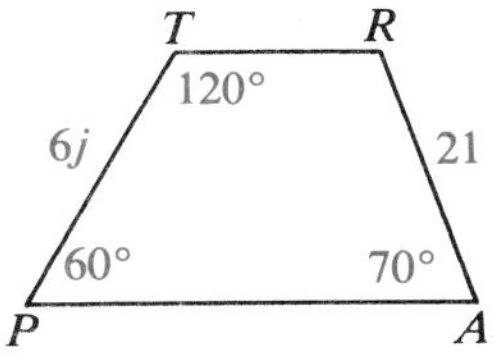

Solution **a.** $\frac{TP}{RA} = \frac{6j}{21} = \frac{2j}{7}$.

The ratio of TP to RA is $2j$ to 7.

b. $\angle R$ has measure $180 - 70$, or 110. Thus $\angle T$ is the largest angle and $\angle P$ is the smallest angle.

$$\frac{m\angle T}{m\angle P} = \frac{120}{60} = \frac{2}{1}$$

The ratio of the measure of the largest angle of the trapezoid to that of the smallest angle is 2 to 1.

Example 2 A poster is 1 m long and 65 cm wide. Find the ratio of the width to the length.

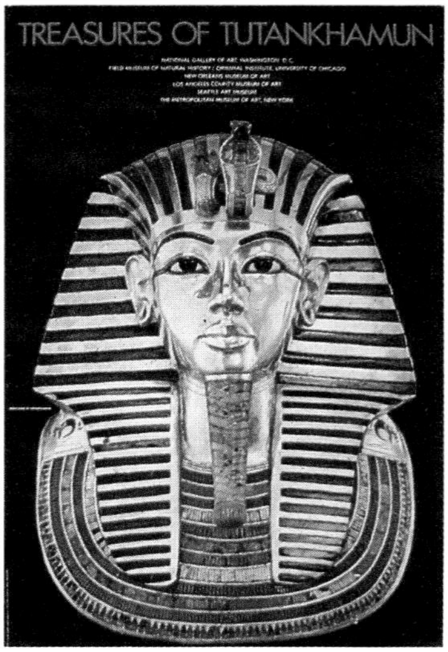

Solution *Method 1*

Use centimeters.

1 m = 100 cm

$$\frac{\text{width}}{\text{length}} = \frac{65}{100} = \frac{13}{20}$$

Method 2

Use meters.

65 cm = 0.65 m

$$\frac{\text{width}}{\text{length}} = \frac{0.65}{1} = \frac{65}{100} = \frac{13}{20}$$

Example 2 shows that the ratio of two quantities is not affected by the unit chosen.

Sometimes the ratio of a to b is written in the form $a:b$. This form can also be used to compare three or more numbers. The statement that three numbers are in the ratio $c:d:e$ (read "c to d to e") means:

(1) The ratio of the first two numbers is $c:d$.
(2) The ratio of the last two numbers is $d:e$.
(3) The ratio of the first and last numbers is $c:e$.

Example 3 The measures of the three angles of a triangle are in the ratio 2:2:5. Find the measure of each angle.

Solution Let $2x$, $2x$, and $5x$ represent the measures.

$$\begin{aligned} 2x + 2x + 5x &= 180 \\ 9x &= 180 \\ x &= 20 \end{aligned}$$

Then $2x = 40$ and $5x = 100$.

The measures of the angles are 40, 40, and 100.

A **proportion** is an equation stating that two ratios are equal. For example,

$$\frac{a}{b} = \frac{c}{d} \quad \text{and} \quad a:b = c:d$$

are equivalent forms of the same proportion. Either form can be read "a is to b as c is to d." The number a is called the first *term* of the proportion. The numbers b, c, and d are the second, third, and fourth terms, respectively.

When three or more ratios are equal, you can write an *extended proportion:*

$$\frac{a}{b} = \frac{c}{d} = \frac{e}{f}$$

Classroom Exercises

Express the ratio in simplest form.

1. $\frac{15}{20}$ **2.** $\frac{4j}{7j}$ **3.** $\frac{4n}{n^2}$ **4.** $\frac{n^2}{4n}$

5. Compare your answers to Exercises 3 and 4. Is the ratio $a:b$ of two numbers always, sometimes, or never the same as the ratio $b:a$?

Express the ratio in simplest form.

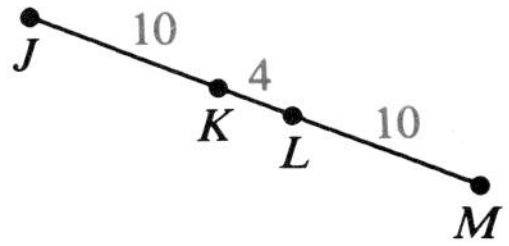

6. $JK:KL$ **7.** $KL:JK$ **8.** $KL:JM$

9. $KM:LK$ **10.** $JL:LM$ **11.** $JK:KL:LM$

12. What is the ratio of 750 milliliters to 1.5 liters?

13. Can you find the ratio of 2 liters to 4 kilometers? Explain.

14. The ratio of the lengths of two segments is 4:3 when they are measured in centimeters. What is their ratio when they are measured in inches?

15. Three numbers aren't known, but the ratio of the numbers is 1:2:5. Is it possible that the numbers are 1, 2, and 5? 10, 20, and 50? 3, 6, and 20? x, $2x$, and $5x$?

16. What is the second term of the proportion $\frac{a}{b} = \frac{x}{y}$?

Written Exercises

***ABCD* is a parallelogram. Find the value of each ratio.**

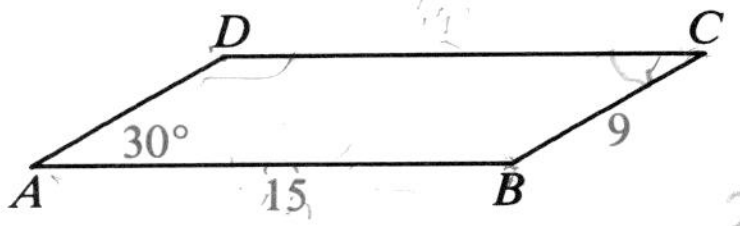

A **1.** $AB:BC$ **2.** $AB:CD$

3. $m\angle C:m\angle D$ **4.** $m\angle B:m\angle C$

5. AD: perimeter of $ABCD$

In Exercises 6–14, $x = 12$, $y = 8$, and $z = 24$. Write each ratio in simplest form.

6. x to y **7.** z to x **8.** $x + y$ to z

9. $\frac{y}{x + z}$ **10.** $\frac{z + x}{z - x}$ **11.** $\frac{x - y}{x + y}$

12. $x:y:z$ **13.** $z:x:y$ **14.** $x:(x + y):(y + z)$

Exercises 15–20 refer to a triangle. Express the ratio of the height to the base in simplest form.

	15.	16.	17.	18.	19.	20.
height	1 m	0.4 km	40 cm	2 cm	3 km	80 mm
base	0.8 m	0.3 km	2 m	5 mm	150 m	0.5 m

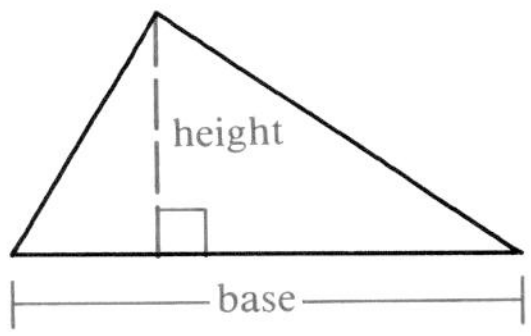

Write the algebraic ratio in simplest form.

21. $\frac{3a}{4ab}$

22. $\frac{2cd}{5c^2}$

23. $\frac{3(x + 4)}{a(x + 4)}$

24. $\frac{10x}{5x}$

25. $\frac{3(x - y)}{(x - y)(x + y)}$

26. $\frac{a + 5}{4a + 20}$

In Exercises 27–32, find the measure of each angle.

B 27. The ratio of the measures of two complementary angles is 4:5. (*Hint:* Let $4x$ and $5x$ represent the measures.)

28. The ratio of the measures of two supplementary angles is 11:4.

29. The measures of the angles of a triangle are in the ratio 3:4:5.

30. The measures of the acute angles of a right triangle are in the ratio 5:7.

31. The measures of the angles of an isosceles triangle are in the ratio 3:3:4.

32. The measures of the angles of a hexagon are in the ratio 4:6:6:7:8:9.

33. The perimeter of a triangle is 96 cm and the lengths of its sides are in the ratio 9:11:12. Find the length of each side.

34. The measures of the consecutive angles of a quadrilateral are in the ratio 6:7:11:12. Find the measure of each angle, draw a quadrilateral that satisfies the requirements, and explain why two sides must be parallel.

35. What is the ratio of the measure of an interior angle to the measure of an exterior angle in a regular decagon? A regular n-gon?

36. A team's best hitter has a lifetime batting average of .320. He has been at bat 325 times.
   a. How many hits has he made?
   b. The player goes into a slump and doesn't get any hits at all in his next ten times at bat. Now what is his batting average to the nearest thousandth?

C 37. A basketball player has made 24 points out of 30 free throws. She hopes to make all her next free throws until her free-throw percentage is 85 or better. How many consecutive free throws will she have to make?

38. Points B and C lie on $\overline{AD}$. $\frac{AB}{BD} = \frac{3}{4}$, $\frac{AC}{CD} = \frac{5}{6}$, and $BD = 66$. Find AC.

39. Find the ratio of x to y: $\frac{4}{y} + \frac{3}{x} = 44$
$\frac{12}{y} - \frac{2}{x} = 44$

5-2 Properties of Proportions

The first and last terms of a proportion are called the *extremes*. The middle terms are the *means*. In the proportions below, the extremes are shown in red. The means are shown in black.

$$a:b = c:d \qquad 6:9 = 2:3 \qquad \frac{6}{9} = \frac{2}{3}$$

Notice that $6 \cdot 3 = 9 \cdot 2$. This illustrates a property of all proportions, called the *means-extremes* property of proportions:

The product of the extremes equals the product of the means.

$\frac{a}{b} = \frac{c}{d}$ is *equivalent* to $ad = bc$.

The two equations are equivalent because we can change either of them into the other by multiplying (or dividing) each side by bd. Try this yourself.

It is often necessary to replace one proportion by an equivalent proportion. When you do so in a proof, you may use the reason "A property of proportions." The following properties will be justified in the exercises.

Properties of Proportions

1. $\frac{a}{b} = \frac{c}{d}$ is equivalent to:

a. $ad = bc$ **b.** $\frac{a}{c} = \frac{b}{d}$ **c.** $\frac{b}{a} = \frac{d}{c}$ **d.** $\frac{a+b}{b} = \frac{c+d}{d}$

2. If $\frac{a}{b} = \frac{c}{d} = \frac{e}{f} = \cdots$, then $\frac{a+c+e+\cdots}{b+d+f+\cdots} = \frac{a}{b} = \cdots$.

Example Use the proportion $\frac{x}{y} = \frac{3}{4}$ to complete each statement.

a. $3y = \underline{\ ?\ }$ **b.** $\frac{x+y}{y} = \frac{?}{?}$

c. $\frac{x}{3} = \frac{?}{?}$ **d.** $\frac{4}{3} = \frac{?}{?}$

Solution **a.** $3y = 4x$ **b.** $\frac{x+y}{y} = \frac{7}{4}$

c. $\frac{x}{3} = \frac{y}{4}$ **d.** $\frac{4}{3} = \frac{y}{x}$

Classroom Exercises

1. If $\frac{e}{f} = \frac{g}{h}$, which equation is correct?

   a. $ef = gh$ b. $eh = fg$ c. $eg = fh$

2. Which proportions are equivalent to $\frac{x}{12} = \frac{3}{4}$?

   a. $\frac{x}{3} = \frac{12}{4}$ b. $\frac{x}{4} = \frac{12}{3}$ c. $\frac{12}{x} = \frac{4}{3}$ d. $\frac{x + 12}{12} = \frac{7}{4}$

Complete the statement.

3. If $\frac{a}{b} = \frac{6}{5}$, then $5a = \underline{\ ?\ }$.

4. If $\frac{c}{d} = \frac{9}{4}$, then $\frac{d}{c} = \frac{?}{?}$.

5. If $\frac{e}{f} = \frac{7}{11}$, then $\frac{e}{7} = \frac{?}{?}$.

6. If $\frac{w}{x} = \frac{y}{z}$, then $\frac{w}{y} = \frac{?}{?}$.

7. If $\frac{a}{b} = \frac{2}{3}$, then $\frac{a + b}{b} = \frac{?}{?}$.

8. If $\frac{a}{b} = \frac{j}{k} = \frac{4}{7}$, then $\frac{a + j + 4}{b + k + 7} = \frac{?}{?}$.

9. a. Apply the means-extremes property of proportions to the proportion $\frac{e}{f} = \frac{g}{5}$ and you get $5e = \underline{\ ?\ }$.

   b. Apply the property to the proportion $\frac{5}{f} = \frac{g}{e}$ and you get $\underline{\ ?\ } = \underline{\ ?\ }$.

   c. Are the proportions $\frac{e}{f} = \frac{g}{5}$ and $\frac{5}{f} = \frac{g}{e}$ equivalent?

10. Explain an easy way to show that the proportions $\frac{x}{7} = \frac{2}{3}$ and $\frac{x}{2} = \frac{3}{7}$ are not equivalent.

What can you conclude from the given information?

11. $\frac{a}{b} = \frac{c}{n}$ and $\frac{b}{a} = \frac{x}{c}$

12. $\frac{3}{4} = \frac{y}{k}$ and $\frac{3}{v} = \frac{4}{k}$

13. Apply the means-extremes property to $\frac{a}{b} = \frac{c}{d}$ and also to $\frac{a}{c} = \frac{b}{d}$. (Note that you have justified Property 1(b) on page 209 by showing that each proportion is equivalent to the same equation.)

14. Explain why $\frac{a}{b} = \frac{c}{d}$ and $\frac{b}{a} = \frac{d}{c}$ are equivalent. (This justifies Property 1(c) on page 209.)

Written Exercises

Complete each statement.

A

1. If $\frac{x}{5} = \frac{3}{4}$, then $4x = \underline{\ ?\ }$.

2. If $\frac{7}{x} = \frac{3}{8}$, then $3x = \underline{\ ?\ }$.

3. If $n:3 = 7:8$, then $8n = \underline{\ ?\ }$.

4. If $4:g = 5:6$, then $5g = \underline{\ ?\ }$.

5. If $\frac{a}{4} = \frac{b}{7}$, then $\frac{a}{b} = \frac{?}{?}$.

6. If $\frac{x}{y} = \frac{3}{8}$, then $\frac{y}{x} = \frac{?}{?}$.

7. If $\frac{x}{2} = \frac{y}{3}$, then $\frac{x+2}{2} = \underline{\ ?\ }$.

8. If $\frac{a}{b} = \frac{5-x}{x}$, then $\frac{a+b}{b} = \underline{\ ?\ }$.

Find the value of *x*.

9. $\frac{x}{3} = \frac{4}{5}$

10. $\frac{x}{7} = \frac{3}{8}$

11. $\frac{2x}{5} = \frac{3}{4}$

12. $\frac{8}{x} = \frac{2}{5}$

13. $\frac{x+5}{4} = \frac{1}{2}$

14. $\frac{x+3}{2} = \frac{4}{3}$

15. $\frac{x+2}{x+3} = \frac{4}{5}$

16. $\frac{2x+1}{4x-1} = \frac{2}{3}$

17. $\frac{x+3}{2} = \frac{2x-1}{3}$

18. $\frac{10}{7x+5} = \frac{7}{6x-2}$

19. $\frac{x+5}{x-5} = \frac{7}{4}$

20. $\frac{4x-5}{4} = \frac{20x+1}{7}$

For the figure shown, it is given that $\frac{KR}{RT} = \frac{KS}{SU}$. Copy and complete the table.

	KR	*RT*	*KT*	*KS*	*SU*	*KU*
21.	12	9	?	16	?	?
22.	8	?	10	12	?	?
23.	16	?	?	?	10	30
24.	?	2	?	9	?	12
B 25.	?	?	12	10	5	?
26.	12	4	?	?	?	20
27.	?	9	36	?	?	48
28.	?	?	30	28	?	42

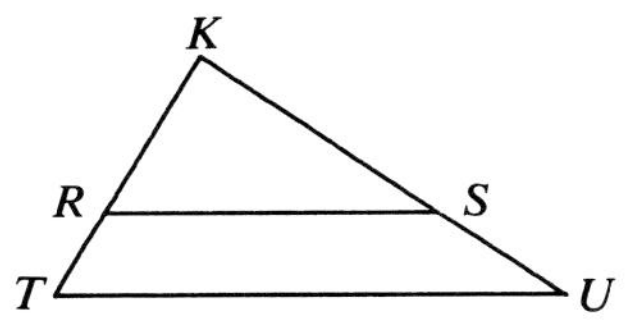

(*Hint for Ex. 25*: Let $KR = x$; then $RT = 12 - x$.)

29. Show that the proportions $\frac{a+b}{b} = \frac{c+d}{d}$ and $\frac{a}{b} = \frac{c}{d}$ are equivalent. (Note that this exercise justifies property 1(d) on page 209.)

30. Given the proportions $\frac{x+y}{y} = \frac{r}{s}$ and $\frac{x-y}{x+y} = \frac{s}{y}$, what can you conclude?

31. Show that the proportions $\frac{a-b}{a+b} = \frac{c-d}{c+d}$ and $\frac{a}{b} = \frac{c}{d}$ are equivalent.

32. Show that the proportions $\frac{a+c}{b+d} = \frac{a-c}{b-d}$ and $\frac{a}{b} = \frac{c}{d}$ are equivalent.

Find the value of *x*.

33. $\frac{x}{x-3} = \frac{x+4}{x}$

34. $\frac{x+2}{x+6} = \frac{x-1}{x+2}$

35. $\frac{x+1}{x-2} = \frac{x+5}{x-6}$

Find the value of x.

C **36.** $\frac{x-2}{x-5} = \frac{2x+1}{x-1}$ **37.** $\frac{x(x+5)}{4x+4} = \frac{9}{5}$ **38.** $\frac{x-1}{x+2} = \frac{10}{3x-2}$

Find the values of x and y.

39. $\frac{x}{y+1} = \frac{3}{2}$

$\frac{x+y}{x-y} = \frac{7}{2}$

40. $\frac{x-3}{4} = \frac{y+2}{2}$

$\frac{x+y-1}{6} = \frac{x-y+1}{5}$

41. Prove: If $\frac{a}{b} = \frac{c}{d} = \frac{e}{f}$, then $\frac{a+c+e}{b+d+f} = \frac{a}{b}$.

(*Hint:* Let $\frac{a}{b} = r$. Then $a = br$, $c = dr$, and $e =$ __?__.)

42. Explain how to extend the proof of Exercise 41 to justify Property 2 on page 209.

43. If $\frac{4a-9b}{4a} = \frac{a-2b}{b}$, find the numerical value of the ratio $a:b$.

5-3 Similar Polygons

When you draw a diagram of a soccer field, you don't need an enormous piece of paper. You use a convenient sheet and draw *to scale*. That is, you show the right shape, but in a convenient size. Two figures, such as those below, that have the same shape are called *similar*.

Two polygons are **similar** if their vertices can be paired so that:

(1) Corresponding angles are congruent.

(2) Corresponding sides are in proportion. (Their lengths have the same ratio.)

When you refer to similar polygons, their corresponding vertices must be in the same order. Given that polygon $PQRST$ is similar to polygon $VWXYZ$, you write:

$$\text{polygon } PQRST \sim \text{polygon } VWXYZ$$

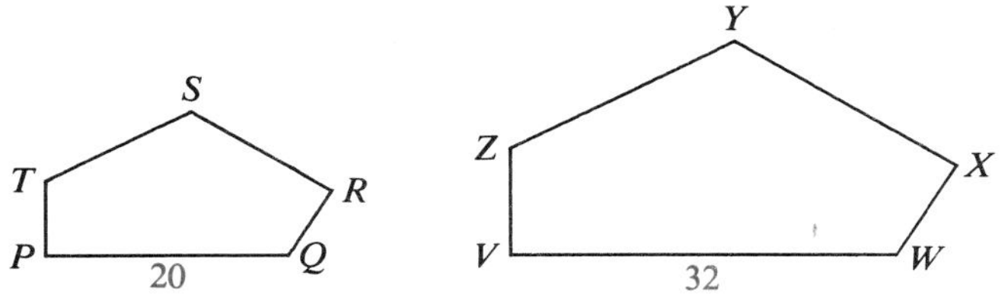

From the definition of similar polygons, we have:

(1) $\angle P \cong \angle V \quad \angle Q \cong \angle W \quad \angle R \cong \angle X \quad \angle S \cong \angle Y \quad \angle T \cong \angle Z$

(2) $\dfrac{PQ}{VW} = \dfrac{QR}{WX} = \dfrac{RS}{XY} = \dfrac{ST}{YZ} = \dfrac{TP}{ZV}$

The ratio of the lengths of two corresponding sides is called the **scale factor** of the similarity. Since $\dfrac{PQ}{VW} = \dfrac{20}{32} = \dfrac{5}{8}$, the scale factor of pentagon $PQRST$ to pentagon $VWXYZ$ is $\dfrac{5}{8}$, or 5 : 8.

The example that follows shows one convenient way to label corresponding vertices: A and A' (read A prime), B and B', and so on.

Example Quad. $ABCD \sim$ quad. $A'B'C'D'$. Find:

a. the scale factor

b. the values of x, y, and z

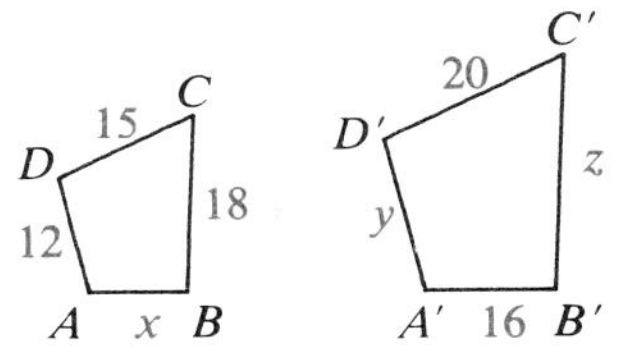

Solution **a.** scale factor $= \dfrac{15}{20} = \dfrac{3}{4}$

b. $\dfrac{x}{16} = \dfrac{3}{4}$, $4x = 48$, $x = 12$ $\qquad$ $\dfrac{12}{y} = \dfrac{3}{4}$, $3y = 48$, $y = 16$ $\qquad$ $\dfrac{18}{z} = \dfrac{3}{4}$, $3z = 72$, $z = 24$

Classroom Exercises

Are the quadrilaterals similar? If they aren't, tell why not.

1. $ABCD$ and $EFGH$
2. $ABCD$ and $JKLM$
3. $ABCD$ and $NOPQ$
4. $JKLM$ and $NOPQ$

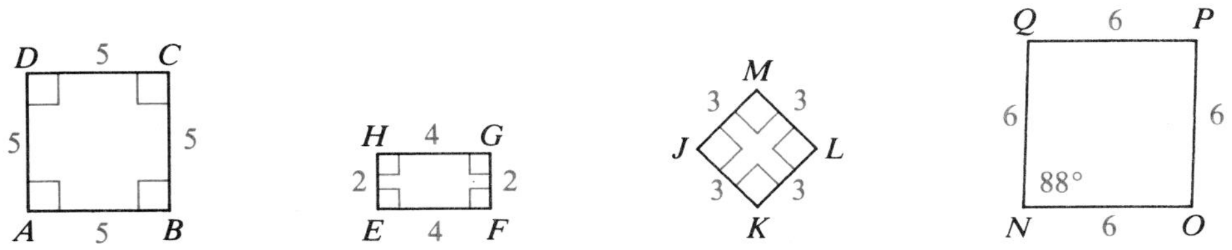

5. If the corresponding angles of two polygons are congruent, must the polygons be similar?
6. If the corresponding sides of two polygons are in proportion, must the polygons be similar?
7. Two polygons are similar. Do they have to be congruent?
8. Two polygons are congruent. Do they have to be similar?
9. Are all regular pentagons similar?
10. Are all isosceles right triangles similar?

11. Quad. $ABCD \sim$ quad. $A'B'C'D'$. Complete.
  a. $m\angle C' = $ _?_ b. $A'D' = $ _?_
  c. $DC = $ _?_ d. Quad. $CBAD \sim$ _?_
  e. Explain why
    quad. $ABCD \sim$ quad. $B'C'D'A'$
   is not a correct statement.

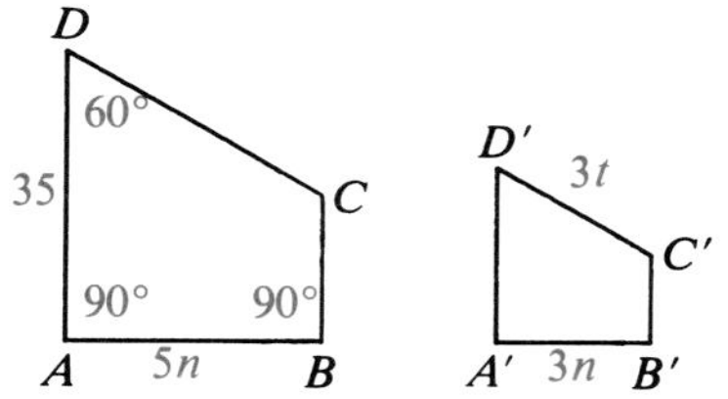

12. The lengths of the sides of a quadrilateral are 4, 6, 6, and 8. The lengths of the sides of a similar quadrilateral are 6, 9, 9, and 12.
  a. What is the scale factor?
  b. What are the perimeters of the two quadrilaterals?
  c. What is the ratio of the perimeters?

13. The triangles are similar. Complete.
  a. $\triangle RST \sim$ _?_
  b. The scale factor is _?_.
  c. $JK = $ _?_ and $KL = $ _?_
  d. The perimeter of $\triangle JKL$ is _?_.
   The perimeter of $\triangle RST$ is _?_.
  e. The ratio of the perimeters is _?_.

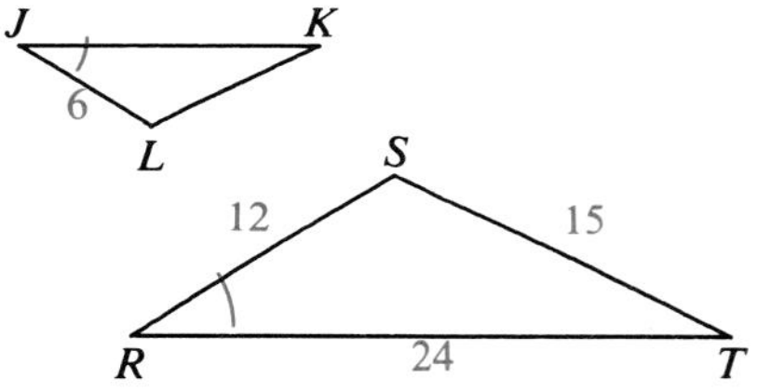

Written Exercises

Tell whether the two polygons are *always, sometimes,* or *never* similar.

A
1. Two equilateral triangles
2. Two right triangles
3. Two isosceles triangles
4. Two scalene triangles
5. Two squares
6. Two rectangles
7. Two rhombuses
8. Two isosceles trapezoids
9. A right triangle and an acute triangle
10. An isosceles triangle and a scalene triangle
11. A right triangle and a scalene triangle
12. An equilateral triangle and an equiangular triangle

In Exercises 13–20, quad. *ABCD* ~ quad. *A′B′C′D′*.

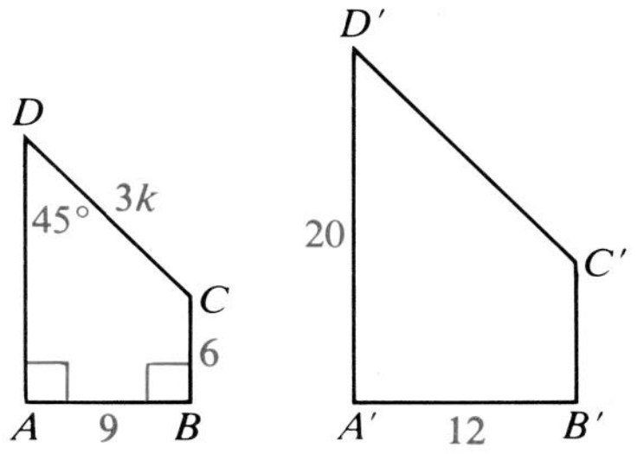

13. What is the scale factor of quad. *ABCD* to quad. *A′B′C′D′*?

14. What special kind of figure must quad. *A′B′C′D′* be? Explain.

15. Find $m\angle D'$.

16. Find $m\angle C'$.

17. Find $B'C'$.

18. Find AD.

19. Find $C'D'$.

20. Find the ratio of the perimeters.

Two similar polygons are shown. Find the values of *x*, *y*, and *z*. (In Exercise 23 find the values of *x* and *y*.)

B **21.**

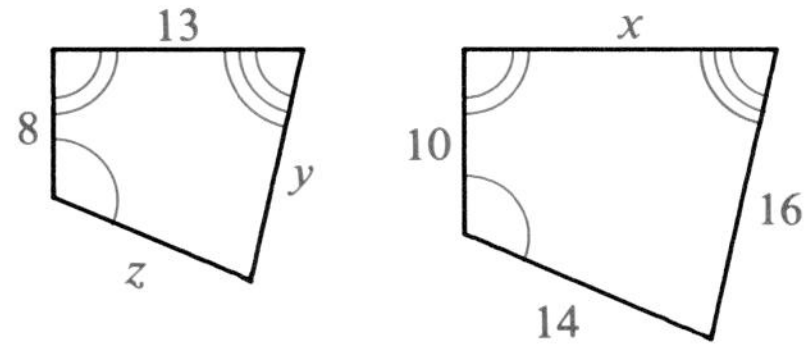

22.

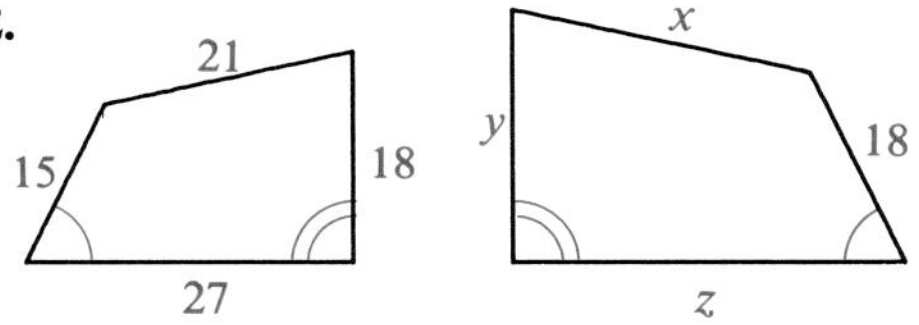

23.

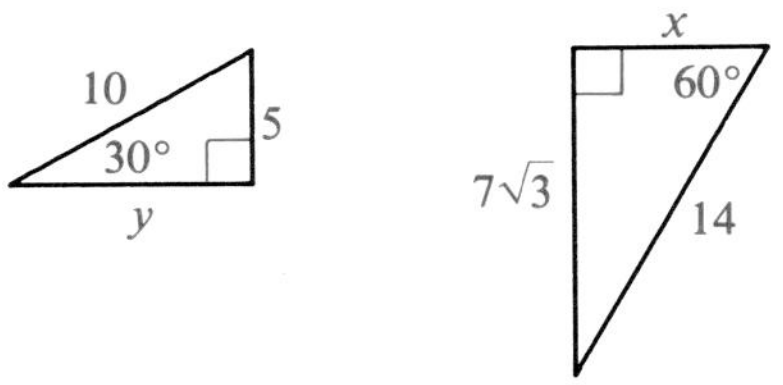

24.

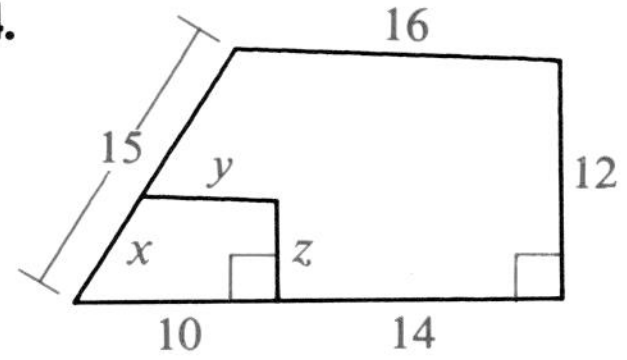

25. Draw two equilateral hexagons that are clearly not similar.

26. Draw two equiangular hexagons that are clearly not similar.

27. If $\triangle ABC \sim \triangle DEF$, express AB in terms of other lengths. (There are two possible answers.)

28. Explain how you can tell at once that quadrilateral *RSWX* is not similar to quadrilateral *RSYZ*.

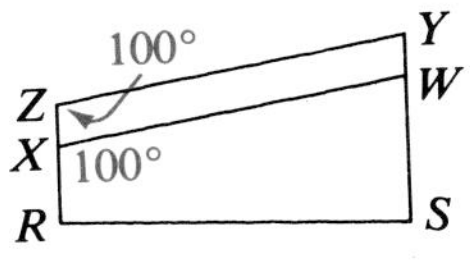

Plot the given points on graph paper. Draw quadrilateral *ABCD* and $\overline{A'B'}$. Locate points *C′* and *D′* so that *A′B′C′D′* is similar to *ABCD*.

29. $A(0, 0)$, $B(4, 0)$, $C(2, 4)$, $D(0, 2)$, $A'(-10, -2)$, $B'(-2, -2)$

30. $A(0, 0)$, $B(4, 0)$, $C(2, 4)$, $D(0, 2)$, $A'(7, 2)$, $B'(7, 0)$

31. The card shown was cut into four congruent pieces with each piece similar to the original. Find the value of *x*.

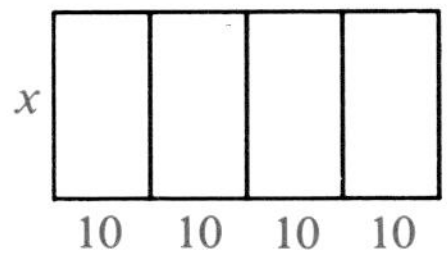

C **32.** What can you deduce from the diagram shown below? Explain.

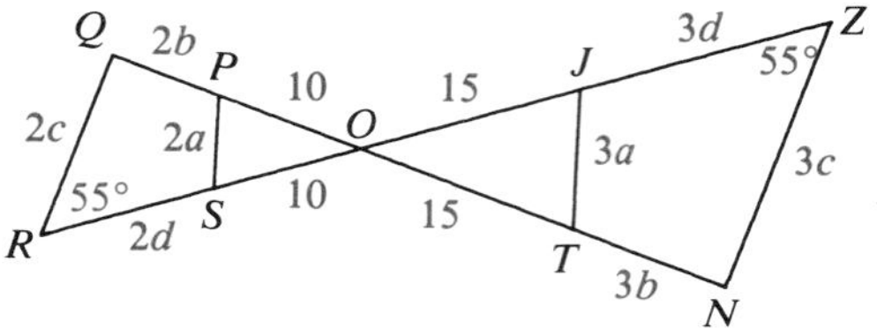

33. The large rectangle shown is a *golden rectangle.* This means that when a square is cut off, the rectangle that remains is similar to the original rectangle.

a. How wide is the original rectangle?

b. The ratio of length to width in a golden rectangle is called the *golden ratio.* Write the golden ratio in simplified radical form. Then use a calculator to find an approximation to the nearest hundredth.

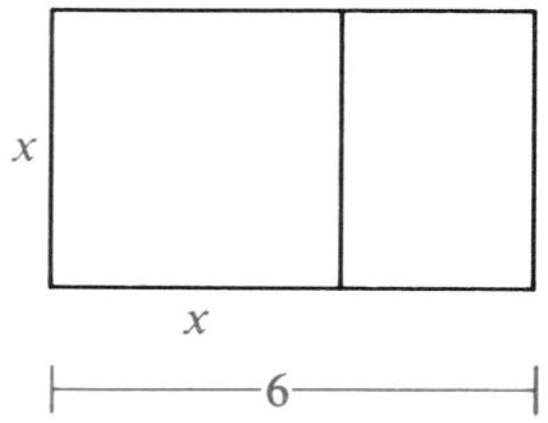

Self-Test 1

Express the ratio in simplest form.

1. 9 : 15 **2.** 60 cm to 2 m **3.** $\frac{4ab}{6b^2}$

Solve for x.

4. $\frac{x}{8} = \frac{9}{12}$ **5.** $\frac{x-2}{2} = \frac{x+6}{4}$ **6.** $\frac{x}{5-x} = \frac{12}{8}$

Tell whether the equation is equivalent to the proportion $\frac{a}{b} = \frac{5}{7}$.

7. $\frac{a}{7} = \frac{b}{5}$ **8.** $7a = 5b$ **9.** $\frac{a+b}{b} = \frac{12}{7}$

10. If $\triangle ABC \sim \triangle RST$, $m\angle A = 45$, and $m\angle C = 60$, then $m\angle R =$ _?_, $m\angle S =$ _?_, and $m\angle T =$ _?_.

The quadrilaterals shown are similar.

11. The scale factor of the smaller quadrilateral to the larger quadrilateral is $\frac{?}{?}$.

12. $x =$ _?_ **13.** $y =$ _?_ **14.** $z =$ _?_

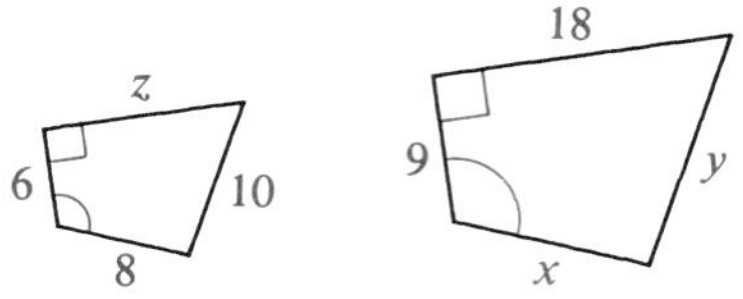

15. The measures of the angles of a hexagon are in the ratio 5 : 5 : 5 : 6 : 7 : 8. Find the measures.

Application

SCALE DRAWINGS

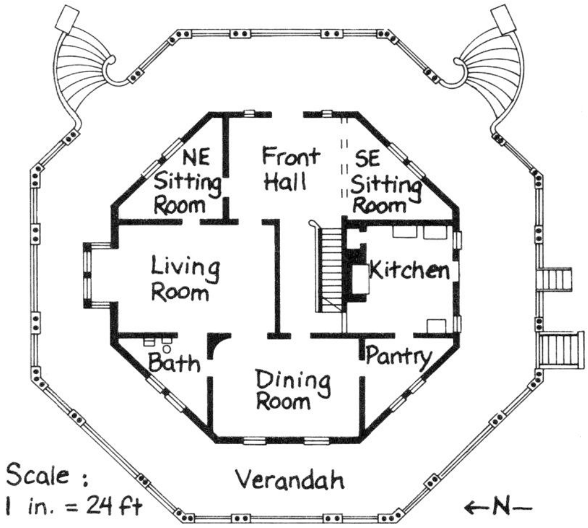

This "octagon house" was built in Irvington, New York, in 1860. The plan shows the rooms on the first floor. The scale on this *scale drawing* tells you that a length of 1 in. on the plan represents a true length of 24 ft.

$$\frac{\text{Plan length in inches}}{\text{True length in feet}} = \frac{1}{24}$$

The following examples show how you can use this formula to find actual dimensions of the house from the plan or to convert dimensions of full-sized objects to plan size.

The verandah measures $\frac{3}{8}$ in. wide on the plan. Find its true width, T.

$\frac{\frac{3}{8}}{T} = \frac{1}{24}$, so $1 \cdot T = \frac{3}{8} \cdot 24$

$T = 9$ The real verandah is 9 ft wide.

A sofa is 6 ft long. Find its plan length, P.

$\frac{P}{6} = \frac{1}{24}$, so $24 \cdot P = 6 \cdot 1$

$P = \frac{1}{4}$ The plan length is $\frac{1}{4}$ in.

Exercises

1. Find the true length and width of the dining room.
2. A rug measures 9 ft by $7\frac{1}{2}$ ft. What would its dimensions be on the floor plan? Would it fit in the northeast sitting room?
3. If a new floor plan is drawn with a scale of 1 in. = 10 ft, how many times longer is each line segment on the new plan than the corresponding segment on the plan shown?
4. Suppose that on the architect's drawings each side of the verandah (the outer octagon) measured 12 in. What was the scale of these drawings?

Independent Reading

When you begin a new chapter or lesson, it is a good idea to have some goals in mind. The objectives listed at the beginning of each main chapter division help you see where you are headed. For example, the title below, "Working with Similar Triangles," and the objectives tell you that pages 218–238 cover several methods of proving that triangles are similar and discuss some important applications of theorems related to similar triangles. You may want to skim all three of the following sections to see what important ideas are covered in each section.

As you read, pay particular attention to new words and phrases, in heavy type, and to important results such as theorems, postulates, and summaries, which are set off by special type and colored rules or boxes. Be sure to study all the worked-out examples. You may want to try solving some of them before you look at the printed solutions. If there is anything you don't understand after you have read and reread it, make a note and ask your teacher about it later.

When you have finished one of the main chapter divisions, look back at the words in heavy type and make sure you know their meanings. When you believe that you understand the material, try the Self-Test. The answers are printed at the back of the book, so you can check your own work. The Chapter Reviews, Chapter Tests, Cumulative Reviews, and, of course, the exercises will also help you check your mastery of a chapter or a group of chapters.

Working with Similar Triangles

Objectives

1. Use the AA Similarity Postulate, the SAS Similarity Theorem, and the SSS Similarity Theorem to prove triangles similar.
2. Deduce information about segments or angles by first proving that two triangles are similar.
3. Apply the Triangle Proportionality Theorem and its corollary.
4. State and apply the Triangle Angle-Bisector Theorem.

5-4 A Postulate for Similar Triangles

You can always prove that two triangles are similar by showing that they satisfy the definition of similar polygons. However, there are simpler methods. For example, the following experiment suggests that two triangles are similar whenever two pairs of angles are congruent.

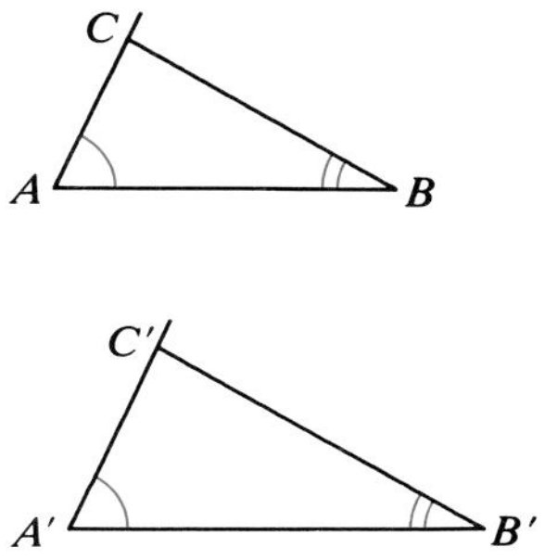

1. Draw any two segments, $\overline{AB}$ and $\overline{A'B'}$.
2. Draw any angle at A and a congruent angle at A'. Draw any angle at B and a congruent angle at B'. Label points C and C' as shown. $\angle ACB \cong \angle A'C'B'$. (Why?)
3. Measure each pair of corresponding sides and compute an approximate decimal value for the ratio of their lengths:

$$\frac{AB}{A'B'} \qquad \frac{BC}{B'C'} \qquad \frac{AC}{A'C'}$$

4. Are the ratios computed in Step 3 approximately equal?

If you worked carefully, your answer in Step 4 was *yes*. Thus, corresponding angles of the two triangles are congruent and corresponding sides are in proportion. By the definition of similar polygons, $\triangle ABC \sim \triangle A'B'C'$.

Whenever you draw two triangles with two angles of one triangle congruent to two angles of the other, you will find that the third angles are also congruent and that corresponding sides are in proportion.

Postulate 15 AA Similarity Postulate

If two angles of one triangle are congruent to two angles of another triangle, then the triangles are similar.

Example

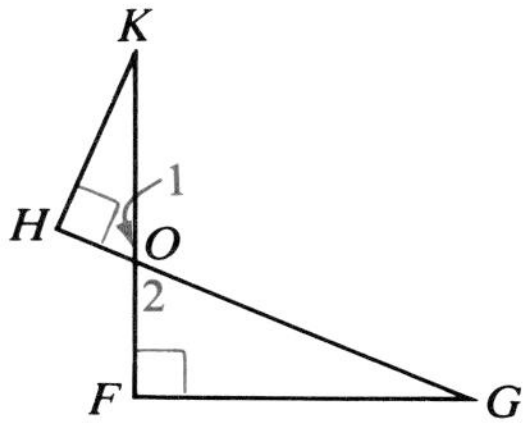

Given: $\angle H$ and $\angle F$ are rt. $\angle$s.

Prove: $HK \cdot GO = FG \cdot KO$

■ ***Plan for Proof:*** We can prove that $HK \cdot GO = FG \cdot KO$ if we can show that $\frac{HK}{FG} = \frac{KO}{GO}$. To do this, we will show that $\triangle HKO \sim \triangle FGO$.

Proof:

Statements	Reasons
1. $\angle 1 \cong \angle 2$	1. Vertical $\angle$s are $\cong$.
2. $\angle H$ and $\angle F$ are rt. $\angle$s.	2. Given
3. $m\angle H = 90 = m\angle F$	3. Def. of a rt. $\angle$
4. $\triangle HKO \sim \triangle FGO$	4. AA Similarity Postulate
5. $\frac{HK}{FG} = \frac{KO}{GO}$	5. Corr. sides of $\sim$ $\triangle$s are in proportion.
6. $HK \cdot GO = FG \cdot KO$	6. A property of proportions

The example shows one way to prove that the product of the lengths of two segments is equal to the product of the lengths of two other segments. You can prove two triangles similar, write a proportion, and then apply the means-extremes property of proportions.

Classroom Exercises

In Exercises 1–8, $\triangle TAN \sim \triangle KEG$. Tell whether each statement must be true.

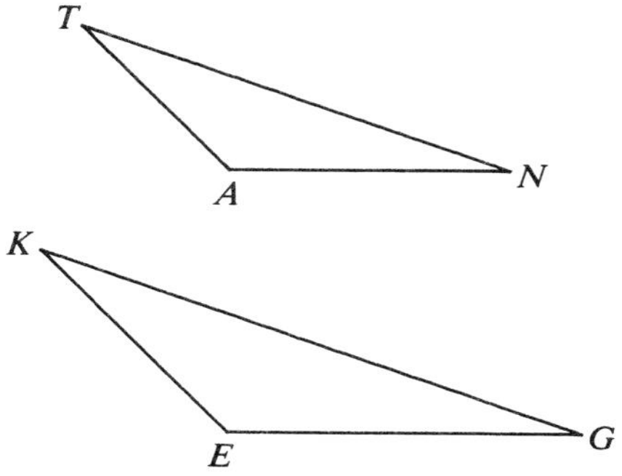

Exs. 1–8

1. $\triangle NTA \sim \triangle GEK$
2. If $m\angle A = 120$, then $m\angle E = 120$.
3. If $m\angle T = 35$, then $m\angle G = 35$.
4. $AT:EK = EG:AN$
5. If $\frac{TA}{KE} = \frac{2}{3}$, then $\frac{TN}{KG} = \frac{2}{3}$.
6. If $\frac{TA}{KE} = \frac{2}{3}$, then $\frac{m\angle T}{m\angle K} = \frac{2}{3}$.
7. If the scale factor of $\triangle TAN$ to $\triangle KEG$ is 4 to 5, then the scale factor of $\triangle KEG$ to $\triangle TAN$ is 5 to 4.
8. If $\overline{KG}$ is twice as long as $\overline{KE}$, then $\overline{TN}$ is twice as long as $\overline{TA}$.

9. Name all pairs of congruent angles in the figure.

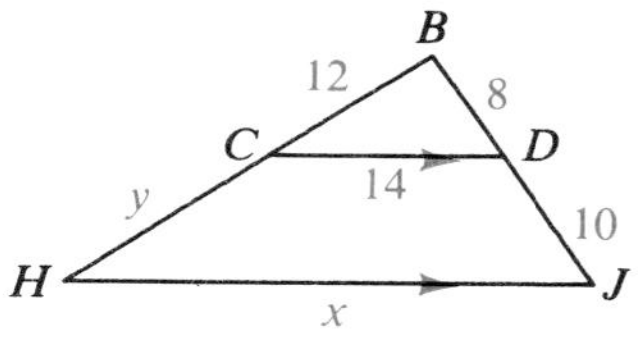

Exs. 9–12

Complete.

10. $\triangle BCD \sim$ _?_
11. $\frac{14}{x} = \frac{8}{?}$ and $x =$ _?_
12. $\frac{12}{12 + y} = \frac{?}{?}$ and $y =$ _?_

13. The diagram shows three triangles. Name two pairs of congruent angles in two of the triangles.
14. Complete: $\triangle RWZ \sim$ _?_
15. Name a third pair of congruent angles.
16. Complete: $\frac{RW}{?} = \frac{WZ}{?} = \frac{RZ}{?}$

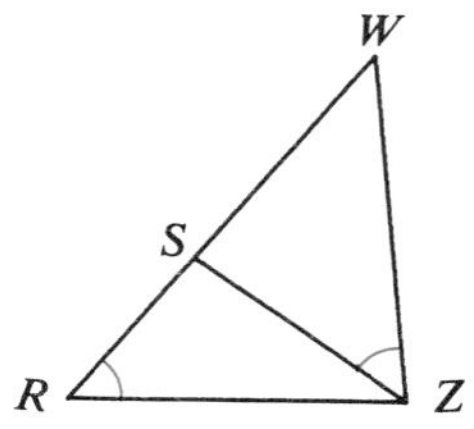

Exs. 13–16

Suppose your goal is to show that $EF \cdot KL = EG \cdot KJ$. Complete each proportion so that it will lead to this equation.

17. $\frac{EF}{KJ} = \frac{?}{?}$

18. $\frac{KL}{EG} = \frac{?}{?}$

Written Exercises

Tell whether the triangles are similar or not similar. If you can't reach a conclusion, write *no conclusion is possible*.

A **1.**
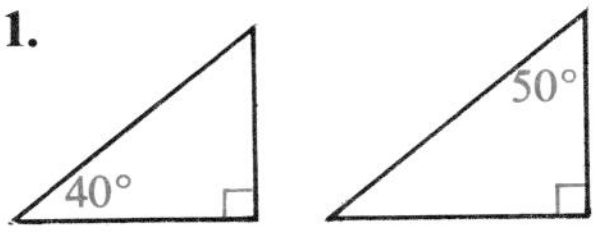

2.
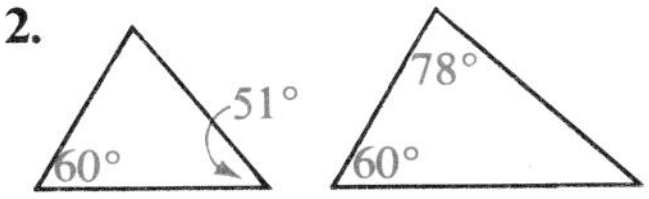

3.
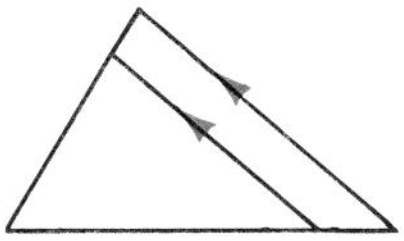

4.
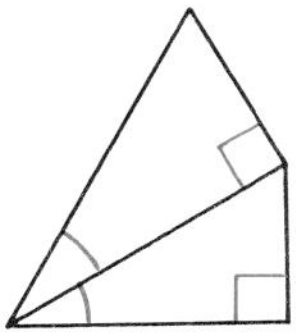

5.
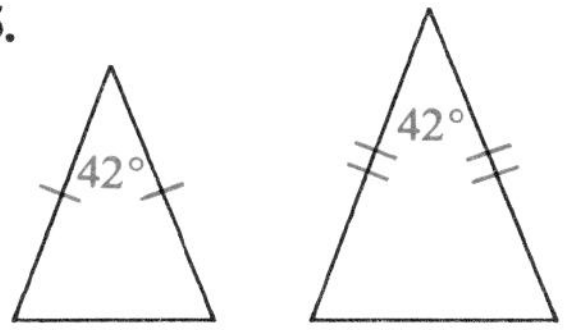

6.
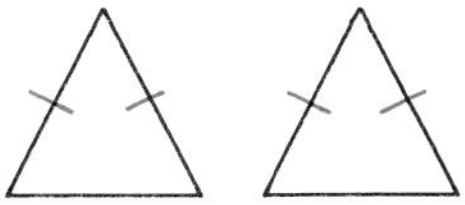

7.
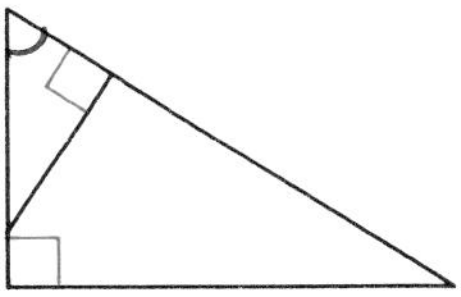

8.
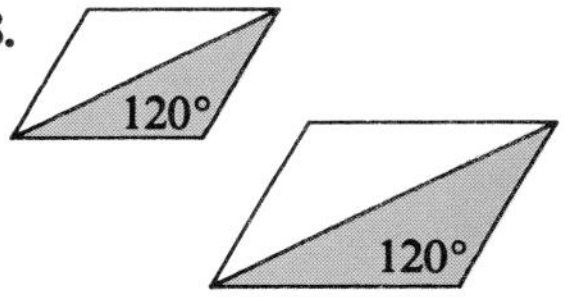

Parallelograms given

9.
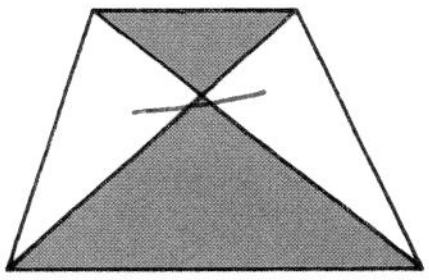

Trapezoid given

Complete.

10. a. $\triangle ABC \sim$ __?__

b. $\frac{AB}{?} = \frac{BC}{?} = \frac{AC}{?}$

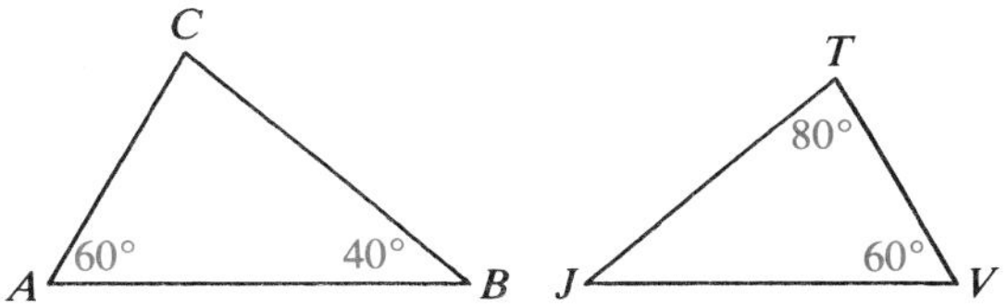

11. a. $\triangle JKN \sim$ __?__

b. $\frac{15}{?} = \frac{18}{?}$ and $\frac{15}{?} = \frac{12}{?}$

c. $x =$ __?__ and $y =$ __?__

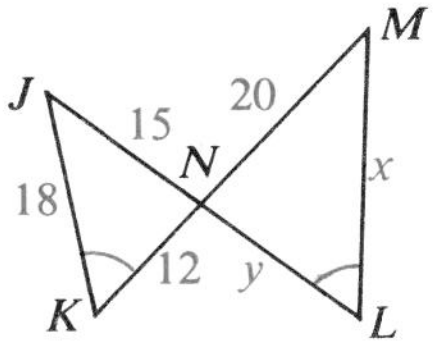

Find the values of *x* and *y*.

12.
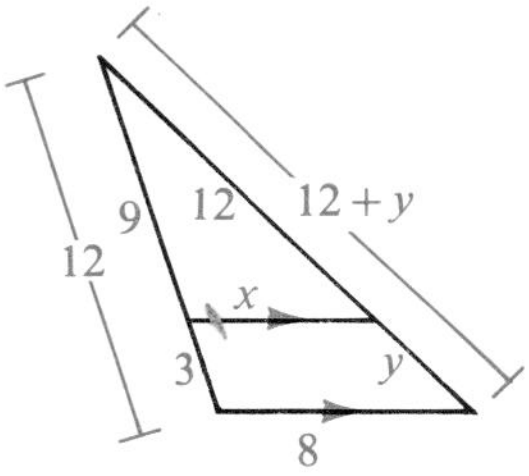

13.
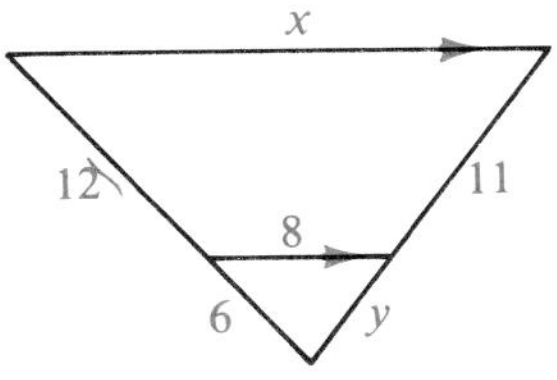

14.
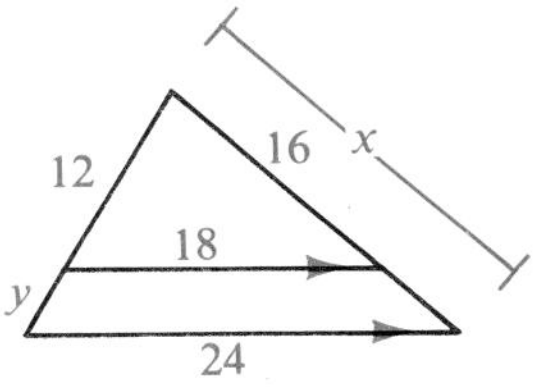

B **15.** **a.** Name two triangles that are similar to $\triangle ABC$.
b. Find the values of x and y.

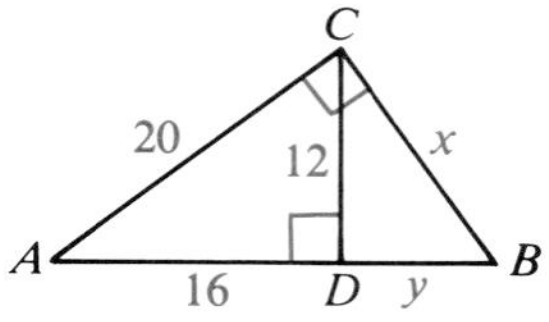

16. To estimate the height of a pole, a basketball player exactly 2 m tall stood so that the ends of the shadows coincided. He found that $\overline{DE}$ and $\overline{DF}$ measured 1.6 m and 4.4 m, respectively. About how tall was the pole?

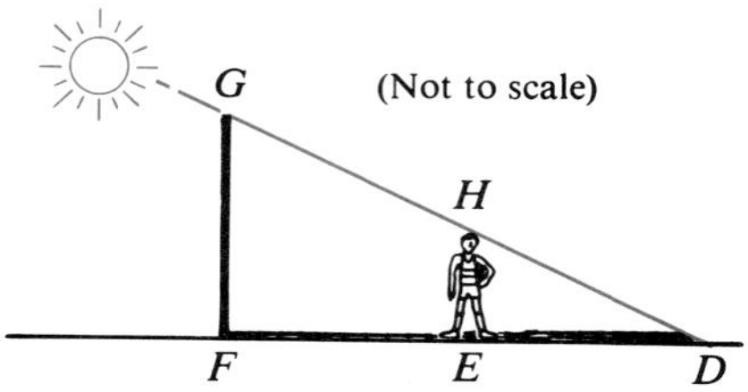

17. The diagram, *not* drawn to scale, shows a film being projected on a screen. $LF = 6$ cm and $LS = 24$ m. The screen image is 2.2 m tall. How tall is the film image?

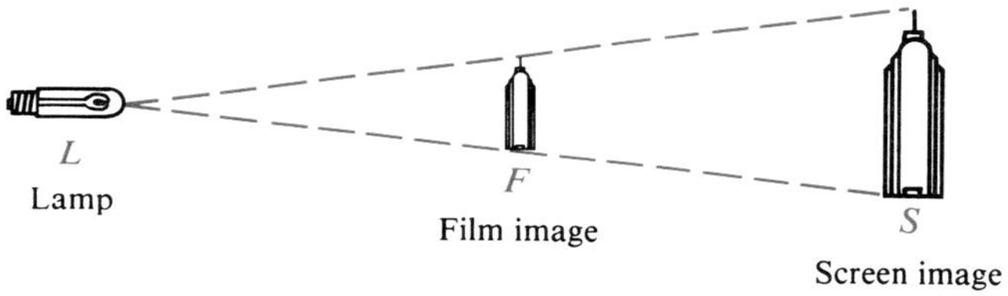

18. If $IV = 36$ m, $VE = 20$ m, and $EB = 15$ m, find the width, RI, of the river.

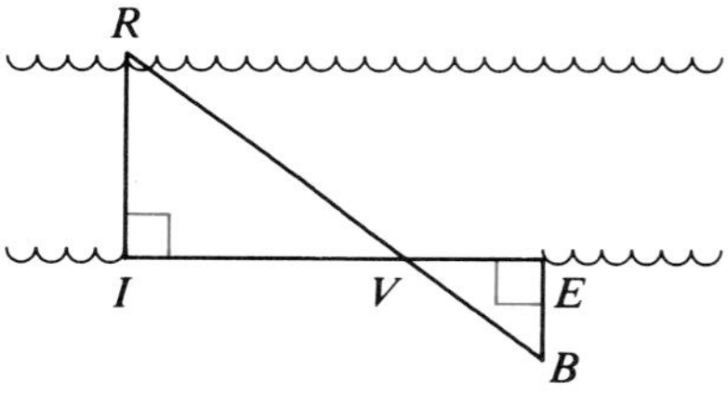

In Exercises 19 and 20, *ABCD* is a parallelogram. Find the values of *x* and *y*.

19.

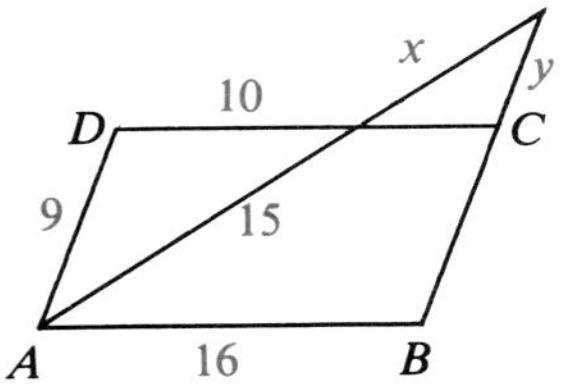

20.

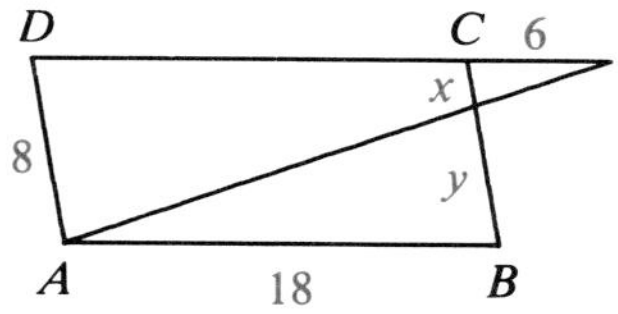

21. You can estimate the height of a flagpole by placing a mirror on level ground so that you see the top of the flagpole in it. The girl shown is 172 cm tall. Her eyes are about 12 cm from the top of her head. By measurement, AM is about 120 cm and $A'M$ is about 4.5 m. From physics it is known that $\angle 1 \cong \angle 2$. Explain why the triangles are similar and find the approximate height of the pole.

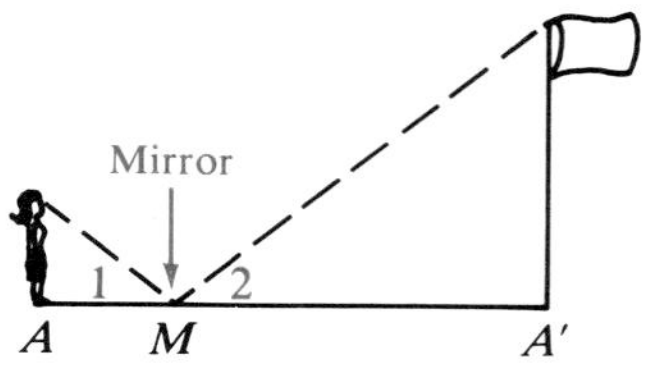

22. Given: $\overline{EF} \parallel \overline{RS}$
Prove: **a.** $\triangle FXE \sim \triangle SXR$
b. $\dfrac{FX}{SX} = \dfrac{EF}{RS}$

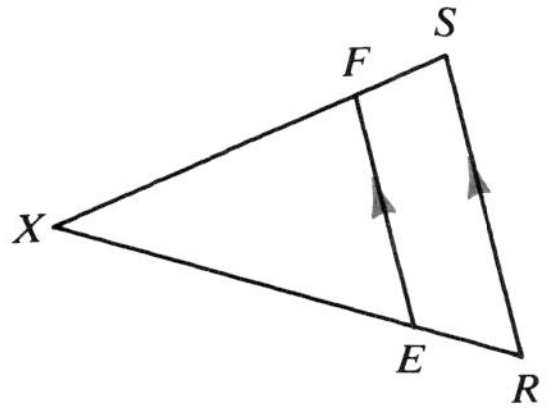

23. Given: $\angle 1 \cong \angle 2$
Prove: **a.** $\triangle JIG \sim \triangle JZY$
b. $\dfrac{JG}{JY} = \dfrac{GI}{YZ}$

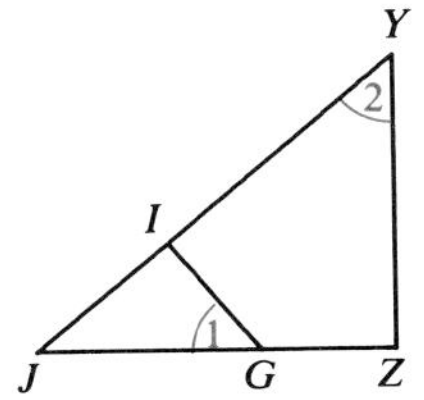

24. Given: $\angle B \cong \angle C$
Prove: $NM \cdot CM = LM \cdot BM$

25. Given: $\overline{BN} \parallel \overline{LC}$
Prove: $BN \cdot LM = CL \cdot NM$

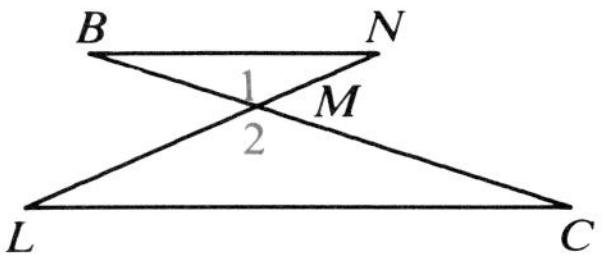

26. Given: $\angle D$ and $\angle AHE$ are right angles.
a. Prove two triangles similar.
b. Prove $AE \cdot DG = AG \cdot HE$

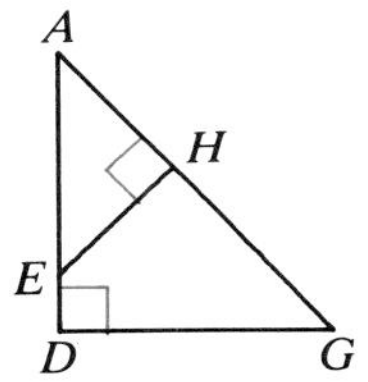

27. Given: $\overline{QT} \parallel \overline{RS}$
Prove: $\dfrac{QU}{RV} = \dfrac{UT}{VS}$

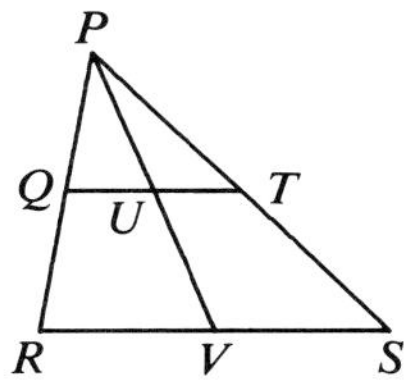

In the diagram for Exercises 28 and 29, the plane of $\triangle A'B'C'$ is parallel to the plane of $\triangle ABC$.

28. $VA' = 15$ and $A'A = 20$
a. If $VC' = 18$, then $VC =$ __?__.
b. If $VB = 49$, then $BB' =$ __?__.
c. If $A'B' = 24$, then $AB =$ __?__.

29. If $VA' = 10$, $VA = 25$, $AB = 20$, $BC = 14$, and $AC = 16$, find the perimeter of $\triangle A'B'C'$.

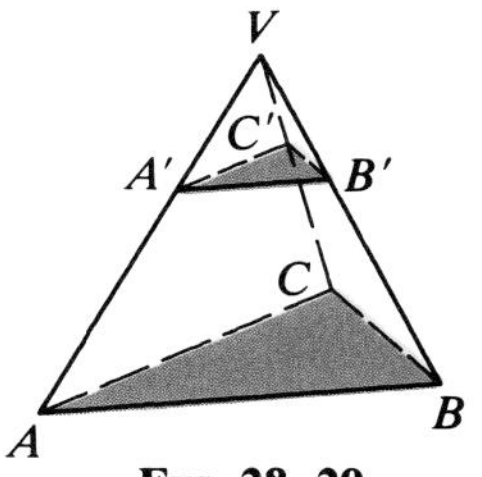

Exs. 28, 29

30. Prove that the lengths of corresponding altitudes of similar triangles have the same ratio as the lengths of corresponding sides.

C **31.** Prove that in any triangle the product of the lengths of one side and the altitude to that side is equal to the product of the lengths of another side and the altitude to that side.

32. Two vertical poles have heights 6 ft and 12 ft. A rope is stretched from the top of each pole to the bottom of the other. How far above the ground do the ropes cross? (*Hint:* The lengths y and z do not affect the answer.)

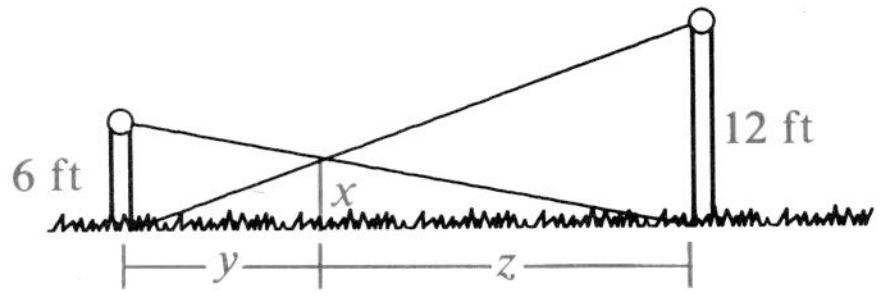

In Exercises 33–36 write a paragraph proof for anything you are asked to prove.

33. Given: $\overline{QN}$ and $\overline{RM}$ are medians of $\triangle QRS$.

Prove: $\frac{QP}{PN} = \frac{2}{1}$ and $\frac{RP}{PM} = \frac{2}{1}$

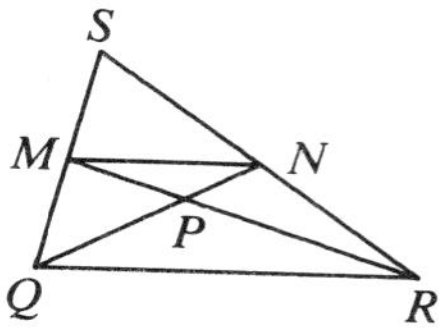

34. Given: Regular pentagon $ABCDE$

a. Make a large copy of the diagram.

b. Write the angle measures on your diagram.

c. Prove that $\frac{DA}{DK} = \frac{DK}{AK}$.

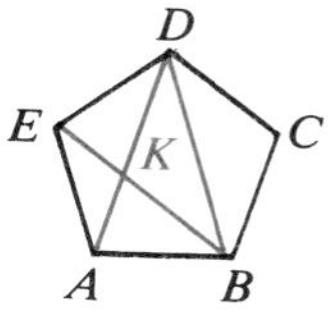

★35. Related to any doubly convex lens there is a focal distance OF. Physicists have determined experimentally that a vertical lens, a vertical object $\overline{JT}$ (with $\overline{JO}$ horizontal), a vertical image $\overline{IM}$, and a focus F are related as shown in the diagram. Once the relationship is known, geometry can be used to establish a lens law:

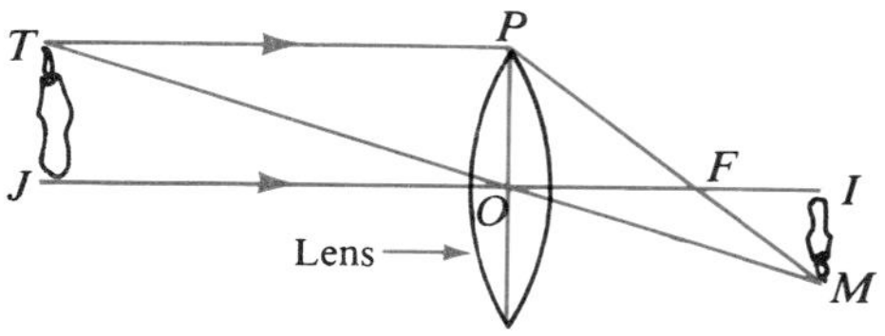

$$\frac{1}{\text{object distance}} + \frac{1}{\text{image distance}} = \frac{1}{\text{focal distance}}$$

a. Prove that $\frac{1}{OJ} + \frac{1}{OI} = \frac{1}{OF}$.

b. Show algebraically that $OF = \frac{OJ \cdot OI}{OJ + OI}$.

★36. Given: rectangle $XYTU$;

$\overline{XV} \perp \overline{UY}$; $\overline{VW} \perp \overline{UX}$

Prove: quad. $XYVW \sim$ quad. $YTUX$

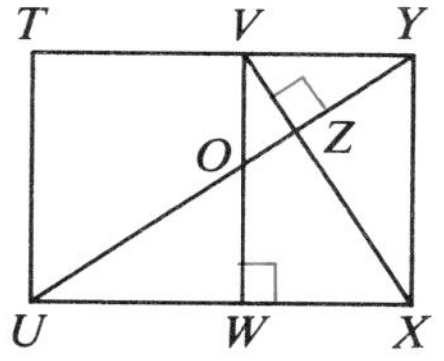

★37. $ABCD$ is a square.

a. Find the distance from H to each side of the square.

b. Find BF, FC, CG, DE, EA, EH, and HF.

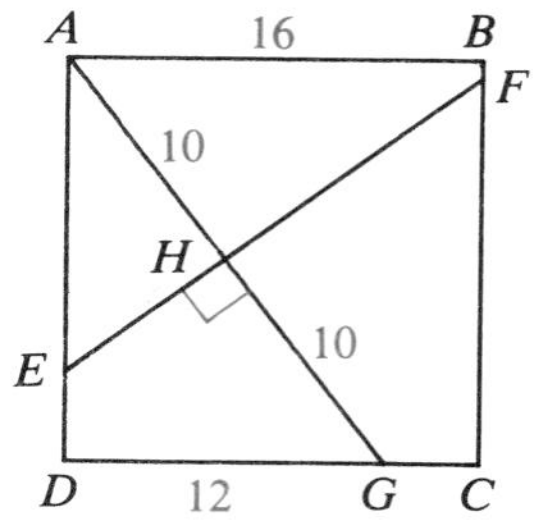

COMPUTER KEY-IN

1. a. Refer to Exercise 35, page 224. Write a computer program that will compute the focal distance when the object distance and image distance are given. Use the formula derived in Exercise 35(b).

   b. Trials with a particular lens resulted in the measurements (given in centimeters) shown in the table below. RUN your program for the values and complete the table.

OJ	60	55	50	45	40	35	30	25	20
OI	20	20.5	21	22.5	24	26	30.5	42	61
OF	?	?	?	?	?	?	?	?	?

   c. Modify your program so that it will compute the *average* focal distance using the nine experimental values of OF found in part (b).

2. a. Let the average focal distance, correct to the nearest tenth, found in Exercise 1(c) be *the* focal distance of the lens. Substitute this value for OF in the equation $\frac{1}{OJ} + \frac{1}{OI} = \frac{1}{OF}$. Then solve for OI in terms of OJ.

   b. Write and RUN a computer program to complete the table below that gives values for OJ, the object distance.

OJ	100	90	80	70	65	15	10	5
OI	?	?	?	?	?	?	?	?

BIOGRAPHICAL NOTE

R. Buckminster Fuller

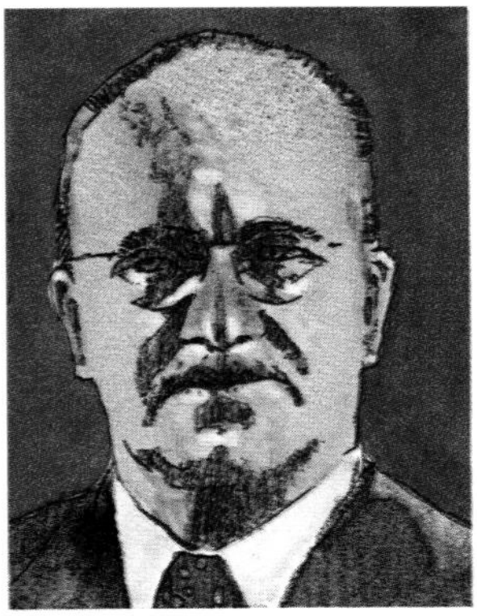

The early curiosity shown by R. Buckminster Fuller (1895–1983) about the world around him led to a life of invention and philosophy. As a mathematician he made many contributions to the fields of engineering, architecture, and cartography. His ultimate goal was always "to do more with less." Thus his discoveries often had economic and ecological implications.

Fuller's inventions include the geodesic dome (see pages 450 and 451), the 3-wheeled Dymaxion car, and the Dymaxion Air-ocean World Map on which he was able to project the spherical earth as a flat surface without any visible distortions. He also designed other structures that were based upon triangles and circles instead of the usual rectangular surfaces.

5-5 Theorems for Similar Triangles

You can prove two triangles similar by using the definition of similar polygons or by using the AA Postulate. Of course, in practice you would always use the AA Postulate instead of the definition. Two additional methods are established in the theorems below. Proofs are left as Exercises 19 and 20.

Theorem 5-1 SAS Similarity Theorem

If an angle of one triangle is congruent to an angle of another triangle and the sides including those angles are in proportion, then the triangles are similar.

Given: $\angle A \cong \angle D$;
$\frac{AB}{DE} = \frac{AC}{DF}$

Prove: $\triangle ABC \sim \triangle DEF$

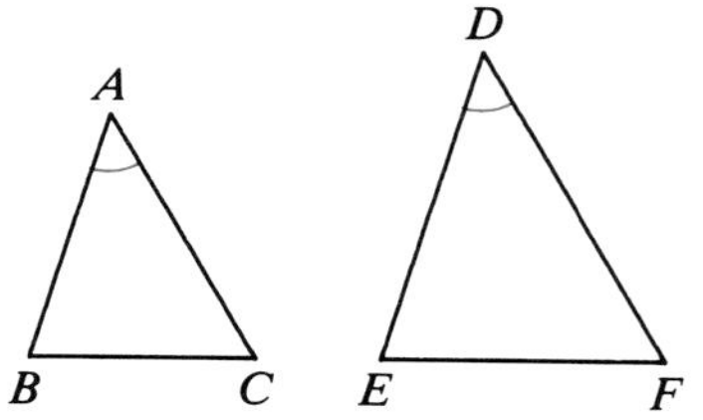

■ ***Plan for Proof:*** Assume $AB < DE$. Take X on $\overline{DE}$ so that $DX = AB$. Through X draw a line parallel to $\overleftrightarrow{EF}$. Then $\triangle DXY \sim \triangle DEF$ and $\frac{DX}{DE} = \frac{DY}{DF}$. Use this proportion, the given proportion, and the fact that $DX = AB$ to show that $DY = AC$. Then $\triangle ABC \cong \triangle DXY$ and $\angle B \cong \angle 1$. But $\angle 1 \cong \angle E$, so $\angle B \cong \angle E$. Also, $\angle A \cong \angle D$. Use the AA Similarity Postulate to show that $\triangle ABC \sim \triangle DEF$.

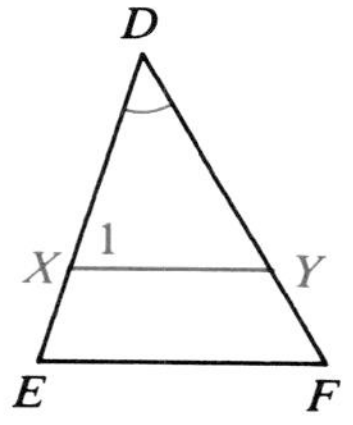

Theorem 5-2 SSS Similarity Theorem

If the sides of two triangles are in proportion, then the two triangles are similar.

Given: $\frac{AB}{DE} = \frac{BC}{EF} = \frac{AC}{DF}$

Prove: $\triangle ABC \sim \triangle DEF$

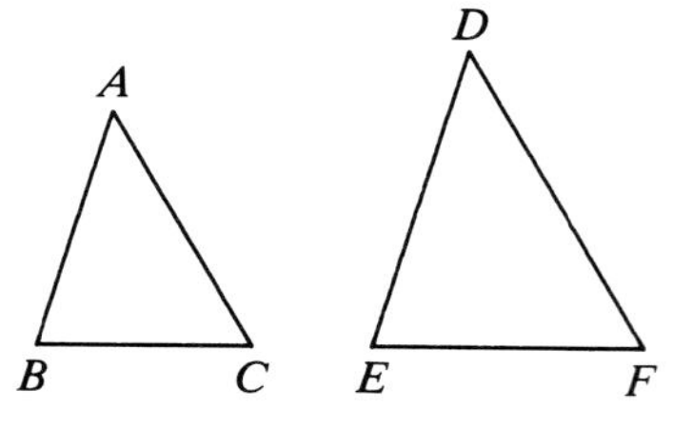

■ ***Plan for Proof:*** Assume $AB < DE$. Take X on $\overline{DE}$ so that $DX = AB$ and draw a line parallel to $\overleftrightarrow{EF}$. $\triangle DXY \sim \triangle DEF$ and $\frac{DX}{DE} = \frac{XY}{EF} = \frac{DY}{DF}$. Use this extended proportion, the given extended proportion, and the fact that $DX = AB$ to show that $BC = XY$ and $AC = DY$. Then $\triangle ABC \cong \triangle DXY$, and $\angle B \cong \angle 1$. Since $\angle 1 \cong \angle E$, $\angle B \cong \angle E$. Finally, apply the SAS Similarity Theorem.

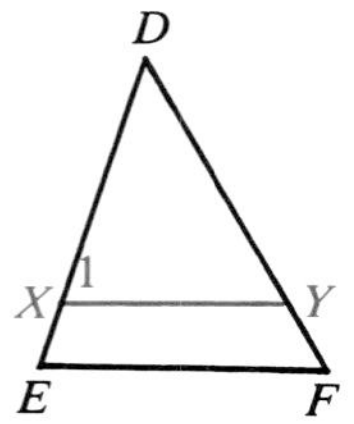

Example Can the given information be used to prove $\triangle RST \sim \triangle WZT$? If so, how?

a. $RS = 18$, $ST = 15$, $RT = 10$, $WT = 6$, $ZT = 9$, $WZ = 10.8$

b. $\overline{ST} \perp \overline{RW}$, $SZ = 8$, $ZT = 24$, $RT = 20$, $WT = 15$

c. $\angle 1 \cong \angle 2$, $\dfrac{ST}{ZT} = \dfrac{RS}{WZ}$

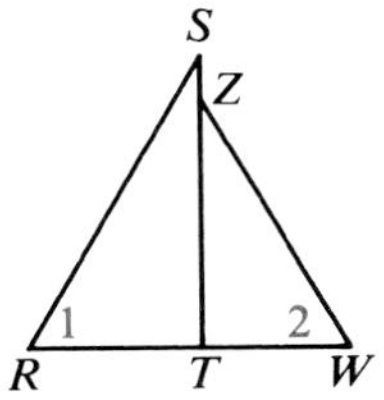

Solution **a.** $\dfrac{RS}{WZ} = \dfrac{18}{10.8} = \dfrac{5}{3}$, $\dfrac{ST}{ZT} = \dfrac{15}{9} = \dfrac{5}{3}$, $\dfrac{RT}{WT} = \dfrac{10}{6} = \dfrac{5}{3}$

Thus $\dfrac{RS}{WZ} = \dfrac{ST}{ZT} = \dfrac{RT}{WT}$.

$\triangle RST \sim \triangle WZT$ by the SSS Similarity Theorem.

b. $\dfrac{ST}{ZT} = \dfrac{32}{24} = \dfrac{4}{3}$ and $\dfrac{RT}{WT} = \dfrac{20}{15} = \dfrac{4}{3}$; $\angle STR \cong \angle ZTW$

$\triangle RST \sim \triangle WZT$ by the SAS Similarity Theorem.

c. The triangles cannot be proved similar. (Notice that $\angle 1$ and $\angle 2$ are not the angles included by the sides that are known to be in proportion.)

Classroom Exercises

Can two triangles shown be proved similar? If so, state the similarity and tell which similarity postulate or theorem you would use.

1.

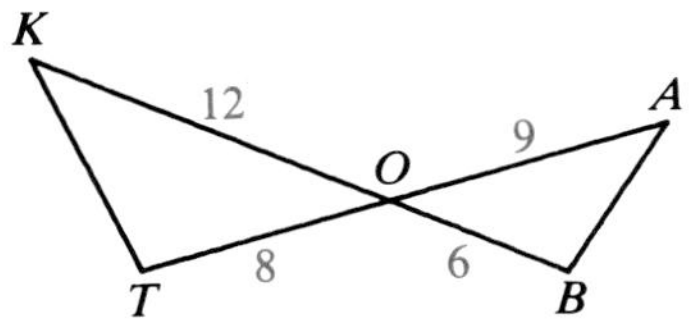

2.

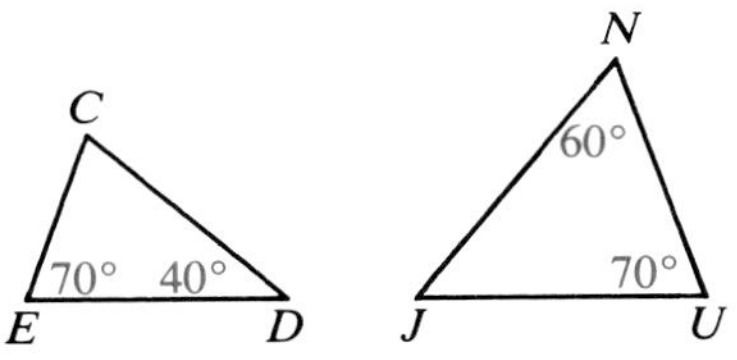

3.

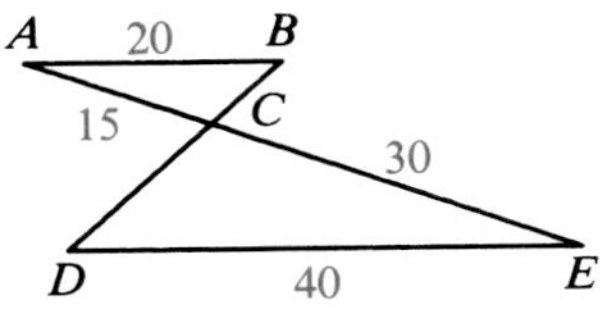

4.

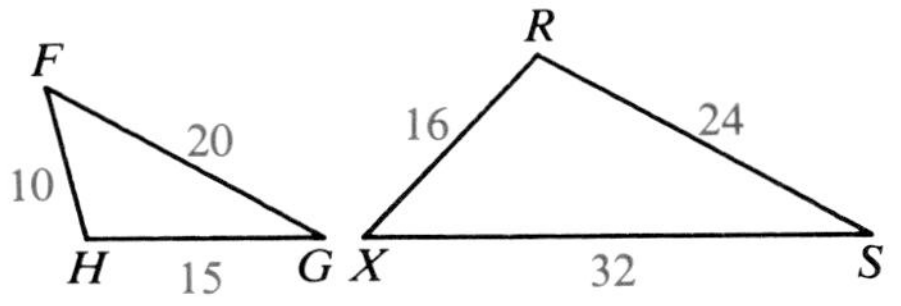

5.

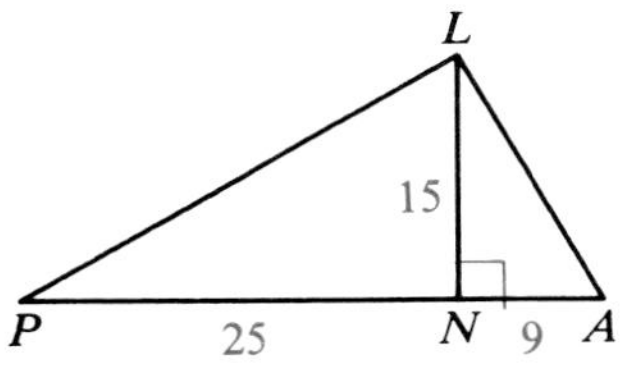

6.

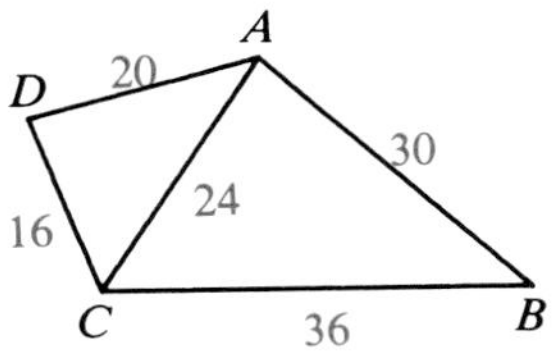

7. Suppose you want to prove that $\triangle RST \sim \triangle XYZ$ by the SSS Similarity Theorem. State the extended proportion you would need to prove first.
8. Suppose you want to prove that $\triangle RST \sim \triangle XYZ$ by the SAS Similarity Theorem. If you know that $\angle R \cong \angle X$, what else would you need to prove?
9. A *pantograph* is a tool for enlarging or reducing maps and drawings. Four bars are pinned together at A, B, C, and D so that $ABCD$ is a parallelogram and points P, D, and E lie on a line. Point P is fixed to the drawing board. To enlarge a figure, the artist inserts a stylus at D and guides the pen or pencil at E so that the stylus traces the original. As E moves, the angles of the parallelogram change, but P, D, and E remain collinear. Suppose PA is 3 units and AB is 7 units.

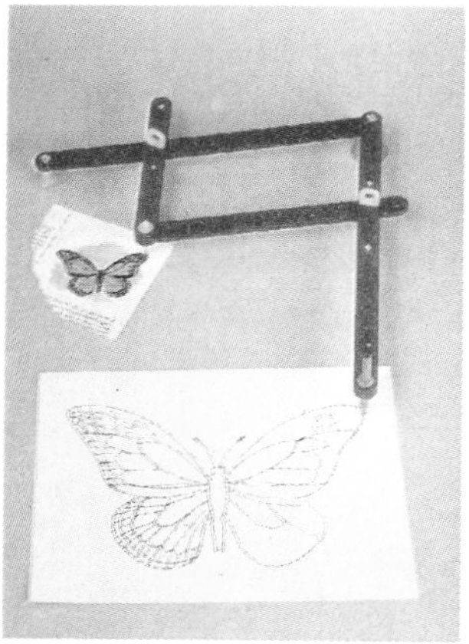

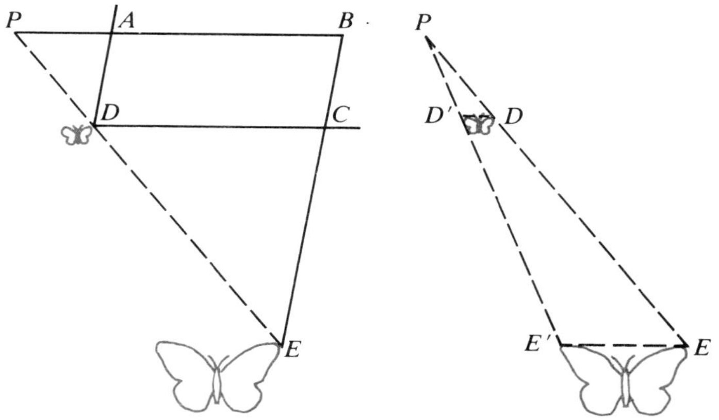

a. Explain why $\triangle PBE \sim \triangle PAD$.
b. What is the ratio of PB to PA?
c. What is the ratio of PE to PD?
d. What is the ratio of the butterfly's wingspan, $E'E$, in the enlargement to its wingspan, $D'D$, in the original?

Written Exercises

Name two similar triangles. Also name the postulate or theorem that justifies your answer.

A 1.

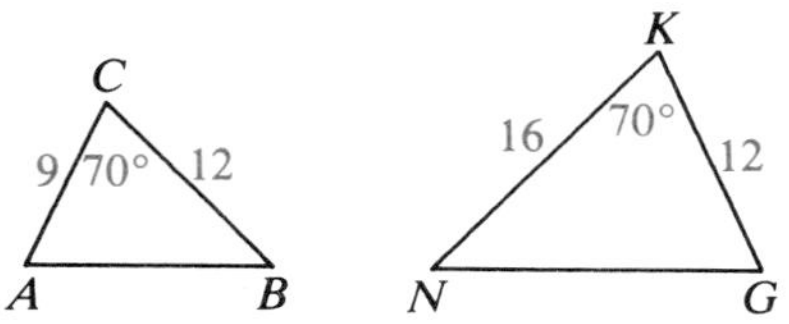

2.

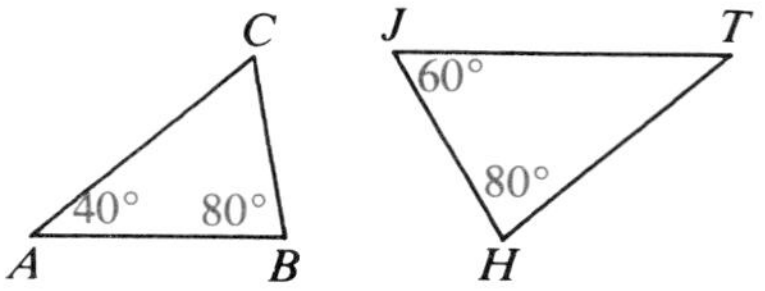

3.
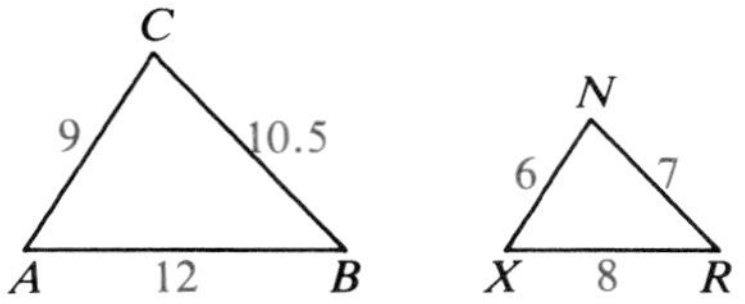

4.
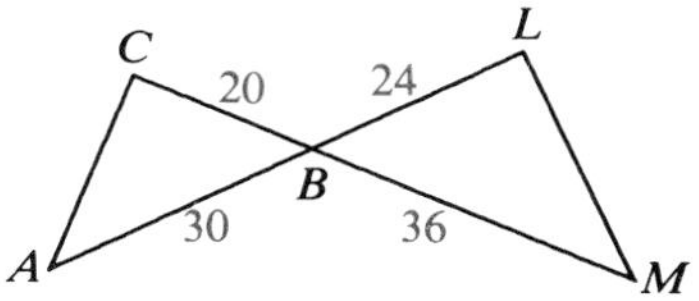

5.
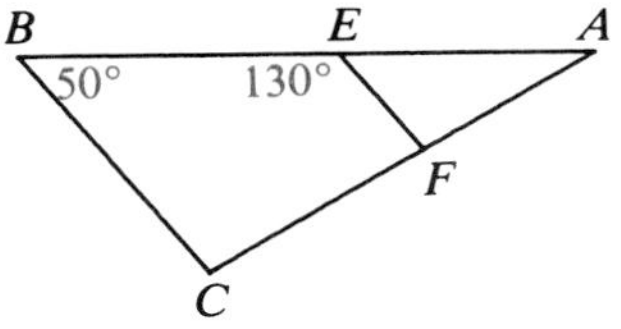

6.
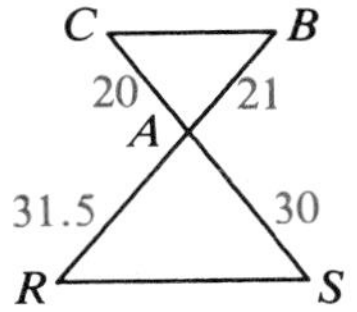

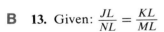

One triangle has vertices A, B, and C. Another triangle has vertices P, K, and N. Are two triangles similar? If so, state the similarity and the scale factor.

	AB	BC	AC	PK	KN	PN
7.	6	8	10	9	12	15
8.	6	8	10	15	9	12
9.	6	8	10	25	20	16
10.	12	16	18	20	22.5	15

11. Given: $\frac{DE}{GH} = \frac{DF}{GI} = \frac{EF}{HI}$

Prove: $\angle E \cong \angle H$

12. Given: $\frac{DE}{GH} = \frac{EF}{HI}$; $\angle E \cong \angle H$

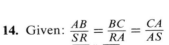

Prove: $\frac{EF}{HI} = \frac{DF}{GI}$

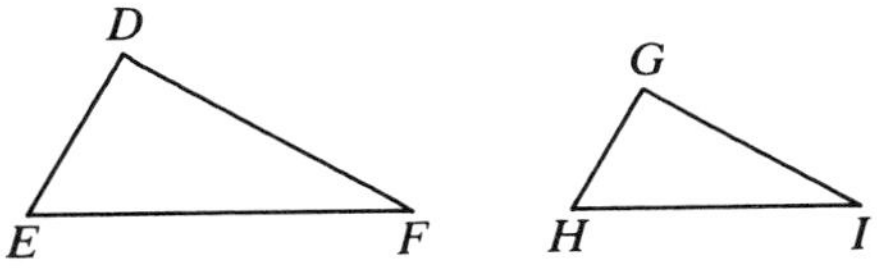

B **13.** Given: $\frac{JL}{NL} = \frac{KL}{ML}$

Prove: $\angle J \cong \angle N$

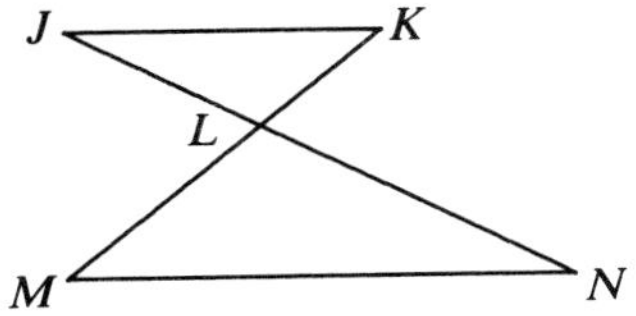

14. Given: $\frac{AB}{SR} = \frac{BC}{RA} = \frac{CA}{AS}$

Prove: $\overline{BC} \parallel \overline{AR}$

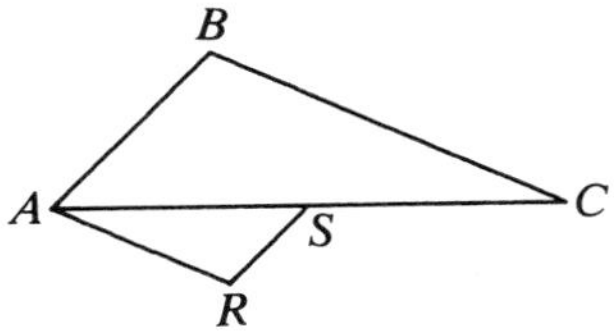

15. Given: $\frac{VW}{VX} = \frac{VZ}{VY}$

Prove: $\overline{WZ} \parallel \overline{XY}$

16. Given: $\frac{VW}{VY} = \frac{VZ}{VX}$

Which one(s) of the following *must* be true?

(1) $\triangle VWZ \sim \triangle VXY$ (2) $\overline{WZ} \parallel \overline{XY}$ (3) $\angle 1 \cong \angle Y$

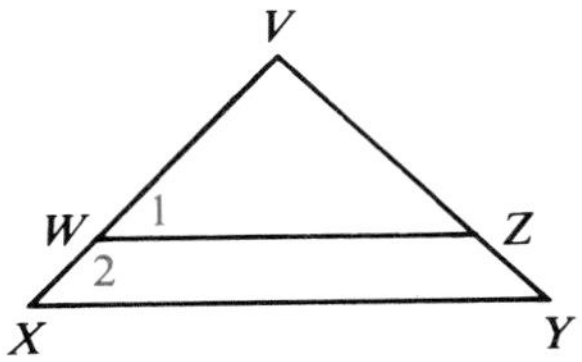

17. The faces of a cube are congruent squares. The cube shown is cut by plane $ABCD$. $VA = VB$ and $VW = 4 \cdot VA$. Find, in terms of AB, the length of the median of trap. $ABCD$.

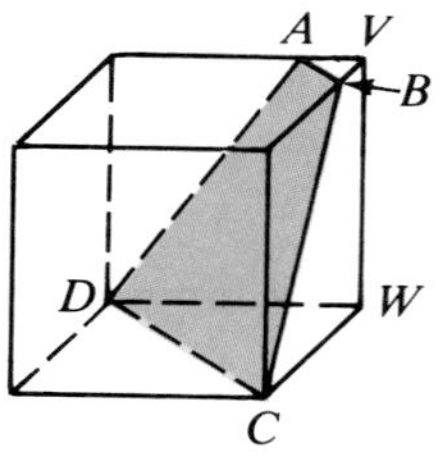

18. Given: $OR' = 2 \cdot OR$;
$OS' = 2 \cdot OS$;
$OT' = 2 \cdot OT$
Prove: $\triangle RST \sim \triangle R'S'T'$

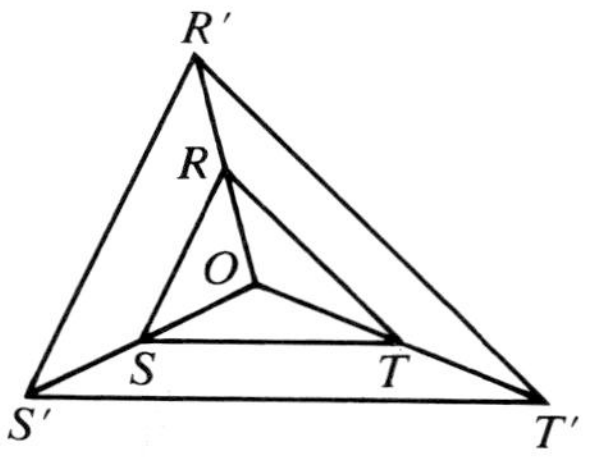

19. Prove the SAS Similarity Theorem.

20. Prove the SSS Similarity Theorem.

21. Prove: If the vertex angle of one isosceles triangle is congruent to the vertex angle of another isosceles triangle, then the triangles are similar.

22. Prove Theorem 4–17 on page 174: The segment that joins the midpoints of two sides of a triangle is parallel to the third side and is half as long as the third side.
Given: M is the midpoint of $\overline{AB}$;
N is the midpoint of $\overline{AC}$.
Prove: $\overline{MN} \parallel \overline{BC}$; $MN = \frac{1}{2}BC$

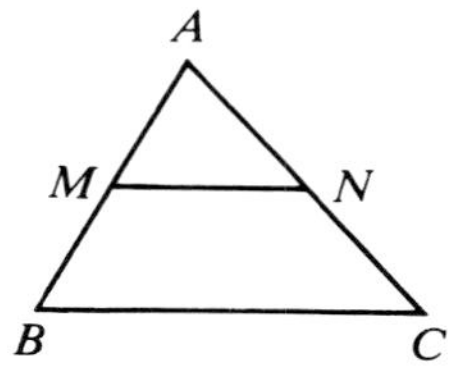

C 23. Prove that the lengths of corresponding medians of similar triangles have the same ratio as the lengths of corresponding sides.

24. Given: $\square WXYZ$
Prove: $\triangle ATB \sim \triangle A'TB'$
(*Hint:* Show that $\frac{AT}{A'T}$ and $\frac{BT}{B'T}$ both equal $\frac{TW}{TY}$.)

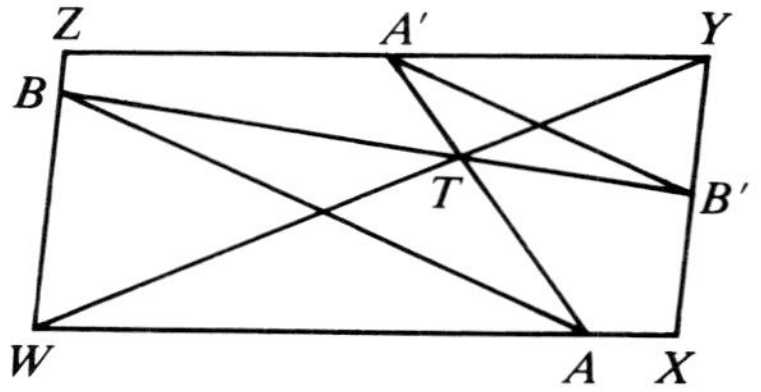

★ 25. In $\triangle CAT$, it is known that $AC = 20$, $AT = 25$, and $\overline{AM}$ is a median. P is any point on $\overline{AT}$. Q lies on $\overline{AC}$ with $AQ = \frac{3}{5}AP$. $\overline{QP}$ and $\overline{AM}$ intersect at K.
a. Draw the diagram for the case when P is point T. Then find the ratio $QK : PK$.
b. Show that the ratio $QK : PK$ is the same for any position of P on $\overline{AT}$.

Challenge

Explain how to pass a plane through a cube in such a way that the intersection is (**a**) an equilateral triangle; (**b**) a trapezoid; (**c**) a pentagon; (**d**) a hexagon.

CALCULATOR KEY-IN

Before the arch on top of the Parthenon in Athens was destroyed, the front of the building fit almost exactly into a *golden rectangle*. A **golden rectangle** is such that its length l and width w satisfy the equation $\frac{l}{w} = \frac{l + w}{l}$. The ratio $\frac{l}{w}$ is called the **golden ratio.**

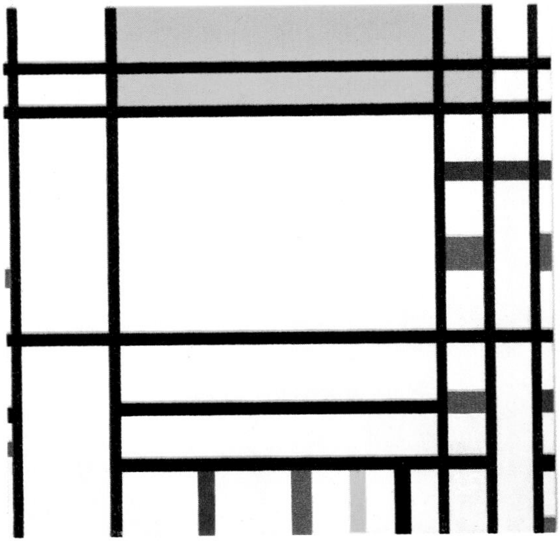

Over the centuries, artists and architects have found the golden rectangle to be especially pleasing to the eye. How many golden rectangles can you find in the painting by Piet Mondrian (1872–1944) that is shown?

Exercises

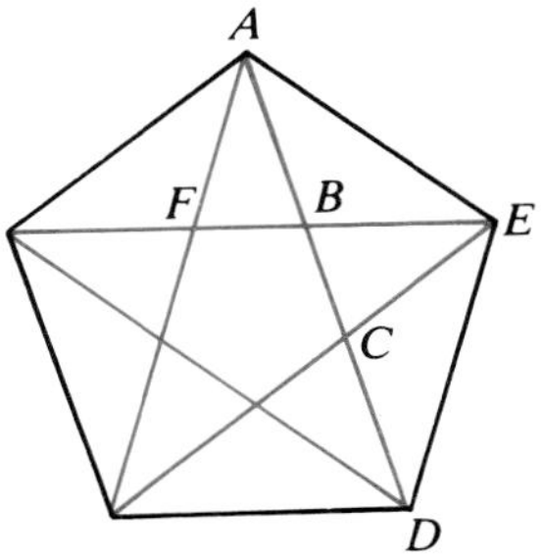

1. A regular pentagon is shown. It happens to be true that $\frac{AD}{AC}$, $\frac{AC}{AB}$, and $\frac{AB}{BC}$ all equal the golden ratio. Measure the appropriate lengths to the nearest millimeter and compute the ratios with a calculator.

2. From the equation $\frac{l}{w} = \frac{l + w}{l}$ it can be shown that the numerical value of $\frac{l}{w}$ is $\frac{1 + \sqrt{5}}{2}$. Express the value of $\frac{l}{w}$, the golden ratio, as a decimal.

3. Sometimes the golden ratio is expressed as $\frac{w}{l}$ rather than $\frac{l}{w}$. From Exercise 2 you see that $\frac{w}{l} = \frac{2}{1 + \sqrt{5}}$. Express $\frac{w}{l}$ as a decimal.

COMPUTER KEY-IN

The sequence 1, 1, 2, 3, 5, 8, 13, 21, . . . is called a *Fibonacci sequence* after its discoverer, Leonardo Fibonacci, a 13th-century mathematician. The first two terms are 1 and 1. You then add two consecutive terms to get the next term.

$$\begin{array}{ccccc} \text{1st term} & + & \text{2nd term} & = & \text{3rd term} \\ 1 & + & 1 & = & 2 \end{array} \qquad \begin{array}{ccccc} \text{2nd term} & + & \text{3rd term} & = & \text{4th term} \\ 1 & + & 2 & = & 3 \end{array}$$

$$\begin{array}{ccccc} \text{3rd} & + & \text{4th} & = & \text{5th} \\ 2 & + & 3 & = & 5 \end{array} \qquad \begin{array}{ccccc} \text{4th} & + & \text{5th} & = & \text{6th} \\ 3 & + & 5 & = & 8 \end{array}$$

The following computer program computes the first twenty-five terms of the Fibonacci sequence shown above and finds the ratio of any term to its preceding term. For example, we want to look at the ratios

$$\frac{1}{1} = 1,\ \frac{2}{1} = 2,\ \frac{3}{2} = 1.5,\ \frac{5}{3} \approx 1.66667, \text{ and so on.}$$

```
10  PRINT "TERM NO.","TERM","RATIO TO PRECEDING TERM"
20  LET A = 1
30  LET B = 1
40  PRINT 1,A,"-"
50  FOR N = 2 TO 25
60  PRINT N,B,B/A
70  LET C = B + A
80  LET A = B
90  LET B = C
100  NEXT N
110  END
```

Exercises

1. RUN the given computer program. As the terms become larger, what happens to the values of the ratios?
2. Suppose another sequence is formed by choosing starting numbers different from 1 and 1. For example, suppose the sequence is 3, 11, 14, 25, 39, . . . , where the pattern for creating the terms of the sequence is still the same. Change lines 20 and 30 to:

   ```
   20  LET A = 3
   30  LET B = 11
   ```

   RUN the modified program. What happens to the values of the ratios as the terms become larger and larger?
3. Modify the program again so that another pair of starting numbers is used and the first fifty terms are computed. RUN the program. What can you conclude from the results?

5-6 Proportional Lengths

Points L and M lie on $\overline{AB}$ and $\overline{CD}$, respectively. If $\frac{AL}{LB} = \frac{CM}{MD}$, we say that $\overline{AB}$ and $\overline{CD}$ are **divided proportionally.**

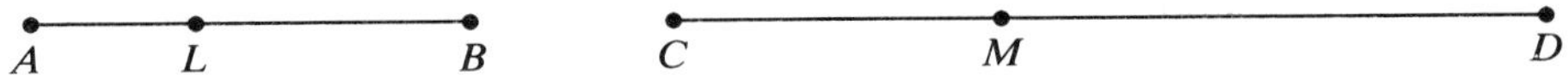

Theorem 5-3 Triangle Proportionality Theorem

If a line parallel to one side of a triangle intersects the other two sides, then it divides those sides proportionally.

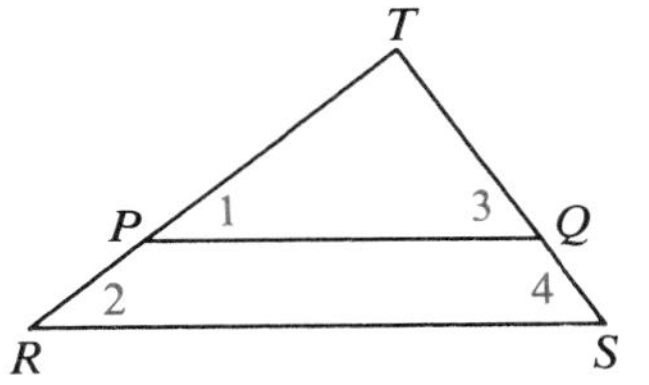

Given: $\triangle RST$; $\overleftrightarrow{PQ} \parallel \overline{RS}$

Prove: $\frac{RP}{PT} = \frac{SQ}{QT}$

Proof:

Statements	Reasons
1. $\overleftrightarrow{PQ} \parallel \overline{RS}$	1. ?
2. $\angle 1 \cong \angle 2$; $\angle 3 \cong \angle 4$	2. ?
3. $\triangle RST \sim \triangle PQT$	3. ?
4. $\frac{RT}{PT} = \frac{ST}{QT}$	4. Corr. sides of $\sim$ ⚠ are in proportion.
5. $RT = RP + PT$; $ST = SQ + QT$	5. ?
6. $\frac{RP + PT}{PT} = \frac{SQ + QT}{QT}$	6. ?
7. $\frac{RP}{PT} = \frac{SQ}{QT}$	7. A property of proportions (Property 1(d), p. 209)

We will use the Triangle Proportionality Theorem to justify any proportion equivalent to $\frac{RP}{PT} = \frac{SQ}{QT}$. For the diagram at the right, some of the proportions that may be justified by the Triangle Proportionality Theorem include:

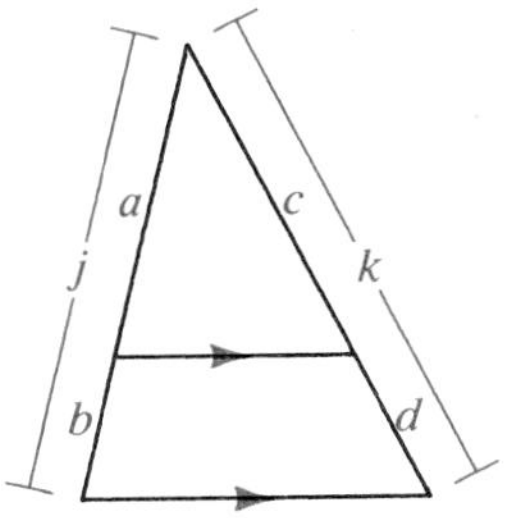

$$\frac{a}{j} = \frac{c}{k} \qquad \frac{a}{c} = \frac{j}{k} \qquad \frac{b}{j} = \frac{d}{k}$$

$$\frac{a}{b} = \frac{c}{d} \qquad \frac{a}{c} = \frac{b}{d} \qquad \frac{b}{d} = \frac{j}{k}$$

Example Find the numerical value.

a. $\frac{TN}{NR}$ **b.** $\frac{TR}{NR}$ **c.** $\frac{RN}{RT}$

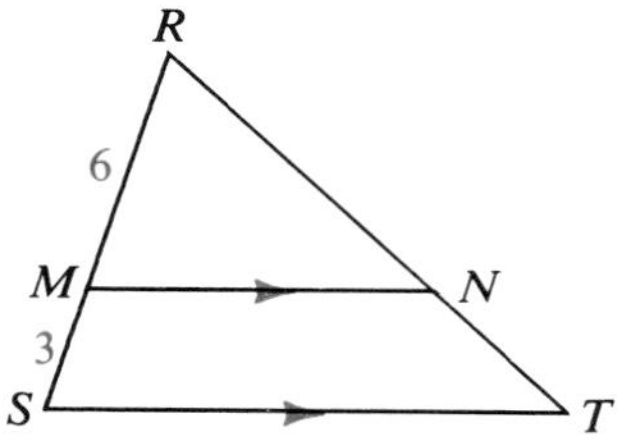

Solution **a.** $\frac{TN}{NR} = \frac{SM}{MR} = \frac{3}{6} = \frac{1}{2}$

b. $\frac{TR}{NR} = \frac{SR}{MR} = \frac{9}{6} = \frac{3}{2}$

c. $\frac{RN}{RT} = \frac{RM}{RS} = \frac{6}{9} = \frac{2}{3}$

Compare the following corollary with Theorem 4–8 on page 164.

Corollary

If three parallel lines intersect two transversals, then they divide the transversals proportionally.

Given: $\overleftrightarrow{RX} \parallel \overleftrightarrow{SY} \parallel \overleftrightarrow{TZ}$

Prove: $\frac{RS}{ST} = \frac{XY}{YZ}$

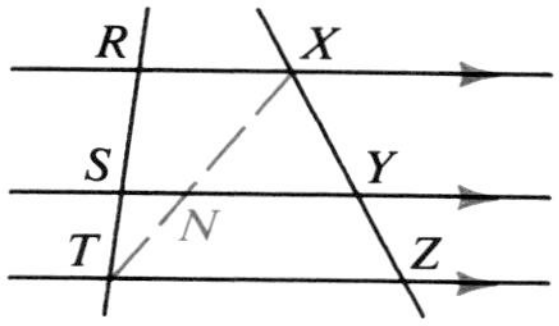

■ ***Plan for Proof:*** Draw $\overline{TX}$, intersecting $\overleftrightarrow{SY}$ at N. Note that $\overleftrightarrow{SY}$ is parallel to one side of $\triangle RTX$, and also to one side of $\triangle TXZ$. You can apply the Triangle Proportionality Theorem to show that $\frac{RS}{ST} = \frac{XN}{NT}$ and $\frac{XY}{YZ} = \frac{XN}{NT}$. Then $\frac{RS}{ST} = \frac{XY}{YZ}$.

Theorem 5–4 Triangle Angle-Bisector Theorem

If a ray bisects an angle of a triangle, then it divides the opposite side into segments proportional to the other two sides.

Given: $\triangle DEF$; $\overrightarrow{DG}$ bisects $\angle FDE$.

Prove: $\frac{GF}{GE} = \frac{DF}{DE}$

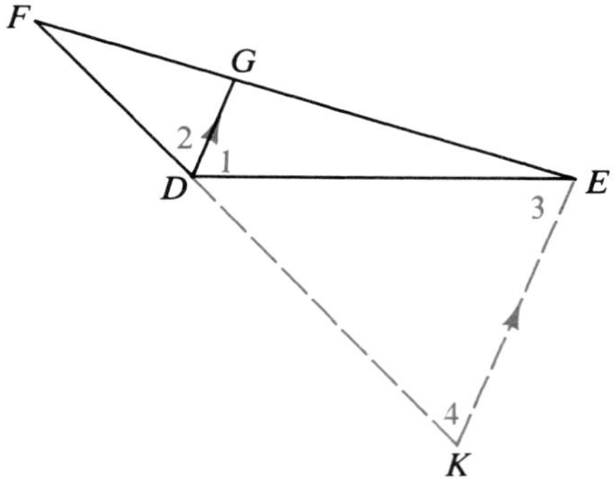

■ ***Plan for Proof:*** By drawing a line through E parallel to $\overrightarrow{DG}$ you can form a triangle to which you can apply the Triangle Proportionality Theorem. Using alternate interior angles, corresponding angles, and what is given, show that $\angle 3 \cong \angle 4$. Then $DK = DE$. But $\frac{GF}{GE} = \frac{DF}{DK}$ by the Triangle Proportionality Theorem, and substitution of DE for DK completes the proof.

Classroom Exercises

1. The two segments are divided proportionally. State several correct proportions.

2. State several proportions informally as shown in the proportion at the left below.

$$\frac{\text{upper left}}{\text{whole left}} = \frac{\text{upper right}}{\text{whole right}} \qquad \frac{\text{lower left}}{\text{upper left}} = \frac{?}{?}$$

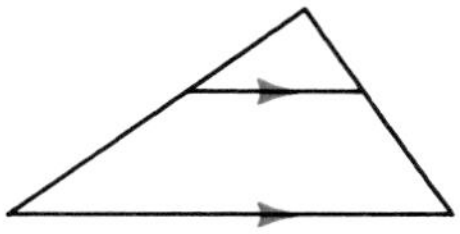

State a proportion for each diagram.

3.

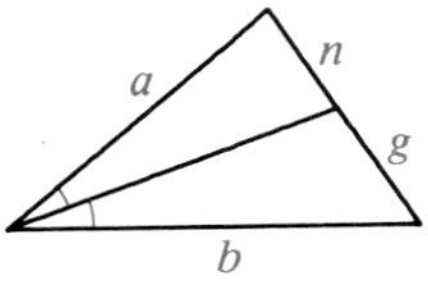

4.

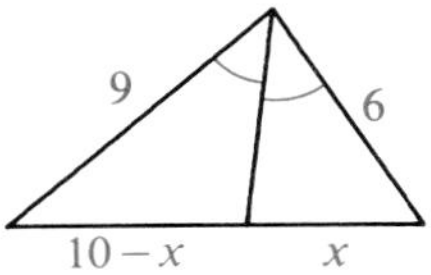

5.

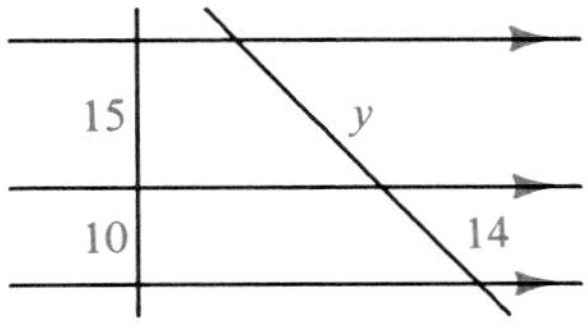

6. Suppose you want to find the length of the segment on the upper left. Three methods are suggested below. Complete each solution.

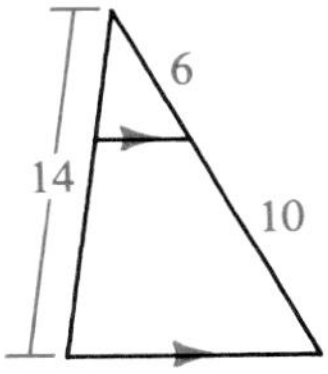

a.

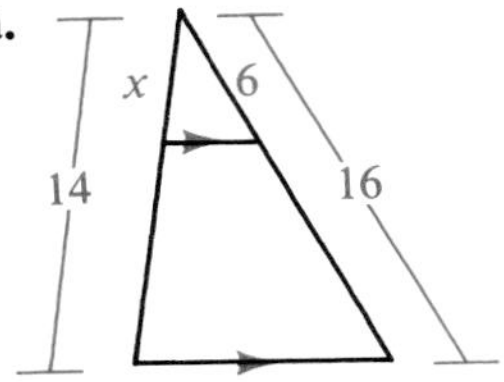

b.

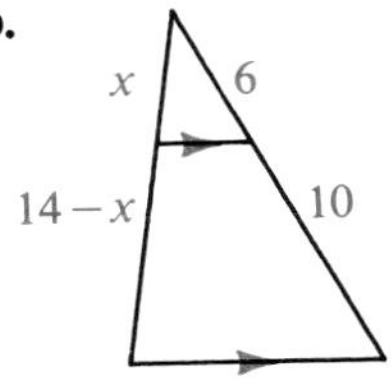

c.

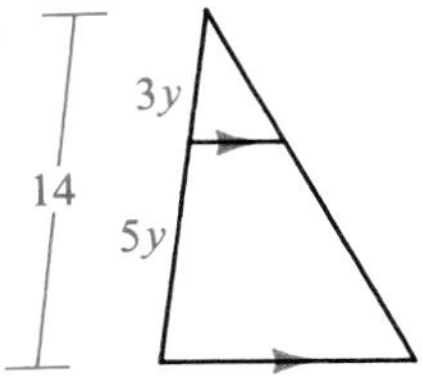

7. Explain why the expressions $3y$ and $5y$ can be used in Exercise 6(c).

8. a. State the converse of the corollary to the Triangle Proportionality Theorem.

b. Is the converse true? (*Hint:* Can you draw a diagram with lengths like those shown below, but in which lines r, s, and t are not parallel?)

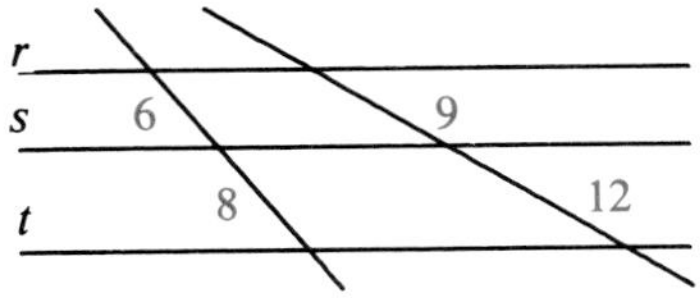

Ex. 8

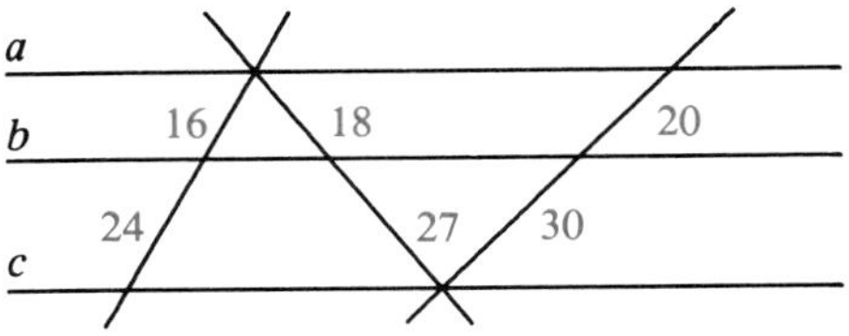

Ex. 9

9. Must lines a, b, and c shown above be parallel? Explain.

Written Exercises

A **1.** Tell whether the proportion is correct.

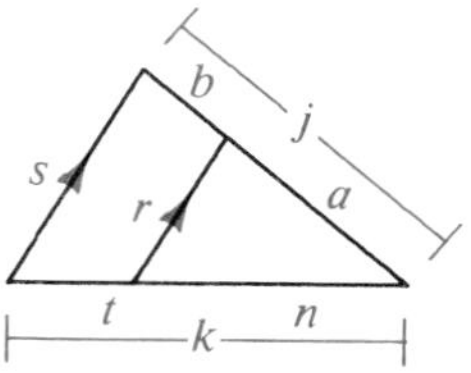

a. $\frac{r}{s} = \frac{a}{b}$ **b.** $\frac{j}{a} = \frac{s}{r}$ **c.** $\frac{a}{b} = \frac{n}{t}$

d. $\frac{t}{k} = \frac{a}{j}$ **e.** $\frac{r}{s} = \frac{n}{k}$ **f.** $\frac{b}{j} = \frac{t}{k}$

2. Tell whether the proportion is correct.

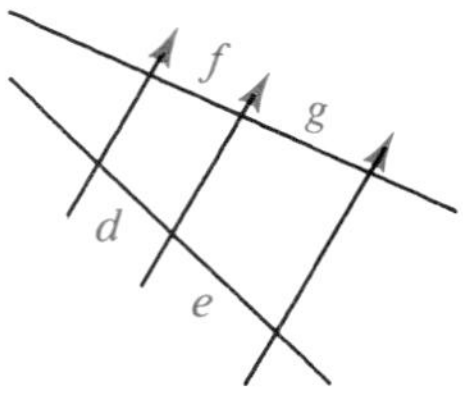

a. $\frac{d}{f} = \frac{g}{e}$ **b.** $\frac{f}{g} = \frac{e}{d}$

c. $\frac{g}{f} = \frac{e}{d}$ **d.** $\frac{d}{f} = \frac{e}{g}$

In Exercises 3–6, $\frac{AB}{BC} = \frac{3}{5}$. Copy and complete the table.

	3.	4.	5.	6.
AB	6	?	?	?
BC	?	25	?	?
AC	?	?	56	100

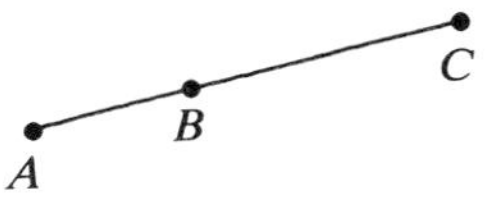

Find the value of x.

7.

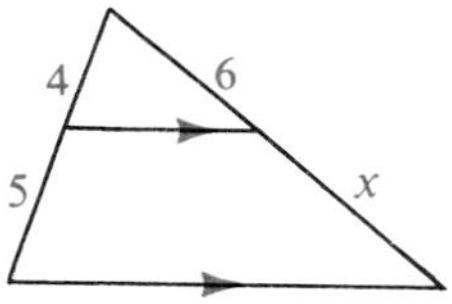

8.

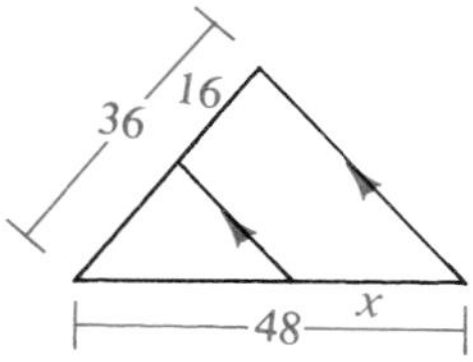

9.

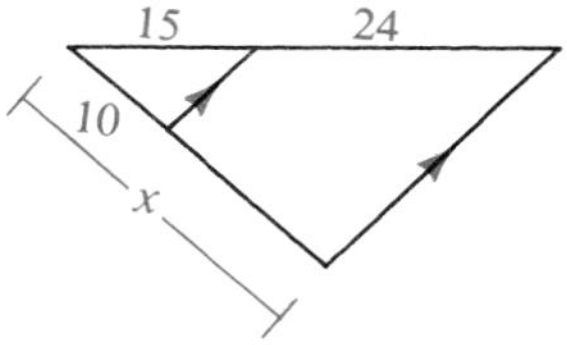

10.

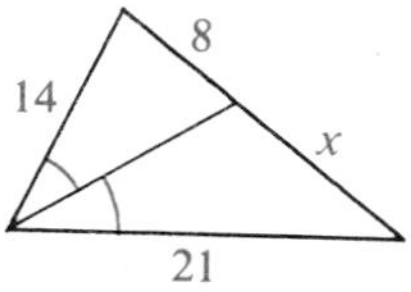

11.

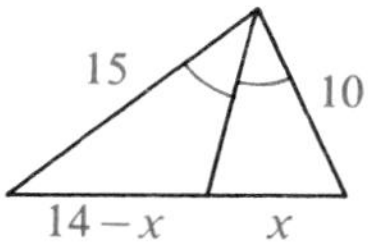

12.

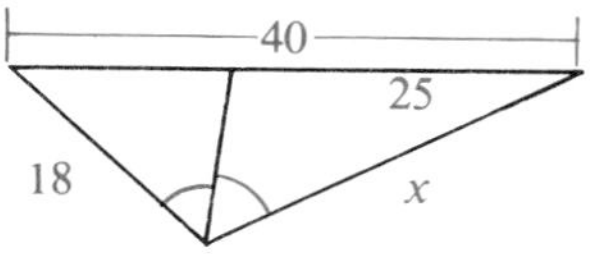

13.

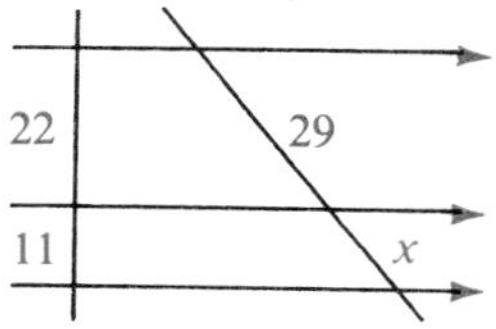

14.

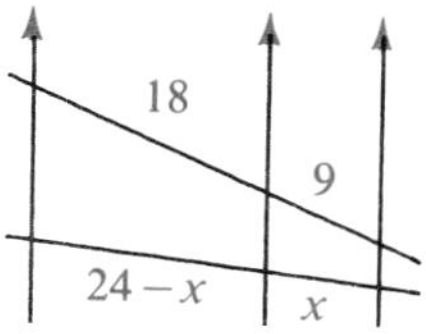

15.

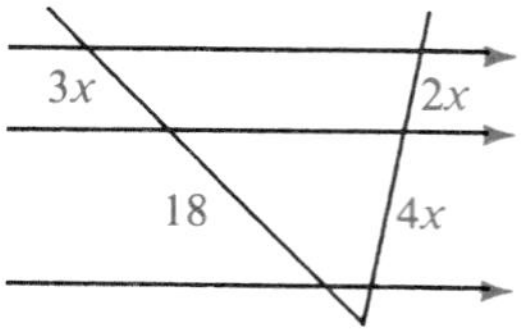

Copy the table and fill in as many spaces *as possible.* It may help to draw a new sketch for each exercise and label lengths as you find them.

	AR	RT	AT	AN	NP	AP	RN	TP
B **16.**	6	4	?	9	?	?	?	15
17.	?	?	?	?	6	16	?	?
18.	18	?	?	?	?	?	30	40
19.	12	?	20	?	?	30	15	?
20.	18	?	?	26	?	?	24	36
21.	?	?	33	24	20	?	?	50

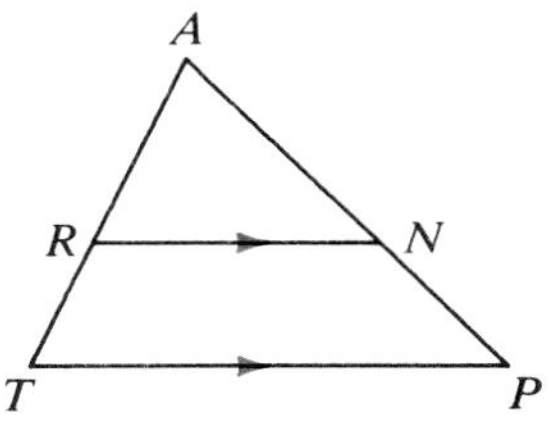

22. Prove the corollary to the Triangle Proportionality Theorem.

23. Prove the Triangle Angle-Bisector Theorem.

Complete.

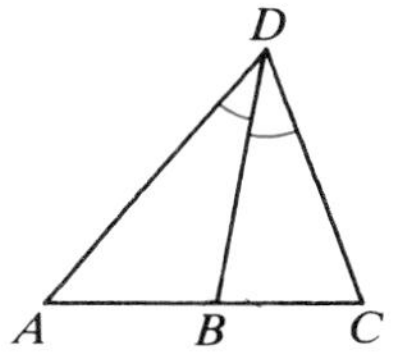

24. $AD = 21$, $DC = 14$, $AC = 25$, $AB =$ _?_

25. $AC = 60$, $CD = 30$, $AD = 50$, $BC =$ _?_

26. $AB = 27$, $BC = x$, $CD = \frac{4}{3}x$, $AD = x$, $AC =$ _?_

27. $AB = 2x - 12$, $BC = x$, $CD = x + 5$, $AD = 2x - 4$, $AC =$ _?_

28. Three lots with parallel side boundaries extend from the avenue to the road as shown. Find, to the nearest tenth of a meter, the frontages of the lots on Martin Luther King Avenue.

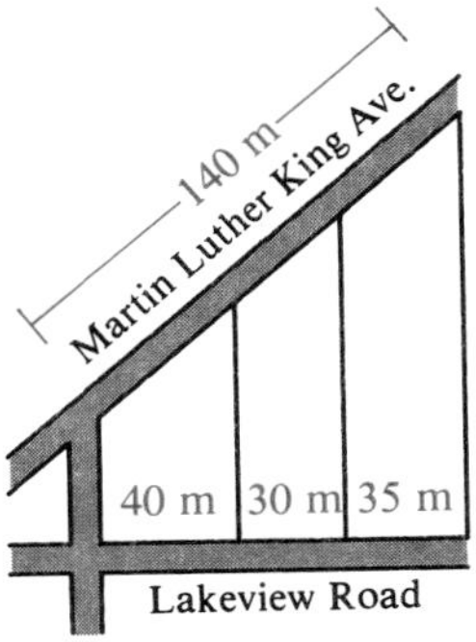

Ex. 28

29. The lengths of the sides of $\triangle ABC$ are $BC = 12$, $CA = 13$, and $AB = 14$. If M is the midpoint of $\overline{CA}$, and P is the point where $\overline{CA}$ is cut by the bisector of $\angle B$, find MP.

30. Prove: If a line bisects both an angle of a triangle and the opposite side, then the triangle is isosceles.

C **31.** Discover and prove a theorem, about planes and transversals, suggested by the corollary to the Triangle Proportionality Theorem.

32. Prove that there cannot be a triangle in which the trisectors of an angle also trisect the opposite side.

33. Can there exist a $\triangle ROS$ in which the trisectors of $\angle O$ intersect $\overline{RS}$ at D and E, with $RD = 1$, $DE = 2$, and $ES = 4$? Explain.

34. Angle E of $\triangle ZEN$ is obtuse. The bisector of $\angle E$ intersects $\overline{ZN}$ at X. J and K lie on $\overline{ZE}$ and $\overline{NE}$ with $ZJ = ZX$ and $NK = NX$. Discover and prove something about quadrilateral $ZNKJ$.

★ **35.** In $\triangle RST$, U lies on $\overline{TS}$ with $TU:US = 2:3$. M is the midpoint of $\overline{RU}$. $\overrightarrow{TM}$ intersects $\overline{RS}$ in V. Find the ratio $RV:VS$.

★ **36.** Prove *Ceva's Theorem:* If P is any point inside $\triangle ABC$, then $\frac{AX}{XB} \cdot \frac{BY}{YC} \cdot \frac{CZ}{ZA} = 1$.

(*Hint:* Draw lines parallel to $\overline{CX}$ through A and B. Apply the Triangle Proportionality Theorem to $\triangle ABM$. Show that $\triangle APN \sim \triangle MPB$, $\triangle BYM \sim \triangle CYP$, and $\triangle CZP \sim \triangle AZN$.)

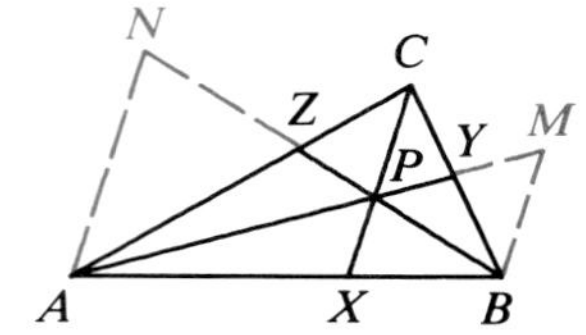

Self-Test 2

State the postulate or theorem you can use to prove that two triangles are similar.

1.

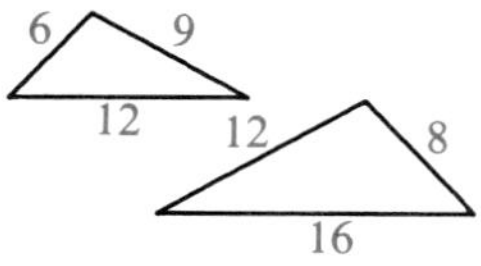

2.

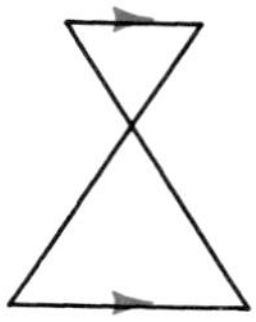

3.

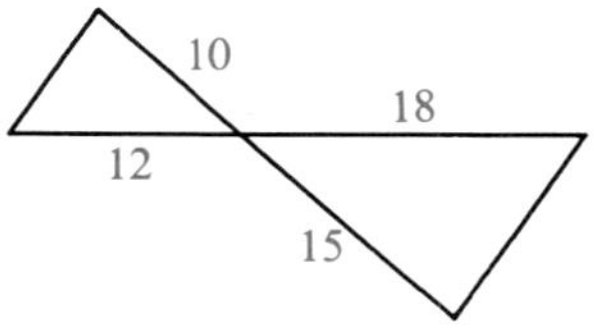

In the figure, it is given that $\overline{RS} \parallel \overline{TQ}$. Complete the proportion.

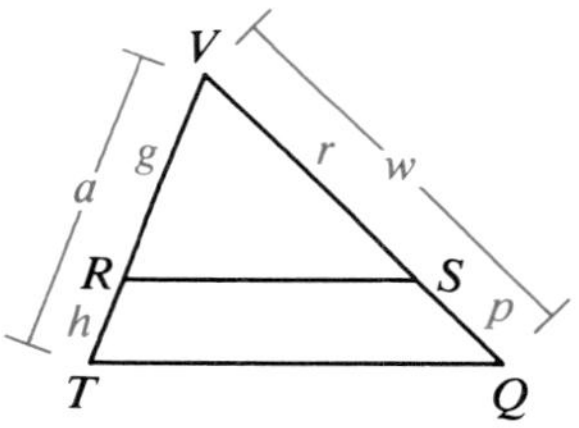

4. $\frac{g}{h} = \frac{?}{p}$

5. $\frac{a}{h} = \frac{w}{?}$

6. $\frac{r}{g} = \frac{p}{?}$

7. $\frac{h}{p} = \frac{?}{w}$

Find the value of x.

8.

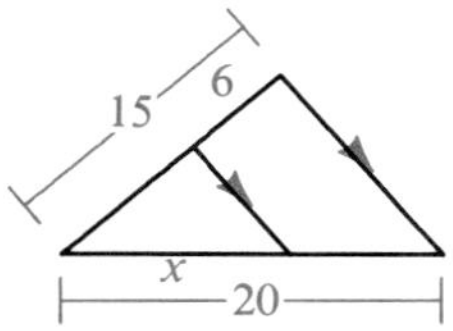

9.

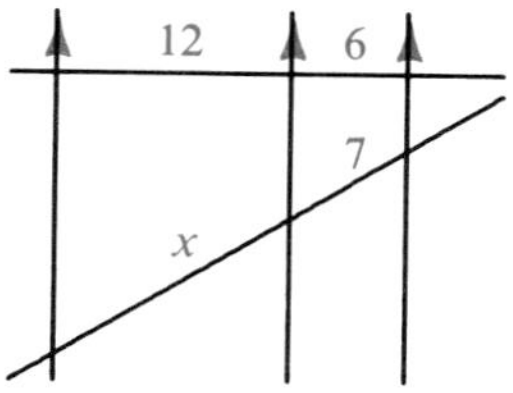

10.

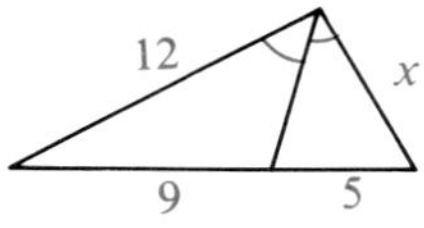

Challenge

Given: $\overline{FD} \parallel \overline{AC}$; $\overline{BD} \parallel \overline{AE}$; $\overline{FB} \parallel \overline{EC}$

Show that B, D, and F are midpoints of $\overline{AC}$, $\overline{CE}$, and $\overline{EA}$.

(*Hint:* If $\frac{AB}{BC} = \frac{BC}{AB}$, then $(AB)^2 = (BC)^2$, and $AB = BC$.)

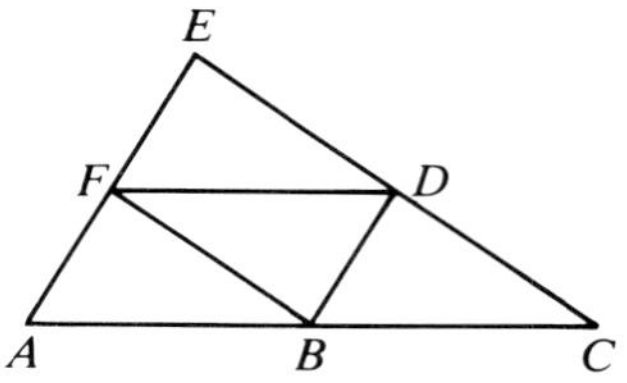

EXTRA

Topology

In the geometry we have been studying, our interest has been in congruent figures and similar figures, that is, figures with the same size and shape or at least the same shape. If we were studying the branch of geometry called *topology,* we would be interested in properties of figures that are even more basic than size and shape. For example, imagine taking a rubber band and stretching it into all kinds of figures.

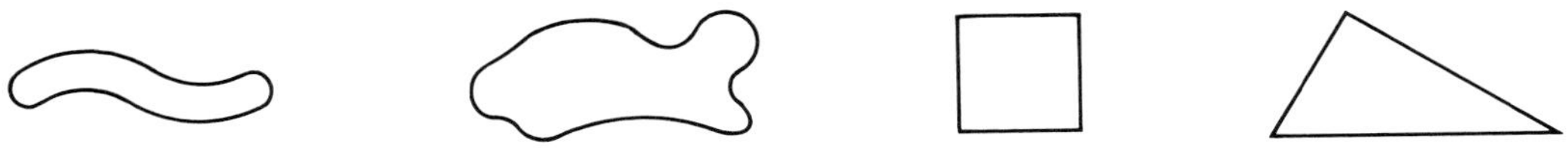

These figures have different sizes and shapes, but they still have something in common: Each one can be turned into any of the others by stretching and bending the rubber band. In topology figures are classified according to this kind of family resemblance. Figures that can be stretched, bent, or molded into the same shape without cutting or puncturing belong to the same family and are called *topologically equivalent.* Thus circles, squares, and triangles are equivalent. Likewise the straight line segment and wiggly curves below are equivalent.

Notice that to make one of these figures out of the rubber band you would have to cut the band, so these two-ended curves are not equivalent to the closed curves in the first illustration. Suppose that in the following plane figures the lines are joined where they cross. Then these figures belong to a third family. They are equivalent to each other but not to any of the figures above.

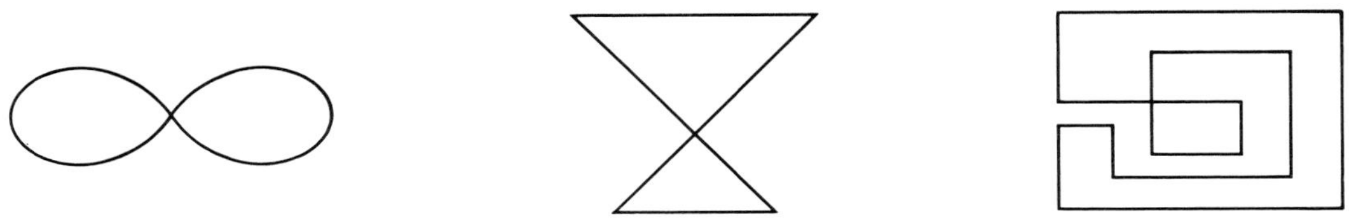

One of the goals of topology is to identify and describe the different families of equivalent figures. A person who studies topology (called a *topologist*) may be interested in classifying solid figures as well as figures in a plane. For example, the topologist would consider an orange, a teaspoon, and a brick equivalent to each other.

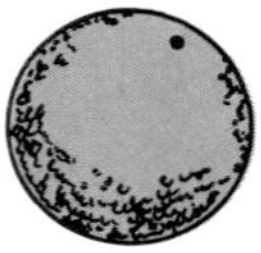

Orange

Teaspoon

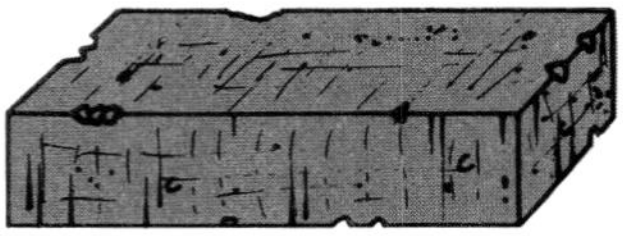

Brick

In fact, a doughnut is topologically equivalent to a coffee cup. (See the diagrams below.) For this reason, a topologist has been humorously described as a mathematician who can't tell the difference between a doughnut and a coffee cup!

Think of the objects as made of modeling clay.

Push thumb into clay to make room for coffee.

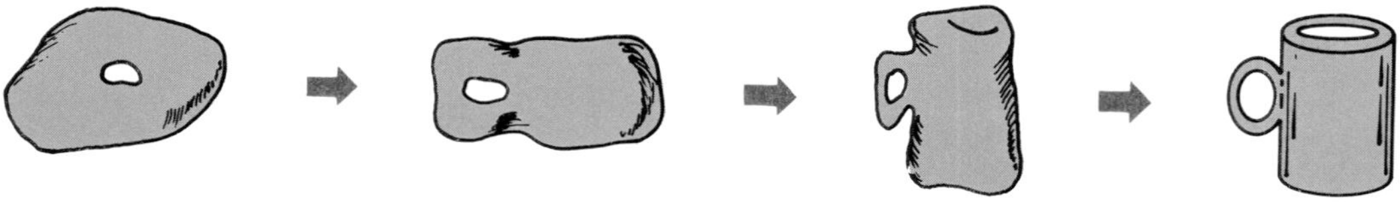

Exercises

In each exercise tell which figure is *not* topologically equivalent to the rest. Exercises 1 and 2 show plane figures.

1. a.  **b.**  **c.**  **d.** 

2. a.  **b.** 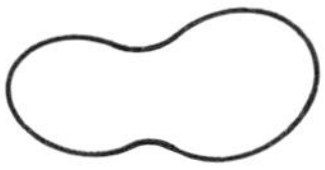 **c.**  **d.** 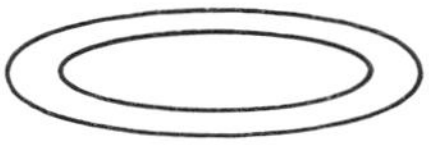

3. a. solid ball **b.** hollow ball **c.** crayon **d.** comb

4. a. saucer **b.** car key **c.** coffee cup **d.** wedding ring

5. a. hammer **b.** screwdriver **c.** thimble **d.** sewing needle

6. Group the block numbers shown into three groups such that the numbers in each group are topologically equivalent to each other.

7. Make a series of drawings showing that the items in each pair are topologically equivalent to each other.

a. a drinking glass and a dollar bill

b. a tack and a paper clip

Chapter Summary

1. The ratio of a to b is the quotient $\frac{a}{b}$ (b cannot be 0). The ratio $\frac{a}{b}$ can also be written $a:b$.
2. A proportion is an equation, such as $\frac{a}{b} = \frac{c}{d}$, stating that two ratios are equal.
3. The properties of proportions (see page 209) are used to change proportions into equivalent equations. For example, the product of the extremes equals the product of the means.
4. Similar figures have the same shape. Two polygons are similar if and only if corresponding angles are congruent and corresponding sides are in proportion.
5. Ways to prove two triangles similar:
   AA Similarity Postulate   SAS Similarity Theorem   SSS Similarity Theorem
6. Ways to show that segments are proportional:
   a. Corresponding sides of similar polygons are in proportion.
   b. If a line is parallel to one side of a triangle and intersects the other two sides, then it divides those sides proportionally.
   c. If three parallel lines intersect two transversals, they divide the transversals proportionally.
   d. If a ray bisects an angle of a triangle, then it divides the opposite side into segments proportional to the other two sides.

Chapter Review

Write the ratio in simplest form.

1. 15:25   **2.** 6:12:9   **3.** $\frac{16xy}{24x^2}$ 5-1

4. The measures of the angles of a triangle are in the ratio 4:4:7. Find the three measures.

Is the equation equivalent to the proportion $\frac{30 - x}{x} = \frac{8}{7}$?

5. $7x = 8(30 - x)$ **6.** $\frac{x}{30 - x} = \frac{7}{8}$ 5-2

7. $8x = 210 - 7x$ **8.** $\frac{30}{x} = \frac{15}{7}$

9. If $\triangle ABC \sim \triangle NJT$, then $\angle B \cong$ _?_. 5-3

10. If quad. $DEFG \sim$ quad. $PQRS$, then $\frac{FG}{RS} = \frac{GD}{?}$.

11. $\triangle ABC \sim \triangle JET$, and the scale factor of $\triangle ABC$ to $\triangle JET$ is $\frac{5}{3}$. If $BC = 20$, then $ET =$ _?_.

12. The quadrilaterals are similar. Find the values of x and y.

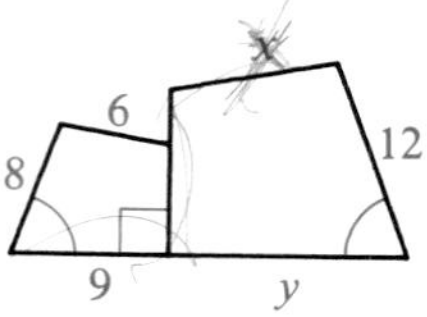

13. a. $\triangle RTS \sim$ _?_ 5-4

b. What postulate or theorem justifies the statement in part (a)?

14. $\frac{RT}{?} = \frac{TS}{?} = \frac{RS}{?}$

15. Suppose you wanted to prove

$$RS \cdot UV = RT \cdot UH.$$

You would first use similar triangles to show that $\frac{RS}{?} = \frac{?}{?}$.

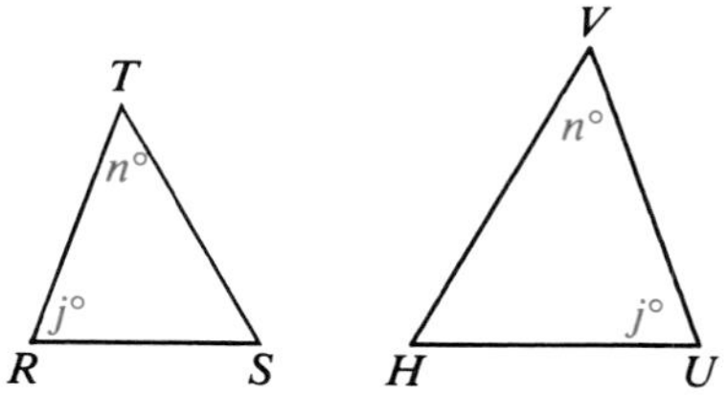

If two triangles shown can be proved similar, state the similarity. If not, write *no*.

16. $\angle A \cong \angle D$

17. $\angle B \cong \angle D$ 5-5

18. $CN = 16$, $ND = 14$, $BN = 7$, $AN = 8$

19. $AN = 7$, $AB = 6$, $DN = 14$, $DC = 12$

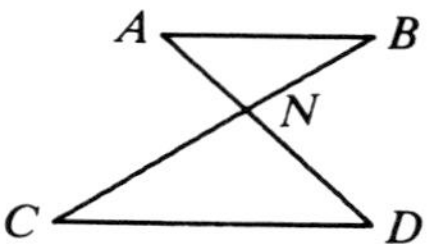

Exs. 16–19

20.

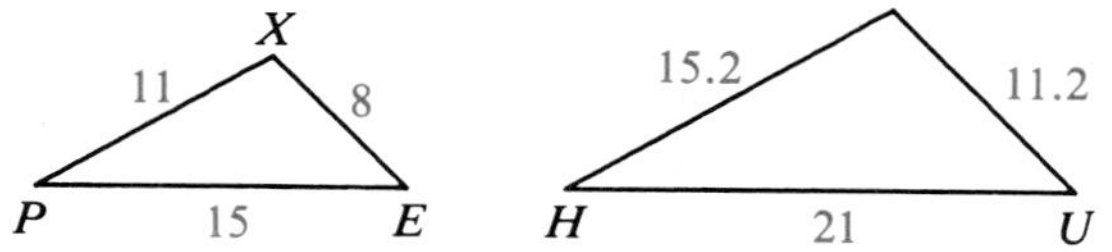

21. Which proportion is *incorrect*? 5-6

(1) $\frac{OS}{ST} = \frac{OV}{VW}$ (2) $\frac{SV}{TW} = \frac{OS}{ST}$ (3) $\frac{OT}{OW} = \frac{OS}{OV}$

22. If $OS = 8$, $ST = 12$, and $OV = 10$, then $OW =$ _?_.

23. If $OS = 8$, $ST = 12$, and $OW = 24$, then $VW =$ _?_.

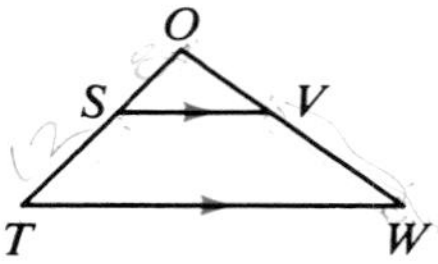

24. In $\triangle ABC$, the bisector of $\angle B$ meets $\overline{AC}$ at K. $AB = 18$, $BC = 24$, and $AC = 28$. Find AK.

Chapter Test

1. Two sides of a rectangle have the lengths 20 and 32. Find, in simplest form, the ratio of:
   a. the length of the shorter side to the length of the longer side
   b. the perimeter to the length of the longer side

2. If quad. $ABCD \sim$ quad. $THUS$, then: **a.** $\angle U \cong$ _?_ **b.** $\frac{BC}{HU} = \frac{AD}{?}$

3. If $x:y:z = 4:6:9$ and $z = 45$, then $x =$ _?_ and $y =$ _?_.

4. If $\frac{8}{9} = \frac{x}{15}$, then $x =$ _?_.

5. If $\frac{a}{b} = \frac{c}{10}$, then $\frac{a+b}{?} = \frac{?}{10}$.

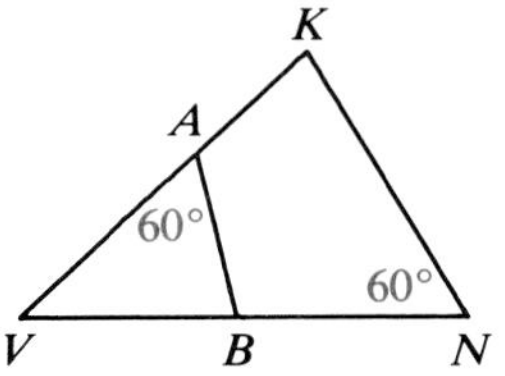

6. What postulate or theorem justifies the statement $\triangle AVB \sim \triangle NVK$?

7. $\frac{AB}{NK} = \frac{VA}{?}$

8. The scale factor of $\triangle AVB$ to $\triangle NVK$ is $\frac{5}{8}$. If $VA = 2.5$ and $VB = 1.7$, then $VN =$ _?_.

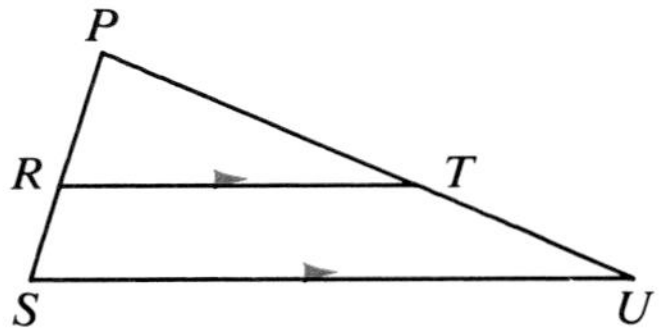

9. If $PR = 10$, $RS = 6$, and $PT = 15$, then $TU =$ _?_.

10. If $PT = 32$, $PU = 48$, and $RS = 10$, then $PR =$ _?_.

11. If $PR = 14$, $RS = 7$, and $RT = 26$, then $SU =$ _?_.

In $\triangle GEB$, the bisector of $\angle E$ meets $\overline{GB}$ at K.

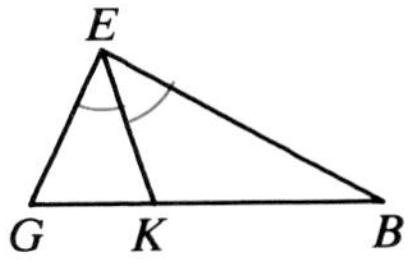

12. If $GK = 5$, $KB = 8$, and $GE = 7$, then $EB =$ _?_.

13. If $GE = 14$, $EB = 21$, and $GB = 30$, then $GK =$ _?_.

14. Given the triangles shown, state a similarity:
    $\triangle$ _?_ $\sim \triangle$ _?_.

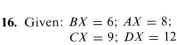

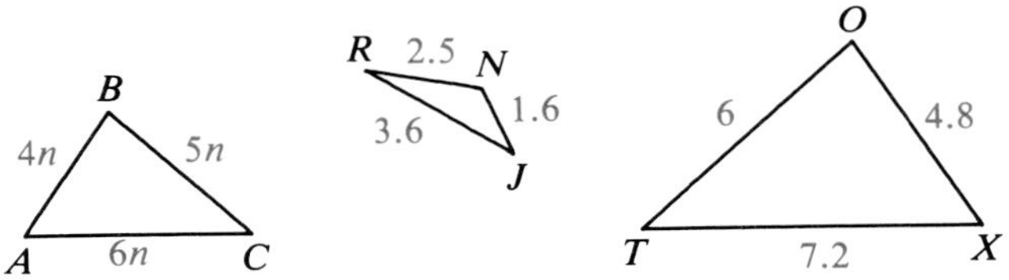

15. Given: $\overleftrightarrow{DE} \parallel \overleftrightarrow{FG} \parallel \overleftrightarrow{HJ}$
    Prove: $DF \cdot GJ = FH \cdot EG$

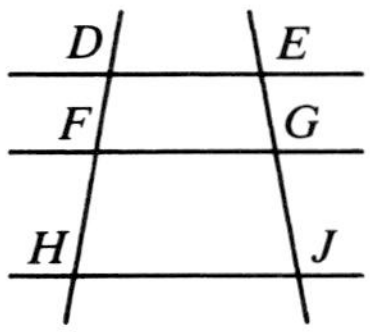

16. Given: $BX = 6$; $AX = 8$;
    $CX = 9$; $DX = 12$
    Prove: $\overline{AB} \parallel \overline{CD}$

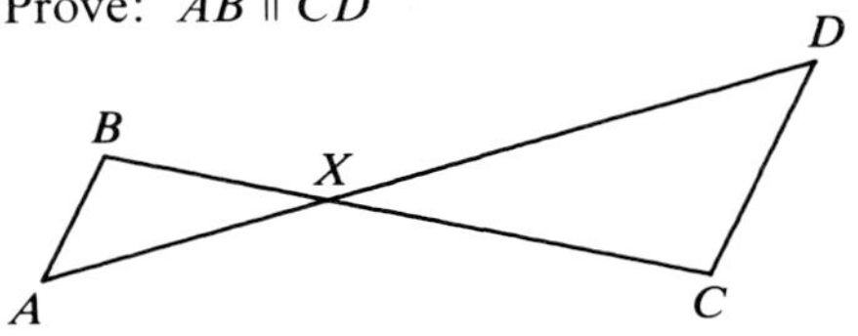

Mixed Review

1. In $\triangle XYZ$, $\angle Y \cong \angle Z$. If $XY = 5x + 3$, $YZ = 4x + 2$, and $XZ = 10x - 32$, find the lengths of $\overline{XY}$, $\overline{YZ}$, and $\overline{XZ}$.

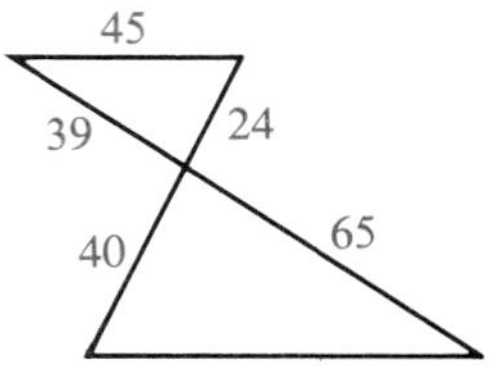

Ex. 2

2. **a.** The _?_ guarantees that the two triangles are _?_.
   b. Find the perimeter of the larger triangle.
3. The coordinates of points P and Q on a number line are -13 and 4. Find the coordinate of the midpoint of $\overline{PQ}$.
4. In $\triangle ABC$, $m\angle A : m\angle B : m\angle C = 3:3:4$.
   a. Is $\triangle ABC$ scalene, isosceles, or equilateral?
   b. Is $\triangle ABC$ acute, right, or obtuse?
   c. Name the longest side of $\triangle ABC$.
5. Describe the possible relationships between two planes.
6. To write an indirect proof, you assume temporarily that the _?_ is not true.
7. If two parallel lines are cut by a transversal, then _?_ angles are congruent, _?_ angles are congruent, and _?_ angles are supplementary.
8. Given: $RSTWYZ$ is a regular hexagon.
   Prove: $RSWY$ is a rectangle.
   (Begin by drawing a diagram.)
9. If $AB = x - 5$, $BC = x - 2$, $CD = x + 4$, and $DA = x$, find the value of x.

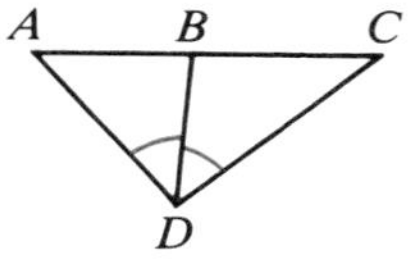

10. Use an indirect proof to show that no triangle has sides of length x, y, and $x + y$.
11. Find the sum of the measures of the angles of an octagon.
12. Given: $\angle WXY \cong \angle XZY$
    Prove: $(XY)^2 = WY \cdot ZY$

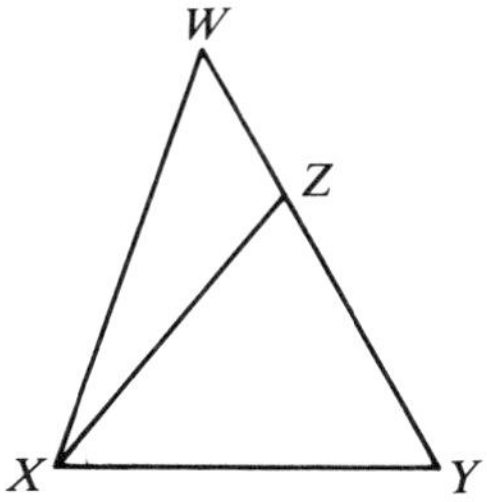

13. The measures of three consecutive angles of a quadrilateral are 58, 122, and 58. Must the diagonals
    a. be perpendicular?
    b. bisect each other?
    c. be congruent?

Algebra Review

Simplify.

Example **a.** $\sqrt{56} = \sqrt{4 \cdot 14} = \sqrt{4} \cdot \sqrt{14} = 2\sqrt{14}$

b. $\sqrt{8} \cdot \sqrt{6} = \sqrt{8 \cdot 6} = \sqrt{48} = \sqrt{16 \cdot 3} = 4\sqrt{3}$

c. $\sqrt{\frac{5}{6}} = \frac{\sqrt{5}}{\sqrt{6}} = \frac{\sqrt{5}}{\sqrt{6}} \cdot \frac{\sqrt{6}}{\sqrt{6}} = \frac{\sqrt{30}}{\sqrt{36}} = \frac{\sqrt{30}}{6}$, or $\frac{1}{6}\sqrt{30}$

d. $\frac{\sqrt{55}}{\sqrt{22}} = \sqrt{\frac{55}{22}} = \sqrt{\frac{5}{2}} = \frac{\sqrt{5}}{\sqrt{2}} \cdot \frac{\sqrt{2}}{\sqrt{2}} = \frac{\sqrt{10}}{2}$

e. $\left(\frac{3\sqrt{6}}{2}\right)^2 = \frac{3\sqrt{6}}{2} \cdot \frac{3\sqrt{6}}{2} = \frac{9 \cdot 6}{4} = \frac{27}{2}$

1. $\sqrt{81}$ **2.** $\sqrt{0}$ **3.** $\sqrt{24}$ **4.** $\sqrt{13^2}$

5. $(\sqrt{7})^2$ **6.** $\sqrt{600}$ **7.** $\sqrt{245}$ **8.** $\frac{1}{\sqrt{5}}$

9. $\frac{12}{\sqrt{2}}$ **10.** $\sqrt{\frac{2}{3}}$ **11.** $\sqrt{4} \cdot \sqrt{7}$ **12.** $\sqrt{8} \cdot \sqrt{15}$

13. $\frac{\sqrt{45}}{\sqrt{5}}$ **14.** $\frac{\sqrt{12}}{\sqrt{24}}$ **15.** $\frac{\sqrt{5}}{\sqrt{3}}$ **16.** $\frac{\sqrt{21}}{\sqrt{18}}$

17. $\frac{12}{\sqrt{15}}$ **18.** $\sqrt{\frac{80}{25}}$ **19.** $\sqrt{\frac{25}{80}}$ **20.** $3\sqrt{27}$

21. $\frac{1}{2}\sqrt{121}$ **22.** $\frac{4\sqrt{125}}{5}$ **23.** $\frac{12}{5\sqrt{6}}$ **24.** $\frac{15\sqrt{2}}{\sqrt{5}}$

25. $(9\sqrt{2})^2$ **26.** $\left(\frac{\sqrt{10}}{2}\right)^2$ **27.** $5(2\sqrt{3})^2$ **28.** $\frac{3}{4}(3\sqrt{8})^2$

Solve for x. Assume that x represents a positive number.

Example **a.** $\frac{x}{6} = \frac{3}{x}$ **b.** $\frac{144}{x} = \frac{x}{50}$ **c.** $x^2 + (3\sqrt{2})^2 = 9^2$

Solution

a.
$$x^2 = 6 \cdot 3$$
$$x = 18$$
$$x = \sqrt{18}$$
$$x = 3\sqrt{2}$$

b.
$$x^2 = 144 \cdot 50$$
$$x = \sqrt{144} \cdot \sqrt{50}$$
$$x = 12 \cdot 5\sqrt{2}$$
$$x = 60\sqrt{2}$$

c.
$$x^2 + 18 = 81$$
$$x = \sqrt{63}$$
$$x = \sqrt{9 \cdot 7}$$
$$x = 3\sqrt{7}$$

29. $\frac{2}{x} = \frac{x}{8}$ **30.** $\frac{1}{x} = \frac{x}{27}$ **31.** $\frac{x}{7} = \frac{5}{x}$

32. $\frac{x}{49} = \frac{100}{x}$ **33.** $\frac{x}{4} = \frac{32}{x}$ **34.** $\frac{30}{x} = \frac{x}{6}$

35. $x^2 = 3^2 + 4^2$ **36.** $x^2 + 3^2 = 4^2$ **37.** $5^2 + x^2 = 9^2$

38. $(4\sqrt{2})^2 + x^2 = (4\sqrt{3})^2$ **39.** $\frac{3x + 6}{x + 4} = \frac{x + 2}{x}$ **40.** $(2x)^2 + 15^2 = (3x)^2$

Right triangles are a prominent feature of the architectural design of this mountain cabin in Colorado. Properties of right triangles and an introduction to right-triangle trigonometry are the topics of this chapter.

Right Triangles

6

The Pythagorean Theorem

Objectives

1. Determine the geometric mean between two numbers.
2. State and apply the relationships that exist when the altitude is drawn to the hypotenuse of a right triangle.
3. State and apply the Pythagorean Theorem.

6-1 Geometric Means

Suppose r, s, and t are positive numbers with $\frac{r}{s} = \frac{s}{t}$. Then s is called the **geometric mean** between r and t.

Example 1 Find the geometric mean between the given numbers.

a. 3 and 7 **b.** 6 and 15

Solution

a. $\frac{3}{x} = \frac{x}{7}$

$x^2 = 21$

$x = \sqrt{21}$

b. $\frac{6}{x} = \frac{x}{15}$

$x^2 = 90$

$x = \sqrt{90} = \sqrt{9 \cdot 10} = \sqrt{9} \cdot \sqrt{10} = 3\sqrt{10}$

The symbol $\sqrt{}$ always indicates the positive square root of a number. In (b) above, the *radical* $\sqrt{90}$ could be *simplified* because the *radicand* 90 has the factor 9, a perfect square. When you write radical expressions you should write them in **simplest form.** This means writing them so that

1. No radicand has a factor, other than 1, that is a perfect square.
2. No radicand is a fraction.
3. No fraction has a denominator that contains a radical.

You should express answers involving radicals in simplest form unless you are asked to use decimal approximations.

Example 2 Simplify the expressions.

a. $3\sqrt{\frac{2}{5}}$ **b.** $\frac{5}{6\sqrt{7}}$

Solution

a. Multiply the numerator and denominator by 5 to obtain a perfect square in the denominator.

$3\sqrt{\frac{2}{5}} = 3\sqrt{\frac{2}{5} \cdot \frac{5}{5}}$

$= 3\sqrt{\frac{10}{25}} = \frac{3\sqrt{10}}{5}$

b. To remove the radical from the denominator, multiply the numerator and denominator by the radical.

$\frac{5}{6\sqrt{7}} = \frac{5}{6\sqrt{7}} \cdot \frac{\sqrt{7}}{\sqrt{7}}$

$= \frac{5\sqrt{7}}{6 \cdot 7} = \frac{5\sqrt{7}}{42}$

The following theorem states a special property of right triangles.

Theorem 6–1

If the altitude is drawn to the hypotenuse of a right triangle, then the two triangles formed are similar to the original triangle and to each other.

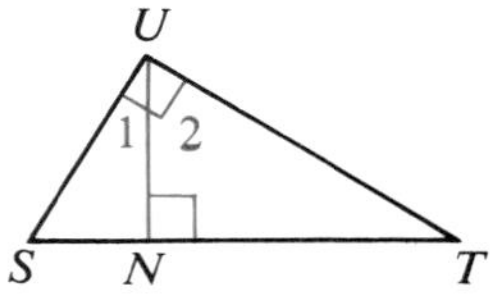

Given: $\triangle SUT$ with rt. $\angle SUT$;
$\overline{UN} \perp \overline{ST}$

Prove: $\triangle SNU \sim \triangle SUT \sim \triangle UNT$

Plan for Proof: Note that $\angle 2$ and $\angle S$ are both complementary to $\angle 1$. So $\angle S \cong \angle S \cong \angle 2$. Also, $\angle SNU \cong \angle SUT \cong \angle UNT$. Therefore, $\triangle SNU \sim \triangle SUT \sim \triangle UNT$.

The altitude to the hypotenuse divides the hypotenuse into two segments. Corollaries 1 and 2 of Theorem 6–1 deal with geometric means involving the lengths of these segments. For proofs of the corollaries, see Classroom Exercises 2–7.

For simplicity in stating theorems about the lengths of sides of triangles, the words *segment, side, leg,* and *hypotenuse* are often used to refer to the *length* of a segment rather than the segment itself. When the context makes this meaning clear, as in the corollaries that follow, we will use this convention.

Corollary 1

When the altitude is drawn to the hypotenuse of a right triangle, the length of the altitude is the geometric mean between the segments of the hypotenuse.

Corollary 2

When the altitude is drawn to the hypotenuse of a right triangle, each leg is the geometric mean between the hypotenuse and the segment of the hypotenuse that is adjacent to that leg.

Example 3 $\angle SUT$ is a right angle. Find the values of h, a, and b.

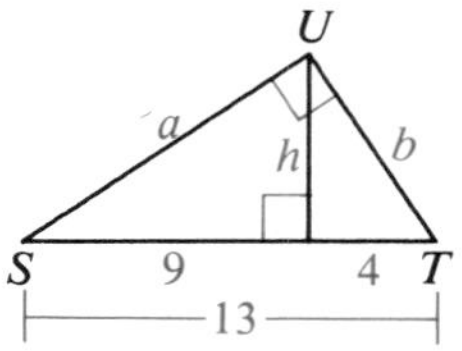

Solution

By Corollary 1:

$\frac{9}{h} = \frac{h}{4}$

$h^2 = 9 \cdot 4 = 36$

$h = \sqrt{36}$

$h = 6$

By Corollary 2:

$\frac{13}{a} = \frac{a}{9}$

$a^2 = 9 \cdot 13$

$a = \sqrt{9 \cdot 13}$

$a = 3\sqrt{13}$

$\frac{13}{b} = \frac{b}{4}$

$b^2 = 4 \cdot 13$

$b = \sqrt{4 \cdot 13}$

$b = 2\sqrt{13}$

Example 4 $\angle PQR$ is a right angle, $RP = 18$, and $RQ = 6$.
Find the values of w, x, y, and z.

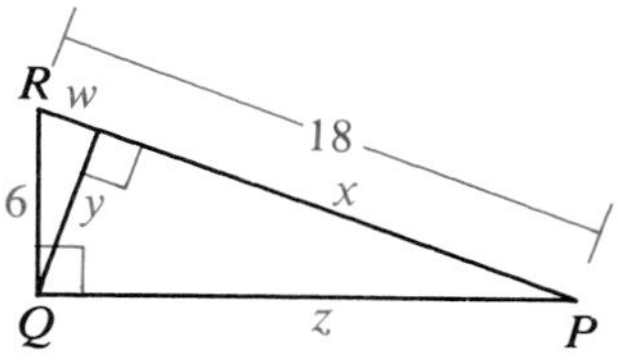

Solution First use Corollary 2 to find w.

$\frac{18}{6} = \frac{6}{w}$

$18w = 36$

$w = 2$

Then $x = 18 - 2 = 16$.

$\frac{16}{y} = \frac{y}{2}$ (Cor. 1)

$y^2 = 16 \cdot 2$

$y = \sqrt{16 \cdot 2}$

$y = 4\sqrt{2}$

$\frac{18}{z} = \frac{z}{16}$ (Cor. 2)

$z^2 = 16 \cdot 18$

$z = \sqrt{16 \cdot 18}$

$z = \sqrt{16 \cdot 9 \cdot 2}$

$z = 4 \cdot 3\sqrt{2} = 12\sqrt{2}$

Classroom Exercises

$\angle ACB$ is a right angle and $\overline{CN} \perp \overline{AB}$.

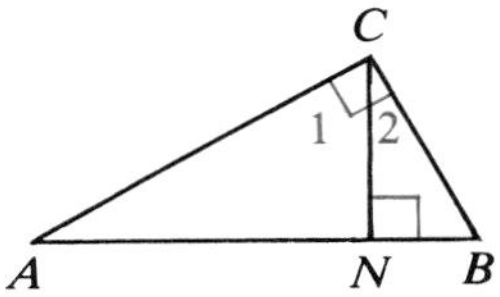

1. If $m\angle 1 = 60$, then $m\angle A =$ _?_, $m\angle 2 =$ _?_, and $m\angle B =$ _?_.
2. If $m\angle 1 = k$, then $m\angle A =$ _?_, $m\angle 2 =$ _?_, and $m\angle B =$ _?_.
3. From Exercise 2 we see that $\angle A \cong$ _?_. Likewise, $\angle B \cong$ _?_.
4. Since $\triangle ANC \sim \triangle CNB$ (Why?), $\frac{AN}{CN} = \frac{CN}{?}$.
5. State in words the corollary proved in Exercises 2–4.
6. **a.** Since $\triangle ACB \sim \triangle ANC$, $\frac{AB}{AC} = \frac{AC}{?}$.

   b. Since $\triangle ACB \sim \triangle CNB$, $\frac{BA}{?} = \frac{?}{BN}$.
7. State in words the corollary proved in Exercise 6.
8. The diagram shows a right triangle with the altitude drawn to the hypotenuse.

   a. p is the geometric mean between _?_ and _?_.

   b. e is the geometric mean between _?_ and _?_.

   c. f is the geometric mean between _?_ and _?_.

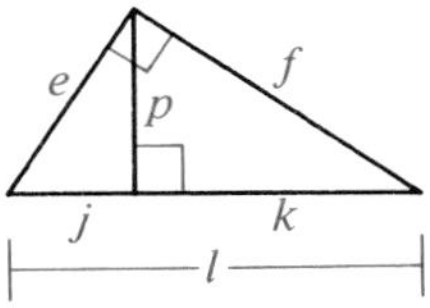

9. Finish simplifying $\sqrt{\frac{7}{50}}$ in the two ways shown.

$$\sqrt{\frac{7}{50}} = \sqrt{\frac{7}{50} \cdot \frac{50}{50}} = \frac{1}{50}\sqrt{7 \cdot 2 \cdot 25} = \frac{?}{?} = \frac{?}{?}$$

$$\sqrt{\frac{7}{50}} = \sqrt{\frac{7}{50} \cdot \frac{2}{2}} = \sqrt{\frac{14}{100}} = \frac{?}{?}$$

10. Finish simplifying $\frac{2}{\sqrt{8}}$ in the two ways shown.

$$\frac{2}{\sqrt{8}} = \frac{2}{\sqrt{8}} \cdot \frac{\sqrt{8}}{\sqrt{8}} = \frac{2\sqrt{8}}{8} = \frac{?}{?} = \frac{?}{?}$$

$$\frac{2}{\sqrt{8}} = \frac{2}{\sqrt{8}} \cdot \frac{\sqrt{2}}{\sqrt{2}} = \frac{2\sqrt{2}}{\sqrt{16}} = \frac{?}{?} = \frac{?}{?}$$

Simplify the radical expressions.

11. $6\sqrt{25}$ **12.** $5\sqrt{18}$ **13.** $\sqrt{\frac{1}{3}}$ **14.** $\frac{15}{\sqrt{3}}$

15. a. $\sqrt{4} \cdot \sqrt{9}$ **16. a.** $\sqrt{7} \cdot \sqrt{16}$ **17. a.** $\frac{\sqrt{4}}{\sqrt{9}}$ **18. a.** $\sqrt{\frac{5}{2}}$

b. $\sqrt{4 \cdot 9}$ **b.** $\sqrt{7 \cdot 16}$ **b.** $\sqrt{\frac{4}{9}}$ **b.** $\frac{\sqrt{5}}{\sqrt{2}}$

Written Exercises

Simplify the expressions.

A **1.** $\sqrt{49}$ **2.** $3\sqrt{64}$ **3.** $\frac{2}{5}\sqrt{9}$ **4.** $\frac{2}{5}\sqrt{25}$

5. $\sqrt{12}$ **6.** $\sqrt{50}$ **7.** $5\sqrt{28}$ **8.** $\frac{1}{2}\sqrt{300}$

9. $\sqrt{\frac{1}{2}}$ **10.** $\frac{1}{\sqrt{2}}$ **11.** $\sqrt{\frac{2}{27}}$ **12.** $6\sqrt{\frac{1}{3}}$

13. $\frac{18}{\sqrt{3}}$ **14.** $\frac{15}{\sqrt{30}}$ **15.** $\frac{3\sqrt{32}}{4}$ **16.** $\frac{5}{2\sqrt{10}}$

Find the geometric mean between the two numbers.

17. 2 and 8 **18.** 3 and 27 **19.** 13 and 25

20. 1 and 50 **21.** 6 and 10 **22.** $\frac{1}{10}$ and 2

Each diagram shows a right triangle with the altitude drawn to the hypotenuse. Find the values of x, y, and z.

B **23.**

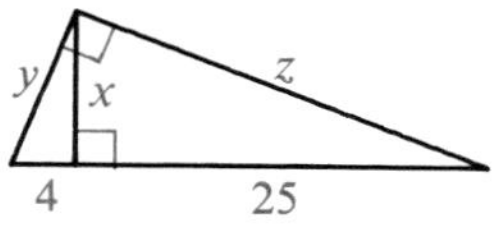

24.

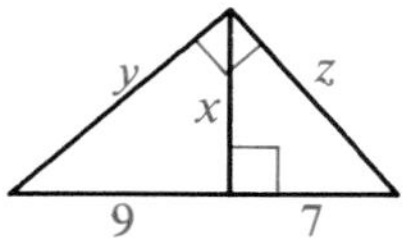

25.

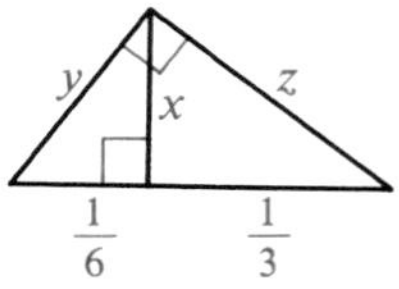

26.

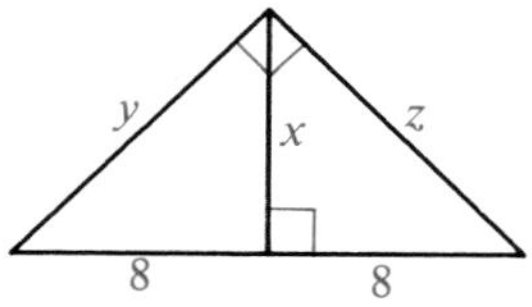

27.

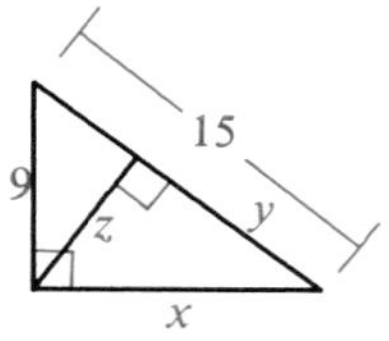

28.

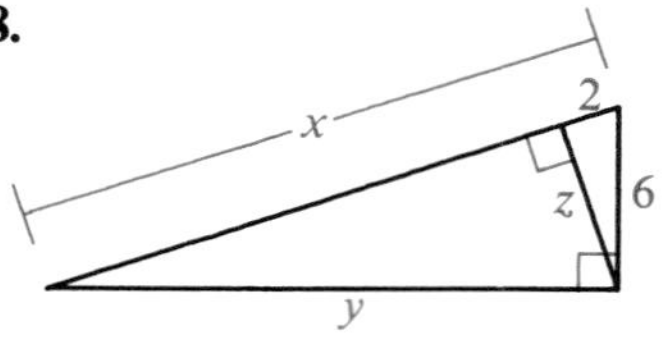

Find the values of x, y, and z.

29. 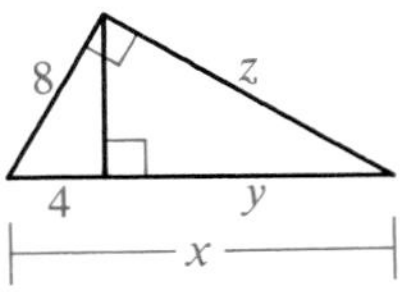

30. 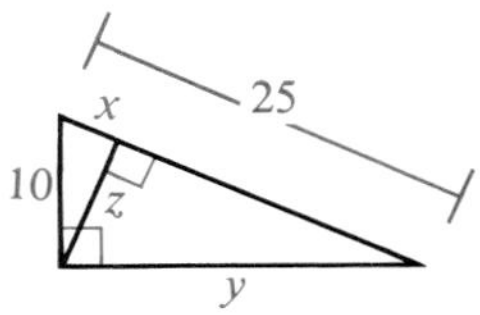

31. 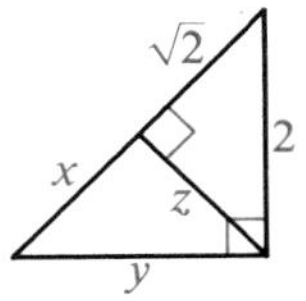

32. Prove Theorem 6–1.

C **33.** Prove: In a right triangle, the product of the hypotenuse and the length of the altitude drawn to the hypotenuse is equal to the product of the two legs.

34. The *arithmetic mean* between two numbers r and s is defined to be the number $\frac{r+s}{2}$.

a. $\overline{CM}$ is the median and $\overline{CH}$ is the altitude to the hypotenuse of right $\triangle ABC$. Show that CM is the arithmetic mean between AH and BH and that CH is the geometric mean between AH and BH. Then use the diagram to show that the arithmetic mean is greater than the geometric mean.

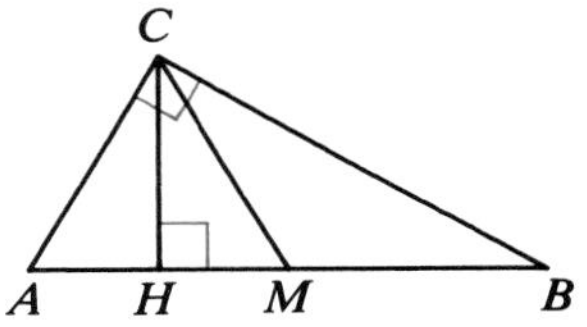

b. Show algebraically that the arithmetic mean between two different numbers r and s is greater than the geometric mean. (*Hint:* The geometric mean is $\sqrt{rs}$. Work backward from $\frac{r+s}{2} > \sqrt{rs}$ to $(r - s)^2 > 0$ and then reverse the steps.)

35. In this exercise p, q, r, s, and t are prime numbers, all different.

a. Note that the number 3 is the geometric mean between two different pairs of integers: 1 and 9, 3 and 3. The number 6 is the geometric mean between five different pairs of integers. List them.

b. The number pq is the geometric mean between five different pairs of integers. List them.

c. The number pqr is the geometric mean between __?__ different pairs of integers. List them.

d. The number $pqrst$ is the geometric mean between __?__ different pairs of integers. (You don't have to list them.)

Challenge

Start with a right triangle. Build a square on each side. Locate the center of the square on the longer leg. Through the center, draw a parallel to the hypotenuse and a perpendicular to the hypotenuse.

Cut out the pieces numbered 1–5. Can you arrange the five pieces to cover exactly the square built on the hypotenuse?

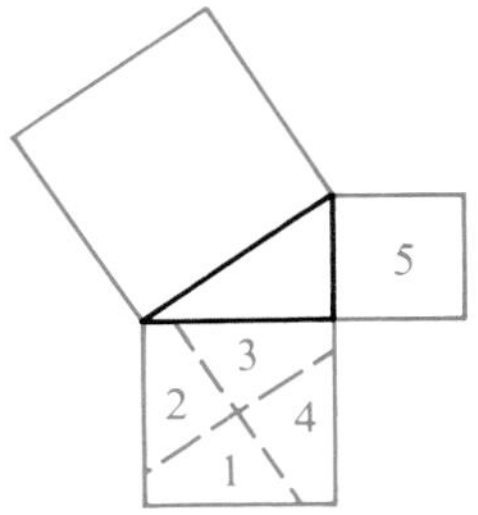

6-2 The Pythagorean Theorem

One of the best known and most useful theorems in all of mathematics is the Pythagorean Theorem. It is believed that Pythagoras, a Greek mathematician and philosopher, proved this theorem about twenty-five hundred years ago. Many different proofs have been discovered since then.

Theorem 6-2 Pythagorean Theorem

In a right triangle, the square of the hypotenuse is equal to the sum of the squares of the legs.

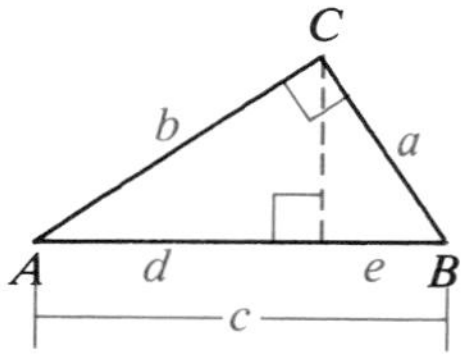

Given: Rt. $\triangle ABC$; $\angle C$ is a rt. $\angle$.

Prove: $c^2 = a^2 + b^2$

Proof:

Statements	Reasons
1. Draw a perpendicular from C to $\overline{AB}$.	1. Through a point outside a line, there is exactly one line _?_.
2. $\frac{c}{a} = \frac{a}{e}$; $\frac{c}{b} = \frac{b}{d}$	2. When the altitude is drawn to the hypotenuse of a rt. $\triangle$, each leg is the geometric mean between _?_.
3. $ce = a^2$; $cd = b^2$	3. A property of proportions
4. $ce + cd = a^2 + b^2$	4. Addition Property of =
5. $c(e + d) = a^2 + b^2$	5. Distributive Property
6. $c^2 = a^2 + b^2$	6. Substitution Property

Example Find the value of x. Keep in mind the fact that the length of a segment must be a positive number.

a.

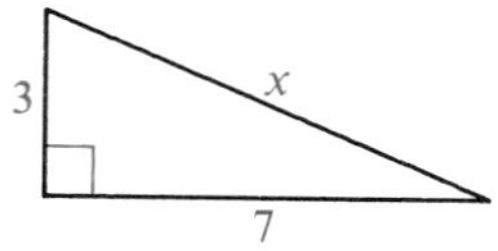

b.

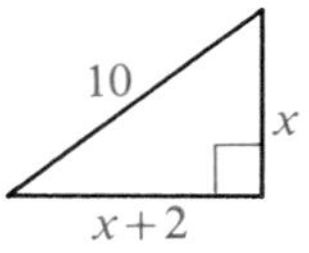

Solution

a.

$$x^2 = 7^2 + 3^2$$
$$x^2 = 49 + 9$$
$$x^2 = 58$$
$$x = \sqrt{58}$$

b.

$$x^2 + (x + 2)^2 = 10^2$$
$$x^2 + x^2 + 4x + 4 = 100$$
$$2x^2 + 4x - 96 = 0$$
$$x^2 + 2x - 48 = 0$$
$$(x + 8)(x - 6) = 0$$

~~$x = -8$~~, $x = 6$

Classroom Exercises

1. The early Greeks thought of the Pythagorean Theorem in this form: *The area of the square on the hypotenuse of a right triangle is equal to the sum of the areas of the squares on the legs.* Draw a diagram to illustrate that interpretation.

2. For the right triangle shown, $t^2 = r^2 + s^2$. Which of the following equations are equivalent to this equation?

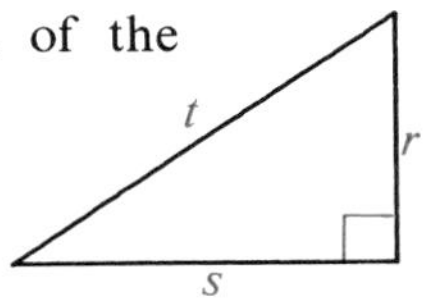

   a. $r^2 = s^2 + t^2$
   b. $r^2 = (t + s)(t - s)$
   c. $s^2 = t^2 - r^2$
   d. $t = r + s$

Complete each simplification.

3. $(\sqrt{3})^2 = \sqrt{3} \cdot$ _?_ = _?_
4. $(3\sqrt{11})^2 =$ _?_ $\cdot$ _?_ $= 9 \cdot$ _?_ = _?_

Simplify each expression.

5. $(\sqrt{5})^2$
6. $(2\sqrt{7})^2$
7. $(7\sqrt{2})^2$
8. $\left(\frac{\sqrt{2}}{2}\right)^2$
9. $\left(\frac{3}{\sqrt{5}}\right)^2$
10. $(2n)^2$
11. $\left(\frac{n}{\sqrt{3}}\right)^2$
12. $\left(\frac{2}{3}\sqrt{6}\right)^2$

State an equation you could use to find the value of x. Then find the value of x.

13.
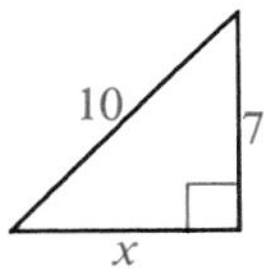

14.
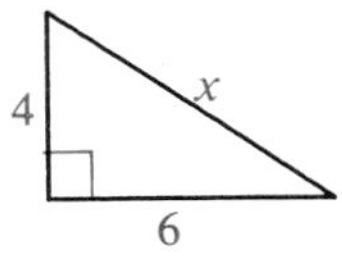

15.
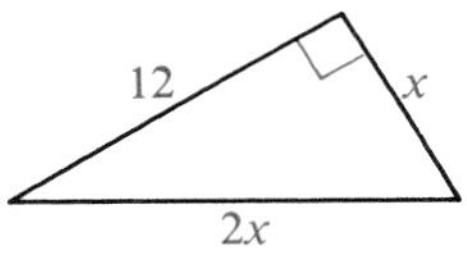

16.
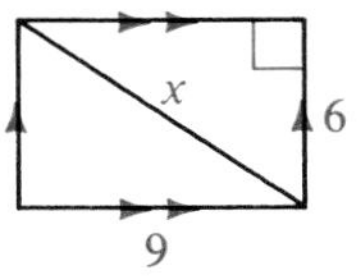

17.
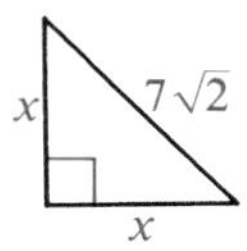

18.
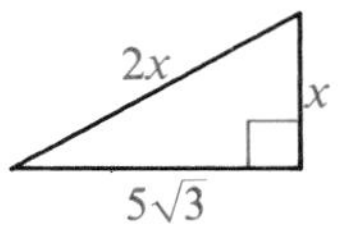

19. In the figure, $m\angle 1 > 90$ and $m\angle 2 = 90$. Two sides of the obtuse triangle are congruent to two sides of the right triangle as shown. State a reason to support each statement.
   a. $k > c$
   b. $k^2 > c^2$
   c. $c^2 = a^2 + b^2$
   d. $k^2 > a^2 + b^2$

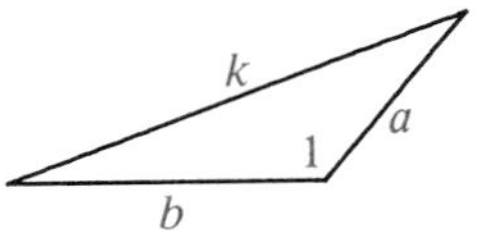

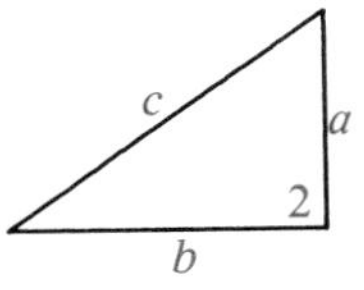

20. Exercise 19 outlines a proof. Complete the statement of what is proved: In an obtuse triangle, the square of the side opposite the obtuse angle is _?_.

Written Exercises

Copy and complete the table.

A

	1.	2.	3.	4.	5.	6.	7.	8.
a	3	?	5	?	11	$3n$	?	6
b	4	8	?	15	?	$4n$	$3\sqrt{2}$	$6\sqrt{3}$
c	?	10	13	17	61	?	6	?

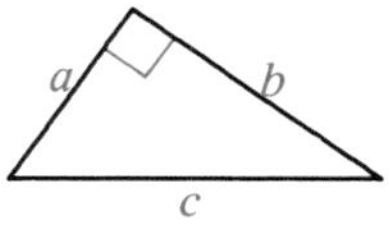

Draw a figure for each exercise. The length and width of a rectangle are given. Find the length of a diagonal.

9. 8 and 6 **10.** 0.8 and 0.6 **11.** 80 and 60 **12.** $\sqrt{3}$ and 1

The length of a diagonal of a square is given. Find the length of a side of the square.

13. 2 **14.** 10 **15.** $20k$ **16.** $7n\sqrt{2}$

Find the value of x in each figure.

B **17.**

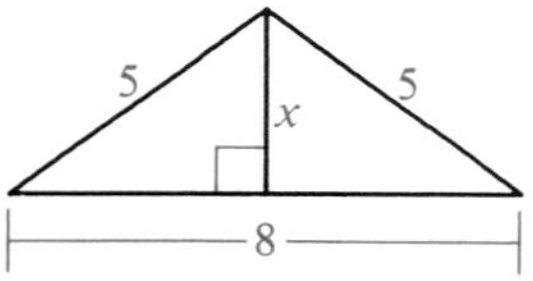

18.

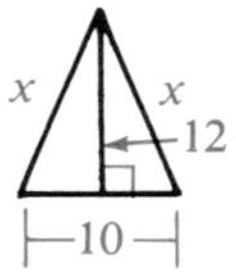

19.

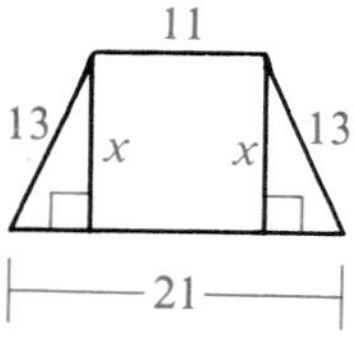

20.

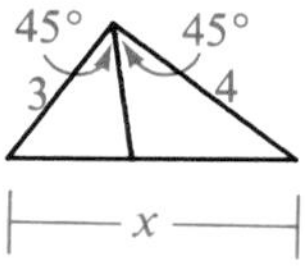

21.

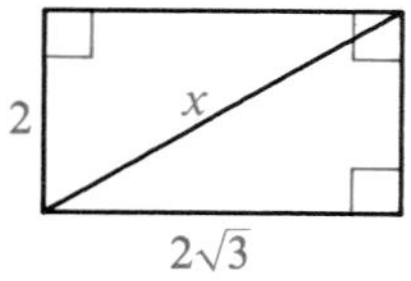

22.

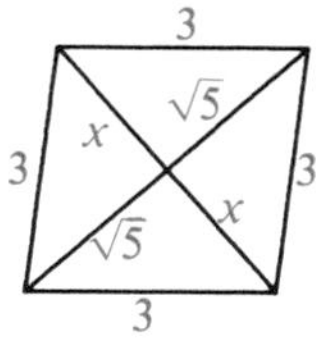

23.

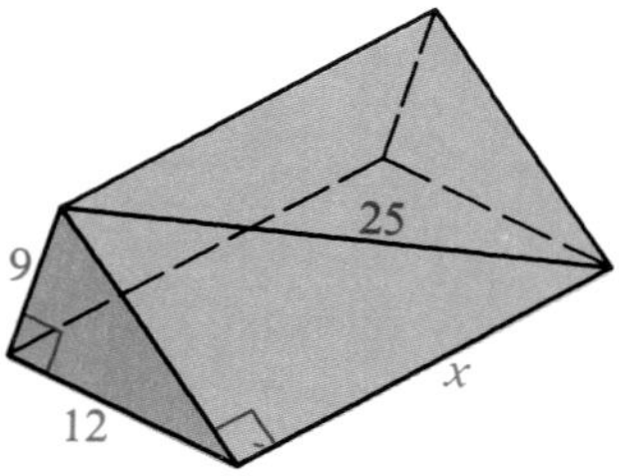

24.

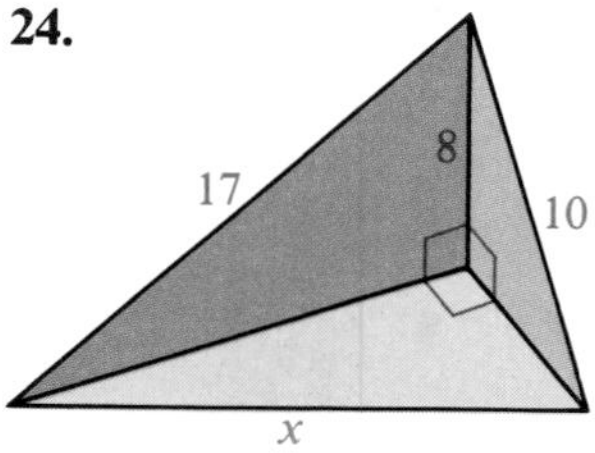

In Exercises 25–30, the dimensions of a rectangular box are given. Find the length of a diagonal of the solid.

Example Dimensions 6, 4, 3

Solution

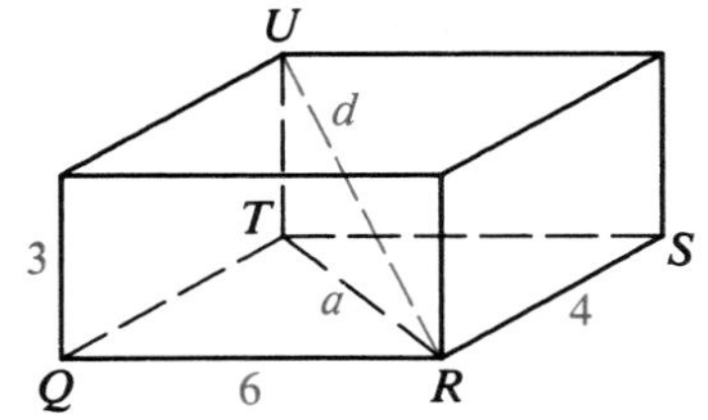

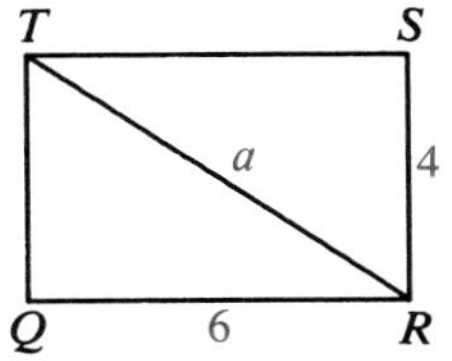

$a^2 = 6^2 + 4^2$
$a^2 = 36 + 16$
$a^2 = 52$

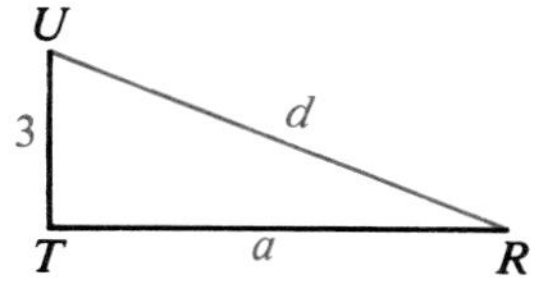

$d^2 = a^2 + 3^2$
$d^2 = 52 + 9$
$d^2 = 61$
$d = \sqrt{61}$

25. 12, 4, 3 **26.** 5, 5, 2 **27.** $\sqrt{7}, \sqrt{6}, \sqrt{5}$

28. e, e, e **29.** l, w, h **30.** $n + 2, \sqrt{2n + 1}, 2$

31. See Classroom Exercise 20. State and prove a theorem about a side opposite an acute angle in any triangle.

C 32. Given: A rt. △ with sides a, b, c;
b is the arithmetic mean of a and c.
Prove: $a:b:c = 3:4:5$

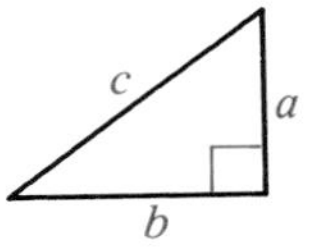

Find the value of *h*.

33.

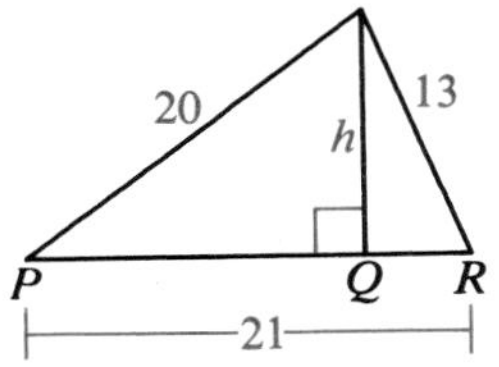

(*Hint:* Let $PQ = x$; $QR = 21 - x$. Use two right triangles.)

34.

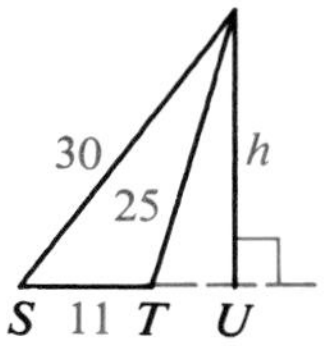

(*Hint:* Let $TU = x$; $SU = x + 11$.)

35. O is the *center* of square $ABCD$ (the point of intersection of the diagonals) and $\overline{VO}$ is perpendicular to the plane of the square. Find OE, the distance from O to the plane of $\triangle VBC$.

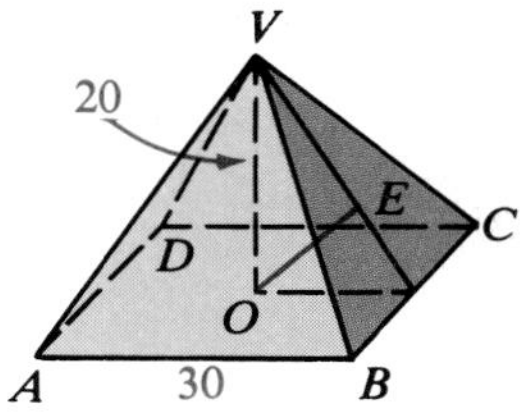

36. Given: $\triangle DEF$ with median $\overline{FG}$

Prove: $m = \frac{1}{2}\sqrt{2r^2 + 2s^2 - t^2}$

(*Hint:* Draw the perpendicular from F to $\overline{DE}$. Let $GH = x$ and $FH = y$.)

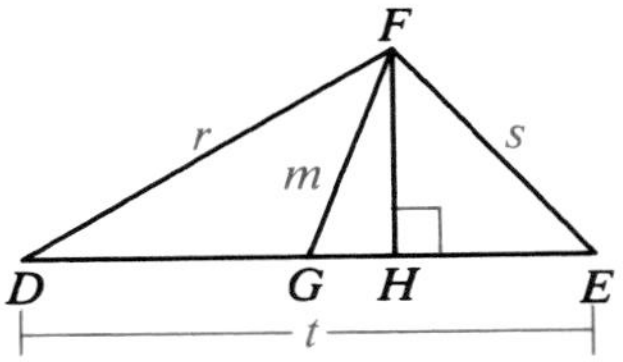

Self-Test 1

1. Simplify the expressions.

a. $\sqrt{48}$ **b.** $\dfrac{7}{\sqrt{2}}$ **c.** $\sqrt{\dfrac{5}{8}}$ **d.** $\dfrac{3\sqrt{20}}{4}$

2. Find the geometric mean between 3 and 15.

The diagram shows the altitude drawn to the hypotenuse of a right triangle.

3. $x =$ __?__

4. $y =$ __?__

5. $z =$ __?__

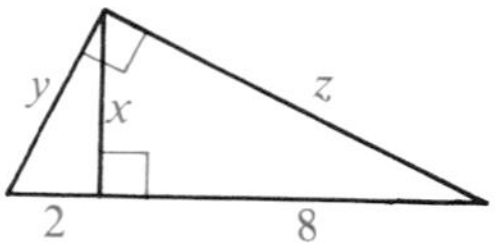

The hypotenuse of a right triangle is *h*. The legs are *f* and *g*. Copy and complete the table.

	6.	7.	8.	9.	10.
f	6	?	?	?	n
g	8	12	$5\sqrt{2}$	$7\sqrt{3}$	n
h	?	13	10	14	?

B I O G R A P H I C A L   N O T E

Pythagoras

Most of the ancient Greeks believed the whole numbers were sufficient to represent all the physical universe. Pythagoras (sixth century B.C.) came to realize that whole numbers and ratios of whole numbers were an inadequate representation of the physical world.

He discovered that the diagonal of a square one unit on each side has a length that can never be represented as the ratio of two integers. Although his best-known work is the Pythagorean Theorem, he also made numerous other contributions to mathematics. Among these is the proof that the sum of a triangle's three angles is equal to two right angles. Pythagoras also discovered the dependence of musical intervals on arithmetical ratios.

Right Triangles

Objectives

1. State and apply the converse of the Pythagorean Theorem and related theorems about obtuse and acute triangles.
2. Determine the lengths of two sides of a 45°–45°–90° or a 30°–60°–90° triangle when the length of the third side is known.

6–3 The Converse of the Pythagorean Theorem

We have seen that the converse of a theorem is not necessarily true. However, the converse of the Pythagorean Theorem *is* true. It is stated below as our next theorem.

Theorem 6–3

If the square of one side of a triangle is equal to the sum of the squares of the other two sides, then the triangle is a right triangle.

Given: $\triangle ABC$ with $c^2 = a^2 + b^2$

Prove: $\triangle ABC$ is a rt. $\triangle$.

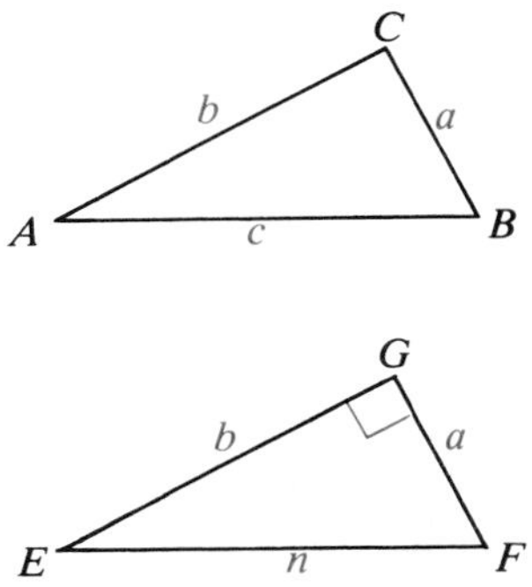

Outline of proof:

1. Draw rt. $\triangle EFG$ with legs a and b.
2. $n^2 = a^2 + b^2$ (Why?)
3. $c^2 = a^2 + b^2$ (Given)
4. $c = n$ (Why?)
5. $\triangle ABC \cong \triangle EFG$ (SSS Postulate)
6. $\angle C$ is a rt. $\angle$. (Why?)
7. $\triangle ABC$ is a rt. $\triangle$. (Why?)

A triangle with sides 3 units, 4 units, and 5 units long is called a 3-4-5 triangle. The numbers 3, 4, and 5 satisfy the equation $a^2 + b^2 = c^2$, so we can apply Theorem 6–3 to conclude that a 3-4-5 triangle is a right triangle. Furthermore, any triangle similar to a 3-4-5 triangle must be a right triangle. The side lengths shown in the table at the top of the next page all satisfy the equation

$$a^2 + b^2 = c^2,$$

so the triangles formed are right triangles.

Some Common Right Triangle Lengths

3, 4, 5	5, 12, 13	8, 15, 17	7, 24, 25
6, 8, 10	10, 24, 26		
9, 12, 15			
12, 16, 20			
15, 20, 25			

When a triangle is not a right triangle, the squares of the sides can be used to determine whether it is obtuse or acute, as shown in Theorems 6–4 and 6–5.

Theorem 6–4

If the square of the longest side of a triangle is greater than the sum of the squares of the other two sides, then the triangle is an obtuse triangle.

Theorem 6–5

If the square of the longest side of a triangle is less than the sum of the squares of the other two sides, then the triangle is an acute triangle.

Example Determine whether a triangle formed with sides having the lengths named is acute, right, or obtuse.

a. 9, 40, 41

b. 6, 7, 8

Solution

a. $9^2 + 40^2 \underline{\ ?\ } 41^2$
$81 + 1600 \underline{\ ?\ } 1681$
$1681 = 1681$
The triangle is right.

b. $6^2 + 7^2 \underline{\ ?\ } 8^2$
$36 + 49 \underline{\ ?\ } 64$
$85 > 64$
The triangle is acute.

Classroom Exercises

If a triangle is formed with sides having the lengths named, is it acute, right, or obtuse? If a triangle can't be formed, say *not possible*.

1. 1, 4, 6
2. 4, 6, 8
3. 6, 8, 10
4. 8, 10, 12
5. $\sqrt{7}$, $\sqrt{7}$, $\sqrt{14}$
6. 4, $4\sqrt{3}$, 8

7. Specify all lengths l that will make the statement true.
a. $\angle 1$ is a right $\angle$.
b. $\angle 1$ is an acute $\angle$.
c. $\angle 1$ is an obtuse $\angle$.
d. The triangle is isosceles.
e. No triangle is possible.

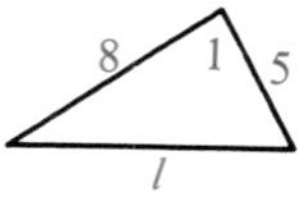

For Exercises 8–10, refer to the figures below.

8. Explain why x must equal 5.

9. Explain why $\angle D$ must be a right angle.

10. Explain why $\angle R'$ must be a right angle.

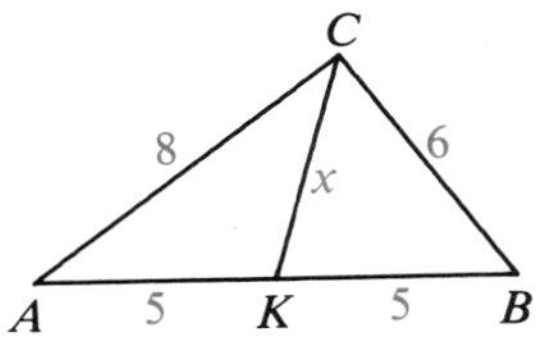

Ex. 8

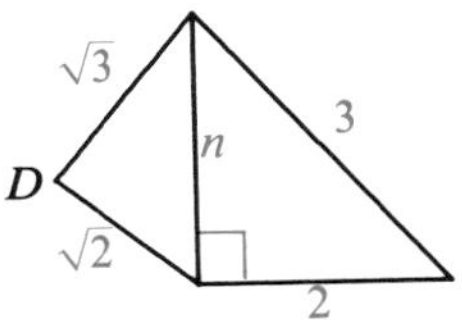

Ex. 9

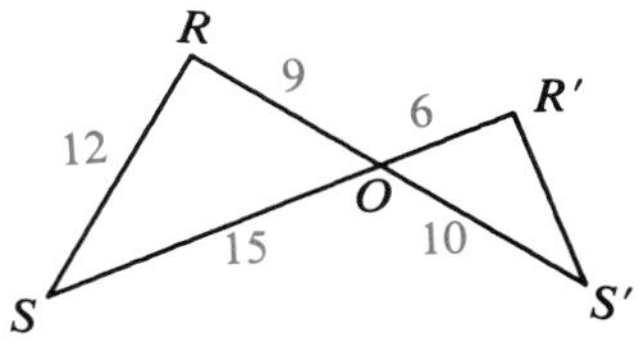

Ex. 10

Written Exercises

Tell whether a triangle formed with sides having the lengths named is acute, right, or obtuse. When it isn't possible to decide, write *can't tell.*

A

1. 11, 11, 15

2. 9, 9, 13

3. 8, $8\sqrt{3}$, 16

4. 0.03, 0.04, 0.05

5. 300, 400, 501

6. 0.6, 0.8, 1

7. $5n$, $12n$, $13n$ where $n > 0$

8. $n + 4$, $n + 5$, $n + 6$ where $n \geq 1$

9. $7 - n$, 7, $7 + n$ where $0 < n \leq 3$

10. Given: $\angle UTS$ is a rt. $\angle$.
Explain why $\triangle TRS$ must be a right triangle.

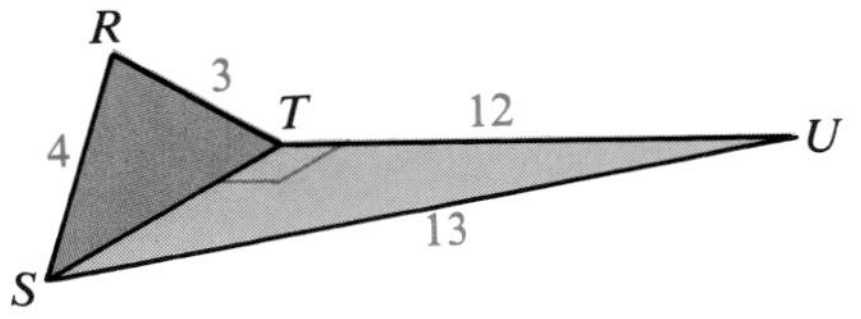

11. Given: $\overline{AC} \perp$ plane P
Explain why $\triangle BCD$ must be an obtuse triangle.

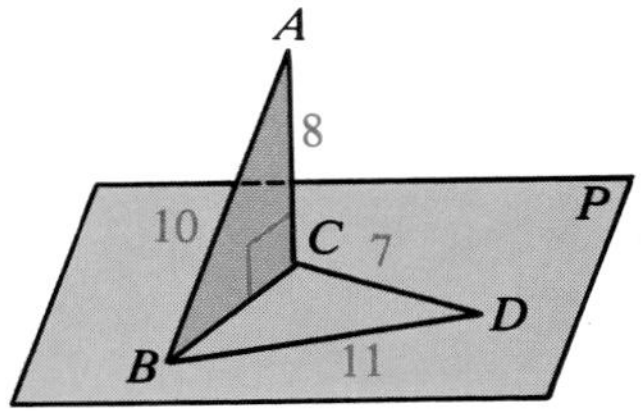

B

12. Sketch $\square ABCD$ with $AB = 13$, $AC = 24$, and $BD = 10$. What special kind of parallelogram is $ABCD$? Explain your answer.

13. Sketch $\square RSTU$, with diagonals intersecting at M. $RS = 9$, $ST = 20$, and $RM = 11$. Which segment is longer, $\overline{SM}$ or $\overline{RM}$? Explain your answer.

14. Given: A triangle with sides a, b, c;
$a = n^2 - 1$; $b = 2n$; $c = n^2 + 1$
Prove: The triangle is a rt. $\triangle$.

15. The sides of a triangle have lengths x, $x + 4$, and 20. Specify those values of x for which the triangle is acute with longest side 20.

16. Is $\triangle ABC$ acute, right, or obtuse? Explain your answer.

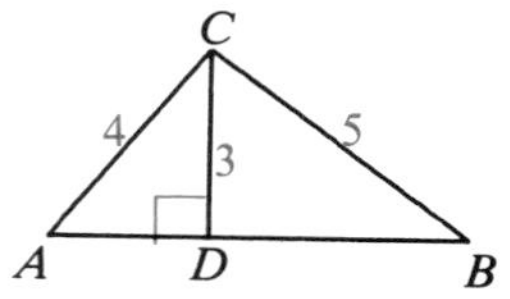

C 17. Write a plan for the proof of Theorem 6–4.
Given: $\triangle RST$; $l^2 > j^2 + k^2$
Prove: $\triangle RST$ is an obtuse triangle.
(*Hint*: Start by drawing right $\triangle UVW$ with legs j and k. Compare lengths l and n.)

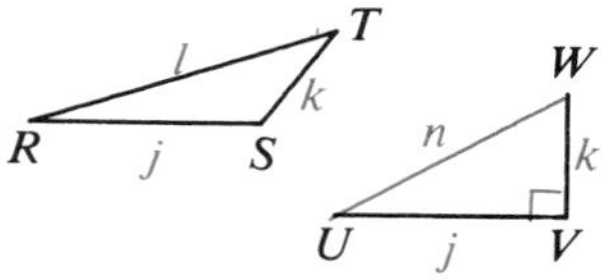

18. Write a plan for the proof of Theorem 6–5.

19. Given: $\triangle ABC$ and $\triangle RST$, with sides having the lengths shown; $\angle C$ is a rt. $\angle$.
Prove: $\triangle RST$ is an acute $\triangle$.

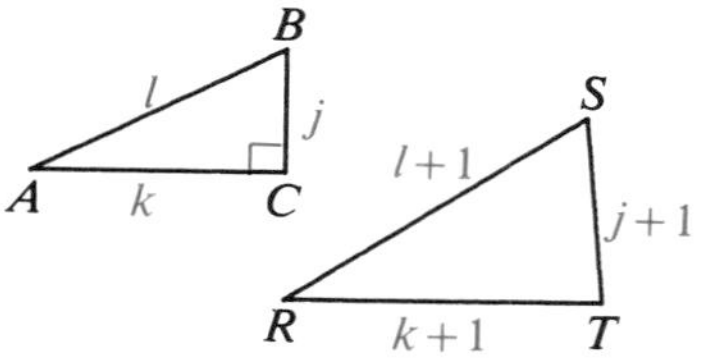

20. Find the values of x and y. *Note:* A frame in this shape, like the simple *scissors truss* shown at the right below, can be used to support a peaked roof. The weight of the roof compresses some parts of the frame (green), while other parts are in tension (blue). A frame made with s segments joined at j points is stable if $s \geq 2j - 3$. In the truss shown, 9 segments connect 6 points. Verify that the truss is stable.

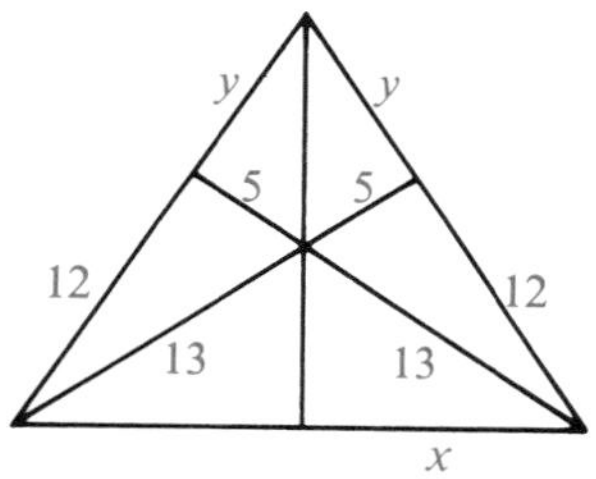

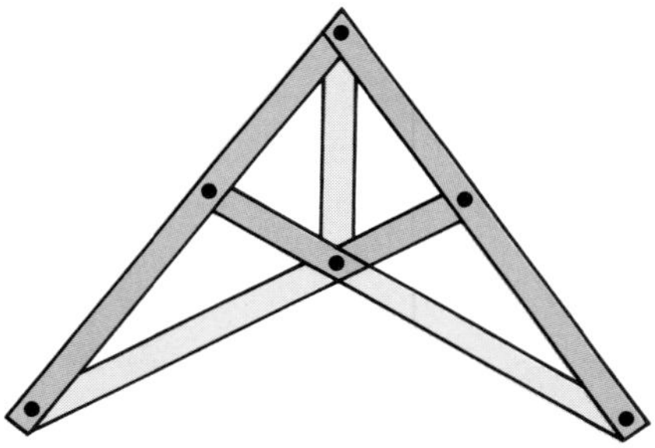

COMPUTER KEY-IN

Suppose a, b, and c are positive integers such that $a^2 + b^2 = c^2$. Then the converse of the Pythagorean Theorem guarantees that a, b, and c are the lengths of the sides of a right triangle. Because of this, any such triple of integers is called a **Pythagorean triple.**

For example, 3, 4, 5 is a Pythagorean triple since $3^2 + 4^2 = 5^2$. Another triple is 6, 8, 10, since $6^2 + 8^2 = 10^2$. The triple 3, 4, 5 is called a *primitive* Pythagorean triple because no factor (other than 1) is common to all three integers. 6, 8, 10 is *not* a primitive triple.

The following program in BASIC lists some Pythagorean triples.

```
10  FOR X = 2 TO 7
20  FOR Y = 1 TO X - 1
30  LET A = 2 * X * Y
40  LET B = X * X  - Y * Y
50  LET C = X * X  + Y * Y
60  PRINT A;",";B;",";C
70  NEXT Y
80  NEXT X
90  END
```

Exercises

1. Type and RUN the program. (If your computer uses a language other than BASIC, write and RUN a similar program.) What Pythagorean triples did it list? Which of these are primitive Pythagorean triples?

2. The program above uses a method for finding Pythagorean triples that was developed by Euclid around 320 B.C. His method can be stated as follows:

If x and y are positive integers with $y < x$, then $a = 2xy$, $b = x^2 - y^2$, and $c = x^2 + y^2$ is a Pythagorean triple.

To verify that Euclid's method is correct, show that the equation below is true.

$$(2xy)^2 + (x^2 - y^2)^2 = (x^2 + y^2)^2$$

3. Explain why, in line 20, the last value for Y must be X − 1. What happens if Y = X?

4. To find several Pythagorean triples containing large numbers change lines 10 and 20 to

```
10  FOR X = 51 TO 55
20  FOR Y = 50 TO X - 1
```

and RUN the program.

5. In Exercise 1 the computer printed 21 Pythagorean triples. Suppose the values of X are the integers from 2 to 10 and the values of Y are the integers from 1 to X − 1. How many Pythagorean triples would you expect the computer to print? RUN the program to verify your answer.

6. Suppose the values of X are the integers from 2 to 100 and the values of Y are the integers from 1 to X − 1. How many Pythagorean triples should the computer print?

Challenge

A room is 30 ft long, 12 ft wide, and 12 ft high. A spider is at the middle of an end wall, 1 ft from the floor. A fly is at the middle of the other end wall, 1 ft from the ceiling, too frightened to move. The spider crawls to the fly. What is the shortest distance?

6-4 Special Right Triangles

An isosceles right triangle is also called a 45°-45°-90° triangle, because the measures of the angles are 45, 45, and 90.

Theorem 6-6 45°-45°-90° Theorem

In a 45°-45°-90° triangle, the hypotenuse is $\sqrt{2}$ times as long as a leg.

Given: A 45°-45°-90° triangle

Prove: hypotenuse $= \sqrt{2} \cdot$ leg

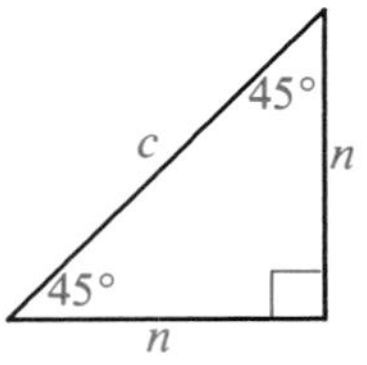

■ ***Plan for Proof:*** Let the sides of the given triangle be n, n, and c. Apply the Pythagorean Theorem and solve for c in terms of n.

Example 1 Find the value of x.

a.

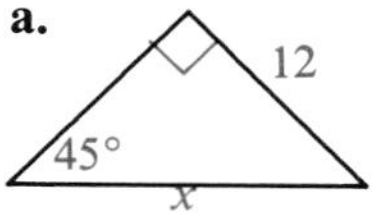

b.

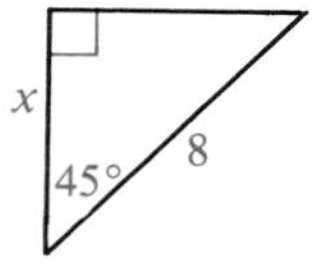

Solution

a.
$$\text{hyp.} = \sqrt{2} \cdot \text{leg}$$
$$x = \sqrt{2} \cdot 12$$
$$x = 12\sqrt{2}$$

b.
$$\text{hyp.} = \sqrt{2} \cdot \text{leg}$$
$$8 = \sqrt{2} \cdot x$$
$$x = \frac{8}{\sqrt{2}} = \frac{8}{\sqrt{2}} \cdot \frac{\sqrt{2}}{\sqrt{2}} = \frac{8\sqrt{2}}{2}$$
$$x = 4\sqrt{2}$$

Another special right triangle has acute angles measuring 30 and 60.

Theorem 6-7 30°-60°-90° Theorem

In a 30°-60°-90° triangle, the hypotenuse is twice as long as the shorter leg, and the longer leg is $\sqrt{3}$ times as long as the shorter leg.

Given: $\triangle ABC$, a 30°-60°-90° triangle

Prove: hypotenuse $= 2 \cdot$ shorter leg
longer leg $= \sqrt{3} \cdot$ shorter leg

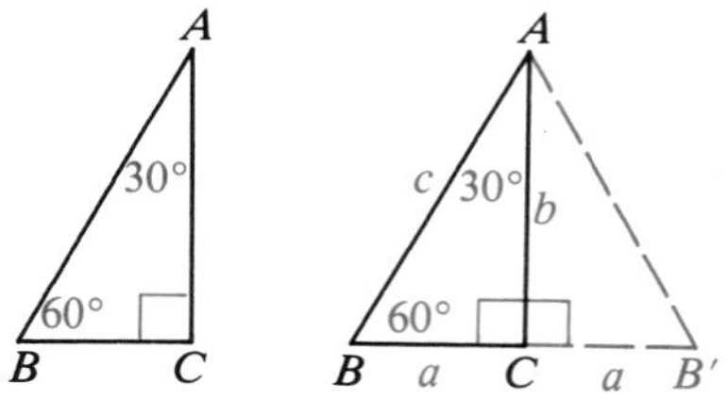

Outline of proof:

1. Build onto $\triangle ABC$ as shown.
2. $\triangle AB'C \cong \triangle ABC$, so $m\angle B' = 60$
3. $\triangle ABB'$ is equiangular. (Why?)
4. $\triangle ABB'$ is equilateral, so $c = 2a$.
5. $a^2 + b^2 = c^2$ (Why?)
6. $a^2 + b^2 = 4a^2$ (Why?)
7. $b^2 = 3a^2$ and $b = a\sqrt{3}$ (Why?)

Example 2 Find the values of x and y.

a.

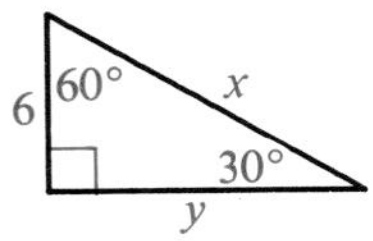

b.

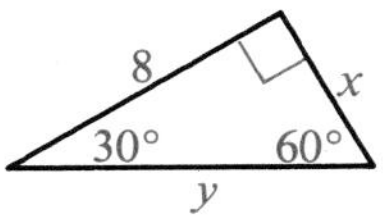

Solution

a. hyp. $= 2 \cdot$ shorter leg

$x = 2 \cdot 6$

$x = 12$

longer leg $= \sqrt{3} \cdot$ shorter leg

$y = 6\sqrt{3}$

b. longer leg $= \sqrt{3} \cdot$ shorter leg

$8 = \sqrt{3}x$

$x = \dfrac{8}{\sqrt{3}} = \dfrac{8\sqrt{3}}{3}$

hyp. $= 2 \cdot$ shorter leg

$y = 2 \cdot \dfrac{8\sqrt{3}}{3} = \dfrac{16\sqrt{3}}{3}$

Classroom Exercises

Find the value of x.

1.

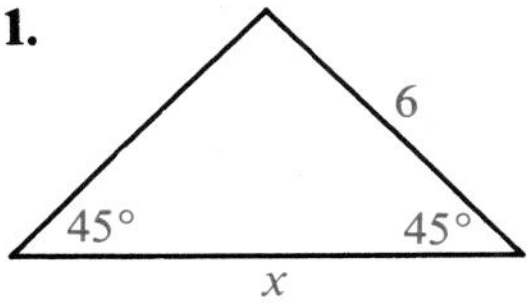

2.

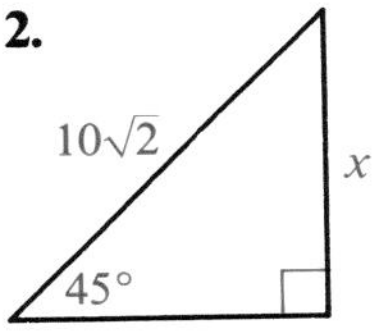

3.

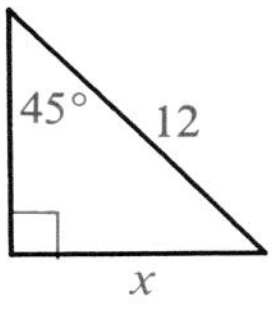

4.

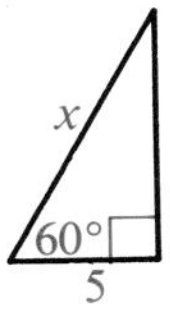

5.

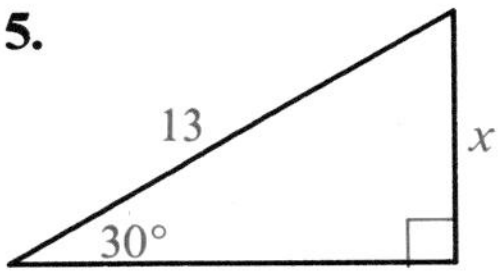

6.

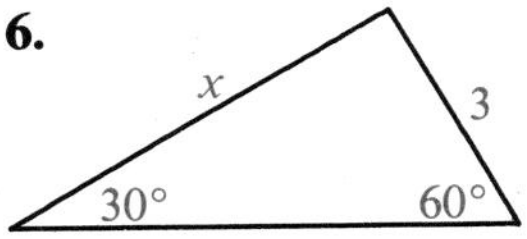

7.

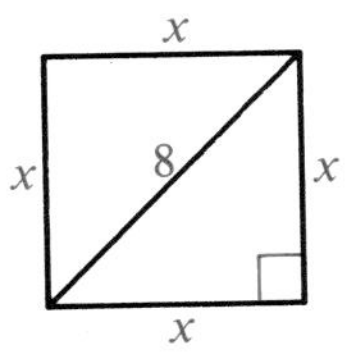

8.

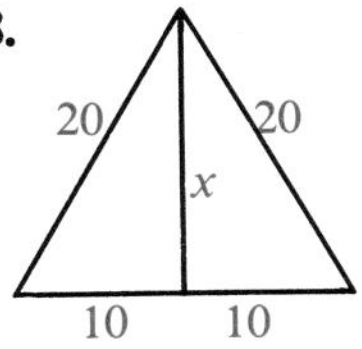

9.

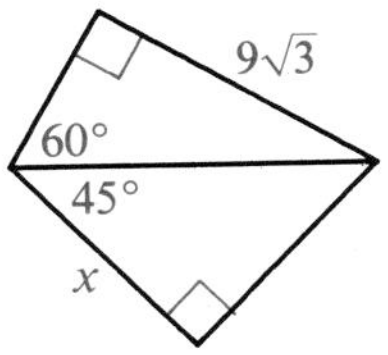

10. In regular hexagon $ABCDEF$, $AB = 8$. Find AC and AD.

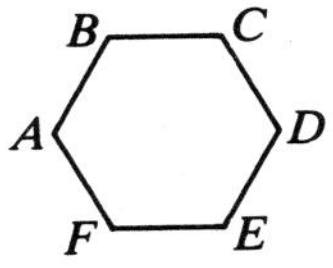

11. Express PQ, PS, and QR in terms of a.

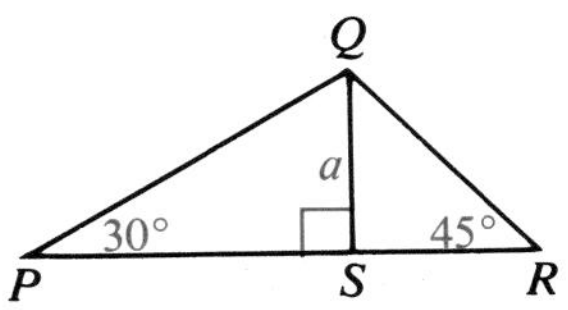

Written Exercises

A **1.** Draw an equilateral triangle with sides 10 units long. Draw an altitude. On your diagram, show the length of each segment of the base.

a. Use Theorem 6–7 to find the length of the altitude.

b. Use the Pythagorean Theorem to find the length of the altitude.

2. Draw a square with a diagonal 13 units long.

a. Use Theorem 6–6 to find the length of a side of the square.

b. Use the Pythagorean Theorem to find the length of a side of the square.

Copy and complete the tables.

	3.	**4.**	**5.**	**6.**	**7.**	**8.**
a	6	?	$\sqrt{5}$	?	?	?
b	?	$\frac{4}{5}$	?	?	?	?
c	?	?	?	$8\sqrt{2}$	6	$\sqrt{22}$

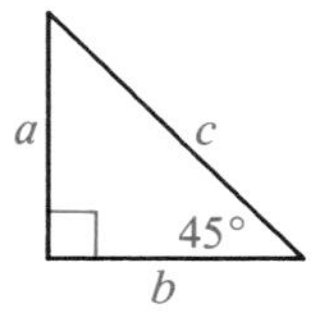

	9.	**10.**	**11.**	**12.**	**13.**	**14.**
d	7	$\frac{1}{5}$	?	?	?	?
e	?	?	$5\sqrt{3}$	6	?	?
f	?	?	?	?	12	5

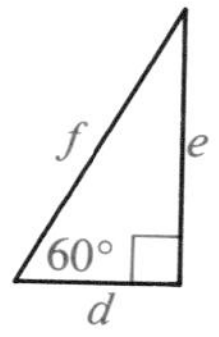

Copy and complete the table. Draw a new sketch for each exercise and label lengths as you find them.

	TU	***UV***	***TV***	***WT***	***WU***	***WV***
B **15.**	7	?	?	?	?	?
16.	?	?	$8\sqrt{3}$	?	?	?
17.	?	?	?	50	?	?
18.	?	?	?	?	7	?

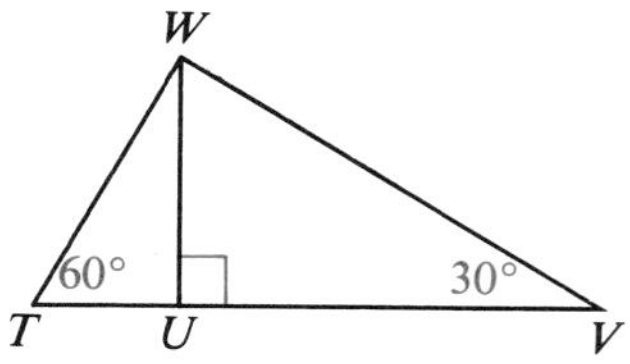

Find the lengths of as many segments as possible.

19.

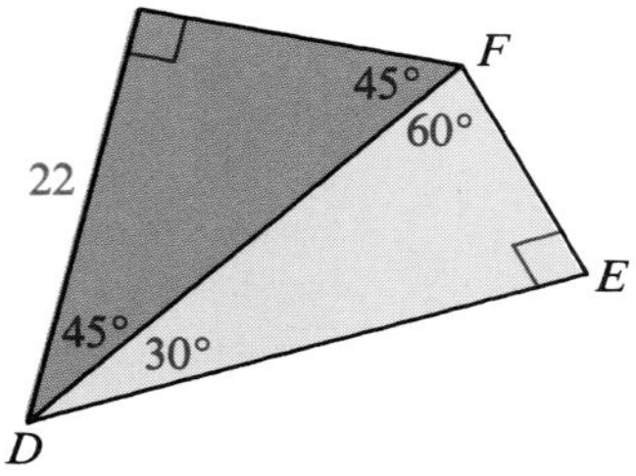

20.

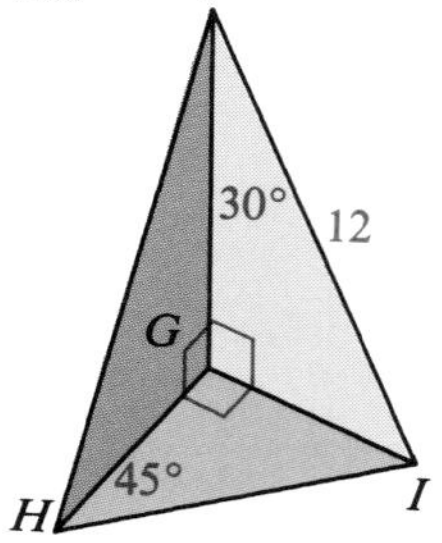

21.

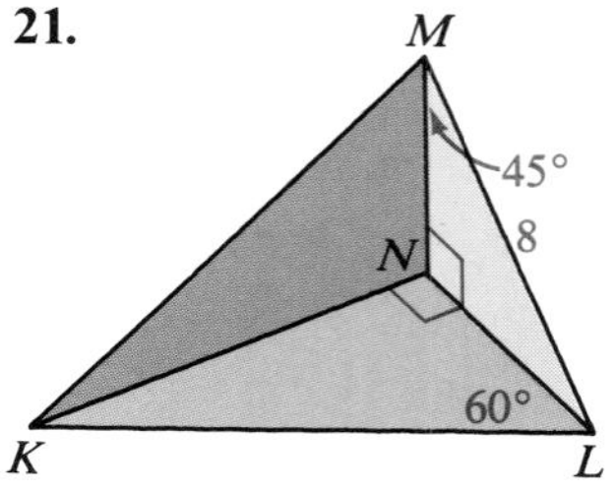

22. The diagonals of a rectangle are 8 units long and intersect at a 60° angle. Find the dimensions of the rectangle.

23. The sum of the lengths of the sides of a rhombus is 64 and one of its angles has measure 120. Find the lengths of the diagonals.

24. Prove Theorem 6–6.

25. Explain why any triangle having sides in the ratio $1 : \sqrt{3} : 2$ must be a 30°-60°-90° triangle.

C 26. In quadrilateral $QRST$: $m\angle R = 60$; $m\angle T = 90$; $QR = RS$; $ST = 8$; $TQ = 8$. How long is the longer diagonal of the quadrilateral?

27. Find the perimeter of the triangle.

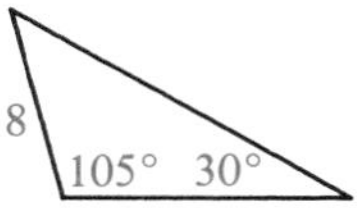

28. Find the length of the median of the trapezoid in terms of j.

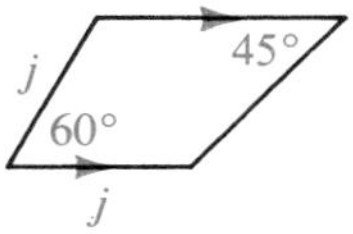

29. If the wrench just fits the hexagonal nut, what is the value of x?

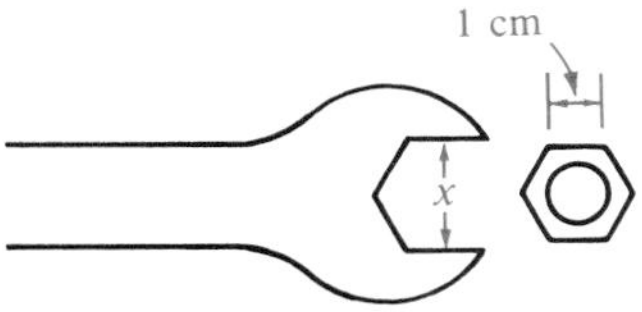

★ 30. The six edges of the solid shown are 8 units long. A and B are midpoints of two edges as shown. Find AB.

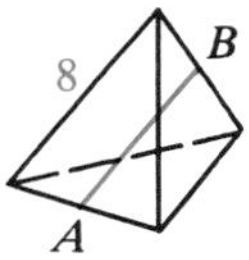

Self-Test 2

Three sides of a triangle are given. Tell whether the triangle is acute, right, or obtuse.

1. 11, 60, 61
2. 7, 9, 11
3. 0.2, 0.3, 0.4

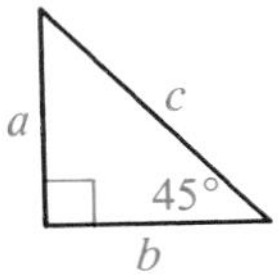

4. If $a = 5$, then $b =$ __?__ and $c =$ __?__.
5. If $c = 12$, then $a =$ __?__ and $b =$ __?__.

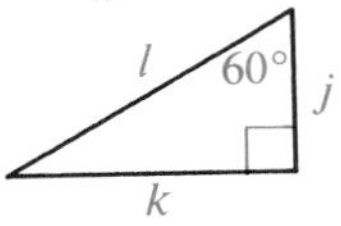

6. If $j = 4$, then $k =$ __?__ and $l =$ __?__.
7. If $l = 10$, then $j =$ __?__ and $k =$ __?__.
8. If $k = 6$, then $j =$ __?__ and $l =$ __?__.
9. The sides of a rhombus are 4 units long, and one diagonal has length 4. How long is the other diagonal?

Trigonometry

Objectives

1. Define the tangent, sine, and cosine ratios for an acute angle.
2. Solve right triangle problems by correct selection and use of the tangent, sine, and cosine ratios.

6-5 The Tangent Ratio

The word *trigonometry* comes from Greek words that mean "triangle measurement." Our study in this book will be limited to the trigonometry of right triangles. In the right triangle shown, one acute angle is marked. The leg opposite this angle and the leg adjacent to this angle are labeled.

The following ratio of the lengths of the legs is called the *tangent ratio.*

tangent of $\angle A = \dfrac{\text{leg opposite } \angle A}{\text{leg adjacent to } \angle A}$

In abbreviated form: $\tan A = \dfrac{\text{opposite}}{\text{adjacent}}$

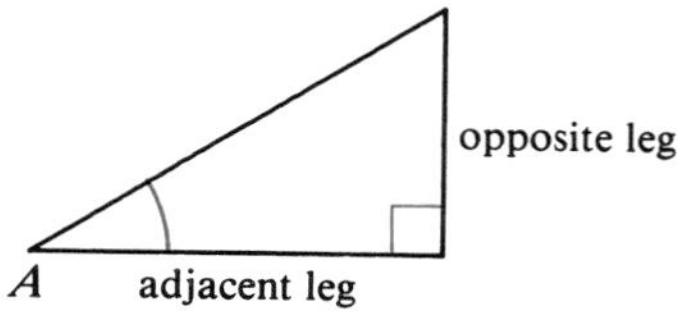

Example 1 Find $\tan X$ and $\tan Y$.

Solution

$$\tan X = \frac{\text{leg opposite } \angle X}{\text{leg adjacent to } \angle X} = \frac{12}{5}$$

$$\tan Y = \frac{\text{leg opposite } \angle Y}{\text{leg adjacent to } \angle Y} = \frac{5}{12}$$

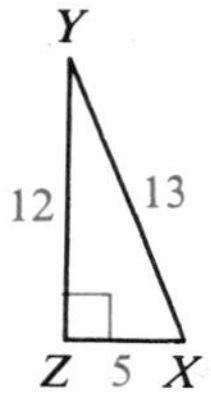

In the right triangles shown below, $m\angle A = m\angle R$. Then by the AA Similarity Postulate, the triangles are similar. We can write these proportions:

$\dfrac{a}{r} = \dfrac{b}{s}$ (Why?)

$\dfrac{a}{b} = \dfrac{r}{s}$ (A property of proportions)

$\tan A = \tan R$ (Def. of tangent ratio)

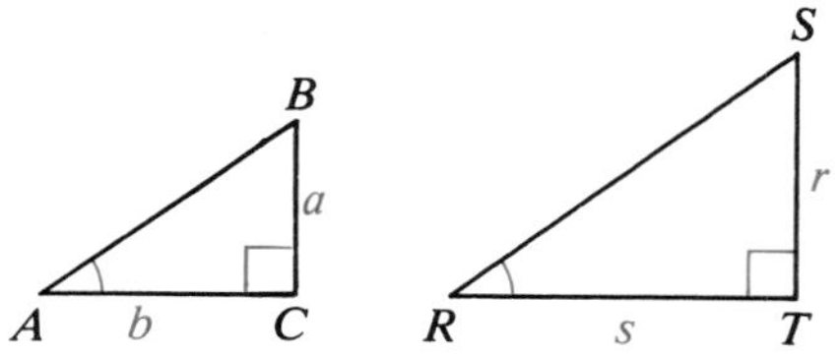

We have shown that if $m\angle A = m\angle R$, then $\tan A = \tan R$. Thus we have shown that the value of the tangent of an angle depends only on the size of the angle, not on the size of the right triangle. It is also true that if $\tan A = \tan R$ for acute angles A and R, then $m\angle A = m\angle R$.

Since the tangent of an angle depends only on the measure of the angle, we can write $\tan 10°$, for example, to stand for the tangent of any angle with a degree measure of 10. The table on page 271 lists the values of the tangents of some angles with measures between 0 and 90. Most of the values are approximations, rounded to four decimal places. Suppose you want the approximate value of tan 33°. Locate 33° in the angle column. Go across to the tangent column. Read .6494. You write $\tan 33° \approx 0.6494$, where the symbol $\approx$ means "is approximately equal to."

Example 2 Find the value of y to the nearest tenth.

Solution

$$\tan 56° = \frac{y}{32}$$

$$1.4826 \approx \frac{y}{32}$$

$$32(1.4826) \approx y$$

$$y \approx 47.4432, \text{ or } 47.4$$

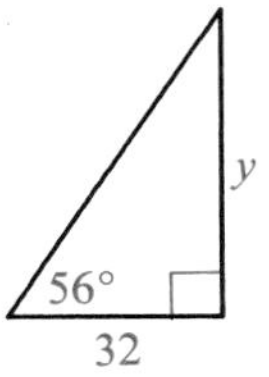

You can find the degree measure of an angle with a given tangent by reading from the tangent column across to the angle column.

Example 3 Find the value of $z°$ to the nearest degree.

Solution

$$\tan z° = \frac{3}{4}$$

$$\tan z° = 0.7500$$

$$z° \approx 37°$$

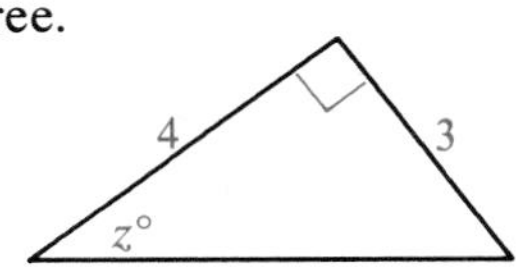

Notice that 0.7500 falls between two values in the tangent column: $\tan 36° \approx 0.7265$ and $\tan 37° \approx 0.7536$. Since 0.7500 is closer to 0.7536, we use 37° as an approximate value for $z°$.

A scientific calculator with keys for computing trigonometric ratios can be used to find the tangent of a given angle. If the calculator has an inverse key the measure of the angle with a given tangent can also be found. Like values found from a table, the values given by a calculator are approximations, usually rounded to seven or eight decimal places.

Classroom Exercises

In Exercises 1–3 express tan *R* as a ratio.

1.

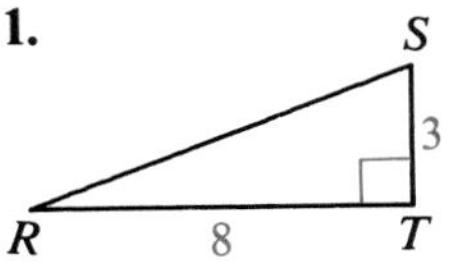

2.

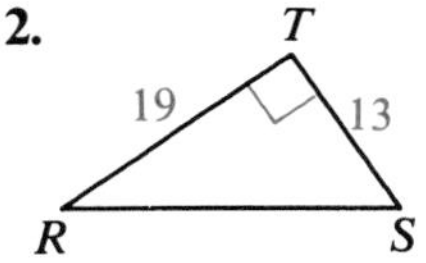

3.

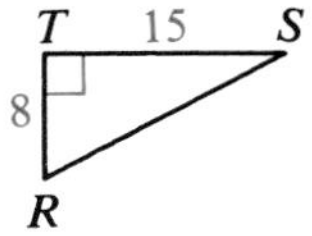

4–6. Express tan S as a ratio for each triangle above.

7. Use the table on page 271 to complete the statements.

a. tan 24° ≈ _?_ b. tan 41° ≈ _?_ c. tan 88° ≈ _?_

d. tan _?_ ≈ 2.4751 e. tan _?_ ≈ 0.3057 f. tan _?_ ≈ 0.8098

8. Three 45°-45°-90° triangles are shown below.

a. In each triangle, express tan 45° in simplified form.

b. See the entry for tan 45° on page 271. Is the entry exact?

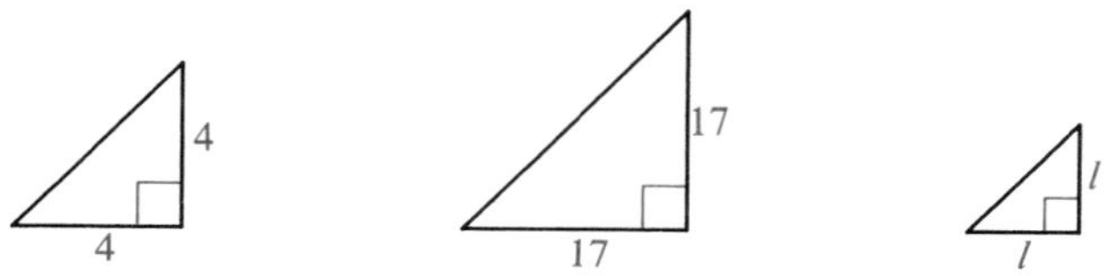

9. Three 30°-60°-90° triangles are shown below.

a. In each triangle, express tan 60° in simplified radical form.

b. Because tan 60° $= \frac{\sqrt{3}}{1}$, and because $\sqrt{3} \approx 1.732051$, you can write tan 60° ≈ _?_.

c. Is the entry for tan 60° on page 271 exact? Is it correct to four decimal places?

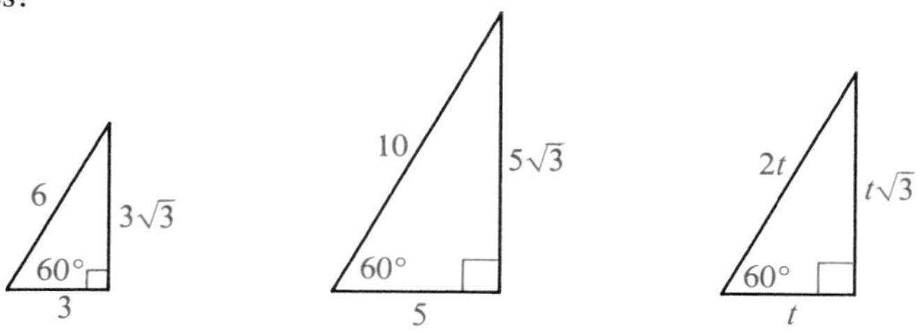

10. Notice that the tangent values increase rapidly toward the end of the table on page 271. Explain how you know that there is some angle with a tangent value equal to 1,000,000. Is there any upper limit to tangent values?

11. Two ways to find the value of x are started below.

Using tan 40°:	Using tan 50°:
$\tan 40° = \frac{27}{x}$	$\tan 50° = \frac{x}{27}$
$0.8391 \approx \frac{27}{x}$	$1.1918 \approx \frac{x}{27}$

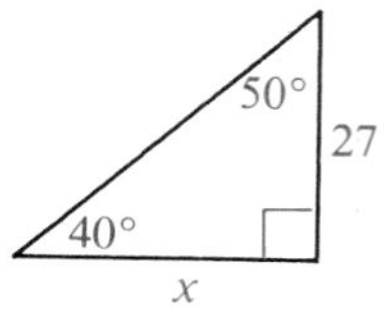

Which of the following statements are correct?

a. $x \approx 27 \cdot 0.8391$

b. $x \approx \frac{27}{0.8391}$

c. $x \approx 27 \cdot 1.1918$

d. $x \approx \frac{27}{1.1918}$

Which method is better if you are not using a calculator for the arithmetic?

1 - 12 odd

Written Exercises

Find x correct to the nearest tenth. Use the table on page 271.

A **1.**

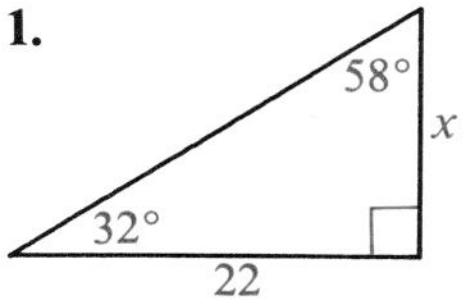

2.

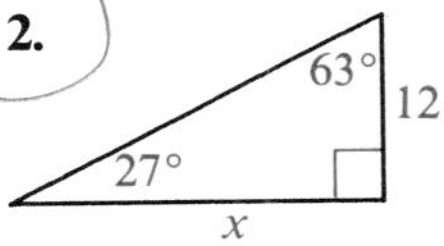

3.

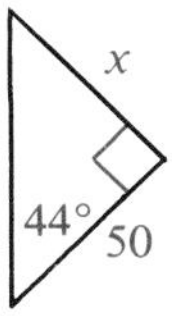

4.

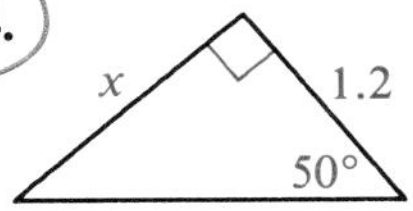

5.

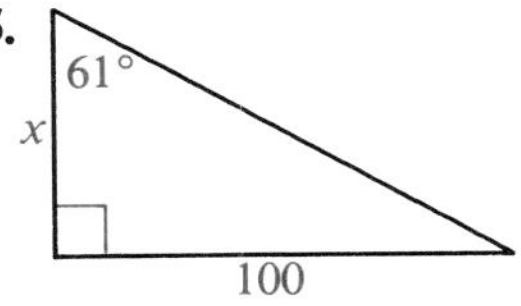

6.

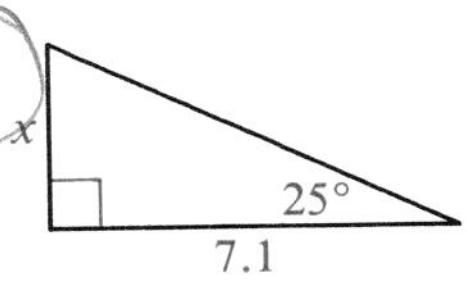

Find $y°$ correct to the nearest degree.

7.

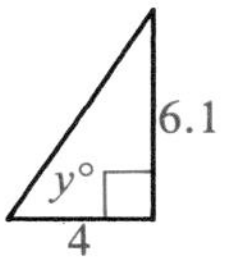

8.

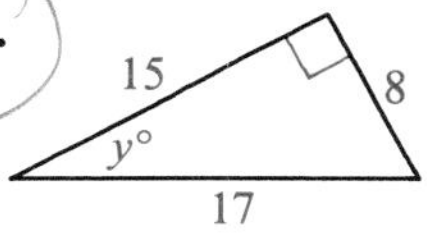

9.

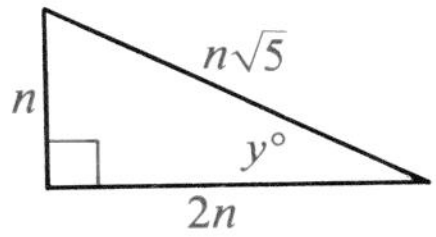

Find w, then z, correct to the nearest integer.

B **10.**

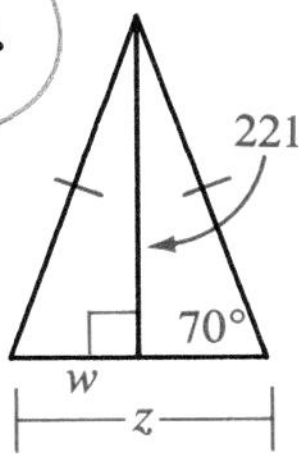

11.

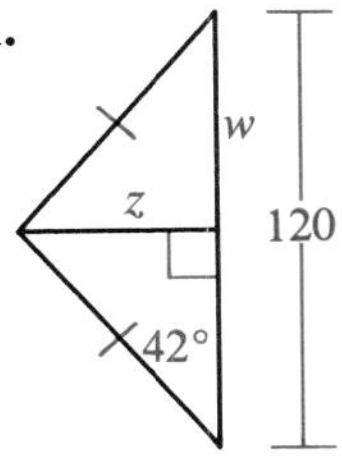

12.

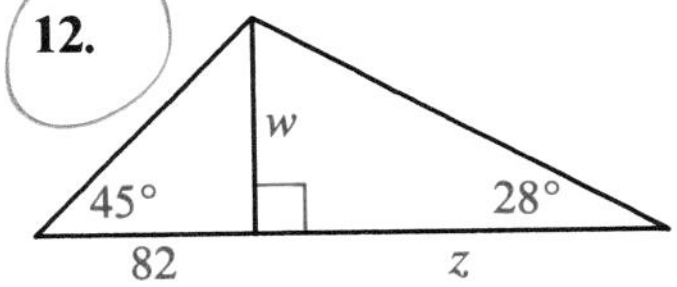

13. A rhombus has diagonals of length 4 and 10. Find the angles of the rhombus to the nearest degree.

14. The sides of a rectangle are 20 and 40. Find, to the nearest degree, the measure of an acute angle formed at the intersection of the diagonals.

15. A natural question to consider is the following:

$$\text{Does } \tan A + \tan B = \tan (A + B)?$$

Try 35 for the measure of A and 25 for the measure of B.

a. $\tan 35° + \tan 25° \approx$ _?_ + _?_ = _?_

b. $\tan (35° + 25°) = \tan$ _?_° $\approx$ _?_

c. What is your answer to the general question raised in this exercise, *yes* or *no*?

16. The shorter diagonal of a rhombus with a 70° angle is 124 cm long. How long (to the nearest centimeter) is the longer diagonal?

17. Complete the proof by supplying reasons and completing statements.
Given: $\angle M$ and $\angle R$ are complementary angles.
Prove: $\tan M \cdot \tan R = 1$

Plan for Proof: Draw right $\triangle MNO$ with $\angle M$ at one vertex and a right angle at O. The other acute angle, $\angle N$, is complementary to $\angle M$, so $\angle N \cong \angle R$. Show that $\tan M \cdot \tan N = 1$ and conclude that $\tan M \cdot \tan R = 1$.

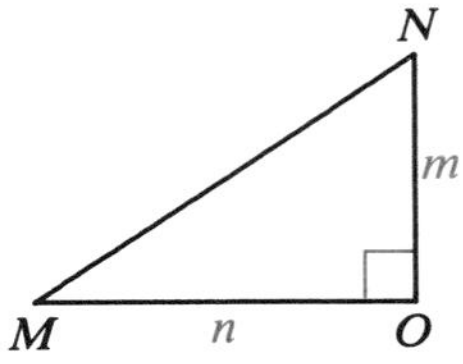

Proof:

1. $\angle N$ is a complement of $\angle M$. (Why?)
2. $\angle R$ is a complement of $\angle M$. (Why?)
3. $m\angle N = m\angle R$ (Why?)
4. $\tan N = \tan R$ (Why?)
5. $\tan M = \frac{m}{n}$ and $\tan N = \underline{\ ?\ }$ (Definition of tangent)
6. $\tan M \cdot \tan N = \frac{m}{n} \cdot \frac{n}{m} = 1$ (Multiplication and Substitution Properties)
7. $\tan M \cdot \tan R = 1$ (Why?)

C **18.** A person at window W, 40 ft above street level, sights points on a building directly across the street. H is chosen so that $\overline{WH}$ is horizontal. T is directly above H, and B is directly below. By measurement, $m\angle TWH = 61$ and $m\angle BWH = 37$. How far above street level is T?

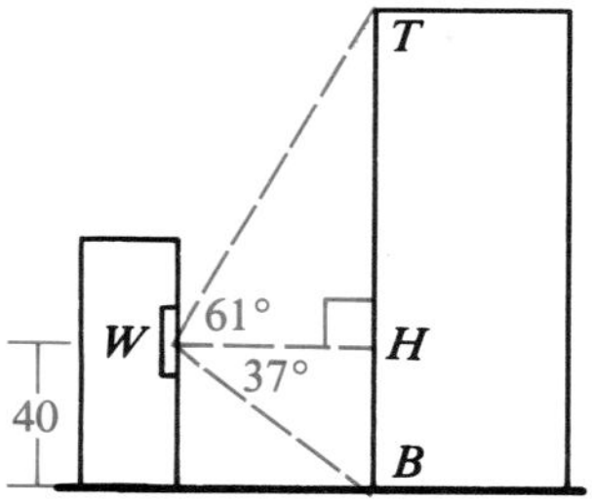

Ex. 18

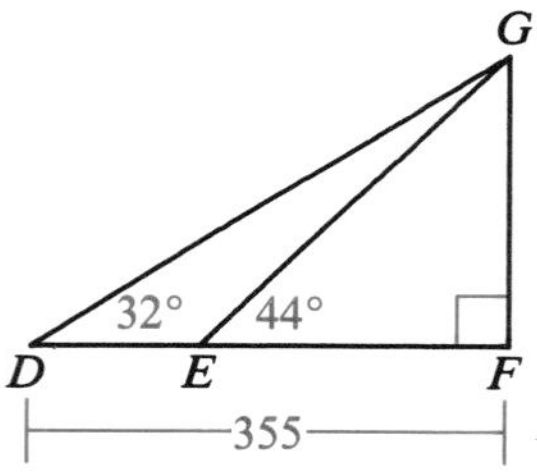

Ex. 19

19. Use the figure to find EF to the nearest integer.

20. In the diagram, half-planes X and Y with the same edge $\overleftrightarrow{FE}$ form **dihedral angle** X-FE-Y. If $\overrightarrow{PR} \perp \overleftrightarrow{FE}$ and $\overrightarrow{PQ} \perp \overleftrightarrow{FE}$, then $\angle RPQ$ is called a *plane angle* of this dihedral angle. The measure of a dihedral angle is defined to be the measure of any of its plane angles, so the measure of this dihedral angle is $m\angle RPQ$.

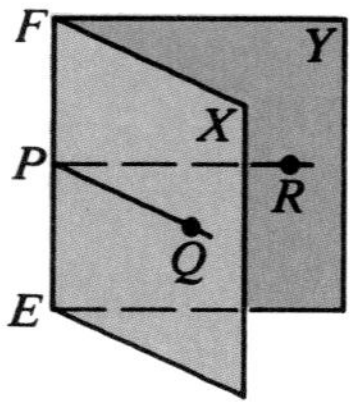

In the cube shown at the right, find, to the nearest degree, the measure of the dihedral angle containing the two shaded triangles.

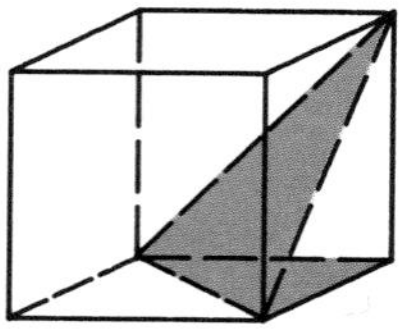

Table of Trigonometric Ratios

Angle	Sine	Cosine	Tangent	Angle	Sine	Cosine	Tangent
1°	.0175	.9998	.0175	**46°**	.7193	.6947	1.0355
2°	.0349	.9994	.0349	**47°**	.7314	.6820	1.0724
3°	.0523	.9986	.0524	**48°**	.7431	.6691	1.1106
4°	.0698	.9976	.0699	**49°**	.7547	.6561	1.1504
5°	.0872	.9962	.0875	**50°**	.7660	.6428	1.1918
6°	.1045	.9945	.1051	**51°**	.7771	.6293	1.2349
7°	.1219	.9925	.1228	**52°**	.7880	.6157	1.2799
8°	.1392	.9903	.1405	**53°**	.7986	.6018	1.3270
9°	.1564	.9877	.1584	**54°**	.8090	.5878	1.3764
10°	.1736	.9848	.1763	**55°**	.8192	.5736	1.4281
11°	.1908	.9816	.1944	**56°**	.8290	.5592	1.4826
12°	.2079	.9781	.2126	**57°**	.8387	.5446	1.5399
13°	.2250	.9744	.2309	**58°**	.8480	.5299	1.6003
14°	.2419	.9703	.2493	**59°**	.8572	.5150	1.6643
15°	.2588	.9659	.2679	**60°**	.8660	.5000	1.7321
16°	.2756	.9613	.2867	**61°**	.8746	.4848	1.8040
17°	.2924	.9563	.3057	**62°**	.8829	.4695	1.8807
18°	.3090	.9511	.3249	**63°**	.8910	.4540	1.9626
19°	.3256	.9455	.3443	**64°**	.8988	.4384	2.0503
20°	.3420	.9397	.3640	**65°**	.9063	.4226	2.1445
21°	.3584	.9336	.3839	**66°**	.9135	.4067	2.2460
22°	.3746	.9272	.4040	**67°**	.9205	.3907	2.3559
23°	.3907	.9205	.4245	**68°**	.9272	.3746	2.4751
24°	.4067	.9135	.4452	**69°**	.9336	.3584	2.6051
25°	.4226	.9063	.4663	**70°**	.9397	.3420	2.7475
26°	.4384	.8988	.4877	**71°**	.9455	.3256	2.9042
27°	.4540	.8910	.5095	**72°**	.9511	.3090	3.0777
28°	.4695	.8829	.5317	**73°**	.9563	.2924	3.2709
29°	.4848	.8746	.5543	**74°**	.9613	.2756	3.4874
30°	.5000	.8660	.5774	**75°**	.9659	.2588	3.7321
31°	.5150	.8572	.6009	**76°**	.9703	.2419	4.0108
32°	.5299	.8480	.6249	**77°**	.9744	.2250	4.3315
33°	.5446	.8387	.6494	**78°**	.9781	.2079	4.7046
34°	.5592	.8290	.6745	**79°**	.9816	.1908	5.1446
35°	.5736	.8192	.7002	**80°**	.9848	.1736	5.6713
36°	.5878	.8090	.7265	**81°**	.9877	.1564	6.3138
37°	.6018	.7986	.7536	**82°**	.9903	.1392	7.1154
38°	.6157	.7880	.7813	**83°**	.9925	.1219	8.1443
39°	.6293	.7771	.8098	**84°**	.9945	.1045	9.5144
40°	.6428	.7660	.8391	**85°**	.9962	.0872	11.4301
41°	.6561	.7547	.8693	**86°**	.9976	.0698	14.3007
42°	.6691	.7431	.9004	**87°**	.9986	.0523	19.0811
43°	.6820	.7314	.9325	**88°**	.9994	.0349	28.6363
44°	.6947	.7193	.9657	**89°**	.9998	.0175	57.2900
45°	.7071	.7071	1.0000				

CAREERS

General Contractor

Every construction project, from remodeling a kitchen to building a skyscraper, requires someone with overall responsibility for getting the job done efficiently and on time. Once the planning phases of a project are completed, the general contractor takes charge.

Construction involves many specialized kinds of work, all of them interrelated. For example, walls cannot be finished until electrical wiring has been completed. The contractor must arrange for the proper personnel to do the work, schedule the sequence of operations carefully, and monitor the pace and quality of work continuously. Effective

management of the project will avoid delays and increases in the cost of the structure. Coordination, scheduling, and supervision are key aspects of the contractor's work. On small projects, the contractor may also do a significant part of the actual construction work in addition to managing the whole project.

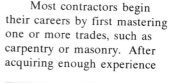

Most contractors begin their careers by first mastering one or more trades, such as carpentry or masonry. After acquiring enough experience

and expertise, a person may decide to go into business as a general contractor. Currently, some colleges are beginning to offer degrees in construction, providing another path for becoming a contractor.

6-6 The Sine and Cosine Ratios

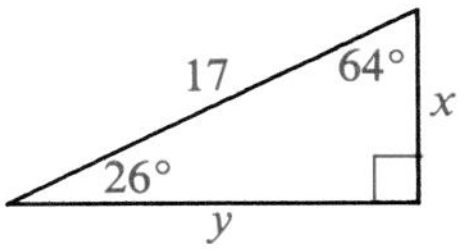

Can you find the values of x and y in the diagram by using tan 64°? by using tan 26°? The answer is *no,* because the only side known, the hypotenuse, is not involved in the definition of tangent. Two ratios that do relate the hypotenuse to the legs are the *sine* and *cosine*.

$$\textbf{sine of } \angle A = \frac{\text{leg opposite } \angle A}{\text{hypotenuse}}$$

$$\textbf{cosine of } \angle A = \frac{\text{leg adjacent to } \angle A}{\text{hypotenuse}}$$

We now have three useful trigonometric ratios:

$$\tan A = \frac{\text{opposite}}{\text{adjacent}}$$

$$\sin A = \frac{\text{opposite}}{\text{hypotenuse}}$$

$$\cos A = \frac{\text{adjacent}}{\text{hypotenuse}}$$

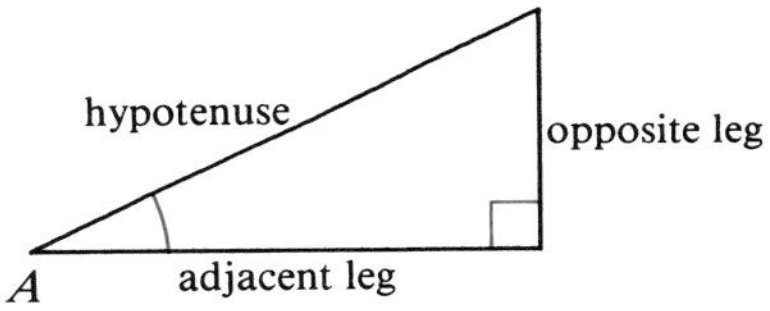

Example 1 Find the values of x and y correct to the nearest integer. Use the table on page 271.

Solution

$$\sin 67° = \frac{x}{120} \qquad \cos 67° = \frac{y}{120}$$

$$0.9205 \approx \frac{x}{120} \qquad 0.3907 \approx \frac{y}{120}$$

$$120 \cdot 0.9205 \approx x \qquad 120 \cdot 0.3907 \approx y$$

$$110.46 \approx x \qquad 46.884 \approx y$$

$$x \approx 110 \qquad y \approx 47$$

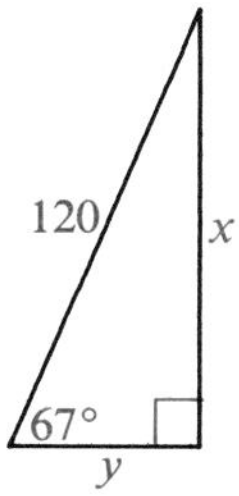

Example 2 Find $n°$ correct to the nearest degree.

Solution

$$\sin n° = \frac{22}{40}$$

$$\sin n° = 0.5500$$

$$n° \approx 33°$$

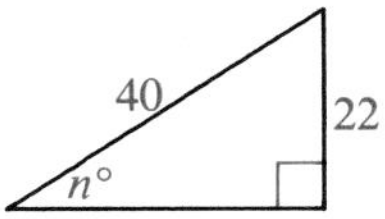

(The two values in the sine column closest to 0.5500 are 0.5446 and 0.5592. The closer is 0.5446.)

Classroom Exercises

In Exercises 1–3, express sin A, cos A, and tan A as fractions.

1.

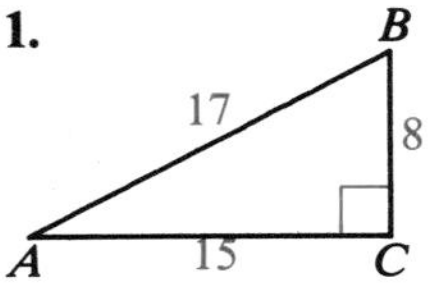

2.

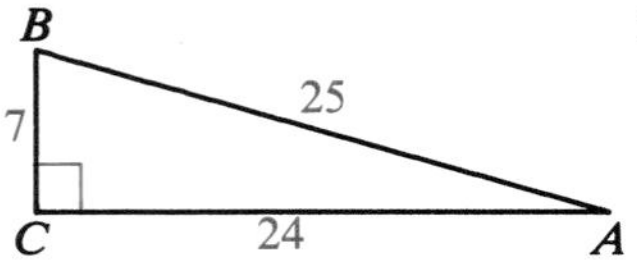

3.

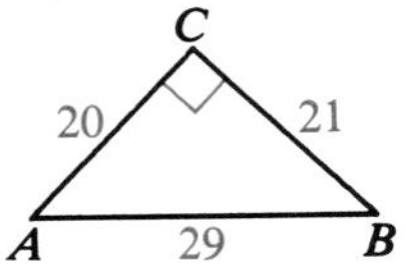

4–6. Using the triangles in Exercises 1–3, express sin B, cos B, and tan B as fractions.

7. Use the table on page 271 to complete the statements.

a. $\sin 13° \approx$ __?__ **b.** $\cos 88° \approx$ __?__

c. sin __?__ ≈ 0.7547 **d.** cos __?__ ≈ 0.9511

State two different equations you could use to find the value of x.

8.

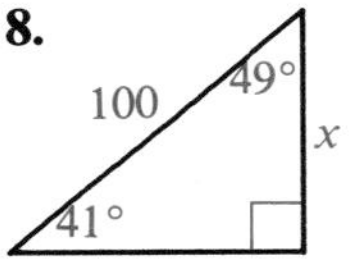

9.

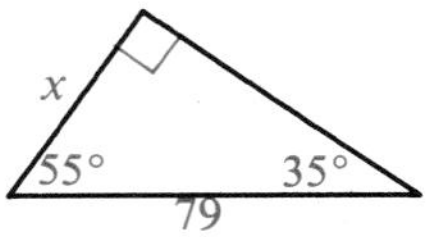

10.

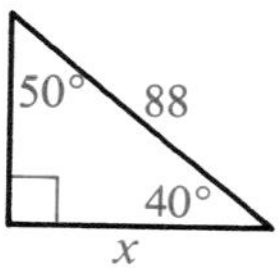

11. State four different equations you could use to find the value of x.

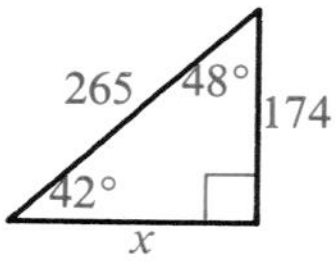

12. The word cosine is related to the phrase complement's sine. Explain the relationship by using the diagram to express the cosine of $\angle A$ and the sine of its complement, $\angle B$.

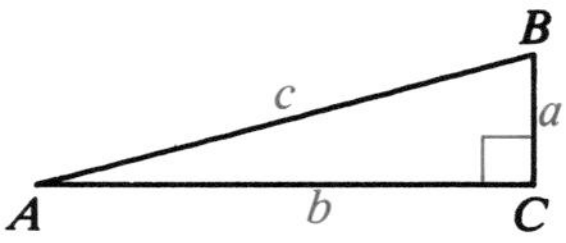

13. The table on page 271 lists 0.5000 as the value of sin 30°. This value is exact. Explain why.

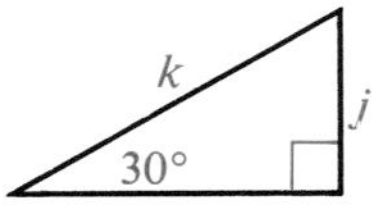

14. Suppose $\sin n° = \frac{15}{17}$. Find $\cos n°$ without using a table.

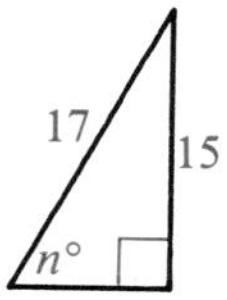

15. Suppose $\sin n° = \frac{a}{t}$. Express $\cos n°$ algebraically.

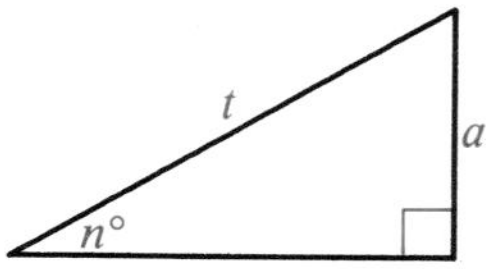

16. According to the table on page 271, sin 1° and tan 1° are both approximately 0.0175. Which is actually larger? How do you know?

Written Exercises

In these exercises, use the table on page 271. Find lengths correct to the nearest integer and angle measures to the nearest degree.

In Exercises 1–9, find the values of the variables.

A **1.**

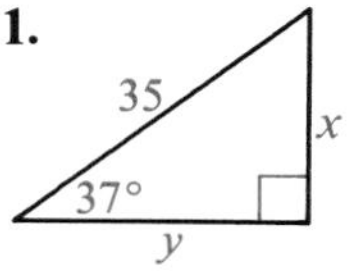

2.

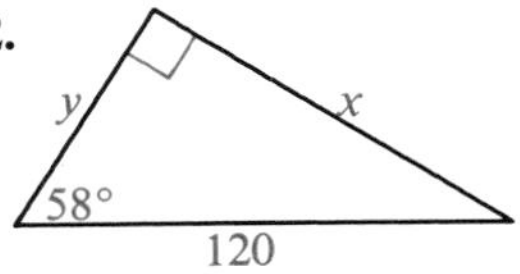

3.

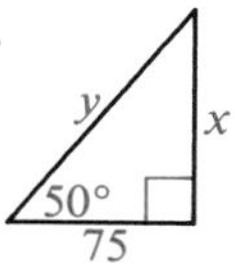

4.

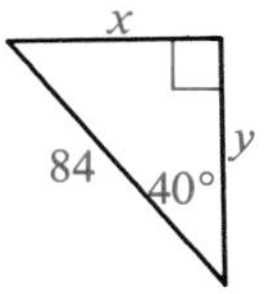

5.

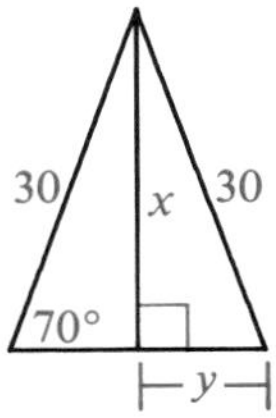

6.

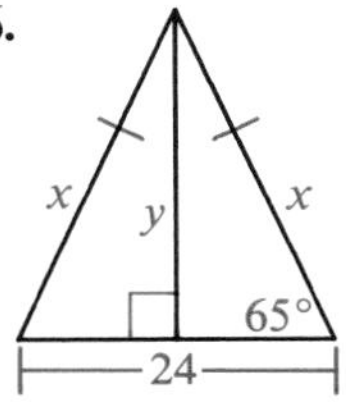

7.

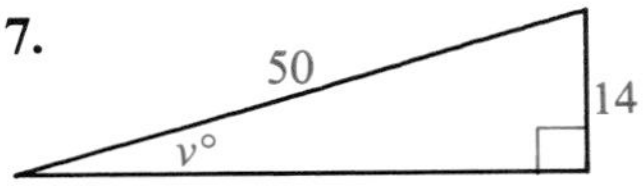

8.

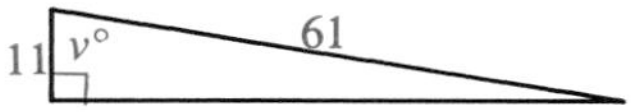

9.

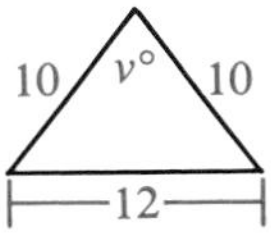

10. **a.** Use the Pythagorean Theorem to find the value of x in radical form.
b. Use the sine table to find $v°$.
c. Use the cosine table and the answer in (b) to find the value of x.
d. Are the x values from (a) and (c) in reasonable agreement?

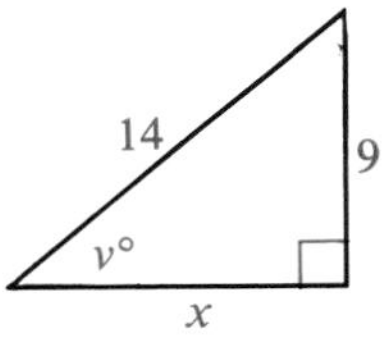

B **11.** The base of an isosceles triangle is 32 cm long and the legs are 24 cm long. Find the measure of a base angle. (*Hint:* Draw the altitude to the base.)

12. The base of an isosceles triangle is 42 cm long and the legs are 25 cm long. Find the measure of the vertex angle.

13. An isosceles triangle with legs 60 cm long has a 42° base angle. Find the lengths of the altitude and the base.

14. Points A, B, and C are three consecutive vertices of a regular decagon whose sides are 16 cm long. How long is diagonal $\overline{AC}$?

15. A guy wire is attached to the top of a 75 m tower and meets the ground at a 65° angle. How long is the wire?

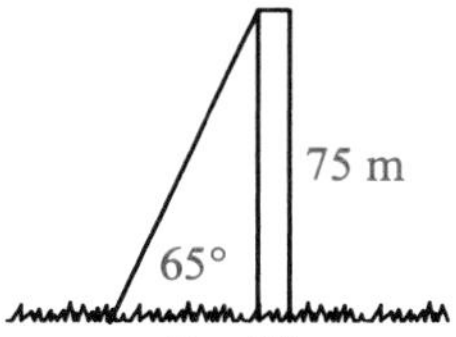

Ex. 15

16. To find the distance from point A on the shore of a lake to point B on an island in the lake, surveyors locate point P with $m\angle PAB = 65$ and $m\angle APB = 25$. They find $PA = 352$ m. Find AB.

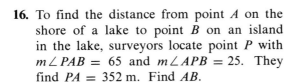

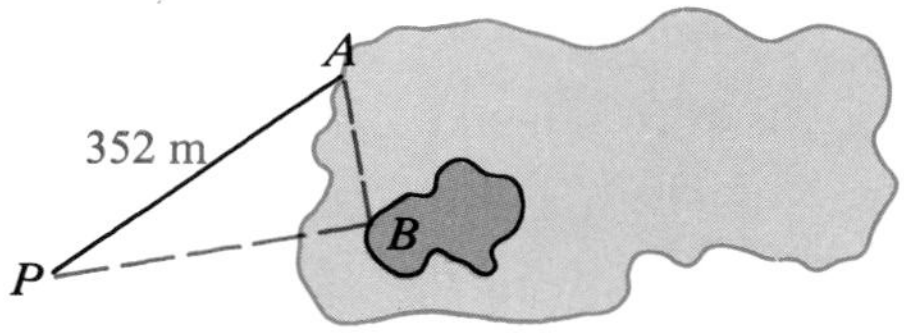

17. A certain jet is capable of a steady 20° climb. How much altitude does the jet gain when it moves 1 km through the air? Answer to the nearest 50 m.

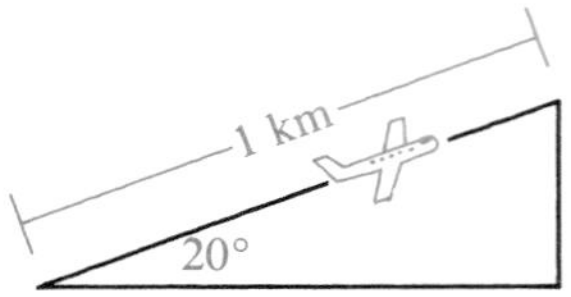

18. A 6 m ladder reaches higher up a wall when placed at a 75° angle than when placed at a 65° angle. How much higher, to the nearest tenth of a meter?

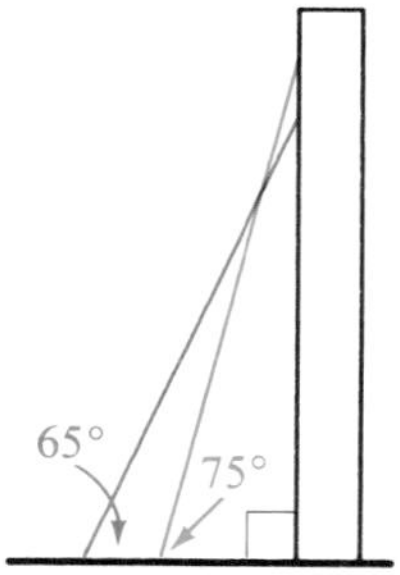

In Exercises 19 and 20, write the statements, but not the supporting reasons, of the proof.

C **19.** Prove that in any triangle with acute angles A and B: $\frac{a}{\sin A} = \frac{b}{\sin B}$. (*Hint:* Draw a perpendicular from the third vertex to $\overline{AB}$. Label it p.)

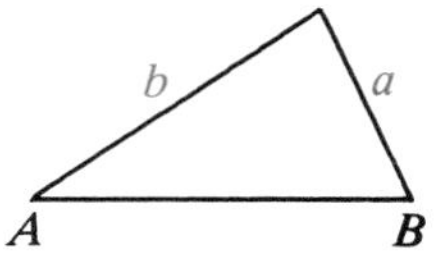

20. Prove: Where E is any acute angle, $(\sin E)^2 + (\cos E)^2 = 1$. (*Hint:* From any point on one side of $\angle E$, draw a perpendicular to the other side.)

21. The diagram in black is given. One way to determine the length x is to draw the red segment and compute the values, in the order named, of p, $q°$, r, and x. Find x.

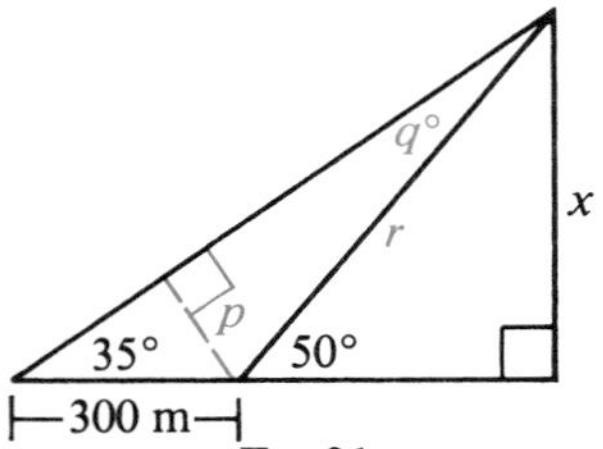

Ex. 21

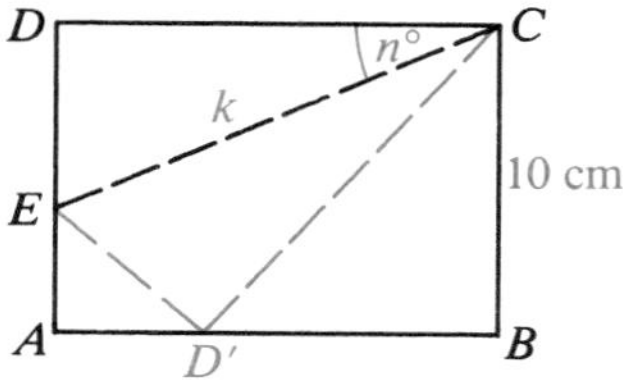

Ex. 22

22. A rectangular card is 10 cm wide. The card is folded so that the vertex D falls at point D' on $\overline{AB}$ as shown. Crease $\overline{CE}$ with length k makes an $n°$ angle with $\overline{CD}$. Show: $k = \frac{10}{\sin (2n)° \cos n°}$

6-7 Using Trigonometric Ratios

Suppose an operator at the top of a lighthouse sights a sailboat on a line that makes a 2° angle with a horizontal line. That angle is called an **angle of depression.** At the same time, a person in the boat must look 2° above the horizontal to see the top of the lighthouse. This is an **angle of elevation.** In this situation, the angle of elevation for one observer has the same measure as the angle of depression for the other observer.

If the top of the lighthouse is 25 m above sea level, the distance x between the boat and the base of the lighthouse can be found in these two ways:

Method 1

$$\tan 2° = \frac{25}{x}$$

$$0.0349 \approx \frac{25}{x}$$

$$0.0349x \approx 25$$
$$x \approx 25 \div 0.0349$$
$$x \approx 716.3$$

Method 2

$$\tan 88° = \frac{x}{25}$$

$$28.6363 \approx \frac{x}{25}$$

$$25 \cdot 28.6363 \approx x$$
$$715.9 \approx x$$

Because the tangent values in the table are approximations, the two methods give slightly different answers. In practice, the angle measurement will not be exact, and the boat may be moving. In a case like this we can not claim high accuracy for our answer. A good answer would be: The boat is roughly 700 m from the lighthouse.

Classroom Exercises

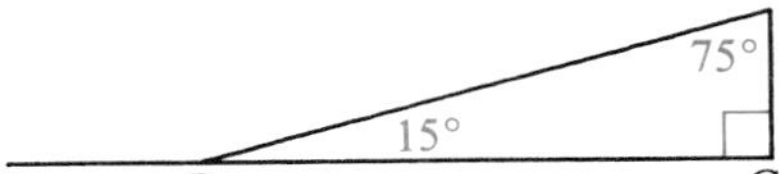

1. $\overleftrightarrow{AC}$ is horizontal.
   a. For an observer at B, what is the angle of elevation of D?
   b. From D, what is the angle of depression of B?
   c. An observer at A measures the angle of elevation of D. Is the measure greater than 15 or less than 15?

The lines shown are horizontal and vertical lines except for $\overleftrightarrow{HT}$ and $\overleftrightarrow{HG}$. Give the number of the angle and its special name when:

2. A person at H sights T.
3. A person at H sights G.
4. A person at T sights H.
5. A person at G sights H.

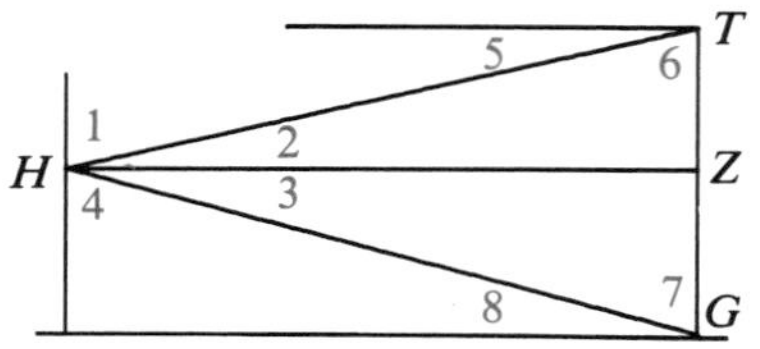

State two equations you could use to find the value of x.

6\.

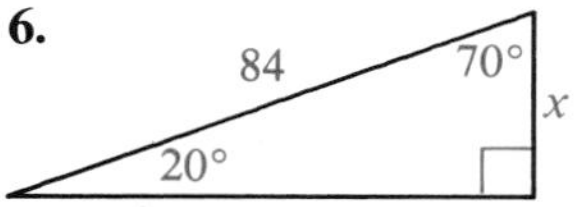

7\.

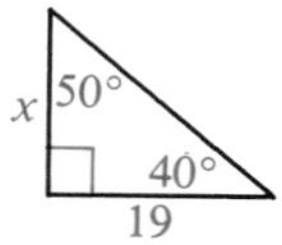

8\.

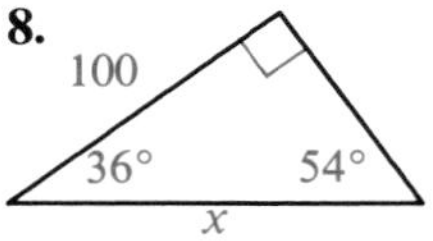

Written Exercises

Express lengths correct to the nearest integer and measures of angles correct to the nearest degree.

A

1. When the sun's angle of elevation is 57°, a building casts a shadow 21 m long. How high is the building?

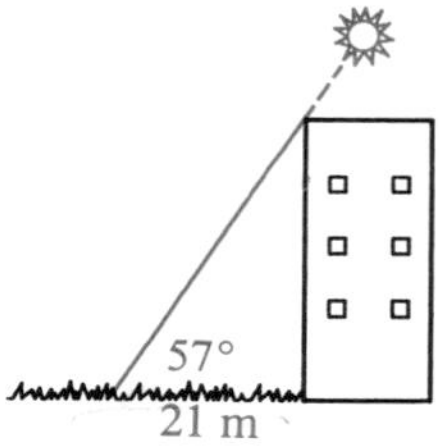

2. At a certain time, a 3 m vertical pole casts a 4 m shadow. What is the angle of elevation of the sun?

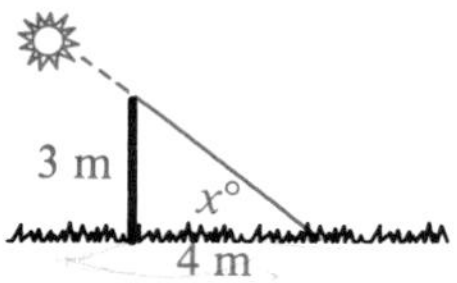

In Exercises 3–6, first draw a diagram.

3. A kite is flying at an angle of elevation of about 40°. All 80 m of string have been let out. Ignoring the sag in the string, find the height of the kite to the nearest 10 m.

4. An advertising blimp hovers over a stadium at an altitude of 125 m. The pilot sights a tennis court at an 8° angle of depression. Find the ground distance in a straight line between the stadium and the tennis court. (*Note:* In an exercise like this one, an answer saying *about . . . hundred meters* is sensible.)

5. A rectangle is 20 m long and 10 m wide.
   - **a.** Find the measure of the angle a diagonal makes with one of the longer sides.
   - **b.** Use trigonometry and your answer from (a) to find the length of a diagonal.
   - **c.** Use the Pythagorean Theorem to find the length of a diagonal.

6. Martha is 180 cm tall and her daughter Heidi is just 90 cm tall. Who casts the longer shadow, Martha when the sun is 70° above the horizon or Heidi when the sun is 35° above the horizon? How much longer?

B

7. Find x, then y and z.

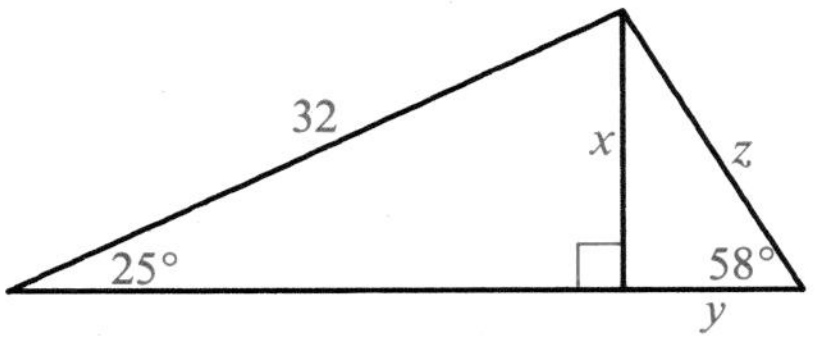

8. Find u, then v and w.

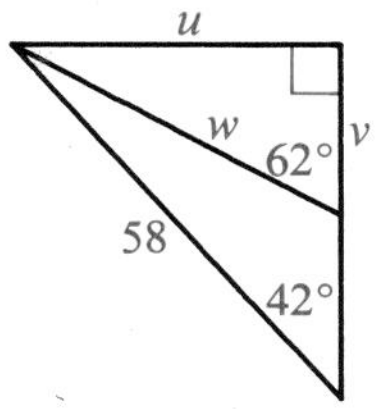

9. Find the length of a diagonal of a regular pentagon with side 20.

10. Points A, B, C, and D are four consecutive vertices of a regular 9-gon (nonagon) with sides 25 mm long. Find the length of $\overline{AD}$.

11. The steepness of a hill is sometimes measured by the grade. A grade of 1 in 4 means that the hill rises one unit for every 4 horizontal units.

   a. For a grade of 1 in 4, what is the measure of $\angle A$, the angle the hill makes with the horizontal?

   b. The force of gravity pulling an object down the hill is its weight multiplied by the sine of $\angle A$. On a 1 in 4 grade, what is the force on a 2500 lb car?

   c. Could you push the car up the hill?

Given the value of one trigonometric ratio, find the exact value of the other two ratios.

Example $\sin A = \frac{7}{9}$

Solution (1) Sketch a right triangle with a leg and hypotenuse in the ratio 7:9.

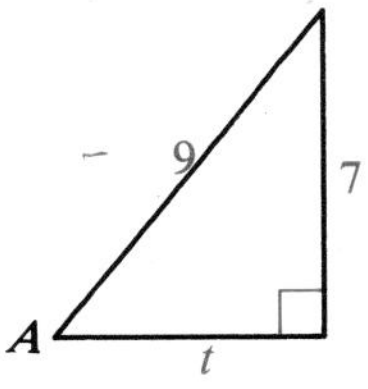

(2) Use the Pythagorean Theorem to find the third side.

$t^2 + 49 = 81$ $\qquad t^2 = 32$ $\qquad t = 4\sqrt{2}$

(3) $\tan A = \frac{7}{t} = \frac{7}{4\sqrt{2}} = \frac{7}{4\sqrt{2}} \cdot \frac{\sqrt{2}}{\sqrt{2}} = \frac{7\sqrt{2}}{8}$

$\cos A = \frac{t}{9} = \frac{4\sqrt{2}}{9}$

12. $\sin A = \frac{11}{61}$

13. $\tan A = \frac{5}{8}$

C

14. $\cos A = \frac{j}{k}$

15. $\tan A = \frac{2uv}{u^2 - v^2}$

16. A surveyor wants to find the distance from points A and B to an inaccessible point C. Point C can be sighted from both A and B. The surveyor measures $\overline{AB}$, $\angle A$, and $\angle B$, with the results shown. Find AC and BC. (*Hint:* Use Exercise 19 on page 276.)

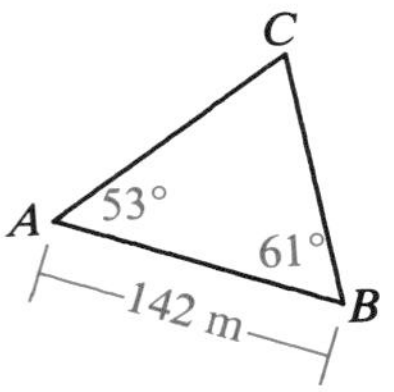

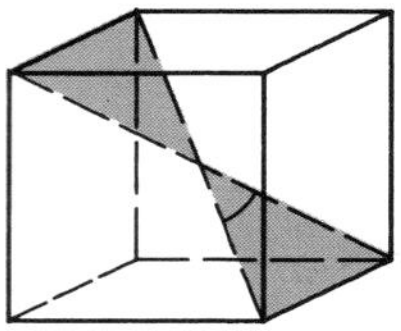

17. Find the acute angle at which two diagonals of a cube intersect.

★18. From the stage of a theater, the angle of elevation of the first balcony is 19°. The angle of elevation of the second balcony, 6.3 m directly above the first, is 29°. How high above stage level is the first balcony? (*Hint:* Use tan 19° and tan 29° to write two equations involving x and d. Solve for d, then find x.)

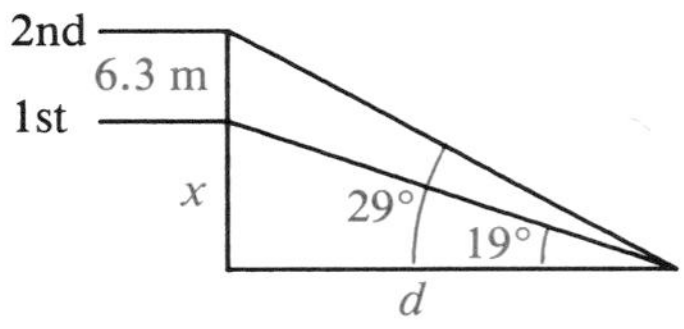

Application

PASSIVE SOLAR DESIGN

Passive solar homes are designed to let the sun heat the house during the winter, but to prevent the sun from heating the house during the summer. Because the Earth's axis is not perpendicular to the *ecliptic* (the plane of the Earth's orbit around the sun), the sun is lower in the sky in the winter than it is in the summer.

From the latitude of the homesite the architect can determine the elevation angle of the sun (the angle at which a person has to look up from the horizontal to see the sun) during the winter and during the summer. The architect can then design an overhang for windows that will let sunlight in the windows during the winter, but will shade the windows during the summer.

The Earth's axis makes an angle of $23\frac{1}{2}°$ with a perpendicular to the ecliptic plane. So for places in the northern hemisphere between the Tropic of Cancer and the Arctic Circle, the angle of elevation of the sun at noon on the longest day of the year, at the summer solstice, is $90°$ – the latitude + $23\frac{1}{2}°$. Its angle of elevation at noon on the shortest day, at the winter solstice, is $90°$ – the latitude – $23\frac{1}{2}°$. For example, in Terre Haute, Indiana, at latitude $39\frac{1}{2}°$ north, the angle of elevation of the sun at noon on the longest day is $74°$ ($90 - 39\frac{1}{2} + 23\frac{1}{2} = 74$), and at noon on the shortest day it is $27°$ ($90 - 39\frac{1}{2} - 23\frac{1}{2} = 27$).

Exercises

Find the angle of elevation of the sun at noon on the longest day and at noon on the shortest day in the following cities. The approximate north latitudes are in parentheses.

1. Seattle, Washington ($47\frac{1}{2}°$)

2. Chicago, Illinois ($42°$)

3. Houston, Texas ($30°$)

4. Los Angeles, California ($34°$)

5. Nome, Alaska ($64\frac{1}{2}°$)

6. Miami, Florida ($26°$)

7. For a city south of the Tropic of Cancer, such as San Juan, Puerto Rico ($18\frac{1}{2}°$N), the formula gives a summer solstice angle greater than $90°$. What does this mean?

8. For a place north of the Arctic Circle, such as Prudhoe Bay, Alaska ($70°$N), the formula gives a negative value for the angle of elevation of the sun at noon at the winter solstice. What does this mean?

9. An architect is designing a passive solar house to be located in Terre Haute, Indiana. The diagram shows a cross-section of a wall that will face south. How long must the overhang x be to shade the entire window at noon at the summer solstice?

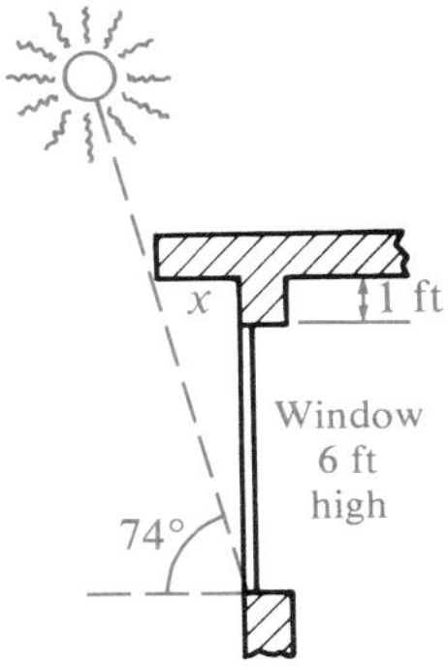

10. If the overhang has the length found in Exercise 9, how much of the window will be in the sun at noon at the winter solstice?

Self-Test 3

Exercises 1–5 refer to the diagram at the right.

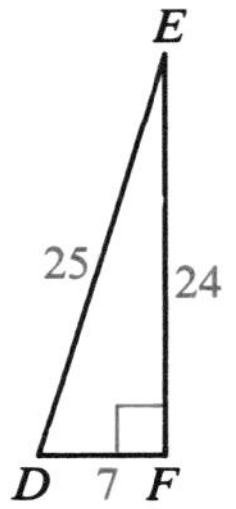

1. $\tan D = \frac{?}{?}$ **2.** $\cos D = \frac{?}{?}$

3. $\sin D = \frac{?}{?}$ **4.** $\tan E = \frac{?}{?}$

5. To the nearest degree, $m\angle D =$ __?__. (Use the table on page 271.)

Find the value of x to the nearest integer. (Use the table on page 271.)

6.

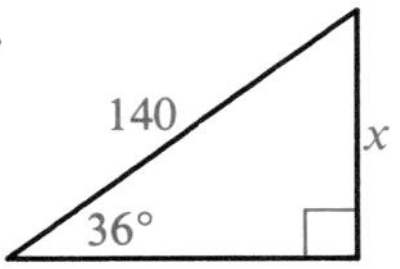

7.

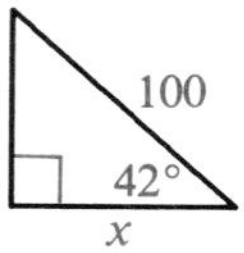

8.

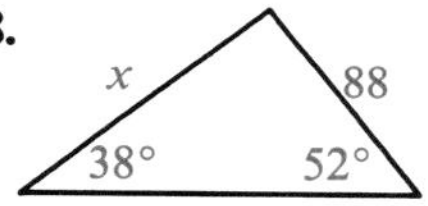

9.

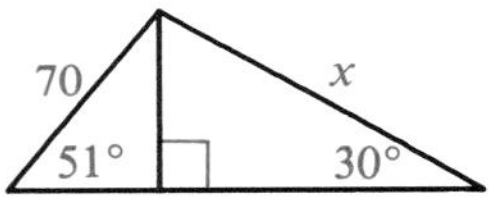

10. A flagpole has a height of 14 m. From a point on the ground 100 m from the foot of the pole, what is the angle of elevation of the top?

Chapter Summary

1. When $\frac{a}{x} = \frac{x}{d}$, x is the geometric mean between a and d.

2. A right triangle is shown with the altitude drawn to the hypotenuse.

a. The two triangles formed are similar to the original triangle and to each other.

$$\frac{x}{h} = \frac{h}{y} \qquad \frac{c}{b} = \frac{b}{x} \qquad \frac{c}{a} = \frac{a}{y}$$

b. Pythagorean Theorem: $c^2 = a^2 + b^2$

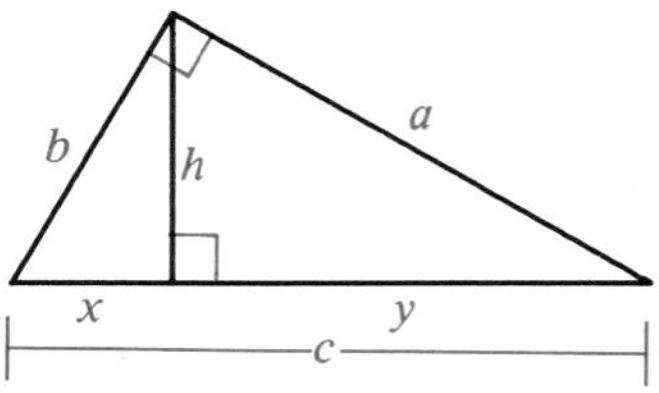

3. The longest side of the triangle shown is t.
If $t^2 = r^2 + s^2$, the triangle is a right triangle.
If $t^2 > r^2 + s^2$, the triangle is obtuse.
If $t^2 < r^2 + s^2$, the triangle is acute.

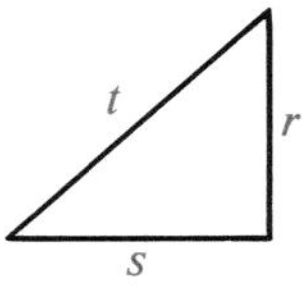

4. The sides of a 45°–45°–90° triangle and the sides of a 30°–60°–90° triangle are related as shown.

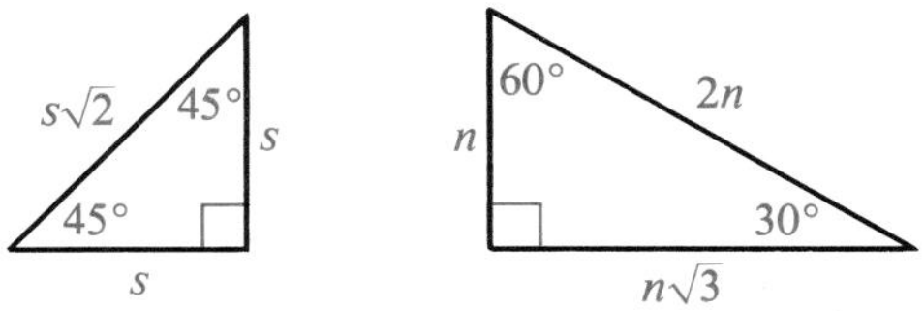

5. In the right triangle shown:

$\tan A = \frac{a}{b}$ $\qquad$ $\sin A = \frac{a}{c}$ $\qquad$ $\cos A = \frac{b}{c}$

The tangent, sine, and cosine ratios are useful in solving problems involving right triangles.

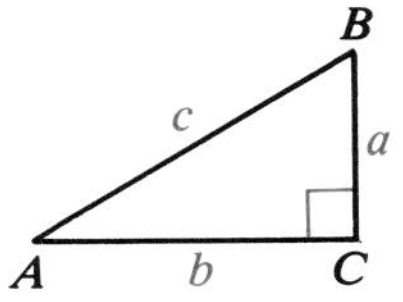

Chapter Review

1. Find the geometric mean between 12 and 3. **6–1**

2. $x =$ __?__

3. $y =$ __?__

4. $z =$ __?__

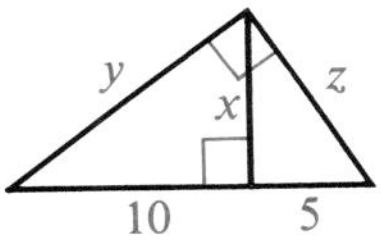

5. The legs of a right triangle are 3 and 6. Find the length of the hypotenuse. **6–2**

6. A rectangle has sides 10 and 8. Find the length of a diagonal.

7. The diagonal of a square has length 14. Find the length of a side.

8. The legs of an isosceles triangle are 10 units long and the altitude to the base is 8 units long. Find the length of the base.

Tell whether a triangle formed with sides having the lengths named is acute, right, or obtuse. If a triangle can't be formed, write *not possible*.

6–3

9. 4, 5, 6

10. 8, 8, 17

11. 11, 60, 61

12. $2\sqrt{3}$, $3\sqrt{2}$, 6

Find the value of *x*.

6–4

13.

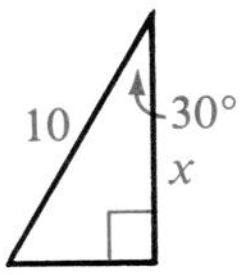

14.

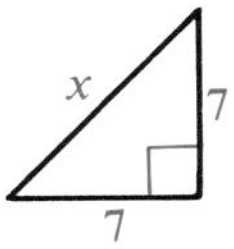

15.

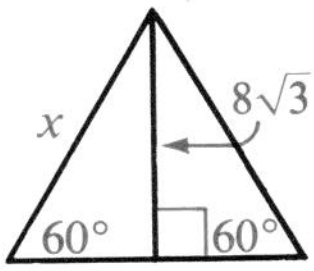

16. Find the value of x.

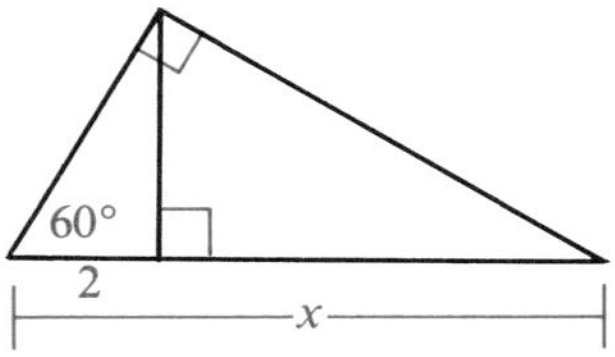

17. Express x in terms of k.

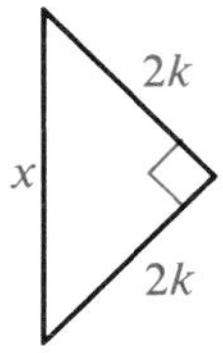

Complete. For Exercises 20, 21, 24, and 25 use the table on page 271.

18. $\tan A =$ ___?___ 6-5

19. $\tan B =$ ___?___

20. $\tan 72° \approx$ ___?___

21. $\tan$ ___?___ ≈ 0.4452

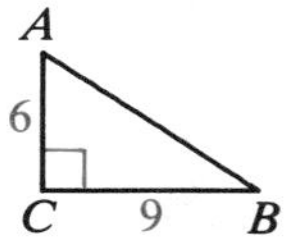

22. $\sin J =$ ___?___ 6-6

23. $\cos K =$ ___?___

24. $\cos$ ___?___ ≈ 0.2588

25. $\sin 43° \approx$ ___?___

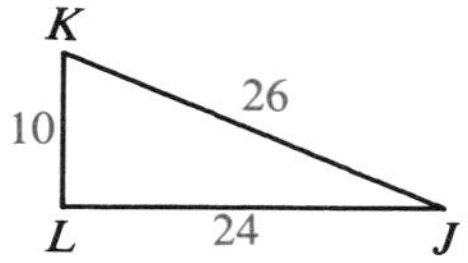

Find x correct to the nearest integer. Find y correct to the nearest degree.

26.

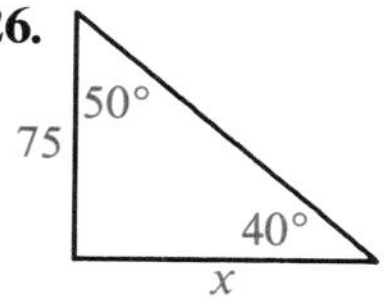

27.

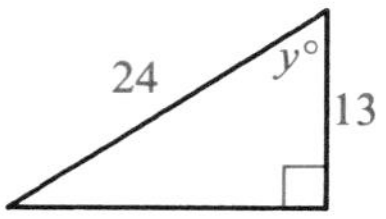

28. 6-7

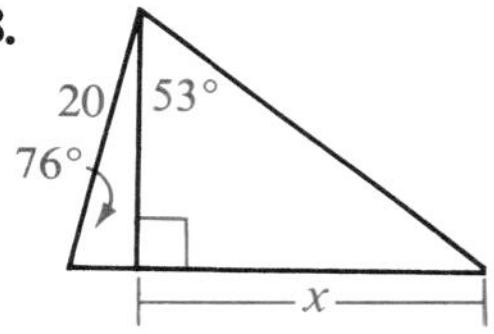

Chapter Test

Find the geometric mean between the numbers.

1. 5 and 20

2. 6 and 8

In the diagram, $\angle DNF$ is a right angle and $\overline{NE} \perp \overline{DF}$.

3. $\triangle DNF \sim \triangle$ ___?___, and $\triangle DNF \sim \triangle$ ___?___.

4. NE is the geometric mean between ___?___ and ___?___.

5. NF is the geometric mean between ___?___ and ___?___.

6. If $DE = 10$ and $EF = 15$, then $ND =$ ___?___.

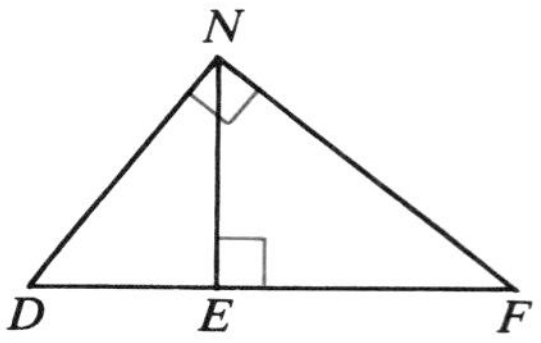

7. Find the values of x and y.

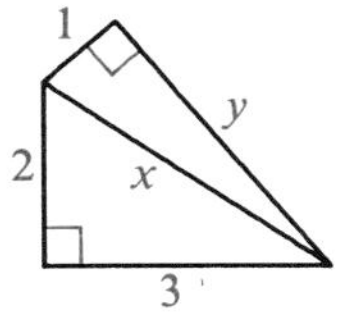

Tell whether a triangle formed with sides having the lengths named is acute, right, or obtuse. If a triangle can't be formed, write *not possible*.

8. 3, 4, 8 **9.** 11, 12, 13 **10.** 7, 7, 10 **11.** $\frac{3}{5}$, $\frac{4}{5}$, 1

Find the value of x.

12.

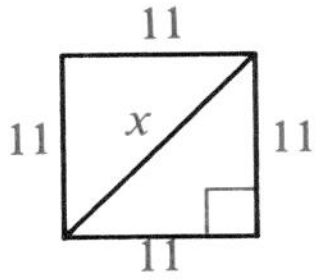

13.

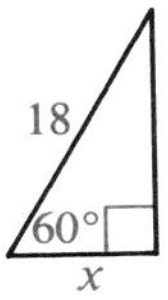

14.

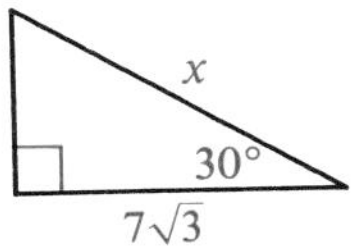

15.

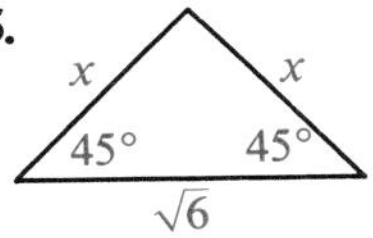

16.

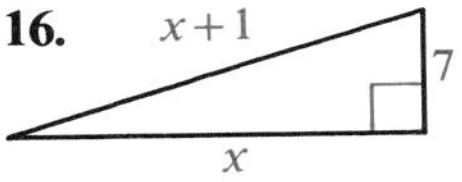

17.

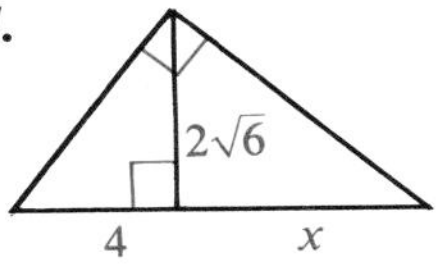

18. In the diagram, $\angle RTS$ is a right angle; $\overline{RT}$, $\overline{RS}$, $\overline{VT}$, and $\overline{VS}$ have the lengths shown.
a. What kind of angle is $\angle V$?
b. Explain your answer to part (a).

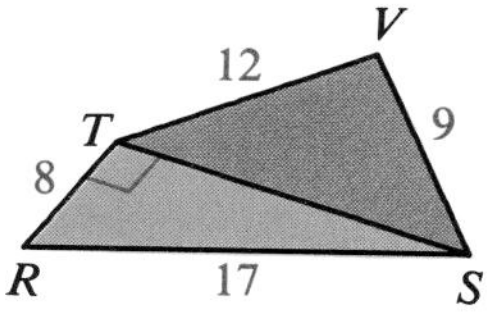

In Exercises 19–24 use the table on page 271. Find lengths correct to the nearest integer and angle measures correct to the nearest degree.

19.

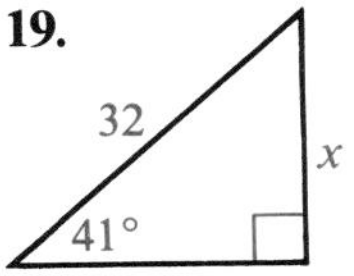

20.

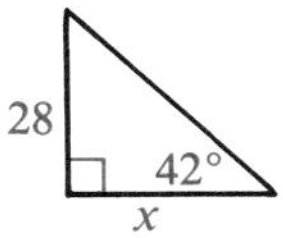

21.

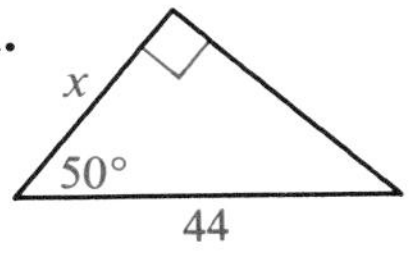

22.

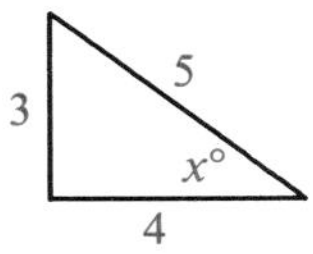

23.

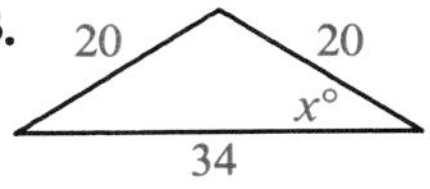

24.

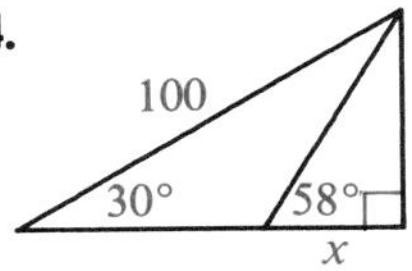

Preparing for College Entrance Exams

Strategy for Success

Problems in college entrance exams often involve right triangles. You can prepare for the exam by learning the common right-triangle lengths listed on page 258. Also, keep in mind that if a, b, and c are the lengths of the sides of a right triangle, then for any $x > 0$, ax, bx, and cx are also lengths of sides of a right triangle.

Indicate the best answer by writing the appropriate letter.

1. In $\triangle ABC$, $m\angle A : m\angle B : m\angle C = 2:5:5$. $m\angle B =$
(A) 75 **(B)** 60 **(C)** 30 **(D)** 40 **(E)** 100

2. The proportion $\frac{t}{z} = \frac{m}{k}$ is *not* equivalent to:
(A) $\frac{t-z}{z} = \frac{m-k}{k}$ **(B)** $\frac{k}{z} = \frac{m}{t}$ **(C)** $\frac{t}{m} = \frac{k}{z}$ **(D)** $tk = mz$ **(E)** $\frac{z}{t} = \frac{k}{m}$

3. If $\triangle ABC \sim \triangle DEF$, which statement is not necessarily true?
(A) $\angle C \cong \angle F$ **(B)** $\overline{BC} \cong \overline{EF}$ **(C)** $\frac{AB}{BC} = \frac{DE}{EF}$
(D) $m\angle A + m\angle E = m\angle B + m\angle D$ **(E)** $AC \cdot DE = DF \cdot AB$

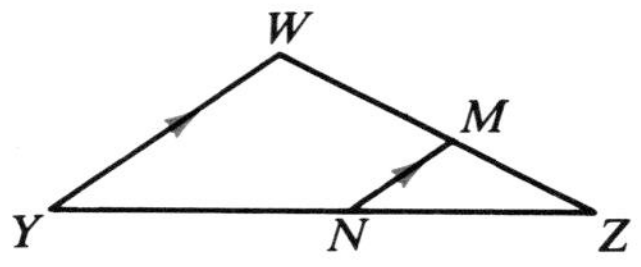

4. If $ZY = 2x + 9$, $ZM = 10$, $ZN = x + 3$, and $MW = x$, then $x =$
(A) $2 + \sqrt{34}$ **(B)** -12 **(C)** 12 **(D)** 5 **(E)** -5

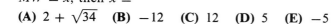

5. $\overrightarrow{BD}$ bisects $\angle ABC$ and D lies on $\overline{AC}$. If $AB = 6$, $BC = 14$, and $AC = 14$, find AD.
(A) 6 **(B)** 8.4 **(C)** 9.8 **(D)** 7 **(E)** 4.2

6. Find the geometric mean of $2x$ and $2y$.
(A) $2\sqrt{xy}$ **(B)** $\sqrt{2xy}$ **(C)** $2\sqrt{x+y}$ **(D)** $\sqrt{2(x+y)}$ **(E)** $4xy$

7. If $XY = 8$, $YZ = 40$, and $XZ = 41$, then:
(A) $\triangle XYZ$ is acute **(B)** $\triangle XYZ$ is right **(C)** $\triangle XYZ$ is obtuse
(D) $m\angle Y < m\angle Z$ **(E)** no $\triangle XYZ$ is possible

8. A rhombus contains a 120° angle. Find the ratio of the length of the longer diagonal to the length of the shorter diagonal.
(A) $\sqrt{3}:1$ **(B)** $\sqrt{3}:3$ **(C)** $\sqrt{2}:1$ **(D)** $\sqrt{2}:2$ **(E)** cannot be determined

9. $k =$

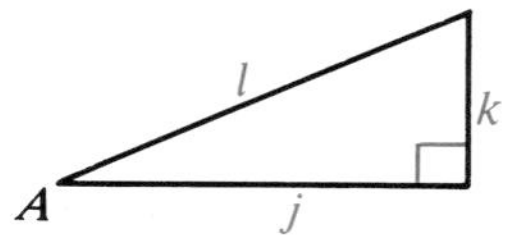

(A) $j \sin A$ **(B)** $j \tan A$ **(C)** $\frac{l}{\sin A}$
(D) $l \cos A$ **(E)** $l \tan A$

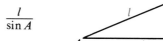

10. The legs of an isosceles triangle have length 4 and the base angles have measure 65. If $\sin 65° \approx 0.91$, $\cos 65° \approx 0.42$, and $\tan 65° \approx 2.14$, then the approximate length of the base of the triangle is:
(A) 1.7 **(B)** 1.9 **(C)** 3.4 **(D)** 3.6 **(E)** 4.4

Cumulative Review: Chapters 1–6

True–False Exercises

Write T or F to indicate your answer.

A
1. Through any three points there is exactly one plane.
2. If $AX = XB$, then X must be the midpoint of $\overline{AB}$.
3. Definitions may be used to justify statements in a proof.
4. If a line and a plane are parallel, then the line is parallel to every line in the plane.
5. Every isosceles trapezoid contains two pairs of congruent angles.
6. When two parallel lines are cut by a transversal, any two angles formed are either congruent or supplementary.
7. If the sides of one triangle are congruent to the corresponding sides of another triangle, then the corresponding angles must also be congruent.
8. The geometric mean of 18 and 50 is 34.

B
9. In quadrilateral $WXYZ$, if $WX = 25$, $XY = 25$, $YZ = 20$, $ZW = 16$, and $WY = 20$, then $\overline{WY}$ divides the quadrilateral into two similar triangles.
10. The bisector of the 60° angle of a 30°–60°–90° triangle separates the opposite side into segments with the ratio 1:2.
11. If a quadrilateral has two pairs of supplementary angles, then it must be a parallelogram.
12. In $\triangle PQR$, $m\angle P = m\angle R = 50$. If T lies on $\overline{PR}$ and $m\angle PQT = 42$, then $PT < TR$.
13. If a line parallel to one side of a triangle intersects the other two sides, then the triangle formed is similar to the given triangle.
14. In any right triangle, the sine of one acute angle is equal to the cosine of the other acute angle.
15. If the diagonals of a quadrilateral bisect each other and are congruent, then the quadrilateral must be a square.

Multiple-Choice Exercises

Indicate the best answer by writing the appropriate letter.

A
1. Which pair of angles must be congruent?
   a. $\angle 1$ and $\angle 4$ b. $\angle 2$ and $\angle 3$
   c. $\angle 2$ and $\angle 4$ d. $\angle 4$ and $\angle 5$
   e. $\angle 2$ and $\angle 8$

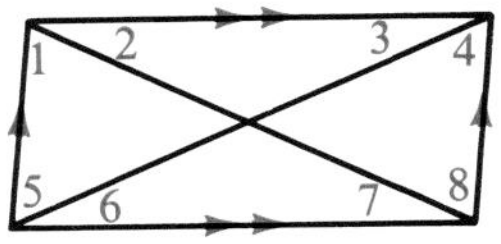

2. Which of the following can be the lengths of the sides of a triangle?
   a. 3, 7, 10 b. 3, 7, 11 c. 0.5, 7, 7 d. $\frac{1}{2}, \frac{1}{4}, \frac{1}{5}$ e. 1, 3, 5

3. Which of the following can be the lengths of the sides of a right triangle?
   a. 2, 3, 4 **b.** 6, 8, $\sqrt{14}$ **c.** $\frac{1}{2}, \frac{1}{2}, 1$ **d.** $\sqrt{3}, \sqrt{5}, \sqrt{15}$ **e.** none of these
4. If $\triangle ABC \cong \triangle NDH$, then it is also true that:
   a. $\angle B \cong \angle H$ **b.** $\angle A \cong \angle H$ **c.** $\overline{AB} \cong \overline{HD}$ **d.** $\overline{CA} \cong \overline{HN}$ **e.** $\triangle CBA \cong \triangle DHN$
5. If $PQRS$ is a parallelogram, which of the following *must* be true?
   a. $PQ = QR$ **b.** $PQ = RS$ **c.** $PR = QS$ **d.** $\overline{PR} \perp \overline{QS}$ **e.** $\angle Q \cong \angle R$

B 6. If the diagonals of a rhombus have lengths 18 and 24, then the sides of the rhombus have length:
   a. 15 **b.** 30 **c.** $6\sqrt{7}$ **d.** 12 **e.** cannot be determined
7. The altitude to the hypotenuse of a 30°–60°–90° triangle has length 6. The longer leg has length:
   a. $2\sqrt{3}$ **b.** $6\sqrt{3}$ **c.** $4\sqrt{3}$ **d.** $8\sqrt{3}$ **e.** 12
8. If a, b, c, and d are coplanar lines such that $a \perp b$, $c \perp d$, and $b \parallel c$, then:
   a. $a \perp d$ **b.** $b \parallel d$ **c.** $a \parallel d$ **d.** $a \parallel c$ **e.** none of these

Always–Sometimes–Never Exercises

Write A, S, or N to indicate your choice.

A 1. An angle ___?___ has a complement.
2. Two vertical angles are ___?___ adjacent.
3. Two parallel lines are ___?___ coplanar.
4. A scalene triangle is ___?___ equiangular.
5. A rectangle is ___?___ a rhombus.
6. A regular polygon is ___?___ equilateral.
7. If a conditional is false, then its converse is ___?___ false.
8. The sine of an acute angle is ___?___ greater than 1.
9. If $\overline{RS} \cong \overline{MN}$, $\overline{ST} \cong \overline{NO}$, and $\angle R \cong \angle M$, then $\triangle RST$ and $\triangle MNO$ are ___?___ congruent.
10. The HL method is ___?___ appropriate for proving that two acute triangles are congruent.
11. If $AX = BX$, $AY = BY$, and points A, B, X, Y are coplanar, then $\overline{AB}$ and $\overline{XY}$ are ___?___ perpendicular.

B 12. The diagonals of a trapezoid are ___?___ perpendicular.
13. If $\triangle JKL \cong \triangle NET$ and $\overline{NE} \perp \overline{ET}$, then it is ___?___ true that $LJ < TE$.
14. Two perpendicular lines are ___?___ both parallel to a third line.
15. If $AB + BC > AC$, then A, B, and C are ___?___ collinear points.
16. Two equilateral octagons are ___?___ similar.
17. Given the lengths of the legs of a right triangle, it is ___?___ possible to find approximations for the measures of the angles of the triangle.
18. A triangle with sides of length $x - 1$, x, and x is ___?___ an obtuse triangle.

Algebraic Exercises

In Exercises 1-12 find the value of x.

A

1. The lengths of the legs of an isosceles triangle are $7x - 13$ and $2x + 17$.
2. An angle and its complement have the measures $x + 38$ and $2x - 5$.
3. Consecutive angles of a parallelogram have the measures $6x$ and $2x + 20$.
4. The measures of the angles of a triangle are $x + 12$, $2x - 7$, and $3x + 1$.
5. On a number line, R and S have coordinates -8 and x, and the midpoint of $\overline{RS}$ has coordinate -1.
6. Two vertical angles have measures $x^2 + 18x$ and $x^2 + 54$.
7. The measures of the angles of a quadrilateral are x, $x + 4$, $x + 8$, and $x + 12$.
8. The hypotenuse of a 30°-60°-90° triangle has length $8\sqrt{3}$ and the longer leg has length $4x$.
9. A trapezoid has bases of length x and $x + 8$ and a median of length 15.
10. $\frac{3x - 1}{4x + 2} = \frac{2}{3}$
11. $\frac{5}{8} = \frac{x - 1}{6}$
12. $\frac{x}{x + 4} = \frac{x + 3}{x + 9}$

B

13. The measure of a supplement of an angle is 8 more than three times the measure of a complement. Find the measure of the angle.
14. A triangle with perimeter 64 cm has sides with lengths in the ratio 4:5:7. Find the length of each side.
15. In $\triangle XYZ$, $XY = YZ$. If $\frac{m\angle X}{m\angle Y} = \frac{5}{2}$, find the numerical measure of $\angle Z$.
16. In a regular polygon, the ratio of the measure of an exterior angle to the measure of an interior angle is 2:13. How many vertices does the polygon have?
17. In $\triangle RST$, $RS = 8$, $ST = 9$, and $RT = 12$. $\overrightarrow{RU}$ bisects $\angle SRT$ and U lies on $\overline{ST}$. $SU = \underline{\ ?\ }$.
18. The sides of an obtuse triangle have lengths x, $2x + 2$, and $2x + 3$. $\underline{\ ?\ } < x < \underline{\ ?\ }$.
19. The sides of a parallelogram have lengths 12 cm and 15 cm. Find the lengths of the sides of a similar parallelogram with perimeter 90 cm.
20. The length of a diagonal of a rectangle is 10, and the perimeter of each right triangle formed is 24. Find the length and width of the rectangle.

In Exercises 21-23 right $\triangle XYZ$ has hypotenuse $\overline{XZ}$.

21. If $\cos X = \frac{7}{10}$ and $XZ = 24$, then to the nearest integer $XY = \underline{\ ?\ }$.
22. If $XY = 10$ and $YZ = 15$, then to the nearest degree $m\angle X = \underline{\ ?\ }$ and $m\angle Z = \underline{\ ?\ }$. (Use the table on page 271.)
23. If $\overline{YM}$ is an altitude of $\triangle XYZ$, $YZ = 18$, and $XZ = 24$, then $XM = \underline{\ ?\ }$.

24. In the diagram, $\overline{AB} \parallel \overline{DC}$ and $\overline{AD} \parallel \overline{GC}$. Find the values of x and y.

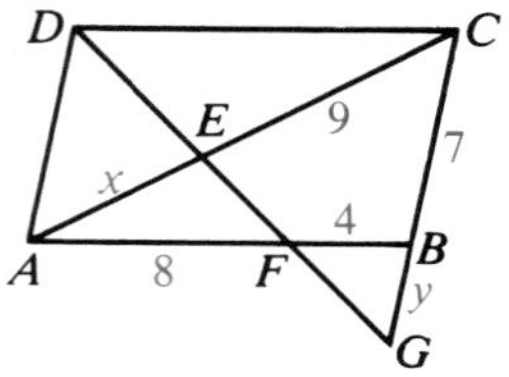

C **25.** The sides of a triangle have lengths $x + y$, $x - y$, and $2\sqrt{xy}$. Is the triangle acute, right, or obtuse?

26. A piece of plywood is in the shape of a right triangle with legs of length 12 cm and 8 cm. A square is cut from the triangle as shown. Find the length of each side of the square.

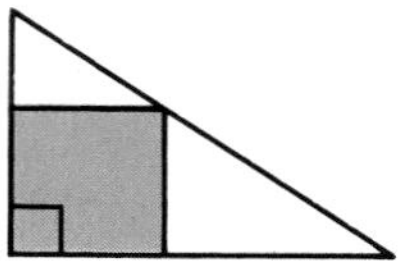

Completion Exercises

Complete each statement in the best way.

A **1.** If $\overrightarrow{YW}$ bisects $\angle XYZ$ and $m\angle WYX = 60$, then $m\angle XYZ =$ _?_.

2. The acute angles of a right triangle are _?_.

3. A supplement of an acute angle is a(n) _?_ angle.

4. Adjacent angles formed by _?_ lines are congruent.

5. The measure of each interior angle of a regular pentagon is _?_.

6. In an isosceles right triangle, the ratio of the length of a leg to the length of the hypotenuse is _?_.

7. In $\triangle ABC$ and $\triangle DEF$, $\angle A \cong \angle D$ and $\angle B \cong \angle E$. $\triangle ABC$ and $\triangle DEF$ must be _?_.

8. If $x = 8$, $y = 10$, and $z = 6$, then $\sin X =$ _?_, $\cos Z =$ _?_, and $\tan Z =$ _?_.

9. If $m\angle X = 50$ and $z = 20$, then to the nearest integer $x \approx$ _?_. (Use the table on page 271.)

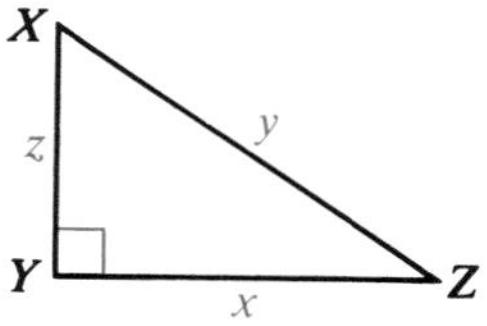

Exs. 8, 9

10. The ratio of the measures of the acute angles of a right triangle is 3:2. The measure of the smaller acute angle is _?_.

B **11.** If $\frac{r}{s} = \frac{t}{u}$, then $\frac{r+s}{t+u} = \frac{?}{?}$.

12. An isosceles triangle has legs of 10 cm and a vertex angle of measure 120. The length of the base is _?_.

13. If $\sin B = \frac{8}{17}$, then $\cos B =$ _?_.

14. When the midpoints of the sides of a rhombus are joined in order, the resulting quadrilateral is best described as a _?_.

15. If a tree is 20 m high and the distance from point P on the ground to the base of the tree is also 20 m, then the angle of elevation of the top of the tree from point P is _?_.

Proof Exercises

A

1. Given: $\overline{AD} \cong \overline{BC}$; $\overline{AD} \parallel \overline{BC}$
Prove: $\angle D \cong \angle B$

2. Given: $ABCD$ is a $\square$; $\overline{CE} \perp \overline{AB}$; $\overline{AF} \perp \overline{CD}$
Prove: $\overline{BE} \cong \overline{DF}$

3. Given: $\triangle DAF \cong \triangle BCE$; $CD = AB$
Prove: $ABCD$ is a $\square$.

4. Given: $\overline{DC} \parallel \overline{AB}$; $\overline{CE} \perp \overline{AB}$; $\overline{AF} \perp \overline{AB}$
Prove: $AECF$ is a rectangle.

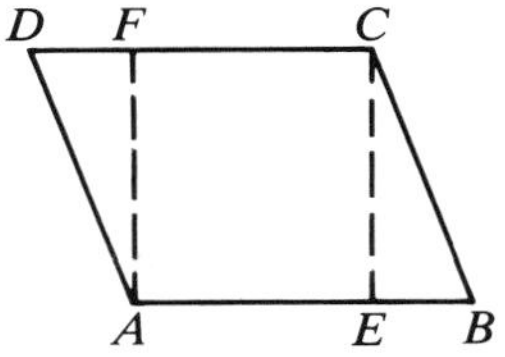

5. Given: $\overline{SU} \cong \overline{SV}$; $\angle 1 \cong \angle 2$
Prove: $\overline{UQ} \cong \overline{VQ}$

6. Given: $\overrightarrow{QS}$ bisects $\angle RQT$; $\angle R \cong \angle T$
Prove: $\overrightarrow{SQ}$ bisects $\angle RST$.

B

7. Given: $\triangle QRU \cong \triangle QTV$; $US = VS$
Prove: $\triangle QRS \cong \triangle QTS$

8. Given: $\overline{QS}$ bisects $\angle UQV$ and $\angle USV$; $\angle R \cong \angle T$
Prove: $\overline{RQ} \cong \overline{TQ}$

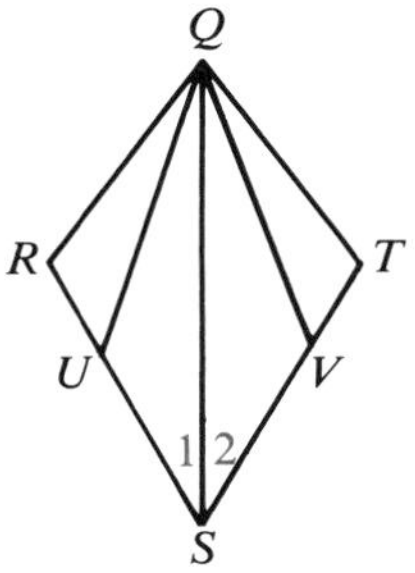

9. Given: $\overline{EF} \parallel \overline{JK}$; $\overline{JK} \parallel \overline{HI}$
Prove: $\triangle EFG \sim \triangle IHG$

10. Given: $\dfrac{JG}{HG} = \dfrac{KG}{IG}$; $\angle 1 \cong \angle 2$
Prove: $\overline{EF} \parallel \overline{HI}$

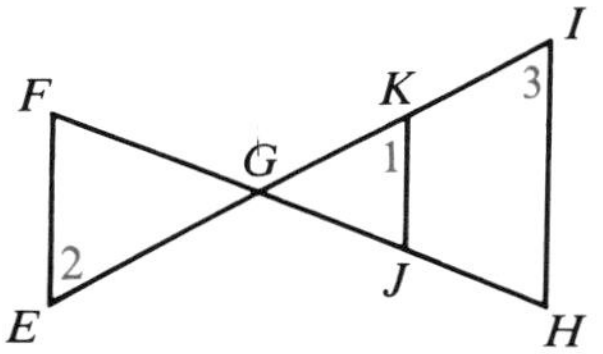

11. Given: $XZ = YW$; $VZ = VW$
Prove: $XW = YZ$

12. Given: $\overline{XW} \cong \overline{YZ}$; $\angle XWZ \cong \angle YZW$
Prove: $\triangle XVW \cong \triangle YVZ$

13. Given: $WXYZ$ is an isosceles trapezoid with $\overline{XW} \cong \overline{YZ}$.
Prove: $\overline{XZ} \cong \overline{YW}$

14. Given: $VW = VZ$; $VX = VY$
Prove: $XY \cdot VW = WZ \cdot VX$

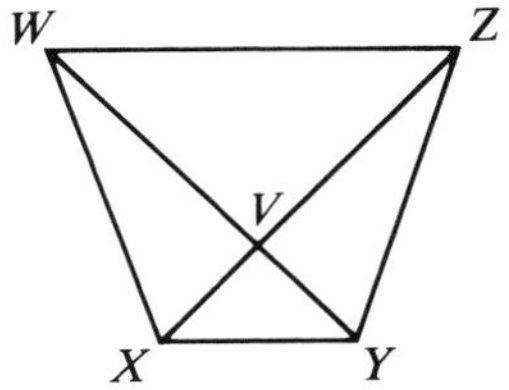

15. Given: $\angle G \cong \angle 1$
Prove: JH is the geometric mean between JK and JG.

16. Given: $\overline{GH} \perp \overline{JH}$; $\overline{HK} \perp \overline{GJ}$
Prove: $(GH)^2 - (GK)^2 = GK \cdot KJ$

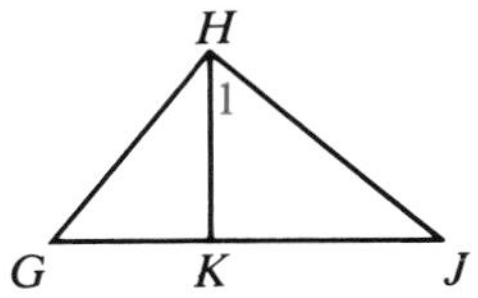

17. Given: Points R, S, and T lie in plane P;
$\overline{QS} \perp$ plane P; $\triangle QRT$ is equilateral.
Prove: $\angle SRT \cong \angle STR$

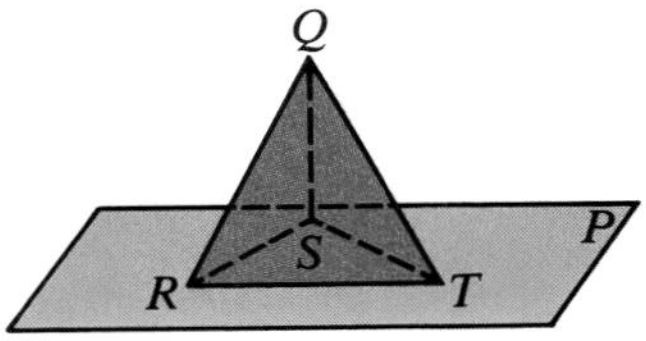

The photograph below, which shows meshed gears, suggests a number of ideas discussed in this chapter—for example, tangent circles, concentric circles, arcs of a circle, and secants of a circle.

Circles

7

Tangents, Arcs, and Chords

Objectives

1. Define a circle, a sphere, and terms related to them.
2. Recognize circumscribed and inscribed polygons and circles.
3. Apply theorems that relate tangents and radii.
4. Define and apply properties of arcs and central angles.
5. Apply theorems about the chords of a circle.

7-1 Basic Terms

A **circle** ($\odot$) is the set of points in a plane that are a given distance from a given point in the plane. The given point is the **center,** and the given distance is the **radius.** A segment that joins the center to a point on the circle is called *a radius.* All radii of a circle are congruent. The circle shown, with center O, is called circle O ($\odot O$).

The radius of $\odot O$ is 8.

$\overline{OS}$ is a radius of $\odot O$.

R lies inside the circle.	$OR < 8$
S lies on the circle.	$OS = 8$
T lies outside the circle.	$OT > 8$

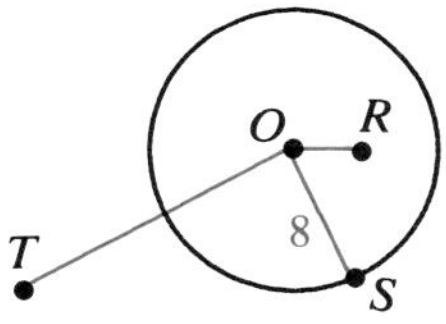

A **chord** is a segment that joins two points on a circle. A **diameter** is a chord that passes through the center. A **secant** is a line that contains a chord.

Chords: $\overline{AB}$, $\overline{CD}$

Diameter: $\overline{CD}$

Secants: $\overleftrightarrow{AB}$, $\overleftrightarrow{CD}$

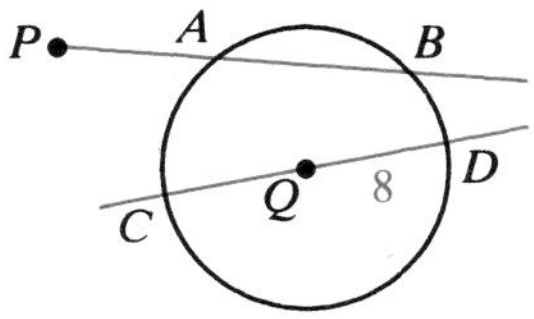

A ray or segment containing a chord is often also called a secant. For example, you can refer to $\overline{PB}$ in the diagram as a secant drawn to $\odot Q$ from external point P.

Like the word *radius,* the word *diameter* can refer to the length of a segment as well as to the segment itself. From the definitions it follows that the diameter of a circle equals twice the radius. Thus $\overline{CD}$ is *a* diameter of $\odot Q$ and *the* diameter of $\odot Q$ is 16.

Congruent circles are circles that have congruent radii. By definition, radii of congruent circles are congruent.

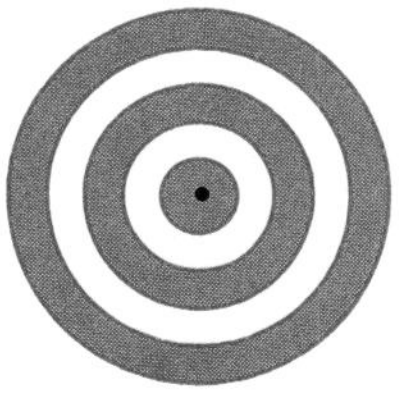

Concentric circles are circles that lie in the same plane and have the same center. The rings of the target illustrate concentric circles.

A polygon is **inscribed in a circle** and the circle is **circumscribed about the polygon** when each vertex of the polygon lies on the circle.

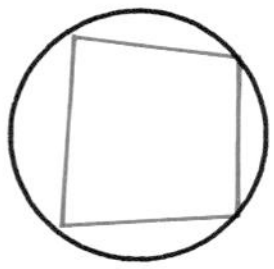

Inscribed polygons

Circumscribed circles

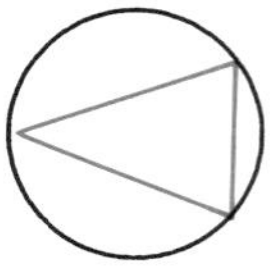

Remove the phrase *in a plane* from the definition of a circle and you have the definition of a *sphere*. A **sphere** is the set of points that are a given distance from a given point. Many of the terms used for circles are also used for spheres.

Classroom Exercises

1. Name two radii, a diameter, three chords, and two secants of $\odot Q$.

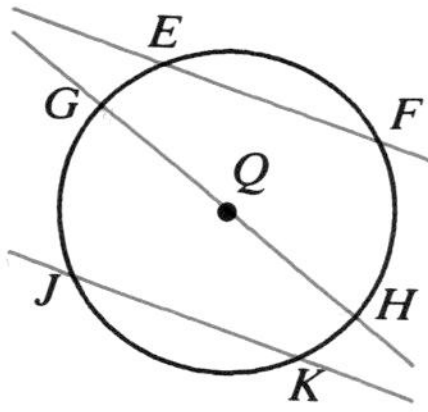

2. Name two radii, a diameter, three chords, and two secants of sphere O.

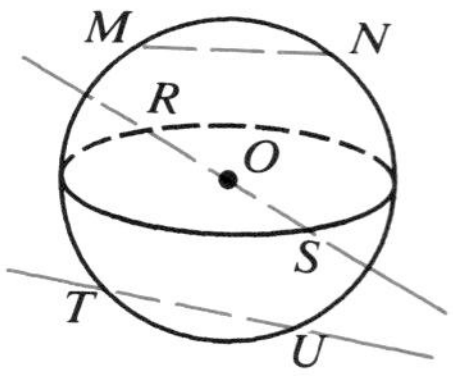

3. See the figure for Exercise 1. Name a secant to $\odot Q$ from external point P.

4. Write a definition of concentric spheres.

5. What is the diameter of a circle with radius 8? 5.2? $4\sqrt{3}$? j?

6. What is the radius of a sphere with diameter 14? 13? 5.6? $6n$?

7. Draw a circle and several parallel chords. What do you think is true of the midpoints of all such chords?

8. Point Z lies on $\odot X$. How many chords can you draw that contain Z? How many diameters? How many radii?

9. Explain why it is not correct to call $ABCD$ an inscribed quadrilateral.

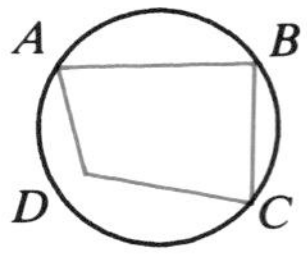

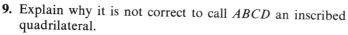

10. Plane Z passes through the center of sphere Q.

a. Explain why $QR = QS = QT$.

b. Explain why the intersection of the plane and the sphere is a circle. (The intersection of a sphere with any plane passing through the center of the sphere is called a **great circle** of the sphere.)

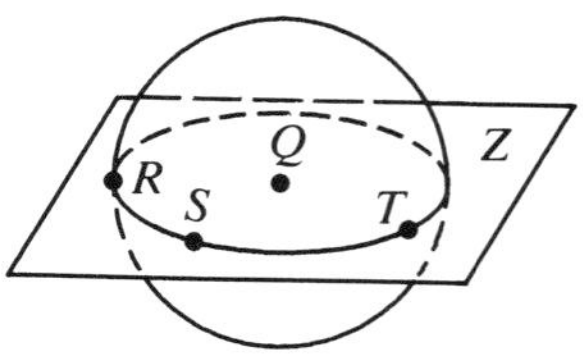

c. On a globe, which of the following are great circles: (1) the equator, (2) the Arctic Circle, and (3) the circle formed by the 0° meridian (also called the prime meridian) and the 180° meridian?

Written Exercises

Point W lies outside $\odot O$ and point X lies inside. In how many points does the figure named intersect the circle?

A 1. $\overline{WX}$ 2. $\overrightarrow{WX}$ 3. $\overrightarrow{XW}$ 4. $\overline{OX}$

Point Y lies outside sphere Q and point Z lies inside. In how many points does the figure named intersect the sphere?

5. $\overline{YZ}$ 6. $\overrightarrow{YQ}$ 7. $\overrightarrow{ZY}$ 8. $\overline{QZ}$

The radius of sphere P is 12. Plane M cuts the sphere in a circle. Tell what you can about the center and radius of the circle if:

9. Plane M passes through P.
10. Plane M does not pass through P.

For each exercise draw a circle and inscribe the figure named in the circle. If a polygon of the type named can't be inscribed, write *not possible.*

11. A rectangle
12. A trapezoid
13. An obtuse triangle
14. A nonrectangular parallelogram
15. An acute isosceles triangle
16. A quadrilateral $PQRS$, with $\overline{PR}$ a diameter

For each exercise draw $\odot O$ with radius 12. Then draw radii $\overline{OA}$ and $\overline{OB}$ to form an angle with the measure named. Find the length of $\overline{AB}$.

B 17. $m\angle AOB = 90$ 18. $m\angle AOB = 180$ 19. $m\angle AOB = 60$ 20. $m\angle AOB = 120$

21. Write a definition of radius of a sphere.
22. Write a definition of congruent spheres.
23. A plane 6 units from the center of a sphere of radius 10 intersects the sphere in a circle. Find the radius of the circle.
24. A plane 15 cm from the center of a sphere cuts the sphere in a circle with a radius of 8 cm. Find the radius of the sphere.

C 25. Two spheres with radii of 6 cm and 4 cm have centers that are 8 cm apart. Find the radius of the circle in which the spheres intersect.
26. Prove: A line intersects a circle in at most two points. (*Hint:* Write an indirect proof.)
27. Exercises 23 and 24 assume the following theorem: If a plane intersects a sphere in more than one point, the intersection is a circle. Prove this theorem. (*Hint:* The case where the plane, Z, passes through the center of the sphere, Q, is covered in Classroom Exercise 10. If Z does not pass through Q, draw a perpendicular from Q to Z, intersecting Z in point P. (You may assume that this is possible.) Let R and S be any two points on the intersection of the sphere and the plane. Show that $\overline{PR} \cong \overline{PS}$.)

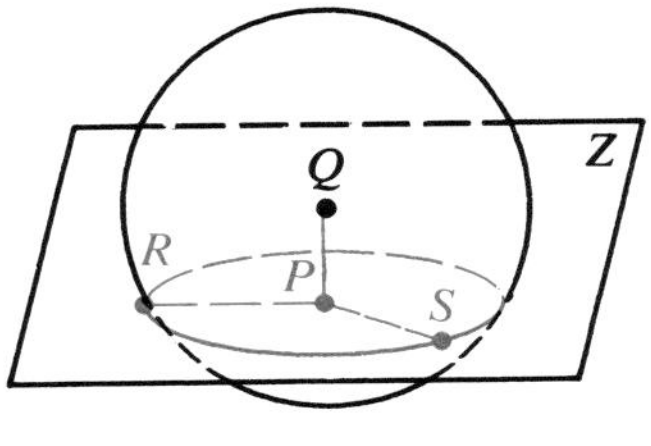

7-2 Tangents

A **tangent** to a circle is a line in the plane of the circle that meets the circle in exactly one point, called the **point of tangency.**

$\overleftrightarrow{AP}$ is tangent to $\odot O$.

$\odot O$ is tangent to $\overleftrightarrow{AP}$.

A is the point of tangency.

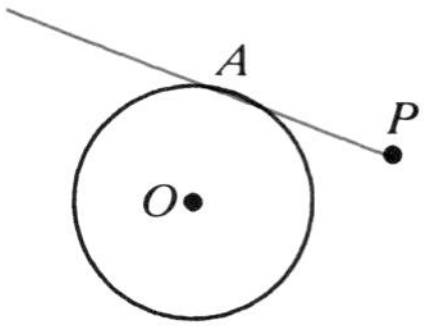

The tangent ray $\overrightarrow{PA}$ and the tangent segment $\overline{PA}$ are often called simply tangents. $\overline{PA}$ is a tangent (or tangent segment) to $\odot O$ from external point P.

Theorem 7-1

If a line is tangent to a circle, then the line is perpendicular to the radius drawn to the point of tangency.

Given: Line m is tangent to $\odot O$ at point T.

Prove: $\overline{OT} \perp m$

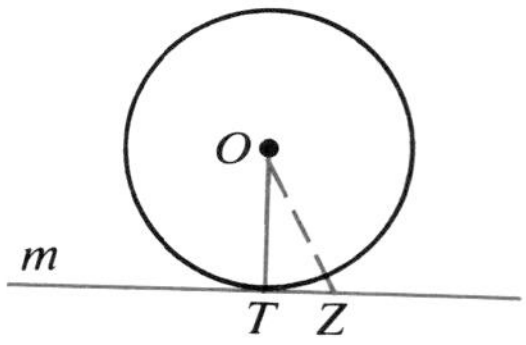

Indirect proof:

Assume temporarily that $\overline{OT}$ is not perpendicular to m. Draw the line through O that is perpendicular to m and call it $\overleftrightarrow{OZ}$. Because the perpendicular segment from O to m is the shortest segment from O to m, we have $OZ < OT$. Since m intersects $\odot O$ only in point T, Z must lie outside $\odot O$, and therefore $OZ > OT$. The statements $OZ < OT$ and $OZ > OT$ are contradictory. This shows that what we temporarily assumed, that $\overline{OT}$ is not perpendicular to m, must be false. We conclude that $\overline{OT} \perp m$.

Corollary

Tangents to a circle from a point are congruent.

In the figure, $\overline{PA}$ and $\overline{PB}$ are tangent to the circle at A and B. The corollary tells us that $\overline{PA} \cong \overline{PB}$. For a proof, see Classroom Exercise 7.

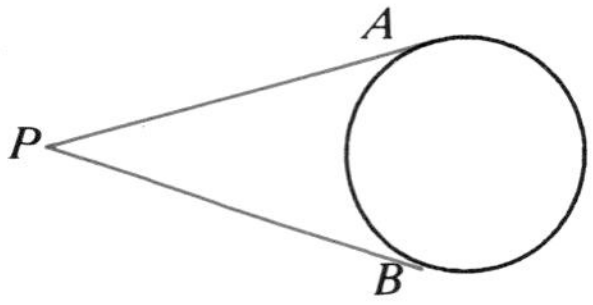

The proof of the following converse to Theorem 7-1 is left as Exercise 23.

Theorem 7-2

If a line in the plane of a circle is perpendicular to a radius at its outer endpoint, then the line is tangent to the circle.

Given: Line l in the plane of $\odot Q$;
$l \perp$ radius $\overline{QR}$ at R

Prove: l is tangent to $\odot Q$.

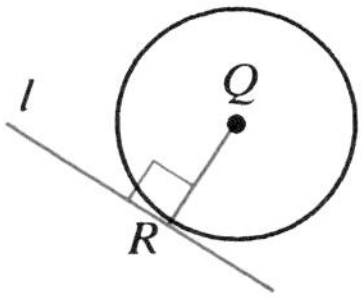

A line that is tangent to each of two coplanar circles is called a **common tangent.**

Common *internal* tangents intersect the segment joining the centers.

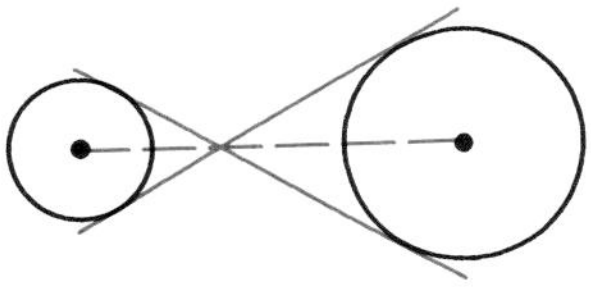

Common *external* tangents do *not* intersect the segment joining the centers.

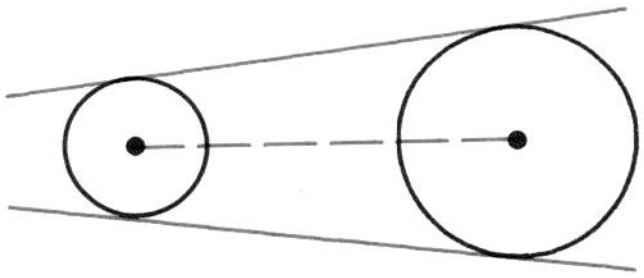

Two *circles are tangent* to each other when they are coplanar and are tangent to the same line at the same point.

$\odot A$ and $\odot B$ are *externally* tangent.

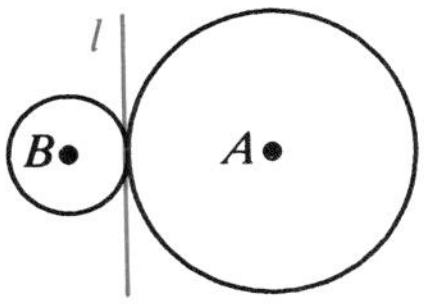

$\odot C$ and $\odot D$ are *internally* tangent.

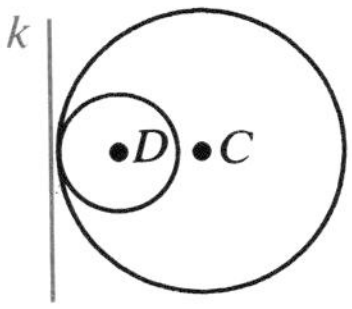

The ends of the plastic industrial pipes shown in the photograph illustrate externally tangent circles. Notice that when a circle is surrounded by tangent circles of the same radius, six of these circles fit exactly around the inner circle.

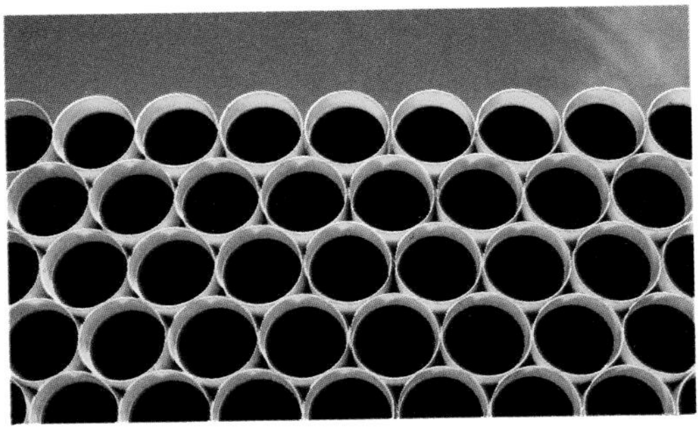

When each side of a polygon is tangent to a circle, the polygon is said to be **circumscribed about the circle.** The circle is **inscribed in the polygon.**

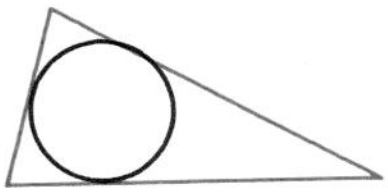

Circumscribed polygons

Inscribed circles

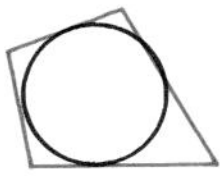

Classroom Exercises

How many common external tangents and how many common internal tangents can be drawn to the two circles?

1.

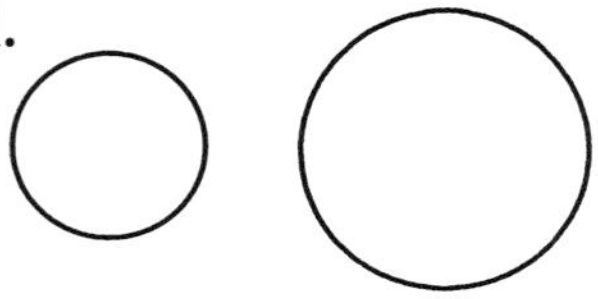

2.

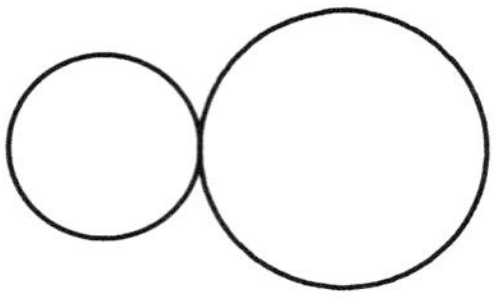

3.

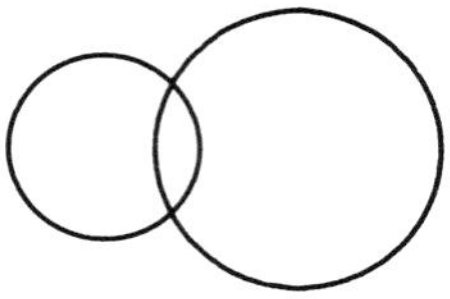

4.

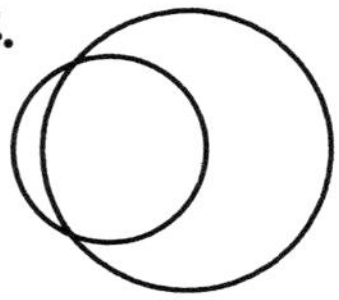

5.

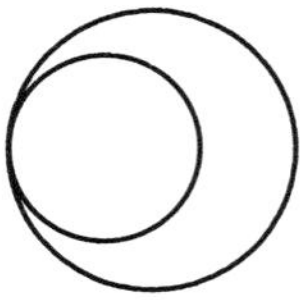

6.

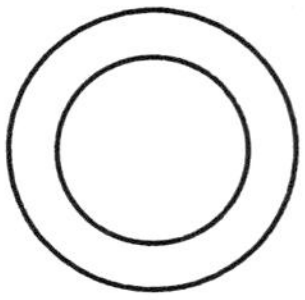

7. Write a proof of the corollary to Theorem 7-1.
Given: $\overline{PA}$ and $\overline{PB}$ are tangents to $\odot O$.
Prove: $\overline{PA} \cong \overline{PB}$

■ ***Plan for Proof:*** Draw $\overline{AO}$, $\overline{BO}$, and $\overline{PO}$. Show that two triangles are congruent and use corresponding parts.

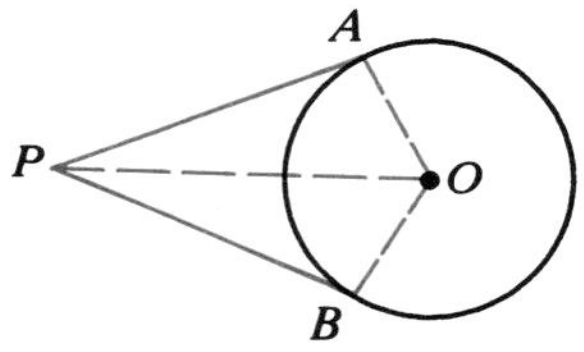

8. State and prove a relation between $\angle APB$ and $\angle AOB$.

Written Exercises

A **1.** Copy and complete this proof that the common external tangent segments to two noncongruent circles are congruent.

Given: Tangents $\overline{AB}$ and $\overline{CD}$

Prove: $\overline{AB} \cong \overline{CD}$

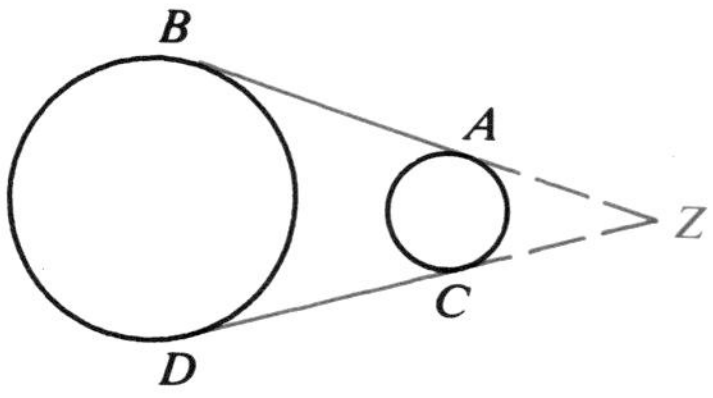

Proof:

Statements	Reasons
1. Draw $\overleftrightarrow{AB}$ and $\overleftrightarrow{CD}$, intersecting at Z.	1. Through any two points there is _?_.
2. $ZA + AB = ZB$; $ZC + CD = ZD$	2. _?_
3. $ZB = ZD$	3. _?_
4. $ZA + AB = ZC + CD$	4. _?_
5. $ZA = ZC$	5. _?_
6. $AB = CD$, or $\overline{AB} \cong \overline{CD}$	6. _?_

2. Suppose, in Exercise 1, that the circles are congruent. Then the tangent lines won't meet. Supply reasons for these key steps of the proof that the common tangent segments are congruent for this special case.

Given: Tangents $\overline{AB}$ and $\overline{CD}$;
$\odot O \cong \odot Q$

Prove: $\overline{AB} \cong \overline{CD}$

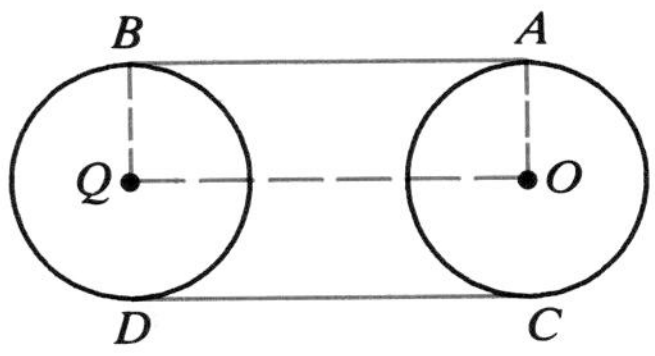

1. $\overline{OA} \perp \overline{AB}$ and $\overline{QB} \perp \overline{AB}$
2. $\overline{OA} \parallel \overline{QB}$
3. $\overline{OA} \cong \overline{QB}$
4. Quad. $AOQB$ is a ▱.
5. $\overline{AB} \cong \overline{OQ}$

By a similar proof, $\overline{OQ} \cong \overline{CD}$.

6. $\overline{AB} \cong \overline{CD}$

Draw two circles, with all their common tangents, so that the number of common tangents is the stated number.

3. one **4.** two **5.** three **6.** four

7. How many circles can be tangent to a given line at a given point on the line?

8. Circles O and Q are tangent at point Z. Suppose you draw a different circle tangent to $\odot O$ at Z. What can you say about the new circle and $\odot Q$?

9. The diagram shows tangent circles and lines.
$PA = 10$ $PB = \underline{\ ?\ }$ $PC = \underline{\ ?\ }$

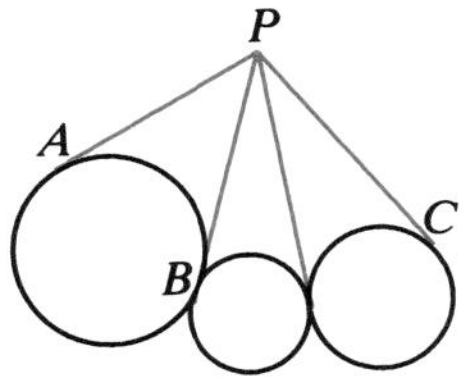

Ex. 9

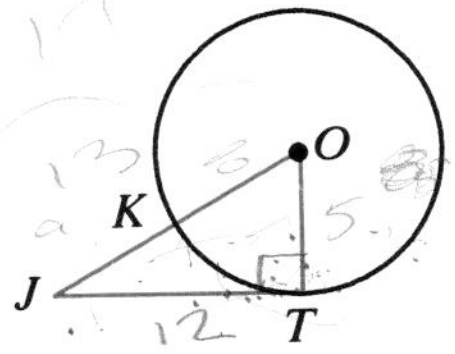

Exs. 10–12

In the diagram for Exercises 10–12, $\overline{JT}$ is tangent to $\odot O$.

10. If $JO = 13$ and $OT = 5$, then $JT = \underline{\ ?\ }$.

11. If $m\angle OJT = 30$ and $JO = 20$, then $JT = \underline{\ ?\ }$.

12. If $JK = 9$ and $KO = 8$, then $JT = \underline{\ ?\ }$.

B **13.** Discover and prove a theorem about two lines tangent to a circle at the endpoints of a diameter.

14. State, without proof, a theorem about spheres related to the theorem in Exercise 13.

15. Quad. $ABCD$ is circumscribed about a circle. Discover and prove a relationship between $AB + DC$ and $AD + BC$.

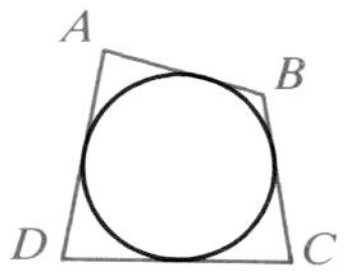

16. Rays $\overrightarrow{AB}$ and $\overrightarrow{AC}$ are tangent to $\odot O$. Discover and prove a theorem about $\overrightarrow{AO}$ and $\angle BAC$.

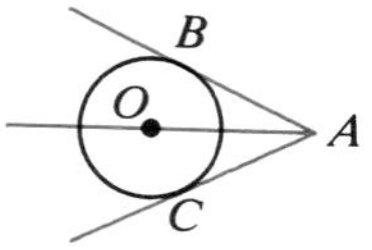

17. $\overline{SR}$ is tangent to $\odot P$ and $\odot Q$.
$QT = 6$; $TR = 8$; $PR = 30$
$PS =$ __?__ $PQ =$ __?__ $ST =$ __?__

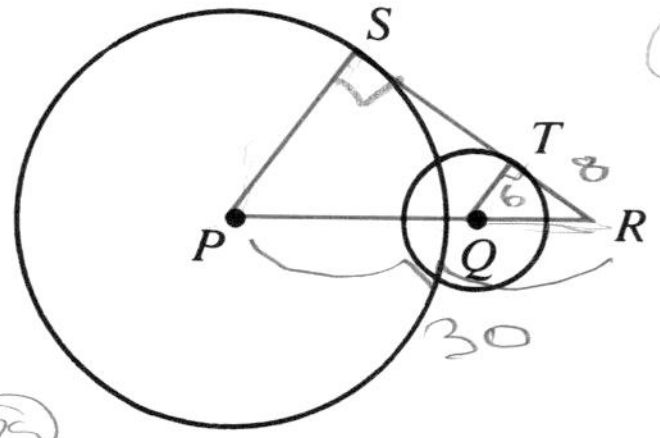

18. $\overline{JK}$ is tangent to $\odot P$ and $\odot Q$. $JK =$ __?__ (*Hint:* What kind of quadrilateral is $JPQK$?)

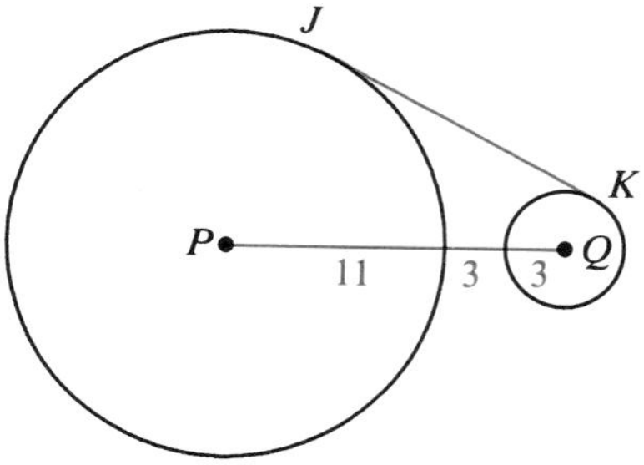

19. Circles P and Q have radii 6 and 2 and are tangent to each other. Find the length of their common external tangent $\overline{AB}$. (*Hint:* Draw $\overline{PQ}$, $\overline{PA}$, and $\overline{QB}$.)

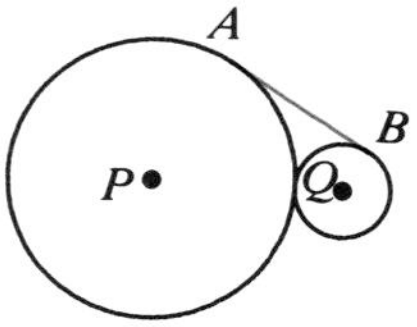

20. Given: Two tangent circles; $\overline{EF}$ is a common external tangent; $\overline{GH}$ is the common internal tangent.
Prove: $\angle EHF$ is a rt. $\angle$.

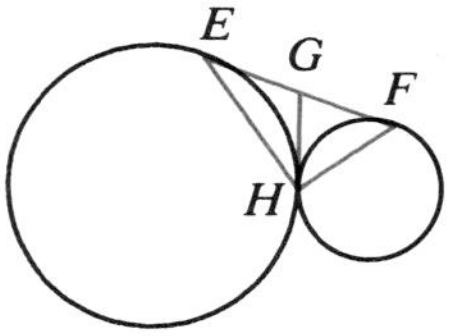

21. Three circles are shown. How many circles tangent to all three of the given circles can be drawn?

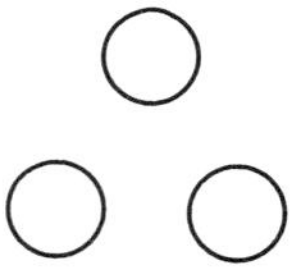

C 22. Suppose the three circles represent three spheres.
a. How many planes tangent to each of the spheres can be drawn?
b. How many spheres tangent to each of the three spheres can be drawn?

23. Prove Theorem 7–2. (*Hint:* Write an indirect proof.)

24. Find the radius of the circle inscribed in the triangle.

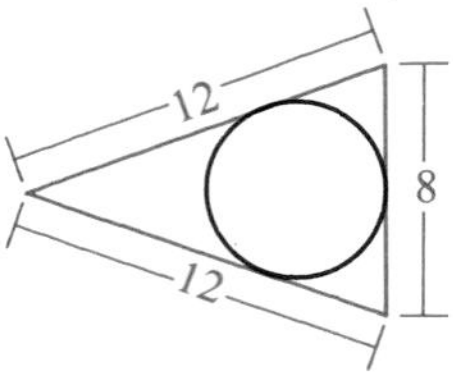

25. A circle inscribed in right $\triangle ABC$ is tangent to hypotenuse $\overline{AB}$ at K. $AK = 20$ and $BK = 6$. Find the sides of $\triangle ABC$.

26. The diameter of the circle inscribed in a certain right triangle is 6 while the diameter of the circumscribed circle is 17. Find the sides of the triangle.

7-3 Arcs and Central Angles

An *arc* is an unbroken part of a circle. Two points Y and Z on a circle O are always the endpoints of two arcs. If $\overline{YZ}$ is a diameter, the two arcs are called **semicircles.** Otherwise, Y and Z and the points of $\odot O$ in the interior of $\angle YOZ$ form a **minor arc.** Y and Z and the remaining points of $\odot O$ form a **major arc.**

Semicircles

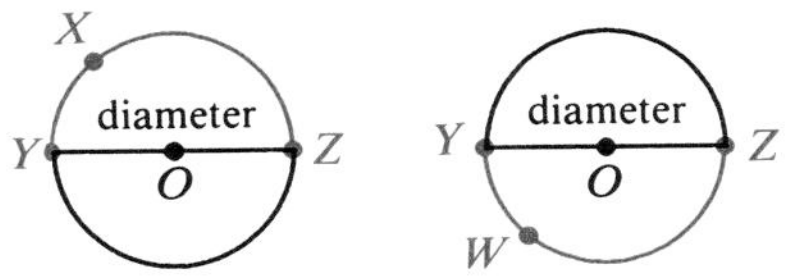

$\overset{\frown}{YXZ}$ $\overset{\frown}{YWZ}$

Minor Arc

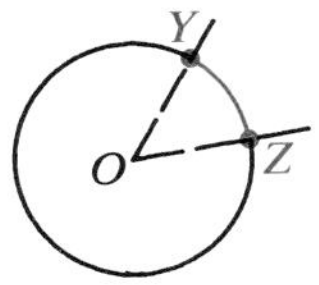

$\overset{\frown}{YZ}$

Major Arc

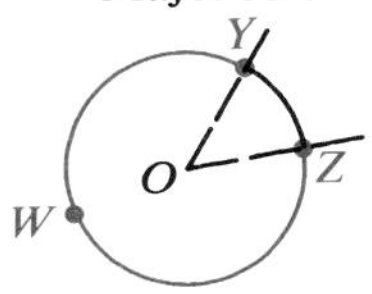

$\overset{\frown}{YWZ}$

A minor arc is named by its endpoints: $\overset{\frown}{YZ}$ is read "arc YZ." You use three letters to name a semicircle or a major arc: $\overset{\frown}{YWZ}$ is read "arc YWZ."

A **central angle** of a circle is an angle with its vertex at the center of the circle. A central angle that intersects a minor arc at its endpoints is called the central angle of that arc. Angle YOZ is the central angle of minor arc YZ shown above.

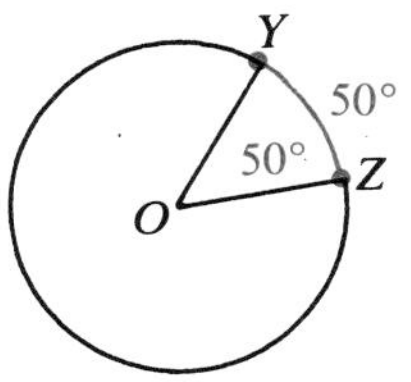

The **measure of a minor arc** is defined to be the measure of its central angle. To indicate that the measure of $\overset{\frown}{YZ}$ equals the measure of central angle YOZ you can write $m\overset{\frown}{YZ} = m\angle YOZ$. For example, if $m\angle YOZ = 50$, then $m\overset{\frown}{YZ} = 50$.

The **measure of a semicircle** is 180. The **measure of a major arc** will always be greater than 180. It is found as shown below.

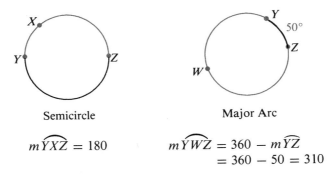

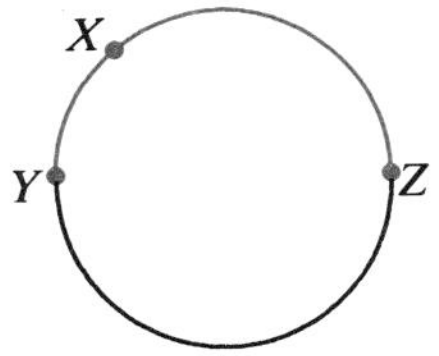

Semicircle

$m\overset{\frown}{YXZ} = 180$

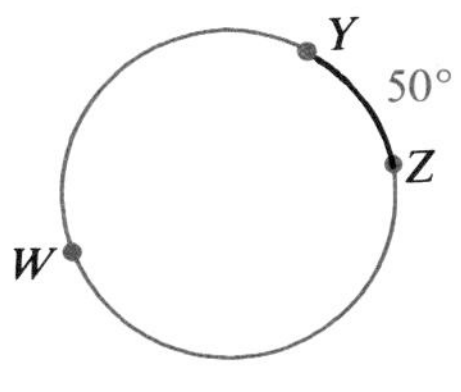

Major Arc

$$m\overset{\frown}{YWZ} = 360 - m\overset{\frown}{YZ}$$
$$= 360 - 50 = 310$$

Adjacent nonoverlapping arcs of a circle are arcs that have exactly one point in common. The following postulate is used when we compute arc measures by adding the measures of adjacent arcs.

Postulate 16   Arc Addition Postulate

The measure of the arc formed by two adjacent nonoverlapping arcs is the sum of the measures of these two arcs.

For example, applying Postulate 16 to the circle shown at the right, we have

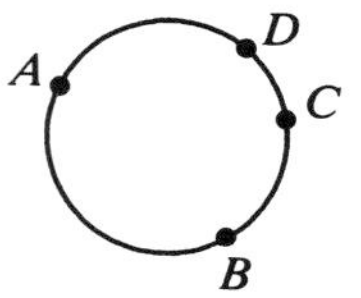

$$m\overset{\frown}{AD} + m\overset{\frown}{DC} = m\overset{\frown}{AC}$$
$$m\overset{\frown}{AB} + m\overset{\frown}{BC} = m\overset{\frown}{ABC}$$
$$m\overset{\frown}{ABC} + m\overset{\frown}{CD} = m\overset{\frown}{ABD}$$

Congruent arcs are arcs, in the same circle or in congruent circles, that have equal measures. In the diagram below, $\odot P$ and $\odot Q$ are congruent circles and $\overset{\frown}{AB} \cong \overset{\frown}{CD} \cong \overset{\frown}{EF}$. However, $\overset{\frown}{EF}$ is not congruent to $\overset{\frown}{RS}$ even though both arcs have the same degree measure, because $\odot Q$ is not congruent to $\odot O$.

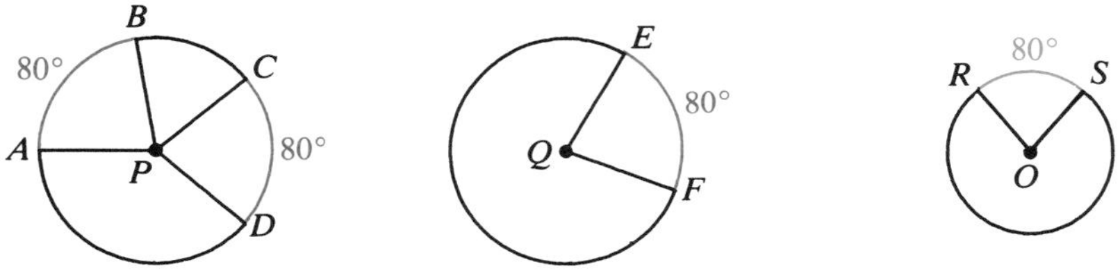

Notice that each of the congruent arcs above has an 80° central angle, so these congruent arcs have congruent central angles. And you can see in the photograph that the congruent central angles formed by adjacent spokes cut off congruent arcs along the rim of the wheel.

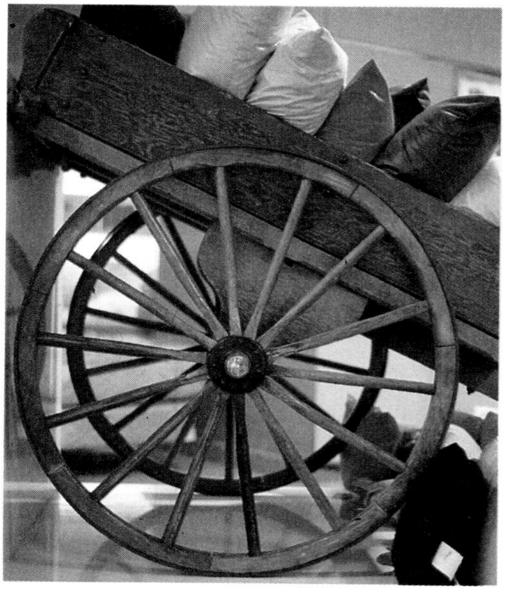

This relationship between congruence of minor arcs and congruence of their central angles is stated by Theorem 7-3 on the next page. The theorem will be proved in Classroom Exercises 14 and 15.

Theorem 7-3

In the same circle or in congruent circles, two minor arcs are congruent if and only if their central angles are congruent.

Classroom Exercises

1. Using the letters shown in the diagram, name:

a. Two central angles **b.** A semicircle

c. Two minor arcs **d.** Two major arcs

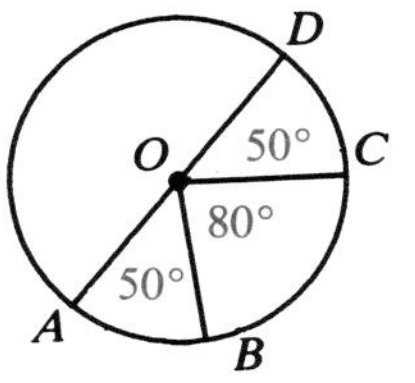

In Exercises 2–7 find the measure of the arc.

2. $\widehat{AB}$ **3.** $\widehat{AC}$ **4.** $\widehat{ABD}$

5. $\widehat{BAD}$ **6.** $\widehat{CDA}$ **7.** $\widehat{CDB}$

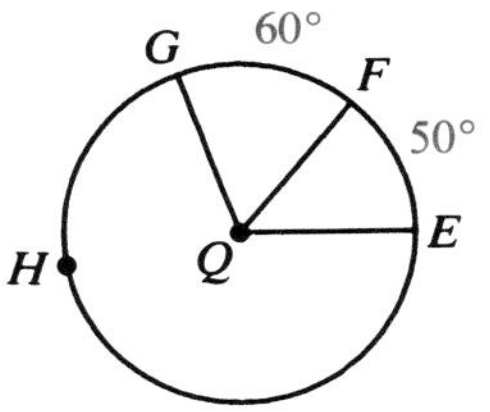

In Exercises 8–13 find the measure of the angle or the arc named.

8. $\angle GQF$ **9.** $\angle EQF$ **10.** $\angle GQE$

11. $\widehat{GE}$ **12.** $\widehat{GHE}$ **13.** $\widehat{EHF}$

14. Complete the proof of the "only if" part of Theorem 7-3 for the case of two congruent circles. That is, prove that in congruent circles two minor arcs are congruent only if their central angles are congruent. Give reasons for each statement.

Given: $\odot P \cong \odot Q$; $\widehat{AB} \cong \widehat{CD}$

Prove: $\angle APB \cong \angle CQD$

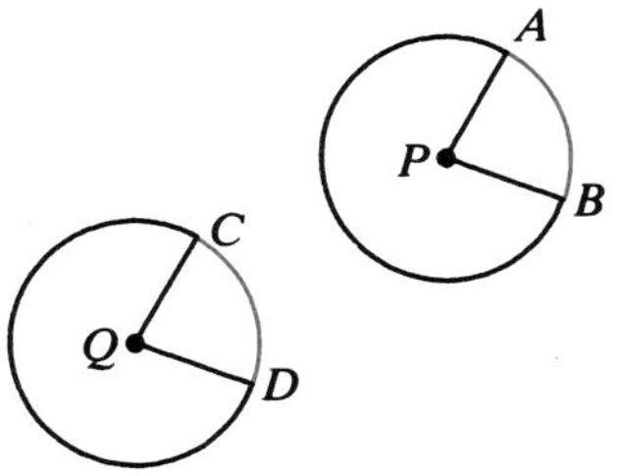

Proof:

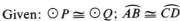

Statements	Reasons
1. $m\widehat{AB} = m\angle APB$; $m\widehat{CD} = m\angle CQD$	1. ?
2. $\widehat{AB} \cong \widehat{CD}$, or $m\widehat{AB} = m\widehat{CD}$	2. ?
3. $m\angle APB = m\angle CQD$, or $\angle APB \cong \angle CQD$	3. ?

15. Write a proof for the "if" part of Theorem 7-3 for congruent angles in the same circle. That is, prove that two minor arcs of a circle are congruent if their central angles are congruent.

Given: $\angle 1 \cong \angle 2$

Prove: $\widehat{TS} \cong \widehat{QR}$

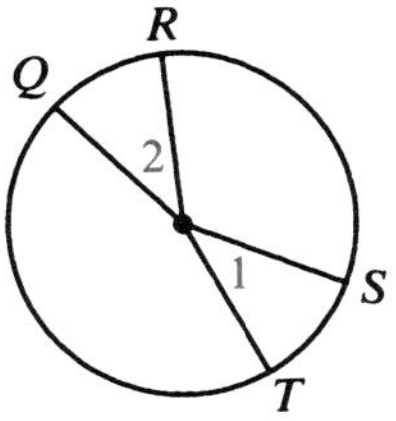

Written Exercises

Find the measure of central $\angle 1$.

A **1.**

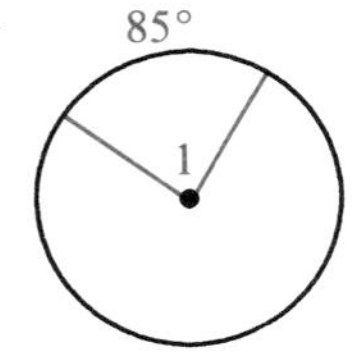

2.

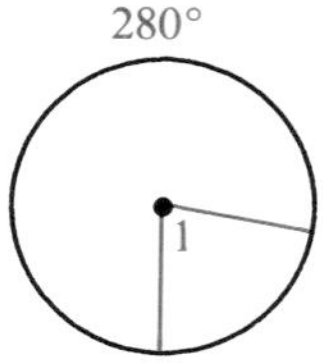

3.

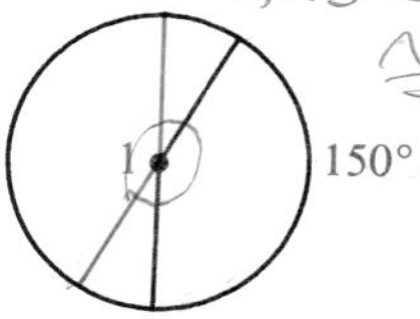

4.

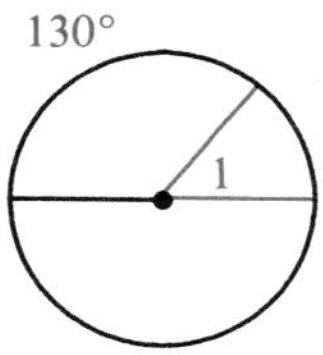

5.

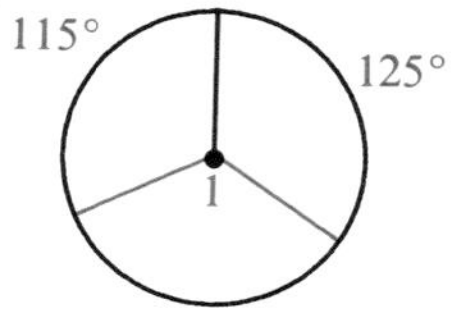

6.

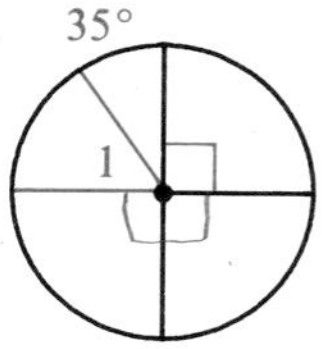

7. At 11 o'clock the hands of a clock form an angle of __?__°.

8. The hands of a clock form a 120° angle at __?__ o'clock and at __?__ o'clock.

9. a. Draw a circle. Place points A, B, and C on it in such positions that $m\widehat{AB} + m\widehat{BC}$ does not equal $m\widehat{AC}$.

b. Does your example in part (a) contradict Postulate 16?

In Exercises 10 and 11, $\overline{AB}$, $\overline{CD}$, and $\overline{EF}$ are diameters.

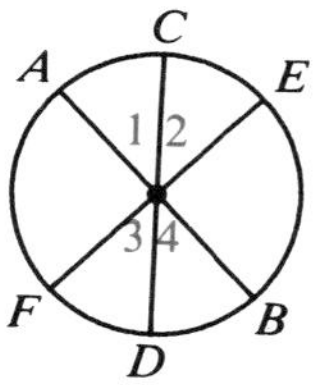

10. Given: $\widehat{AC} \cong \widehat{CE}$
Prove: $\angle 3 \cong \angle 4$

11. Given: $\angle 1 \cong \angle 2$
Prove: $\widehat{BD} \cong \widehat{DF}$

12. Given: $\odot O$ with $m\angle 5 = m\angle 7$
Prove: $\widehat{RT} \cong \widehat{SU}$

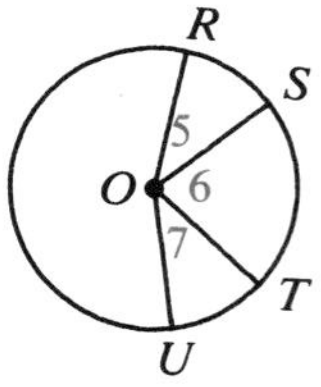

B **13.** Given: $\overline{WZ}$ is a diameter of $\odot O$; $\overline{OX} \parallel \overline{ZY}$
Prove: $\widehat{WX} \cong \widehat{XY}$
(*Hint:* Draw $\overline{OY}$.)

14. Given: $\overline{WZ}$ is a diameter of $\odot O$; $m\widehat{WX} = m\widehat{XY} = n$
Prove: $m\angle Z = n$

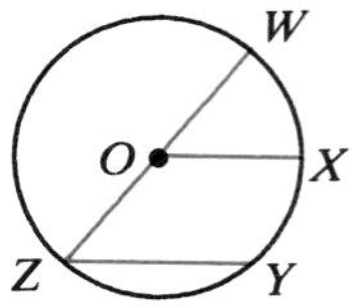

15. $\overline{AC}$ is a diameter of $\odot O$.

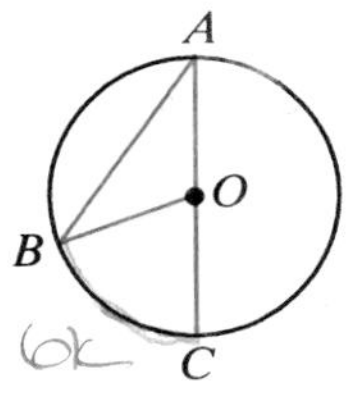

a. If $m\angle A = 35$, then $m\angle B =$ _?_, $m\angle BOC =$ _?_, and $m\widehat{BC} =$ _?_.

b. If $m\angle A = n$, then $m\widehat{BC} =$ _?_.

c. If $m\widehat{BC} = 6k$, then $m\angle A =$ _?_.

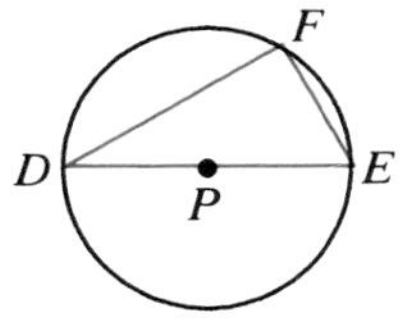

16. $\overline{DE}$ is a diameter of $\odot P$ and $m\widehat{EF} = n$. $m\angle DEF =$ _?_. (*Hint:* Draw $\overline{PF}$.)

The diagram, not drawn to scale, shows satellite S above the Earth, represented as sphere E. All lines tangent to the Earth from S touch the Earth at points on a circle with center C. Any two points on the Earth's surface on or above that circle can communicate with each other via S. X and Y are as far apart as two communication points can be. The Earth distance between X and Y equals the length of $\widehat{XTY}$, which equals $\frac{n}{360} \cdot$ circumference of the Earth. That circumference is approximately 40,200 km and the radius of the Earth is approximately 6400 km.

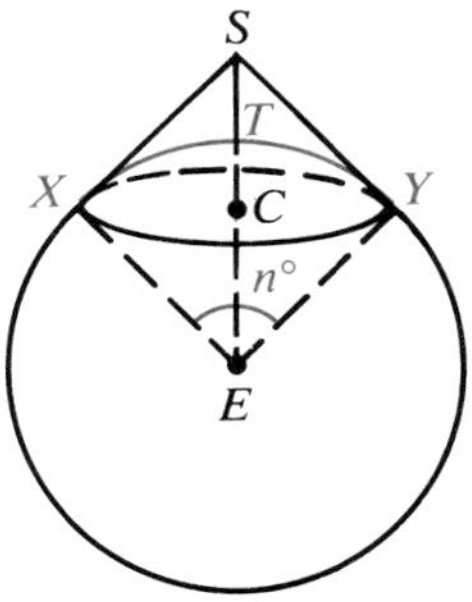

C 17. The photograph above shows the view from Gemini V looking north over the Gulf of California toward Los Angeles. The orbit of Gemini V ranged from 160 km to 300 km above the Earth. Take S to be 300 km above the Earth. That is, $ST = 300$ km. Find the Earth distance, rounded to the nearest 100 km, between X and Y. (*Hint:* Since you can find the value of $\cos \frac{n^\circ}{2}$ you can determine n°.)

18. Repeat Exercise 17, but with S twice as far from the Earth. Note that the distance between X and Y is not twice as great as before.

19. Given: $\odot O$ and $\odot Q$ intersect at R and S;
$m\overset{\frown}{RVS} = 60$; $m\overset{\frown}{RUS} = 120$
Prove: $\overline{OR}$ is tangent to $\odot Q$;
$\overline{QR}$ is tangent to $\odot O$.

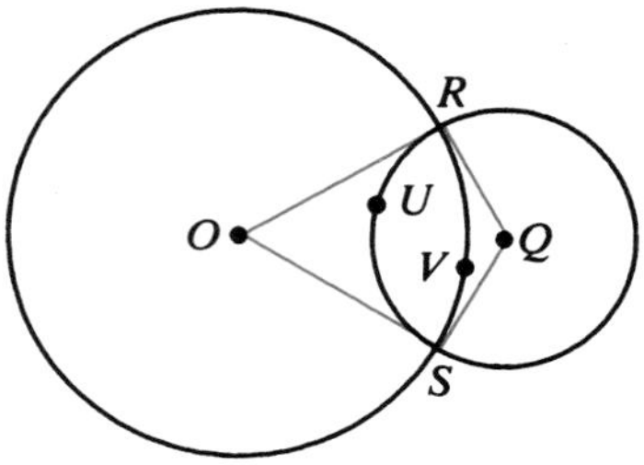

20. Given: $\overline{AB}$ is a diameter of $\odot Z$; points J and K lie on $\odot Z$ with $m\overset{\frown}{AJ} = m\overset{\frown}{BK}$. Discover and prove something about $\overline{JK}$. (*Hint:* There are two possibilities, depending on whether $\overset{\frown}{AJ}$ and $\overset{\frown}{BK}$ lie on the same side of $\overline{AB}$ or on opposite sides. So your statement will be of the *either . . . or* type.)

7-4 Arcs and Chords

In $\odot O$, $\overline{RS}$ cuts off two arcs, $\overset{\frown}{RS}$ and $\overset{\frown}{RTS}$. We speak of $\overset{\frown}{RS}$, the minor arc, as being *the arc of chord RS.*

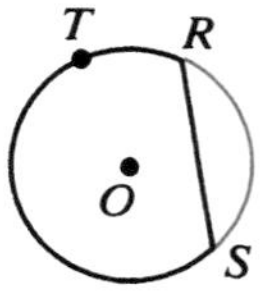

Theorem 7-4

In the same circle or in congruent circles:

(1) Congruent chords have congruent arcs.

(2) Congruent arcs have congruent chords.

We outline the proof of part (1) for one circle.

Given: $\odot O$; $\overline{RS} \cong \overline{TU}$

Prove: $\overset{\frown}{RS} \cong \overset{\frown}{TU}$

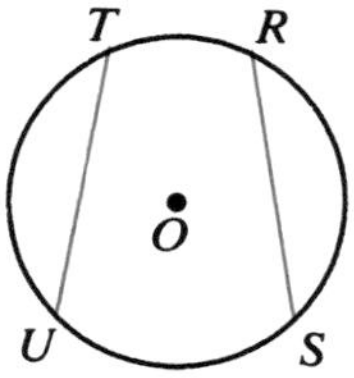

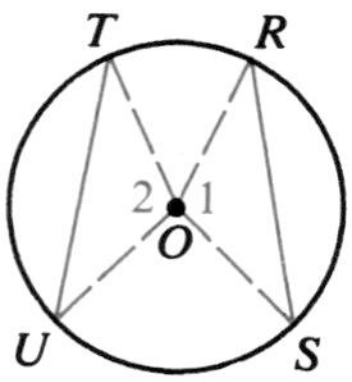

Outline of proof:

1. Draw radii $\overline{OR}$, $\overline{OS}$, $\overline{OT}$, and $\overline{OU}$.
2. $\overline{OR} \cong \overline{OT}$; $\overline{OS} \cong \overline{OU}$ (Why?)
3. $\overline{RS} \cong \overline{TU}$ (Given)
4. $\triangle ROS \cong \triangle TOU$ (Why?)
5. $\angle 1 \cong \angle 2$ (Corres. __?__.)
6. $\overset{\frown}{RS} \cong \overset{\frown}{TU}$ (Why?)

The next theorem involves the idea of bisecting an arc. Y is called the midpoint of $\widehat{XYZ}$ if $\widehat{XY} \cong \widehat{YZ}$. Any line, segment, or ray that contains Y bisects $\widehat{XYZ}$.

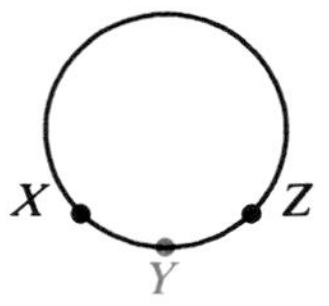

Theorem 7-5

A diameter that is perpendicular to a chord bisects the chord and its arc.

Given: $\odot O$; $\overline{CD} \perp \overline{AB}$

Prove: $\overline{AZ} \cong \overline{BZ}$; $\widehat{AD} \cong \widehat{BD}$

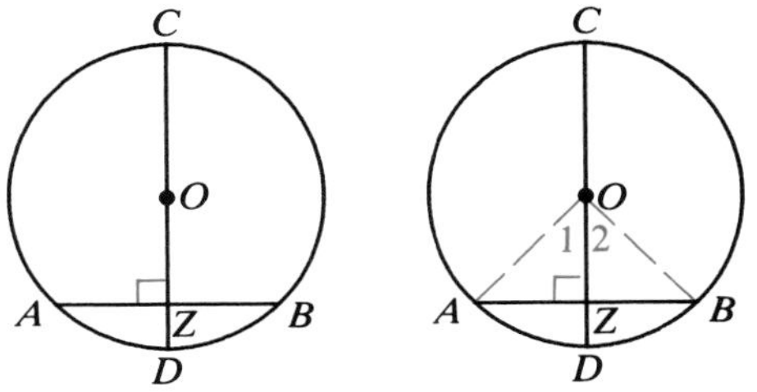

Plan for Proof: Draw $\overline{OA}$ and $\overline{OB}$. Use the HL Theorem to prove that right triangles OZA and OZB are congruent. Then use corresponding parts of congruent triangles to show that $\overline{AZ} \cong \overline{BZ}$ and $\angle 1 \cong \angle 2$. Finally, apply the theorem that congruent central angles have congruent arcs.

You will prove part (1) of the next theorem as Classroom Exercise 2.

Theorem 7-6

In the same circle or in congruent circles:

(1) Congruent chords are equally distant from the center (or centers).

(2) Chords equally distant from the center (or centers) are congruent.

Example 1 Find the values of x and y.

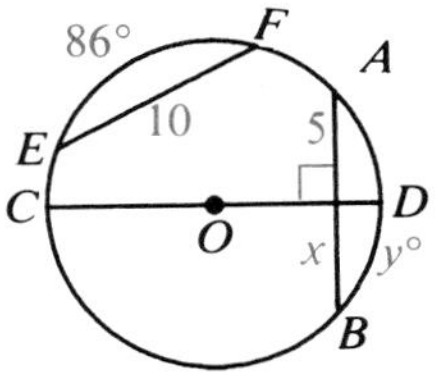

Solution Diameter $\overline{CD}$ bisects chord $\overline{AB}$, so $x = 5$. (Theorem 7-5)
$\overline{AB} \cong \overline{EF}$, so $m\widehat{AB} = 86$. (Theorem 7-4)
Diameter $\overline{CD}$ bisects $\widehat{AB}$, so $y = 43$. (Theorem 7-5)

Example 2 Find the value of x.

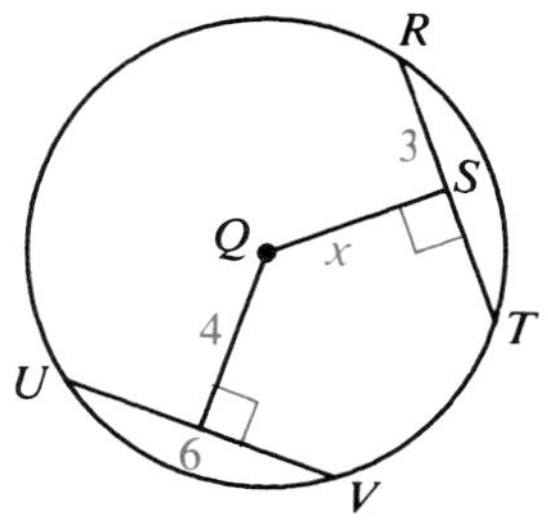

Solution S is the midpoint of $\overline{RT}$, so $RT = 6$. (Theorem 7-5)
$\overline{RT} \cong \overline{UV}$, so $x = 4$. (Theorem 7-6)

Classroom Exercises

1. See Theorem 7–4 on page 306. Which statements in the outline of the proof would you modify to get a proof of part (1) for congruent circles?

2. Supply reasons to complete a proof of Theorem 7–6, part (1), for one circle.
   Given: $\odot O$; $\overline{AB} \cong \overline{CD}$;
   $\overline{OY} \perp \overline{AB}$; $\overline{OZ} \perp \overline{CD}$
   Prove: $OY = OZ$

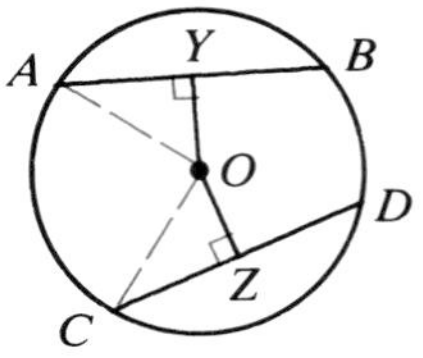

Proof:

Statements	Reasons
1. Draw radii $\overline{OA}$ and $\overline{OC}$.	1. ___?___
2. $\overline{AB} \cong \overline{CD}$, or $AB = CD$	2. ___?___
3. $\frac{1}{2}AB = \frac{1}{2}CD$	3. ___?___
4. $AY = \frac{1}{2}AB$; $CZ = \frac{1}{2}CD$	4. ___?___
5. $AY = CZ$	5. ___?___
6. $OA = OC$	6. ___?___
7. rt. $\triangle OYA \cong$ rt. $\triangle OZC$	7. ___?___
8. $OY = OZ$	8. ___?___

3. In $\odot E$, diameter $\overline{TU} \perp \overline{RS}$; $VS = 4$; $m\widehat{US} = n$.
   a. $RV =$ ___?___ and $m\widehat{RU} =$ ___?___.
   b. In terms of n, $m\widehat{ST} =$ ___?___ and $m\widehat{RT} =$ ___?___.
   c. From part (b) we conclude that $\widehat{RT} \cong$ ___?___.
   d. Suppose $TU = 10$. Then $EV =$ ___?___.

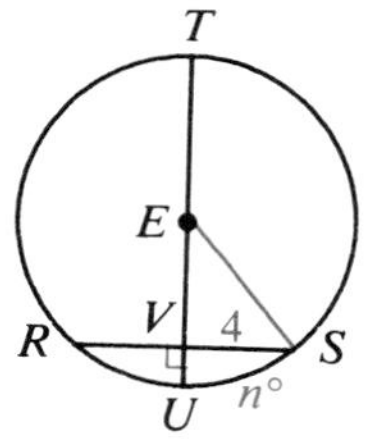

4. In $\odot O$, the chords have unequal lengths, with $2j > 2k$.
   a. From $m^2 + j^2 = r^2$ we get $m = \sqrt{\underline{\ ?\ } - \underline{\ ?\ }}$.
   b. We also have $n = \sqrt{\underline{\ ?\ } - \underline{\ ?\ }}$.
   c. Because $j > k$, $\sqrt{r^2 - j^2}$ ___?___ (</=/>) $\sqrt{r^2 - k^2}$.
   d. By substitution in (c) we get m ___?___ (</=/>) n.

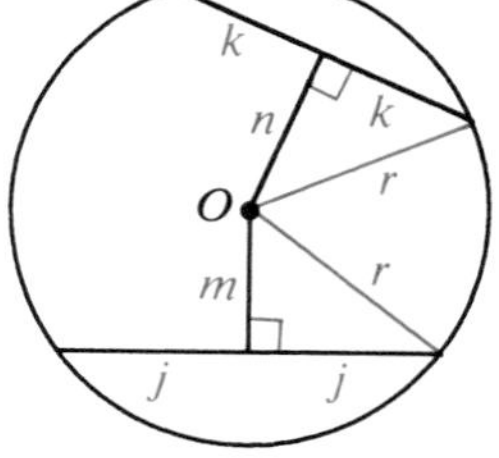

5. Exercise 4 provides a condensed proof of a theorem not stated in this textbook. Complete the statement: If two chords of a circle have unequal lengths, then the longer chord is ___?___.

Written Exercises

In the diagrams that follow, *O* is the center of each circle.

A

1.

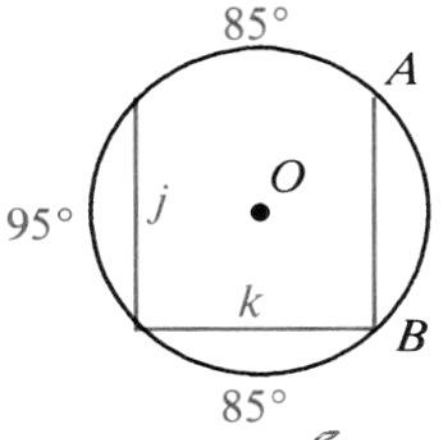

$AB =$ ___?___

2.

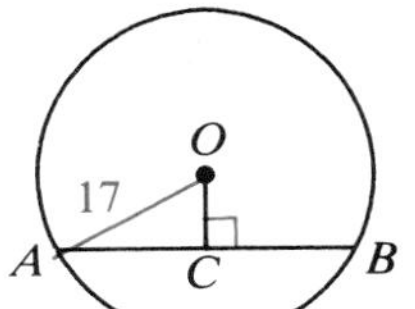

$AB = 30$; $OC =$ ___?___

3.

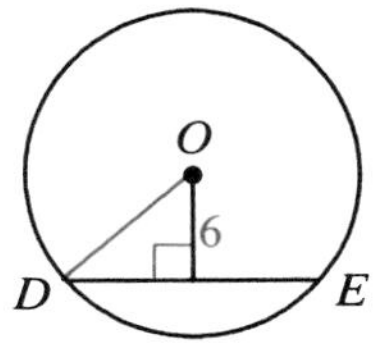

$DE = 16$; $OD =$ ___?___

4.

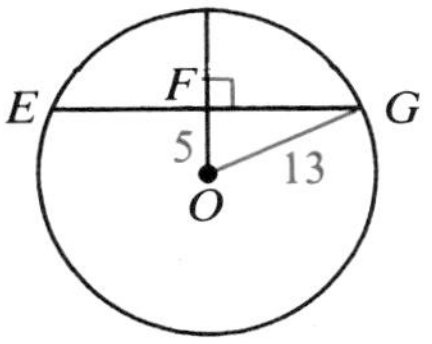

$EG =$ ___?___

5.

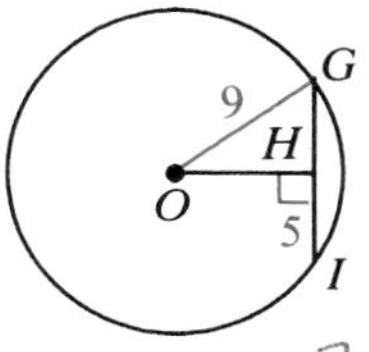

$OH =$ ___?___

6.

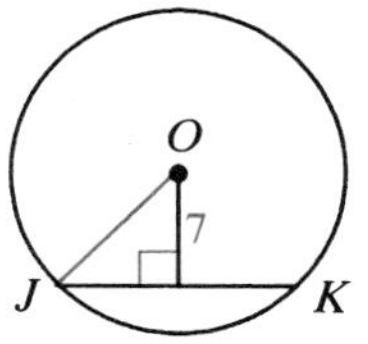

$JK = 14$; $OJ =$ ___?___

7. Prove part 2 of Theorem 7–4 for congruent circles. First list what is given and what you are to prove.

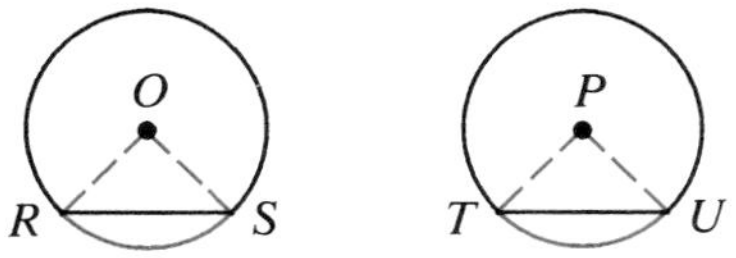

8. a. Given: $\overset{\frown}{JZ} \cong \overset{\frown}{KZ}$
Prove: $\angle J \cong \angle K$

b. Is the converse of part (a) true?

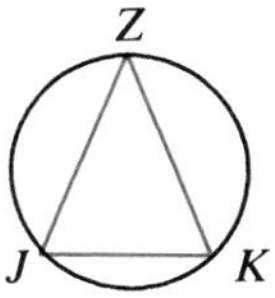

B

9. Given: $\overline{RS} \cong \overline{UT}$
Prove: $\overline{RT} \cong \overline{US}$
(*Hint:* Apply Theorem 7–4 and the Arc Addition Postulate.)

10. Given: $\overset{\frown}{RS} \cong \overset{\frown}{UT}$; $\angle R \cong \angle U$
Prove: $\overline{VS} \cong \overline{VT}$ and $\overline{RV} \cong \overline{UV}$

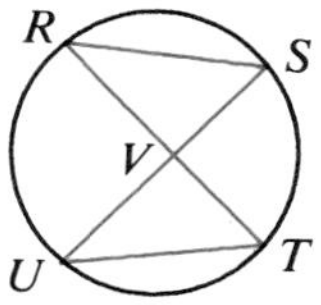

11. The informal statement "When you double the length of an arc you double the length of the chord" may seem at first glance to be true. But use the figure, in which $m\overset{\frown}{AC} = 2 \cdot m\overset{\frown}{AB}$, to show that $AC \neq 2 \cdot AB$.

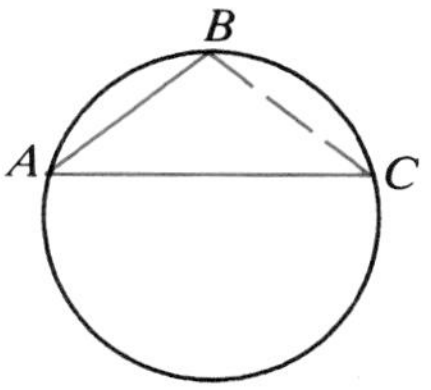

12. a. Draw three generous-sized circles and inscribe a different-shaped quadrilateral *ABCD* in each.
b. Use a protractor to measure all the angles.
c. Compare $\angle A$ and $\angle C$, $\angle B$ and $\angle D$.
d. Although you haven't proved anything in this exercise, you should wonder about a possible theorem. State the theorem.

13. 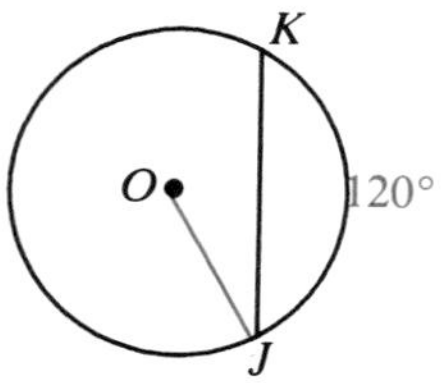

14. 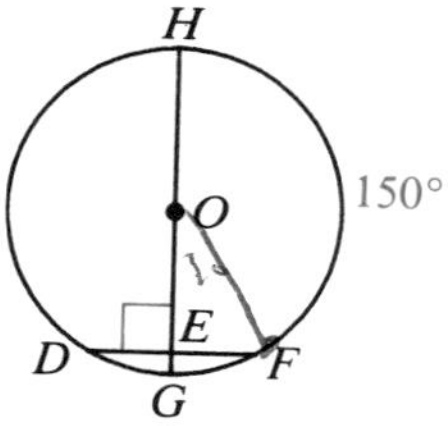

15. 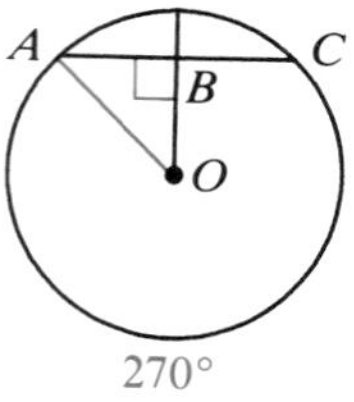

If $OJ = 12$, $JK =$ __?__. If $OE = 8\sqrt{3}$, $HG =$ __?__. If $OA = 9$, $BC =$ __?__.

16. The radius of a sphere is j. The distance from the center of the sphere to a certain chord is k. How long is the chord? Answer in terms of j and k.

State and prove a theorem suggested by the figure.

C **17.** 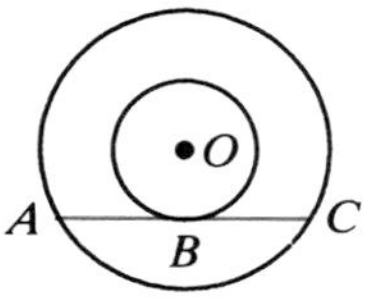

18. 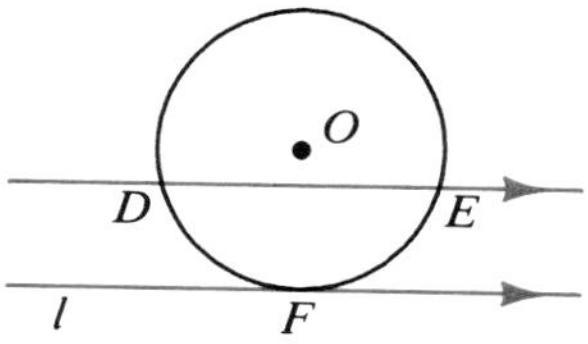

19. Investigate the possibility, given a circle, of drawing two chords whose lengths are in the ratio 1:2 and whose distances from the center are in the ratio 2:1. If the chords can be drawn, find the length of each in terms of the radius. If not, prove that the figure is impossible.

20. Three parallel chords of ⊙O are drawn as shown. Their lengths are 20, 16, and 12 cm. Find, to the nearest tenth of a centimeter, the length of chord $\overline{XY}$ (not shown).

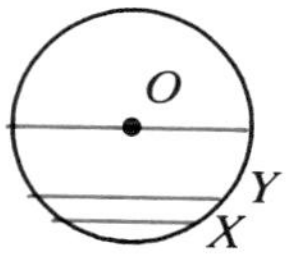

Self-Test 1

1. Sketch a triangle inscribed in one circle and sketch a quadrilateral circumscribed about another circle.
2. Circles O and Q are congruent circles. The radius of ⊙O is 8. The diameter of ⊙Q is __?__.
3. Two circles intersect in two points. How many common tangents can be drawn to the circles?
4. A plane passes through the common center of two concentric spheres. Describe the intersection of the plane and the two spheres.

Points *E*, *F*, *G*, *H*, and *J* lie on ⊙*O*.

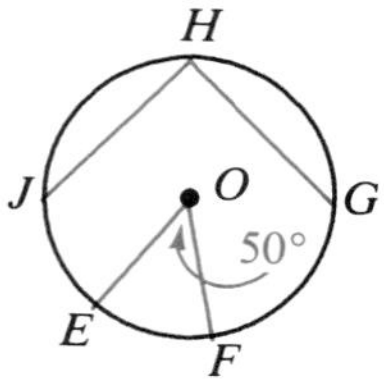

5. $m\widehat{EF} =$ __?__ 6. $m\widehat{EHF} =$ __?__
7. Suppose $\overline{JH} \cong \overline{HG}$. State the theorem that supports the conclusion that $\widehat{JH} \cong \widehat{HG}$.

In $\odot Q$, diameter $\overline{US} \perp \overline{RT}$.

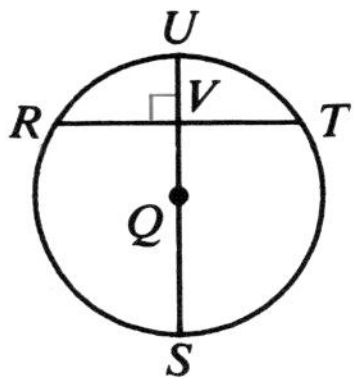

8. If $m\widehat{RST} = 220$, $m\widehat{UT} =$ _?_.

9. If $RT = 16$ and $QS = 10$, then $QV =$ _?_.

10. Given: Tangent circles O and Q with tangents $\overline{PA}$, $\overline{PB}$, and $\overline{PC}$ as shown.
Prove: $\overline{PA} \cong \overline{PC}$

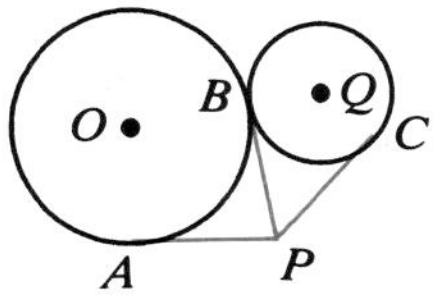

Ex. 10

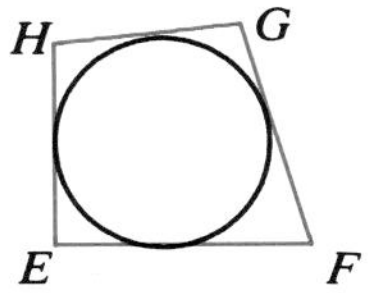

Ex. 11

11. In the circumscribed quadrilateral, $EF = 16$, $FG = 15$, and $GH = 12$. $HE =$ _?_.

Angles and Segments

Objectives

1. Solve problems and prove statements involving inscribed angles.
2. Solve problems and prove statements involving angles formed by chords, secants, and tangents.
3. Solve problems involving lengths of chords, secant segments, and tangent segments.

7-5 Inscribed Angles

In the figures below, we say that the angles *intercept* the arcs shown in color.

An **inscribed angle** is an angle whose vertex is on a circle and whose sides contain chords of the circle. In the diagrams above, only $\angle 1$ and $\angle 2$ are inscribed angles. $\angle 1$ intercepts a minor arc, and $\angle 2$ intercepts a major arc. Some inscribed angles intercept semicircles. $\angle 3$ intercepts two arcs and $\angle 4$ intercepts one arc.

The next theorem compares the measure of an inscribed angle with the measure of its intercepted arc.

Theorem 7-7

The measure of an inscribed angle is equal to half the measure of its intercepted arc.

Given: $\angle ABC$ inscribed in $\odot O$

Prove: $m\angle ABC = \frac{1}{2}m\widehat{AC}$

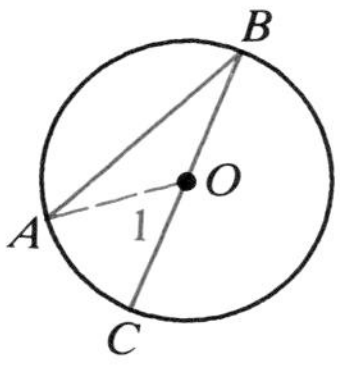

Case I:
Point O lies on $\angle ABC$.

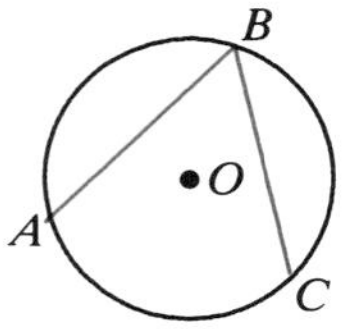

Case II:
Point O lies inside $\angle ABC$.

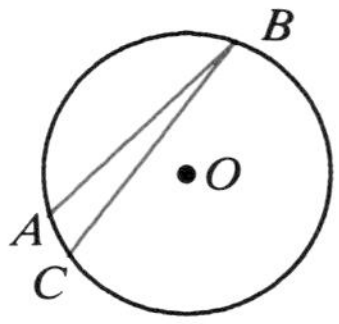

Case III:
Point O lies outside $\angle ABC$.

Outline of Proof of Case I:

1. Draw radius $\overline{OA}$. (Through any two points there is exactly one line.)
2. $\overline{OB} \cong \overline{OA}$ (All radii of a circle are congruent.)
3. $m\angle A = m\angle B$ (If two sides of a $\triangle$ are $\cong$, _?_.) angles opp are ≅
4. $m\angle A + m\angle B = m\angle 1$ (The measure of an exterior angle of a $\triangle$ = _?_.) the sum of 2 measures of 2 remote ∠s.
5. $m\angle B + m\angle B = m\angle 1$ (Why?) Substitution
6. $m\angle B = \frac{1}{2}m\angle 1$ (Division Property of =)
7. $m\widehat{AC} = m\angle 1$ (Why?) Def of minor arc.
8. $m\angle B = \frac{1}{2}m\widehat{AC}$ (Substitution Property)

The proofs of Cases II and III of Theorem 7-7 are left as Classroom Exercises 14 and 15. Proofs of the following three corollaries will be considered in Classroom Exercises 3-5.

Corollary 1

If two inscribed angles intercept the same arc, then the angles are congruent.

Corollary 2

If a quadrilateral is inscribed in a circle, then its opposite angles are supplementary.

Corollary 3

An angle inscribed in a semicircle is a right angle.

The following diagrams illustrate Corollaries 1–3.

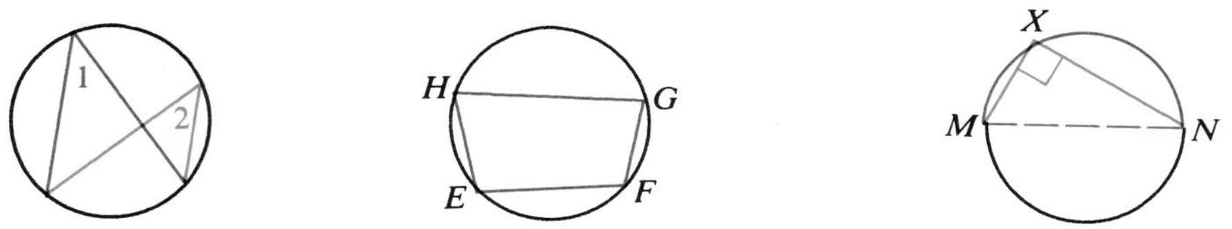

Corollary 1	Corollary 2	Corollary 3
$\angle 1 \cong \angle 2$	$\angle E$ is supp. to $\angle G$. $\angle F$ is supp. to $\angle H$.	If $\widehat{MXN}$ is a semicircle, then $\angle X$ is a right angle.

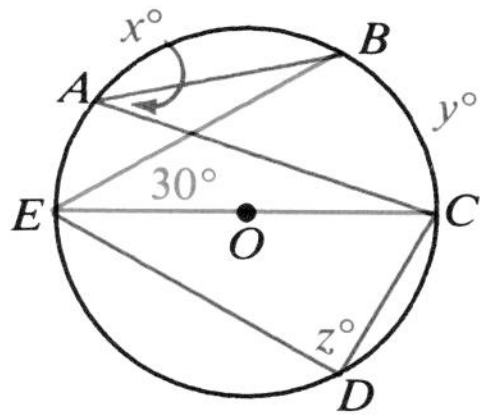

Example Find the values of x, y, and z in $\odot O$.

Solution $\angle BAC \cong \angle BEC$ since they intercept the same arc, so $x = 30$.

$30 = \frac{1}{2}m\widehat{BC}$, so $y = 60$.

$\angle EDC$ is inscribed in a semicircle, so $z = 90$.

Study the diagrams below from left to right. Point B moves along the circle closer and closer to point T. Finally, in diagram (4), point B has merged with T, and one side of $\angle T$ has become a tangent.

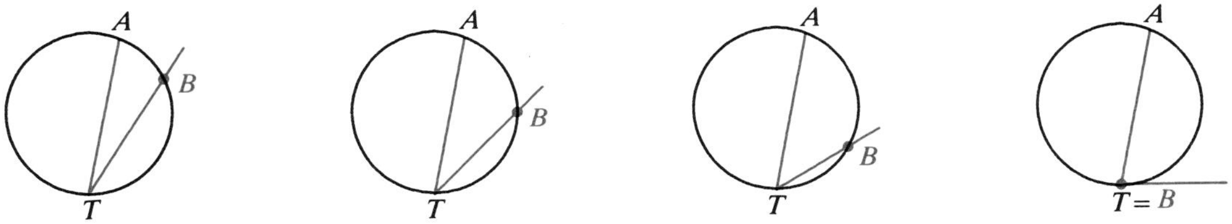

(1) (2) (3) (4)

Apply Theorem 7–7 to diagrams (1), (2), and (3) and you have $m\angle T = \frac{1}{2}m\widehat{AB}$. As you might expect, this equation applies to diagram (4), too. Diagram (4) suggests Theorem 7–8. In Exercises 9–11 you will prove the three cases of the theorem.

Theorem 7–8

The measure of an angle formed by a chord and a tangent is equal to half the measure of the intercepted arc.

For example, if $\overline{PT}$ is tangent to the circle and $m\widehat{AT} = 140$, then $m\angle ATP = 70$.

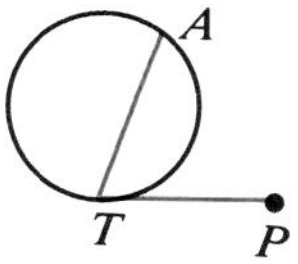

Classroom Exercises

1. A regular hexagon is inscribed in a circle. What is the measure of each arc?
2. A regular 15-gon is inscribed in a circle. What is the measure of each arc?
3. Explain why Corollary 1 of Theorem 7-7 is true. That is, explain why $\angle 1 \cong \angle 2$.

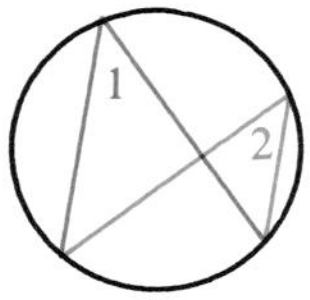

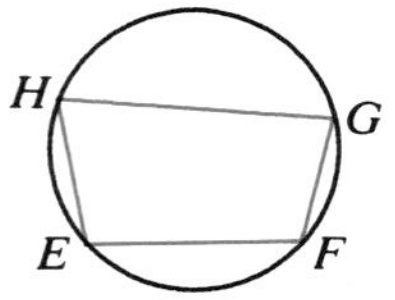

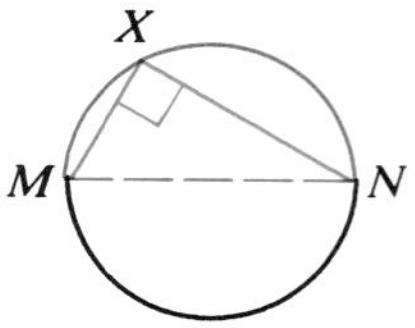

4. Explain why Corollary 2 is true. That is, explain why $\angle E$ and $\angle G$ are supplementary. (*Hint:* Let $m\widehat{FGH} = n$. Express $m\angle E$ and $m\angle G$ in terms of n.)
5. Explain why Corollary 3 is true. That is, explain how the fact that $\widehat{MXN}$ is a semicircle leads to the conclusion that $\angle X$ is a right angle.

Find the values of x and y. In Exercise 8, O is the center of the circle.

6.

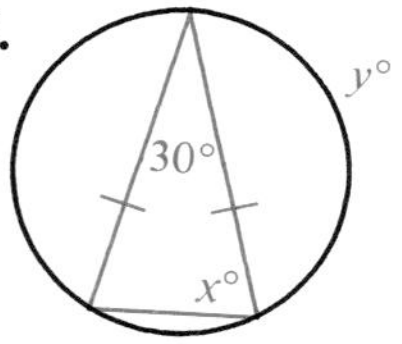

7.

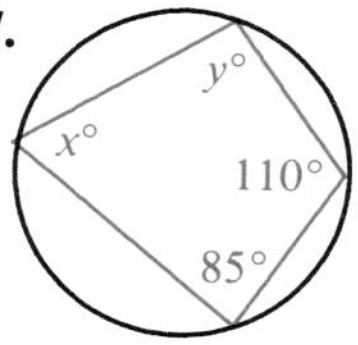

8.

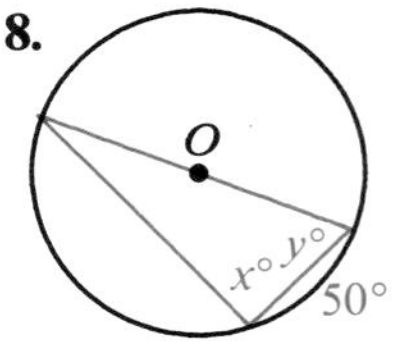

9. In quadrilateral $RSTU$, it is known that $m\angle R = 80$, $m\angle S = 100$, $m\angle T = 110$, and $m\angle U = 70$. Explain how you know that it isn't possible to draw a circle through points R, S, T, and U.
10. Suppose a parallelogram is inscribed in a circle. What special kind of parallelogram must it be? Explain.
11. In the figure, $m\angle AKB = m\angle CKD = n$. $m\widehat{AB} =$ __?__ and $m\widehat{CD} =$ __?__. State a theorem suggested by this exercise.

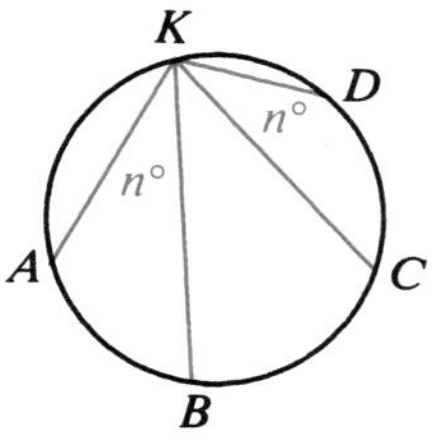

Ex. 11

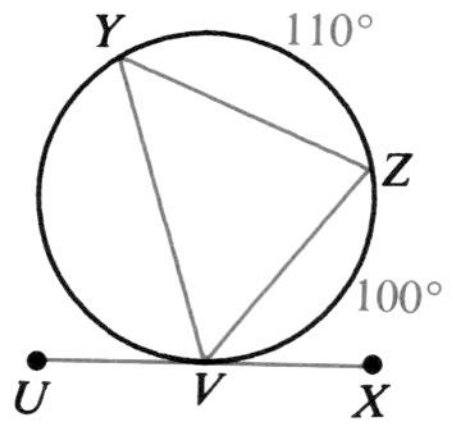

Ex. 12

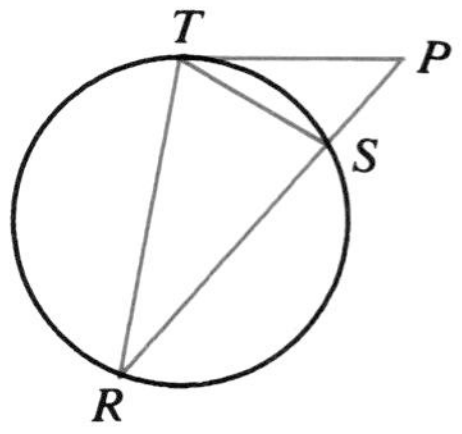

Ex. 13

12. $\overline{VX}$ is tangent to the circle. State the degree measures of as many angles as possible.
13. $\overline{PT}$ is tangent to the circle. Compare the measures of $\angle PRT$ and $\angle PTS$.

14. Outline a proof of Case II of Theorem 7–7. Use the diagram on page 312. (*Hint:* Draw the diameter from B and apply Case I.)

15. Repeat Exercise 14 for Case III.

Written Exercises

When the letter O is used in a diagram in these exercises, point O is the center of the circle. In Exercises 1–6, find the values of x, y, and z.

A

1.

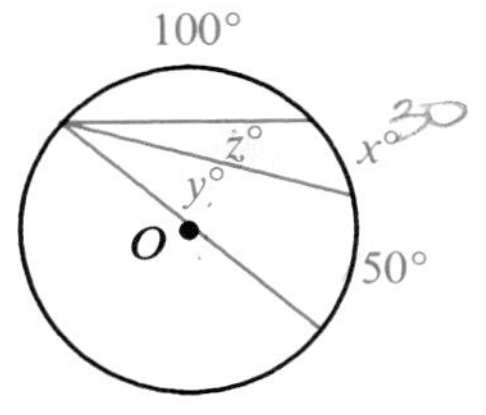

2.

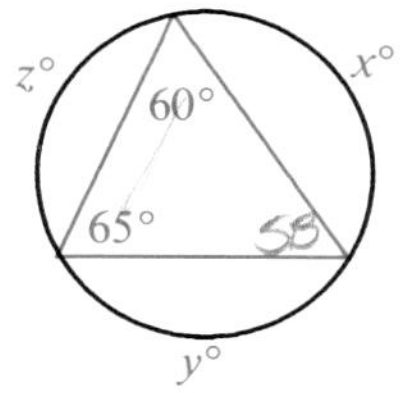

3.

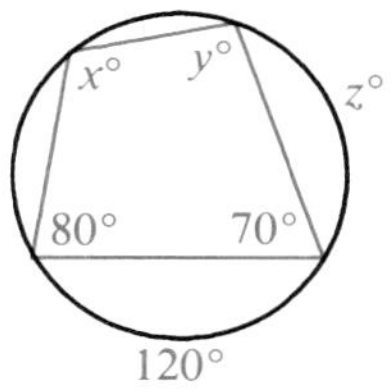

4.

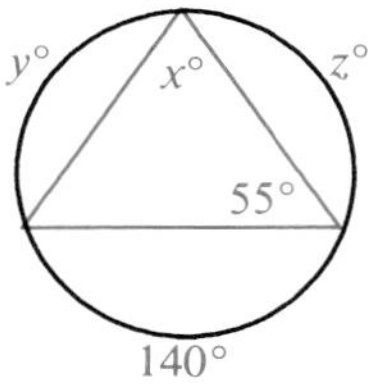

5.

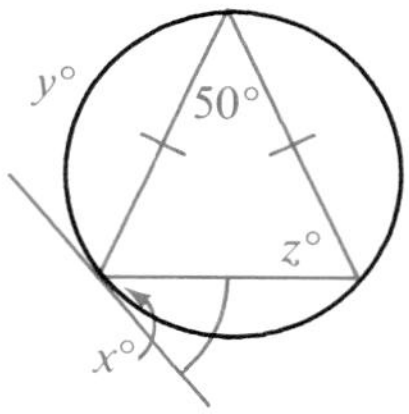

6.

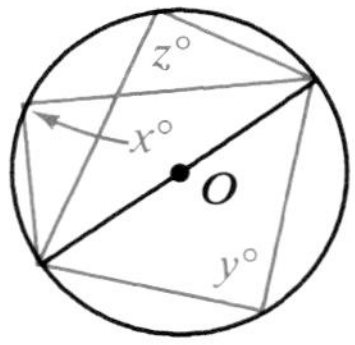

7. Prove: If two chords of a circle are parallel, the two arcs between the chords are congruent.

   Given: $\overline{AB} \parallel \overline{CD}$

   Prove: $\overset{\frown}{AC} \cong \overset{\frown}{BD}$

   (*Hint:* Draw $\overline{BC}$.)

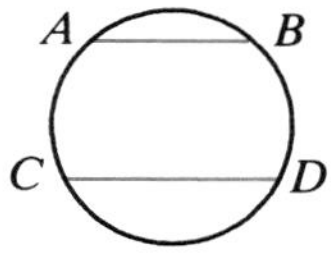

8. Prove: $\triangle UXZ \sim \triangle YVZ$

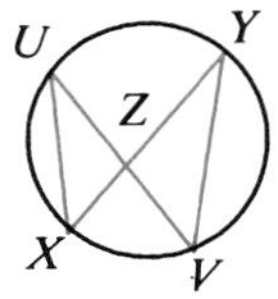

Exercises 9–11 prove the three possible cases of Theorem 7–8. In each case you are given chord $\overline{TA}$ and tangent $\overline{TP}$ of $\odot O$.

9. Supply reasons for the key steps of the proof that $m\angle ATP = \frac{1}{2}m\overset{\frown}{ANT}$ in Case I.

   Case I: O lies on $\angle ATP$.

   1. $\overline{TP} \perp \overline{TA}$ and $m\angle ATP = 90$.
   2. $\overset{\frown}{ANT}$ is a semicircle and $m\overset{\frown}{ANT} = 180$.
   3. $\frac{1}{2}m\overset{\frown}{ANT} = 90$
   4. $m\angle ATP = \frac{1}{2}m\overset{\frown}{ANT}$

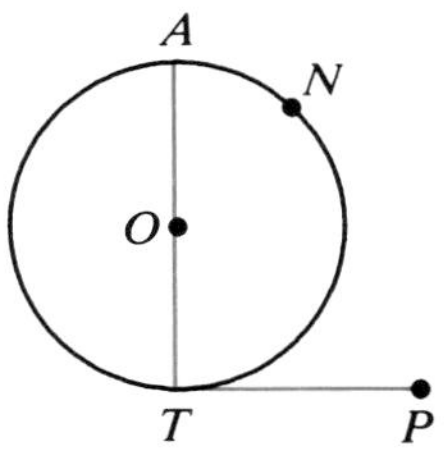

In Case II and Case III, $\overline{AT}$ is not a diameter. You can draw diameter $\overline{TZ}$ and then use Case I, Theorem 7-7, and the Angle Addition and Arc Addition Postulates.

B **10.** Prove $m\angle ATP = \frac{1}{2}m\widehat{ANT}$ in Case II.

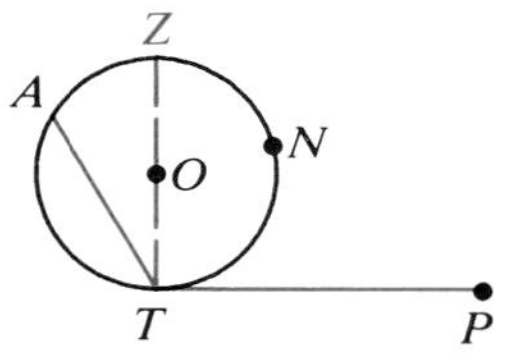

Case II. O lies inside $\angle ATP$.

11. Prove $m\angle ATP = \frac{1}{2}m\widehat{ANT}$ in Case III.

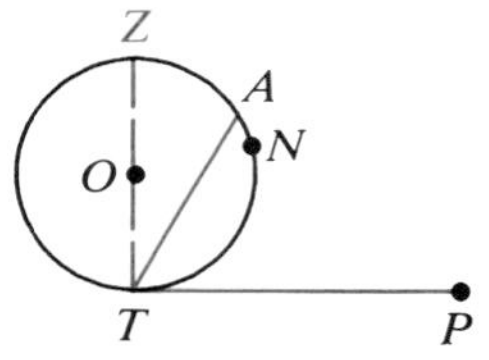

Case III. O lies outside $\angle ATP$.

12. Points A, B, C, D, and E are five consecutive vertices of a regular inscribed 15-gon. Chord $\overline{BE}$ is drawn. $m\angle ABE =$ __?__

In Exercises 13 and 14, quadrilateral *FGHJ* is inscribed in a circle. Give numerical answers.

13. $m\angle F = x$, $m\angle G = x$, and $m\angle H = x + 20$. $m\angle J =$ __?__

14. $m\angle F = x^2$, $m\angle G = 9x - 2$, $m\angle H = 11x$, and $m\angle J = x^2 + 20$. The measure of the largest angle of the quadrilateral is __?__.

15. The diagram at the right shows a regular polygon with 7 sides.
 a. Explain why the numbered angles are all congruent. (*Hint:* You may assume that a circle can be circumscribed about any regular polygon.)
 b. Will your reasoning apply to a regular polygon with any number of sides?

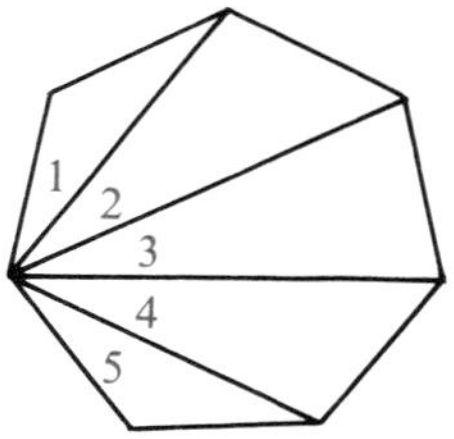

C **16.** Given: Vertices A, B, and C of quadrilateral $ABCD$ lie on $\odot O$;
 $m\angle A + m\angle C = 180$; $m\angle B + m\angle D = 180$.
 Prove: D lies on $\odot O$.
 (*Hint:* Use an indirect proof. Assume temporarily that D is not on $\odot O$. You must then treat two cases: (1) D is inside $\odot O$, and (2) D is outside $\odot O$. In each case let X be the point where $\overrightarrow{AD}$ intersects $\odot O$ and draw $\overline{CX}$. Show that what you can conclude about $\angle AXC$ contradicts the given information.)

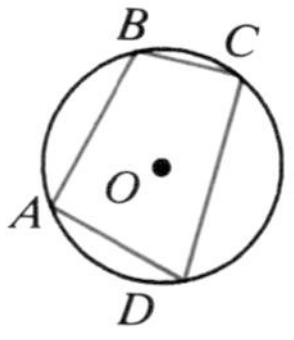

17. Given: $\overline{PT}$ is a tangent; $\overline{TU} \parallel \overline{PS}$.
 Find three similar triangles and prove them similar. Write a paragraph proof.

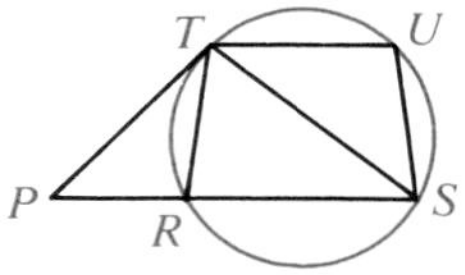

18. Circle I is inscribed in $\triangle FGH$ and $\odot O$ is circumscribed about $\triangle FGH$. $\overrightarrow{FI}$ intersects $\odot O$ in a point K. Discover and prove a relationship between $\overline{KG}$, $\overline{KH}$, and $\overline{KI}$.

★ **19.** Angle C of $\triangle ABC$ is a right angle. The sides of the triangle have the lengths shown. The smallest circle (not shown) through C that is tangent to $\overline{AB}$ intersects $\overline{AC}$ at J and $\overline{BC}$ at K. Express the distance JK in terms of a, b, and c.

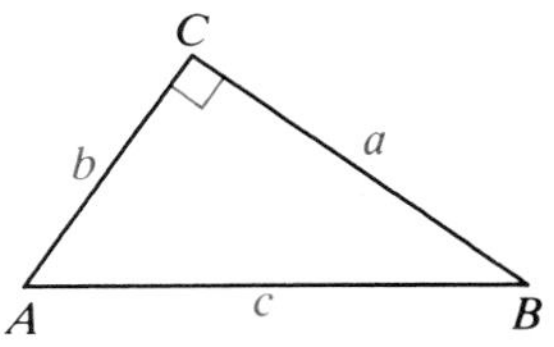

★ **20.** Prove: When a circle is circumscribed about an equilateral triangle, and chords are drawn from any point on the circle to the three vertices of the triangle, then the length of the longest chord is equal to the sum of the lengths of the other two chords.

★ **21.** Prove: The product of the lengths of the diagonals of an inscribed quadrilateral is equal to the sum of the products of the lengths of the opposite sides. (*Hint:* If the quadrilateral is $ABCD$, draw a segment, intersecting $\overline{AC}$ at X, such that $\angle ADX \cong \angle BDC$.)

7-6 Other Angles

The preceding section dealt with angles that have their vertices on a circle. Theorem 7-9 deals with the angle formed by two chords that intersect inside a circle. Such an angle and its vertical angle intercept two arcs.

Theorem 7-9

The measure of an angle formed by two chords that intersect inside a circle is equal to half the sum of the measures of the intercepted arcs.

Given: Chords $\overline{AB}$ and $\overline{CD}$ intersect inside a circle.

Prove: $m\angle 1 = \frac{1}{2}(m\widehat{AC} + m\widehat{BD})$

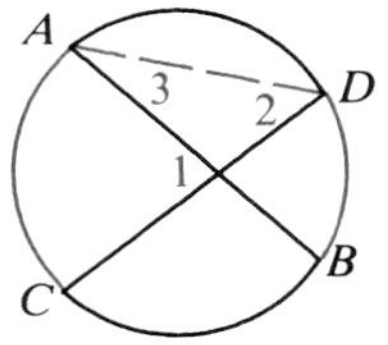

Proof:

Statements	Reasons
1. Draw chord $\overline{AD}$.	1. Through any two points there is exactly one line.
2. $m\angle 1 = m\angle 2 + m\angle 3$	2. The measure of an exterior $\angle$ of a $\triangle$ = the sum of the measures of the two remote interior $\angle$s.
3. $m\angle 2 = \frac{1}{2}m\widehat{AC}$; $m\angle 3 = \frac{1}{2}m\widehat{BD}$	3. The measure of an inscribed angle is equal to half the measure of its intercepted arc.
4. $m\angle 1 = \frac{1}{2}m\widehat{AC} + \frac{1}{2}m\widehat{BD}$, or $m\angle 1 = \frac{1}{2}(m\widehat{AC} + m\widehat{BD})$	4. Substitution (Step 3 in Step 2)

One case of the next theorem will be proved in Classroom Exercise 10, the other two cases in Exercises 22 and 23.

Theorem 7–10

The measure of an angle formed by two secants, two tangents, or a secant and a tangent drawn from a point outside a circle is equal to half the difference of the measures of the intercepted arcs.

Case I: Two secants

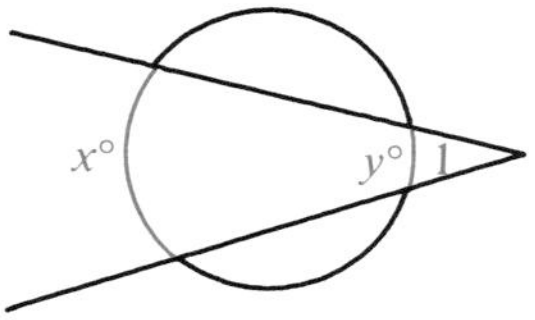

$m\angle 1 = \frac{1}{2}(x - y)$

Case II: Two tangents

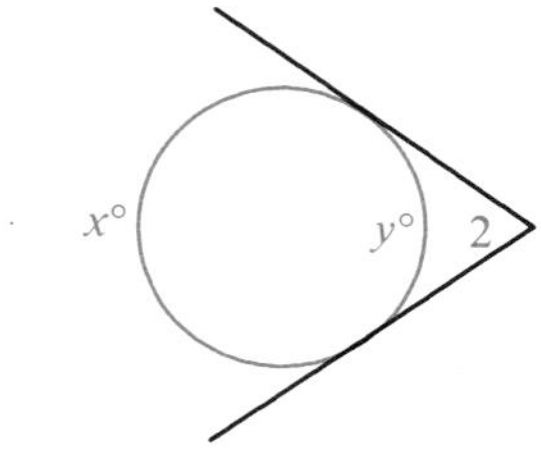

$m\angle 2 = \frac{1}{2}(x - y)$

Case III: A secant and a tangent

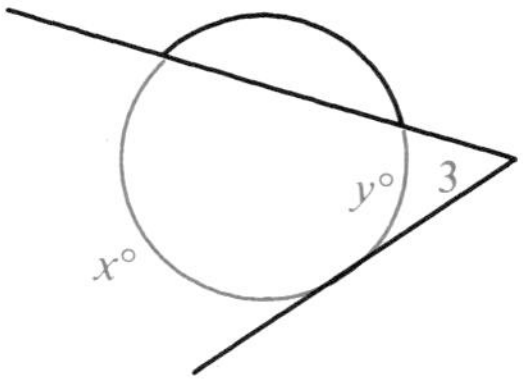

$m\angle 3 = \frac{1}{2}(x - y)$

Example 1 $m\widehat{CE} = 70$, $m\widehat{FD} = 50$, and $m\widehat{DB} = 26$. Find the measures of $\angle 1$ and $\angle 2$.

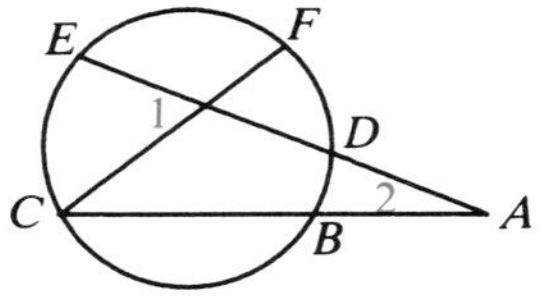

Solution

$m\angle 1 = \frac{1}{2}(m\widehat{CE} + m\widehat{FD})$ (Theorem 7–9)
$m\angle 1 = \frac{1}{2}(70 + 50) = 60$
$m\angle 2 = \frac{1}{2}(m\widehat{CE} - m\widehat{BD})$ (Theorem 7–10)
$m\angle 2 = \frac{1}{2}(70 - 26) = 22$

Example 2 $\overline{BA}$ is a tangent. Find $m\widehat{BD}$ and $m\widehat{BC}$.

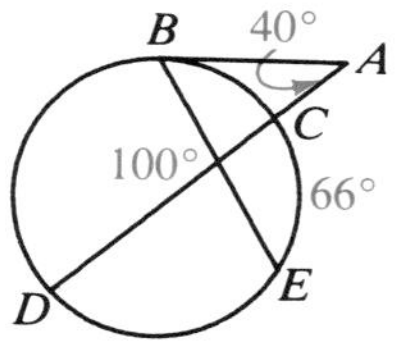

Solution

$100 = \frac{1}{2}(m\widehat{BD} + 66)$ (Theorem 7–9)
$200 = m\widehat{BD} + 66$, so $m\widehat{BD} = 134$
$40 = \frac{1}{2}(m\widehat{BD} - m\widehat{BC})$ (Theorem 7–10)
$80 = 134 - m\widehat{BC}$, so $m\widehat{BC} = 54$

Classroom Exercises

Find the measure of each numbered angle.

1.

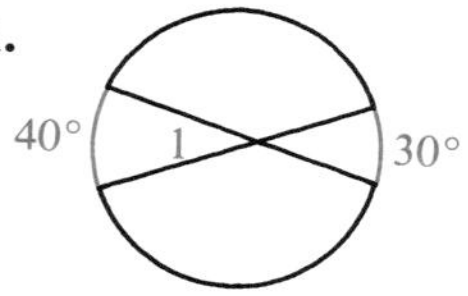

2.

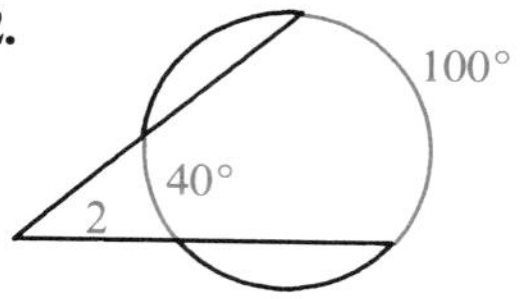

3.

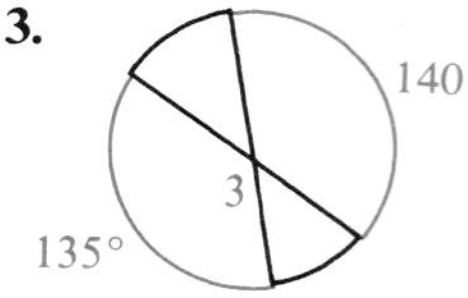

4.

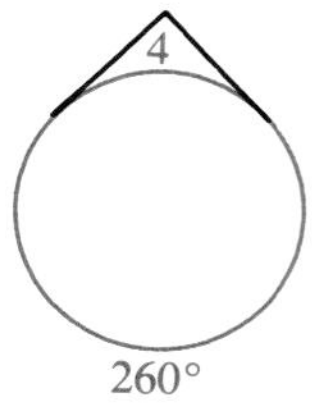

5.

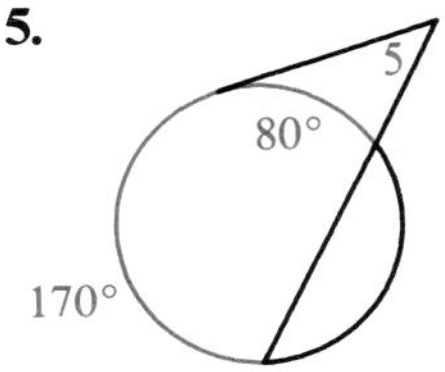

6.

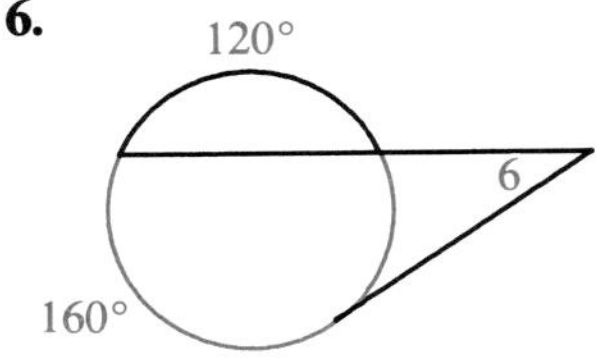

State an equation you can use to find x. Then find the value of x.

7.

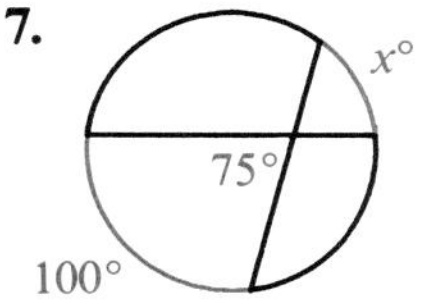

8.

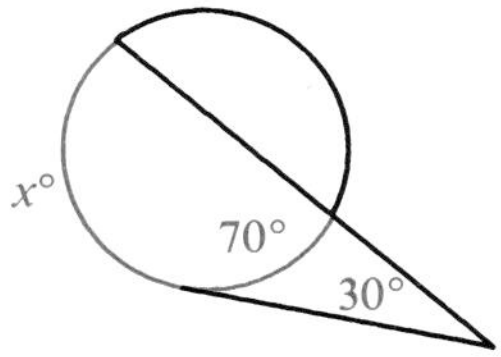

9.

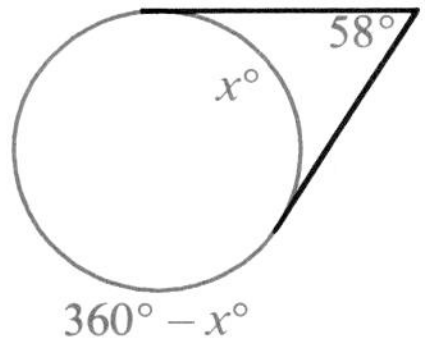

10. Supply reasons to complete a proof of the first case of Theorem 7–10.

Given: Secants $\overline{PA}$ and $\overline{PC}$

Prove: $m\angle 1 = \frac{1}{2}(m\widehat{AC} - m\widehat{BD})$

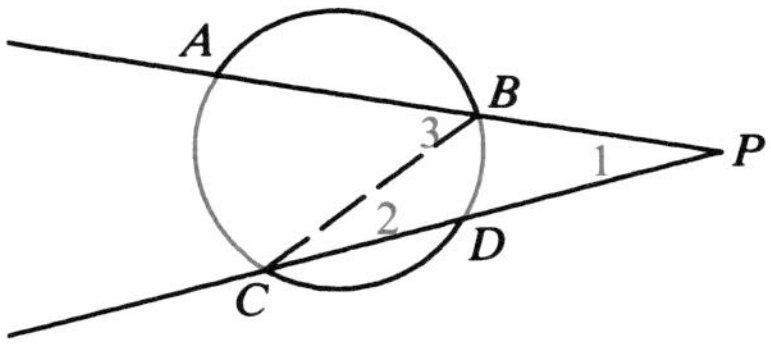

Proof:

1. Draw chord $\overline{BC}$.
2. $m\angle 1 + m\angle 2 = m\angle 3$
3. $m\angle 1 = m\angle 3 - m\angle 2$
4. $m\angle 3 = \frac{1}{2}m\widehat{AC}$; $m\angle 2 = \frac{1}{2}m\widehat{BD}$
5. $m\angle 1 = \frac{1}{2}m\widehat{AC} - \frac{1}{2}m\widehat{BD}$, or
   $m\angle 1 = \frac{1}{2}(m\widehat{AC} - m\widehat{BD})$

Written Exercises

A **1–10.** $\overleftrightarrow{BZ}$ is tangent to $\odot O$; $\overline{AC}$ is a diameter; $m\widehat{BC} = 90$; $m\widehat{CD} = 30$; $m\widehat{DE} = 20$
Draw your own large diagram so that you can write arc measures alongside the arcs. Find the measure of each numbered angle.

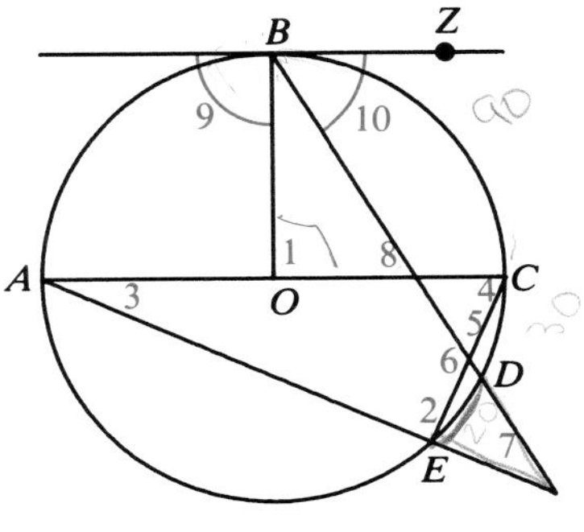

11. If $m\widehat{RT} = 80$ and $m\widehat{US} = 40$, then $m\angle 1 = \underline{\ ?\ }$.

12. If $m\widehat{RU} = 130$ and $m\widehat{TS} = 100$, then $m\angle 1 = \underline{\ ?\ }$.

13. If $m\angle 1 = 50$ and $m\widehat{RT} = 70$, then $m\widehat{US} = \underline{\ ?\ }$.

14. If $m\angle 1 = 52$ and $m\widehat{US} = 36$, then $m\widehat{RT} = \underline{\ ?\ }$.

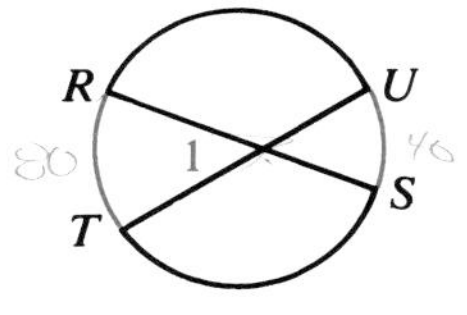

In Exercises 15–17, $\overline{PX}$ and $\overline{PY}$ are tangents.

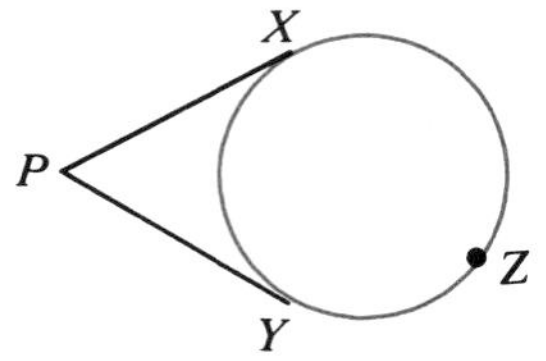

15. If $m\widehat{XZY} = 250$, then $m\angle P =$ __?__.
16. If $m\widehat{XY} = 90$, then $m\angle P =$ __?__.
17. If $m\angle P = 85$, then $m\widehat{XY} =$ __?__.

In Exercises 18–20, $\overline{AT}$ is a tangent.

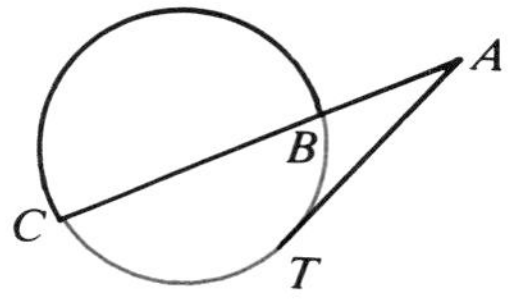

18. If $m\widehat{CT} = 110$ and $m\widehat{BT} = 50$, then $m\angle A =$ __?__.
19. If $m\angle A = 40$ and $m\widehat{BT} = 40$, then $m\widehat{CT} =$ __?__.
20. If $m\angle A = 35$ and $m\widehat{CT} = 110$, then $m\widehat{BT} =$ __?__.

B 21. A quadrilateral circumscribed about a circle has angles of 80°, 90°, 94°, and 96°. Find the measures of the four nonoverlapping arcs determined by the points of tangency.

22. Prove Case II of Theorem 7-10. (*Hint:* See Classroom Exercise 10. In a figure like the second one shown below the theorem on page 318, draw the chord joining the points of tangency.)

23. Prove Case III of Theorem 7-10.

24. Write an equation involving a, b, and c.

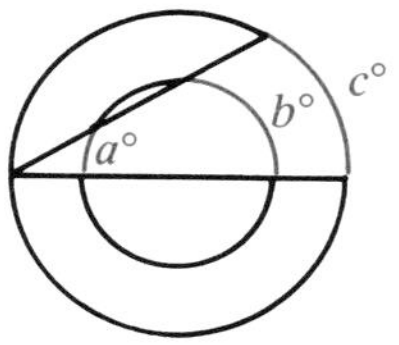

25. Find the ratio $x:y$.

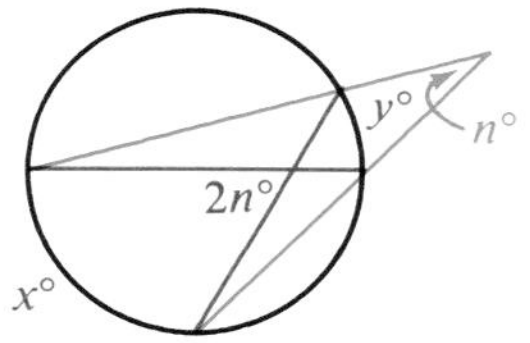

C 26. $\overline{PT}$ is a tangent. It is known that $80 < m\widehat{RS} < m\widehat{ST} < 90$. State as much as you can about the measure of $\angle P$.

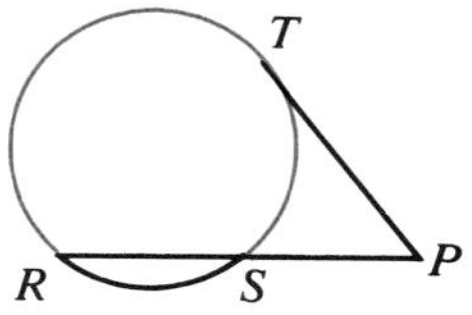

27. $\overline{AC}$ and $\overline{AE}$ are secants of $\odot O$. It is given that $\overline{AB} \cong \overline{OB}$. Discover and prove a relation between the measures of $\widehat{CE}$ and $\widehat{BD}$.

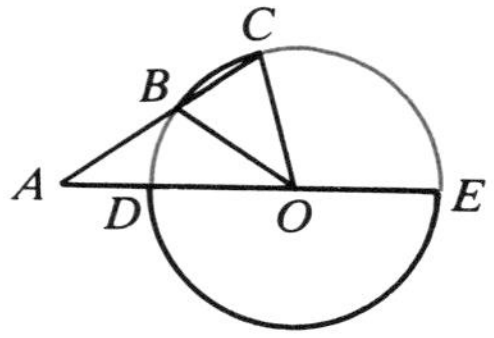

28. Take any point P outside a circle. Draw a tangent segment $\overline{PT}$ and a secant $\overline{PBA}$ with A and B points on the circle. Take K on $\overline{PA}$ so that $PK = PT$. Draw $\overrightarrow{TK}$. Let the intersection of $\overrightarrow{TK}$ with the circle be point X. Discover and prove a relationship between $\widehat{AX}$ and $\widehat{XB}$.

7-7 Circles and Lengths of Segments

You can use similar triangles to prove that lengths of chords, secants, and tangents are related in interesting ways.

Theorem 7-11

When two chords intersect inside a circle, the product of the lengths of the segments of one chord equals the product of the lengths of the segments of the other chord.

Given: $\overline{AB}$ and $\overline{CD}$ intersect in P.

Prove: $r \cdot s = t \cdot u$

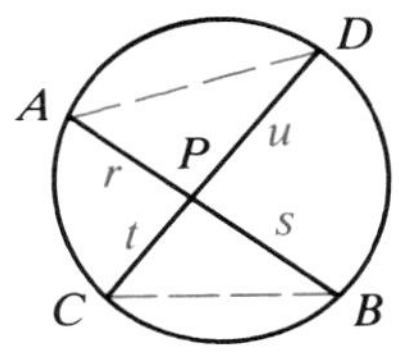

Proof:

Statements	Reasons
1. Draw chords $\overline{AD}$ and $\overline{CB}$.	1. Through any two points there is exactly one line.
2. $\angle A \cong \angle C$; $\angle D \cong \angle B$	2. If two inscribed angles intercept $\underline{\ ?\ }$.
3. $\triangle APD \sim \triangle CPB$	3. Why?
4. $\frac{r}{t} = \frac{u}{s}$	4. Why?
5. $r \cdot s = t \cdot u$	5. A property of proportions

For a proof of the following theorem, see Classroom Exercise 7. In the diagram for the theorem, $\overline{AP}$ and $\overline{CP}$ are *secant segments.* $\overline{BP}$ and $\overline{DP}$ are exterior to the circle and are referred to as *external segments.*

Theorem 7-12

When two secant segments are drawn to a circle from an external point, the product of the lengths of one secant segment and its external segment equals the product of the lengths of the other secant segment and its external segment.

Given: $\overline{PA}$ and $\overline{PC}$ drawn to the circle from point P

Prove: $r \cdot s = t \cdot u$

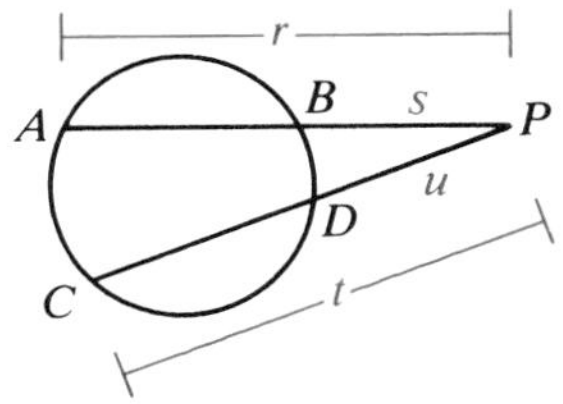

Study the diagrams below from left to right. As $\overline{PC}$ approaches a position of tangency, C and D move closer together until they merge, $\overline{PC}$ is a tangent, and $t = u$.

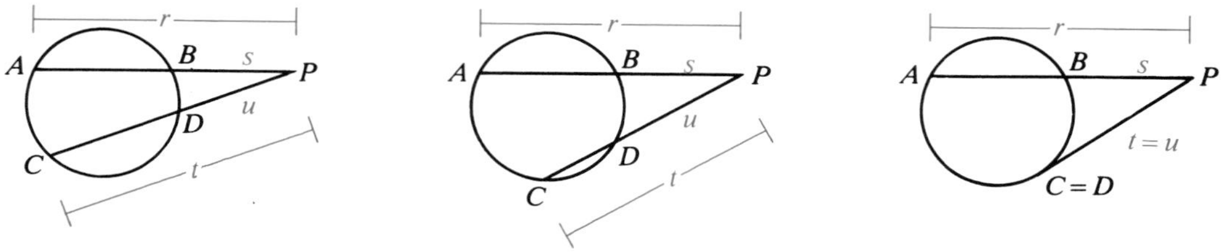

In the first two diagrams we know that $r \cdot s = t \cdot u$. In the third figure, u and t both become equal to the length of the tangent segment, and the equation becomes $r \cdot s = t^2$. This result, stated below, will be proved in Exercise 10.

Theorem 7-13

When a secant segment and a tangent segment are drawn to a circle from an external point, the product of the lengths of the secant segment and its external segment is equal to the square of the length of the tangent segment.

Example 1 Find the value of x.

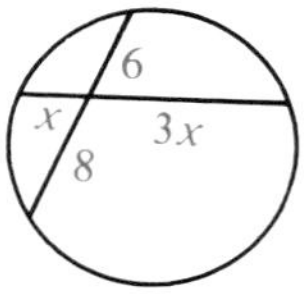

Solution $3x \cdot x = 6 \cdot 8$ (Theorem 7-11)
$3x^2 = 48$, $x^2 = 16$, and $x = 4$

Example 2 Find the values of x and y.

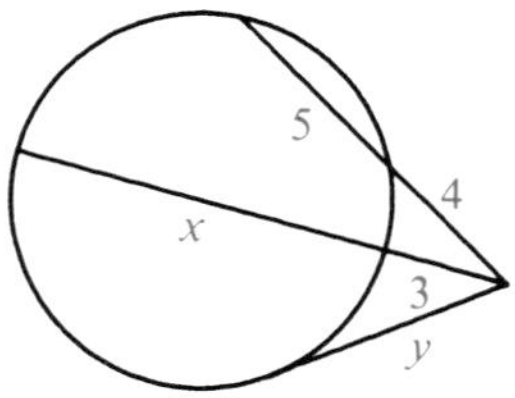

Solution $4 \cdot 9 = 3(3 + x)$ (Theorem 7-12)
$36 = 3(3 + x)$, $12 = 3 + x$, and $x = 9$
$4 \cdot 9 = y^2$ (Theorem 7-13)
$36 = y^2$, so $y = 6$

Classroom Exercises

Chords, secants, and tangents are shown. State the equation you would use to find x. Then solve for x.

1.

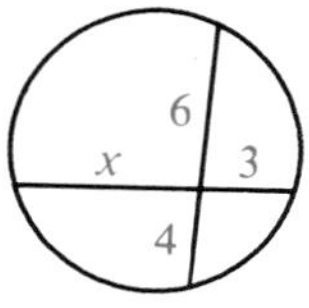

2.

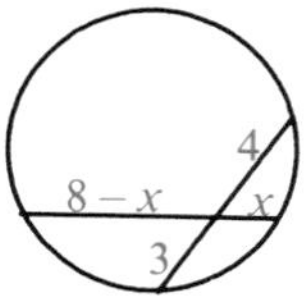

3.

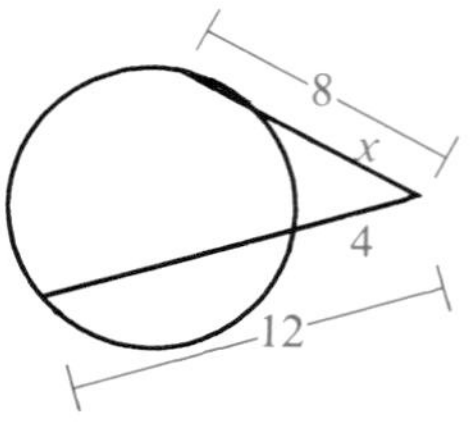

4.

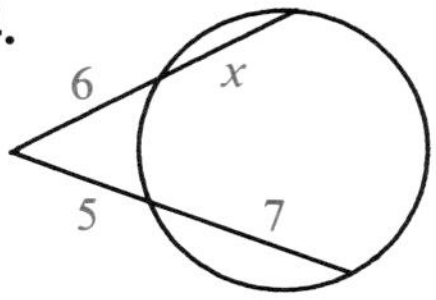

5.

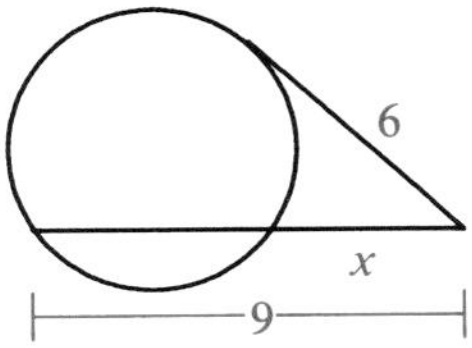

6.

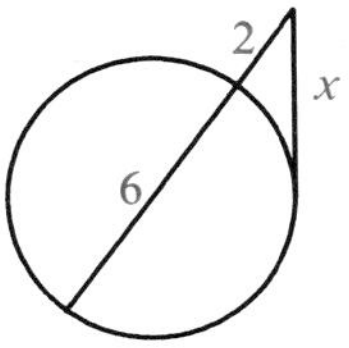

7. Supply reasons to complete the proof of Theorem 7–12.
Given: $\overline{PA}$ and $\overline{PC}$ drawn to the circle from point P
Prove: $r \cdot s = t \cdot u$

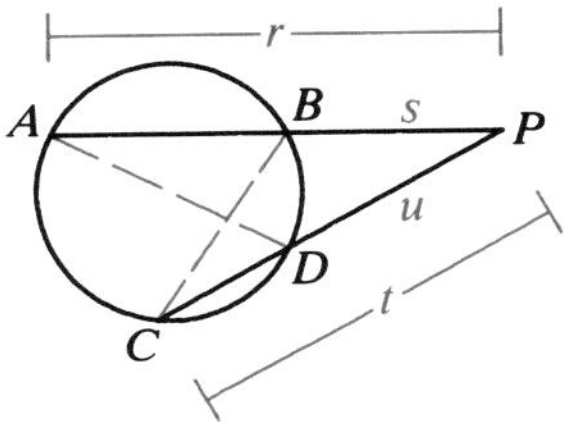

Proof:

1. Draw chords $\overline{AD}$ and $\overline{BC}$.
2. $\angle A \cong \angle C$
3. $\angle P \cong \angle P$
4. $\triangle APD \sim \triangle CPB$
5. $\frac{r}{t} = \frac{u}{s}$
6. $r \cdot s = t \cdot u$

Written Exercises

Chords, secants, and tangents are shown. Find the value of *x*.

A **1.**

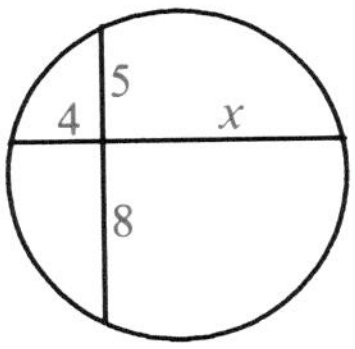

2.

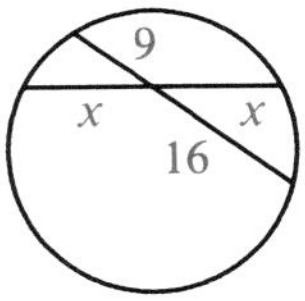

3.

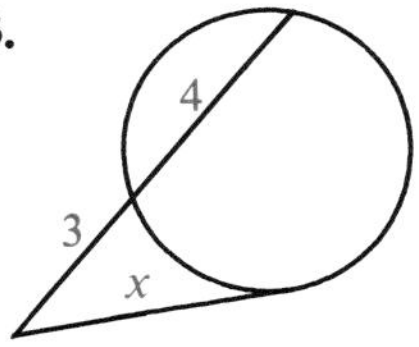

4.

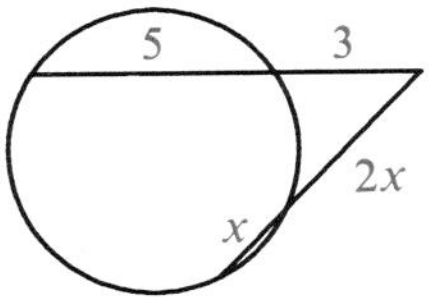

5.

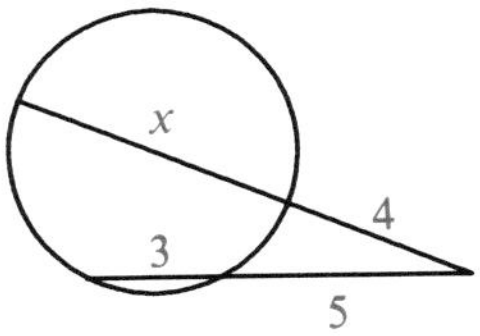

6.

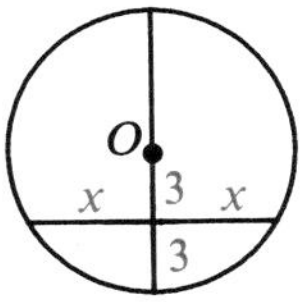

7.

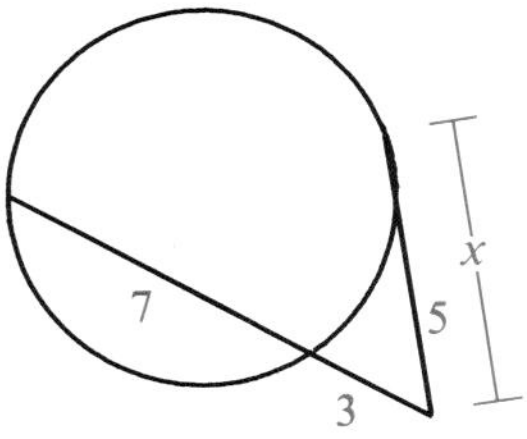

8.

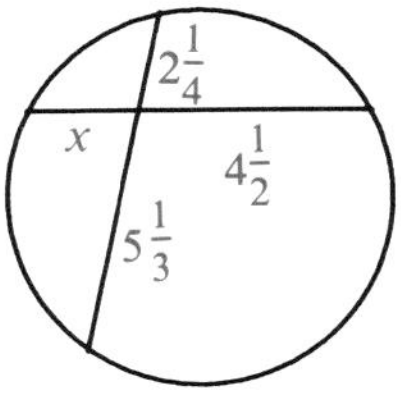

9.

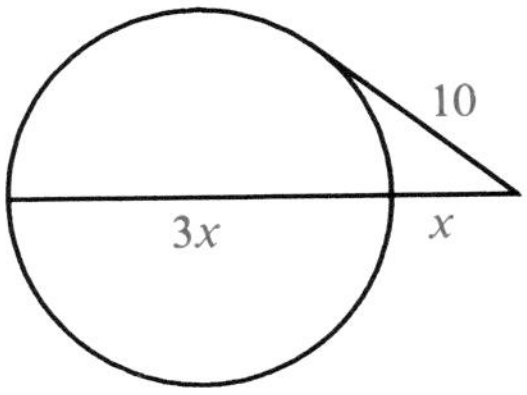

10. Copy and complete the proof of Theorem 7–13.

Given: Secant segment $\overline{PA}$ and tangent segment $\overline{PC}$ drawn to the circle from P.

Prove: $r \cdot s = t^2$

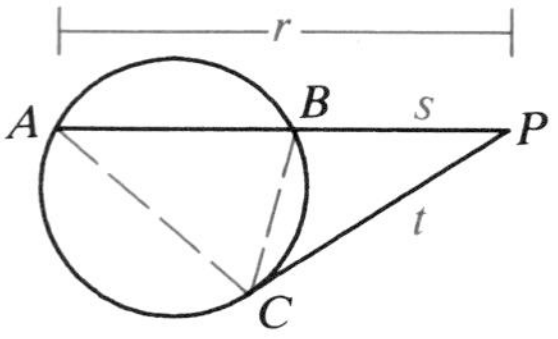

Proof:

Statements	Reasons
1. Draw chords $\overline{AC}$ and $\overline{BC}$.	1. __?__
2. $m\angle A = \frac{1}{2}m\widehat{BC}$	2. __?__
3. $m\angle BCP = \frac{1}{2}m\widehat{BC}$	3. The measure of an angle formed by a chord and a tangent __?__.
4. $\angle A \cong \angle BCP$	4. __?__
5. $\angle P \cong \angle P$	5. __?__

(*Hint:* You need three more steps. Apply similar triangles as in Classroom Exercise 7.)

B **11.** Given: $\odot O$ and $\odot P$ are tangent at T.
Prove: $UV \cdot UW = UX \cdot UY$

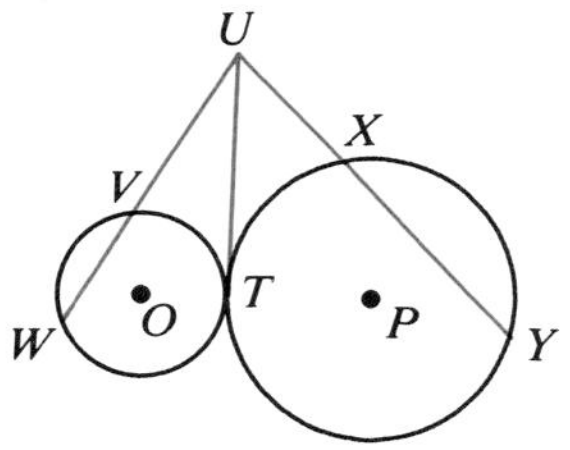

12. Given: $\overline{AB}$ is tangent to $\odot Q$;
$\overline{AC}$ is tangent to $\odot S$.
Prove: $\overline{AB} \cong \overline{AC}$

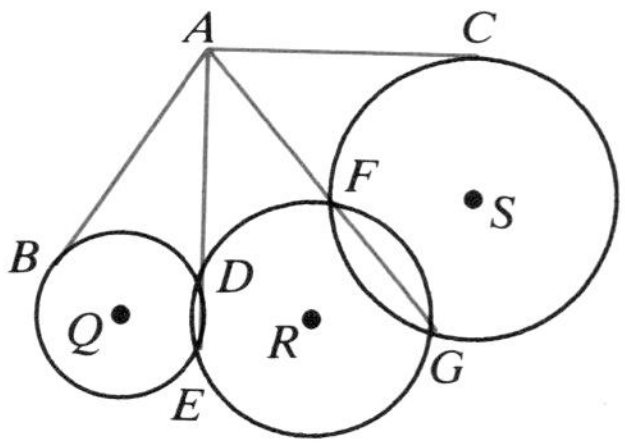

Chords $\overline{AB}$ and $\overline{CD}$ intersect at P. Find the lengths indicated.

Example: $AP = 5$; $BP = 4$; $CD = 12$; $CP =$ __?__

Solution: Let $CP = x$. Then $DP = 12 - x$.

$x(12 - x) = 5 \cdot 4$
$12x - x^2 = 20$
$x^2 - 12x + 20 = 0$
$(x - 2)(x - 10) = 0$
$x = 2$ or $x = 10$
$CP = 2$ or 10

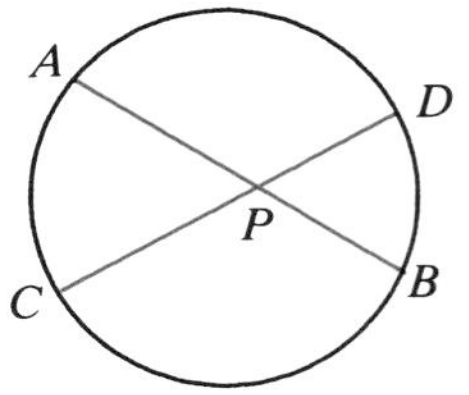

13. $AP = 6$; $BP = 8$; $CD = 16$; $DP =$ __?__

14. $CD = 10$; $CP = 6$; $AB = 11$; $AP =$ __?__

15. $AB = 12$; $CP = 9$; $DP = 4$; $BP =$ __?__

16. $AP = 6$; $BP = 5$; $CP = 3 \cdot DP$; $DP =$ __?__

$\overline{PT}$ is tangent to the circle. Find the lengths indicated.

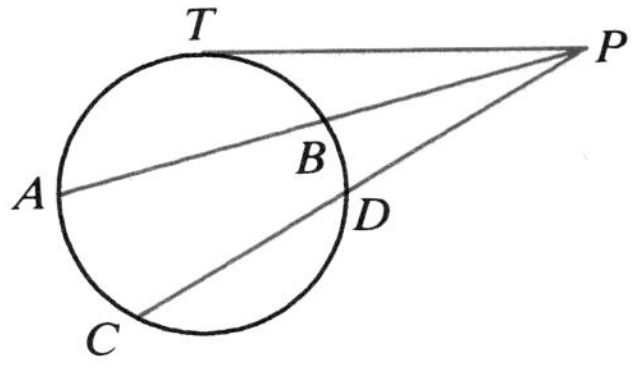

17. $PT = 6$; $PB = 3$; $AB =$ ___?___

18. $PT = 12$; $CD = 18$; $PC =$ ___?___

19. $PD = 5$; $CD = 7$; $AB = 11$; $PB =$ ___?___

20. $PB = AB = 5$; $PD = 4$; $PT =$ ___?___ and $PC =$ ___?___

21. A circle can be drawn through points X, Y, and Z.

a. What is the radius of the circle?

b. How far is the center of the circle from point W?

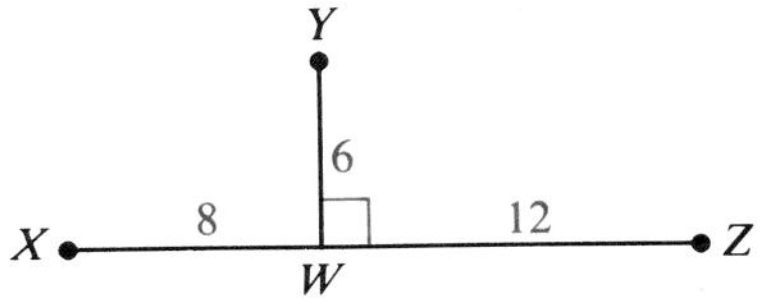

C **22.** $\overline{PT}$ is tangent to $\odot O$ and $\overline{PN}$ intersects $\odot O$ at J. Find the radius of the circle.

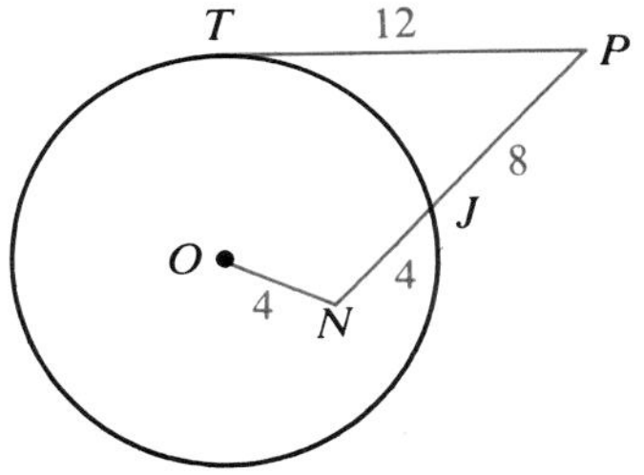

Ex. 22

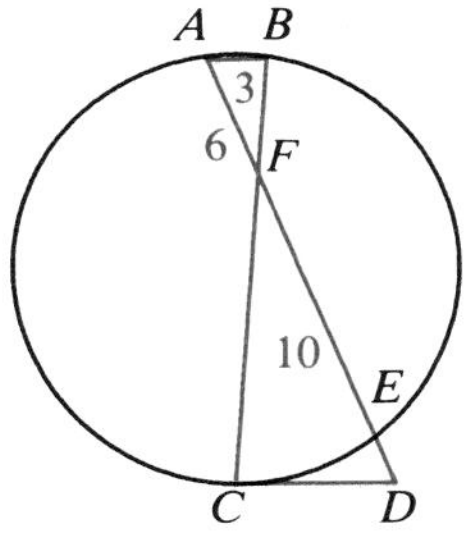

Ex. 23

★**23.** In the diagram at the right above, $\overline{CD}$ is a tangent, $\widehat{AC} \cong \widehat{BC}$, $AB = 3$, $AF = 6$, and $FE = 10$. Find ED.

Application

DISTANCE TO THE HORIZON

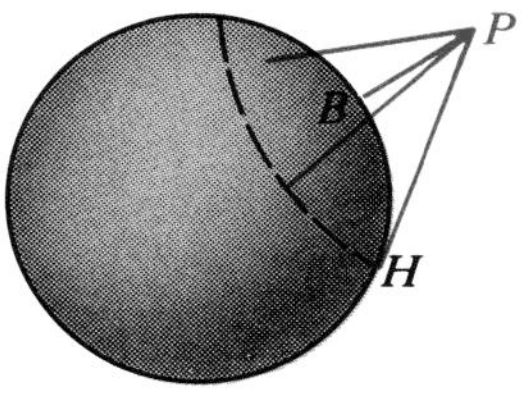

If you look out over the surface of the Earth from a position at P, directly above point B on the surface, you see the horizon wherever your line of sight is tangent to the surface of the Earth. If the surface around B is smooth (say you are on the ocean on a calm day), the horizon will be a circle, and the higher your lookout is, the farther away this horizon circle will be.

You can use Theorem 7–13 to derive a formula that tells how far you can see from any given height. As shown on the following page, the picture is simpler if you imagine a section through the Earth containing P, H, and O, the center of the Earth.

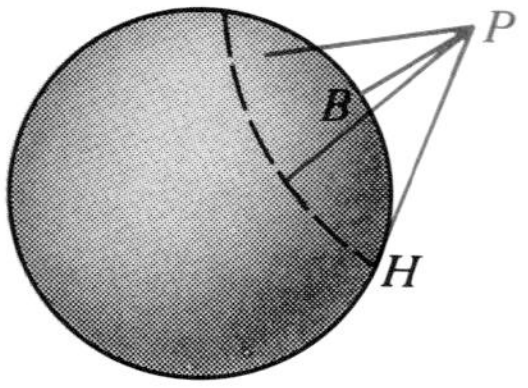

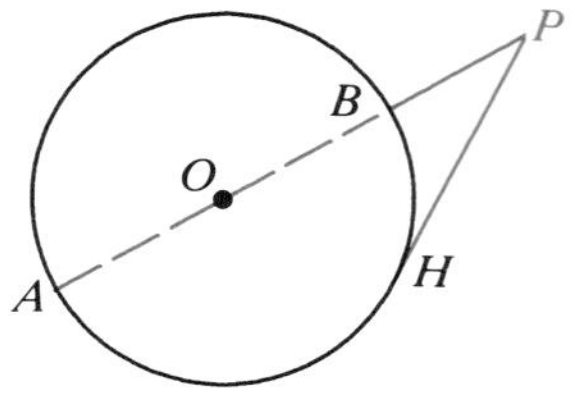

In the diagram at the right above, $\overline{PH}$ is tangent to circle O at H. $\overline{PA}$ is a secant passing through the center O. By Theorem 7–13:

$$(PH)^2 = AP \cdot BP$$

PH is the distance from the observer to the horizon, and BP is the observer's height above the surface of the Earth. If this height is small compared to the diameter, AB, of the Earth, then $AP \approx AB$. Using 12,800,000 m for AB, you can rewrite the formula as:

$$(\text{distance})^2 \approx (12{,}800{,}000)(\text{height})$$

Taking square roots, you get:

$$\text{distance} \approx \sqrt{12{,}800{,}000} \cdot \sqrt{\text{height}} \approx 3600\sqrt{\text{height}}$$

So the approximate distance (in meters) to the horizon is 3600 times the square root of your height (in meters) above the surface of the Earth. If your height is less than 400 km, the error in this approximation will be less than one percent.

Exercises

In Exercises 1 and 2 give your answer to the nearest kilometer, in Exercises 3 and 5 to the nearest 10 km, and in Exercise 4 to the nearest meter.

1. If you stand on a dune with your eyes about 16 m above sea level, how far out to sea can you look?
2. A lookout climbs high in the rigging of a sailing ship to a point 36 m above the water line. About how far away is the horizon?
3. From a balloon floating 10 km above the ocean, how far away is the farthest point you can see on the Earth's surface?
4. If you want to have a horizon of 8 km, how high a lookout must you have?
5. You are approaching the coast of Japan in a small sailboat. The highest point on the central island of Honshu is the cone of Mount Fuji, 3776 m above sea level. Roughly how far away from the mountain will you be when you can first see the top? (Assume that the sky is clear!)

Self-Test 2

$\overleftrightarrow{MD}$ is tangent to the circle.

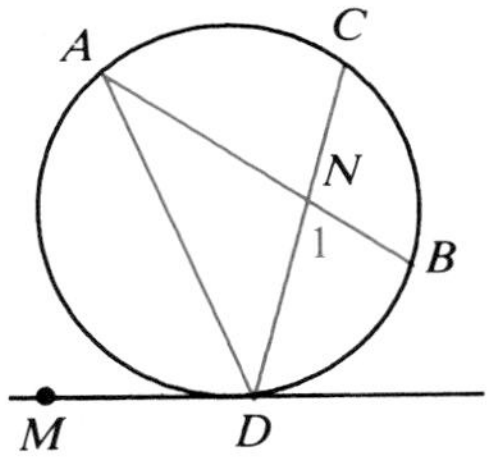

1. If $m\widehat{BD} = 80$, then $m\angle A =$ __?__.

2. If $m\angle ADM = 75$, then $m\widehat{AD} =$ __?__.

3. If $m\widehat{BD} = 80$ and $m\angle 1 = 81$, then $m\widehat{AC} =$ __?__.

4. If $AN = 12$, $BN = 6$, and $CN = 8$, then $DN =$ __?__.

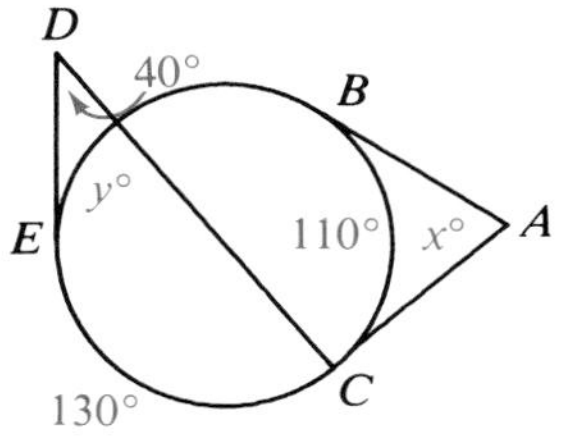

5. $\overline{AB}$, $\overline{AC}$, and $\overline{DE}$ are tangents. Find the values of x and y.

$\overline{PE}$ and $\overline{PF}$ are secants.

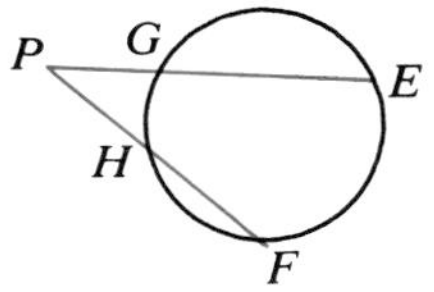

6. If $m\widehat{EF} = 100$ and $m\widehat{GH} = 30$, then $m\angle P =$ __?__.

7. If $PG = 4$, $PE = 15$, and $PH = 6$, then $PF =$ __?__.

8. If $PG = 8$, $GE = 12$, and $HF = 6$, then $PH =$ __?__.

9. Given: Chords as shown.
Prove: $\dfrac{AK}{CK} = \dfrac{AD}{CB}$

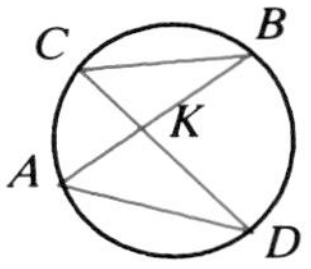

CALCULATOR KEY-IN

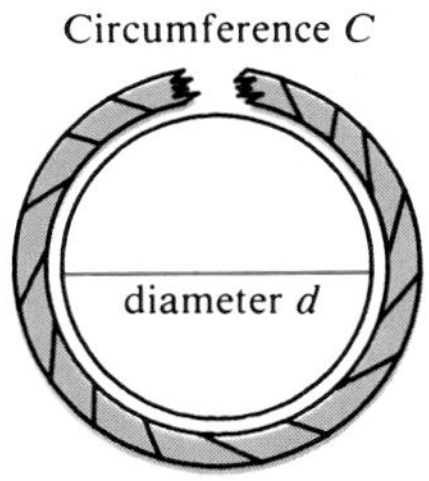

The ratio of the circumference of a circle to the diameter is a constant for all circles. The ratio is denoted by the Greek letter π (*pi*).

$$\pi = \frac{C}{d}$$

If you were to wrap a string around a circle to measure the circumference, C, and if you then measured the diameter, d, you would find that $\frac{C}{d} \approx 3$.

π is an irrational number that cannot be expressed exactly as the ratio of two integers. Decimal approximations of π have been computed to thousands of decimal places. (The value of π to 5 places is 3.14159.) We can easily look up values in reference books, but such was not always the case. In the past, mathematicians had to rely on their cleverness to compute an approximate value of π. One of the earliest approximations was that of Archimedes, who found that $3\frac{1}{7} > \pi > 3\frac{10}{71}$.

Exercises

1. Find decimal approximations of $3\frac{1}{7}$ and $3\frac{10}{71}$. Did Archimedes approximate π correct to hundredths?

In Exercises 2–5, find approximations for π. The more terms or factors you use, the better your approximations will be. But you can't use them all!

2. $\pi \approx 2\sqrt{3}\left(1 - \frac{1}{3 \cdot 3} + \frac{1}{3^2 \cdot 5} - \frac{1}{3^3 \cdot 7} + \frac{1}{3^4 \cdot 9} - \frac{1}{3^5 \cdot 11} + \cdots\right)$ (Sharpe, 18th century)

3. $\pi \approx 2 \cdot \frac{2}{1} \cdot \frac{2}{3} \cdot \frac{4}{3} \cdot \frac{4}{5} \cdot \frac{6}{5} \cdot \frac{6}{7} \cdot \frac{8}{7} \cdot \frac{8}{9} \cdot \cdots$ (Wallis, 17th century)

4. This exercise is for calculators that have a square root function and a memory.
$\pi \approx 2 \div \left(\sqrt{0.5} \cdot \sqrt{0.5 + 0.5\sqrt{0.5}} \cdot \sqrt{0.5 + 0.5\sqrt{0.5 + 0.5\sqrt{0.5}}} \cdot \cdots\right)$
(Vieta, 16th century)

5. $\pi \approx 4 \cdot \left(1 - \frac{1}{3} + \frac{1}{5} - \frac{1}{7} + \frac{1}{9} - \frac{1}{11} + \cdots\right)$ (Leibniz, 17th century)

Note: Although the expression here is simple in appearance, your approximations will move *very* slowly toward π. If you used one hundred terms within the parentheses, your approximation would not be correct to more than one decimal place.

Chapter Summary

1. If a line is tangent to a circle, then the line is perpendicular to the radius drawn to the point of tangency. The converse is also true.
2. Tangents to a circle from a point are congruent.
3. In the same circle or in congruent circles:
   - **a.** Congruent minor arcs have congruent central angles. Congruent central angles have congruent arcs.
   - **b.** Congruent chords have congruent arcs. Congruent arcs have congruent chords.
   - **c.** Congruent chords are equally distant from the center. Chords equally distant from the center are congruent.
4. A diameter that is perpendicular to a chord bisects the chord and its arc.
5. If two inscribed angles intercept the same arc, then the angles are congruent.
6. If a quadrilateral is inscribed in a circle, then opposite angles are supplementary.
7. An angle inscribed in a semicircle is a right angle.

8. Relationships expressed by formulas:

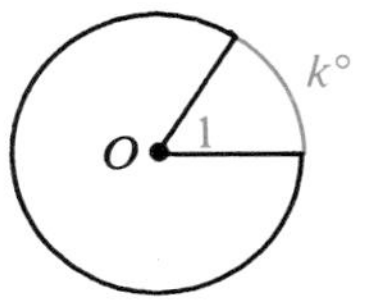

$m\angle 1 = k$

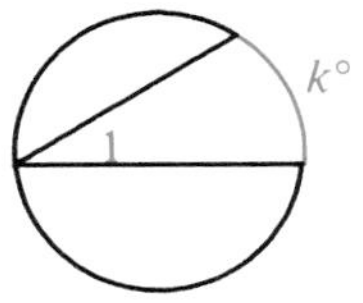

$m\angle 1 = \frac{1}{2}k$

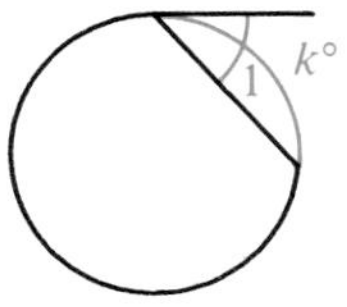

$m\angle 1 = \frac{1}{2}k$

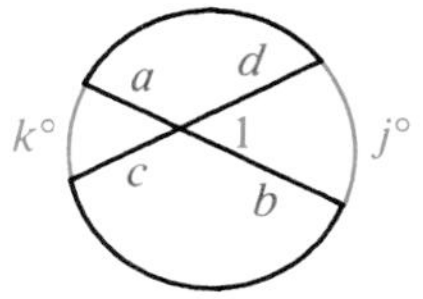

$m\angle 1 = \frac{1}{2}(k + j)$
$a \cdot b = c \cdot d$

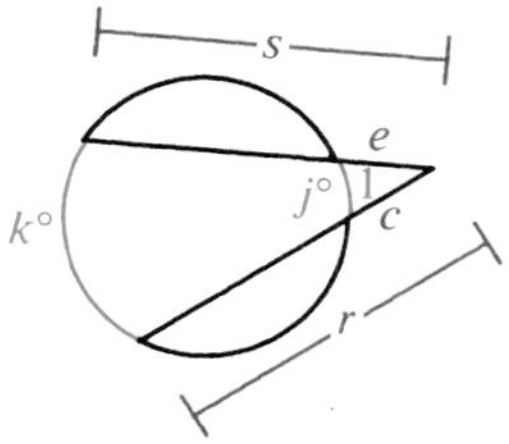

$m\angle 1 = \frac{1}{2}(k - j)$
$s \cdot e = r \cdot c$

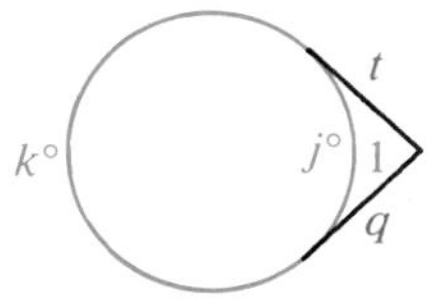

$m\angle 1 = \frac{1}{2}(k - j)$
$t = q$

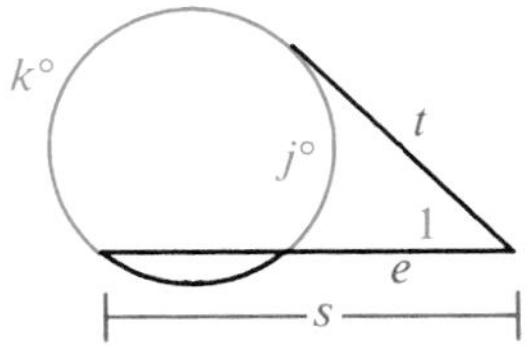

$m\angle 1 = \frac{1}{2}(k - j)$
$s \cdot e = t^2$

Chapter Review

Points A, B, and C lie on $\odot O$.

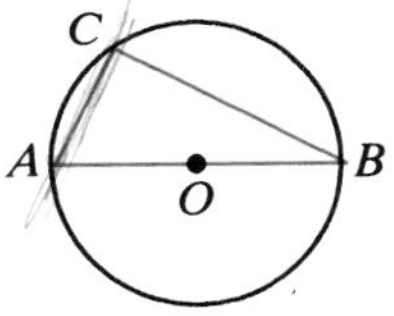

7–1

1. $\overline{AC}$ is called a __?__, while $\overleftrightarrow{AC}$ is called a __?__.
2. $\overline{OB}$ is called a __?__.
3. The best name for $\overline{AB}$ is __?__.
4. $\triangle ABC$ is ______?______ $\odot O$.
   (inscribed in/circumscribed about)

Lines $\overleftrightarrow{ZX}$ and $\overleftrightarrow{ZY}$ are tangent to $\odot P$.

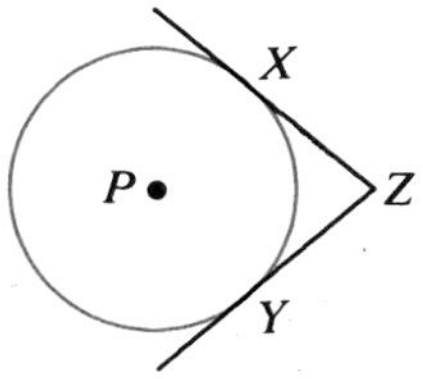

7–2

5. $\overline{PX}$, if drawn, would be __?__ to $\overleftrightarrow{XZ}$.
6. If the radius of $\odot P$ is 6 and if $XZ = 8$, the distance between points P and Z is __?__.
7. If $m\angle Z = 90$ and if $XZ = 13$, the distance between points X and Y is __?__.

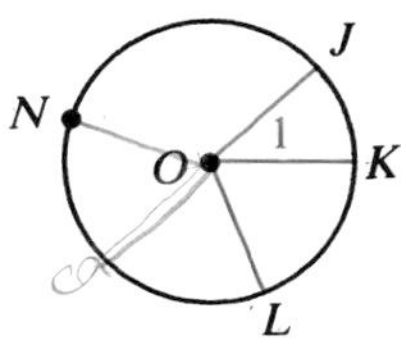

7–3

8. If $m\angle 1 = 42$, then $m\widehat{JK} =$ __?__.
9. If $m\widehat{JN} = 120$ and $m\widehat{NL} = 130$, then $m\angle JOL =$ __?__.
10. Suppose $\overrightarrow{JO}$ intersects $\odot O$ at G.
    a. $m\widehat{JLG} =$ __?__
    b. If $\angle NOG \cong \angle KOL$, then $\widehat{NG}$ __?__ $\widehat{KL}$.

In $\odot X$, $m\widehat{AC} = 120$.

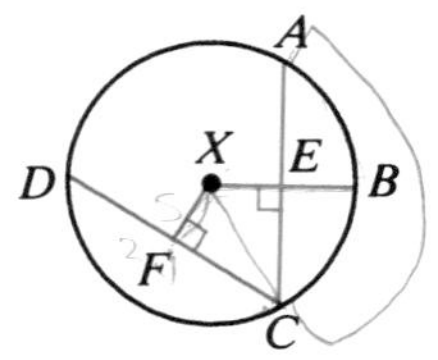

11. $m\widehat{AB} =$ _?_. 7–4
12. If $\overline{AC} \cong \overline{CD}$, then $m\widehat{CD} =$ _?_.
13. If $CD > AC$, then XF _?_ XE. ($</=/>$)
14. If $DC = 24$ and $XF = 5$, the radius of $\odot X =$ _?_.

$\overleftrightarrow{RS}$ **is tangent to the circle at** N.

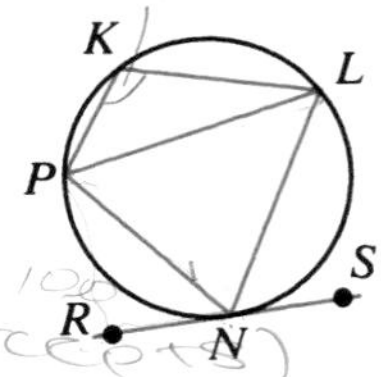

15. If $m\angle K = 105$, then $m\angle PNL =$ _?_. 7–5
16. If $m\widehat{PN} = 100$, then $m\angle PLN =$ _?_ and $m\angle PNR =$ _?_.
17. If $m\angle K = 110$, then $m\widehat{PNL} =$ _?_ and $m\widehat{PL} =$ _?_.

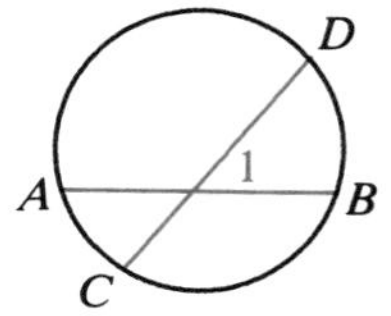

18. If $m\widehat{AC} = 40$ and $m\widehat{BD} = 60$, then $m\angle 1 =$ _?_. 7–6
19. If $m\widehat{AC} = 44$ and $m\angle 1 = 55$, then $m\widehat{BD} =$ _?_.

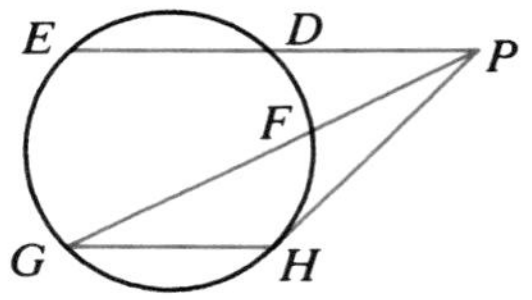

20. If $m\widehat{EG} = 100$ and $m\widehat{DF} = 40$, then $m\angle EPG =$ _?_.
21. If $\overline{PH}$ is a tangent, $m\widehat{GH} = 90$ and $m\angle GPH = 25$, then $m\widehat{FH} =$ _?_.

Chords, secants, and a tangent are shown. Find x.

22.

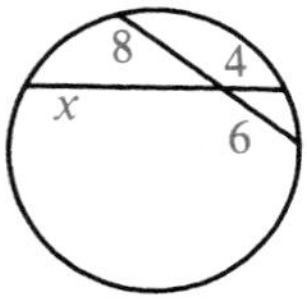

23.

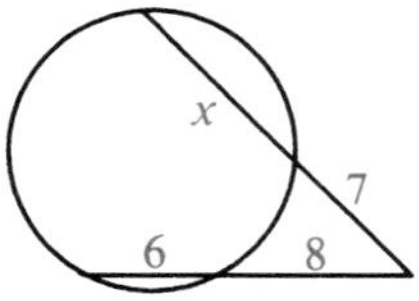

24. 7–7

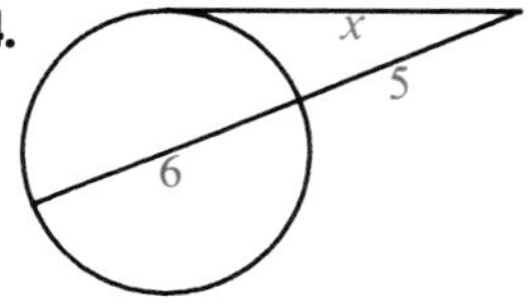

Chapter Test

Tell whether the statement is true or false.

1. It is possible to place points A, B, and C on a circle so that $m\widehat{AB} + m\widehat{BC} > m\widehat{AC}$.
2. If two circles are congruent, their diameters are congruent.
3. If a chord in one circle is congruent to a chord in another circle, the arcs of these chords must have congruent central angles.
4. Opposite angles of an inscribed quadrilateral must be congruent.

5. If a diameter is perpendicular to a chord, the diameter must bisect the chord.
6. If a line bisects a chord, that line must pass through the center of the circle.
7. If $\overrightarrow{GM}$ intersects a circle in just one point, $\overline{GM}$ must be tangent to the circle.
8. It is possible to draw two circles so that no common tangents can be drawn.
9. An angle inscribed in a semicircle must be a right angle.
10. When one chord is farther from the center of a circle than another chord, the chord farther from the center is the longer of the two chords.

11. In $\odot O$, if $m\widehat{AB} = 100°$, then $m\widehat{AC} = $ _?_.
12. If the radius of $\odot O$ is 17 and $AB = 30$, then $OE = $ _?_.

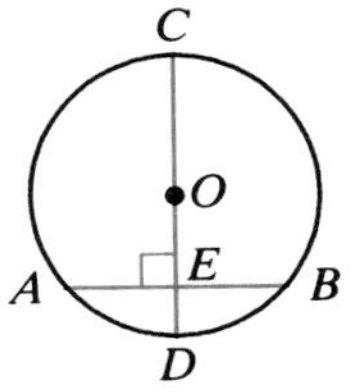

$\overline{AD}$ and $\overline{DB}$ are tangent to the circle.

13. If $AB = BC$ and $m\widehat{BC} = 80$, then $m\angle ABC = $ _?_.
14. If $m\angle D = 110$, then $m\angle BCA = $ _?_.

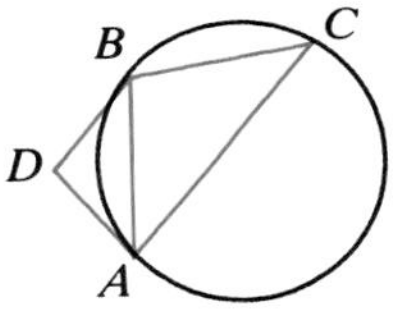

15. If $m\widehat{AC} = 50$ and $m\widehat{BD} = 38$, then $m\angle AEC = $ _?_.
16. If $AE = 10$, $EB = 9$, and $CE = 15$, then $ED = $ _?_.

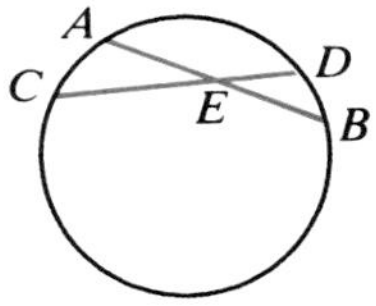

$\overline{PT}$ is a tangent to the circle.

17. If $m\widehat{RS} = 120$ and $m\widehat{ST} = 160$, then $m\angle P = $ _?_.
18. If $PT = 12$ and $PS = 18$, then $PR = $ _?_.

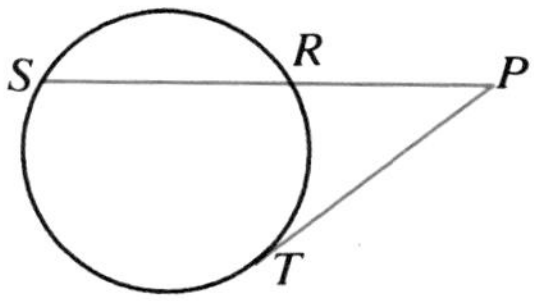

19. Given: $\odot O$; $\overline{TK} \cong \overline{KQ}$
Prove: $\overline{TQ} \perp \overline{OK}$

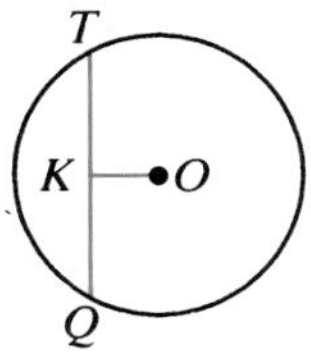

20. Given: $\overline{AD}$ is tangent to $\odot P$.
Prove: $\triangle BAD \sim \triangle ACD$

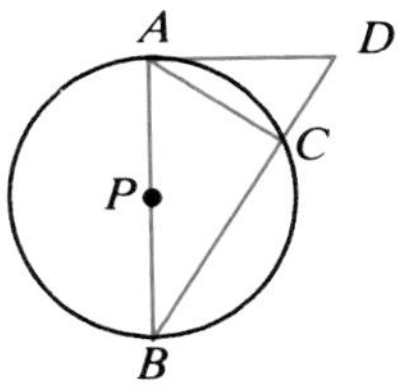

Mixed Review

1. If x, $x + 3$, and y are the lengths of the sides of a triangle, then
$\underline{\ ?\ } < y < \underline{\ ?\ }$.

2. Find the measure of an angle if the measures of a supplement and a complement of the angle have the ratio 5:2.

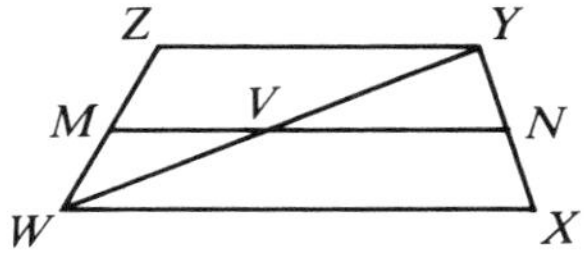

3. Given: $\overline{MN}$ is the median of trapezoid $WXYZ$.
Prove: $\overline{MN}$ bisects $\overline{WY}$.

4. Prove: The diagonals of a rhombus divide the rhombus into four congruent triangles.

5. A 30°-60°-90° triangle is inscribed in a circle of radius 7. Find the length of each leg of the triangle.

6. Must three parallel lines be coplanar? Draw a diagram to illustrate your answer.

7. If a regular polygon has 18 sides, find the measure of each interior angle and the measure of each exterior angle.

8. If $ABCD$ is a square and $AC = 4$, find AB.

9. If the lengths of two sides of a right triangle are 6 and 10, find two possible lengths for the third side.

10. Given: If Pete is particular, then Julie is jovial.
Pete is not particular.
What, if anything, can you conclude?

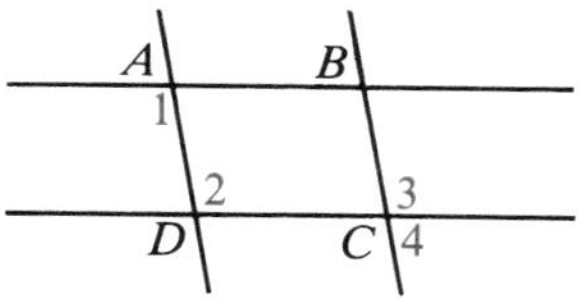

11. Given: $\angle 1 \cong \angle 2$; $\angle 2 \cong \angle 3$
Prove: $\overline{AB} \cong \overline{DC}$

12. When the altitude to the hypotenuse of a certain right triangle is drawn, the altitude divides the hypotenuse into segments of lengths 8 and 10. Find the length of the shorter leg.

13. Complete with *outside, inside,* or *on*: In a right triangle, (a) the medians intersect $\underline{\ ?\ }$ the triangle, (b) the altitudes intersect $\underline{\ ?\ }$ the triangle, and (c) the perpendicular bisectors of the sides intersect $\underline{\ ?\ }$ the triangle.

14. If $\overrightarrow{OB}$ bisects $\angle AOC$, $m\angle AOB = 5t - 7$, and $m\angle AOC = 8t + 10$, find the numerical measure of $\angle BOC$.

15. Two chords of a circle intersect inside a circle, dividing one chord into segments of length 15 and 12 and the other chord into segments of length 9 and t. Find the value of t.

16. The sine of any acute angle must be greater than $\underline{\ ?\ }$ and less than $\underline{\ ?\ }$.

17. If points R and S on a number line have coordinates -11 and 3, and $\overline{RS}$ has midpoint T, find RS and ST.

Algebra Review

Evaluate each expression for the given values of the variables.

Example $\frac{1}{2}bh$ when $b = 3.4$ and $h = 4.5$ **Solution** $\frac{1}{2}(3.4)(4.5) = 7.65$

1. Area of a square: s^2 when $s = 1.3$
2. Length of hypotenuse of a right triangle: $\sqrt{a^2 + b^2}$ when $a = 20$ and $b = 21$
3. Perimeter of parallelogram: $2(x + y)$ when $x = \frac{5}{3}$ and $y = \frac{3}{2}$
4. Perimeter of triangle: $a + b + c$ when $a = 11.5$, $b = 7.2$, and $c = 9.9$
5. Area of a rectangle: lw when $l = 2\sqrt{6}$ and $w = 3\sqrt{3}$
6. Perimeter of isosceles trapezoid: $2r + s + t$ when $r = \frac{4}{7}$, $s = 1$, and $t = \frac{13}{7}$
7. πr^2 when $r = 30$ (Use 3.14 for π.)
8. lwh when $l = 8$, $w = 6\frac{1}{4}$, and $h = 3\frac{1}{2}$
9. $2(lw + wh + lh)$ when $l = 4.5$, $w = 3$, and $h = 1$
10. $\frac{x - 3}{y + 2}$ when $x = 3$ and $y = -4$
11. $\frac{x + 5}{y - 2}$ when $x = -2$ and $y = -4$
12. $mx + b$ when $x = -6$, $m = \frac{5}{2}$, and $b = -2$
13. $6t^2$ when $t = 3$
14. $(6t)^2$ when $t = 3$
15. $\frac{1}{2}h(a + b)$ when $h = 3$, $a = 3\sqrt{2}$, and $b = 7\sqrt{2}$
16. $\sqrt{(x - 5)^2 + (y - 3)^2}$ when $x = 1$ and $y = 0$
17. $\frac{1}{3}x^2h$ when $x = 4\sqrt{3}$ and $h = 6$
18. $2s^2 + 4sh$ when $s = \sqrt{6}$ and $h = \frac{5}{2}\sqrt{6}$

Use the given information to rewrite each expression.

Example Bh when $B = \frac{1}{2}rs$ **Solution** $Bh = \left(\frac{1}{2}rs\right)h = \frac{1}{2}rsh$

19. $c(x + y)$ when $x + y = d$
20. $\frac{1}{3}Bh$ when $B = \pi r^2$
21. $\frac{1}{2}pl$ when $p = 2\pi r$
22. $2(l + w)$ when $l = s$ and $w = s$
23. $4\pi r^2$ when $r = \frac{1}{2}d$
24. $n\left(\frac{1}{2}sa\right)$ when $ns = p$

Solve each formula for the variable shown in color.

Example $y = mx + b$ **Solution** $x = \frac{y - b}{m}$; $m \neq 0$

25. $ax + by = c$
26. $C = \pi d$
27. $S = (n - 2)180$
28. $x^2 + y^2 = r^2$
29. $\frac{x}{h} = \frac{h}{y}$
30. $a^2 + b^2 = (a\sqrt{2})^2$
31. $A = \frac{1}{2}bh$
32. $m = \frac{y + 4}{x - 2}$

Most quilt patterns are designed by using geometric patterns. The quilt shown consists of concentric circles, squares, and triangles. The endless combinations of geometric designs used for creating patterns make quilting a beautiful and challenging craft.

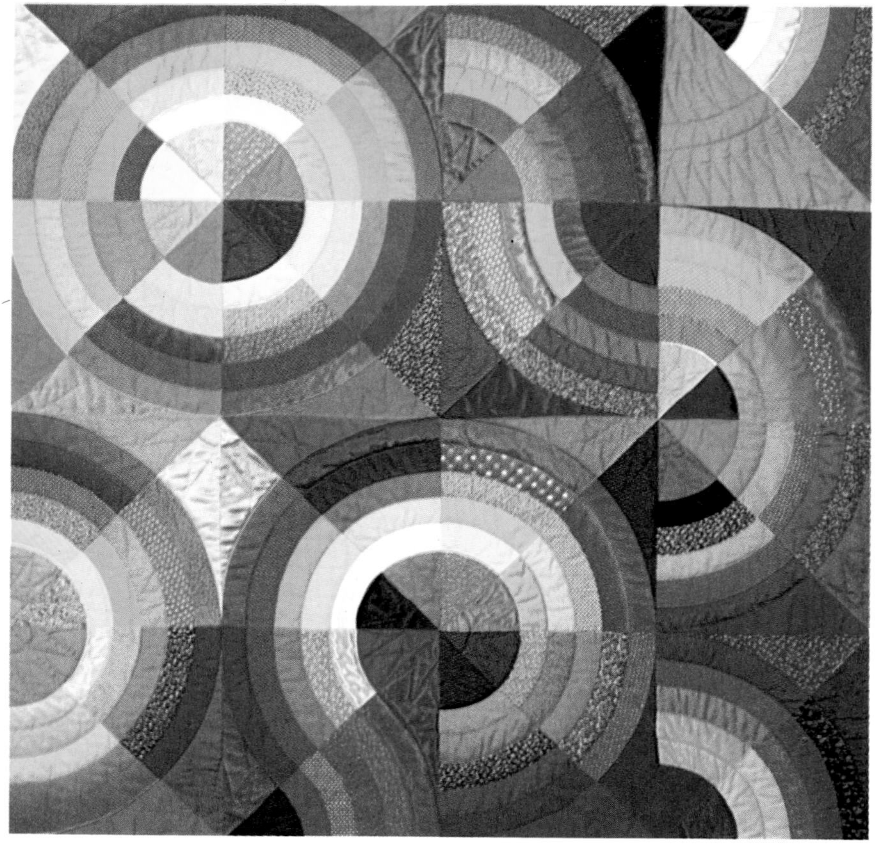

Constructions and Loci

8

Basic Constructions

Objectives

1. Perform seven basic constructions.
2. Use these basic constructions in original construction exercises.
3. State and apply theorems involving concurrent lines.

8-1 What Construction Means

In Chapters 1–7 we have used rulers and protractors to draw segments with certain lengths and angles with certain measures. In this chapter we will *construct* geometric figures using only two instruments, a *straightedge* and a *compass*. (You may use a ruler as a straightedge as long as you do not use the marks on the ruler.)

Using a Straightedge in Constructions

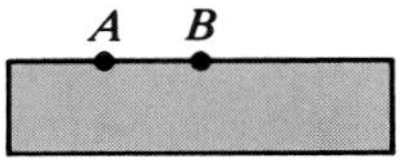

Given two points A and B, we know from Postulate 6 that there is exactly one line through A and B. We agree that we can use a straightedge to draw $\overleftrightarrow{AB}$ or parts of the line, such as $\overline{AB}$ and $\overrightarrow{AB}$.

Using a Compass in Constructions

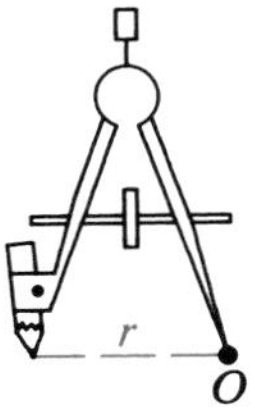

Given a point O and a length r, we know from the definition of a circle that there is exactly one circle with center O and radius r. We agree that we can use a compass to draw this circle or arcs of the circle.

Construction 1

Given a segment, construct a segment congruent to the given segment.

Given: $\overline{AB}$

Construct: A segment congruent to $\overline{AB}$

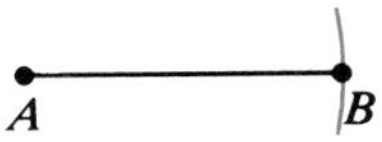

Procedure:

1. Use a straightedge to draw a line. Call it l.
2. Choose any point on l and label it X.
3. Set your compass for radius AB. Using X as center, draw an arc intersecting line l. Label the point of intersection Y.

$\overline{XY}$ is congruent to $\overline{AB}$.

Justification: Since AB was used for the radius of $\odot X$, $\overline{XY}$ is congruent to $\overline{AB}$.

Construction 2

Given an angle, construct an angle congruent to the given angle.

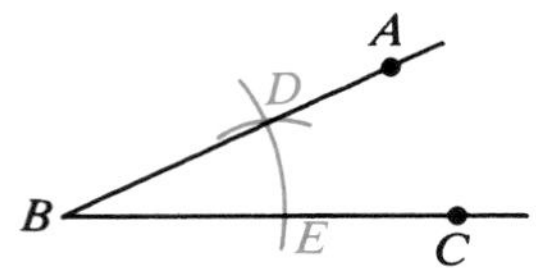

Given: $\angle ABC$

Construct: An angle congruent to $\angle ABC$

Procedure:

1. Draw a ray. Label it $\overrightarrow{RY}$.
2. Using B as center and any radius, draw an arc intersecting $\overrightarrow{BA}$ and $\overrightarrow{BC}$. Label the points of intersection D and E.
3. Using R as center and the same radius as before, draw an arc intersecting $\overrightarrow{RY}$. Label the arc $\overset{\frown}{XS}$, with S the point where the arc intersects $\overrightarrow{RY}$.
4. Using S as center and a radius equal to DE, draw an arc that intersects $\overset{\frown}{XS}$ at a point Q.
5. Draw $\overrightarrow{RQ}$.

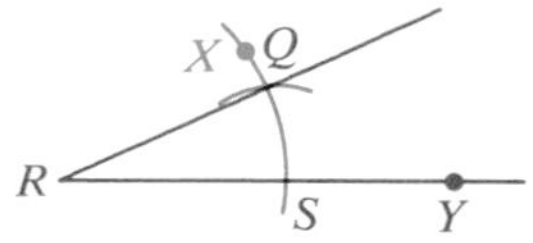

$\angle QRS$ is congruent to $\angle ABC$.

Justification: If $\overline{DE}$ and $\overline{QS}$ are drawn, $\triangle DBE \cong \triangle QRS$ (SSS Postulate). Then $\angle QRS \cong \angle ABC$.

Construction 3

Given an angle, construct the bisector of the angle.

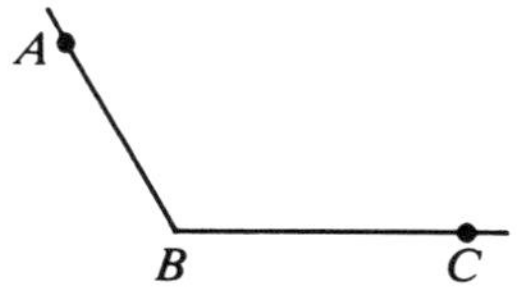

Given: $\angle ABC$

Construct: The bisector of $\angle ABC$

Procedure:

1. Using B as center and any radius, draw an arc that intersects $\overrightarrow{BA}$ at X and $\overrightarrow{BC}$ at Y.
2. Using X as center and a suitable radius, draw an arc. Using Y as center and the same radius, draw an arc that intersects the first arc at Z.
3. Draw $\overrightarrow{BZ}$.

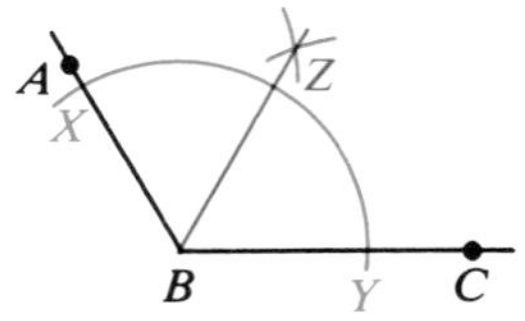

$\overrightarrow{BZ}$ bisects $\angle ABC$.

Justification: Draw $\overline{XZ}$ and $\overline{YZ}$. Then $\triangle XBZ \cong \triangle YBZ$ (SSS Postulate). Thus $\angle XBZ \cong \angle YBZ$ and $\overrightarrow{BZ}$ bisects $\angle ABC$.

Example Given $\angle 1$ and $\angle 2$, construct an angle whose measure is equal to $m\angle 1 + m\angle 2$.

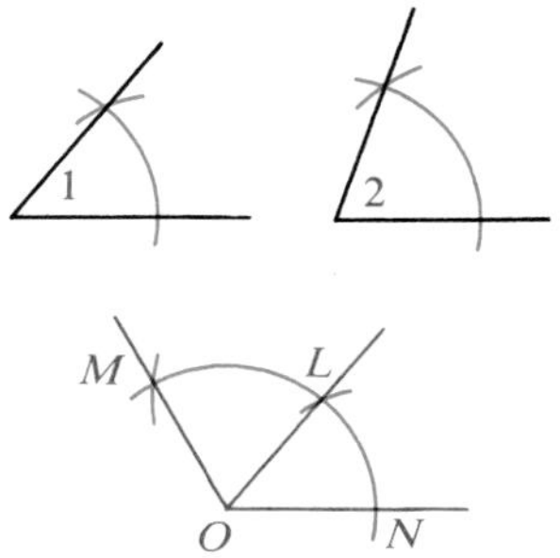

Solution First use Construction 2 to construct $\angle LON$ congruent to $\angle 1$. Then use the same method to construct $\angle MOL$ congruent to $\angle 2$ (as shown) so that $m\angle MON = m\angle 1 + m\angle 2$.

In construction exercises, you won't ordinarily have to write out the procedure and the justification. However, you should be able to supply them when asked to do so.

Classroom Exercises

1. Given: $\triangle JKM$
Explain how to construct a triangle that is congruent to $\triangle JKM$.

2. Draw any $\overline{AB}$.
  a. Construct $\overline{XY}$ so that $XY = AB$.
  b. Using X and Y as centers, and a radius equal to AB, draw arcs that intersect. Label the point of intersection Z.
  c. Draw $\overline{XZ}$ and $\overline{YZ}$.
  d. What kind of triangle is $\triangle XYZ$?

3. Explain how you could construct a 30° angle.

4. Exercise 3 suggests that you could construct other angles with certain measures. Name some.

5. Suppose you are given the three lengths shown and are asked to construct a triangle whose sides have lengths r, s, and t. Can you do so? State the theorem from Chapter 4 that applies.

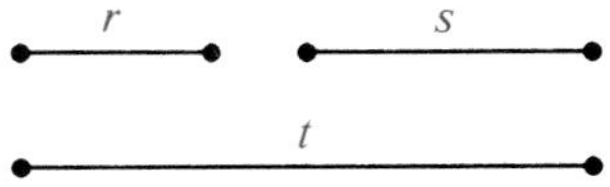

6. $\angle 1$ and $\angle 2$ are given. You see two attempts at constructing an angle whose measure is equal to $m\angle 1 + m\angle 2$. Are both constructions satisfactory?

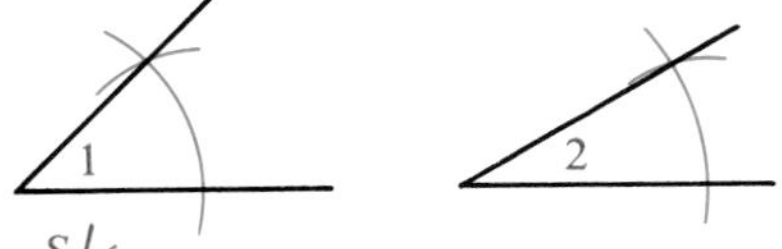

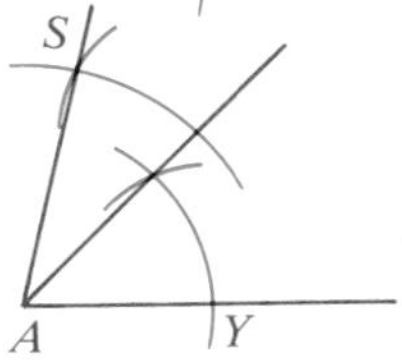

$m\angle SAY = m\angle 1 + m\angle 2$

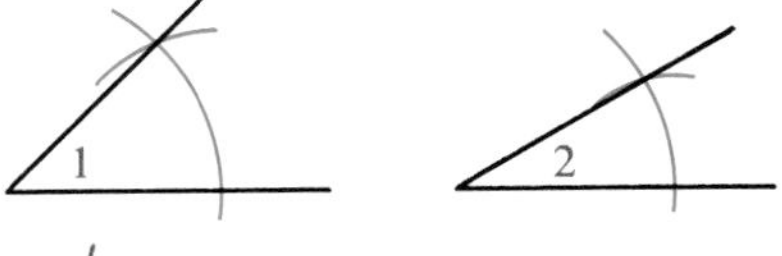

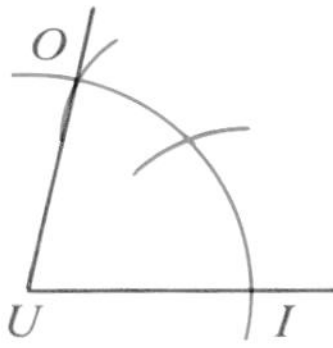

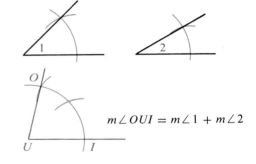

$m\angle OUI = m\angle 1 + m\angle 2$

Written Exercises

On your paper, draw two segments roughly like those shown. Use these lengths in Exercises 1–4 to construct a segment having the indicated length.

A **1.** $a + b$ **2.** $b - a$ **3.** $3a - b$ **4.** $a + 2b$

5. Using any convenient length for a side, construct an equilateral triangle.

6. **a.** Construct a 30° angle. **b.** Construct a 15° angle.

7. Draw any acute $\triangle ACU$. Use a method based on the SSS Postulate to construct a triangle congruent to $\triangle ACU$.

8. Draw any obtuse $\triangle OBT$. Use the SSS method to construct a triangle congruent to $\triangle OBT$.

9. Repeat Exercise 7, but use the SAS method.

10. Repeat Exercise 8, but use the ASA method.

On your paper, draw two angles roughly like those shown. Then for Exercises 11–14 construct an angle having the indicated measure.

11. $x + y$ **12.** $x - y$ **13.** $\frac{3}{4}x$ **14.** $180 - 2y$

B **15.** **a.** Draw any acute triangle. Bisect each of the three angles.
b. Draw any obtuse triangle. Bisect each of the three angles.
c. What do you notice about the points of intersection of the bisectors in parts (a) and (b)?

16. Construct a six-pointed star using the following procedure.
1. Draw a ray, $\overrightarrow{AB}$. On $\overrightarrow{AB}$ mark off, in order, points C and D such that $AB = BC = CD$.
2. Construct equilateral $\triangle ADG$.
3. On $\overline{AG}$ mark off points E and F such that $AE = EF = FG$. (Note that $AE = AB$.)
4. On $\overline{GD}$ mark off points H and I such that $GH = HI = ID$.
5. To complete the star, draw the three lines $\overleftrightarrow{FH}$, $\overleftrightarrow{EB}$, and $\overleftrightarrow{CI}$.

Construct an angle having the indicated measure.

17. 120 **18.** 150 **19.** 165 **20.** 45

21. Draw any $\triangle ABC$. Construct $\triangle DEF$ so that $\triangle DEF \sim \triangle ABC$ and $DE = 2AB$.

22. Construct a $\triangle RST$ such that $RS:ST:TR = 4:6:7$.

On your paper draw figures roughly like those shown. Use them in constructing the figures described in Exercises 23–25.

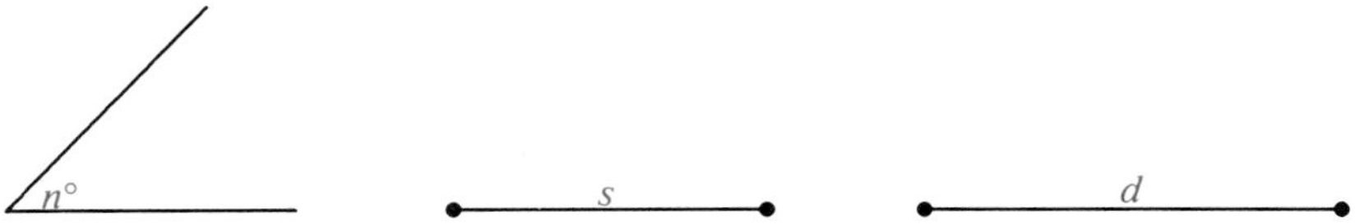

23. An isosceles triangle with a vertex angle of $n°$ and legs d

24. An isosceles triangle with a vertex angle of $n°$ and base s

C 25. A parallelogram with an $n°$ angle, longer side s, and longer diagonal d

★ 26. On your paper draw figures roughly like the ones shown. Then construct a triangle whose three angles are congruent to $\angle 1$, $\angle 2$, and $\angle 3$, and whose circumscribed circle has radius r.

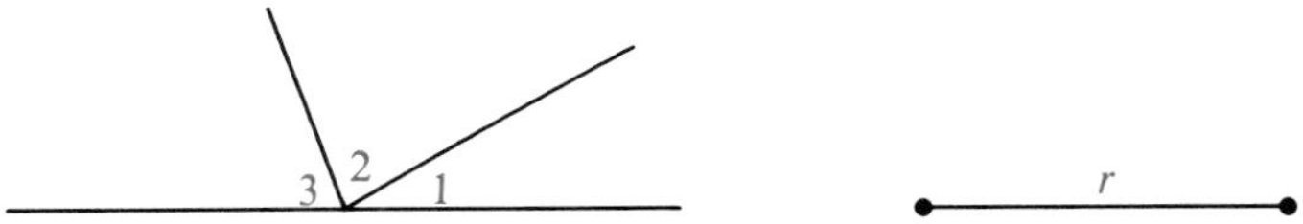

8-2 Perpendiculars and Parallels

The next three constructions are based on the following theorem and postulate.

(1) If a point is equidistant from the endpoints of a segment, then the point lies on the perpendicular bisector of the segment.

(2) Through any two points there is exactly one line.

Construction 4

Given a segment, construct the perpendicular bisector of the segment.

Given: $\overline{AB}$

Construct: The perpendicular bisector of $\overline{AB}$

Procedure:

1. Using any radius greater than $\frac{1}{2}AB$, draw four arcs of equal radii, two with center A and two with center B. Label the points of intersection of these arcs X and Y.
2. Draw $\overleftrightarrow{XY}$.

$\overleftrightarrow{XY}$ is the perpendicular bisector of $\overline{AB}$.

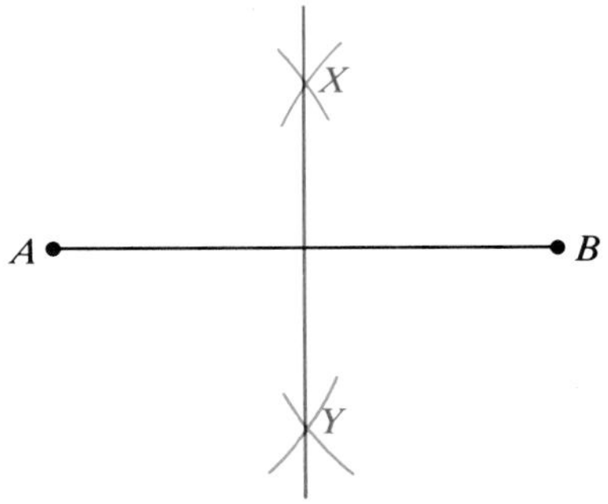

Justification: Points X and Y are equidistant from A and B. Thus $\overleftrightarrow{XY}$ is the perpendicular bisector of $\overline{AB}$.

Construction 5

Given a point on a line, construct the perpendicular to the line at the given point.

Given: Point C on line k

Construct: The perpendicular to k at C

Procedure:

1. Using C as center and any radius, draw arcs intersecting k at X and Y.
2. Using X as center and a radius greater than CX, draw an arc. Using Y as center and the same radius, draw an arc intersecting the first arc at Z.
3. Draw $\overleftrightarrow{CZ}$.

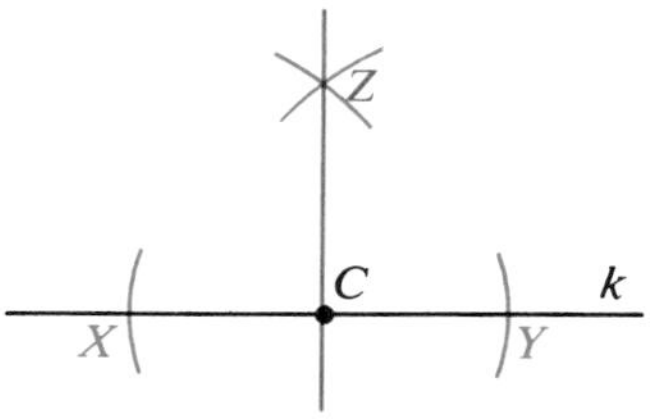

$\overleftrightarrow{CZ}$ is perpendicular to k at C.

Justification: Points X and Y were constructed so that C is equidistant from X and Y. Then point Z was constructed so that Z is equidistant from X and Y. Since $\overleftrightarrow{CZ}$ is the perpendicular bisector of $\overline{XY}$, $\overleftrightarrow{CZ}$ is perpendicular to k at C.

Construction 6

Given a point outside a line, construct the perpendicular to the line from the given point.

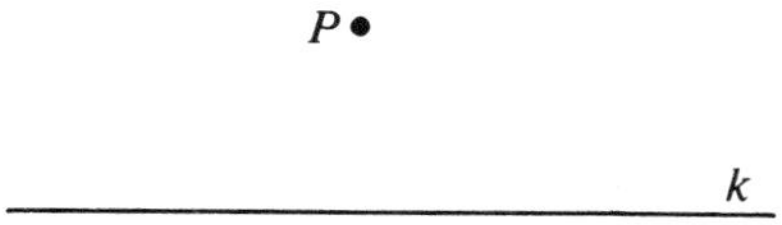

Given: Point P outside line k

Construct: The perpendicular to k from P

Procedure:

1. Using P as center, draw two arcs of equal radii that intersect k at points X and Y.
2. Using X and Y as centers and a suitable radius, draw arcs that intersect at a point Z.
3. Draw $\overleftrightarrow{PZ}$.

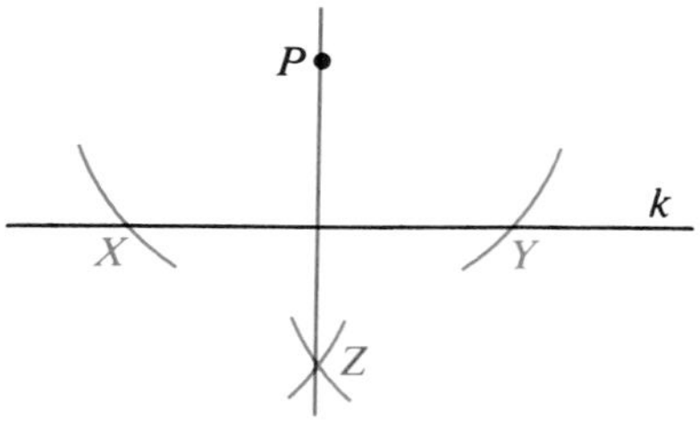

$\overleftrightarrow{PZ}$ is perpendicular to k.

Justification: Both P and Z are equidistant from X and Y. Thus $\overleftrightarrow{PZ}$ is the perpendicular bisector of $\overline{XY}$, and $\overleftrightarrow{PZ} \perp k$.

Construction 7

Given a point outside a line, construct the parallel to the given line through the given point.

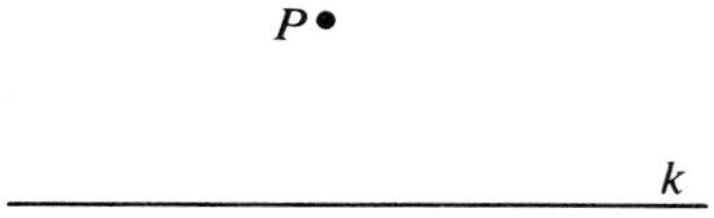

Given: Point P outside line k

Construct: The line through P parallel to k

Procedure:

1. Let A and B be two points on line k. Draw $\overleftrightarrow{PA}$.
2. At P, construct $\angle 1$ so that $\angle 1$ and $\angle PAB$ are congruent corresponding angles. Let l be the line containing the ray you just constructed.

l is the line through P parallel to k.

Justification: If two lines are cut by a transversal and corresponding angles are congruent, then the lines are parallel. (Postulate 11)

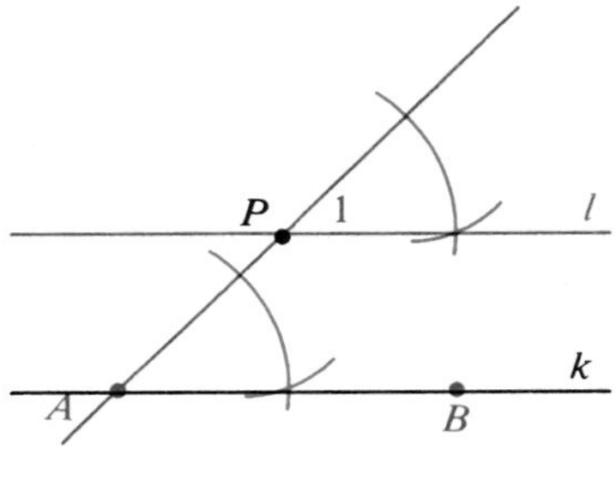

Classroom Exercises

1. Suggest an alternative procedure for Construction 7 that uses Constructions 5 and 6.

Describe how you would construct each of the following.

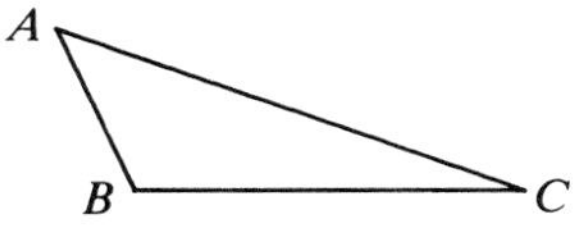

2. The median of $\triangle ABC$ that contains vertex B

3. The altitude of $\triangle ABC$ that contains vertex B

4. The altitude of $\triangle ABC$ that contains vertex A

5. The perpendicular to $\overline{BC}$ at C

6. A square whose sides each have length AC

7. A square whose perimeter equals AC

8. A right triangle with hypotenuse and one leg equal to AC and BC, respectively

9. A triangle whose sides are in the ratio $1:2:\sqrt{5}$

10. Suppose you want to construct a circle that is tangent to l at X, and that passes through point Y.

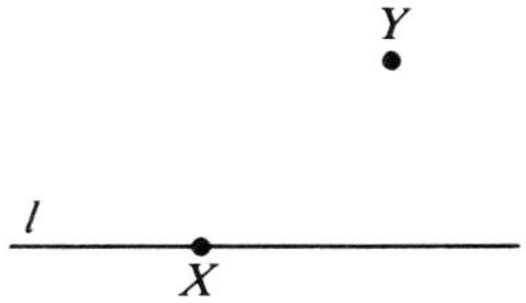

a. Where must the center lie with respect to line l and point X?

b. Where must the center lie with respect to points X and Y?

c. Explain how you could carry out the construction.

Written Exercises

Draw a figure roughly like the one shown, but larger. Do the indicated construction clearly enough so that your method can be understood easily.

A **1.** The perpendicular to l at P

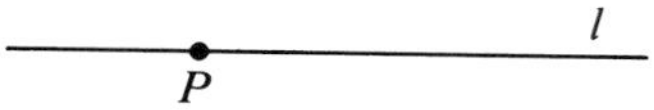

2. The perpendicular to l from S

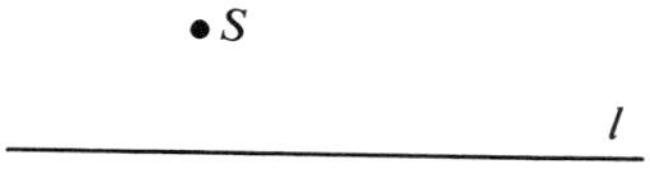

3. The perpendicular bisector of $\overline{JK}$

4. The parallel to l through T

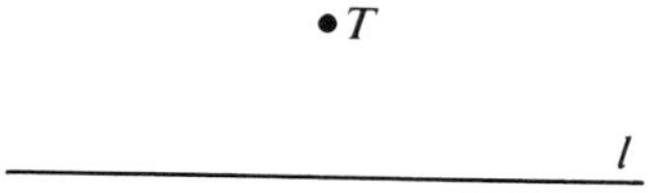

5. The perpendicular to $\overleftrightarrow{BA}$ at A

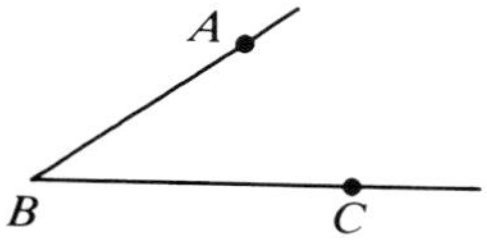

6. The parallel to $\overleftrightarrow{ED}$ through F

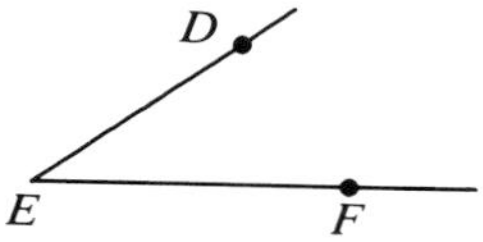

7. The perpendicular to $\overleftrightarrow{HJ}$ from G

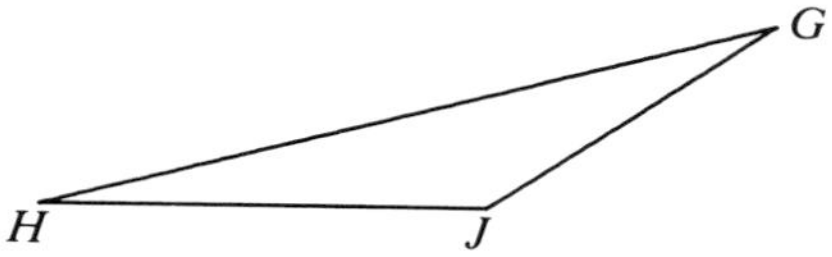

8. A complement of $\angle KMN$

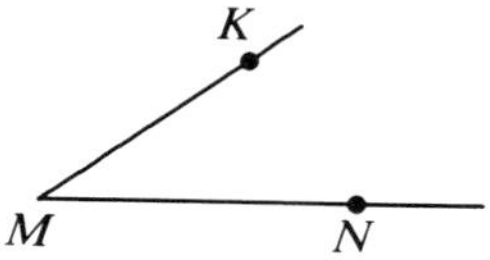

Construct an angle with the indicated measure.

9. 45 **10.** 135 **11.** $22\frac{1}{2}$ **12.** 105

13. Draw a segment $\overline{AB}$. Construct a segment $\overline{XY}$ whose length equals $\frac{3}{4}AB$.

B **14. a.** Draw an acute triangle. Construct the perpendicular bisector of each side.
b. Do the perpendicular bisectors intersect in one point?
c. Repeat parts (a) and (b) using an obtuse triangle.

15. a. Draw an acute triangle. Construct the three altitudes.
b. Do the lines that contain the altitudes intersect in one point?
c. Repeat parts (a) and (b) using an obtuse triangle.

16. a. Draw a very large acute triangle. Construct the three medians.
b. Do the lines that contain the medians intersect in one point?
c. Repeat parts (a) and (b) using an obtuse triangle.

On your paper draw figures roughly like those shown. Use them in constructing the figures described in Exercises 17–24.

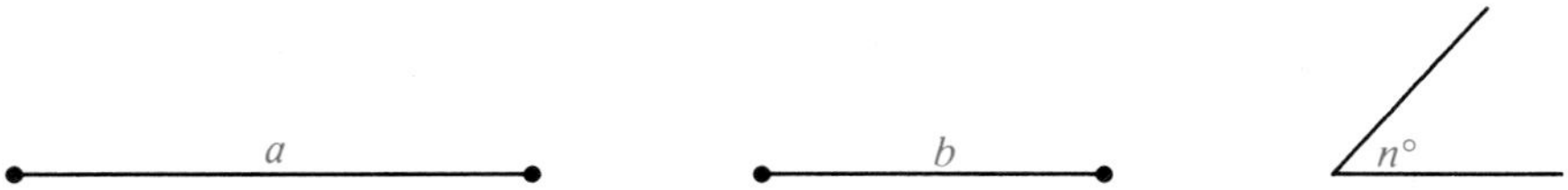

17. A parallelogram with sides a and b and an $n°$ angle

18. A rectangle with sides a and b

19. A square with perimeter $2a$

20. A rhombus with diagonals a and b

21. A square with diagonal b

22. A segment with length $\sqrt{a^2 + b^2}$

23. A square with diagonals $b\sqrt{2}$

24. A right triangle with hypotenuse a and one leg b

C **25.** Draw a segment and let its length be s. Construct a segment whose length is $s\sqrt{3}$.

26. Draw a diagram roughly like the one shown. Without laying your straightedge across any part of the lake, construct more of $\overrightarrow{RS}$.

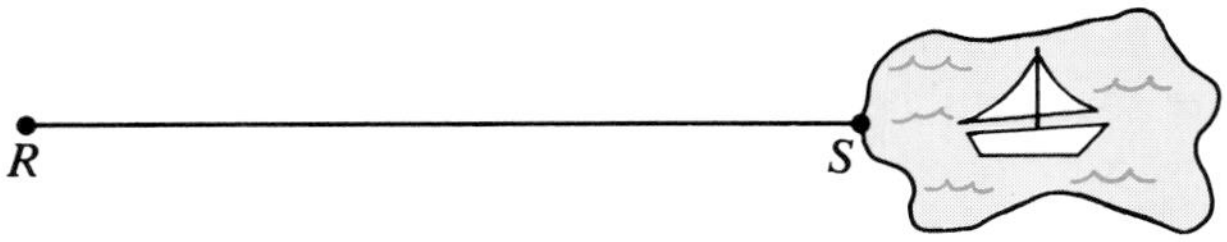

27. Draw three noncollinear points R, S, and T. Construct a triangle whose sides have R, S, and T as midpoints. (*Hint:* How is $\overline{RT}$ related to the side of the triangle that has S as midpoint?)

28. Draw a segment and let its length be 1.

a. Construct a segment with length $\sqrt{5}$.

b. Construct a segment with length $\frac{1}{2} + \frac{\sqrt{5}}{2}$, or $\frac{1 + \sqrt{5}}{2}$.

c. Construct a *golden rectangle* whose sides are in the ratio $1 : \frac{1 + \sqrt{5}}{2}$.

Challenge

Given $\overline{AB}$, its midpoint M, and a point Z outside $\overline{AB}$, use only a straightedge (and *no* compass) to construct a line through Z parallel to $\overleftrightarrow{AB}$. (*Hint:* Use Ceva's Theorem, Exercise 36, page 238.)

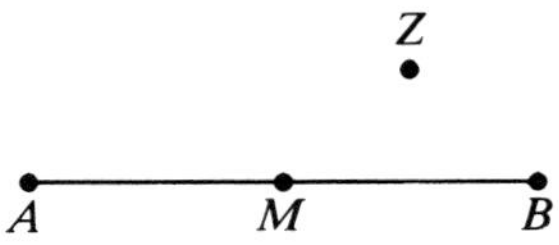

8-3 Concurrent Lines

When two or more lines intersect in one point, the lines are said to be **concurrent.** As you saw in Exercise 15, page 338, the bisectors of the angles of a triangle are concurrent.

Theorem 8-1

The bisectors of the angles of a triangle intersect in a point that is equidistant from the three sides of the triangle.

Given: $\triangle ABC$; the bisectors of $\angle A$, $\angle B$, and $\angle C$

Prove: The angle bisectors intersect in a point; that point is equidistant from $\overline{AB}$, $\overline{BC}$, and $\overline{AC}$.

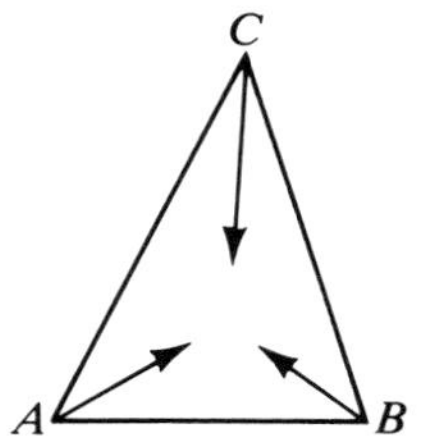

Proof:

Label the point where the bisectors of $\angle A$ and $\angle B$ intersect as I. We will show that point I also lies on the bisector of $\angle C$ and that I is equidistant from $\overline{AB}$, $\overline{BC}$, and $\overline{AC}$.

Draw perpendiculars from I intersecting $\overline{AB}$, $\overline{BC}$, and $\overline{AC}$ at R, S, and T, respectively. Since any point on the bisector of an angle is equidistant from the sides of the angle (Theorem 3-7, page 139), $IT = IR$ and $IR = IS$. Thus $IT = IS$. Since any point equidistant from the sides of an angle is on the bisector of the angle (Theorem 3-8, page 139), I is on the bisector of $\angle C$. Since $IR = IS = IT$, point I is equidistant from $\overline{AB}$, $\overline{BC}$, and $\overline{AC}$.

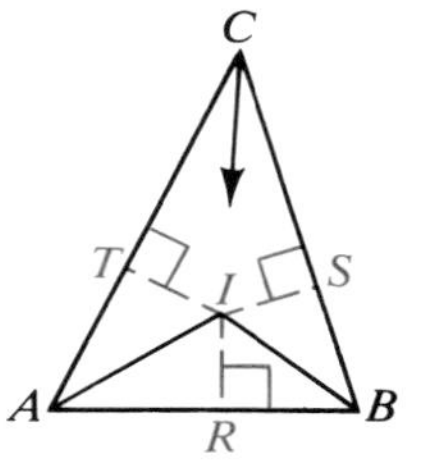

In Exercises 14–16, page 342, you discovered three other sets of concurrent lines related to triangles: the perpendicular bisectors of the sides, the lines containing the altitudes, and the medians. As you can see in the diagrams below, concurrent lines may intersect in a point outside the triangle. The intersection point may also lie on the triangle (see Classroom Exercise 4, page 346).

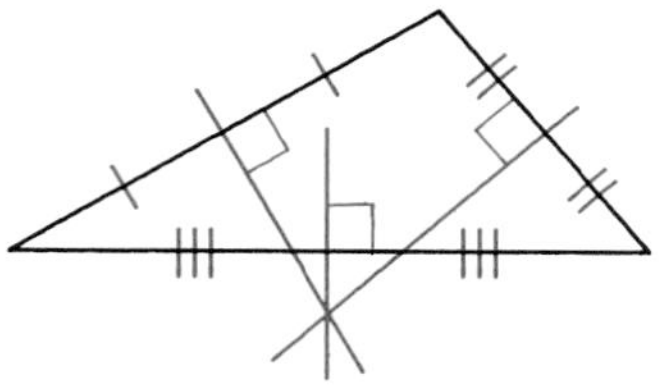

Perpendicular bisectors

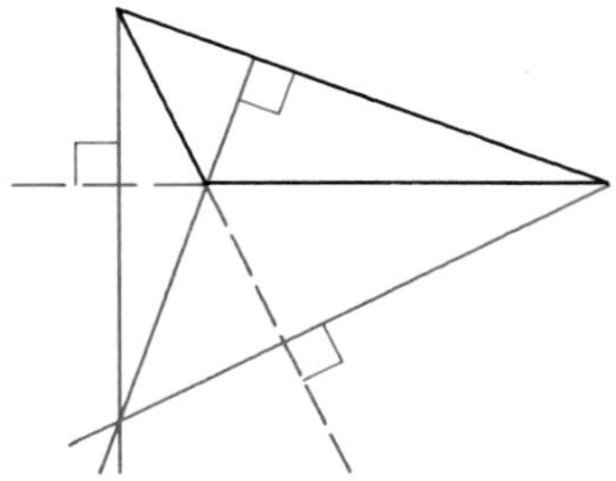

Lines containing altitudes

Theorem 8-2

The perpendicular bisectors of the sides of a triangle intersect in a point that is equidistant from the three vertices of the triangle.

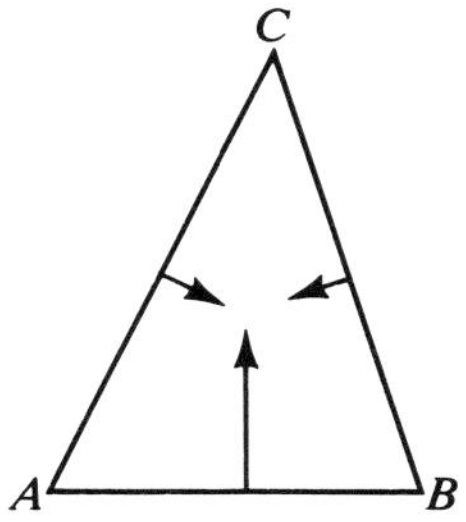

Given: $\triangle ABC$; the $\perp$ bisectors of $\overline{AB}$, $\overline{BC}$, and $\overline{AC}$

Prove: The $\perp$ bisectors intersect in a point; that point is equidistant from A, B, and C.

Proof:

Label the point where the perpendicular bisectors of $\overline{AC}$ and $\overline{BC}$ intersect as O. We will show that point O lies on the perpendicular bisector of $\overline{AB}$ and is equidistant from A, B, and C.

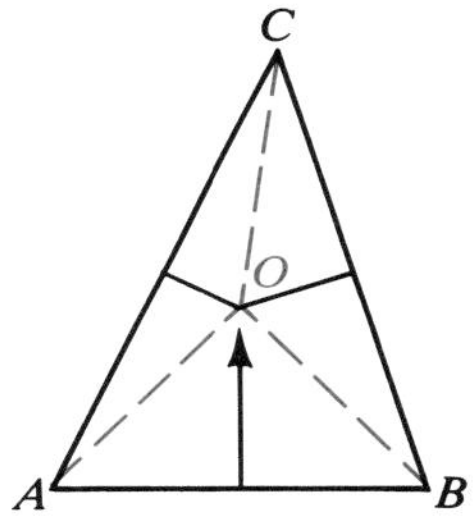

Draw $\overline{OA}$, $\overline{OB}$, and $\overline{OC}$. Since any point on the perpendicular bisector of a segment is equidistant from the endpoints of the segment (Theorem 3-5, page 138), $OA = OC$ and $OC = OB$. Thus $OA = OB$. Since any point equidistant from the endpoints of a segment lies on the perpendicular bisector of the segment (Theorem 3-6, page 138), O is on the perpendicular bisector of $\overline{AB}$. Since $OA = OB = OC$, point O is equidistant from A, B, and C.

The following theorems will be proved in Chapter 11.

Theorem 8-3

The lines that contain the altitudes of a triangle intersect in a point.

Theorem 8-4

The medians of a triangle intersect in a point that is two thirds of the distance from each vertex to the midpoint of the opposite side.

According to Theorem 8-4, if $\overline{AM}$, $\overline{BN}$, and $\overline{CO}$ are medians of $\triangle ABC$, then:

$$AX = \frac{2}{3}AM$$

$$XN = \frac{1}{3}BN$$

$$CX : XO : CO = 2 : 1 : 3$$

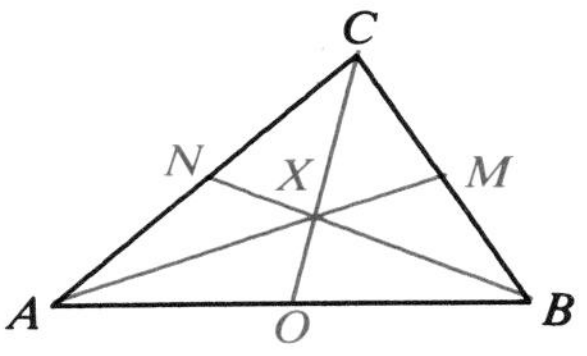

Classroom Exercises

1. Draw, if possible, a triangle in which the perpendicular bisectors of the sides intersect in the point described.

a. A point inside the triangle **b.** A point outside the triangle
c. A point on the triangle

2. Repeat Exercise 1, but work with angle bisectors.

3. Is there some kind of triangle such that the perpendicular bisector of each side is also an angle bisector, a median, and an altitude?

4. $\triangle JAM$ is a right triangle.

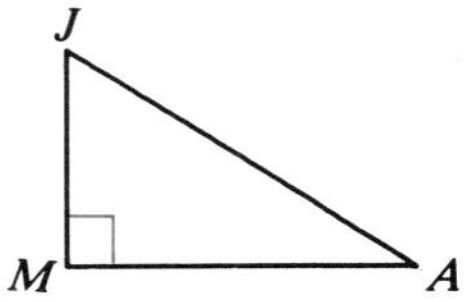

a. Is $\overline{JM}$ an altitude of $\triangle JAM$?
b. Name another altitude shown.
c. In what point do the three altitudes of $\triangle JAM$ meet?
d. Where do the perpendicular bisectors of the sides of $\triangle JAM$ meet?
e. Does your answer to (d) agree with Theorem 8-2?

5. The medians of $\triangle DEF$ are shown. Find the lengths indicated.

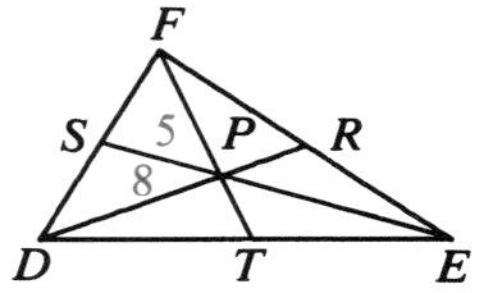

a. $EP =$ _?_ **b.** $PR =$ _?_
c. If $FT = 9$, then $PT =$ _?_ and $FP =$ _?_.

6. Given: $\overline{RJ}$ and $\overline{SK}$ are medians of $\triangle RST$;
X and Y are the midpoints of $\overline{RG}$ and $\overline{SG}$.

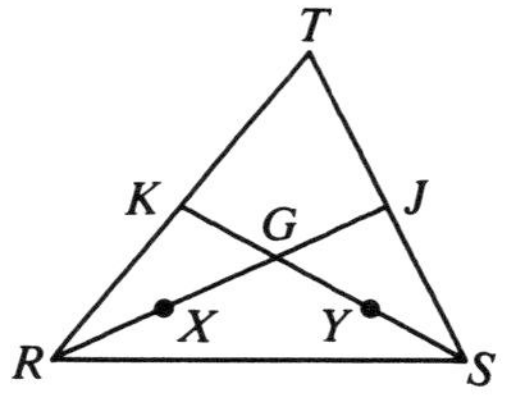

a. How are $\overline{XY}$ and $\overline{RS}$ related? Why?
b. How are $\overline{KJ}$ and $\overline{RS}$ related? Why?
c. How are $\overline{KJ}$ and $\overline{XY}$ related? Why?
d. What special kind of quadrilateral is $XYJK$? Why?
e. Why does $XG = GJ$?
f. Explain why $RG = \frac{2}{3}RJ$.

Written Exercises

A **1.** Draw a triangle such that the lines containing the three altitudes intersect in the point described.

a. A point inside the triangle **b.** A point outside the triangle
c. A point on the triangle

Exercises 2–5 refer to the diagram.

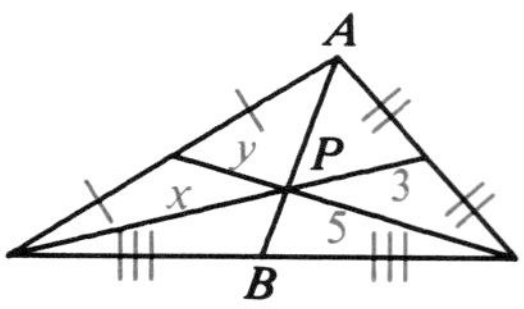

2. Find the values of x and y.

3. If $AB = 6$, then $BP =$ _?_ and $AP =$ _?_.

4. If $AB = 7$, then $BP =$ _?_ and $AP =$ _?_.

5. If $PB = 1.9$, then $AP =$ _?_ and $AB =$ _?_.

6. Use a ruler and a protractor to draw a regular pentagon. Then construct the perpendicular bisectors of the five sides.

7. Draw a regular pentagon as in Exercise 6. Construct the angle bisectors.

B **8.** Three towns, located as shown, plan to build one recreation center to serve all three towns. They decide that the fair thing to do is to build the hall equidistant from the three towns. Comment about the wisdom of the plan.

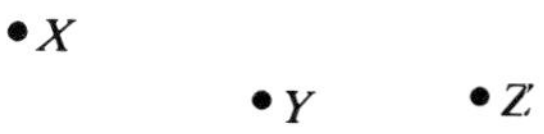

9. See Exercise 8. Locate three towns so that it isn't possible to find a spot equidistant from the three towns.

10. In the figure, $\overline{AD}$ and $\overline{BE}$ are congruent medians of $\triangle ABC$.

a. Explain why $GD = GE$.

b. $GA = \underline{\ ?\ }$

c. Name three angles congruent to $\angle GAB$.

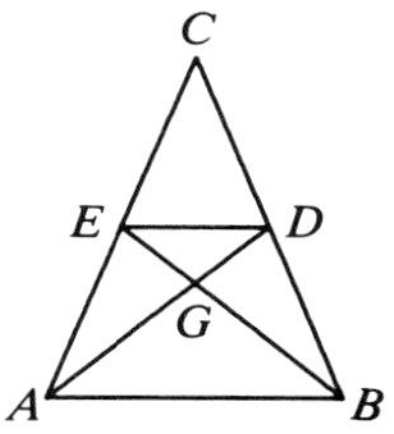

$\overline{AU}$, $\overline{BV}$, and $\overline{CW}$ are the medians of $\triangle ABC$.

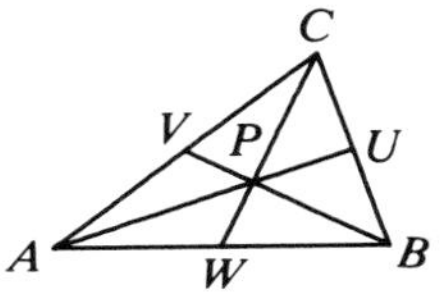

11. If $AP = x^2$ and $PU = 2x$, then $x = \underline{\ ?\ }$.

12. If $BP = y^2 + 1$ and $PV = y + 2$, then $y = \underline{\ ?\ }$ or $y = \underline{\ ?\ }$.

13. If $CW = 2z^2 - 5z - 12$ and $CP = z^2 - 15$, then $z = \underline{\ ?\ }$ and $PW = \underline{\ ?\ }$.

14. $ABCD$ is a parallelogram with M the midpoint of $\overline{CD}$. If $\overline{BM}$ intersects $\overline{AC}$ at X, prove that $CX = \frac{1}{3}AC$. (*Hint:* Draw $\overline{BD}$.)

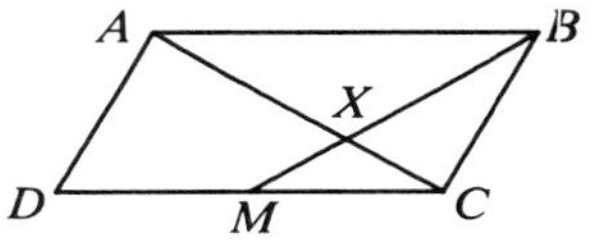

C **15.** In the plane figure, point P is equidistant from R, S, and T. Describe the location of the following points in the plane.

a. Points farther from both R and S than from T

b. Points closer to both R and S than to T

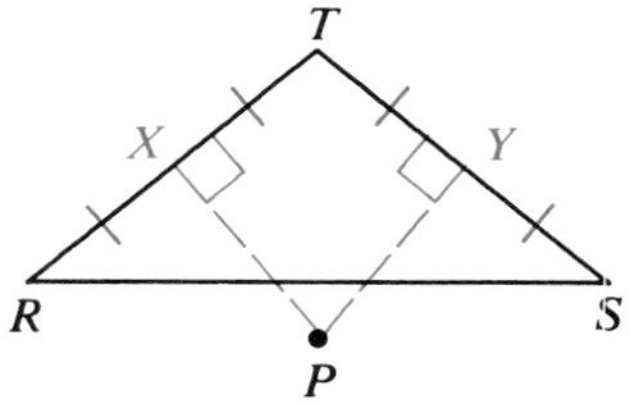

Ex. 15

16. Prove: If two of the medians of a triangle are congruent, then the triangle is isosceles.

Application

CENTER OF GRAVITY

The *center of gravity* of an object is the point where the weight of the object is focused. If you lift or support an object, you can do this most easily under its center of gravity.

A mobile is either hung or supported at its center of gravity. In planning a mobile, a sculptor must take into account the centers of gravity of the component parts.

If an object is not supported under its center of gravity, it becomes unstable. Suppose you hold a heavy bar in one hand. If you support it near the center of gravity, it will be easy to hold (Figure 1). To support it at one end requires more effort (Figure 2), since the pole tends to turn until the center of gravity is directly below the point of support (Figure 3).

The center of gravity may be inside an object or outside of it. The center of gravity of an ice cube is in the middle of the ice, but the center of gravity of an automobile tire is not in a part of the tire itself.

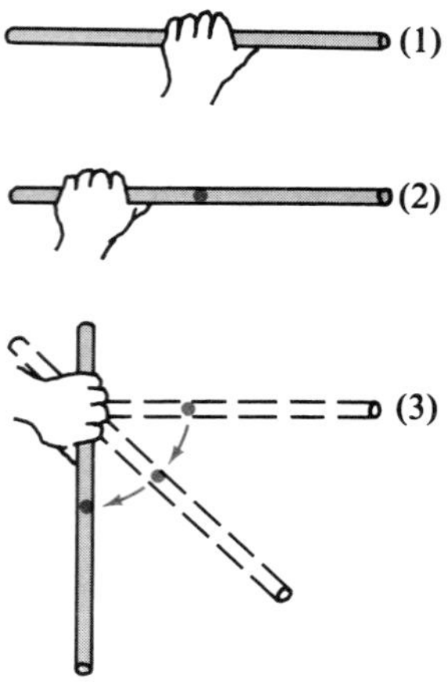

Exercises

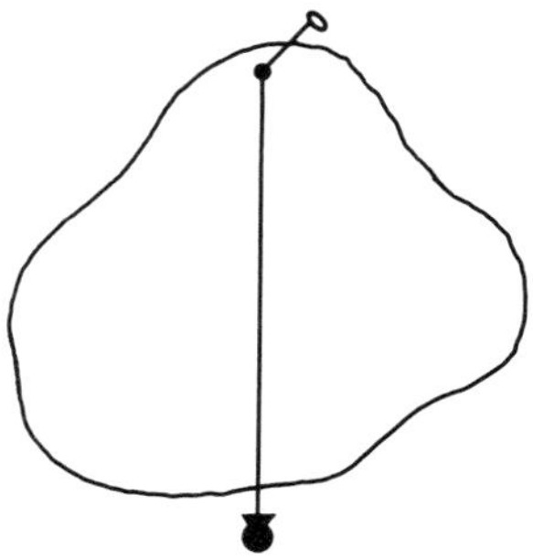

1. For this experiment, cut out a large, irregularly shaped piece of cardboard.
   a. Near the edge, poke a hole just large enough to allow the cardboard to rotate freely when pinned through the hole.
   b. Pin the cardboard through the hole to a suitable wall surface. The piece will position itself so that its center of gravity is as low as possible. This means that it will lie on a vertical line through the point of suspension. To find this line, tie a weighted string to the pin. Then draw on the cardboard the line determined by the string.
   c. Repeat (a) and (b) but use a different hole. The center of gravity of the cardboard ought to lie on both of the lines you have drawn and should therefore be their point of intersection. The cardboard should balance if supported at this point.

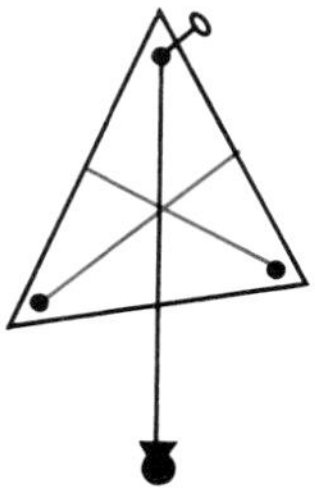

2. Cut out a piece of cardboard in the shape of a large scalene triangle.
   a. Follow the steps of Exercise 1 using three holes, one near each of the three vertices.
   b. If you worked carefully, all three lines drawn intersect in one point, the center of gravity of the cardboard. This point is also referred to as the *center of mass* or the *centroid* of the cardboard. Study the lines you have drawn and explain why in geometry the point of intersection of the medians of a triangle is called the *centroid of the triangle*.

3. Do you think that the center of gravity of a parallelogram is the point where the diagonals intersect? Use the technique of Exercise 1 to test this idea.

Self-Test 1

1. Draw any $\overline{CD}$. Construct the perpendicular bisector of $\overline{CD}$.
2. Construct a 60° angle, $\angle RST$, and its bisector, $\overrightarrow{SQ}$.
3. Draw a large acute $\triangle ABC$. Then construct altitude $\overline{AD}$ from vertex A.
4. Draw a line t and choose any point P that is not on line t. Construct $\overleftrightarrow{PQ} \parallel t$.
5. Draw any $\overline{AB}$. Construct rectangle $JKLM$ so that $JK = 2AB$ and $KL = AB$.
6. Name four types of concurrent lines, rays, or segments that are associated with triangles.
7. The perpendicular bisectors of the sides of a right triangle intersect in a point located at _?_.
8. The medians of equilateral $\triangle ABC$ intersect at point X. If $\overline{AD}$ is a median and $AB = 12$, then $AX =$ _?_ and $XD =$ _?_.

More Constructions

Objectives

1. Perform seven additional basic constructions.
2. Use the basic constructions in original construction exercises.

8-4 Circles

Construction 8

Given a point on a circle, construct the tangent to the circle at the given point.

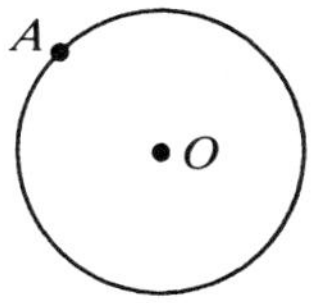

Given: Point A on $\odot O$

Construct: The tangent to $\odot O$ at A

Procedure:

1. Draw $\overrightarrow{OA}$.
2. Construct the line perpendicular to $\overrightarrow{OA}$ at A. Call it t.

Line t is tangent to $\odot O$ at A.

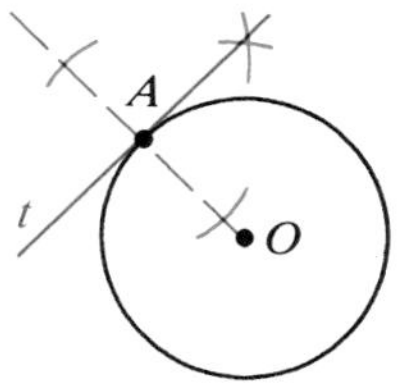

Justification: Because t is perpendicular to radius $\overline{OA}$ at A, t is tangent to $\odot O$.

Construction 9

Given a point outside a circle, construct a tangent to the circle from the given point.

Given: Point P outside $\odot O$

Construct: A tangent to $\odot O$ from P

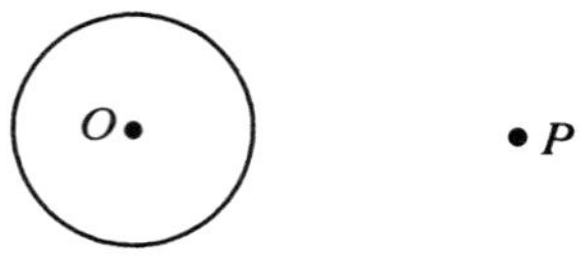

Procedure:

1. Draw $\overline{OP}$.
2. Find the midpoint M of $\overline{OP}$ by constructing the perpendicular bisector of $\overline{OP}$.
3. Using M as center and MP as radius, draw a circle that intersects $\odot O$ in a point X.
4. Draw $\overrightarrow{PX}$.

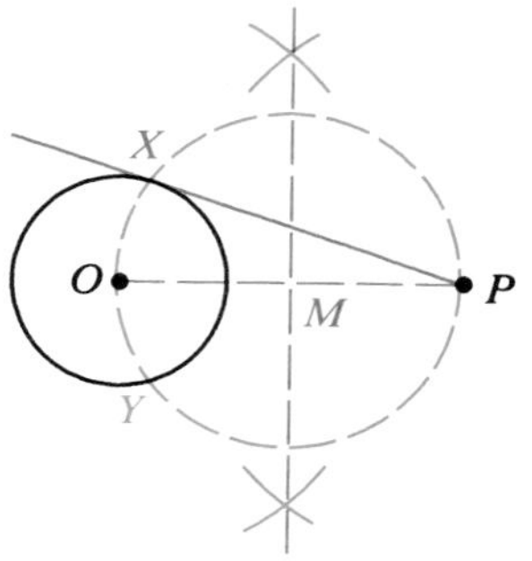

$\overrightarrow{PX}$ is tangent to $\odot O$ from P. $\overrightarrow{PY}$, not drawn, is the other tangent from P.

Justification: Draw $\overline{OX}$. Because $\angle OXP$ is inscribed in a semicircle, $\angle OXP$ is a right angle and $\overrightarrow{PX} \perp \overline{OX}$. Because $\overrightarrow{PX}$ is perpendicular to radius $\overline{OX}$ at its outer endpoint, $\overrightarrow{PX}$ is tangent to $\odot O$.

Construction 10

Given a triangle, circumscribe a circle about the triangle.

Given: $\triangle ABC$

Construct: A circle passing through A, B, and C

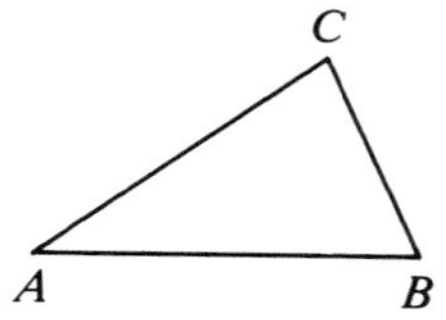

Procedure:

1. Construct the perpendicular bisectors of any two sides of $\triangle ABC$. Label the point of intersection O.
2. Using O as center and OA as radius, draw a circle.

Circle O passes through A, B, and C.

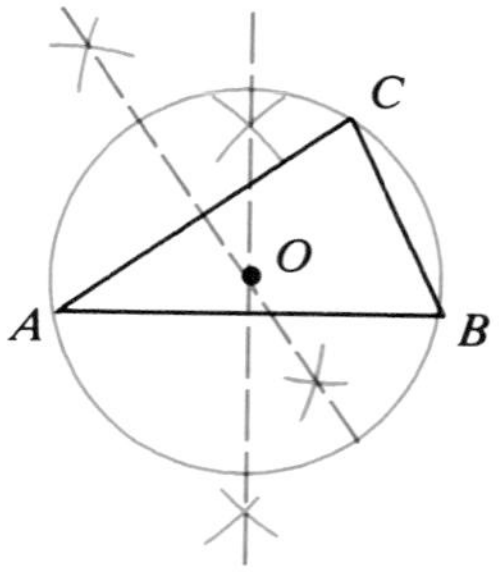

Justification: See Theorem 8-2 on page 345.

Construction 11

Given a triangle, inscribe a circle in the triangle.

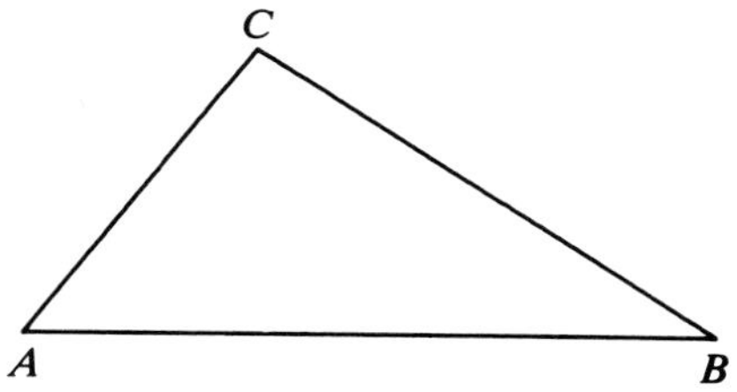

Given: $\triangle ABC$

Construct: A circle tangent to $\overline{AB}$, $\overline{BC}$, and $\overline{AC}$

Procedure:

1. Construct the bisectors of $\angle A$ and $\angle B$. Label the point of intersection I.
2. Construct a perpendicular from I to $\overline{AB}$. It intersects $\overline{AB}$ at a point R.
3. Using I as center and IR as radius, draw a circle.

Circle I is tangent to $\overline{AB}$, $\overline{BC}$, and $\overline{AC}$.

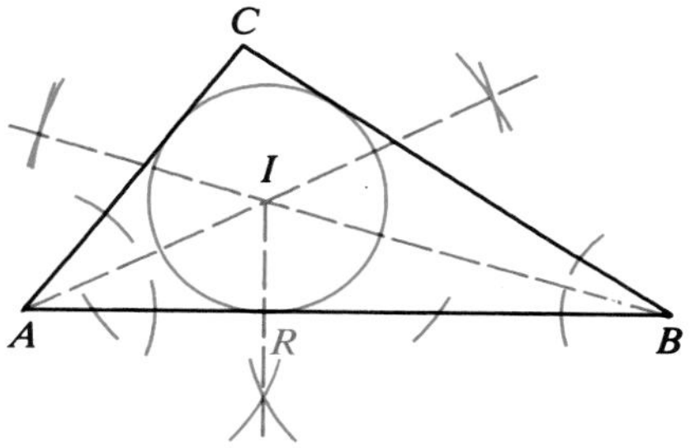

Justification: See Theorem 8-1 on page 344.

Classroom Exercises

1. Explain how to find the midpoint of $\widehat{AB}$.
2. Explain how to construct the center of the circle containing points A, B, and C.

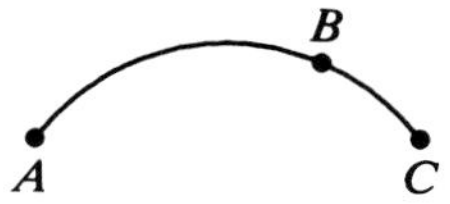

3. Explain how to find a line that is parallel to $\overline{RS}$ and tangent to $\odot P$.

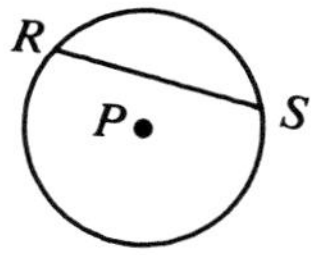

4. Here you see a common method for using just one compass setting for drawing a circle and dividing the circle into six congruent arcs. Explain how the method works.
5. Suppose a circle is given. Explain how you can use the method of Exercise 4 to inscribe an equilateral triangle in the circle.
6. Suppose the construction of Exercise 4 has been carried out. Explain how you can then inscribe a regular twelve-sided polygon in the circle.

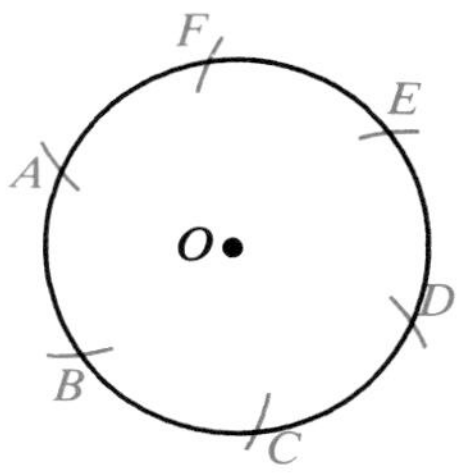

7. A student intends to inscribe a circle in $\triangle RST$. The center I has been found as shown. How should the student find the radius needed?

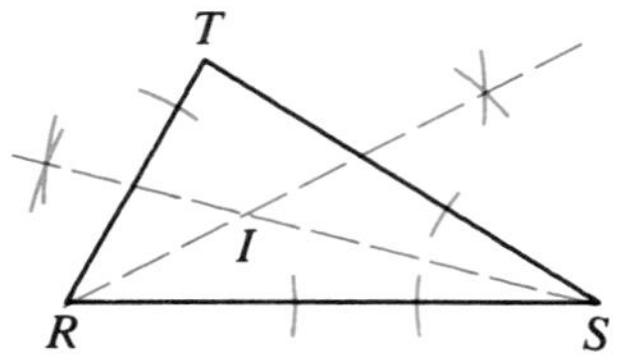

Written Exercises

In Exercises 1 and 2, draw a diagram similar to the one shown, but larger.

A 1. Construct a tangent at A.

2. Construct two tangents from P.

3. Draw a large acute triangle. Construct the circumscribed circle.

4. Construct a large right triangle. Construct the circumscribed circle.

5. Draw a large obtuse triangle. Construct the circumscribed circle.

6. Draw a large acute triangle. Construct the inscribed circle.

7. Construct a large right triangle. Construct the inscribed circle.

8. Draw a large obtuse triangle. Construct the inscribed circle.

B 9. Draw a circle. Inscribe an equilateral triangle in the circle.

10. Draw a circle. Inscribe a square in the circle.

11. Draw a circle. Circumscribe a square about the circle.

12. Construct a square. Circumscribe a circle about the square.

13. Construct a square. Inscribe a circle in the square.

14. Draw a circle. Circumscribe an equilateral triangle about the circle.

In each of Exercises 15 and 16 begin with a diagram roughly like the one shown, but larger.

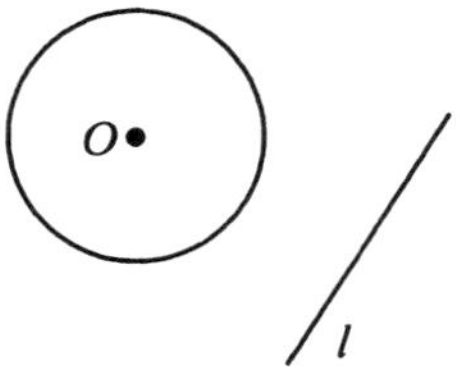

15. Construct a line that is parallel to line l and tangent to $\odot O$.

16. Construct a line that is perpendicular to line l and tangent to $\odot O$.

C 17. Construct three congruent circles, each tangent to the other two circles. Then construct an equilateral triangle, each side of which is tangent to two of the three circles.

In Exercises 18–20 begin with two circles P and Q such that $\odot P$ and $\odot Q$ do not intersect and $\odot P$ is larger than $\odot Q$. Let the radii of $\odot P$ and $\odot Q$ be p and q.

18. Construct a circle, with radius equal to PQ, that is tangent to $\odot P$ and $\odot Q$.

19. Construct a common external tangent to $\odot P$ and $\odot Q$. One method is suggested on the next page.

1. Draw a circle with center P and radius $p - q$.
2. Construct a tangent to this circle from Q, and call the point of tangency Z.
3. Draw $\overrightarrow{PZ}$. $\overrightarrow{PZ}$ intersects $\odot P$ in a point X.
4. With center X and radius ZQ, draw an arc that intersects $\odot Q$ in a point Y.
5. Draw $\overleftrightarrow{XY}$.

If you draw $\overline{QY}$, you can show that $XZQY$ is a rectangle. The rest of a justification is easy.

20. Construct a common internal tangent to $\odot P$ and $\odot Q$. (*Hint:* Draw a circle with center P and radius $p + q$.)

21. Draw any acute angle. Call its measure n. Construct two congruent circles, each tangent to the other circle. Describe clearly a procedure for constructing an isosceles triangle with the following properties.
(1) The vertex angle has measure n.
(2) The base is tangent to both of the circles.
(3) Each leg is tangent to one of the circles.

8-5 Special Segments

Construction 12

Given a segment, divide the segment into a given number of congruent parts. (3 shown)

Given: $\overline{AB}$

Construct: Points X and Y on $\overline{AB}$ so that $AX = XY = YB$

Procedure:

1. Choose any point Z not on $\overleftrightarrow{AB}$. Draw $\overrightarrow{AZ}$.
2. Using any radius, start with A as center and mark off R, S, and T so that $AR = RS = ST$.
3. Draw $\overline{TB}$.
4. At R and S construct lines parallel to $\overline{TB}$ and intersecting $\overline{AB}$ in X and Y.

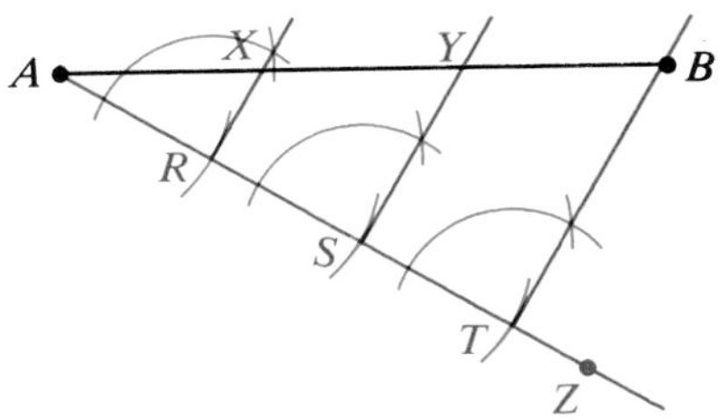

$\overline{AX}$, $\overline{XY}$, and $\overline{YB}$ are congruent parts of $\overline{AB}$.

Justification: Since the lines constructed parallel cut off congruent segments on transversal $\overleftrightarrow{AZ}$, they cut off congruent segments on transversal $\overleftrightarrow{AB}$. (It may help you to think of the parallel to $\overline{TB}$ through A.)

Construction 13

Given three segments, construct a fourth segment so that the four segments are in proportion.

Given: Segments with lengths a, b, and c

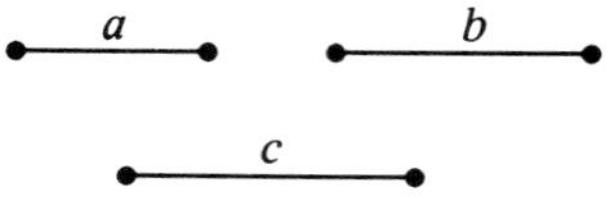

Construct: A segment of length x such that $\frac{a}{b} = \frac{c}{x}$

Procedure:

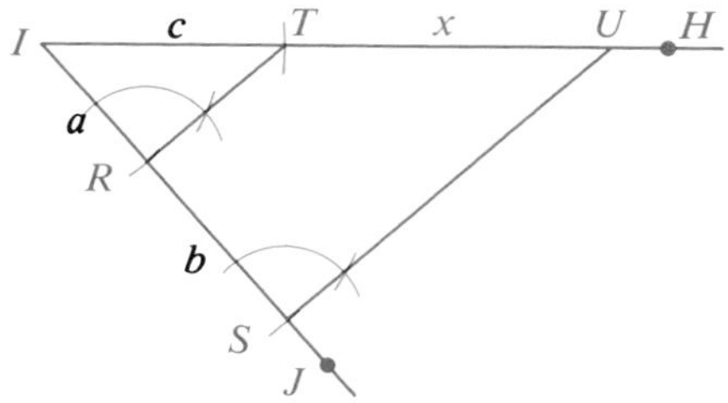

1. Draw an $\angle HIJ$.
2. On $\overrightarrow{IJ}$, mark off $IR = a$ and $RS = b$.
3. On $\overrightarrow{IH}$, mark off $IT = c$.
4. Draw $\overline{RT}$.
5. At S, construct a parallel to $\overline{RT}$, intersecting $\overrightarrow{IH}$ in a point U.

$\overline{TU}$ has length x such that $\frac{a}{b} = \frac{c}{x}$.

Justification: In $\triangle ISU$, $\overline{RT} \parallel \overline{SU}$. Therefore, $\frac{a}{b} = \frac{c}{x}$.

Construction 14

Given two segments, construct their geometric mean.

Given: Segments with lengths a and b

Construct: A segment of length x such that $\frac{a}{x} = \frac{x}{b}$ (or $x = \sqrt{ab}$)

Procedure:

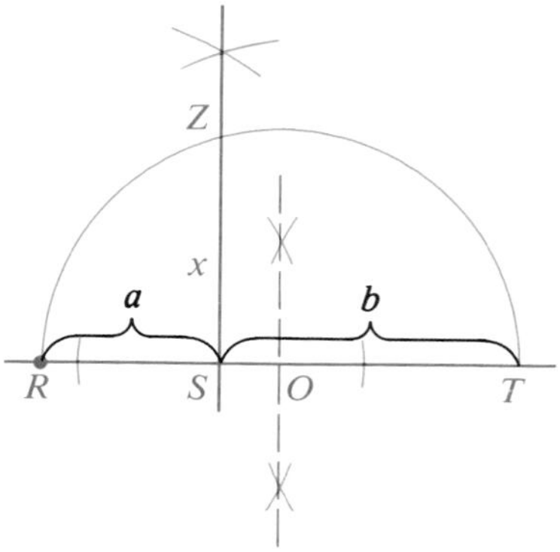

1. Draw a line and mark off $RS = a$ and $ST = b$.
2. Locate the midpoint O of $\overline{RT}$ by constructing the perpendicular bisector of $\overline{RT}$.
3. Using O as center draw a semicircle with a radius equal to OR.
4. At S, construct a perpendicular to $\overline{RT}$. The perpendicular intersects the semicircle at a point Z.

ZS, or x, is the geometric mean between a and b.

Justification: Draw $\overline{RZ}$ and $\overline{ZT}$. Since $\overparen{RZT}$ is a semicircle, $\triangle RZT$ is a right triangle. Since $\overline{ZS}$ is the altitude to the hypotenuse of rt. $\triangle RZT$, $\frac{a}{x} = \frac{x}{b}$.

Classroom Exercises

1. Given a segment, tell how to construct an equilateral triangle whose perimeter equals the length of the given segment.

Draw three segments and label their lengths *a*, *b*, and *c*.

2. Construct a segment of length x such that $\frac{c}{a} = \frac{b}{x}$.
3. Describe how to construct a segment of length x such that $x = \sqrt{2ab}$.
4. Describe how to construct a segment of length x such that $x = \sqrt{5ab}$.
5. Describe how to construct a segment of length x such that $x = \sqrt{4ab}$.

Exercises 6–11 will analyze the following problem.

Given: Line t; points A and B

Construct: A circle through A and B and tangent to t

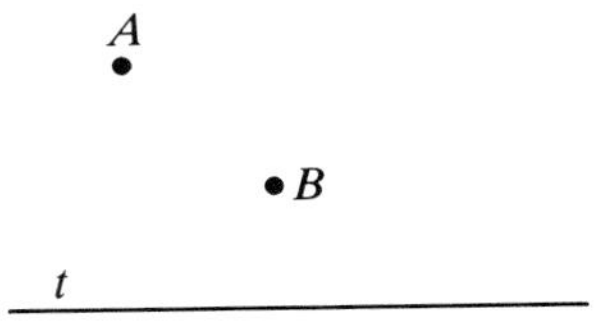

If the problem had been solved, we would have a diagram something like the one shown.

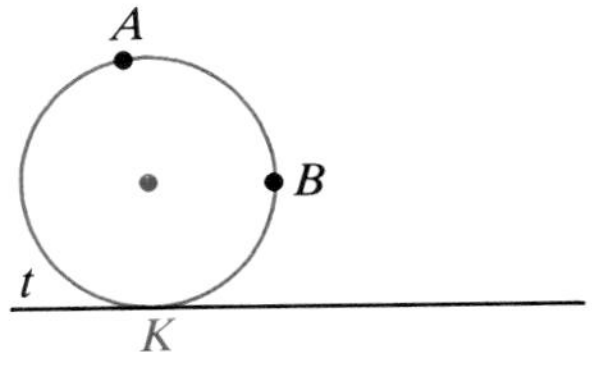

6. Where does the center of the circle lie with respect to $\overline{AB}$?
7. Where does the center of the circle lie with respect to line t and K, the point of tangency?

Note that we don't have point K located in the given diagram. Hunting for ideas, we draw $\overleftrightarrow{AB}$. We now have a point J, which we can locate in the given diagram.

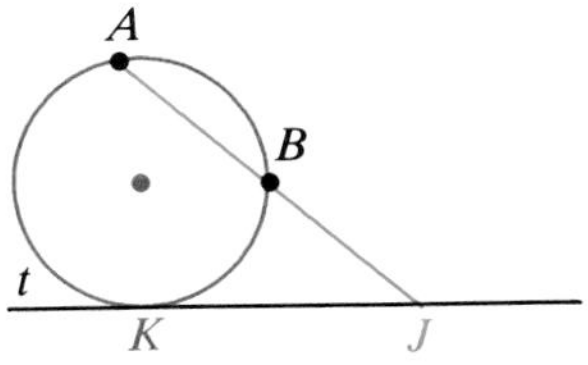

8. State an equation that relates JK to JA and JB.
9. Rewrite your equation in the form $\frac{?}{JK} = \frac{JK}{?}$.
10. What construction can we use to get the length JK?

In a *separate* diagram we can mark off the lengths JA and JB on some line l and then use Construction 14 to find x such that $\frac{JA}{x} = \frac{x}{JB}$. Once we have x, which equals JK, we return to the given diagram and draw an arc to locate K.

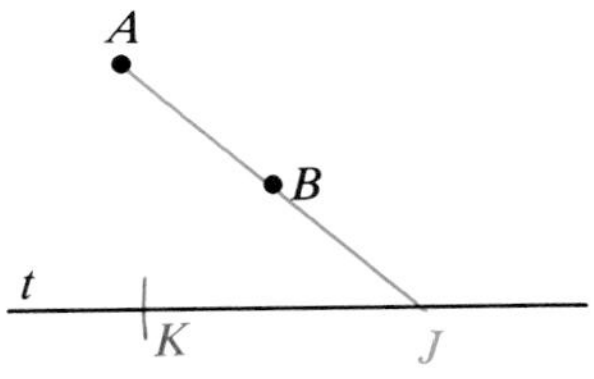

11. Explain how to complete the construction of the circle.

Written Exercises

In each of Exercises 1–4 begin by drawing $\overline{AB}$, roughly 15 cm long.

A **1.** Divide $\overline{AB}$ into three congruent segments.

2. a. Use Construction 12 to divide $\overline{AB}$ into four congruent segments.

b. Use Construction 4 to divide $\overline{AB}$ into four congruent segments.

3. a. Use Construction 12 to divide $\overline{AB}$ into five congruent segments.

b. Can Construction 4 be used to divide $\overline{AB}$ into five congruent segments?

c. Divide $\overline{AB}$ into two segments that have the ratio 2:3.

4. Divide $\overline{AB}$ into two segments that have the ratio 3:4.

On your paper draw four segments roughly as long as those shown below. Use your segments in Exercises 5–14. In each exercise construct a segment that has length x.

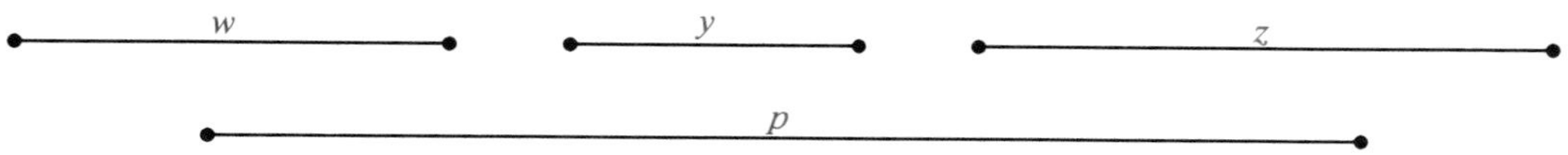

5. $\frac{y}{w} = \frac{z}{x}$ **6.** $\frac{w}{x} = \frac{x}{y}$ **7.** $x = \sqrt{yp}$ **8.** $3x = w + 2y$

B **9.** $zx = wy$ (*Hint:* First write a proportion that is equivalent to the given equation and has x as the last term.)

10. $x = \frac{yp}{z}$ **11.** $x = \frac{1}{3}\sqrt{yp}$ **12.** $x = \sqrt{3wz}$ **13.** $x = \sqrt{6yz}$

14. Construct $\overline{AB}$, with $AB = p$. Divide $\overline{AB}$ into two parts that have the ratio $w:y$.

15. Draw a segment like the one shown and let its length be 1. Use the segment to construct a segment of length $\sqrt{15}$.

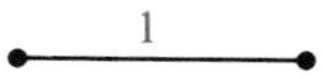

C **16.** Draw a segment about 20 cm long. Label the endpoints C and D. Construct a triangle whose perimeter is equal to CD and whose sides are in the ratio 2:2:3.

★17. To trisect a general angle G, a student tried this procedure:

1. Mark off $\overline{GA}$ congruent to $\overline{GB}$.
2. Draw $\overline{AB}$.
3. Divide $\overline{AB}$ into three congruent parts using Construction 12.
4. Draw $\overline{GX}$ and $\overline{GY}$.

Show that the student did not trisect $\angle G$. That is, show that $m\angle 2 \neq m\angle 1$. (*Hint:* Show that $GA > GY$ and use an indirect proof.)

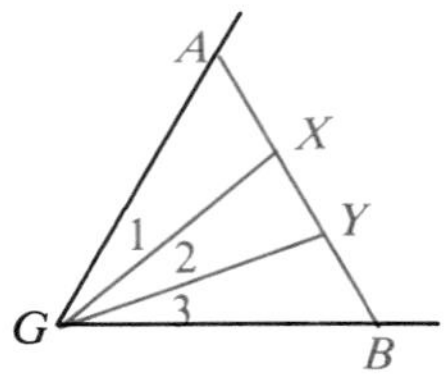

CAREERS

Cartographer

When you think of a map, do you think of a piece of paper with colored areas and lines? Surprisingly, some maps now consist of thousands, or even millions, of numbers stored on computer tapes. Obviously *cartography,* or map-making, is changing.

Several technical advances have led to changes in mapping. Space satellites carrying scanners produce extremely detailed images of the entire world at regular intervals. Besides conventional photographs, these scanners also record images using infrared and other wavelengths beyond the range of visible light. After processing by computer, such images provide many more kinds of information than the traditional political boundaries and topographic features of conventional maps. For example, they can map soil types and land use, distinguishing among farm fields, forests, and urban areas. In fact, they can even differentiate a corn field from a soybean field, or a freshly plowed field from a field with a mature crop.

In the false-color map shown below of Oregon and Washington, vegetation appears as red, dry regions are blue-green, water is black, and snow on the Cascade Mountains and the Olympic Mountains is white.

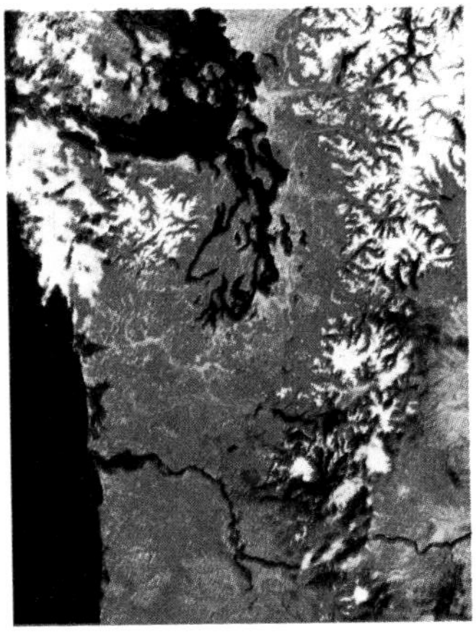

Both new images and conventional map data are being digitized, that is, converted to numerical codes and stored on computer tape. As new images are received, changes in physical features are coded and recorded. Map users can thus be provided with maps that are constantly being revised and kept up to date.

Self-Test 2

1. Draw a large $\odot O$. Choose a point A that is outside $\odot O$. Construct the two tangents to $\odot O$ from point A.
2. Draw a very large obtuse triangle. Construct the inscribed circle.
3. Draw a segment about half as long as the width of your paper. Then divide the segment by construction into two segments whose lengths have the ratio 2:1.
4. Draw a large $\triangle ABC$. Then construct $\overline{DE}$ such that $\frac{AB}{BC} = \frac{AC}{DE}$.
5. Use $\triangle ABC$ drawn in Exercise 4 to construct a segment, $\overline{PQ}$, whose length is the geometric mean of AB and AC.
6. You are given $\odot S$ and diameter $\overline{FG}$. To construct parallel tangents to $\odot S$, you could construct a line that is __?__ to $\overline{FG}$ at __?__ and a line that is __?__ to $\overline{FG}$ at __?__.
7. You are given $\triangle TRI$. Describe the three steps you would use to circumscribe a circle about $\triangle TRI$.

Locus

Objectives

1. Describe the locus that satisfies a given condition.
2. Describe the locus that satisfies more than one given condition.
3. Apply the concept of locus in the solution of construction exercises.

8-6 The Meaning of Locus

A radar system is used to determine the position, or *locus,* of airplanes relative to an airport. In geometry **locus** means a figure that is the set of all points, and only those points, that satisfy one or more conditions.

Suppose we have a line k in a plane and wish to picture the locus of points in the plane that are 1 cm from k. Several points are shown in the first diagram below.

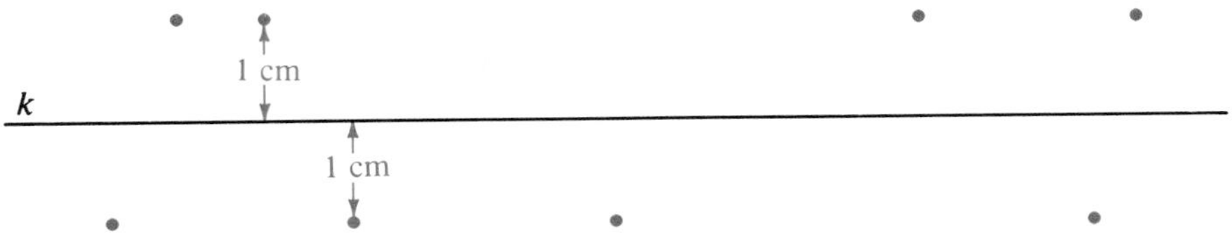

All the points satisfying the given conditions are indicated in the next diagram. You see that the required locus is a pair of lines parallel to, and 1 cm from, k.

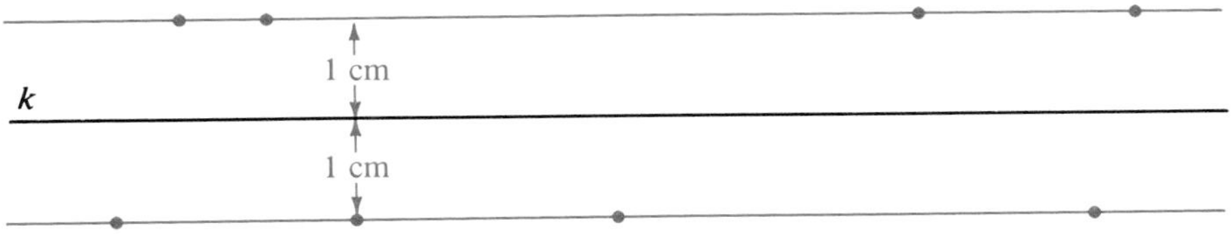

Suppose we wish to picture the locus of points 1 cm from k without requiring the points to be *in a plane*. The problem changes. Now you need to consider all the points in space that are 1 cm from line k. The required locus is a cylindrical surface with axis k and a 1 cm radius, as shown below. Of course, the surface will extend in both directions without end, just as line k does.

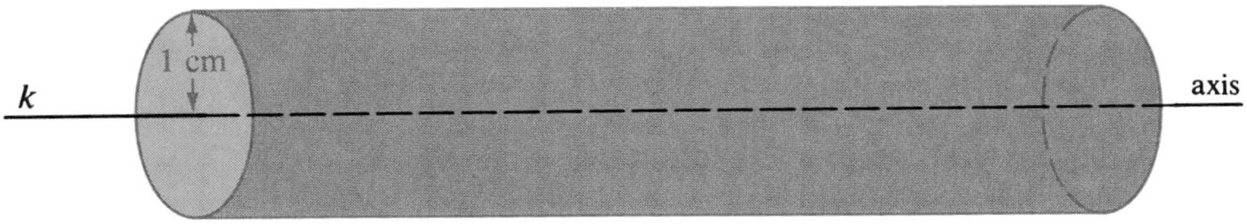

When you are solving a locus problem, always think in terms of three dimensions unless the statement of the problem restricts the locus to a plane.

Classroom Exercises

1. Draw a point A on the chalkboard.
 a. Draw several points on the chalkboard that are 20 cm from A.
 b. Draw all the points on the chalkboard that are 20 cm from A.
 c. Complete: The locus of all points on the chalkboard that are 20 cm from point A is _?_.
 d. Remove the restriction that the points must lie in the plane of the chalkboard. Now describe the locus.

2. Draw two parallel lines k and l.
   a. Draw several points that are in the plane containing k and l and are equidistant from k and l.
   b. Draw all the points that are in the plane containing k and l and are equidistant from k and l.
   c. Describe the locus of points that are in the plane of two parallel lines and equidistant from them.
   d. Remove the restriction that the points must lie in the plane of the two lines. Now describe the locus.
3. Draw an angle.
   a. Draw several points in the plane of the angle that are equidistant from the sides of the angle.
   b. Draw all the points in the plane of the angle that are equidistant from the sides of the angle.
   c. Describe the locus of points in the plane of a given angle that are equidistant from the sides of the angle.
4. What is the locus of points in your classroom that are equidistant from the ceiling and floor?
5. What is the locus of points in your classroom that are 1 m from the floor?
6. Choose a point P on the floor of the classroom.
   a. What is the locus of points, on the floor, that are 1 m from P?
   b. What is the locus of points, in the room, that are 1 m from P?
7. What is the locus of points in your classroom that are equidistant from the ceiling and floor and are also equidistant from the two side walls?
8. Draw a circle with radius 6 cm. Use the following definition of *distance from a circle:* A point P is x cm from a circle if there is a point of the circle that is x cm from P and no point of the circle is less than x cm from P.
   a. Draw all the points in the plane of the circle that are 2 cm from the circle.
   b. Complete: Given a circle with a 6 cm radius, the locus of all points in the plane of the circle and 2 cm from the circle is __?__.
   c. Remove the restriction that the points must lie in the plane of the circle. Now describe the locus.
9. Make up a locus problem for which the locus doesn't contain any points.

Written Exercises

Exercises 1–4 deal with figures in a plane. Draw a diagram showing the locus. Then write a description of the locus.

A

1. Given two points A and B, what is the locus of points equidistant from A and B?
2. Given parallel lines j and k, what is the locus of points equidistant from j and k?
3. Given a point O, what is the locus of points 2 cm from O?
4. Given a line h, what is the locus of points 2 cm from h?

In Exercises 5–8, begin each exercise with a square $ABCD$ that has sides 4 cm long. Draw a diagram showing the locus of the points inside the square that satisfy the given conditions. Then write a description of the locus.

5. Equidistant from $\overline{AB}$ and $\overline{CD}$
6. Equidistant from points B and D
7. Equidistant from $\overline{AB}$ and $\overline{BC}$
8. Equidistant from all four sides

Exercises 9–12 deal with figures in space.

9. Given two parallel planes, what is the locus of points equidistant from the two planes?
10. Given a plane, what is the locus of points 5 cm from the plane?
11. Given point E, what is the locus of points 3 cm from E?
12. Given points C and D, what is the locus of points equidistant from C and D?

Exercises 13–17 deal with figures in a plane. (*Note:* If a point in a segment or an arc is not included in the locus, indicate the point by a small circle.)

B 13. a. Draw an angle HEX. Construct the locus of points equidistant from the sides of $\angle HEX$.
b. Draw two intersecting lines j and k. Construct the locus of points equidistant from j and k.

14. Draw a segment $\overline{DE}$ and a line n. Construct the locus of points whose distance from n is DE.
15. Draw a segment $\overline{AB}$. Construct the locus of points P such that $\angle APB$ is a right angle.
16. Draw a segment $\overline{CD}$. Construct the locus of points Q such that $\triangle CQD$ is isosceles with base $\overline{CD}$.
17. Draw a circle. Construct the locus of the midpoints of all radii of the circle.

Exercises 18–20 deal with figures in space.

18. Given a sphere, what is the locus of the midpoints of the radii of the sphere?
19. Given a square, what is the locus of points equidistant from the sides?
20. Given a scalene triangle, what is the locus of points equidistant from the vertices?

C 21. A ladder leans against a house. As A moves up or down on the wall, B moves along the ground. What path is followed by midpoint M? (*Hint:* Experiment with a meter stick, a wall, and the floor.)

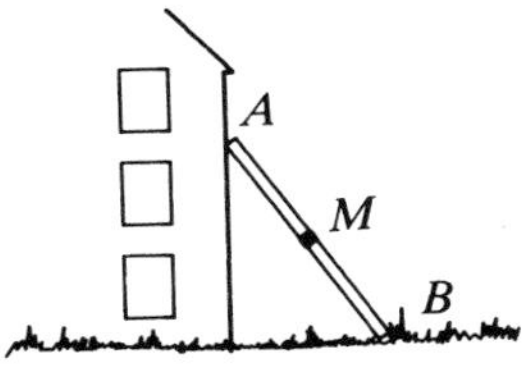

22. Given a segment $\overline{CD}$, what is the locus in space of points P such that $m\angle CPD = 90$?

23. A goat is tied to a square shed as shown. Using the scale 1:100, carefully draw a diagram that shows the region over which the goat can graze.

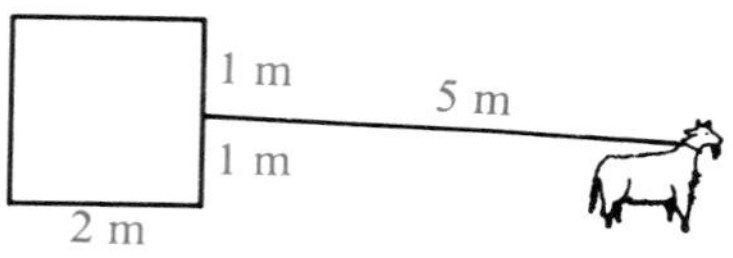

24. A tight wire $\overline{AC}$ is stretched between the tops of two vertical posts $\overline{AB}$ and $\overline{CD}$ that are 5 m apart and 2 m high. A ring, at one end of a 6 m leash, can slide along $\overline{AC}$. A dog is tied to the other end of the leash. Draw a diagram that shows the region over which the leashed dog can roam. Use the scale 1:100.

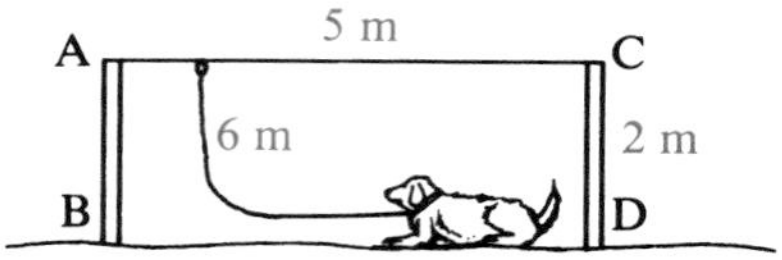

8-7 Locus Problems

The plural of *locus* is *loci*. The following problem involves intersections of loci.

Suppose you are given three noncollinear points *A*, *B*, and *C*. In the plane of *A*, *B*, and *C*, what is the locus of points that are 1 cm from *A* and are, at the same time, equidistant from *B* and *C*?

You can analyze one part of the problem at a time.

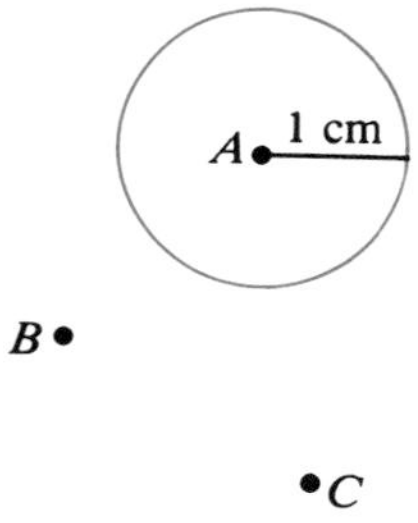

The locus of points 1 cm from *A* is ⊙*A* with radius 1 cm.

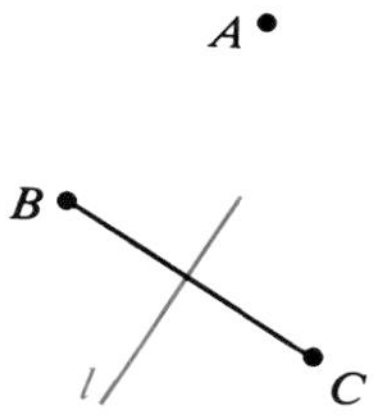

The locus of points equidistant from *B* and *C* is *l*, the perpendicular bisector of $\overline{BC}$.

The locus of points satisfying *both* conditions must lie on both circle *A* and line *l*. There are three possibilities, depending on the positions of *A*, *B*, and *C*, as shown below.

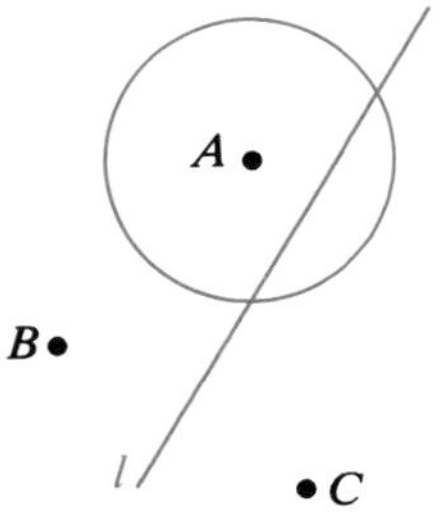

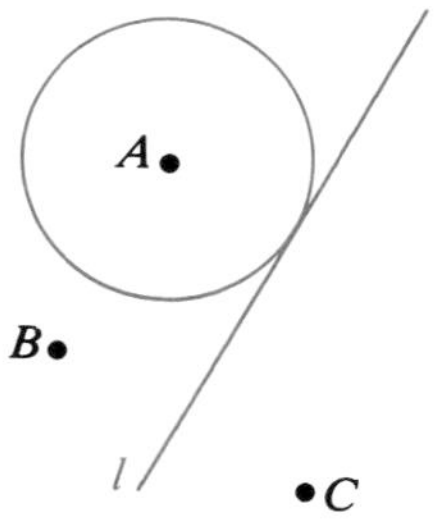

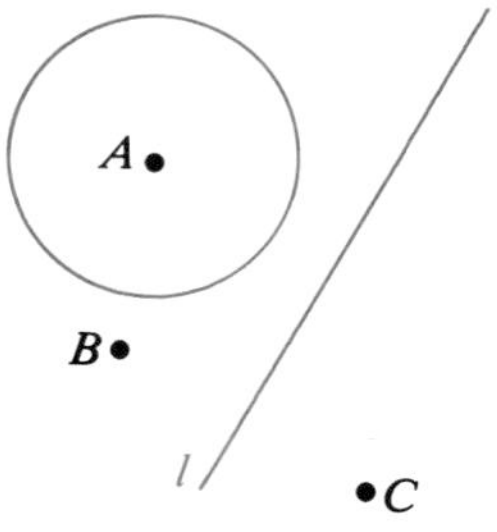

All three can be described in one sentence:

The locus is two points, one point, or no points, depending on the intersection of the circle with center A and radius 1 cm and the line that is the perpendicular bisector of $\overline{BC}$.

The example that follows deals with the corresponding problem in three dimensions.

Example Given three noncollinear points A, B, and C, what is the locus of points 1 cm from A and equidistant from B and C?

Solution

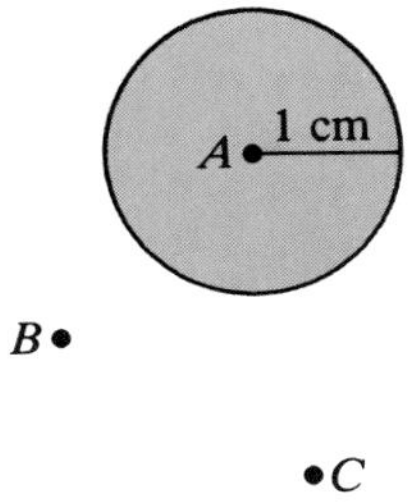

The first locus is sphere A with radius 1 cm.

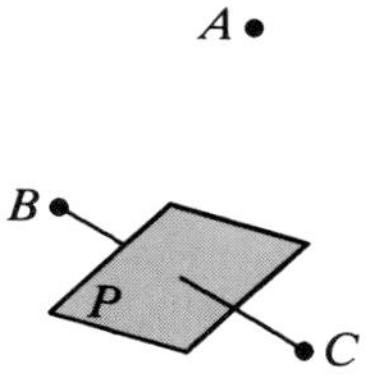

The second locus is plane P, the perpendicular bisector of $\overline{BC}$.

Possibilities:

The plane might cut the sphere in a circle.
The plane might be tangent to the sphere.
The plane might not have any points in common with the sphere.

Thus, the locus is a circle, one point, or no points, depending on the intersection of the sphere with center A and radius 1 cm and the plane which is the perpendicular bisector of $\overline{BC}$.

Classroom Exercises

Exercises 1–4 refer to coplanar figures. Describe the intersections that are possible.

1. A line and a circle

2. Two circles

3. Two parallel lines and a circle

4. Two perpendicular lines and a circle

5. Consider the following problem: In a plane, what is the locus of points that are equidistant from the sides of $\angle A$ and are equidistant from two points B and C?

a. The locus of points equidistant from the sides of $\angle A$ is __?__.

b. The locus of points equidistant from B and C is __?__.

c. Draw diagrams to show three possibilities with regard to points that satisfy both conditions (a) and (b).

d. Describe the locus.

Exercises 6–9 refer to figures in space. Describe the intersections that are possible.

6. A line and a plane
7. A line and a sphere
8. Two spheres
9. A plane and a sphere

10. Let C be the point in the center of your classroom (*not* the center of the floor). Describe the locus of points in the room that satisfy the given conditions.
    a. 3 m from C
    b. 3 m from C and equidistant from the ceiling and the floor
    c. 3 m from C and 1 m from either the ceiling or the floor

Written Exercises

Exercises 1–4 refer to plane figures.

A

1. Draw a new $\odot O$ for each part. Then place two points A and B outside $\odot O$ so that the locus of points on $\odot O$ and equidistant from A and B is:
   a. 2 points **b.** 0 points **c.** 1 point
2. Draw two parallel lines m and n. Then place two points R and S so that the locus of points equidistant from m and n and also equidistant from R and S is:
   a. 1 point **b.** 1 line **c.** 0 points
3. Consider the following problem: Given two points D and E, what is the locus of points 1 cm from D and 2 cm from E?
   a. The locus of points 1 cm from D is __?__.
   b. The locus of points 2 cm from E is __?__.
   c. Draw diagrams to show three possibilities with regard to points that satisfy both conditions (a) and (b).
   d. Give a one-sentence solution to the problem.
4. Consider the following problem: Given a point A and a line k, what is the locus of points 3 cm from A and 1 cm from k?
   a. The locus of points 3 cm from A is __?__.
   b. The locus of points 1 cm from k is __?__.
   c. Draw diagrams to show five possibilities with regard to points that satisfy both conditions (a) and (b).
   d. Give a one-sentence solution to the problem.

Exercises 5–10 refer to plane figures. Draw a diagram of the locus. Then write a description of the locus.

5. Point P lies on line l. What is the locus of points on l and 3 cm from P?
6. Point Q lies on line l. What is the locus of points 5 cm from Q and 3 cm from l?

7. Points A and B are 3 cm apart. What is the locus of points 2 cm from both A and B?

8. Lines j and k intersect in point P. What is the locus of points equidistant from j and k, and 2 cm from P?

9. Given $\angle A$, what is the locus of points equidistant from the sides of $\angle A$ and 2 cm from vertex A?

10. Given $\triangle RST$, what is the locus of points equidistant from $\overline{RS}$ and $\overline{RT}$ and also equidistant from R and S?

In Exercises 11–14 draw diagrams to show the possibilities with regard to points in a plane.

B 11. Given points C and D, what is the locus of points 2 cm from C and 3 cm from D?

12. Given point E and line k, what is the locus of points 3 cm from E and 2 cm from k?

13. Given a point A and two parallel lines j and k, what is the locus of points 30 cm from A and equidistant from j and k?

14. Given four points P, Q, R, and S, what is the locus of points that are equidistant from P and Q and equidistant from R and S?

Exercises 15–18 refer to figures in space. In each exercise tell what the locus is. You need not draw the locus or describe it precisely.

Example Given two parallel planes and a point A, what is the locus of points equidistant from the planes and 3 cm from A?

Solution The locus is a circle, a point, or no points.

15. Given plane Z and point B outside Z, what is the locus of points in Z that are 3 cm from B?

16. Given $\overleftrightarrow{AB} \perp$ plane Q, what is the locus of points 2 cm from $\overleftrightarrow{AB}$ and 2 cm from Q?

17. Given square $ABCD$, what is the locus of points equidistant from the vertices of the square?

18. Given point A in plane Z, what is the locus of points 5 cm from A and d cm from Z? (More than 1 possibility)

19. Points R, S, T, and W are not coplanar and no three of them are collinear.
    a. The locus of points equidistant from R and S is __?__.
    b. The locus of points equidistant from R and T is __?__.
    c. The loci found in (a) and (b) intersect in a __?__, and all points in this __?__ are equidistant from points R, S, and T.
    d. The locus of points equidistant from R and W is __?__.
    e. The intersection of the figures found in (c) and (d) is a __?__. This __?__ is equidistant from the four given points.

C **20.** Can you locate four points R, S, T, and W so that the locus of points equidistant from R, S, T, and W is named below? If the answer is *yes*, describe the location of the points R, S, T, and W.

a. a point **b.** a line **c.** a plane **d.** no points

21. Assume that the Earth is a sphere. How many points are there on the Earth's surface that are equidistant from

a. Houston and Toronto?

b. Houston, Toronto, and Los Angeles?

c. Houston, Toronto, Los Angeles, and Mexico City?

22. A mini-radio transmitter has been secured to a bear. Rangers at D, E, and F are studying the bear's movements. Rangers D and E can receive the bear's beep at distances up to 10 km, ranger F at distances up to 15 km.

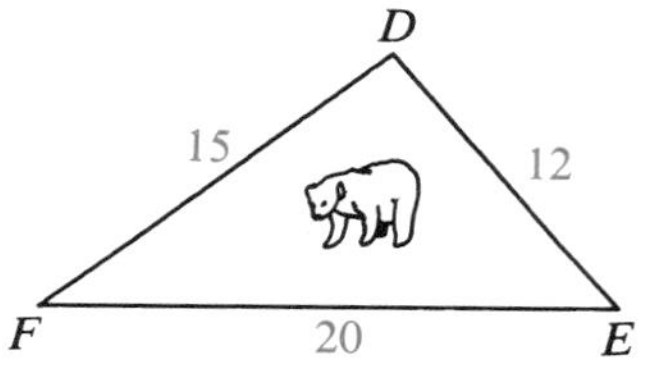

Draw a diagram showing where the bear might be at these times:

a. When all three rangers can receive the signal

b. When F suddenly detects the signal after a period of time during which only D and E could receive the signal

c. When station D is shut down, and F begins to detect the signal just as E loses it

8-8 Locus and Construction

Sometimes the solution to a construction problem depends on finding a point that satisfies more than one condition. To locate the point, you may have to construct the locus of points satisfying one of the conditions.

Example Given the angle and the segments shown, construct $\triangle ABC$ with $m\angle A = n$, $AB = r$, and the altitude to $\overleftrightarrow{AB}$ having length s.

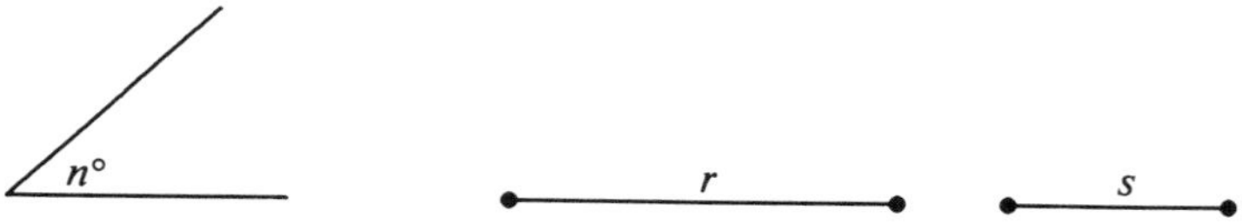

Solution It is easy to construct $\angle A$ and side $\overline{AB}$. Point C must satisfy two conditions: C must lie on $\overrightarrow{AZ}$, and C must be s units from $\overleftrightarrow{AB}$. The locus of points s units from $\overleftrightarrow{AB}$ is a pair of parallel lines. Only the upper parallel will intersect $\overrightarrow{AZ}$. We construct that parallel to $\overleftrightarrow{AB}$ as follows:

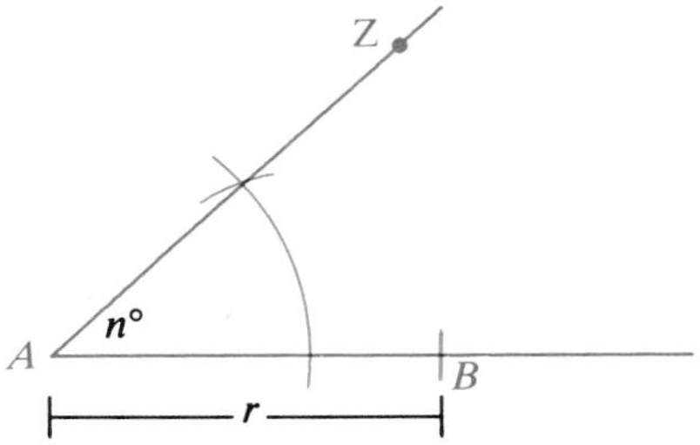

1. Construct the perpendicular to $\overleftrightarrow{AB}$ at any convenient point X.
2. Mark off s units on the perpendicular to locate point Y.
3. Construct the perpendicular to $\overrightarrow{XY}$ at Y. Call it $\overleftrightarrow{YW}$.

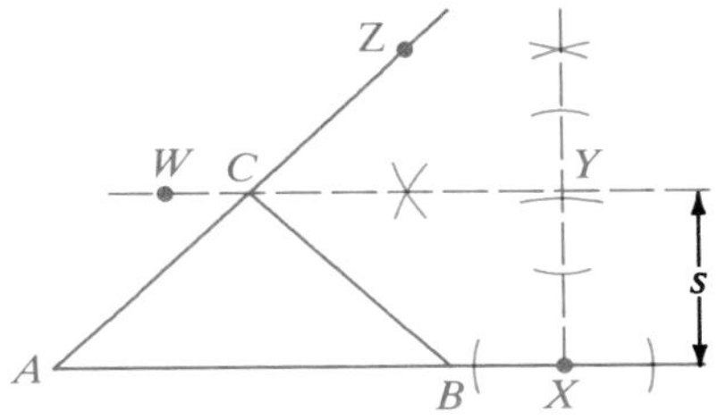

Note that all points on $\overleftrightarrow{YW}$ are s units from $\overleftrightarrow{AB}$. Thus the intersection of $\overleftrightarrow{YW}$ and $\overrightarrow{AZ}$ is the desired point C. To complete the solution, we simply draw $\overline{CB}$.

Classroom Exercises

1. The purpose of this exercise is to analyze the following construction problem:

Given a circle and a segment with length k, inscribe in the circle an isosceles triangle RST with base $\overline{RS}$ k units long.

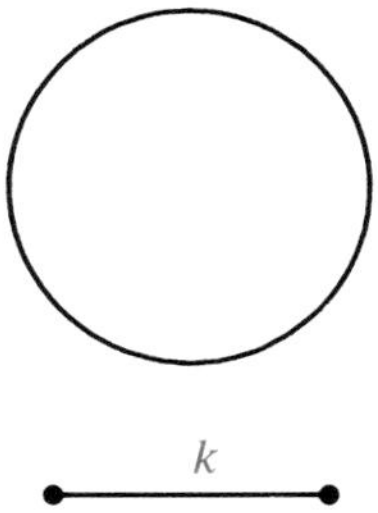

a. Suppose R has been chosen. Where must S lie so that RS equals k? (In other words, what is the locus of points k units from R?)

b. Suppose $\overline{RS}$ is fixed. Where must T lie so that $RT = ST$? (In other words, what is the locus of points equidistant from R and S?)

c. Explain the steps of the construction shown.

(1)

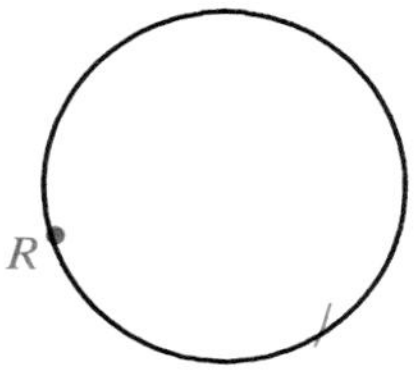

(2)

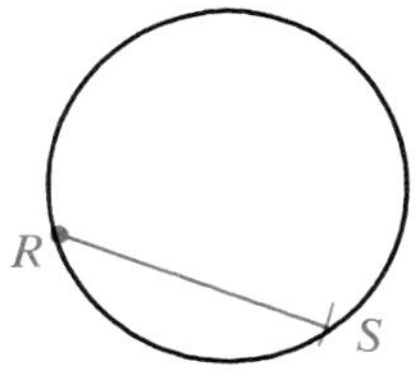

(3)

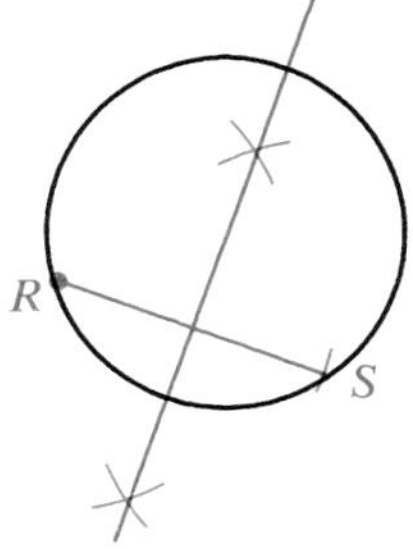

d. Explain two different ways to finish the construction.

2. Two different solutions, both correct, are shown for the following construction problem. Analyze the diagrams and explain the solutions.

Given segments with lengths r and s, construct $\triangle ABC$ with $m\angle C = 90$, $AC = r$, and $AB = s$.

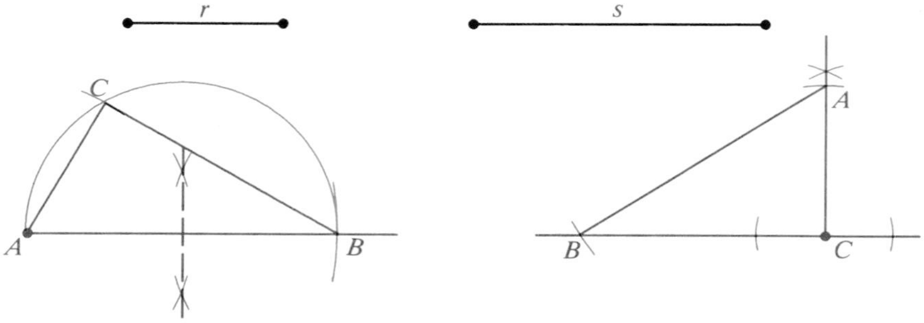

First solution Second solution

Written Exercises

A 1. Draw any $\overline{AB}$ and a segment with length h. Construct the locus of points P such that for every $\triangle APB$ the altitude from P to $\overleftrightarrow{AB}$ would equal h.

2. Begin each part of this exercise by drawing any $\overline{CD}$. Then construct the locus of points P that meet the given condition.
   a. $\angle CDP$ is a right angle.
   b. $\angle CPD$ is a right angle. (*Hint:* See Classroom Exercise 2.)

On your paper draw a segment roughly as long as the one shown. Use it in Exercises 3 and 4.

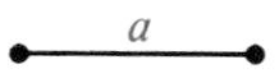

3. Draw an angle XYZ. Construct a circle, with radius a, that is tangent to the sides of $\angle XYZ$.

4. Draw a figure roughly like the one shown. Then construct a circle, with radius a, that passes through N and is tangent to line k. (*Hint:* Construct the locus of points that would, as centers, be the correct distance from k. Also construct the locus of points that would, as centers, be the correct distance from N.)

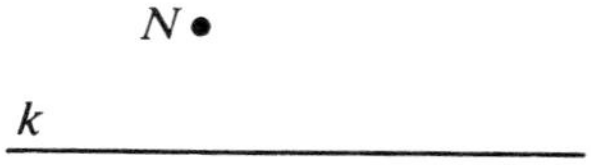

On your paper draw an angle and three segments roughly like those shown. Use them in Exercises 5–19. You may find it helpful to begin with a sketch.

5. Construct $\overline{AB}$ so that $AB = t$. Then construct the locus of all points C so that in $\triangle ABC$ the altitude from C has length r.

6. Construct $\overline{AB}$ so that $AB = t$. Then construct the locus of all points C so that in $\triangle ABC$ the median from C has length s.

B 7. Construct isosceles $\triangle ABC$ so that $AB = AC = t$ and so that the altitude from A has length s.

8. Construct an isosceles trapezoid $ABCD$ with $\overline{AB}$ the shorter base, with $AB = AD = BC = t$, and with an altitude having length r.

9. Construct $\triangle ABC$ so that $AB = t$, $AC = s$, and the median to $\overline{AB}$ has length r.

10. Construct $\triangle ABC$ with $m\angle A = m\angle B = n$ and the altitude to $\overline{AB}$ having length s.

11. Construct $\triangle ABC$ with $m\angle C = 90$, $m\angle A = n$, and the altitude to $\overline{AB}$ having length s.

12. Construct $\triangle ABC$ with $AB = s$, $AC = t$, and the altitude to $\overline{AB}$ having length r.

13. Construct $\triangle ABC$ with $AB = t$, the median to $\overline{AB}$ and the altitude to $\overline{AB}$ having lengths s and r, respectively.

14. Construct a right triangle with the altitude to the hypotenuse and the median to the hypotenuse having lengths r and s, respectively.

15. Construct both an acute isosceles triangle and an obtuse isosceles triangle such that each leg has length s and each altitude to a leg has length r.

C 16. Construct a square whose sides each have length $4s$. A segment of length $3s$ moves so that its endpoints are always on the sides of the square. Construct the locus of the midpoint of the moving segment.

17. Construct a right triangle such that the bisector of the right angle divides the hypotenuse into segments whose lengths are r and s.

18. Construct an isosceles right triangle such that the radius of the inscribed circle is r.

19. Construct $\overline{AB}$ so that $AB = t$. Then construct the locus of points P such that $m\angle APB = n$.

Challenge

Given $\overline{AB}$, it is possible to construct the midpoint M of $\overline{AB}$ using only a compass (and *no* straightedge). Study the diagram until you understand the procedure. Then draw $\overline{AB}$, about 10 cm long, construct its midpoint M as shown, and prove that M is the midpoint.

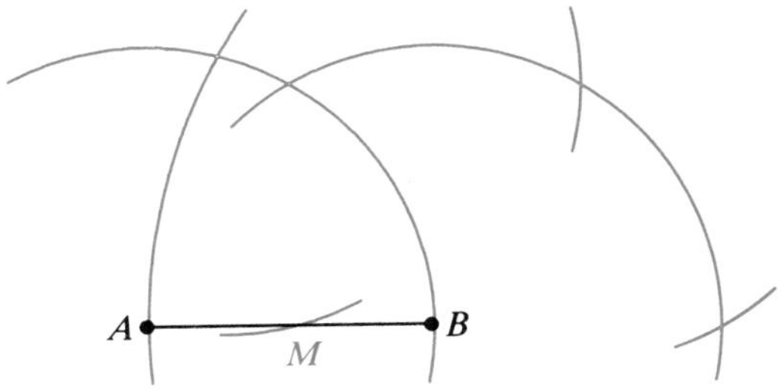

EXTRA

The Nine-Point Circle

Given any $\triangle ABC$, let H be the intersection of the three altitudes. There is a circle that passes through these nine special points:

midpoints L, M, N of the three sides

points R, S, T, where the three altitudes of the triangle meet the sides

midpoints X, Y, Z of $\overline{HA}$, $\overline{HB}$, $\overline{HC}$

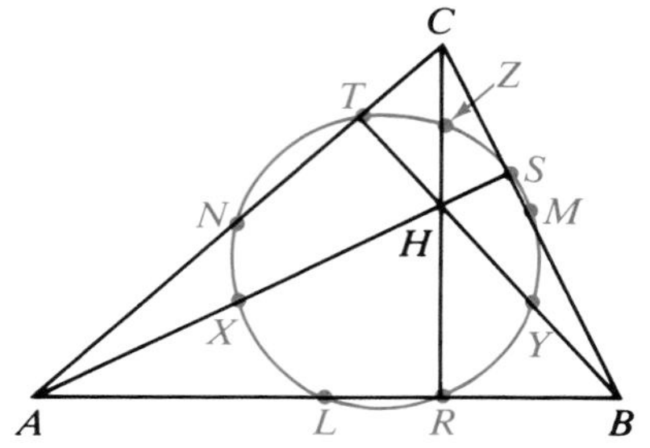

Outline of proof:

1. $XYMN$ is a rectangle.
2. The circle circumscribed about $XYMN$ has diameters $\overline{MX}$ and $\overline{NY}$.
3. Because $\angle XSM$ and $\angle YTN$ are right angles, the circle contains points S and T as well as X, Y, M, and N.
4. $XLMZ$ is a rectangle.
5. The circle circumscribed about $XLMZ$ has diameters $\overline{MX}$ and $\overline{LZ}$.
6. Because $\angle XSM$ and $\angle ZRL$ are right angles, the circle contains points S and R as well as X, L, M, and Z.
7. The circle of Steps 1–3 and the circle of Steps 4–6 must be the same circle, because $\overline{MX}$ is a diameter of both circles.
8. There is a circle that passes through the nine points, L, M, N, R, S, T, X, Y, and Z. (See Steps 3 and 6.)

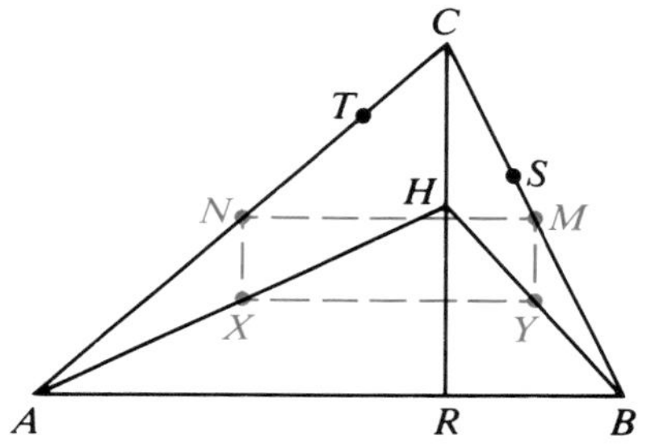

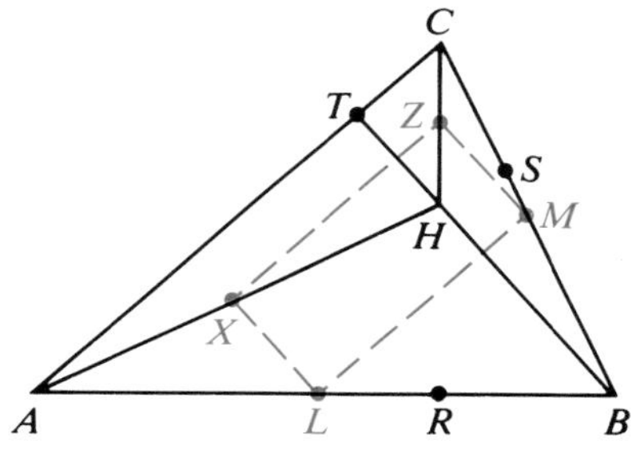

One way to locate the center of the circle is to locate points X and M, then the midpoint of $\overline{XM}$.

Exercises

1. Test your mechanical skill by constructing the nine-point circle for an acute triangle. (The larger the figure, the better.)
2. Repeat Exercise 1, but use an obtuse triangle.
3. Repeat Exercise 1, but use an equilateral triangle. What happens to some of the nine points?

4. Repeat Exercise 1, but use a right triangle. How many of the nine points are at the vertex of the right angle?
5. Prove that $XYMN$ is a rectangle. Use the diagram shown for Steps 1–3 of the outline of proof. (*Hint:* Compare $\overline{NX}$ with $\overline{CR}$ and $\overline{NM}$ with $\overline{AB}$.)
6. What is the ratio of the radius of the nine-point circle to the radius of the circumscribed circle?

Self-Test 3

Describe briefly the locus of points that satisfy the conditions.

1. In the plane of two intersecting lines j and k, and equidistant from the lines
2. In space and t units from point P
3. In space and equidistant from points W and X that are 10 cm apart
4. In the plane of $\angle DEF$, equidistant from the sides of the angle, and 4 cm from $\overrightarrow{EF}$
5. In the plane of two parallel lines s and t, equidistant from s and t, and 4 cm from a particular point A in the plane (three possibilities)
6. Construct a large isosceles $\triangle RST$. Then construct the locus of points that are equidistant from the vertices of $\triangle RST$.
7. Draw a long segment, $\overline{BC}$, and an acute angle, $\angle 1$. Construct a right triangle with an acute angle congruent to $\angle 1$ and hypotenuse congruent to $\overline{BC}$.

Chapter Summary

1. Geometric constructions are diagrams that are drawn using only a straightedge and a compass.
2. Basic constructions:
   (1) A segment congruent to a given segment, page 335
   (2) An angle congruent to a given angle, page 336
   (3) The bisector of a given angle, page 336
   (4) The perpendicular bisector of a given segment, page 339
   (5) A line perpendicular to a given line at a given point on the line, page 340
   (6) A line perpendicular to a given line from a given point outside the line, page 340
   (7) A line parallel to a given line through a given point outside the line, page 341
   (8) A tangent to a given circle at a given point on the circle, page 349
   (9) A tangent to a given circle from a given point outside the circle, page 350

(10) A circle circumscribed about a given triangle, page 350
(11) A circle inscribed in a given triangle, page 351
(12) Division of a given segment into any number of congruent parts, page 353
(13) A segment of length x such that $\frac{a}{b} = \frac{c}{x}$ when segments of length a, b, and c are given, page 354
(14) A segment whose length is the geometric mean between the lengths of two given segments, page 354

3. Every triangle has these concurrency properties:
   (1) The bisectors of the angles intersect in a point that is equidistant from the three sides of the triangle.
   (2) The perpendicular bisectors of the sides intersect in a point that is equidistant from the three vertices of the triangle.
   (3) The lines that contain the altitudes intersect in a point.
   (4) The medians intersect in a point that is two thirds of the distance from each vertex to the midpoint of the opposite side.
4. A locus is the set of all points, and only those points, that satisfy one or more conditions.
5. A locus that satisfies more than one condition is found by considering all possible intersections of the loci for the separate conditions.

Chapter Review

In Exercises 1–3 draw a diagram that is similar to, but larger than, the one shown. Then do the constructions.

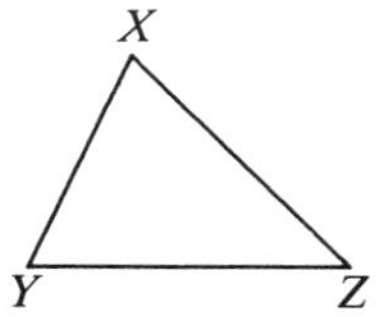

1. Draw any line m. On m construct $\overline{ST}$ such that $ST = 3XY$. 8–1
2. Construct an angle with measure equal to $m\angle X + m\angle Z$.
3. Bisect $\angle Y$.

Use a diagram like the one below for Exercises 4–7.

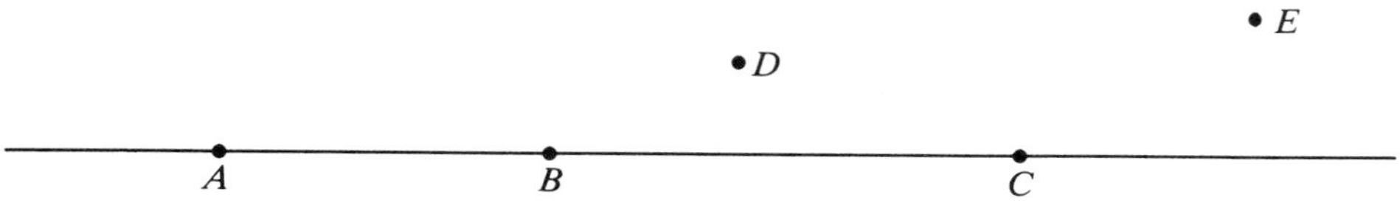

4. Construct the perpendicular bisector of $\overline{AB}$. 8–2
5. Construct the perpendicular to $\overleftrightarrow{AC}$ at C.
6. Construct the perpendicular to $\overleftrightarrow{AC}$ from D.
7. Construct the parallel to $\overleftrightarrow{AC}$ through E.

8. The ___?___ of a triangle intersect in a point that is equidistant from the vertices of the triangle. 8–3

9. The ___?___ of a triangle intersect in a point that is equidistant from the sides of the triangle.

10. If $MR = 12$, then $MP =$ ___?___.

11. $QR:RO =$ ___?___ (numerical answer)

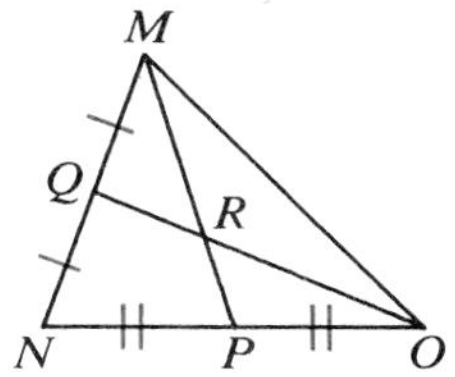

Exs. 10, 11

Draw a large $\odot O$. Label a point F on $\odot O$ and a point G outside $\odot O$.

12. Construct the tangent to $\odot O$ at F. 8–4

13. Construct a tangent to $\odot O$ from G.

14. Draw a large acute triangle. Find, by construction, the center of the circle that could be inscribed in the triangle.

15. Draw a large obtuse triangle. Construct a circle that circumscribes the triangle.

Draw segments about as long as those shown below. In each exercise, construct a segment with the required length t.

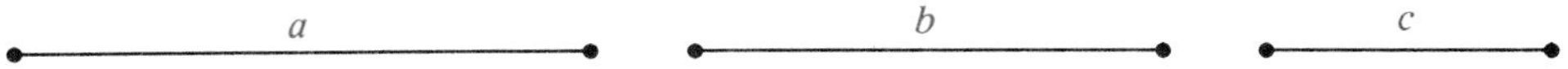

16. $t^2 = bc$ 17. $at = bc$ 18. $t = \frac{1}{3}(a + b)$ 8–5

19. Given two parallel lines l and m, what is the locus of points in their plane and equidistant from them? 8–6

20. Given two points A and B, what is the locus of points, in space, equidistant from A and B?

21. What is the locus of points in space equidistant from two parallel planes?

22. What is the locus of points in space that are equidistant from the vertices of equilateral $\triangle HJK$?

23. Points P and Q are 6 cm apart. What is the locus of points in a plane that are equidistant from P and Q and are 8 cm from P? Sketch the locus. 8–7

24. Point R is on line l. What is the locus in space of points that are 8 cm from l and 8 cm from R?

25. What is the locus of points in space that are 1 m from plane Q and 2 m from point Z not in Q? (There is more than one possibility.)

Use the segments with lengths a, b, and c that you drew for Exercises 16–18.

26. Construct an isosceles right triangle with hypotenuse of length a. 8–8

27. Construct a $\triangle RST$ with $RS = a$, $RT = c$, and the median to $\overline{RS}$ of length b.

Chapter Test

Begin by drawing segments and an angle roughly like those shown.

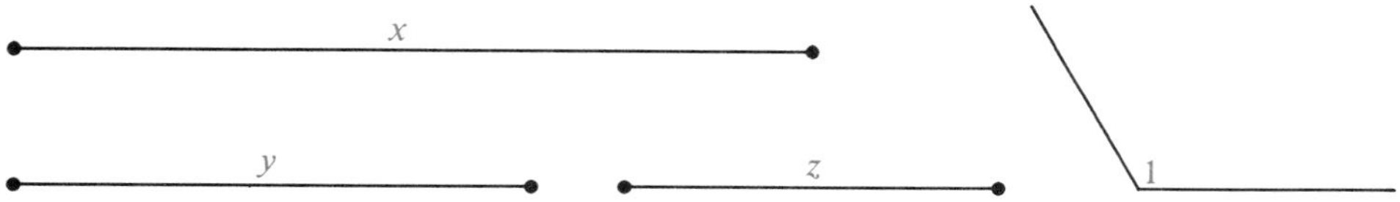

1. Construct an isosceles triangle with vertex angle congruent to $\angle 1$ and legs of length z.
2. Construct a 30°–60°–90° triangle with shorter leg of length y.
3. Construct a segment of length $\sqrt{xy}$.
4. Construct a segment of length $\frac{2}{3}(y + 2z)$.
5. Construct a segment of length n such that $\frac{x}{z} = \frac{y}{n}$.
6. Draw a large circle and a point K not on the circle. Using K as one vertex, construct any triangle that is circumscribed about the circle.
7. Draw a large triangle and construct the circle inscribed in the triangle.
8. In a right triangle **(a)** the _?_ of the triangle intersect at a point on the hypotenuse, **(b)** the _?_ intersect at a point inside the triangle, and **(c)** the altitudes of the triangle intersect at a _?_ of the triangle.
9. An isosceles triangle has sides of length 5, 5, and 8.
   a. What is the length of the median to the base?
   b. When the three medians are drawn, the median to the base is divided into segments with lengths _?_ and _?_.
10. Given points R and S in plane Z, what is the locus of points **(a)** in Z and equidistant from R and S and **(b)** in space and equidistant from R and S?
11. Given points T and U 8 units apart, what is the locus of points, in space, that are 6 units from T and 4 units from U?
12. Draw a line l and a point A on it. Using y and z from Exercises 1–5, construct the locus of points z units from l and y units from A.

Preparing for College Entrance Exams

Strategy for Success

Often the answer to a question can be found by writing an equation or inequality and solving it. When a complete solution is time-consuming, you may find that the fastest way to answer the question is to test the suggested answers in your equation or inequality.

Indicate the best answer by writing the appropriate letter.

1. $\overline{AB}$ and $\overline{AC}$ are tangent to $\odot O$ at B and C. If $m\widehat{BC} = x$, then $m\angle BAC =$
(A) x **(B)** $180 - x$ **(C)** $360 - x$ **(D)** $180 + x$ **(E)** $\frac{1}{2}x$

2. If quadrilateral $JKLM$ is inscribed in a circle and $\angle J$ and $\angle K$ are supplementary angles, then $\angle J$:
(A) must be congruent to $\angle L$
(B) must be a right angle
(C) must be congruent to $\angle M$
(D) must be an acute angle
(E) must be supplementary to $\angle M$

3. In $\odot M$, chords $\overline{RS}$ and $\overline{TU}$ intersect at X. If $RX = 15$, $XS = 18$, and $TX:XU = 3:10$, then $XU =$
(A) 3 **(B)** 9 **(C)** $20\frac{10}{13}$ **(D)** $25\frac{5}{13}$ **(E)** 30

4. If $m\widehat{XW} = 60$, $m\widehat{WZ} = 70$, and $m\widehat{ZY} = 70$, then $m\angle 1 =$
(A) 45 **(B)** 50 **(C)** 60 **(D)** 65 **(E)** 70

5. If $VW = 10$, $WX = 6$, and $VZ = 8$, then $ZY =$
(A) 4.8 **(B)** 12 **(C)** 7.5 **(D)** 20 **(E)** 16

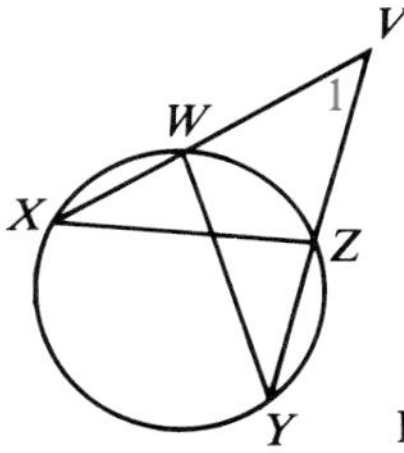

Exs. 4, 5

6. Given $\triangle ABC$, you can construct the locus of points in the plane of $\triangle ABC$ and equidistant from $\overline{AB}$, $\overline{BC}$, and $\overline{AC}$ by constructing:
(A) two medians **(B)** two altitudes **(C)** two angle bisectors
(D) the perpendicular bisectors of two sides **(E)** the circumscribed circle

7. To construct a tangent to $\odot R$ from a point S outside $\odot R$, you need to construct:
(A) the perpendicular bisector of $\overline{RS}$
(B) a perpendicular to $\overline{RS}$ at the point where $\overline{RS}$ intersects $\odot R$
(C) a diameter that is perpendicular to $\overline{RS}$
(D) a perpendicular to $\overleftrightarrow{RS}$ through point S
(E) a 30°-60°-90° triangle with vertex S

8. The locus of points 6 cm from plane P and 10 cm from a given point J *cannot* be:
(A) no points **(B)** one point **(C)** a line **(D)** a circle **(E)** two circles

9. The locus of the centers of all 8 cm chords in a circle of radius 5 cm is:
(A) a point **(B)** a segment **(C)** a line **(D)** a ray **(E)** a circle

Cumulative Review: Chapters 1–8

Write *always, sometimes,* or *never* to complete each statement.

A

1. Two lines that are not parallel are __?__ intersecting lines.
2. A supplement of an acute angle is __?__ an acute angle.
3. Three given points are __?__ collinear and __?__ coplanar.
4. When two parallel lines are cut by a transversal, two exterior angles on the same side of the transversal are __?__ complementary.
5. A quadrilateral __?__ has four obtuse angles.
6. A true conditional __?__ has a true converse.
7. Two isosceles right triangles with congruent hypotenuses are __?__ congruent.
8. A diagonal of an isosceles trapezoid __?__ divides the trapezoid into two congruent triangles.
9. A triangle with two congruent angles is __?__ equilateral.
10. If $\widehat{AC}$ on $\odot O$ and $\widehat{BD}$ on $\odot P$ have the same measure, then $\widehat{AC}$ is __?__ congruent to $\widehat{BD}$.
11. If $ABCD$ is a parallelogram and M is the midpoint of $\overline{AC}$, then M is also __?__ the midpoint of $\overline{BD}$.
12. If two consecutive sides of a parallelogram are perpendicular, then the diagonals are __?__ perpendicular.
13. If two sides and one angle of one triangle are congruent to the corresponding parts of another triangle, then the triangles are __?__ congruent.
14. Two equiangular hexagons are __?__ similar.
15. If the lengths of the sides of two triangles are in proportion, then the corresponding angles are __?__ congruent.
16. The tangent of an angle is __?__ greater than 1.
17. A triangle with sides of length $2x$, $3x$, and $4x$, with $x > 0$, is __?__ acute.
18. The altitude to the hypotenuse of a $30°$-$60°$-$90°$ triangle __?__ divides the hypotenuse into segments with lengths in the ratio $1 : \sqrt{3}$.
19. When a tangent segment and a secant segment are drawn to a circle from an external point, the square of the length of the tangent segment is __?__ equal to the product of the lengths of the secant segment and its external segment.
20. Given two segments with lengths r and s, it is __?__ possible to construct a segment of length $\frac{3}{4}\sqrt{2rs}$.
21. Given a plane containing points A and B, the locus of points in the plane that are equidistant from A and B and are 10 cm from A is __?__ one point.

Complete each statement in Exercises 22–25.

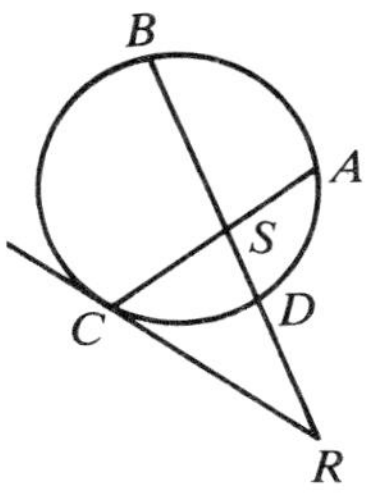

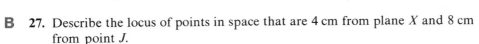

22. If $m\widehat{AB} = 80$, $m\widehat{CD} = 66$, and $m\widehat{DA} = 70$, $m\angle ASD = \underline{\ ?\ }$.

23. If $BS = 12$, $SD = 6$, and $AS = 8$, then $SC = \underline{\ ?\ }$.

24. If $RD = 9$ and $DB = 16$, then $RC = \underline{\ ?\ }$.

25. If $m\widehat{AB} = 80$, $m\widehat{CD} = 66$, and $m\widehat{DA} = 70$, then $m\angle R = \underline{\ ?\ }$.

26. Draw a large $\triangle RST$. Construct a $\triangle XYZ$ congruent to $\triangle RST$.

B **27.** Describe the locus of points in space that are 4 cm from plane X and 8 cm from point J.

28. If 4, 7, and x are the lengths of the sides of a triangle and x is an integer, list the possible values for x.

29. $\triangle DEF$ is a right triangle with hypotenuse $\overline{DF}$. $DE = 6$ and $EF = 8$.

a. If $\overline{EX} \perp \overline{DF}$ at X, find DX.

b. If Y lies on $\overline{DF}$ and $\overrightarrow{EY}$ bisects $\angle DEF$, find DY.

30. $\tan \underline{\ ?\ }° = \sqrt{3}$

31. If each interior angle of a regular polygon has measure 160, how many sides does the polygon have?

32. Prove: If the diagonals of rhombus $PQRS$ intersect at M, then $\triangle PQM \cong \triangle RQM$.

33. Given: $\odot O$; $m\angle 1 = 45$
Prove: $\triangle OPQ$ is a 45°–45°–90° $\triangle$.

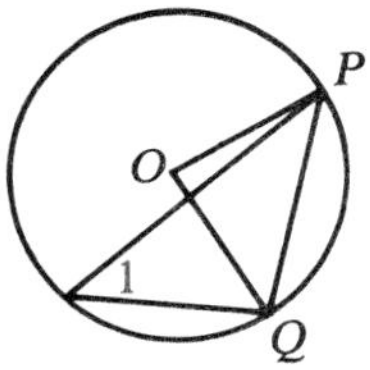

34. Use the given diagram to prove that $WX \cdot YV = XV \cdot ZY$.

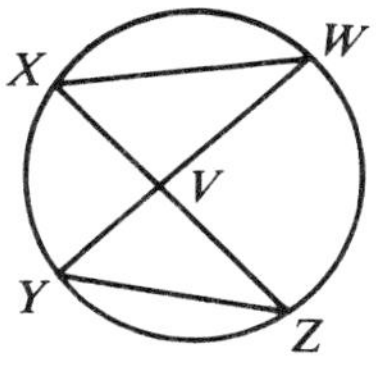

Find the value of x. (In Exercise 36, $\overrightarrow{AB}$ and $\overrightarrow{AC}$ are tangents.)

35.

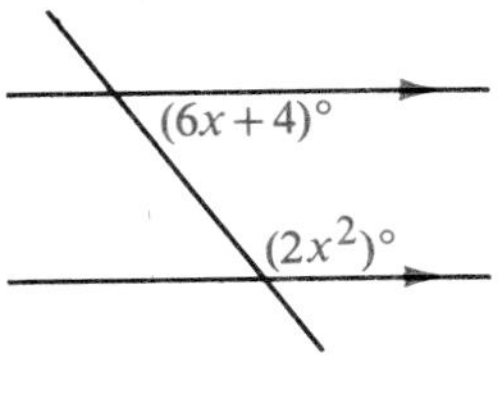

36.

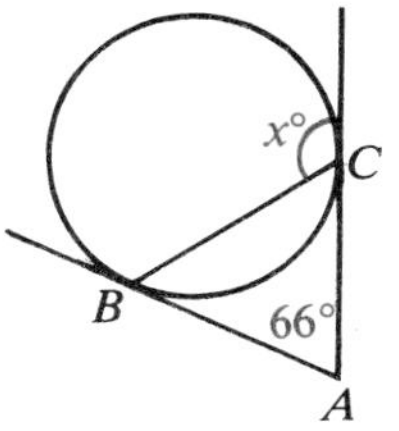

37.

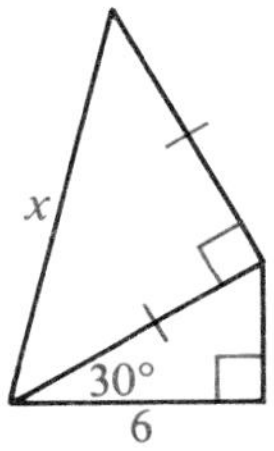

38. Draw a circle. Construct a regular inscribed octagon.

39. Draw $\overline{XY}$. Construct any rectangle with a diagonal congruent to $\overline{XY}$.

The aerial photograph below shows clearly outlined regions of land. In this chapter you will learn how to find the area of a number of different types of regions, including those bounded by rectangles, triangles, and trapezoids.

Areas of Plane Figures

9

Areas of Polygons

Objectives

1. Understand what is meant by the area of a polygon.
2. Understand the area postulates.
3. Know and use the formulas for the areas of rectangles, parallelograms, triangles, trapezoids, and regular polygons.

9-1 Areas of Rectangles

In everyday conversation people often refer to the *area* of a rectangle when what they really mean is the area of a rectangular region.

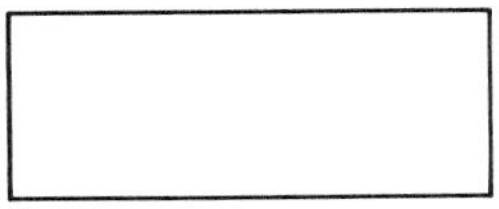

Rectangle

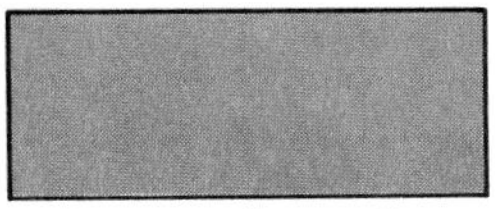

Rectangular region

For the sake of simplicity, we will continue this common practice. Thus, when we speak of the area of a triangle, we will mean the area of the triangular region that includes the triangle *and* its interior.

In Chapter 1 we accepted postulates that enable us to express the lengths of segments and the measures of angles as positive numbers. Similarly, the areas of figures are positive numbers with properties given by the following area postulates.

Postulate 17

The area of a square is the square of the length of a side. ($A = s^2$)

Length: 1 unit

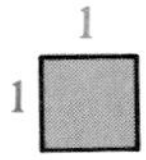

Area: 1 square unit

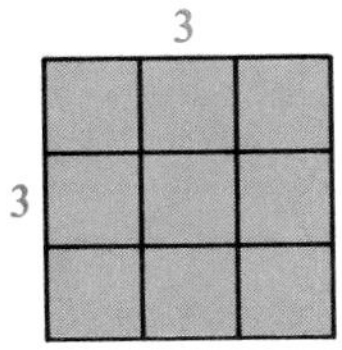

Area: 3^2, or 9, square units

Postulate 18 Area Congruence Postulate

If two figures are congruent, then they have the same area.

Postulate 19 Area Addition Postulate

The area of a region is the sum of the areas of its non-overlapping parts.

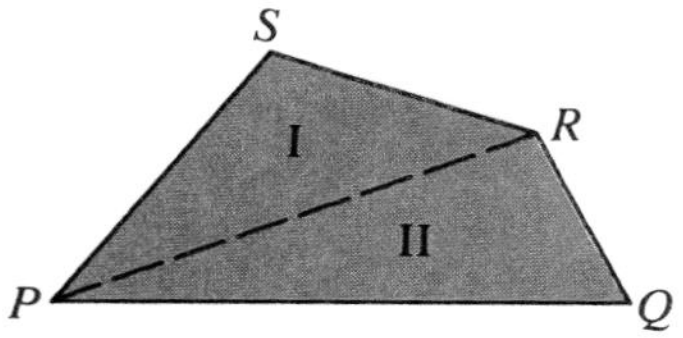

Area of $PQRS$:
Area I + Area II

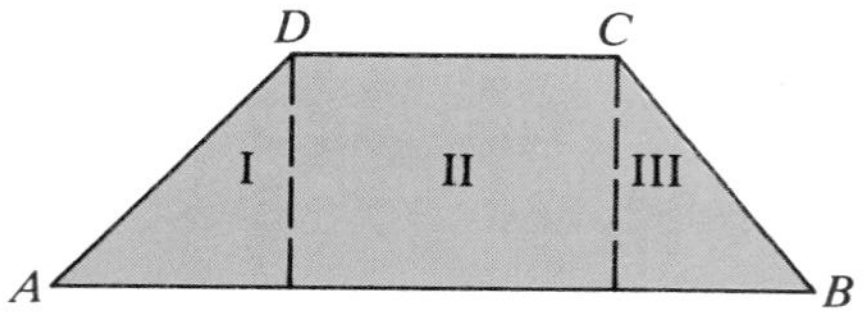

Area of $ABCD$:
Area I + Area II + Area III

Any side of a rectangle or other parallelogram can be considered to be a **base.** The length of a base will be denoted by b. In this text the term *base* will be used to refer either to the line segment or to its length. An **altitude** to a base is any segment perpendicular to the line containing the base from any point on the opposite side. The length of an altitude is called the **height** (h). All the altitudes to a particular base have the same length.

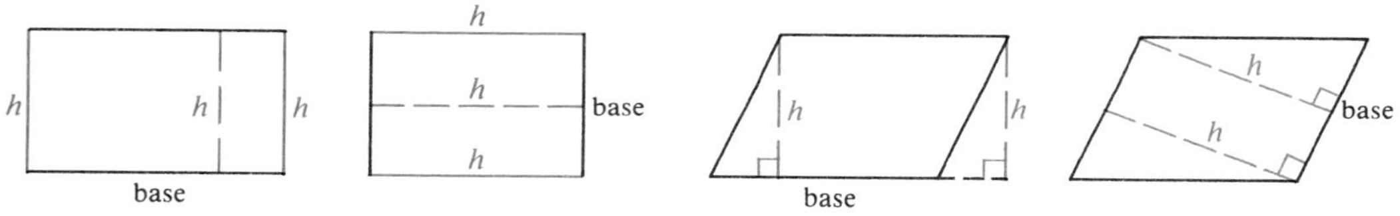

Theorem 9-1

The area of a rectangle equals the product of its base and height. ($A = bh$)

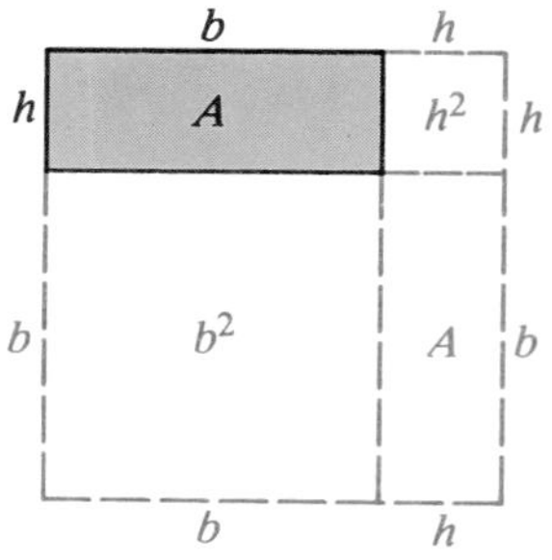

Given: A rectangle with base b and height h

Prove: $A = bh$

Proof:

Building onto the given rectangle, we can draw a large square consisting of these non-overlapping parts:
the given rectangle with area A;
a congruent rectangle with area A;
a square with area b^2;
a square with area h^2.

Area of big square $= 2A + b^2 + h^2$	(Area Addition Postulate)
Area of big square $= (b + h)^2 = b^2 + 2bh + h^2$	($A = s^2$)
$2A + b^2 + h^2 = b^2 + 2bh + h^2$	(Substitution Property)
$2A = 2bh$	(Subtraction Property)
$A = bh$	(Division Property)

Some common units of area are the square centimeter (cm^2) and the square meter (m^2).

Example Find the area of each figure.

a.

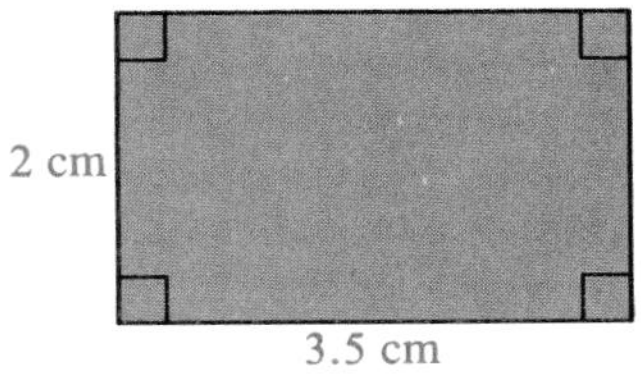

b.

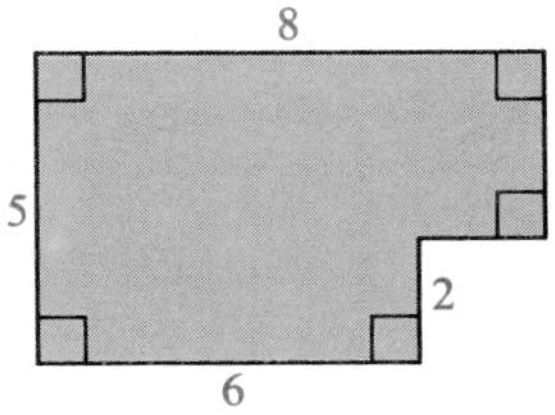

Solution a. $A = 3.5 \cdot 2$
$= 7\ (cm^2)$

b. $A = (8 \cdot 5) - (2 \cdot 2) = 40 - 4$
$= 36$

or

$A = (8 \cdot 3) + (6 \cdot 2) = 24 + 12$
$= 36$

Notice that in part (b) of the example above, the unit of length and the unit of area are understood to be "units" and "square units," respectively. It is important to remember that the implied units for length and area are different.

Classroom Exercises

1. Tell what each letter represents in the formula $A = s^2$.
2. Tell what each letter represents in the formula $A = bh$.
3. Find the area and perimeter of a square with sides 5 cm long.
4. The perimeter of a square is 28 cm. What is the area?
5. The area of a square is 64 cm^2. What is the perimeter?

Exercises 6–13 refer to rectangles. Complete the table.

	6.	7.	8.	9.	10.	11.	12.	13.
b	8 cm	4 cm	12 m	?	$3\sqrt{2}$	$4\sqrt{2}$	$5\sqrt{3}$	$x + 3$
h	3 cm	1.2 cm	?	5 cm	2	$\sqrt{2}$	$2\sqrt{3}$	x
A	?	?	36 m^2	55 cm^2	?	?	?	?

14. a. What is the converse of the Area Congruence Postulate?
    b. Is this converse true or false? Explain.
15. a. Draw three noncongruent rectangles, each with perimeter 20 cm. Find the area of each rectangle.
    b. Of all rectangles having perimeter 20 cm, which one do you think has the greatest area? (Give its length and width.)

Written Exercises

Complete the tables. Exercises 1–16 refer to rectangles. p is the perimeter.

A

	1.	2.	3.	4.	5.	6.	7.	8.
b	12 cm	8.2 cm	16 cm	?	$3\sqrt{2}$	$\sqrt{6}$	$2x$	$4k - 1$
h	5 cm	4 cm	?	8 m	$4\sqrt{2}$	$\sqrt{2}$	$x - 3$	$k + 2$
A	?	?	80 cm²	120 m²	?	?	?	?

	9.	10.	11.	12.	13.	14.	15.	16.
b	9 cm	10 cm	16 cm	$x + 5$	$a + 3$	$k + 7$	x	?
h	4 cm	?	?	x	$a - 3$	?	?	y
A	?	?	?	?	?	?	$x^2 - 3x$	$y^2 + 7y$
p	?	30 cm	42 cm	?	?	$4k + 20$	?	?

Consecutive sides of the figures below are perpendicular. Find the area of each figure.

B **17.**

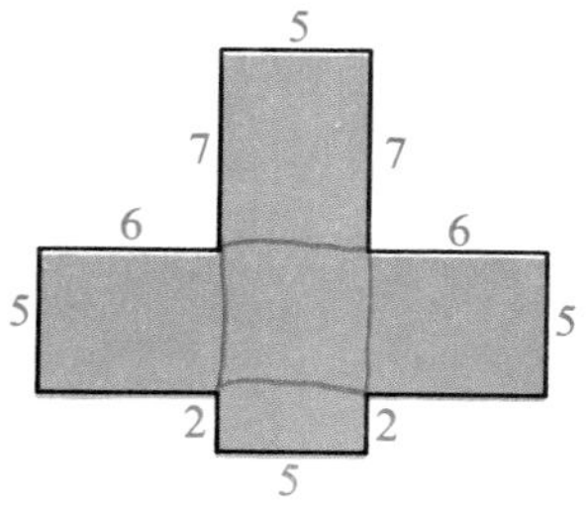

18.

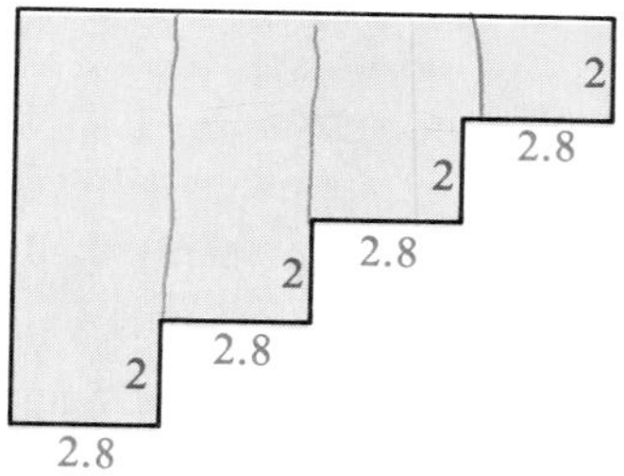

19.

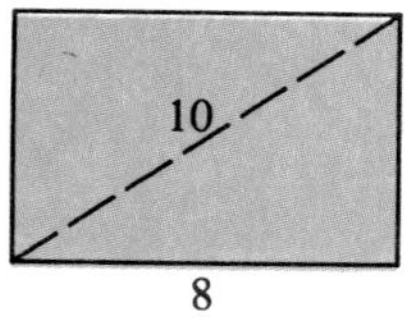

20.

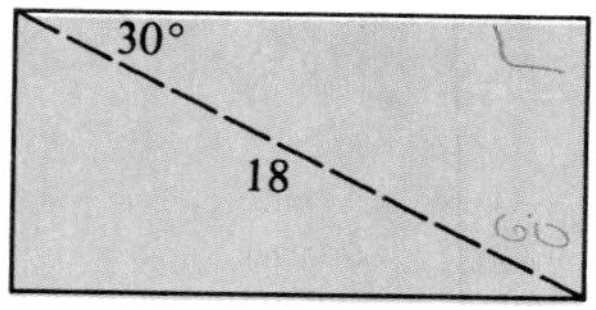

21.

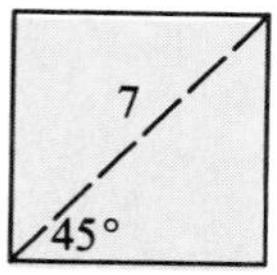

22.

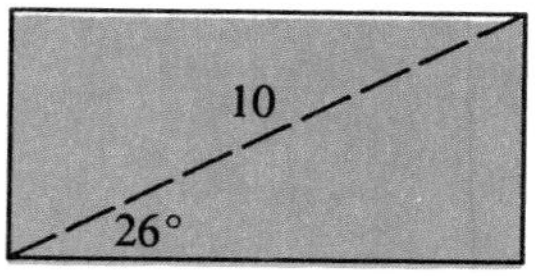

(Give answer correct to the nearest tenth.)

If the program on page 384 is run, the computer will print

```
AREA IS APPROXIMATELY 0.385
```

A better approximation can be found by using 100 smaller rectangles with base vertices at 0, 0.01, 0.02, 0.03, . . . , 1.00. Change lines 10 and 30 as follows:

```
10 FOR X = 0.01 TO 1.00 STEP 0.01
30 LET A = A + Y * 0.01
```

With this change the computer will print

```
AREA IS APPROXIMATELY 0.33835
```

Exercises

1. Modify the given program so that it will use 1000 rectangles with base vertices at 0, 0.001, 0.002, 0.003, . . . , 1.000 to approximate the area of the shaded region.

Write and RUN a program using ten rectangles to find the approximate area of each region described.

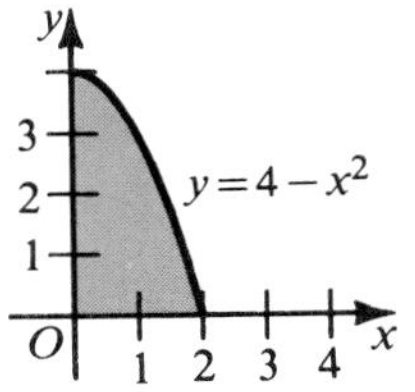

2. The region below the graph of $y = x^3$, above the x-axis, and between the vertical lines $x = 0$ and $x = 2$
3. The shaded region shown at the right

9-2 Areas of Parallelograms and Triangles

Detailed formal proofs of most area theorems are lengthy and time consuming. For that reason, we will show outlines of proofs.

Theorem 9-2

The area of a parallelogram equals the product of its base and height. ($A = bh$)

Given: $\square PQRS$

Prove: $A = bh$

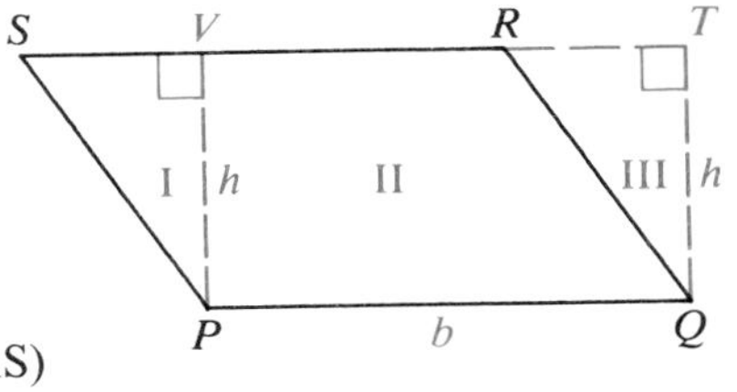

Outline of proof:

1. Draw altitudes $\overline{PV}$ and $\overline{QT}$, forming two rt. $\triangle$s.
2. Area I = Area III ($\triangle PSV \cong \triangle QRT$ by HL or AAS)
3. Area of $\square PQRS$ = Area II + Area I
   = Area II + Area III
   = Area of rect. $PQTV$
   = bh

Theorem 9-3

The area of a triangle equals half the product of its base and height. $(A = \frac{1}{2}bh)$

Given: $\triangle XYZ$

Prove: $A = \frac{1}{2}bh$

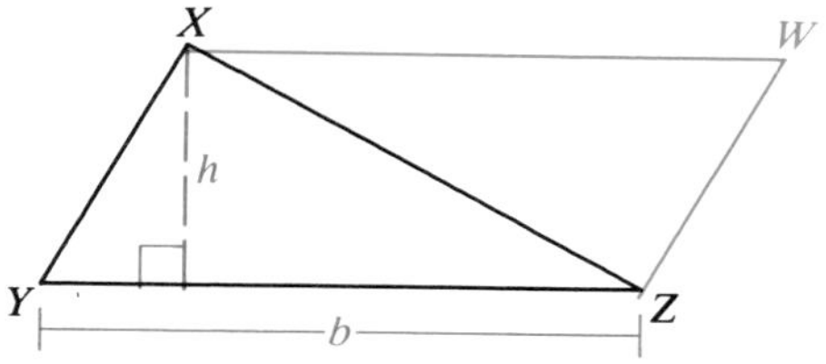

Outline of proof:

1. Draw $\overline{XW} \parallel \overline{YZ}$ and $\overline{ZW} \parallel \overline{YX}$, forming $\square XYZW$.
2. $\triangle XYZ \cong \triangle ZWX$ (SAS or SSS)
3. Area of $\triangle XYZ = \frac{1}{2}$ Area of $\square XYZW$
$= \frac{1}{2}bh$

Corollary

The area of a rhombus equals half the product of its diagonals. $(A = \frac{1}{2}d_1d_2)$

The proof of the corollary is left as Exercise 29.

Example 1 Find the area of a triangle with sides 8, 8, and 6.

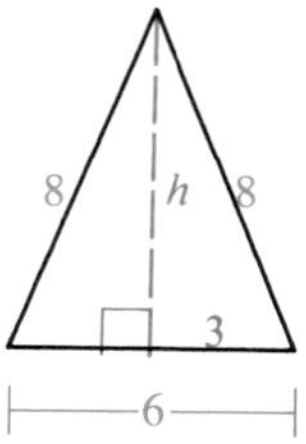

Solution Draw the altitude to the base shown. Since the triangle is isosceles, this altitude bisects the base.

$h^2 + 3^2 = 8^2$ (Pythagorean Theorem)

$h^2 = 8^2 - 3^2 = 55$

$h = \sqrt{55}$

$A = \frac{1}{2}bh = \frac{1}{2} \cdot 6 \cdot \sqrt{55} = 3\sqrt{55}$

Example 2 Find the area of an equilateral triangle with side 6.

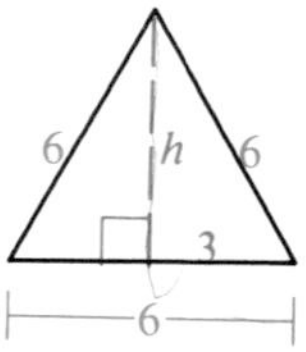

Solution Draw an altitude. Two 30°-60°-90° triangles are formed.

$h = 3\sqrt{3}$

$A = \frac{1}{2}bh = \frac{1}{2} \cdot 6 \cdot 3\sqrt{3} = 9\sqrt{3}$

Classroom Exercises

1. The area of the parallelogram can be found in two ways:

a. $A = 8 \cdot$ _?_ $=$ _?_

b. $A = 4 \cdot$ _?_ $=$ _?_

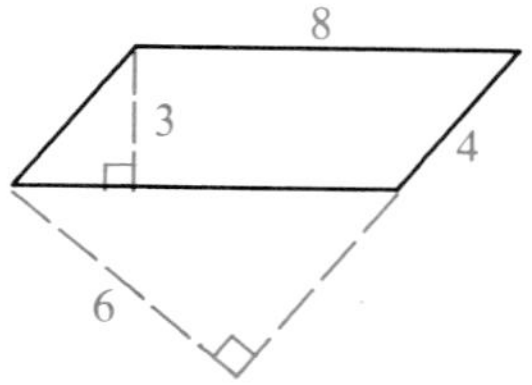

2. Find the area of $\triangle ABC$.
3. Find the area of $\triangle DBC$.
4. Find the area of $\triangle EBC$.

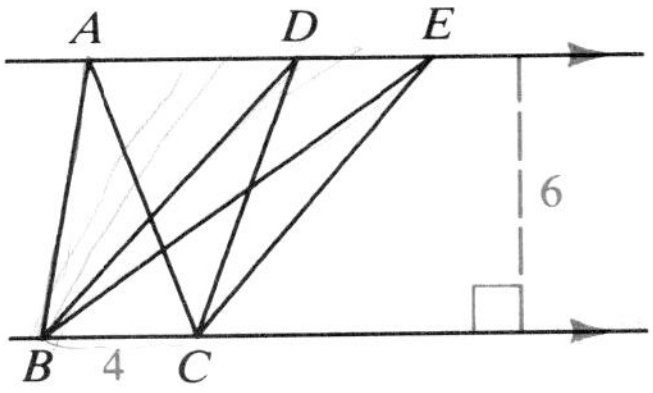

Find the area of each figure.

5.
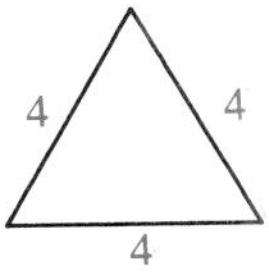

6.
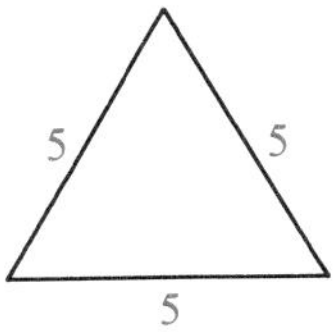

7.
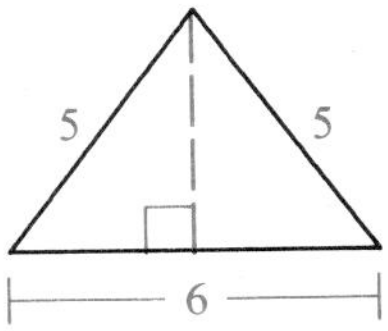

8.
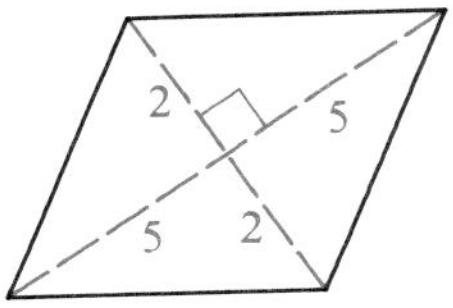

9.
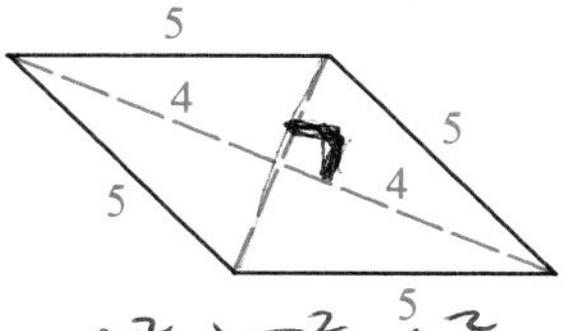

10.
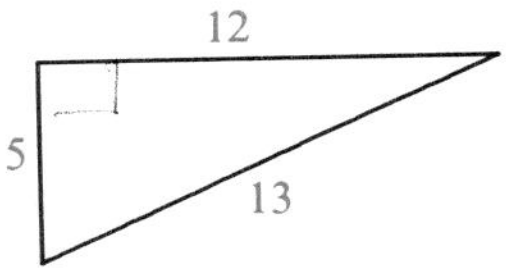

Written Exercises

Exercises 1–8 refer to triangles. Complete the table.

A

	1.	2.	3.	4.	5.	6.	7.	8.
b	8 cm	5.2 m	18	?	$3\sqrt{2}$	$6\sqrt{3}$	$5x$	?
h	7 cm	11.5 m	?	14	$2\sqrt{2}$	$3\sqrt{6}$	?	$3\sqrt{2}$
A	?	?	108	56	?	?	$15xy$	$24\sqrt{2}$

Find the area of each figure.

9.
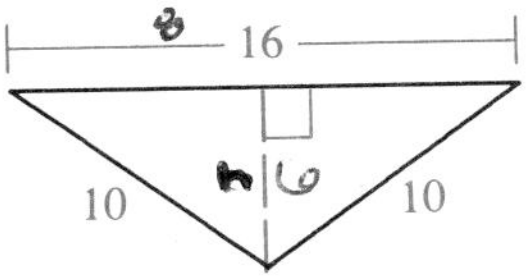

10.
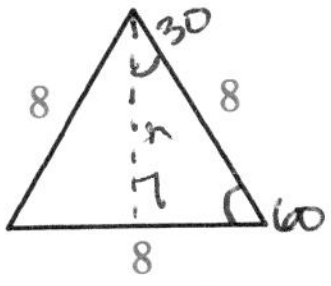

11.
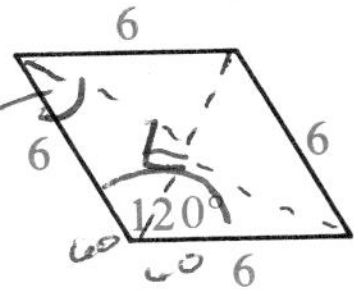

12.
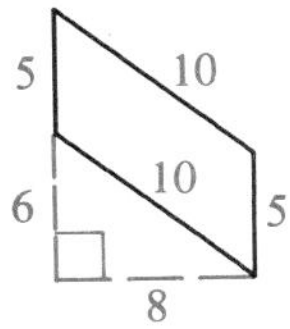

13.
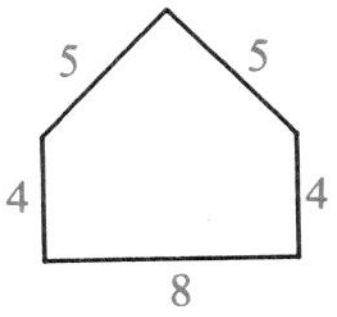

14.
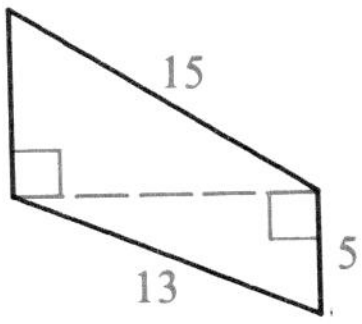

15. A parallelogram has sides 12 cm and 20 cm long. If the shorter altitude is 6 cm long, how long is the other altitude?

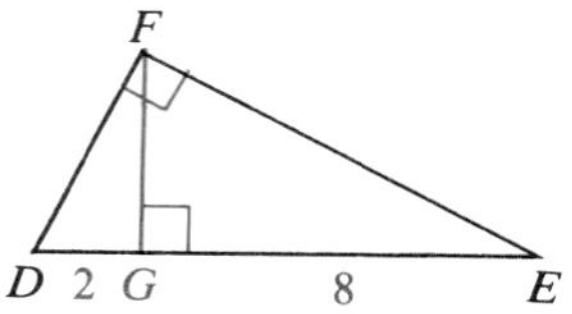

16. $\overline{FG}$ is the altitude to the hypotenuse of $\triangle DEF$. Name three similar triangles and find their areas. (*Hint:* See Theorem 6-1 and Corollary 1 on page 248.)

B 17. a. Let $\overline{AM}$ be a median of $\triangle ABC$. If $BC = 16$ and $h = 5$, find the areas of $\triangle ABC$ and $\triangle ABM$.

b. Write an outline of a proof that if $\overline{AM}$ is a median of $\triangle ABC$, then

$$\text{Area of } \triangle ABM = \tfrac{1}{2} \cdot \text{Area of } \triangle ABC.$$

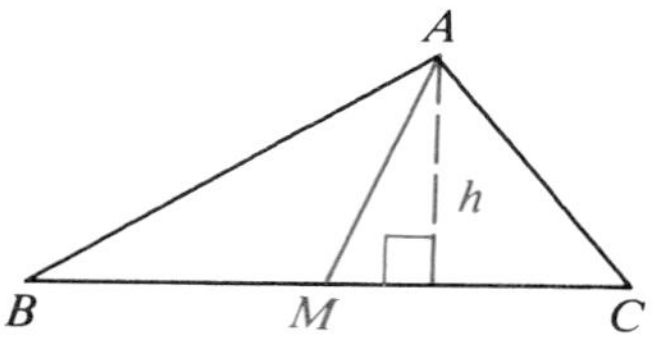

18. An isosceles triangle has sides 5 cm, 5 cm, and 8 cm long.
a. Find its area.
b. Find the lengths of the three altitudes.

19. If the area of parallelogram *PQRS* is 36, find the area of $\triangle TRS$.

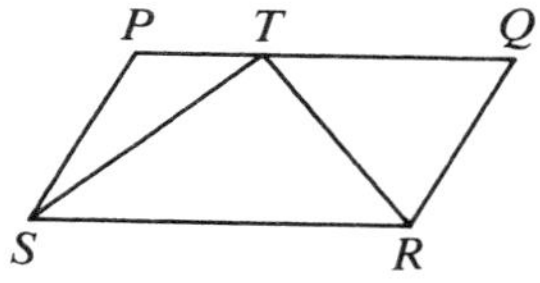

20. Find the ratio of the areas of $\triangle ABD$ and $\triangle ADC$.

21. If the area of $\triangle ABC$ is 240, find the length of the altitude from *C* to $\overleftrightarrow{AB}$.

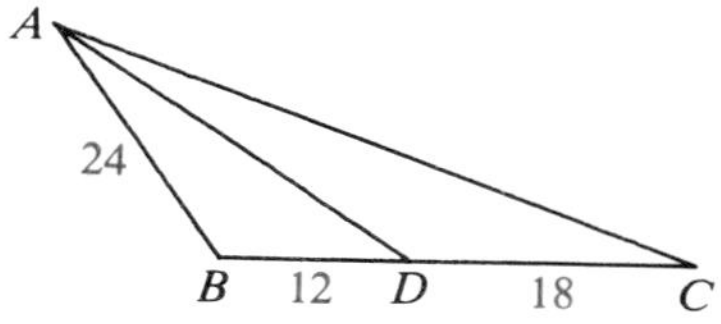

Find the area of each figure.

22. A rhombus with perimeter 40 and one diagonal 12

23. A 30°-60°-90° triangle with hypotenuse 8

24. An isosceles right triangle with hypotenuse *x*

25. An equilateral triangle with height 12

26. A regular hexagon with perimeter 60

27. A rectangle with length 24 inscribed in a circle with radius 13

28. Use the diagram shown at the right.
a. Find the area of ▱*PQRS*.
b. Find the area of $\triangle PSR$.
c. Find the area of $\triangle OSR$. (*Hint:* Refer to $\triangle PSR$ and use Exercise 17.)
d. What is the area of $\triangle PSO$?
e. What must the area of $\triangle POQ$ be? Why? What must the area of $\triangle OQR$ be?
f. State what you have shown in parts (a)–(e) about how the diagonals divide a parallelogram.

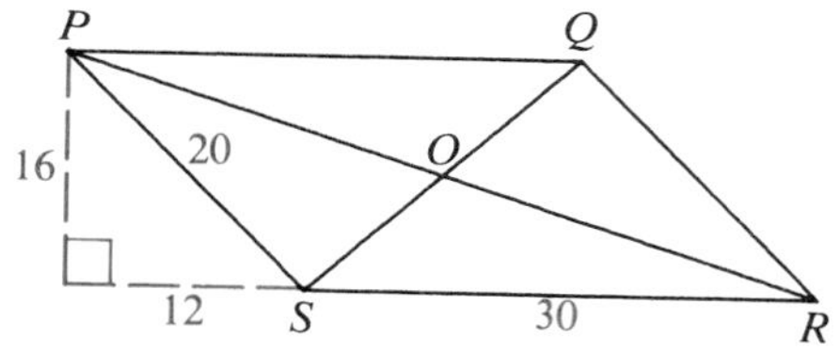

29. Using the diagram given below, write an outline of a proof of the corollary to Theorem 9–3.

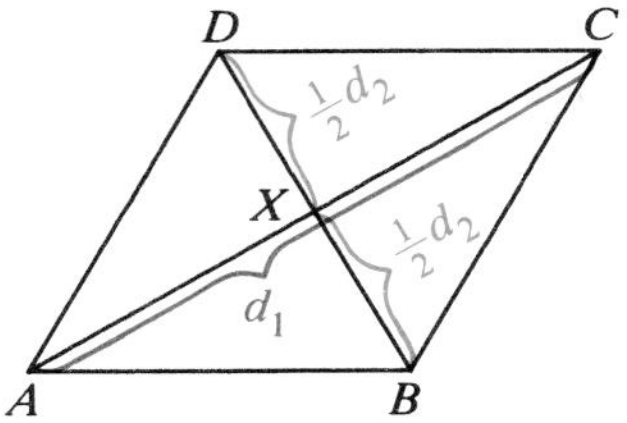

Ex. 29

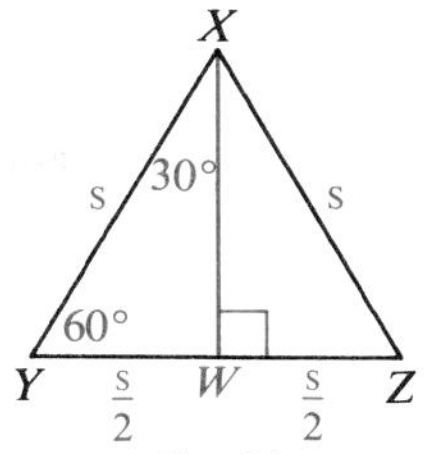

Ex. 30

30. The area of an equilateral triangle with side s can be found by using the formula $A = \frac{s^2\sqrt{3}}{4}$. Using the diagram given above, prove that the formula is correct.

31. Think of a parallelogram made with cardboard strips and hinged at each vertex so that the measure of $\angle C$ will vary. Find the area of the parallelogram for each measure of $\angle C$ given in parts (a)–(e).

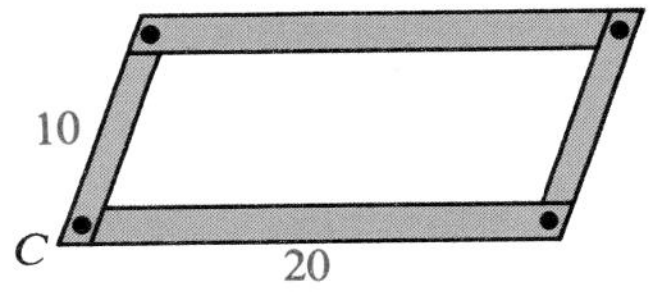

a. 30° **b.** 45° **c.** 60° **d.** 90° **e.** 120°

f. Approximate your answers to parts (b), (c), and (e) by using $\sqrt{2} \approx 1.4$ and $\sqrt{3} \approx 1.7$. Then record your answers to parts (a)–(e) on a set of axes like the one shown below.

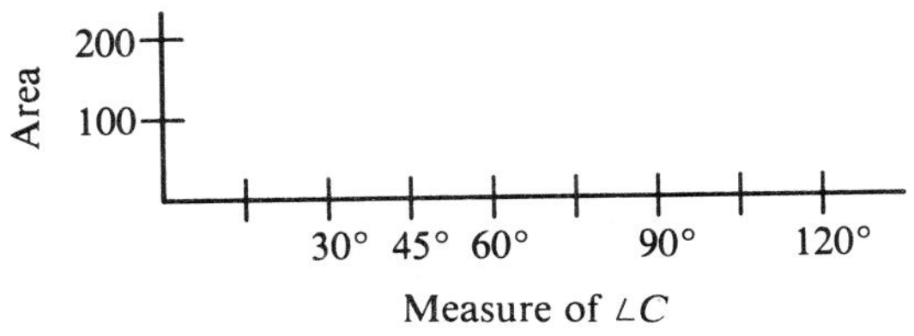

32. The base of a triangle is 1 cm longer than its altitude. If the area of the triangle is 210 cm², how long is the altitude?

C **33.** The diagonals of a parallelogram are 82 cm and 30 cm. One altitude is 18 cm long. Find the two possible values for the area.

For Exercises 34–36, draw a scalene triangle *ABC*.

34. Construct an isosceles triangle whose area is equal to the area of $\triangle ABC$.

35. Construct an isosceles right triangle whose area is equal to the area of $\triangle ABC$.

36. Construct an equilateral triangle whose area is equal to the area of $\triangle ABC$.

37. **a.** Accurately draw or construct a large equilateral triangle. Choose any point inside the triangle and carefully measure the distances x, y, and z. Find $x + y + z$.

b. Now choose another point on or inside the triangle and find $x + y + z$. What do you notice? Why does this happen?

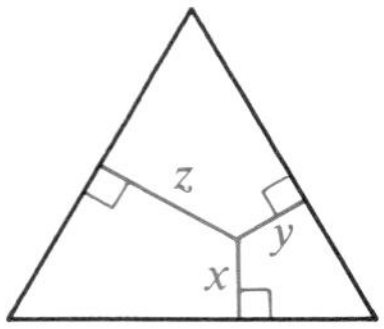

38. Two squares each with sides 12 cm are placed so that a vertex of one lies at the center of the other. Find the area of the shaded region.

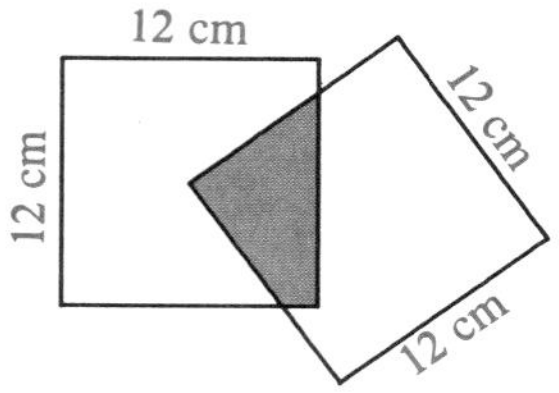

CALCULATOR KEY-IN

A formula for finding the area of a triangle, given the lengths of its sides, has been known for over two thousand years.

Let a, b, and c represent the lengths of the sides of a triangle and let:

$$s = \frac{a + b + c}{2}$$

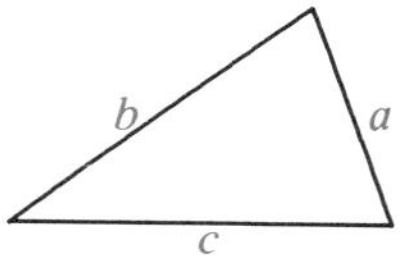

The area K is then given by the following expression, known as Heron's Formula:

$$K = \sqrt{s(s - a)(s - b)(s - c)}$$

If $a = 6$, $b = 8$, and $c = 10$, we can calculate s and K as follows.

$$s = \frac{6 + 8 + 10}{2} = 12 \qquad K = \sqrt{12(12 - 6)(12 - 8)(12 - 10)} = \sqrt{576} = 24$$

Since 6, 8, and 10 are the sides of a right triangle with base 6 and height 8, we can check the value we found for K by using another method:

$$K = \tfrac{1}{2}bh = \tfrac{1}{2}(6)(8) = 24$$

Exercises

1. A triangle has sides 3, 13, and 14.
   a. Show that Heron's Formula gives the area as $6\sqrt{10}$.
   b. If 3 is taken as the base, then $6\sqrt{10} = \frac{1}{2} \cdot 3h$, and the height, h, is __?__.
   c. If 13 is taken as the base, then the height is __?__.
   d. If 14 is taken as the base, then the height is __?__.

For Exercises 2–7, the lengths of the sides of a triangle are given. Find the area and the three heights of each triangle, correct to the nearest thousandth.

2. 11, 13, 15
3. 8, 8, 10
4. 12, 18, 27
5. 6.3, 7.2, 10.1
6. 68, 77, 105
7. 5.5, 6.5, 10

9-3 Areas of Trapezoids

An altitude of a trapezoid is any segment perpendicular to a line containing the base from a point on the opposite base. Since the bases are parallel, all altitudes have the same length, called the *height* (h) of the trapezoid.

Theorem 9-4

The area of a trapezoid equals half the product of the height and the sum of the bases. $A = \frac{1}{2}h(b_1 + b_2)$

Outline of proof:

1. Draw diagonal $\overline{BD}$ of trap. $ABCD$, forming two triangular regions, I and II, each with height h.
2. Area of trapezoid = Area I + Area II

$$= \tfrac{1}{2}b_1h + \tfrac{1}{2}b_2h$$
$$= \tfrac{1}{2}h(b_1 + b_2)$$

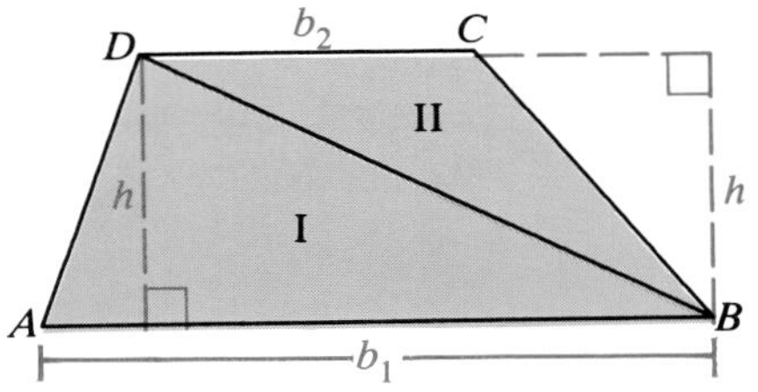

Example 1 Find the area of a trapezoid with height 7 and bases 12 and 8.

Solution $A = \frac{1}{2}h(b_1 + b_2) = \frac{1}{2} \cdot 7 \cdot (12 + 8) = 70$

Example 2 Find the area of an isosceles trapezoid with legs 5 and bases 6 and 10.

Solution When you draw the two altitudes shown, you get a rectangle and two congruent right triangles. The segments of the lower base must have lengths 6, 2, and 2. First find h:

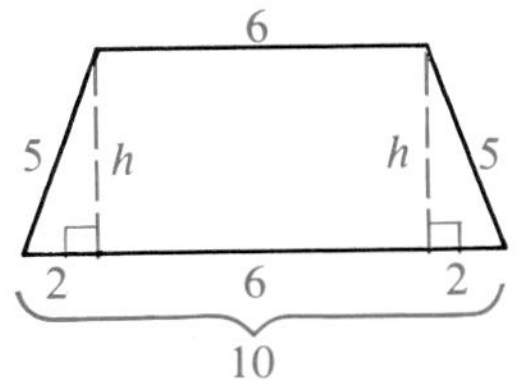

$$h^2 + 2^2 = 5^2$$
$$h^2 = 21$$
$$h = \sqrt{21}$$

Then find the area: $A = \frac{1}{2}h(b_1 + b_2)$

$$= \tfrac{1}{2}\sqrt{21}(10 + 6) = 8\sqrt{21}$$

Classroom Exercises

Find the area of each trapezoid.

1.

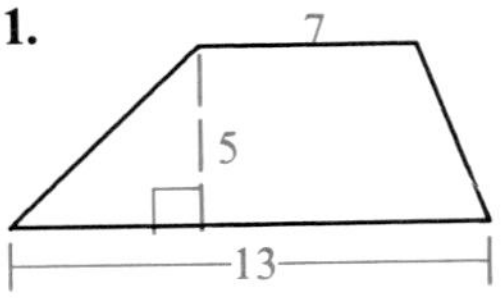

2.

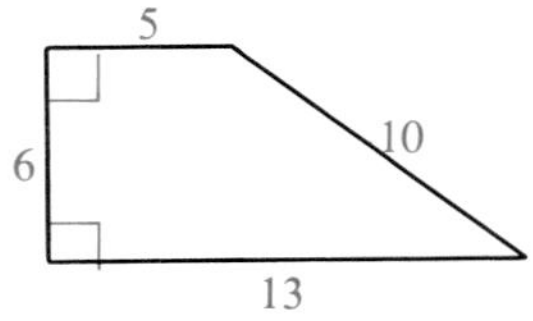

3.

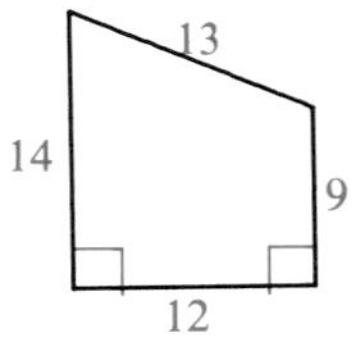

Find the area of each trapezoid.

4.

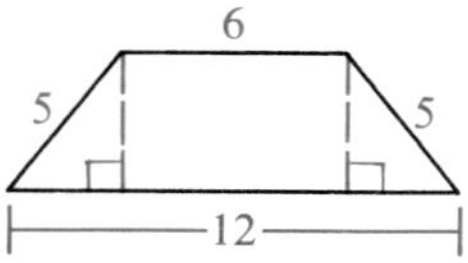

5.

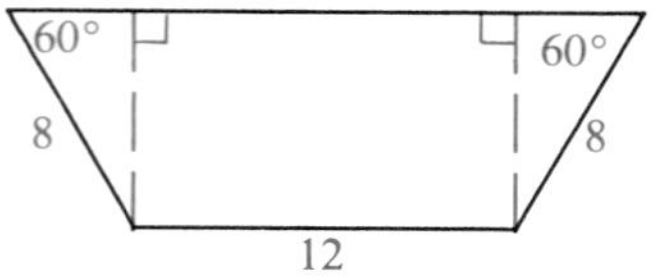

6.

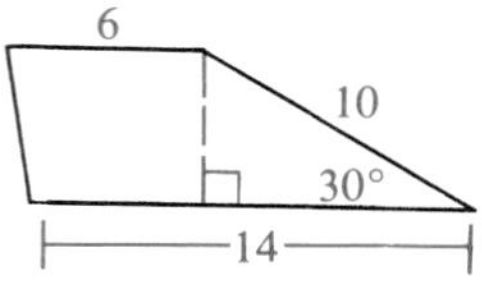

7. a. Find the lengths of the medians of the trapezoids in Exercises 1–3.

b. Explain why the area of a trapezoid can also be given by the formula *Area* = *height* × *median*.

8. a. If the congruent trapezoids shown are slid together, what special quadrilateral is formed?

b. Use your answer to derive the formula $A = \frac{1}{2}h(b_1 + b_2)$.

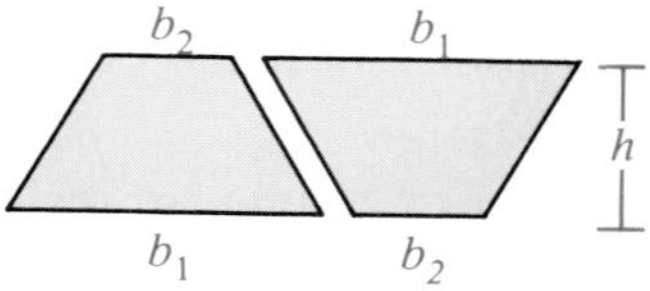

Written Exercises

Exercises 1–8 refer to trapezoids. Complete the table.

A

	1.	**2.**	**3.**	**4.**	**5.**	**6.**	**7.**	**8.**
b_1	12	6.8	$3\frac{1}{6}$	45	27	3	7	?
b_2	8	3.2	$4\frac{1}{3}$	15	9	?	?	$3k$
h	7	6.1	$1\frac{3}{5}$	?	?	3	$9\sqrt{2}$	$5k$
A	?	?	?	300	90	12	$36\sqrt{2}$	$45k^2$

9. Find the lengths of the medians of the trapezoids in Exercises 1–3.

10. A trapezoid has area 54 and height 6. How long is its median?

Find the area of each trapezoid.

11.

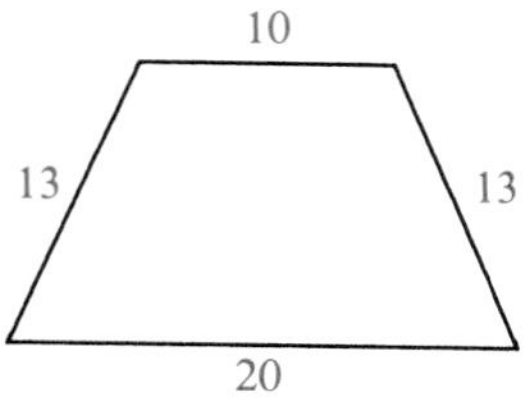

12.

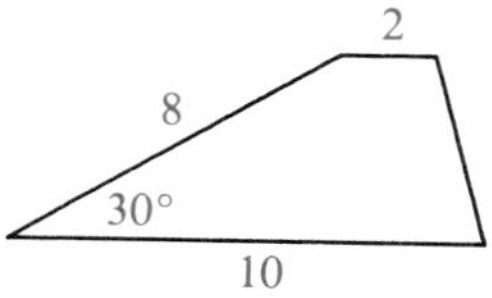

13.

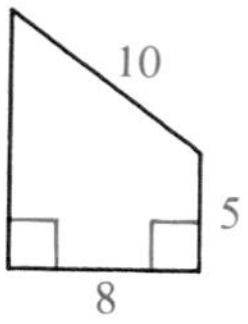

14.

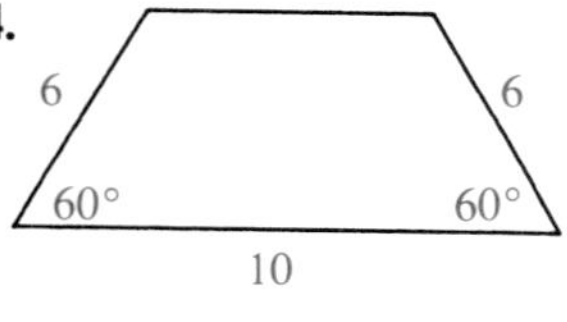

15.

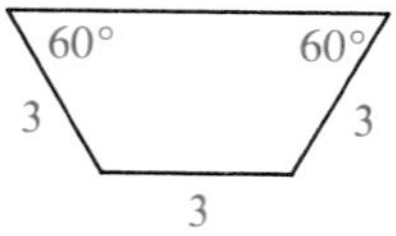

16.

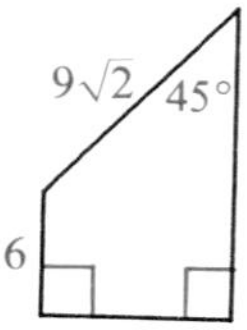

17. An isosceles trapezoid with 45° base angles has bases 8 and 20. Find its area.

B **18.** **a.** The legs of an isosceles trapezoid are 10 cm and the bases are 9 cm and 21 cm. Find the area of the trapezoid.
b. Find the lengths of the diagonals.

19. An isosceles trapezoid has bases 12 and 28. The area is 300.
a. Find the height. **b.** Find the perimeter.

20. The bases of trapezoid $RSTV$ are $\overline{RS}$ and $\overline{VT}$. $RS = 5$, $VT = 11$, $RV = 4$, and $m\angle V = 37$. Find the area of the trapezoid, correct to the nearest tenth. (Use the trigonometry table on page 271 or a calculator.)

21. In trapezoid $ABCD$, $m\angle A = m\angle B = 90$, $AB = 16$, $BC = 8$, and $AD = 12$. If M is the midpoint of $\overline{AB}$, find the area of $\triangle DMC$.

22. $ABCD$ is a trapezoid with bases 4 cm and 12 cm, as shown. Find the ratio of the areas of:
a. $\triangle ABD$ and $\triangle ABC$
b. $\triangle AOD$ and $\triangle BOC$
c. $\triangle ABD$ and $\triangle ADC$

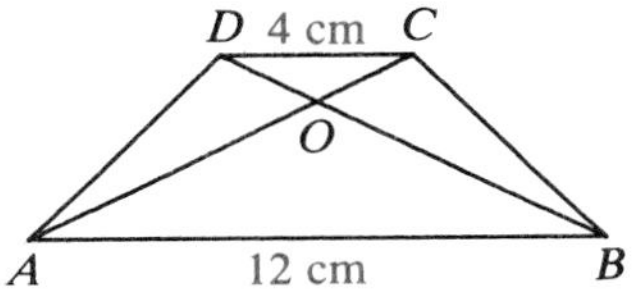

23. $ABCDEF$ is a regular hexagon with side 12. Find the areas of the three regions formed when diagonals $\overline{AC}$ and $\overline{AD}$ are drawn.

24. An isosceles trapezoid with bases 12 and 16 is inscribed in a circle of radius 10. The center of the circle lies in the interior of the trapezoid. Find the area of the trapezoid.

C **25.** Draw a non-isosceles trapezoid, then construct an isosceles trapezoid with equal area.

For Exercises 26 and 27, find the area of the trapezoid shown.

26.

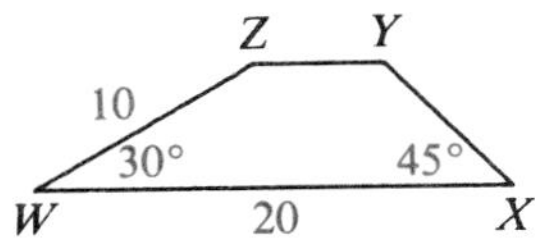

27.

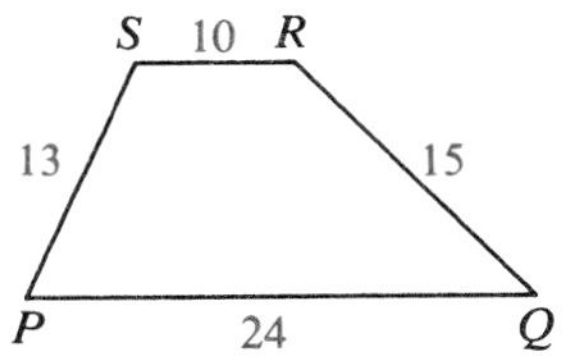

28. In the figure given below, prove that the area of square $ABCD$ equals the area of rectangle $EFGD$.

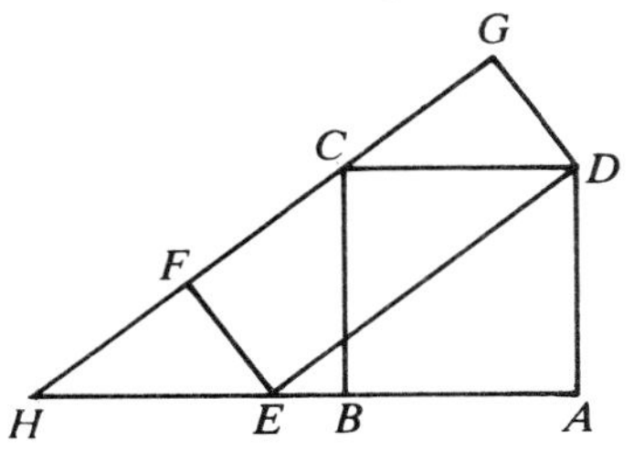

Ex. 28

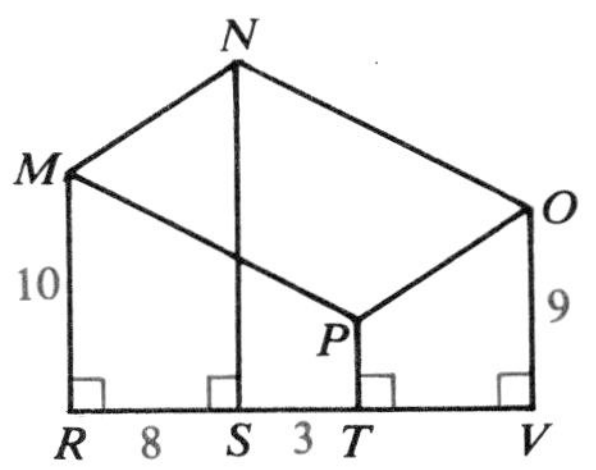

Ex. 29

★**29.** If $NS = 16$, find the area of ▱$MNOP$ shown above.

CALCULATOR KEY-IN

A plot of land is described as follows:

From point A proceed 178 m east to point B, then 195 m north to point C, then 132 m northwest to point D (northwest is a 45° turn toward west from the north direction), then 200 m south to point E, then west to point F that is located due north of point A, then south to point A.

Find the area of this plot of land, correct to the nearest square meter.

COMPUTER KEY-IN

The shaded region shown is bounded by the graph of $y = x^2$, the x-axis, and the vertical lines $x = 1$ and $x = 2$. The area of this region can be approximated by constructing rectangles. (See Computer Key-In, pages 384–385.) This area can also be approximated by constructing trapezoids. For illustration purposes the curve $y = x^2$ has been exaggerated slightly to better show the constructed trapezoids. As the diagrams below show, you can obtain a closer approximation by using trapezoids than by using rectangles.

Let us compare the approximations computed when five rectangles and five trapezoids are used.

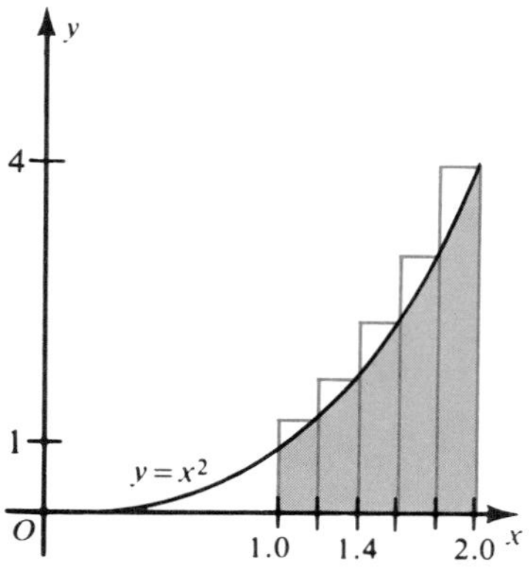

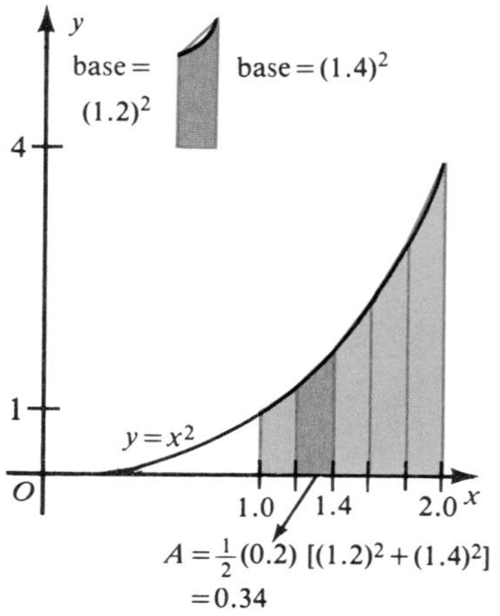

The base of each rectangle is 0.2 and the height is given by $y = x^2$. For any trapezoid in the diagram at the right above, the parallel bases are vertical segments from the x-axis to the curve $y = x^2$. The altitude is a horizontal segment with length 0.2. For example, in the second trapezoid, the bases are $(1.2)^2$ and $(1.4)^2$, respectively, and the height is 0.2.

On the next page the area of the shaded region is first approximated by the sum of the areas of the five rectangles and then by the sum of the areas of the five trapezoids. Compare the two approximations. (The exact area is $\frac{7}{3}$.)

Area approximated by rectangles:

$$A \approx (1.2)^2(0.2) + (1.4)^2(0.2) + (1.6)^2(0.2) + (1.8)^2(0.2) + (2.0)^2(0.2) \approx 2.64$$

Area approximated by five trapezoids:

$$A \approx \frac{1}{2}(0.2)[(1.0)^2 + (1.2)^2] + \frac{1}{2}(0.2)[(1.2)^2 + (1.4)^2] + \frac{1}{2}(0.2)[(1.4)^2 + (1.6)^2]$$
$$+ \frac{1}{2}(0.2)[(1.6)^2 + (1.8)^2] + \frac{1}{2}(0.2)[(1.8)^2 + (2.0)^2] \approx 2.34$$

The following computer program will compute and add the areas of the five trapezoids shown above.

```
10 FOR X = 1 TO 1.8 STEP 0.2
20 LET B1 = X↑2
30 LET B2 = (X + 0.2)↑2
40 LET A = A + 0.5 * 0.2 * (B1 + B2)
50 NEXT X
60 PRINT "AREA IS APPROXIMATELY ";A
70 END
```

If the program above is run for 100 smaller trapezoids with vertices on the x-axis at $x = 1.00, 1.01, 1.02, \ldots, 1.99, 2.00$, then a better approximation can be found. Change lines 10, 30, and 40 to

```
10 FOR X = 1 TO 1.99 STEP 0.01
30 LET B2 = (X + 0.01)↑2
40 LET A = A + 0.5 * 0.01 * (B1 + B2)
```

With this change the computer will print

```
AREA IS APPROXIMATELY 2.33335
```

Exercises

1. Modify the given computer program so that it will use 1000 trapezoids with base vertices at 1.000, 1.001, 1.002, . . . , 2.000 to approximate the area of the shaded region.
2. Modify the given computer program so that it will use ten trapezoids to approximate the area of the region that is bounded by the graph of $y = x^2$, the x-axis, and *the vertical lines* $x = 0$ *and* $x = 1$. Compare your answer with that obtained on page 385, where ten rectangles were used. (*Note:* Calculus can be used to prove that the exact area is $\frac{1}{3}$.)
3. Modify the given computer program so that it will use ten trapezoids to approximate the area of the region that is bounded by the graph of $y = 4 - x^2$, the x-axis, and *the vertical lines* $x = 0$ *and* $x = 2$. Compare your answer with that obtained on page 385, where ten rectangles were used. (*Note:* Calculus can be used to prove that the exact area is $\frac{16}{3}$.)

9-4 Areas of Regular Polygons

The beautifully symmetrical designs of kaleidoscopes are produced by mirrors that reflect light through loose particles of colored glass. Since the body of a kaleidoscope is a tube, the designs always appear to be inscribed in a circle. The photograph of a kaleidoscope pattern at the right suggests a regular hexagon.

Given any circle, you can inscribe in it a regular polygon of any number of sides. The diagrams below show how this can be done.

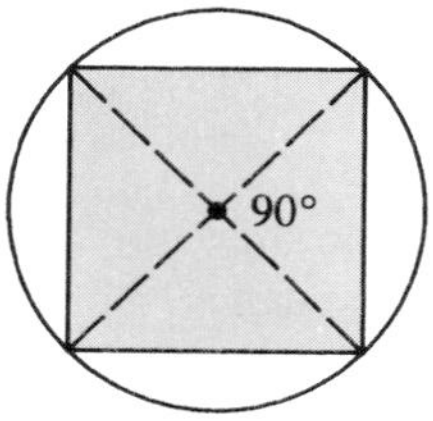

Square in circle: Draw four 90° central angles.

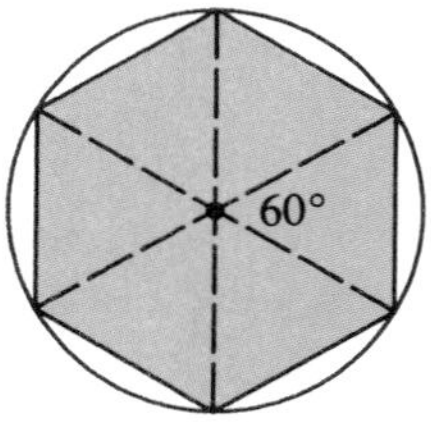

Regular hexagon in circle: Draw six 60° central angles.

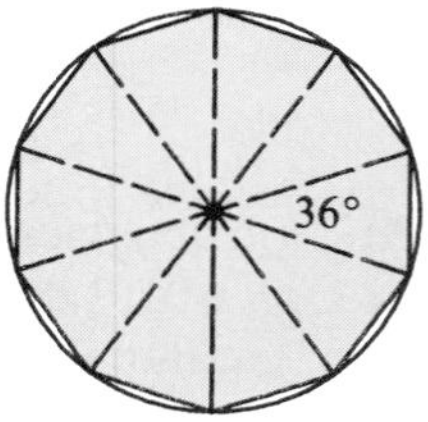

Regular decagon in circle: Draw ten 36° central angles.

It is also true that if you are given any regular polygon, you can circumscribe a circle about it. This relationship between circles and regular polygons leads us to the following definitions:

The **center of a regular polygon** is the center of the circumscribed circle.

The **radius of a regular polygon** is the distance from the center to a vertex.

A **central angle of a regular polygon** is an angle formed by two radii drawn to consecutive vertices.

The **apothem of a regular polygon** is the (perpendicular) distance from the center of the polygon to a side.

Center of regular octagon: O

Radius: OA, OB, OC, and so on

Central angle: $\angle AOB$, $\angle BOC$, and so on

Measure of central angle: $\frac{360}{8} = 45$

Apothem: OX

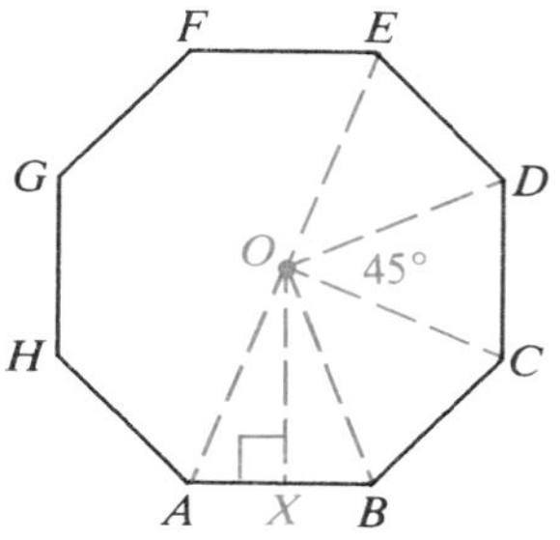

If you know the apothem and perimeter of a regular polygon, you can use the next theorem to find the area of the polygon.

Theorem 9-5

The area of a regular polygon is equal to half the product of the apothem and the perimeter. ($A = \frac{1}{2}ap$)

Given: Regular n-gon $TUVW\ldots$; apothem a; side s; perimeter p; area A

Prove: $A = \frac{1}{2}ap$

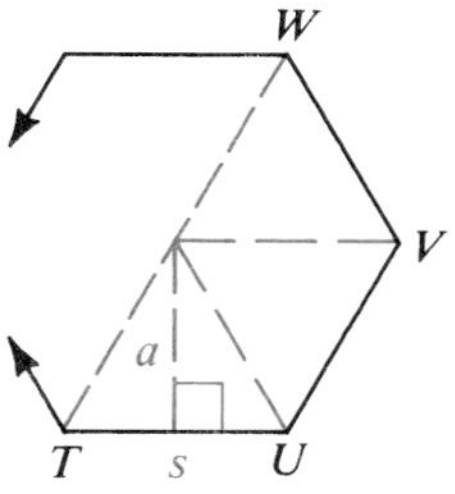

Outline of proof:

1. If all radii are drawn, n congruent triangles are formed.
2. Area of each $\triangle = \frac{1}{2}sa$
3. $A = n(\frac{1}{2}sa) = \frac{1}{2}a(ns)$
4. Since $ns = p$, $A = \frac{1}{2}ap$.

Example 1 Find the radius and apothem of an equilateral triangle with side 6.

Solution Use 30°-60°-90° $\triangle$ relationships.

$$a = \frac{3}{\sqrt{3}} = \sqrt{3}$$
$$r = 2a = 2\sqrt{3}$$

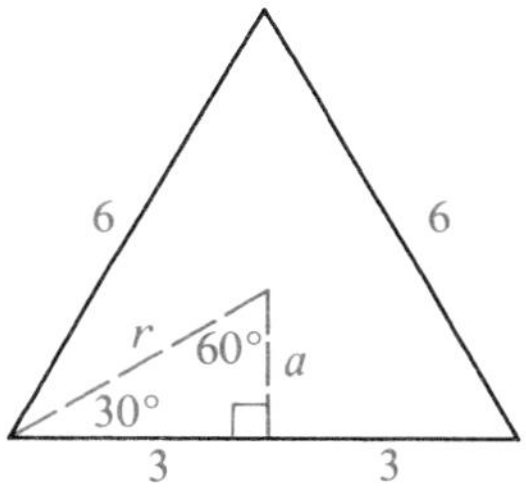

Example 2 Find the area of a regular hexagon with apothem 9.

Solution Use 30°-60°-90° $\triangle$ relationships.

$$\frac{1}{2}s = \frac{9}{\sqrt{3}} = 3\sqrt{3}$$
$$s = 6\sqrt{3};\ p = 36\sqrt{3}$$
$$A = \frac{1}{2}ap = \frac{1}{2}\cdot 9 \cdot 36\sqrt{3}$$
$$= 162\sqrt{3}$$

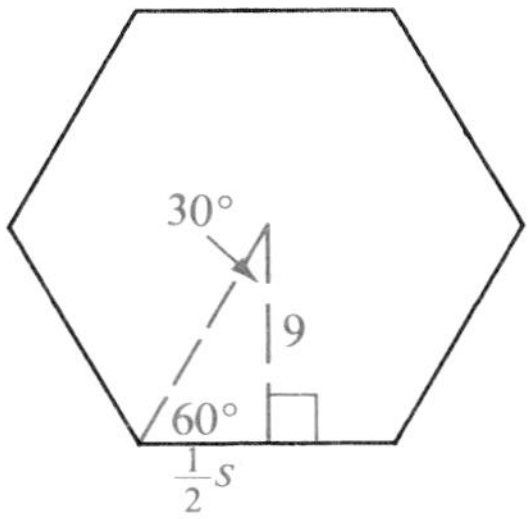

Classroom Exercises

For each regular polygon shown, find:
a. the measure of a central angle
b. the apothem a **c. the radius r**

1. 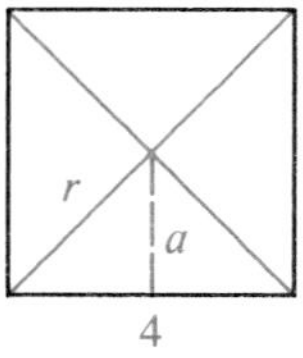

2. 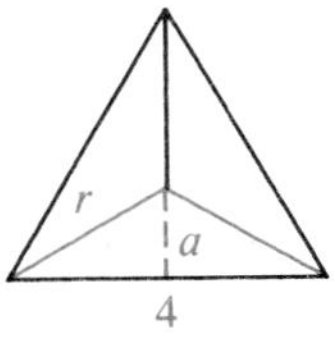

3. 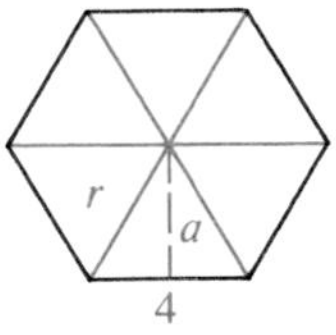

4. Complete the table below.

Number of sides of regular polygon	9	10	360	?	?
Measure of central angle (in degrees)	?	?	?	30	20

Find the area of each regular polygon described.

5. A regular octagon with side 4 and apothem a.
6. A regular pentagon with side s and apothem 3.
7. A regular decagon with side s and apothem a.
8. $ABCDE$ is a regular pentagon with radius 10.
   a. $m\angle AOB = \underline{\ ?\ }$
   b. Explain why $m\angle AOX = 36$.
   Note: For parts (c)–(e), use the table on page 271 or a calculator.
   c. $\cos 36° = \frac{a}{10}$. To the nearest tenth, $a \approx \underline{\ ?\ }$.
   d. $\sin 36° = \frac{\frac{1}{2}s}{?}$. To the nearest tenth, $s \approx \underline{\ ?\ }$.
   e. Find the perimeter and area of the pentagon.

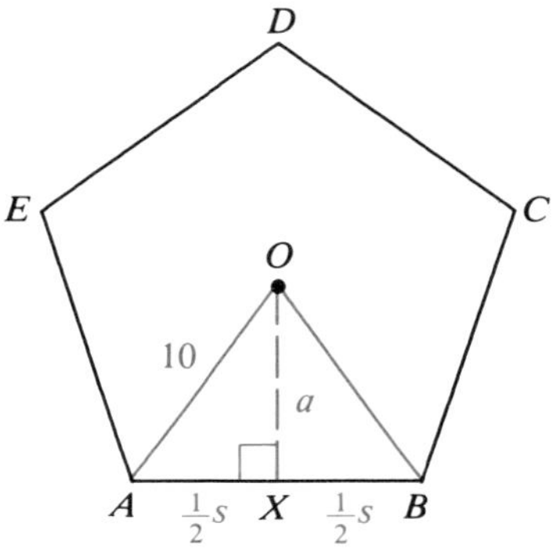

Written Exercises

Copy and complete the tables for the regular polygons shown. In these tables, *p* represents the perimeter and *A* represents the area.

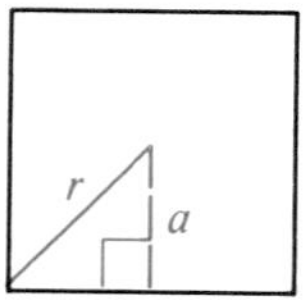

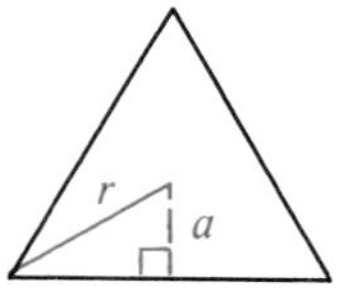

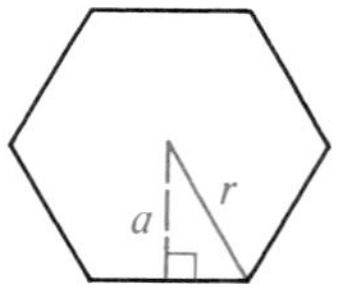

A

	r	a	A
1.	$8\sqrt{2}$	?	?
2.	?	5	?
3.	?	?	49
4.	?	$\sqrt{6}$	?

	r	a	p	A
5.	6	?	?	?
6.	?	4	?	?
7.	?	?	12	?
8.	?	?	$9\sqrt{3}$	?

	r	a	p	A
9.	4	?	?	?
10.	?	$5\sqrt{3}$	?	?
11.	?	6	?	?
12.	?	?	$12\sqrt{3}$	?

Find the area of each polygon.

B 13. Equilateral triangle with radius $4\sqrt{3}$
14. Square with radius $8k$
15. Regular hexagon with perimeter 72
16. Regular hexagon with apothem 4

Three regular polygons are inscribed in circles with radii 1. Find the apothem, the perimeter, and the area of each polygon. Use $\sqrt{3} \approx 1.73$ and $\sqrt{2} \approx 1.41$.

17.

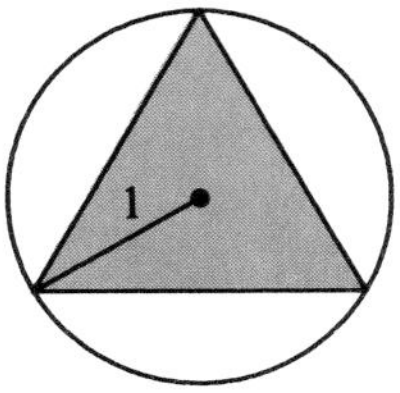

18.

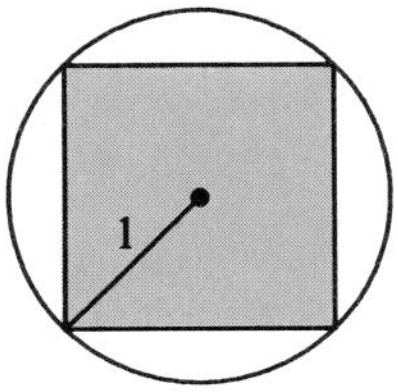

19.

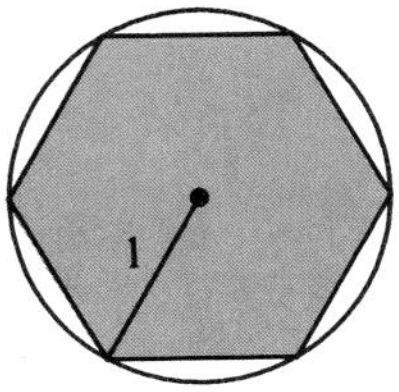

20. Let s be the length of the side of a square that is inscribed in a circle with radius r.

a. Find s and the perimeter, p, in terms of r.

b. Express an approximation to the perimeter found in part (a) by using $\sqrt{2} \approx 1.414$.

21. A regular decagon is shown inscribed in a circle with radius 1.

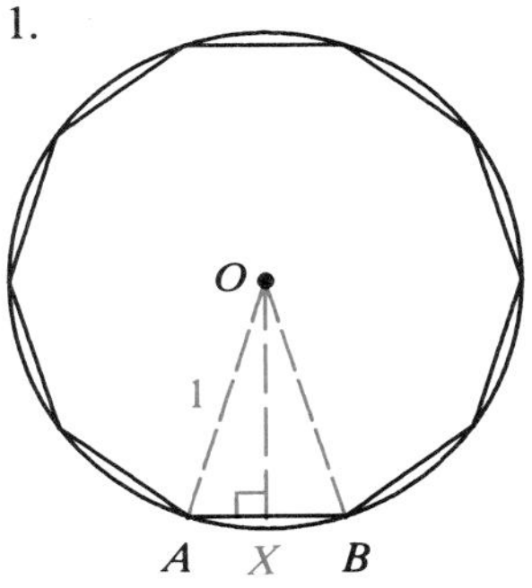

a. Explain why $m\angle AOX = 18$.

b. Use a calculator or the table on page 271 to evaluate OX and AX below.

$\sin 18° = \frac{AX}{1}$, so $AX \approx$ _?_

$\cos 18° = \frac{OX}{1}$, so $OX \approx$ _?_

c. Perimeter of decagon $\approx$ _?_

d. Area of $\triangle AOB \approx$ _?_

e. Area of decagon $\approx$ _?_

C **22.** Find the area and perimeter of a regular dodecagon (12 sides) inscribed in a circle with radius 1. Use the procedure suggested by Exercise 21.

23. A regular polygon with n sides is inscribed in a circle with radius 1.

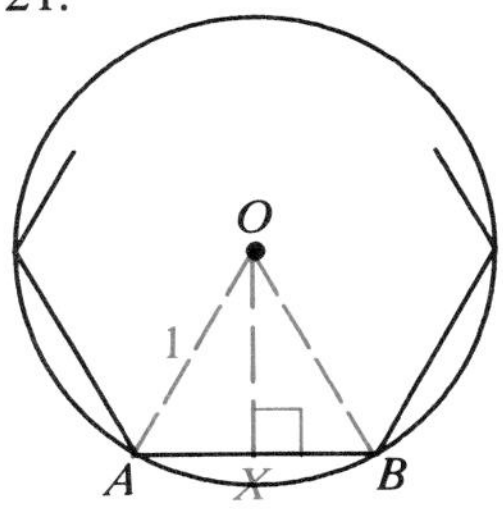

a. Explain why $m\angle AOX = \frac{180}{n}$.

b. Show that $AX = \sin\left(\frac{180}{n}\right)^\circ$.

c. Show that $OX = \cos\left(\frac{180}{n}\right)^\circ$.

d. Show that the perimeter of the polygon is $p = 2n \cdot \sin\left(\frac{180}{n}\right)^\circ$.

e. Show that the area of the polygon is $A = n \cdot \sin\left(\frac{180}{n}\right)^\circ \cdot \cos\left(\frac{180}{n}\right)^\circ$.

COMPUTER KEY-IN

If a regular n-sided polygon is inscribed in a circle with radius 1, then its perimeter and area are given by the formulas previously derived in Exercise 23.

$$\text{Perimeter} = 2n \cdot \sin\left(\frac{180}{n}\right)^\circ \qquad \text{Area} = n \cdot \sin\left(\frac{180}{n}\right)^\circ \cdot \cos\left(\frac{180}{n}\right)^\circ$$

The following computer program uses these formulas to find the perimeter and area of any regular *N*-sided polygon inscribed in a circle with radius 1. Most computer languages require that angle measures be given in *radians* instead of degrees. 180° is approximately the same as 3.14159 radians. Thus the formulas on page 399 can be rewritten in computer format as

```
P = 2 * N * SIN(3.14159/N)
A = N * SIN(3.14159/N) * COS(3.14159/N).
```

```
10  PRINT "HOW MANY SIDES";
20  INPUT N
30  LET P = 2 * N * SIN(3.14159/N)
40  LET A = N * SIN(3.14159/N) * COS(3.14159/N)
50  PRINT "PERIMETER IS ";P
60  PRINT "AREA IS ";A
70  END
```

Exercises

1. Run the given computer program using different values for the number of sides, *N*.
2. Use the given program to complete the table at the right.
3. Use your answers in Exercise 2 to suggest an approximation to the perimeter and area of a circle with radius 1.

Number of sides	Perimeter	Area
18	?	?
180	?	?
1800	?	?
18000	?	?

Self-Test 1

Find the area of each polygon.

1. A square with diagonal $9\sqrt{2}$
2. A rectangle with base 12 and diagonal 13
3. A parallelogram with sides 8 and 10 and an angle of measure 60
4. An equilateral triangle with perimeter 12 cm
5. An isosceles triangle with sides 7 cm, 7 cm, and 12 cm
6. A rhombus with diagonals 8 and 10
7. An isosceles trapezoid with legs 5 and bases 9 and 17
8. A regular hexagon with sides 10
9. Find the area of the quadrilateral shown.

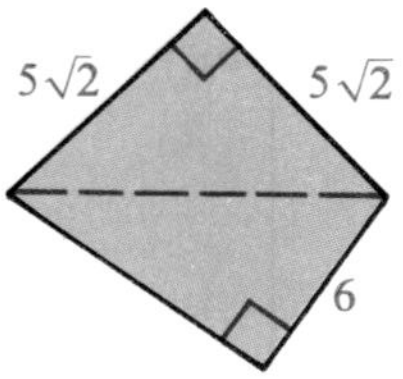

CAREERS

Carpenter

Carpenters work in all parts of the construction industry. A self-employed carpenter may work on relatively small-scale projects—for example, remodeling rooms or making other

alterations in existing houses or even building new single-family houses. As an employee of a large building contractor, a carpenter may be part of the work force building apartment or office complexes, stores, factories, and other major projects. Some carpenters are employed solely to provide maintenance to a large structure, where they do repairs and upkeep and make any alterations in the structure that are required.

Carpenters with adequate experience and expertise may become specialists in some skill of their own choice, for example, framing, interior finishing, or cabinet making. A carpenter who learns all aspects of the building industry thoroughly may decide to go into business as a general contractor, responsible for all work on an entire project.

Although some carpenters learn the trade through four-year apprenticeships, most learn on the job. These workers begin as laborers or as carpenters' helpers. While they work in these jobs they gradually acquire the skills necessary to become carpenters themselves. Carpenters must be able to measure accurately and to apply their knowledge of arithmetic, geometry, and informal algebra. They also benefit from being able to read and understand plans, blueprints, and charts.

Circles and Similar Figures

Objectives

1. Understand how the area and perimeter formulas for regular polygons relate to the area and circumference formulas for circles.
2. Compute the circumferences and areas of circles.
3. Compute arc lengths and the areas of sectors of a circle.
4. Understand and apply the relationships between scale factors, perimeters, and areas of similar figures.

9-5 Circumference and Area of a Circle

When you think of the perimeter of a figure, you probably think of the distance around the figure. Since the word "around" is not mathematically precise, perimeter is usually defined in other ways. For example, the perimeter of a polygon is defined as the sum of the lengths of its sides, which are segments. Since a circle has no sides, the perimeter of a circle must be defined differently. The preceding Computer Key-In suggests the following approach.

First consider a sequence of regular polygons inscribed in a circle with radius r. Four such polygons are shown below. Imagine that we can keep increasing the number of sides of the regular polygons. As you can see in the diagrams, the more sides there are, the closer the regular polygon approximates (or "fits") the curve of the circle.

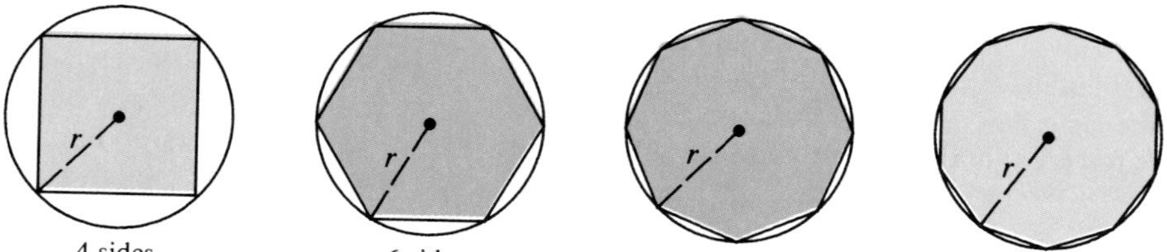

4 sides — 6 sides — 8 sides — 10 sides

Now consider the perimeters and the areas of many different regular polygons. The following table contains values that are approximations (using trigonometry) of the perimeters and the areas of regular polygons in terms of the radius, r.

Number of Sides of Polygon	Perimeter	Area
4	$5.66r$	$2.00r^2$
6	$6.00r$	$2.60r^2$
8	$6.12r$	$2.83r^2$
10	$6.18r$	$2.93r^2$
20	$6.26r$	$3.09r^2$
30	$6.27r$	$3.12r^2$
100	$6.28r$	$3.14r^2$

As the table suggests, these perimeters give us a sequence of numbers that get closer and closer to a limiting number. This limiting number is defined to be the perimeter, or **circumference,** of the circle.

The area of a circle is defined in a similar way. The areas of the inscribed regular polygons get closer and closer to a limiting number, and this limit is defined to be the **area** of the circle.

The results in the table suggest that the circumference and the area of a circle with radius r are approximately $6.28r$ and $3.14r^2$. The exact values are given by the formulas below. (Proofs are suggested in Classroom Exercise 13 and Written Exercise 31.)

Circumference of circle:	$2\pi \cdot$ radius	$C = 2\pi r$
Area of circle:	$\pi \cdot$ radius squared	$A = \pi r^2$

These formulas involve a famous number denoted by the Greek letter π (*pi*). There isn't any decimal or fraction that expresses π exactly. Here are some common approximations for π:

$$3.14 \qquad \frac{22}{7} \qquad 3.1416 \qquad 3.14159$$

When you calculate the circumference and area of a circle, leave your answers in terms of π unless you are told to replace π by one of its approximations.

Example 1 Find the circumference and area of a circle with radius 6 cm.

Solution $C = 2\pi r = 2\pi(6) = 12\pi$ (cm)
$A = \pi r^2 = \pi(6^2) = 36\pi$ (cm^2)

Example 2 Find the circumference of a circle with radius 10 cm. Use $\pi \approx 3.14$.

Solution $C = 2\pi r \approx 2(3.14)(10) \approx 62.8$ (cm)

Example 3 Find the circumference of a circle if the area is 25π.

Solution From $\pi r^2 = 25\pi$, we get $r^2 = 25$ and $r = 5$.
Thus, $C = 2\pi r = 2\pi(5) = 10\pi$.

Classroom Exercises

Complete the table.

	1.	2.	3.	4.	5.	6.	7.	8.
Radius	3	4	8	?	?	?	?	?
Circumference	?	?	?	10π	18π	?	?	?
Area	?	?	?	?	?	36π	49π	144π

Find the circumference and area to the nearest tenth. Use $\pi \approx 3.14$.

9. $r = 2$ **10.** $r = 6$ **11.** $r = \frac{1}{2}$ **12.** $r = 0.3$

13. The number π is defined to be the ratio of the circumference of a circle to the diameter. This ratio is constant from circle to circle. Supply the missing reasons in the outline of proof below.

Given: $\odot O$ and $\odot O'$ with circumferences C and C' and diameters d and d'

Prove: $\frac{C}{d} = \frac{C'}{d'}$

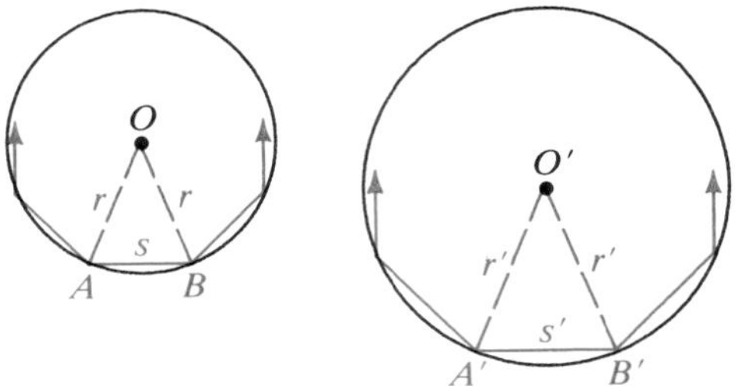

Outline of proof:

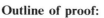

Inscribe in each circle a regular polygon of n sides. Let p and p' be the perimeters.

1. $p = ns$ and $p' = ns'$ (Why?)
2. $\frac{p}{p'} = \frac{ns}{ns'} = \frac{s}{s'}$ (Why?)
3. $\triangle AOB \sim \triangle A'O'B'$ (Why?)
4. $\frac{s}{s'} = \frac{r}{r'} = \frac{d}{d'}$ (Why?)
5. Thus, $\frac{p}{p'} = \frac{d}{d'}$ (Steps 2 and 4)
6. Steps 1–4 hold for any number of sides n. We can let n be so large that p is practically the same as C, and p' is practically the same as C'. In advanced courses, you learn that C and C' can be substituted for p and p' in Step 5. This gives $\frac{C}{C'} = \frac{d}{d'}$, or $\frac{C}{d} = \frac{C'}{d'}$.

14. The photograph shows a piece of land that is supplied with water by a circular irrigation system. This system consists of a moving arm that sprinkles water over a circular region. If the arm is 430 m long, what is the area, correct to the nearest thousand square meters, of the region being irrigated? (Use $\pi \approx 3.14$)

Written Exercises

Complete the table. Leave answers in terms of π.

A

	1.	**2.**	**3.**	**4.**	**5.**	**6.**	**7.**	**8.**
Radius	7	120	$\frac{5}{2}$	$6\sqrt{2}$	?	?	?	?
Circumference	?	?	?	?	20π	12π	?	?
Area	?	?	?	?	?	?	25π	50π

Find the circumference and area. Use $\pi \approx \frac{22}{7}$.

9. $r = 42$ **10.** $d = \frac{7}{2}$ **11.** $d = 2\frac{6}{11}$ **12.** $r = 7k$

Find the circumference and area, correct to the nearest tenth. Use $\pi \approx 3.14$.

13. $r = 10$ **14.** $d = 3$ **15.** $d = 0.5$ **16.** $r = 1.1$

17. Which is the better buy, a 10-inch round pizza costing \$4 or a 15-inch round pizza costing \$7?

B **18.** A target consists of four concentric circles with radii 1, 2, 3, and 4. Find the area of the bull's-eye and of each ring of the target. (What would be the area of the *n*th ring?)

19. Semicircles are constructed on the sides of the right triangle shown at the right. Show that

Area I + Area II = Area III.

20. Suppose that in Exercise 19 the lengths of the legs of the right triangle are a and b and the length of the hypotenuse is c. Show that

Area I + Area II = Area III.

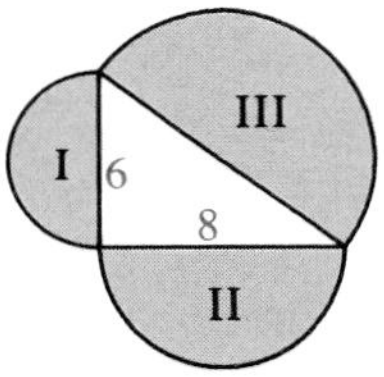

Exs. 19, 20

Find the area of each shaded region in terms of *r*.

21.

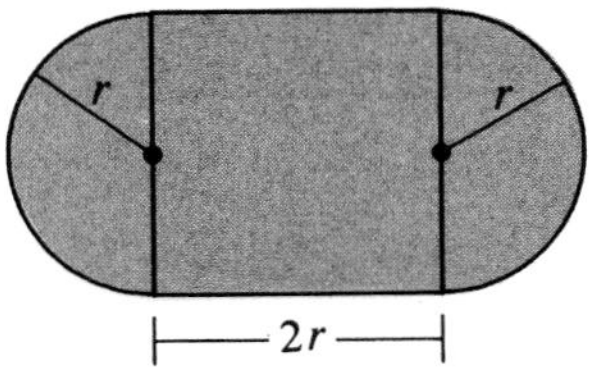

22.

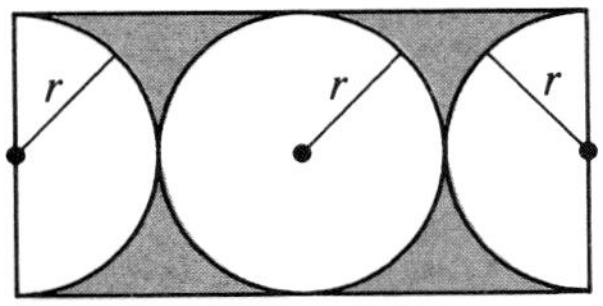

23.

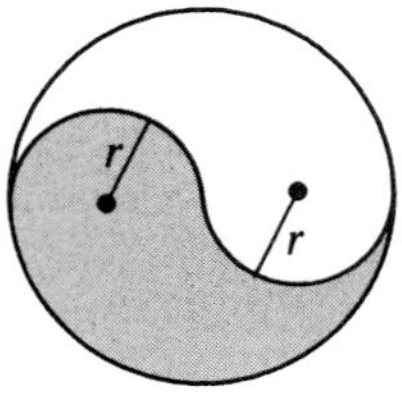

24.

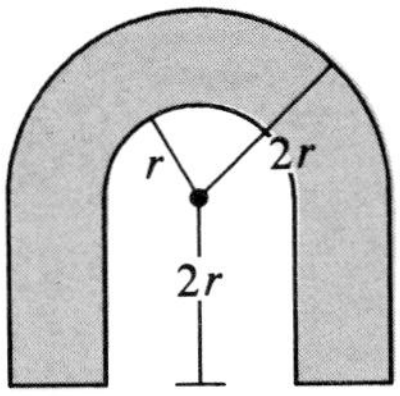

In Exercises 25 and 26, the tires of a racing bike are approximately 70 cm in diameter.

25. How far does the bike travel if the tires make 10 revolutions? Use $\pi \approx \frac{22}{7}$.

26. About how many revolutions will the wheel make in a 22 km race? Recall that 1 km = 1000 m = 100,000 cm. Use $\pi \approx \frac{22}{7}$.

Each diagram shows a regular polygon with sides 6 cm long. Also shown are the polygon's inscribed and circumscribed circles. Find the area of each circle.

27.

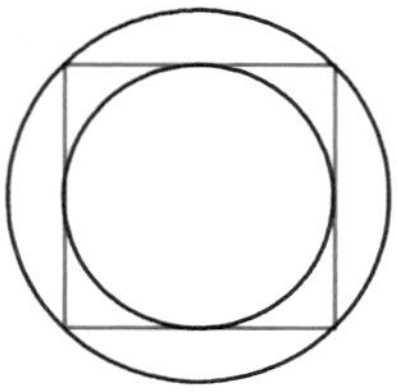

28.

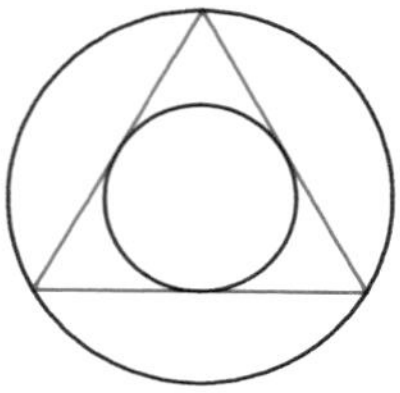

29.

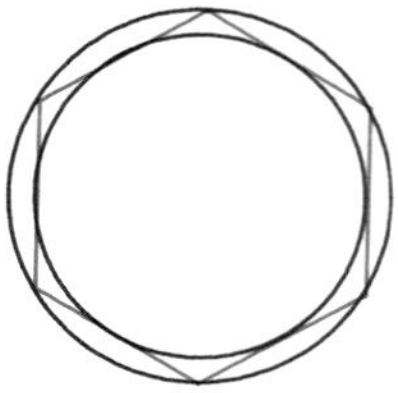

C **30.** A regular polygon with 12 sides is inscribed in a circle with radius 6. Find the area enclosed between the circle and the polygon.

31. A regular polygon with apothem a is inscribed in a circle with radius r.

a. As the number of sides increases, the value of a gets nearer to __?__ and the perimeter of the polygon gets nearer to $2\pi r$.

b. In the formula $A = \frac{1}{2}ap$, replace a by r, and p by $2\pi r$. What formula do you get?

32. Find the circumference of a circle inscribed in a rhombus with diagonals 12 cm and 16 cm.

33. Write a formula giving the area (A) of a circle in terms of the circumference (C) and the number π.

34. A circle is inscribed in a right triangle with sides 6, 8, and 10. Find the area of the circle.

35. Draw any circle O and any circle P. Construct a circle whose area equals the sum of the areas of circle O and circle P.

9-6 Areas of Sectors and Arc Lengths

A *pie chart* is often used to analyze data or to help plan business strategy. The radii of a pie chart divide the interior of the circle into regions called *sectors,* whose areas represent the relative size of particular items. A **sector of a circle** is a region bounded by two radii and an arc of the circle. The shaded region of the diagram at the right below is called sector AOB. The unshaded region is also a sector.

The length of $\widehat{AB}$ in circle O is part of the circumference of the circle. Since $m\widehat{AB} = 60$ and $\frac{60}{360} = \frac{1}{6}$,

$$\text{Length of } \widehat{AB} = \frac{1}{6}(2\pi \cdot 5) = \frac{5}{3}\pi.$$

Similarly, the area of sector AOB is $\frac{1}{6}$ of the area of the circle. Thus,

$$\text{Area of sector } AOB = \frac{1}{6}(\pi \cdot 5^2) = \frac{25}{6}\pi.$$

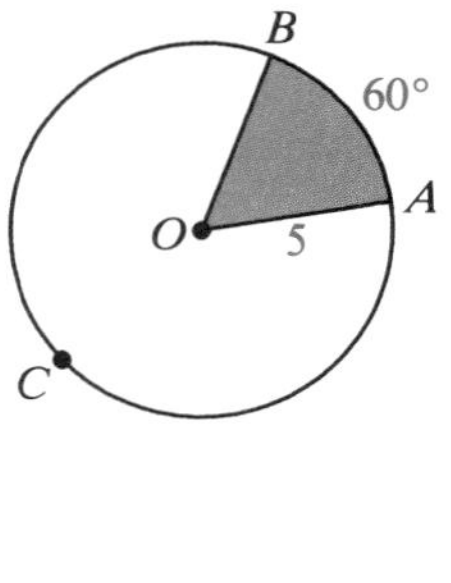

In general, if $m\widehat{AB} = x$:

$$\text{Length of } \widehat{AB} = \frac{x}{360} \cdot 2\pi r$$

$$\text{Area of sector } AOB = \frac{x}{360} \cdot \pi r^2$$

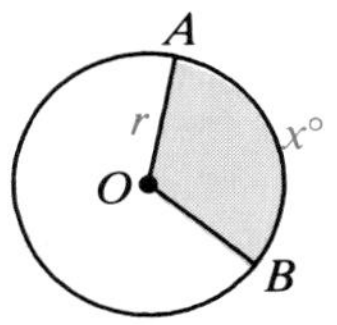

Example 1 In $\odot O$ with radius 9, $m\angle AOB = 120$.
Find the lengths of the arcs $\widehat{AB}$ and $\widehat{ACB}$ and the areas of the two sectors shown.

Solution Make a sketch as shown.
$m\widehat{AB} = 120$, and $m\widehat{ACB} = 240$.

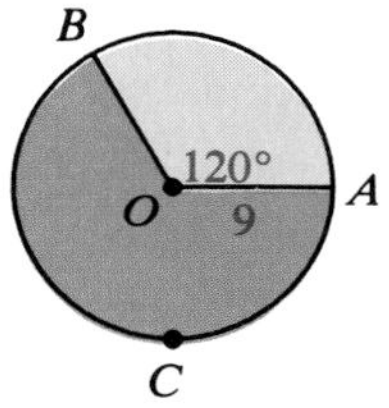

Minor arc $\widehat{AB}$:

$$\text{Arc length} = \frac{120}{360} \cdot (2\pi \cdot 9) = \frac{1}{3}(18\pi) = 6\pi$$

$$\text{Area of sector} = \frac{120}{360} \cdot (\pi \cdot 9^2) = \frac{1}{3}(81\pi) = 27\pi$$

Major arc $\widehat{ACB}$:

$$\text{Arc length} = \frac{240}{360} \cdot (2\pi \cdot 9) = \frac{2}{3}(18\pi) = 12\pi$$

$$\text{Area of sector} = \frac{240}{360} \cdot (\pi \cdot 9^2) = \frac{2}{3}(81\pi) = 54\pi$$

Example 2 Find the area of the shaded region bounded by $\overline{XY}$ and $\widehat{XY}$.

Solution

$$\text{Area of sector } XOY = \frac{90}{360} \cdot \pi \cdot 10^2 = 25\pi$$

$$\text{Area of } \triangle XOY = \frac{1}{2} \cdot 10 \cdot 10 = 50$$

$$\text{Area of shaded region} = 25\pi - 50$$

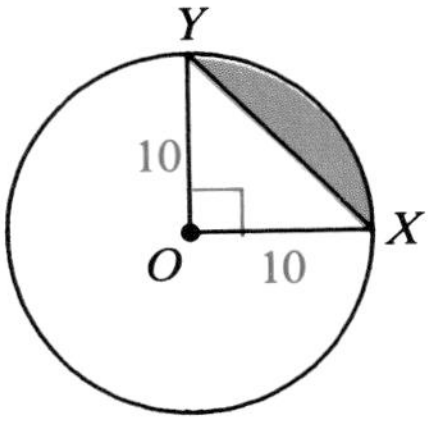

Classroom Exercises

Find the arc length and area of each shaded sector.

1.

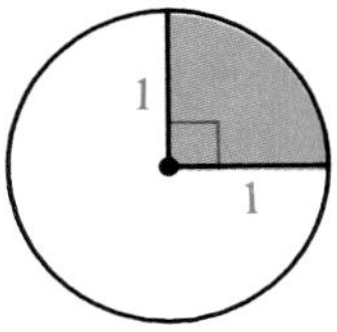

2.

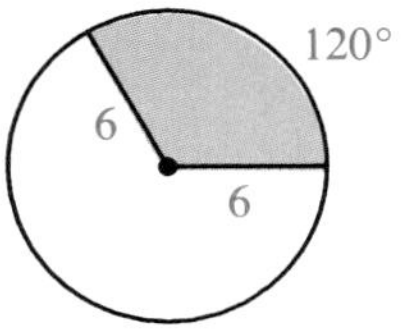

3.

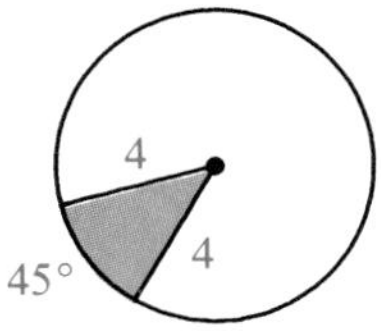

4.

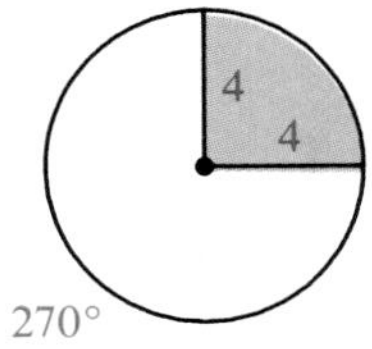

5. In a circle with radius 6, $m\widehat{AB} = 60$. Make a sketch and find the area of the region bounded by $\overline{AB}$ and $\widehat{AB}$.

Written Exercises

Sector AOB is described by giving $m\angle AOB$ and the radius of circle O. Make a sketch and find the length of $\widehat{AB}$ and the area of sector AOB.

A

	1.	2.	3.	4.	5.	6.	7.	8.	9.	10.	11.	12.
$m\angle AOB$	90	60	30	45	120	240	180	270	40	320	108	200
radius	10	12	12	4	3	3	5	8	6	6	25	3

13. The area of sector AOB is 10π and $m\angle AOB = 100$. Find the radius of circle O.

14. The area of sector AOB is $\frac{7\pi}{2}$ and $m\angle AOB = 315$. Find the radius of circle O.

Find the area of each shaded region. Point O marks the center of a circle.

B **15.**

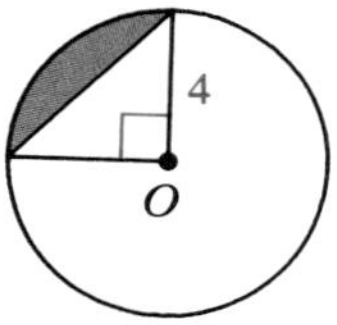

16.

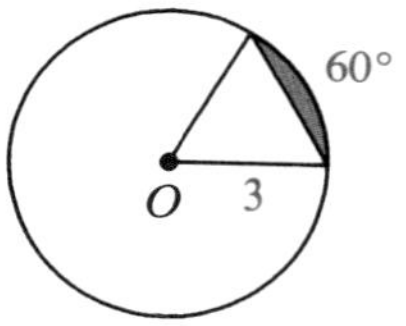

17.

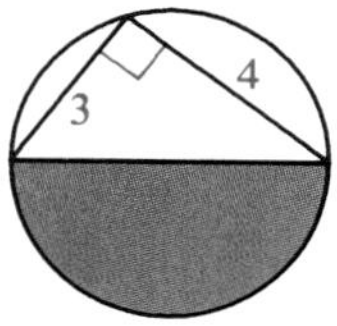

18.

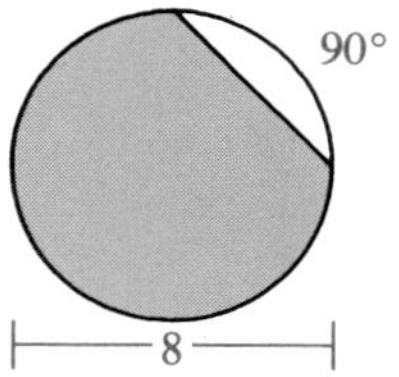

19.

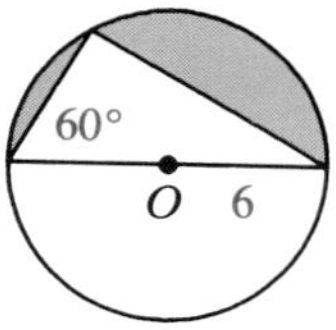

20.

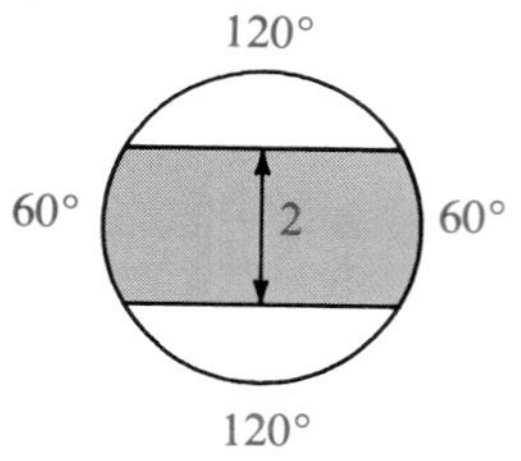

21. A rectangle with length 16 cm and width 12 cm is inscribed in a circle. Find the area of the region inside the circle but outside the rectangle.

22. From point P, $\overline{PA}$ and $\overline{PB}$ are drawn tangent to circle O at points A and B. If the radius of the circle is 6 and $m\angle APB = 60$, find the area of the region outside the circle but inside quadrilateral $AOBP$.

23. $\overline{AB}$ is a chord of a circle with radius 10. If $m\widehat{AB} = 72$, find the area of the region bounded by $\overline{AB}$ and $\widehat{AB}$, correct to the nearest tenth. (Use trigonometry.)

24. $ABCD$ is a square with sides 8 cm long. Two circles each with radius 8 cm are drawn, one with center A and the other with center C. Find the area of the region inside both circles.

25. **a.** Draw a square, then construct the figure shown at the right.
b. If the radius of the square is 2, find the area of the shaded region.

C 26. **a.** Using only a compass, construct the six-pointed figure shown at the right.
b. If the radius of the circle is 6, find the area of the shaded region.

27. Three circles with radii 6 are tangent to each other. Find the area of the region enclosed between them.

★28. Circles X and Y, with radii 6 and 2, are tangent to each other. $\overline{AB}$ is a common external tangent. Find the area of the shaded region. (*Hint:* What kind of figure is $AXYB$? What is the measure of $\angle AXY$?)

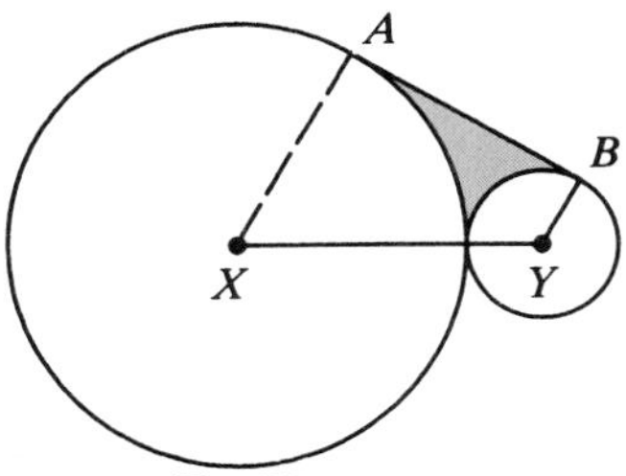

Ex. 28

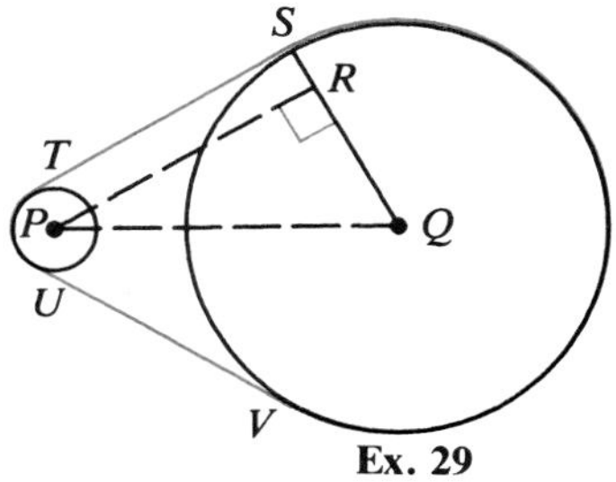

Ex. 29

★29. The diagram at the right above shows a belt tightly stretched over two wheels with radii 5 cm and 25 cm. The distance between the centers of the wheels is 40 cm. Find the length of the belt.

★30. Given: $\overset{\frown}{ACB}$, $\overset{\frown}{AXC}$, and $\overset{\frown}{CYB}$ are semicircles; $AC = CB$
a. Show that Area I + Area II = Area of $\triangle ABC$.
b. Is the statement in part (a) true if $AC \neq CB$?

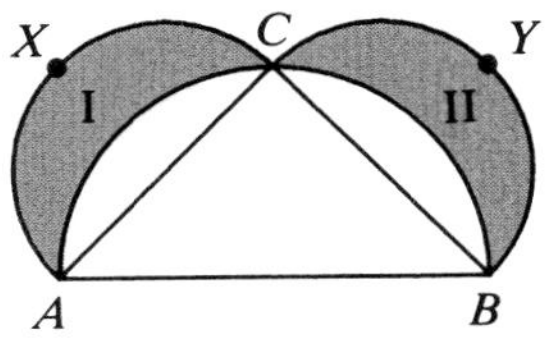

Challenge

Here $\overline{XY}$ has been divided into 5 congruent segments and semicircles have been drawn. But suppose $\overline{XY}$ were divided into millions of congruent segments and semicircles were drawn. What would the sum of the lengths of the arcs be?

Sarah says, "XY, because all the points would be so close to $\overline{XY}$." Mike says, "A really large number, because there would be so many arc lengths to add up." What do you say?

CALCULATOR KEY-IN

If your calculator does not have a π key, use $\pi \approx 3.14159$.

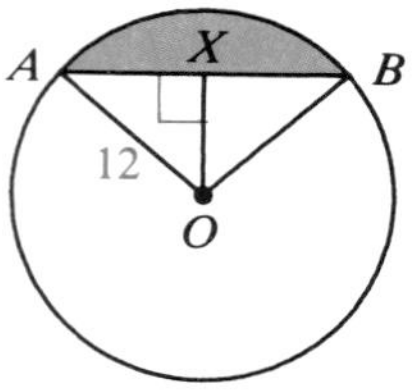

1. Chord AB is 18 cm long and the radius of the circle is 12 cm.
 a. Use trigonometry to find the measures of $\angle AOX$ and $\angle AOB$, correct to the nearest degree.
 b. Find the area of the shaded region, correct to the nearest hundredth of a square centimeter.

2. A cow is tied to a 25 m rope that is tied to the corner of a barn as shown. A fence keeps the cow out of the garden. Find, to the nearest square meter, the grazing area.

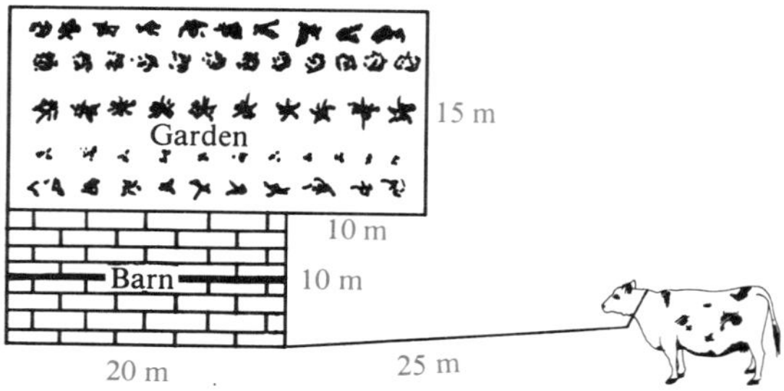

9-7 Areas of Similar Figures

The triangles shown below are similar. Notice that corresponding sides are in a 2:1 ratio and corresponding altitudes are in a 2:1 ratio also. What do you think the ratio of the perimeters is? the ratio of the areas is?

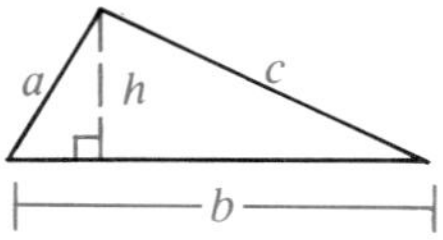

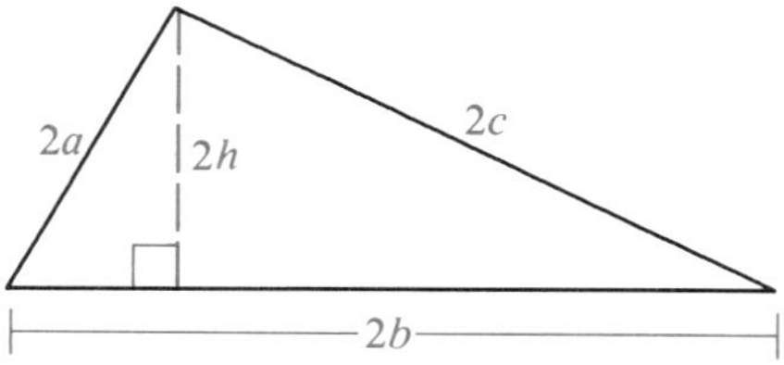

If you thought that the ratio of the perimeters is 2:1, you were right.

$$\frac{\text{Perimeter of larger triangle}}{\text{Perimeter of smaller triangle}} = \frac{2a + 2b + 2c}{a + b + c} = \frac{2(a + b + c)}{a + b + c} = \frac{2}{1}$$

Many people think that the ratio of the areas is also 2:1, but it is 4:1.

$$\frac{\text{Area of larger triangle}}{\text{Area of smaller triangle}} = \frac{\frac{1}{2}(2b)(2h)}{\frac{1}{2}bh} = \frac{4bh}{bh} = \frac{4}{1}$$

Theorem 9-6

If the scale factor of two similar figures is $a:b$, then:
(1) The ratio of the perimeters is $a:b$.
(2) The ratio of the areas is $a^2:b^2$.

Example 1 Find the ratio of the perimeters and the ratio of the areas of the two similar figures.

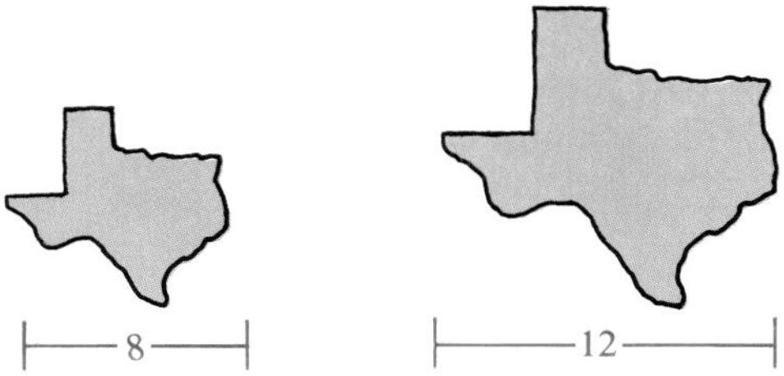

Solution The scale factor is 8:12, or 2:3.
Therefore, the ratio of the perimeters is 2:3.
The ratio of the areas is $2^2:3^2$, or 4:9.

Example 2 *ABCD* is a trapezoid.

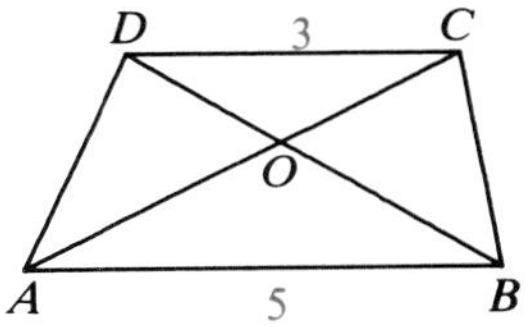

a. Find the ratio of the areas of $\triangle COD$ and $\triangle AOB$.
b. Find the ratio of the areas of $\triangle COD$ and $\triangle DOA$.

Solution **a.** $\triangle COD \sim \triangle AOB$ (AA Similarity Postulate) with scale factor 3:5.

Then $\dfrac{\text{Area of } \triangle COD}{\text{Area of } \triangle AOB} = \dfrac{3^2}{5^2} = \dfrac{9}{25}$

b. $\triangle COD$ is *not* similar to $\triangle DOA$, so we cannot use Theorem 9-6. Instead, we consider $\overline{CO}$ and $\overline{AO}$ as bases of the two triangles, which then have the same height.

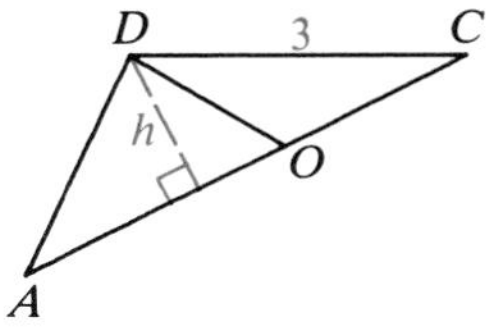

Then $\dfrac{\text{Area of } \triangle COD}{\text{Area of } \triangle DOA} = \dfrac{\frac{1}{2} \cdot CO \cdot h}{\frac{1}{2} \cdot AO \cdot h} = \dfrac{CO}{AO}$

To find $\dfrac{CO}{AO}$, we note that $\overline{CO}$ and $\overline{AO}$ are corresponding sides of similar triangles *COD* and *AOB* in the given trapezoid. Thus, $\dfrac{CO}{AO} = \dfrac{3}{5}$.

Classroom Exercises

The table refers to similar figures. Complete the table.

	1.	2.	3.	4.	5.	6.	7.	8.
Scale factor	1:3	1:5	3:4	6:9	?	?	?	?
Perimeter ratio	?	?	?	?	4:5	12:20	?	?
Area ratio	?	?	?	?	?	?	16:49	36:25

9. a. Are all circles similar?
   b. If two circles have radii 9 and 12, what is the ratio of the circumferences? of the areas?

10. The areas of two circles are 25π and 81π. What is the ratio of the circumferences?

11. *ABCD* is a parallelogram. Find each ratio.

    a. $\frac{\text{Area of } \triangle DEF}{\text{Area of } \triangle ABF}$ b. $\frac{\text{Area of } \triangle DEF}{\text{Area of } \triangle CEB}$

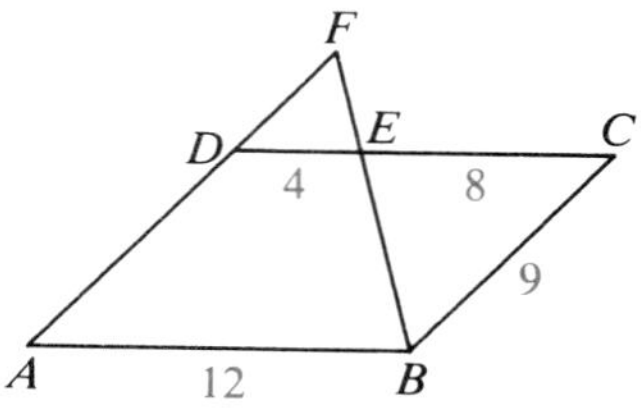

12. Consider triangles I, II, and III.
    a. Are any of these triangles similar?
    b. Can Theorem 9-6 be used?
    c. Find the ratio of the areas of triangles I and II.
    d. Find the ratio of the areas of triangles II and III.

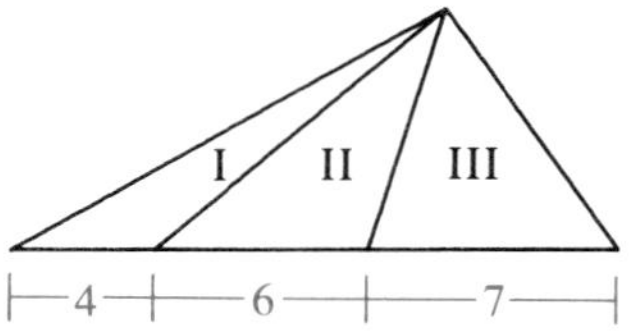

13. The figure is a trapezoid. Find the ratio of the areas of each pair of triangles.
    a. I and III   b. I and II
    c. I and IV   d. II and IV

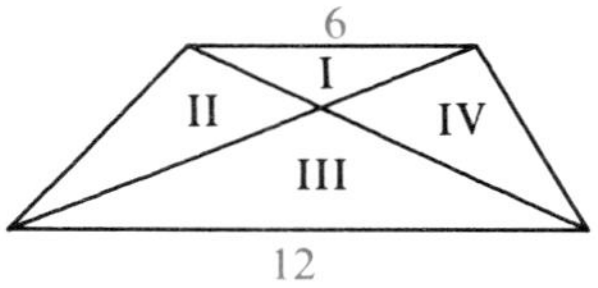

14. On a map of California, 1 cm corresponds to 50 km. Find the ratio of the map's area to the actual area of California.

Written Exercises

The table refers to similar figures. Copy and complete the table.

A

	1.	2.	3.	4.	5.	6.	7.	8.
Scale factor	1:4	3:2	6:7	?	?	?	?	?
Ratio of perimeters	?	?	?	9:5	3:13	?	?	?
Ratio of areas	?	?	?	?	?	25:1	9:64	2:1

9. Two circles have radii 7 and 11. What is the ratio of the areas?

10. The areas of two circles are 36π and 64π. What is the ratio of the circumferences?

11. *L*, *M*, and *N* are the midpoints of the sides of $\triangle ABC$. Find the ratio of the perimeters and the ratio of the areas of $\triangle LMN$ and $\triangle ABC$.

12. $\triangle ABC \sim \triangle XYZ$, $AB = 6$, $BC = 8$, $AC = 9$, and $XY = 10$. Find the ratio of the perimeters and the ratio of the areas.

13. The lengths of two similar rectangles are x^2 and xy, respectively. What is the ratio of the areas?

14. In the diagram below, $PQRS$ is a parallelogram. Find the ratio of the areas for each pair of triangles.
 a. $\triangle TOS$ and $\triangle QOP$ **b.** $\triangle TOS$ and $\triangle TQR$

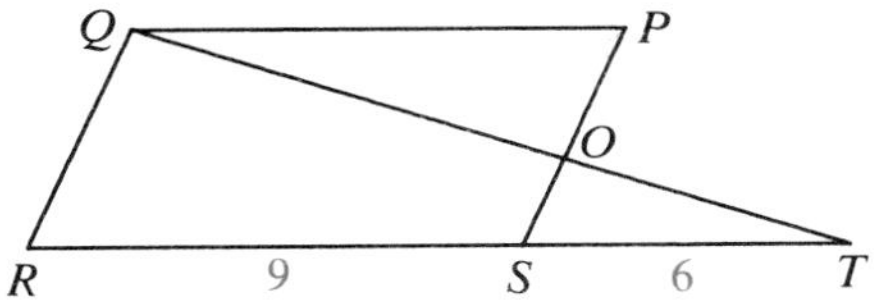

Ex. 14

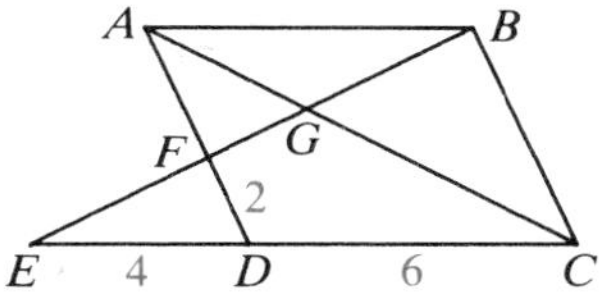

Ex. 15

B 15. In the diagram above, $ABCD$ is a parallelogram. Name *four pairs* of similar triangles and give the ratio of the areas for each pair.

The figures in Exercises 16 and 17 are trapezoids. Find the ratio of the areas of each pair of triangles.

a. I and III **b. I and II** **c. I and IV** **d. II and IV**

16.

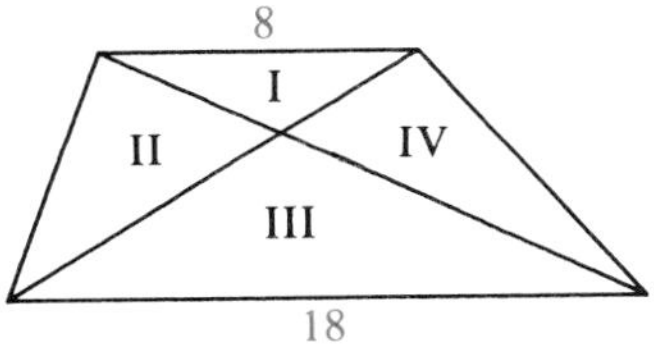

17.

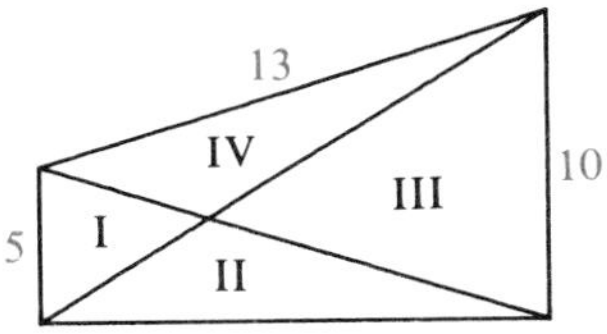

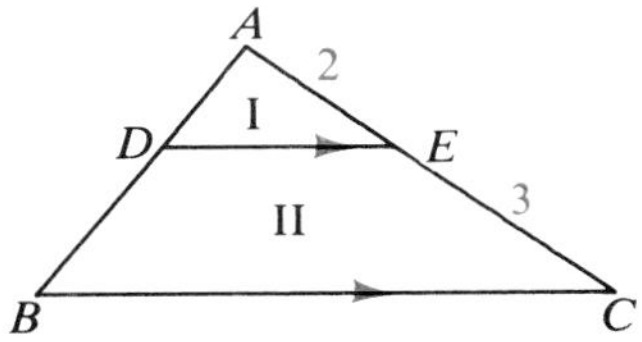

18. **a.** Are figures I and II similar?
 b. Name two similar triangles.
 c. What is the ratio of their areas?
 d. What is the ratio of the areas of figures I and II?

For Exercises 19 and 20, find the ratio of the areas of figures I and II. Note that these figures are not similar.

19.

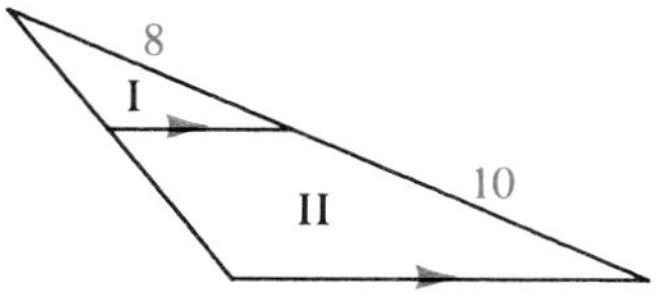

20.

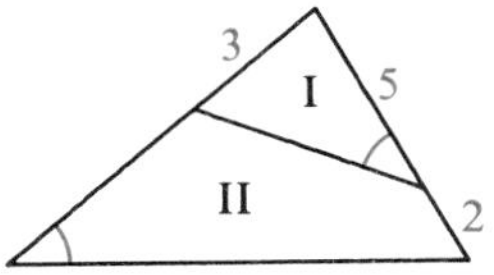

21. A square is inscribed in a 30°–60°–90° triangle. Find the ratio of the areas of regions I and II.

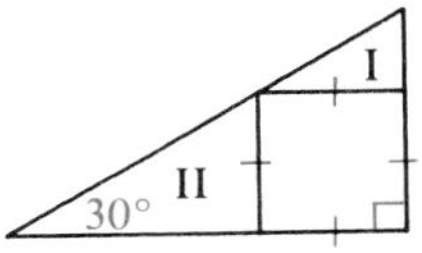

22. The area of parallelogram $ABCD$ is 48 cm^2 and $DE = 2 \cdot EC$. Find the area of each triangle.
 a. $\triangle ABE$ **b.** $\triangle BEC$ **c.** $\triangle ADE$ **d.** $\triangle CEF$ **e.** $\triangle DEF$

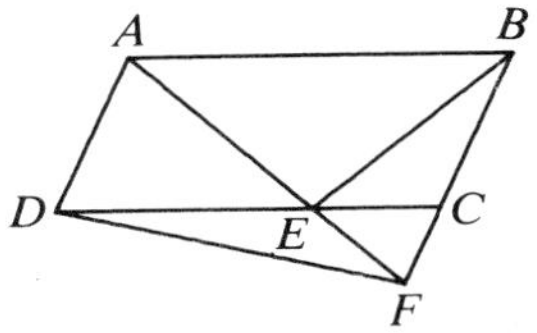

C★ 23. If you draw the three medians of a triangle, six small triangles are formed. Prove whatever you can about the areas of these six triangles.

★ 24. D, E, F, and G are the midpoints of the sides of square $PQRS$. Show that the area of the shaded region is one fifth of the area of $PQRS$.

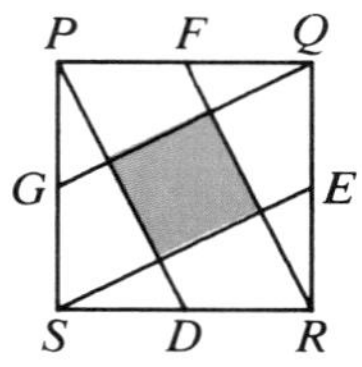

Self-Test 2

Leave your answers in terms of π unless you are told to use an approximation.

1. Find the circumference and area of a circle with radius 14. Use $\pi \approx \frac{22}{7}$.
2. The circumference of a circle is 18π. What is its area?
3. In $\odot O$ with radius 12, $m\widehat{AB} = 90$.
   a. Find the length of $\widehat{AB}$.
   b. Find the area of sector AOB.
   c. Find the area of the region bounded by $\overline{AB}$ and $\widehat{AB}$.
4. Find the ratio of the areas of two circles with radii 4 and 7.
5. The areas of two similar triangles are 36 and 81. Find the ratio of their perimeters.
6. $PQRS$ is a parallelogram. Find the ratio of the areas of $\triangle PTO$ and $\triangle SRO$.

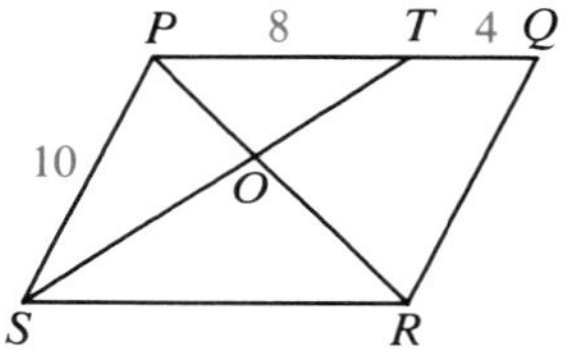

Each polygon is a regular polygon. Find the area of the shaded region.

7. 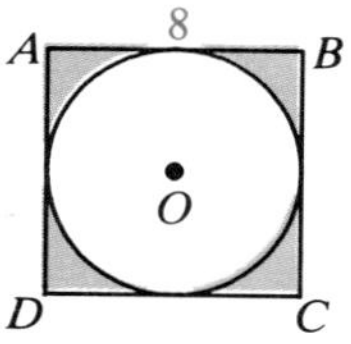

$\odot O$ is inscribed in square $ABCD$.

8. 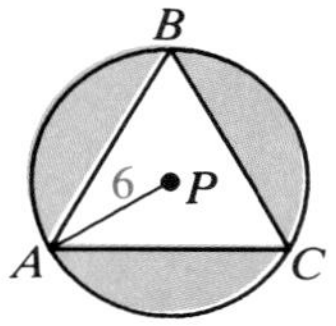

$\triangle ABC$ is inscribed in $\odot P$.

Challenge

Imagine that the Earth is a perfectly smooth sphere without mountains and valleys. A band fits snugly around the Earth at the equator. The band is stretched 1 m and is placed so that its points are equidistant around the Earth. Which of the following animals could crawl under the band—a flea, a mouse, a kitten, a large dog?

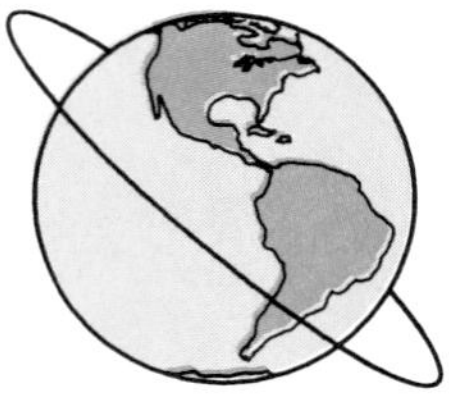

Application

SPACE SHUTTLE LANDINGS

The space shuttle is launched vertically as a rocket but lands horizontally as a glider with no power and no second chance at the runway. NASA studied many different guidance systems for the final portion of entry and landing. The system that was selected for the first shuttle flights used a cylinder called the Heading Alignment Cylinder (HAC), shown in the diagram below. Notice that the projection of the flight path onto the Earth's surface is called the *ground track.*

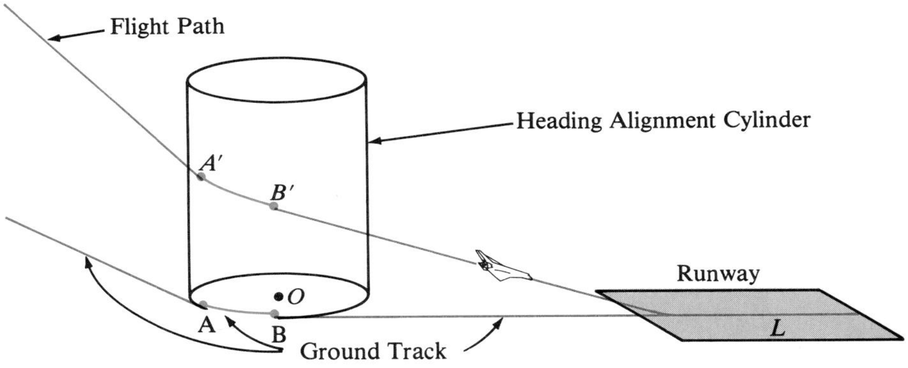

The shuttle followed a straight-line flight path to A' and then it followed a curved path along the Heading Alignment Cylinder to point B'. The shuttle continued to lose altitude so that it was closer to the Earth's surface at B' than at A'. From B' to landing at L the shuttle followed a straight path aligned with the center of the runway.

The points A' and B' where the shuttle's turn begins and ends can be determined by looking at the ground track. The figure at the right shows what you would see if you were high above the ground looking straight down on the ground track and runway. A' and B' are directly above A and B, which are located as follows: Extend the HAC acquisition line and the center line of the runway to meet at M. Bisect the angle at M and choose point O on the bisector so that a circle with center O and radius 20,000 ft will be tangent to the sides of the angle. Call the points of tangency A and B. $\odot O$ is the base of the Heading Alignment Cylinder and A' and B' are on the cylinder directly above A and B.

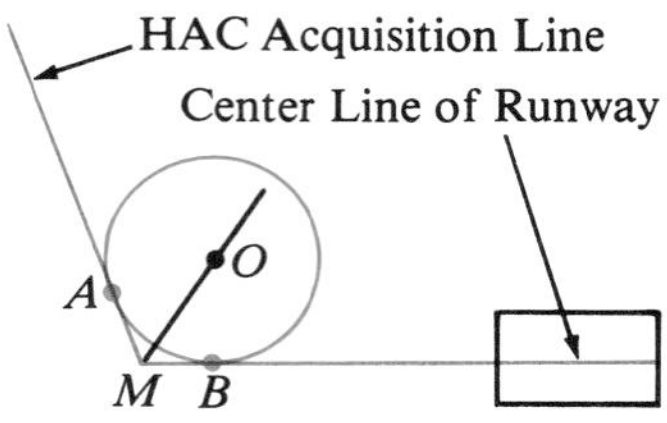

Normally the shuttle approached the cylinder at 800 ft/s and turned along its surface by lowering one wing tip so that the wings formed an angle of about 45° with the horizontal (called the *bank angle*). Under high-speed conditions the shuttle would have approached the cylinder at 1000 ft/s. This would have required a bank angle of about 57° to follow the surface of the cylinder. Unfortunately, at that bank angle, the shuttle would have lost lift, and the astronauts would have lost some of their control capability.

NASA has refined the guidance system so that the shuttle can be safely landed even under these adverse circumstances. Now, instead of following the surface of a cylinder, it spirals along the surface of a cone called the Heading Alignment Cone. Once every second during this part of the landing the shuttle's computers recompute the radius of turn necessary to keep the shuttle on the surface of the cone. Now even under most high-speed conditions the bank angle will not exceed approximately 42°. At point Q the shuttle is heading directly toward the runway; it leaves the cone and continues along a straight course to touchdown.

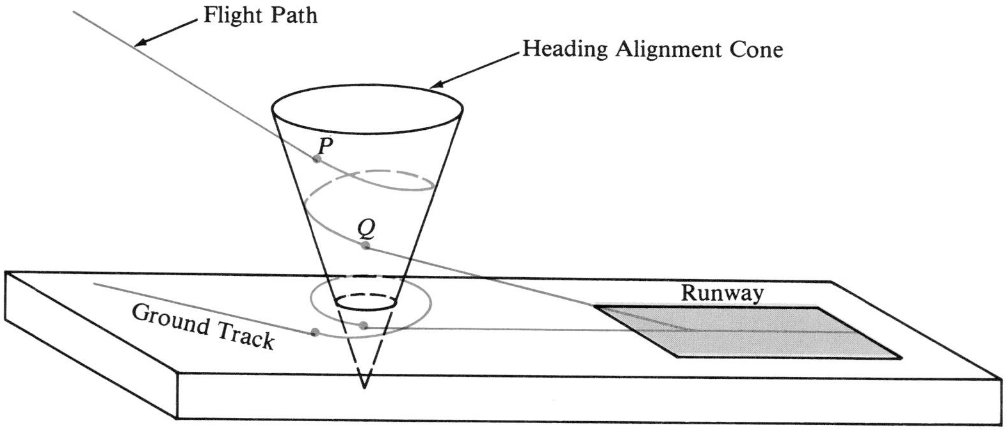

Exercises

1. Let T be any point on the bisector of $\angle AMB$. Show that if $\odot T$ is drawn tangent to $\overleftrightarrow{MA}$ it will also be tangent to $\overleftrightarrow{MB}$.
2. The radius of the Heading Alignment Cylinder was 20,000 ft, and a typical value for $m\angle AMB$ was 120. How long was the curved portion of the ground track, $\widehat{AB}$? (Use 3.1416 for π and give your answer to the nearest foot.)
3. The shuttle's turning radius changes as it moves along the surface of the Heading Alignment Cone. Is the radius larger near P or near Q?
4. A good approximation of the detailed landing procedure uses a Heading Alignment Cone with vertex below the surface of the Earth. A typical radius of the cone at a height of 30,000 ft above the Earth's surface is 20,000 ft. At a height of 12,000 ft, which is a typical height for Q, the radius of the cone is 14,000 ft.
   - **a.** How far below the surface of the Earth is the vertex of the cone?
   - **b.** What is the radius of the cone at a height of 15,000 ft?
   - **c.** At what height is the radius of the cone equal to 12,000 ft?

Chapter Summary

1. If two figures are congruent, then they have the same area.
2. The area of a region is the sum of the areas of its non-overlapping parts.
3. The list below gives the formulas for areas of polygons.

   Square: $A = s^2$
   Rectangle: $A = bh$
   Parallelogram: $A = bh$
   Triangle: $A = \frac{1}{2}bh$
   Rhombus: $A = \frac{1}{2}d_1d_2$
   Trapezoid: $A = \frac{1}{2}h(b_1 + b_2)$
   Regular polygon: $A = \frac{1}{2}ap$, where a is the apothem and p is the perimeter
4. The list below gives the formulas related to circles.

   $C = 2\pi r$ Length of arc $= \frac{x}{360} \cdot 2\pi r$

   $A = \pi r^2$ Area of sector $= \frac{x}{360} \cdot \pi r^2$
5. If the scale factor of two similar figures is $a:b$, then:
   (1) The ratio of the perimeters is $a:b$.
   (2) The ratio of the areas is $a^2:b^2$.

Chapter Review

1. The perimeter of a square is 32. Find the area. 9–1
2. Find the area of a rectangle with length 4 and diagonal 6.
3. Find the area of a square with side $3\sqrt{2}$ cm.
4. Find the area of a rhombus with side 17 and longer diagonal 30. 9–2
5. A parallelogram has sides 8 and 12. The shorter altitude is 6. Find the length of the other altitude.
6. Find the perimeter and the area of the triangle shown.

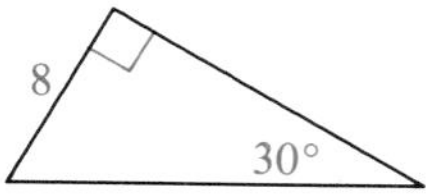

7. Find the height of a trapezoid with median 12 and area 84. 9–3
8. Find the area of an isosceles trapezoid with legs 5 and bases 4 and 12.
9. Find the perimeter and the area of the figure shown.

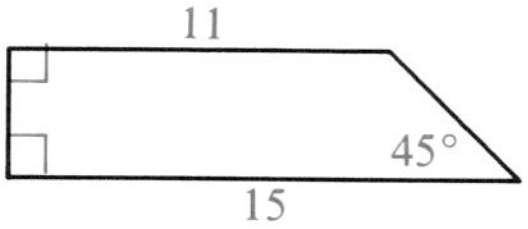

Find the area of each regular polygon.

10. A square with apothem 3 m 9–4
11. An equilateral triangle with radius $2\sqrt{3}$
12. A regular hexagon with perimeter 12 cm
13. Find the circumference and area of a circle with radius 30. Use $\pi \approx 3.14$. 9–5
14. The area of a circle is 121π cm^2. Find the diameter.
15. A square with side 8 is inscribed in a circle. Find the circumference of the circle.
16. Find the length of a 135° arc in a circle with radius 24. 9–6

Find the area of each shaded region.

17. 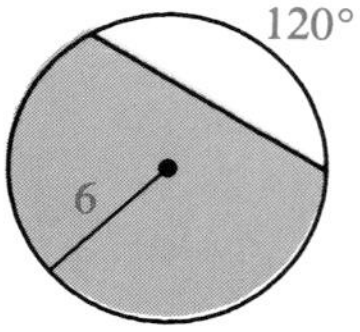

18. 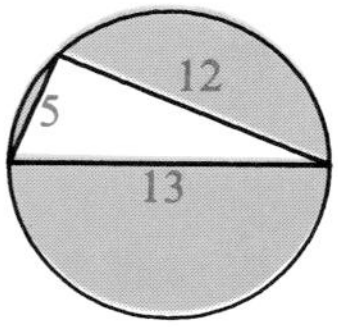

19. If $AB = 9$ and $CD = 12$, find the ratio of the areas of: 9–7
    a. $\triangle AEB$ and $\triangle DEC$
    b. $\triangle AED$ and $\triangle DEC$

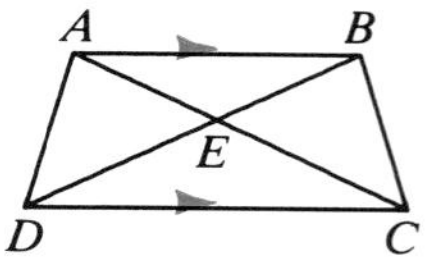

20. Two regular octagons have perimeters 16 cm and 32 cm, respectively. What is the ratio of their areas?
21. Two similar polygons have the scale factor 7:5. The area of the large polygon is 147. Find the area of the smaller polygon.

Chapter Test

Find the area of each figure described.

1. A circle with diameter 10
2. A square with diagonal 4 cm
3. An isosceles right triangle with hypotenuse $6\sqrt{2}$
4. A circle with circumference 30π m
5. A rhombus with diagonals 5 and 4
6. An isosceles trapezoid with legs 10 and bases 6 and 22
7. A parallelogram with sides 6 and 10 that form a 30° angle
8. A regular hexagon with apothem $2\sqrt{3}$ cm
9. Sector AOB of $\odot O$ with radius 4 and $m\widehat{AB} = 45$
10. A rectangle with length 12 inscribed in a circle with radius 7.5
11. A sector of a circle with radius 12 and arc length 10π
12. A square with radius 9

Find the area of each shaded region.

13.

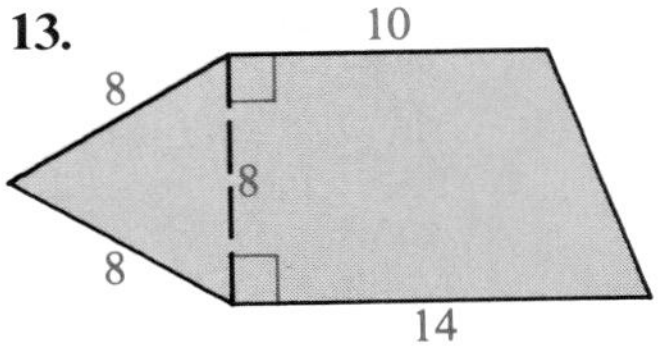

14.

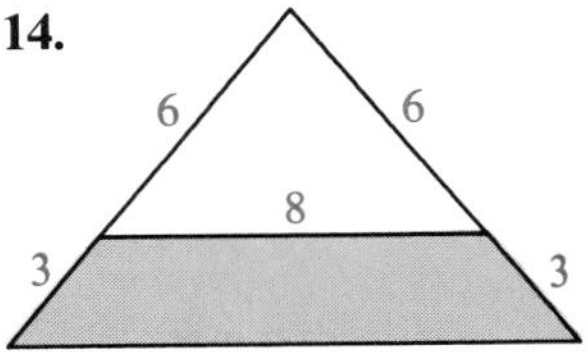

15.

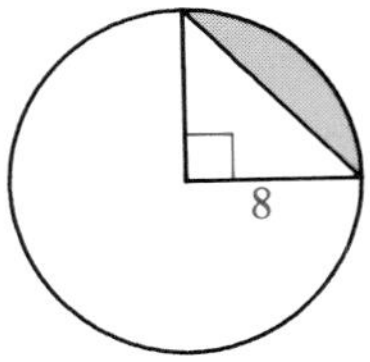

16. The areas of two circles are 100π and 36π. Find the ratio of their radii and the ratio of their circumferences.
17. Two regular pentagons have sides of 14 m and 3.5 m, respectively. Find their scale factor and the ratio of their areas.
18. In $\odot Q$, $m\widehat{ABC} = 288$ and $QA = 10$. Find the length of $\widehat{AC}$.

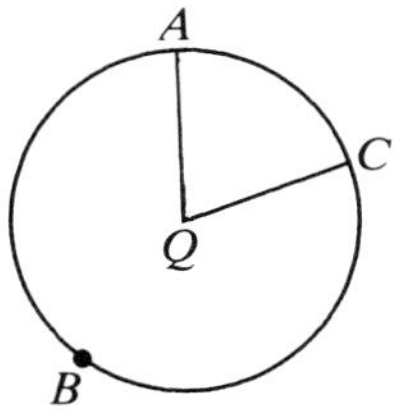

Mixed Review

1. Given: $\overline{AB} \perp \overline{BC}$; $\overline{DC} \perp \overline{BC}$; $\overline{AC} \cong \overline{BD}$
   Prove: $\triangle BCE$ is isosceles.

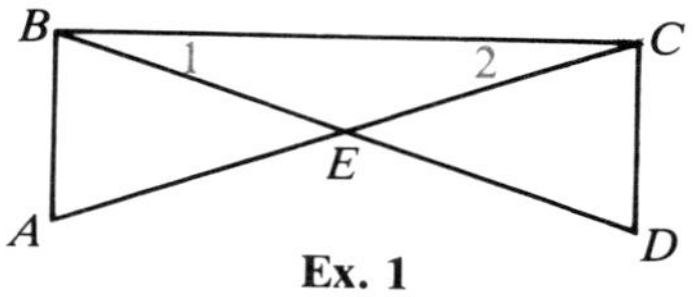

Ex. 1

2. If a 45°–45°–90° triangle has legs of length $5\sqrt{2}$, find the length of the altitude to the hypotenuse.
3. In $\triangle BEV$, $m\angle B = 53$ and $m\angle V = 64$. Name **(a)** the longest and **(b)** the shortest side of $\triangle BEV$.
4. Find the value of x in the diagram.

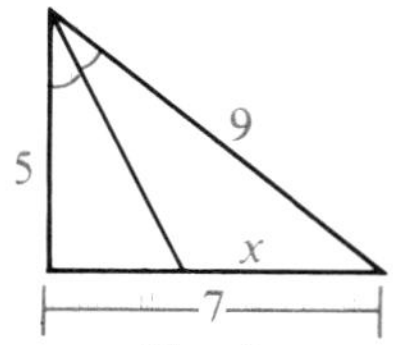

Ex. 4

5. An equilateral triangle has perimeter 12 cm. Find its area.
6. The legs of a right triangle are 4 cm and 8 cm long. What is the length of the median to the hypotenuse?
7. If $\overline{PQ}$ and $\overline{PR}$ are tangents to the circle and $m\angle 1 = 58$, find $m\angle 2$.

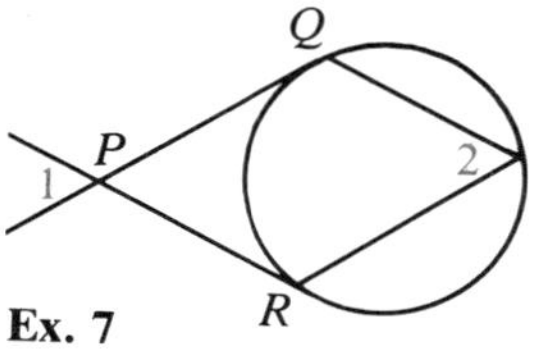

Ex. 7

8. Draw two segments and let their lengths be x and y. Construct a segment of length t such that $t = \dfrac{2x^2}{y}$.
9. Given: Quadrilateral $EFGH$; $\overline{EF} \cong \overline{HG}$; $\overline{EF} \parallel \overline{HG}$
   Prove: $\angle EHF \cong \angle GFH$
10. For $\triangle JKL$ and $\triangle XYZ$ use the statement "If $\angle J \cong \angle X$ and $\angle K \cong \angle Y$, then $\triangle JKL \sim \triangle XYZ$."
    a. Name the postulate or theorem that justifies the statement.
    b. Write the converse of the statement and classify the converse as true or false.
11. **a.** Find the length of a 200° arc in a circle with diameter 24.
    b. Find the area of the sector determined by this arc.
12. Two similar rectangles have diagonals of $6\sqrt{3}$ and 9. Find the ratio of their perimeters and the ratio of their areas.
13. Find the sum of the measures of the angles of a convex decagon.
14. Two lines that do not intersect are either __?__ or __?__.
15. Describe the locus of points in space that are no more than 4 cm from a given point P.
16. A trapezoid has bases with lengths $x + 3$ and $3x - 1$ and a median of length 11. Find the value of x.
17. A tree 5 m tall casts a shadow 8 m long. Use the table on page 271 to find the angle of elevation of the sun to the nearest degree.
18. Use inductive thinking to guess the next number: 10, 9, 5, -4, -20, __?__

Algebra Review

Find the value of each expression using the given values of the variables. Give your answer in terms of π.

Example $2\pi r$ when $r = \frac{5}{4}$

Solution $2 \cdot \pi \cdot \frac{5}{4} = \left(2 \cdot \frac{5}{4}\right)\pi = \frac{5}{2}\pi$

1. $2\pi r$ when $r = \frac{9}{2}$

2. πd when $r = 2\sqrt{3}$ and $d = 2r$

3. πr^2 when $r = 5\sqrt{3}$

4. πr^2 when $r = \frac{2}{3}\sqrt{3}$

5. $4\pi r^2$ when $r = \frac{10}{3}$

6. πrl when $r = 4\frac{1}{5}$ and $l = 15$

7. $2\pi rh$ when $r = h = 1\frac{3}{4}$

8. $\frac{x}{360} \cdot 2\pi r$ when $x = 270$ and $r = 2$

9. $\frac{x}{360} \cdot \pi r^2$ when $x = 180$ and $r = 6\sqrt{2}$

10. $\pi r^2 h$ when $r = \frac{1}{4}$ and $h = 4$

11. $\frac{1}{3}\pi r^2 h$ when $r = 2\sqrt{6}$ and $h = 4$

12. $\frac{4}{3}\pi r^3$ when $r = 6$

13. $\pi r\sqrt{r^2 + h^2}$ when $r = h = \sqrt{5}$

14. $2\pi r^2 + 2\pi rh$ when $r = 10$ and $h = 6$

15. $\pi r^2 + \pi r\sqrt{r^2 + h^2}$ when $r = 2$ and $h = 2\sqrt{3}$

16. $\pi({r_1}^2 - {r_2}^2)$ when $r_1 = 6$ and $r_2 = 3\sqrt{2}$

Find an approximation of the value of each expression using the given approximation for π.

Example $2\pi rh$ when $r = 28$ and $h = 20$; $\pi \approx \frac{22}{7}$

Solution $2\pi rh \approx 2 \cdot \frac{22}{7} \cdot 28 \cdot 20 = 2 \cdot 22 \cdot 4 \cdot 20 = 3520$

17. πr^2 when $r = 20$; $\pi \approx 3.14$

18. πr^2 when $r = \sqrt{21}$; $\pi \approx \frac{22}{7}$

19. $\frac{4}{3}\pi r^3$ when $r = 3$; $\pi \approx \frac{22}{7}$

20. $4\pi r^2$ when $r = \sqrt{6}$; $\pi \approx 3.1416$

Solve for the variable shown in color. Assume that all variables represent positive numbers.

Example **a.** $4\pi x^2 = 25\pi$ **b.** $C = 2\pi r$

Solution $x^2 = \frac{25\pi}{4\pi} = \frac{25}{4}$, so $x = \frac{5}{2}$ $r = \frac{C}{2\pi}$

21. $\pi x^2 = 121\pi$

22. $\frac{32\pi}{3} = \frac{4}{3}\pi x^3$

23. $A = 2\pi rh$

24. $V = \frac{1}{3}\pi r^2 h$

25. $\frac{1}{3}\pi r^2 \cdot 10 = 3000\pi$

26. $\frac{x}{360} \cdot 2\pi \cdot 10 = \frac{\pi}{9}$

The bundles of pipes in the photograph suggest hexagonal prisms. Each individual pipe is in the shape of a circular cylinder. Prisms and cylinders are two of the geometric solids you will study in this chapter.

Areas and Volumes of Solids

10

Important Solids

Objectives

1. Identify the parts of prisms, pyramids, cylinders, and cones.
2. Find the lateral area, total area, and volume of a right prism or regular pyramid.
3. Find the lateral area, total area, and volume of a right cylinder or cone.

10-1 Prisms

In this chapter you will be calculating surface areas and volumes of special solids. It is possible to begin with some postulates and then prove as theorems the formulas for areas and volumes of solids, as we did for plane figures. Instead, informal arguments will be given to show you that the formulas for solids are reasonable.

The first solid we will study is the **prism.** The two shaded faces of the prism shown are its **bases.** Notice that the bases are congruent polygons lying in parallel planes. An **altitude** of a prism is a segment joining the two base planes and perpendicular to both. The length of an altitude is the *height* (h) of the prism.

The faces of a prism that are not its bases are called **lateral faces.** Adjacent lateral faces intersect in parallel segments called **lateral edges.**

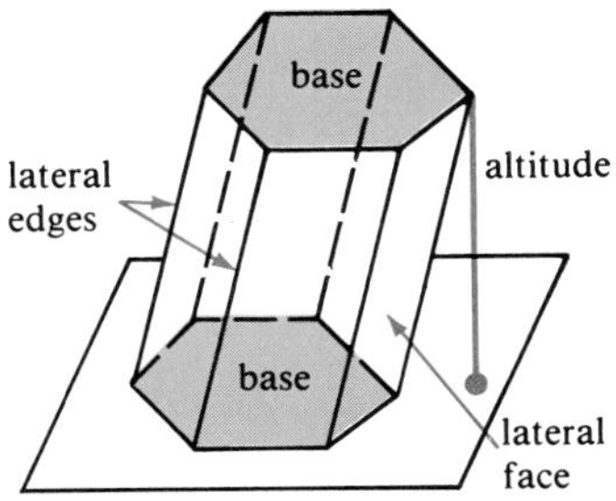

The lateral faces of a prism are parallelograms. If they are rectangles, the prism is a **right prism.** Otherwise the prism is an **oblique prism.** The diagrams below show that a prism is also classified by the shape of its base.

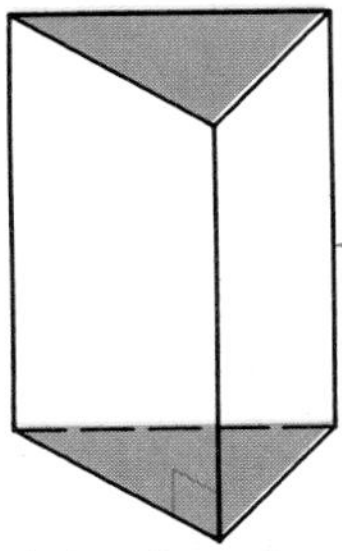

Right triangular prism

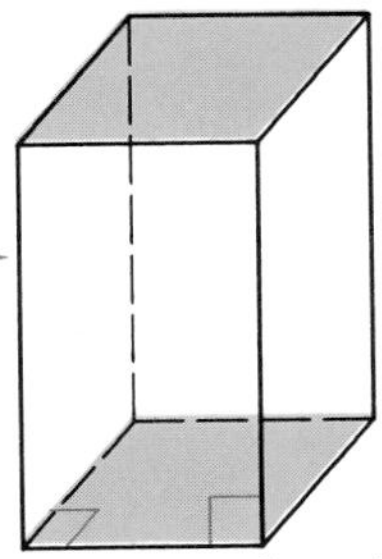

Right rectangular prism (Rectangular solid)

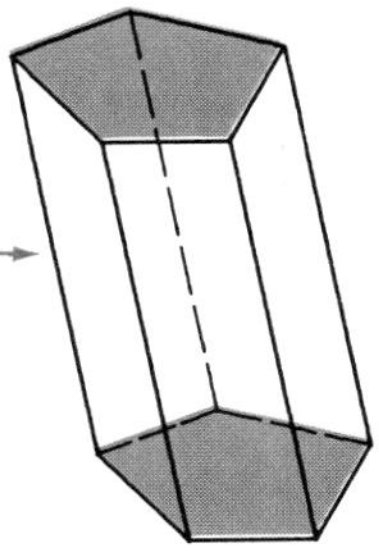

Oblique pentagonal prism

Note that in a right prism, the lateral edges are also altitudes.

The **lateral area** (L.A.) of a prism is the sum of the areas of its lateral faces. The **total area** (T.A.) is the sum of the areas of all its faces. Using B to denote the area of a base, we have the formula:

$$\textbf{T.A.} = \textbf{L.A.} + 2B$$

If a prism is a right prism, the next theorem gives us an easy way to get the lateral area.

Theorem 10-1

The lateral area of a right prism equals the perimeter of a base times the height of the prism. (L.A. = ph)

The formula for lateral area applies to any right prism. We'll use a right pentagonal prism to illustrate the development of the formula.

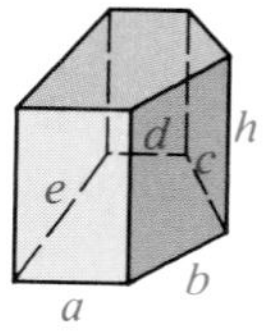

$$\begin{aligned}\text{L.A.} &= ah + bh + ch + dh + eh \\ &= (a + b + c + d + e)h \\ &= \text{perimeter} \cdot h \\ &= ph\end{aligned}$$

Prisms have *volume* as well as area. A rectangular solid with square faces is a cube. Since each edge of the shaded cube shown is 1 unit long, the cube is said to have a volume of 1 cubic unit. The larger rectangular solid has 3 layers of cubes, each layer containing (4 · 2) cubes. Hence its volume is (4 · 2) · 3, or 24 cubic units.

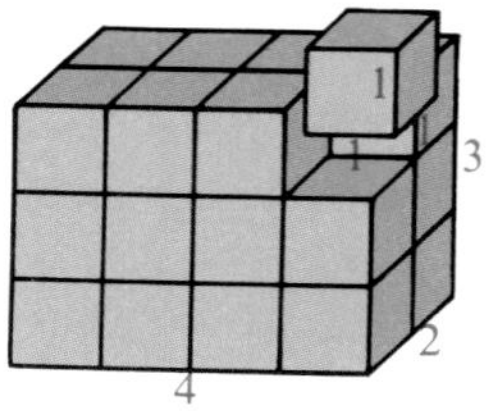

$$\begin{aligned}\text{Volume} &= \text{Base area} \times \text{height} \\ &= (4 \cdot 2) \cdot 3 \\ &= 24 \text{ cubic units}\end{aligned}$$

The same sort of reasoning is used to find the volume of any right prism. We will list the result as a theorem without giving the details of the proof.

Theorem 10-2

The volume of a right prism equals the area of a base times the height of the prism. ($V = Bh$)

Some common units for measuring volume are the cubic centimeter (cm^3) and the cubic meter (m^3).

Example 1 A right trapezoidal prism is shown. Find the (a) lateral area, (b) total area, and (c) volume.

Solution

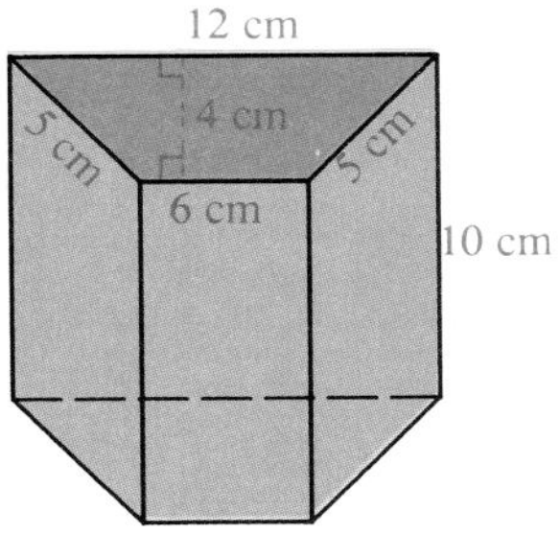

a. Lateral area
First find the perimeter of a base.
$p = 5 + 6 + 5 + 12 = 28$ (cm)
Now use the formula for lateral area.
$\text{L.A.} = ph = 28 \cdot 10 = 280$ (cm^2)

b. Total area
First find the area of a base.
$B = \frac{1}{2} \cdot 4 \cdot (12 + 6) = 36$ (cm^2)
Now use the formula for total area.
$\text{T.A.} = \text{L.A.} + 2B = 280 + 2 \cdot 36 = 352$ (cm^2)

c. Volume
$V = Bh = 36 \cdot 10 = 360$ (cm^3)

Example 2 A right triangular prism is shown. The volume is 315. Find the total area.

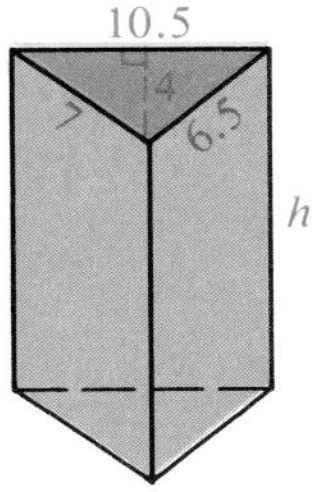

Solution First find the height of the prism.

$$V = Bh$$
$$315 = \tfrac{1}{2} \cdot 10.5 \cdot 4 \cdot h$$
$$315 = 21h$$
$$15 = h$$

Second, find the lateral area.
$\text{L.A.} = ph = (10.5 + 6.5 + 7) \cdot 15 = 24 \cdot 15 = 360$
Now use the formula for total area.
$\text{T.A.} = \text{L.A.} + 2B = 360 + 2 \cdot 21 = 402$

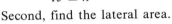

Classroom Exercises

Exercises 1–8 refer to the right prism shown.

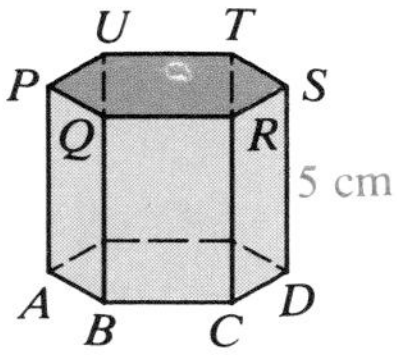

1. What kind of polygons are the bases?
2. The prism is called a right __?__ prism.
3. How many lateral faces are there?
4. What kind of figure is each lateral face?
5. Name two lateral edges.
6. Name an altitude.
7. The length of an altitude is called the __?__ of the prism.
8. Suppose the bases are regular hexagons with 4 cm edges.
   a. Find the perimeter of a base.
   b. Given that the height of the prism is 5 cm, find the lateral area.
   c. Find the base area.
   d. Find the total area.
   e. Find the volume.

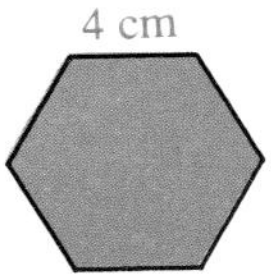

Written Exercises

Exercises 1–6 refer to rectangular solids with dimensions *l*, *w*, and *h*. Complete the table.

	l	*w*	*h*	L.A.	T.A.	*V*
A **1.**	6	4	2	?	?	?
2.	50	30	15	?	?	?
3.	6	3	?	?	?	54
4.	?	8	5	?	?	360
5.	9	?	2	60	?	?
6.	$5x$	$4x$	$3x$	?	?	?

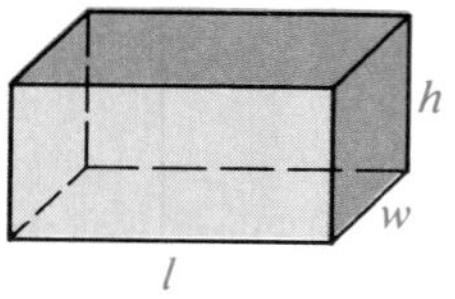

Exercises 7–12 refer to cubes with edges of length *e*. Complete the table.

	7.	**8.**	**9.**	**10.**	**11.**	**12.**
e	3	6	?	?	?	$2x$
T.A.	?	?	?	?	150	?
V	?	?	1000	64	?	?

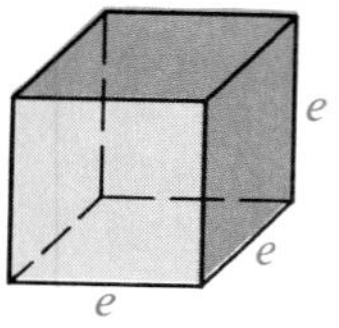

13. If the edge of a cube is doubled, the total area is multiplied by __?__ and the volume is multiplied by __?__.

14. Find the lateral area of a right pentagonal prism with height 13 and base edges 3.2, 5.8, 6.9, 4.7, and 9.4.

Facts about the base of a right prism and the height of the prism are given. Sketch each prism and find its lateral area, total area, and volume.

15. Equilateral triangle with side 8; $h = 10$

16. Triangle with sides 9, 12, 15; $h = 10$

B **17.** Isosceles triangle with sides 13, 13, 10; $h = 7$

18. Isosceles trapezoid with sides 10, 5, 4, 5; $h = 20$

19. Regular hexagon with side 8; $h = 12$

20. Rhombus with diagonals 6 and 8; $h = 9$

21. The container shown has the shape of a rectangular solid. When a rock is submerged, the water level rises 2 cm. Find the volume of the rock.

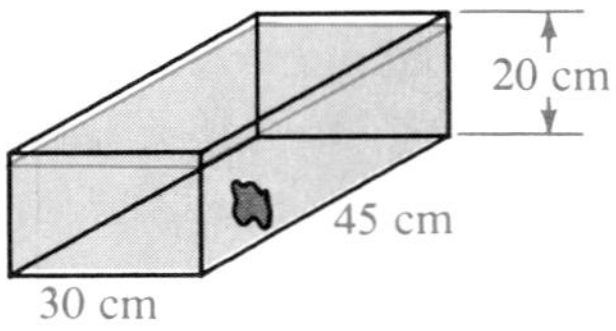

22. A driveway 30 m long and 5 m wide is to be paved with blacktop 3 cm thick. How much will the blacktop cost if it is sold at the price of $42 per cubic meter?

23. A brick with dimensions 20 cm, 10 cm, and 5 cm weighs 1.2 kg. A second brick of the same material has dimensions 25 cm, 15 cm, and 4 cm. What is its weight?

24. A drinking trough for horses is a right trapezoidal prism with dimensions shown below. If it is filled with water, how much will the water weigh? (1 m^3 of water weighs 1 metric ton.)

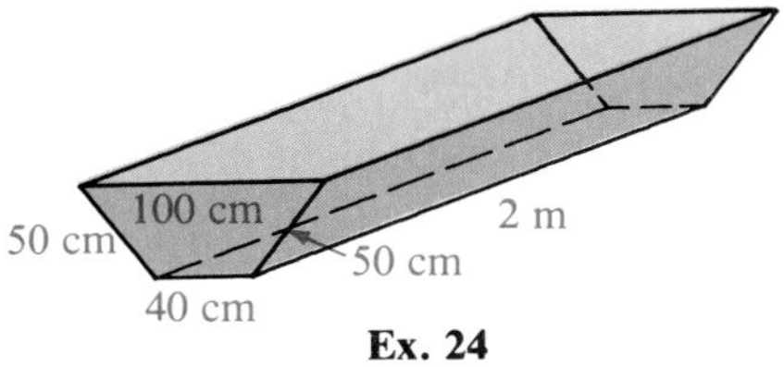

Ex. 24

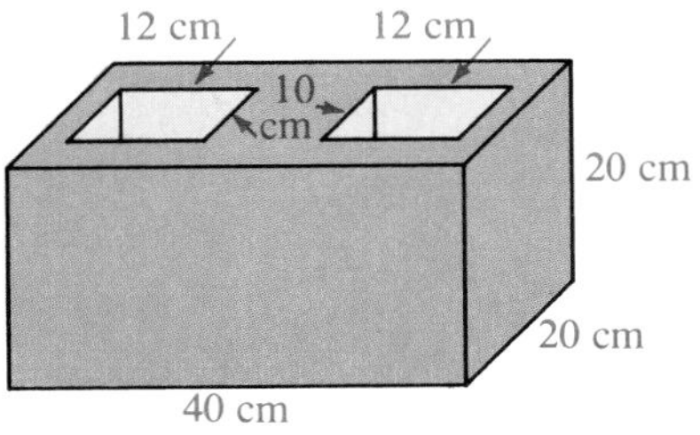

Ex. 25

25. Find the weight of the cement block shown. Cement weighs 1700 kg/m^3.

26. Find the weight of the steel **I**-beam shown below. Steel weighs 7860 kg/m^3.

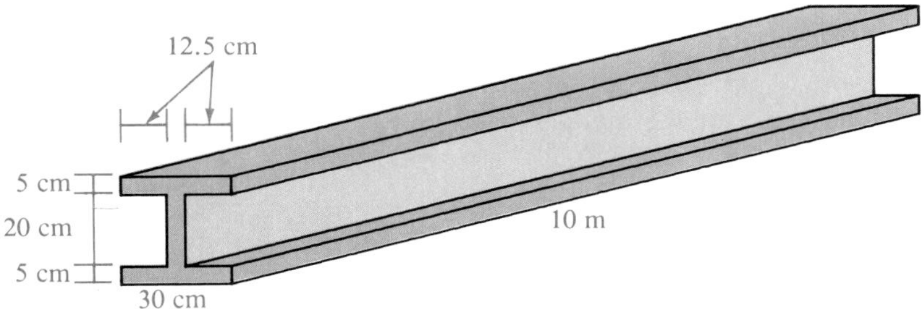

For Exercises 27 and 28 find the volume and total surface area of each solid in terms of the given variables.

27.

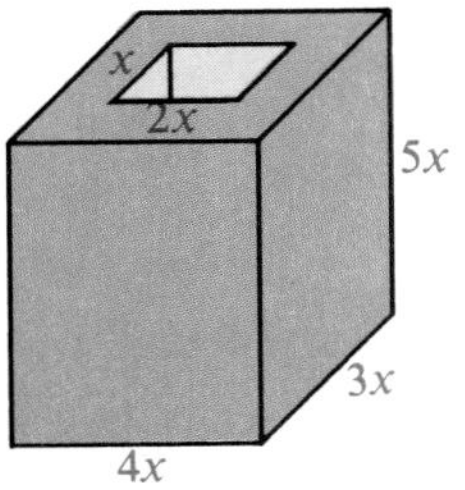

28.

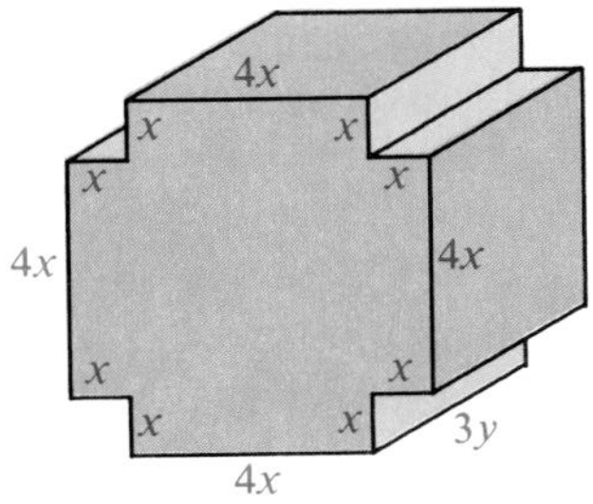

29. The length of a rectangular solid is twice the width, and the height is three times the width. If the volume is 162 cm^3, what are the dimensions of the solid?

30. A diagonal of a box forms a 35° angle with a diagonal of the base, as shown below. Use trigonometry to find the volume of the box.

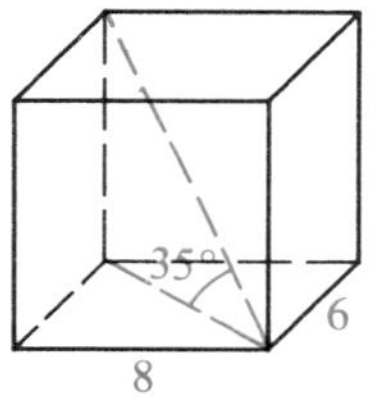

Ex. 30

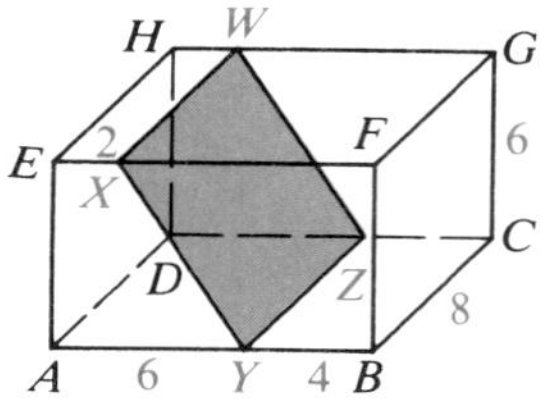

Ex. 31

C 31. The rectangular solid shown above has length 10, width 8, and height 6. Plane $WXYZ$ intersects the solid as shown, forming two trapezoidal prisms, one with base $AYXE$ and the other with base $BYXF$.

a. Find the volumes of the two trapezoidal prisms.

b. Find the total surface area of the prism with base $BYXF$.

32. A rectangular beam of wood 3 m long is cut into six pieces, as shown in the diagram. Find the volume of each piece.

33. A diagonal of a cube joins two vertices not in the same face. If the diagonals are $4\sqrt{3}$ cm long, what is the volume?

34. All nine edges of a right triangular prism are equal. Find the length of these edges if the volume is $54\sqrt{3}\text{ cm}^3$.

35. A right prism has height h and has bases which are regular hexagons with sides s. Show that the volume is $V = \frac{3}{2}\sqrt{3}\, s^2 h$.

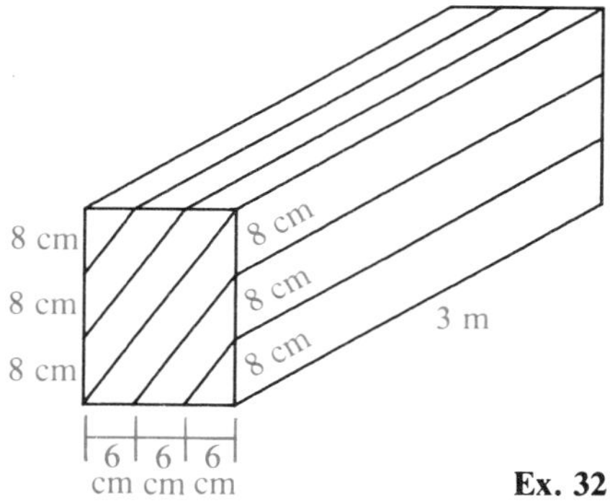

Ex. 32

COMPUTER KEY-IN

A manufacturing company produces metal boxes of different sizes by cutting out square corners from rectangular pieces of metal that measure 9 in. by 12 in. The metal is then folded along the dashed lines to form a box without a top. If a customer requests the box with the greatest possible volume, what dimensions should be used?

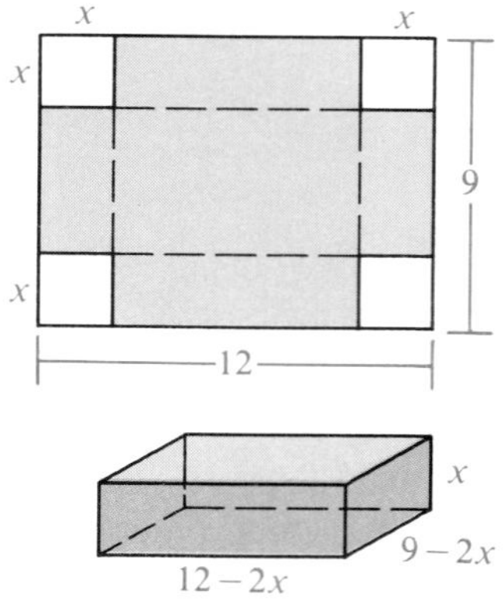

The volume, V, of the box can be expressed in terms of x.

$$V = \text{length} \cdot \text{width} \cdot \text{height}$$
$$= (12 - 2x) \cdot (9 - 2x) \cdot x$$

To form a box, the possible values for x are $0 < x < \frac{9}{2}$.

The following computer program finds the volumes of the boxes produced for values of x from 0 to 4.5.

```
10  PRINT "X", "VOLUME"
20  FOR X = 0 TO 4.5 STEP 0.5
30  LET V = (12 - 2 * X)*(9 - 2 * X)* X
40  PRINT X, V
50  NEXT X
60  END
```

```
RUN

X              VOLUME

 0              0
 .5             44
 1              70
 1.5            81
 2              80
 2.5            70
 3              54
 3.5            35
 4              16
 4.5            0
```

The print-out shows that the maximum volume of the box probably occurs when the value of x is between 1 and 2.

Exercises

1. To find a more accurate value for x, change line 20 to:

```
FOR X = 1 TO 2 STEP 0.1
```

Between what values of x will the maximum volume occur?

2. Modify line 20 so that you find the maximum volume, correct to the nearest tenth of a cubic inch. What are the approximate dimensions of this box?

3. Suppose the manufacturing company cuts square corners out of pieces of metal 8 in. by 15 in.

a. Express the volume in terms of x.

b. Find the maximum volume, correct to the nearest tenth of a cubic inch.

c. What are the approximate dimensions of the box that has maximum volume?

Challenge

A cube with sides n cm long is painted on all faces. It is then cut into cubes with sides 1 cm long. If $n = 4$, as the diagram at the right illustrates, how many of these smaller cubes will have paint on

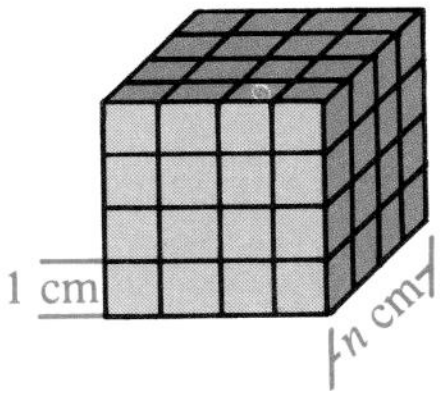

a. 3 surfaces? **b.** 2 surfaces?
c. 1 surface? **d.** 0 surfaces?

Answer the questions above for the case of any positive integer n.

10-2 Pyramids

The diagram shows the pentagonal **pyramid** V-$ABCDE$. Point V is the **vertex** of the pyramid and pentagon $ABCDE$ is the **base.** The segment from the vertex perpendicular to the base is the **altitude** and its length is the *height* (h) of the pyramid.

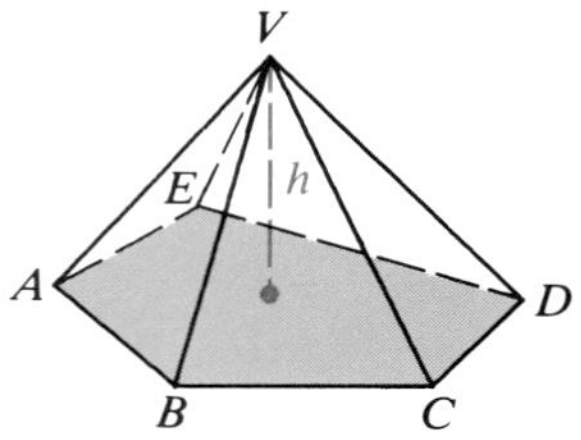

The five triangular faces with V in common, such as $\triangle VAB$, are **lateral faces.** These faces intersect in segments called **lateral edges.**

Most of the pyramids you'll study will be **regular pyramids.** These are pyramids having the following properties:

(1) The base is a regular polygon.
(2) All lateral edges are congruent.
(3) All lateral faces are congruent isosceles triangles. The height of a lateral face is called the **slant height** of the pyramid. It is denoted by l.
(4) The altitude meets the base at its center, O.

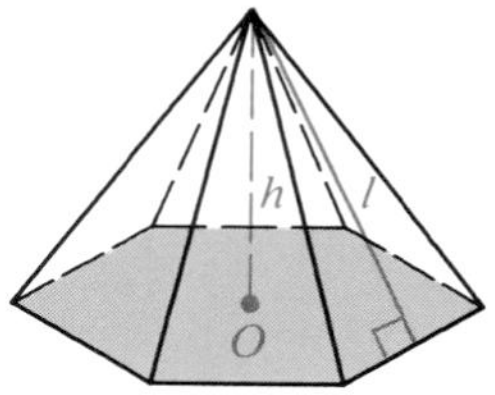

Regular hexagonal pyramid

To find the lateral area of a regular pyramid, you may use either of the following methods:

Method 1 Find the area of one lateral face.
Multiply this area by the number of lateral faces.

Method 2 Use the following formula, in which p denotes perimeter of the base. (See Classroom Exercise 1.)

$$\textbf{L.A.} = \tfrac{1}{2}pl$$

The prism and pyramid below have congruent bases and equal heights. Since the volume of the prism is Bh, the volume of the pyramid must be less than Bh. In fact, it is exactly $\frac{1}{3}Bh$.

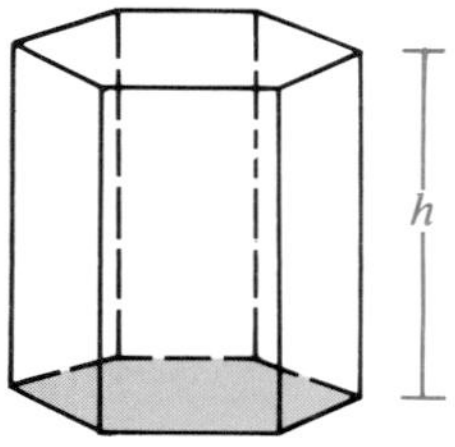

$V = Bh$

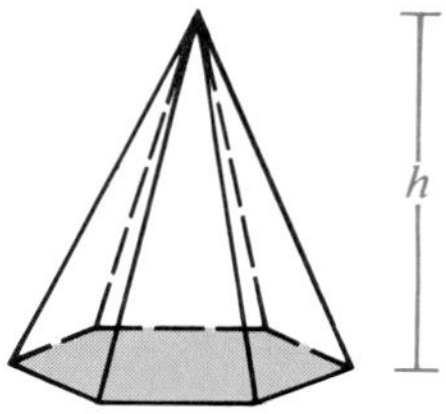

$V = \frac{1}{3}Bh$

Example 1 Given a regular square pyramid with base edge 10 and lateral edge 13, find the (a) lateral area, (b) total area, and (c) volume.

Solution   **a.** $l = \sqrt{13^2 - 5^2} = \sqrt{144} = 12$

Method 1:
Area of one lateral face $= \frac{1}{2} \cdot 10 \cdot 12 = 60$
Area of four lateral faces $= 4 \cdot 60 = 240$

Method 2:
Perimeter of base $= p = 40$
L.A. $= \frac{1}{2}pl = \frac{1}{2} \cdot 40 \cdot 12 = 240$

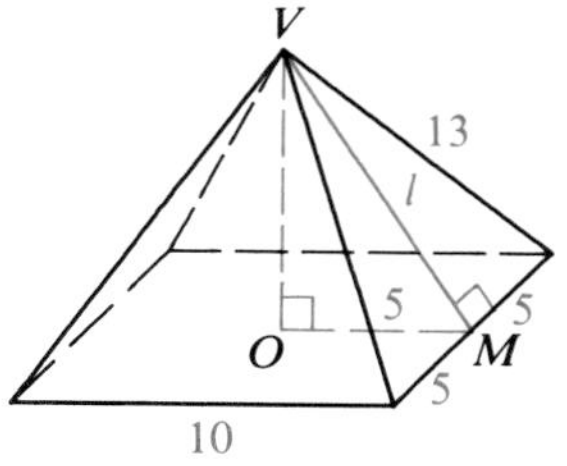

b. Area of base $= B = 10^2 = 100$
T.A. $=$ L.A. $+ B = 240 + 100 = 340$

c. In rt. $\triangle VOM$, $h = \sqrt{l^2 - 5^2}$
$= \sqrt{144 - 25} = \sqrt{119}$

$$V = \frac{1}{3}Bh = \frac{1}{3} \cdot 100 \cdot \sqrt{119} = \frac{100\sqrt{119}}{3}$$

Example 2   Given a regular triangular pyramid with lateral edge 10 and height 6, find the (a) lateral area and (b) volume.

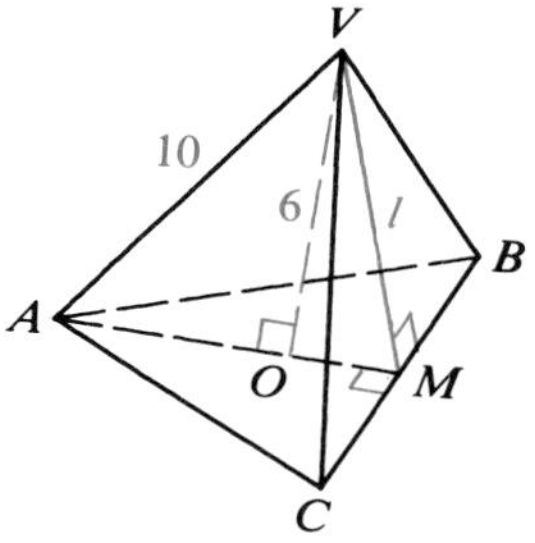

Solution   **a.** In rt. $\triangle VOA$, $AO = \sqrt{10^2 - 6^2}$
$= \sqrt{100 - 36}$
$= \sqrt{64} = 8$

Since $AO = \frac{2}{3}AM$ (why?), $\frac{2}{3}AM = 8$, $AM = 12$, and $OM = 4$.

$l = \sqrt{6^2 + 4^2} = \sqrt{52} = 2\sqrt{13}$

In 30°–60°–90° $\triangle AMC$, $CM = \dfrac{12}{\sqrt{3}} = 4\sqrt{3}$

Base edge $= AC = BC = 2 \cdot 4\sqrt{3} = 8\sqrt{3}$

L.A. $= \frac{1}{2}pl = \frac{1}{2} \cdot 24\sqrt{3} \cdot 2\sqrt{13} = 24\sqrt{39}$

b. Area of base $= B = \frac{1}{2} \cdot 8\sqrt{3} \cdot 12 = 48\sqrt{3}$

$V = \frac{1}{3}Bh = \frac{1}{3} \cdot 48\sqrt{3} \cdot 6 = 96\sqrt{3}$

Classroom Exercises

1. The base of the regular pyramid shown is an n-sided polygon with edge a. The slant height is l.
   a. Perimeter of base $= p =$ _?_
   b. Area of one lateral face $=$ _?_
   c. Area of n lateral faces $=$ _?_
   d. Express your answer to (c) in terms of p and l. What formula have you developed?

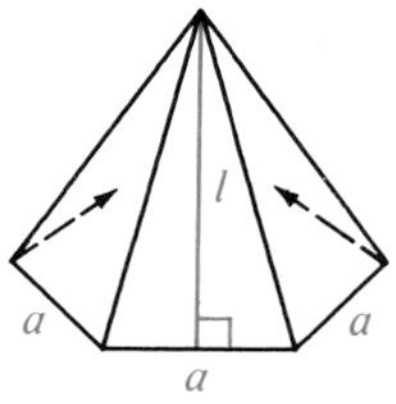

***V-ABCD* is a regular square pyramid. Find numerical answers.**

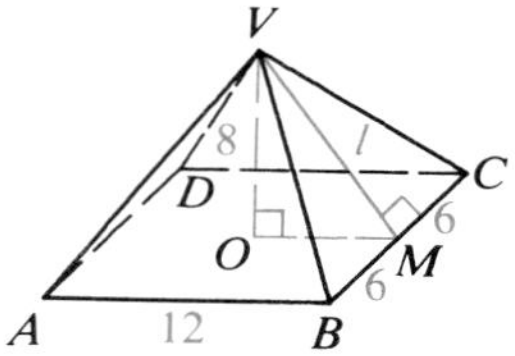

2. $OM =$ _?_
3. $l =$ _?_
4. Area of $\triangle VBC =$ _?_
5. L.A. = _?_
6. Volume = _?_
7. $VC =$ _?_

All edges of regular pyramid *V-XYZ* are 6 cm long. Find numerical answers.

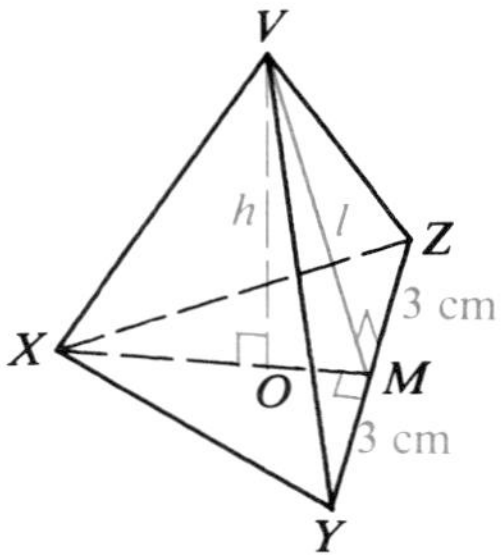

8. $XM =$ _?_
9. $XO =$ _?_
10. $h =$ _?_
11. Base area = _?_
12. Volume = _?_
13. Slant height = _?_

Written Exercises

You can use the following three steps to sketch a square pyramid.

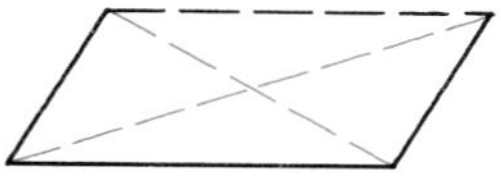

(1) Draw a parallelogram for the base and sketch diagonals.

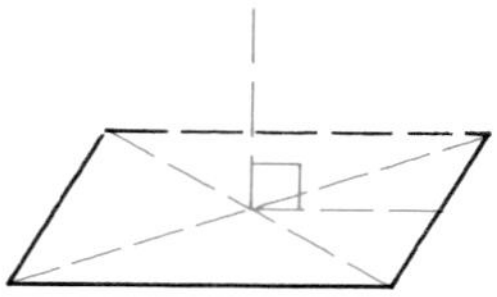

(2) Draw a vertical line segment from the point where the diagonals intersect.

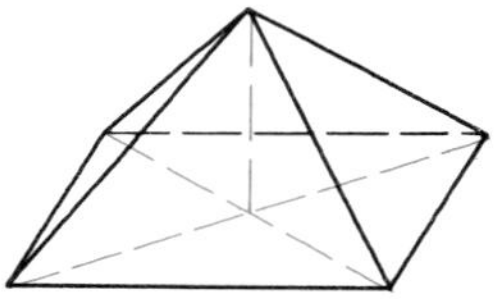

(3) Join the vertex to the base vertices.

Sketch each pyramid. Then find its lateral area.

A

1. A regular square pyramid with base edge 1.5 and slant height 9
2. A regular triangular pyramid with base edge 4 and slant height 6
3. A regular square pyramid with base edge 12 and lateral edge 10
4. A regular hexagonal pyramid with base edge 10 and lateral edge 13

Copy and complete the table below for the regular square pyramid shown.

	5.	**6.**	**7.**	**8.**	**9.**	**10.**
height, *h*	4	12	24	?	?	15
slant height, *l*	5	13	?	12	5	?
base edge	?	?	14	?	8	?
lateral edge	?	?	?	15	?	17

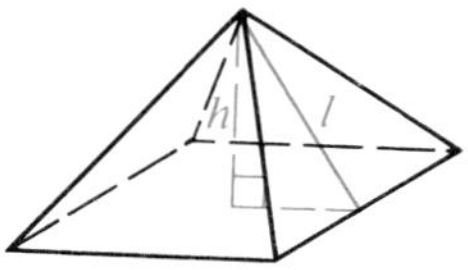

For Exercises 11–14 sketch each square pyramid described. Then find its lateral area, total area, and volume.

11. base edge = 6, height = 4

12. base edge = 16, slant height = 10

13. base edge = 16, lateral edge = 17

14. height = 12, slant height = 13

B **15.** *V-ABCD* is a pyramid with a rectangular base 18 cm long and 10 cm wide. *O* is the center of the rectangle. The height, *VO*, of the pyramid is 12 cm.

a. Find *VX* and *VY*.

b. Find the lateral area of the pyramid. (Why can't you use the formula L.A. = $\frac{1}{2}pl$?)

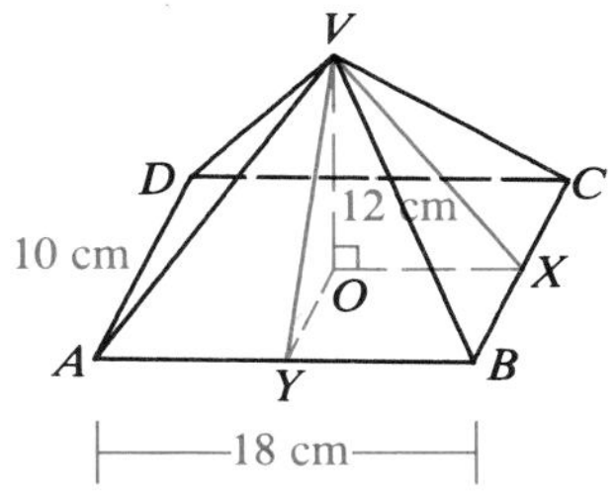

16. A pyramid and a prism both have height 8.2 cm and congruent hexagonal bases with area 22.3 cm^2. Give the ratio of the volumes. (You do not *need* to calculate their volumes.)

17. The shaded pyramid is cut from a rectangular solid. How does the volume of the pyramid compare with the volume of the rectangular solid?

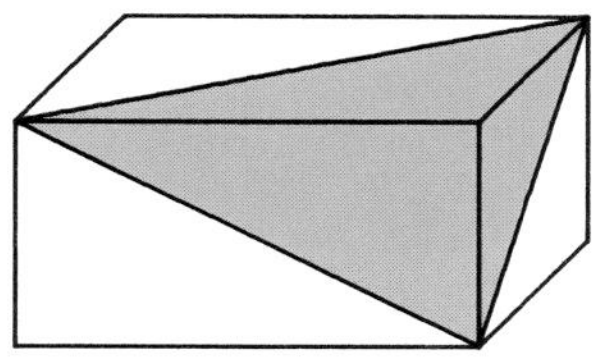

Ex. 17

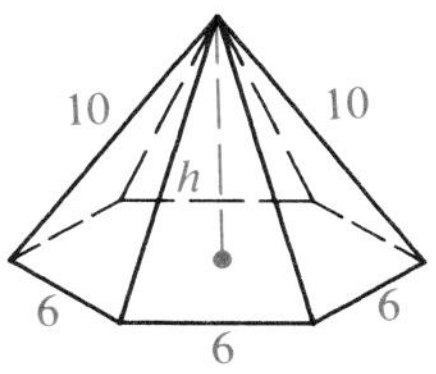

Ex. 18

18. Find the height and the volume of a regular hexagonal pyramid with lateral edges 10 units and base edges 6 units. (*Hint:* The diagonals of the base form six equilateral triangles.)

For Exercises 19–25 refer to the regular triangular pyramid shown below.

19. If $AM = 9$ and $VA = 10$, find h and l.

20. a. If $BC = 6$, find AM and AO.

b. If $BC = 6$ and $VA = 4$, find h and l.

21. a. If $h = 4$ and $l = 5$, find OM, OA, and BC.

b. Find the lateral area and the volume.

22. If $VA = 5$ and $h = 3$, find the slant height, the lateral area, and the volume.

23. If $AB = 12$ and $VA = 10$, find the lateral area and the volume.

24. a. If all edges of the pyramid are 6, show that $h = \sqrt{24}$, or $2\sqrt{6}$.

b. Find the total area and the volume.

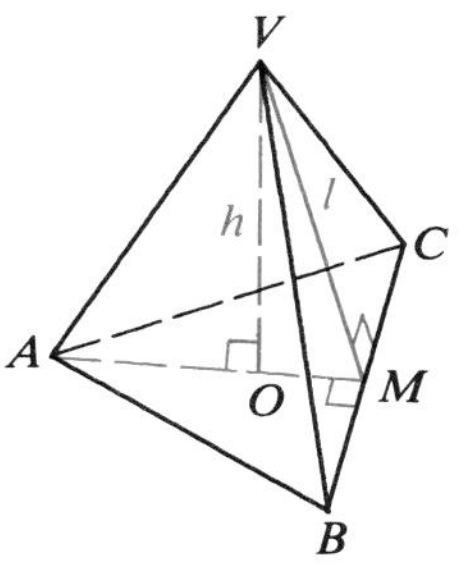

Exs. 19–25

C **25.** Suppose all edges of the pyramid shown above are *e* units long. Find the volume in terms of *e*.

26. The base of a pyramid is a regular hexagon with sides x cm long. The lateral edges are $2x$ cm long. Find the volume of the pyramid in terms of x.

27. Show that the ratio of the volumes of the two regular square pyramids shown is $\frac{\tan 40°}{\tan 80°}$.

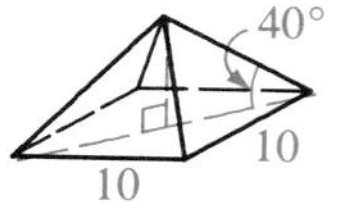

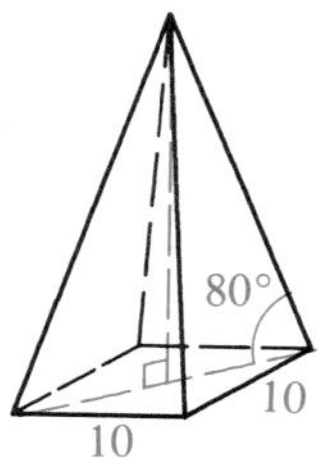

★ 28. Different pyramids are inscribed in two identical cubes, as shown below.
 a. Which pyramid has the greater volume?
 b. Which pyramid has the greater total area?

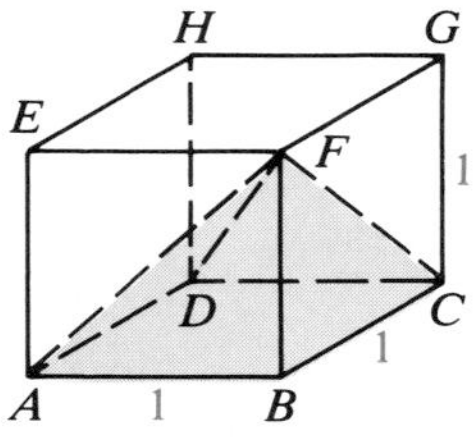

Pyramid *F-ABCD*

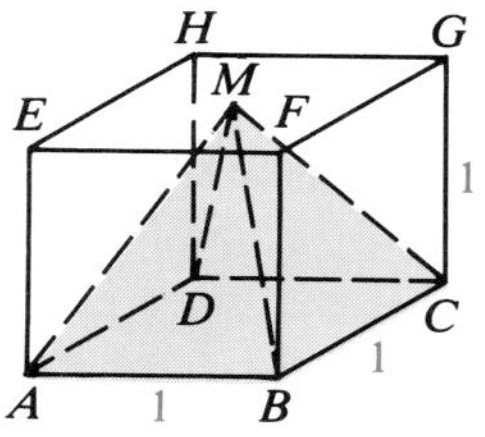

Pyramid *M-ABCD* has vertex *M* at center of square *EFGH*.

COMPUTER KEY-IN

The earliest pyramids, which were built about 2750 B.C., are called *step pyramids* because the lateral faces are not really triangles but a series of great stone steps. To find the volume of such a pyramid it is only necessary to find the sum of the volumes of the steps, or layers. Each layer is a rectangular solid with a square base.

Let us consider a pyramid with base edges 10 and height 10. Suppose that this pyramid is made up of 10 steps with equal heights. The top layer is a cube (base edges equal the height), and the base edge for each succeeding layer increases by an amount equal to the height of a layer. As the left side of the diagram at the top of the next page shows, the height of each step is $\frac{10}{10} = 1$, and the volume of the top layer is $V_1 = Bh = (1^2) \cdot 1 = 1$. The volumes of the second and third layers are $V_2 = (2^2) \cdot 1 = 4$ and $V_3 = (3^2) \cdot 1 = 9$. Continuing in this way, the total volume for the pyramid is:

$$V = 1^2 \cdot 1 + 2^2 \cdot 1 + 3^2 \cdot 1 + 4^2 \cdot 1 + \cdots + 9^2 \cdot 1 + 10^2 \cdot 1 = 385$$

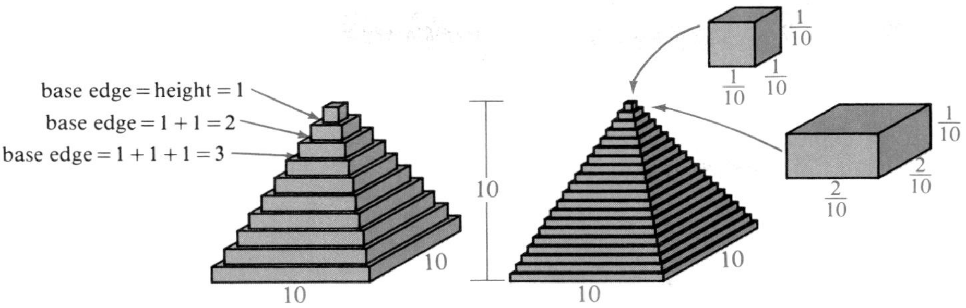

Now consider another pyramid with the same base and height but having 100 steps instead of 10 steps. The height of each layer is $\frac{10}{100} = \frac{1}{10}$, and the volume of each layer is as follows:

$$\text{Volume of each layer} = B \cdot h$$

$$\text{Volume of top layer} = \left(\frac{1}{10}\right)^2 \cdot \frac{1}{10}$$

$$\text{Volume of second layer} = \left(2 \cdot \frac{1}{10}\right)^2 \cdot \frac{1}{10} = \left(\frac{2}{10}\right)^2 \cdot \frac{1}{10}$$

$$\text{Volume of third layer} = \left(3 \cdot \frac{1}{10}\right)^2 \cdot \frac{1}{10} = \left(\frac{3}{10}\right)^2 \cdot \frac{1}{10}$$

$$\vdots \qquad\qquad \vdots$$

$$\text{Volume of 99th layer} = \left(99 \cdot \frac{1}{10}\right)^2 \cdot \frac{1}{10} = \left(\frac{99}{10}\right)^2 \cdot \frac{1}{10}$$

$$\text{Volume of 100th layer} = \left(100 \cdot \frac{1}{10}\right)^2 \cdot \frac{1}{10} = \left(\frac{100}{10}\right)^2 \cdot \frac{1}{10}$$

Thus, the volume of the pyramid is:

$$V = \left(\frac{1}{10}\right)^2 \cdot \frac{1}{10} + \left(\frac{2}{10}\right)^2 \cdot \frac{1}{10} + \left(\frac{3}{10}\right)^2 \cdot \frac{1}{10} + \cdots + \left(\frac{99}{10}\right)^2 \cdot \frac{1}{10} + \left(\frac{100}{10}\right)^2 \cdot \frac{1}{10}$$

The following computer program finds the total volume for the given pyramid having N steps.

```
10  LET V = 0
20  PRINT "HOW MANY STEPS ARE THERE";
30  INPUT N
40  LET H = 10/N
50  FOR X = 1 TO N
60  LET V = V + (X * H)↑2 * H
70  NEXT X
80  PRINT "VOLUME OF PYRAMID WITH ";N;"STEPS IS ";V
90  END
```

Exercises

1. RUN the given program to verify the volume of the 10-step pyramid and to find the volume of the 100-step pyramid.
2. a. Suppose that another pyramid with the same base and height has 1000 steps. RUN the program to find the volume.
   b. Complete the chart.

Number of steps	Volume
10	?
100	?
1000	?
5000	?
10,000	?
15,000	?

   c. As the number of steps increases, what value do the volumes seem to be getting close to?
   d. What is the volume of a regular square pyramid with base edge of length 10 and height 10?
   e. What can you conclude from comparing the answers to parts (b)–(d)?

CALCULATOR KEY-IN

The base of a pyramid is a triangle with sides 7 cm, 8 cm, and 9 cm long. The height of the pyramid is 11 cm. Find its volume. (*Hint:* Use Heron's Formula, page 390.)

Challenge

Given any two rectangles, draw one line that divides each into two parts of equal area.

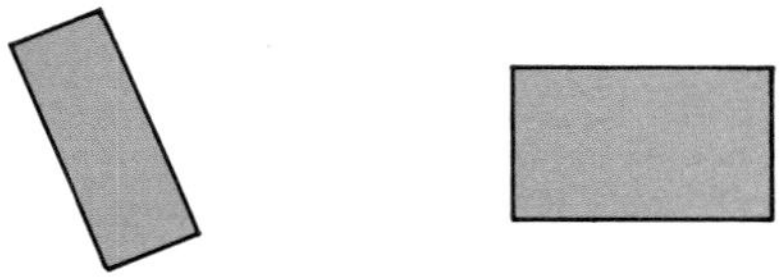

10-3 Cylinders and Cones

A **cylinder** is like a prism except that its bases are circles instead of polygons. In a **right cylinder,** the segment joining the centers of the circular bases is an **altitude.** The length of an altitude is called the *height* (h) of the cylinder. A radius of a base is also called a **radius** r of the cylinder.

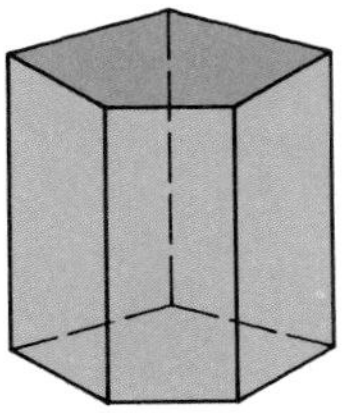

Right prism

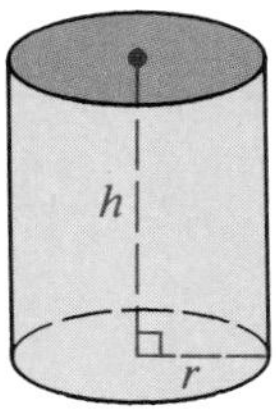

Right cylinder

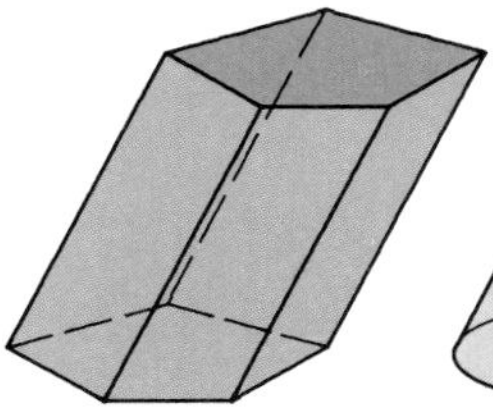

Oblique prism

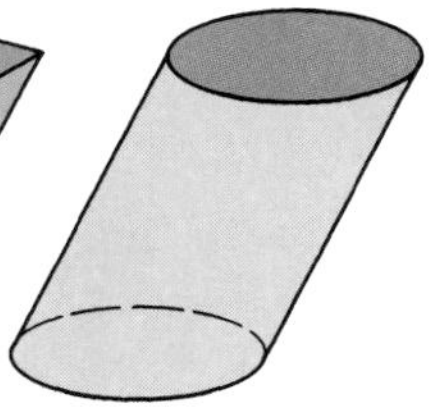

Oblique cylinder

The diagrams above show the relationship between prisms and cylinders. The relationship between pyramids and **cones** is shown in the diagrams below.

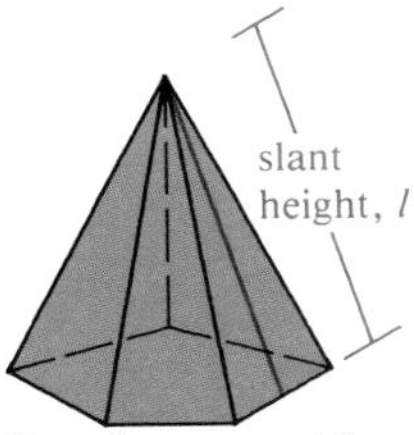

Regular pyramid

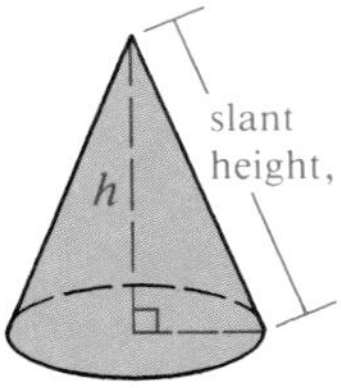

Right cone

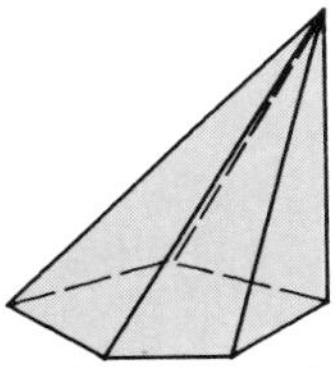

Oblique pyramid

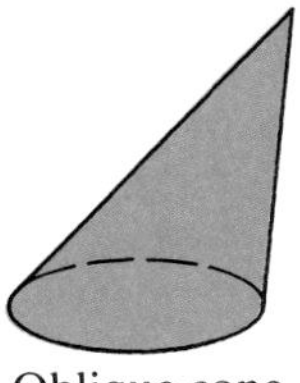

Oblique cone

In the discussion and exercises that follow, the words "cylinder" and "cone" will always refer to a right cylinder and a **right cone.** Note that "slant height" applies only to a regular pyramid and a right cone.

The formulas for cylinders are related to those for prisms, and the formulas for cones are related to those for pyramids. Classify these solids and their corresponding formulas as follows: Prisms and cylinders have parallel bases ($V = Bh$) whereas pyramids and cones have one base and a pointed end, or vertex ($V = \frac{1}{3}Bh$). Since cylinders and cones have circular bases, use πr^2 for B and $2\pi r$ for p.

Prism	*Cylinder*
$V = Bh$	$V = Bh = \pi r^2 h$
$\text{L.A.} = ph$	$\text{L.A.} = ph = 2\pi rh$
Pyramid	*Cone*
$V = \frac{1}{3}Bh$	$V = \frac{1}{3}Bh = \frac{1}{3}\pi r^2 h$
$\text{L.A.} = \frac{1}{2}pl$	$\text{L.A.} = \frac{1}{2}pl = \frac{1}{2}(2\pi r)l = \pi rl$

So far our study of solids has not included formulas for oblique solids. The volume formulas shown above, but not the area formulas, can be used for the corresponding oblique solids. (See the Extra on pages 460–461.)

Example 1 Given a cylinder with radius 5 cm and height 4 cm, find the (a) lateral area, (b) total area, and (c) volume.

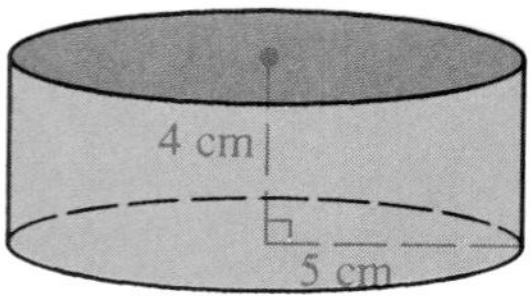

Solution

a. L.A. $= 2\pi rh = 2\pi \cdot 5 \cdot 4 = 40\pi$ (cm^2)
b. T.A. $=$ L.A. $+ 2B$
$= 40\pi + 2(\pi \cdot 5^2) = 90\pi$ (cm^2)
c. $V = \pi r^2 h = \pi \cdot 5^2 \cdot 4 = 100\pi$ (cm^3)

Example 2 Find the (a) lateral area, (b) total area, and (c) volume for the cone shown.

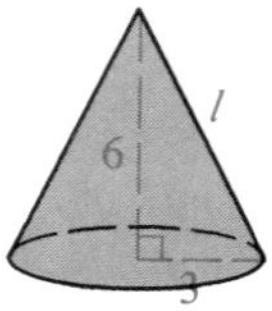

Solution

a. To find L.A., first find l.
$l = \sqrt{6^2 + 3^2} = \sqrt{45} = 3\sqrt{5}$
L.A. $= \pi rl = \pi \cdot 3 \cdot 3\sqrt{5} = 9\pi\sqrt{5}$
b. T.A. $=$ L.A. $+ B = 9\pi\sqrt{5} + \pi \cdot 3^2 = 9\pi\sqrt{5} + 9\pi$
c. $V = \frac{1}{3}\pi r^2 h = \frac{1}{3}\pi \cdot 3^2 \cdot 6 = 18\pi$

Classroom Exercises

1. a. When the label of a soup can is cut off and laid flat, it is a rectangular piece of paper. (See diagram below.) How are the length and width of this rectangle related to r and h?
b. What is the area of this rectangle?

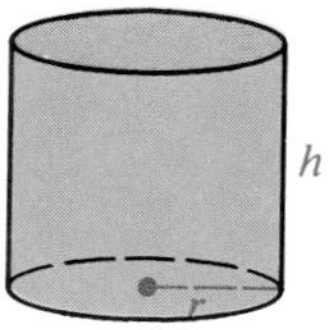

2. a. Find the lateral areas of cylinders I, II, and III.
b. Notice that the height of II is twice the height of I. Is the lateral area of II twice the lateral area of I?
c. Notice that the radius of III is twice the radius of I. Is the lateral area of III twice the lateral area of I?

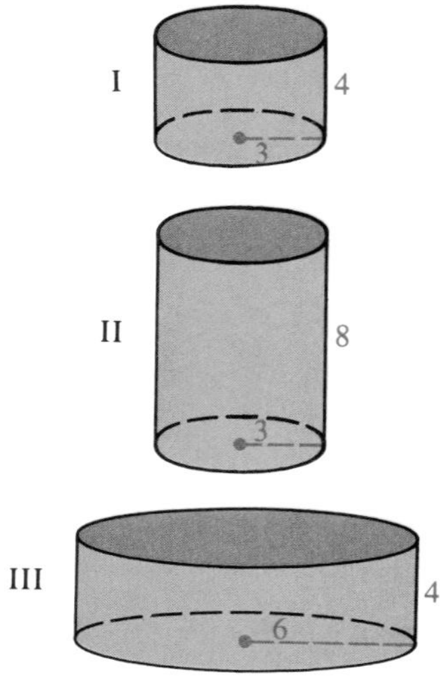

3. a. Find the volumes of cylinders I, II, and III.
b. Notice that the height of II is twice the height of I. Is the volume of II twice the volume of I?
c. Notice that the radius of III is twice the radius of I. Is the volume of III twice the volume of I?

4. Suppose the radius and height of a cylinder are both doubled.
a. What happens to the lateral area?
b. What happens to the volume?

5. A manufacturer needs to decide which container to use for packaging a product. One container is twice as wide as another but only half as tall. Which container holds more, or do they hold the same amount? Guess first and then calculate the ratio of their volumes.

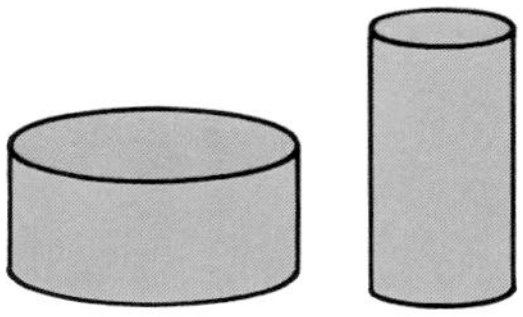

Complete the table for the cone shown.

	r	h	l	L.A.	T.A.	V
6.	3	4	?	?	?	?
7.	5	12	?	?	?	?
8.	6 cm	?	10 cm	?	?	?

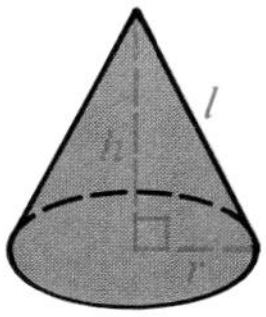

Written Exercises

Find the lateral area, total area, and volume of each cylinder.

A **1.** $r = 4$, $h = 5$ **2.** $r = 8$, $h = 10$ **3.** $r = 4$, $h = 3$ **4.** $r = 8$, $h = 6$

5. The volume of a cylinder is 64π. If $r = h$, find r.

6. The lateral area of a cylinder is 18π. If $h = 6$, find r.

7. The volume of a cylinder is 72π. If $h = 8$, find the lateral area.

8. The total area of a cylinder is 100π. If $r = h$, find r.

Copy and complete the table below for the cone shown.

	r	h	l	L.A.	T.A.	V
9.	4	3	?	?	?	?
10.	8	6	?	?	?	?
11.	12	?	13	?	?	?
12.	?	2	6	?	?	?
13.	?	?	15	180π	?	?
14.	15	?	?	?	?	600π

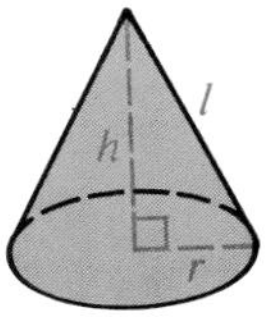

15. In the first two rows of the preceding table, the ratio of the radii is $\frac{4}{8}$, or $\frac{1}{2}$, and the ratio of the heights is $\frac{3}{6}$, or $\frac{1}{2}$. Use your answers from these two rows of the table to determine the ratios of the following:
a. lateral areas **b.** total areas **c.** volumes

16. A cone and a cylinder both have height 48 and radius 15. Give the ratio of their volumes without actually calculating the two volumes.

B **17.** **a.** Guess which contains more, the can or the bottle. (Assume that the top part of the bottle is a complete cone.)
b. See if your guess is right by finding the volumes of both.

Ex. 17

18. A solid metal cylinder with radius 6 cm and height 18 cm is melted down and recast as a solid cone with radius 9. Find the height of the cone.

19. A pipe is 2 m long and has inside radius 5 cm and outside radius 6 cm. How many cubic centimeters of metal are in the pipe?

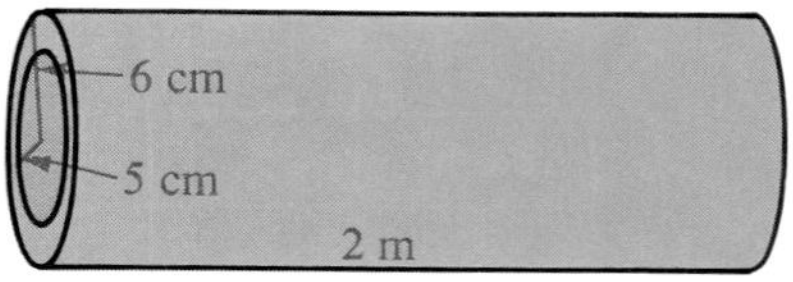

20. Two water pipes of the same length have diameters 6 cm and 8 cm. These two pipes are replaced by a single pipe of the same length, which has the same capacity as the smaller pipes combined. What should the diameter of the new pipe be?

21. If the radius and height of a cylinder are both multiplied by 3, the lateral area is multiplied by __?__ and the volume is multiplied by __?__.

22. The total area of a cylinder is 40π. If $h = 8$, find r.

23. The total area of a cylinder is 90π. If $h = 12$, find r.

24. In rectangle $ABCD$, $AB = 10$ and $AD = 6$.
a. If the rectangle is revolved in space about $\overline{AB}$, what is the volume of the space through which it moves?
b. Answer part (a) if the rectangle is revolved about $\overline{AD}$.

25. **a.** The segment joining (0, 0) and (4, 3) is rotated about the x-axis, forming the lateral surface of a cone. Find the lateral area and the volume of this cone.
b. Make a sketch showing the cone that would be formed if the segment had been rotated about the y-axis. Find the lateral area and the volume of this cone.
c. Are your answers to parts (a) and (b) the same?

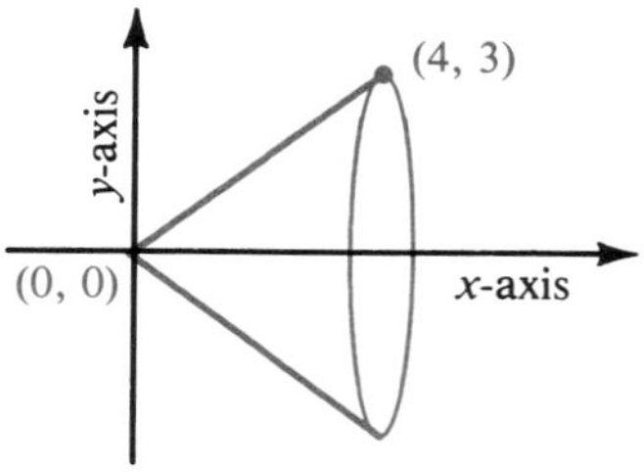

26. Each of the prisms shown below is inscribed in a cylinder with height 10 and radius 6. Find the volume and lateral area of each prism.

a.
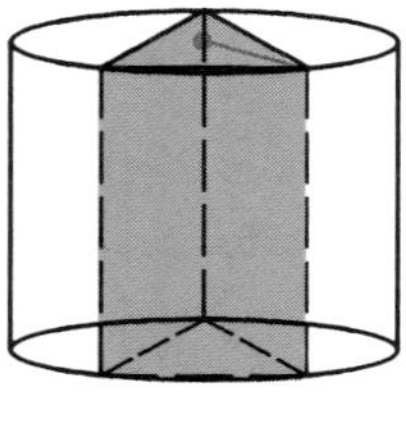

Base is an equilateral triangle.

b.
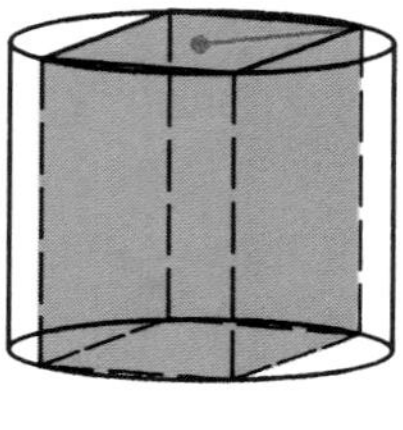

Base is a square.

c.
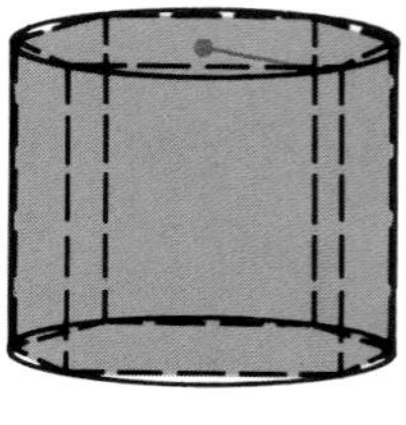

Base is a regular hexagon.

27. A square pyramid with base edge 4 is inscribed in a cone with height 6. What is the volume of the cone?

28. A square pyramid is inscribed in a cone with radius 4 and height 4.
   a. What is the volume of the pyramid?
   b. Find the slant heights of the cone and the pyramid.

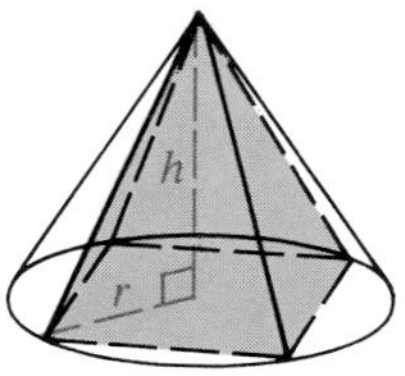

Exs. 27, 28

C 29. A regular hexagonal pyramid with base edge 6 and height 8 is inscribed in a cone. Show that the lateral area of the cone is 60π and the lateral area of the pyramid is $18\sqrt{91}$.

30. In $\triangle ABC$, $AB = 15$, $AC = 20$, and $BC = 25$. If the triangle is rotated in space about $\overline{BC}$, what is the volume of the space through which it moves?

31. A 120° sector is cut out of a circular piece of tin with radius 6 and bent to form the lateral surface of a cone. What is the volume of the cone?

CALCULATOR KEY-IN

1. Water is pouring into a conical reservoir at the rate of 1.8 m^3 per minute. How long will it take to fill the reservoir?

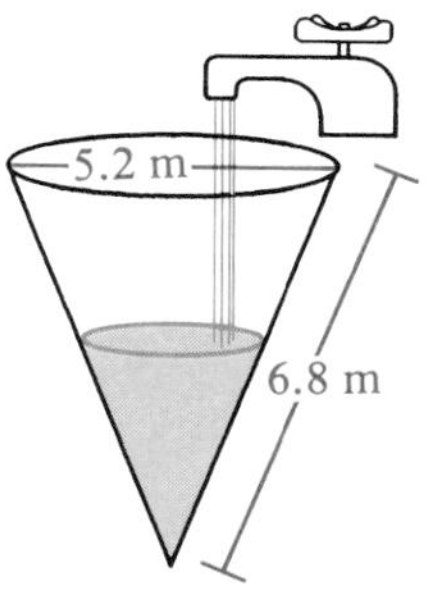

Ex. 1

2. Given a cylinder with radius 10 and height 12, suppose that the lateral surface of the cylinder is covered with a thin coat of paint having thickness 0.1. The volume of the paint can be calculated approximately or exactly.
   a. Use the diagrams below to explain the following formula.

   Approximate volume = (lateral area of cylinder) · (thickness of paint)

   $$V \approx (2\pi rh) \cdot (t)$$

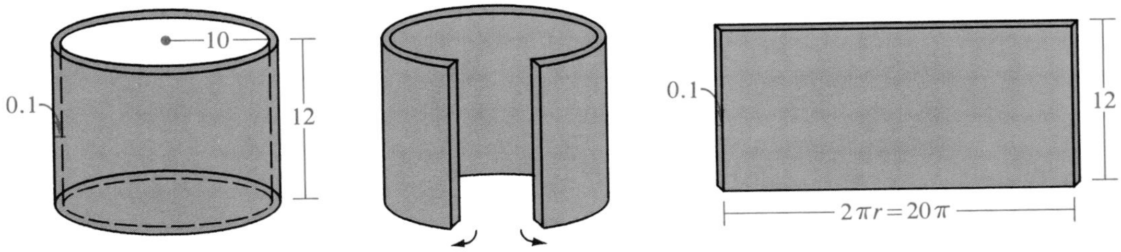

   Why is this formula only an approximation of the volume?
   b. Use the formula above to find the approximate volume of the paint.
   c. Find the exact volume of paint by subtracting the volume of the inner cylinder (the given cylinder) from the volume of the outer cylinder (the given cylinder plus paint).

COMPUTER KEY-IN

Refer to Exercise 2 of the Calculator Key-In.

1. Write a computer program that will find the approximate volume of paint for each thickness: 0.1, 0.01, 0.001
2. Write a computer program that will find the exact volume of paint for each thickness: 0.1, 0.01, 0.001
3. Complete the table at the right.
   What can you conclude from the values listed in the table?

Paint thickness	Approximate volume	Exact volume
0.1	?	?
0.01	?	?
0.001	?	?

Self-Test 1

For Exercises 1–5 find the lateral area, the total area, and the volume of each solid.

1. A rectangular solid with length 10, width 8, and height 4.5
2. A regular square pyramid with base edge 24 and slant height 13
3. A cylinder with radius 10 and height 7
4. A right hexagonal prism with height 5 cm and base edge 6 cm
5. A cone with height 12 and radius 9
6. The total area of a cube is 2400 m^2. Find the volume.
7. A solid metal cylinder with radius 2 and height 2 is recast as a solid cone with radius 2. Find the height of the cone.
8. A prism with height 2 and a pyramid with height 5 have congruent triangular bases. Which solid has the greater volume?

Challenge

A piece of wood contains a square hole, a circular hole, and a triangular hole as illustrated. Explain how one block of wood in the shape of a cube with a 2 cm edge can be cut down so that it will pass through, but will plug, each of the holes in turn.

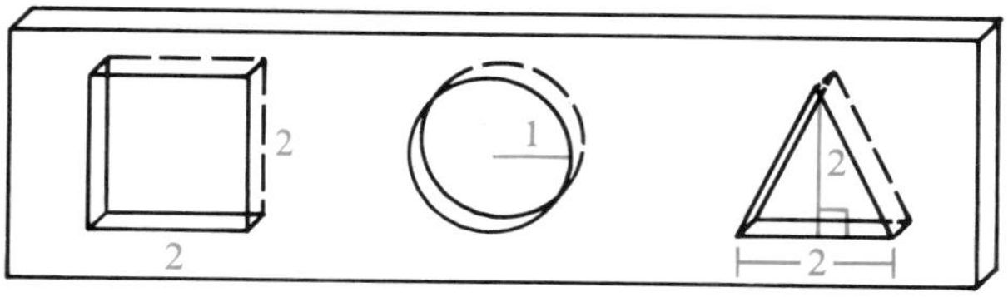

Similar Solids

Objectives

1. Find the area and the volume of a sphere.
2. State and apply the properties of similar solids.

10-4 Spheres

The sphere has many useful applications. One recent application is the development of a spherical blimp. An experimental model of the blimp is shown in the photograph and drawing. A spherical shape was selected for this blimp because a sphere gives excellent maneuverability, stability, hovering capabilities, and lift. The rotation of the top of a sphere away from the direction in which the sphere is traveling provides lifting power.

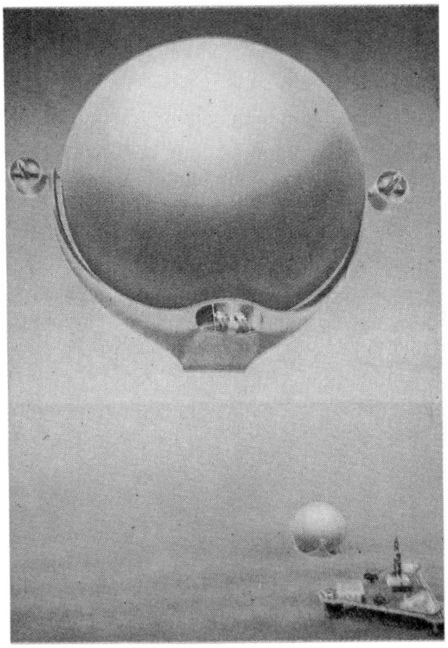

The area and the volume of a sphere are given by the formulas below. After some examples showing how these formulas are used, we will see how they may be derived.

$$\mathbf{A = 4\pi r^2} \qquad \mathbf{V = \tfrac{4}{3}\pi r^3}$$

Example 1 Find the area and the volume of a sphere with radius 2.

Solution $A = 4\pi r^2 = 4\pi \cdot 2^2 = 16\pi$

$$V = \frac{4}{3}\pi r^3 = \frac{4}{3}\pi \cdot 2^3 = \frac{32\pi}{3}$$

Example 2 The area of a sphere is 256π. Find the volume.

Solution To find the volume, we must first find the radius.

(1) $A = 256\pi = 4\pi r^2$
$64 = r^2$
$8 = r$

(2) $V = \frac{4}{3}\pi r^3 = \frac{4}{3}\pi \cdot 8^3$
$= \frac{2048\pi}{3}$

The next example shows how to find the area of the circle formed when a sphere is cut by a plane.

Example 3 A plane passes 4 cm from the center of a sphere with radius 7 cm. Find the area of the circle of intersection.

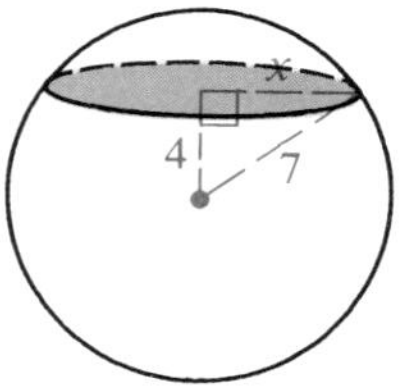

Solution Let x = radius of the circle.

$x = \sqrt{7^2 - 4^2} = \sqrt{33}$

$A = \pi x^2 = \pi(\sqrt{33})^2 = 33\pi \text{ (cm}^2\text{)}$

Deriving the Volume Formula (Optional)

Any solid can be approximated by a stack of thin circular discs of equal thickness, as shown by the sphere drawn at the right. Each disc is actually a cylinder with height h.

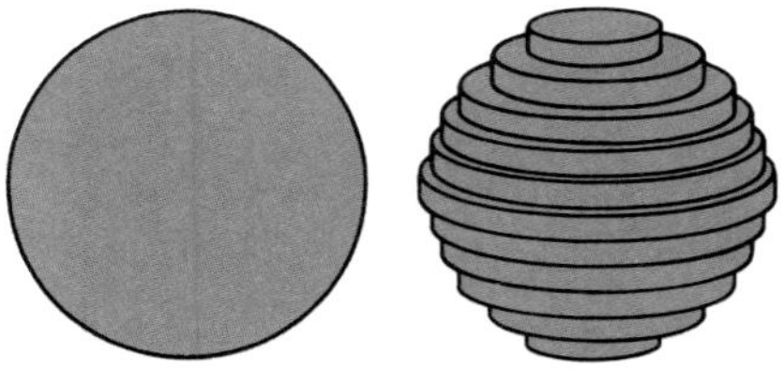

The sphere, the cylinder, and the double cone below all have radius r and height $2r$. Look at the disc that is x units above the center of each solid.

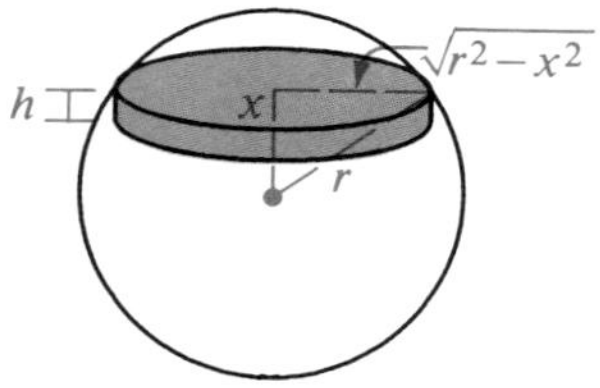

Disc volume:
$\pi(\sqrt{r^2 - x^2})^2 h = \pi(r^2 - x^2)h$
$= \pi r^2 h - \pi x^2 h$

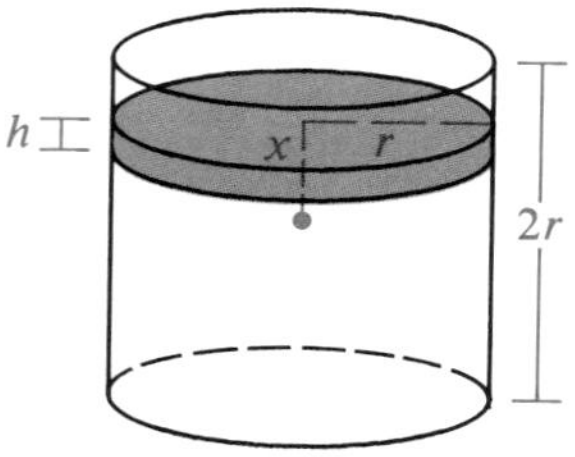

Disc volume: $\pi r^2 h$

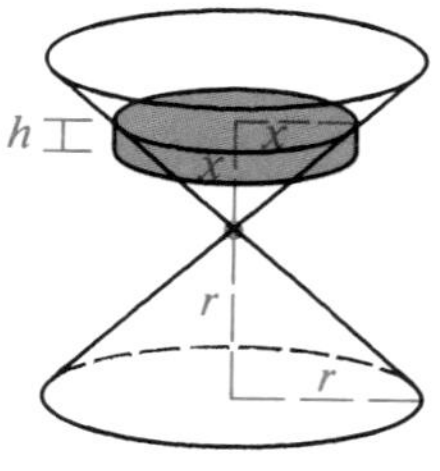

Disc volume: $\pi x^2 h$

Study the calculations shown for the figures above. Notice that no matter what the distance x is, the volume of the first disc equals the difference between the volumes of the other two discs. This means that if you take many discs of the same thickness in the three solids (as in the diagrams on the next page) the total volume of all the discs in the first solid will equal the difference between the total volumes of all the discs in each of the other two solids.

Total volume of discs in sphere = Total volume of discs in cylinder − Total volume of discs in double cone

The equation on the preceding page holds if there are just a few discs approximating each solid or very many discs. If there are very many discs, their total volume will be practically the same as the volume of the solid. It follows that we can find the volume of the sphere by subtracting the volume of the double cone from the volume of the cylinder.

$$\begin{aligned}\text{Volume of sphere} &= \pi r^2 \cdot 2r - 2(\tfrac{1}{3}\pi r^2 \cdot r)\\ &= 2\pi r^3 \quad - \tfrac{2}{3}\pi r^3\\ &= \tfrac{4}{3}\pi r^3\end{aligned}$$

Deriving the Area Formula (Optional)

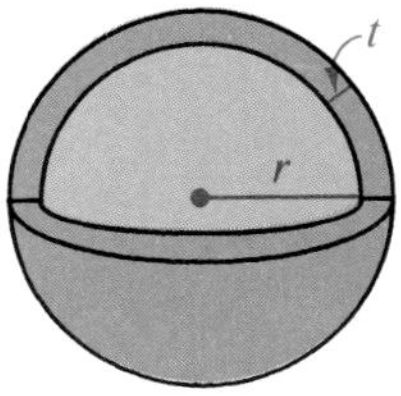

Imagine a rubber ball with inner radius r and rubber thickness t. To find the volume of the rubber, we can use the formula for the volume of a sphere. We just subtract the volume of the inner sphere from the volume of the outer sphere.

$$\begin{aligned}\text{Exact vol. of rubber} &= \tfrac{4}{3}\pi(r+t)^3 - \tfrac{4}{3}\pi r^3\\ &= \tfrac{4}{3}\pi[(r+t)^3 - r^3]\\ &= \tfrac{4}{3}\pi[r^3 + 3r^2t + 3rt^2 + t^3 - r^3]\\ &= 4\pi r^2 t + 4\pi r t^2 + \tfrac{4}{3}\pi t^3\end{aligned}$$

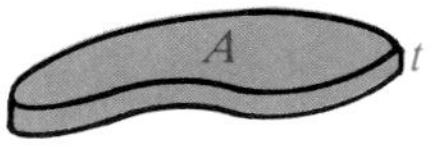

The volume of the rubber can be found in another way as well. If we think of a small piece of the rubber ball, its approximate volume would be its outer area A times its thickness t. The same thing is true for the whole ball.

$$\text{Volume of rubber} \approx \text{Surface area} \times \text{thickness}$$
$$V \approx At$$

Now let us use both results for the volume of the rubber.

$$At \approx 4\pi r^2 t + 4\pi r t^2 + \tfrac{4}{3}\pi t^3$$

If we divide both sides of the equation by t, we get the result:

$$A \approx 4\pi r^2 + 4\pi r t + \tfrac{4}{3}\pi t^2$$

This approximation for A gets better and better as the layer of rubber gets thinner and thinner. As t gets near zero, the last two terms in the formula above also get near zero. The limiting result is the formula

$$A = 4\pi r^2.$$

This is exactly what we would expect, since the surface area of a ball clearly does not depend at all on the thickness of the rubber, but only on the size of the radius.

Classroom Exercises

Copy and complete the table for spheres.

	1.	2.	3.	4.	5.	6.
Radius	1	2	4	?	?	?
Area	?	?	?	36π	100π	?
Volume	?	?	?	?	?	$\frac{4000\pi}{3}$

A plane passes h cm from the center of a sphere with radius r cm. Find the area of the circle of intersection, shaded in the diagram, for the given values.

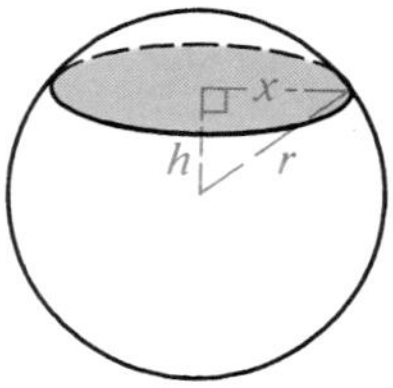

7. $r = 5$, $h = 3$

8. $r = 17$, $h = 8$

9. $r = 7$, $h = 6$

Written Exercises

Copy and complete the table for spheres.

A

	1.	2.	3.	4.	5.	6.	7.	8.
Radius	3	5	$\frac{1}{2}$	$\frac{3}{4}$	?	?	$\sqrt{2}$	?
Area	?	?	?	?	64π	324π	?	?
Volume	?	?	?	?	?	?	?	288π

9. If you double the radius of a sphere, the area of the sphere is multiplied by _?_ and the volume is multiplied by _?_.

10. If you triple the radius of a sphere, the area of the sphere is multiplied by _?_ and the volume is multiplied by _?_.

11. Find the area of the circle formed when a plane passes 2 cm from the center of a sphere with radius 5 cm.

12. Find the area of the circle formed when a plane passes 7 cm from the center of a sphere with radius 8 cm.

13. A sphere has radius 2 and a hemisphere has radius 4. Compare their volumes.

14. A scoop of ice cream with diameter 6 cm is placed in an ice-cream cone with diameter 5 cm and height 10 cm. Is the cone big enough to hold all the ice cream if it melts?

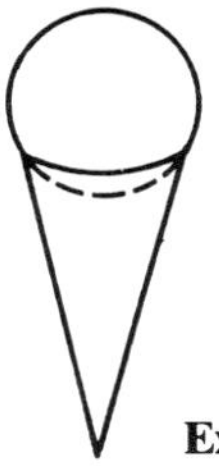
Ex. 14

B 15. An experimental one-room house is a hemisphere with a floor. If three cans of paint are needed to cover the floor, how many cans will be needed to paint the ceiling? (Ignore door and windows.)

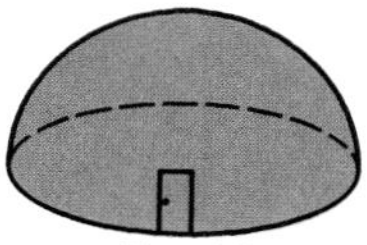
Ex. 15

16. A silo of a barn consists of a cylinder capped by a hemisphere as shown. Find the volume of the silo.

17. Two cans of paint cover the hemispherical dome of the silo shown. Approximately how many cans are needed to paint the rest of the silo's exterior?

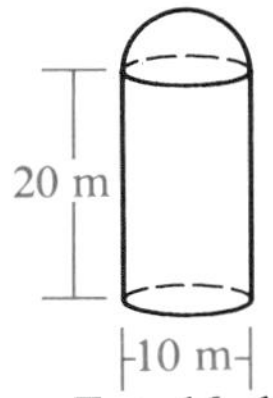

Exs. 16, 17

18. A hemispheric bowl with radius 25 contains water whose depth is 10. What is the area of the water's surface?

19. The circle containing points midway between the Earth's equator and the North Pole is at latitude 45°N. What is the ratio of the area of this circle to the area of the circle at the equator?

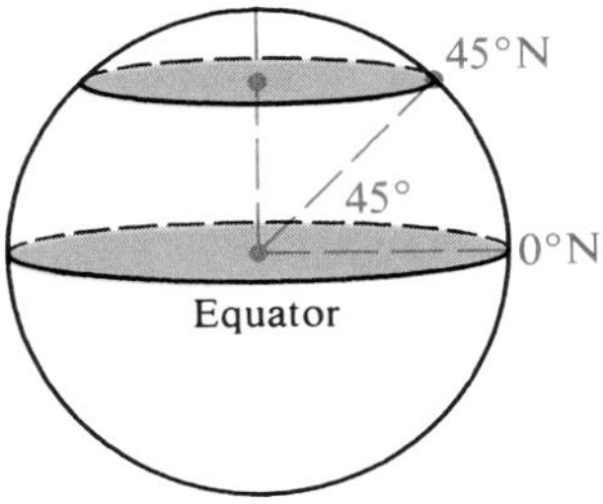

20. A metal ball with radius 8 cm is melted down and recast as a cone with the same radius. What is the height of the cone?

21. Four metal balls fit snugly inside a cylindrical can. A geometry student claims that two extra balls of the same size can be put into the can, provided all six balls can be melted down and the molten liquid poured into the can. Is the student correct? (*Hint:* Let the radius of the balls be r.)

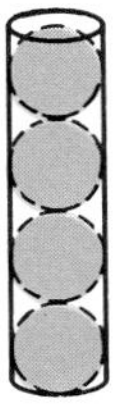

22. A sphere with radius r is inscribed in a cylinder. Find the volume of the cylinder in terms of r.

23. A sphere is inscribed in a cylinder. Show that the area of the sphere equals the lateral area of the cylinder.

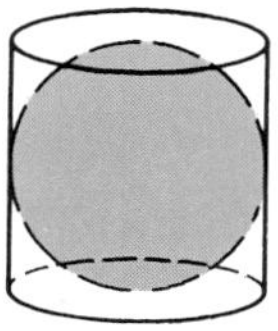

Exs. 22, 23

24. A double cone is inscribed in the cylinder shown. Find the volume of the space inside the cylinder but outside the double cone.

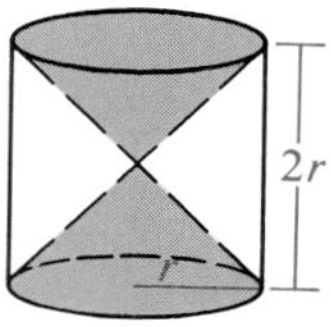

Ex. 24

25. A hollow rubber ball has outer radius 11 cm and inner radius 10 cm.

a. Find the exact volume of the rubber.

b. The volume of the rubber can be approximated by the formula:

$$V \approx \text{inner surface area} \times \text{thickness of rubber}$$

Use this formula to approximate V and compare your answer with part (a).

c. Is the approximation method used in part (b) better for a ball with a thick layer of rubber or a ball with a thin layer?

Ex. 25

26. A cylinder with height 12 is inscribed in a sphere with radius 10 Find the volume of the cylinder.

C **27.** A cylinder with height $2x$ is inscribed in a sphere with radius 10.

a. Show that the volume of the cylinder, V, is

$$2\pi x(100 - x^2).$$

b. By using calculus, one can show that V is maximum when $x = \frac{10\sqrt{3}}{3}$. If you substitute this value for x, the maximum volume $V = \underline{\quad ? \quad}$.

c. (Optional exercise) Use a calculator or a computer to evaluate $V = 2\pi x(100 - x^2)$ for various values of x between 0 and 10. Show that the maximum volume V occurs when x is approximately 5.77.

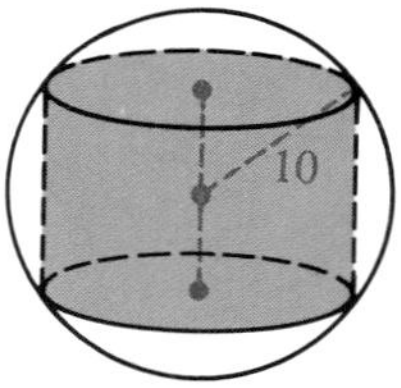

Exs. 26, 27

28. A cone is inscribed in a sphere with radius 10 as shown.

a. Show that the volume of the cone, V, is

$$\tfrac{1}{3}\pi(100 - x^2)(10 + x).$$

b. By using calculus, one can show that V is maximum when $x = \frac{10}{3}$. If you substitute this value for x, the maximum volume $V = \underline{\quad ? \quad}$.

c. (Optional) Use a calculator or a computer to evaluate $V = \frac{1}{3}\pi(100 - x^2)(10 + x)$ for various values of x between 0 and 10. Show that the maximum volume V occurs when x is $\frac{10}{3}$.

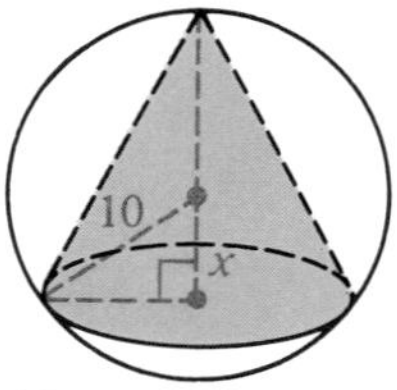

29. Sketch two intersecting spheres with radii 15 cm and 20 cm, respectively. The centers of the spheres are 25 cm apart. Find the area of the circle that is formed by the intersection. (*Hint:* Use Exercise 33 on page 251.)

30. A sphere is inscribed in a cone whose radius is 6 and whose height is 8. Find the radius of the sphere.

COMPUTER KEY-IN

The volume of a sphere with radius 10 can be approximated by cylindrical discs with equal heights. It is convenient to work with the upper half of the sphere, then double the result.

Suppose you use ten discs to approximate the upper hemisphere, as shown at the left below.

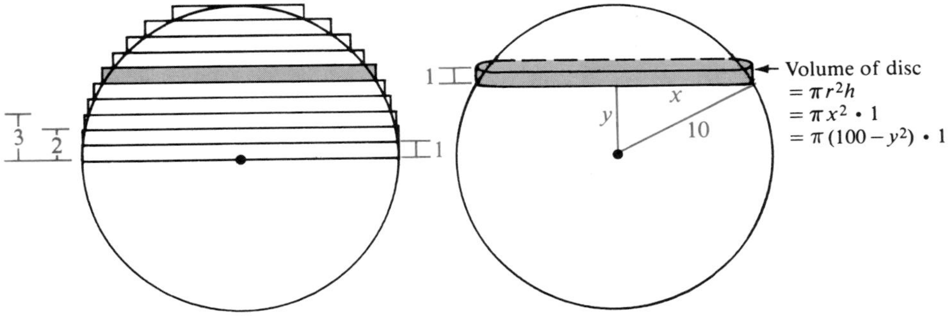

The diagram at the right above shows that the volume of a disc y units from the center of the sphere is $V = \pi(100 - y^2)$. Substitute $y = 0, 1, 2, \ldots,$ 9 to get the volumes of the ten discs.

Suppose you use n discs to approximate the upper hemisphere. Then the altitude of each disc equals $\frac{10}{n}$, and the volume of a disc y units from the center of the sphere is $V = \pi(100 - y^2) \cdot \frac{10}{n}$. The following computer program totals the volumes of n discs, then doubles the result. Note that line 80 calculates the volume of the sphere using the formula $V = \frac{4}{3}\pi r^3$.

```
10  LET Y = 0
15  LET V = 0
20  PRINT "HOW MANY DISCS";
25  INPUT N
30  FOR I = 1 TO N
40  LET Y = (I - 1)*10/N
50  LET V = V + 3.14159*(100 - Y↑2)*(10/N)
60  NEXT I
70  PRINT "VOLUME OF DISCS IS ";2*V
80  PRINT "VOLUME OF SPHERE IS ";4/3*3.14159*10↑3
90  END
```

Exercises

1. Use 10 for N and RUN the program. By approximately what percent does the disc method overcompute the volume of the sphere?
2. RUN the program several times and then complete the chart on the next page.

3. Our discs, outside the sphere, have yielded approximations greater than the true volume. Replace line 40 with `LET Y = I*(10/N)` to obtain a set of discs inside the sphere. RUN the new program for `N = 100`. Find the average of this result and the result listed for `N = 100` in Exercise 2.

Number of Discs	Height of Each Disc	Total Volume of Discs
20	?	?
50	?	?
100	?	?
1000	?	?

CALCULATOR KEY-IN

A 1. Approximately 70% of the Earth's surface is covered by water. How many square kilometers is this area? (The radius of the Earth is approximately 6380 km.)

2. a. Find the volume, correct to the nearest tenth of a cubic centimeter, of a sphere inscribed in a cube with edges 6 cm long.
   b. Show that the volume of the sphere is a little more than half the volume of the cube.

B 3. A solid metal sphere with radius 3 is melted and recast as a cone with radius 3. Show that the lateral area of the cone is about 3% more than the area of the sphere.

Application

GEODESIC DOMES

A spherical dome is an efficient way of enclosing space, since a sphere holds a greater volume than any other container with the same surface area. In 1947, R. Buckminster Fuller patented the *geodesic dome,* a framework made by joining straight pieces of steel or aluminum tubing in a network of triangles. A thin cover of aluminum or plastic is then attached to the tubing.

The segments forming the network are of various lengths, but the vertices are all equidistant from the center of the dome, so that they lie on a sphere. When we follow a chain of segments around the dome, we find that they approximate a circle on this sphere, often a great circle. It is this property that gives the dome design its name: A *geodesic* on any surface is a path of minimum length between two points on the surface, and on a sphere these shortest paths are arcs of great circles.

Though the geodesic dome is very light and has no internal supports, it is very strong, and standardized parts make construction of the dome relatively easy. Domes have been used with success for theaters, exhibition halls, sports arenas, and greenhouses.

The United States Pavilion that Fuller designed for Expo '67 in Montreal uses two domes linked together. The design of this structure is illustrated at the right. The red triangular network is the outer dome, the black hexagons form the inner dome, and the blue segments represent the trusses that tie the two domes together. The arrows mark one of the many chains of segments that form arcs of circles on the dome. You can see all of these features of the structure in the photograph at the right, which shows a view from inside the dome.

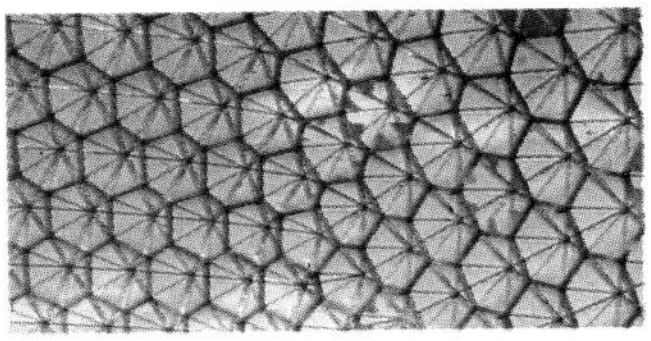

Exercises

1. In this exercise you will compare the volume of a sphere with the volume of a cube having about the same surface area as the sphere.
- **a.** Find the volume and surface area of a sphere of radius 7. (Use 3.14 for π.)
- **b.** Find the volume and surface area of a cube with edge 10.
- **c.** What is the approximate ratio of the sphere volume to the cube volume?

A cube has 8 vertices, 6 faces, and 12 edges. $8 + 6 - 12 = 2$. Leonard Euler (1707–1783) showed that for *any* solid with polygons for faces, vertices + faces − edges = 2. You can use Euler's formula to test possible frame patterns for a geodesic dome that completely encloses a volume. The pieces of tubing are the edges of the frame, and the points where they join are the vertices of the frame. Suppose you try to build a frame using n hexagons in the pattern shown in black in the diagram. Each hexagon contributes 6 vertices, 6 edges, and 1 face to the total, but 3 hexagon vertices combine at each vertex of the frame, and 2 hexagon edges combine at each edge of the frame.

2. The frame must have $6n \div 3$ or ____ vertices, ____ faces, and ____ edges.

3. What is vertices + faces − edges for this frame? Does Euler's formula hold? Can the frame be built?

4. Show that if you change exactly 12 hexagons to pentagons, the frame will satisfy Euler's formula. (*Hint:* 12 hexagon vertices are lost, but since they combine in threes, the frame loses 4 vertices. How many edges does it lose?)

5. One of these pentagons appears in the photograph above. Can you find it?

10-5 Areas and Volumes of Similar Solids

One of the best-known attractions in The Hague, the Netherlands, is a unique miniature city, Madurodam, consisting of five acres of carefully crafted reproductions done on a scale of 1:25. Everything in this model city works, including the two-mile railway network, the canal locks, the harbor fireboats, and the nearly 50,000 tiny lights that come on at dusk. In this section you will learn about the relationship between scale factors of *similar solids* and their areas and volumes.

Similar solids are solids that have the same shape but not necessarily the same size. It's easy to see that all spheres are similar. To decide whether two other solids are similar, determine whether bases are similar and corresponding lengths are proportional.

Right cylinders

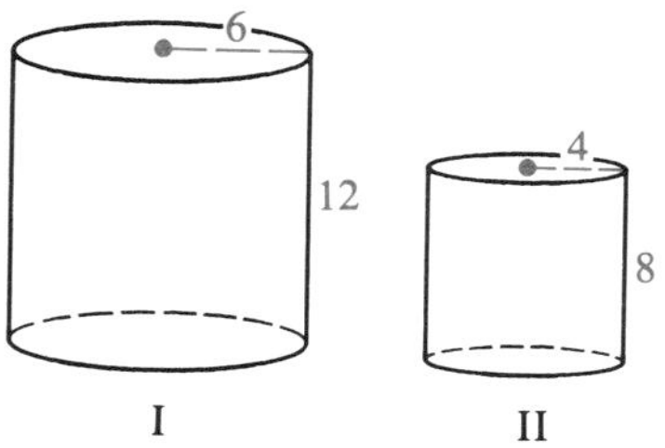

Similar because $\frac{6}{4} = \frac{12}{8}$

Scale factor: $\frac{3}{2}$

Regular square pyramids

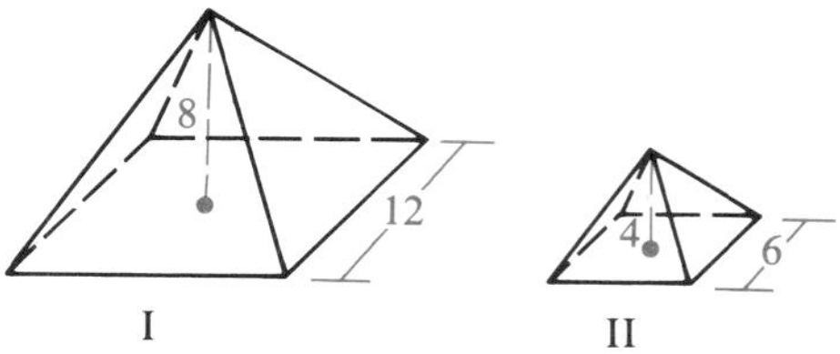

Similar because $\frac{12}{6} = \frac{8}{4}$

Scale factor: $\frac{2}{1}$

The table below shows the ratios of the perimeters, areas, and volumes for both pairs of similar solids shown on page 452. What do you notice about the ratios for the cylinders? for the pyramids?

	Cylinders I and II	**Pyramids I and II**
Scale factor	$\frac{3}{2}$	$\frac{2}{1}$
$\frac{\text{Base perimeter (I)}}{\text{Base perimeter (II)}}$	$\frac{2\pi \cdot 6}{2\pi \cdot 4} = \frac{6}{4}$, or $\frac{3}{2}$	$\frac{4 \cdot 12}{4 \cdot 6} = \frac{12}{6}$, or $\frac{2}{1}$
$\frac{\text{L.A. (I)}}{\text{L.A. (II)}}$	$\frac{2\pi \cdot 6 \cdot 12}{2\pi \cdot 4 \cdot 8} = \frac{9}{4}$, or $\frac{3^2}{2^2}$	$\frac{\frac{1}{2} \cdot 48 \cdot 10}{\frac{1}{2} \cdot 24 \cdot 5} = \frac{4}{1}$, or $\frac{2^2}{1^2}$
$\frac{\text{Volume (I)}}{\text{Volume (II)}}$	$\frac{\pi \cdot 6^2 \cdot 12}{\pi \cdot 4^2 \cdot 8} = \frac{27}{8}$, or $\frac{3^3}{2^3}$	$\frac{\frac{1}{3} \cdot 12^2 \cdot 8}{\frac{1}{3} \cdot 6^2 \cdot 4} = \frac{8}{1}$, or $\frac{2^3}{1^3}$

The results shown in the table above are generalized in the following theorem. (See Exercises 20–25 for proofs.)

Theorem 10-3

If the scale factor of two similar solids is $a:b$, then:
(1) The ratio of corresponding perimeters is $a:b$.
(2) The ratios of the base areas, of the lateral areas, and of the total areas are $a^2:b^2$.
(3) The ratio of the volumes is $a^3:b^3$.

Example For the similar solids shown, find the ratios of (a) base perimeters, (b) lateral areas, and (c) volumes.

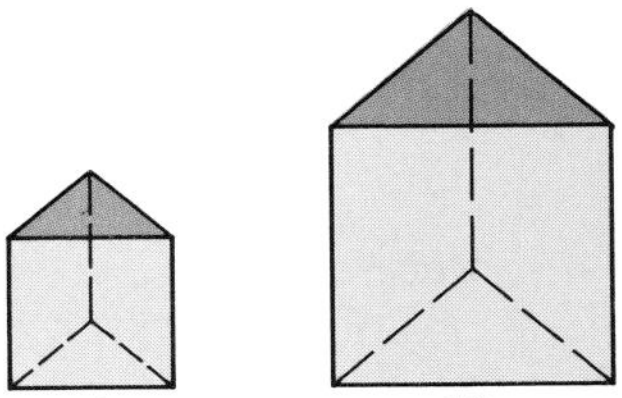

Solution The scale factor is 6:10, or 3:5.
a. Ratio of base perimeters = 3:5.
b. Ratio of lateral areas = $3^2:5^2 = 9:25$
c. Ratio of volumes = $3^3:5^3 = 27:125$

Theorem 10-3 above is the three-dimensional counterpart of Theorem 9-6 on page 410. (Take a minute to compare these theorems.) There is a similar relationship between the two cases shown below.

In two dimensions:
If $\overline{XY} \parallel \overline{AB}$, then $\triangle VXY \sim \triangle VAB$.

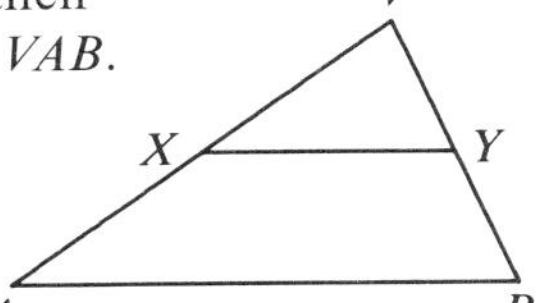

In three dimensions:
If plane $XYZ \parallel$ plane ABC, then $V\text{-}XYZ \sim V\text{-}ABC$.

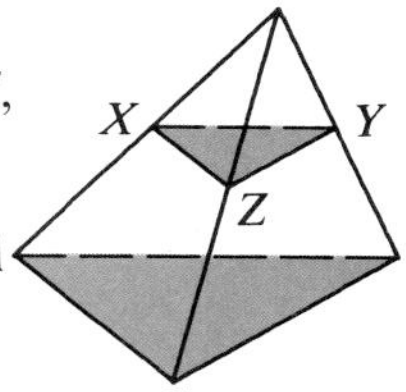

Classroom Exercises

Tell whether the solids in each pair are similar.

1.

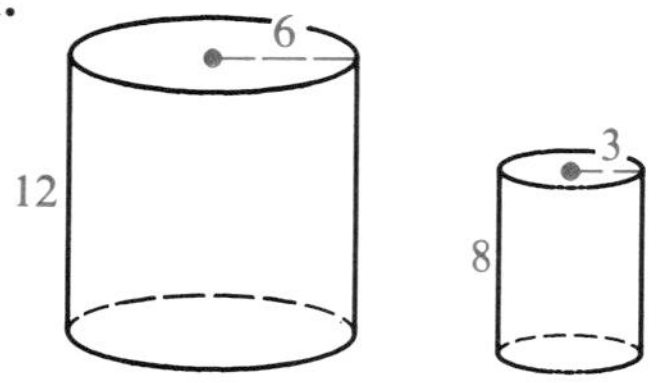

Right cylinders

2.

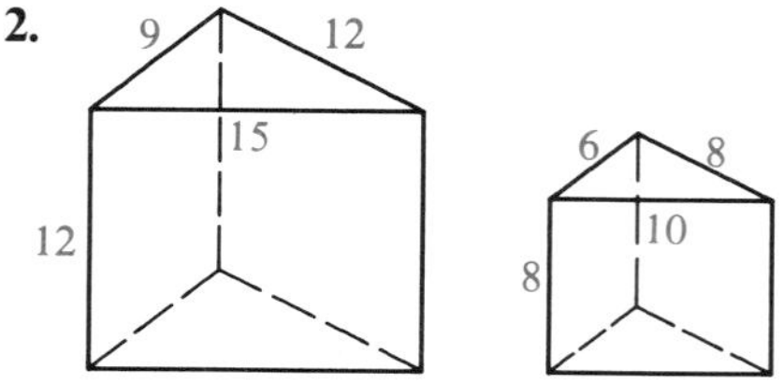

Right prisms

3. a. What is the ratio of the lateral areas of the prisms in Exercise 2?
b. What is the ratio of the total areas of the prisms?
c. What is the ratio of the volumes of the prisms?

4. Two spheres have diameters 24 and 36.
a. What is the ratio of the areas?
b. What is the ratio of the volumes?

5. Two spheres have volumes 2π and 16π. Find the ratios of the following:
a. volumes **b.** radii **c.** areas

Complete the table below, which refers to two similar pyramids.

	6.	**7.**	**8.**	**9.**	**10.**	**11.**
Scale factor	3:4	5:7	?	?	?	?
Ratio of base perimeters	?	?	2:1	?	?	?
Ratio of slant heights	?	?	?	4:9	?	?
Ratio of lateral areas	?	?	?	?	4:9	?
Ratio of total areas	?	?	?	?	?	?
Ratio of volumes	?	?	?	?	?	8:125

12. Plane PQR is parallel to the base of the pyramid and bisects the altitude. Find the following ratios.
a. The perimeter of $\triangle PQR$ to the perimeter of $\triangle ABC$
b. The lateral area of the top part of the pyramid to the lateral area of the whole pyramid
c. The lateral area of the top part of the pyramid to the lateral area of the bottom part
d. The volume of the top part of the pyramid to the volume of the bottom part

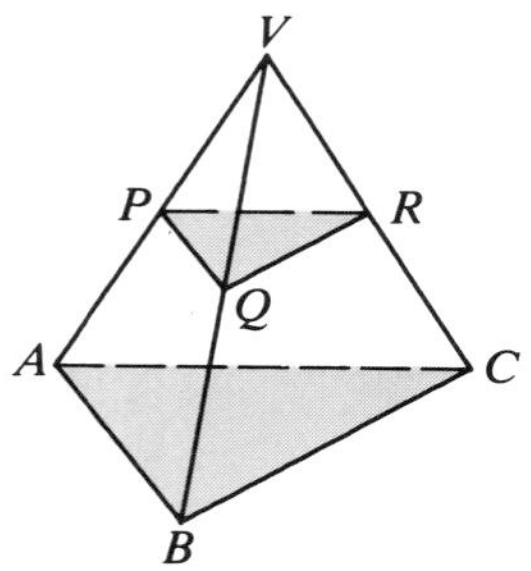

13. Find the ratios in Exercise 12 (a)–(d) if the height of the top pyramid is 3 and the height of the whole pyramid is 5.

Written Exercises

A **1.** Two cones have radii 6 and 9. The heights are 10 and 15. Are the cones similar?

2. The heights of two right prisms are 18 and 30. The bases are squares with sides 8 and 15. Are the prisms similar?

3. Two similar cylinders have radii 3 and 4. Find the ratios of the following:
a. heights **b.** base circumferences **c.** lateral areas **d.** volumes

4. Two similar pyramids have heights 12 and 18. Find the ratios of the following:
a. base areas **b.** lateral areas **c.** total areas **d.** volumes

5. Assume that the Earth and the moon are smooth spheres with diameters 12,800 km and 3,200 km, respectively. Find the ratios of the following:
a. lengths of their equators **b.** areas **c.** volumes

6. The package of a model airplane kit states that the scale is 1:200. Compare the amounts of paint required to cover the model and the actual airplane. (Assume the paint on the model is as thick as that on the actual plane.)

7. The scale for a certain model freight train is 1:48. If the model hopper car (usually used for carrying coal) will hold 90 in.^{3} of coal, what is the capacity of the actual hopper car?

8. Two similar cylinders have lateral areas 81π and 144π. Find the ratios of the following:
a. heights **b.** total areas **c.** volumes

9. Two similar cones have volumes 8π and 27π. Find the ratios of the following:
a. radii **b.** slant heights **c.** lateral areas

10. Two similar pyramids have volumes 3 and 375. Find the ratios of the following:
a. heights **b.** base areas **c.** total areas

B **11.** A certain kind of string is sold in a ball 6 cm in diameter and in a ball 12 cm in diameter. The smaller ball costs $1.00 and the larger one costs $6.50. Which is the better buy?

12. Two balls made of the same metal have radii 6 cm and 10 cm. If the smaller ball weighs 4 kg, how much does the larger ball weigh?

13. A snow man is made using three balls of snow with diameters 30 cm, 40 cm, and 50 cm. If the head weighs roughly 6 kg, find the total weight of the snow man. (Ignore arms, eyes, nose, and mouth.)

14. Construction engineers know that the strength of a column is proportional to the area of its cross section. Suppose that the larger of two similar columns is three times as high as the smaller column.

a. The larger column is __?__ times as strong as the smaller column.

b. The larger column is __?__ times as heavy as the smaller column.

c. Which can support more, *per pound of column material,* the larger or the smaller column?

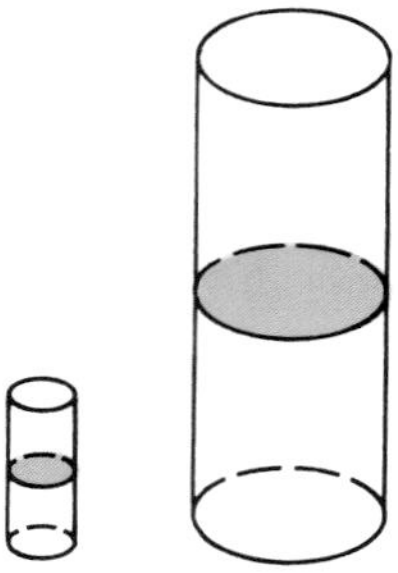

15. Two similar pyramids have lateral areas 8 and 18. If the volume of the smaller pyramid is 32, what is the volume of the larger?

16. Two similar cones have volumes 12π and 96π. If the lateral area of the smaller cone is 15π, what is the lateral area of the larger?

17. A plane parallel to the base of a cone divides the cone into two pieces. Find the ratios of the following:

a. The areas of the shaded circles

b. The lateral area of the top piece to that of the whole cone

c. The lateral area of the top piece to that of the bottom piece

d. The volume of the top piece to that of the whole cone

e. The volume of the top piece to that of the bottom piece

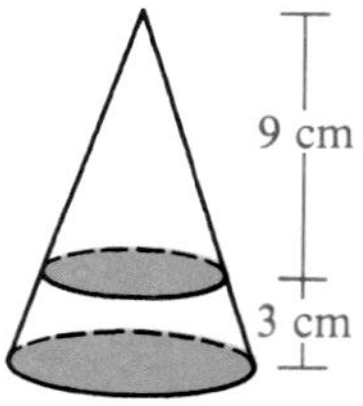

18. Redraw the figure for Exercise 17, changing the 9 cm and 3 cm dimensions to 10 cm and 4 cm. Then find the five ratios described in Exercise 17.

19. A pyramid with height 15 cm is separated into two pieces by a plane parallel to the base and 6 cm above it. What are the volumes of these two pieces if the volume of the original pyramid is 250 cm^3?

The purpose of Exercises 20–25 is to prove Theorem 10-3 for some similar solids.

20. Two spheres have radii a and b. Prove that the ratio of the areas is $a^2:b^2$.

21. Two spheres have radii a and b. Prove that the ratio of the volumes is $a^3:b^3$.

22. Two similar cones have radii r_1 and r_2 and heights h_1 and h_2. Prove that the ratio of the volumes is $h_1^3:h_2^3$.

23. Two similar cones have radii r_1 and r_2 and lateral heights l_1 and l_2. Prove that the ratio of the lateral areas is $r_1^2:r_2^2$.

24. The bases of two similar prisms are regular pentagons with base edges e_1 and e_2 and base areas B_1 and B_2. The heights are h_1 and h_2. Prove that the ratio of the lateral areas is $e_1^2:e_2^2$.

25. Refer to Exercise 24 and prove that the ratio of the volumes of the prisms is $e_1^3:e_2^3$.

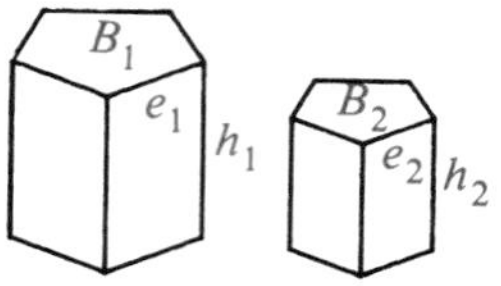

C **26.** The purpose of this exercise is to prove that if plane $XYZ \parallel$ plane ABC, then $V\text{-}XYZ \sim V\text{-}ABC$. To do this, suppose that $VA = k \cdot VX$ and show that every edge of $V\text{-}ABC$ is k times as long as the corresponding edge of $V\text{-}XYZ$. (*Hint:* Use Theorem 2–1.)

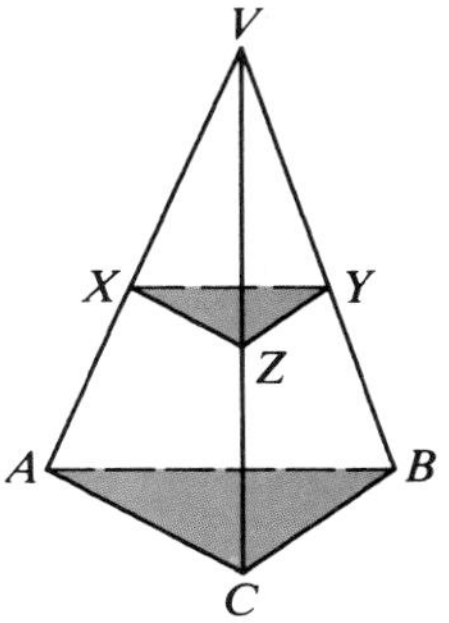

27. A plane parallel to the base of a pyramid separates the pyramid into two pieces with equal volumes. If the height of the pyramid is 12, find the height of the top piece.

CALCULATOR KEY-IN

The Great Pyramid of King Cheops was a true pyramid, not a step pyramid. The pyramid has a square base with sides 755 feet long. The original height was 481 feet, but the top part of the pyramid, which was 31 feet in height, has been destroyed. Approximately what percent of the original volume remains?

Self-Test 2

1. Find the area and volume of a sphere with diameter 6 cm.
2. The volume of a sphere is $\frac{32}{3}\pi$. Find the area.
3. The students of a school decide to bury a time capsule consisting of a cylinder capped by two hemispheres. Find the volume of the time capsule shown.

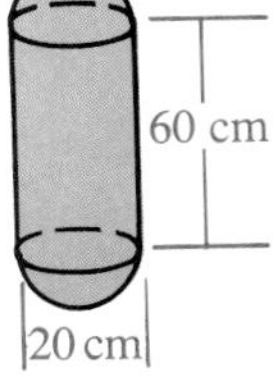

4. Find the area of the circle formed when a plane passes 12 cm from the center of a sphere with radius 13 cm.
5. One regular triangular pyramid has base edge 8 and height 6. A similar pyramid has height 4.
   a. Find the base edge of the smaller pyramid.
   b. Find the ratio of the total areas.
6. The base areas of two similar prisms are 32 and 200, respectively.
   a. Find the ratio of their heights.
   b. Find the ratio of their volumes.

CALCULATOR KEY-IN

The diagrams show two rectangles inscribed in an isosceles triangle with legs 5 and base 6. There are many more such rectangles. The question is, which one has the greatest area?

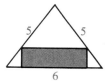

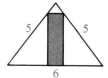

To solve the problem, let *CDEF* represent any rectangle inscribed in isosceles $\triangle ABV$ with legs 5 and base 6. If we let $OD = x$ and $ED = y$, the area of the rectangle is $2xy$. Our goal is to express this area in terms of x alone. Then we can find out how the area changes as x changes.

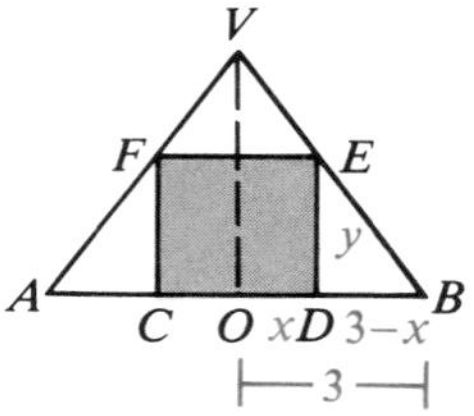

1. In right $\triangle VOB$, $OB = 3$ and $VB = 5$. Thus $VO = 4$.
2. $\triangle EDB \sim \triangle VOB$ (Why?)
3. $\frac{ED}{VO} = \frac{DB}{OB}$ (Why?)
4. $\frac{y}{4} = \frac{3 - x}{3}$ (By substitution in Step 3)
5. $y = \frac{4}{3}(3 - x)$ (Multiplication Postulate)
6. Area of rectangle: $A = 2xy = 2x \cdot \frac{4}{3}(3 - x) = \frac{8x(3 - x)}{3}$

We can use the formula found in Step 6 and a calculator to find the area for many values of x. It is easiest to calculate $3 - x$ first, then multiply by x, then multiply by 8, and finally divide by 3.

x	**Area**
0	0
0.25	1.83333
0.5	3.33333
0.75	4.5
1	5.33333
1.25	5.83333
1.5	6
1.75	5.83333
2	5.33333
2.25	4.5
2.50	3.33333
2.75	1.83333
3	0

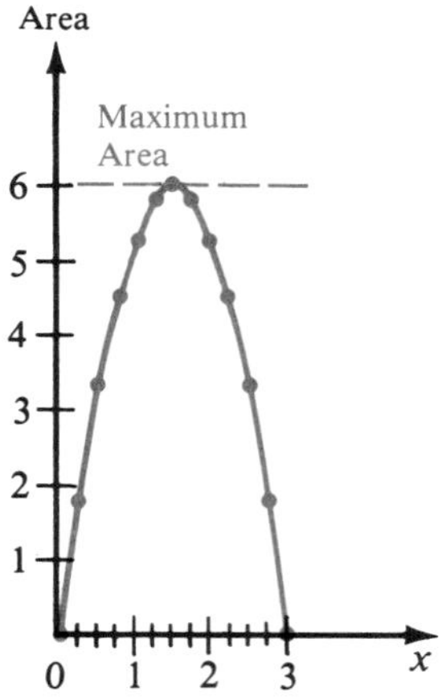

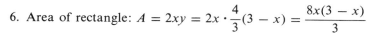

The table was used to make a graph showing how the area varies with x. Both the table and the graph suggest that the greatest area, 6, occurs when $x = 1.5$.

Exercises

Suppose the original triangle had sides 5, 5, and 8 instead of 5, 5, and 6.

1. Draw a diagram like the third diagram on page 458 and show that $A = \dfrac{3x(4 - x)}{2}$.
2. Find the value of x for which the greatest area occurs.

COMPUTER KEY-IN

The diagrams at the left below show two cylinders inscribed in a cone with diameter 6 and lateral height 5. There are many more such cylinders. The question is, Which one has the greatest volume?

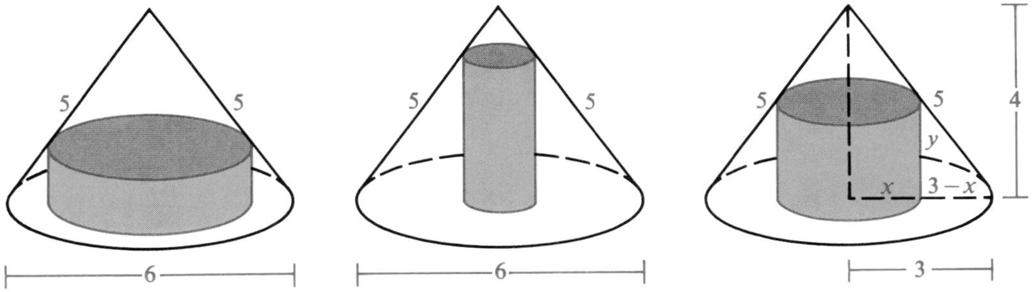

The diagram at the right above shows a typical inscribed cylinder. By using similar triangles, we can write the proportion $\dfrac{y}{4} = \dfrac{3 - x}{3}$. Thus $y = \dfrac{4}{3}(3 - x)$. The volume of the cylinder is as follows:

$$V = \pi x^2 y = \pi x^2 \cdot \frac{4}{3}(3 - x) \approx \frac{4}{3}(3.14159)x^2(3 - x)$$

The program in BASIC below will evaluate V for various values of x.

```
10  PRINT "X", "VOLUME"
20  FOR X=0 TO 3 STEP 0.25
30  LET V=4/3 * 3.14159 * X↑2 * (3-X)
40  PRINT X, V
50  NEXT X
60  END
```

Exercises

1. If your computer uses a language other than BASIC, write a similar program in that language to evaluate V for various values of x.
2. RUN the program. Make a graph that shows how the volume varies with x. For what value of x did you find the greatest volume?

EXTRA

Cavalieri's Principle

Suppose you have a right rectangular prism and divide it horizontally into thin rectangular slices. The base of each rectangular slice has the same area as the base of the prism. If you rearrange the slices, the total volume of the slices does not change.

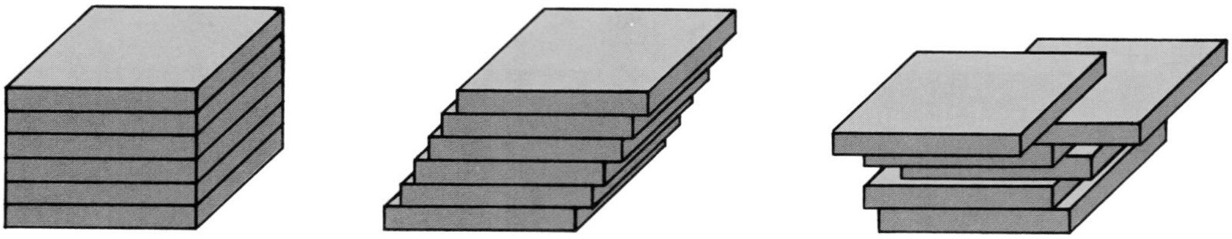

Bonaventura Cavalieri (1598–1647), an Italian mathematician, used this idea to compare the volumes of solids. His conclusion is known as Cavalieri's Principle.

Cavalieri's Principle

If two solids lying between parallel planes have equal heights and all cross-sections at equal distances from their bases have equal areas, then the solids have equal volumes.

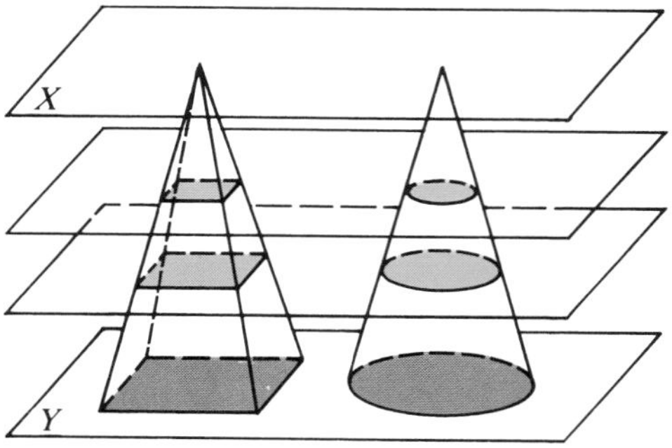

Using Cavalieri's Principle you can find the volume of an oblique prism. Consider a right triangular prism and an oblique prism that have the same base and height.

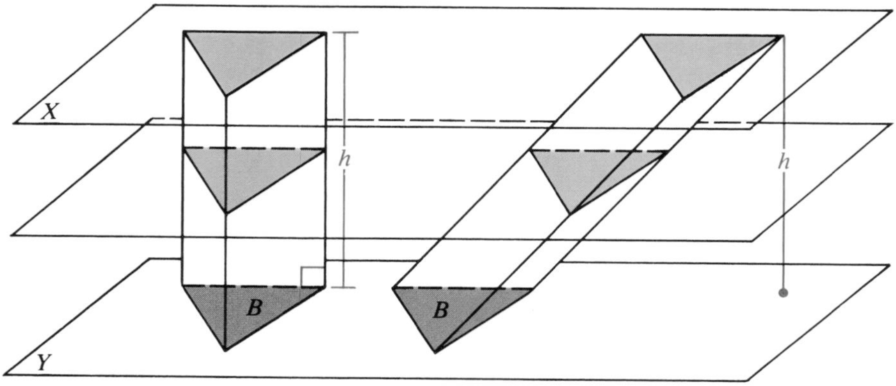

The volume of the right prism is $V = Bh$ (Theorem 10-2). Every cross-section of each prism has the same area as that prism's base. Since the base areas are equal, the corresponding cross-sections of the two prisms have equal areas. Therefore by Cavalieri's Principle, the volume of the oblique prism also is $V = Bh$.

Similarly, you can show that the volume formulas given for a regular pyramid, right cylinder, and right cone hold true for the corresponding oblique solids.

$$V = Bh \quad \text{for } any \text{ prism or cylinder}$$

$$V = \frac{1}{3}Bh \text{ for } any \text{ pyramid or cone}$$

Exercises

Find the volume of the solid shown with the given altitude.

1.

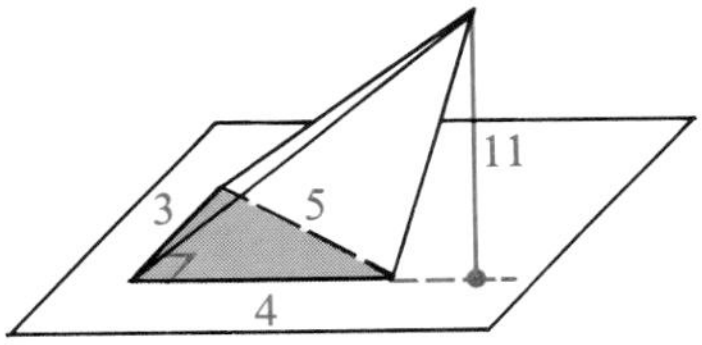

2.

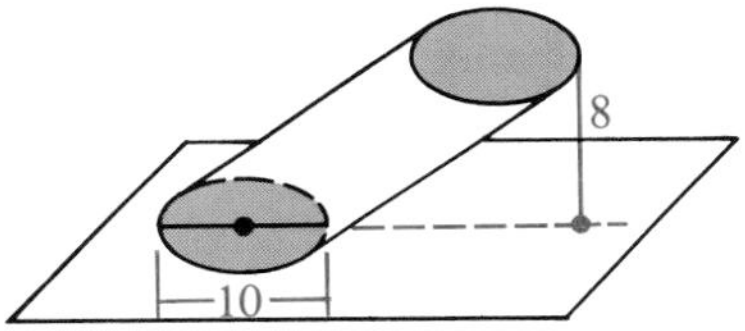

3. Find the volume of an oblique cone with radius 4 and height 3.5.
4. The oblique square prism shown below has base edge 3. A lateral edge that is 15 makes a 60° angle with the plane containing the base. Find the volume.

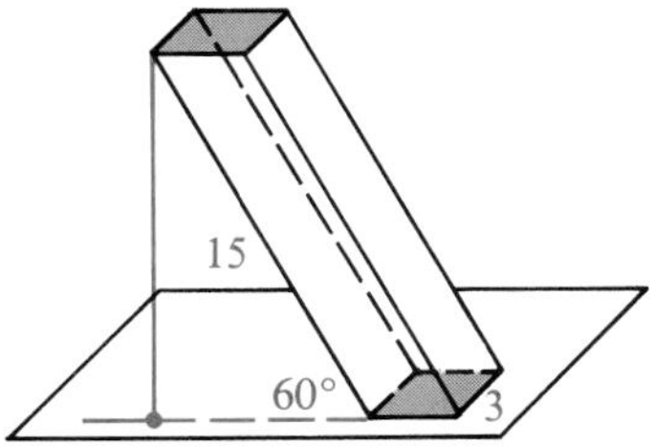

Ex. 4

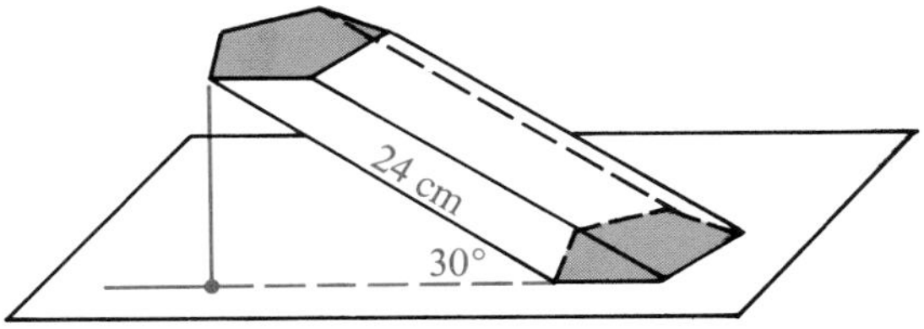

Ex. 5

5. The volume of the oblique pentagonal prism shown above is 96 cm^3. A lateral edge that is 24 cm makes a 30° angle with the plane containing the base. Find the area of the base.
6. Refer to the derivation of the formula for the volume of a sphere given on pages 444–445. How does Cavalieri's Principle justify the statement that the volume of the sphere is equal to the difference between the volumes of the cylinder and the double cone?

READING GEOMETRY

Three-dimensional Figures

When a three-dimensional figure is pictured on paper, you may find it difficult to picture in your mind segments not shown in the drawing. The use of actual three-dimensional models often makes the relationships clearer. Look at the rectangular solid with dimensions l, w, and h shown on page 426. Can you picture a diagonal of the solid? Try using your geometry book as a model.

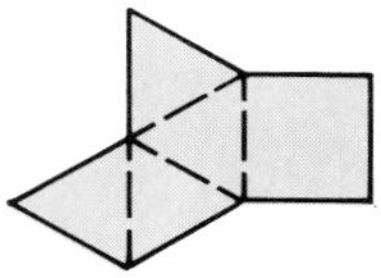

You probably have models at hand for many situations. To picture perpendicular or parallel planes, consider the floor, ceiling, and walls of a room. You can use pencils for lines, a ball for a sphere, and a can for a cylinder. If a set of building blocks is available, you may find cubes, cones, pyramids, and rectangular prisms as well. If not, you can make your own models.

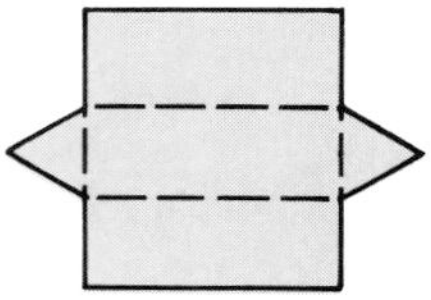

Patterns for three figures are shown. Can you tell what they are? To build them, make large copies of the patterns. Trace them onto cardboard or stiff paper. Cut along the solid lines, fold along the dashed lines, and tape the edges together.

If you want to make a pattern for a figure, think about the number of faces, their shapes, and how the edges are related. Try a triangular pyramid.

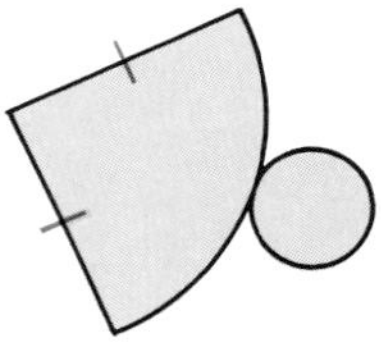

Chapter Summary

1. You should know the following formulas for areas and volumes.

Solid	L.A.	T.A.	V
Right prism	L.A. $= ph$	T.A. = L.A. $+ 2B$	$V = Bh$
Regular pyramid	L.A. $= \frac{1}{2}pl$	T.A. = L.A. $+ B$	$V = \frac{1}{3}Bh$
Right cylinder	L.A. $= 2\pi rh$	T.A. = L.A. $+ 2B$	$V = \pi r^2 h$
Right cone	L.A. $= \pi rl$	T.A. = L.A. $+ B$	$V = \frac{1}{3}\pi r^2 h$

Sphere: $A = 4\pi r^2$ and $V = \frac{4}{3}\pi r^3$

2. If the scale factor of two similar solids is $a:b$, then:
   (a) The ratio of corresponding perimeters is $a:b$.
   (b) The ratio of corresponding areas is $a^2:b^2$.
   (c) The ratio of the volumes is $a^3:b^3$.

Chapter Review

1. In a right prism, each __?__ is also an altitude. 10-1
2. Find the lateral area of a right octagonal prism with height 12 and base edge 7.
3. Find the total area and volume of a rectangular solid with dimensions 8, 6, and 5.
4. A right square prism has base edge 9 and volume 891. Find the total area.
5. Find the volume of a regular triangular pyramid with base edge 8 and height 10. 10-2
6. A regular pentagonal pyramid has base edge 6 and lateral edge 5. Find the slant height and the lateral area.

A regular square pyramid has base edge 30 and total area 1920.

7. Find the area of the base, the lateral area, and the slant height.
8. Find the height and the volume of the pyramid.
9. Find the lateral area and the total area of a cylinder with radius 4 and height 3. 10-3
10. Find the lateral area, total area, and volume of a cone with radius 6 cm and slant height 10 cm.
11. A cone has volume 8π and height 6. Find its slant height.
12. The radius of a cylinder is doubled and its height is halved. How does the volume change?
13. A sphere has radius 7. Use $\pi \approx \frac{22}{7}$ to find the approximate area of the sphere. 10-4
14. Find, in terms of π, the volume of a sphere with diameter 12 ft.
15. A plane intersects a sphere with radius 29, forming a circle with radius 21. Find the distance from the center of the sphere to the plane.
16. Find the volume of a sphere with area 484π cm^2.

Plane $RST \parallel$ plane XYZ and $VS : VY = 1:3$. 10-5

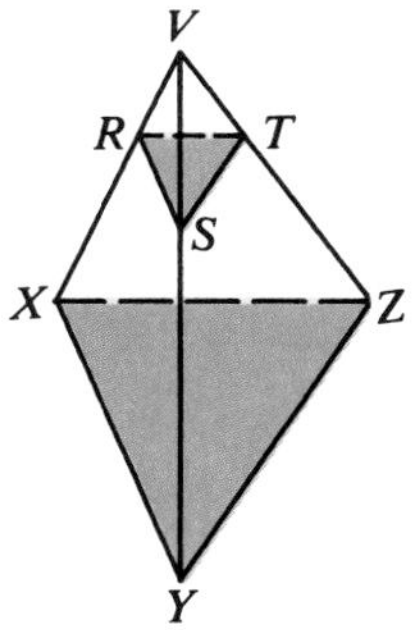

17. $\dfrac{\text{perimeter of } \triangle RST}{\text{perimeter of } \triangle XYZ} = $ __?__

18. $\dfrac{\text{total area of small pyramid}}{\text{total area of large pyramid}} = $ __?__

19. $\dfrac{\text{volume of small pyramid}}{\text{volume of bottom part}} = $ __?__

20. Two similar cylinders have lateral areas 48π and 27π. Find the ratio of their volumes.

Chapter Test

1. Find the volume and the total area of a cube with edge $2k$.
2. A pyramid has a rectangular base 10 cm long and 6 cm wide. The pyramid's height is 4 cm. Find the volume.
3. A cone has radius 8 and height 6. Find the volume.
4. Find the lateral area and the total area of the cone in Exercise 3.
5. A right triangular prism has height 20 and base edges 5, 12, and 13. Find the total area.
6. Find the volume of the prism in Exercise 5.
7. A cylinder has radius 6 cm and height 4 cm. Find the lateral area.
8. Find the volume of the cylinder in Exercise 7.
9. A regular square pyramid has lateral area 60 m^2 and base edge 6 m. Find the volume.
10. A sphere has radius 6 cm. Find the area and the volume.
11. Two cones have radii 12 cm and 18 cm and have slant heights 18 cm and 24 cm. Are the cones similar?
12. A regular pyramid has height 18 and total area 648. A similar pyramid has height 6. Find the total area of the smaller pyramid.
13. The volumes of two similar rectangular solids are 1000 cm^3 and 64 cm^3. What is the ratio of their lateral areas?
14. A cone and a cylinder each have radius 3 and height 4. Find the ratio of their volumes and of their lateral areas.
15. Find the volume of a sphere with area 9π.
16. A cylinder with radius 7 has total area 168π cm^2. Find its height.

Preparing for College Entrance Exams

Strategy for Success

Questions on college entrance exams often require knowledge of areas and volumes. Be sure that you know all the important formulas developed in Chapters 9 and 10. To save time in doing unnecessary calculations, be sure to read the directions to find out whether answers may be expressed in terms of π.

Indicate the best answer by writing the appropriate letter.

1. A cone has volume 320π and height 15. Find the total area.
(A) 200π **(B)** 368π **(C)** 264π **(D)** 136π **(E)** 320π

2. Two equilateral triangles have perimeters 6 and $9\sqrt{3}$. The ratio of their areas is:
(A) $2:3\sqrt{3}$ **(B)** $2\sqrt{3}:9$ **(C)** $4:27$ **(D)** $4:9$ **(E)** $8:81\sqrt{3}$

3. A sphere has volume 288π. Its diameter is:
(A) $12\sqrt{6}$ **(B)** $6\sqrt{2}$ **(C)** 12 **(D)** $12\sqrt{2}$ **(E)** 6

4. *RSTW* is a rhombus with $m\angle R = 60$ and $RS = 4$. If X is the midpoint of $\overline{RS}$, find the area of trapezoid *SXWT*.
(A) 12 **(B)** 16 **(C)** $6\sqrt{3}$ **(D)** $8\sqrt{3}$ **(E)** $16 - 2\sqrt{2}$

5. If *ABCD* is a square and $AE = y$, the area of *ABCDE* is
(A) $\frac{5}{4}y^2$ **(B)** $\frac{5}{2}y^2$ **(C)** $3y^2$
(D) $(4 + \frac{1}{2}\sqrt{3})y^2$ **(E)** $(\frac{1}{2} + \sqrt{2})y^2$

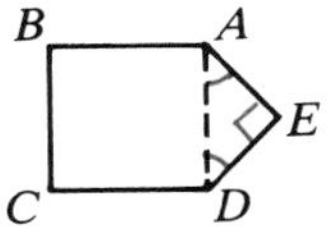

Compare the quantity in Column A with that in Column B. Select:
(A) if the quantity in Column A is greater;
(B) if the quantity in Column B is greater;
(C) if the two quantities are equal;
(D) if the relationship cannot be determined from the information given.

Column A	**Column B**
6. volume of square pyramid	volume of square prism

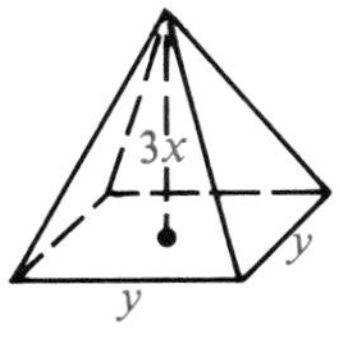

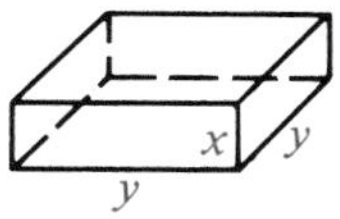

7. area of triangle	area of sector

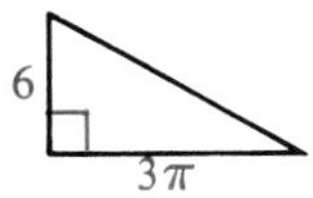

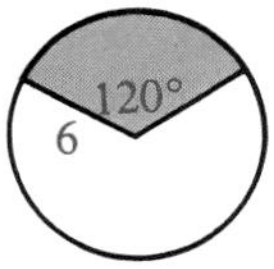

Cumulative Review: Chapters 1–10

Classify each statement as true or false.

A **1.** If point A lies on $\overrightarrow{BC}$, but not on $\overline{BC}$, then B is between A and C.

2. The statement "If $ac = bc$, then $a = b$" is true for all real numbers a, b, and c.

3. The conditional "p only if q" is equivalent to "if p, then q."

4. If two parallel lines are cut by transversal t and t is perpendicular to one of the lines, then t must also be perpendicular to the other line.

5. If $\triangle ABC \cong \triangle DEF$ and $\angle A \cong \angle B$, then $\overline{DE} \cong \overline{EF}$.

6. If the vertex angle of an isosceles triangle has measure j, then the measure of a base angle is $180 - 2j$.

7. If the opposite sides of a quadrilateral are congruent and the diagonals are perpendicular, then the quadrilateral must be a square.

8. In $\triangle RST$, if $m\angle R = 48$ and $m\angle S = 68$, then $RT > RS$.

9. If $\triangle GBS \sim \triangle JFK$, then $\frac{JF}{JK} = \frac{GB}{GS}$.

10. The length of the altitude to the hypotenuse of a right triangle is always the geometric mean between the lengths of the legs.

11. A triangle with sides of length $\sqrt{3}$, 2, and $\sqrt{7}$ is a right triangle.

12. If right $\triangle JEH$ has hypotenuse $\overline{JE}$, then $\tan J = \frac{JH}{EH}$.

13. If an angle inscribed in a circle intercepts a major arc, then the measure of the angle must be between 180 and 360.

14. If $JK = 10$, then the locus of points in space that are 4 units from J and 5 units from K is a circle.

15. It is possible to construct an angle of measure 105.

16. The area of a triangle with sides 3, 3, and 2 is $4\sqrt{2}$.

17. When a square is circumscribed about a circle, the ratio of the areas is $4:\pi$.

18. The volume of both a pyramid and a cone can be computed by using the formula $V = \frac{1}{3}Bh$.

19. B and E are the respective midpoints of $\overline{AC}$ and $\overline{AD}$. Given that $AB = 9$, $BE = 6$, and $AE = 8$, find:

a. the perimeter of $\triangle ACD$

b. the ratio of the areas of $\triangle ABE$ and $\triangle ACD$

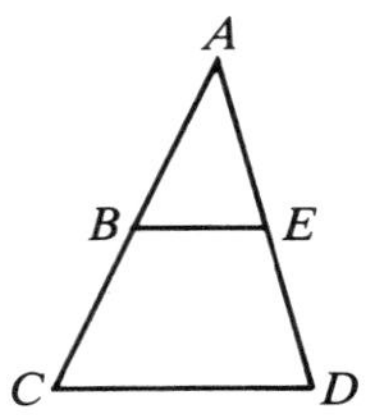

20. An equilateral triangle has sides 4 cm long. Find its altitude and its area.

21. Given: $\overline{WZ} \perp \overline{ZY}$; $\overline{WX} \perp \overline{XY}$; $\overline{WX} \cong \overline{YZ}$
Prove: $\overline{WZ} \parallel \overline{XY}$

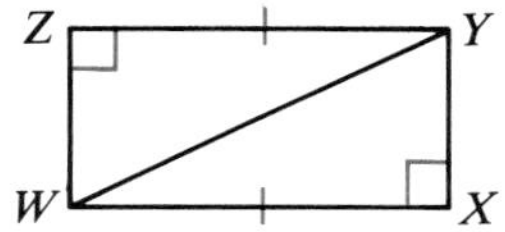

22. A sphere has a diameter of 1.8 cm. Find its surface area to the nearest square centimeter. (Use $\pi \approx 3.14$.)

23. A regular square pyramid has base edge 10 and height 12. Find its total area and volume.

B 24. In $\triangle RST$, $m\angle R = 2x + 10$, $m\angle S = 3x - 10$, and $m\angle T = 4x$.
   a. Find the numerical measure of each angle.
   b. Is $\triangle RST$ a scalene triangle, an isosceles triangle, or a right triangle? Why?

25. In $\square JKLM$, $m\angle J = \frac{3}{2}x$ and $m\angle L = x + 17$.
Find the numerical measure of $\angle K$.

26. Find the value of x in the diagram.

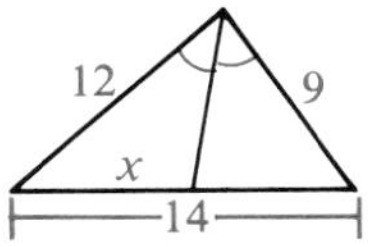

27. A cylinder has a radius equal to its height. The total area of the cylinder is 100π cm^2. Find its volume.

28. $\widehat{AB}$ lies on $\odot O$ with $m\widehat{AB} = 60$. $\odot O$ has radius 8. Find AB.

29. Prove: If the diagonals of a parallelogram are perpendicular, then the parallelogram must be a rhombus.

30. $\triangle ABC$ is an isosceles right triangle with hypotenuse $\overline{AC}$ of length $2\sqrt{2}$. If medians $\overline{AD}$ and $\overline{BE}$ intersect at M, find AD and AM.

31. Draw two segments with lengths y and z. Construct a segment of length t such that $t = \frac{y^2}{z}$.

32. Describe each possibility for the locus of points in space that are equidistant from the sides of a $\triangle ABC$ and 4 cm from A.

33. In $\triangle DEF$, $m\angle F = 42$, $m\angle E = 90$, and $DE = 12$. Find EF to the nearest integer. (Use the table on page 271.)

34. Find the area of a trapezoid with legs 7 and bases 11 and 21.

35. In $\odot O$, $m\widehat{AB} = 90$ and $OA = 6$.
   a. Find the perimeter of sector AOB.
   b. Find the area of the region bounded by $\overline{AB}$ and $\widehat{AB}$.

36. $\overline{AB}$ and $\overline{CD}$ are chords of $\odot P$ intersecting at X. If $AX = 7.5$, $BX = 3.2$, $CD = 11$, and $CX > DX$, find CX.

Satellites have a variety of commercial applications, ranging from communication systems to oil exploration. Satellite tracking systems determine the coordinates of a satellite's position. A portion of the visual display from a tracking system is shown below.

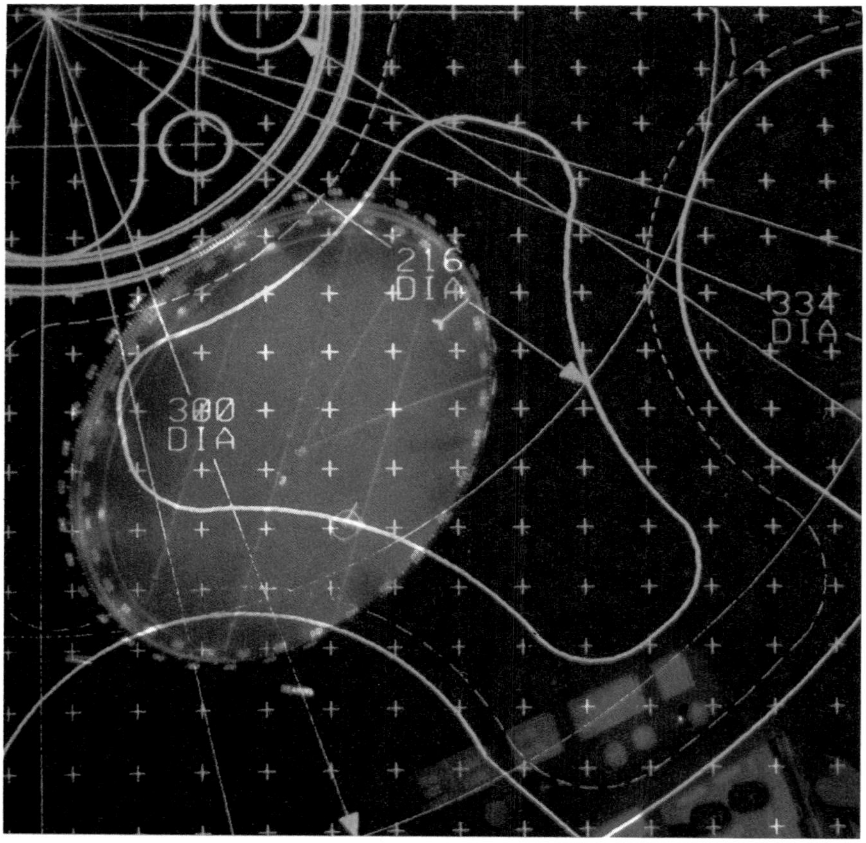

Coordinate Geometry

11

Using the Distance Formula

Objectives

1. Specify points in the coordinate plane by means of their coordinates.
2. State and apply the distance formula.
3. State and apply the general equation of a circle.
4. State and apply the midpoint formula.

11-1 The Distance Formula

Some of the terms you have used in your study of graphs are reviewed below.

Origin: Point O

Axes: x-axis and y-axis

Quadrants: Regions I, II, III, and IV

Coordinate plane: The plane of the x-axis and the y-axis

The arrowhead on each axis shows the positive direction.

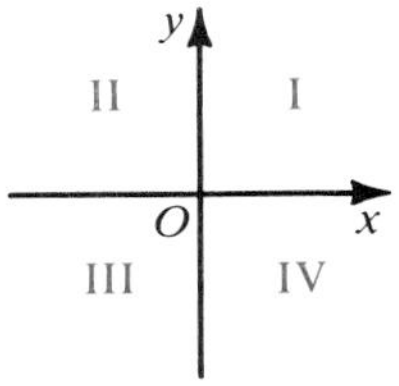

The **x-coordinate** of P is 3.

The **y-coordinate** of P is 2.

The **coordinates** of P are 3 and 2. Point P is the **graph** of the *ordered pair* (3, 2). We often denote this as $P(3, 2)$.

Note that the order of the coordinates has significance in locating points in the coordinate plane. For example, points $(-2, 2)$ and $(2, -2)$ are the points R and W, respectively. The other points shown are:

$O(0, 0)$ $\quad$ $S(-3, 0)$ $\quad$ $T(-2, -3)$ $\quad$ $V(0, -2)$

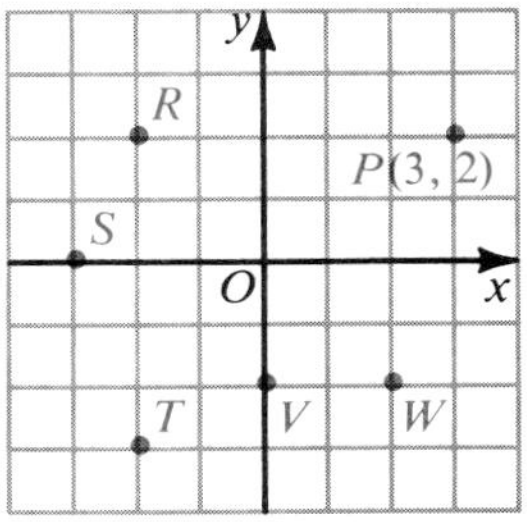

You can easily find the distance between two points that lie on a horizontal line or on a vertical line.

The distance between A and B is 4.
Using the x-coordinates of A and B:
$|3 - (-1)| = 4$, or $|(-1) - 3| = 4$

The distance between C and D is 3.
Using the y-coordinates of C and D:
$|1 - (-2)| = 3$, or $|-2 - 1| = 3$

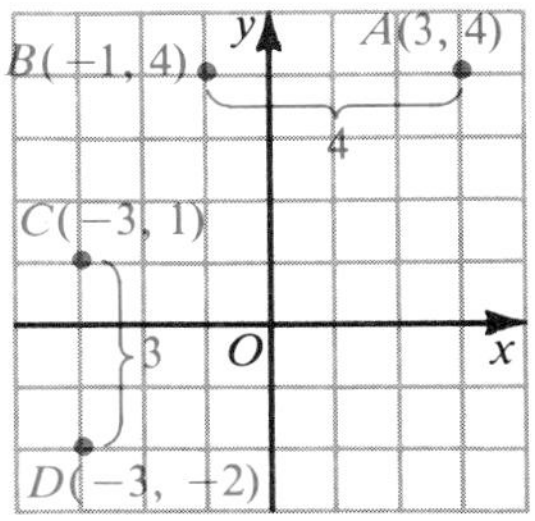

When two points do not lie on a horizontal or vertical line, you can find the distance between the points by using the Pythagorean Theorem.

Example 1 Find the distance between points $A\ (4, -2)$ and $B(1, 2)$.

Solution Draw the horizontal and vertical segments shown. The coordinates of T are $(1, -2)$. Then $AT = 3$, $BT = 4$, $(AB)^2 = 3^2 + 4^2 = 25$, and $AB = 5$.

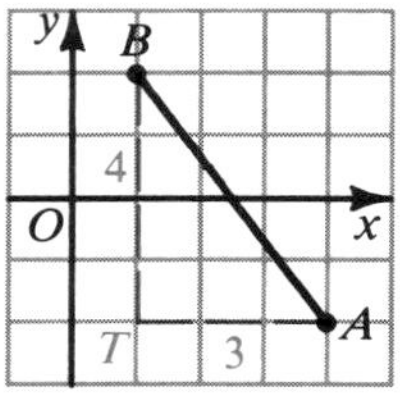

Using a method suggested by Example 1, you can find a formula for the distance between points $P_1(x_1, y_1)$ and $P_2(x_2, y_2)$. First draw a right triangle as shown. The coordinates of T are (x_2, y_1).

$$P_1T = |x_2 - x_1|;\ P_2T = |y_2 - y_1|$$

$$\begin{aligned} d^2 &= (P_1T)^2 + (P_2T)^2 \\ &= |x_2 - x_1|^2 + |y_2 - y_1|^2 \\ &= (x_2 - x_1)^2 + (y_2 - y_1)^2 \\ d &= \sqrt{(x_2 - x_1)^2 + (y_2 - y_1)^2} \end{aligned}$$

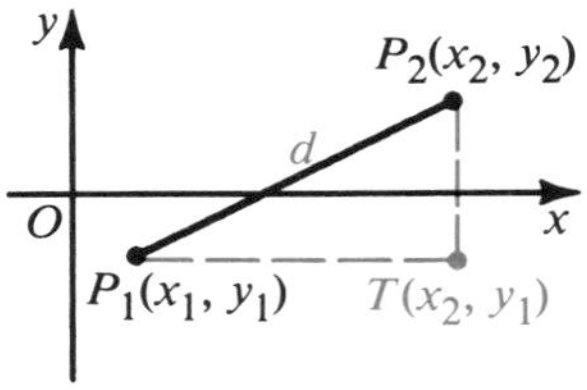

Since d represents distance, d must be nonnegative.

Theorem 11-1 The Distance Formula

The distance d between points (x_1, y_1) and (x_2, y_2) is given by:

$$d = \sqrt{(x_2 - x_1)^2 + (y_2 - y_1)^2}$$

Example 2 Find the distance between points $(-4, 2)$ and $(2, -1)$.

Solution 1 Draw a right triangle. The legs have lengths 6 and 3.

$$d^2 = 6^2 + 3^2 = 36 + 9 = 45$$
$$d = \sqrt{45} = \sqrt{9}\sqrt{5} = 3\sqrt{5}$$

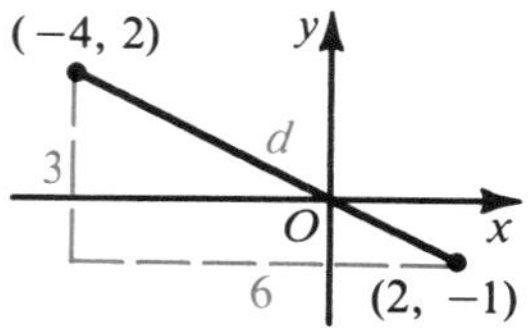

Solution 2 Let (x_1, y_1) be $(-4, 2)$ and (x_2, y_2) be $(2, -1)$.

$$\begin{aligned} \text{Then } d &= \sqrt{(x_2 - x_1)^2 + (y_2 - y_1)^2} \\ &= \sqrt{(2 - (-4))^2 + ((-1) - 2)^2} \\ &= \sqrt{6^2 + (-3)^2} = \sqrt{36 + 9} = \sqrt{45} = 3\sqrt{5} \end{aligned}$$

Classroom Exercises

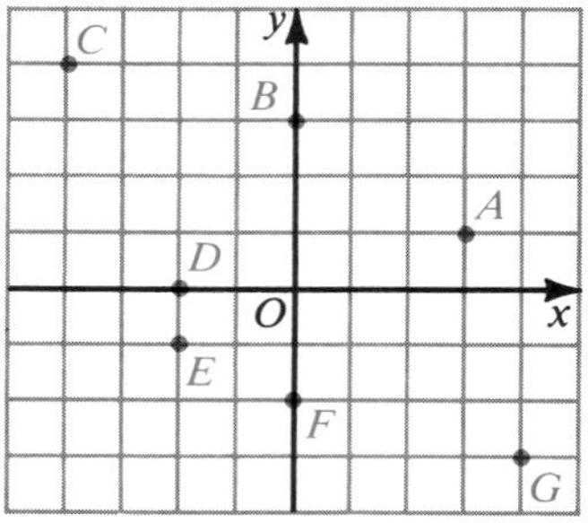

1. State the coordinates of A, B, C, D, E, F, G, and O.
2. What is the x-coordinate of every point that lies on a vertical line through C?
3. Which of the following points lie on a horizontal line through C?

(2, 4)	(2, −4)	(0, 4)
(4, 3)	(15, 4)	(−4, 3)

In Exercises 4-9 state: a. the coordinates of T
b. the lengths of the legs of the right triangle
c. the length of the segment shown

4.

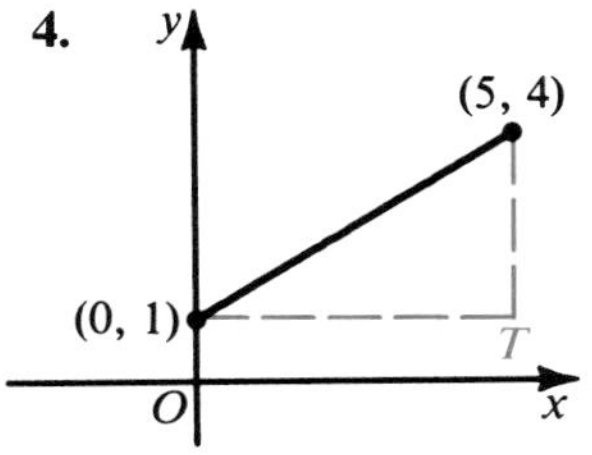

5.

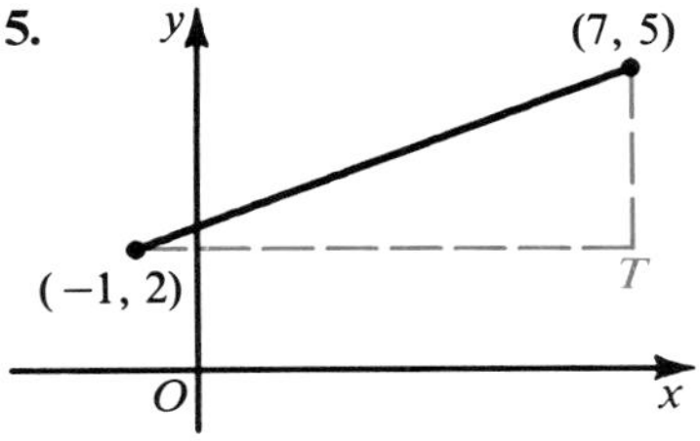

6.

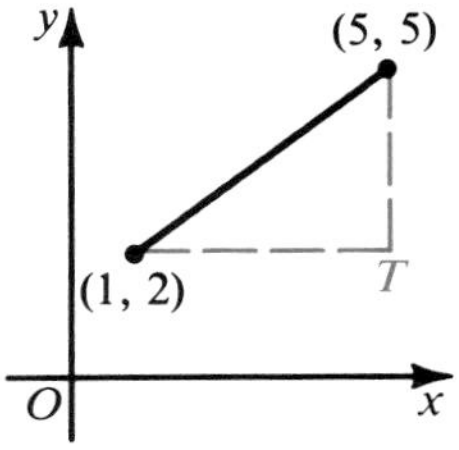

7.

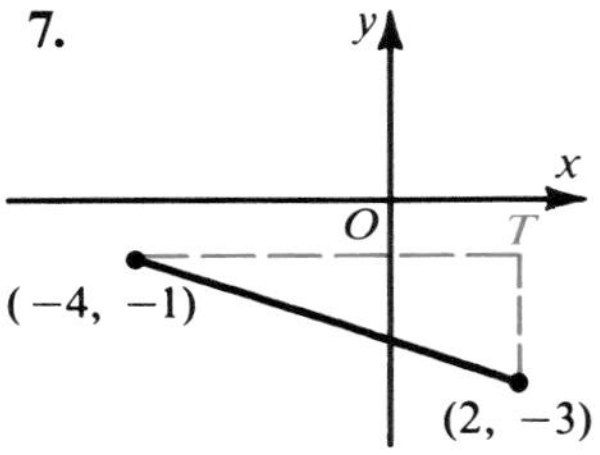

8.

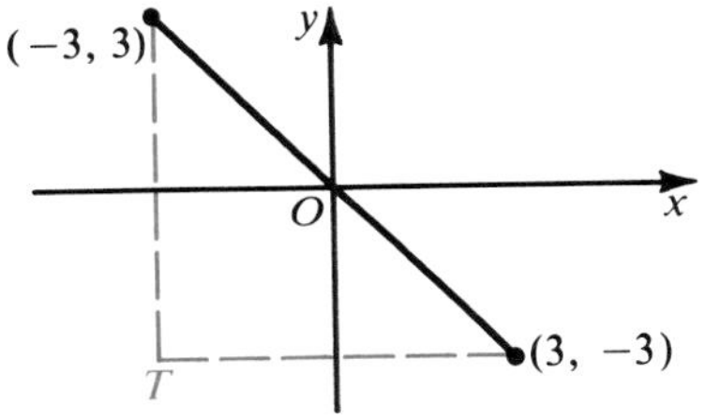

9.

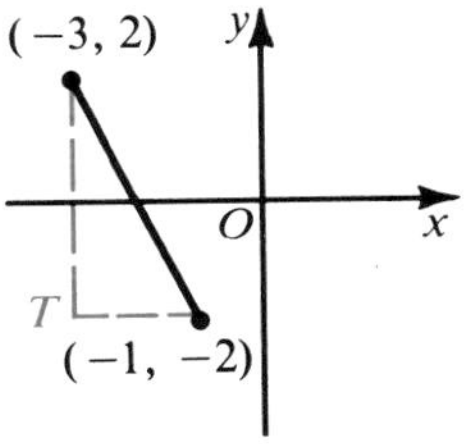

10. a. To use the distance formula to find the distance between points (2, 3) and (4, 7), you can let P_1 be (2, 3) and P_2 be (4, 7). Then:
$(x_2 - x_1)^2 = \underline{\ ?\ }$; $(y_2 - y_1)^2 = \underline{\ ?\ }$; $d = \underline{\ ?\ }$

b. On the other hand, you can let $P_1 = (4, 7)$ and $P_2 = (2, 3)$. Then:
$(x_2 - x_1)^2 = \underline{\ ?\ }$; $(y_2 - y_1)^2 = \underline{\ ?\ }$; $d = \underline{\ ?\ }$

11. Find the distance between the points named. Write all answers in simplest form.
a. (0, 0) and (5, −3) **b.** (3, −2) and (−5, −2) **c.** (4, 4) and (−3, −3)

Written Exercises

Find the distance between the two points. If necessary, you may draw graphs, but you shouldn't need to use the distance formula.

A

1. (0, 2) and (0, −5)
2. (−2, −3) and (−2, 4)
3. (3, 3) and (−2, 3)
4. (3, −4) and (−1, −4)
5. (1, −2) and (5, −2)
6. (0, 0) and (3, 4)

Use the distance formula to find the distance between the two points.

7. (−6, −2) and (−7, −5)
8. (5, 4) and (1, −2)
9. (−1, −1) and (3, 3)
10. (−8, 6) and (0, 0)
11. (3, 2) and (5, −2)
12. (0, 0) and (3, 4)

Find the distance between the points named. Use any method you choose.

13. $(-2, -2)$ and $(5, 7)$

14. $(-4, -1)$ and $(-4, 3)$

15. $(-6, 0)$ and $(0, 8)$

16. $(-2, 3)$ and $(3, -2)$

Write an expression for the distance between the points named.

17. (a, b) and (c, d)

18. $(e, -f)$ and $(-g, -h)$

B **19.** There are twelve points, each with integer coordinates, that are 10 units from the origin. List the points. (*Hint:* Recall the 6, 8, 10 right triangle.)

20. List twelve points, each with integer coordinates, that are 5 units from $(-8, 1)$.

In Exercises 21–32 you will want to find and then compare lengths of segments.

21. Show that the triangle with vertices $A(-3, 4)$, $M(3, 1)$, and $Y(0, -2)$ is isosceles.

22. Quadrilateral $TAUL$ has vertices $T(4, 6)$, $A(6, -4)$, $U(-4, -2)$, and $L(-2, 4)$. Show that the diagonals are congruent.

23. Show that the triangle with vertices $A(3, -1)$, $B(5, 1)$, and $C(-1, 1)$ is a scalene triangle.

24. Find the length of the median of the trapezoid with vertices $B(-2, -4)$, $C(-2, 6)$, $D(3, 1)$, and $E(3, 5)$. Note that the bases are $\overline{BC}$ and $\overline{DE}$.

25. Triangles JAN and RFK have vertices $J(-2, -2)$, $A(4, -2)$, $N(2, 2)$, $R(8, 1)$, $F(8, 4)$, and $K(6, 3)$. Show that $\triangle JAN$ is similar to $\triangle RFK$.

26. Discover and prove something about the quadrilateral with vertices $R(-1, -6)$, $A(1, -3)$, $Y(11, 1)$, and $J(9, -2)$.

27. Find the area of the rectangle with vertices $B(8, 0)$, $T(2, -9)$, $R(-1, -7)$, and $C(5, 2)$.

28. The vertices of $\triangle KAT$ and $\triangle IES$ are $K(3, -1)$, $A(2, 6)$, $T(5, 1)$, $I(-4, 1)$, $E(-3, -6)$, and $S(-6, -1)$. What word best describes the relationship between $\triangle KAT$ and $\triangle IES$?

C **29.** It is known that $\triangle GHM$ is isosceles. G is point $(-2, -3)$; H is point $(-2, 7)$; the x-coordinate of M is 4. Find all five possible values for the y-coordinate of M.

30. Discover and prove two things about the triangle with vertices $K(-3, 4)$, $M(3, 1)$, and $J(-6, -2)$.

31. The point (a, b) is equidistant from $(-2, 5)$, $(8, 5)$, and $(6, 7)$. Find the values of a and b.

32. Use an indirect proof to show that there is no point equidistant from $(2, 2)$, $(-1, 8)$, and $(1, 4)$.

Application

VECTORS

Air shows sometimes include performances of precision flying, as shown in the photograph. The pilots of these jets demonstrate their skill in flying at the same speed in close formation.

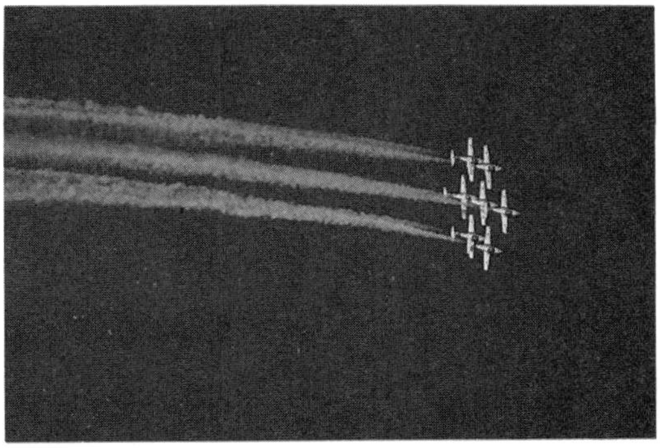

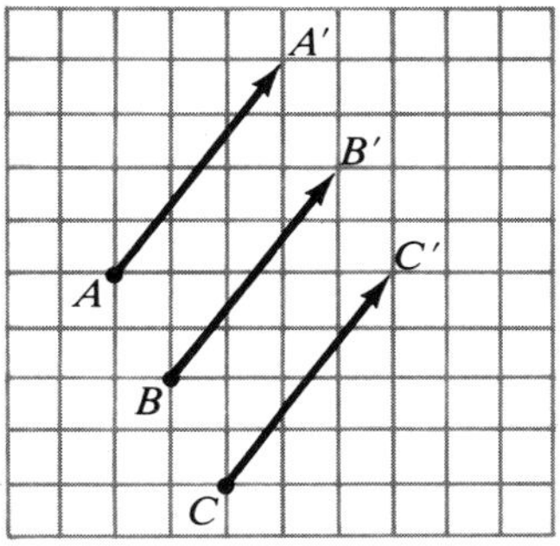

The diagram at the right above illustrates three jets cruising on the same course and at the same speed. In one hour, the jets travel from points A, B, and C to points A', B', and C'. The flight path of the jet traveling from A to A' is represented by the arrow from A to A' and is denoted $\overrightarrow{AA'}$ (not to be confused with ray $\overrightarrow{AA'}$).

The flight path $\overrightarrow{AA'}$ can also be represented by the pair of numbers (3, 4) because the jet has traveled 3 units east and 4 units north. Thus, we can write $\overrightarrow{AA'} = (3, 4)$. Since the other two jets are flying on the same course and at the same speed, we can also write $\overrightarrow{BB'} = (3, 4)$ and $\overrightarrow{CC'} = (3, 4)$. When we recognize the familiar 3–4–5 right triangle, we see that each jet has traveled 5 units.

If positive numbers are used for travel east and north, negative numbers are used for movement west and south. Thus, referring to the diagram, $\overrightarrow{EE'} = (-3, -1)$. This jet has traveled $\sqrt{(-3)^2 + (-1)^2}$ units, or $\sqrt{10}$ units. What ordered pairs describe $\overrightarrow{FF'}$ and $\overrightarrow{GG'}$? How far has each of these jets traveled?

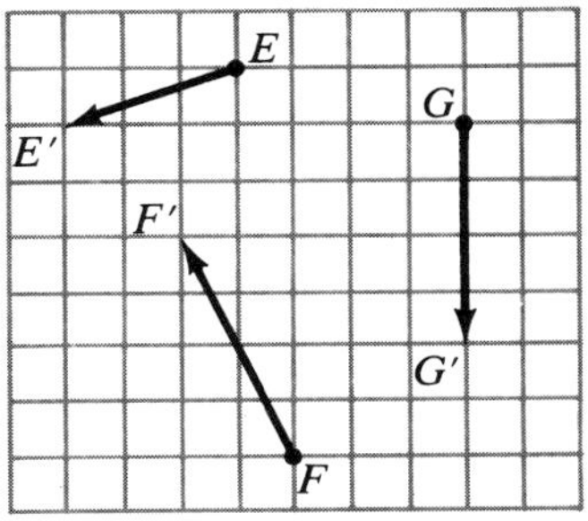

When an ordered pair is used to represent a flight path, or any quantity having both *magnitude* and *direction,* the ordered pair is called a **vector.** Examples of such quantities that can be represented by vectors are velocity, acceleration, and force. A vector may also represent a *translation* of the plane. In Chapter 12 you will learn that translations move all points of the plane the same distance in the same direction; thus, a translation has magnitude and direction.

Vectors can be added by the following simple rule:

$$(a, b) + (c, d) = (a + c, b + d)$$

To see an application of adding vectors, suppose that a jet travels from P to Q and then changes its course to travel from Q to R. A succession of these two flight paths will move the jet to the same position as the single flight path from P to R. We abbreviate this fact by writing

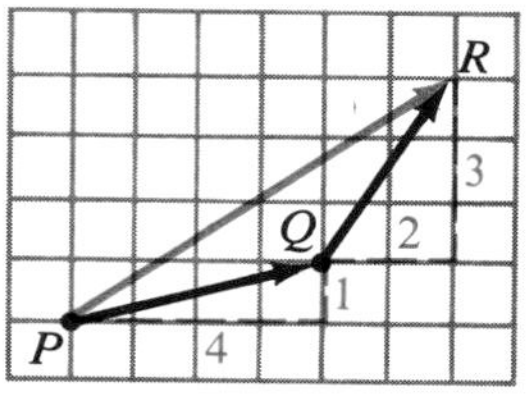

$$\overrightarrow{PQ} + \overrightarrow{QR} = \overrightarrow{PR}$$
$$(4, 1) + (2, 3) = (6, 4)$$

To see another application of adding vectors, suppose a heavy crate is being pulled along the floor by two people using ropes. One person pulls with a force of 40 lb in a northerly direction (represented by $\overrightarrow{KX}$) while the other person pulls with a force of 50 lb in a southeasterly direction (represented by $\overrightarrow{KY}$).

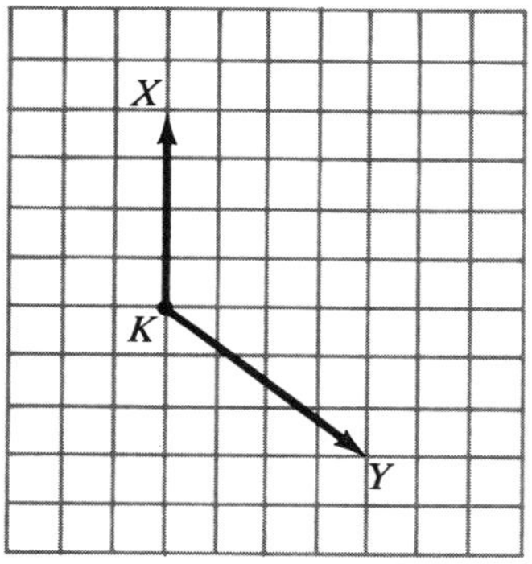

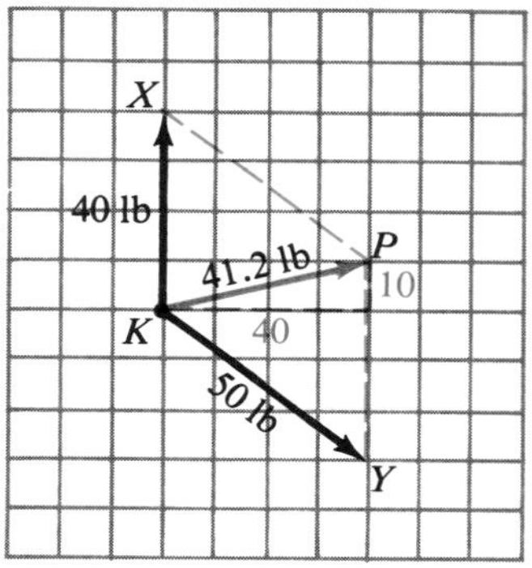

You can think of $\overrightarrow{KX}$ and $\overrightarrow{KY}$ as two forces simultaneously acting on an object at point K. The single force that has the same effect as these two forces can be found by adding vectors:

$$\overrightarrow{KX} + \overrightarrow{KY} = (0, 40) + (40, -30) = (40, 10) = \overrightarrow{KP}.$$

The magnitude of $\overrightarrow{KP}$ is $\sqrt{(40)^2 + (10)^2} \approx 41.2$ lb, and this force is acting in a direction between north and southeast. Notice in the diagram at the right above that this vector sum is represented by the diagonal, $\overrightarrow{KP}$, of the parallelogram with sides $\overrightarrow{KX}$ and $\overrightarrow{KY}$. This rule for the addition of vectors is sometimes called the *parallelogram law.*

Exercises

1. Use a grid and any starting points you choose. Draw arrows to represent the following vectors:

a. $(3, 1)$ **b.** $(4, -4)$ **c.** $(-5, 0)$ **d.** $(-3, -4)$

2. Find the magnitude of each vector in Exercise 1.
3. Find the following vector sums:
   a. (2, 1) + (4, 3) **b.** (−4, 7) + (3, −2) **c.** (4, −9) + (−4, 6)
4. A girl rides her bicycle from *A* to *B* and then from *B* to *C*. Show her trip on a coordinate grid if $\overrightarrow{AB} = (2, 9)$ and $\overrightarrow{BC} = (6, -3)$. How far is *C* from *A*?
5. The vector (−5, 5) represents a force given in pounds. In what compass direction is the force acting? What is the magnitude of the force?
6. Make a drawing showing an object being pulled by the two forces $\overrightarrow{KX} = (-1, 5)$ and $\overrightarrow{KY} = (7, 3)$. What single force has the same effect as the two forces acting together? What is the magnitude of this force?
7. Repeat Exercise 6 for the forces $\overrightarrow{KX} = (2, -3)$ and $\overrightarrow{KY} = (-2, 3)$.
8. If *ABCD* is a parallelogram, does $\overrightarrow{AB} = \overrightarrow{DC}$? Does $\overrightarrow{BC} = \overrightarrow{AD}$?

11-2 Circles

You can use the distance formula to develop an equation of a circle with center at the origin and with radius 6.

Let $P(x, y)$ represent any point on the circle. The distance between $O(0, 0)$ and $P(x, y)$ is 6.

$$\sqrt{(x-0)^2 + (y-0)^2} = 6$$
$$\sqrt{x^2 + y^2} = 6$$
$$x^2 + y^2 = 36$$

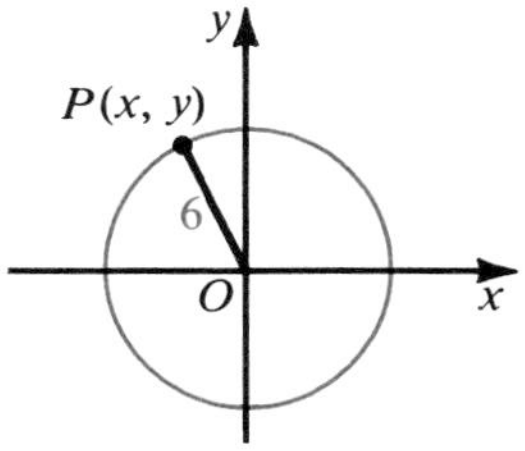

You can proceed in much the same way to develop an equation of a circle with center $C(a, b)$ and radius r.

Let $P(x, y)$ represent any point on the circle. The distance between $C(a, b)$ and $P(x, y)$ is r.

$$\sqrt{(x-a)^2 + (y-b)^2} = r$$
$$(x-a)^2 + (y-b)^2 = r^2$$

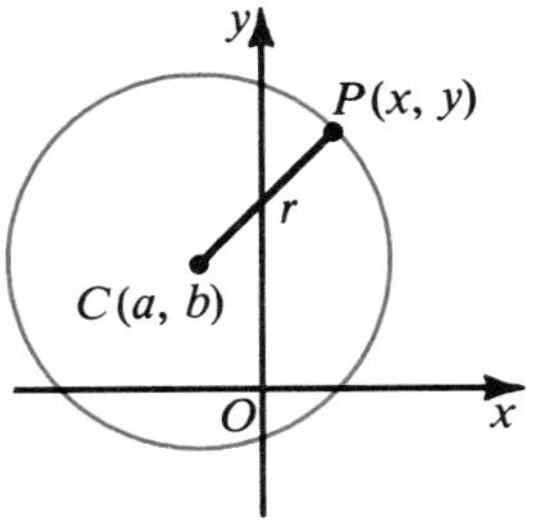

This development proves Theorem 11-2.

Theorem 11-2

An equation of the circle with center (a, b) and radius r is

$$(x-a)^2 + (y-b)^2 = r^2.$$

Example 1 Find an equation of the circle with center $(1, -2)$ and radius 3. Sketch the graph.

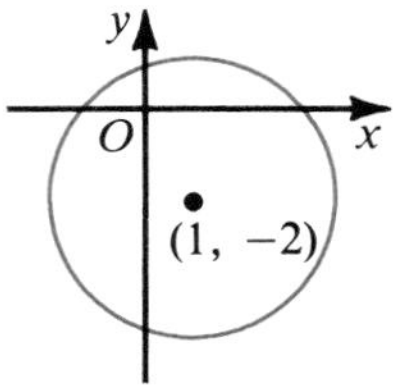

Solution $(x-1)^2 + (y-(-2))^2 = 3^2$, or

$$(x-1)^2 + (y+2)^2 = 9$$

Example 2 Find the center and the radius of the circle with equation

$$(x+4)^2 + (y-2)^2 = 64$$

Solution Rewrite the equation in the form $(x-a)^2 + (y-b)^2 = r^2$.

$$(x-(-4))^2 + (y-2)^2 = 8^2$$

The center is point $(-4, 2)$ and the radius is 8.

Classroom Exercises

1. Find the center and the radius of each circle.

a. $(x-4)^2 + (y-2)^2 = 7^2$

b. $(x-3)^2 + (y-0)^2 = 8^2$

c. $(x-2)^2 + y^2 = 1$

d. $(x+2)^2 + (y-8)^2 = 16$

e. $x^2 + (y+5)^2 = 112$

f. $(x+3)^2 + (y+7)^2 = 14$

2. Find an equation of the circle that has the given center and radius.

a. Center $(2, 5)$; radius 3

b. Center $(-2, 0)$; radius 5

c. Center $(-2, 3)$; radius r

d. Center (j, k); radius n

3. Write an inequality that describes the points (x, y) that are less than 5 units from the origin.

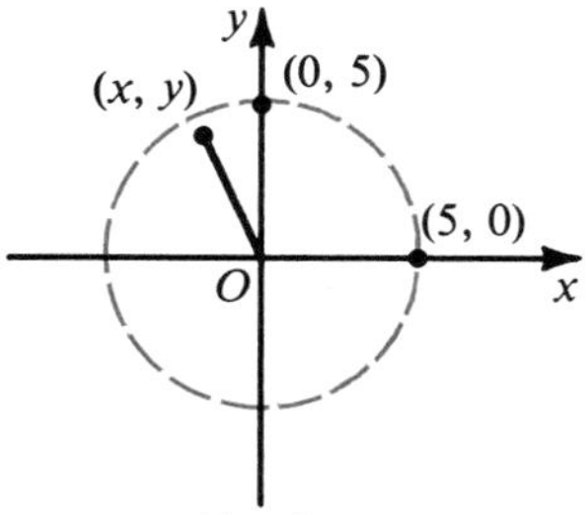

Ex. 3

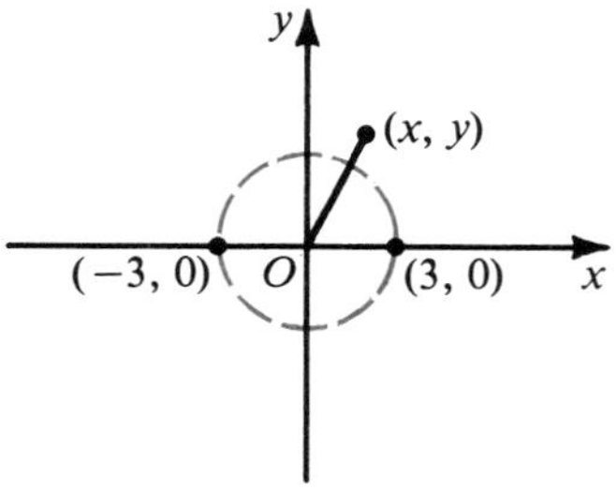

Ex. 4

4. Write an inequality that describes the points (x, y) that are more than 3 units from the origin.

5. Write an inequality that describes the points (x, y) that are less than or equal to 4 units from the point $(3, 3)$.

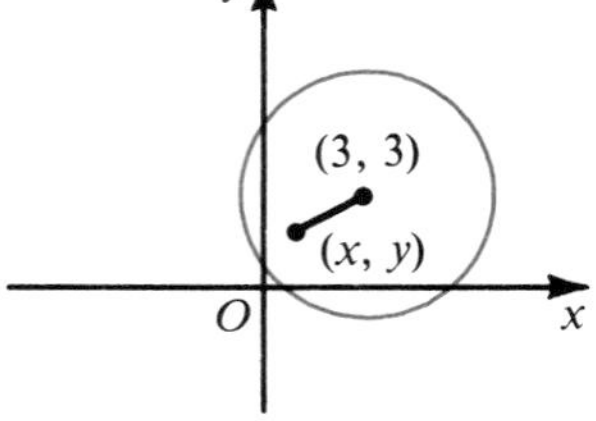

6. Refer to the diagram. Complete the statement to describe the points (x, y) that are more than 2 units but less than 6 units from the point $(-1, 3)$.

$$2^2 < (\underline{\ ?\ })^2 + (\underline{\ ?\ })^2 < 6^2$$

7. Describe the locus of points whose coordinates satisfy both statements: $x^2 + y^2 = 64$ and $x \geq 0$.

8. Describe the locus of points whose coordinates satisfy both statements: $x^2 + y^2 = 9$ and $y \leq 0$.

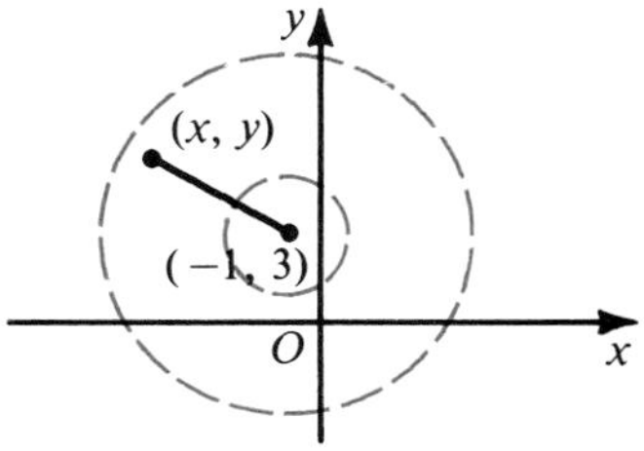

Ex. 6

Written Exercises

Find the center and the radius of each circle.

A 1. $(x - 4)^2 + (y - 3)^2 = 9^2$

2. $(x - 87)^2 + (y - 94)^2 = 6^2$

3. $(x + 3)^2 + y^2 = 49$

4. $(x + 7)^2 + (y + 8)^2 = \frac{36}{25}$

5. $(x - j)^2 + (y + 14)^2 = 17$

6. $(x + a)^2 + (y - b)^2 = c^2$

Write an equation of the circle that has the center and radius named.

	7.	8.	9.	10.	11.	12.
Center	(0, 0)	(3, 0)	(2, −1)	(−2, 5)	(−4, −7)	(p, q)
Radius	2	8	1	$\frac{1}{3}$	g	t

13. Sketch the graph of $(x - 3)^2 + (y + 4)^2 = 36$.

14. Sketch the graph of $(x - 2)^2 + (y - 5)^2 < 9$.

In Exercises 15–20, find an equation of the circle described.

B 15. The circle has center (5, 5) and is tangent to both axes.

16. The circle has center (p, q) and is tangent to the x-axis.

17. The circle has center (0, 6) and passes through point (6, 14).

18. The circle has center (−2, −4) and passes through point (3, 8).

19. The circle has diameter $\overline{PD}$ where P is (0, 0) and D is (0, 4).

20. The circle has diameter $\overline{RS}$ where R is (−3, 2) and S is (3, 2).

21. Two points on the circle $(x - 2)^2 + (y - 4)^2 = 25$ both have y-coordinate 7. What are the x-coordinates of those two points?

22. Find an equation of the locus of the centers of all circles with radius 4 that pass through (−3, 2).

C 23. Given $O(0, 0)$ and $A(0, 6)$, describe the locus of all points K such that $\angle OKA$ is a right angle.

24. Given $O(0, 0)$ and $M(0, 4)$, describe the locus of all points N such that $\triangle OMN$ is a right triangle. (*Hint:* There are three possibilities to consider for the right angle: O, M, and N.)

25. Find the center and the radius of the circle $x^2 + 4x + y^2 - 8y = 16$. (*Hint:* Fill in the blanks in $(x^2 + 4x + \underline{\ ?\ }) + (y^2 - 8y + \underline{\ ?\ }) = 16 + \underline{\ ?\ }$ in such a way that you can proceed to an equation in the form $(x - a)^2 + (y - b)^2 = r^2$.)

26. **a.** Draw the four circles obtained by using all combinations of signs in the equation $(x \pm 4)^2 + (y \pm 4)^2 = 16$.

b. Write an equation of a circle that surrounds the four given circles and is tangent to each of them.

c. Write an equation of another circle that is tangent to each of the four given circles.

COMPUTER KEY-IN

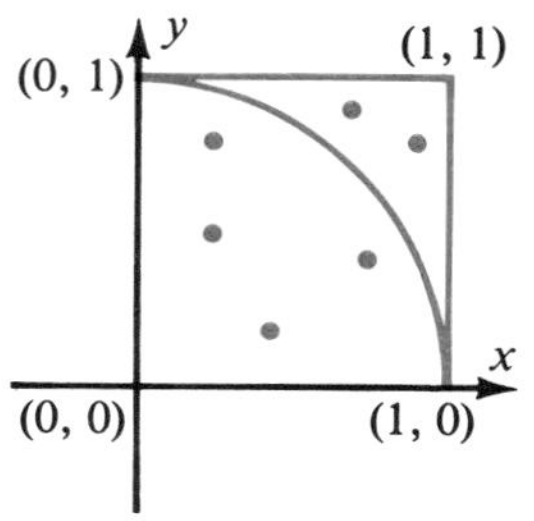

The graph shows a quarter-circle inscribed in a square with area 1. If points are picked at random inside the square, some of them will also be inside the quarter-circle. Let n be the number of points picked inside the square and let q be the number of these points that fall inside the quarter-circle. If many, many points are picked at random inside the square, the following ratios are approximately equal:

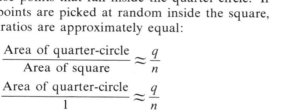

$$\frac{\text{Area of quarter-circle}}{\text{Area of square}} \approx \frac{q}{n}$$

$$\frac{\text{Area of quarter-circle}}{1} \approx \frac{q}{n}$$

$$\text{Area of whole circle} \approx 4 \times \frac{q}{n}$$

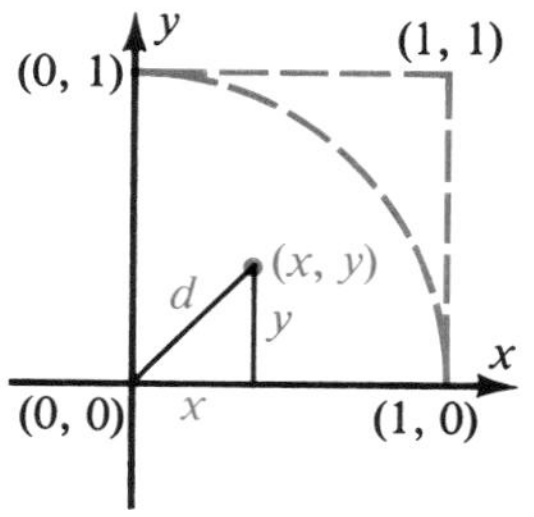

Any point (x, y) in the square region has coordinates such that $0 < x < 1$ and $0 < y < 1$. (Note that this restriction excludes points on the boundaries of the square.) A computer can pick a random point inside the unit square by choosing two random numbers x and y between 0 and 1. We let d be the distance from O to the point (x, y). By the Pythagorean Theorem, $d = \sqrt{x^2 + y^2}$. Do you see that if $d < 1$, the point lies inside the quarter-circle?

Exercises

1. Write a computer program to do all of the following:
   a. Choose n random points (x, y) inside the unit square.
   b. Using the distance formula test each point chosen to see whether it lies inside the quarter-circle.
   c. Count the number of points (q) which *do* lie inside the quarter-circle.
   d. Print out the value of $4 \times \frac{q}{n}$.

2. RUN your program for $n = 100$, $n = 500$, and $n = 1000$.
3. Calculate the area of the circle, using the formula given on page 403. Compare this result with your computer approximations.

11-3 The Midpoint Formula

You can see that the midpoint M of $\overline{AB}$ has coordinate 1, the average of -1 and 3. In Exercise 25, page 27, you learned that if A and B have coordinates a and b, then the midpoint of $\overline{AB}$ has coordinates $\frac{a+b}{2}$.

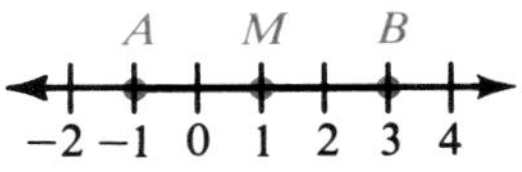

$$\frac{(-1)+3}{2} = 1$$

It is easy to find the midpoint of two points that lie on a horizontal line or a vertical line.

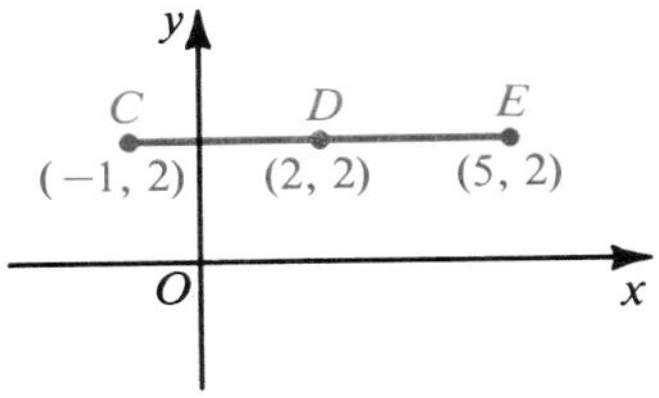

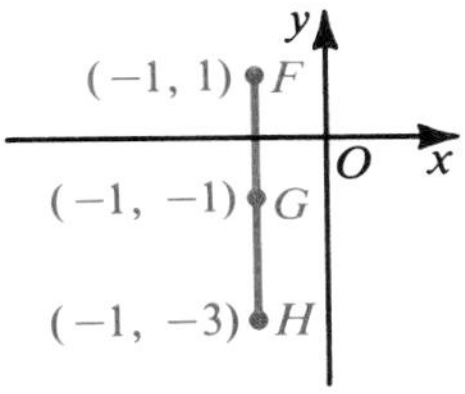

The midpoint of $\overline{CE}$ is $D(2, 2)$; the midpoint of $\overline{FH}$ is $G(-1, -1)$.

You can use this idea twice to find the coordinates of the midpoint of a slanting segment with endpoints $P_1(x_1, y_1)$ and $P_2(x_2, y_2)$. In the diagram below, M is the midpoint of $\overline{P_1P_2}$. Horizontal and vertical segments are drawn as shown. Then T is point (x_2, y_1).

Since $\overline{MR} \parallel \overline{P_2T}$, the corollary to Theorem 4-8, page 165, tells us that R is the midpoint of $\overline{P_1T}$. Thus the x-coordinate of R is $\frac{x_1 + x_2}{2}$, and M also has x-coordinate $\frac{x_1 + x_2}{2}$.

Similarly, $\overline{MS} \parallel \overline{P_1T}$, and S is the midpoint of $\overline{P_2T}$. Both S and M have y-coordinate $\frac{y_1 + y_2}{2}$.

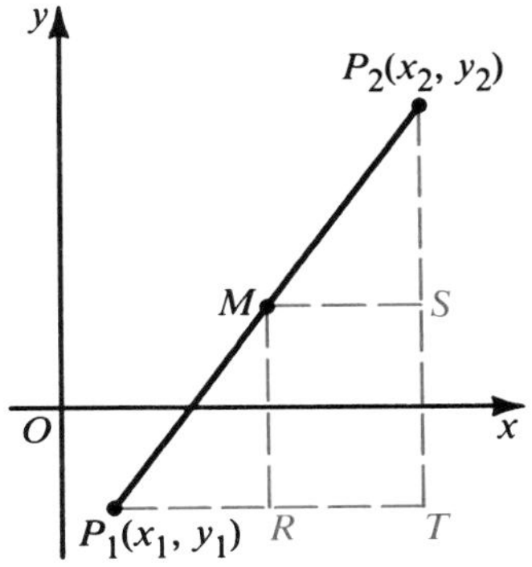

Theorem 11-3 The Midpoint Formula

The midpoint of the segment that joins points (x_1, y_1) and (x_2, y_2) is the point

$$\left(\frac{x_1 + x_2}{2}, \frac{y_1 + y_2}{2}\right).$$

Example 1 Find the midpoint of the line segment that joins $(-11, 3)$ and $(8, -7)$.

Solution The x-coordinate of the midpoint is

$$\frac{x_1 + x_2}{2} = \frac{-11 + 8}{2} = \frac{-3}{2}, \text{ or } -\frac{3}{2}$$

The y-coordinate of the midpoint is

$$\frac{y_1 + y_2}{2} = \frac{3 - 7}{2} = \frac{-4}{2} = -2$$

The midpoint is $(-\frac{3}{2}, -2)$.

Example 2 $M(2, -3)$ is the midpoint of $\overline{AB}$, where A has coordinates $(x_1, y_1) = (-5, 1)$. Find the coordinates of B.

Solution Let the coordinates of B be (x_2, y_2). Since M is the midpoint:

The x-coordinate of M is:

$$\frac{x_1 + x_2}{2} = \frac{-5 + x_2}{2} = 2$$
$$-5 + x_2 = 4$$
$$x_2 = 9$$

The y-coordinate of M is:

$$\frac{y_1 + y_2}{2} = \frac{1 + y_2}{2} = -3$$
$$1 + y_2 = -6$$
$$y_2 = -7$$

Thus B has coordinates $(9, -7)$

Classroom Exercises

Find the coordinates of the midpoint of the segment that joins the given points.

1. $(3, 5)$ and $(7, 5)$
2. $(0, 4)$ and $(4, 3)$
3. $(-2, 2)$ and $(6, 4)$
4. $(-3, 7)$ and $(-7, -5)$
5. $(-1, -3)$ and $(-3, 6)$
6. (a, n) and (d, y)
7. $(t, 2)$ and $(t + 4, -4)$
8. $(2b, 3)$ and $(4, -5)$

9. $M(3, 5)$ is the midpoint of $\overline{P_1P_2}$, where P_1 has coordinates $(0, 1)$. Find the coordinates of P_2.
10. Point $(1, -1)$ is the midpoint of $\overline{AB}$, where A has coordinates $(-1, 3)$. Find the coordinates of B.

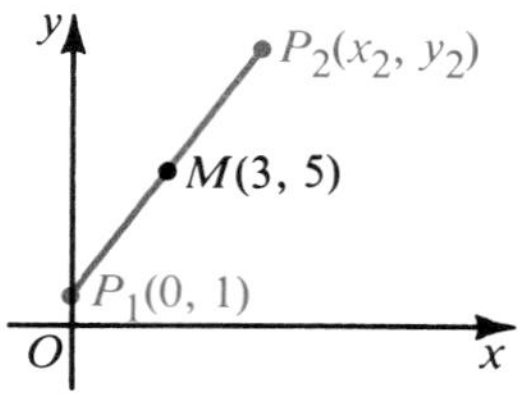

Written Exercises

Find the coordinates of the midpoint of the segment that joins the given points.

A

1. $(0, 2)$ and $(6, 4)$
2. $(-2, 6)$ and $(4, 3)$
3. $(5, 3)$ and $(3, 7)$
4. $(a, 4)$ and $(a + 2, 0)$
5. $(2.3, 3.7)$ and $(1.5, -2.9)$
6. (a, b) and (c, d)

In Exercises 7-10, M is the midpoint of $\overline{AB}$, where the coordinates of A are given. Find the coordinates of B.

7. $A(4, -2)$; $M(4, 4)$ **8.** $A(1, -3)$; $M(5, 1)$

9. $A(5, 2)$; $M(-4, b)$ **10.** $A(r, s)$; $M(t, v)$

B **11.** Using the midpoint formula, find the length of the median of the trapezoid with vertices $C(-4, -3)$, $D(-1, 4)$, $E(4, 4)$, and $F(7, -3)$.

12. Repeat Exercise 11 for the trapezoid with vertices $W(-3, -1)$, $R(-3, 2)$, $S(1, 4)$, and $T(1, -10)$.

13. The vertices of quadrilateral $KOSE$ have the coordinates shown.

a. Show that $OK = SE$ and $OS = KE$.

b. What special kind of quadrilateral is $KOSE$?

c. The midpoint of $\overline{OE}$ has coordinates (_?_, _?_).

d. The midpoint of $\overline{KS}$ has coordinates (_?_, _?_).

e. Note that $\overline{OE}$ and $\overline{KS}$ have the same midpoint. State a theorem, from Chapter 4, that is suggested by this exercise.

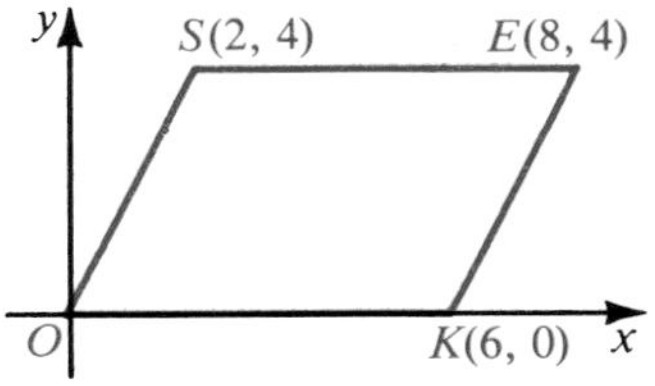

14. The vertices of right triangle OAT have the coordinates shown. M is the midpoint of $\overline{AT}$.

a. M has coordinates (_?_, _?_).

b. Find, and compare, the lengths MA, MT, and MO.

c. State a theorem, from Chapter 4, suggested by this exercise.

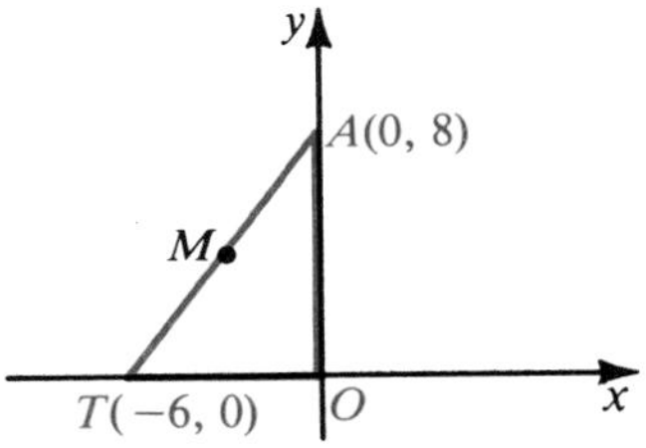

15. The vertices of quadrilateral $JOFL$ have the coordinates shown. P, G, R, and T are the midpoints of the sides of quadrilateral $JOFL$.

a. Find the midpoints of $\overline{RP}$ and $\overline{GT}$.

b. What special kind of quadrilateral is $PGRT$? State a theorem, from Chapter 4, to justify your answer.

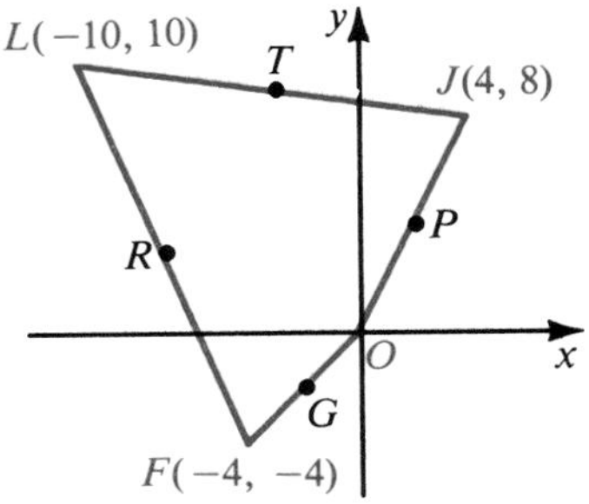

C **16.** According to Theorem 11-3, the midpoint of $\overline{AC}$ has coordinates $\left(\frac{a+c}{2}, \frac{b+d}{2}\right)$. Verify that this point is the midpoint of $\overline{AC}$ by using the distance formula and showing that $AM = MC = \frac{1}{2}AC$.

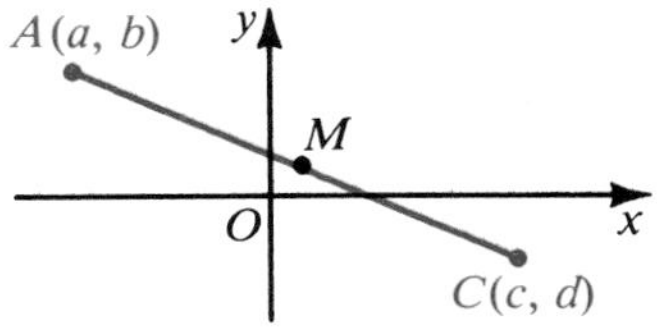

17. Given points $P(2, 1)$ and $D(7, 11)$, find the coordinates of a point T on $\overline{PD}$ such that $\frac{PT}{TD} = \frac{2}{3}$.

Self-Test 1

For each pair of points find (a) the distance between the two points and (b) the midpoint of the segment that joins the two points.

1. (5, 1) and (3, 1)

2. (8, −6) and (0, 0)

3. (−2, 7) and (8, −3)

4. (−3, 2) and (−5, 7)

Write an equation of the circle described.

5. Center at the origin; radius 9

6. Center (−1, 2); radius 5

7. Find the center and the radius of the circle $(x + 2)^2 + (y - 3)^2 = 36$.

Lines

Objectives

1. Find the slope of the line containing two given points.
2. Determine whether two lines are parallel, perpendicular, or neither.
3. Draw the graph of a line specified by a given equation.
4. Write an equation of a line when given either one point and the slope of the line or two points on the line.
5. Given an equation of a line, identify its slope and y-intercept.
6. Determine the intersection of two lines.

11-4 Slope of a Line

The effect of steepness, or slope, must be taken into consideration in a variety of everyday situations. Some examples are the grade of a road, the pitch of a roof, the incline of a wheelchair ramp, and the tilt of an unloading platform, such as the one at a paper mill in Maine shown in the photograph at the right. In this section, you will see that the intuitive idea of steepness is generalized and made precise by the mathematical concept of *slope of a line through two points.*

Refer to the diagrams below. In each of these diagrams, you can think of getting from one point to another by moving horizontally and then vertically. Informally, the slope is the quotient:

$$\frac{\text{change in } y}{\text{change in } x}, \text{ or } \frac{\text{rise}}{\text{run}}$$

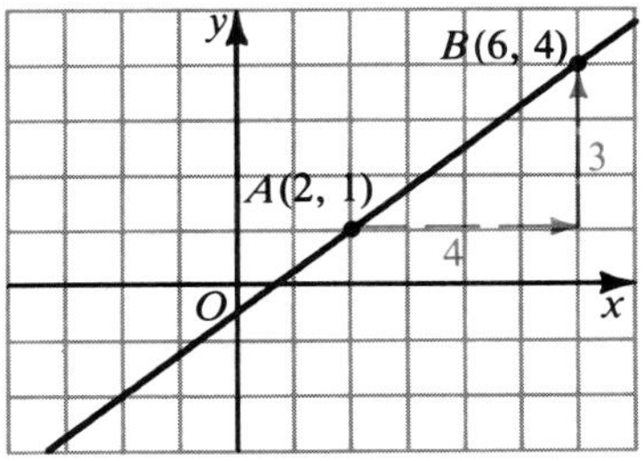

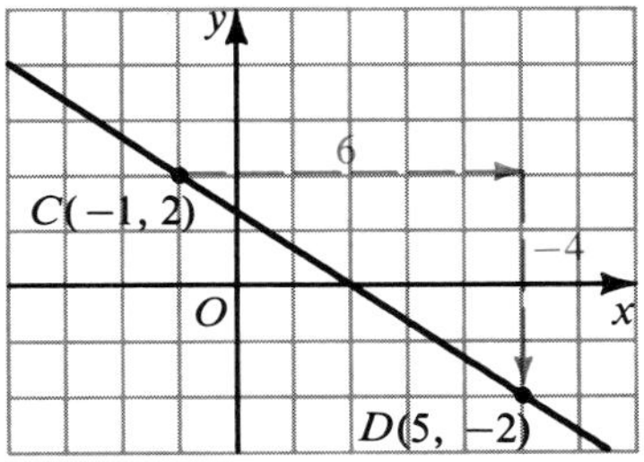

Slope of $\overleftrightarrow{AB} = \dfrac{\text{change in } y}{\text{change in } x} = \dfrac{3}{4}$ Slope of $\overleftrightarrow{CD} = \dfrac{\text{change in } y}{\text{change in } x} = \dfrac{-4}{6} = -\dfrac{2}{3}$

You see that the slope of $\overleftrightarrow{AB}$ is positive and that the slope of $\overleftrightarrow{CD}$ is negative. In general, lines that rise to the right have positive slope. Lines that fall to the right have negative slope.

The *slope m of a line* through $P_1(x_1, y_1)$ and $P_2(x_2, y_2)$, where $x_1 \neq x_2$, is defined as follows:

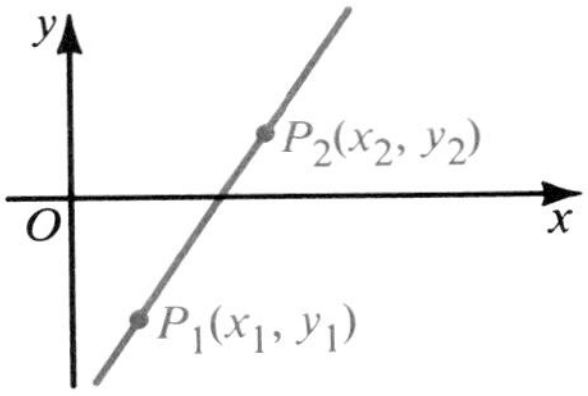

$$m = \frac{y_2 - y_1}{x_2 - x_1}$$

This is also the slope of $\overline{P_1P_2}$, a segment of $\overleftrightarrow{P_1P_2}$.

When you are given several points on a line, you can use any two of them to compute the slope. See Classroom Exercise 6. Furthermore, you can choose either point to be (x_1, y_1), as the two solutions to the following example suggest.

Example Find the slope of the segment that joins points $(-1, 3)$ and $(2, 5)$.

Solution 1 Let $(-1, 3)$ be (x_1, y_1) and $(2, 5)$ be (x_2, y_2).

$$m = \frac{y_2 - y_1}{x_2 - x_1} = \frac{5 - 3}{2 - (-1)} = \frac{2}{3}$$

Solution 2 Let $(2, 5)$ be (x_1, y_1) and $(-1, 3)$ be (x_2, y_2).

$$m = \frac{y_2 - y_1}{x_2 - x_1} = \frac{3 - 5}{-1 - 2} = \frac{-2}{-3} = \frac{2}{3}$$

Consider any horizontal line. Note that y_1 and y_2 must be equal. Thus $m = \dfrac{y_2 - y_1}{x_2 - x_1} = \dfrac{0}{x_2 - x_1} = 0$.

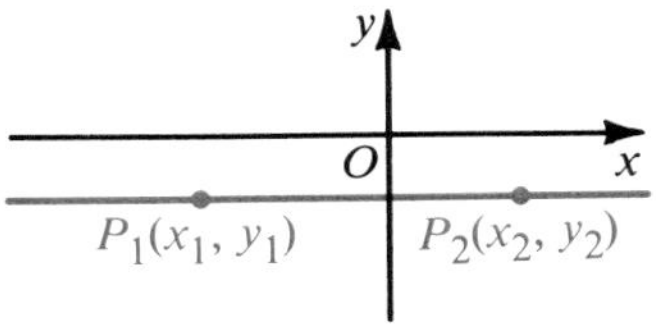

The slope of any horizontal line equals zero.

Consider any vertical line. Note that x_1 and x_2 must be equal. Since the denominator of $\frac{y_2 - y_1}{x_2 - x_1}$ is then 0, the expression for m doesn't mean anything.

Slope is not defined for vertical lines.

Notice that the definition of slope includes the phrase "where $x_1 \neq x_2$"; that is, the definition excludes vertical lines.

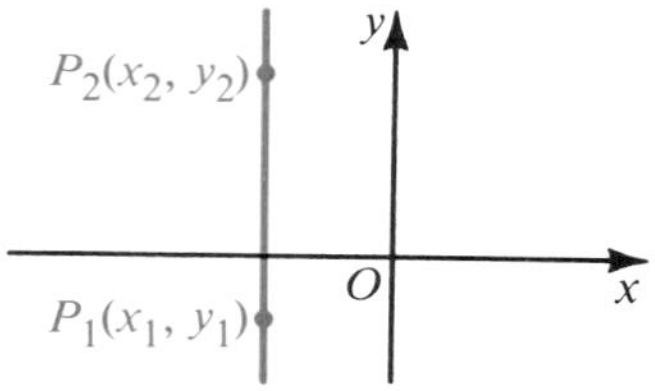

Classroom Exercises

1. Find the slope of the line.

a.
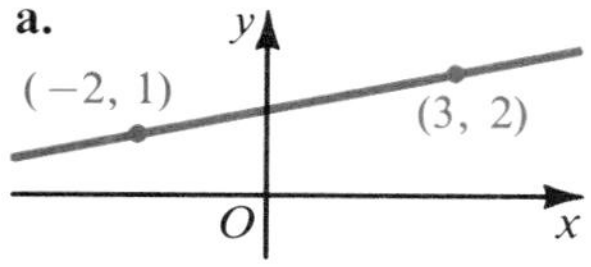

b.
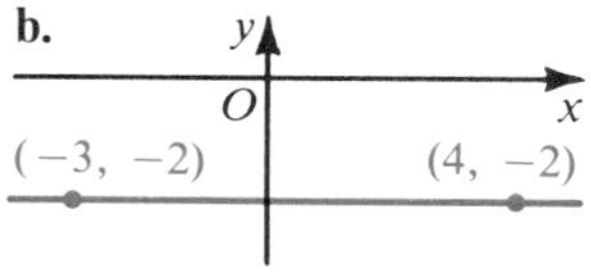

c.
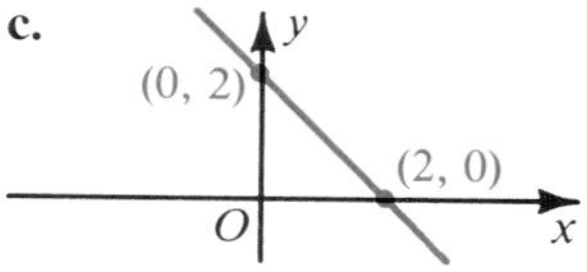

Tell whether each expression is positive or negative for the line shown in Exercises 2 and 3.

a. $y_2 - y_1$ **b.** $x_2 - x_1$ **c.** $\frac{y_2 - y_1}{x_2 - x_1}$

2.
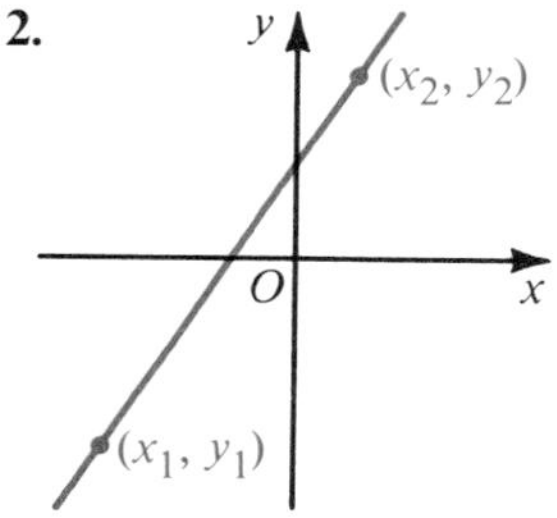

3.
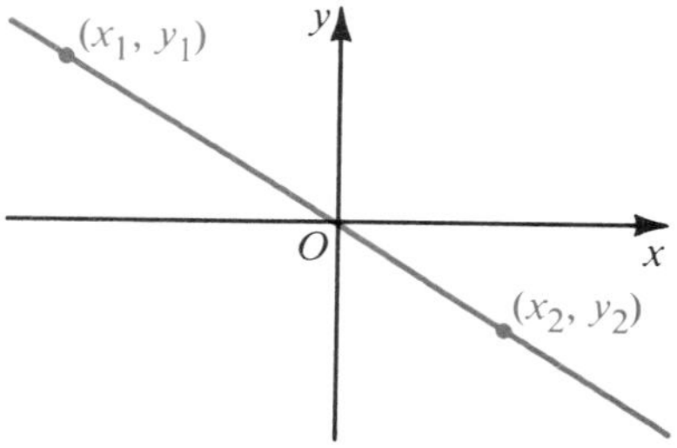

4. Does the slope of the line appear to be positive, negative, zero, or not defined?

a.

b.

c.

d.

5. a. What value does the slope of $\overleftrightarrow{AB}$ have?
   b. What value does $\tan n°$ have?
   c. Consider the statement: If a line with positive slope makes an acute angle of $n°$ with the x-axis, then the slope of the line is $\tan n°$. Do you think this statement is true or false? Explain.

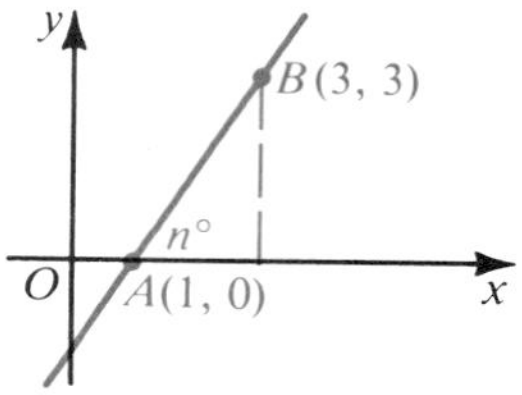

6. In this exercise you will prove that you can use any two points on a line to determine the slope of the line. Horizontal and vertical segments have been drawn as shown. Supply the reason for each step.

Outline of proof:

1. $\angle I \cong \angle A$
2. $\angle 1 \cong \angle 2$
3. $\triangle LIN \sim \triangle DAJ$
4. $\frac{IN}{AJ} = \frac{LI}{DA}$, or $\frac{IN}{LI} = \frac{AJ}{DA}$
5. The slope of $\overline{LN}$ equals $\frac{IN}{LI}$, and the slope of $\overline{DJ}$ equals $\frac{AJ}{DA}$.
6. Slope of $\overline{LN}$ = slope of $\overline{DJ}$

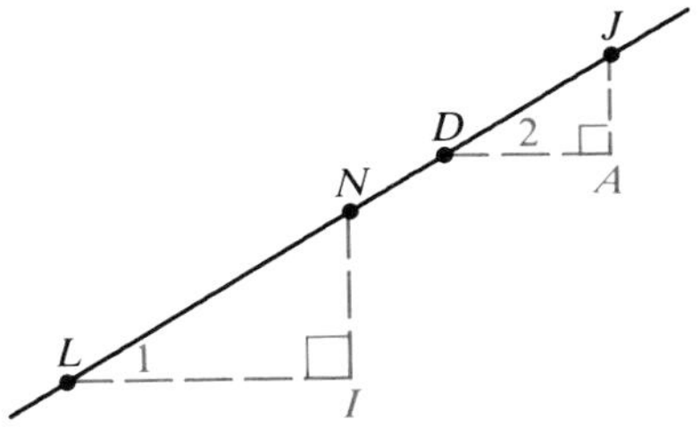

Written Exercises

Find the slope of the line through the points named. If a line is vertical, write *slope not defined*.

A

1. (1, 2); (3, 4)

2. (1, 2); (−2, −5)

3. (1, 2); (−2, 5)

4. (0, 0); (5, 1)

5. (7, 2); (2, 7)

6. (3, 3); (3, 7)

7. (6, −6); (−6, −6)

8. (6, −6); (4, 3)

9. $(0, a)$; $(b, 0)$

10. (p, q); $(-m, n)$

11. (r, s); (t, v)

12. $(-e, f)$; $(-d, f)$

Find the missing coordinate.

B

13. A line with slope $\frac{3}{4}$ passes through points (2, 3) and (10, __?__).

14. A line with slope $-\frac{5}{2}$ passes through points (7, −4) and (__?__, 6).

15. A line with slope m passes through points (p, q) and $(r,$ __?__$)$. Answer in terms of p, q, r, and m.

For each line described below, find three other points on the line.

16. **a.** The line with slope $\frac{2}{3}$ that passes through (0, 0)

b. The line with slope $-\frac{5}{2}$ that passes through (−3, 7)

17. **a.** The line with slope 2 that passes through (1, 1)

b. The line with slope −3 that passes through (−2, −1)

In Exercises 18–21, R, S, and T are vertices of a right triangle with right angle at S. Find the slopes of the legs. What is the relationship between these slopes?

18. $R(4, 3)$, $S(2, 1)$, $T(-3, 6)$

19. $R(-1, 1)$, $S(2, 4)$, $T(5, 1)$

20. $R(3, 6)$, $S(5, 2)$, $T(1, 0)$

21. $R(-3, -4)$, $S(2, 2)$, $T(14, -8)$

22. According to national guidelines for wheelchair ramps, if a ramp has a rise greater than 6 in., then it must have handrails on both sides. Suppose the slope of a ramp is $\frac{1}{12}$ and the bottom of the ramp is 10 ft from the base of a building. Find the rise of the ramp. Should handrails be installed?

23. Given $D(-5, -3)$, $E(-2, 4)$, $F(8, 9)$, and $G(14, 23)$, Jackie decided, after computing the slopes of $\overline{DE}$ and $\overline{FG}$, that the four given points were collinear. Was the conclusion correct? Explain.

C **24.** A line passes through points $(-2, -1)$ and $(4, 3)$. Where does the line intersect the x-axis? the y-axis?

25. A line through $H(3, 1)$ and $J(5, a)$ has positive slope and makes a $60°$ angle with the x-axis. Find the value of a.

26. Find two values of k such that the points $(-3, 4)$, $(0, k)$, and $(k, 10)$ are collinear.

BIOGRAPHICAL NOTE

Maria Gaetana Agnesi

Maria Gaetana Agnesi (1718–1799) was born in Milan, Italy. A child prodigy, she had mastered seven languages by the age of thirteen. Between the ages of twenty and thirty she compiled the works of the mathematicians of her time into two volumes on calculus, called *Analytical Institutions.* This was an enormous task, since the mathematicians had originally published their results in different languages and had used a variety of methods of approach.

Her volumes were praised as clear, methodical, and comprehensive. They were translated into English and French and were widely used as textbooks. Agnesi was elected to the Bologna Academy of Sciences and was appointed honorary lecturer in mathematics at the University of Bologna.

11-5 Parallel and Perpendicular Lines

When you look at two parallel lines, you probably believe that the lines have equal slopes. This idea is illustrated by the photograph below. The parallel beams shown are needed to support a roof with a fixed pitch.

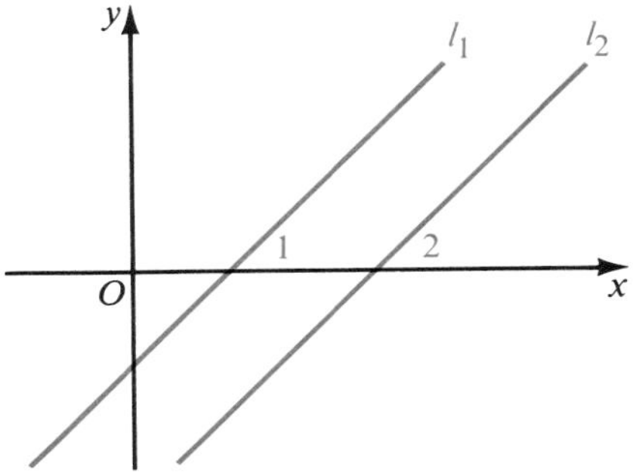

You can use trigonometry and properties of parallel lines to show the following for two nonvertical lines l_1 and l_2 (see the diagram at the right above):

1. $l_1 \parallel l_2$ if and only if $\angle 1 \cong \angle 2$
2. $\angle 1 \cong \angle 2$ if and only if $\tan \angle 1 = \tan \angle 2$
3. $\tan \angle 1 = \tan \angle 2$ if and only if slope of l_1 = slope of l_2

Therefore $l_1 \parallel l_2$ if and only if slope of l_1 = slope of l_2.

Although the diagram shows two lines with positive slope, this result can also be proved for two lines with negative slope. When the lines are parallel to the x-axis, both have slope zero.

Theorem 11-4

Two nonvertical lines are parallel if and only if their slopes are equal.

In Exercises 18–21 of the preceding section, you may have noticed that perpendicular lines, too, have slopes that are related in a special way. See Classroom Exercise 11 and Written Exercise 22 for proofs of the following theorem.

Theorem 11-5

Two nonvertical lines are perpendicular if and only if the product of their slopes is -1.

$$m_1 \cdot m_2 = -1, \quad \text{or} \quad m_1 = -\frac{1}{m_2}$$

Classroom Exercises

1. Line j has slope $\frac{4}{5}$. State the slope of every line that is:
   a. parallel to j **b.** perpendicular to j
2. Line k has slope -3, or $-\frac{3}{1}$. State the slope of every line that is:
   a. parallel to k **b.** perpendicular to k

Are two lines with the given slopes parallel, perpendicular, or neither?

3. $m_1 = \frac{3}{4}$; $m_2 = \frac{12}{16}$
4. $m_1 = 1$; $m_2 = -1$
5. $m_1 = 3$; $m_2 = -3$
6. $m_1 = -\frac{3}{4}$; $m_2 = -\frac{4}{3}$
7. $m_1 = 3$; $m_2 = \frac{-1}{3}$
8. $m_1 = \frac{-2}{3}$; $m_2 = \frac{2}{-3}$
9. $m_1 = 0$; $m_2 = -1$
10. $m_1 = 0$; m_2 is not defined.
11. In this exercise you will prove the statement: If two nonvertical lines are perpendicular, then the product of their slopes is -1. Supply the reason for each step.

Given: l_1 has slope m_1;
l_2 has slope m_2;
$l_1 \perp l_2$

Prove: $m_1 \cdot m_2 = -1$

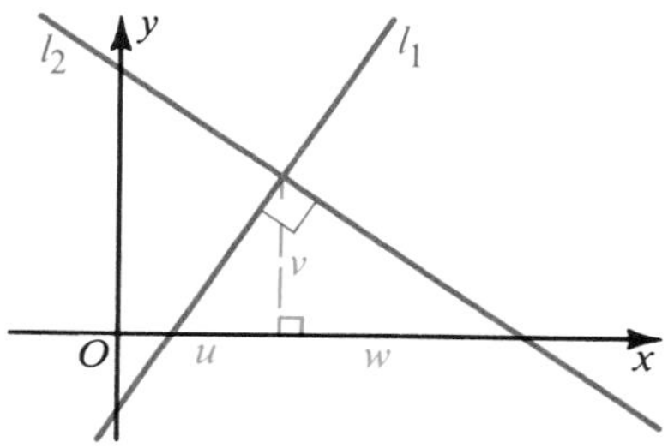

Outline of proof:

1. Draw the vertical segment shown.
2. $m_1 = \frac{v}{u}$
3. $m_2 = -\frac{v}{w}$
4. $m_1 \cdot m_2 = \left(\frac{v}{u}\right) \cdot \left(-\frac{v}{w}\right)$
5. $\frac{u}{v} = \frac{v}{w}$
6. $m_1 \cdot m_2 = \left(\frac{v}{u}\right) \cdot \left(-\frac{u}{v}\right)$, or -1

Written Exercises

A 1. Given $A(-2, 0)$ and $B(4, 3)$, find the slope of each line described.
   a. $\overleftrightarrow{AB}$ **b.** any line parallel to $\overleftrightarrow{AB}$
   c. any line perpendicular to $\overleftrightarrow{AB}$

2. Given $C(-3, 1)$ and $D(2, -1)$, find the slope of each line described.
   a. $\overleftrightarrow{CD}$ **b.** any line parallel to $\overleftrightarrow{CD}$
   c. any line perpendicular to $\overleftrightarrow{CD}$

3. In the diagram at the left below, $OEFG$ is a parallelogram. What is the slope of $\overline{OE}$? of $\overline{OG}$? of $\overline{GF}$? of $\overline{EF}$?

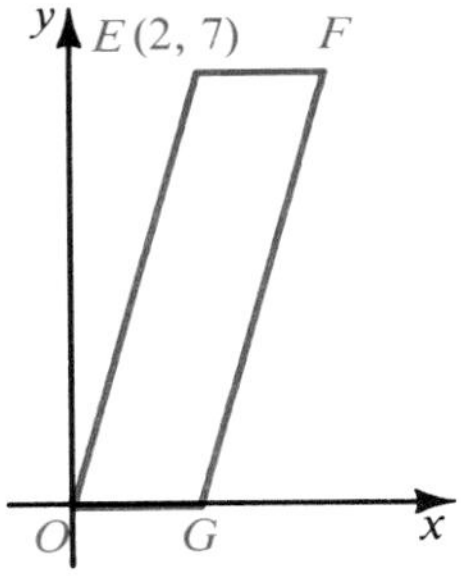

Ex. 3

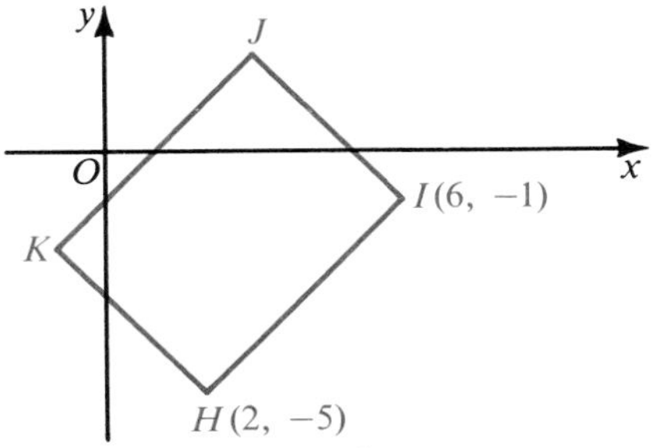

Ex. 4

4. In the diagram at the right above, $HIJK$ is a rectangle. What is the slope of $\overline{HI}$? of $\overline{IJ}$? of $\overline{JK}$? of $\overline{KH}$?

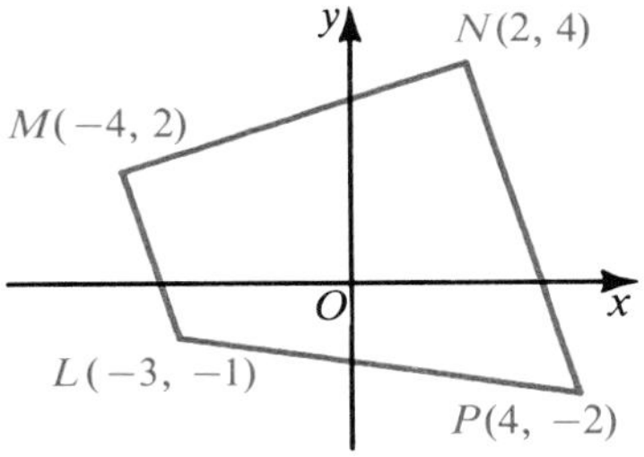

5. **a.** What is the slope of $\overline{LM}$? of $\overline{PN}$?
   b. Why is $\overline{LM} \parallel \overline{PN}$?
   c. What is the slope of $\overline{MN}$? of $\overline{LP}$?
   d. Why is $\overline{MN}$ not parallel to $\overline{LP}$?
   e. What special kind of quadrilateral is $LMNP$?

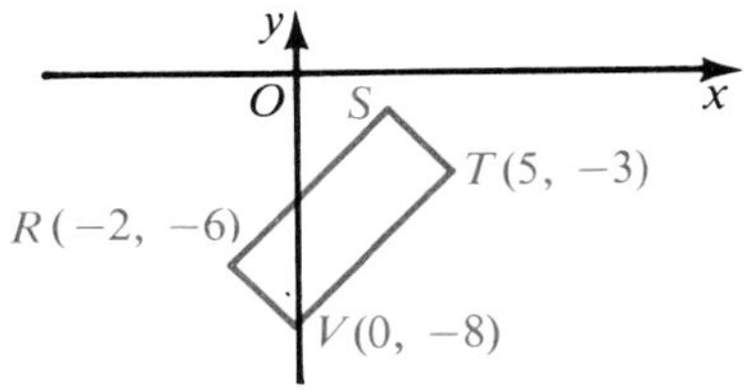

6. Quadrilateral $RSTV$ is known to be a parallelogram.
   a. What is the slope of $\overline{RV}$? of $\overline{TV}$?
   b. Why is $\overline{RV} \perp \overline{TV}$?
   c. Why is $\square RSTV$ a rectangle?
   d. Find the coordinates of S.

Find the slope of each side and each altitude of $\triangle ABC$.

7. $A(0, 0)$	$B(7, 3)$	$C(2, -5)$
8. $A(1, 4)$	$B(-1, -3)$	$C(4, -5)$

Find the slope of each side and each median of $\triangle DEF$.

B

9. $D(0, 0)$	$E(8, 0)$	$F(6, 6)$
10. $D(3, -5)$	$E(9, 1)$	$F(5, 5)$

If $\overline{AB} \parallel \overline{CD}$, find the value of t.

11. $A(-1, 5)$	$B(-3, 9)$	$C(4, -9)$	$D(0, t)$
12. $A(4, -3)$	$B(13, 9)$	$C(t, -2)$	$D(5, t)$

Decide what special type of quadrilateral *HIJK* is. Then prove that your answer is correct.

13. $H(0, 0)$	$I(5, 0)$	$J(7, 9)$	$K(1, 9)$
14. $H(0, 1)$	$I(2, -3)$	$J(-2, -1)$	$K(-4, 3)$
15. $H(7, 5)$	$I(8, 3)$	$J(0, -1)$	$K(-1, 1)$
16. $H(-3, -3)$	$I(-5, -6)$	$J(4, -5)$	$K(6, -2)$

17. Point $N(3, -4)$ lies on the circle $x^2 + y^2 = 25$. What is the slope of the line that is tangent to the circle at N? (*Hint:* Draw a diagram; also draw $\overline{ON}$.)

18. Point $P(6, 7)$ lies on the circle $(x + 2)^2 + (y - 1)^2 = 100$. What is the slope of the line that is tangent to the circle at P?

In Chapter 2 parallel lines are defined as coplanar lines that do not intersect. It is also possible to define parallel lines *algebraically* as follows:

Lines a and b are *parallel* if and only if slope of a = slope of b (or both a and b are vertical).

19. Use the algebraic definition to classify each statement as true or false.
a. For any line l in a plane, $l \parallel l$.
b. For any lines l and m in a plane, if $l \parallel m$, then $m \parallel l$.
c. For any lines l, m, and n in a plane, if $l \parallel m$ and $m \parallel n$, then $l \parallel n$.

20. Refer to Exercise 19. Is parallelism of lines an equivalence relation? (See Exercise 15, page 22.) Explain.

C **21.** In $\triangle TAY$, the midpoint of $\overline{TA}$ is $R(8, 4)$; the midpoint of $\overline{AY}$ is $C(15, 7)$; and the midpoint of $\overline{YT}$ is $S(7, 3)$. Find the coordinates of T, A, and Y.

22. Here is another way to prove Theorem 11-5.
a. Use the Pythagorean Theorem to prove:
If $\overleftrightarrow{TU} \perp \overleftrightarrow{US}$, then the product of the slopes of $\overleftrightarrow{TU}$ and $\overleftrightarrow{US}$ equals -1. That is, prove $\left(-\frac{c}{a}\right) \cdot \left(-\frac{c}{b}\right) = -1$.
b. Use the converse of the Pythagorean Theorem to prove:
If $\left(-\frac{c}{a}\right) \cdot \left(-\frac{c}{b}\right) = -1$, then $\overleftrightarrow{TU} \perp \overleftrightarrow{US}$.

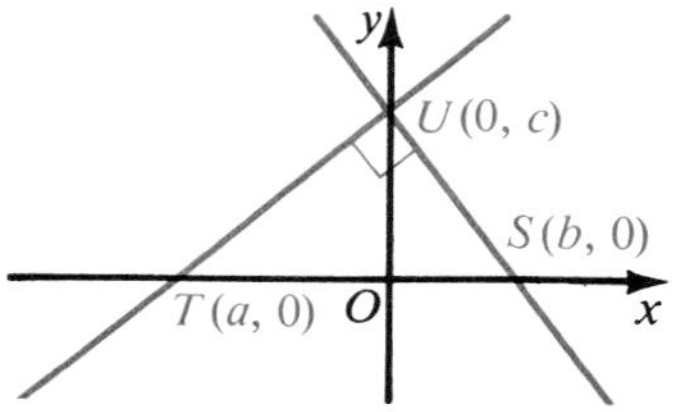

11-6 Equation of a Line

A **linear equation** is an equation whose graph is a line. Linear equations can be written in different forms: *standard form, point-slope form,* and *slope-intercept form.* We state a theorem for the standard form, but omit the proof.

Theorem 11–6 Standard Form

The graph of any equation that can be written in the form $ax + by = c$, where a and b are not both zero, is a line.

The advantage of the standard form is that it is easy to determine the point where the line crosses the x-axis (the x-coordinate of this point is called the *x-intercept*) and the point where the line crosses the y-axis (the y-coordinate of this point is called the *y-intercept*).

Example 1 Draw the graph of $2x - 3y = 12$.

Solution Since two points determine a line, begin by plotting two convenient points, such as the points where the line crosses the axes. Then draw the line.

To find the x-intercept, let $y = 0$.

$$2x - 3(0) = 12$$
$$2x = 12$$
$$x = 6$$

Thus $(6, 0)$ is a point on the line.

To find the y-intercept, let $x = 0$.

$$2(0) - 3y = 12$$
$$-3y = 12$$
$$y = -4$$

Thus $(0, -4)$ is a point on the line.

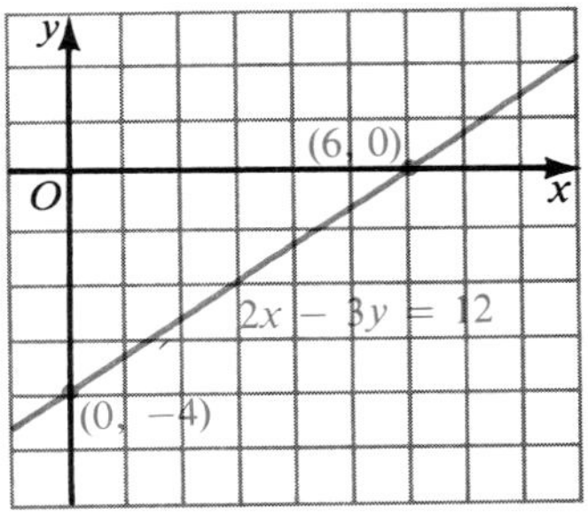

It can be proved that the converse of Theorem 11–6 is also true. That is, given any line, you can write an equation for that line in the form $ax + by = c$.

Suppose a line passes through a known point (x_1, y_1). Let (x, y) represent any other point on the line. Then the slope of the line is by definition

$$m = \frac{y - y_1}{x - x_1}.$$

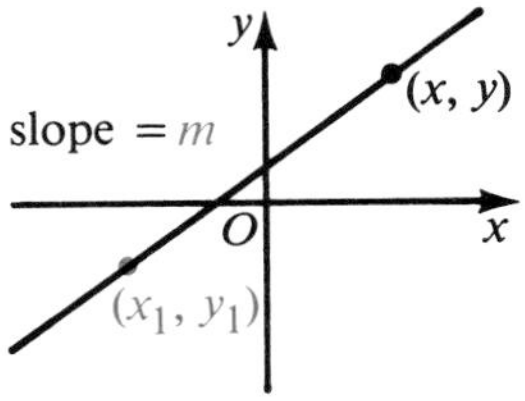

Multiplying both sides of the equation above by $(x - x_1)$, you get

$$y - y_1 = m(x - x_1).$$

We have just proved the next theorem.

Theorem 11–7 Point-Slope Form

An equation of the line that passes through point (x_1, y_1) and has slope m is $y - y_1 = m(x - x_1)$.

Example 2 Find an equation of the line containing points (1, −4) and (3, 1).

Solution Let (1, −4) be (x_1, y_1) and (3, 1) be (x_2, y_2).

First find the slope: $m = \frac{1 - (-4)}{3 - 1} = \frac{5}{2}$

An equation of the line through (1, −4) with slope $\frac{5}{2}$ is

$$y - (-4) = \frac{5}{2}(x - 1), \text{ or } y + 4 = \frac{5}{2}(x - 1).$$

In Example 2, if you let (3, 1) be (x_1, y_1), you get the equation $y - 1 = \frac{5}{2}(x - 3)$. This equation and $y + 4 = \frac{5}{2}(x - 1)$ are equivalent. See Exercise 21.

Example 3 Find an equation of the line with slope $\frac{1}{2}$ that passes through (0, −3).

Solution Using the point-slope form,

$$y - (-3) = \frac{1}{2}(x - 0)$$

$$y + 3 = \frac{1}{2}x$$

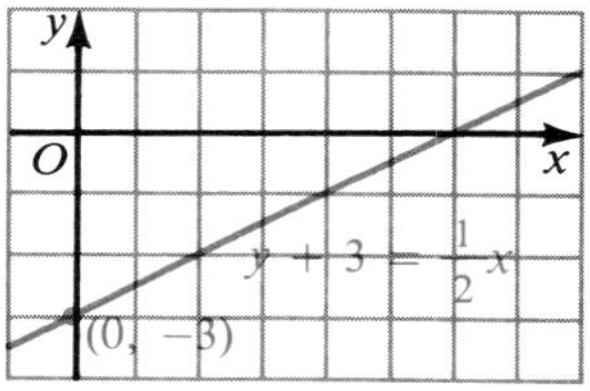

In Example 3, notice that $\frac{1}{2}$ is the slope and −3 is the y-intercept of the line. If you solve the equation found in the example for y, you get

$$y = \frac{1}{2}x - 3$$

Example 3 suggests a way to prove the following theorem.

Theorem 11-8 Slope-Intercept Form

An equation of the line that has slope m and y-intercept b is $y = mx + b$.

The advantage of this form is that you can tell at a glance what the slope and y-intercept of a line are.

Example 4 Find the slope and y-intercept of the line.

a. $y = \frac{2}{3}x + 4$

b. $5x + 6y = -12$

Solution a. $y = \frac{2}{3}x + 4$ is in slope-intercept form.

Slope $= \frac{2}{3}$ and y-intercept $= 4$

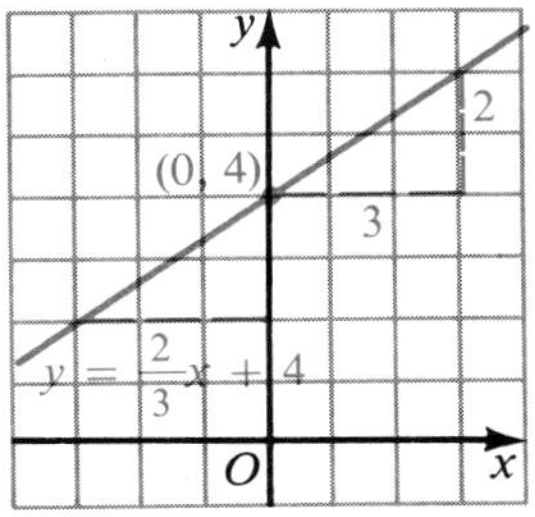

b. To get the equation in slope-intercept form, solve for y.

$$5x + 6y = -12$$
$$6y = -5x - 12$$
$$y = -\frac{5}{6}x - 2$$

Slope $= -\frac{5}{6}$ and y-intercept $= -2$

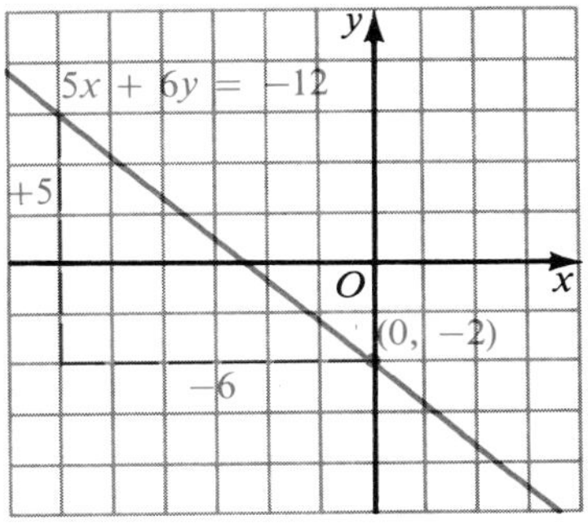

When you are given the equations of two lines, you can graph each line and find the approximate coordinates of the point of intersection. On the other hand, you can use algebra to find the exact coordinates.

Example 5 Find the point of intersection of the lines $7x + 2y = -4$ and $2x + y = 1$.

Solution 1 *Addition or Subtraction Method:* First we choose a letter to eliminate. Here we choose y:

$7x + 2y = -4$	(First equation)
$4x + 2y = 2$	(Second equation $\times$ 2)
$3x = -6$	(Subtract to eliminate y.)
$x = -2$	
$2(-2) + y = 1$	(Substitute in second equation.)
$y = 5$	

The point of intersection is $(-2, 5)$.

Solution 2 *Substitution Method:* First we choose one equation to solve for one letter.

$2x + y = 1$	(Second equation)
$y = 1 - 2x$	(Solve for y.)
$7x + 2(1 - 2x) = -4$	(Substitute in the other equation.)
$3x = -6$	
$x = -2$	
$y = 1 - 2(-2) = 5$	

The point of intersection is $(-2, 5)$.

Classroom Exercises

Complete the table of values for each equation. Then draw the graph.

1. $x + y = 4$
2. $x + 2y = 6$
3. $2x - 3y = 12$
4. $3x + 2y = 6$
5. $y = 3x$
6. $y - 2 = 3(x - 1)$

x	y
0	?
?	0
?	1

Find the slope and y-intercept of each line.

7. $y = -\frac{3}{4}x + 5$
8. $y = \frac{5}{9}x + 2$
9. $3x + 5y = 15$
10. $y - x = 12$

State an equation of the line described.

11. slope $= 3$; y-intercept $= 14$
12. slope $= -\frac{2}{5}$; y-intercept $= -8$
13. slope $= -1$; y-intercept $= 0$
14. slope $= \frac{11}{6}$; y-intercept $= \frac{5}{6}$

State an equation of the line through point P and having slope m.

15. $P(8, 2)$ $m = \frac{3}{5}$
16. $P(0, 3)$ $m = -2$
17. $P(5, 0)$ $m = \frac{1}{4}$
18. $P(-2, 4)$ $m = -\frac{3}{2}$

Solve each pair of equations by the Addition or Subtraction Method.

19. $3x + y = 5$
$2x + 3y = 8$
20. $4x + 5y = -7$
$2x - 3y = 13$
21. $4x - 3y = 3$
$7x + y = -1$

22. It is easy to see that the coordinates of all points on line k satisfy the equation:

$$0x + y = 1, \text{ or } y = 1$$

What is the equation of each line shown below?

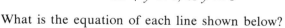

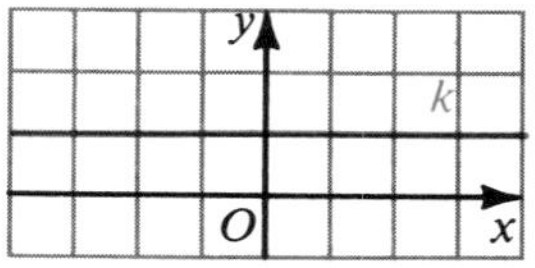

a.

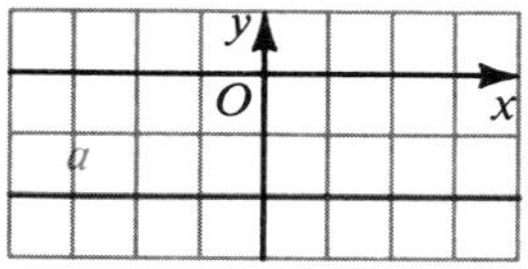

b.

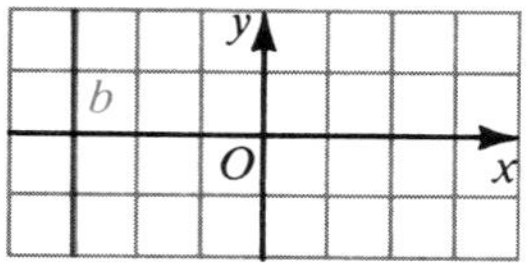

c.

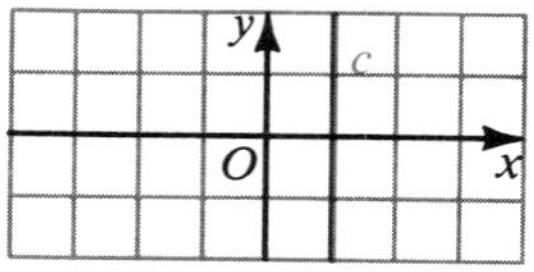

Written Exercises

Draw the graph of each equation.

A

1. $x + 3y = 6$
2. $x - y = 3$
3. $2x + 3y = 6$
4. $2x + y = 4$
5. $y = \frac{1}{3}x + 5$
6. $3x = 2y$

Write, in point-slope form, an equation of the line that passes through *P* and has slope *m*. (See Example 3.)

7. $P(2, 3);\ m = \frac{3}{4}$

8. $P(4, -3);\ m = -\frac{3}{5}$

9. $P(0, 7);\ m = 4$

10. $P(-5, 0);\ m = \frac{7}{4}$

Write an equation of the line that contains the given points. (See Example 2.)

11. $(1, 2)$ and $(5, 3)$

12. $(2, 0)$ and $(4, 3)$

13. $(7, -2)$ and $(-3, 2)$

14. $(0, 5)$ and $(-5, 6)$

Write an equation of the line described. (See Example 4(a).)

15. slope $= \frac{5}{7}$; y-intercept $= -9$

16. slope $= -\frac{1}{2}$; y-intercept $= 3$

Find the point of intersection of the two lines. (See Example 5.)

17. $x + y = 3$
    $x - y = -1$

18. $2x + 3y = 1$
    $3x + y = 12$

19. $3x + 2y = 8$
    $-x + 3y = 12$

20. $7x - 4y = -14$
    $3x + 2y = -6$

Show that the equations are equivalent by writing each in the form $ax + by = c$.

B 21. $y - 1 = \frac{5}{2}(x - 3)$
    $y + 4 = \frac{5}{2}(x - 1)$

22. $y + 2 = -\frac{4}{3}(x - 3)$
    $y - 2 = -\frac{4}{3}x$

Find an equation of the line described.

23. y-intercept $= 8$; x-intercept $= 4$

24. y-intercept $= 6$; x-intercept $= -2$

25. Find an equation of the line that contains the median of trapezoid $AODH$, with vertices $A(4, 2)$, $O(0, 0)$, $D(7, 0)$, and $H(5, 2)$.

26. Quadrilateral $BECK$ is known to be a rhombus. Two of the vertices are $B(3, 5)$ and $C(7, -3)$.
    a. Find the slope of diagonal $\overline{EK}$.
    b. Find an equation of $\overleftrightarrow{EK}$.

Write an equation of the line described.

27. The line through $(5, 8)$ and parallel to the x-axis

28. The x-axis

29. The line through $(7, -9)$ and parallel to the y-axis

30. The y-axis

In Exercises 31–34 write an equation of the line described.

31. The line through (4, 1) and parallel to $\overleftrightarrow{OM}$

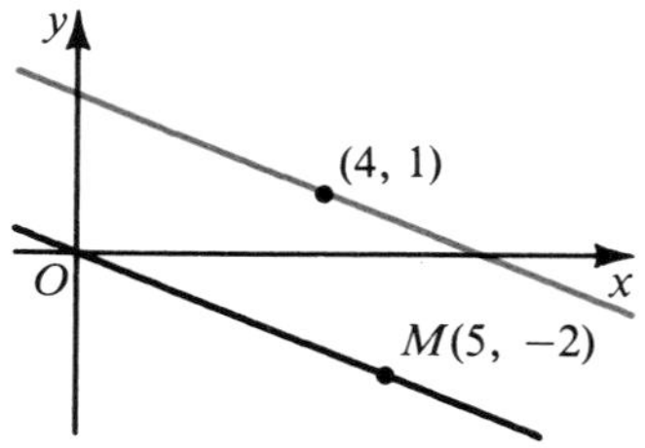

Ex. 31

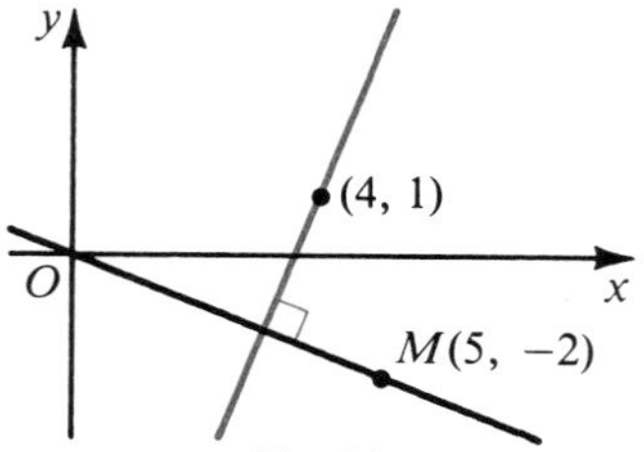

Ex. 32

32. The line through (4, 1) and perpendicular to $\overleftrightarrow{OM}$

33. The line through D that is perpendicular to side $\overline{AN}$

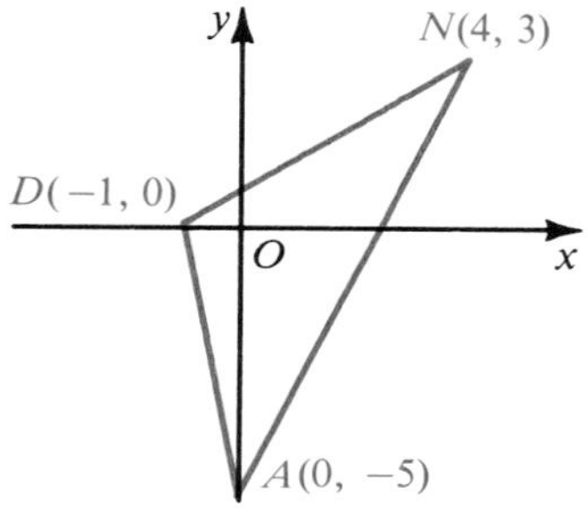

Ex. 33

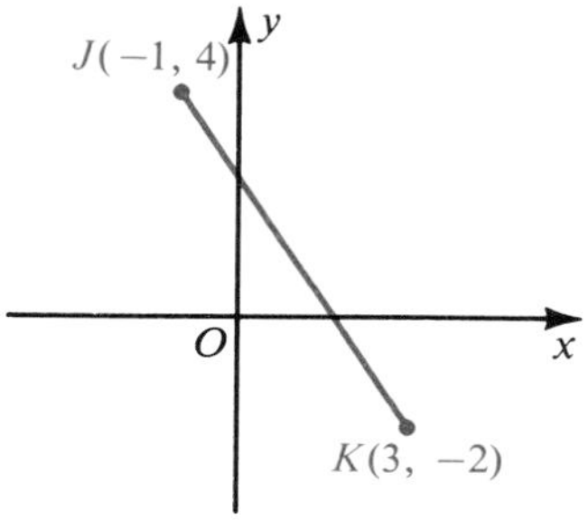

Ex. 34

34. The perpendicular bisector of $\overline{JK}$

C **35.** Write an equation of the line that is tangent to the circle $(x + 4)^2 + (y - 7)^2 = 25$ at (0, 4).

36. The vertices of $\triangle LOE$ are $L(14, 0)$, $O(0, 0)$, and $E(6, 8)$.

a. Find the equations of the three lines that contain the three altitudes of $\triangle LOE$.

b. Find, algebraically, the intersection point, S, of two of these lines.

c. Show that the third altitude also passes through S. You can do this by showing that the coordinates of S satisfy the equation of the line containing this altitude.

37. The vertices of $\triangle EOF$ are $E(18, 0)$, $O(0, 0)$, and $F(6, 6)$. The medians of $\triangle EOF$ are $\overline{EA}$, $\overline{OB}$, and $\overline{FC}$.

a. Find the equations of $\overleftrightarrow{EA}$, $\overleftrightarrow{OB}$, and $\overleftrightarrow{FC}$.

b. Find, algebraically, the intersection point, P, of two of the medians.

c. Show that the third median passes through P.

d. Find FP and FC and show that $FP = \frac{2}{3}FC$.

38. Find an equation of the circle circumscribed about the triangle whose vertices are $K(-4, -4)$, $A(4, 4)$, and $I(2, 8)$.

CAREERS

Accountant

An accountant is a financial expert. Accountants study and analyze a company's or an organization's overall financial dealings. They prepare many different kinds of financial reports. These include profit-and-loss statements, which summarize the company's earnings for a given period of time, and balance sheets, which state the current net worth of the company. Another important accounting function is the preparation of tax reports and statements. A company's owners or managers rely on the accountant's reports

to determine whether the company is operating efficiently and profitably, and where improvements can be made.

Accountants are usually college graduates with a major in accounting. They often begin work as junior accountants. After sufficient work experience they take an examination in order to become certified. Upon passing this examination, an accountant becomes a certified public accountant, or CPA. Further career advancement may lead to promotion to senior accountant, specializing in areas such as cost accounting or auditing. Advancement in a different direction may lead to a financial policy position such as that of controller or to starting an independent accounting company.

Self-Test 2

Find the slope of the line through the points named.

1. (0, 0) and (7, 4)

2. (−4, 2) and (1, −1)

3. For which is slope *not* defined, a horizontal line or a vertical line?

4. The slope of line l is $-\frac{5}{3}$. Find the slope of any line that is:

a. parallel to l **b.** perpendicular to l

5. A line has the equation $y - 4 = \frac{2}{3}(x + 3)$.

a. Find the slope of the line.
b. Find the coordinates of a point on the line.

6. Draw the graph of the line $x - 3y = 6$.

7. What are the slope and y-intercept of the line $-2x + 5y = -30$?

8. Find an equation of the line through points (−5, −2) and (3, 4).

9. Find the point of intersection of the lines $x + y = 7$ and $3x - y = 5$.

Coordinate Geometry Proofs

Objectives

1. Given a polygon, choose a convenient placement of the coordinate axes and assign appropriate coordinates.
2. Prove statements by using coordinate geometry methods.

11-7 Organizing Coordinate Proofs

We will illustrate coordinate geometry methods by proving Theorem 4-12: *The midpoint of the hypotenuse of a right triangle is equidistant from the three vertices.*

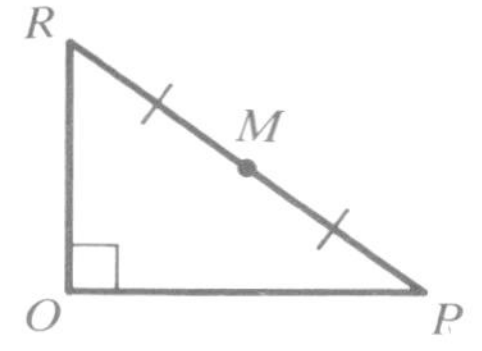

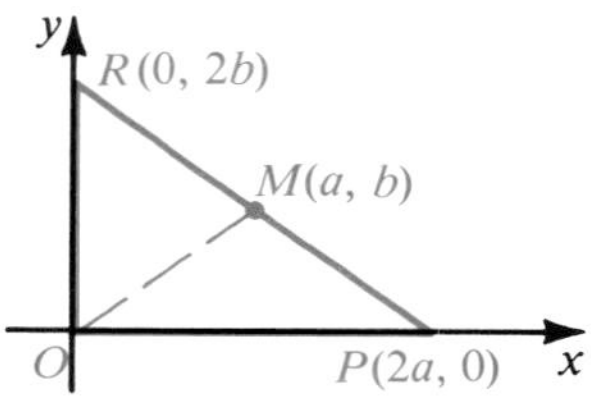

Proof:

Let $\overleftrightarrow{OP}$ and $\overleftrightarrow{OR}$ be the x-axis and y-axis.
Let P and R have the coordinates shown.
Then the coordinates of M are (a, b).

$MO = \sqrt{(a-0)^2 + (b-0)^2} = \sqrt{a^2 + b^2}$

$MP = \sqrt{(a-2a)^2 + (b-0)^2} = \sqrt{a^2 + b^2}$

Thus $MO = MP$.

By the definition of midpoint, $MP = MR$.
Hence $MO = MP = MR$.

Notice that $2a$ and $2b$ are convenient choices for coordinates since they lead to expressions that do not contain fractions for the coordinates of M.

If you have a right triangle, such as $\triangle POR$ on page 498, the most convenient place to put the x-axis and y-axis is usually along the legs of the triangle. If a triangle is not a right triangle, the two most convenient ways to place your axes are shown below. Notice that these locations for the axes maximize the number of times zero is a coordinate of a vertex.

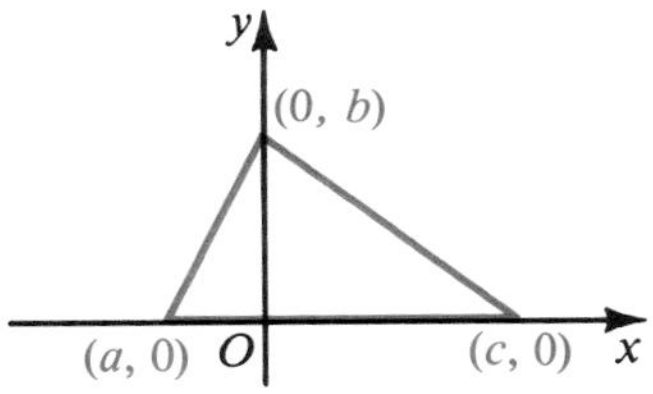

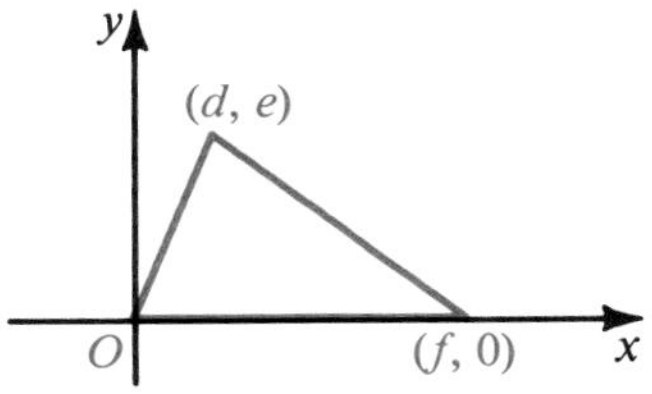

Some common ways of placing coordinate axes on other special figures are shown below.

$\triangle COD$ is isosceles; $CO = CD$.
Then C can be labeled (a, b).

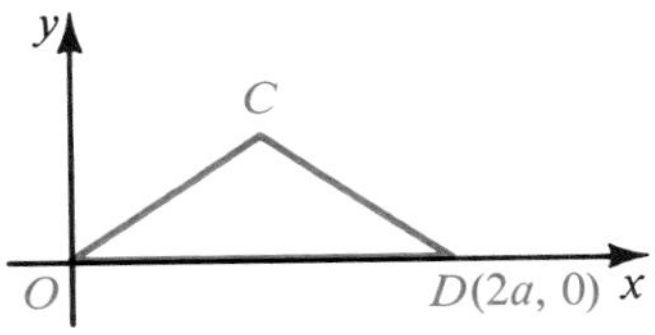

$\triangle EFG$ is isosceles; $EF = EG$.
Then F can be labeled $(-a, 0)$.

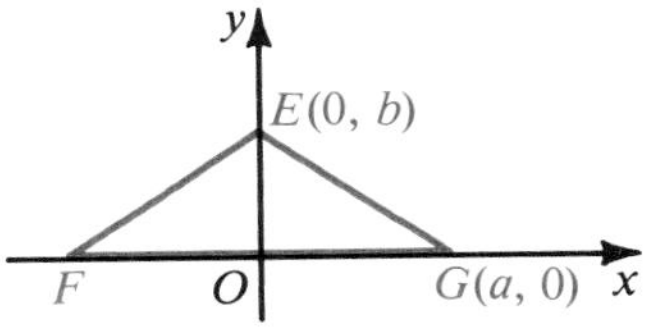

$HOJK$ is a rectangle.
Then K can be labeled (a, b).

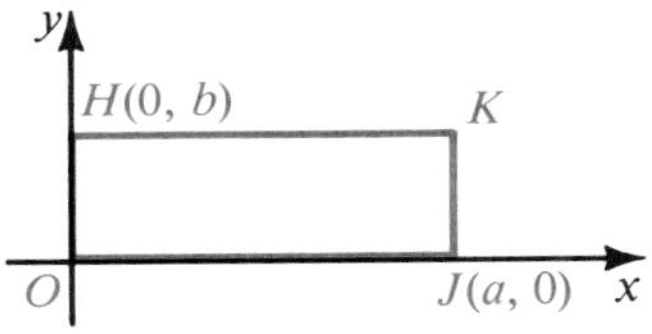

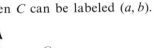

$MONP$ is a parallelogram.
Then P can be labeled $(a + b, c)$.

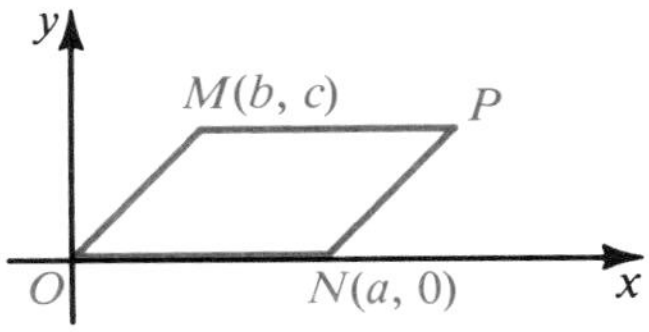

$ROST$ is a trapezoid.
Then T can be labeled (d, c).

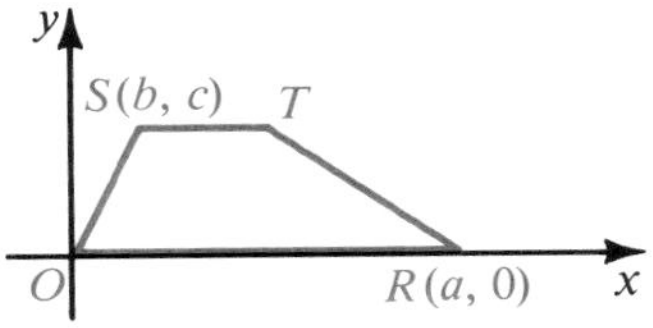

$UOVW$ is an isosceles trapezoid.
Then W can be labeled $(a - b, c)$.

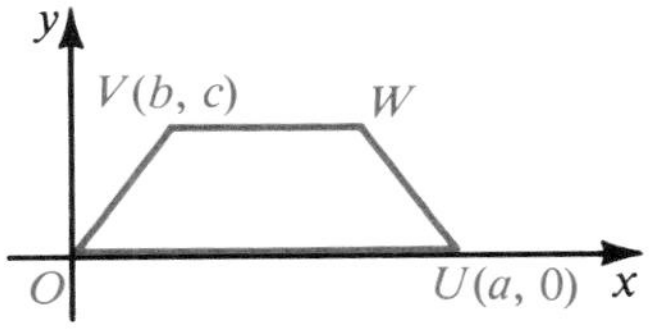

Classroom Exercises

Supply the missing coordinates without introducing any new letters.

1. *POST* is a square.

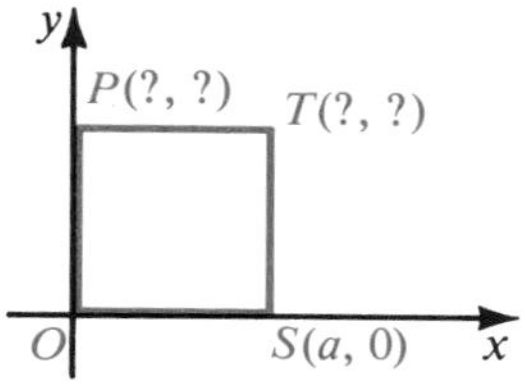

2. △*MON* is isosceles.

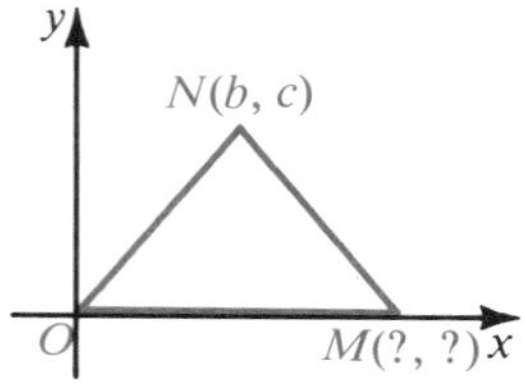

3. *JOKL* is a trapezoid.

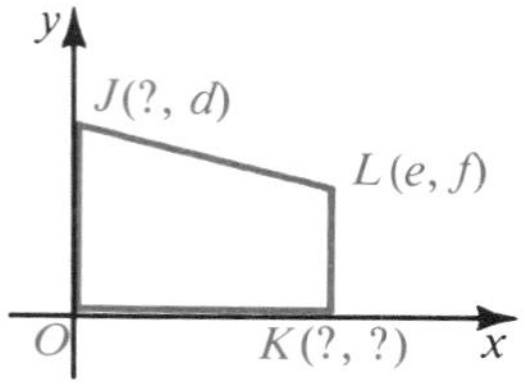

4. *GEOM* is a parallelogram.

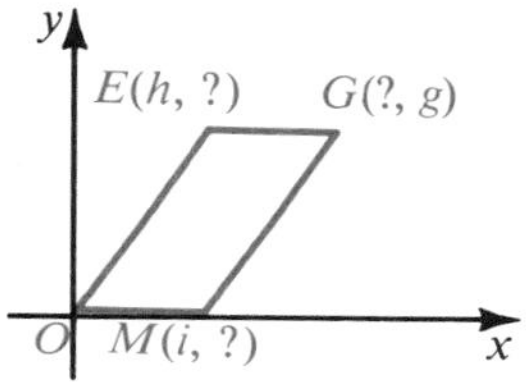

5. *GOLD* is a rectangle.

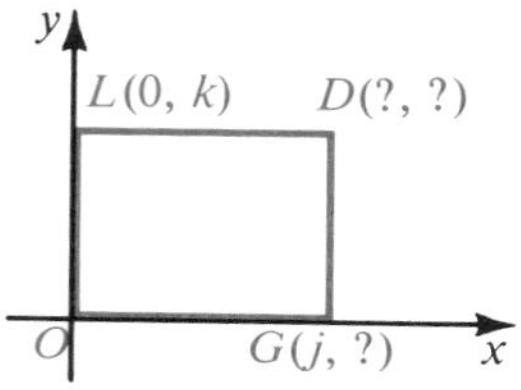

6. Rt. △*TOP* is isosceles.

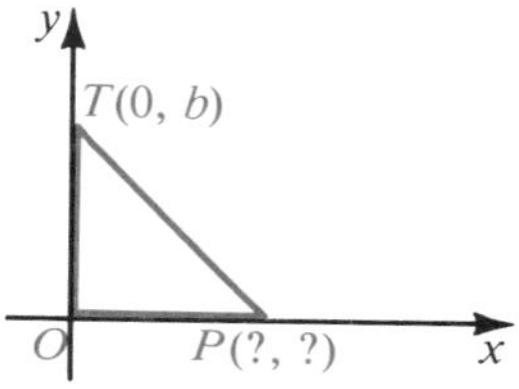

Written Exercises

Copy the figure. Supply the missing coordinates without introducing any new letters.

A **1.** Rectangle

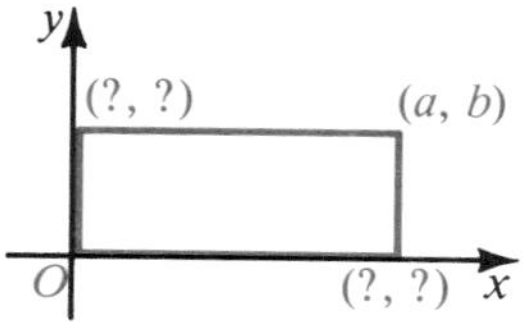

2. Parallelogram

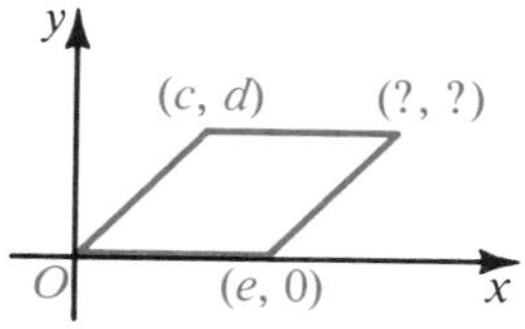

3. Square

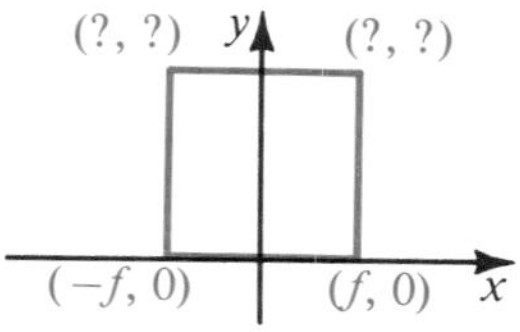

4. Isosceles triangle

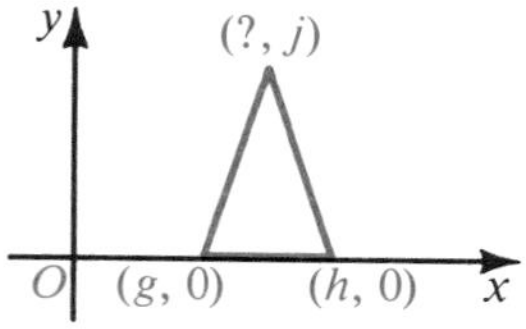

5. Parallelogram

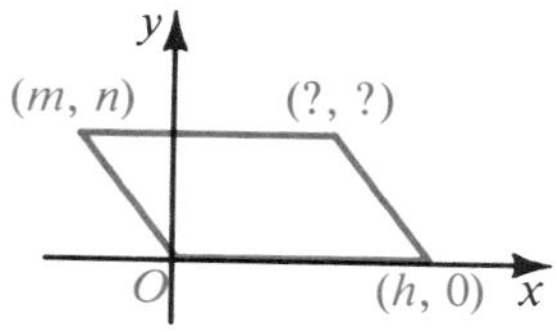

6. Isosceles trapezoid

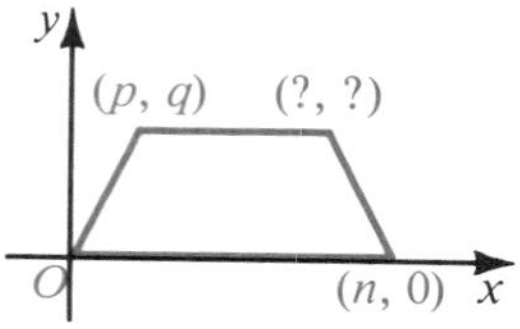

B 7. An equilateral triangle is shown below. Express the missing coordinates in terms of s.

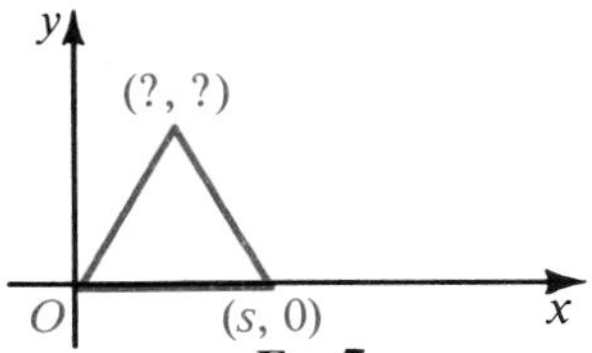

Ex. 7

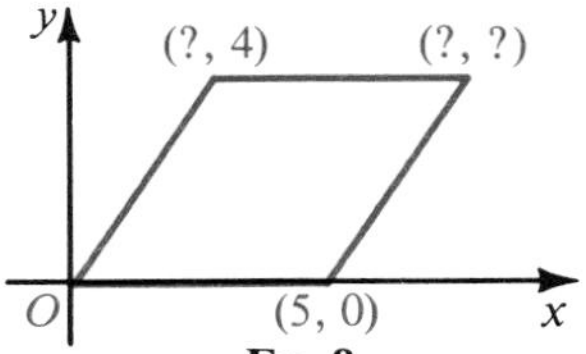

Ex. 8

8. A rhombus is shown above. Find the missing coordinates.

9. Rhombus $OABC$ is shown at the right. Express the missing coordinates in terms of a and b. (*Hint:* See Exercise 8.)

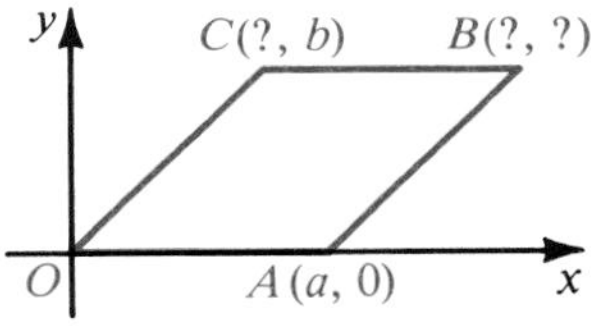

10. Supply the missing coordinates to prove: The segments that join the midpoints of opposite sides of any quadrilateral bisect each other. Let H, E, A, and R be the midpoints of the sides of quadrilateral $SOMK$. Choose axes and coordinates as shown.

   a. R has coordinates (__?__, __?__).
   b. E has coordinates (__?__, __?__).
   c. The midpoint of $\overline{RE}$ has coordinates (__?__, __?__).
   d. A has coordinates (__?__, __?__).
   e. H has coordinates (__?__, __?__).
   f. The midpoint of $\overline{AH}$ has coordinates (__?__, __?__).
   g. Because (__?__, __?__) is the midpoint of both $\overline{RE}$ and $\overline{AH}$, $\overline{RE}$ and $\overline{AH}$ bisect each other.

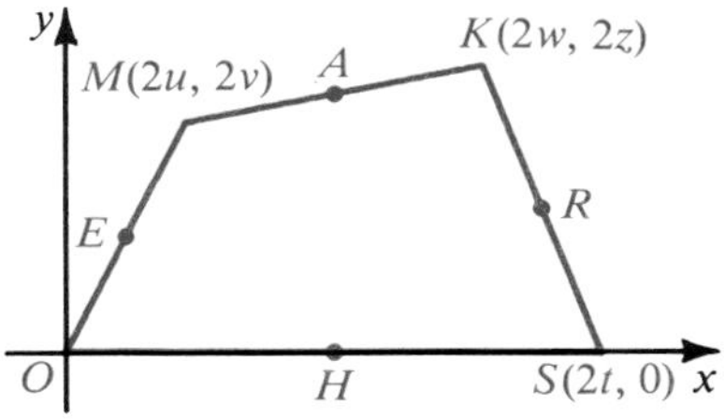

Draw the figure named. Select axes and label the coordinates of the vertices in terms of a single letter.

C 11. a regular hexagon

12. a regular octagon

13. Given isosceles trapezoid $HOJK$ and the axes and coordinates shown, use the definition of an isosceles trapezoid to prove that $e = c$ and $d = a - b$.

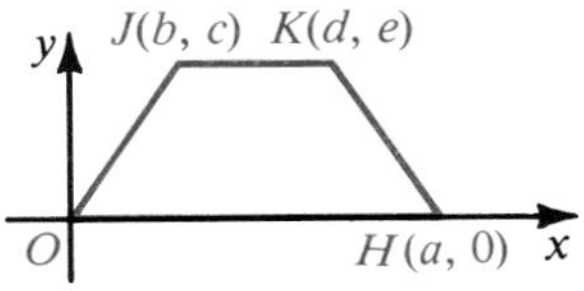

Challenge

Draw segments that divide an obtuse triangle into acute triangles.

11-8 Using Coordinate Geometry in Proofs

You can now use coordinate geometry to prove some theorems stated in previous chapters.

Example Prove that the median of a trapezoid
(1) is parallel to the bases;
(2) has a length equal to half the sum of the lengths of the bases.
(Theorem 4-16)

Solution **Proof:**

Given trapezoid *OMNP*, we choose convenient axes and coordinates as shown.

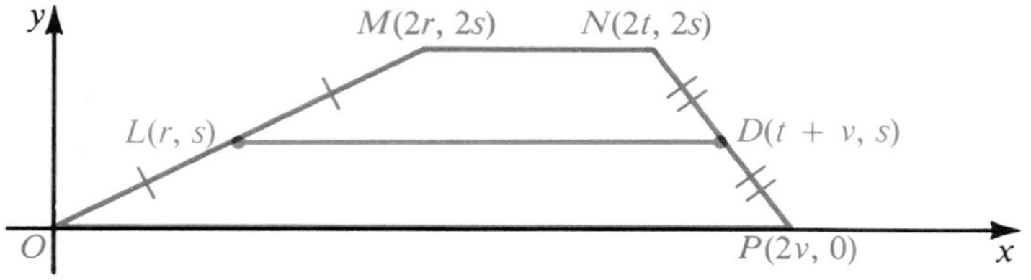

Midpoint L has coordinates (r, s).
Midpoint D has coordinates $(t + v, s)$.
Slope of $\overline{OP}$ = slope of $\overline{MN} = 0$

Slope of $\overline{LD} = \dfrac{s - s}{(t + v) - r} = 0$

Since $\overline{OP}$, $\overline{LD}$, and $\overline{MN}$ have equal slopes, $\overline{OP} \parallel \overline{LD} \parallel \overline{MN}$.
$MN = 2t - 2r$; $LD = t + v - r$; $OP = 2v$
$\frac{1}{2}(MN + OP) = \frac{1}{2}[(2t - 2r) + 2v] = \frac{1}{2}(2t + 2v - 2r) = t + v - r$
$\frac{1}{2}(MN + OP) = LD$

Classroom Exercises

The purpose of Exercises 1–8 is to prove that the lines that contain the altitudes of a triangle intersect in a point.

Given $\triangle ROM$, with lines j, k, and l containing the altitudes, we choose axes and coordinates as shown.

1. The equation of line k is __?__.
2. Since the slope of $\overline{MR}$ is $\dfrac{c}{b - a}$, the slope of line l is __?__.
3. The equation of line l is $y = \left(\dfrac{a - b}{c}\right)x$. (Explain.)
4. Lines k and l intersect where $x = b$ and $y = \left(\dfrac{a - b}{c}\right)b$, or $y = \dfrac{ab - b^2}{c}$. (Explain.)

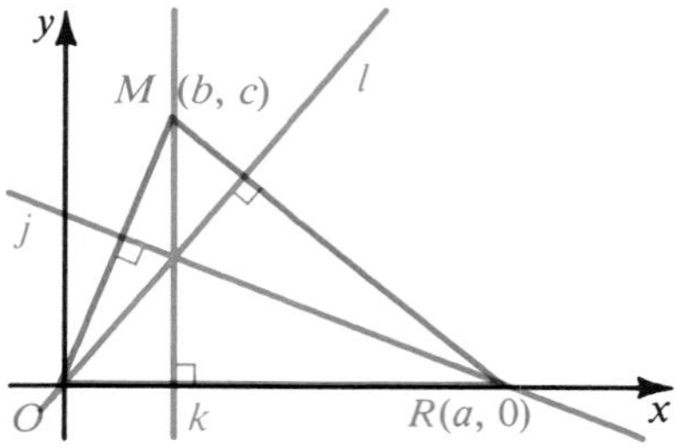

5. Since the slope of $\overline{OM} = \frac{c}{b}$, the slope of line j is __?__.

6. The equation of line j is $y = -\frac{b}{c}(x - a)$. (Explain.)

7. Lines k and j intersect where $x = b$ and $y = -\frac{b}{c}(b - a)$, or $y = \frac{ab - b^2}{c}$. (Explain.)

8. The three altitude lines intersect in a point. Name the coordinates of that point.

Written Exercises

In Exercises 1–8, use coordinate geometry to prove the statement.

A 1. If a point lies on the perpendicular bisector of a segment, then the point is equidistant from the endpoints of the segment.

2. The diagonals of a parallelogram bisect each other.

3. The diagonals of a rectangle are congruent.

4. The diagonals of a square are perpendicular.

B 5. The diagonals of an isosceles trapezoid are congruent.

6. The medians drawn to the legs of an isosceles triangle are congruent.

7. The quadrilateral formed by joining the midpoints of the sides of a rectangle is a rhombus.

8. A triangle is formed by joining the midpoints of the sides of an isosceles triangle. What can you prove about the smaller triangle?

C 9. The segment that joins the midpoints of two sides of a triangle
(1) is parallel to the third side;
(2) has a length equal to half the length of the third side.
Write a coordinate geometry proof.

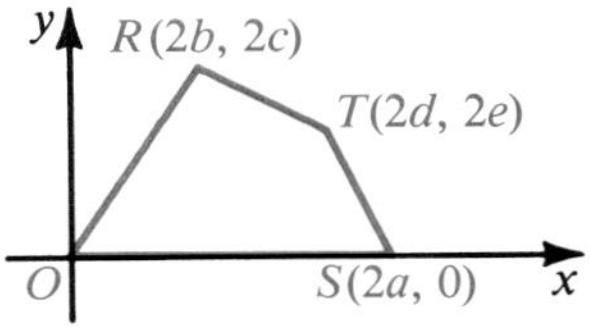

10. a. Prove that the quadrilateral formed by joining, in order, the midpoints of the sides of quadrilateral $ROST$ is a parallelogram.
b. Find an expression for c (in terms of a, b, d, and e) for which the parallelogram is a rectangle.

11. What two properties can you prove for the segment that joins the midpoints of the diagonals of a trapezoid? Write coordinate geometry proofs.

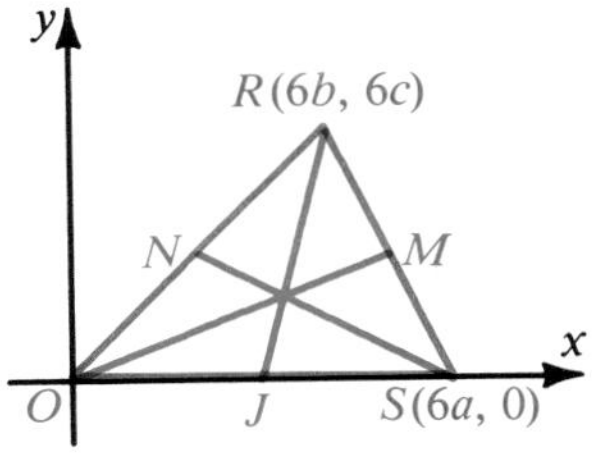

★12. Use axes and coordinates as shown to prove: The medians of a triangle intersect in a point that is two thirds of the distance from each vertex to the midpoint of the opposite side. (*Hint:* Find the coordinates of the midpoints; then the slopes of the medians; then the equations of the lines containing the medians.)

★**13.** The diagram shows axes and coordinates that can be used for any triangle *ROS*.
C, the intersection point of the perpendicular bisectors of the sides, has coordinates $\left(3a, \frac{3b^2 + 3c^2 - 3ab}{c}\right)$.
G, the intersection point of the medians, has coordinates $(2a + 2b, 2c)$. (See Exercise 12.)
H, the intersection point of the altitudes, has coordinates $\left(6b, \frac{6ab - 6b^2}{c}\right)$. (See the Classroom Exercises.)

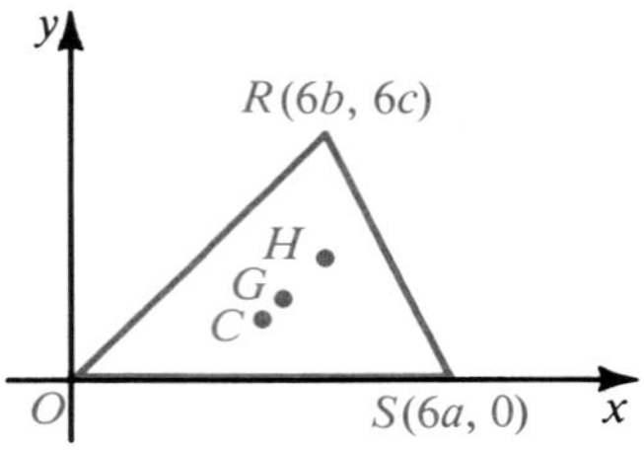

Prove each statement.

a. Points *C*, *G*, and *H* are collinear. The line containing these points is called *Euler's Line.* (*Hint:* One way to prove this is to show that slope of $\overline{CG}$ = slope of $\overline{GH}$.)

b. $CG = \frac{1}{3}CH$

Self-Test 3

State the coordinates of point *J* without introducing any new letters.

1. Isosceles triangle

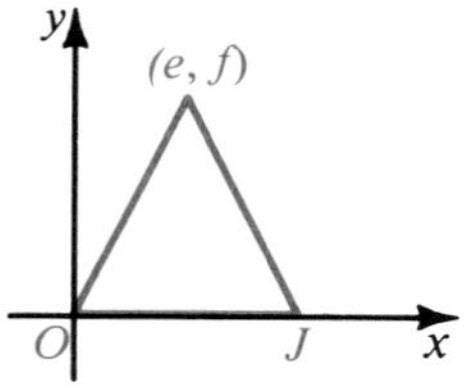

2. Parallelogram

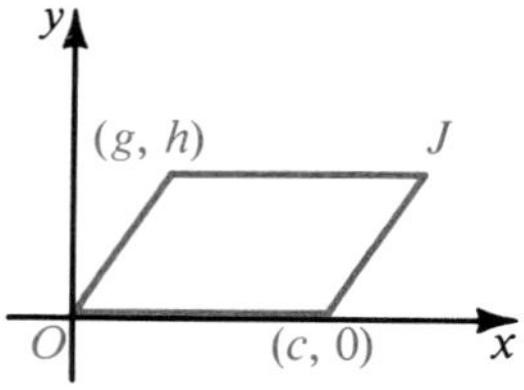

3. Isosceles trapezoid

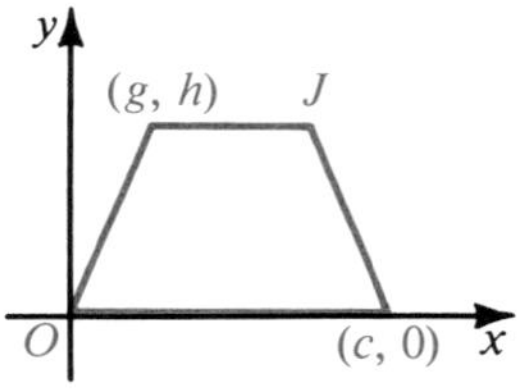

4. Parallelogram *QRST* is shown at the right. Supply the missing coordinates without introducing any new letters.

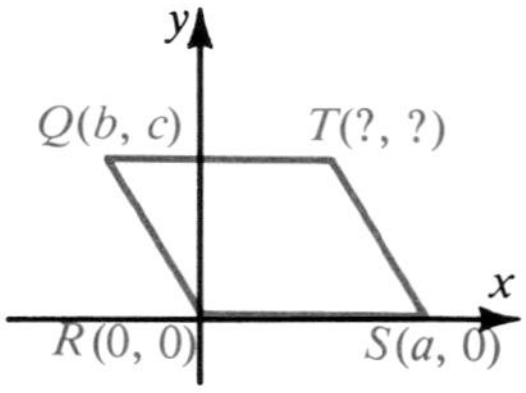

5. The vertices of a quadrilateral are $G(4, -1)$, $O(0, 0)$, $L(2, 6)$, and $D(6, 5)$. Show that quadrilateral *GOLD* is a parallelogram.

6. Use coordinate geometry to prove that the medians of an equilateral triangle are congruent. (*Hint:* Use the diagram given at the right.)

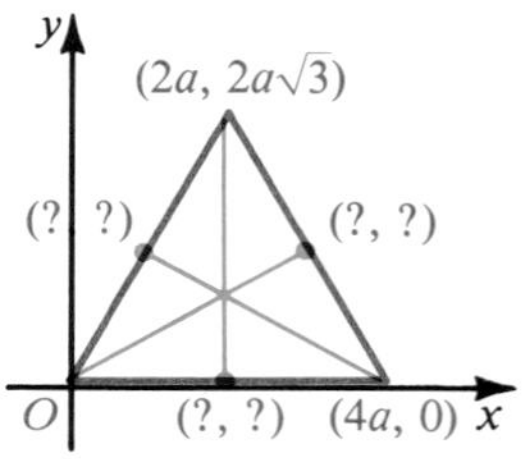

Chapter Summary

1. You can represent ordered pairs of numbers by points located with respect to the x- and y-axes in a coordinate plane. For any ordered pair (a, b), a is the x-coordinate and b is the y-coordinate.
2. The distance between points (x_1, y_1) and (x_2, y_2) is
$$\sqrt{(x_2 - x_1)^2 + (y_2 - y_1)^2}.$$
The midpoint of the segment joining these points is the point
$$\left(\frac{x_1 + x_2}{2}, \frac{y_1 + y_2}{2}\right).$$
3. The circle with center (a, b) and radius r has the equation
$$(x - a)^2 + (y - b)^2 = r^2.$$
4. The slope m of a line through two points (x_1, y_1) and (x_2, y_2), $x_1 \neq x_2$, is defined as follows: $m = \dfrac{y_2 - y_1}{x_2 - x_1}$. The slope of a horizontal line is zero. Slope is not defined for vertical lines.
5. Two nonvertical lines with slopes m_1 and m_2 are:
   a. parallel if and only if $m_1 = m_2$.
   b. perpendicular if and only if $m_1 \cdot m_2 = -1$.
6. The graph of any equation that can be written in the form $ax + by = c$, with a and b not both zero, is a line. An equation of the line through point (x_1, y_1) with slope m is $y - y_1 = m(x - x_1)$. An equation of the line with slope m and y-intercept b is $y = mx + b$.
7. To prove theorems using coordinate geometry, proceed as follows:
   a. Place x- and y-axes in a convenient position with respect to a figure.
   b. Use known properties to assign coordinates to points of the figure.
   c. Use the distance formula, the midpoint formula, and the slope properties of parallel and perpendicular lines to prove theorems.

Chapter Review

Exercises 1–3 refer to points $X(-2, -4)$, $Y(2, 4)$, and $Z(2, -6)$.

1. Graph X, Y, and Z on one set of axes. 11-1
2. Find XY, YZ, and XZ.
3. Use your answers to Exercise 2 to show that $\triangle XYZ$ is a right triangle.

Find the center and radius of each circle.

4. $(x + 3)^2 + y^2 = 100$

5. $(x - 5)^2 + (y + 1)^2 = 49$ 11-2

Write an equation of the circle described.

6. The circle has center $(2, -3)$ and is tangent to the x-axis.
7. The circle has center $(-6, -1)$ and passes through point $(-3, 3)$.

Find the coordinates of the midpoint of the segment that joins the given points.

8. $(7, -2)$ and $(1, -1)$ 9. $(-4, 5)$ and $(2, -5)$ 10. (a, b) and $(-a, b)$ 11-3

11. $M(0, 5)$ is the midpoint of $\overline{RS}$. If S has coordinates $(11, -1)$, then R is point $(\underline{\ ?\ }, \underline{\ ?\ })$.

12. Find the slope of the line through $(-5, -1)$ and $(15, -6)$. 11-4
13. A line with slope $\frac{2}{3}$ passes through $(9, -13)$ and $(0, \underline{\ ?\ })$.
14. A line with slope 5 passes through $(0, -2)$. Find three other points on the line.

15. What is the slope of a line that is parallel to the x-axis? 11-5
16. Show that $QRST$ is a trapezoid.
17. Since the slope of $\overline{QT}$ is $\underline{\ ?\ }$, the slope of an altitude to $\overline{QT}$ is $\underline{\ ?\ }$.
18. If U is a point on $\overline{QT}$ such that $\overline{UR} \parallel \overline{ST}$, then U has coordinates $(\underline{\ ?\ }, \underline{\ ?\ })$.

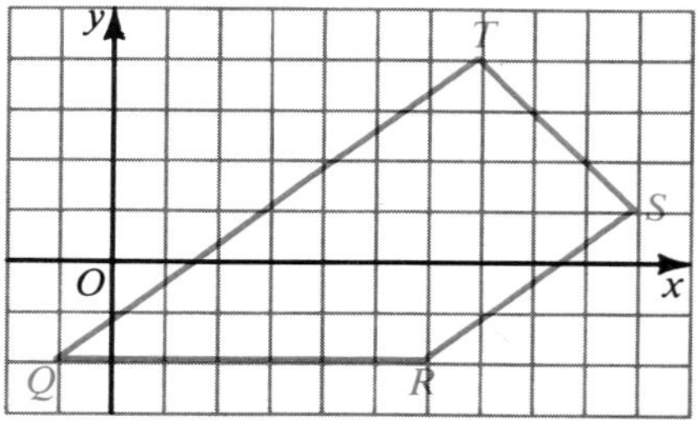

19. Draw the graph of $5x + 2y = 10$. 11-6
20. Find an equation of the line that passes through $(5, 2)$ and has slope -2.
21. Write an equation of the line through $(-2, -3)$ and $(-8, -1)$.
22. Find the slope and y-intercept of the line whose equation is $4x - 6y = 12$.
23. Find the point of intersection of the lines $x + 5y = 8$ and $4x - 3y = 9$.

24. A rectangle is shown. Find the coordinates of points A and C without introducing any new letters. 11-7
25. Let equilateral $\triangle RST$ have vertices $R(0, 0)$ and $T(2a, 0)$. Then S has coordinates $(\underline{\ ?\ }, \underline{\ ?\ })$.

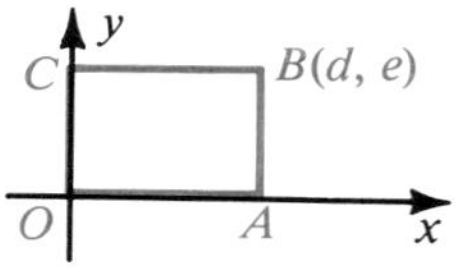

Use coordinate geometry to prove each statement.

26. If rectangle $DOEF$ has $OE = 2 \cdot EF$ and G is the midpoint of $\overline{DF}$, then $\overline{OG}$ is perpendicular to $\overline{EG}$. 11-8
27. In rectangle $DOEF$, if G is the midpoint of $\overline{DF}$, then $\triangle OEG$ is isosceles.

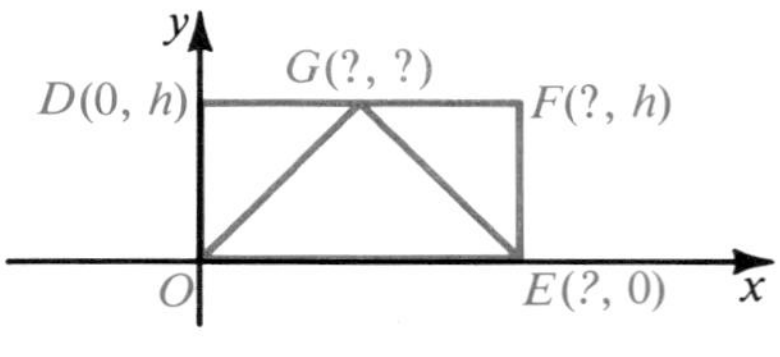

28. The segment joining the midpoints of two opposite sides of a parallelogram is parallel to the other two sides.

Chapter Test

Given points $M(-2, 1)$ and $N(2, 4)$, complete the statements.

1. The x-coordinate of point N is _?_.
2. Point M lies in Quadrant _?_.
3. The distance between M and N equals _?_.
4. The midpoint of $\overline{MN}$ is (_?_, _?_).
5. The circle that has center N and radius MN has equation _?_.
6. An equation of $\overleftrightarrow{MN}$ is _?_.
7. If $\overleftrightarrow{LM} \perp \overleftrightarrow{MN}$, then $\overleftrightarrow{LM}$ has slope _?_.
8. An equation of a line through $K(1, -2)$ and perpendicular to $\overleftrightarrow{MN}$ is _?_.
9. A line parallel to $\overleftrightarrow{MN}$ contains points (3, 1) and (−1, _?_).
10. An equation of a line that is parallel to $\overleftrightarrow{MN}$ and has y-intercept −4 is _?_.
11. If M is the midpoint of $\overline{NZ}$, then Z has coordinates (_?_, _?_).
12. Find the point of intersection of the lines $5x - 3y = -1$ and $2x - y = -1$.
13. Find the radius and the center of the circle with equation $(x + 6)^2 + y^2 = 9$.
14. An isosceles trapezoid is shown. Give the missing coordinates without introducing any new letters.

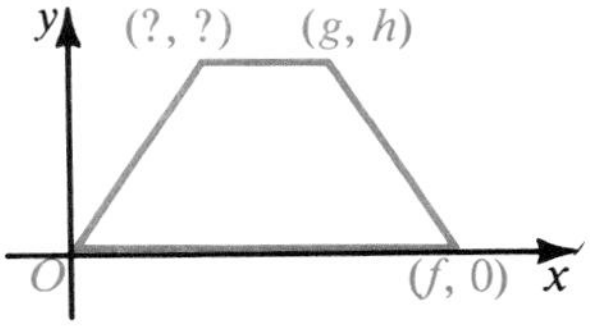

Use points $J(-12, 0)$, $K(0, 6)$, and $L(-3, -3)$.

15. Show that $\triangle JKL$ is isosceles.
16. Use slopes to show that $\triangle JKL$ is a right triangle.

Use coordinate geometry to prove each statement.

17. The diagonals of a square bisect each other.
18. The segments joining the midpoints of consecutive sides of a rectangle form a rhombus.

Mixed Review

1. Refer to the diagram.
   a. Show that $\angle B \cong \angle D$.
   b. Find the value of x.
   c. Find the ratio of the areas of the triangles.

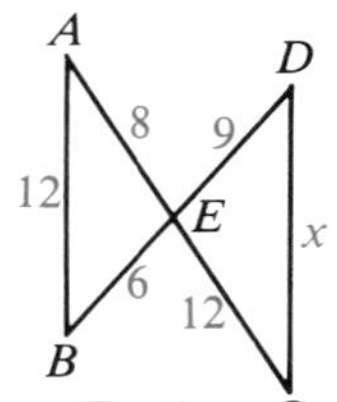

Ex. 1

2. If F is the point $(-3, 5)$ and G is the point $(0, -4)$, graph $\overleftrightarrow{FG}$ and find an equation for $\overleftrightarrow{FG}$.
3. Is a triangle with sides of lengths 12, 35, and 37 acute, right, or obtuse?
4. Find the total area and the volume of a cylinder with radius 10 and height 8.2.
5. If x is the length of a tangent segment in the diagram, find the values of x and y.

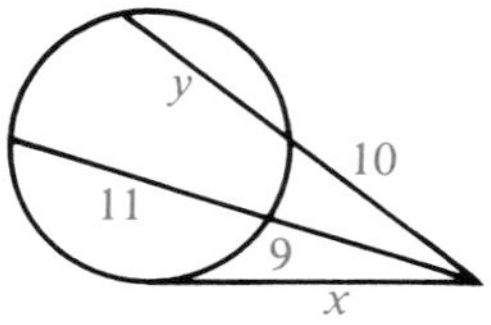

6. If the diagonals of a quadrilateral are congruent and perpendicular, must the quadrilateral be a square? a rhombus? Draw a diagram to illustrate your answer.
7. Prove: If the ray that bisects an angle of a triangle is perpendicular to the side that it intersects, then the triangle is an isosceles triangle.
8. Find the perimeter and area of a square with apothem $\sqrt{2}$ cm.

Find the value of x.

9.

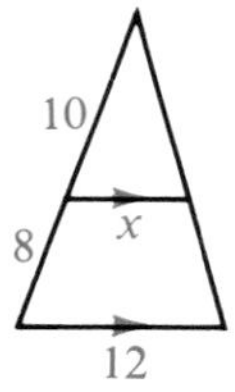

10.

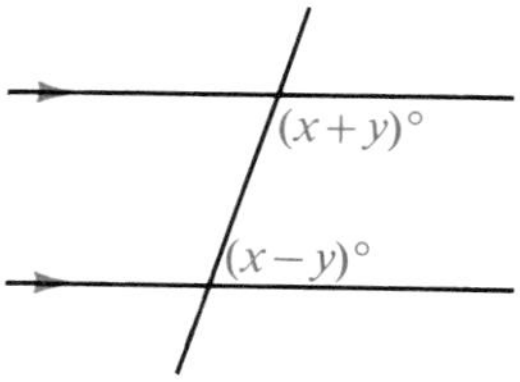

11.

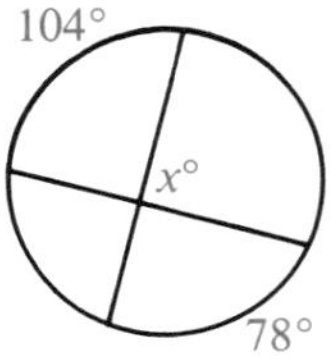

12. Draw a large obtuse triangle. Construct a circumscribed circle about the triangle.
13. Use coordinate geometry to prove that the median of a trapezoid is parallel to each base.
14. Write "$x = 1$ only if $x \neq 0$" in if-then form. Then write the contrapositive and classify the contrapositive as true or false.
15. In $\triangle ABC$, $\overline{AB} \perp \overline{BC}$, $AB = 1$, and $AC = 3$. Find:
    a. $\cos A$ **b.** $\sin C$ **c.** $\tan A$ **d.** $\cos C$
16. A sphere has surface area 144π cm^2. Find its volume.

Algebra Review

Rules of Exponents

When a and b are nonzero real numbers and m and n are integers:

1. $a^0 = 1$
2. $a^m \cdot a^n = a^{m+n}$
3. $\frac{a^m}{a^n} = a^{m-n}$
4. $(a^m)^n = a^{mn}$
5. $(ab)^m = a^m b^m$
6. $\left(\frac{a}{b}\right)^m = \frac{a^m}{b^m}$
7. $a^{-m} = \frac{1}{a^m}$

Simplify. Use only positive exponents in your answers.

Examples

1. $(-2)^4 = (-2)(-2)(-2)(-2) = 16$

2. $4^{-2} = \frac{1}{4^2} = \frac{1}{16}$

3. $x^2 \cdot x^{-2} = x^2 \cdot \frac{1}{x^2} = 1$ or $x^2 \cdot x^{-2} = x^0 = 1$

4. $a^{-5} \cdot a = a^{-5+1} = a^{-4} = \frac{1}{a^4}$

5. $(3xy^2)^2 = 3xy^2 \cdot 3xy^2 = 9x^2y^4$ or
$(3xy^2)^2 = 3^2x^2(y^2)^2 = 9x^2y^4$

6. $\left(\frac{3}{2}\right)^{-2} = \frac{1}{\left(\frac{3}{2}\right)^2} = \frac{1}{\frac{9}{4}} = 1 \cdot \frac{4}{9} = \frac{4}{9}$

1. $(-6)^3$ **2.** 15^0 **3.** $2^3 \cdot 2^2$ **4.** $4^2 \cdot 3^3$

5. 2^{-4} **6.** $r^5 \cdot r^8$ **7.** $\frac{r^9}{r^4}$ **8.** $\frac{t^3}{t^4}$

9. $a \cdot a^{-1}$ **10.** $a^3 \cdot a^{-5}$ **11.** $x^{-1} \cdot x^{-2}$ **12.** $(-3)^{-2}$

13. $(3^2)^2$ **14.** $\left(\frac{7}{9}\right)^0$ **15.** $(-1)^8$ **16.** $(-1)^{99}$

17. $\left(\frac{1}{2}\right)^{-3}$ **18.** $(2x)^3$ **19.** $(3y^2)(2y^4)$ **20.** $(x^2y)^5$

21. $(m^{-5})(m^5)$ **22.** $\frac{d^{10}}{d^{-10}}$ **23.** $7^5 \cdot 7^{-11} \cdot 7^7$ **24.** $x^{-6} \cdot x^{-2} \cdot x^3$

25. $(x^2)^{-2}$ **26.** $(x^{-2})^2$ **27.** $(4x^3y^2)^2$ **28.** $(3x)^2(2xy^3)$

29. $(-2s^5t)(-4st)^2$ **30.** $4^{-1} \cdot 3^{-2}$ **31.** $\left(\frac{2}{3}\right)^{-3}$ **32.** $\left(\frac{7}{5}\right)^{-2}$

The Buckeye butterfly shown in the photograph is native to Florida. Notice that the wings on the right side are mirror images of those on the left. This kind of symmetry is very common in living things.

Transformations

12

Some Basic Mappings

Objectives

1. Recognize and use the terms *mapping, image, preimage, transformation,* and *isometry.*
2. Locate images of figures by reflection, translation, glide reflection, rotation, and dilation.
3. Recognize the properties of the basic mappings.

12-1 Isometries

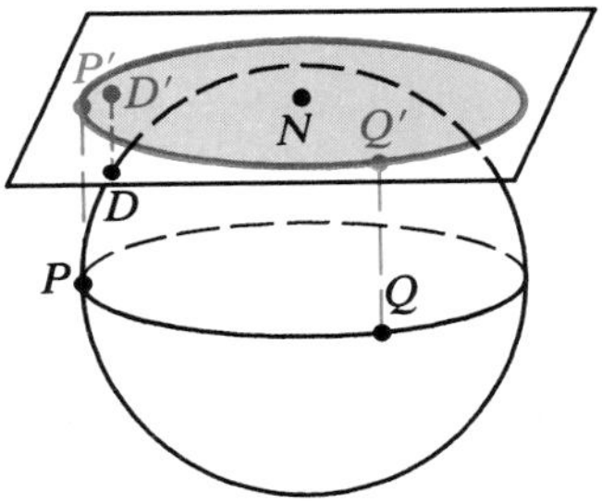

Have you ever wondered how maps of the round Earth can be made on flat paper? The diagram illustrates the idea behind a *polar map* of the northern hemisphere. A plane is placed tangent to a globe of the Earth at its North Pole N. Every point P of the globe is projected straight upward to a point P' in the plane. P' is called the **image** of P, and P is called the **preimage** of P'. The diagram shows the images of two points P and Q on the globe's equator. It also shows D', the image of a point D not on the equator.

This correspondence between points of the globe's northern hemisphere and points in the plane is an example of a **mapping.** If A and B are two sets of points, a mapping from A to B is called **one-to-one** when every point of A has exactly one image in B and every point of B has exactly one preimage in A. The polar projection illustrated above is a one-to-one mapping of the northern hemisphere of the globe onto a disc in the tangent plane (the shaded disc in the figure). The projection of a plane to a line, described in Example 1, is *not* one-to-one.

Example 1 Given a plane and a line l in the plane, the projection of the plane onto l maps each point P in the plane to the intersection point of l and the perpendicular from P to l. Any point on l is mapped to itself. Let P' be the image of P. Find all the preimages of P'.

Solution Every point on $\overleftrightarrow{PP'}$ maps to P', and any point not on $\overleftrightarrow{PP'}$ (point S, for example) maps to a point different from P'. So the preimages of P' are the points of $\overleftrightarrow{PP'}$. Since P' has many preimages, the projection is not one-to-one.

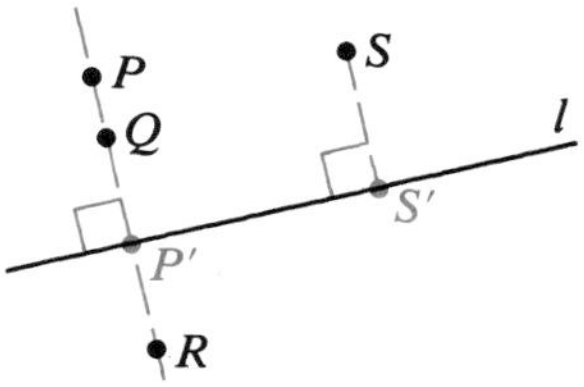

We are now ready to define the special kind of mapping we will study in this chapter. A **transformation** is a one-to-one mapping from the whole plane to the whole plane. Transformations are named by capital letters and are often described in terms of coordinates. Suppose a transformation R maps each point with coordinates (x, y) to an image point with coordinates $(-x, y)$. We write $R:(x, y) \rightarrow (-x, y)$ and we say that R maps (x, y) to $(-x, y)$.

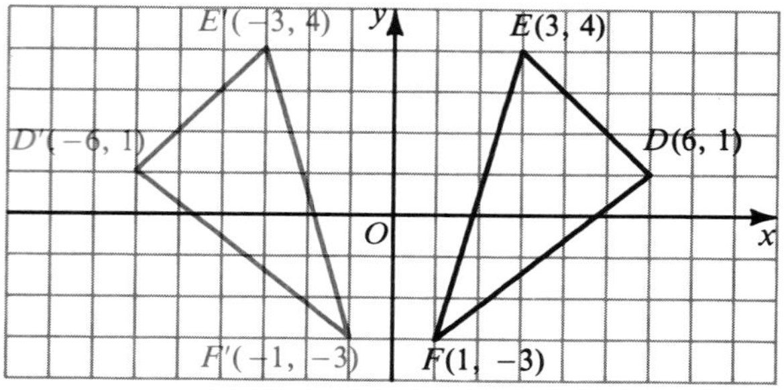

The diagram shows the effect of transformation R on the points of $\triangle DEF$. $R:\triangle DEF \rightarrow \triangle D'E'F'$. We say that R maps $\triangle DEF$ to $\triangle D'E'F'$ and that $\triangle D'E'F'$ is the image of $\triangle DEF$ *by* R or *under* R.

Notice that the distance between D and E is equal to the distance between D' and E' (both are equal to $3\sqrt{2}$). Exercise 17 outlines a proof that under transformation R the distance between *any* two points is equal to the distance between their images. We say that such a mapping *preserves distance*. A transformation that preserves distance is called an **isometry.** Thus an isometry is a transformation with the property that $PQ = P'Q'$ for any two points P and Q with images P' and Q'.

Example 2 $M:(x, y) \rightarrow (4x, y + 1)$

a. Find the images of $P(1, 2)$ and $Q(2,5)$.

b. Is M an isometry?

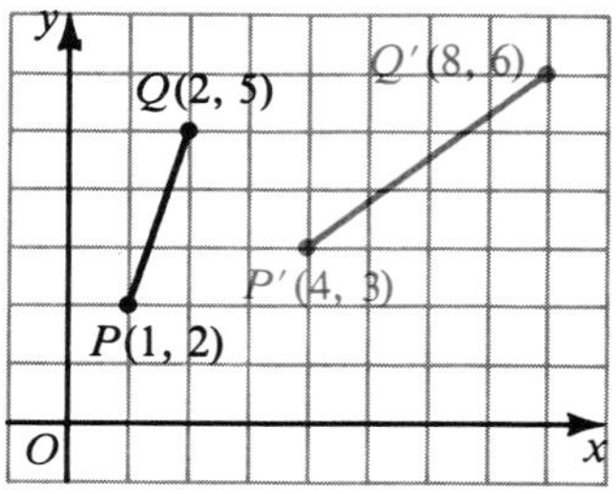

Solution

a. P' has coordinates $(4 \cdot 1, 2 + 1)$, or $(4, 3)$.
Q' has coordinates $(4 \cdot 2, 5 + 1)$, or $(8, 6)$.

b. $PQ = \sqrt{10}$ and $P'Q' = 5$
Since $PQ \neq P'Q'$, M is not an isometry.

We have defined isometries as distance-preserving transformations. Using this definition it can be shown that isometries preserve many other properties of geometric figures. For example, under any isometry the image of a line is a line, the image of a segment is a congruent segment, and the image of an angle is a congruent angle.

Classroom Exercises

1. Explain why each of the mappings pictured below is not a one-to-one mapping from set A to set B.

a.

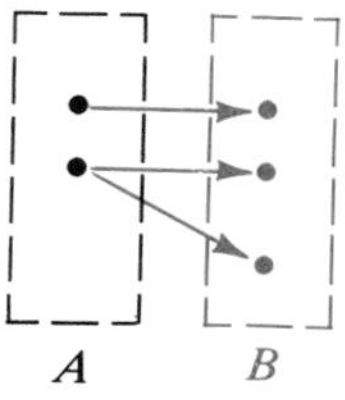

b.

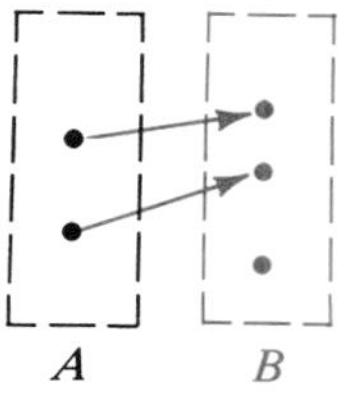

c.

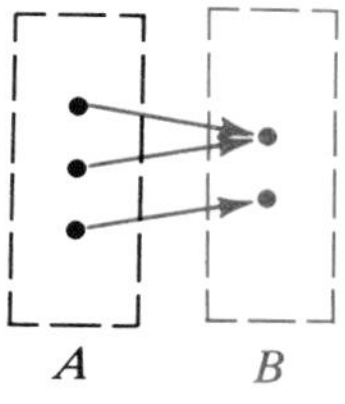

d.

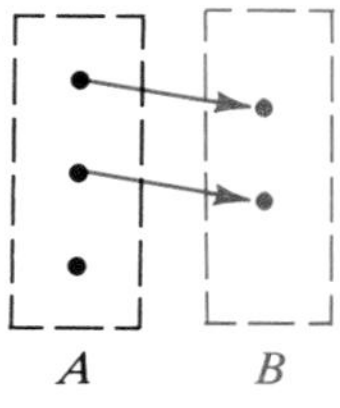

Transformation S maps $\triangle ABC$ to $\triangle A'B'C'$.

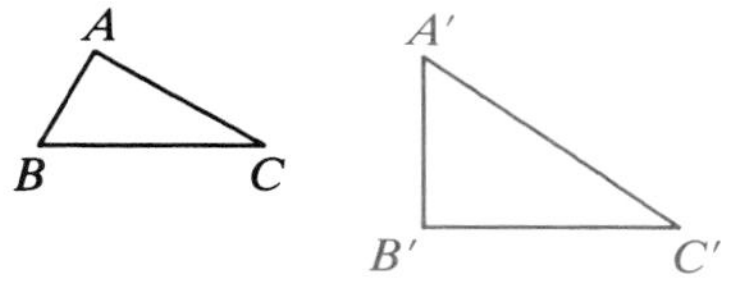

Exs. 2–4

2. What is the image of A? of B?
3. Is the transformation S an isometry? Explain your answer.
4. What is the preimage of C'?
5. Use the transformation $T:(x, y) \rightarrow (x + 1, y + 2)$ in this exercise.
   a. Plot the following points and their images on the chalkboard: $A(0, 0)$, $B(3, 4)$, $C(5, 1)$, $D(-1, -3)$
   b. Find the distances AB and $A'B'$.
   c. Find the distances CD and $C'D'$.
   d. Is this transformation an isometry?
   e. What is the preimage of $(0, 0)$? of $(4, 5)$?

Exercises 6–8 refer to the globe shown on page 511.

6. What is the image of point N?
7. Is the distance between N and P on the globe the same as the corresponding distance on the polar map?
8. Does the polar map preserve distance?
9. Does the projection of a plane onto a line shown in Example 1 preserve distance? Explain your answer.
10. This exercise illustrates why isometries are also called *congruence mappings*. Suppose an isometry maps $\triangle PQR$ to $\triangle P'Q'R'$.
   a. Which distances must be equal?
   b. Which segments must be congruent?
   c. Why is $\triangle PQR \cong \triangle P'Q'R'$?
   d. Which angles must be congruent? (We say that an isometry *preserves angle measure*.)

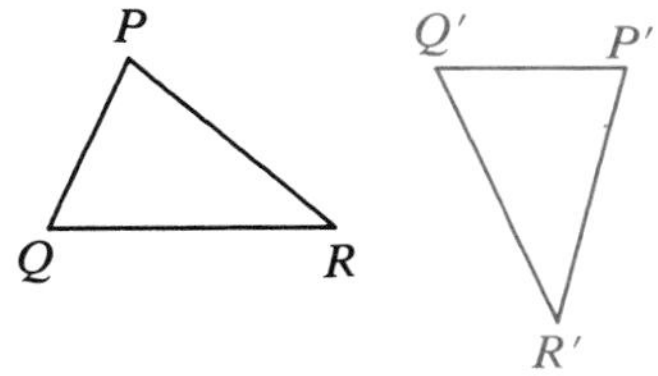

11. Use the mapping $(x, y) \rightarrow (x, 0)$ in this exercise.
   a. Find the images of $(2, 4)$ and $(2, 5)$.
   b. Name several preimages of $(2, 0)$. Is this mapping a transformation? Explain your answer.

Written Exercises

For each transformation given in Exercises 1–6:
a. Plot the three points $A(0, 4)$, $B(4, 6)$, and $C(2, 0)$ and their images A', B', and C' under the transformation.
b. State whether or not the transformation appears to be an isometry.
c. Find the preimage of $(12, 6)$.

A 1. $T:(x, y) \rightarrow (x + 4, y - 2)$

2. $S:(x, y) \rightarrow (2x + 4, 2y - 2)$
3. $D:(x, y) \rightarrow (3x, 3y)$
4. $H:(x, y) \rightarrow (-x, -y)$
5. $M:(x, y) \rightarrow (12 - x, y)$
6. $G:(x, y) \rightarrow (-\frac{1}{2}x, -\frac{1}{2}y)$

7. O is a point equidistant from parallel lines l_1 and l_2. A mapping M maps each point P of l_1 to the point P' where $\overrightarrow{PO}$ intersects l_2.
   a. Is the mapping a one-to-one mapping of l_1 onto l_2?
   b. Does this mapping preserve distance?
   c. If l_1 and l_2 were not parallel, would the mapping preserve distance? Illustrate your answer with a sketch.

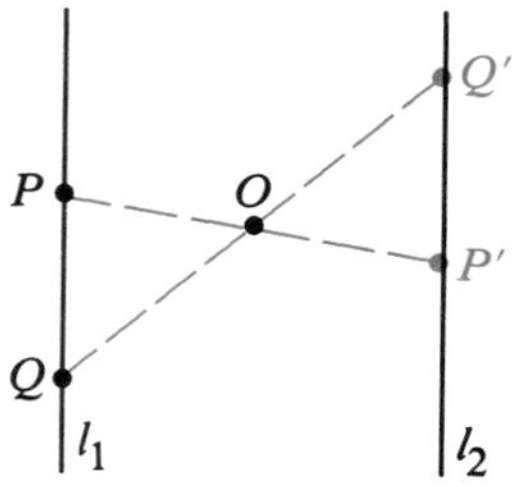

8. $\triangle XYZ$ is isosceles with $\overline{XY} \cong \overline{XZ}$. Describe a way of mapping each point of $\overline{XY}$ to a point of $\overline{XZ}$ so that the mapping is one-to-one and preserves distance.

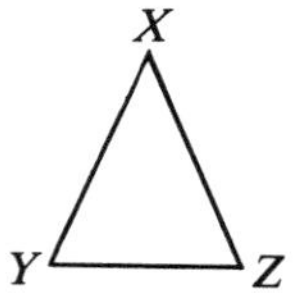

B 9. $ABCD$ is a trapezoid. Describe a way of mapping each point of $\overline{DC}$ to a point of $\overline{AB}$ so that the mapping is one-to-one. Does your mapping preserve distance?

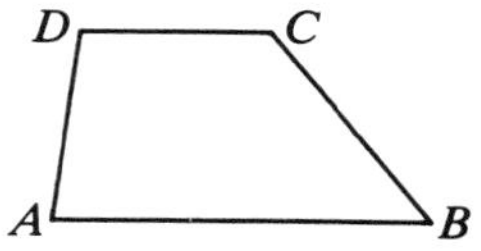

10. The red and blue squares are congruent and have the same center O. A mapping maps each point P of the red square to the point P' where $\overrightarrow{OP}$ intersects the blue square.
   a. Is this mapping one-to-one?
   b. Copy the diagram and locate a point X that is its own image.
   c. Locate two points R and S on the red square and their images R' and S' on the blue square that have the property that $RS \neq R'S'$.
   d. Does this mapping preserve distance?
   e. Describe a mapping from the red square onto the blue square that *does* preserve distance.

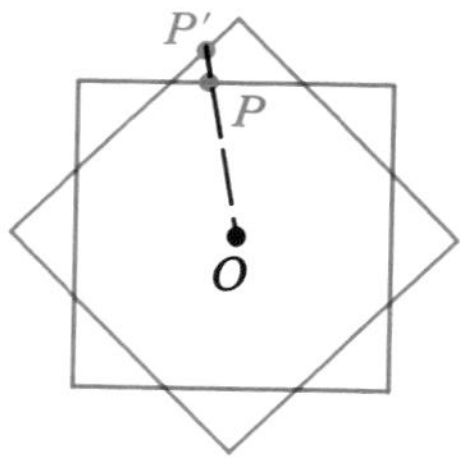

11. The transformation $T:(x, y) \rightarrow (x + y, y)$ preserves areas of figures even though it does not preserve distances. Illustrate this by drawing a square with vertices $A(2, 3)$, $B(4, 3)$, $C(4, 5)$, and $D(2, 5)$ and its image $A'B'C'D'$. Find the area and perimeter of each figure.

A piece of paper is wrapped around a globe of the Earth to form a cylinder as shown. O is the center of the Earth and a point P of the globe is projected along $\overrightarrow{OP}$ to a point P' of the cylinder.

12. Describe the image of the globe's equator.
13. Is the image of the Arctic Circle the same size as the image of the equator?
14. Are distances near the equator distorted more than or less than distances near the Arctic Circle?
15. Does the North Pole (point N) have an image?

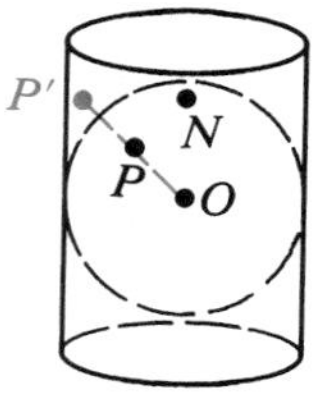

16. Points A and B have the same image under a mapping U. Explain why the mapping U does not preserve distance.

C **17.** Prove that the transformation R discussed on page 512 is an isometry. Let $P(x_1, y_1)$ and $Q(x_2, y_2)$ be any two points. Find the coordinates of their images P' and Q'. Then show $PQ = P'Q'$ by using the distance formula.

18. Isometry I maps points A, B, and C to A', B', and C'. Use the definition of isometry to show that if B is between A and C, then B' is on $\overleftrightarrow{A'C'}$. (*Hint:* If A', B', and C' are *not* collinear, they form a triangle. Show that this is impossible.)

12-2 Reflections

When you stand before a mirror, your image appears to be as far behind the mirror as you are in front of it. The diagram shows a transformation in which a line acts like a mirror. Points P and Q are reflected in line j to their images P' and Q'. This transformation is called a *reflection.* Line j is called the *line of reflection.*

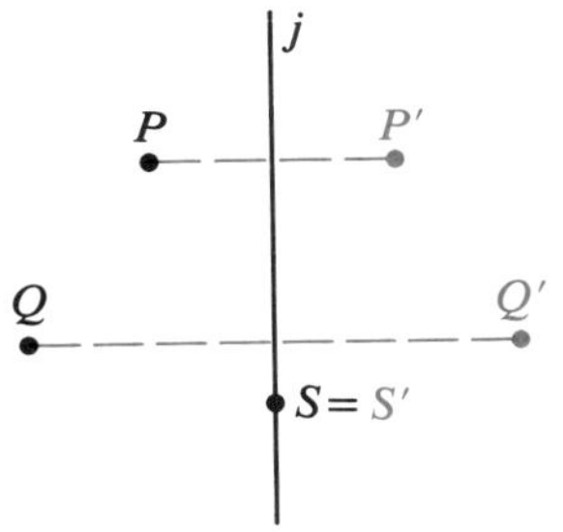

A **reflection** in line j maps every point P to a point P' such that:
(1) If P is not on the line j, then j is the perpendicular bisector of $\overline{PP'}$.
(2) If P is on line j, then $P' = P$.

To abbreviate *reflection in line j,* we write R_j. To abbreviate the statement *R_j maps P to P',* we write $R_j: P \rightarrow P'$. This may also be read as *P is reflected in line j to P'.*

Theorem 12-1

A reflection in a line is an isometry.

Theorem 12–1 can be proved with or without the use of coordinates. If coordinates are not used, we must show that $PQ = P'Q'$ for all choices of P and Q. Four of the possible cases are shown below. In Written Exercises 19–21 you will prove Theorem 12–1 for Cases 2–4, using the fact that the line of reflection is the perpendicular bisector of $\overline{PP'}$ and $\overline{QQ'}$.

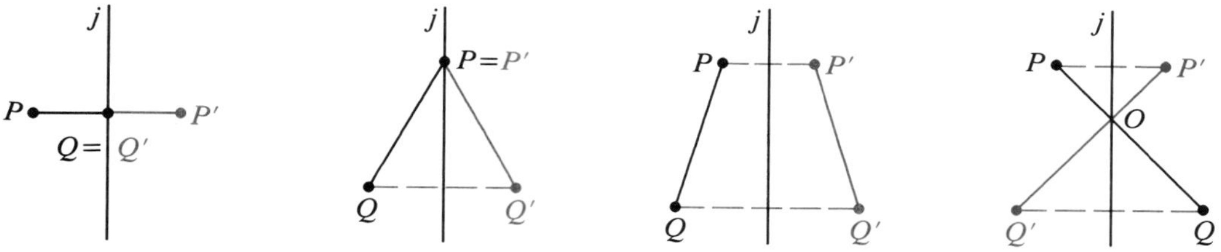

Case 1 Case 2 Case 3 Case 4

To prove the theorem by using coordinates, we assign coordinates in the plane so that the line of reflection becomes the y-axis. Then in coordinate terms the reflection is $R:(x, y) \to (-x, y)$. This transformation was introduced in the preceding section, and in Exercise 17 on page 515 the distance formula was used to prove that $PQ = P'Q'$. Although the diagram shows P and Q on the same side of the y-axis, you should realize that the coordinates x_1, y_1, x_2, and y_2 can be positive, negative, or zero, thereby covering all cases.

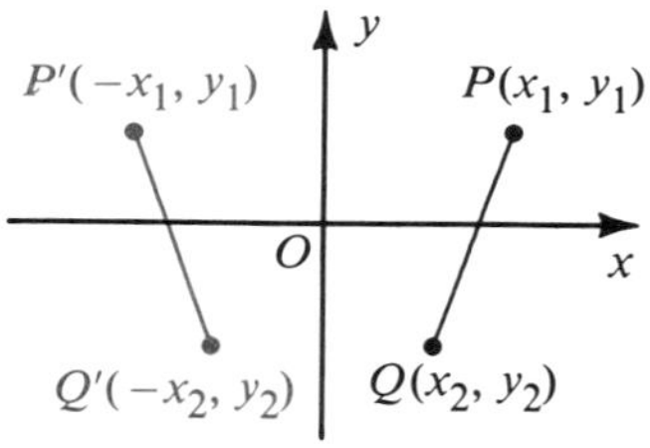

Since a reflection is an isometry, it preserves distance and angle measure. The image of a triangle is a congruent triangle, and in general any polygon is reflected to a congruent polygon.

Example Find the image of point $P(2, 4)$ and $\triangle ABC$ under each reflection.

a. The line of reflection is the x-axis.

b. The line of reflection is the line $y = x$.

Solution The images are shown in red.

a.

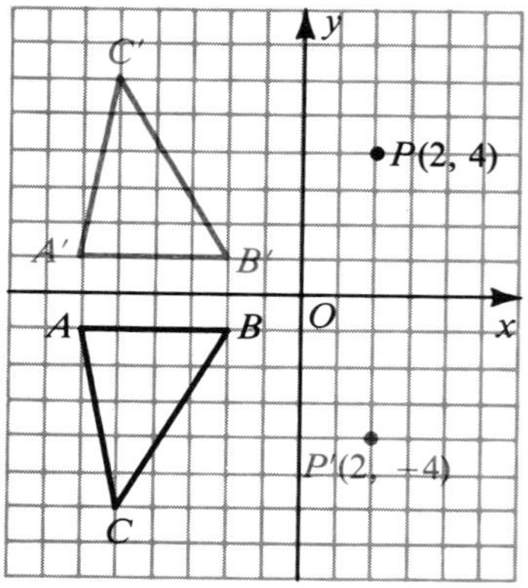

b.

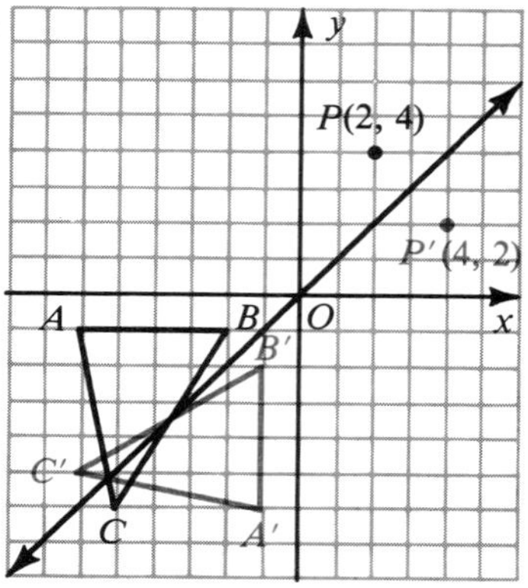

Classroom Exercises

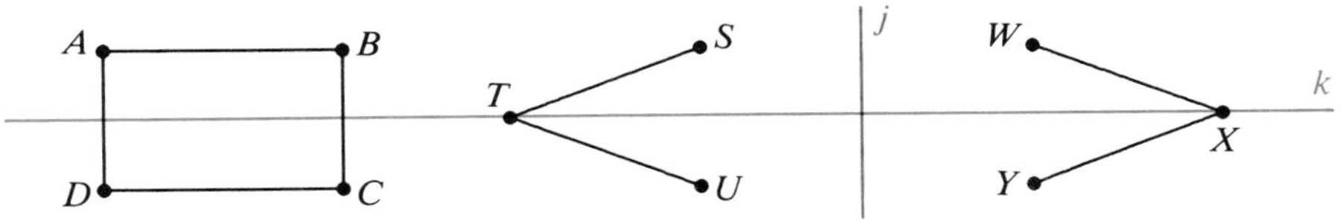

Complete the following.

1. R_k stands for __?__.
2. $R_k: A \rightarrow$ __?__
3. $R_k: \overline{AB} \rightarrow$ __?__
4. $R_k: B \rightarrow$ __?__
5. $R_k: C \rightarrow$ __?__
6. $R_k: \overline{BC} \rightarrow$ __?__
7. $R_k: T \rightarrow$ __?__
8. $R_k: \angle STU \rightarrow$ __?__
9. $R_j: S \rightarrow$ __?__
10. $R_j: \overline{ST} \rightarrow$ __?__
11. $R_j:$ __?__ $\rightarrow \overline{XY}$
12. $R_j:$line $k \rightarrow$ __?__

Points A–D are reflected in the x-axis. Points E–H are reflected in the y-axis. State the coordinates of the images.

13.

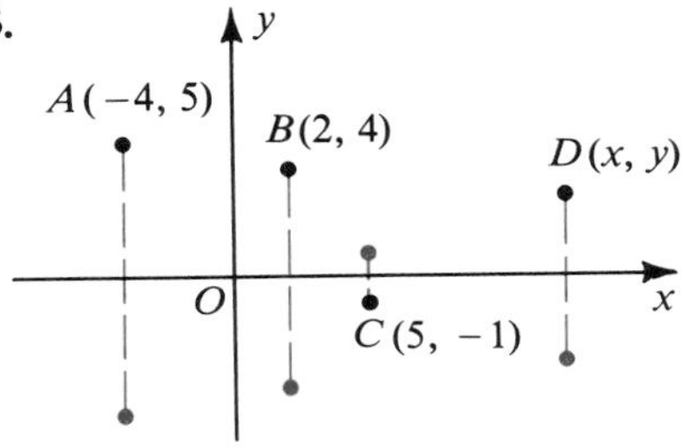

14.

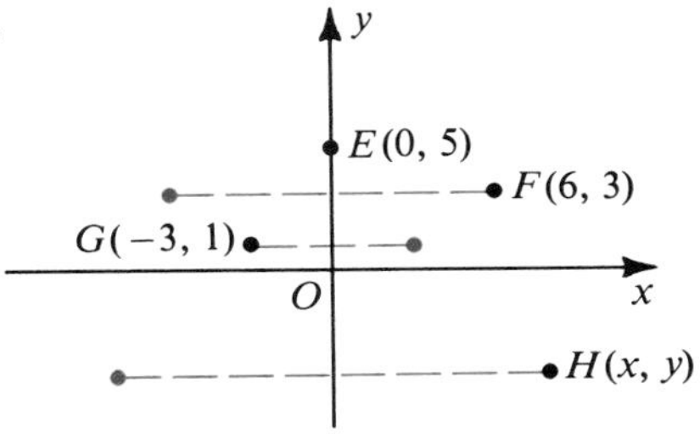

Sketch each figure on the chalkboard. With a different color, sketch its image, using the dashed line as the line of reflection.

15.

16.

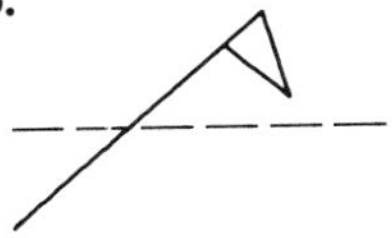

17.

18.

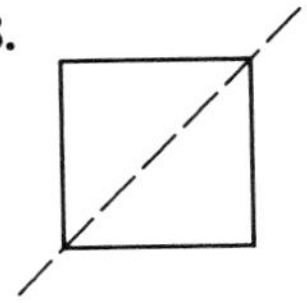

19.

20.

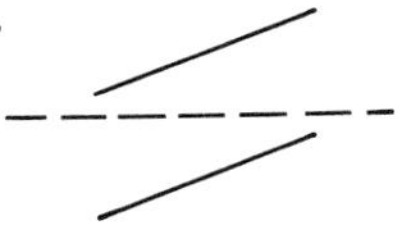

21.

22.

23. When the word MOM is reflected in a vertical line, the image is still MOM. Can you think of other words that are unchanged when reflected in a vertical line?
24. When the word DECK is reflected in a horizontal line, the image is still DECK. Can you think of other words that are unchanged when reflected in a horizontal line?

Written Exercises

Copy each figure on squared paper. Then draw the image by reflection in line *k*.

A **1.**

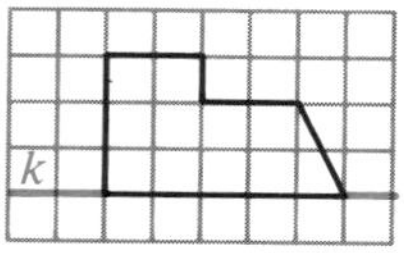

2.

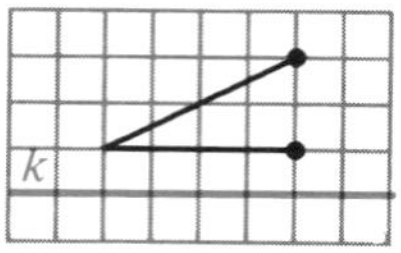

3.

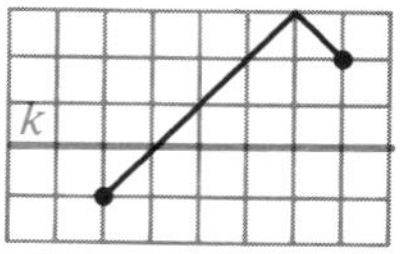

4.

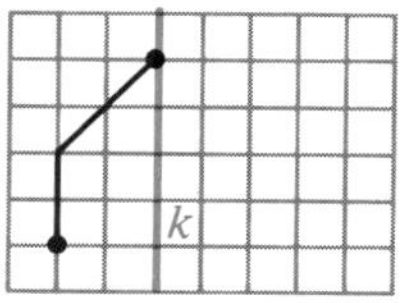

5.

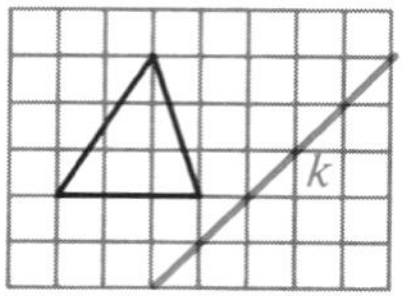

6.

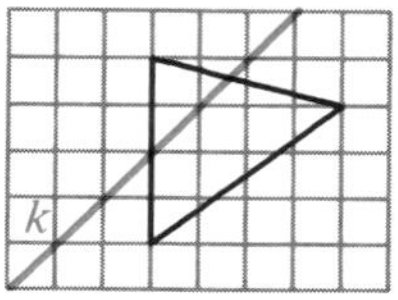

Write the coordinates of the image of each point by reflection in (a) the *x*-axis and (b) the *y*-axis.

7. *A* **8.** *B* **9.** *C*

10. *D* **11.** *E* **12.** *O*

B **13–18.** Write the coordinates of the image of each point in Exercises 7–12 by reflection in the line $y = x$.

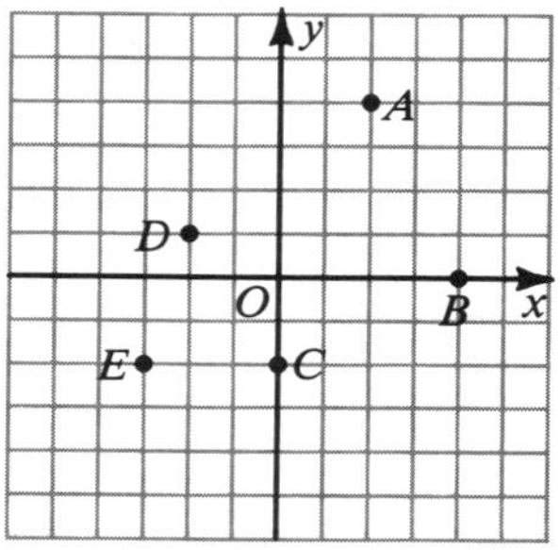

In Exercises 19–21, refer to the diagrams on page 516. Given the reflection $R_j : \overline{PQ} \rightarrow \overline{P'Q'}$, write the key steps of a proof that $PQ = P'Q'$ for each case.

19. Case 2 **20.** Case 3 **21.** Case 4

Copy the figures shown. Use a straightedge and compass.

22. Construct the image of *A* under R_t.

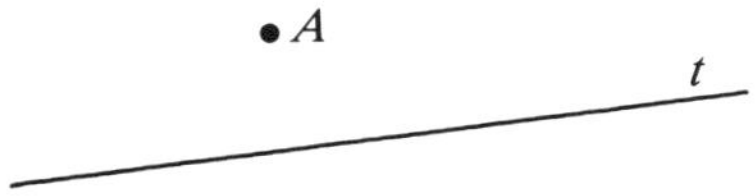

23. Construct the line of reflection *t* so that $R_t : B \rightarrow B'$.

The photograph shows a reflected beam of laser light. Exercises 24–28 deal with the similar reflected path of a golf ball bouncing off the walls of a miniature golf layout. These exercises show how the geometry of reflections can be used to solve the problem of aiming a reflected path at a particular target.

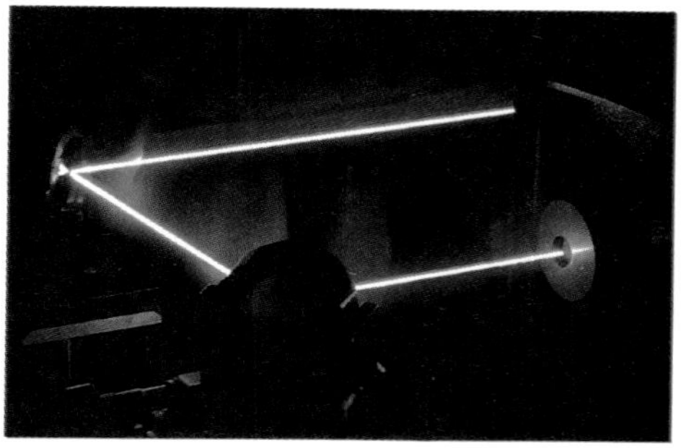

24. A ball that does not have much spin will roll off a wall so that the two angles that the path forms with the wall are congruent. Thus, to roll the ball from B off the wall shown and into hole H, you need to aim the ball so that $\angle 1 \cong \angle 2$.

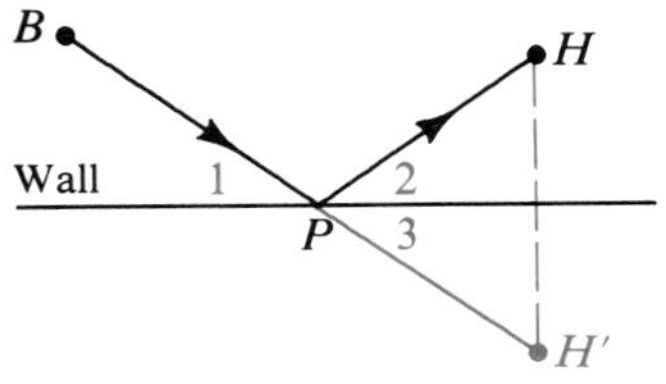

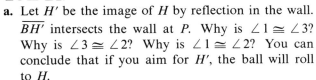

a. Let H' be the image of H by reflection in the wall. $\overline{BH'}$ intersects the wall at P. Why is $\angle 1 \cong \angle 3$? Why is $\angle 3 \cong \angle 2$? Why is $\angle 1 \cong \angle 2$? You can conclude that if you aim for H', the ball will roll to H.

b. Show that the distance traveled by the ball equals the distance BH'.

25. In the two-wall shot illustrated at the right, a reflection in one wall maps H to H', and a reflection in a second wall (extended) maps H' to H''. To roll the ball from B to H, you aim for H''. Show that the total distance traveled by the ball equals the distance BH''.

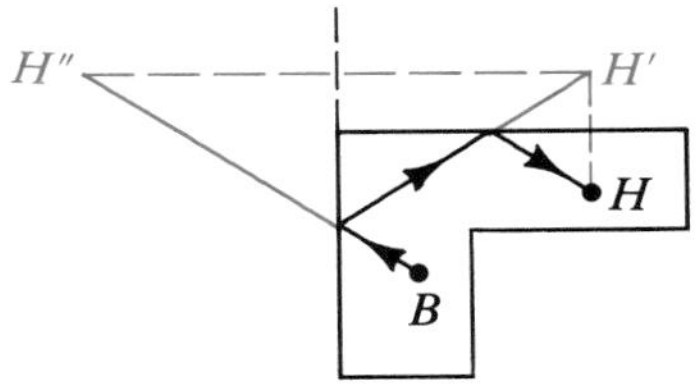

26. Show how to score a hole in one on the fifth hole of the golf course shown by rolling the ball off one wall.

27. Repeat Exercise 26 but roll the ball off two walls.

28. Repeat Exercise 26 but roll the ball off three walls.

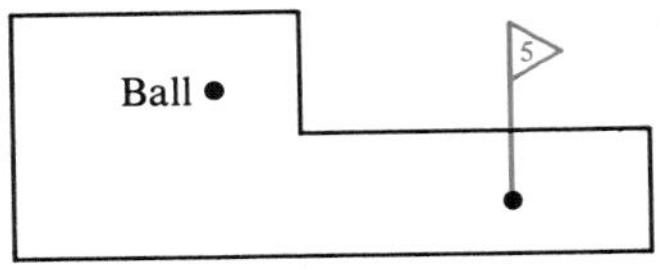

29. A ball rolls at a 45° angle away from one side of a billiard table that has a coordinate grid on it. If the ball starts at the point (0, 1) it will eventually return to its starting point. Would this happen if the ball started from other points on the y-axis between (0, 0) and (0, 4)?

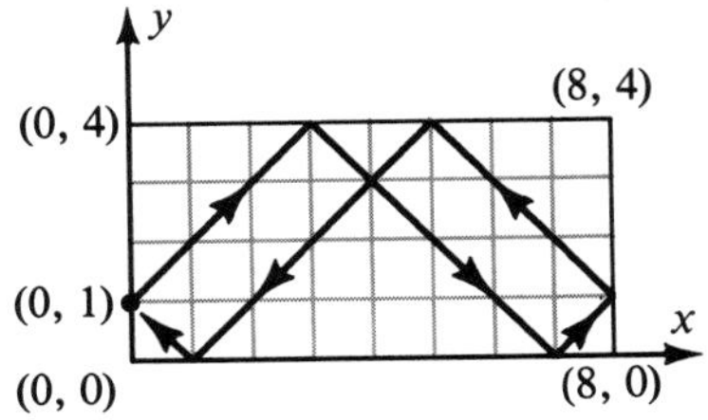

30. The line with equation $y = 2x + 3$ is reflected in the y-axis. Find an equation of the image line.

31. The line with equation $y = x + 5$ is reflected in the x-axis. Find an equation of the image line.

C 32. Draw the x- and y-axes and the line l with equation $y = -x$. Plot several points and their images under R_l. What is the image of (x, y)?

33. Draw the x- and y-axes and the vertical line j with equation $x = 5$. Find the images under R_j of the following points.

a. (4, 3) **b.** (0, −2) **c.** (−3, 1) **d.** (x, y)

34. Repeat Exercise 33 letting j be the horizontal line with equation $y = 6$.

For each exercise make a sketch showing A' as the mirror image of point A in a line k. Then find an equation of line k.

	35.	**36.**	**37.**	**38.**	**39.**	**40.**	**41.**
A	(2, 3)	(5, 0)	(1, 4)	(4, 0)	(5, 1)	(0, 2)	(−1, 2)
A'	(−2, 3)	(9, 0)	(3, 4)	(4, 6)	(1, 5)	(4, 6)	(4, 5)

Application

MIRRORS

If a ray of light strikes a mirror at an angle of 40°, it will be reflected off the mirror at an angle of 40° also. The angle between the mirror and the reflected ray is always congruent to the angle between the mirror and the initial light ray. In the diagram at the left below, $\angle 2 \cong \angle 1$.

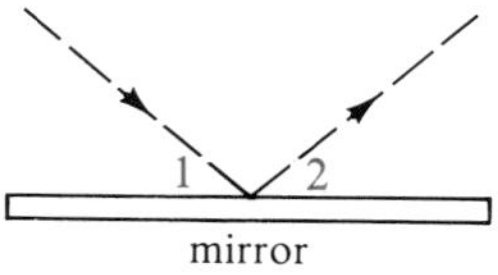

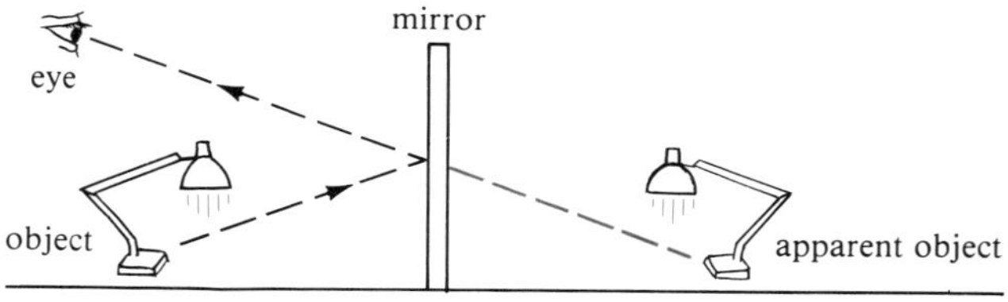

We see objects in a mirror when the reflected light ray reaches the eye. The object appears to lie behind the mirror as shown in the diagram at the right above.

You can see all of yourself in a mirror that is only half as tall as you are if the mirror is in a position as shown. You see the top of your head at the top of the mirror and your feet at the bottom of the mirror. If the mirror is too high or too low, you will not see your entire body.

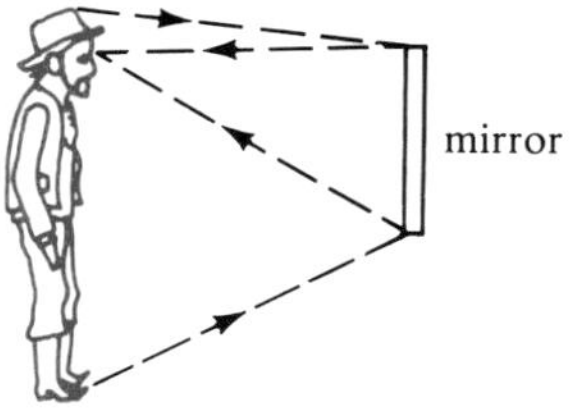

A periscope uses mirrors to enable a viewer to see above the line of sight. The diagram at the right is a simple illustration of the principle used in a periscope. It has two mirrors, parallel to each other, at the top and at the bottom. The mirrors are placed at an angle of 45° with the horizontal. Horizontal light rays from an object entering at the top are reflected down to the mirror at the bottom. They are then reflected to the eye of the viewer.

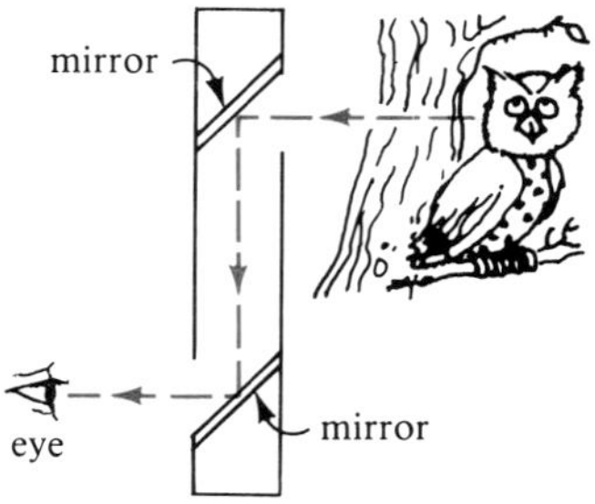

Exercises

1. What are the measures of the angles that the initial light ray and the reflected light rays make with the mirrors in the diagram of the periscope shown on page 520?
2. If you can see the eyes of someone when you look into a mirror, can the other person see your eyes in that same mirror?
3. A person with eyes at A, 150 cm above the floor, faces a mirror 1 m away. The mirror extends 30 cm above eye level. How high can the person see on a wall 2 m behind point A?
4. Prove that the height of your image in a mirror is one-half your height when you are just able to see the top of your head at the top of a mirror and your feet at the bottom.
5. Prove that the point D which is as far behind the mirror as the object A is in front of the mirror lies on $\overleftrightarrow{BC}$. (*Hint:* Show that $\angle CBE$ and $\angle EBD$ are supplementary.)
6. Show that the light ray follows the shortest possible path from A to C via the mirror by proving that for any point E on the mirror (other than B) $AE + EC > AB + BC$. (*Hint:* $AE + EC =$ __?__ $+ EC$)

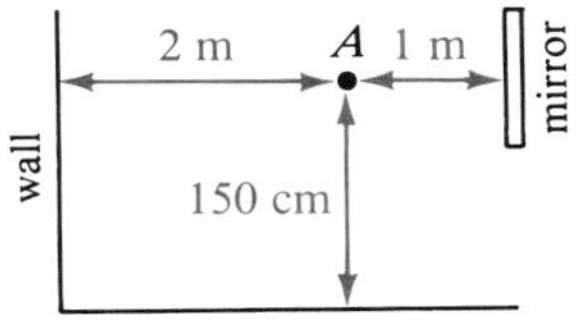

Ex. 3

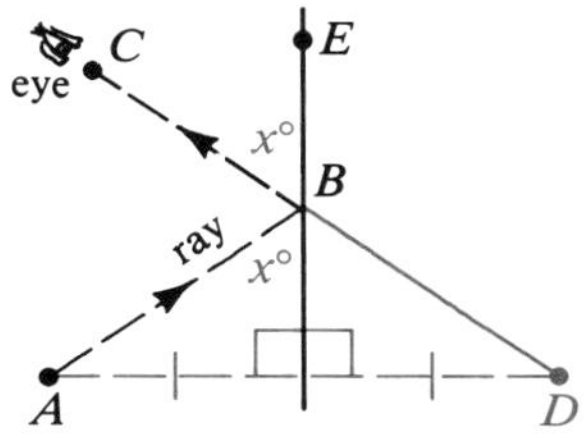

Exs. 5, 6

12-3 Translations and Glide Reflections

The photograph suggests the transformation called a *translation,* or *glide.* A *translation* can be thought of as a transformation that glides all points of the plane the same distance in the same direction.

If a translation maps A to A', B to B', and C to C', points A, B, and C glide along parallel or collinear segments so that $AA' = BB' = CC'$.

While the idea of gliding points the same distance in the same direction may be clear to you, it is difficult to state precisely what is meant by "same direction" until your next mathematics course. Therefore we adopt the slightly different but equivalent definition of translation that is given on the following page.

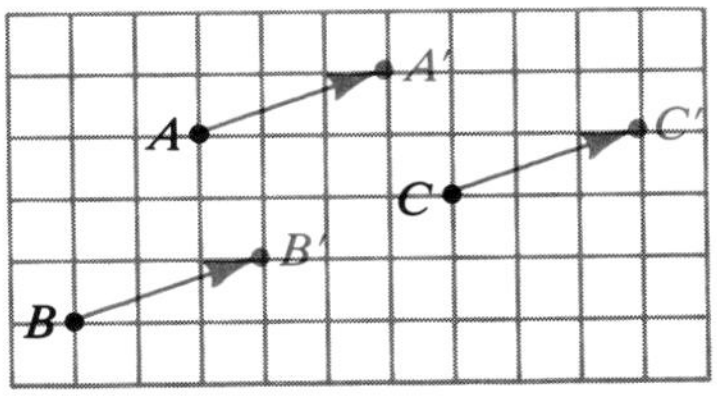

A **translation,** or glide, is an isometry that glides all points the same distance. That is, a transformation T is a translation if for any two points P and Q with images P' and Q' two properties hold:

(1) $PQ = P'Q'$ (T is an isometry.)
(2) $PP' = QQ'$ (P and Q glide the same distance.)

The easiest way to describe a translation is to use coordinates. For instance, consider a translation in which every point glides 8 units right and 2 units up. To indicate this we write $T:(x, y) \to (x + 8, y + 2)$. The following diagram shows how $\triangle PQR$ is mapped by T to $\triangle P'Q'R'$. You can use the distance formula to check that $PQ = P'Q'$ and that $PP' = QQ'$.

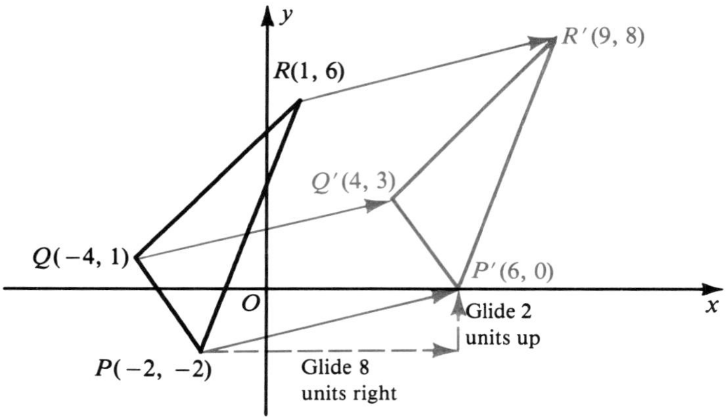

The distance formula can be used to prove that any coordinate transformation of this kind is a translation.

Theorem 12-2

If a transformation T maps any point (x, y) to $(x + a, y + b)$, then T is a translation.

■ ***Plan for Proof:*** Label two points P and Q and their images P' and Q' as shown in the diagram. To show that T is a translation we need to show that $PQ = P'Q'$ and $PP' = QQ'$. Use the distance formula to find these four distances and compare them.

PQ and $P'Q'$ both equal $\sqrt{(x_2 - x_1)^2 + (y_2 - y_1)^2}$.
PP' and QQ' both equal $\sqrt{a^2 + b^2}$.

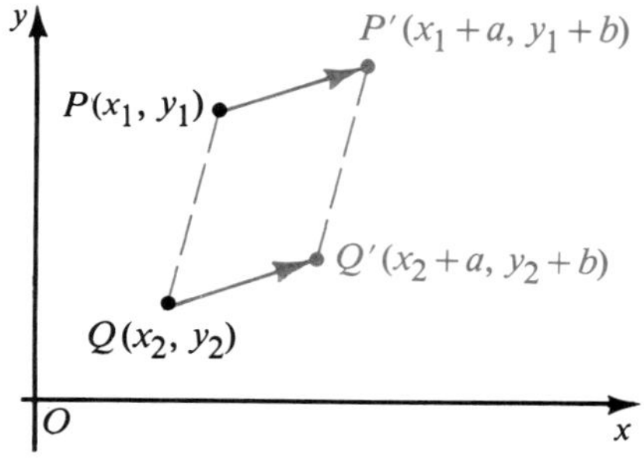

The converse of this theorem is also true: Every translation is a coordinate mapping of the form $(x, y) \to (x + a, y + b)$.

A glide and a reflection can be carried out one after the other to produce a transformation known as a *glide reflection.* The succession of footprints shown illustrates a glide reflection. Note that the reflection line is parallel to the direction of the glide. This is an important part of the definition that follows.

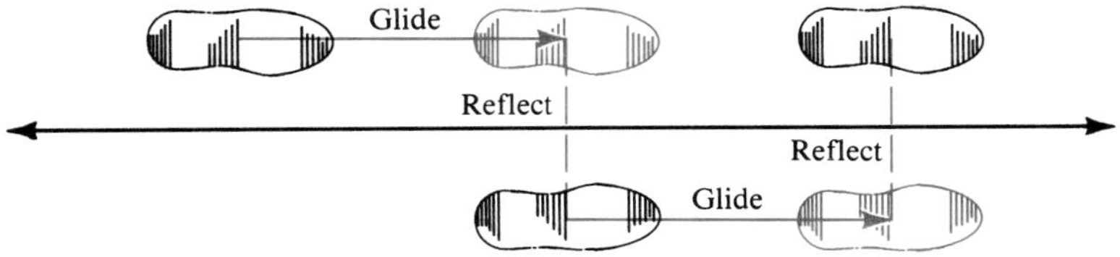

A **glide reflection** is a transformation in which every point P is mapped to a point P'' by the following procedure:

First, a glide maps P to P'. Second, a reflection in a line parallel to the glide line maps P' to P''.

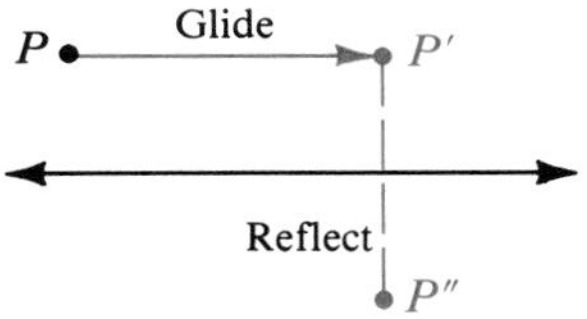

A glide reflection combines two isometries to produce a new transformation. As you would expect, this new transformation is itself an isometry. We will look further at such combinations, or *products,* of mappings in Section 12-6.

Classroom Exercises

1. Complete each statement for the translation $T:(x, y) \rightarrow (x + 3, y - 1)$.

a. T glides points __?__ units right and 1 unit __?__.

b. The image of (4, 6) is (__?__, __?__).

c. The preimage of (2, 3) is (__?__, __?__).

Describe each translation in words, as in Exercise 1(a), and give the image of (4, 6) and the preimage of (2, 3).

2. $T:(x, y) \rightarrow (x - 5, y + 4)$

3. $T:(x, y) \rightarrow (x + 1, y)$

Each diagram shows a point *P* on the coordinate plane and its image *P′* under a translation *T*. Complete the statement $T:(x, y) \rightarrow$ (__?__, __?__).

4.

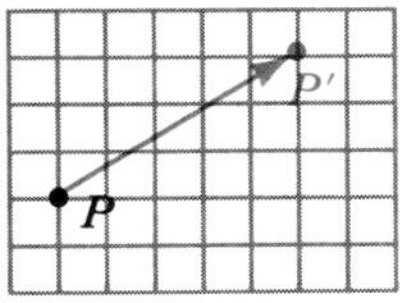

5.

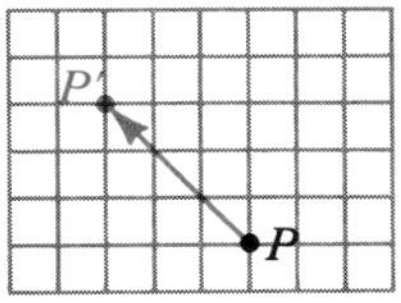

6.

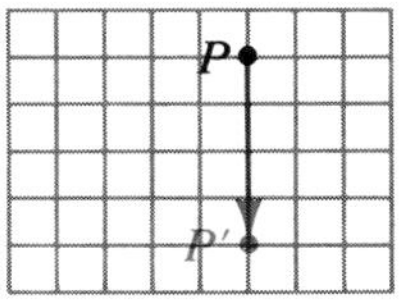

7. For a given translation, the image of the origin is (5, 7). What is the preimage of the origin?

8. Describe a transformation that will map figure F_1 to figure F_2.

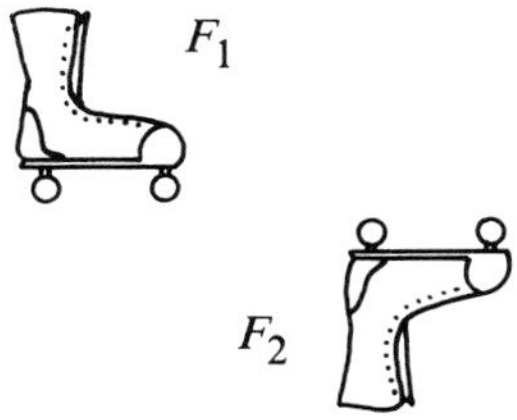

Written Exercises

In each exercise a translation T is described. For each T:
a. Make a drawing showing $\triangle ABC$ and its image $\triangle A'B'C'$.
b. In color, draw arrows from A to A', B to B', and C to C'.
c. Are your colored arrows the same length? Are they parallel?

A

1. $T:(x, y) \to (x + 8, y + 6)$
   $A(-2, 0)$, $B(0, 4)$, $C(3, -1)$

2. $T:(x, y) \to (x - 3, y - 6)$
   $A(3, 6)$, $B(-3, 6)$, $C(-1, -2)$

3. A translation maps the origin to (5, 1). Where does it map (3, 3)?

4. A translation maps the point (1, 1) to (3, 0). Where does it map (0, 0)?

5. A translation maps the point (−2, 3) to (2, 6). What point gets mapped to the origin?

6. The image of $P(-1, 5)$ under a translation is $P'(5, 7)$. What is the preimage of P?

In each exercise a glide reflection is described. Make a diagram showing $\triangle ABC$ and its image under the glide, $\triangle A'B'C'$. Also show $\triangle A''B''C''$, the image of $\triangle A'B'C'$ under the reflection.

7. Glide: All points move up 4 units.
   Reflection: All points are reflected in the y-axis.
   $A(1, 0)$, $B(4, 2)$, and $C(5, 6)$

8. Glide: All points move left 7 units.
   Reflection: All points are reflected in the x-axis.
   $A(4, 2)$, $B(7, 0)$, and $C(9, -3)$

B

9. Where does the glide reflection in Exercise 7 map (x, y)?

10. Where does the glide reflection in Exercise 8 map (x, y)?

11. Given $\odot A$ and $\odot B$ and $\overline{CD}$, construct a segment $\overline{XY}$ parallel to and congruent to $\overline{CD}$ and having X on $\odot A$ and Y on $\odot B$. (*Hint:* Translate $\odot A$ along a path parallel to and congruent to $\overline{CD}$.)

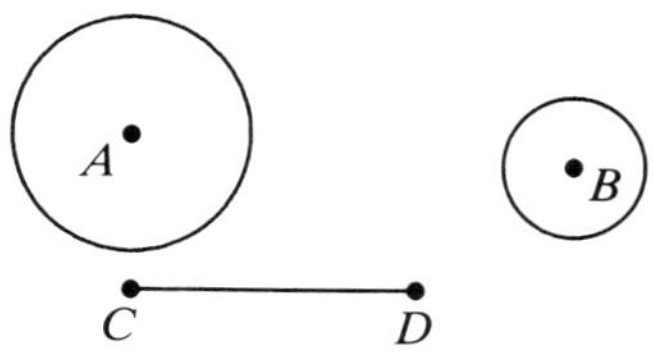

12. Describe how you would construct points X and Y, one on each of the lines shown, so that $\overline{XY}$ is parallel to and congruent to $\overline{EF}$.

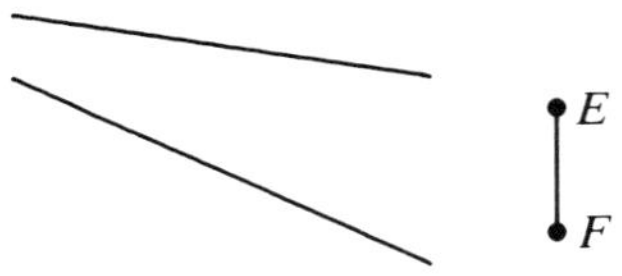

In Exercises 13 and 14, translations R and S are described. R maps point P to P' and S maps P' to P''. Find T, the translation that maps P directly to P''.

13. $R:(x, y) \to (x + 1, y + 2)$
    $S:(x, y) \to (x - 5, y + 7)$
    $T:(x, y) \to (\underline{\ ?\ }, \underline{\ ?\ })$

14. $R:(x, y) \to (x - 5, y - 3)$
    $S:(x, y) \to (x + 4, y - 6)$
    $T:(x, y) \to (\underline{\ ?\ }, \underline{\ ?\ })$

15. A glide reflection maps $\triangle ABC$ to $\triangle A'B'C'$. Copy the diagram and locate the midpoints of $\overline{AA'}$, $\overline{BB'}$, and $\overline{CC'}$. What seems to be true about these midpoints? Try to prove your conjecture.

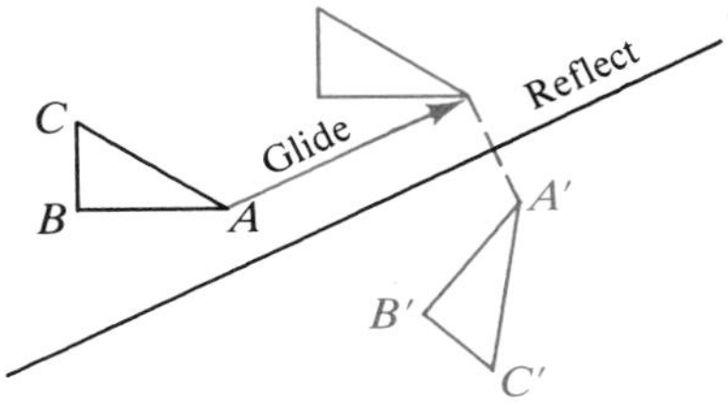

16. Copy the figure and use the result of Exercise 15 to find the reflecting line of the glide reflection that maps $\triangle ABC$ to $\triangle A'B'C'$. Also draw the glide image of $\triangle ABC$.

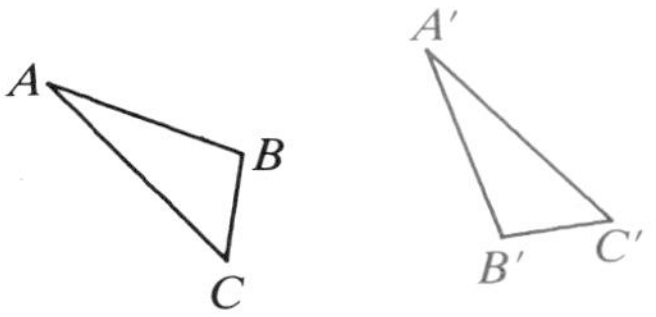

C 17. Prove that a glide reflection is an isometry.

18. Consider the translation $T:(x, y) \to (x + a, y + b)$. Let $P(x_1, y_1)$ and $Q(x_2, y_2)$ be any two points, with images P' and Q'.
   a. Find the coordinates of P' and Q' and use the distance formula to show that the values for PQ, $P'Q'$, PP', and QQ' given in the plan for the proof of Theorem 12–2 are correct. This proves that $PQ = P'Q'$ and $PP' = QQ'$.
   b. Prove that $\overline{PP'}$ and $\overline{QQ'}$ have equal slopes.
   c. Prove that $\overline{PQ}$ and $\overline{P'Q'}$ have equal slopes.
   d. What kind of quadrilateral is $PP'Q'Q$?

12–4 Rotations

A *rotation* is a transformation suggested by a rotating paddle wheel. When the wheel moves, each paddle rotates to a new position. When the wheel stops, the position of a paddle (P') can be referred to mathematically as the image of the initial position of the paddle (P).

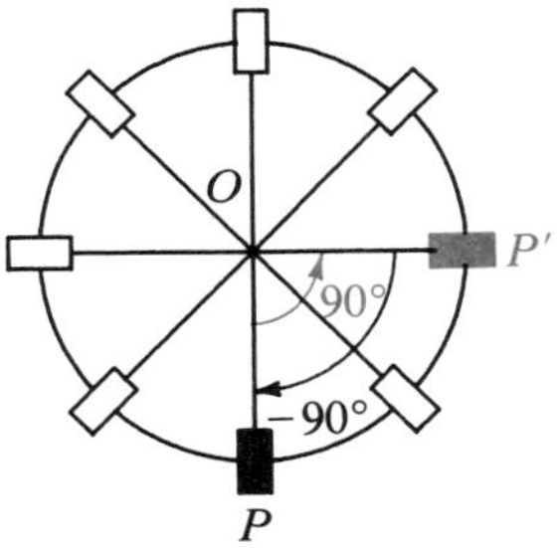

For the counterclockwise rotation shown about point O through $90°$, we write $\mathcal{R}_{O,90}$. A counterclockwise rotation is considered positive, and a clockwise rotation is considered negative. If the red paddle is rotated about O clockwise until it moves into the position of the black paddle, the rotation is denoted by $\mathcal{R}_{O,-90}$. (Note that to avoid confusion with the R used for reflections we use a script $\mathcal{R}$ for rotations.)

A 360° rotation about point O will rotate any point P around to itself so that $P' = P$. The diagram at the left below shows a rotation of 390° about O. Since 390° is 30° more than one full revolution, the image of any point P under a 390° rotation is the same as its image under a 30° rotation, and the two rotations are said to be equal. Thus, $\mathcal{R}_{O,30}$ is another name for $\mathcal{R}_{O,390}$, and we write $\mathcal{R}_{O,390} = \mathcal{R}_{O,30}$. Similarly, the diagram at the right below shows that a 90° counterclockwise rotation is equal to a 270° clockwise rotation because both have the same effect on any point P.

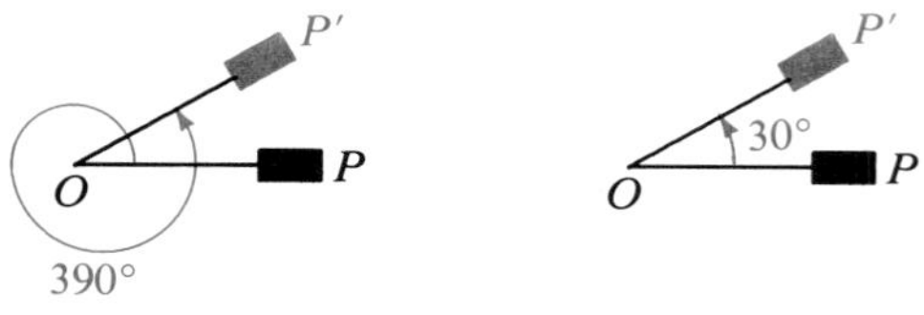

$\mathcal{R}_{O,390} = \mathcal{R}_{O,30}$
Notice: $390 - 360 = 30$

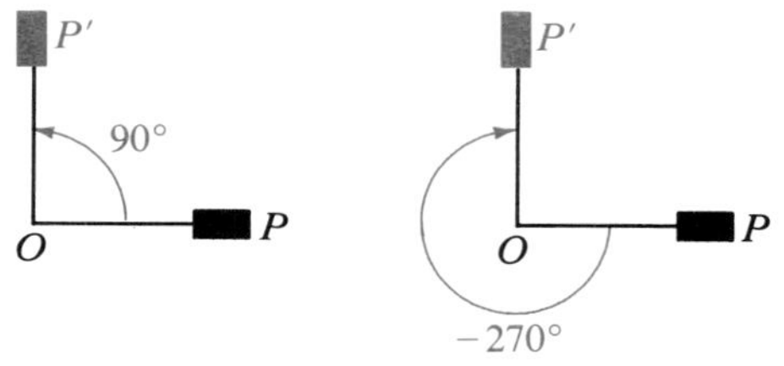

$\mathcal{R}_{O,90} = \mathcal{R}_{O,-270}$
Notice: $90 - 360 = -270$

In the following definition of a rotation, the angle measure x can be positive or negative and can be more than 180 in absolute value.

A **rotation** about point O through $x°$ is a transformation such that:

(1) If point P is different from O, then $OP' = OP$ and $m\angle POP' = x$.
(2) If point P is the same as O, then $P' = O$.

Theorem 12-3

A rotation is an isometry.

Given: $\mathcal{R}_{O,x}$ maps P to P' and Q to Q'.
Prove: $PQ = P'Q'$

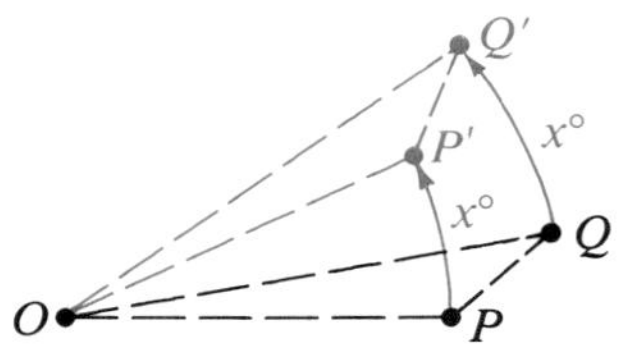

Outline of proof:

1. $OP = OP'$, $OQ = OQ'$ (Definition of rotation)
2. $m\angle POP' = m\angle QOQ' = x$ (Definition of rotation)
3. $m\angle POQ = m\angle P'OQ'$ (Subtraction Property of =; subtract $m\angle QOP'$.)
4. $\triangle POQ \cong \triangle P'OQ'$ (SAS Postulate)
5. $PQ = P'Q'$ (Corr. parts of $\cong$ ⚠ are $\cong$.)

A rotation about point O through 180° is called a **half-turn** about O and is usually denoted by H_O. The diagram shows $\triangle PQR$ and its image $\triangle P'Q'R'$ by H_O. Notice that O is the midpoint of $\overline{PP'}$, $\overline{QQ'}$, and $\overline{RR'}$.

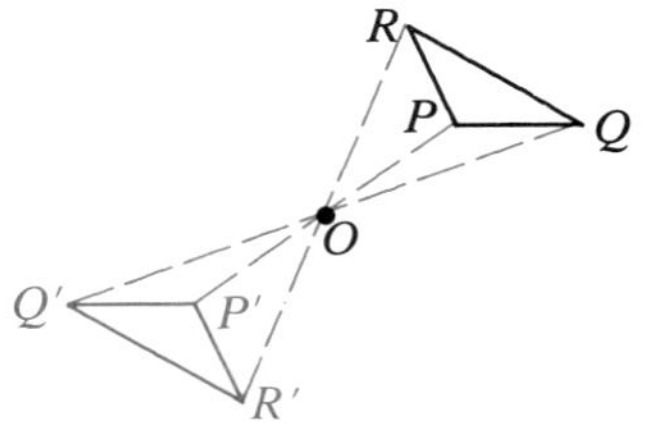

Using coordinates, a half-turn H_O about the origin can be written

$$H_O:(x, y) \rightarrow (-x, -y).$$

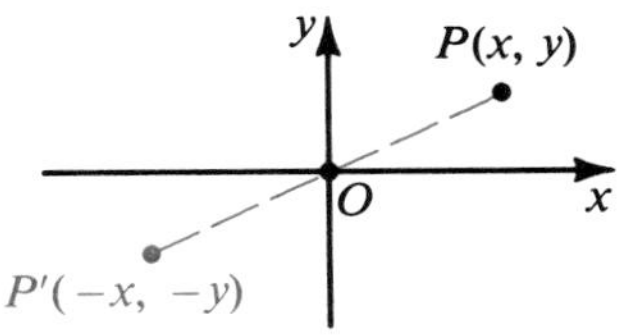

Classroom Exercises

State another name for each rotation.

1. $\mathcal{R}_{O,10}$ **2.** $\mathcal{R}_{O,50}$ **3.** $\mathcal{R}_{O,-40}$ **4.** $\mathcal{R}_{O,-90}$

5. $\mathcal{R}_{O,380}$ **6.** $\mathcal{R}_{O,400}$ **7.** $\mathcal{R}_{O,-180}$ **8.** $\mathcal{R}_{O,360}$

In the diagram for Exercises 9–13, *O* is the center of equilateral $\triangle PST$. State the images of points *P*, *S*, and *T* for each rotation.

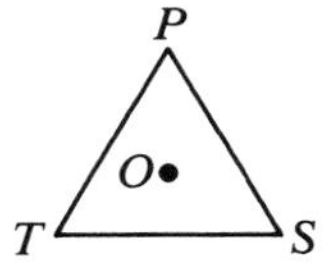

Exs. 9–13

9. $\mathcal{R}_{O,120}$ **10.** $\mathcal{R}_{O,-120}$ **11.** $\mathcal{R}_{O,360}$

12. What is the image of S by $\mathcal{R}_{T,60}$?

13. What is the image of P by $\mathcal{R}_{T,-60}$?

Draw a coordinate grid on the chalkboard and plot the following points: $A(2, 0)$, $B(2, 1)$, $C(4, 1)$, $D(4, -2)$.

14. Plot the images of these points by a half-turn about the origin O.

15. Plot the images of these points by $\mathcal{R}_{O,90}$.

16. Plot the images of these points by $\mathcal{R}_{O,-90}$.

Written Exercises

State another name for each rotation.

A **1.** $\mathcal{R}_{O,80}$ **2.** $\mathcal{R}_{O,-15}$ **3.** $\mathcal{R}_{A,450}$ **4.** $\mathcal{R}_{B,-720}$ **5.** H_O

The diagonals of regular hexagon *ABCDEF* form six equilateral triangles as shown. Complete each statement below.

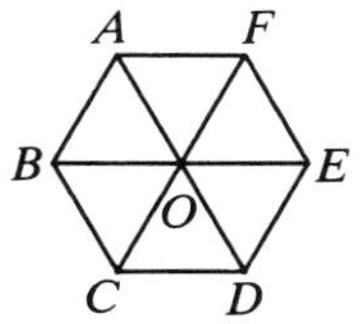

Exs. 6–13

6. $\mathcal{R}_{O,60}:E \rightarrow$ _?_ **7.** $\mathcal{R}_{O,-60}:D \rightarrow$ _?_

8. $\mathcal{R}_{O,120}:F \rightarrow$ _?_ **9.** $H_O:A \rightarrow$ _?_

10. $\mathcal{R}_{B,-60}:O \rightarrow$ _?_ **11.** $\mathcal{R}_{D,60}:$ _?_ $\rightarrow O$

12. If a translation maps A to B, then it also maps O to _?_ and E to _?_.

13. A reflection in $\overleftrightarrow{FC}$ maps B to _?_.

Complete each statement below by using one of the words *reflection, translation, rotation,* or *half-turn.*

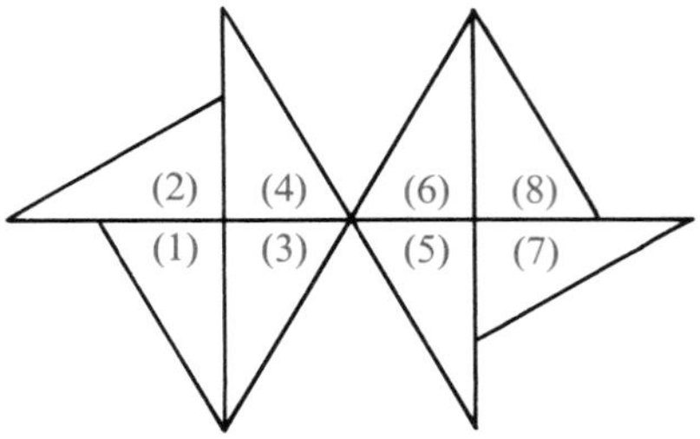

14. A _?_ maps triangle (1) to (2).

15. A _?_ maps triangle (1) to (3).

16. A _?_ maps triangle (1) to (4).

17. A _?_ maps triangle (1) to (5).

18. A _?_ maps triangle (2) to (4).

19. A _?_ maps triangle (2) to (7).

20. A _?_ maps triangle (4) to (6).

21. A _?_ maps triangle (4) to (8).

B **22.** In the diagram above there is a glide reflection mapping triangle (1) to triangle (_?_).

23. Name another pair of triangles for which the first is mapped to the second by a glide reflection.

Copy the figure on squared paper. Draw the image of the figure by the rotation specified.

24. $\mathcal{R}_{O,90}$

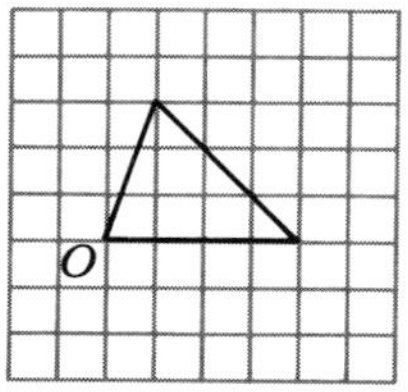

25. $\mathcal{R}_{O,-90}$

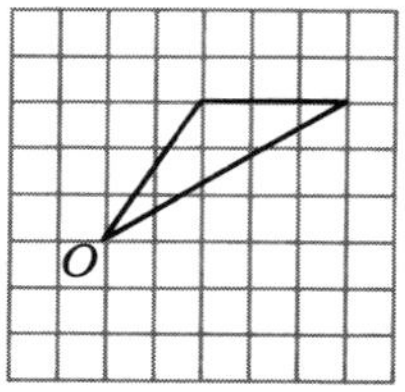

26. H_O

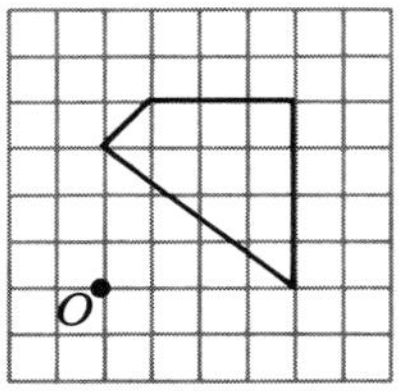

27. If H_C:(1, 1) → (7, 3), find the coordinates of C.

28. A rotation maps A to A' and B to B'. Construct the center of the rotation. (*Hint:* If the center is O, then $OA = OA'$ and $OB = OB'$.)

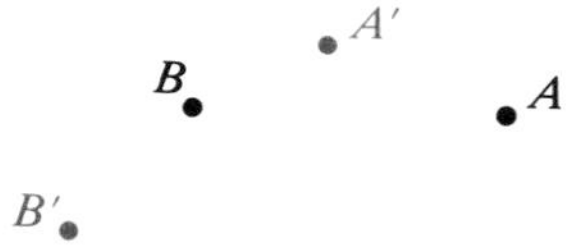

29. **a.** Draw a coordinate grid with origin O and plot the points $A(0, 3)$ and $B(4, 1)$.

b. Plot A' and B', the images of A and B by $\mathcal{R}_{O,90}$.

c. Compare the slopes of $\overleftrightarrow{AB}$ and $\overleftrightarrow{A'B'}$. What does this tell you about these lines?

d. Without using the distance formula, you know that $A'B' = AB$. What theorem tells you this?

30. A half-turn about (3, 2) maps P to P'. Where does this half-turn map the following points?

a. P' **b.** (0, 0) **c.** (3, 0)
d. (1, 4) **e.** (−2, 1) **f.** (x, y)

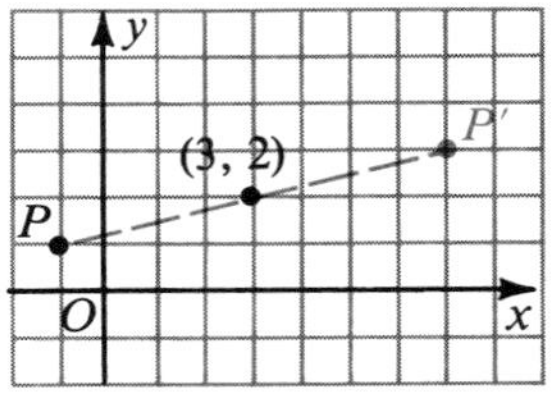

31. The rotation $\mathcal{R}_{O,x}$ maps line l to line l'. (You can think of rotating $\overline{OF}$, the perpendicular from O to l, through $x°$. Its image will be $\overline{OF'}$.) Prove that one of the angles between l and l' has measure x.

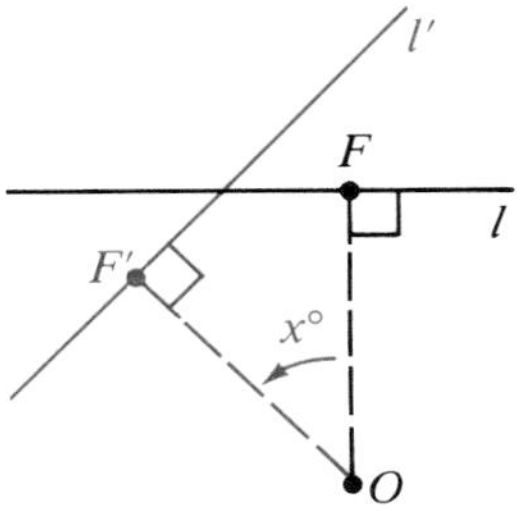

32. $\triangle ABC$ and $\triangle DCE$ are equilateral.

a. What rotation maps A to B and D to E?

b. $AD = BE$ because a rotation is an __?__.

c. Find the measure of an acute angle between $\overleftrightarrow{AD}$ and $\overleftrightarrow{BE}$. (*Hint:* See Exercise 31.)

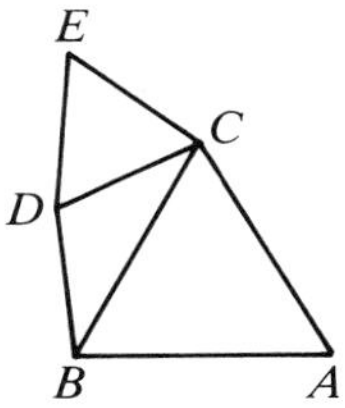

33. $\triangle ABC$ and $\triangle DEC$ are isosceles right triangles.

a. What rotation maps B to A and E to D?

b. Why does $AD = BE$?

c. Explain why $\overline{AD} \perp \overline{BE}$. (*Hint:* See Exercise 31.)

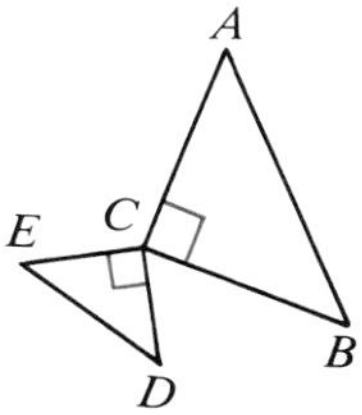

C **34.** Given: Parallel lines l and k and point A

a. Construct an equilateral $\triangle ABC$ with B on k and C on l using the following method.

Step 1. Rotate l through $60°$ about A and let B be the point on k where the image of l intersects k. (The diagram for Exercise 31 may be helpful in rotating l.)

Step 2. Let point C on l be the preimage of B.

b. Explain why $\triangle ABC$ is equilateral.

c. Are there other equilateral triangles with vertices at A and on l and k?

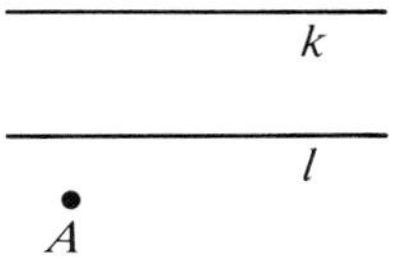

35. Given the figure for Exercise 34, construct a square $AXYZ$ with X on k and Z on l.

Challenge

A mouse moves along $\overline{AJ}$. For any position M of the mouse, X and Y are such that $\overline{AX} \perp \overline{AJ}$ with $AX = AM$, and $\overline{JY} \perp \overline{AJ}$ with $JY = JM$. The cat is at C, the midpoint of $\overline{XY}$. Describe the locus of the cat as the mouse moves from A to J.

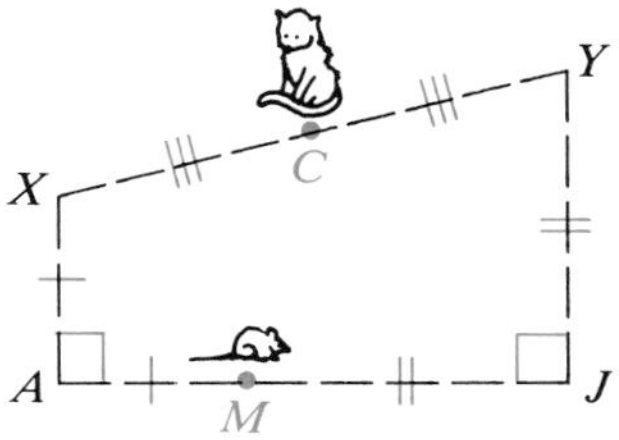

12-5 Dilations

Reflections, translations, glide reflections, and rotations are isometries, or *congruence* mappings. In this section we consider a transformation related to *similarity* rather than congruence. It is called a **dilation.** The dilation $D_{O,k}$ has *center* O and nonzero *scale factor* k. $D_{O,k}$ maps any point P to a point P' determined as follows:

(1) If $k > 0$, P' lies on $\overrightarrow{OP}$ and $OP' = k \cdot OP$.
(2) If $k < 0$, P' lies on the ray opposite $\overrightarrow{OP}$ and $OP' = |k| \cdot OP$.
(3) The center O is its own image.

If $|k| > 1$, the dilation is called an **expansion.**
If $|k| < 1$, the dilation is called a **contraction.**

Example 1 Find the image of $\triangle ABC$ under the expansion $D_{O,2}$.

Solution $D_{O,2}: \triangle ABC \to \triangle A'B'C'$
$OA' = 2 \cdot OA$
$OB' = 2 \cdot OB$
$OC' = 2 \cdot OC$

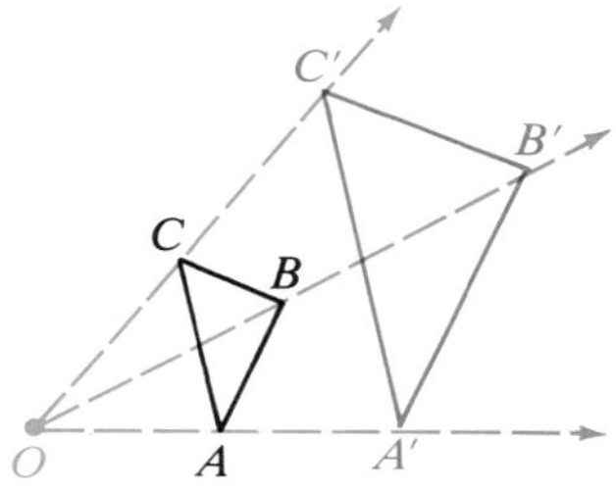

Example 2 Find the image of $\triangle RST$ under the contraction $D_{O,\frac{2}{3}}$.

Solution $D_{O,\frac{2}{3}}: \triangle RST \to \triangle R'S'T'$
$OR' = \frac{2}{3} \cdot OR$
$OS' = \frac{2}{3} \cdot OS$
$OT' = \frac{2}{3} \cdot OT$

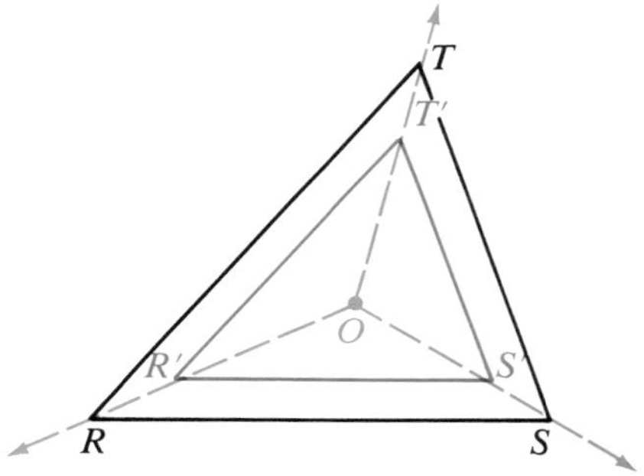

Example 3 Find the image of figure F under the contraction $D_{O,-\frac{1}{2}}$.

Solution $D_{O,-\frac{1}{2}}$:figure $F \to$ figure F'
$\overrightarrow{OP}$ is opposite to $\overrightarrow{OP'}$.
$OP' = |-\frac{1}{2}| \cdot OP = \frac{1}{2} \cdot OP$

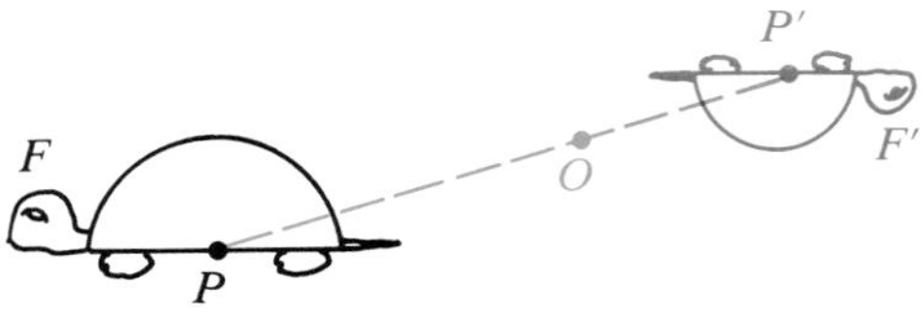

As these examples illustrate, dilations do not preserve distance (unless the scale factor is 1 or -1). But a dilation always preserves angle measure, and so it maps any geometric figure to a similar figure. In the examples above, $\triangle ABC \sim \triangle A'B'C'$, $\triangle RST \sim \triangle R'S'T'$, and the figure F is similar to the figure F'.

A developing leaf undergoes a dilation with scale factor greater than 1, keeping approximately the same shape as it grows in size.

Theorem 12-4

The dilation $D_{O,k}$ maps every line segment to a parallel segment that is $|k|$ times as long.

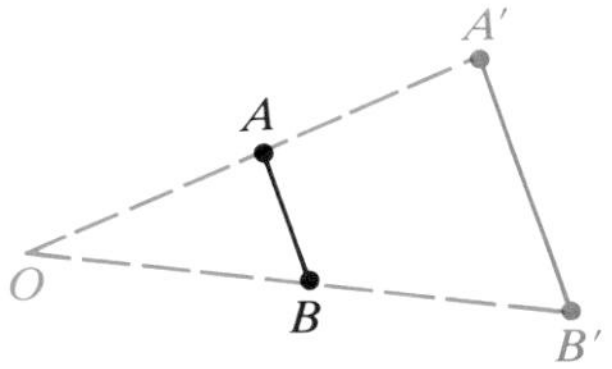

Given: $D_{O,k}: \overline{AB} \to \overline{A'B'}$

Prove: $\overline{A'B'} \parallel \overline{AB}$;
$A'B' = |k| \cdot AB$

Proof:

Statements	Reasons
1. $OA' = \|k\| \cdot OA$, $OB' = \|k\| \cdot OB$	1. Definition of dilation
2. $\triangle OA'B' \sim \triangle OAB$	2. SAS Similarity Theorem
3. $\frac{A'B'}{AB} = \frac{OA'}{OA} = \|k\|$ (so that $A'B' = \|k\| \cdot AB$)	3. Corr. sides of ~ $\triangle$s are proportional.
4. $\angle OA'B' \cong \angle OAB$	4. Corr. $\angle$s of ~ $\triangle$s are $\cong$.
5. $\overline{A'B'} \parallel \overline{AB}$	5. ?

The diagram for the proof of Theorem 12-4 shows the case when $k > 0$. You should draw the diagram for $k < 0$ and convince yourself that the proof is the same.

A dilation with its center at the origin of the coordinate plane can be described easily in terms of coordinates. If the scale factor is k, the image of any point (x, y) is the point (kx, ky) and we can write $D_{O,k}:(x, y) \to (kx, ky)$. The dilation $D_{O,3}$ is illustrated at the right.

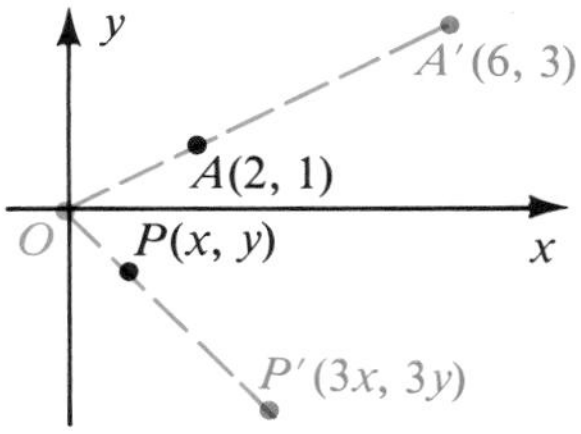

Classroom Exercises

Sketch each triangle on the chalkboard. Then sketch its image under the given dilation.

1. $D_{O,3}$

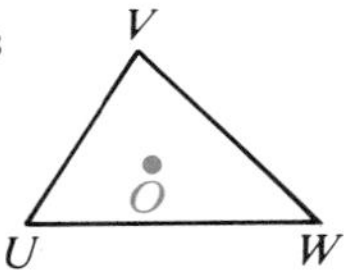

2. $D_{S,\frac{1}{2}}$

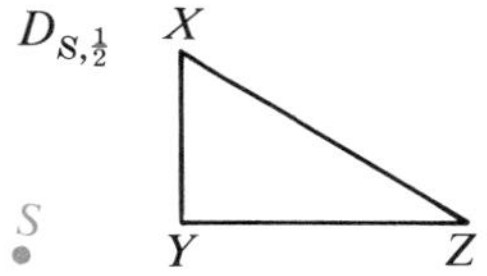

3. $D_{E,-2}$

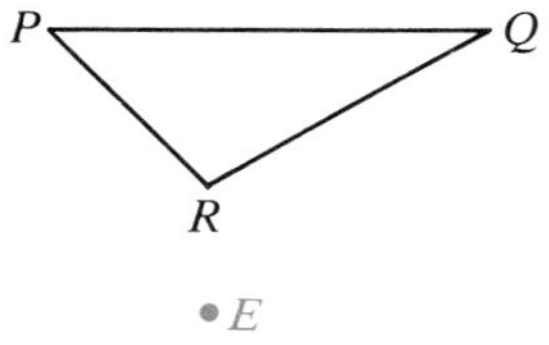

4. $D_{N,-\frac{1}{3}}$

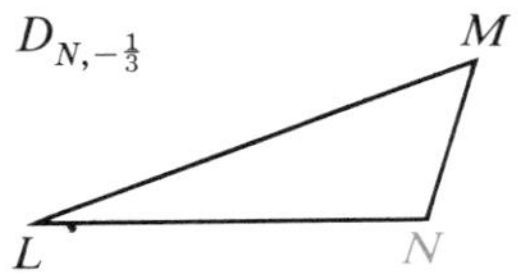

5. Find the coordinates of the images of points A, B, and C under the dilation $D_{O,2}$.
6. Find the image of (x, y) under $D_{O,2}$.
7. What dilation maps A to B?
8. What dilation maps C to the point $(-6, 0)$?

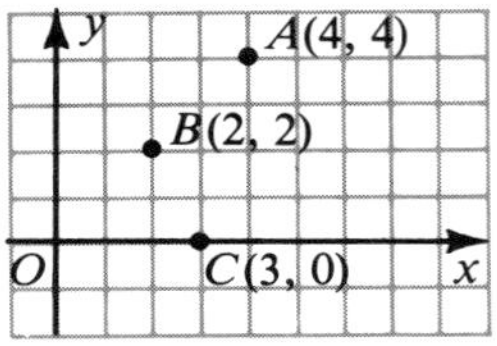

9. Match each scale factor in the first column with the name of the corresponding dilation in the second column.

Scale factor	*Transformation*
$\frac{2}{5}$	Half-turn
-4	Expansion
-1	Contraction

10. How would you describe the dilation $D_{O,1}$?
11. If $\odot S$ has radius 4, describe the image of $\odot S$ under $D_{S,5}$ and under $D_{S,-1}$.
12. If point A is on line k, what is the image of line k under $D_{A,2}$?
13. The dilation $D_{O,3}$ maps P to P' and Q to Q'.
    - **a.** If $OQ = 2$, find OQ'.
    - **b.** If $PQ = 7$, find $P'Q'$.
    - **c.** If $PP' = 10$, find OP.

Written Exercises

Find the coordinates of the images of A, B, and C by the given dilation.

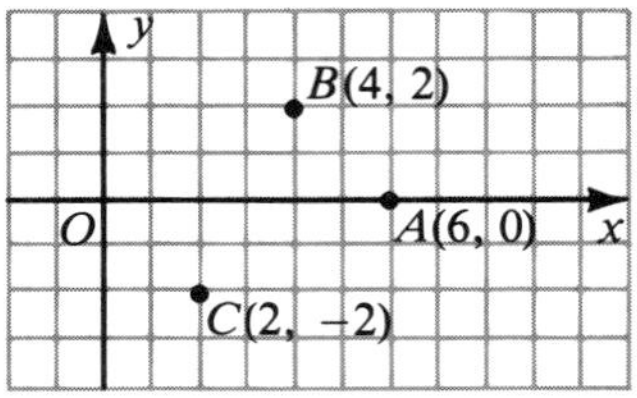

A

1. $D_{O,2}$	2. $D_{O,3}$	3. $D_{O,\frac{1}{2}}$	4. $D_{O,-\frac{1}{2}}$
5. $D_{O,-2}$	6. $D_{O,1}$	7. $D_{A,-\frac{1}{2}}$	8. $D_{A,2}$

A dilation with the origin, O, as center maps the given point to the image point named. Find the scale factor of the dilation. Is the dilation an expansion or a contraction?

9. $(2, 0) \to (8, 0)$ **10.** $(2, 3) \to (4, 6)$ **11.** $(3, 9) \to (1, 3)$

12. $(4, 10) \to (-2, -5)$ **13.** $(0, \frac{1}{6}) \to (0, \frac{2}{3})$ **14.** $(-6, 2) \to (18, -6)$

Draw quadrilateral $PQRS$ in black. Then draw in color the image, quadrilateral $P'Q'R'S'$, by the dilation given.

B **15.** $P(-1, 1)$ $Q(0, -1)$ $R(4, 0)$ $S(2, 2)$; $D_{O,3}$

16. $P(12, 0)$ $Q(0, 15)$ $R(-9, 6)$ $S(3, -9)$; $D_{O,\frac{2}{3}}$

17. $P(3, 0)$ $Q(3, 4)$ $R(6, 6)$ $S(5, -1)$; $D_{O,-2}$

18. $P(-2, -2)$ $Q(0, 0)$ $R(4, 0)$ $S(6, -2)$; $D_{O,-\frac{1}{2}}$

19. $D_{O,3}$ maps $\triangle ABC$ to $\triangle A'B'C'$.
  a. What is the ratio of the perimeters of $\triangle A'B'C'$ and $\triangle ABC$?
  b. What is the ratio of the areas of $\triangle A'B'C'$ and $\triangle ABC$?

20. The diagram illustrates a dilation of three-dimensional space. $D_{O,2}$ maps the smaller cube to the larger cube.
  a. What is the ratio of the surface areas of these cubes?
  b. What is the ratio of the volumes of these cubes?

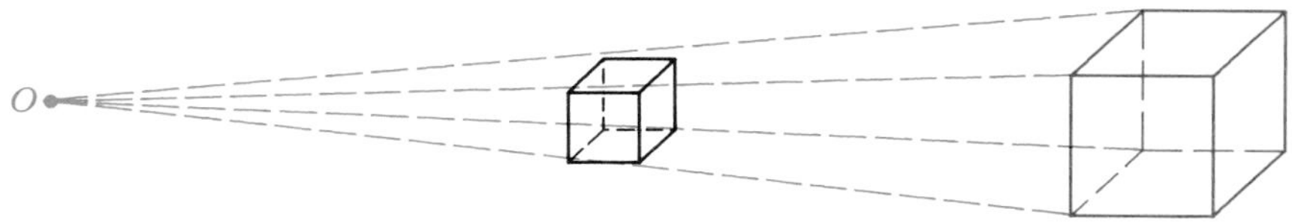

21. G is the intersection of the medians of $\triangle XYZ$. Complete the following statements. (*Hint:* Use Theorem 8–4 on page 345.)

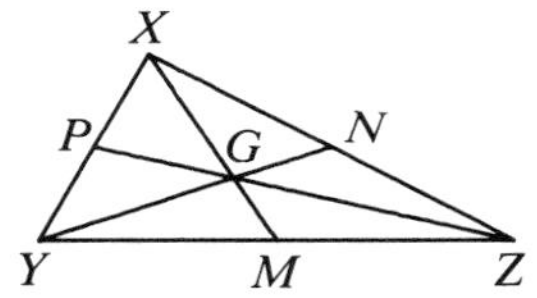

  a. $\frac{XG}{XM} = ?$ **b.** $\frac{GM}{GX} = ?$
  c. What dilation maps X to M?
  d. What is the image under this dilation of Y? of Z?

22. $D_{O,k}$ maps $\overline{PQ}$ to $\overline{P'Q'}$.
  a. Show that the slopes of $\overline{PQ}$ and $\overline{P'Q'}$ are equal.
  b. Part (a) proves that $\overline{PQ}$ and $\overline{P'Q'}$ are __?__.

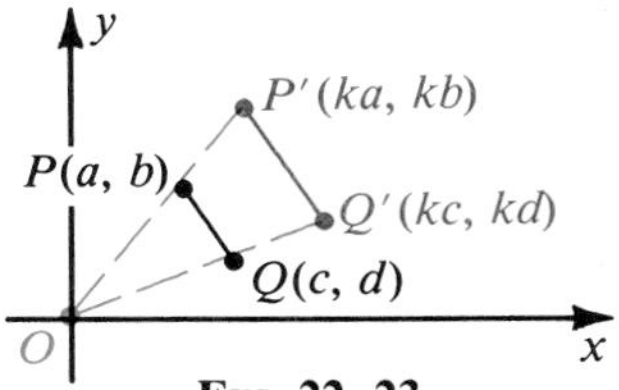

Exs. 22, 23

C **23.** Use the distance formula to show that
$P'Q' = |k|\sqrt{(a-c)^2 + (b-d)^2} = |k| \cdot PQ.$

24. A dilation with center (a, b) and scale factor k maps $A(3, 4)$ to $A'(1, 8)$ and $B(3, 2)$ to $B'(1, 2)$. Find the coordinates of the center (a, b) and the value of k.

Self-Test 1

Transformation S maps $\triangle PQR$ to $\triangle P'Q'R'$.

1. The image of Q is _?_.
2. The preimage of R' is _?_.
3. If S is an isometry, then $PQ =$ _?_.
4. What are the coordinates of the image of (2, 3) by the translation $T:(x, y) \rightarrow (x + 1, y - 2)$?
5. A reflection in line k maps A to A'. If A is not on line k, then k is the perpendicular bisector of _?_.
6. Give two other names for the rotation $\mathcal{R}_{O,-30}$.

Complete.

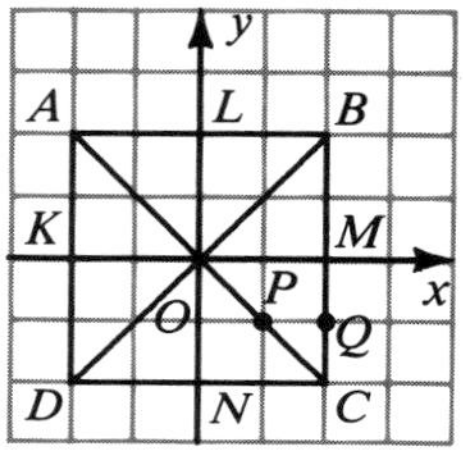

7. $R_y:A \rightarrow$ _?_
8. $R_x:B \rightarrow$ _?_
9. $R_x:\overline{DC} \rightarrow$ _?_
10. $R_y:$ _?_ $\rightarrow \overline{OA}$
11. $H_O:K \rightarrow$ _?_
12. $H_O:$ _?_ $\rightarrow \overline{CO}$
13. $\mathcal{R}_{O,90}$ maps M to _?_.
14. $\mathcal{R}_{O,-90}$ maps $\triangle MCO$ to $\triangle$ _?_.
15. $D_{O,2}$ maps P to _?_.
16. $D_{M,-\frac{1}{2}}$ maps B to _?_.
17. A translation that maps A to L maps N to _?_.
18. A translation that maps D to O maps _?_ to B.
19. $\overleftrightarrow{DB}$ is a glide line for a glide reflection mapping $\triangle DON$ to $\triangle$ _?_.

Products and Symmetry

Objectives

1. Locate the images of figures by products of mappings.
2. Recognize and use the terms *identity* and *inverse* in relation to mappings.
3. Describe the symmetry of figures.

12-6 Products of Mappings

Suppose a transformation T maps point P to P' and a transformation S maps P' to P''. Then T and S can be combined to produce a new transformation that maps P to P''. This new transformation is called the **product** of S and T. The product is written $S \circ T$, which is read "S circle T" or "T followed by S." Notice that when we write $S \circ T:P \rightarrow P''$, the transformation that acts *first* (T) is written on the right, nearest P.

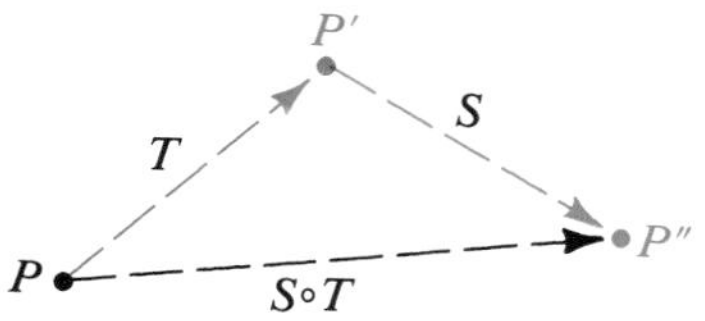

As you work with transformations, you will see that their products behave very much like products of numbers. There is one important exception, however, where the product of transformations is not at all like the product of numbers. For numbers, the products *ab* and *ba* are equal. But for transformations, the products $S \circ T$ and $T \circ S$ are usually not equal. The following example shows this.

Example Show that $H_O \circ R_j \neq R_j \circ H_O$.

Solution Study the two diagrams below and notice that the two products map P to different points.

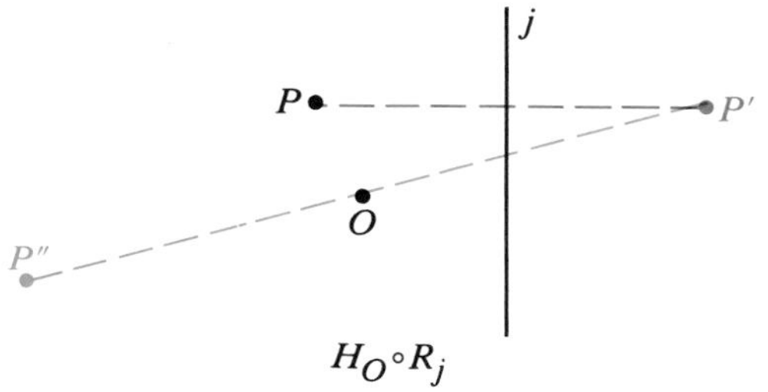

Here R_j, the reflection of P in line j, is carried out first, mapping P to P'. Then H_O maps P' to P''. Thus P'' is the image of P under the product $H_O \circ R_j$.

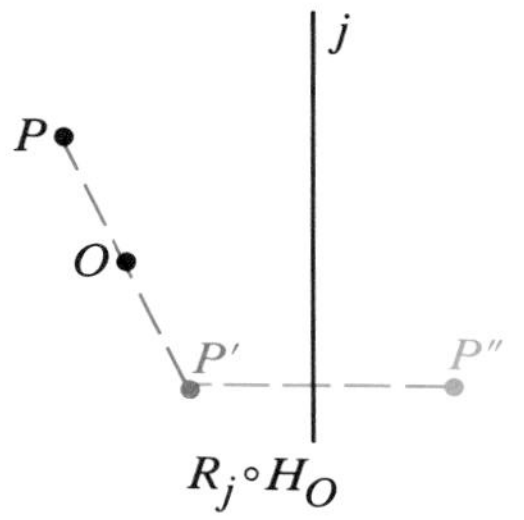

With the order changed in the product, the half-turn is carried out first, followed by the reflection in line j. The image point P'' is now in a different place.

This example shows that the order in a product of transformations can be very important, but this is not always the case. For example, if S and T are two translations, then order is not important, since $S \circ T = T \circ S$. See Exercise 8.

The example above shows the effect of a product of mappings on a single point P. The diagram below shows a product of reflections acting on a whole figure, F. F is reflected in line j to F', and F' is reflected in line k to F''. The product $R_k \circ R_j$ maps F to F''. Again notice that the first reflection, R_j, is written on the right.

The final image F'' is the same size and shape as F but has been translated to a new position on the plane. This illustrates our next two theorems. First, the product of any two isometries is an isometry. Second, the product of reflections in two parallel lines is a translation.

Theorem 12–5

The product of two isometries is an isometry.

Proof: Let S and T be isometries. To show that $S \circ T$ is an isometry we must show that this mapping preserves the distance between any two points P and Q. Suppose T maps P and Q to P' and Q', and that S maps P' and Q' to P'' and Q''. Then the images of P and Q under the product $S \circ T$ are P'' and Q''. $PQ = P'Q'$ because T is an isometry. $P'Q' = P''Q''$ (Why?) Thus $PQ = P''Q''$ (why?) and therefore $S \circ T$ preserves distance.

Theorem 12–6

A product of reflections in two parallel lines is a translation. The translation glides all points through twice the distance between the lines.

Given: $j \parallel k$

Prove: $R_k \circ R_j$ is a translation through twice the distance between j and k.

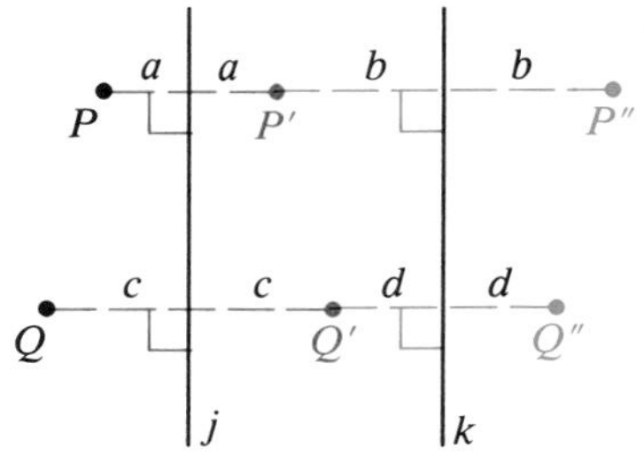

Proof:

To prove that the product is a translation we must prove two things: (1) $R_k \circ R_j$ is an isometry, and (2) $PP'' = QQ''$ for any two points P and Q with images P'' and Q'' under $R_k \circ R_j$.

(1) Since R_j and R_k are isometries (Theorem 12–1), so is $R_k \circ R_j$ (Theorem 12–5).

(2) R_j maps P and Q to P' and Q'; R_k maps P' and Q' to P'' and Q''. The letters a, b, c, and d in the diagram label pairs of distances that are equal according to the definition of a reflection. $\overline{PP'} \perp j$ and $\overline{P'P''} \perp k$. (Why?) Since j and k are parallel, P, P', and P'' are collinear and

$$PP'' = 2a + 2b = 2(a + b)$$

Similarly, $$QQ'' = 2c + 2d = 2(c + d)$$

$a + b = c + d$, since the distance between the parallel lines j and k is constant. Therefore $PP'' = QQ''$ = twice the distance between j and k.

You should make diagrams for the case when P is on j or k, when P is located between j and k, and when P is to the right of k. Convince yourself that $PP'' = 2(a + b)$ in these cases also. In every case, the glide is perpendicular to j and k and goes in the direction from j to k (from left to right in the diagram).

Theorem 12–6 shows that when lines j and k are parallel, $R_k \circ R_j$ translates points through twice the distance between the lines. If j and k intersect, $R_k \circ R_j$ rotates points through twice the angle between the lines. This is our next theorem.

Theorem 12-7

A product of reflections in two intersecting lines is a rotation about the point of intersection of the two lines. The measure of the angle of rotation is twice the measure of the angle from the first line of reflection to the second.

Given: j intersects k, forming an angle of $y°$ at O.

Prove: $R_k \circ R_j = \mathcal{R}_{O,2y}$

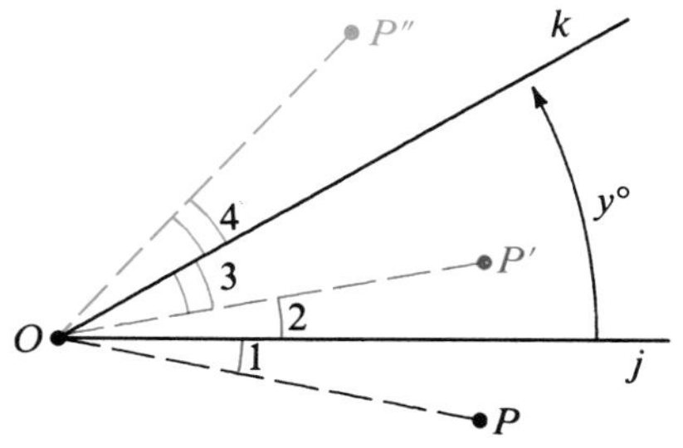

Proof:

The diagram shows an arbitrary point P and its image P' by reflection in j. The image of P' by reflection in k is P''. According to the definition of a rotation we must prove that $OP = OP''$ and $m\angle POP'' = 2y$.

R_j and R_k are isometries, so they preserve both distance and angle measure. Therefore $OP = OP'$, $OP' = OP''$, $m\angle 1 = m\angle 2$, and $m\angle 3 = m\angle 4$. Thus $OP = OP''$ and the measure of the angle of rotation equals

$$m\angle 1 + m\angle 2 + m\angle 3 + m\angle 4 = 2m\angle 2 + 2m\angle 3 = 2y.$$

Corollary

A product of reflections in perpendicular lines is a half turn about the point where the lines intersect.

Classroom Exercises

Copy the figure on the chalkboard and find its image by $R_k \circ R_j$. Then copy the figure again and find its image by $R_j \circ R_k$.

1.

2.

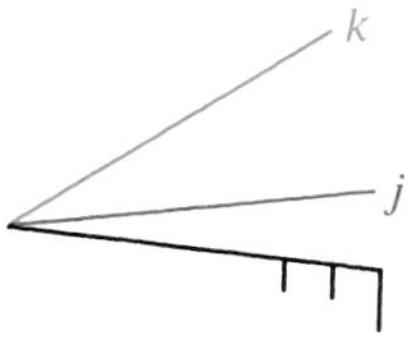

Complete the following.

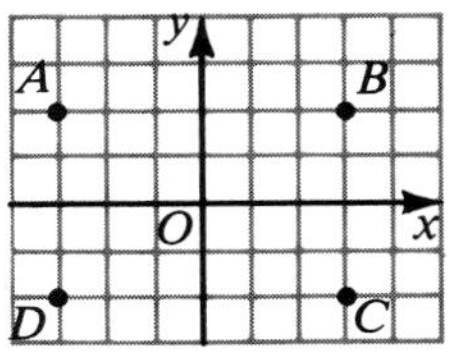

3. $R_x \circ R_y : A \to$ __?__

4. $R_x \circ R_y : D \to$ __?__

5. $H_O \circ R_y : B \to$ __?__

6. $R_y \circ H_O : B \to$ __?__

7. $H_O \circ H_O : A \to$ __?__

8. $R_y \circ R_y : C \to$ __?__

In the diagram, line l is the line $y = x$, which makes a 45° angle with the x- and y-axes. Complete each statement.

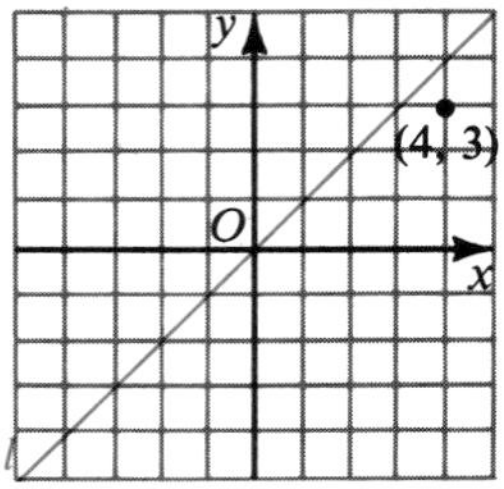

9. $R_l:(4, 3) \rightarrow (\underline{\ ?\ }, \underline{\ ?\ })$
10. $R_y \circ R_l:(4, 3) \rightarrow (\underline{\ ?\ }, \underline{\ ?\ })$
11. $R_y \circ R_l = \mathcal{R}_{?,?}$
12. $R_l \circ R_y:(4, 3) \rightarrow (\underline{\ ?\ }, \underline{\ ?\ })$
13. $R_l \circ R_y = \mathcal{R}_{?,?}$

Written Exercises

Copy each figure and show F', the image of figure F under the reflection R_j. Also show F'', the image of F under the product $R_k \circ R_j$.

A 1.

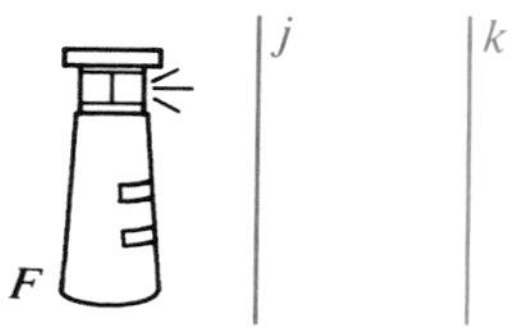

2.

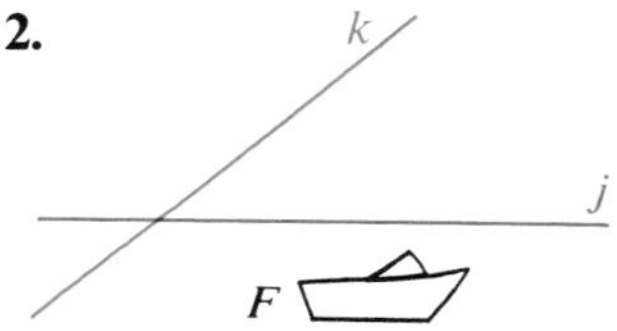

3. Redraw the figure for Exercise 1 and show F and its images F' and F'' under the mappings R_k and $R_j \circ R_k$.
4. Redraw the figure for Exercise 2 and show F and its images F' and F'' under the mappings R_k and $R_j \circ R_k$.

Copy each figure twice and show the image of the red flag under each of the products given.

5. a. $H_B \circ H_A$
   b. $H_A \circ H_B$

6. a. $R_j \circ H_C$
   b. $H_C \circ R_j$

7. a. $H_E \circ D_{E,\frac{1}{3}}$
   b. $D_{E,\frac{1}{3}} \circ H_E$

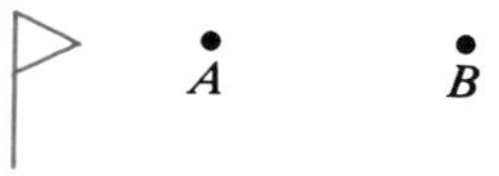

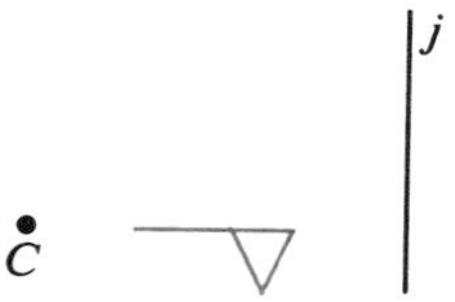

8. $A = (4, 1)$, $B = (1, 5)$, and $C = (0, 1)$. S and T are translations. $S:(x, y) \rightarrow (x + 1, y + 4)$ and $T:(x, y) \rightarrow (x + 3, y - 1)$. Draw $\triangle ABC$ and its images under $S \circ T$ and $T \circ S$.
   a. Does $S \circ T$ appear to be a translation?
   b. Is $S \circ T$ equal to $T \circ S$?
   c. $S \circ T:(x, y) \rightarrow (\underline{\ ?\ }, \underline{\ ?\ })$
   d. $T \circ S:(x, y) \rightarrow (\underline{\ ?\ }, \underline{\ ?\ })$

9. L, M, and N are midpoints of the sides of $\triangle QRS$. Complete the following.

a. $H_N \circ H_M : S \rightarrow$ _?_

b. $H_M \circ H_N : Q \rightarrow$ _?_

c. $D_{S,\frac{1}{2}} \circ H_N : Q \rightarrow$ _?_

d. $H_N \circ D_{S,2} : M \rightarrow$ _?_

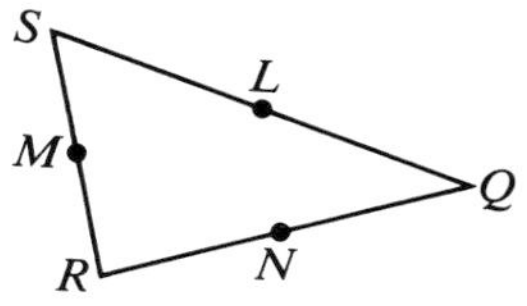

Exs. 9–11

B 10. If T is a translation that maps R to N, then

a. $T:M \rightarrow$ _?_ b. $T \circ D_{S,\frac{1}{2}} : R \rightarrow$ _?_ c. $T \circ T : R \rightarrow$ _?_

11. $H_L \circ H_M \circ H_N : Q \rightarrow$ _?_

For each exercise draw a grid and find the coordinates of the image point. O is the origin and A is the point (3, 1). R_x and R_y are reflections in the x- and y-axes.

12. $R_x \circ R_y : (3, 1) \rightarrow$ (_?_, _?_)

13. $R_y \circ H_O : (1, -2) \rightarrow$ (_?_, _?_)

14. $H_A \circ H_O : (3, 0) \rightarrow$ (_?_, _?_)

15. $H_O \circ H_A : (1, 1) \rightarrow$ (_?_, _?_)

16. $R_x \circ D_{O,2} : (2, 4) \rightarrow$ (_?_, _?_)

17. $\mathcal{R}_{O,90} \circ R_y : (-2, 1) \rightarrow$ (_?_, _?_)

18. $\mathcal{R}_{A,90} \circ \mathcal{R}_{O,-90} : (-1, -1) \rightarrow$ (_?_, _?_)

19. $D_{O,-\frac{1}{3}} \circ D_{A,4} : (3, 0) \rightarrow$ (_?_, _?_)

20. Let R_l be a reflection in the line $y = x$ and R_y a reflection in the y-axis. Draw a grid and label the origin O.

a. Plot the point $P(5, 2)$ and its image Q under the mapping $R_y \circ R_l$.

b. According to Theorem 12-7, $m\angle POQ =$ _?_.

c. Use the slopes of $\overline{OP}$ and $\overline{OQ}$ to verify that $\overline{OP} \perp \overline{OQ}$.

21. Copy the figure, which shows four parallel lines a, b, c, and d, with the distance between a and b equal to the distance between c and d.

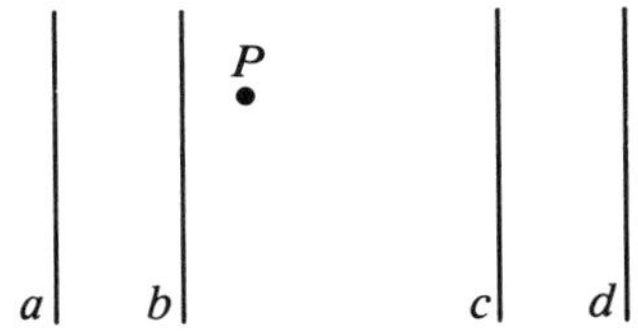

a. Carefully find the image of P under the product $R_a \circ R_b$.

b. Carefully find the image of P under the product $R_c \circ R_d$.

c. The image points in parts (a) and (b) should coincide. Explain why.

22. Explain how you would construct line j so that $R_k \circ R_j : A \rightarrow B$.

C 23. The figure shows that $H_B \circ H_A : P \rightarrow P''$.

a. Copy the figure and verify by measuring that $PP'' = 2 \cdot AB$. What theorem about the midpoints of the sides of a triangle does this suggest?

b. Choose another point Q and carefully locate Q'', the image of Q under $H_B \circ H_A$. Does $QQ'' = 2 \cdot AB$?

c. Measure PQ and $P''Q''$. Are they equal? What kind of transformation does $H_B \circ H_A$ appear to be?

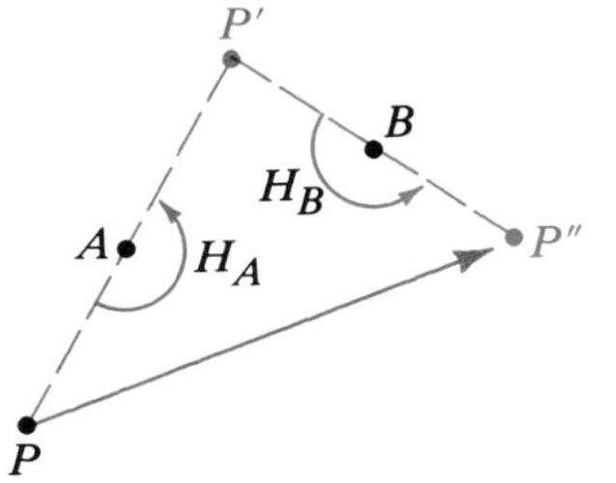

24. $D_{A,2}: \overline{PQ} \to \overline{P'Q'}$ and $D_{B,\frac{1}{2}}: \overline{P'Q'} \to \overline{P''Q''}$. What kind of transformation is the product $D_{B,\frac{1}{2}} \circ D_{A,2}$? Explain.

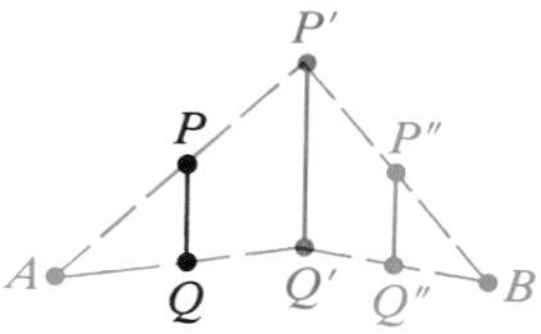

25. The point P is called a *fixed point* of the transformation T if $T: P \to P$.

a. How many fixed points does each of the following have: $\mathcal{R}_{O,90}$? R_y? $D_{O,3}$? The translation $T:(x, y) \to (x - 3, y + 2)$?

b. O is the origin and A is the point $(1, 0)$. Find the coordinates of a fixed point of the product $D_{O,2} \circ D_{A,\frac{1}{4}}$.

12-7 Inverses and the Identity

Suppose that the pattern below continues indefinitely to both the left and the right. The translation T glides each runner one place to the right. The translation that glides each runner one place to the *left* is called the *inverse* of T, and is denoted T^{-1}. Notice that T followed by T^{-1} keeps *all* points fixed:

$$T^{-1} \circ T: P \to P$$

The product $T \circ T$, usually denoted T^2, glides each runner two places to the right.

The mapping that maps every point to itself is called the **identity** transformation I. The words "identity" and "inverse" are used for mappings in much the same way that they are used for numbers.

Relating Geometry and Algebra

For products of numbers, 1 is the identity. $a \cdot 1 = a$ and $1 \cdot a = a$

For products of mappings, I is the identity. $S \circ I = S$ and $I \circ S = S$

For numbers, the inverse of a is written a^{-1} $\left(\text{or } \frac{1}{a}\right)$. $a \cdot a^{-1} = 1$ and $a^{-1} \cdot a = 1$

For mappings, the inverse of S is written S^{-1}. $S \circ S^{-1} = I$ and $S^{-1} \circ S = I$

In general, the **inverse** of a transformation T is defined as the transformation S such that $S \circ T = I$. If we apply this definition to the figure at the right we see that $R_j \circ R_j = I$, so that the inverse of R_j is R_j itself. In symbols, $R_j^{-1} = R_j$. Do you see that the inverse of any reflection is that same reflection? The inverses of translations, rotations, and dilations are illustrated below.

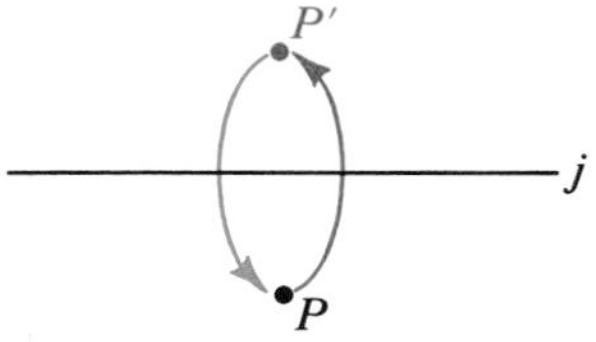

Translation T

$T:(x, y) \rightarrow (x + 5, y - 4)$
$T^{-1}:(x, y) \rightarrow (x - 5, y + 4)$

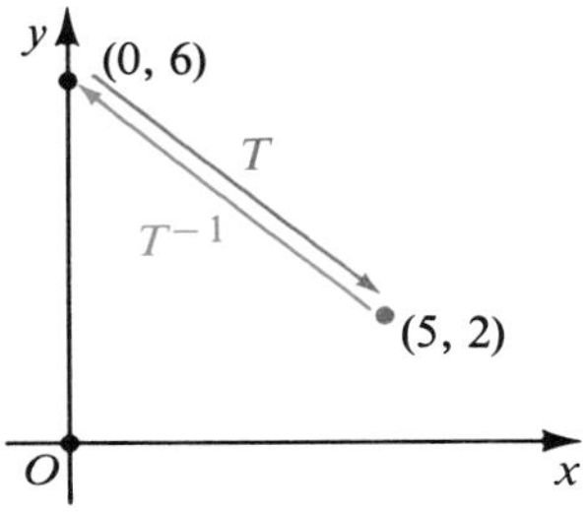

$T:(0, 6) \rightarrow (5, 2)$
$T^{-1}:(5, 2) \rightarrow (0, 6)$

Rotation $\mathcal{R}$

The inverse of $\mathcal{R}_{O,x}$ is $\mathcal{R}_{O,-x}$.

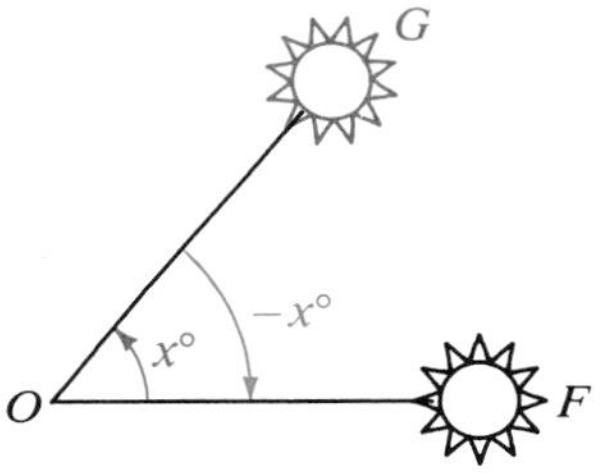

$\mathcal{R}_{O,x}:F \rightarrow G$
$\mathcal{R}_{O,-x}:G \rightarrow F$

Dilation D

The inverse of $D_{O,2}$ is $D_{O,\frac{1}{2}}$.

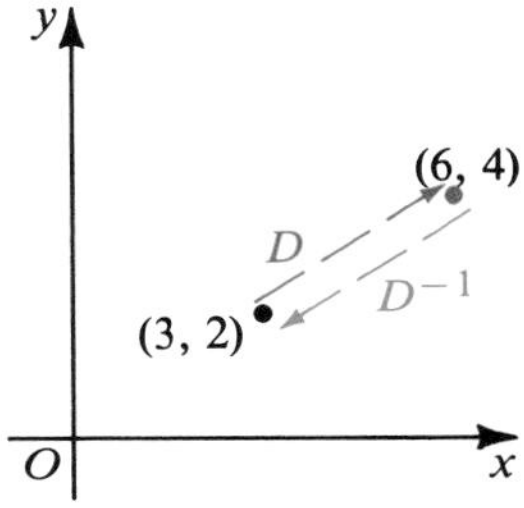

$D_{O,2}:(3, 2) \rightarrow (6, 4)$
$D_{O,\frac{1}{2}}:(6, 4) \rightarrow (3, 2)$

Classroom Exercises

The symbol 2^{-1} stands for the multiplicative inverse of 2, or $\frac{1}{2}$. Give the value of each of the following.

1. 3^{-1} **2.** 7^{-1} **3.** $(\frac{4}{5})^{-1}$ **4.** $(2^{-1})^{-1}$

The translation T maps all points five units right. Describe each of the following transformations.

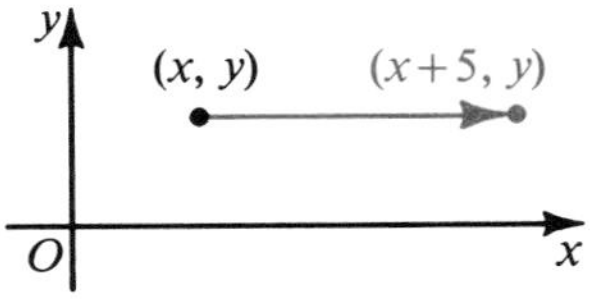

5. T^2 **6.** T^3

7. T^{-1} **8.** T^{-2}

9. $T \circ T^{-1}$ **10.** $(T^{-1})^{-1}$

The rotation $\mathcal{R}$ maps all points 120° about G, the center of equilateral $\triangle ABC$. Give the image of A by each of the following.

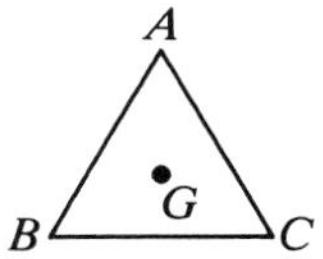

11. $\mathcal{R}$ **12.** $\mathcal{R}^2$ **13.** $\mathcal{R}^3$

14. $\mathcal{R}^6$ **15.** $\mathcal{R}^{-1}$ **16.** $\mathcal{R}^{-2}$

17. $\mathcal{R}^2 \circ \mathcal{R}^{-2}$ **18.** $\mathcal{R}^2 \circ \mathcal{R}^{-3}$ **19.** $\mathcal{R}^{100}$

20. What number is the identity for multiplication?
21. The product of any number t and the identity for multiplication is __?__.
22. The product of any transformation T and the identity transformation is __?__.
23. Write the inverse of each transformation.
  a. R_l b. $\mathcal{R}_{O,30}$ c. $T:(x, y) \to (x - 4, y + 1)$

Written Exercises

Give the value of each of the following.

A 1. 4^{-1} 2. 9^{-1} 3. $(\frac{2}{3})^{-1}$ 4. $(5^{-1})^{-1}$

The rotation $\mathcal{R}$ maps all points 90° about O, the center of square $ABCD$. Give the image of A by each of the following.

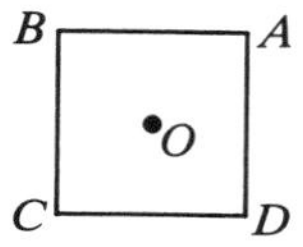

5. $\mathcal{R}^2$ 6. $\mathcal{R}^3$ 7. $\mathcal{R}^4$
8. $\mathcal{R}^{-1}$ 9. $\mathcal{R}^{-2}$ 10. $\mathcal{R}^{-3}$
11. $\mathcal{R}^{-3} \circ \mathcal{R}^3$ 12. $\mathcal{R}^5$ 13. $\mathcal{R}^{50}$

Complete.

14. By definition, the identity mapping I maps every point P to __?__.
15. $H_O{}^2$ is the same as the mapping __?__.
16. The inverse of H_O is __?__.
17. $H_O{}^3$ is the same as the mapping __?__.
18. If $T:(x, y) \to (x + 2, y)$, then $T^2:(x, y) \to$ (__?__, __?__).
19. If $T:(x, y) \to (x + 3, y + 4)$, then $T^2:(x, y) \to$ (__?__, __?__).

In each exercise, a rule is given for a mapping S. Write the rule for S^{-1}.

B 20. $S:(x, y) \to (x + 5, y + 2)$ 21. $S:(x, y) \to (x - 3, y - 1)$
22. $S:(x, y) \to (3x, 3y)$ 23. $S:(x, y) \to (\frac{1}{4}x, \frac{1}{4}y)$
24. $S:(x, y) \to (x + 4, 4y)$ 25. $S:(x, y) \to (y, x)$

26. If $S:(x, y) \to (x + 12, y + 3)$, find a translation T such that $T^6 = S$.
27. Find a transformation S (other than the identity) for which $S^5 = I$.

C 28. a. j and k are vertical lines 1 unit apart. According to Theorem 12-6, $R_k \circ R_j$ and $R_j \circ R_k$ are both translations. Describe in words the distance and direction of each translation.
  b. Show that $R_k \circ R_j$ and $R_j \circ R_k$ are inverses by showing that their product is I. *Note:* Forming products of transformations is an associative operation, so $(R_k \circ R_j) \circ (R_j \circ R_k) = R_k \circ (R_j \circ R_j) \circ R_k$.

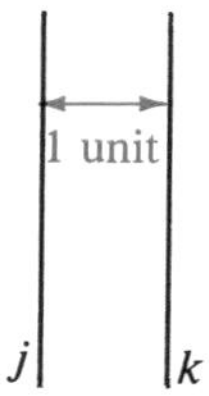

29. The blue lines in the diagram illustrate the statement $H_B \circ H_A$ = translation T. The red lines show that $H_A \circ H_B$ = translation S.

a. How is translation S related to translation T?

b. Prove your answer correct by showing that $(H_A \circ H_B) \circ (H_B \circ H_A) = I$.

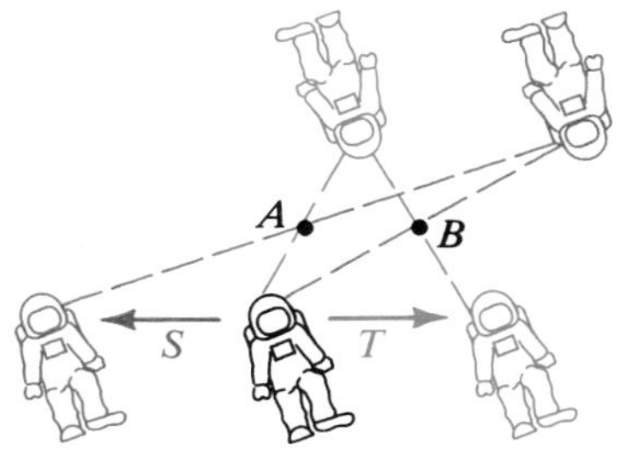

30. Complete the proof by giving a reason for each step.

Given: $l_1 \perp l_2$; $l_3 \perp l_2$; R_1, R_2, and R_3 denote reflections in l_1, l_2, and l_3.

Prove: $H_B \circ H_A$ is a translation.

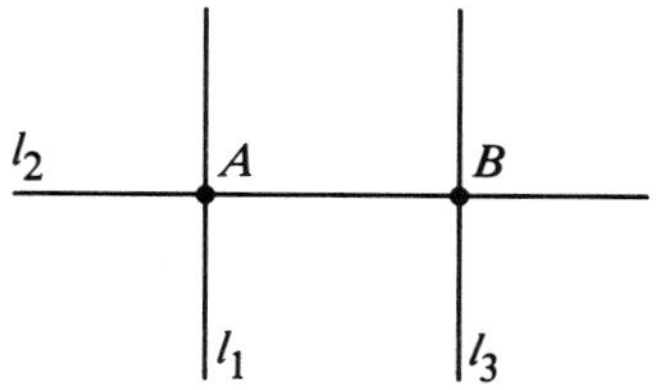

Proof:

Statements	Reasons
1. $H_A = R_2 \circ R_1$	1. ?
2. $H_B = R_3 \circ R_2$	2. ?
3. $H_B \circ H_A = (R_3 \circ R_2) \circ (R_2 \circ R_1)$	3. ?
4. $H_B \circ H_A = (R_3 \circ (R_2 \circ R_2)) \circ R_1$	4. Forming products of transformations is an associative operation.
5. $H_B \circ H_A = (R_3 \circ I) \circ R_1$	5. ?
6. $H_B \circ H_A = R_3 \circ R_1$	6. ?
7. $H_B \circ H_A$ is a translation.	7. ?

12-8 Symmetry

A figure in the plane has **symmetry** if there is an isometry, other than the identity, that maps the figure onto itself. We call such an isometry *a symmetry* of the figure.

Each of the figures below has **line symmetry.** This means that for each figure there is a symmetry line k such that the reflection R_k maps the figure onto itself. (Though the objects shown are three-dimensional, we can consider their pictures as plane figures. The reflection symmetry of living creatures like the lizard and butterfly is often called *bilateral symmetry.*)

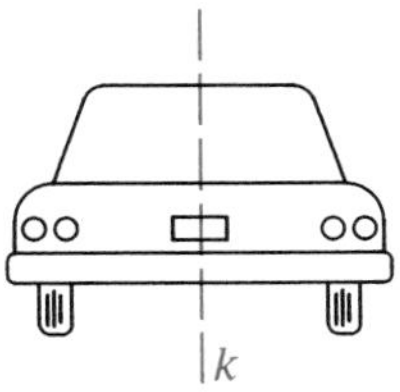

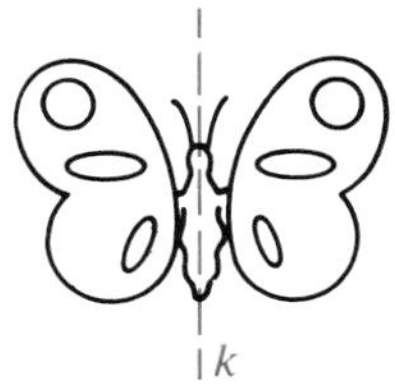

Each of the figures below has **point symmetry.** This means that for each figure there is a symmetry point O such that the half-turn H_O maps the figure onto itself.

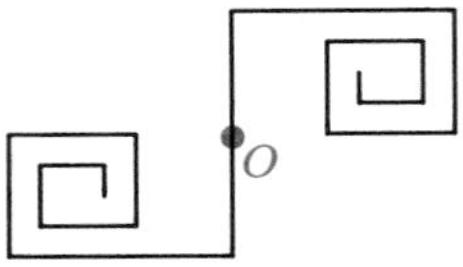

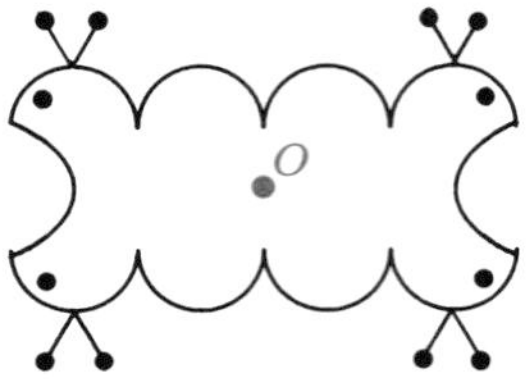

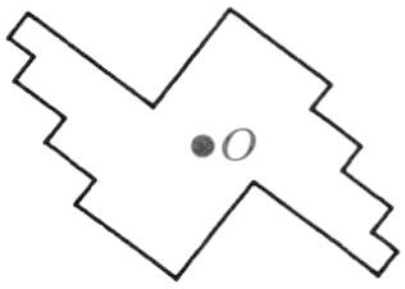

Besides having a symmetry point, the middle figure above has a vertical symmetry line and a horizontal symmetry line.

A third kind of symmetry is **rotational symmetry.** The figure at the right has the four rotational symmetries listed. Each symmetry has center O and rotates the figure onto itself. Note that 180° rotational symmetry is another name for point symmetry.

(1) 90° rotational symmetry; $\mathcal{R}_{O,90}$
(2) 180° rotational symmetry; $\mathcal{R}_{O,180}$ (or H_O)
(3) 270° rotational symmetry; $\mathcal{R}_{O,270}$
(4) 360° rotational symmetry; the identity I

The identity mapping always maps a figure onto itself, and we usually include the identity when listing the symmetries of a figure. (The Extra beginning on page 548 explores some properties of such complete lists of symmetries.) However, we do not call a figure *symmetric* if the identity is its only symmetry.

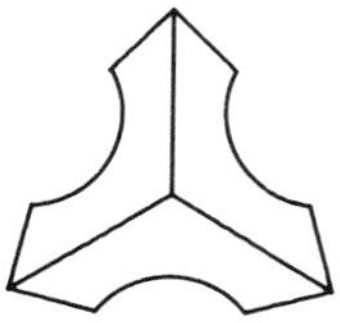

This symmetric figure has a total of six symmetries. Can you name them?

A figure can also have **translational symmetry** if there is a translation that maps the figure onto itself. For example, imagine that the design at the right extends in all directions to fill the plane. If you consider the distance between the eyes of adjacent blue fish as a unit, then a translation through one or more units right, left, up, or down maps the whole pattern onto itself. Do you see that you can also translate the pattern along diagonal lines?

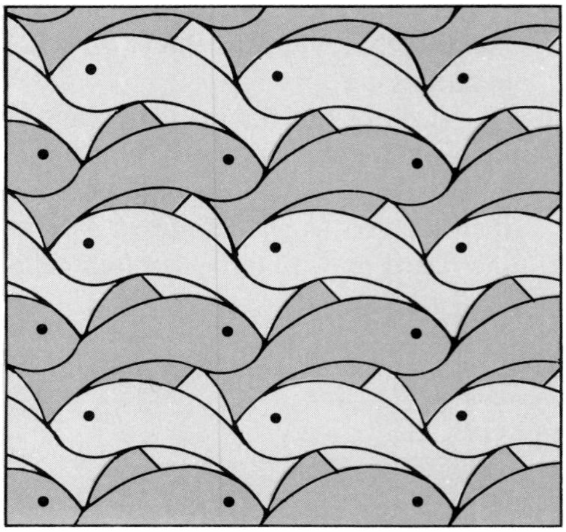

It is also possible to map the blue fish, which all face to the left, onto the right-facing green fish by translating the whole pattern a half unit up and then reflecting it in a vertical line. Thus, if we ignore color differences, the pattern has glide reflection symmetry.

A design like this pattern of fish, in which congruent copies of a figure completely fill the plane without overlapping, is called a *tessellation*. Tessellations can have any of the kinds of symmetry we have discussed. Here are two more examples.

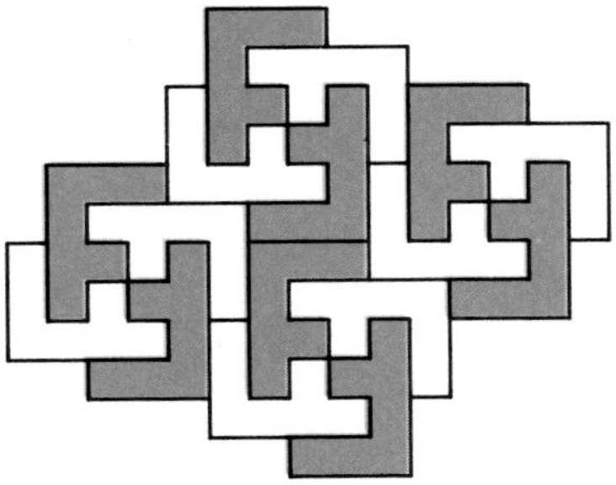

A tessellation of the letter F. This pattern has point symmetry and translational symmetry.

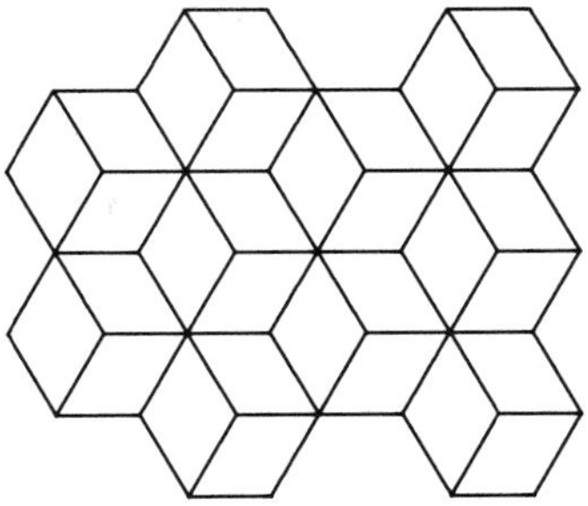

This tessellation has line, point, rotational, translational, and glide reflection symmetry.

Notice that the congruent figures in a tessellation may be given different colors, as in the fish design and the tessellation made from the letter F. Coloring a pattern in this way often changes its symmetries.

Classroom Exercises

Tell how many symmetry lines each figure has. In Exercise 2, *O* is the center of the equilateral triangle.

1. 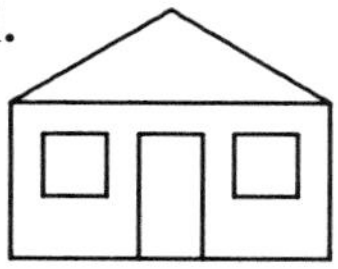

2. 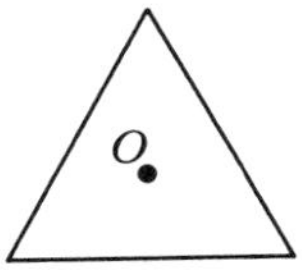

3. 

4. 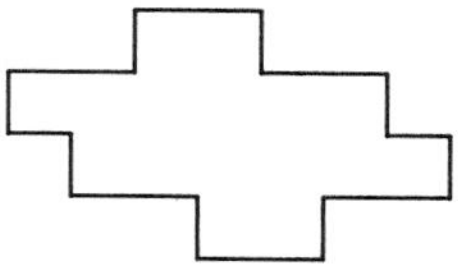

5. Which figures above have point symmetry?
6. Describe all of the rotational symmetries of the figure in Exercise 2.
7. Describe all of the rotational symmetries of the figure in Exercise 3.

Draw each figure on the chalkboard and describe all of its symmetries.

8. isosceles triangle
9. parallelogram
10. rectangle
11. rhombus

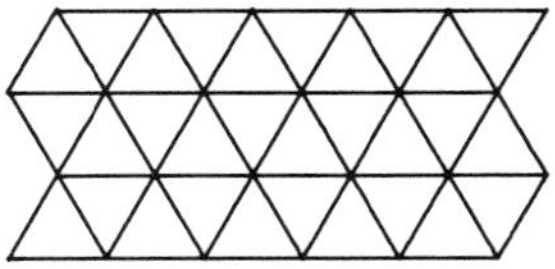

12. Imagine that the pattern shown fills the entire plane. Does the pattern have the symmetry named?
    a. translational symmetry
    b. line symmetry
    c. point symmetry
    d. rotational symmetry

Written Exercises

Consider the object shown in each photograph as a plane figure.
a. State how many symmetry lines each figure has.
b. State whether or not the figure has a symmetry point.
c. List all the rotational symmetries of each figure between 0° and 360°.

A **1.**

2.

3.

4.

5. Which capital letters of the alphabet have just one line of symmetry? (One answer is "D".)
6. Which capital letters of the alphabet have two lines of symmetry?
7. Which capital letters of the alphabet have a point of symmetry?

Make a tessellation of the given figure.

8.

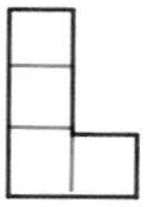

9.

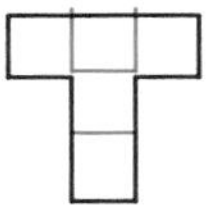

10.

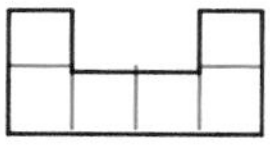

11.

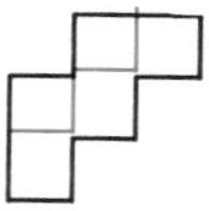

Copy the figure shown. Then complete the figure so that it has the specified symmetries.

12.

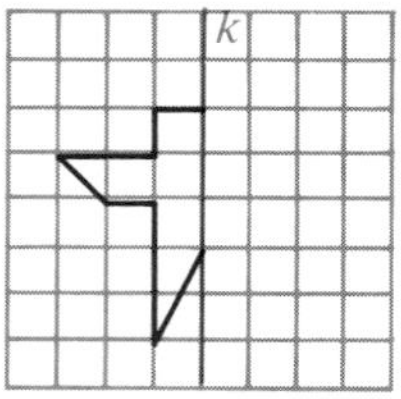

symmetry in line *k*

13.

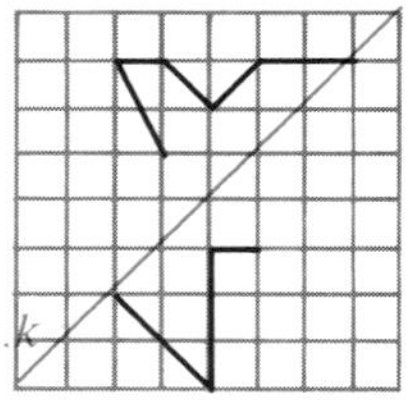

symmetry in line *k*

14.

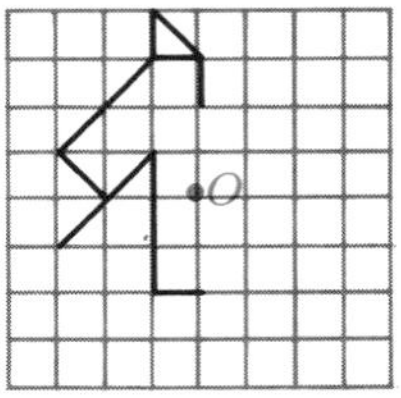

symmetry in point *O*

Copy the figure shown. Then complete the figure so that it has the specified symmetries.

B **15.**

60°, 120°, and 180° rotational symmetry

16.

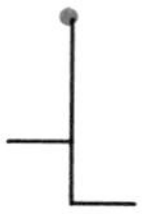

90°, 180°, and 270° rotational symmetry

17.

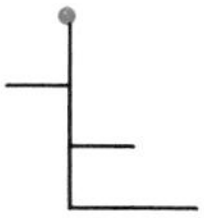

2 symmetry lines and 1 symmetry point

18. Tell whether or not a tessellation can be made with the given figure.
a. A regular hexagon
b. A scalene triangle
c. A regular pentagon
d. A nonisosceles trapezoid

In Exercises 19–21, draw the figure if there is one that meets the conditions. Otherwise write *not possible.*

19. A trapezoid with **(a)** no symmetry, **(b)** one symmetry line, **(c)** a symmetry point.

20. A parallelogram with **(a)** 4 symmetry lines, **(b)** just two symmetry lines, **(c)** just one symmetry line.

21. An octagon with **(a)** 8 rotational symmetries, **(b)** just 4 rotational symmetries, **(c)** only point symmetry.

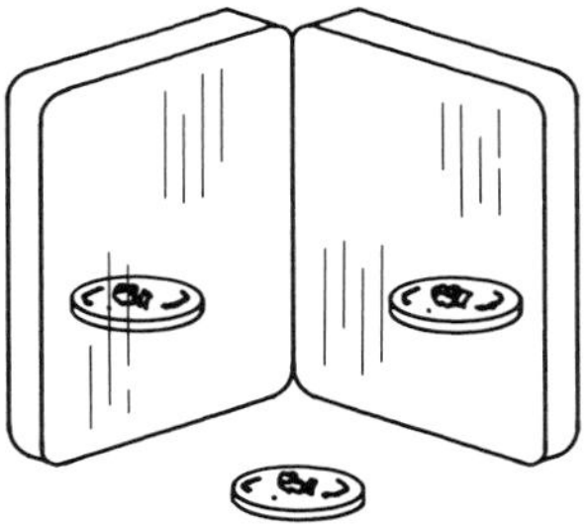

22. If you use tape to hinge together two pocket mirrors as shown and place the mirrors at a 120° angle, then a coin placed between the mirrors will be reflected, giving a pattern with 120° and 240° rotational symmetry.
a. What kinds of symmetries occur when the mirrors are at a right angle?
b. Experiment by forming various angles with two mirrors. Be sure to try 60°, 45°, and 30° angles. Record the number of coins you see, including the actual coin.

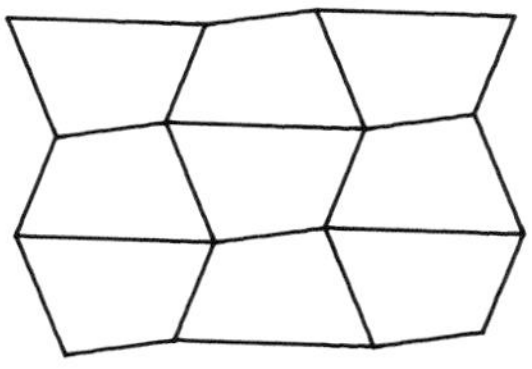

23. You can make a tessellation by tracing around *any* quadrilateral, placing copies of the quadrilateral systematically as shown.
a. The tessellation shown has many symmetry points but none of these are at vertices of the quadrilateral. Where are they?
b. What other kind of symmetry does this mosaic have?

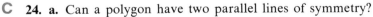

C **24.** **a.** Can a polygon have two parallel lines of symmetry?
b. Can a tessellation have two parallel lines of symmetry?

25. Show that if a hexagon has point symmetry, then its opposite sides must be parallel.

26. A figure has 60° rotational symmetry. What other rotational symmetries *must* it have? Explain your answer.

27. A figure has 50° rotational symmetry. What other rotational symmetries *must* it have? Explain your answer.

Self-Test 2

For Exercises 1–6, refer to the figure.

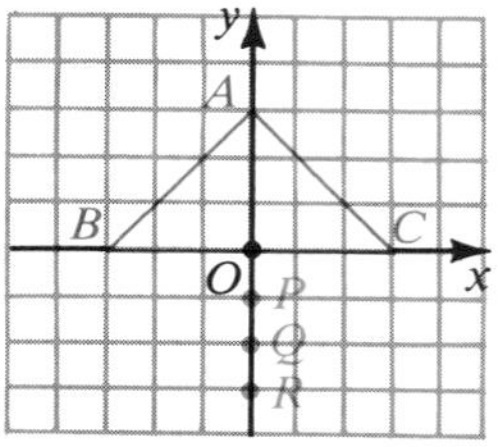

1. $R_x \circ \mathcal{R}_{O,90}: B \to$ ___?___
2. $R_x \circ H_O: A \to$ ___?___
3. $\mathcal{R}_{O,110} \circ \mathcal{R}_{O,70}: C \to$ ___?___
4. $D_{O,\frac{1}{2}} \circ D_{R,\frac{1}{2}}: P \to$ ___?___
5. What is the symmetry line of $\triangle ABC$?
6. Does $\triangle ABC$ have point symmetry?
7. For any transformation T, $T^{-1} \circ T: P \to$ ___?___.
8. The product of any transformation T and the identity is ___?___.
9. If line a is parallel to line b, then the product $R_a \circ R_b$ is a ___?___.
10. Give the inverse of each transformation.
    a. $S:(x, y) \to (x + 2, y - 3)$ **b.** $D_{O,5}$
    c. $\mathcal{R}_{O,-70}$ **d.** R_y
11. How many lines of symmetry does a regular hexagon have?

EXTRA

Symmetry Groups

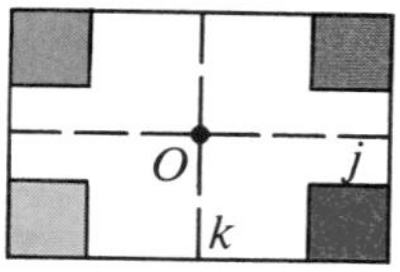

Cut out a cardboard or paper rectangle and color each corner with a color of its own on both front and back. Also on the front and back draw symmetry lines j and k and label symmetry point O. The rectangle has four symmetries: I, R_j, R_k, and H_O. The effect of each of these on the original rectangle is shown below.

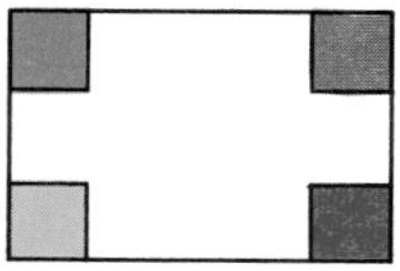

Effect of I:
Rectangle unchanged

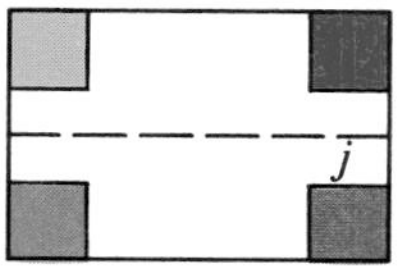

Effect of R_j

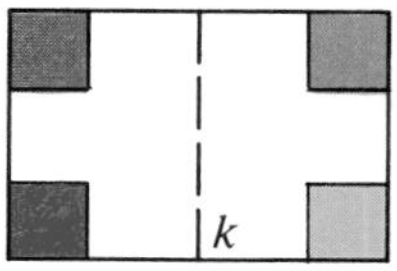

Effect of R_k

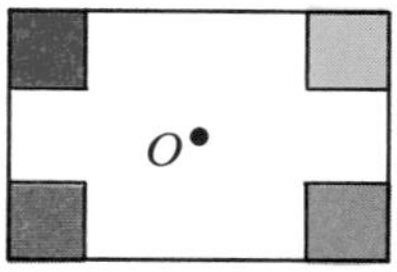

Effect of H_O

Our goal is to see how the four symmetries of the rectangle combine with each other. For example, if the original rectangle is mapped first by R_j and then by H_O, the images look like this:

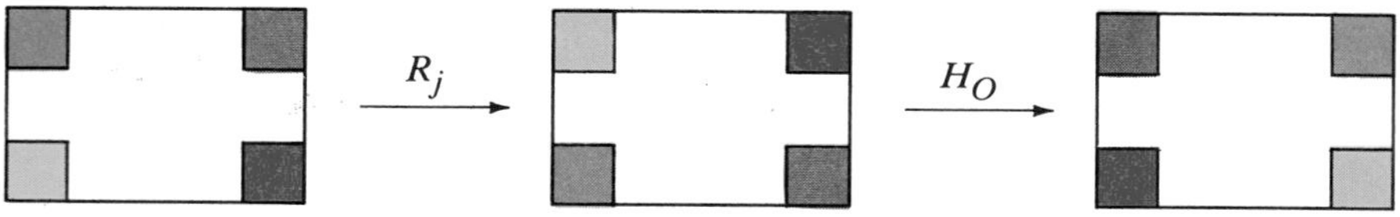

Mapping the rectangle by R_j and then by H_O has the same effect as the single symmetry R_k, so $H_O \circ R_j = R_k$. We can record this fact in a table resembling a multiplication table. Where the row for R_j meets the column for H_O, we enter the product $H_O \circ R_j$, which is R_k.

$\circ$	I	R_j	R_k	H_O
I				
R_j				R_k
R_k				
H_O				

We can determine other products of symmetries in the same way, but sometimes shortcuts can be used. For example, we know that

(1) $R_j \circ R_j = I$ and $R_k \circ R_k = I$ (Why?)
(2) $H_O \circ H_O = I$ (Why?)
(3) $R_j \circ R_k = H_O$ and $R_k \circ R_j = H_O$
(Corollary to Theorem 12–7)

Also we know that the product of any symmetry and the identity is that same symmetry. The completed table is shown at the right.

$\circ$	I	R_j	R_k	H_O
I	I	R_j	R_k	H_O
R_j	R_j	I	H_O	R_k
R_k	R_k	H_O	I	R_j
H_O	H_O	R_k	R_j	I

By studying the table you can see that the symmetries of the rectangle have these four properties:

(1) The product of two symmetries is another symmetry.
(2) The set of symmetries contains the identity.
(3) Each symmetry has an inverse which is also a symmetry. (In this example each symmetry is its own inverse.)
(4) Forming products of transformations is an associative operation: $A \circ (B \circ C) = (A \circ B) \circ C$ for any three symmetries A, B, and C.

A set of symmetries with these four properties is called a symmetry *group.* Symmetry groups are used in crystallography, and more general groups are important in physics and advanced mathematics. The exercises that follow illustrate the fact that the symmetries of any figure form a group.

Exercises

1. An isosceles triangle has just two symmetries, including the identity. Make a 2 by 2 group table showing how these symmetries combine.

2. a. List the four symmetries of the rhombus shown. (Include the identity.)

b. Make a group table showing all products of two symmetries.

c. Is your table in part (b) identical to the table of symmetries for the rectangle?

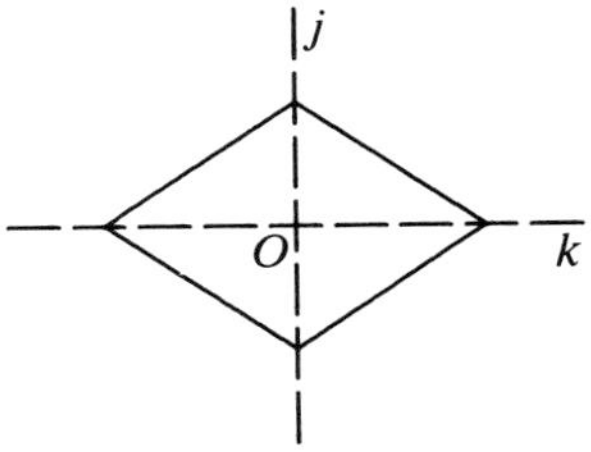

3. Make a group table for the three symmetries of this figure.

4. Make a group table for the four symmetries of this figure.

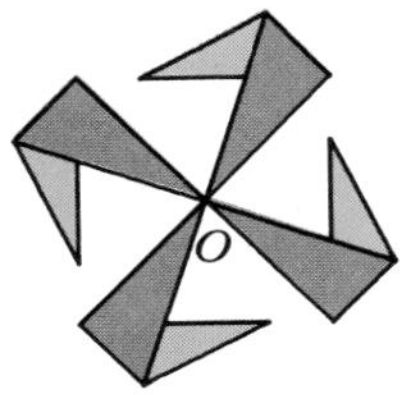

5. A transformation that is its own inverse is called a *self-inverse.*
   a. How many of the four symmetries of the figure in Exercise 4 are self-inverses?
   b. How many of the four symmetries of the rectangle are self-inverses?

6. A symmetry group is called commutative if $A \circ B = B \circ A$ for every pair of symmetries A and B in the group. The symmetry group of the rectangle is *commutative,* as you can see from the completed table. (For example, $H_O \circ R_j$ and $R_j \circ H_O$ are both equal to R_k.) Tell whether the groups in Exercises 3 and 4 are commutative or not.

7. An equilateral triangle has three rotational symmetries (I, $\mathcal{R}_{O,120}$, and $\mathcal{R}_{O,240}$) and three line symmetries (R_j, R_k, and R_l).
   a. Make a group table for these six symmetries.
   b. Give an example which shows that this group is *not* commutative.

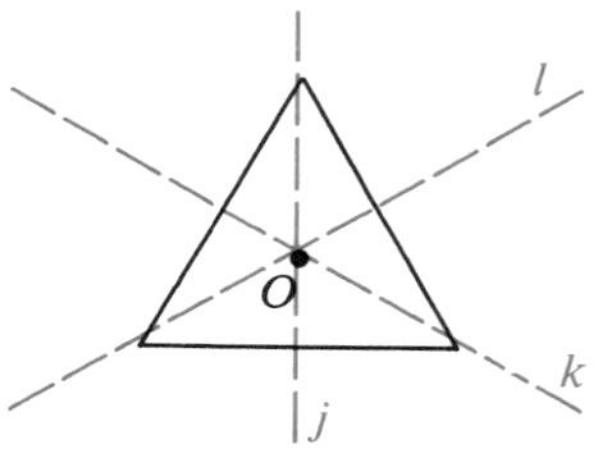

8. A square has four rotational symmetries (including the identity) and four line symmetries. Make a group table for these symmetries. Is this a commutative group?

9. The four rotational symmetries of the square satisfy the four requirements for a group, and so they are called a *subgroup* of the full symmetry group. (Notice that the identity is one of these rotational symmetries and that the product of two rotations is another rotation in the subgroup.)
   a. Do the four line symmetries of the square form a subgroup?
   b. Does the symmetry group of the equilateral triangle have a subgroup?
   c. Which two symmetries of the figure in Exercise 4 form a subgroup?

10. The tessellation with fish on page 544 has translational symmetry. Let S be the horizontal translation mapping each fish to the fish of the same color to its right, and let T be the vertical translation mapping each fish to the fish of the same color directly above.
   a. Describe the mapping S^3. Is it a symmetry of the pattern?
   b. Describe T^{-1}. Is it a symmetry?
   c. Describe $S \circ T$. Is it a symmetry?
   d. How many symmetries does the tessellation have?
   e. Does this set of symmetries satisfy the four requirements for a group?

Chapter Summary

1. A transformation is a one-to-one mapping from the whole plane to the whole plane. If the transformation S maps P to P', we write $S:P \rightarrow P'$.
2. An isometry is a transformation that preserves distance. An isometry maps any figure to a congruent figure.
3. Some basic isometries are:

   Reflection in a line. R_j is a reflection in line j.

   Translation or glide. The coordinate mapping $T:(x, y) \rightarrow (x + a, y + b)$ is a translation.

   Rotation about a point. $\mathcal{R}_{O,x}$ is a rotation counterclockwise about O through $x°$. H_O is a half-turn about O.

   Glide reflection. A glide followed by a reflection in a line parallel to the glide yields a glide reflection.
4. A dilation maps any figure to a similar figure. $D_{O,k}$ is a dilation with center O and nonzero scale factor k.
5. The combination of one mapping followed by another is called a product of mappings. The mapping A followed by B is written $B \circ A$.
6. A product of isometries is an isometry.
   A product of reflections in two parallel lines is a translation.
   A product of reflections in two intersecting lines is a rotation.
   A product of reflections in two perpendicular lines is a half-turn.
7. The identity transformation I keeps all points fixed. A transformation S followed by its inverse S^{-1} is equal to the identity.
8. A symmetry of a figure is an isometry that maps the figure onto itself. Figures can have line symmetry, point symmetry, and rotational symmetry. A tessellation, or covering of the plane with congruent figures, may also have translational and glide reflection symmetry.

Chapter Review

1. If isometry S maps A to A' and B to B', then $\overline{AB}$ _?_ $\overline{A'B'}$. 12-1

The transformation $S:(x, y) \rightarrow (2x, y - 2)$ is given.

2. The image of (3, 3) is (_?_, _?_).
3. The preimage of (3, 3) is (_?_, _?_).
4. Is S an isometry?

Copy each figure on squared paper. Then draw the image by reflection in line *k*.

5. 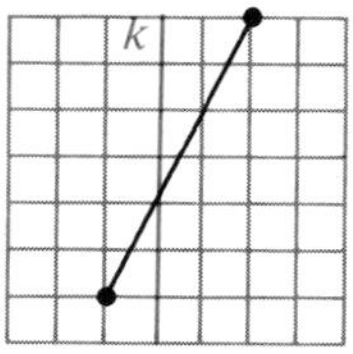

6. 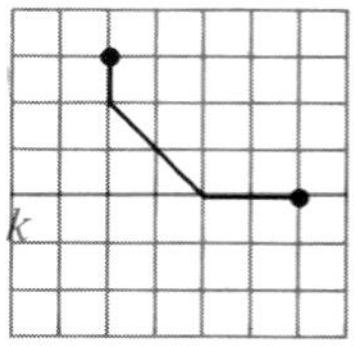

7. 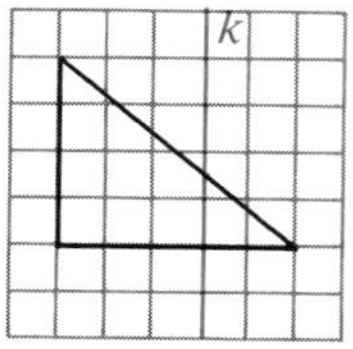

12–2

8. Under R_y the image of $(-7, 5)$ is (__?__, __?__).
9. Under R_x the image of $(5, 0)$ is (__?__, __?__).

10. The translation T maps the point $(5, 5)$ to $(7, 1)$ 12–3
    a. T:(0, 0) → (__?__, __?__) **b.** T:(x, y) → (__?__, __?__)
    c. T:(__?__, __?__) → (3, 2)
11. Find the image of $(7, -2)$ under the glide reflection that moves all points 5 units to the right and then reflects all points in the x-axis.

12. Plot the following points on a coordinate grid: $A(3, 2)$, $B(-1, 1)$, and $C(1, -3)$. Label the origin O. Draw $\triangle ABC$ and its images under 12–4
    a. $\mathcal{R}_{O,90}$ **b.** H_O
13. Which of the given rotations are equal to $\mathcal{R}_{O,140}$?
    a. $\mathcal{R}_{O,500}$ **b.** $\mathcal{R}_{O,-140}$ **c.** $\mathcal{R}_{O,-220}$

Give the coordinates of the image of *A* by the dilation specified.

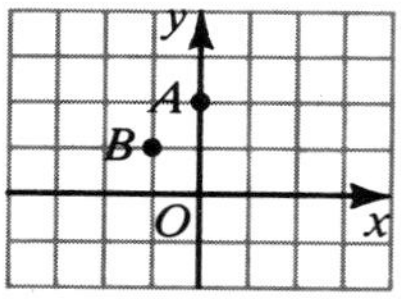

12–5

14. $D_{O,2}$
15. $D_{O,-\frac{1}{2}}$
16. $D_{B,3}$

Find the image of (3, 1) under each mapping.

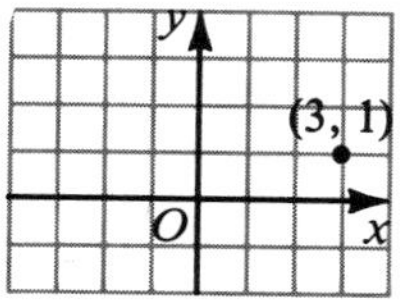

12–6

17. $R_x \circ R_y$:(3, 1) → (__?__, __?__)
18. $R_y \circ H_O$:(3, 1) → (__?__, __?__)
19. $R_x \circ \mathcal{R}_{O,-90}$:(3, 1) → (__?__, __?__)

Complete.

20. If T:$(x, y) \to (x - 1, y + 6)$, then T^{-1}:(x, y) → (__?__, __?__) 12–7
21. The inverse of $D_{O,5}$ is $D_{?,?}$.
22. $R_j \circ R_j$ = __?__ **23.** $\mathcal{R}_{O,75} \circ \mathcal{R}_{O,?} = I$
24. Does a scalene triangle have line symmetry? 12–8
25. Does a rectangle have point symmetry?
26. Does a regular octagon have 90° rotational symmetry?
27. Name a figure that has 72° rotational symmetry.

Chapter Test

State whether the transformation mapping the black triangle to the red triangle is a reflection, a translation, a glide reflection, or a rotation.

1. 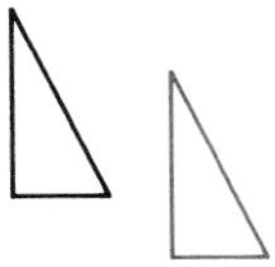

2. 

3. 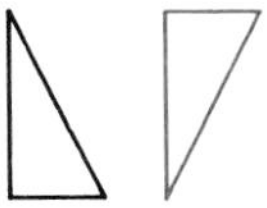

4. 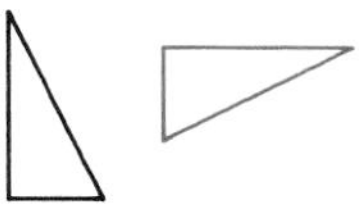

Give the coordinates of the image of point *P* under the transformation specified.

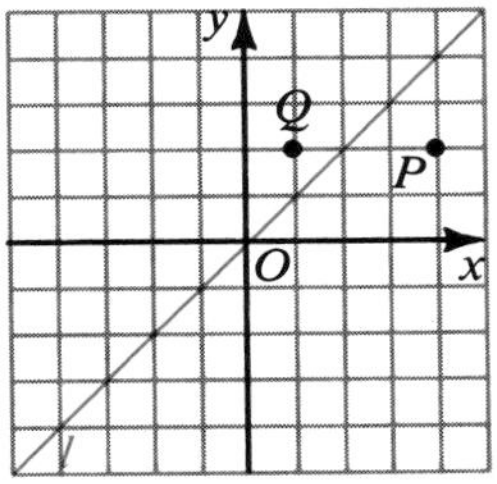

5. R_l

6. $\mathcal{R}_{O,-90}$

7. $D_{O,\frac{1}{2}}$

8. $H_O \circ R_x$

9. $\mathcal{R}_{O,90} \circ \mathcal{R}_{O,90}$

10. $D_{Q,\frac{1}{3}}$

11. $\mathcal{R}_{O,180} \circ H_O$

12. $R_l \circ R_y$

13. $R_l \circ D_{Q,-1}$

14. $R_l \circ (R_y \circ R_x)$

Give the inverse of each transformation.

15. H_O

16. R_x

17. $D_{O,-1}$

***T* is the translation mapping (4, 1) to (6, 2). Find the coordinates of the image of the origin under each mapping.**

18. T

19. T^3

20. T^{-1}

21. $I \circ T$

Classify each statement as true or false.

22. All regular polygons have rotational symmetry.

23. A parallelogram need not have any symmetry except the identity.

24. 180° rotational symmetry is the same as point symmetry.

25. All regular n-gons have exactly n symmetry lines.

26. A figure that has two intersecting lines of symmetry must have rotational symmetry.

Preparing for College Entrance Exams

Strategy for Success

Try to work quickly and accurately on exam questions. Do not take time to double-check your answers unless you finish all the questions before the deadline. Skip questions that are too difficult for you, and spend no more than a few minutes on each question.

Indicate the best answer by writing the appropriate letter.

1. Find an equation of the perpendicular bisector of the segment joining $(3, -1)$ and $(-1, 7)$.
(A) $x + 2y = 7$ **(B)** $x - 2y = -5$ **(C)** $2x + y = -5$
(D) $2x + y = 5$ **(E)** $2x - y = -1$

2. A circle has a diameter with endpoints $(0, -8)$ and $(-6, -16)$. An equation of the circle is:
(A) $(x + 3)^2 + (y + 12)^2 = 25$ **(B)** $(x + 3)^2 + (y + 12)^2 = 100$
(C) $(x - 3)^2 + (y - 12)^2 = 25$ **(D)** $(x - 3)^2 + (y - 12)^2 = 100$
(E) $(x + 6)^2 + (x + 24)^2 = 100$

3. The point $(\frac{1}{2}, -\frac{1}{2})$ lies on line t. Which of the following allow you to find an equation for t?
I. slope of t is -3 II. x-intercept of t is 7 III. t is parallel to $4x - 5y = 7$
(A) I only **(B)** III only **(C)** I and III only **(D)** II only **(E)** I, II, and III

4. Given $A(-3, 5)$, $B(0, -4)$, $C(2, 5)$, and $D(-6, -1)$, find the intersection point of $\overleftrightarrow{AB}$ and $\overleftrightarrow{CD}$.
(A) $(6, 23)$ **(B)** $(2, -10)$ **(C)** $(-2, 2)$ **(D)** $(-18, 5)$ **(E)** cannot be determined

5. What is the best name for quadrilateral $WXYZ$ with vertices $W(-3, -2)$, $X(-5, 2)$, $Y(1, 5)$, and $Z(3, 1)$?
(A) isosceles trapezoid **(B)** parallelogram **(C)** rectangle
(D) rhombus **(E)** square

6. Two vertices of an isosceles right triangle are $(0, 0)$ and $(j, 0)$. The third vertex cannot be:
(A) $(0, j)$ **(B)** $(0, -j)$ **(C)** (j, j) **(D)** $\left(\frac{j}{2}, \frac{j}{2}\right)$ **(E)** $\left(\frac{j}{2}, j\right)$

7. What is the image of $(-2, 3)$ under reflection in the line $y = x$?
(A) $(3, -2)$ **(B)** $(2, 3)$ **(C)** $(-2, -3)$ **(D)** $(2, -3)$ **(E)** $(-3, 2)$

8. Find the preimage of $(0, 0)$ under $D_{P,\frac{1}{4}}$, where P is the point $(-1, 1)$.
(A) $(-4, 4)$ **(B)** $(-\frac{3}{4}, \frac{3}{4})$ **(C)** $(-\frac{1}{4}, \frac{1}{4})$ **(D)** $(4, -4)$ **(E)** $(3, -3)$

9. A regular pentagon does *not* have:
(A) line symmetry **(B)** point symmetry **(C)** 360° rotational symmetry
(D) 216° rotational symmetry **(E)** 72° rotational symmetry

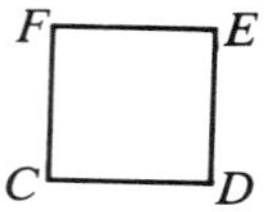

10. If $CDEF$ is a square, then $\mathcal{R}_{C,-450}: \overline{CF} \to$ ___?___.
(A) $\overline{FE}$ **(B)** $\overline{ED}$ **(C)** $\overline{CF}$ **(D)** $\overline{CD}$ **(E)** $\overline{CE}$

Cumulative Review: Chapters 1–12

True-False Exercises

Write T or F to indicate your answer.

A **1.** The line through $(-2, -5)$ and $(1, 1)$ passes through the origin.

2. If two rectangles have equal perimeters, then they have equal areas.

3. If a cylinder and a right prism have equal base areas and equal heights, then they have equal volumes.

4. An acute angle inscribed in a circle must intercept a minor arc.

5. If the slopes of two lines have opposite signs, the lines are perpendicular.

6. If two quadrilaterals are congruent, then they are also similar.

7. Three given points are always coplanar.

8. A triangle with sides of length $2x$, $3x$, and $4x$ must be obtuse.

9. Some rectangles are equilateral.

10. The contrapositive of a true conditional is sometimes false.

11. If $\triangle RST \cong \triangle TSR$, then $\angle R \cong \angle T$.

12. In a plane the locus of points equidistant from M and N is the midpoint of $\overline{MN}$.

13. Corresponding parts of similar triangles must be congruent.

14. $R_k \circ R_k = I$

15. All cylinders are similar.

16. Given a segment of length t, it is possible to construct a segment of length $t\sqrt{3}$.

17. A point lies on the bisector of $\angle ABC$ if and only if it is equidistant from A and C.

B **18.** The lateral area of a cone can be equal to the area of the base of the cone.

19. A triangle with vertices $(a, 0)$, $(-a, 0)$, and $(0, a)$ is equilateral.

20. If an equilateral triangle and a regular hexagon are inscribed in a circle, then the ratio of their areas is 1:2.

21. In $\triangle RST$, if $RS < ST$, then $\angle R$ must be the largest angle of the triangle.

22. The angle bisectors of an obtuse triangle intersect at a point that is equidistant from the three vertices.

23. Each interior angle of a regular n-gon has measure $\frac{(n-2)180}{n}$.

24. If a figure has 90° rotational symmetry, then it also has point symmetry.

25. The circle $(x + 3)^2 + (y - 2)^2 = 4$ is tangent to the line $x = -1$.

26. In a right triangle, the altitude to the hypotenuse is always the shortest altitude.

Multiple-Choice Exercises

Write the letter that indicates the best answer.

A

1. In a plane, the locus of points equidistant from two parallel lines is:
a. a point **b.** a line **c.** a pair of lines **d.** a plane

2. For every acute angle X:
a. $\cos X < \sin X$ **b.** $\cos X > \tan X$ **c.** $\tan X > 1$ **d.** $\cos X < 1$

3. The line through $(-7, 0)$ and $(3, -5)$ has slope:
a. $-\frac{1}{2}$ **b.** $\frac{1}{2}$ **c.** 2 **d.** -2

4. The median to the hypotenuse of a right triangle divides the triangle into two triangles that are both:
a. similar **b.** right **c.** scalene **d.** isosceles

5. If O is the origin, then $H_O:(-5, 3) \rightarrow$ _?_.
a. $(-5, -3)$ **b.** $(3, -5)$ **c.** $(5, -3)$ **d.** $(-3, 5)$

6. Which of the following is not a method for proving two triangles congruent?
a. HL **b.** AAS **c.** SSA **d.** SAS

7. What is the total area of a cone with radius 8 and height 6?
a. 96π **b.** 128π **c.** 160π **d.** 144π

8. Which group of numbers can be the lengths of the sides of an obtuse triangle?
a. 3, 5, 7 **b.** 2, 4, 6 **c.** 0.5, 0.6, 0.7 **d.** $\frac{3}{4}$, 1, $\frac{5}{4}$

9. If A, B, and C are points on $\odot O$, $\overline{AC}$ is a diameter, and $m\angle AOB = 60$, then $m\angle ACB =$
a. 30 **b.** 60 **c.** 90 **d.** 120

10. Two regular octagons have sides of length $6\sqrt{3}$ and 9. The ratio of their areas is:
a. $2\sqrt{3}:3$ **b.** $4:3$ **c.** $2:3$ **d.** $8\sqrt{3}:9$

B

11. Which proportion is *not* equivalent to $\frac{a}{b} = \frac{c}{d}$?
a. $\frac{a}{c} = \frac{b}{d}$ **b.** $\frac{b}{a + b} = \frac{d}{c + d}$ **c.** $\frac{b}{a} = \frac{d}{c}$ **d.** $\frac{a}{d} = \frac{c}{b}$

12. A regular hexagon with perimeter 24 has area:
a. $48\sqrt{3}$ **b.** $16\sqrt{3}$ **c.** $24\sqrt{3}$ **d.** $32\sqrt{3}$

13. A product of reflections in intersecting lines must be:
a. a translation **b.** a rotation **c.** a half-turn **d.** a dilation

14. In $\triangle RST$, X and Y are the midpoints of $\overline{RS}$ and $\overline{RT}$, respectively. $\overline{XT}$ and $\overline{YS}$ intersect at Z. Then

a. Z is equidistant from $\overline{RS}$, $\overline{ST}$, and $\overline{RT}$.
b. Z is equidistant from R, S, and T.
c. $TZ = 2 \cdot XZ$
d. Point Z is outside $\triangle RST$ when $\triangle RST$ is obtuse.

15. A rectangle with perimeter 27 and area 44 has length:
a. $2\sqrt{11}$ **b.** 8 **c.** 11 **d.** 10

Completion Exercises

Write the correct word, number, phrase, or expression.

A **1.** If $j \perp k$ and line j has slope $\frac{2}{3}$, then k has slope _?_.

2. The measures of two angles of a triangle are 56 and 62. The measure of the largest exterior angle of the triangle is _?_.

3. The slope of a _?_ line is not defined.

4. A cube with side of length 11 has volume _?_.

5. If A is $(-8, 3)$ and B is $(-4, -1)$, then the midpoint of $\overline{AB}$ is (_?_, _?_).

6. If l is the x-axis, then R_l:(4, 1) $\rightarrow$ (_?_, _?_).

7. An angle inscribed in a semicircle must be a(n) _?_ angle.

8. The distance between $(-5, -2)$ and $(1, -6)$ is _?_.

9. In a convex n-gon, _?_ diagonals can be drawn from a particular vertex.

10. Two circles can have at most _?_ common tangents.

11. If the vertices of a triangle are $(-2, 0)$, $(9, 0)$, and $(3, 6)$, then its area is _?_.

12. A circle with area 100π has circumference _?_.

13. A regular square pyramid with base edge 16 cm and height 6 cm has volume _?_.

14. If $(4, -3)$ and $(7, -6)$ are opposite vertices of a square, then the perimeter of the square is _?_ and the area is _?_.

15. If $\overline{RS} \cong \overline{ST}$, and $\overrightarrow{SZ}$ bisects $\angle RST$ and intersects $\overline{RT}$ at Z, then the _?_ methods can be used to prove that $\triangle RSZ \cong \triangle TSZ$.

16. If $5x - 1 = 14$, then the statement $5x = 15$ is justified by the _?_.

17. If the longest possible chord of a circle has length 12, then the radius is _?_.

18. A rhombus with diagonals of 16 and 12 has area _?_.

19. A regular hexagon with radius 12 has apothem _?_.

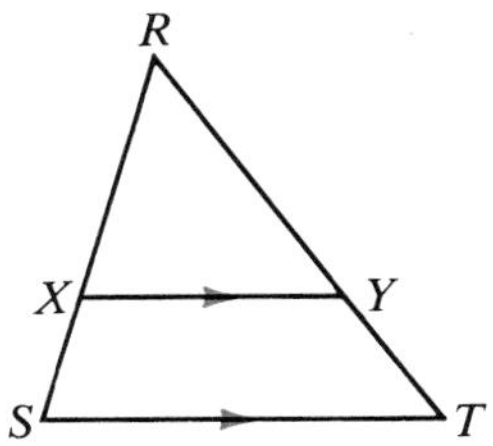

Exs. 20, 21

B **20.** If $RX = 18$, $XS = 10$, and $RT = 35$, then $YT =$ _?_.

21. If $RX = 16$, $XS = 8$, and $XY = 15$, then $ST =$ _?_.

22. A cone with radius 9 and slant height 12 has volume _?_ and lateral area _?_.

23. The shortest chord that can be drawn through a point 5 cm from the center of a circle of radius 10 cm has a length of _?_.

24. A sector with an arc of measure 72 in a circle of radius 4 m has an area of _?_.

25. In $\triangle ABC$, $\overline{AB} \perp \overline{BC}$, $AB = 15$, and $BC = 8$. Then the exact value of $\sin C$ is _?_.

26. A circle with center $(-2, 3)$ and radius 5 contains points $(-5,$ _?_$)$ and $(-5,$ _?_$)$.

27. A trapezoid with sides 8, 8, 8, and 10 has area _?_.

28. If each edge of a regular triangular pyramid is 6 cm, then the pyramid has total area __?__ and volume __?__.

29. If R_x is reflection in the x-axis and O is the origin, then

$$R_x \circ \mathcal{R}_{O,90}:(-2, 5) \rightarrow (?, ?).$$

30. A plane parallel to the base of a cone and bisecting the altitude divides the cone into two parts whose volumes have the ratio __?__.

Always-Sometimes-Never Exercises

Write A, S, or N to indicate your answer.

A 1. The lengths of the sides of a 45°-45°-90° triangle are __?__ in the ratio 1:1:2.

2. A lateral edge of a regular pyramid is __?__ longer than the slant height.

3. Two similar triangles are __?__ congruent.

4. A conclusion based on deductive reasoning is __?__ correct.

5. A conclusion based on inductive reasoning is __?__ correct.

6. Under a half-turn about point O, point O is __?__ mapped into itself.

7. Transformations are __?__ isometries.

8. Given any three lengths, it is __?__ possible to construct a triangle with sides of these lengths.

9. If J is a point outside $\odot P$ and $\overline{JA}$ and $\overline{JB}$ are tangent to $\odot P$, then $\triangle JAB$ is __?__ scalene.

10. Vertical angles are __?__ adjacent angles.

11. Two right triangles with congruent hypotenuses are __?__ congruent.

12. If the diagonals of a quadrilateral are perpendicular bisectors of each other, then the quadrilateral __?__ is a rhombus.

13. In a regular polygon, an interior angle and an exterior angle are __?__ complementary.

14. If $\triangle RST$ is a right triangle with hypotenuse $\overline{RS}$, then sin R and cos S are __?__ equal.

15. The locus of points in space equidistant from two parallel lines is __?__ a line.

B 16. The center of the circle that can be circumscribed about a given triangle is __?__ outside the triangle.

17. A circle __?__ contains three collinear points.

18. Given a square, it is __?__ possible to construct a square with area twice as large as the given square.

19. A product of reflections in two lines is __?__ a translation.

20. A triangle with sides of length x, $x + 2$, and $x + 4$ is __?__ an acute triangle.

21. A median of a triangle __?__ separates the triangle into two triangles with equal areas.

22. If $\widehat{RS}$ and $\widehat{XY}$ are arcs of $\odot O$ and $m\widehat{RS} < m\widehat{XY}$, then RS and XY are __?__ equal.

23. If A and B are transformations, then the image of a point under $A \circ B$ is __?__ the same as the image under $B \circ A$.

Construction Exercises

A 1. Draw a large obtuse triangle. Inscribe a circle in the triangle.

2. Draw a circle O and choose a point T on $\odot O$. Construct the tangent to $\odot O$ at T.

Draw two long segments. Let their lengths be x and y, with $x > y$.

3. Construct an isosceles triangle with legs of length x and with vertex angle of measure 30.

4. Construct a segment of length $\frac{1}{2}(3x - y)$.

5. Construct a rectangle with width y and diagonal x.

B 6. Construct a segment with length $\sqrt{3xy}$.

7. Construct any triangle with area xy.

8. Construct a rhombus with sides of length y and a diagonal of length x.

9. Draw a very long $\overline{AB}$. Construct a parallelogram with perimeter AB and consecutive sides in the ratio 3:2.

10. Construct an angle of measure $157\frac{1}{2}$.

Proof Exercises

A 1. Given: $\overline{OP} \cong \overline{OQ}$; $\overline{OS} \cong \overline{OR}$
Prove: $\overline{PS} \cong \overline{QR}$

2. Given: $\overline{PQ} \parallel \overline{RS}$
Prove: $\dfrac{PO}{RO} = \dfrac{PQ}{RS}$

3. Given: $\overline{PR} \perp \overline{QS}$; $\overline{PS} \cong \overline{QR}$; $\overline{OS} \cong \overline{OR}$
Prove: $\angle PSO \cong \angle QRO$

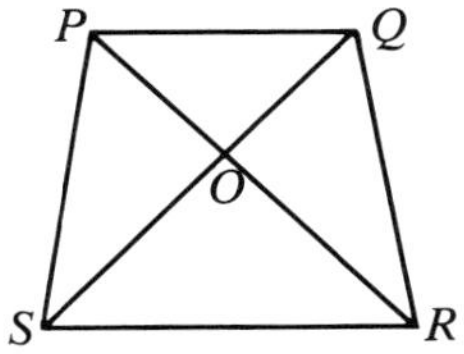

Exs. 1–4

B 4. Given: $\angle OSR \cong \angle ORS$; $\angle OPQ \cong \angle OQP$
Prove: $\triangle PSR \cong \triangle QRS$

5. Prove: The diagonals of a rectangle intersect to form four congruent segments.

6. Use coordinate geometry to prove that the triangle formed by joining the midpoints of the sides of an isosceles triangle is an isosceles triangle.

C 7. Use an indirect proof to show that a trapezoid cannot have two pairs of congruent sides.

8. Prove: If two coplanar circles intersect in two points, then the line joining those points bisects a common tangent segment.

Examinations

Chapter 1

Indicate the best answer by writing the appropriate letter.

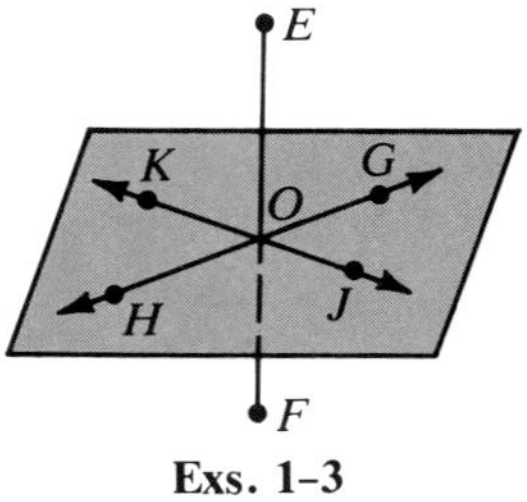

Exs. 1–3

1. Which of the following sets of points are *not* coplanar?
   a. E, H, O, G **b.** K, O, G, E
   c. E, O, F, J **d.** H, K, O, J
2. Which of the following sets of points are contained in *more* than one plane?
   a. G, O, J **b.** E, O, G
   c. H, E, G **d.** G, O, H
3. If $m\angle KOG = 140$, which of the following *must* be true?
   a. $m\angle EOG = 90$ **b.** $m\angle EOJ = 140$
   c. $m\angle HOJ = 140$ **d.** $m\angle HOF = 40$
4. Which of the following *cannot* be used as a reason in a proof?
   a. a definition **b.** a postulate
   c. yesterday's theorem **d.** tomorrow's theorem
5. Points A, B, and C are collinear, but they do not necessarily lie on a line in the order named. If $AB = 5$ and $BC = 3$, what is the length of $\overline{AC}$?
   a. either 2 or 8 **b.** either 2 or 4 **c.** 2 **d.** 8
6. If $\angle 1$ and $\angle 2$ are complements, $\angle 2$ and $\angle 3$ are complements, and $\angle 3$ and $\angle 4$ are supplements, what are $\angle 1$ and $\angle 4$?
   a. supplements **b.** complements **c.** congruent angles **d.** can't tell
7. The statement "If $m\angle A = m\angle B$ and $m\angle D = m\angle A + m\angle C$, then $m\angle D = m\angle B + m\angle C$" is justified by what property?
   a. Transitive **b.** Substitution **c.** Symmetric **d.** Reflexive

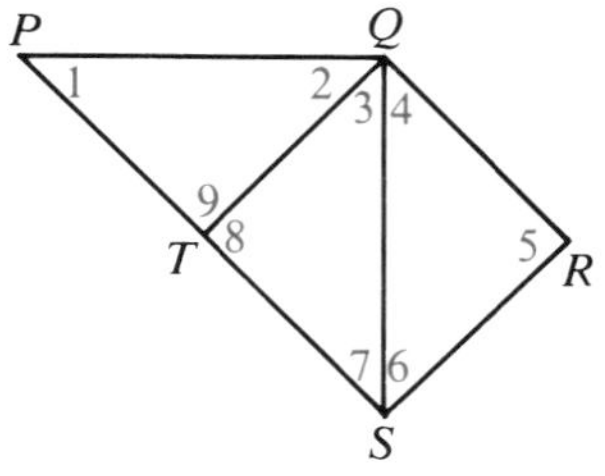

Exs. 8–11

8. If $\overline{TQ} \perp \overline{QR}$, which angles *must* be complementary angles?
   a. $\angle 2$ and $\angle 3$ **b.** $\angle 3$ and $\angle 4$
   c. $\angle 5$ and $\angle 8$ **d.** $\angle 3$ and $\angle 7$
9. If $m\angle 8 = x + 80$, what is the measure of $\angle 9$?
   a. $100 - x$ **b.** $100 + x$
   c. $x - 80$ **d.** $x - 180$
10. Which angle is obtuse?
    a. $\angle 7$ **b.** $\angle PQR$ **c.** $\angle 9$ **d.** $\angle R$
11. If $\overrightarrow{SQ}$ bisects $\angle RST$, which angles *must* be congruent?
    a. $\angle 6$ and $\angle 7$ **b.** $\angle 3$ and $\angle 4$
    c. $\angle 2$, $\angle 3$, and $\angle 4$ **d.** $\angle 4$ and $\angle 6$

Chapter 2

Indicate the best answer by writing the appropriate letter.

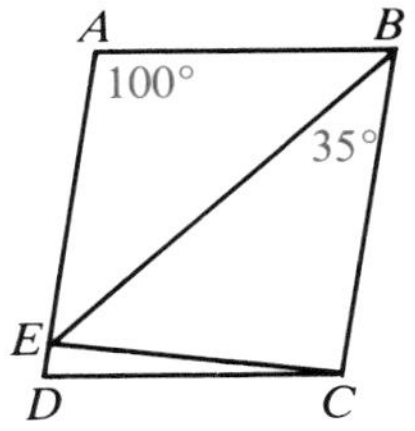

Exs. 1–4

1. If $\overrightarrow{BE}$ bisects $\angle ABC$, what does $m\angle AEB$ equal?
   a. 30 **b.** 35 **c.** 40 **d.** 45

2. If $m\angle ABE = 40$, what does $m\angle BED$ equal?
   a. 140 **b.** 40 **c.** 75 **d.** 135

3. If $\overline{AB} \parallel \overline{DC}$, what does $m\angle D$ equal?
   a. 70 **b.** 80 **c.** 90 **d.** 100

4. Which of the following would allow you to conclude that $\overline{AD} \parallel \overline{BC}$?
   a. $\angle DEC \cong \angle BCE$ **b.** $\angle ABE \cong \angle BEC$
   c. $\angle BEC \cong \angle BCE$ **d.** $m\angle A + m\angle AEC = 180$

5. Given: (1) If A is white, then B is red.
   (2) B is not red.
   Which of the following *must* be true?
   a. B is white. **b.** B is not white. **c.** A is not white. **d.** A is red.

6. If a statement is known to be true, then which of the following *must* also be true?
   a. its converse **b.** its contrapositive
   c. its inverse **d.** none of these

7. What is the measure of each angle of a regular octagon?
   a. 150 **b.** 144 **c.** 140 **d.** 135

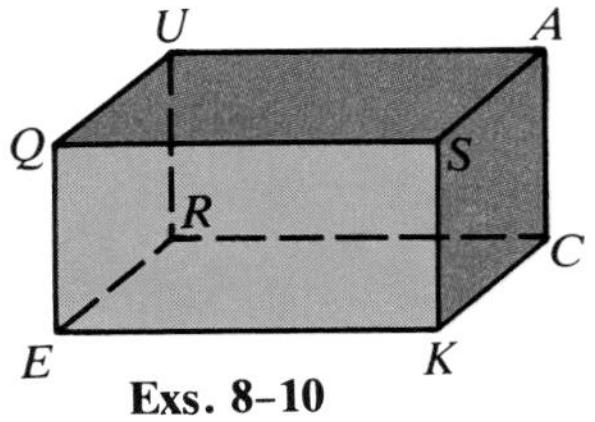

Exs. 8–10

8. The plane containing S, A, C, K appears to be parallel to the plane containing which points?
   a. Q, E, K, S **b.** Q, U, R, E
   c. A, S, Q, U **d.** U, R, C, A

9. Which of the following appear to be skew lines?
   a. $\overleftrightarrow{QE}$ and $\overleftrightarrow{AC}$ **b.** $\overleftrightarrow{QU}$ and $\overleftrightarrow{KC}$
   c. $\overleftrightarrow{AC}$ and $\overleftrightarrow{UR}$ **d.** $\overleftrightarrow{QS}$ and $\overleftrightarrow{AC}$

10. $\overleftrightarrow{EK}$ does *not* appear to be parallel to the plane containing which points?
    a. U, A, C **b.** Q, U, A **c.** Q, U, R **d.** Q, S, C

11. The sum of the interior angles of a certain polygon is the same as the sum of its exterior angles. How many sides does the polygon have?
    a. four **b.** six **c.** eight **d.** ten

12. What is the sum of the measures of the interior angles of a pentagon?
    a. 180 **b.** 360 **c.** 540 **d.** 900

Chapter 3

In Exercises 1–8 write a method (SSS, SAS, ASA, AAS, or HL) that can be used to prove the two triangles congruent.

1.

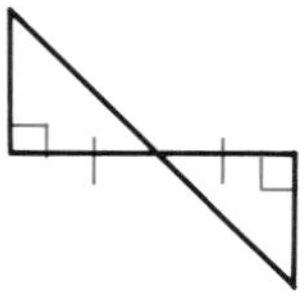

2.

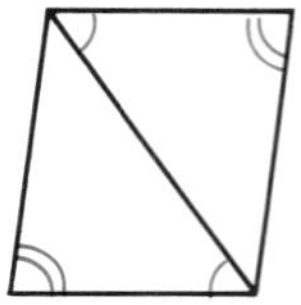

3.

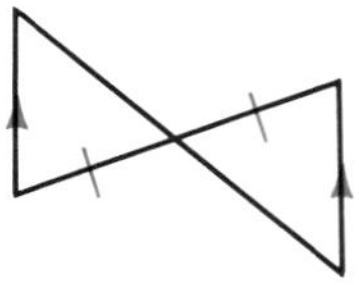

4.

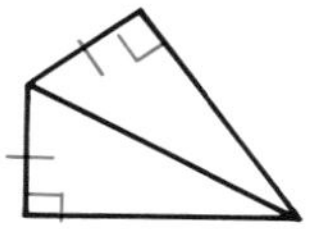

5.

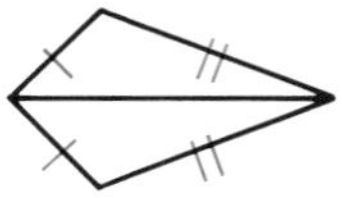

6.

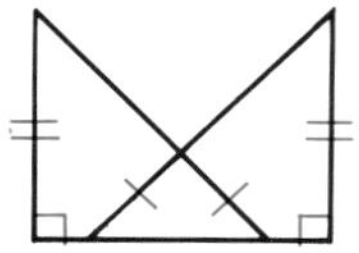

7. Given: $\overline{PO} \perp$ plane X; $OT = OS$
8. Given: $\overline{PO} \perp$ plane X; $PT = PS$

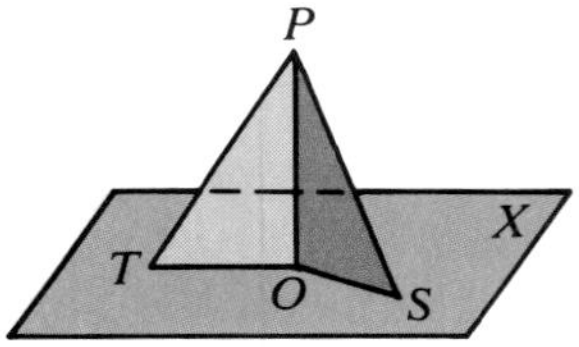

Exs. 7, 8

Indicate the best answer by writing the appropriate letter.

9. An equiangular triangle can *not* be which of the following?
   a. equilateral **b.** isosceles **c.** scalene **d.** acute
10. $\overline{RT}$ is the base of $\triangle RXT$. $\angle R \cong \angle T$, $RT = 2x + 5$, $RX = 5x - 7$, and $TX = 2x + 8$. What is the perimeter of $\triangle RXT$?
    a. 5 **b.** 15 **c.** 18 **d.** 51
11. If $\triangle DEF \cong \triangle PRS$, which of these congruences must be true?
    a. $\overline{DF} \cong \overline{PS}$ **b.** $\overline{EF} \cong \overline{PR}$ **c.** $\angle E \cong \angle S$ **d.** $\angle F \cong \angle R$
12. What is the *principal* basis for inductive reasoning?
    a. definitions **b.** previously proved theorems
    c. postulates **d.** past observations
13. Point X is equidistant from vertices T and N of $\triangle TEN$. Point X must lie on which of the following?
    a. bisector of $\angle E$ **b.** perpendicular bisector of $\overline{TN}$
    c. median to $\overline{TN}$ **d.** none of these
14. $\overline{AC}$ is a diagonal of regular pentagon $ABCDE$. What does $m\angle ACD$ equal?
    a. 36 **b.** 54 **c.** 72 **d.** 108
15. In $\triangle ABC$, $AB = AC$, $m\angle A = 46$, and $\overline{BD}$ is an altitude. What does $m\angle CBD$ equal?
    a. 23 **b.** 44 **c.** 67 **d.** 134

Chapter 4

Indicate the best answer by writing the appropriate letter.

1. Both pairs of opposite sides of a quadrilateral are parallel. Which special kind of quadrilateral *must* it be?
   a. parallelogram **b.** rectangle **c.** rhombus **d.** trapezoid
2. The diagonals of a certain quadrilateral are congruent. Which term could *not* be used to describe the quadrilateral?
   a. isosceles trapezoid **b.** rectangle
   c. rhombus **d.** parallelogram with a $60°$ angle
3. M is the midpoint of hypotenuse $\overline{TK}$ of right $\triangle TAK$. $AM = 13$. What is the length of $\overline{TK}$?
   a. 26 **b.** $19\frac{1}{2}$ **c.** 13 **d.** none of these
4. You don't need a figure to do this exercise. Given that $m\angle A = m\angle B$, $AC = BD$, and $l \parallel m$, you want to prove that $m\angle 3 = m\angle 4$. To write an indirect proof, you should begin by temporarily assuming what?
   a. $m\angle A \neq m\angle B$ **b.** $AC \neq BD$
   c. $l \perp m$ **d.** $m\angle 3 \neq m\angle 4$
5. A diagonal of a parallelogram bisects one of its angles. Which special kind of parallelogram *must* it be?
   a. rectangle **b.** rhombus
   c. square **d.** parallelogram with a $60°$ angle
6. The lengths of the bases of a trapezoid are 18 and 26. What is the length of the median?
   a. 8 **b.** 22 **c.** 44 **d.** none of these
7. In $\triangle RST$, $RS = 8$ and $ST = 10$. Which of these *must* be true?
   a. $RT > 2$ **b.** $RT < 2$
   c. $RT > 10$ **d.** $RT < 10$
8. In $\triangle ABC$, $AB = 8$, $BC = 10$, and $AC = 12$. M is the midpoint of $\overline{AB}$, and N is the midpoint of $\overline{BC}$. What is the length of $\overline{MN}$?
   a. 4 **b.** 5 **c.** 6 **d.** 9
9. If $EFGH$ is a parallelogram, which of the following *must* be true?
   a. $\angle E \cong \angle F$ **b.** $\angle F \cong \angle H$
   c. $\overline{FG} \parallel \overline{GH}$ **d.** $m\angle E + m\angle G = 180$
10. Which information does *not* prove that quad. $ABCD$ is a parallelogram?
    a. $\overline{AC}$ and $\overline{BD}$ bisect each other. **b.** $\overline{AD} \parallel \overline{BC}$; $\overline{AD} \cong \overline{BC}$
    c. $\overline{AB} \parallel \overline{CD}$; $\overline{AD} \cong \overline{BC}$ **d.** $\angle A \cong \angle C$; $\angle B \cong \angle D$
11. In quadrilateral $RSTU$, $RS = 5$, $ST = 6$, $TU = 7$, and $UR = 9$. Which of the following might possibly be the length of $\overline{SU}$?
    a. 12.5 **b.** 14 **c.** 2 **d.** all of these

Chapter 5

Indicate the best answer by writing the appropriate letter.

1. If the measures of the angles of a triangle are in the ratio 3:3:4, what is the measure of the largest angle of the triangle?
   a. 40 **b.** 54 **c.** 72 **d.** 90

2. If $\triangle ABC \sim \triangle JOT$, which of these is a correct proportion?
   a. $\frac{BC}{AC} = \frac{JT}{OT}$ **b.** $\frac{AB}{JT} = \frac{AC}{JO}$ **c.** $\frac{AB}{BC} = \frac{OT}{JT}$ **d.** $\frac{AC}{JT} = \frac{BC}{OT}$

3. If $\frac{a}{b} = \frac{x}{y}$, what does $\frac{y}{b}$ equal?
   a. $\frac{x}{a}$ **b.** $\frac{a}{x}$ **c.** $\frac{y}{x}$ **d.** $\frac{b}{y}$

4. $\triangle ABC \sim \triangle DEF$, $AB = 8$, $BC = 12$, $AC = 16$, and $DE = 12$. What is the perimeter of $\triangle DEF$?
   a. 36 **b.** 40 **c.** 48 **d.** 54

5. Which of the following pairs of polygons *must* be similar?
   a. two rectangles **b.** two regular hexagons
   c. two isosceles triangles **d.** two parallelograms with a 60° angle

6. Quad. $GHJK \sim$ quad. $RSTU$, $GH = JK = 10$, $HJ = KG = 14$, and $RS = TU = 16$. What is the scale factor of quad. $GHJK$ to quad. $RSTU$?
   a. $\frac{5}{7}$ **b.** $\frac{5}{8}$ **c.** $\frac{7}{8}$ **d.** $\frac{16}{10}$

7. Which of the following can you use to prove that the two triangles are similar?
   a. SAS Similarity Theorem **b.** AA Postulate
   c. SSS Similarity Theorem **d.** def. of similar triangles

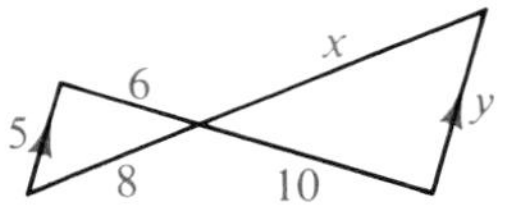

Exs. 7, 8

8. Which statement is correct?
   a. $\frac{6}{10} = \frac{8}{x}$ **b.** $\frac{6}{8} = \frac{x}{10}$ **c.** $6 \cdot 10 = 8x$ **d.** $\frac{5}{y} = \frac{8}{10}$

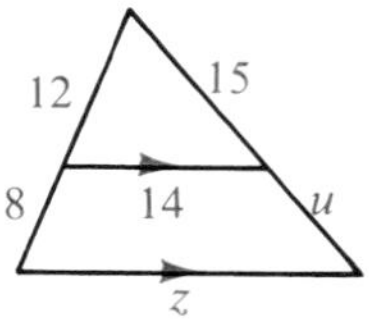

9. What is the value of u?
   a. 8 **b.** 10 **c.** 16 **d.** 25

10. What is the value of z?
    a. 25 **b.** 28 **c.** $\frac{28}{3}$ **d.** $\frac{70}{3}$

11. In $\triangle APC$, the bisector of $\angle P$ meets $\overline{AC}$ at B. $PA = 30$, $PC = 50$, and $AB = 12$. What is the length of $\overline{BC}$?
    a. $\frac{36}{5}$ **b.** 12 **c.** 20 **d.** 32

Chapter 6

Indicate the best answer by writing the appropriate letter.

1. The shorter leg of a 30°–60°–90° triangle is 7. Find the hypotenuse.
a. 14 **b.** $7\sqrt{2}$ **c.** $7\sqrt{3}$ **d.** $\sqrt{14}$

2. What is the simplified form of the number $3\sqrt{60}$?
a. $5\sqrt{15}$ **b.** $6\sqrt{15}$ **c.** $7\sqrt{15}$ **d.** $12\sqrt{15}$

3. The altitude to the 55 cm hypotenuse of a right triangle divides the hypotenuse into segments 25 cm and 30 cm long. How long is the altitude?
a. $15\sqrt{3}$ cm **b.** $15\sqrt{5}$ cm **c.** $5\sqrt{30}$ cm **d.** $5\sqrt{55}$ cm

4. The legs of a right triangle are 4 and 7. Find the hypotenuse.
a. $2\sqrt{7}$ **b.** $\sqrt{28}$ **c.** $\sqrt{33}$ **d.** $\sqrt{65}$

5. The hypotenuse and one leg of a right triangle are 61 and 11. Find the other leg.
a. 36 **b.** $5\sqrt{2}$ **c.** 60 **d.** $\sqrt{3842}$

6. Each side of an equilateral triangle is 10. Find an altitude.
a. 5 **b.** 10 **c.** $5\sqrt{2}$ **d.** $5\sqrt{3}$

7. One side of a square is s. Find a diagonal.
a. $2\sqrt{s}$ **b.** $s\sqrt{2}$ **c.** $\frac{s}{2}\sqrt{3}$ **d.** $s\sqrt{3}$

8. What kind of triangle is a triangle whose sides are 12, 13, and 18?
a. an obtuse triangle **b.** a right triangle
c. an acute triangle **d.** an impossibility

9. In $\triangle RST$, $m\angle S = 90$. What does sin T equal?
a. $\frac{ST}{RT}$ **b.** $\frac{RS}{ST}$ **c.** $\frac{RS}{RT}$ **d.** $\frac{RT}{RS}$

10. What is the geometric mean between $\frac{5}{7}$ and 2?
a. $\frac{19}{14}$ **b.** $2\sqrt{\frac{5}{7}}$ **c.** $\frac{1}{10}\sqrt{70}$ **d.** $\frac{1}{7}\sqrt{70}$

11. One acute angle of a certain right triangle has measure n. If $\sin n° = \frac{3}{5}$, what does $\tan n°$ equal?
a. $\frac{4}{3}$ **b.** $\frac{4}{5}$ **c.** $\frac{3}{4}$ **d.** none of these

12. Which equation could be used to find the value of x?
a. $\cos 58° = \frac{x}{18.9}$ **b.** $\sin 32° = \frac{x}{16}$
c. $\cos 44° = \frac{x}{10.4}$ **d.** $\tan 46° = \frac{x}{10.4}$

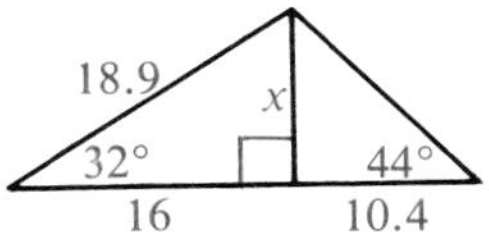

Chapter 7

Indicate the best answer by writing the appropriate letter.

In Exercises 1–3, $\overline{PT}$ is tangent to $\odot M$ at T.

Exs. 1–3

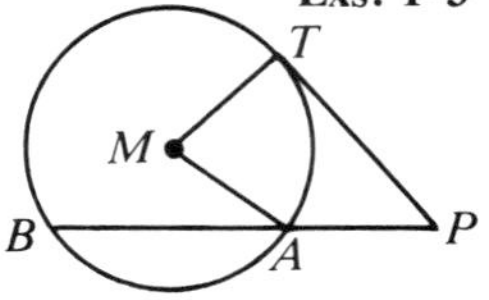

1. If $m\angle TMA = 70$, what is the measure of $\widehat{TBA}$?
   a. 35 **b.** 70 **c.** 290 **d.** 145
2. If $m\angle M = 70$ and $m\angle P = 50$, what is the measure of $\angle MAP$?
   a. 140 **b.** 150 **c.** 160 **d.** 170
3. If $PA = 9$ and $AB = 16$, what does PT equal?
   a. 12 **b.** $\frac{25}{2}$ **c.** 15 **d.** 20
4. In a plane, two circles intersect in two points. What is the number of common tangents that can be drawn to the circles?
   a. zero **b.** one **c.** two **d.** none of these
5. Points A, B, and C lie on a circle in the order named. $m\widehat{AB} = 110$ and $m\widehat{BC} = 120$. What is the measure of $\angle BAC$?
   a. 130 **b.** 65 **c.** 60 **d.** 55
6. In Exercise 5 point D lies on $\widehat{AC}$. What is the sum of the measures of $\angle ABC$ and $\angle ADC$?
   a. 180 **b.** 170 **c.** 160 **d.** can't tell

Exs. 7–9

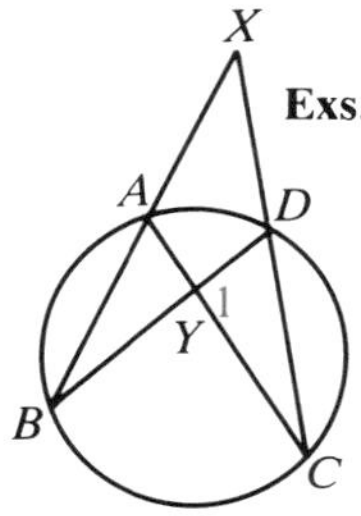

7. If $m\widehat{BC} = 120$ and $m\widehat{AD} = 50$, what does $m\angle X$ equal?
   a. 25 **b.** 35 **c.** 60 **d.** 70
8. If $m\widehat{BC} = 120$ and $m\widehat{AD} = 50$, what does $m\angle 1$ equal?
   a. 60 **b.** 85 **c.** 90 **d.** 95
9. If $AY = j$, $YC = k$, and $YD = 7$, what does BY equal?
   a. $\frac{jk}{7}$ **b.** $\frac{7j}{k}$ **c.** $\frac{7k}{j}$ **d.** $\frac{k}{7j}$
10. R and S are points on a circle. $\overline{RS}$ could be which of these?
    a. radius **b.** diameter **c.** secant **d.** tangent

In Exercises 11–13, $\overleftrightarrow{XA}$ is tangent to $\odot O$ at X.

Exs. 11–13

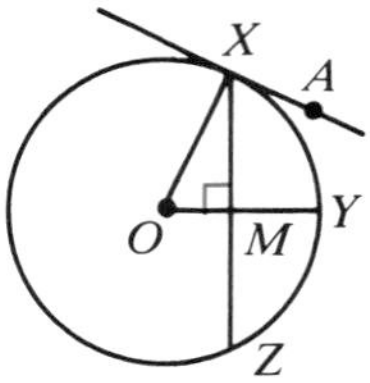

11. Which of these equals $m\angle AXZ$?
    a. $m\widehat{XYZ}$ **b.** $m\angle OXM$ **c.** $\frac{1}{2}m\widehat{XY}$ **d.** $\frac{1}{2}m\widehat{XZ}$
12. If the radius of $\odot O$ is 13 and $XZ = 24$, what is the distance from O to chord $\overline{XZ}$?
    a. 5 **b.** 8 **c.** 11 **d.** $\sqrt{407}$
13. If $OM = 8$ and $MY = 9$, what does XZ equal?
    a. $6\sqrt{2}$ **b.** $2\sqrt{17}$ **c.** $\sqrt{145}$ **d.** 30

Chapter 8

Indicate the best answer by writing the appropriate letter.

1. In a plane, what is the locus of points equidistant from two given points?
a. a point **b.** a circle **c.** a line **d.** a pair of lines

2. Point P lies on line l in a plane. What is the locus of points, in that plane, that lie 8 cm from P and 2 cm from l?
a. no points **b.** two points **c.** three points **d.** four points

3. To inscribe a circle in a triangle, what should you construct first?
a. two medians **b.** two angle bisectors
c. two altitudes **d.** two perpendicular bisectors

4. The lengths of two segments are r and s, with $r > s$. It is *not* possible to construct a segment with which of these lengths?
a. $\frac{1}{5}(r + s)$ **b.** $s - r$ **c.** $\sqrt{rs}$ **d.** $\sqrt{r^2 + s^2}$

5. It *is* possible to construct an angle with which of these measures?
a. 10 **b.** 20 **c.** 30 **d.** 40

6. You are to construct a tangent to a given $\odot O$ from a point P outside the circle. In the process, it would be *useless* to construct which of these?
a. $\overline{OP}$ **b.** the perpendicular bisector of $\overline{OP}$
c. a circle with O and P on it **d.** a line parallel to $\overline{OP}$

7. Which of these could *not* be the intersection of a plane and a sphere?
a. a pair of circles **b.** a circle **c.** a point **d.** the empty set

8. Where *must* the perpendicular bisectors of the sides of a triangle meet?
a. inside the triangle **b.** on the triangle
c. outside the triangle **d.** none of these

9. In space what is the locus of points 3 cm from a given point A?
a. a line **b.** a plane
c. a cylindrical surface **d.** a sphere

10. In a plane, what is the locus of points equidistant from the sides of a square?
a. a square **b.** a pair of squares
c. a circle **d.** a point

11. Three segments related to a triangle intersect in a point that divides each segment in the ratio 2:1. What are those segments?
a. medians **b.** altitudes
c. angle bisectors **d.** perpendicular bisectors

12. You are to construct a perpendicular to a line l at a given point X on l. In how many places on l will you need to position the point of your compass in order to do this construction?
a. one **b.** two **c.** three **d.** four

Chapter 9

Indicate the best answer by writing the appropriate letter.

1. One side of a rectangle is 14 and the perimeter is 44. What is the area?
a. 112 **b.** 210 **c.** 224 **d.** 420

2. A square is inscribed in a circle with radius 8. What is the area of the square?
a. 32 **b.** 64 **c.** $64\sqrt{2}$ **d.** 128

3. The area of a circle is 25π. What is its circumference?
a. 5π **b.** 10π **c.** 12.5π **d.** 50π

4. What is the area of a trapezoid with bases 7 and 8 and height 6?
a. 90 **b.** 336 **c.** 45 **d.** 168

5. A parallelogram and a triangle have equal areas. The base and height of the parallelogram are 12 and 9. If the base of the triangle is 36, find its height.
a. 3 **b.** 6 **c.** 9 **d.** 12

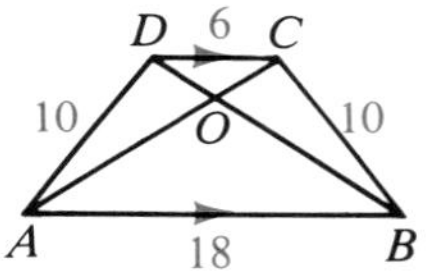

Exs. 6, 7

6. What is the area of trapezoid $ABCD$?
a. 96 **b.** 120 **c.** 144 **d.** 192

7. What is the ratio of the areas of $\triangle AOB$ and $\triangle DOC$?
a. $\sqrt{3}:1$ **b.** $\sqrt{3}:3$ **c.** $3:1$ **d.** $9:1$

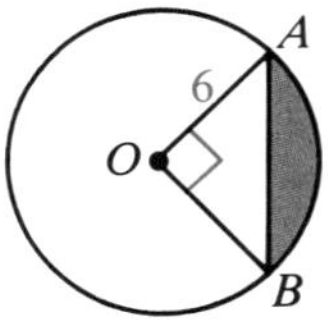

Exs. 8, 9

8. In the diagram, what is the length of $\widehat{AB}$?
a. $6\sqrt{2}$ **b.** 6π **c.** 3π **d.** 36π

9. In the diagram, what is the area of the shaded region?
a. $9\pi - 36$ **b.** $12\pi - 36$ **c.** $9\pi - 18$ **d.** $12\pi - 18$

10. A rhombus has diagonals 6 and 8. What is its area?
a. 12 **b.** 24 **c.** 36 **d.** 48

11. What is the area of a circle with diameter 12?
a. $24\pi^2$ **b.** 12π **c.** 144π **d.** 36π

12. What is the area of an equilateral triangle with perimeter 24?
a. $64\sqrt{3}$ **b.** $32\sqrt{3}$ **c.** $\frac{32\sqrt{3}}{3}$ **d.** $16\sqrt{3}$

13. What is the area of a triangle with sides 15, 15, and 24?
a. 54 **b.** 108 **c.** 180 **d.** 216

Chapter 10

Indicate the best answer by writing the appropriate letter.

1. What is the volume of a rectangular solid with dimensions 12, 9, and 6?
a. 108 **b.** 216 **c.** 432 **d.** 648

2. What is the total surface area of the solid in Exercise 1?
a. 234 **b.** 468 **c.** 252 **d.** 360

3. Two similar cones have heights 5 and 20. What is the ratio of their volumes?
a. 1:64 **b.** 1:4 **c.** 1:16 **d.** 4:16

4. What is the volume of a regular square pyramid with base edge 16 and height 6?
a. 128 **b.** 256 **c.** 512 **d.** 1536

5. What is the lateral area of the pyramid in Exercise 4?
a. 256 **b.** 320 **c.** 576 **d.** 640

6. A sphere has area 16π. What is its volume?
a. $\frac{8\pi}{3}$ **b.** $\frac{32\pi}{3}$ **c.** $\frac{64\pi}{3}$ **d.** $\frac{256\pi}{3}$

7. A cone has radius 5 and height 12. A cylinder with radius 10 has the same volume as the cone. What is the cylinder's height?
a. 1 **b.** 2 **c.** 3 **d.** 4

8. A square prism is inscribed in a cylinder with radius 6 and height 10. What is the volume of the prism?
a. 1080 **b.** 720 **c.** 360 **d.** 240

9. A plane passes 2 cm from the center of a sphere with radius 4 cm. What is the area of the circle of intersection?
a. 12π cm^2 **b.** 16π cm^2 **c.** 18π cm^2 **d.** 20π cm^2

10. Find the total surface area of a cylinder with radius 4 and height 6.
a. 16π **b.** 32π **c.** 48π **d.** 80π

11. Two similar pyramids have volumes 27 and 125. If the smaller has lateral area 18, what is the lateral area of the larger?
a. 30 **b.** $83\frac{1}{3}$ **c.** 50 **d.** 25

12. The base of a right prism is a regular hexagon with side 4. The height of the prism is 6. What is the volume of the prism?
a. $144\sqrt{3}$ **b.** $72\sqrt{3}$ **c.** $48\sqrt{3}$ **d.** $36\sqrt{3}$

13. What is the lateral area of the prism in Exercise 12?
a. 24 **b.** 36 **c.** 72 **d.** 144

Chapter 11

Indicate the best answer by writing the appropriate letter.

1. What is the distance between points $(-2, 0)$ and $(2, 5)$?
a. 5 **b.** 3 **c.** $\sqrt{29}$ **d.** $\sqrt{41}$

2. A line with slope $\frac{2}{5}$ passes through point $(1, 4)$. What is an equation of the line?
a. $y - 4 = \frac{2}{5}(x - 1)$ **b.** $y - 4 = \frac{5}{2}(x - 1)$
c. $y + 4 = \frac{2}{5}(x + 1)$ **d.** $y - 1 = \frac{5}{2}(x - 4)$

3. The midpoint of $\overline{AB}$ is $(3, 4)$. If the coordinates of B are $(6, 6)$, what are the coordinates of A?
a. $(9, 10)$ **b.** $(4.5, 5)$ **c.** $(0, 2)$ **d.** $(9, 10)$

4. The slope of line l is $\frac{2}{7}$. What is the slope of any line perpendicular to l?
a. $\frac{2}{7}$ **b.** $-\frac{2}{7}$ **c.** $\frac{7}{2}$ **d.** $-\frac{7}{2}$

5. Which point lies on the line $3x + 2y = 12$?
a. $(0, 4)$ **b.** $(\frac{14}{3}, 1)$ **c.** $(\frac{10}{3}, -1)$ **d.** $(\frac{10}{3}, 1)$

6. What is an equation of the circle with center $(3, 0)$ and radius 8?
a. $x^2 + y^2 = 64$ **b.** $(x - 3)^2 + y^2 = 64$
c. $(x + 3)^2 + y^2 = 8$ **d.** $(x - 3)^2 + y^2 = 8$

7. What is the slope of the segment that joins points $(-3, 5)$ and $(2, 8)$?
a. -1 **b.** $-\frac{3}{5}$ **c.** $\frac{5}{3}$ **d.** $\frac{3}{5}$

8. What is the *best* term for a quadrilateral with vertices $(-5, 0)$, $(3, 6)$, $(6, 2)$, and $(-2, -4)$?
a. trapezoid **b.** parallelogram **c.** rectangle **d.** rhombus

9. Three consecutive vertices of a parallelogram are $(j, 5)$, $(0, 0)$, and $(7, 0)$. Which is the fourth vertex?
a. $(7, 5)$ **b.** $(5, 7)$ **c.** $(j + 7, 5)$ **d.** $(j + 5, 7)$

10. Points $(2, 2)$ and $(8, v)$ lie on a line whose slope is $\frac{1}{2}$. What is the value of v?
a. -10 **b.** -1 **c.** 5 **d.** 14

11. What is the *best* term for a triangle with vertices $(1, -3)$, $(6, 2)$, and $(0, 4)$?
a. isosceles triangle **b.** equilateral triangle
c. right triangle **d.** none of these

12. Which point is the intersection of lines $3x + 2y = 17$ and $x - 4y = 1$?
a. $(1, 5)$ **b.** $(5, 1)$ **c.** $(-1, 5)$ **d.** $(\frac{33}{5}, \frac{7}{5})$

Chapter 12

Indicate the best answer by writing the appropriate letter.

1. A regular hexagon does *not* have which symmetry?
a. line **b.** point **c.** 30° rotational **d.** 120° rotational

2. What is the image of the point (2, 3) by reflection in the x-axis?
a. (3, 2) **b.** $(-2, 3)$ **c.** $(2, -3)$ **d.** $(-2, -3)$

3. $T:(x, y) \to (x, y - 2)$. What is the preimage of (3, 5)?
a. (5, 7) **b.** (3, 7) **c.** (3, 3) **d.** (5, 3)

4. If O is the point (0, 0), what is the image of (3, 6) by $D_{O,\frac{1}{3}}$?
a. (9, 18) **b.** (2, 4) **c.** (1, 2) **d.** $(-1, -2)$

5. What is the image of $(-1, 3)$ by a half-turn about (1, 2)?
a. (3, 1) **b.** $(1, -3)$ **c.** $(-1, -2)$ **d.** $(3, -1)$

6. T: $(x, y) \to (x, y - 2)$. What is the image of (5, 3) by T^{-1}?
a. (3, 3) **b.** (5, 1) **c.** (3, 5) **d.** (5, 5)

7. Isometry S: ▱ $ABCD \to$ ▱$JKLM$. Which statement *must* be true?
a. $\angle DAB \cong \angle JKL$ **b.** $AC = JL$
c. $S: C \to M$ **d.** $CD = MJ$

8. What is the line of reflection for a transformation that maps $(-2, 1)$ to (2, 1)?
a. the x-axis **b.** the line $y = x$
c. the y-axis **d.** the origin

9. How many lines of symmetry does a rhombus with a 60° angle have?
a. none **b.** one **c.** two **d.** four

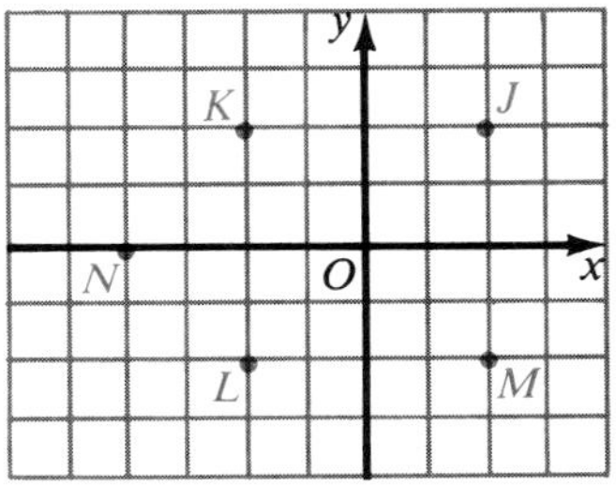

Exs. 10–13

10. What is the image of J by $R_x \circ R_y$?
a. J **b.** K **c.** L **d.** M

11. T is a translation that maps K to N. What is the image of J under T?
a. K **b.** O **c.** N **d.** L

12. What is the image of J under $R_x \circ H_O$?
a. J **b.** K **c.** L **d.** M

13. What is the image of $\triangle LMJ$ by $\mathcal{R}_{O,90}$?
a. $\triangle JKL$ **b.** $\triangle KLM$ **c.** $\triangle LMJ$ **d.** $\triangle MJK$

College Entrance Exams

If you are planning to attend college, you will probably be required to take college entrance exams. Some of these exams test your knowledge of specific subject areas; others are more general exams that attempt to measure the extent to which your verbal and mathematical reasoning abilities have been developed. These abilities are ones that can be improved through study and practice. Generally the best preparation for college entrance exams is to follow a strong academic program in high school and to read as extensively as possible.

The following test-taking strategies may be useful:

- Familiarize yourself with the test you will be taking well in advance of the test date. Sample tests, with accompanying explanatory material, are available for many standardized tests. By working through this sample material, you become comfortable with the types of questions and directions that will appear on the test and you develop a feeling for the pace at which you must work in order to complete the test.
- Find out how the test is scored so that you know whether it is advantageous to guess.
- Skim sections of the test before starting to answer the questions, to get an overview of the questions. You may wish to answer the easiest questions first. In any case, do not waste time on questions you do not understand; go on to those that you do.
- Mark your answer sheet carefully, checking the numbering on the answer sheet about every five questions to avoid errors caused by misplaced answer markings.
- Write in the test booklet if it is helpful; for example, cross out incorrect alternatives and do mathematical calculations.
- Work carefully, but do not take time to double-check your answers unless you finish before the deadline and have extra time.
- Arrive at the test center early and come well prepared with any necessary supplies such as sharpened pencils and a watch.

College entrance exams that test general reasoning abilities, such as the Scholastic Aptitude Test, frequently include questions relating to basic geometric concepts and skills. The following topics often appear on such exams. For each topic, page references have been given to the places in your textbook where the topic is discussed.

Properties of Parallel and Perpendicular Lines (pages 35–36, 55–56, 60, 487)

If two parallel lines are cut by a transversal, then alternate interior angles are congruent, corresponding angles are congruent, and same-side interior angles are supplementary.

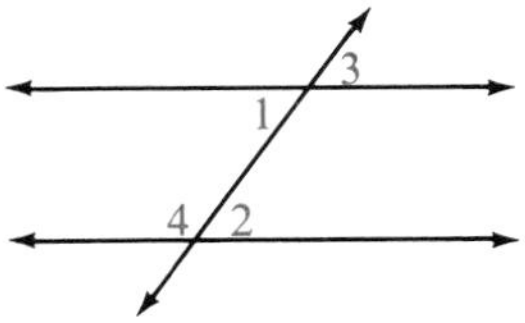

$$m\angle 1 = m\angle 2$$
$$m\angle 3 = m\angle 2$$
$$m\angle 1 + m\angle 4 = 180$$

Angle Measure Relationships (pages 31, 73–74, 81)

Vertical angles are congruent.

$$m\angle 3 = m\angle 5$$

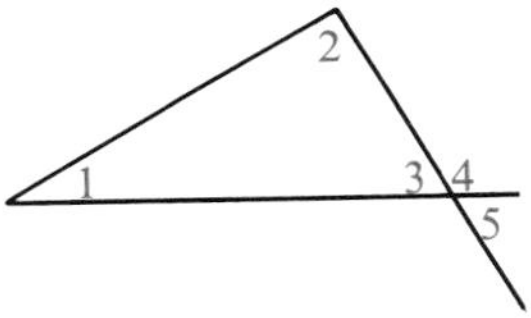

The sum of the measures of the angles of a triangle is 180.

$$m\angle 1 + m\angle 2 + m\angle 3 = 180$$

The measure of an exterior angle of a triangle equals the sum of the measures of the two remote interior angles.

$$m\angle 4 = m\angle 1 + m\angle 2$$

The sum of the measures of the angles of a convex polygon with n sides is $(n - 2)180$.

For example, the sum of the measures of the angles of the pentagon at right is $3 \cdot 180 = 540$.

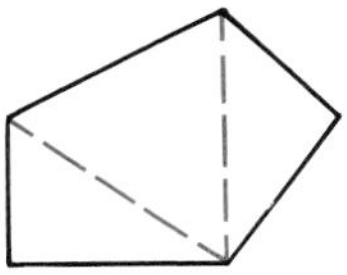

Triangle Side Relationships (pages 183–184)

The sum of the lengths of any two sides of a triangle is greater than the length of the third side.

For example, $AB + BC > AC$.

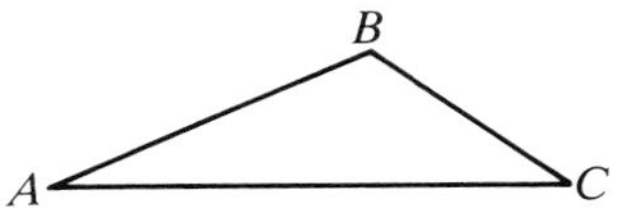

If one side of a triangle is longer than a second side, then the angle opposite the first side is larger than the angle opposite the second side, and conversely.

For example, if $AC > BC$, then $m\angle B > m\angle A$;
if $m\angle C < m\angle B$, then $AB < AC$.

Special Triangle Relationships (pages 72, 252, 257, 262)

Isosceles Triangle

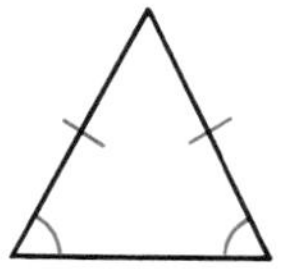

At least 2 sides are congruent.
Angles opposite congruent sides are congruent.

Equilateral Triangle

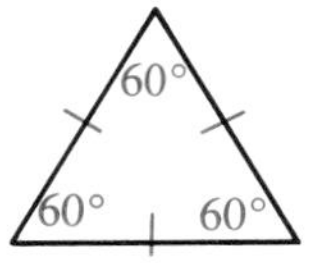

All sides are congruent.
All angles are congruent.

By the Pythagorean Theorem, in $\triangle ABC$

$$c^2 = a^2 + b^2.$$

Since $\angle C$ is a right angle,

$$m\angle A + m\angle B = 90.$$

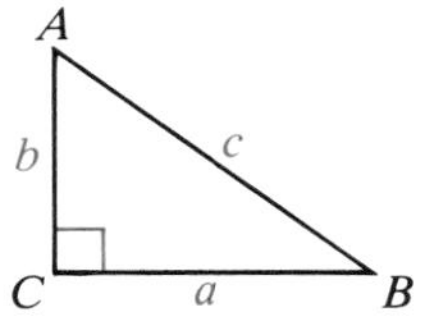

45°-45°-90° Triangle

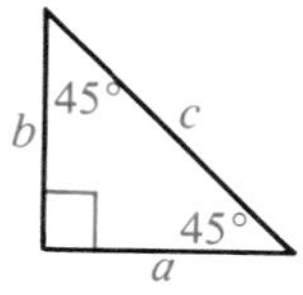

$a = b$
$c = \sqrt{2}\,a$
$\quad = \sqrt{2}\,b$

Legs are congruent.
Hypotenuse $= \sqrt{2} \cdot$ leg

30°-60°-90° Triangle

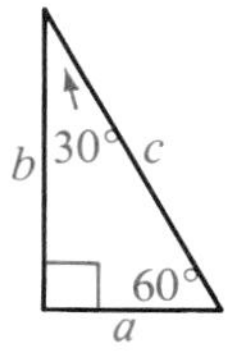

$c = 2a$
$b = \sqrt{3}\,a$

Hypotenuse $= 2 \cdot$ shorter leg
Longer leg $= \sqrt{3} \cdot$ shorter leg

Perimeter, Area, and Volume Formulas (pages 380, 386, 402–403, 417, 424, 462)

Rectangle

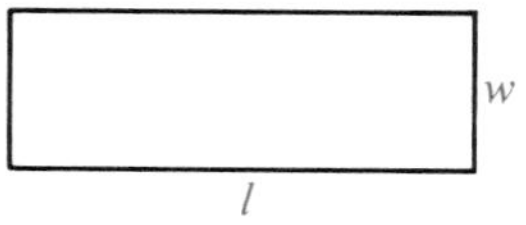

Perimeter $= 2l + 2w$
Area $= lw$

Triangle

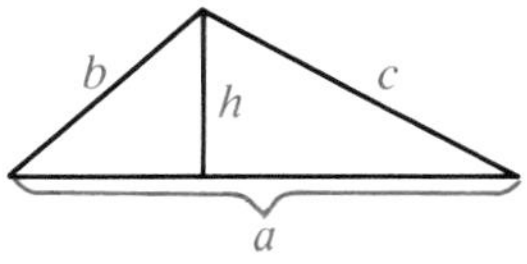

Perimeter $= a + b + c$
Area $= \frac{1}{2}(\text{base} \times \text{height})$
$\quad = \frac{1}{2}ah$

Circle

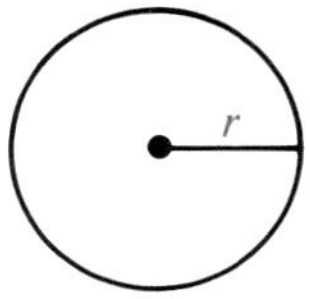

Circumference $= 2\pi r$
Area $= \pi r^2$

Rectangular Solid

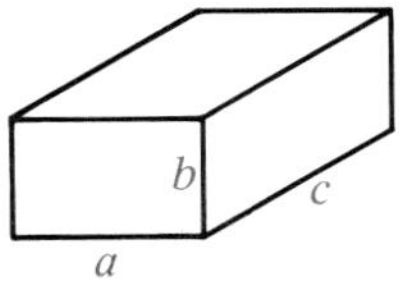

Total area $= 2ab + 2bc + 2ac$
Volume $= abc$

Locating Points on a Grid (page 469)

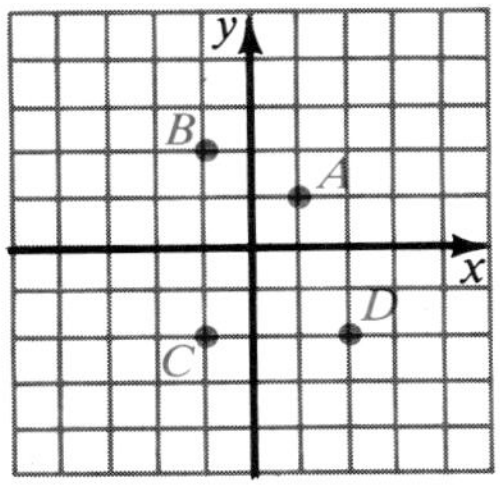

The points shown are $A(1, 1)$, $B(-1, 2)$, $C(-1, -2)$, and $D(2, -2)$.

Logic

Statements and Truth Tables

In algebra, you have used letters to represent numbers. In logic, letters are used to represent simple statements that are either true or false. For example, p might represent the statement "Paris is the capital city of France," and q might represent the statement "The moon is made of green cheese."

Simple statements can be joined to form **compound statements.** Two important compound statements are defined below.

A **conjunction** is a compound statement composed of two simple statements joined by the word "and." The symbol $\wedge$ is used to represent the word "and."

A **disjunction** is a compound statement composed of two simple statements joined by the word "or." The symbol $\vee$ is used to represent the word "or."

Example

Simple statements:	p	Mom plays the guitar.
	q	Dad plays the piano.
Conjunction:	$p \wedge q$	Mom plays the guitar and Dad plays the piano.
Disjunction:	$p \vee q$	Mom plays the guitar or Dad plays the piano.

The table at the right is called a **truth table.** It tells you the conditions under which a conjunction is a true statement. "T" stands for "true" and "F" for "false." The first row of the table shows that when statement p is true and statement q is true, the conjunction $p \wedge q$ is true. The other rows of the table show that $p \wedge q$ is false when either of its component statements is false.

Truth table for conjunction

p	q	$p \wedge q$
T	T	T
T	F	F
F	T	F
F	F	F

In everyday speech, the word "or" is used in two different ways.

Inclusive use of "or": Rosa will go or Mary will go. This statement is true if Rosa goes, or if Mary goes, or if both Rosa and Mary go.

Exclusive use of "or": Jake will be elected class president or Carol will be elected class president. This statement is true if either Jake or Carol is elected class president. Obviously they can't both be elected. The statement means that Jake will be chosen or else Carol will be.

We shall deal only with the inclusive use of "or" in this course. The first row of the truth table for disjunction shows that when both p and q are true, $p \vee q$ is true. The next two rows show that the compound statement $p \vee q$ is true when either of its components is true. The last row shows that a disjunction is false when both of its components are false.

Truth table for disjunction

p	q	$p \vee q$
T	T	T
T	F	T
F	T	T
F	F	F

In addition to the words "and" and "or," the word "not" is an important word in logic. If p is a statement, then the statment "p is not true," usually shortened to "not p" and written $\sim p$, is called the **negation** of p.

Example

Statement: p Sam is sleeping in class.

Negation: $\sim p$ It is not true that Sam is sleeping in class.

or $\sim p$ Sam is not sleeping in class.

The truth table for negation shows that when p is true, $\sim p$ is false. When p is false, $\sim p$ is true.

Truth table for negation

p	$\sim p$
T	F
F	T

An example will show how to make truth tables for some other compound statements.

Example Make a truth table for the statement $\sim p \vee \sim q$.

Solution

1. Make a column for p and a column for q. Write all possible combinations of T and F in the standard pattern shown.
2. Since $\sim p$ is a part of the given statement, add a column for $\sim p$. To fill out this column, use the first column and refer to the truth table for negation above. Similarly, add a column for $\sim q$.
3. Using the columns for $\sim p$ and $\sim q$, refer to the truth table for disjunction above in order to fill out the column for $\sim p \vee \sim q$. Remember that a disjunction is false only when both of its components are false.

p	q	$\sim p$	$\sim q$	$\sim p \vee \sim q$
T	T	F	F	F
T	F	F	T	T
F	T	T	F	T
F	F	T	T	T

To make a truth table for a compound statement involving three simple statements p, q, and r, you would need an eight-row table to show all possible combinations of T and F. The standard pattern across the three columns headed p, q, and r is as follows: TTT, TTF, TFT, TFF, FTT, FTF, FFT, FFF.

Exercises

Suppose p stands for "I like the city," and q stands for "You like the country." Express in words each of the following statements.

1. $p \wedge q$
2. $\sim p$
3. $\sim q$
4. $p \vee q$
5. $p \vee \sim q$
6. $\sim(p \wedge q)$
7. $\sim p \vee \sim q$
8. $\sim p \wedge q$
9. $\sim(p \vee q)$
10. $\sim p \wedge \sim q$

Suppose p stands for "Hawks swoop," and q stands for "Gulls glide." Express in symbolic form each of the following statements.

11. Hawks swoop or gulls glide.

12. Gulls do not glide.

13. It is not true that "Hawks swoop or gulls glide."

14. Hawks do not swoop and gulls do not glide.

15. It is not true that "Hawks swoop and gulls glide."

16. Hawks do not swoop or gulls do not glide.

17. Do the statements in Exercises 13 and 14 mean the same thing?

18. Do the statements in Exercises 15 and 16 mean the same thing?

Make a truth table for each of the following statements.

19. $p \lor \sim q$

20. $\sim p \lor q$

21. $\sim(\sim p)$

22. $\sim(p \land q)$

23. $p \lor \sim p$

24. $p \land \sim p$

25. $p \land (q \lor r)$

26. $(p \land q) \lor (p \land r)$

Truth Tables for Conditionals

The conditional statement "If p then q," which is discussed in Section 2–6, is symbolized as $p \rightarrow q$. This is also read as "p implies q" and as "q follows from p." The truth table for $p \rightarrow q$ is shown at the right. Notice that the only time a conditional is false is when the hypothesis p is true and the conclusion q is false. The example below will show why this is a reasonable way to make out the truth table.

Truth table for conditionals

p	q	$p \rightarrow q$
T	T	T
T	F	F
F	T	T
F	F	T

Example Mom promises, "If I catch the early train home, I'll take you swimming." Consider the four possibilities of the truth table.

1. Mom catches the early train home and takes you swimming. She kept her promise; her statement was *true*.
2. Mom catches the early train home but does not take you swimming. She broke her promise; her statement was *false*.
3. Mom does not catch the early train home but still takes you swimming. She has not broken her promise; her statement was *true*.
4. Mom does not catch the early train home and does not take you swimming. She has not broken her promise; her statement was *true*.

The tables below show the converse and contrapositive of $p \to q$. Make sure that you understand how these tables were made. Notice that the last column of the table for the contrapositive $\sim q \to \sim p$ is identical with the last column of the table for the conditional on page 578. In other words, the contrapositive of a statement is true (or false) exactly when the statement itself is true (or false). This is what we mean when we say that a statement and its contrapositive are logically equivalent. On the other hand, a statement and its converse are not logically equivalent. The last columns in their truth tables are different.

Converse of $p \to q$

p	q	$q \to p$
T	T	T
T	F	T
F	T	F
F	F	T

Contrapositive of $p \to q$

p	q	$\sim q$	$\sim p$	$\sim q \to \sim p$
T	T	F	F	T
T	F	T	F	F
F	T	F	T	T
F	F	T	T	T

Exercises

Suppose p represents "You like to paint," q represents "You are an artist," and r represents "You draw landscapes." Express in words each of the following statements.

1. $p \to q$ **2.** $q \to r$ **3.** $\sim q \to \sim r$ **4.** $\sim(p \to q)$

5. $(p \wedge q) \to r$ **6.** $p \wedge (q \to r)$ **7.** $(r \vee q) \to p$ **8.** $r \vee (q \to p)$

Let *b*, *s*, and *w* represent the following statements.
***b:* Bonnie bellows.** ***s:* Sheila shouts** ***w:* Wilbur whispers.**
Express in symbolic form each of the following statements.

9. If Bonnie bellows, then Wilbur whispers.

10. If Wilbur whispers, then Sheila does not shout.

11. If Bonnie does not bellow or Wilbur does not whisper, then Sheila shouts.

12. Sheila shouts, and if Bonnie bellows, then Wilbur whispers.

13. It is not true that Sheila shouts if Bonnie bellows.

14. If Bonnie does not bellow, then Wilbur whispers and Sheila shouts.

15. **a.** Make a truth table for $\sim p \to \sim q$ (the inverse of $p \to q$). Your first two columns should be the same as the first two columns of the table for $p \to q$. The last columns of the two tables should be different. Are they? Is $\sim p \to \sim q$ logically equivalent to $p \to q$?

b. Compare the truth table for $\sim p \to \sim q$ (the inverse of $p \to q$) with the truth table for $q \to p$ (the converse of $p \to q$). Are the last columns the same? Are the inverse and the converse logically equivalent?

Make truth tables for the following statements.

16. $p \rightarrow \sim q$ **17.** $\sim(p \rightarrow q)$ **18.** $p \wedge \sim q$

19. By comparing the truth tables in Exercises 16–18, you should find that two of the three statements are logically equivalent. Which two?

20. The statement "p if and only if q" is defined as $(p \rightarrow q) \wedge (q \rightarrow p)$. Make a truth table for this statement.

Some Rules of Inference

Four rules for making logical inferences are symbolized below. A horizontal line separates the given information, or premises, from the conclusion. If you accept the given statement or statements as true, then you must accept as true the conclusion shown.

1. Modus Ponens

$p \rightarrow q$
p
―――――
Therefore, q

2. Modus Tollens

$p \rightarrow q$
$\sim q$
―――――
Therefore, $\sim p$

3. Simplification

$p \wedge q$
―――――
Therefore, p

4. Disjunctive Syllogism

$p \vee q$
$\sim p$
―――――
Therefore, q

You should convince yourself that these rules make good sense. For example, Rule 4 says that if you know that "p or q" is true and then you find out that p is not true, you must conclude that q is true.

Example 1 If today is Tuesday, then tomorrow is Wednesday.
Today is Tuesday.
―――――
Therefore, tomorrow is Wednesday. (Rule 1)

Example 2 If a figure is a triangle, then it is a polygon.
This figure is not a polygon.
―――――
Therefore, this figure is not a triangle. (Rule 2)

Example 3 It is Tuesday and it is April.
―――――
Therefore, it is Tuesday. (Rule 3)

Example 4 It is a square or it is a trapezoid.
It is not a square.
―――――
Therefore, it is a trapezoid. (Rule 4)

Rules 1–4 can be used to prove more complicated arguments as in the following example.

Example 5

Given: $p \rightarrow q$; $p \vee r$; $\sim q$

Prove: r

Proof

Statements	Reasons
1. $p \rightarrow q$	1. Given
2. $\sim q$	2. Given
3. $\sim p$	3. Steps 1 and 2 and Modus Tollens
4. $p \vee r$	4. Given
5. r	5. Steps 3 and 4 and Disjunctive Syllogism

Exercises

Supply the reasons to complete each proof.

1. Given: $p \wedge q$; $p \rightarrow s$
Prove: s

Statements
1. $p \wedge q$
2. p
3. $p \rightarrow s$
4. s

2. Given: $r \rightarrow s$; r; $s \rightarrow t$
Prove: t

Statements
1. $r \rightarrow s$
2. r
3. s
4. $s \rightarrow t$
5. t

Write two-column proofs for the following.

3. Given: $p \vee q$; $\sim p$; $q \rightarrow s$
Prove: s

4. Given: $a \rightarrow b$; $a \vee c$; $\sim b$
Prove: c

5. Given: $a \wedge b$; $a \rightarrow \sim c$; $c \vee d$
Prove: d

6. Given: $p \wedge q$; $p \rightarrow \sim s$; $r \rightarrow s$
Prove: $\sim r$

Symbolize the statements, accept them as true, and write two-column proofs.

7. If Jim jogs, then the dog barks.
If the dog barks, then the cat scats.
Jim jogs and Willa runs.
Prove that the cat scats. (Use the letters *j*, *d*, *c*, and *w*.)

8. Alice is on the team and Rachel is also.
Barb is on the team or Carol is.
If Alice is on the team, then Barb is not.
Prove that Carol is on the team. (Use the letters *a*, *r*, *b*, and *c*.)

Symbolize the statements, accept them as true, and write a two-column proof.

9. Turner will play fullback or Rizzo will play fullback.
If Turner plays fullback, than Packard will play halfback.
If Rizzo plays fullback, then Sullivan will be quarterback.
Packard will not play halfback.
Prove that Sullivan will be quarterback. (Use the letters t, r, p, and s.)

Some Rules of Replacement

The symbol $\equiv$ means "is logically equivalent to." Thus Rule 5 below states that the conditional statement $p \rightarrow q$ is logically equivalent to its contrapositive, $\sim q \rightarrow \sim p$. Rules 6–10 give other logical equivalences. These can be verified by comparing the truth tables of the statements on both sides of the $\equiv$ sign.

5. Contrapositive Rule

$$p \rightarrow q \equiv \sim q \rightarrow \sim p$$

6. Double Negation

$$\sim(\sim p) \equiv p$$

7. Commutative Rules

$$p \wedge q \equiv q \wedge p$$
$$p \vee q \equiv q \vee p$$

8. Associative Rules

$$(p \wedge q) \wedge r \equiv p \wedge (q \wedge r)$$
$$(p \vee q) \vee r \equiv p \vee (q \vee r)$$

9. Distributive Rules

$$p \wedge (q \vee r) \equiv (p \wedge q) \vee (p \wedge r)$$
$$p \vee (q \wedge r) \equiv (p \vee q) \wedge (p \vee r)$$

10. DeMorgan's Rules

$$\sim(p \wedge q) \equiv \sim p \vee \sim q$$
$$\sim(p \vee q) \equiv \sim p \wedge \sim q$$

Any logically equivalent expressions can replace each other wherever they occur in a proof.

Example

Given: $p \wedge q$; $q \rightarrow \sim(r \vee s)$

Prove: $\sim r \wedge \sim s$

Proof:

Statements	Reasons
1. $p \wedge q$	1. Given
2. $q \wedge p$	2. Step 1 and Commutative Rule
3. q	3. Step 2 and Simplification
4. $q \rightarrow \sim(r \vee s)$	4. Given
5. $\sim(r \vee s)$	5. Steps 3 and 4 and Modus Ponens
6. $\sim r \wedge \sim s$	6. Step 5 and DeMorgan's Rule

Exercises

Supply the reasons to complete each proof.

1. Given: $a \to \sim b$; b
Prove: $\sim a$

Statements
1. b
2. $\sim(\sim b)$
3. $a \to \sim b$
4. $\sim a$

2. Given: $a \vee (b \wedge c)$; $\sim b$
Prove: a

Statements
1. $a \vee (b \wedge c)$
2. $(a \vee b) \wedge (a \vee c)$
3. $a \vee b$
4. $b \vee a$
5. $\sim b$
6. a

Write two-column proofs for the following.

3. Given: $a \wedge (b \wedge c)$
Prove: c

4. Given: $(p \wedge q) \to s$; $\sim s$
Prove: $\sim p \vee \sim q$

5. Given: $p \vee (\sim q)$; q
Prove: p

6. Given: $\sim q \to \sim p$; $q \to r$; p
Prove: r

7. Given: $p \vee (q \wedge s)$
Prove: $p \vee s$

8. Given: $\mathrm{t} \vee (r \vee s)$; $\sim r \wedge \sim s$
Prove: t

Symbolize the statements and write two-column proofs.

9. Observers agree on three things:
If the strike is settled by Monday, then the workers will return to work on Tuesday.
If the strike is not settled by Monday, then the picket lines will be crossed.
The picket lines will never be crossed.
Prove that the workers will return to work on Tuesday. (Let s = strike is settled by Monday, p = picket lines will be crossed, and w = workers will return to work on Tuesday.)

10. A detective has established these facts:
Lady Eastwick stole the jewels or Sir Castleton stole them.
The butler is telling the truth or the maid is lying.
If the butler is telling the truth, then the safe was left open.
If the maid is lying, then Lady Eastwick did not steal the jewels.
The safe was not left open.
Prove that Sir Castleton stole the jewels. (Let e = Lady Eastwick stole the jewels, c = Sir Castleton stole the jewels, b = butler is telling truth, m = maid is lying, and s = safe was left open.)

Application of Logic to Electrical Circuits

The diagram at the right represents part of an electrical circuit. When switch p is open, the electricity that is flowing from A will not reach B. When switch p is closed, as in the second diagram, the electricity flows through the switch to B.

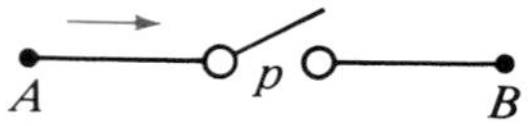

Open switch

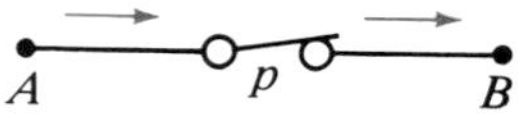

Closed switch

The diagram at the left below represents two switches p and q that are *connected in series*. Notice that current will flow if and only if both switches are closed. The diagram at the right represents the switches p and q *connected in parallel*. Notice that current will flow if either switch is closed or if both switches are closed. If switches p and q are both open, the current cannot flow.

Series circuit

Parallel circuit

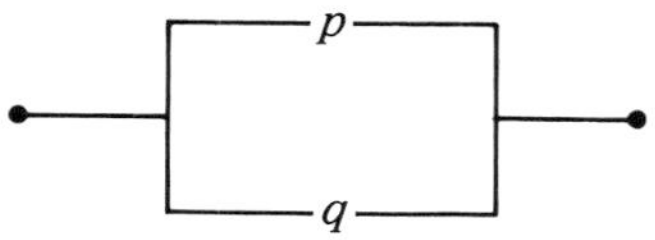

In order to understand how circuits are related to truth tables, let us do the following:

1. If a switch is closed, label it T. If it is open, label it F.
2. If current will flow in a circuit, label the circuit T. If the current will not flow, label the circuit F.

With these agreements, we can use truth tables to show what happens in the series circuit and the parallel circuit illustrated above.

Series circuit

p	q	Circuit
T	T	T
T	F	F
F	T	F
F	F	F

Parallel circuit

p	q	Circuit
T	T	T
T	F	T
F	T	T
F	F	F

Notice that the truth table for the series circuit is just like the truth table for $p \wedge q$. Also, the truth table for the parallel circuit is just like the table for $p \vee q$.

Now study the circuit shown at the right. Notice that one of the switches is labeled $\sim q$. This means that this switch is open if switch q is closed, and vice versa. (In many electrical circuits, switch $\sim q$ will open automatically when switch q is closed.)

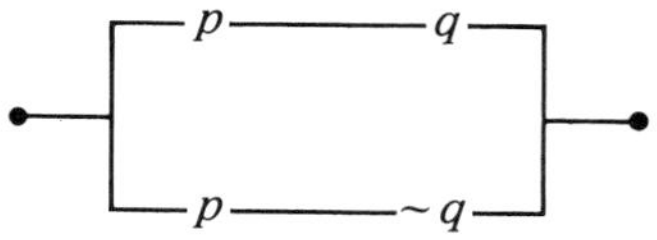

$(p \wedge q) \vee (p \wedge \sim q)$

The circuit shown is basically a parallel circuit, but in each branch of the circuit there are two switches connected in series. This explains why the circuit is labeled $(p \wedge q) \vee (p \wedge \sim q)$. A truth table for this circuit is given on the next page.

p	q	$\sim q$	$p \wedge q$	$p \wedge \sim q$	$(p \wedge q) \vee (p \wedge \sim q)$
T	T	F	T	F	T
T	F	T	F	T	T
F	T	F	F	F	F
F	F	T	F	F	F

Notice that the first and last columns of the truth table are identical. This means that the complicated circuit shown can be replaced by a simpler circuit that contains just switch p! In other words, logic can be used to replace a complex electrical circuit by a simpler one.

Exercises

Symbolize each circuit, using $\wedge$, $\vee$, $\sim$, and letters given for the switches in each diagram.

1. 

2. 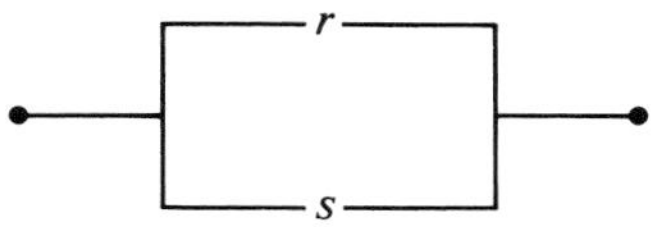

3. 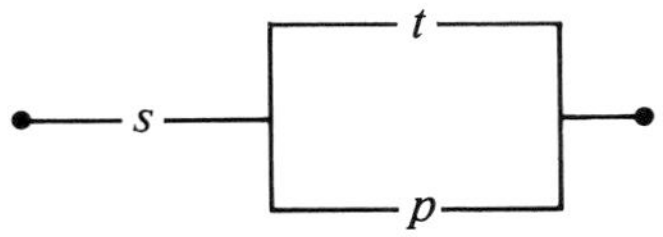

4. 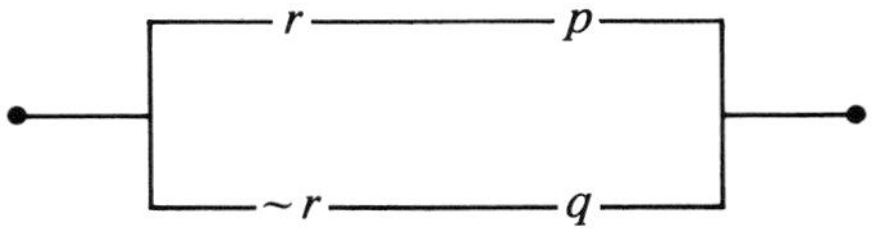

5. 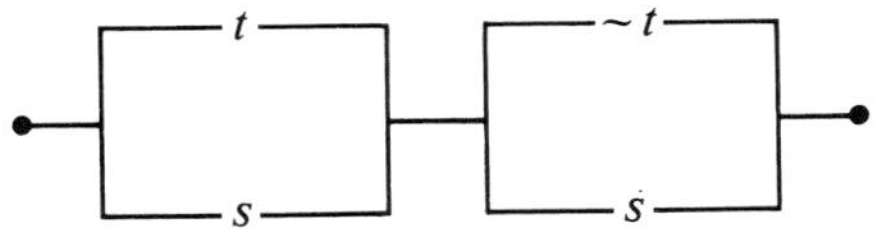

6. 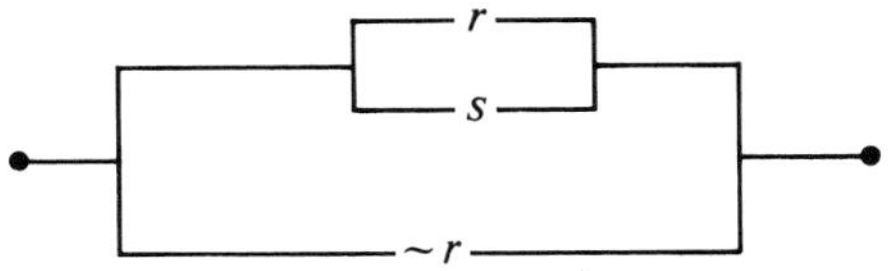

7. Draw a diagram for the circuit $p \wedge \sim p$; also for the circuit $p \vee \sim p$. Electricity can always pass through one of these circuits and can never pass through the other. Which is which?

8. According to the commutative rule, $p \wedge q \equiv q \wedge p$. This means that the circuit $p \wedge q$ does the same thing as the circuit $q \wedge p$. Make a diagram of each circuit.

9. According to the associative rule, $(p \vee q) \vee r \equiv p \vee (q \vee r)$. Draw diagrams for each circuit.

10. According to the distributive rule, $p \wedge (q \vee r) \equiv (p \wedge q) \vee (p \wedge r)$. Draw diagrams for each circuit.

11. Make both a diagram and a truth table for the circuit $(p \vee q) \vee \sim q$. Notice that the last column of your table is always T so that current always flows. This means that all of the switches could be eliminated.

12. Make both a diagram and a truth table for the circuit $(p \vee q) \wedge (p \vee \sim q)$. Describe a simpler circuit equivalent to this circuit.

Projections; Dihedral Angles

Projections

The purpose of this section is to help you develp skill in working with three-dimensional figures. As a start, you should read the first paragraph on page 511 carefully. On that page, point P' is called the image of point P. You can also say that P' is the **projection** of P into the plane.

In the figure below, each point of $\triangle ABC$ is mapped into a point on plane M by dropping a perpendicular from the point of $\triangle ABC$ to plane M.

Original figure	Projection into plane M
Point A	Point A'
$\overline{AB}$	$\overline{A'B'}$
$\overleftrightarrow{AB}$	$\overleftrightarrow{A'B'}$
$\angle CAB$	$\angle C'A'B'$
$\triangle CAB$	$\triangle C'A'B'$

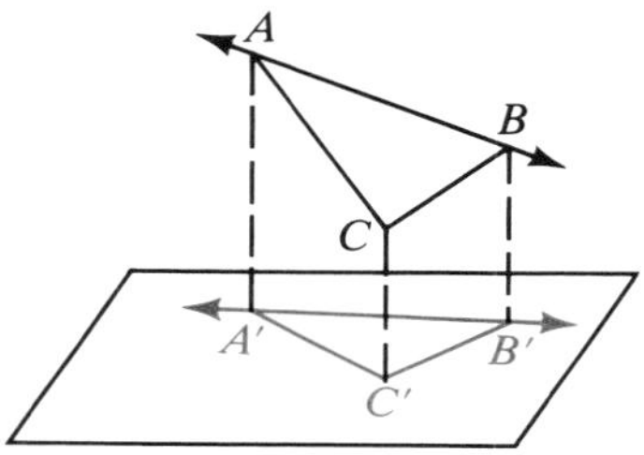

Exercises

Exercises 1–10 deal with a square $ABCD$, whose sides are 20 units long, and the projection of that square into a plane F. You can improve your thinking by cutting a square out of cardboard and holding your model in various positions with respect to the floor. Tell whether it is possible for the square to be in such position that its projection into F becomes the figure named.

A

1. A square

2. A 10 by 20 rectangle

3. A 1 by 20 rectangle

4. A 20-unit segment

5. A 30-unit segment

6. A 10-unit segment

7. A $20\sqrt{2}$-unit segment

8. A trapezoid

B

9. A nonsquare rhombus

10. A parallelogram that is neither a rectangle nor a rhombus

For each exercise, draw a diagram to show that the projection of an acute angle *RUZ* into a plane *Q* is the figure named.

11. A ray

12. A line

13. An angle congruent to $\angle RUZ$

14. An obtuse angle

The figure named is to be projected into some plane. Describe the various projections that can be formed.

15. Two parallel lines
16. A circle
17. An equilateral triangle
18. A regular hexagon
19. An ordinary parallelogram
20. An obtuse angle

Dihedral Angles

Your answers to Exercises 4 and 7 on page 586 should have been *yes.* You might naturally say, in each case, that the plane of square $ABCD$ is perpendicular to plane F. To describe the ways in which two planes can intersect, we use the terms *half-plane* and *dihedral angle.* Any line in a plane separates the plane into two **half-planes,** and all the points on one side of the line lie in the same half-plane. The diagrams below suggest that a **dihedral angle** is the figure formed by two half-planes that have the same edge.

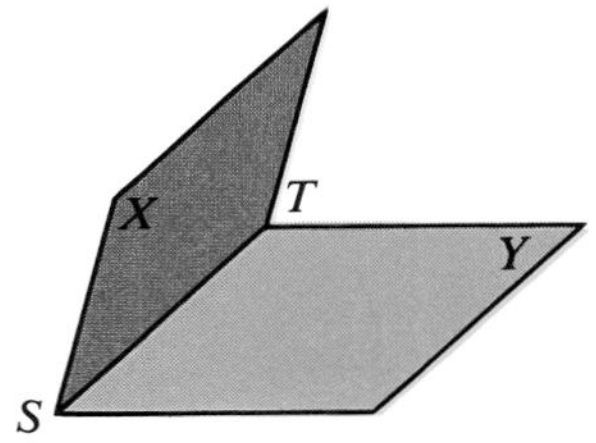

Dihedral angle X-ST-Y

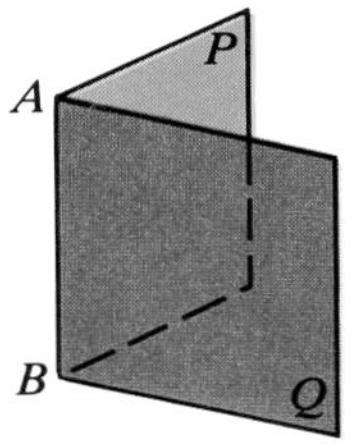

Dihedral angle P-AB-Q

The diagram at the left above suggests an *acute dihedral angle.* But just what does this phrase mean? You measure a dihedral angle as shown below.

In X, $\overrightarrow{EJ} \perp \overleftrightarrow{ST}$.
In Y, $\overrightarrow{EK} \perp \overleftrightarrow{ST}$.
$\angle JEK$ is a **plane angle** of dihedral angle X-ST-Y.

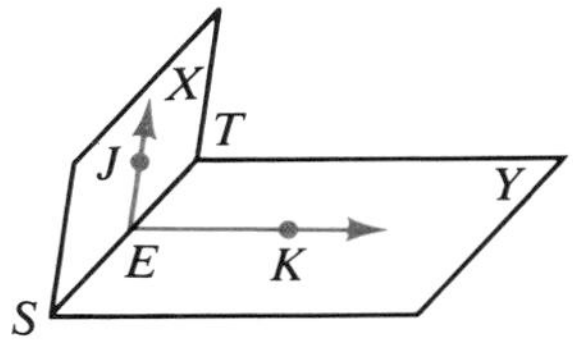

Dihedral angle X-ST-Y is measured in terms of one of its plane angles. In Exercise 13 you will prove that any two plane angles of a dihedral angle are congruent. Suppose $m\angle JEK = 80$. Then dihedral angle X-ST-Y is an 80° dihedral angle, and we say that planes X and Y meet at an 80° angle.

The diagram shows how to measure the angle that a line makes with a plane.

$\overrightarrow{AL}$ is the projection of $\overrightarrow{AP}$ into plane M.

$m\angle PAL = 30$

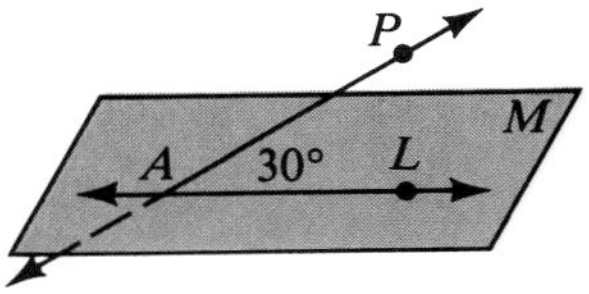

Noting that $\overleftrightarrow{AL}$ is a very special line, namely the projection of $\overleftrightarrow{AP}$ into the plane, we say that $\overleftrightarrow{AP}$ makes a 30° angle with the plane and that $\overleftrightarrow{AP}$ is inclined at a 30° angle to the plane.

Exercises

Draw a diagram that represents the figure named.

A

1. An acute dihedral angle
2. A right dihedral angle
3. An obtuse dihedral angle
4. Adjacent dihedral angles

B

5. Write your own definition of perpendicular planes. (*Hint:* Use the idea of a plane angle of a dihedral angle.)

6. **a.** Draw a plane X and a line k that is perpendicular to X.
   b. Draw a plane Y that contains k.
   c. What appears to be true about plane Y with respect to plane X?
   d. Complete this statement suggested by parts (a)–(c):
   If a line is perpendicular to a plane, and a second plane . . .

7. State theorems dealing with planes and dihedral angles suggested by the three theorems found on pages 60 and 61.

In Exercises 8–10 use a regular square pyramid each of whose edges is 6 units long.

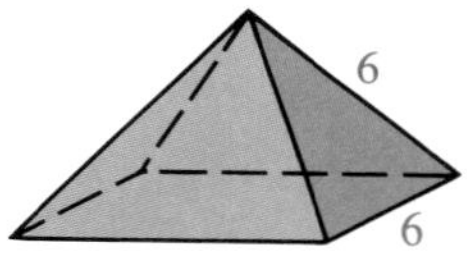

Exs. 8–10

C

8. **a.** Estimate the angle a lateral edge makes with the base of the pyramid.
   b. Compute to the nearest degree the angle named in part (a). (*Hint:* Draw your own figure, including the altitude and a segment joining the foot of the altitude to a vertex of the base.)

9. **a.** Estimate the measure of the dihedral angle formed by a lateral face and the base.
   b. Compute to the nearest degree the angle described in part (a). You will need to use a trigonometry table.

10. Repeat Exercise 9, but use the dihedral angle formed by two intersecting lateral faces.

11, 12. Using a regular triangular pyramid, repeat Exercises 8 and 9.

13. To show that it is reasonable to speak of *the* plane angle of a dihedral angle, you can show that any two plane angles of a dihedral angle are congruent. Explain how you can show this. You will probably need to use the theorem: Two lines parallel to a third line are parallel to each other.

14. *Challenge problem:* In the diagram, $\overleftrightarrow{AX}$ is oblique to plane ABC, $m\angle CAB = 50$, $m\angle XAB = 65$, $m\angle XAC = 70$, $AX = 10$, $\overline{XP} \perp$ plane ABC, and all right angles are as shown. Find:
    a. AQ, AR, and XR
    b. The angle that AX makes with plane ABC

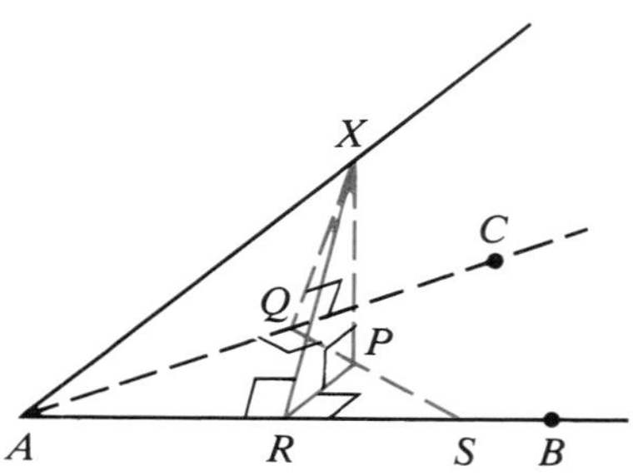

Postulates

Postulate 1 **(Ruler Postulate)** The points on a line can be paired with the real numbers in such a way that:

1. Any two desired points can have coordinates 0 and 1.
2. The distance between any two points equals the absolute value of the difference of their coordinates. (p. 7)

Postulate 2 **(Segment Addition Postulate)** If B is between A and C, then

$$AB + BC = AC. \qquad \text{(p. 7)}$$

Postulate 3 **(Protractor Postulate)** On $\overleftrightarrow{AB}$ in a given plane, choose any point O between A and B. Consider $\overrightarrow{OA}$ and $\overrightarrow{OB}$ and all the rays that can be drawn from O on one side of $\overleftrightarrow{AB}$. These rays can be paired with the real numbers from 0 to 180 in such a way that:

a. $\overrightarrow{OA}$ is paired with 0, and $\overrightarrow{OB}$ with 180.

b. If $\overrightarrow{OP}$ is paired with x, and $\overrightarrow{OQ}$ with y, then $m\angle POQ = |x - y|$. (p. 11)

Postulate 4 **(Angle Addition Postulate)** If point B lies in the interior of $\angle AOC$, then $m\angle AOB + m\angle BOC = m\angle AOC$. If $\angle AOC$ is a straight angle and B is any point not on $\overleftrightarrow{AC}$, then $m\angle AOB + m\angle BOC = 180$. (p. 12)

Postulate 5 A line contains at least two points; a plane contains at least three points not all in one line; space contains at least four points not all in one plane. (p. 46)

Postulate 6 Through any two points there is exactly one line. (p. 46)

Postulate 7 Through any three points there is at least one plane, and through any three noncollinear points there is exactly one plane. (p. 46)

Postulate 8 If two points are in a plane, then the line that contains the points is in that plane. (p. 46)

Postulate 9 If two planes intersect, then their intersection is a line. (p. 46)

Postulate 10 If two parallel lines are cut by a transversal, then corresponding angles are congruent. (p. 60)

Postulate 11 If two lines are cut by a transversal and corresponding angles are congruent, then the lines are parallel. (p. 64)

Postulate 12 **(SSS Postulate)** If three sides of one triangle are congruent to three sides of another triangle, then the triangles are congruent. (p. 110)

Postulate 13 **(SAS Postulate)** If two sides and the included angle of one triangle are congruent to two sides and the included angle of another triangle, then the triangles are congruent. (p. 110)

Postulate 14 **(ASA Postulate)** If two angles and the included side of one triangle are congruent to two angles and the included side of another triangle, then the triangles are congruent. (p. 111)

Postulate 15 **(AA Similarity Postulate)** If two angles of one triangle are congruent to two angles of another triangle, then the triangles are similar. (p. 219)

Postulate 16 **(Arc Addition Postulate)** The measure of the arc formed by two adjacent non-overlapping arcs is the sum of the measures of these two arcs. (p. 302)

Postulate 17 The area of a square is the square of the length of a side ($A = s^2$). (p. 379)

Postulate 18 **(Area Congruence Postulate)** If two figures are congruent, then they have the same area. (p. 379)

Postulate 19 **(Area Addition Postulate)** The area of a region is the sum of the areas of its non-overlapping parts. (p. 380)

Theorems

Points, Lines, Planes, and Angles

1-1 **(Midpoint Theorem)** If M is the midpoint of $\overline{AB}$, then: $2AM = AB$ and $AM = \frac{1}{2}AB$; $2MB = AB$ and $MB = \frac{1}{2}AB$. (p. 23)

1-2 **(Angle Bisector Theorem)** If $\overrightarrow{BX}$ is the bisector of $\angle ABC$, then:
$2m\angle ABX = m\angle ABC$ and $m\angle ABX = \frac{1}{2}m\angle ABC$;
$2m\angle XBC = m\angle ABC$ and $m\angle XBC = \frac{1}{2}m\angle ABC$. (p. 24)

1-3 Vertical angles are congruent. (p. 31)

1-4 Adjacent angles formed by perpendicular lines are congruent. (p. 36)

1-5 If two lines form congruent adjacent angles, then the lines are perpendicular. (p. 36)

1-6 If the exterior sides of two adjacent acute angles are perpendicular, then the angles are complementary. (p. 36)

1-7 If two angles are supplements of congruent angles (or of the same angle), then the two angles are congruent. (p. 41)

1-8 If two angles are complements of congruent angles (or of the same angle), then the two angles are congruent. (p. 41)

1-9 If two lines intersect, then they intersect in exactly one point. (p. 46)

1-10 If there is a line and a point not in the line, then exactly one plane contains them. (p. 46)

1-11 If two lines intersect, then exactly one plane contains them. (p. 46)

Parallel Lines and Planes

2-1 If two parallel planes are cut by a third plane, then the lines of intersection are parallel. (p. 56)

2-2 If two parallel lines are cut by a transversal, then alternate interior angles are congruent. (p. 60)

2-3 If two parallel lines are cut by a transversal, then same-side interior angles are supplementary. (p. 60)

2-4 If a transversal is perpendicular to one of two parallel lines, then it is perpendicular to the other one also. (p. 61)

2-5 If two lines are cut by a transversal and alternate interior angles are congruent, then the lines are parallel. (p. 65)

2-6 If two lines are cut by a transversal and same-side interior angles are supplementary, then the lines are parallel. (p. 65)

2-7 In a plane, two lines perpendicular to the same line are parallel. (p. 65)

2-8 Through a point outside a line, there is exactly one line parallel to the given line. (p. 66)

2-9 Through a point outside a line, there is exactly one line perpendicular to the given line. (p. 66)

2-10 Two lines parallel to a third line are parallel to each other. (p. 67)

2-11 The sum of the measures of the angles of a triangle is 180. (p. 73)

Corollary 1 If two angles of one triangle are congruent to two angles of another triangle, then the third angles are congruent. (p. 73)

Corollary 2 Each angle of an equiangular triangle has measure 60. (p. 73)

Corollary 3 In a triangle, there can be at most one right angle or obtuse angle. (p. 73)

Corollary 4 The acute angles of a right triangle are complementary. (p. 73)

2-12 The measure of an exterior angle of a triangle equals the sum of the measures of the two remote interior angles. (p. 74)

2-13 The sum of the measures of the angles of a convex polygon with n sides is $(n - 2)180$. (p. 81)

2-14 The sum of the measures of the exterior angles of any convex polygon, one angle at each vertex, is 360. (p. 81)

Congruent Triangles

3-1 If two sides of a triangle are congruent, then the angles opposite those sides are congruent. (p. 124)

Corollary 1 An equilateral triangle is also equiangular. (p. 125)

Corollary 2 An equilateral triangle has three 60° angles. (p. 125)

Corollary 3 The bisector of the vertex angle of an isosceles triangle is perpendicular to the base at its midpoint. (p. 125)

3-2 If two angles of a triangle are congruent, then the sides opposite those angles are congruent. (p. 125)

Corollary An equiangular triangle is also equilateral. (p. 125)

3-3 **(AAS Theorem)** If two angles and a non-included side of one triangle are congruent to the corresponding parts of another triangle, then the triangles are congruent. (p. 129)

3-4 **(HL Theorem)** If the hypotenuse and a leg of one right triangle are congruent to the corresponding parts of another right triangle, then the triangles are congruent. (p. 130)

3-5 If a point lies on the perpendicular bisector of a segment, then the point is equidistant from the endpoints of the segment. (p. 138)

3-6 If a point is equidistant from the endpoints of a segment, then the point lies on the perpendicular bisector of the segment. (p. 138)

3-7 If a point lies on the bisector of an angle, then the point is equidistant from the sides of the angle. (p. 139)

3-8 If a point is equidistant from the sides of an angle, then the point lies on the bisector of the angle. (p. 139)

Using Congruent Triangles

4-1 Opposite sides of a parallelogram are congruent. (p. 159)

Corollary If two lines are parallel, then all points on one line are equidistant ı the other line. (p. 159)

4-2 Opposite angles of a parallelogram are congruent. (p. 160)

4-3 The diagonals of a parallelogram bisect each other. (p. 160)

4-4 If both pairs of opposite sides of a quadrilateral are congruent, then the quadrilateral is a parallelogram. (p. 163)

4-5 If one pair of opposite sides of a quadrilateral are both congruent and parallel, then the quadrilateral is a parallelogram. (p. 163)

4-6 If both pairs of opposite angles of a quadrilateral are congruent, then the quadrilateral is a parallelogram. (p. 164)

4-7 If the diagonals of a quadrilateral bisect each other, then the quadrilateral is a parallelogram. (p. 164)

4-8 If three parallel lines cut off congruent segments on one transversal, then they cut off congruent segments on every transversal. (p. 164)

Corollary A line that contains the midpoint of one side of a triangle and is parallel to another side bisects the third side. (p. 165)

4-9 The diagonals of a rectangle are congruent. (p. 169)

4-10 The diagonals of a rhombus are perpendicular. (p. 169)

4-11 Each diagonal of a rhombus bisects two angles of the rhombus. (p. 169)

4-12 The midpoint of the hypotenuse of a right triangle is equidistant from the three vertices. (p. 169)

4-13 If an angle of a parallelogram is a right angle, then the parallelogram is a rectangle. (p. 169)

4-14 If two consecutive sides of a parallelogram are congruent, then the parallelogram is a rhombus. (p. 169)

4-15 Base angles of an isosceles trapezoid are congruent. (p. 173)

4-16 The median of a trapezoid
(1) is parallel to the bases;
(2) has a length equal to half the sum of the lengths of the bases. (p. 174)

4-17 The segment that joins the midpoints of two sides of a triangle
(1) is parallel to the third side;
(2) has a length equal to half the length of the third side. (p. 174)

4-18 If one side of a triangle is longer than a second side, then the angle opposite the first side is larger than the angle opposite the second side. (p. 183)

4-19 If one angle of a triangle is larger than a second angle, then the side opposite the first angle is longer than the side opposite the second angle. (p. 183)

Corollary 1 The perpendicular segment from a point to a line is the shortest segment from the point to the line. (p. 183)

Corollary 2 The perpendicular segment from a point to a plane is the shortest segment from the point to the plane. (p. 183)

-20 **(The Triangle Inequality)** The sum of the lengths of any two sides of a triangle is greater than the length of the third side. (p. 184)

4-21 **(SAS Inequality Theorem)** If two sides of one triangle are congruent to two sides of another triangle, but the included angle of the first triangle is greater than the included angle of the second, then the third side of the first triangle is longer than the third side of the second triangle. (p. 189)

4-22 **(SSS Inequality Theorem)** If two sides of one triangle are congruent to two sides of another triangle, but the third side of the first triangle is longer than the third side of the second, then the included angle of the first triangle is larger than the included angle of the second. (p. 190)

Similar Polygons

5-1 **(SAS Similarity Theorem)** If an angle of one triangle is congruent to an angle of another triangle and the sides including those angles are in proportion, then the triangles are similar. (p. 226)

5-2 **(SSS Similarity Theorem)** If the sides of two triangles are in proportion, then the two triangles are similar. (p. 226)

5-3 **(Triangle Proportionality Theorem)** If a line parallel to one side of a triangle intersects the other two sides, then it divides those sides proportionally. (p. 233)

Corollary If three parallel lines intersect two transversals, then they divide the transversals proportionally. (p. 234)

5-4 **(Triangle Angle-Bisector Theorem)** If a ray bisects an angle of a triangle, then it divides the opposite side into segments proportional to the other two sides. (p. 234)

Right Triangles

6-1 If the altitude is drawn to the hypotenuse of a right triangle, then the two triangles formed are similar to the original triangle and to each other. (p. 248)

Corollary 1 When the altitude is drawn to the hypotenuse of a right triangle, the length of the altitude is the geometric mean between the segments of the hypotenuse. (p. 248)

Corollary 2 When the altitude is drawn to the hypotenuse of a right triangle, each leg is the geometric mean between the hypotenuse and the segment of the hypotenuse that is adjacent to that leg. (p. 248)

6-2 **(Pythagorean Theorem)** In a right triangle, the square of the hypotenuse is equal to the sum of the squares of the legs. (p. 252)

6-3 If the square of one side of a triangle is equal to the sum of the squares of the other two sides, then the triangle is a right triangle. (p. 257)

6-4 If the square of the longest side of a triangle is greater than the sum of the squares of the other two sides, then the triangle is an obtuse triangle. (p. 258)

6-5 If the square of the longest side of a triangle is less than the sum of the squares of the other two sides, then the triangle is an acute triangle. (p. 258)

6-6 **(45°-45°-90° Theorem)** In a 45°-45°-90° triangle, the hypotenuse is $\sqrt{2}$ times as long as a leg. (p. 262)

6-7 **(30°-60°-90° Theorem)** In a 30°-60°-90° triangle, the hypotenuse is twice as long as the shorter leg, and the longer leg is $\sqrt{3}$ times as long as the shorter leg. (p. 262)

Circles

7-1 If a line is tangent to a circle, then the line is perpendicular to the radius drawn to the point of tangency. (p. 296)

Corollary Tangents to a circle from a point are congruent. (p. 296)

7-2 If a line in the plane of a circle is perpendicular to a radius at its outer endpoint, then the line is tangent to the circle. (p. 297)

7-3 In the same circle or in congruent circles, two minor arcs are congruent if and only if their central angles are congruent. (p. 303)

7-4 In the same circle or in congruent circles:
(1) Congruent chords have congruent arcs.
(2) Congruent arcs have congruent chords. (p. 306)

7-5 A diameter that is perpendicular to a chord bisects the chord and its arc. (p. 307)

7-6 In the same circle or in congruent circles:
(1) Congruent chords are equally distant from the center (or centers).
(2) Chords equally distant from the center (or centers) are congruent. (p. 307)

7-7 The measure of an inscribed angle is equal to half the measure of its intercepted arc. (p. 312)

Corollary 1 If two inscribed angles intercept the same arc, then the angles are congruent. (p. 312)

Corollary 2 If a quadrilateral is inscribed in a circle, then its opposite angles are supplementary. (p. 312)

Corollary 3 An angle inscribed in a semicircle is a right angle. (p. 312)

7-8 The measure of an angle formed by a chord and a tangent is equal to half the measure of the intercepted arc. (p. 313)

7-9 The measure of an angle formed by two chords that intersect inside a circle is equal to half the sum of the measures of the intercepted arcs. (p. 317)

7-10 The measure of an angle formed by two secants, two tangents, or a secant and a tangent drawn from a point outside a circle is equal to half the difference of the measures of the intercepted arcs. (p. 318)

7-11 When two chords intersect inside a circle, the product of the lengths of the segments of one chord equals the product of the lengths of the segments of the other. (p. 321)

7-12 When two secant segments are drawn to a circle from an external point, the product of the lengths of one secant segment and its external segment equals the product of the lengths of the other secant segment and its external segment. (p. 321)

7-13 When a secant segment and a tangent segment are drawn to a circle from an external point, the product of the lengths of the secant segment and its external segment is equal to the square of the length of the tangent segment. (p. 322)

Constructions and Loci

8-1 The bisectors of the angles of a triangle intersect in a point that is equidistant from the three sides of the triangle. (p. 344)

8-2 The perpendicular bisectors of the sides of a triangle intersect in a point that is equidistant from the three vertices of the triangle. (p. 345)

8-3 The lines that contain the altitudes of a triangle intersect in a point. (p. 345)

8-4 The medians of a triangle intersect in a point that is two thirds of the distance from each vertex to the midpoint of the opposite side. (p. 345)

Areas of Plane Figures

9-1 The area of a rectangle equals the product of its base and height. ($A = bh$) (p. 380)

9-2 The area of a parallelogram equals the product of its base and height. ($A = bh$) (p. 385)

9-3 The area of a triangle equals half the product of its base and height. ($A = \frac{1}{2}bh$) (p. 386)

Corollary The area of a rhombus equals half the product of its diagonals. ($A = \frac{1}{2}d_1d_2$) (p. 386)

9-4 The area of a trapezoid equals half the product of the height and the sum of the bases. $A = \frac{1}{2}h(b_1 + b_2)$ (p. 391)

9-5 The area of a regular polygon is equal to half the product of the apothem and the perimeter. ($A = \frac{1}{2}ap$) (p. 397)

Related formulas In a circle: $C = 2\pi r$ $A = \pi r^2$ (p. 403)

9-6 If the scale factor of two similar figures is $a:b$, then:
(1) The ratio of the perimeters is $a:b$.
(2) The ratio of the areas is $a^2:b^2$. (p. 410)

Areas and Volumes of Solids

10-1 The lateral area of a right prism equals the perimeter of a base times the height of the prism. (L.A. $= ph$) (p. 424)

10-2 The volume of a right prism equals the area of a base times the height of the prism. ($V = Bh$) (p. 424)

Related formulas

In a pyramid:	L.A. $= \frac{1}{2}pl$	$V = \frac{1}{3}Bh$
In a cylinder:	L.A. $= 2\pi rh$	$V = \pi r^2h$
In a cone:	L.A. $= \pi rl$	$V = \frac{1}{3}\pi r^2h$
In a sphere:	$A = 4\pi r^2$	$V = \frac{4}{3}\pi r^3$

10-3 If the scale factor of two similar solids is $a:b$, then:
(1) The ratio of corresponding perimeters is $a:b$.
(2) The ratios of the base areas, of the lateral areas, and of the total areas are $a^2:b^2$.
(3) The ratio of the volumes is $a^3:b^3$. (p. 453)

Coordinate Geometry

11-1 **(The Distance Formula)** The distance d between points (x_1, y_1) and (x_2, y_2) is given by $d = \sqrt{(x_2 - x_1)^2 + (y_2 - y_1)^2}$. (p. 470)

11-2 An equation of the circle with center (a, b) and radius r is $(x - a)^2 + (y - b)^2 = r^2$. (p. 475)

11-3 **(The Midpoint Formula)** The midpoint of the segment that joins points (x_1, y_1) and (x_2, y_2) is the point $\left(\frac{x_1 + x_2}{2}, \frac{y_1 + y_2}{2}\right)$. (p. 479)

11-4 Two nonvertical lines are parallel if and only if their slopes are equal. (p. 487)

11-5 Two nonvertical lines are perpendicular if and only if the product of their slopes is -1. (p. 487)

$$m_1 \cdot m_2 = -1, \quad \text{or} \quad m_1 = -\frac{1}{m_2}$$

11-6 **(Standard Form)** The graph of any equation that can be written in the form $ax + by = c$, with a and b not both zero, is a line. (p. 491)

11-7 **(Point-Slope Form)** The equation of the line that passes through point (x_1, y_1) and has slope m is $y - y_1 = m(x - x_1)$. (p. 491)

11-8 **(Slope-Intercept Form)** An equation of the line that has slope m and y-intercept b is $y = mx + b$. (p. 492)

Transformations

12-1 A reflection in a line is an isometry. (p. 515)

12-2 If a transformation T maps any point (x, y) to $(x + a, y + b)$, then T is a translation. (p. 522)

12-3 A rotation is an isometry. (p. 526)

12-4 The dilation $D_{O,k}$ maps every line segment to a parallel segment that is $|k|$ times as long. (p. 531)

12-5 The product of two isometries is an isometry. (p. 536)

12-6 A product of reflections in two parallel lines is a translation. The translation glides all points through twice the distance between the lines. (p. 536)

12-7 A product of reflections in two intersecting lines is a rotation about the point of intersection of the two lines. The measure of the angle of rotation is twice the measure of the angle from the first line of reflection to the second. (p. 537)

Corollary A product of reflections in perpendicular lines is a half turn about the point where the lines intersect. (p. 537)

Constructions

Construction 1	Given a segment, construct a segment congruent to the given segment. (p. 335)
Construction 2	Given an angle, construct an angle congruent to the given angle. (p. 336)
Construction 3	Given an angle, construct the bisector of the angle. (p. 336)
Construction 4	Given a segment, construct the perpendicular bisector of the segment. (p. 339)
Construction 5	Given a point on a line, construct the perpendicular to the line at the given point. (p. 340)
Construction 6	Given a point outside a line, construct the perpendicular to the line from the given point. (p. 340)
Construction 7	Given a point outside a line, construct the parallel to the given line through the given point. (p. 341)
Construction 8	Given a point on a circle, construct the tangent to the circle at the given point. (p. 349)
Construction 9	Given a point outside a circle, construct a tangent to the circle from the given point. (p. 350)
Construction 10	Given a triangle, circumscribe a circle about the triangle. (p. 350)
Construction 11	Given a triangle, inscribe a circle in the triangle. (p. 351)
Construction 12	Given a segment, divide the segment into a given number of congruent parts. (p. 353)
Construction 13	Given three segments, construct a fourth segment so that the four segments are in proportion. (p. 354)
Construction 14	Given two segments, construct their geometric mean. (p. 354)

Answers for Self-Tests

Chapter 1

SELF-TEST 1, PAGE 16

1. T **2.** F **3.** F **4.** T **5.** T **6.** $\overrightarrow{EG}$ **7.** 11 **8.** vertex **9.** $\overrightarrow{SW}$, $\angle RST$ **10.** 34

SELF-TEST 2, PAGE 35

1. given, def., theorem, postulate **2.** Subtr. Prop. = **3.** $\angle$ Bis. Th. **4.** Trans. Prop. **5.** Subst. Prop. **6.** Vert. $\angle$s are $\cong$. **7.** $\angle 1$, $\angle DXB$ **8.** 60, 30, 120

SELF-TEST 3, PAGE 49

1. Adj. $\angle$s formed by $\perp$ lines are $\cong$. **2.** Def. $\perp$ lines **3.** If the ext. sides of 2 adj. acute $\angle$s are $\perp$, then the $\angle$s are comp. **4.** the lines are $\perp$ **5.** statement, diagram, given, prove, series of statements and reasons **6.** a line **7.** It lies in *M*. **8.** at least **9.** 3 noncollinear pts.; 2 intersecting lines; a line and a pt. not on the line **10.** $\angle AOC \cong \angle EOC$; $m\angle AOB = \frac{1}{2}m\angle AOC$, $m\angle DOE = \frac{1}{2}m\angle EOC$; $m\angle AOB = m\angle DOE$; $\angle AOB \cong \angle DOE$

Chapter 2

SELF-TEST 1, PAGE 70

1. sometimes **2.** never **3.** always **4.** sometimes **5.** sometimes **6.** $\angle 4$, $\angle 3$ **7.** $\angle 8$ and $\angle 2$, $\angle 7$ and $\angle 4$ **8.** $\angle 2$, $\angle 8$ **9.** $\overline{EB} \parallel \overline{DC}$ **10.** none **11.** $\overline{EA} \parallel \overline{DB}$

SELF-TEST 2, PAGE 86

1. acute **2.** scalene **3.** 105, 35 **4.** 19 **5.** equilateral, equiangular **6.** 360, 144 **7.** 60

SELF-TEST 3, PAGE 96

1. $\triangle ABC$ is rt. $\triangle$ **2.** $\angle C$ is rt. $\angle$. **3.** If $\angle C$ is rt. $\angle$, then $\triangle ABC$ is rt. $\triangle$; true. **4.** If $\triangle ABC$ is not rt. $\triangle$, then $\angle C$ is not rt. $\angle$; true. **5.** If $\angle C$ is not rt. $\angle$, then $\triangle ABC$ is not rt. $\triangle$; false. **6. a.** Valerie can go. **b.** no conc. **c.** no conc. **d.** Dan can go. **7. a.** If a figure is a square, then it is a rectangle.
b.

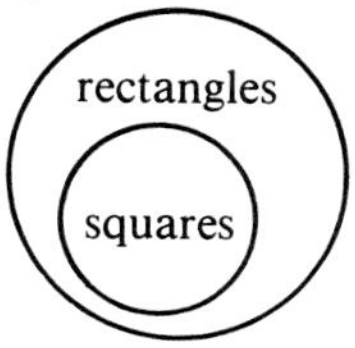

8. If a $\triangle$ is equilateral, then all sides $\cong$. If all sides of a $\triangle$ are $\cong$, then it is equilateral.

Chapter 3

SELF-TEST 1, PAGES 122–123

1. $\angle R \cong \angle C$; corr. parts $\cong$ $\triangle$s **2.** $\overline{RE}$, $\overline{CA}$; $\overline{ED}$, $\overline{AB}$; $\overline{RD}$, $\overline{CB}$ **3.** $\triangle BAE \cong \triangle DCE$, ASA **4.** $\triangle XWY \cong \triangle ZWY$, SSS **5.** no $\cong$ $\triangle$s **6.** 1. $\overline{QU} \cong \overline{AD}$ (Given) 2. $\overline{QU} \parallel \overline{DA}$ (Given) 3. $\angle 2 \cong \angle 3$ (If lines $\parallel$, alt. int. $\angle$s $\cong$.) 4. $\overline{QA} \cong \overline{QA}$ (Reflex.) 5. $\triangle AQU \cong \triangle QAD$ (SAS) 6. $\overline{UA} \cong \overline{DQ}$ (corr. parts $\cong$ $\triangle$s) **7.** 1. $\overline{UA} \perp \overline{QA}$, $\overline{DQ} \perp \overline{QA}$ (Given) 2. $\angle 4 \cong \angle 1$ (Def. of $\perp$ lines, rt. $\angle$s, and $\cong$ $\angle$s) 3. $\angle 2 \cong \angle 3$ (Given) 4. $\overline{QA} \cong \overline{QA}$ (Reflex.) 5. $\triangle AUQ \cong \triangle QDA$ (ASA) 6. $\angle U \cong \angle D$ (corr. parts $\cong$ $\triangle$s)

SELF-TEST 2, PAGE 136

1. $x = 55$ **2.** $x = 7$ **3.** no; cannot prove $\triangle WXY \cong \triangle WZY$ **4.** 1. $\angle X \cong \angle Z$ (Given) 2. $\overline{WY} \perp P$ (Given) 3. $\overline{WY} \perp \overline{YX}$, $\overline{WY} \perp \overline{YZ}$ (Def. line $\perp$ plane) 4. $m\angle WYX = m\angle WYZ = 90$ (Def. $\perp$ lines, rt. $\angle$) 5. $\overline{WY} \cong \overline{WY}$ (Reflex.) 6. $\triangle WXY \cong \triangle WZY$ (AAS) **5.** 1. $\overline{WX} \cong \overline{WZ}$ (Given) 2. $\overline{WY} \perp P$ (Given) 3. $\overline{WY} \perp \overline{XY}$, $\overline{WY} \perp \overline{ZY}$ (Def. line $\perp$ plane) 4. $\triangle WXY$ and $\triangle WZY$ rt. $\triangle$s (Def. $\perp$, rt. $\triangle$) 5. $\overline{WY} \cong \overline{WY}$ (Reflex.) 6. $\triangle WXY \cong \triangle WZY$ (HL) 7. $\overline{XY} \cong \overline{ZY}$ (corr. parts $\cong$ $\triangle$s) **6.** Since $\overline{AB} \cong \overline{AC}$, $\angle A \cong \angle A$, and $\angle ANB \cong \angle AMC$, $\triangle ABN \cong \triangle ACM$ by AAS.

SELF-TEST 3, PAGE 152

1. altitude **2.** median **3.** $\perp$ bis. of $\overline{BD}$ **4.** $\overline{AX}$, $\overline{EX}$ **5.** distance; X; $\overline{AE}$ **6.** (Answers will vary.) Since $\angle A \cong \angle E$, $\overline{AX} \cong \overline{EX}$. $\triangle XAB \cong \triangle XED$ by ASA. Corr. parts $\overline{BX}$ and $\overline{DX}$ are $\cong$. Since $\overline{XC} \cong \overline{XC}$ and $\angle 2 \cong \angle 3$, $\triangle BXC \cong \triangle DXC$ by SAS. $\overline{BC} \cong \overline{DC}$, and $\overline{XC}$ is a median of $\triangle BXD$. **7.** 31 **8.** 16

Chapter 4

SELF-TEST 1, PAGE 178

1. may be **2.** must be **3.** must be **4.** cannot be **5.** parallelogram **6.** trap. **7.** rect. **8.** square **9.** $JQUK$, $JQRL$, $TJLS$, $TQRS$ **10.** 8, 2 **11.** 1. $\overline{EH} \cong \overline{FG}$, $\overline{EH} \parallel \overline{FG}$ (Given) 2. $EFGH$ is ▱ (If one pair of opp. sides of a quad. are both $\parallel$ and $\cong$, then the quad is a ▱.) 3. $\overline{EF} \parallel \overline{HG}$ (Def. ▱) 4. $\overline{EF} \cong \overline{HG}$ (Opp. sides of a ▱ are $\cong$.) **12.** 1. $\angle 1 \cong \angle 2 \cong \angle 3 \cong \angle 4$ (Given) 2. $\overline{EH} \parallel \overline{FG}$, $\overline{EF} \parallel \overline{HG}$ (If alt. int. $\angle$s $\cong$, lines $\parallel$.) 3. $EFGH$ is ▱ (Def. ▱) 4. $\overline{EH} \cong \overline{HG}$ (If 2 $\angle$s of a $\triangle$ are $\cong$, then the sides opp. those $\angle$s are $\cong$.) 5. $EFGH$ is rhombus (If 2 consec. sides of a ▱ are $\cong$, then the ▱ is a rhombus.)

SELF-TEST 2, PAGE 195

1. d, b, a, c **2.** $<$ **3.** $=$ **4.** $>$ **5.** $<$ **6.** 1, 11 **7.** $\overline{DO}$ **8.** cannot be **9.** must be **10.** may be

Chapter 5

SELF-TEST 1, PAGE 216

1. 3 : 5 **2.** 3 : 10 **3.** $\frac{2a}{3b}$ **4.** $x = 6$ **5.** $x = 10$ **6.** $x = 3$ **7.** no **8.** yes **9.** yes **10.** 45; 75; 60 **11.** $\frac{2}{3}$ **12.** 12 **13.** 15 **14.** 12 **15.** 100; 100; 100; 120; 140; 160

SELF-TEST 2, PAGE 238

1. SSS **2.** AA **3.** SAS **4.** r **5.** p **6.** h **7.** a **8.** $x = 12$ **9.** $x = 14$ **10.** $x = 6\frac{2}{3}$

Chapter 6

SELF-TEST 1, PAGE 256

1. a. $4\sqrt{3}$ **b.** $\frac{7\sqrt{2}}{2}$ **c.** $\frac{\sqrt{10}}{4}$ **d.** $\frac{3\sqrt{5}}{2}$ **2.** $3\sqrt{5}$ **3.** 4 **4.** $2\sqrt{5}$ **5.** $4\sqrt{5}$ **6.** 10 **7.** 5 **8.** $5\sqrt{2}$ **9.** 7 **10.** $n\sqrt{2}$

SELF-TEST 2, PAGE 265

1. rt. **2.** acute **3.** obtuse **4.** 5; $5\sqrt{2}$ **5.** $6\sqrt{2}$; $6\sqrt{2}$ **6.** $4\sqrt{3}$; 8 **7.** 5; $5\sqrt{3}$ **8.** $2\sqrt{3}$; $4\sqrt{3}$ **9.** $4\sqrt{3}$

SELF-TEST 3, PAGE 282

1. $\frac{24}{7}$ **2.** $\frac{7}{25}$ **3.** $\frac{24}{25}$ **4.** $\frac{7}{24}$ **5.** 74° **6.** 82 **7.** 74 **8.** 113 **9.** 109 **10.** 8°

Chapter 7

SELF-TEST 1, PAGES 310–311

1. 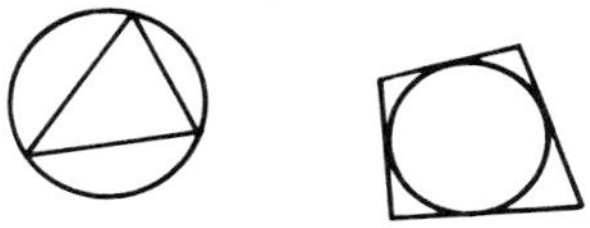

2. 16 **3.** 2 **4.** 2 concentric circles **5.** 50 **6.** 310 **7.** In the same ⊙, ≅ chords have ≅ arcs. **8.** 70 **9.** 6 **10.** 1. $\overline{PA}$ and $\overline{PB}$ are tan. to ⊙O; $\overline{PB}$ and $\overline{PC}$ are tan. to ⊙Q (Given); 2. $\overline{PA} \cong \overline{PB}$; $\overline{PB} \cong \overline{PC}$ (Tan. to a ⊙ from a pt. are ≅.); 3. $\overline{PA} \cong \overline{PC}$ (Trans. Prop.) **11.** 13

SELF-TEST 2, PAGE 327

1. 40 **2.** 150 **3.** 82 **4.** 9 **5.** 70; 50 **6.** 35 **7.** 10 **8.** 10 **9.** 1. $\angle B \cong \angle D$; $\angle C \cong \angle A$ (If 2 inscrib. ∠s int. the same arc, the ∠s are ≅.); 2. $\triangle KAD \sim \triangle KCB$ (AA); 3. $\frac{AK}{CK} = \frac{AD}{CB}$ (Corr. sides of ~ △s are in prop.)

Chapter 8

SELF-TEST 1, PAGE 349

1. Use Const. 4 **2.** Draw $\overline{ST}$. With ctrs. S and T, and $r = ST$, draw arcs int. at R, Draw $\overrightarrow{SR}$; $m\angle RST = 60$. Use Const. 3 **3.** Use Const. 6 **4.** Use Const. 7 **5.** Use Const. 1 to const. $\overline{JK}$ such that $JK = 2AB$. Const. lines ⊥ $\overline{JK}$ at J and K. Const. $\overline{JM} \cong \overline{KL} \cong \overline{AB}$. Draw $\overline{ML}$. (Other methods are poss.) **6.** ∠ bisectors; ⊥ bisectors of the sides; altitudes; medians **7.** the midpt. of the hyp. **8.** $(AD)^2 = 12^2 - 6^2 = 144 - 36 = 108$; $AD = \sqrt{108} = 6\sqrt{3}$; $AX = \frac{2}{3}\cdot 6\sqrt{3} = 4\sqrt{3}$; $XD = \frac{1}{3}\cdot 6\sqrt{3} = 2\sqrt{3}$

SELF-TEST 2, PAGE 358

1. Use Const. 9 **2.** Use Const. 11 **3.** Use Const. 12, but you don't need to const. the ∥ at R. **4.** Use Const. 13 **5.** Use Const. 14 **6.** ⊥; F; ⊥; G **7.** Const. the ⊥ bis. of 2 sides of $\triangle TRI$, int. at O. Draw a ⊙ with ctr. O, $r = OT$.

SELF-TEST 3, PAGE 371

1. the bisectors of the vert. ∠s formed by j and k (2 lines) **2.** a sphere with ctr. P and with radius t **3.** a plane ⊥ $\overline{WX}$ that bis. $\overline{WX}$ **4.** the int. of the bis. of $\angle DEF$ with one of the lines that are ∥ to $\overline{EF}$ and are 4 cm from $\overline{EF}$ (1 pt.) **5.** 0, 1, or 2 points formed by the int. of the line ∥ to s and t and midway between them with a circle having center A and radius 4 **6.** Construct the ⊥ bisectors of 2 (or 3) sides of the △. The pt. where they intersect is the locus. (This pt. may be inside, outside, or on the △, depending on the particular △ you have drawn.)

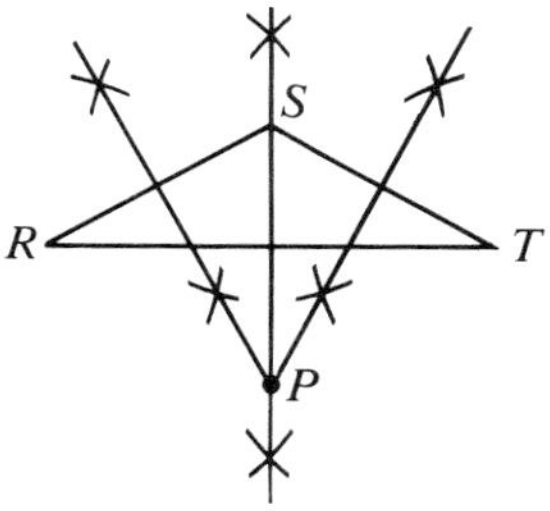

Ex.6

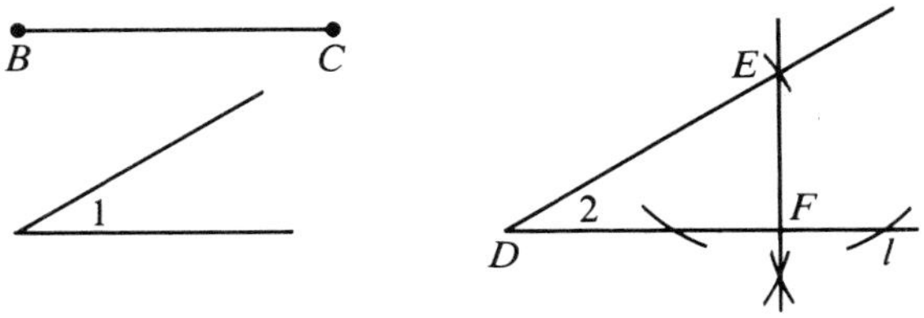

Ex.7

7. Const. $\angle 2 \cong \angle 1$, $\overline{DE} \cong \overline{BC}$, and the ⊥ to l from E, int. l at F. (Other constructions are poss.)

Chapter 9

SELF-TEST 1, PAGE 400

1. 81 **2.** 60 **3.** $40\sqrt{3}$ **4.** $4\sqrt{3}$ **5.** $6\sqrt{13}$ **6.** 40 **7.** 39 **8.** $150\sqrt{3}$ **9.** 49

SELF-TEST 2, PAGE 414

1. 88; 616 **2.** 81π **3. a.** 6π **b.** 36π **c.** $36\pi - 72$ **4.** 16 : 49 **5.** 2 : 3 **6.** 4 : 9 **7.** $64 - 16\pi$ **8.** $36\pi - 27\sqrt{3}$

Chapter 10

SELF-TEST 1, PAGE 442

1. 162; 322; 360 **2.** 624; 1200; 960 **3.** 140π; 340π; 700π **4.** 180; $180 + 108\sqrt{3}$; $270\sqrt{3}$ **5.** 135π; 216π; 324π **6.** 8000 m^3 **7.** 6 **8.** prism

SELF-TEST 2, PAGE 457

1. 36π; 36π **2.** 16π **3.** $\frac{22{,}000\pi}{3}$ cm^3 **4.** 25π cm^2 **5. a.** $5\frac{1}{3}$ **b.** 9 : 4 **6. a.** 2 : 5 **b.** 8 : 125

Chapter 11

SELF-TEST 1, PAGE 482

1. a. 2 **b.** (4, 1) **2. a.** 10 **b.** (4, −3) **3. a.** $10\sqrt{2}$ **b.** (3, 2) **4. a.** $\sqrt{29}$ **b.** $(-4, \frac{9}{2})$ **5.** $x^2 + y^2 = 81$ **6.** $(x + 1)^2 + (y - 2)^2 = 25$ **7.** (−2, 3); 6

SELF-TEST 2, PAGE 498

1. $\frac{4}{7}$ **2.** $-\frac{3}{5}$ **3.** vert. line **4. a.** $-\frac{5}{3}$ **b.** $\frac{3}{5}$ **5. a.** $\frac{2}{3}$ **b.** Answers may vary; e.g. (0, 6)

6.

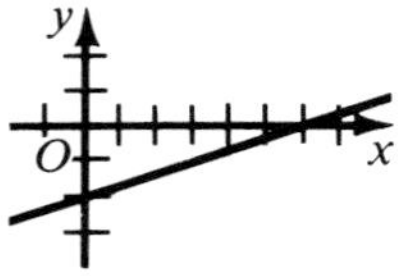

7. $\frac{2}{5}$; −6 **8.** Answers may vary; e.g., $-3x + 4y = 7$ **9.** (3, 4)

SELF-TEST 3, PAGE 504

1. (2e, 0) **2.** $(c + g, h)$ **3.** $(c - g, h)$ **4.** $(a + b, c)$ **5.** Slope of $\overline{GO} = -\frac{1}{4} =$ slope of $\overline{LD}$, slope of $\overline{OL} = 3 =$ slope of $\overline{GD}$; $\overline{GO} \parallel \overline{LD}$ and $\overline{OL} \parallel \overline{GD}$ **6.** Midpts. are $(a, a\sqrt{3})$, $(2a, 0)$, and $(3a, a\sqrt{3})$. Using the distance formula, the length of each median is $\sqrt{12a^2}$, or $2a\sqrt{3}$.

Chapter 12

SELF-TEST 1, PAGE 534

1. Q' **2.** R **3.** $\overline{P'Q'}$ **4.** (3, 1) **5.** $\overline{AA'}$ **6.** $\mathcal{R}_{O,330}$, $\mathcal{R}_{O,690}$ **7.** B **8.** C **9.** $\overline{AB}$ **10.** $\overline{OB}$ **11.** M **12.** $\overline{AO}$ **13.** L **14.** $\triangle NDO$ **15.** C **16.** Q **17.** C **18.** O **19.** OBL

SELF-TEST 2, PAGE 548

1. A **2.** A **3.** B **4.** P **5.** y-axis **6.** no **7.** P **8.** T **9.** translation **10. a.** $S^{-1}:(x, y) \rightarrow (x - 2, y + 3)$ **b.** $D_{O,\frac{1}{5}}$ **c.** $\mathcal{R}_{O,70}$ **d.** R_y **11.** 6

Technical Drawing

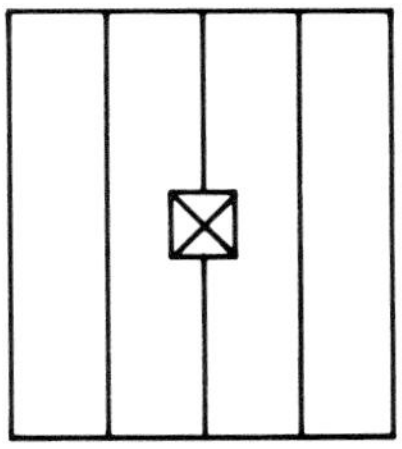

Can the shape of a three-dimensional object be determined from a single two-dimensional image? Usually the shape cannot be so determined. For example, if you photographed the barn shown on page 57 from a point directly above, then your photograph might look something like the sketch shown at the right. You cannot tell from this one photograph whether the roof slopes or whether the sides of the barn slope in or out. You would have a much better idea of the shape of the barn if you could also see it from the front and one side. The following figures show the actual shape of the building much more clearly.

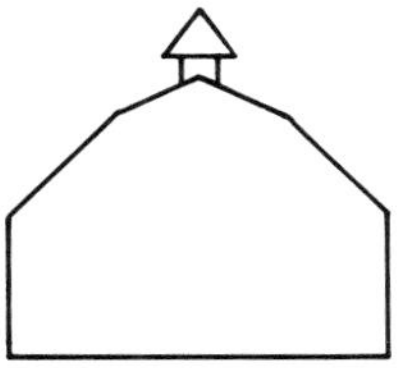

Front view

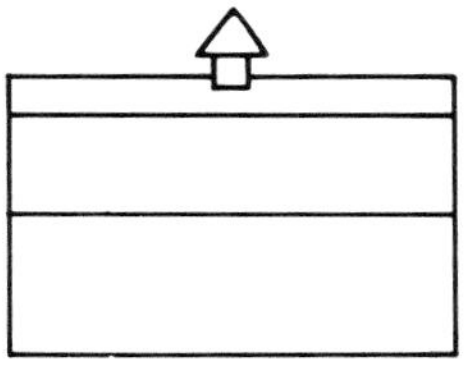

Side view

These three views of the barn illustrate *orthographic projection,* a set of projections of an object onto three planes perpendicular to one another.

To make an orthographic projection of an object, draw a top view, a front view, and a side view. In each projection, all visible edges, contours, and intersections of surfaces are shown as solid lines. Hidden edges and contours are shown as dashed lines. Arrange the three views in an ''L'' shaped pattern as illustrated in the figure below. Some corresponding vertices have been connected with red lines.

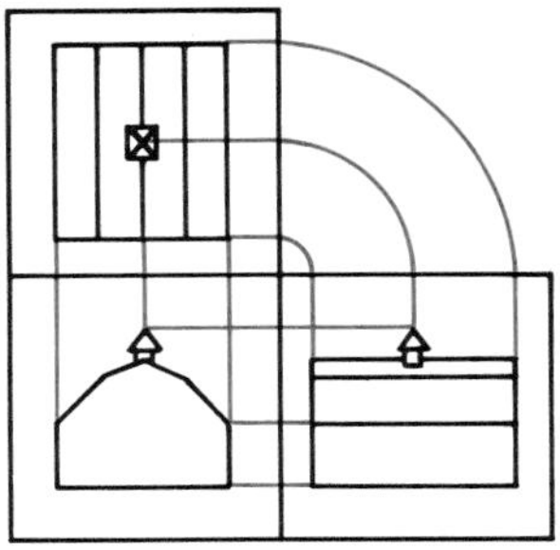

Another method of representing a three-dimensional object by means of a two-dimensional image is an *isometric drawing*. In this type of representation the object is viewed at an angle that allows simultaneous vision of the top, front, and one side. Unlike a photograph, however, an isometric drawing does not show perspective. Rather, congruent sides appear to be congruent. Because we are accustomed to seeing objects in perspective, an isometric representation often appears distorted to us. The following figures illustrate the difference between a perspective drawing (left) and an isometric drawing (right).

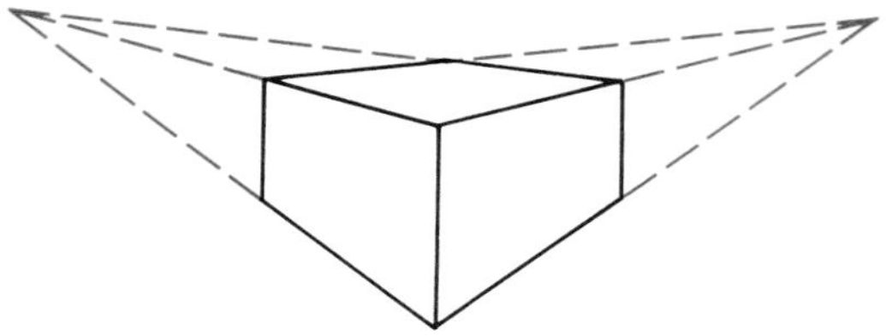

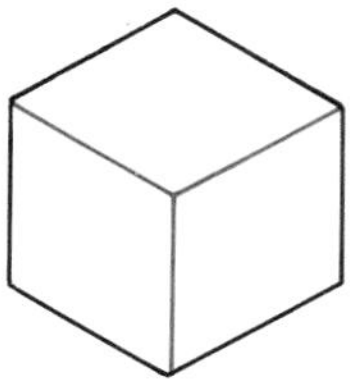

To illustrate how to make an isometric drawing, we shall make one of the solid whose orthographic projection is shown below.

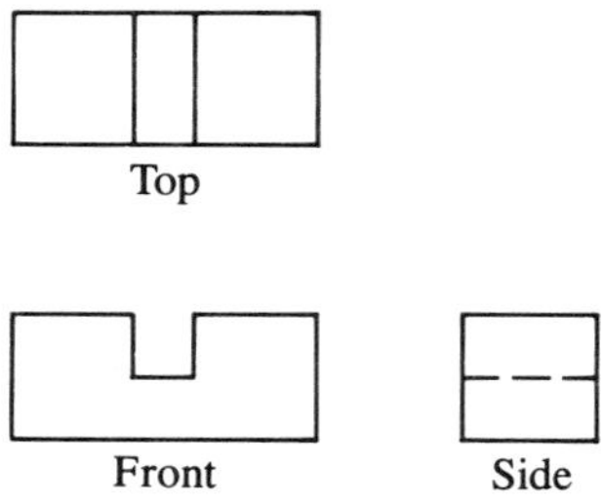

Begin by drawing three rays that intersect at their endpoints and make 120° angles with one another. These are the isometric axes. Mark off the intersection of the front and the side along one axis. Along the other two axes, mark off two adjacent edges of the top. By constructing congruent segments and by showing parallel edges as parallel edges in the drawing, you can finish the isometric drawing. The figures that follow suggest the procedure.

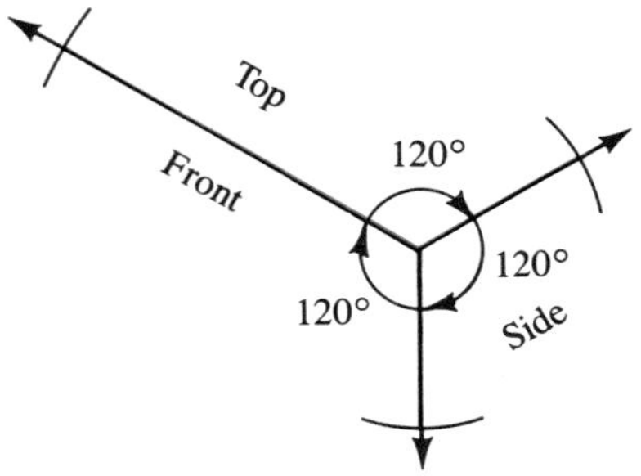

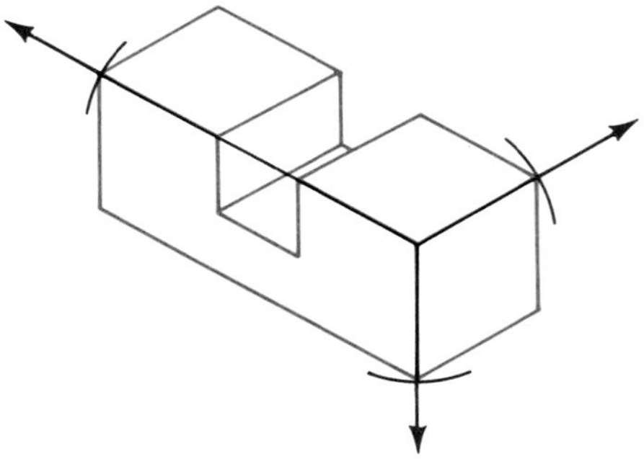

As shown in the preceding figures, a compass, a protractor, and a ruler are very useful tools for making isometric drawings. In particular, the compass can be used to copy congruent edges.

Exercises

Trace each figure. Then make an orthographic projection of the figure.

1.

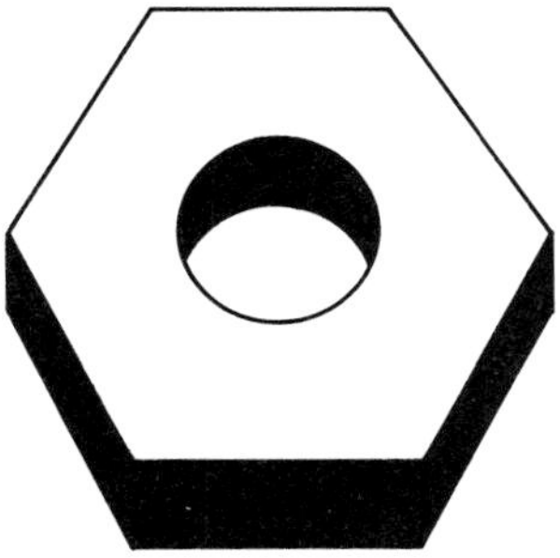

2.

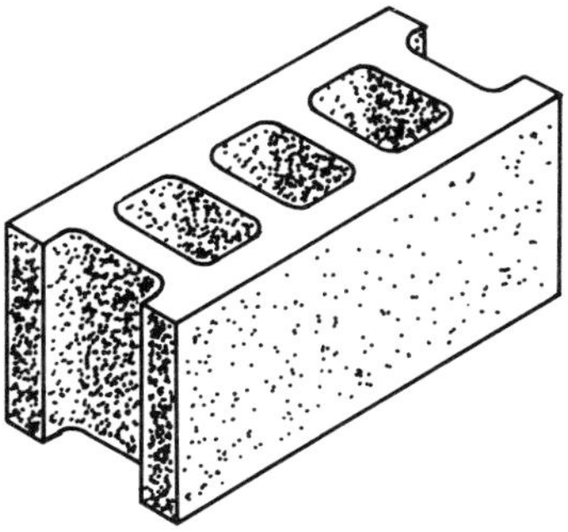

Trace each given orthographic projection. Then make an isometric drawing of each solid.

3.

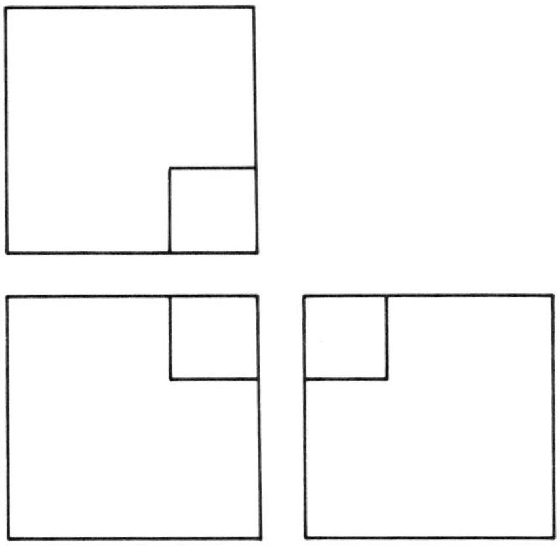

4.

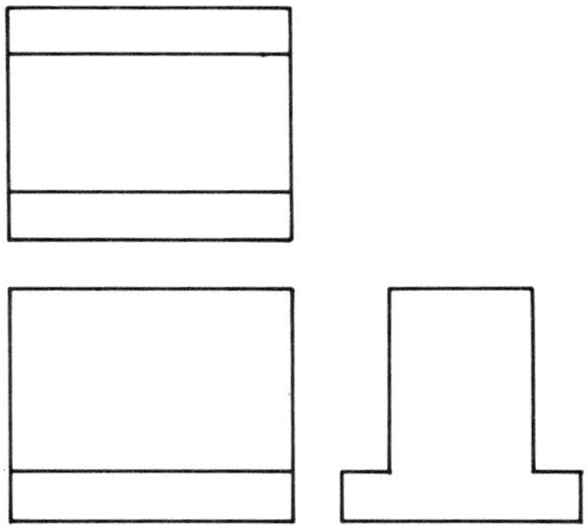

Glossary

acute angle: An angle with measure between 0 and 90. (p. 12)
acute triangle: A triangle having three acute angles. (p. 72)
adjacent angles: Two angles in a plane that have a common vertex and a common side but no common interior points. (p. 12)
alternate interior angles: Two nonadjacent interior angles on opposite sides of a transversal. ∠s 1 and 2 are alternate interior angles. (p. 56)

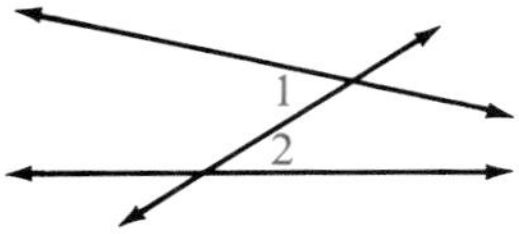

altitude to a base of a parallelogram: Any segment perpendicular to the line containing the base from any point on the opposite side. The length of an altitude is called the *height*. (p. 380)
altitude of a cone: *See* cone.
altitude of a prism: *See* prism.
altitude of a pyramid: *See* pyramid.
altitude of a right cylinder: *See* cylinder.
altitude of a trapezoid: Any segment perpendicular to a line containing the base from a point on the opposite base. (p. 391)
altitude of a triangle: The perpendicular segment from a vertex to the line containing the opposite side. $\overline{BD}$ and $\overline{AD}$ are altitudes. (pp. 137, 138)

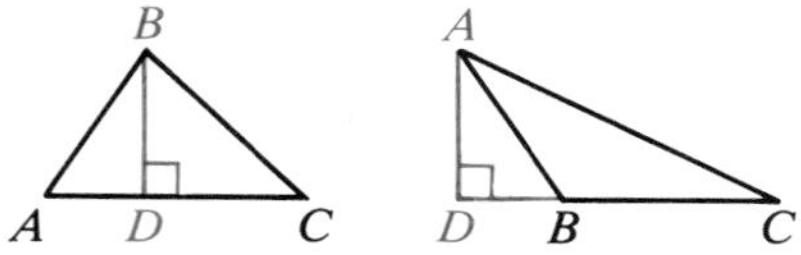

angle: A figure formed by two rays that have the same endpoint. The two rays are called the *sides* of the angle. Their common endpoint is the *vertex*. (p. 10)
angle of depression: When a point B is viewed from a higher point A, as shown by the diagram in the next column, ∠1 is the angle of depression. (p. 277)
angle of elevation: When a point A is viewed from a lower point B, ∠2 is the angle of elevation. (p. 277)

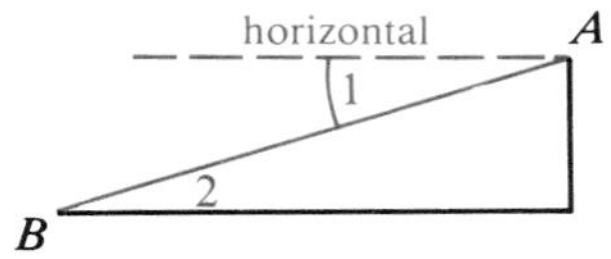

apothem of a regular polygon: The (perpendicular) distance from the center of the polygon to a side. (p. 396)
arc: An unbroken part of a circle. (p. 301)
area of a circle: The limiting number approached by the areas of a sequence of inscribed regular polygons. For radius r, $A = \pi r^2$. (pp. 402, 403)
auxiliary line: A line (or ray or segment) added to a diagram to help in a proof. (p. 73)
axes: Usually the horizontal and vertical axes of a coordinate system. (p. 469)
axioms: Assumptions accepted without proof. (p. 7)

base of a parallelogram: Any side of the figure can be considered its base. The term *base* may refer to the line segment or its length. (p. 380)
base of a pyramid: *See* pyramid.
bases of a prism: *See* prism.
bisector of an angle: A ray that divides the angle into two congruent adjacent angles. (p. 13)
bisector of a segment: A line, segment, ray, or plane that intersects the segment at its midpoint. (p. 7)

center of a regular polygon: The center of the circumscribed circle. (p. 396)
central angle of a circle: An angle with its vertex at the center of the circle. (p. 300)
central angle of a regular polygon: An angle formed by two radii drawn to consecutive vertices. (p. 396)

chord of a circle: A segment that joins two points on the circle. (p. 293)

circle: The set of points in a plane that are a given distance from a given point in the plane. The given point is the *center,* and the given distance is the *radius.* (p. 293)

circumference of a circle: The limiting number approached by the perimeters of a sequence of regular inscribed polygons. For radius r, $C = 2\pi r$. (pp. 402, 403)

circumscribed circle: A circle is circumscribed about a polygon when each vertex of the polygon lies on the circle. (p. 294)

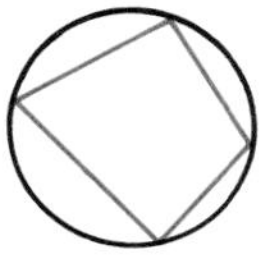

circumscribed polygon: Each side of the polygon is tangent to a circle. The polygon is said to be *circumscribed* about the circle. The circle is *inscribed* in the polygon. (p. 297)

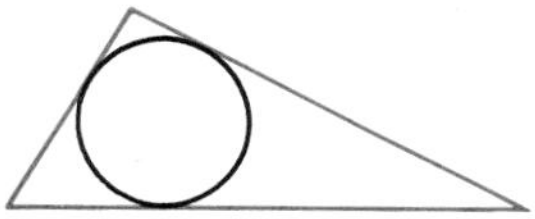

collinear points: Points all in one line. (p. 2)

common tangent: A line that is tangent to each of two coplanar circles. (p. 297)

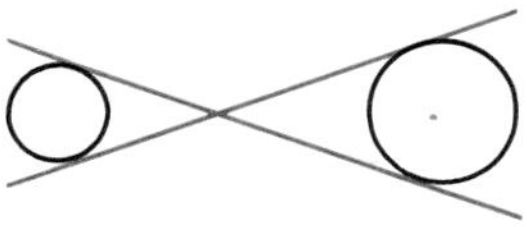

Common *internal* tangents

Common *external* tangents

complementary angles: Two angles whose measures have the sum 90. (p. 30)

concentric circles: Circles that lie in the same plane and have the same center. (p. 293)

conclusion: *See* if-then statement.

concurrent lines: Two or more lines that intersect in one point. (p. 344)

conditional (conditional statement): A statement that is, or may be expressed as, an if-then statement. (p. 87)

cone: The diagrams illustrate a right cone and an oblique cone. Both have circular bases and a vertex V. In the right cone, h is the *altitude* and l is the *slant height.* (p. 437)

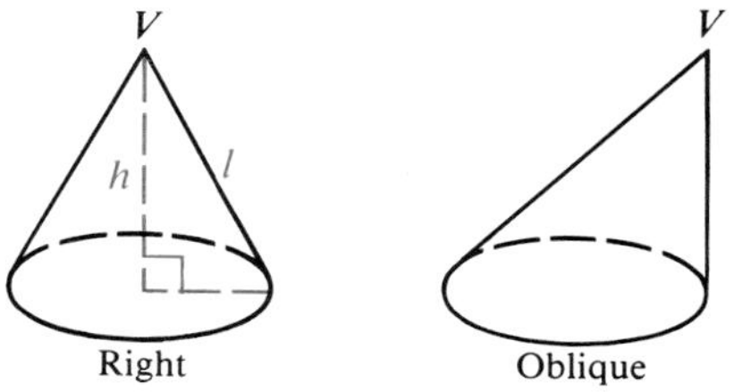

congruent angles: Angles that have equal measures. (p. 12)

congruent circles: Circles that have congruent radii. (p. 293)

congruent figures: Figures having the same size and shape. (p. 105)

congruent segments: Segments that have equal lengths. (p. 7)

contraction: *See* dilation.

contrapositive of a conditional: The contrapositive of the statement *If p, then q* is *If not q, then not p.* (p. 92)

converse: A statement in which the hypothesis and conclusion of a given statement are interchanged. (p. 87)

convex polygon: A polygon such that no line containing a side of the polygon contains a point in the interior of the polygon. (p. 80)

coordinate plane: The plane of the x-axis and y-axis. (p. 469)

coordinates: For point $P(3, 2)$ the coordinates are 3 and 2. (3, 2) is called an *ordered pair.* The *x-coordinate* is 3 and the *y-coordinate* is 2. Point P is the *graph* of (3, 2). (p. 469)

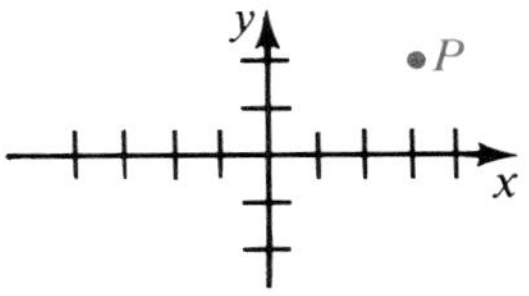

coplanar points: Points all in one plane. (p. 2)

corollary of a theorem: A statement that can be easily proved by applying the theorem. (p. 73)

corresponding angles: Two angles in corresponding positions relative to two lines. $\angle$s 2 and 6 are corresponding angles. (p. 56)

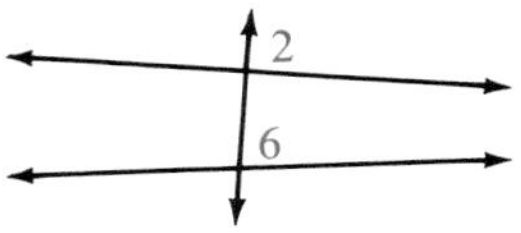

cosine ratio:

cosine of $\angle A = \frac{AC}{AB}$

or $\cos A = \frac{\text{adjacent}}{\text{hypotenuse}}$

(p. 273)

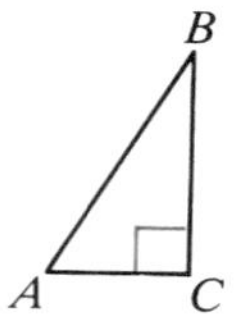

counterexample: An example used to prove an if-then statement false. For that counterexample, the hypothesis is true and the conclusion is false. (p. 87)

cube: A rectangular solid with square faces. (p. 424)

cylinder: Diagram (A) shows a *right cylinder;* (B) shows an *oblique cylinder.* In a right cylinder, the segment joining the centers of the circular bases is an *altitude.* The length of an altitude is the *height* (h) of the cylinder. A radius of a base is a *radius* r of the cylinder. (p. 437)

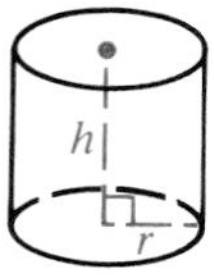

Right
(A)

Oblique
(B)

decagon: A 10-sided polygon. (p. 80)

deductive reasoning: Proving statements by reasoning from accepted statements. (p. 24)

diagonal of a polygon: A segment joining two nonconsecutive vertices of a polygon. (p. 80)

diameter: A chord that passes through the center of a circle. (p. 293)

dilation: A dilation with center O and nonzero scale factor k maps any point P to a point P' determined as follows: (1) If $k > 0$, P' lies on $\overrightarrow{OP}$ and $OP' = k \cdot OP$. (2) If $k < 0$, P' lies on the ray opposite $\overrightarrow{OP}$ and $OP' = |k| \cdot OP$. (3) The center O is its own image. If $|k| > 1$, the dilation is an *expansion;* if $|k| < 1$, it is a *contraction.* (p. 530)

distance between two points: The absolute value of the difference of the coordinates of two points on a number line. (p. 6)

distance from a point to a line (or plane): The length of the perpendicular segment from the point to the line (or plane). (p. 139)

equiangular triangle: A triangle with all angles congruent. (p. 72)

equilateral triangle: A triangle with all sides congruent. (p. 72)

equivalence relation: A relation that is reflexive, symmetric, and transitive. (p. 22)

expansion: *See* dilation.

exterior angle of a triangle: The angle formed when one side of the triangle is extended. $\angle CAD$ is an exterior angle of $\triangle ABC$. The term is also applied to other polygons. (p. 74)

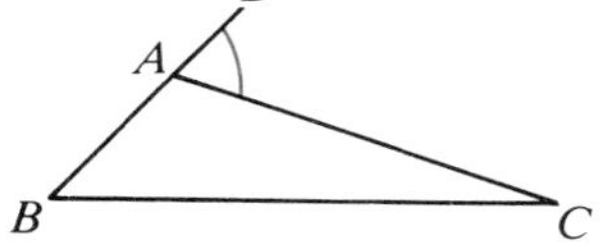

geometric mean: If r, s, and t are positive numbers with $\frac{r}{s} = \frac{s}{t}$, then s is the geometric mean between r and t. (p. 247)

glide reflection: A transformation in which every point P is mapped to a point P'' by this procedure: (1) a glide maps P to P'; (2) a reflection in a line parallel to the glide line maps P' to P''. (p. 523)

golden rectangle: A rectangle such that its length l and width w satisfy the equation $\frac{l}{w} = \frac{l + w}{l}$. (pp. 216, 231)

graph of a point: *See* coordinates.

half-turn: A rotation about O through 180°. (p. 526)

height of a prism: *See* prism.

height of a pyramid: *See* pyramid.

height of a right cylinder: *See* cylinder.

hexagon: A 6-sided polygon. (p. 80)

hypotenuse: In a right triangle the side opposite the right angle. The other two sides are called *legs.* (p. 130)

hypothesis: *See* if-then statement.

identity transformation: The mapping that maps every point to itself. (p. 540)

if-then statement: A statement whose basic form is "If p, then q." "p" is the *hypothesis* and "q" is the *conclusion.* (p. 86)

image: *See* mapping.

indirect proof: A proof that begins by assuming temporarily that the conclusion is not true. (p. 179)

inductive reasoning: A kind of reasoning in which the conclusion is based on several past observations. (pp. 147–148)

inscribed angle: An angle whose vertex is on a circle and whose sides contain chords of the circle. (p. 311)

inscribed polygon: Each vertex of the polygon lies on the circle. (p. 294)

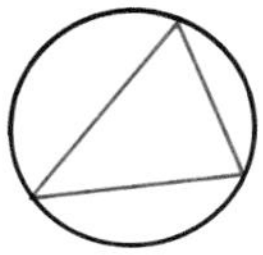

intersection of two figures: The set of points that are in both figures. (p. 2)

inverse of a conditional: The inverse of the statement *If p, then q* is the statement *If not p, then not q.* (p. 92)

inverse of a transformation: The inverse of T is the transformation S such that $S \circ T = I$. (p. 541)

isometry: A transformation that preserves distance. (p. 512)

isosceles triangle: A triangle having at least two sides congruent. (p. 72)

lateral area of a prism: The sum of the areas of its lateral faces. (p. 424)

lateral edges of a prism: *See* prism.

lateral edges of a pyramid: *See* pyramid.

lateral faces of a prism: *See* prism.

lateral faces of a pyramid: *See* pyramid.

legs of an isosceles triangle: The two congruent sides. The third side is the *base*. (p. 124)

length of a segment: The distance between its endpoints. (p. 6)

line symmetry: A figure has line symmetry if there is a symmetry line k such that the reflection R_k maps the figure onto itself. (p. 543)

linear equation: An equation whose graph is a line. (p. 490)

locus: A figure that is the set of all points, and only those points, that satisfy one or more conditions. (p. 358)

mapping: A correspondence between points. Each point P in a given set is *mapped* to some point P' in the same or a different set. P' is called the *image* of P, and P is called the *preimage* of P'. (p. 511)

median of a trapezoid: The segment that joins the midpoints of the legs. (p. 174)

median of a triangle: A segment from a vertex to the midpoint of the opposite side. (p. 137)

midpoint of a segment: The point that divides the segment into two congruent segments. (p. 7)

minor and major arcs: $\overparen{YZ}$ is a minor arc of $\odot O$. $\overparen{YXZ}$ is a major arc. The measure of a minor arc is the measure of its central angle, here $\angle YOZ$. The measure of a major arc is found by subtracting the measure of the minor arc from 360. (p. 301)

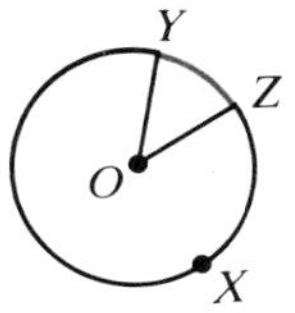

***n*-gon:** A polygon of n sides. (p. 80)

oblique cone: *See* cone.

oblique cylinder: *See* cylinder.

obtuse angle: An angle with measure between 90 and 180. (p. 12)

obtuse triangle: A triangle with one obtuse angle. (p. 72)

octagon: An 8-sided polygon. (p. 80)

one and only one: Exactly one. (p. 46)

opposite rays: Given three collinear points R, S, and T: If S is between R and T, then $\overrightarrow{SR}$ and $\overrightarrow{ST}$ are opposite rays. (p. 6)

ordered pair: *See* coordinates.

origin: The intersection point, denoted $O(0, 0)$, of two perpendicular number lines at their zero points. (p. 469)

parallel lines: Lines that do not intersect and are coplanar. (p. 55)

parallel planes: Planes that do not intersect. (p. 55)

parallelogram: A quadrilateral with both pairs of opposite sides parallel. (p. 159)

pentagon: A 5-sided polygon. (p. 80)

perimeter of a polygon: The sum of the lengths of its sides. (p. 402)

perpendicular bisector of a segment: A line (or ray or segment) that is perpendicular to the segment at its midpoint. (p. 138)

perpendicular line and plane: A line and a plane are perpendicular if and only if they intersect and the line is perpendicular to all lines in the plane that pass through the point of intersection. (p. 117)

perpendicular lines: Two lines that form right angles. (p. 36)

point symmetry: A figure has point symmetry if there is a symmetry point O such that the half-turn H_O maps the figure onto itself. (p. 544)

polygon: A plane figure formed by coplanar segments (sides) such that (1) each segment intersects exactly two other segments, one at each endpoint; and (2) no two points with a common endpoint are collinear. A *regular* polygon is both equiangular and equilateral. (pp. 80, 82)

postulates: Assumptions accepted without proof. (p. 7)

prism: The solids shown are *prisms*. The shaded faces are the *bases* (congruent polygons lying in parallel planes). The other faces are *lateral faces* and all are parallelograms. Adjacent lateral faces intersect in parallel segments called *lateral edges*. An *altitude* of a prism is a segment joining the two base planes and perpendicular to both. The length of an altitude is the *height* (h) of the prism. Figure (A), in which the lateral faces are rectangles, is called a *right prism*. Figure (B) is an *oblique prism*. (p. 423)

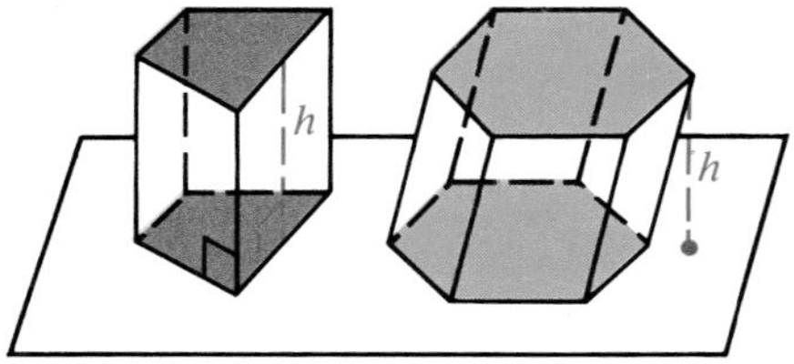

(A) (B)

proportion: An equation stating that two ratios are equal. The first and last terms are the *extremes;* the middle terms are the *means*. (p. 206)

protractor: An instrument for finding the measure in degrees of an angle. (p. 11)

pyramid: The diagram shows a pyramid. Point V is its *vertex*; the pentagon $ABCDE$ is its *base*. The five triangular faces meeting at V are *lateral faces;* they intersect in segments called *lateral edges*.

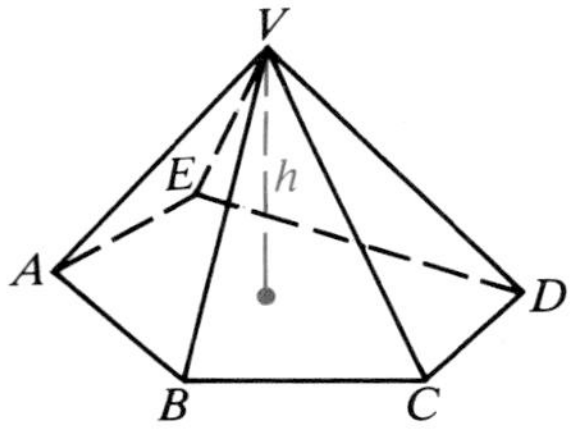

In a *regular pyramid,* the base is a regular polygon, all lateral edges are congruent, all lateral faces are congruent isosceles triangles, and the altitude meets the base at its center. The height of a lateral face is the *slant height* of the pyramid. The segment from the vertex perpendicular to the base is the *altitude,* and its length (h) is the *height* of the pyramid. (p. 430)

quadrants: The four regions into which the plane is divided by the coordinate axes. (p. 469)

quadrilateral: A 4-sided polygon. (p. 80)

radius of a regular polygon: The distance from the center to a vertex. (p. 396)

radius of a right cylinder: *See* cylinder.

ratio: The ratio of x to y ($y \neq 0$) is $\frac{x}{y}$. (p. 205)

ray: The ray AC ($\overrightarrow{AC}$) consists of segment $\overline{AC}$ and all other points P such that C is between A and P. The point named first, here A, is the *endpoint* of ray $\overrightarrow{AC}$. (p. 6)

rectangle: A quadrilateral with four right angles. (p. 168)

rectangular solid: A right rectangular prism. (p. 423) *See also* prism.

reflection: A transformation in which a line acts like a mirror, reflecting points to their images. A reflection in a line j maps every point P to a point P' such that: (1) If P is not on line j, then j is the perpendicular bisector of PP'. (2) If P is on line j, then $P' = P$. (p. 515)

regular pyramid: *See* pyramid.

remote interior angles: $\angle 1$ is an exterior angle of $\triangle ABC$, and $\angle$s B and C are remote interior angles with respect to $\angle 1$. (p. 74)

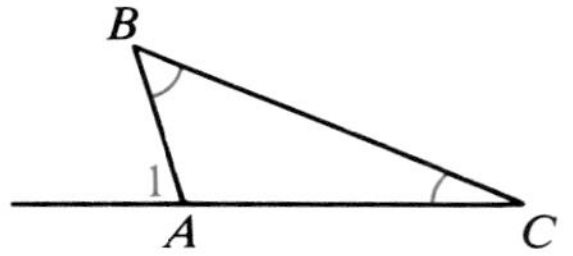

rhombus: A quadrilateral with four congruent sides. (p. 168)

right angle: An angle with measure 90. (p. 12)

right cone: *See* cone.

right cylinder: *See* cylinder.

right triangle: A triangle with one right angle. (p. 72)

rotation: A rotation about point O through x is a transformation such that: (1) If point P is different from O, then $OP' = OP$ and $m\angle POP' = x$. (2) If point P is the same as O, then $P' = O$. (p. 525)

rotational symmetry: A figure has rotational symmetry if there is a rotation that maps the figure onto itself. (p. 544)

same-side interior angles: Two interior angles on the same side of a transversal. (p. 56)

scale factor: For similar polygons, the ratio of the lengths of two corresponding sides. (p. 213)

scalene triangle: A triangle with no sides congruent. (p. 72)

secant of a circle: A line that contains a chord. (p. 293)

sector of a circle: A region bounded by two radii and an arc of the circle. (p. 406)

segment of a line: Two points on the line and all points between them. The two points are called the *endpoints* of the segment. (p. 6)

semicircles: The two arcs of a circle that are cut off by a diameter. The *measure* of a semicircle is 180. (p. 301)

sides of an angle: *See* angle.

sides of a triangle: *See* triangle.

similar figures: Figures that have the same shape but not necessarily the same size. (p. 212)

similar solids: Solids that have the same shape but not necessarily the same size. (p. 452)

sine ratio:

sine of $\angle A = \dfrac{BC}{AB}$

or $\sin A = \dfrac{\text{opposite}}{\text{hypotenuse}}$

(p. 266)

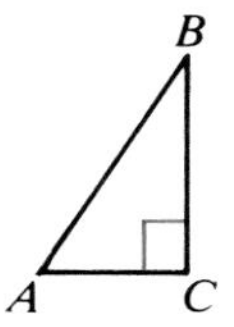

skew lines: Lines that do not intersect and are not coplanar. (p. 55)

slant height of a regular cone: *See* cone.

slant height of a regular pyramid: *See* pyramid.

slope of a line: The steepness of a nonvertical line, defined by

$$\frac{y_2 - y_1}{x_2 - x_1}, \quad x_1 \neq x_2,$$

where $P_1(x_1, y_1)$ and $P_2(x_2, y_2)$ are two points on the line. The slope of any horizontal line is zero. (p. 483)

space: The set of all points. (p. 2)

sphere: The set of points that are a given distance from a given point. (p. 294)

square: A quadrilateral with four right angles and four congruent sides. (p. 168)

straight angle: Its measure is 180. (p. 12)

supplementary angles: Two angles whose measures have the sum 180. (p. 30)

symmetry: A figure in the plane has symmetry if there is an isometry, other than the identity, that maps the figure onto itself. (p. 543)

tangent to a circle: A line in the plane of the circle that meets the circle in exactly one point, called the *point of tangency*. (p. 296)

tangent circles: Two circles that are coplanar and tangent to the same line at the same point. (p. 297)

tangent ratio:

tangent of $\angle A = \dfrac{BC}{AC}$

or $\tan A = \dfrac{\text{opposite}}{\text{adjacent}}$

(p. 266)

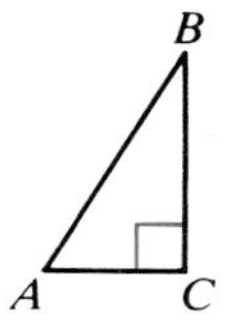

tessellation: A pattern in which congruent copies of a figure completely fill the plane without overlapping. (p. 545)

theorems: Statements that can be proved. (p. 23)

transformation: A one-to-one mapping from the whole plane to the whole plane. (p. 512)

translation: An isometry that glides all points the same distance. (p. 522)

transversal: A line that intersects two or more coplanar lines in different points. (p. 56)

trapezoid: A quadrilateral with exactly one pair of parallel sides. An *isosceles* trapezoid has congruent legs. (p. 173)

triangle: The figure formed by three segments joining three noncollinear points. Each of the three points is a *vertex* of the triangle and the segments are the *sides*. (p. 72)

vertex of an angle: *See* angle.

vertex of a pyramid: *See* pyramid.

vertex of a triangle: *See* triangle.

vertical angles: Two angles whose sides form two pairs of opposite rays. $\angle$s 1 and 2 are vertical angles, as are $\angle$s 3 and 4. (p. 31)

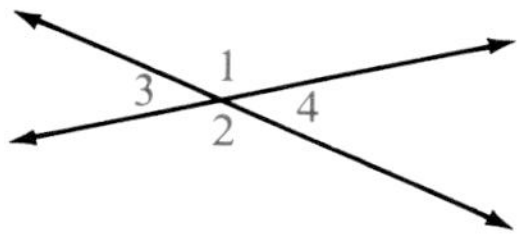

Index

Credits

Cover concept by Kirchoff/Wohlberg, Inc.

Cover: Sculpture by Morton C. Bradley, Jr., Arlington, MA 02174, © 1979; Photo by Gardiner Hutchinson.

Technical art by ANCO/Boston.

Illustrations by Jerry Malone.

Photographs

xii James H. Karales/Peter Arnold, Inc. **1** Barbara Burten. **5** Lou Jones. **6** Greig Cranna. **22** Malcolm Kirk/Peter Arnold, Inc. **35** Harvey Lloyd/Peter Arnold, Inc. **54** Alex S. MacLean/Landslides. **57** Grant Heilman. **71** Yoram Kahana/Peter Arnold, Inc.; © David Meunch 1984. **80** What's Up. **83** Grant Heilman. **92** Focus on Sports. **94** Paul Slaughter/The Image Bank. **104** John Running/Black Star. **116** Dick Luria/FPG; Tom Tracy/FPG; Michal Heron. **117** Thomas Russell. **123** Fred Zimmerman/The Phelps Agency; Stanley Rowin/The Picture Cube; Swanke, Hayden, Connell Architects. **124** Fred Morales/FPG. **129** Barbara Burten. **158** © 1981 Harold and Erica Van Pelt, Photographers L.A. **161** Mike Mazzaschi/Stock, Boston. **168** Ed Brunette. **173** Cary Wolinsky/Stock, Boston. **187** D. Bartruff/FPG. **204** Aquatic Assemblage by Joey Skaggs; photo by Irene Stern. **206** Poster by Lee Boltin; photo by Ed Brunette. **208** Martin T. Rotker/Taurus Photos. **212** Thomas Russell. **217** Eric Kroll/Taurus Photos. **231** The Dallas Museum of Art, Foundation for the Arts Collection, gift of the James H. and Lillian Clark Foundation; Gail Page. **246** Arley Rinehart and Associates. **265** Ed Brunette. **272** Mimi Levine/Folio, Inc.; John Running/Stock, Boston; Eric Anderson/Stock, Boston; Ken Lax/The Stock Shop. **276** Dominique Berretty/Black Star. **277** Karl Hentz/The Image Bank. **280** Deck House, Inc.; Milton and Joan Mann/Cameramann, International. **281** Charles F. Norton/Deck House, Inc. **292** Peter Garfield/Folio, Inc. **297** Jonathan Barkan/The Picture Cube. **302** Ed Brunette. **305** NASA. **326** Michael Rothwell/FPG. **334** Michael James. **347** Martine Franck/Magnum. **357** NASA; Ken Lax/The Stock Shop. **358** Jon Riley/The Stock Shop. **366** Animals, Animals/Brian Milne. **378** Grant Heilman. **396** © 1983 Wayne Sorce. **401** Barbara Burten; Freddie Lieberman/Folio, Inc.; Eric Kroll/Taurus Photos. **404** Grant Heilman. **406** Lou Jones. **415** NASA. **422** Jon Riley/Folio, Inc. **434** Christopher Crowley/Tom Stack & Associates. **443** Van Dusen Corporation. **447** Judy Gibbs. **450** Owen Franken/Stock, Boston. **451** City of Montreal, Archives Division. **452** Vance Henry/Taurus Photos. **455** Barbara Burten. **457** M. Timothy O'Keefe/Tom Stack & Associates. **468** Gabe Palmer/Palmer/Kane, Inc. **473** B. Staley/FPG. **482** Story Litchfield/Stock, Boston. **486** Ken Lax/The Stock Shop/Medichrome. **487** Harold Sund/The Image Bank. **497** Ellis Herwig/The Picture Cube; Tom Tracy/The Stock Shop; Sepp Seitz/Woodfin Camp and Associates. **510** James H. Carmichael/The Image Bank. **515** Larry Sutton/FPG. **518** Robert Herko/The Image Bank. **521** Focus on Sports. **525** Braniff/FPG. **531** Eric Neurath/Stock, Boston. **546** Animals, Animals/Carl Roessler; Clyde H. Smith/Peter Arnold, Inc.; Jack Baker/The Image Bank; David Stone.

Selected Answers

The answers in this answer section have been written in abbreviated form. Check with your teacher regarding the form in which you should give answers in your own work.

Chapter 1

WRITTEN EXERCISES, PAGES 4–5

1. T **3.** T **5.** T **7.** T **9.** F **13.** No **15.** *WRS, VST* **17.** $\overleftrightarrow{VS}$ and *WRS,* for example **21.** $\overleftrightarrow{ER}$, $\overleftrightarrow{EA}$, $\overleftrightarrow{EH}$ **23. a.** No **b.** Yes **25.** Answers may vary. **a.** *EAB, HDC* **b.** *EAD, FBC*

WRITTEN EXERCISES, PAGES 8–10

1. 15 **3.** 4.5 **5.** T **7.** T **9.** F **11.** F **13.** T **15.** T **17.** *B* **19.** *G, C* **21.** *D* **23.** -1 **29. a.** 5 **b.** 10 **c.** 10 **d.** 6 **31.** $x = 6$ **33.** $x = 3$ **35.** $y = 6$ **37.** $z = 8$; $GE = 10$, $EH = 10$; yes **39. a.** 1 **b.** 2 **41.** $\overline{HN}$ **43.** *M* **45.** $\overline{GT}$ **47.** $-2 \le x \le 8$ **49. a.** 37°C, −40°C **b.** $F = \frac{9}{5}C + 32$ **c.** 212°F, 14°F

WRITTEN EXERCISES, PAGES 14–15

1. *E* Answers may vary in Exs. 3–7. **3.** $\angle ELS$ **5.** $\angle AEL$ **7.** $\angle 7$ **9.** acute **11.** rt. **13.** straight **15.** *LKP* **17.** $\overrightarrow{KN}$, $\angle LKP$ **19.** 180 **27.** $\angle JZK$, $\angle LZK$, $\angle JZM$, $\angle LZM$ **29.** $m\angle JKL = 100$, $m\angle KLM = 120$, $m\angle LMJ = 100$ **33.** 31 **35.** $x = 30$ **37.** $x = 6$ **39. a.** 6, 10 **b.** 15 **c.** $\frac{n(n-1)}{2}$

WRITTEN EXERCISES, PAGES 20–22

1. Given; Add. Prop. =; Div. Prop. = **3.** Given; Mult. Prop. =; Subtr. (or Add.) Prop. = **5.** Given; Mult. Prop. =; Dist. Prop.; Add. Prop. =; Div. Prop. = **7.** 1. ∠ Add. Post. 2. ∠ Add. Post. 3. $m\angle AOD = m\angle 1 + m\angle 2 + m\angle 3$; Subst. Prop. **9.** 1. Given 2. $OW + WN$; Seg. Add. Post. 3. $DO + OW = OW + WN$ 4. Ref. Prop. 5. $DO = WN$; Subtr. Prop. = **11.** 1. $m\angle 1 = m\angle 2$; $m\angle 3 = m\angle 4$ (Given) 2. $m\angle 1 + m\angle 3 = m\angle 2 + m\angle 4$ (Add. Prop. =) 3. $m\angle SRT = m\angle 1 + m\angle 3$; $m\angle STR = m\angle 2 + m\angle 4$ (∠ Add. Post.) 4. $m\angle SRT = m\angle STR$ (Subst. Prop.) **13.** 1. $RQ = TP$; $ZQ = ZP$ (Given) 2. $RQ = RZ + ZQ$; $TP = TZ + ZP$ (Seg. Add. Post.) 3. $RZ + ZQ = TZ + ZP$ (Subst. Prop.) 4. $RZ = TZ$ (Subtr. Prop. =) **15.** *b*

WRITTEN EXERCISES, PAGES 26–28

1. Def. of midpt. **3.** ∠ Bis. Th. **5.** Def. of midpt. **7.** ∠ Add. Post. **9.** Def. ∠ bis. **11.** 75 **13.** 8.5 **15.** −20 **17.** 38 **19. a.** $\overline{LM} \cong \overline{MK}$, $\overline{GN} \cong \overline{NH}$ **b.** $LK = GH$ **21.** $AC = DB$ **23. a.** 12 **b.** 6 **c.** 22 **d.** 28 **25.** 1. Given 2. Ruler Post. 3. Given 4. Def. of midpt. 5. Subst. Prop. 6. $a + b$; Add. Prop. = 7. Div. Prop. = **27. a.** $x = 45$ **b.** $0 < x < 45$ **c.** $45 < x < 90$ **29.** 1. *M* is midpt. of $\overline{LK}$; *N* is midpt. of $\overline{GH}$ (Given) 2. $MK = \frac{1}{2}LK$; $NH = \frac{1}{2}GH$ (Midpt. Th.) 3. $LK = GH$ (Given) 4. $\frac{1}{2}LK = \frac{1}{2}GH$ (Mult. Prop. =) 5. $MK = NH$ (Subst. Prop.) **31. a.** $2x$ **b.** $3x$ **c.** $8x$ **d.** $3x$

WRITTEN EXERCISES, PAGES 32–34

1. 35, 125 **3.** 17.5, 107.5 **5.** 45, 45 **7.** $\angle LAE$ **9.** $\angle LAN$, $\angle NAK$; $\angle BAE$, $\angle EAN$; $\angle LAB$, $\angle BAK$; $\angle LAB$, $\angle LAN$; $\angle NAK$, $\angle KAB$ **11.** $\angle LAB$, $\angle BAE$; $\angle BAE$, $\angle NAK$ **13.** 35 **15.** 25 **17.** 60 **19.** $x = 25$ **21.** $x = 10$ or $x = -10$ **23.** $x = 65$, $m\angle A = 130$, $m\angle B = 50$ **25.** $y = 17$, $m\angle C = 56$, $m\angle D = 34$ **27.** $180 - x = 2x$; 60, 120 **29.** $90 - x = 2x - 6$; 32, 58 **31.** 72, 18, 108 **33.** 1. Th. 1-3 2. Given 3. Th. 1-3 4. $\angle 1 \cong \angle 4$ **35.** $x = 47$, $y = 75$ **37.** $x = 65$, $y = 30$ **39.** $90 - x = \frac{1}{2}(180 - x)$ has no solution except $x = 0$

WRITTEN EXERCISES, PAGES 37-39

1. Def. $\perp$ lines **3.** Th. 1-6 **5.** Def. rt. $\angle$ **7.** Def. $\perp$ lines **9.** 1. $\angle$ Add. Post. 2. Given 3. Subst. Prop. 4. $m\angle 1 = 90$ 5. Def. rt. $\angle$ 6. $a \perp b$; Def. $\perp$ lines **11.** 1. $\overline{SW} \perp \overline{RT}$ (Given) 2. $m\angle 1 = m\angle 2$ (Th. 1-4) **13.** $90 + x$ **15.** $x + y$ **17.** Yes **19.** No **21.** No **23.** Yes **25.** $\angle RSX$ and $\angle YST$ are comp. **27.** $\overrightarrow{AD} \perp \overleftrightarrow{AC}$, $\overrightarrow{CE} \perp \overleftrightarrow{AC}$ **29.** 1. $\angle 1$ and $\angle 2$ are comp. $\angle$s (Given) 2. $m\angle 1 + m\angle 2 = 90$ (Def. comp. $\angle$s) 3. $m\angle 1 + m\angle 2 = m\angle AOC$ ($\angle$ Add. Post.) 4. $m\angle AOC = 90$ (Subst. Prop.) 5. $\angle AOC$ is a rt. $\angle$ (Def. rt. $\angle$) 6. $\overleftrightarrow{AO} \perp \overleftrightarrow{CO}$ (Def. $\perp$ lines)

WRITTEN EXERCISES, PAGES 43-45

1. a. $\angle 1$ **b.** $\angle 4$ **c.** $\angle$ Add. Post. **d.** Th. 1-7 **3.** Seg. Add. Post. **5.** Add. Prop. = **7.** Def. $\angle$ bis. **9.** Def. $\perp$ lines, rt. $\angle$ **11.** Def. comp. $\angle$s **13.** Midpt. Th. **15.** Th. 1-4 **17.** Def. rt. $\angle$, $\perp$ lines **19. a.** 1. Given 2. Th. 1-6 3. Given 4. Th. 1-8 **b.** Use Th. 1-7 **21.** $\angle 2 \cong \angle 3$ (Given); $\angle 1$ and $\angle 2$ are supp.; $\angle 3$ and $\angle 4$ are supp. ($\angle$ Add. Post., Def. supp. $\angle$s); $\angle 1 \cong \angle 4$ (Th. 1-7) **23.** $\angle 1$ and $\angle 2$ are comp., $\angle 3$ and $\angle 4$ are comp. (Given); $m\angle 1 + m\angle 2 = 90$, $m\angle 3 + m\angle 4 = 90$ (Def. comp. $\angle$s); $m\angle 1 + m\angle 2 = m\angle 3 + m\angle 4$ (Subst. Prop.); $m\angle 2 = m\angle 4$ (Given); $m\angle 1 = m\angle 3$ (Subtr. Prop. =) **25. a.** $\angle 3$ **b.** $\angle 4$ **c.** $\angle 7$ **d.** $\angle 8$ **27.** $\angle 1 \cong \angle 2$ (Th. 1-3); $\angle 2 \cong \angle 3$ (Given); $\angle 3 \cong \angle 4$ (Th. 1-3); $\angle 4 \cong \angle 5$ (Given); $\angle 1 \cong \angle 5$ (Trans. Prop. used several times); $m\angle 5 + m\angle 6 = 180$ ($\angle$ Add. Post.); $m\angle 1 + m\angle 6 = 180$ (Subst. Prop.); $\angle 1$ is supp. to $\angle 6$ (Def. supp. $\angle$s) **29.** $\overrightarrow{OC} \perp \overleftrightarrow{AE}$ (Given); $\angle AOC \cong \angle EOC$ (Th. 1-4); $m\angle AOC = m\angle 1 + m\angle 3$, $m\angle EOC = m\angle 2 + m\angle 4$ ($\angle$ Add. Post.); $m\angle 1 + m\angle 3 = m\angle 2 + m\angle 4$ (Subst. Prop.); $\overrightarrow{OC}$ bis. $\angle BOD$ (Given); $m\angle 1 = m\angle 2$ (Def. $\angle$ bis.); $m\angle 3 = m\angle 4$ (Subtr. Prop. =) **31.** $\overrightarrow{QX}$ bis. $\angle PQR$ (Given); $\angle PQX \cong \angle RQX$ (Def. $\angle$ bis.); $\angle PQY$ is supp. to $\angle PQX$, $\angle RQY$ is supp. to $\angle RQX$ ($\angle$ Add. Post., Def. supp. $\angle$s); $\angle PQY \cong \angle RQY$ (Th. 1-7)

WRITTEN EXERCISES, PAGES 48-49

1. always **3.** sometimes **5.** sometimes **7.** always **9.** always **11.** never **13.** An angle has a bisector. An angle has no more than one bisector. **15. a.** Post. 7 **b.** Post. 8 **c.** Post. 6 **d.** Post. 8 **17. a.** 3 **b.** 6 **c.** 10 **d.** 15 **e.** 21 **f.** $\frac{n(n-1)}{2}$

CHAPTER REVIEW, PAGES 50-51

4. U, V **5. a.** 3, 3 **b.** $\cong$ **6.** $x = 9$ **7.** $\angle ADC$, $\angle 1$, $\angle 2$; $\angle 1$ and $\angle 2$ **8. a.** 92 **b.** $\angle$ Add. Post. **c.** obtuse **9.** $x = 7$ **10.** Trans. Prop. **11.** Add. Prop. = **12.** Subst. Prop. **13.** Midpt. Th. **14.** Def. $\angle$ bis. **15.** $\angle$ Bis. Th. **16.** 45 **17.** 90; Th. 1-3 **18.** Answers may vary. $\angle 1$ and $\angle 2$; $\angle PXQ$ and $\angle QXT$ **19.** Th. 1-4 **20.** Th. 1-5 **21.** $t = 14$ **22.** $\overleftrightarrow{GI} \perp \overleftrightarrow{GF}$ **23.** $\angle 2 \cong \angle 4$ **24.** $\angle 4 \cong \angle 6$, $\angle 6$ supp. to $\angle 3$, $\angle 4$ supp. to $\angle 1$. **25.** F **26.** F **27.** F **28.** T

ALGEBRA REVIEW, PAGE 53

1. $a = 6$ **3.** $x = 8$ **5.** $x = 60$ **7.** $n = 18$ **9.** $k = 0$ **11.** $m = -2.5$ **13.** $x = 5, y = 4$ **15.** $x = -10, y = -4$ **17.** $x = 9, y = 1$ **19.** $x = 7, y = -6$ **21.** $x = 3, y = -5$ **23.** $x = 1, y = -\frac{5}{4}$ **25.** $x = -10, y = -8$ **27.** $x = 5, y = -6$ **29.** $x = -14, y = 5$

Chapter 2

WRITTEN EXERCISES, PAGES 58-59

1. alt. int. **3.** s.-s. int. **5.** corr. **7.** $\overleftrightarrow{PQ}$ and $\overleftrightarrow{SR}$; $\overleftrightarrow{SQ}$ **9.** $\overleftrightarrow{PQ}$ and $\overleftrightarrow{SR}$; $\overleftrightarrow{PS}$ **11.** Answers may vary. $\overleftrightarrow{SK}$ and $\overleftrightarrow{AN}$; $\overleftrightarrow{RN}$ **13.** corr. **15.** alt. int. **17.** s.-s. int. **21.** $\angle$s are $\cong$ **25.** $\overleftrightarrow{GH}$, $\overleftrightarrow{KJ}$, $\overleftrightarrow{ED}$ **27.** Answers may vary. DCI, KLG **29.** 4 **31.** always **33.** sometimes **35.** sometimes **37.** sometimes

WRITTEN EXERCISES, PAGES 62-64

1. $\angle 3$, $\angle 6$, $\angle 8$ **3.** $\angle 2$, $\angle 5$, $\angle 7$, $\angle 10$, $\angle 12$, $\angle 13$, $\angle 15$ **5.** 50, 130 **7.** $x = 60, y = 61$ **9.** $x = 60, y = 18$ **11.** $x = 65, y = 105$ **13.** 1. Given 2. If lines $\parallel$, corr. $\angle$s $\cong$ 3. $\angle$ Add. Post. 4. Subst. Prop. 5. Def. supp. $\angle$s **15. a.** $m\angle DAB = 64$, $m\angle KAB = 32$, $m\angle DKA = 32$ **b.** more information needed **17.** 1. $k \parallel l$ (Given) 2. $\angle 5 \cong \angle 7$ (If lines $\parallel$, corr. $\angle$s $\cong$) 3. $\angle 2 \cong \angle 5$

(Vert. ∠s ≅) 4. $\angle 2 \cong \angle 7$ (Trans. Prop.) **19.** 1. $m\angle 4 + m\angle 2 = 180$ (∠ Add. Post.) 2. $k \parallel m$ (Given) 3. $m\angle 1 = m\angle 2$ (If lines ∥, alt. int. ∠s ≅) 4. $m\angle 4 + m\angle 1 = 180$ (Subst. Prop.) 5. $\angle 1$ is supp. to $\angle 4$ (Def. supp. ∠s)

WRITTEN EXERCISES, PAGES 68-70

1. $\overline{RU}, \overline{AT}$ **3.** none **5.** $\overline{AU}, \overline{NT}$ **7.** $\overline{RU}, \overline{AT}; \overline{RN}, \overline{OT}$ **9.** none **11.** 1. Given 2. Vert. ∠s are ≅ 3. Given 4. Trans. Prop. 5. Post. 11 **13.** $\overleftrightarrow{FD} \parallel \overleftrightarrow{AC}$ ($\angle FDC$ and $\angle DCA$ are supp.), $\overleftrightarrow{AG} \parallel \overleftrightarrow{CL}$ ($\angle BAG \cong \angle ACL$) **15.** $x = 90, y = 130$ **17.** $\overline{PQ} \parallel \overline{RS}$ **19.** 1. Trans. t cuts lines k and n; $\angle 1$ supp. to $\angle 2$ (Given) 2. $\angle 2$ is supp. $\angle 3$ (∠ Add. Post., Def. supp. ∠s) 3. $\angle 1 \cong \angle 3$ (supp.'s of same ∠ are ≅) 4. $k \parallel n$ (Theorem 2-5) **23.** 80 **25.** 1. $m\angle 1 = m\angle 4$; $\overline{BC} \parallel \overline{ED}$ (Given) 2. $m\angle 2 = m\angle 3$ (Th. 2-2) 3. $m\angle 1 + m\angle 2 = m\angle 3 + m\angle 4$ (Add. Prop. =) 4. $m\angle ABD = m\angle 1 + m\angle 2, m\angle BDF = m\angle 3 + m\angle 4$ (∠ Add. Post.) 5. $m\angle ABD = m\angle BDF$ (Subst. Prop.) 6. $\overline{AB} \parallel \overline{DF}$ (Theorem 2-5) **27.** $x = 50, y = 20$ **29.** $x = 12$

WRITTEN EXERCISES, PAGES 76-78

3. not possible **5.** 100 **7.** 115 **9.** 80 **11.** $x = 27$ **13.** $x = 30, y = 75$ **15.** $x = 130, y = 60$ **17.** $x = 90, y = 25$ **19.** Yes; $n = 5$ **21.** $60 < m\angle C < 120$ **23. a.** 22 **b.** 23 **25.** 1. $\angle A \cong \angle A$ (Refl. Prop.) 2. $\angle ABD \cong \angle AED$ (Given) 3. $\angle C \cong \angle F$ (Th. 2-11 Cor. 1) **27.** 360 **29.** 720 **31.** 1. $m\angle JGI = m\angle H + m\angle I$ (Th. 2-12) 2. $m\angle H = m\angle I$ (Given) 3. $m\angle JGI = 2\,m\angle H$ (Subst. Prop.) 4. $\frac{1}{2}m\angle JGI = m\angle H$ (Mult. Prop. =) 5. $\overrightarrow{GK}$ bis. $\angle JGI$ (Given) 6. $m\angle 1 = \frac{1}{2}m\angle JGI$ (∠ Bis. Th.) 7. $m\angle 1 = m\angle H$, $\angle 1 \cong \angle H$ (Subst. Prop.) 8. $\overline{GK} \parallel \overline{HI}$ (Post. 11) **33.** $x = 8, y = 12$ **35.** $\angle 7 \cong \angle 8$, $\angle 11 \cong \angle 12$ **37.** Sum of measures of ∠s is 360. Opposite ∠s are supp.

WRITTEN EXERCISES, PAGES 83-84

1. 540, 360 **3.** 360, 360 **5.** 1440, 360 **7.** 60, 45, 24, 180; 40, 24, 12, 15, 2; 140, 156, 168, 174, 172 **13.** not possible **15.** 10 **19. a.** Yes **21.** 108 **23. a.** $m\angle R = 60, m\angle S = m\angle T = 120$ **b.** $\overline{QR} \parallel \overline{TS}$ **25.** 180 **27. b.** Yes

WRITTEN EXERCISES, PAGES 89-91

1. a. $3x - 7 = 32$ **b.** $x = 13$ **c.** If $x = 13$, then $3x - 7 = 32$ **3. a.** You will. **b.** I'll try. **c.** If I'll try, then you will. **5. a.** $|x| = 0$ **b.** $x = 0$ **c.** If $x = 0$, then $|x| = 0$. **7. a.** If a person is an Olympic competitor, then the person is an athlete. True. **b.** If a person is an athlete, then the person is an Olympic competitor. False. **9. a.** If $x^2 = 0$, then $x = 0$. True. **b.** If $x = 0$, then $x^2 = 0$. True. **11. a.** If two integers are odd, then their product is odd. True. **b.** If the product of two integers is odd, then the integers are odd. True. **13. a.** If $-2x < 2$, then $x > -1$. True. **b.** If $x > -1$, then $-2x < 2$. True. **15. a.** If a polygon is regular, then it is equiangular. True. **b.** If a polygon is equiangular, then it is regular. False. **17.** If 2 ∠s are ≅, then their measures are =. If the measures of 2 ∠s are =, then the ∠s are ≅. **19.** If $ab > 0$, then a and b are both pos. or both neg. If a and b are both pos. or both neg., then $ab > 0$. **21.** If corr. ∠s formed by 2 lines and a transversal are ≅, then the 2 lines are ∥. True. **23. a.** If $a - c = b - d$, then $a = b$ and $c = d$. **b.** Answers may vary. $a = 4, b = 7, c = 2, d = 5$. **25.** True **27.** True **29.** False **31.** True **33.** necessary and sufficient **35.** sufficient **37.** necessary and sufficient

WRITTEN EXERCISES, PAGES 94-96

1. a. If $5n - 5 \neq 100$, then $n \neq 21$. **b.** If $5n - 5 = 100$, then $n = 21$. **c.** If $n \neq 21$, then $5n - 5 \neq 100$. **3. a.** If $x + 1$ is odd, then x is even. **b.** If $x + 1$ is not odd, then x is not even. **c.** If x is even, then $x + 1$ is odd. **9.** If $\angle 1 \cong \angle 2$, then $\angle 1$ and $\angle 2$ are vert. ∠s; F. If $\angle 1$ and $\angle 2$ are not vert. ∠s, then $\angle 1$ is not $\cong \angle 2$; F. If $\angle 1$ and $\angle 2$ are vert. ∠s, then $\angle 1 \cong \angle 2$; T. If $\angle 1$ is not $\cong \angle 2$, then $\angle 1$ and $\angle 2$ are not vert. ∠s; T. **11.** If $AM = MB$, then M is midpt. of $\overline{AB}$; F. If M is not the midpt. of $\overline{AB}$, then $AM \neq MB$; F. If M is the midpt. of $\overline{AB}$, then $AM = MB$; T. If $AM \neq MB$, then M is not midpt. of $\overline{AB}$; T. **13.** If a not neg., then $|a| = a$; T. If $|a| \neq a$, then a is neg.; T. If $|a| = a$, then a not neg.; T. If a is neg., then $|a| \neq a$; T. **15.** If $x^2 > y^2$, then $x > y$; F. If $x \leq y$, then $x^2 \leq y^2$; F. If $x > y$, then $x^2 > y^2$; F. If $x^2 \leq y^2$, then $x \leq y$; F. **17. a.** $\overline{AC} \cong \overline{BD}$ **b.** no conc. **c.** no conc. **d.** $EFGH$ is not equiangular. **19. a.** Stu loves geometry. **b.** no conc. **c.** no conc. **d.** George is not my student.

CHAPTER REVIEW, PAGES 99–100

1. 2 **2.** corr. **3.** alt. int. **4.** no **5.** 105, 105 **6.** $x = 12$ **7.** $y = 20$ **8.** $b \perp c$; Th. 2-4 **9.** $\overleftrightarrow{DF}$; $\angle A$ and $\angle ADF$ are supp. s.-s. int. $\angle$s **10.** $\overleftrightarrow{BH} \parallel \overleftrightarrow{CI}$; Th. 2-7 **11.** 80 **12.** $\cong$ alt. int. $\angle$s, $\cong$ corr. $\angle$s, supp. s.-s. int. $\angle$s, 2 lines $\parallel$ third line, 2 coplanar lines $\perp$ same line **13.** $x = 35$ **14.** 180 **15.** 100 **16.** $\angle 3 \cong \angle 6$ (supp.'s of $\cong$ $\angle$s); $\angle 2 \cong \angle 8$ (Th. 2-12 or Cor. 1 to Th. 2-11) **17. b.** 720 **c.** 360 **18.** 160 **19.** 15 **20.** 15 **21.** If a quad. is a square, then it is equilateral. **22.** Hyp.: a quad. is a square Conc.: it is equilateral **23.** If a quad. is equilateral, then it is a square; false. **24.** If 2 segments are $\cong$, then their lengths are $=$. If the lengths of 2 segments are $=$, then the segments are $\cong$. **25.** Toddie is an amphibian. **26.** nothing **27.** A dog isn't a toad. **28.** nothing

PREPARING FOR COLLEGE ENTRANCE EXAMS, PAGE 102

1. C **2.** C **3.** D **4.** C **5.** D **6.** E **7.** B **8.** B

CUMULATIVE REVIEW: CHAPTERS 1–2, PAGE 103

1. $\cong$ **3.** 45 **5.** $\perp$ **7.** a line **9.** acute **11.** exactly one plane **13.** $m\angle 1 = 75$; $m\angle 2 = 40$; $m\angle 3 = 40$; $m\angle 4 = 65$ **15.** Yes; $c \parallel d$ **17.** Yes; $a \parallel b$ **19.** 180; Th. 2-11 **21.** 360; Th. 2-14 **23.** $\angle 1$; If lines $\parallel$, corr. $\angle$s $\cong$

Chapter 3

WRITTEN EXERCISES, PAGES 108–109

1. $\angle T$ **3.** CA **5.** $\triangle ATC$ **7.** $\angle I$, $\angle G$, $\angle A$, $\angle T$ **9.** $\angle L \cong \angle F$, $\angle X \cong \angle N$, $\angle R \cong \angle E$, $\overline{LX} \cong \overline{FN}$, $\overline{XR} \cong \overline{NE}$, $\overline{LR} \cong \overline{FE}$ **11.** $\angle K$, corr. parts of $\cong$ $\triangle$s are $\cong$ **13.** TO, $\overline{TR}$ **15.** $\triangle RLA$ **17. a.** $\angle 4$, corr. parts of $\cong$ $\triangle$s are $\cong$ **b.** $\overline{PL} \parallel \overline{AR}$; alt. int. $\angle$s are $\cong$ **19.** $C(7, 2)$ **21.** $\triangle FDE$ **23.** $\triangle FED$ **25.** (8, 5), (4, 5) **27.** $MARO$ **29. b.** Yes **31.** $\triangle RST \cong \triangle XYZ$

WRITTEN EXERCISES, PAGES 113–115

1. $\triangle ABC \cong \triangle NPY$, ASA **3.** $\triangle ABC \cong \triangle CKA$, SSS **5.** no $\cong$ **7.** $\triangle ABC \cong \triangle PQC$, SAS **9.** $\triangle ABC \cong \triangle AGC$, ASA **11.** $\triangle ABC \cong \triangle BST$, ASA **13.** no $\cong$ **15.** $\triangle ABC \cong \triangle MNC$, ASA **17.** 1. Given 2. T 3. Given 4. $\overline{VT}$; Def. midpt. 5. UVT; vert. $\angle$s are $\cong$ 6. RSV, UTV; ASA **19.** 1. M is midpt. of $\overline{AB}$; M is midpt. of $\overline{CD}$ (Given) 2. $\overline{AM} \cong \overline{BM}$, $\overline{DM} \cong \overline{CM}$ (Def. midpt.) 3. $\angle AMD \cong \angle BMC$ (Vert. $\angle$s $\cong$) 4. $\triangle MAD \cong \triangle MBC$ (SAS) **21.** 1. Plane M bisects $\overline{AB}$ (Given) 2. $\overline{AO} \cong \overline{BO}$ (Def. bis.) 3. $\overline{PO} \perp \overline{AB}$ (Given) 4. $\angle POA \cong \angle POB$ ($\perp$ lines form $\cong$ adj. $\angle$s) 5. $\overline{PO} \cong \overline{PO}$ (Refl.) 6. $\triangle POA \cong \triangle POB$ (SAS) **23.** SAS

WRITTEN EXERCISES, PAGES 119–122

1. 1. Given 2. Given 3. Def. midpt. 4. Vert. $\angle$s $\cong$ 5. ASA 6. Corr. parts $\cong$ $\triangle$s 7. Def. midpt. **3.** 1. $\overline{WO} \cong \overline{ZO}$, $\overline{XO} \cong \overline{YO}$ (Given) 2. $\angle WOX \cong \angle ZOY$ (Vert. $\angle$s $\cong$) 3. $\triangle WOX \cong \triangle ZOY$ (SAS) 4. $\angle W \cong \angle Z$ (Corr. parts $\cong$ $\triangle$s) **5.** 1. $\overline{SK} \parallel \overline{NR}$, $\overline{SN} \parallel \overline{KR}$ (Given) 2. $\angle 1 \cong \angle 3$, $\angle 2 \cong \angle 4$ (If lines $\parallel$, alt. int. $\angle$s $\cong$) 3. $\overline{SR} \cong \overline{SR}$ (Refl.) 4. $\triangle SRK \cong \triangle RSN$ (ASA) 5. $\overline{SK} \cong \overline{NR}$, $\overline{SN} \cong \overline{KR}$ (Corr. parts $\cong$ $\triangle$s) **7. a.** If opp. sides of quad. $\cong$, then they are also $\parallel$. **b.** Yes **11.** $\angle Q \cong \angle S$ not needed. 1. $\overline{PQ} \cong \overline{PS}$, $\overline{QR} \cong \overline{SR}$ (Given) 2. $\overline{PR} \cong \overline{PR}$ (Refl.) 3. $\triangle QPR \cong \triangle SPR$ (SSS) 4. $\angle QPR \cong \angle SPR$ (Corr. parts $\cong$ $\triangle$s) **15.** 1 and 2 **17.** 1. $\overline{PA}$ and $\overline{QB} \perp$ plane X (Given) 2. $\overline{PA} \perp \overline{AB}$, $\overline{QB} \perp \overline{AB}$ (Def. of line $\perp$ plane) 3. $m\angle A = m\angle B = 90°$ (Def. $\perp$ lines, rt. $\angle$s) 4. O is midpt. of $\overline{AB}$ (Given) 5. $\overline{AO} \cong \overline{BO}$ (Def. midpt.) 6. $\angle POA \cong \angle QOB$ (Vert. $\angle$s $\cong$) 7. $\triangle POA \cong \triangle QOB$ (ASA) 8. $\overline{PO} \cong \overline{QO}$ (Corr. parts $\cong$ $\triangle$s) 9. O is midpt. of $\overline{PQ}$ (Def. midpt.) **21.** $\triangle BEG$ is equilateral. In $\triangle EAB$, $\triangle BCG$, and $\triangle EFG$, $\overline{EA}$, $\overline{AB}$, $\overline{BC}$, $\overline{GC}$, $\overline{EF}$, and $\overline{FG}$ are $\cong$. $\angle EAB$, $\angle BCG$, and $\angle EFG$ are rt. $\angle$s and thus $\cong$. $\triangle EAB \cong \triangle BCG \cong \triangle EFG$ by SAS, and $\overline{BE} \cong \overline{BG} \cong \overline{EG}$ by corr. parts $\cong$ $\triangle$s.

WRITTEN EXERCISES, PAGES 126–128

1. $x = 5$ **3.** $x = 41$ **5.** 1. Given 2. $\angle B$; Th. 3-1 3. Given 4. $\angle A$; $\angle KQP$; If lines $\parallel$, corr. $\angle$s $\cong$. 5. $\angle KPQ \cong \angle KQP$ 6. Th. 3-2 **7.** 1. $\overline{TU} \cong \overline{TV}$ (Given) 2. $\angle 1 \cong \angle 2$ (Th. 3-1)

3. $\overline{UV} \parallel \overline{RS}$ (Given) 4. $\angle R \cong \angle 1$, $\angle 2 \cong \angle S$ (If lines $\parallel$, corr. $\angle$s $\cong$) 5. $\angle R \cong \angle S$ (Trans.)
9. 1. $\overline{XY} \cong \overline{XZ}$ (Given) 2. $m\angle XYZ = m\angle XZY$ (Th. 3–1) 3. $\overrightarrow{YO}$ bis. $\angle XYZ$, $\overrightarrow{ZO}$ bis. $\angle XZY$ (Given) 4. $m\angle XYZ = 2m\angle 2$, $m\angle XZY = 2m\angle 3$ ($\angle$ Bis. Th.) 5. $2m\angle 2 = 2m\angle 3$ (Subst.) 6. $m\angle 2 = m\angle 3$ (Div. Prop. $=$) 7. $\overline{YO} \cong \overline{ZO}$ (Th. 3–2) **11.** 1 and 2 **13.** 1. $\angle 4 \cong \angle 7$, $\angle 1 \cong \angle 3$ (Given) 2. $\angle B \cong \angle C$ (If 2 $\angle$s of a $\triangle \cong$ 2 $\angle$s of another $\triangle$, 3rd $\angle$s $\cong$) 3. $\overline{AB} \cong \overline{AC}$ (Th. 3–2) 4. $\triangle ABC$ is isos. (Def. isos. $\triangle$)
15. a. $m\angle 2 = 40$, $m\angle 7 = 100$, $m\angle 5 = m\angle 6 = 40°$; $\overline{PQ} \parallel \overline{SR}$ **b.** $m\angle 2 = x$, $m\angle 7 = 180 - 2x$, $m\angle 5 = m\angle 6 = x$. Yes. **17. a.** 40, 40, 60 **b.** $2x$, $2x$, $3x$ **19. a.** 90 **b.** 90 **21.** $x = 40$, $y = 20$
23. $\triangle DAC \cong \triangle DAB$. 1. $\angle ACB \cong \angle ABC$, $\angle DCB \cong \angle DBC$ (Given) 2. $\overline{AB} \cong \overline{AC}$, $\overline{DB} \cong \overline{DC}$ (Th. 3–2) 3. $\overline{DA} \cong \overline{DA}$ (Refl.) 4. $\triangle DAC \cong \triangle DAB$ (SSS) **25.** It is isos.

WRITTEN EXERCISES, PAGES 132–135

1. 1. Given 2. Def. rt. $\triangle$ 3. Given 4. $\overline{XZ} \cong \overline{XZ}$ 5. $\triangle YXZ$; HL 6. $\overline{WZ} \cong \overline{YZ}$; corr. parts $\cong$ $\triangle$s
3. 1. $\overline{EF} \perp \overline{EG}$, $\overline{HG} \perp \overline{EG}$ (Given) 2. $\triangle EGH$ and $\triangle GEF$ are rt. $\triangle$s (Def. $\perp$ lines, rt. $\triangle$) 3. $\overline{EH} \cong \overline{GF}$ (Given) 4. $\overline{EG} \cong \overline{EG}$ (Refl.) 5. $\triangle EGH \cong \triangle GEF$ (HL) 6. $\angle H \cong \angle F$ (Corr. parts $\cong$ $\triangle$s) **5.** 1. Given 2. $\angle PRQ$; Th. 3–1 3. Given 4. Refl. 5. $\triangle QRT$; SAS 6. Corr. parts $\cong$ $\triangle$s **7.** 1. $\overline{RT} \cong \overline{AS}$, $\overline{RS} \cong \overline{AT}$ (Given) 2. $\overline{ST} \cong \overline{ST}$ (Refl.) 3. $\triangle TSA \cong \triangle STR$ (SSS) 4. $\angle TSA \cong \angle STR$ (Corr. parts $\cong$ $\triangle$s) **9.** Line $\perp$ plane if and only if they intersect and line $\perp$ all lines in pl. that pass through pt. of intersection. **11.** AAS
13. a. Yes **b.** No **15.** $\angle 1 \cong \angle 2$, $\angle 3 \cong \angle 4$, $\overline{PQ} \cong \overline{PT}$; ASA **17. b.** $\triangle RMS$, $\triangle RNT$; $\triangle MNS$, $\triangle NMT$; $\triangle ONS$, $\triangle OMT$; $\triangle NST$, $\triangle MTS$ **19.** 1. $AB = AC$ (Given) 2. $m\angle ABC = m\angle ACB$ (Th. 3–1) 3. $\overrightarrow{BX}$ bis. $\angle ABC$, $\overrightarrow{CY}$ bis. $\angle ACB$ (Given) 4. $m\angle XBC = \frac{1}{2}m\angle ABC$, $m\angle YCB = \frac{1}{2}m\angle ACB$ ($\angle$ Bis. Th.) 5. $\frac{1}{2}m\angle ABC = \frac{1}{2}m\angle ACB$ (Mult. Prop. $=$) 6. $m\angle XBC = m\angle YCB$ (Subst.) 7. $\overline{BC} \cong \overline{BC}$ (Refl.) 8. $\triangle XCB \cong \triangle YBC$ (ASA) 9. $BX = CY$ (Corr. parts $\cong$ $\triangle$s) **21.** 1. $FL = AK$ (Given) 2. $FL + LA = LA + AK$ (Add. Prop. $=$) 3. $FA = FL + LA$; $LK = LA + AK$ (Seg. Add. Post.) 4. $FA = LK$ (Subst.) 5. $SF = SK$ (Given) 6. $\angle F \cong \angle K$ (Th. 3–1) 7. $\frac{1}{2}SF = \frac{1}{2}SK$ (Mult. Prop. $=$) 8. M midpt. of $\overline{SF}$, N midpt. of $\overline{SK}$ (Given) 9. $MF = \frac{1}{2}SF$, $NK = \frac{1}{2}SK$ (Midpt. Th.) 10. $MF = NK$ (Subst.) 11. $\triangle MFA \cong \triangle NKL$ (SAS) 12. $\overline{AM} \cong \overline{LN}$ (Corr. parts $\cong$ $\triangle$s)

WRITTEN EXERCISES, PAGES 140–142

1. b. no **5.** Yes; midpt. of hypotenuse **7.** $\overline{NS}$, $\overline{NK}$ **9.** L, A **11.** $\perp$ bis. of $\overline{LF}$ **13.** 1. $\overleftrightarrow{AX}$ is $\perp$ bis. of $\overline{BC}$ (Given) 2. $\overline{XB} \cong \overline{XC}$ (Def. $\perp$ bis.) 3. $\angle AXB \cong \angle AXC$ (Th. 1–4) 4. $\overline{AX} \cong \overline{AX}$ (Refl.) 5. $\triangle AXB \cong \triangle AXC$ (SAS) 6. $AB = AC$ (Corr. parts $\cong$ $\triangle$s) **17. a.** 1. $\overrightarrow{BZ}$ bis. $\angle ABC$ (Given) 2. $\angle PBX \cong \angle PBY$ (Def. $\angle$ bis.) 3. $\overline{PX} \perp \overrightarrow{BA}$, $\overline{PY} \perp \overrightarrow{BC}$ (Given) 4. $m\angle PXB = m\angle PYB = 90$ (Def. $\perp$, rt. $\angle$) 5. $\overline{PB} \cong \overline{PB}$ (Refl.) 6. $\triangle PXB \cong \triangle PYB$ (AAS) 7. $PX = PY$ (Corr. parts $\cong$ $\triangle$s) **b.** 1. $\overline{PX} \perp \overrightarrow{BA}$, $\overline{PY} \perp \overrightarrow{BC}$ (Given) 2. $\triangle PXB$ and $\triangle PYB$ rt. $\triangle$s (Def. $\perp$, rt. $\angle$, rt. $\triangle$) 3. $PB = PB$ (Refl.) 4. $PX = PY$ (Given) 5. $\triangle PXB \cong \triangle PYB$ (HL) 6. $\angle PBX \cong \angle PBY$ (Corr. parts $\cong$ $\triangle$s) 7. $\overrightarrow{BP}$ bis. $\angle ABC$ (Def. $\angle$ bis.) **19. b.** Alt. to legs of isos. $\triangle$ are $\cong$. **25.** $\triangle NQR \cong \triangle MRQ$ (AAS), $\triangle NOQ \cong \triangle MOR$ (AAS), $\overline{NO} \cong \overline{MO}$ (Corr. parts $\cong$ $\triangle$s), $\triangle MNO$ is isos.

WRITTEN EXERCISES, PAGES 144–147

1. a. SSS **b.** Corr. parts $\cong$ $\triangle$s **c.** SAS **d.** Corr. parts $\cong$ $\triangle$s **3. a.** AAS **b.** Corr. parts $\cong$ $\triangle$s **c.** SAS **d.** Corr. parts $\cong$ $\triangle$s **5. a.** SAS **b.** Corr. parts $\cong$ $\triangle$s **c.** HL **d.** Corr. parts $\cong$ $\triangle$s **7.** 1. $\overline{FL} \cong \overline{FK}$, $\overline{LA} \cong \overline{KA}$ (Given) 2. $\overline{FA} \cong \overline{FA}$ (Refl.) 3. $\triangle FLA \cong \triangle FKA$ (SSS) 4. $\angle 1 \cong \angle 2$ (or $\angle 5 \cong \angle 6$) (Corr. parts $\cong$ $\triangle$s) 5. $\overline{FJ} \cong \overline{FJ}$ (or $\overline{JA} \cong \overline{JA}$) (Refl.) 6. $\triangle FLJ \cong \triangle FKJ$ (or $\triangle LAJ \cong \triangle KAJ$) (SAS) 7. $\overline{LJ} \cong \overline{KJ}$ (Corr. parts $\cong$ $\triangle$s) **13.** 1. $\overline{DE} \cong \overline{FG}$, $\overline{GD} \cong \overline{EF}$ (Given) 2. $\overline{GE} \cong \overline{GE}$ (Refl.) 3. $\triangle GDE \cong \triangle EFG$ (SSS) 4. $\angle DEH \cong \angle FGK$ (Corr. parts $\cong$ $\triangle$s) 5. $\angle HDE$ and $\angle KFG$ rt. $\angle$s (Given) 6. $m\angle HDE = m\angle KFG = 90$ (Def. rt. $\angle$) 7. $\triangle DEH \cong \triangle FGK$ (ASA) 8. $\overline{DH} \cong \overline{FK}$ (Corr. parts $\cong$ $\triangle$s) **19.** isosceles; $\triangle XAZ \cong \triangle YAZ$, and $\overline{XZ} \cong \overline{YZ}$

WRITTEN EXERCISES, PAGES 149–150

1. 256; 1024 **3.** $\frac{1}{81}$; $\frac{1}{243}$ **5.** 17; 23 **7.** 15; 4 **9.** Chan is older than Sarah. **11.** G has 7 sides.
13. none **15.** No; deductively **17.** $9876 \times 9 + 4 = 88{,}888$ **19. a.** 16 **b.** Guess: 32; actual: 31
21. a. 13, 17, 23, 31, 41, 53, 67, 83, 101 **b.** prime **c.** 121, 143; they are not prime **23.** 5-pt.: 180; 6-pt.: 360; n-pt.: $180(n - 4)$

EXTRA, PAGE 151

3. one **5, 7.** Two linked non-Möbius (2-sided) bands are formed.

CHAPTER REVIEW, PAGES 153–154

1. $\triangle RSQ$ **2.** $\angle QRS$ **3.** QS; midpt. of $\overline{PQ}$ **4.** 48 **5.** $\triangle ABC \cong \triangle DEC$, SAS **6.** no **7.** $\triangle ABC \cong \triangle DEC$, ASA **8.** no **9.** 1. $\overline{JM} \cong \overline{LM}$, $\overline{JK} \cong \overline{LK}$ (Given) 2. $\overline{MK} \cong \overline{MK}$ (Refl.) 3. $\triangle MJK \cong \triangle MLK$ (SSS) 4. $\angle MJK \cong \angle MLK$ (Corr. parts $\cong$ $\triangle$s) **10.** 1. $\angle JMK \cong \angle LMK$, $\overline{MK} \perp P$ (Given) 2. $\overline{MK} \perp \overline{JK}$, $\overline{MK} \perp \overline{LK}$ (Def. line $\perp$ plane) 3. $m\angle MKJ = m\angle MKL = 90$ (Def. $\perp$ lines, rt. $\angle$) 4. $\overline{MK} \cong \overline{MK}$ (Refl.) 5. $\triangle MKJ \cong \triangle MKL$ (ASA) 6. $\overline{JK} \cong \overline{LK}$ (Corr. parts $\cong$ $\triangle$s) **11.** $\angle D \cong \angle F$; if 2 sides of $\triangle$ $\cong$, opp. $\angle$s $\cong$ also. **12.** $t = 3$, $DF = 9$ **13.** $x = 4$, $y = 8$ **14.** 1. $\overline{GH} \perp \overline{HJ}$, $\overline{KJ} \perp \overline{HJ}$ (Given) 2. $m\angle GHJ = m\angle KJH = 90$ (Def. $\perp$ lines, rt. $\angle$) 3. $\angle G \cong \angle K$ (Given) 4. $\overline{HJ} \cong \overline{HJ}$ (Refl.) 5. $\triangle GHJ \cong \triangle KJH$ (AAS) **15.** 1. $\overline{GH} \perp \overline{HJ}$, $\overline{KJ} \perp \overline{HJ}$ (Given) 2. $\triangle GHJ$ and $\triangle KJH$ rt. $\triangle$s (Def. $\perp$ lines, rt. $\triangle$) 3. $\overline{GJ} \cong \overline{KH}$ (Given) 4. $\overline{HJ} \cong \overline{HJ}$ (Refl.) 5. $\triangle GHJ \cong \triangle KJH$ (HL) 6. $\overline{GH} \cong \overline{KJ}$ (Corr. parts $\cong$ $\triangle$s) **17.** $\overline{QP} \perp \overline{AB}$, $\overline{AP} \cong \overline{BP}$ (P midpt. of $\overline{AB}$), $\overline{AQ} \cong \overline{BQ}$, $\triangle QAB$ isos., $\triangle QAP \cong \triangle QBP$, etc. **18.** bis. of $\angle AQB$ **19.** altitude **20.** 1. ASA 2. Corr. parts $\cong$ $\triangle$s 3. Refl. 4. HL 5. Corr. parts $\cong$ $\triangle$s **21. a.** Answers may vary. She may conclude next toss is tails. She may conclude next toss is heads. **b.** Yes. Conclusion reached by inductive reasoning. **22.** 125; 216 **23.** $\frac{1}{100}$, $-\frac{1}{1000}$

MIXED REVIEW, PAGE 156

1. 171 **3.** 92, 141 **5.** SSS **7.** 36, 36 **9.** 1. $\overline{RU} \parallel \overline{ST}$ (Given) 2. $\angle RUS \cong \angle TSU$ (If 2 $\parallel$ lines are cut by a trans., alt. int. $\angle$s are $\cong$.) 3. $\angle R \cong \angle T$ (Given) 4. $\angle RSU \cong \angle TUS$ (If 2 $\angle$s of one $\triangle$ are $\cong$ 2 $\angle$s of another $\triangle$, then the third $\angle$s are $\cong$.) 5. $\overline{RS} \parallel \overline{UT}$ (If lines are cut by a trans. so alt. int $\angle$s are $\cong$, then the lines are $\parallel$.) **11.** obtuse **13.** postulate **15.** parallel **17.** Segment Addition Postulate

ALGEBRA REVIEW, PAGE 157

1. $-6, 1$ **3.** $-2, 9$ **5.** $0, 5$ **7.** $\pm 2\sqrt{41}$ **9.** $-6, 6$ **11.** $-5, 2$ **13.** $-\frac{4}{9}, 1$ **15.** $\frac{-7 \pm \sqrt{37}}{2}$ **17.** $-8, 3$ **19.** $-1, 4$ **21.** $2, 18$ **23.** $-1, \frac{7}{5}$ **25.** $-1, 2$ **27.** $-5, 2$ **29.** $2, 6$ **31.** $4, 6$ **33.** $2, 5$ **35.** $3, \frac{5}{4}$ **37.** 12

Chapter 4

WRITTEN EXERCISES, PAGES 161–163

1. 10 **3.** 100 **5.** 110 **7.** 100 **9.** $x = 6$ **11.** $x = 8$ **13.** $y = 25$ **15.** $y = 10$ **17.** 1. Draw diagonal $\overline{EG}$ (through 2 pts. there is a line) 2. $EFGH$ is ▱ (Given) 3. $\overline{HG} \parallel \overline{EF}$, $\overline{HE} \parallel \overline{GF}$ (Def. ▱) 4. $\angle 1 \cong \angle 2$, $\angle 3 \cong \angle 4$ (If lines $\parallel$, alt. int. $\angle$s $\cong$) 5. $\overline{EG} \cong \overline{EG}$ (Reflex. Prop.) 6. $\triangle EFG \cong \triangle GHE$ (ASA) 7. $\overline{EF} \cong \overline{HG}$, $\overline{FG} \cong \overline{EH}$ (Corr. parts $\cong$ $\triangle$s) **19.** 1. ▱$QRST$, diagonals $\overline{QS}$ and $\overline{TR}$ (Given) 2. $\overline{QR} \parallel \overline{TS}$ (Def. ▱) 3. $\angle 1 \cong \angle 2$, $\angle 3 \cong \angle 4$ (If lines $\parallel$, alt. int. $\angle$s $\cong$) 4. $\overline{QR} \cong \overline{TS}$ (Th. 4–1) 5. $\triangle QMR \cong \triangle SMT$ (ASA) 6. $\overline{QM} \cong \overline{MS}$, $\overline{TM} \cong \overline{MR}$ (Corr. parts $\cong$ $\triangle$s) 7. $\overline{QS}$ and $\overline{TR}$ bis. each other (Def. seg. bis.) **21.** $x = 5$, $y = 2$ **23.** $x = 10$, $m\angle CED = 70$ **25.** 1. ▱$PQRS$; $\overline{PJ} \cong \overline{RK}$ (Given) 2. $\angle P \cong \angle R$ (Th. 4–2) 3. $\overline{SP} \cong \overline{RQ}$ (Th. 4–1) 4. $\triangle SPJ \cong \triangle QRK$ (SAS) 5. $\overline{SJ} \cong \overline{QK}$ (Corr. parts $\cong$ $\triangle$s) **27.** 1. $ABCD$ is ▱, $\overline{CD} \cong \overline{CE}$ (Given) 2. $\overline{AB} \parallel \overline{CD}$ (Def. ▱) 3. $\angle A \cong \angle CDE$ (If lines $\parallel$, corr. $\angle$s $\cong$) 4. $\angle CDE \cong \angle E$ (Isos. $\triangle$ Th.) 5. $\angle A \cong \angle E$ (Trans.) **31. a.** No. If $GFIJ$ and $EGJH$ were parallelograms, $GF = JI$ and $GE = JH$.

WRITTEN EXERCISES, PAGES 166–168

1. Def. ▱ **3.** Th. 4–5 **5.** Th. 4–6 **7.** 15 **9.** $x = 4$ **11.** $x = 10$, $y = 5$ **13.** $\overline{NC} \parallel \overline{AM}$; $\overline{AB} \cong \overline{DC}$, so $\overline{NC} \cong \overline{AM}$. $AMCN$ is ▱ by Th. 4–5 **15.** $OD = OB$, $OA = OC$ (diag. ▱ bis. each other); $OX = \frac{1}{2}OB = \frac{1}{2}OD = OZ$, $OW = \frac{1}{2}OA = \frac{1}{2}OC = OY$. $\overline{XZ}$ and $\overline{WY}$ bis. each other, so $WXYZ$ is ▱ by Th. 4–7 **17.** $\overline{AB} \parallel \overline{CD} \parallel \overline{EF}$, so $\overline{AB} \parallel \overline{EF}$; $\overline{AB} \cong \overline{CD} \cong \overline{EF}$, so $\overline{AB} \cong \overline{EF}$. $ABEF$ is ▱ by Th. 4–5

23. $x = 22, y = 11$ **25.** 1. $\angle DBC \cong \angle C$ (Given) 2. $\overline{CD} \cong \overline{BD}$ (Th. 3–2) 3. $\overline{AE} \cong \overline{CD}$ (Given) 4. $\overline{AE} \cong \overline{BD}$ (Trans. Prop.) 5. $\angle A \cong \angle DBC$ (Given) 6. $\overline{AE} \parallel \overline{BD}$ (If corr. $\angle$s $\cong$, lines are $\parallel$) 7. $ABDE$ is ▱ (Th. 4–5)

WRITTEN EXERCISES, PAGES 171–173

1. all **3.** all **5.** all **7.** rh., sq. **9.** rect., sq. **11.** Sum of meas. of $\angle$s $= 360$. Each $\angle$ $= \frac{360}{4} = 90$, so all $\angle$s are rt. $\angle$s. **13.** 13 **15.** 20 **17.** 1. $WXYZ$ is ▱ (Given) 2. $m\angle 1 = 90$ (Given) 3. $m\angle 2 = m\angle 1 = 90$ (vert. $\angle$s $\cong$) 4. $\angle 2$ is rt. $\angle$ (Def. rt. $\angle$) 5. $WXYZ$ is rect. (Th. 4–13) **19.** 1. $ABCD$ is rhombus (Given) 2. $\angle DAC \cong \angle CAB$ (Th. 4–11, Def. $\angle$ bis.) 3. $\angle 1 \cong \angle CAB$, $\angle DAC \cong \angle 2$ (vert. $\angle$s $\cong$) 4. $\angle 1 \cong \angle 2$ (Trans. Prop.) **21.** 1. $QRST$ is rect., $RKST$ is ▱ (Given) 2. $\overline{SK} \cong \overline{TR}$ (Th. 4–1) 3. $\overline{TR} \cong \overline{SQ}$ (Th. 4–9) 4. $\overline{SK} \cong \overline{SQ}$ (Trans. Prop.) 5. $\triangle QSK$ is isos. (Def. isos. $\triangle$) **27.** $m\angle 1 = m\angle 4 = 30$; $m\angle 2 = m\angle 3 = 15$ **29. a.** Yes. It has 2 $\cong$ consecutive sides. **b.** Yes. It is a rhombus. **31.** $\triangle RYZ$ is equilateral.

WRITTEN EXERCISES, PAGES 175–177

1. 12, 12 **3.** 4 **5.** 11 **7.** 4.3 **9.** $x = 11$ **11.** $x = 3\frac{1}{3}$ **13.** $BE = 2\,AD$, $CF = 3\,AD$ **15.** 13, 39 **17.** $x = 9, y = 15$ **19.** $CF = 3\,AD$, but $17 \neq 3 \cdot 5$ **21.** rect. **23.** ▱ **27.** 1. Draw $\overline{DP}$ and $\overline{CQ} \perp \overline{AB}$ (Th. 2–9) 2. $\overline{AB} \parallel \overline{DC}$ (Def. Trap.) 3. $\overline{DP} \cong \overline{CQ}$ (Th. 4–1 Cor.) 4. $\triangle APD$, $\triangle BQC$ rt. $\triangle$s (Def. $\perp$ lines, rt. $\triangle$) 5. $ABCD$ is trap., $\overline{AD} \cong \overline{BC}$ (Given) 6. $\triangle APD \cong \triangle BQC$ (HL) 7. $\angle A \cong \angle B$ (Corr. parts $\cong$ $\triangle$s) 8. $\angle ADC$ supp. $\angle A$, $\angle BCD$ supp. $\angle B$ (If lines $\parallel$, s–s. int. $\angle$s supp.) 9. $\angle ADC \cong \angle BCD$ (Th. 1–7) **29.** $\overline{PM} \parallel \overline{WK} \parallel \overline{ON}$ and $PM = \frac{1}{2}WK = ON$ (Th. 4–17). Thus $\overline{PM} \parallel \overline{ON}$ and $PM = ON$. $PMNO$ is ▱ (Th. 4–5). **33.** $x = 3\frac{1}{2}$ **35.** If diag. of trap. $\cong$, then trap. is isos.

WRITTEN EXERCISES, PAGES 181–182

1. Suppose $\angle Y$ is not acute. Then $m\angle Y \geq 90$ and the sum of $\angle$s of $\triangle XYZ \geq 190$. Contradicts fact that sum of $\angle$s of $\triangle = 180$. Assumption that $\angle Y$ is not acute is false, so $\angle Y$ must be acute. **3.** Suppose $\overrightarrow{OE}$ bis. $\angle JOK$. Then $\angle 1 \cong \angle 2$, $\overline{OJ} \cong \overline{OK}$, $\overline{OE} \cong \overline{OE}$. $\triangle OJE \cong \triangle OKE$ (SAS), and $\overline{JE} \cong \overline{KE}$ (Corr. parts $\cong$ $\triangle$s). Contradicts $\overline{JE} \ncong \overline{KE}$ (Given). Assumption that $\overrightarrow{OE}$ bis. $\angle JOK$ false, so $\overrightarrow{OE}$ doesn't bis. $\angle JOK$. **5.** Suppose pl. P and Q don't intersect. Then $P \parallel Q$, and $\overline{AB} \parallel \overline{CD}$ by Th. 2–1. Contradicts $\overline{AB} \nparallel \overline{CD}$ (Given). Assumption that P and Q don't intersect is false; P and Q intersect. **9.** Suppose $n \parallel m$. Then $n \parallel l$ (2 lines $\parallel$ 3rd line $\parallel$ each other). Contradicts n intersects l (Given). Assumption that $n \parallel m$ false; n must intersect m. **13.** Given: trap. $ABCD$. Suppose $\overline{AC}$ and $\overline{BD}$ bis. each other. Then $ABCD$ is ▱ (Th. 4–7). Contradicts fact that $ABCD$ is trap. Assumption that diag. bis. each other is false; diagonals do not bis. each other.

WRITTEN EXERCISES, PAGES 185–187

1. 3; 15 **3.** 0; 200 **5.** 5; $2k + 5$ **7.** $\angle 2$ **9.** $\angle 3$ **11.** $\overline{WT}$ **13.** $\overline{WY}$ **15.** c, d, e, b, a **17.** $m\angle 2$, $m\angle X$, $m\angle XZY$, $m\angle Y$, $m\angle 1$ **19.** 1. $EFGH$ ▱, $EF > FG$ (Given) 2. $HG > EH$ (Th. 4–1, Subst. Prop.) 3. $m\angle 1 > m\angle 2$ (Th. 4–18)

WRITTEN EXERCISES, PAGES 191–193

1. always **3.** never **5.** always **7.** sometimes **9.** always **11.** $\overline{XY}$ **13.** $\angle VRT$ **15.** $\angle VRS$ **17.** 1. $AB > AC$ (Given) 2. $m\angle ACB > m\angle ABC$ (Th. 4–18) 3. $BD = EC$ (Given) 4. $BC = BC$ (Reflex. Prop.) 5. $BE > CD$ (Th. 4–21) **19.** $CA > BA$. Proof: $\angle BAC$ is rt. $\angle$, $\triangle BAC$ is rt. $\triangle$. M is midpt. of $\overline{BC}$ (Given) so $BM = MC = AM$ (midpt. of hyp. equidistant from vertices). $\triangle BMA$ isos., $m\angle B = m\angle MAB = 46$; $\triangle CMA$ isos., $m\angle C = m\angle MAC = 44$. In $\triangle BAC$, $m\angle B > m\angle C$, so $CA > BA$ (Th. 4–19)

CHAPTER REVIEW, PAGES 198–199

1. 110 **2.** 28 **3.** 38 **4.** 6 **5.** $GS = 5$ **6.** $\angle SGN \cong \angle SAN$ **7.** $\overline{GZ} \cong \overline{AZ}$ **8.** $GN = 17$ **9.** parallelogram **10.** square **11.** rhombus **12.** rectangle **13.** $\overline{ZO}, \overline{DI}$ **14.** 14 **15.** 4 **16.** 100 **17.** c, d, a, b **18.** X **19.** BN, BA **20.** 3, 21 **21.** $>$ **22.** $<$ **23.** $=$ **24.** $>$

PREPARING FOR COLLEGE ENTRANCE EXAMS, PAGE 201

1. B **2.** C **3.** A **4.** D **5.** A **6.** B **7.** C **8.** E **9.** D **10.** B

CUMULATIVE REVIEW: CHAPTERS 1-4, PAGES 202-203

1. -1 **3.** bisects, $\perp$ **5. a.** Yes; skew **b.** No **7.** $j \parallel k$ **9.** 540 **11.** False **13. a.** RTA **b.** $m\angle E$ **c.** $\overline{AR}$ **15. a.** A and B **b.** $\overrightarrow{SR}$ and $\overrightarrow{ST}$ **17. a.** sometimes **b.** sometimes **c.** never **d.** always **e.** always **f.** always **19. a.** $\overline{RS}$ **b.** Th. 4-21 **21.** $x = 4$ **23.** $z > 3$ **25.** No; since $AB + BC = AC$, $y = 2.5$ but $AB = 3.2(2.5) = 8$ and $BC = 2(2.5) + 1 = 6$ **27.** 1. $\overline{AD} \cong \overline{BC}$; $\overline{AD} \parallel \overline{BC}$ (Given) 2. $ABCD$ is ▱ (Th. 4-5) 3. $\overline{AB} \parallel \overline{CD}$ (Def. ▱) 4. $\angle FAE \cong \angle FCG$; $\angle AEF \cong \angle CGF$ (If lines $\parallel$, alt. int. ∠s $\cong$) 5. $\overline{AF} \cong \overline{CF}$ (Th. 4-3) 6. $\triangle FAE \cong \triangle FCG$ (AAS) 7. $\overline{EF} \cong \overline{FG}$ (Corr. parts $\cong$ △s)

Chapter 5

WRITTEN EXERCISES, PAGES 207-208

1. 5:3 **3.** 1:5 **5.** 3:16 **7.** 2:1 **9.** $\frac{2}{9}$ **11.** $\frac{1}{5}$ **13.** 6:3:2 **15.** $\frac{5}{4}$ **17.** $\frac{1}{5}$ **19.** $\frac{20}{1}$ **21.** $\frac{3}{4b}$ **23.** $\frac{3}{a}$ **25.** $\frac{3}{x+y}$ **27.** 40; 50 **29.** 45; 60; 75 **31.** 54; 54; 72 **33.** 27 cm; 33 cm; 36 cm **35.** 4:1; $(n-2)$:2 **37.** 10 or more **39.** 5:8

WRITTEN EXERCISES, PAGES 210-212

1. 15 **3.** 21 **5.** $\frac{4}{7}$ **7.** $\frac{y+3}{3}$ **9.** $x = 2\frac{2}{5}$ **11.** $x = 1\frac{7}{8}$ **13.** $x = -3$ **15.** $x = 2$ **17.** $x = 11$ **19.** $x = 18\frac{1}{3}$ **21.** 21; 12; 28 **23.** 8; 24; 20 **25.** 8; 4; 15 **27.** 27; 36; 12 **29.** $\frac{a+b}{b} = \frac{c+d}{d}$; $d(a+b) = b(c+d)$; $ad + bd = bc + bd$; $ad = bc$; $\frac{a}{b} = \frac{c}{d}$ **31.** $(a-b)(c+d) = (a+b)(c-d)$; $ac + ad - bc - bd = ac - ad + bc - bd$; $2ad = 2bc$; $ad = bc$; $\frac{a}{b} = \frac{c}{d}$ **33.** $x = 12$ **35.** $x = \frac{1}{2}$ **37.** $x = 4$, $x = -\frac{9}{5}$ **39.** $x = 9$, $y = 5$ **43.** 3:2

WRITTEN EXERCISES, PAGES 214-216

1. always **3.** sometimes **5.** always **7.** sometimes **9.** never **11.** sometimes **13.** $\frac{3}{4}$ **15.** 45 **17.** 8 **19.** $4k$ **21.** $x = 16\frac{1}{4}$, $y = 12\frac{4}{5}$, $z = 11\frac{1}{5}$ **23.** $x = 7$, $y = 5\sqrt{3}$ **27.** $AB = \frac{BC \cdot DE}{EF}$ or $AB = \frac{AC \cdot DE}{DF}$ **29.** $C'(-6, -10)$, $D'(-10, -6)$ or $C'(-6, 6)$, $D'(-10, 2)$ **31.** $x = 20$ **33. a.** $-3 + 3\sqrt{5}$ **b.** $\frac{1+\sqrt{5}}{2} \approx 1.62$

WRITTEN EXERCISES, PAGES 221-224

1. sim. **3.** sim. **5.** sim. **7.** sim. **9.** sim. **11. a.** $\triangle MLN$ **b.** 20; x; 20; y **c.** 24; 16 **13.** $x = 24$, $y = 5\frac{1}{2}$ **15. a.** $\triangle ACD$, $\triangle CBD$ **b.** $x = 15$, $y = 9$ **17.** 0.55 cm **19.** $x = 9$, $y = 5.4$ **21.** ∠s A and A' are rt. ∠s, $\angle 1 \cong \angle 2$; △s $\sim$ by AA; height ≈ 6 m **23. a.** 1. $\angle 1 \cong \angle 2$ (Given) 2. $\angle J \cong \angle J$ (Reflex.) 3. $\triangle JIG \sim \triangle JZY$ (AA) **b.** 4. $\frac{JG}{JY} = \frac{GI}{YZ}$ (Corr. sides of $\sim$ △s are in prop.) **25.** 1. $\overline{BN} \parallel \overline{LC}$ (Given) 2. $\angle B \cong \angle C$, $\angle N \cong \angle L$ (If lines $\parallel$, alt. int. ∠s $\cong$) 3. $\triangle MLC \sim \triangle MNB$ (AA) 4. $\frac{BN}{CL} = \frac{NM}{LM}$ (Corr. sides of $\sim$ △s are in prop.) 5. $BN \cdot LM = CL \cdot NM$ (means-ext. prop.) **27.** 1. $\overline{QT} \parallel \overline{RS}$ (Given) 2. $\angle PQU \cong \angle R$; $\angle PUQ \cong \angle PVR$; $\angle PUT \cong \angle PVS$; $\angle PTU \cong \angle S$ (If lines $\parallel$, corr. ∠s $\cong$) 3. $\triangle PQU \sim \triangle PRV$, $\triangle PUT \sim \triangle PVS$ (AA) 4. $\frac{QU}{RV} = \frac{PU}{PV}$, $\frac{UT}{VS} = \frac{PU}{PV}$ (Corr. sides of $\sim$ △s are in prop.) 5. $\frac{QU}{RV} = \frac{UT}{VS}$ (Subst.) **29.** 20 **37. a.** dist. of H to: $\overline{AB}$, 8; $\overline{BC}$, 10; $\overline{DC}$, 8; $\overline{AD}$, 6 **b.** 0.5; 15.5; 4; 3.5; 12.5; 7.5; 12.5

WRITTEN EXERCISES, PAGES 228–230

1. $\triangle ABC \sim \triangle GNK$, SAS **3.** $\triangle ABC \sim \triangle XRN$, SSS **5.** $\triangle ABC \sim \triangle AEF$, AA **7.** $\triangle ABC \sim \triangle PKN$, 2:3 **9.** No **11.** 1. $\frac{DE}{GH} = \frac{DF}{GI} = \frac{EF}{HI}$ (Given) 2. $\triangle DEF \sim \triangle GHI$ (SSS) 3. $\angle E \cong \angle H$ (Corr. $\angle$s of $\sim$ $\triangle$s are $\cong$) **15.** 1. $\frac{VW}{VX} = \frac{VZ}{VY}$ (Given) 2. $\angle V \cong \angle V$ (Reflex.) 3. $\triangle VWZ \sim \triangle VXY$ (SAS) 4. $\angle 1 \cong \angle 2$ (Corr. $\angle$s of $\sim$ $\triangle$s are $\cong$) 5. $\overline{WZ} \parallel \overline{XY}$ (If corr. $\angle$s $\cong$, lines $\parallel$) **17.** $\frac{5}{2}AB$ **21.** Given: Isos. $\triangle ABC$, $AB = AC$; isos. $\triangle DEF$, $DE = DF$; $\angle A \cong \angle D$. 1. $AB = AC$, $DE = DF$ (Given) 2. $\frac{AB}{DE} = \frac{AC}{DF}$ (Div. prop. =) 3. $\angle A \cong \angle D$ (Given) 4. $\triangle ABC \sim \triangle DEF$ (SAS) **25. a.** 3:4

WRITTEN EXERCISES, PAGES 236–238

1. a. No **b.** Yes **c.** Yes **d.** No **e.** Yes **f.** Yes **3.** 10; 16 **5.** 21; 35 **7.** $x = 7\frac{1}{2}$ **9.** $x = 26$ **11.** $x = 5\frac{3}{5}$ **13.** $x = 14\frac{1}{2}$ **15.** $x = 3$ **17.** $AN = 10$ **19.** 8; 18; 12; 25 **21.** 18; 15; 44; $\frac{300}{11}(= 27\frac{3}{11})$ **23.** 1. $\overrightarrow{KE} \parallel \overrightarrow{DG}$ (Given) 2. $\frac{GF}{GE} = \frac{DF}{DK}$ (Th. 5-3) 3. $\overrightarrow{DG}$ bis. $\angle FDE$ (Given) 4. $\angle 1 \cong \angle 2$ (Def. $\angle$ bis.) 5. $\angle 3 \cong \angle 1$ (If lines $\parallel$, alt. int. $\angle$s $\cong$) 6. $\angle 2 \cong \angle 4$ (If lines $\parallel$, corr. $\angle$s $\cong$) 7. $\angle 3 \cong \angle 4$ (Trans.) 8. $DK = DE$ (If 2 $\angle$s of $\triangle$ are $\cong$, sides opp. the $\angle$s are $\cong$) 9. $\frac{GF}{GE} = \frac{DF}{DE}$ (Subst.) **25.** 22.5 **27.** 78 **29.** 0.5

EXTRA, PAGE 240

1. *c* **3.** *b* **5.** *d*

CHAPTER REVIEW, PAGES 241–242

1. 3:5 **2.** 2:4:3 **3.** $\frac{2y}{3x}$ **4.** 48; 48; 84 **5.** No **6.** Yes **7.** Yes **8.** Yes **9.** $\angle J$ **10.** SP **11.** 12 **12.** $x = 9$, $y = 13\frac{1}{2}$ **13. a.** $\triangle UVH$ **b.** AA **14.** UV, VH, UH **15.** $\frac{RS}{UH} = \frac{RT}{UV}$ **16.** $\triangle NCD \sim \triangle NBA$ **17.** $\triangle NCD \sim \triangle NAB$ **18.** $\triangle NCD \sim \triangle NAB$ **19.** No **20.** No **21.** (2) **22.** 25 **23.** 14.4 **24.** 12

MIXED REVIEW, PAGE 244

1. 38, 30, 38 **3.** −4.5 **5.** Two planes must be parallel or intersecting. **7.** corresponding, alternate interior, same-side interior **9.** 20 **11.** 1080 **13. a.** No **b.** Yes **c.** No

ALGEBRA REVIEW, PAGE 245

1. 9 **3.** $2\sqrt{6}$ **5.** 7 **7.** $7\sqrt{5}$ **9.** $6\sqrt{2}$ **11.** $2\sqrt{7}$ **13.** 3 **15.** $\frac{\sqrt{15}}{3}$ **17.** $\frac{4\sqrt{15}}{5}$ **19.** $\frac{\sqrt{5}}{4}$ **21.** $\frac{11}{2}$ **23.** $\frac{2\sqrt{6}}{5}$ **25.** 162 **27.** 60 **29.** 4 **31.** $\sqrt{35}$ **33.** $8\sqrt{2}$ **35.** 5 **37.** $2\sqrt{14}$ **39.** 2

Chapter 6

WRITTEN EXERCISES, PAGES 250–251

1. 7 **3.** $\frac{6}{5}$ **5.** $2\sqrt{3}$ **7.** $10\sqrt{7}$ **9.** $\frac{\sqrt{2}}{2}$ **11.** $\frac{\sqrt{6}}{9}$ **13.** $6\sqrt{3}$ **15.** $3\sqrt{2}$ **17.** 4 **19.** $5\sqrt{13}$ **21.** $2\sqrt{15}$ **23.** $x = 10$, $y = 2\sqrt{29}$, $z = 5\sqrt{29}$ **25.** $x = \frac{\sqrt{2}}{6}$, $y = \frac{\sqrt{3}}{6}$, $z = \frac{\sqrt{6}}{6}$ **27.** $x = 12$, $y = 9.6$, $z = 7.2$ **29.** $x = 16$, $y = 12$, $z = 8\sqrt{3}$ **31.** $x = \sqrt{2}$, $y = 2$, $z = \sqrt{2}$ **35. a.** 6, 6; 4, 9; 3, 12; 2, 18; 1, 36 **b.** pq, pq; p^2, q^2; p, pq^2; q, p^2q; 1, p^2q^2 **c.** 14; pqr, pqr; 1, $p^2q^2r^2$; p, pq^2r^2; q, p^2qr^2; r, p^2q^2r; pq, pqr^2; pr, pq^2r; qr, p^2qr; p^2q, qr^2; p^2r, q^2r; pq^2, pr^2; p^2, q^2r^2; q^2, p^2r^2; r^2, p^2q^2 **d.** 122

WRITTEN EXERCISES, PAGES 254-255

1. 5 **3.** 12 **5.** 60 **7.** $3\sqrt{2}$ **9.** 10 **11.** 100 **13.** $\sqrt{2}$ **15.** $10k\sqrt{2}$ **17.** $x = 3$ **19.** $x = 12$ **21.** $x = 4$ **23.** $x = 20$ **25.** 13 **27.** $3\sqrt{2}$ **29.** $\sqrt{l^2 + w^2 + h^2}$ **31.** In a $\triangle$ the sq. of the side opp. an acute $\angle$ is $<$ sum of squares of the other 2 sides. Given: $\triangle ABC$, $\angle C$ acute, $BC = a$, $AC = b$, $AB = c$. Prove: $c^2 < a^2 + b^2$. 1. $\triangle ABC$, $\angle C$ acute (Given) 2. Draw rt. $\triangle A'B'C'$ with $B'C' = a$, $A'C' = b$, $\angle C'$ rt. $\angle$ (Post. 1, 3, 6) 3. Let $A'B' = k$; then $k^2 = a^2 + b^2$ (Pythagorean Th.) 4. $c < k$ (SAS Ineq. Th.), so $c^2 < k^2$ (algebra) 5. $c^2 < a^2 + b^2$ (Subst.) **33.** $h = 12$ **35.** 12

WRITTEN EXERCISES, PAGES 259-260

1. acute **3.** rt. **5.** obt. **7.** rt. **9.** can't tell **11.** $BC = 6$; $6^2 + 7^2 < 11^2$ **13.** $\overline{RM}$. $\angle U$ and $\angle S$ are obt., $\angle R$ and $\angle T$ are acute. **15.** $12 < x \le 16$ **17.** $l^2 > j^2 + k^2 = n^2$ (Given, Pythagorean Th.); $l > n$; $m\angle S > m\angle V = 90$ (SSS Ineq. Th.); $\angle S$ and $\triangle RST$ obt.

WRITTEN EXERCISES, PAGES 264-265

1. a. $5\sqrt{3}$ **b.** $5\sqrt{3}$ **3.** 6; $6\sqrt{2}$ **5.** $\sqrt{5}$; $\sqrt{10}$ **7.** $3\sqrt{2}$; $3\sqrt{2}$ **9.** $7\sqrt{3}$; 14 **11.** 5; 10 **13.** 6; $6\sqrt{3}$ **15.** 21; 28; 14; $7\sqrt{3}$; $14\sqrt{3}$ **17.** 25; 75; 100; $25\sqrt{3}$; $50\sqrt{3}$ **19.** $GF = 22$, $DF = 22\sqrt{2}$, $FE = 11\sqrt{2}$, $DE = 11\sqrt{6}$ **21.** $MN = NL = 4\sqrt{2}$; $KL = 8\sqrt{2}$; $KN = 4\sqrt{6}$ **23.** 16; $16\sqrt{3}$ **25.** 30°-60°-90° $\triangle$ with hyp. 2 has legs 1 and $\sqrt{3}$. Any $\triangle$ with sides in ratio $1:\sqrt{3}:2$ is $\sim$ this 30°-60°-90° $\triangle$, and is thus a 30°-60°-90° $\triangle$. **27.** $8 + 12\sqrt{2} + 4\sqrt{6}$ **29.** $\sqrt{3}$ cm

WRITTEN EXERCISES, PAGES 269-270

1. 13.7 **3.** 48.3 **5.** 55.4 **7.** 57° **9.** 27° **11.** $w = 60$, $z \approx 54$ **13.** 44°, 136° **15. a.** 0.7002; 0.4663; 1.1665 **b.** 60; 1.7321 **c.** No **17.** 1. Acute $\angle$s in rt. $\triangle$ are comp. 2. Given 3. Comp.'s of same $\angle$ are $\cong$ 4. If $m\angle A = m\angle B$, then $\tan A = \tan B$ 5. $\frac{n}{m}$ 7. Subst. **19.** 230

WRITTEN EXERCISES, PAGES 275-276

1. $x \approx 21$, $y \approx 28$ **3.** $x \approx 89$, $y \approx 117$ **5.** $x \approx 28$, $y \approx 10$ **7.** $v \approx 16$ **9.** $v \approx 74$ **11.** 48° **13.** alt., 40 cm; base, 89 cm **15.** 83 m **17.** 350 m **21.** 509 m

WRITTEN EXERCISES, PAGES 278-280

1. 32 m **3.** 50 m **5. a.** 27° **b.** 22 m **c.** $10\sqrt{5} \approx 22$ m **7.** $x \approx 14$; $y \approx 8$; $z \approx 16$ **9.** 32 **11. a.** 14° **b.** 605 lb **c.** No **13.** $\sin A = \frac{5\sqrt{89}}{89}$; $\cos A = \frac{8\sqrt{89}}{89}$ **15.** $\sin A = \frac{2uv}{u^2 + v^2}$; $\cos A = \frac{u^2 - v^2}{u^2 + v^2}$ **17.** $70\frac{1}{2}°$

CHAPTER REVIEW, PAGES 283-284

1. 6 **2.** $5\sqrt{2}$ **3.** $5\sqrt{6}$ **4.** $5\sqrt{3}$ **5.** $3\sqrt{5}$ **6.** $2\sqrt{41}$ **7.** $7\sqrt{2}$ **8.** 12 **9.** acute **10.** not possible **11.** rt. **12.** obtuse **13.** $5\sqrt{3}$ **14.** $7\sqrt{2}$ **15.** 16 **16.** 8 **17.** $2k\sqrt{2}$ **18.** 1.5 **19.** $\frac{2}{3}$ **20.** 3.0777 **21.** 24° **22.** $\frac{5}{13}$ **23.** $\frac{5}{13}$ **24.** 75° **25.** 0.6820 **26.** 89 **27.** 57° **28.** 26

PREPARING FOR COLLEGE ENTRANCE EXAMS, PAGE 286

1. A **2.** C **3.** B **4.** C **5.** E **6.** A **7.** C **8.** A **9.** B **10.** C

CUMULATIVE REVIEW: CHAPTERS 1-6, PAGES 287-291

True-False Exercises **1.** F **3.** T **5.** T **7.** T **9.** T **11.** F **13.** T **15.** F **Multiple-Choice Exercises** **1.** d **3.** e **5.** b **7.** e **Always-Sometimes-Never Exercises** **1.** S **3.** A **5.** S **7.** S **9.** S **11.** A **13.** N **15.** S **17.** A **Algebraic Exercises** **1.** 6 **3.** 20 **5.** 6 **7.** 84 **9.** 11 **11.** $4\frac{3}{4}$ **13.** 49 **15.** 75 **17.** $3\frac{3}{5}$ **19.** 20 cm, 25 cm **21.** 17 **23.** 10.5 **25.** rt. $\triangle$

Completion Exercises **1.** 120 **3.** obtuse **5.** 108 **7.** similar **9.** 24 **11.** $\frac{s}{u}$ **13.** $\frac{15}{17}$ **15.** 45°

Proof Exercises **1.** 1. $\overline{AD} \cong \overline{BC}$; $\overline{AD} \parallel \overline{BC}$ (Given) 2. $ABCD$ is $\square$ (Th. 4-5) 3. $\angle D \cong \angle B$ (Th. 4-2) **3.** 1. $\triangle DAF \cong \triangle BCE$ (Given) 2. $AD = CB$ (Corr. parts $\cong$ $\triangle$s are $\cong$) 3. $CD = AB$ (Given) 4. $ABCD$ is $\square$ (Th. 4-4) **5.** 1. $\overline{SU} \cong \overline{SV}$; $\angle 1 \cong \angle 2$ (Given) 2. $\overline{QS} \cong \overline{QS}$ (Reflex.) 3. $\triangle QUS \cong \triangle QVS$ (SAS) 4. $\overline{UQ} \cong \overline{VQ}$ (Corr. parts $\cong$ $\triangle$s are $\cong$) **9.** 1. $\overline{EF} \parallel \overline{JK}$; $\overline{JK} \parallel \overline{HI}$ (Given) 2. $\overline{EF} \parallel \overline{HI}$ (Th. 2-10) 3. $\angle 2 \cong \angle 3$; $\angle F \cong \angle H$ (If lines $\parallel$, alt. int. $\angle$s $\cong$) 4. $\triangle EFG \sim \triangle IHG$ (AA Sim. Post.) **13.** 1. $WXYZ$ is isos. trap. with

$\overline{XW} \cong \overline{YZ}$ (Given) 2. $\angle WXY \cong \angle ZYX$ (Th. 4-15) 3. $\overline{XY} \cong \overline{XY}$ (Reflex.) 4. $\triangle WXY \cong \triangle ZYX$ (SAS) 5. $\overline{XZ} \cong \overline{YW}$ (Corr. parts $\cong$ $\triangle$s are $\cong$)

Chapter 7

WRITTEN EXERCISES, PAGE 295

1. 1 **3.** 1 **5.** 1 **7.** 1 **9.** Ctr. P, $r = 12$ **17.** $12\sqrt{2}$ **19.** 12 **21.** the seg. from ctr. to any pt. on sphere. **23.** 8 **25.** $\frac{3\sqrt{15}}{4}$

WRITTEN EXERCISES, PAGES 298-301

1. 1. exactly one line 2. Seg. Add. Post. 3. tan. to a $\odot$ from a pt. are $\cong$ 4. Subst. 5. tan. to a $\odot$ from a pt. are $\cong$ 6. Subtr. Prop. = **7.** infinitely many **9.** 10; 10 **11.** $10\sqrt{3}$ **13.** Th.: 2 lines tan. to a $\odot$ at the endpts. of a diam. are $\parallel$. Given: $\overline{AB}$, a diam. of $\odot O$; j tan. to $\odot O$ at A, k tan. to $\odot O$ at B. Prove: $j \parallel k$. Proof: 1. j is tan. to $\odot O$ at A; k is tan to $\odot O$ at B (Given) 2. $j \perp \overline{AB}$; $k \perp \overline{AB}$ (If a line is tan. to a $\odot$, it is $\perp$ to the radius drawn to the pt. of tangency.) 3. $j \parallel k$ (In a plane, 2 lines $\perp$ to the same line are $\parallel$.) **17.** 18; 20; 16 **19.** $4\sqrt{3}$ **21.** 8 **25.** 10; 24; 26

WRITTEN EXERCISES, PAGES 304-306

1. 85 **3.** 150 **5.** 120 **7.** 30 **9. b.** No **11.** 1. $\angle 4 \cong \angle 1$; $\angle 2 \cong \angle 3$ (Vert. $\angle$s $\cong$) 2. $\angle 1 \cong \angle 2$ (Given) 3. $\angle 4 \cong \angle 3$ (Trans.) 4. $\widehat{BD} \cong \widehat{DF}$ (In the same $\odot$, minor arcs are $\cong$ if and only if their central $\angle$s are $\cong$.) **15. a.** 35; 70; 70 **b.** $2n$ **c.** $3k$ **17.** 3800 km

WRITTEN EXERCISES, PAGES 309-311

1. j **3.** 10 **5.** $2\sqrt{14}$ **7.** Given: $\odot O \cong \odot P$; $\widehat{RS} \cong \widehat{TU}$. Prove: $\overline{RS} \cong \overline{TU}$. Proof: 1. Draw $\overline{OR}$, $\overline{OS}$, $\overline{PT}$, and $\overline{PU}$ (Through any 2 pts. there is exactly one line) 2. $\overline{OR} \cong \overline{OS} \cong \overline{PT} \cong \overline{PU}$ (All radii of $\cong$ $\odot$s are $\cong$) 3. $\widehat{RS} \cong \widehat{TU}$ (Given) 4. $\angle ROS \cong \angle TPU$ (In $\cong$ $\odot$s, 2 minor arcs are $\cong$ if and only if their central $\angle$s are $\cong$) 5. $\triangle ROS \cong \triangle TPU$ (SAS) 6. $\overline{RS} \cong \overline{TU}$ (Corr. parts $\cong$ $\triangle$s) **11.** $\widehat{AB} \cong \widehat{BC}$ so $AB = BC$ and $2AB = AB + BC > AC$ by $\triangle$ Ineq. Th. **13.** $12\sqrt{3}$ **15.** $\frac{9\sqrt{2}}{2}$

WRITTEN EXERCISES, PAGES 315-317

1. $x = 30$; $y = 25$; $z = 15$ **3.** $x = 110$; $y = 100$; $z = 100$ **5.** $x = 50$; $y = 130$; $z = 65$ **7.** 1. $\overline{AB} \parallel \overline{CD}$ (Given) 2. Draw $\overline{BC}$ (Through any 2 pts. there is exactly one line) 3. $\angle ABC \cong \angle BCD$ so $m\angle ABC = m\angle BCD$ (If lines $\parallel$, alt. int. $\angle$s $\cong$) 4. $m\angle ABC = \frac{1}{2}m\widehat{AC}$; $m\angle BCD = \frac{1}{2}m\widehat{BD}$ (The meas. of an inscribed $\angle$ is $\frac{1}{2}$ the meas. of the int. arc) 5. $\frac{1}{2}m\widehat{AC} = \frac{1}{2}m\widehat{BD}$ (Subst.) 6. $m\widehat{AC} = m\widehat{BD}$ (Mult. Prop. =) **9.** 1. The tan. line is $\perp$ to the radius drawn to the pt. of tan.; Def. $\perp$ lines 2. Def. semicircle 3. Mult. Prop. = 4. Subst. **13.** 100 **15. a.** In circum. circle, all $\angle$s intercept $\cong$ arcs. **b.** Yes **19.** $JK = \frac{ab}{c}$

WRITTEN EXERCISES, PAGES 319-320

1. 90 **3.** 25 **5.** 55 **7.** 35 **9.** 90 **11.** 60 **13.** 30 **15.** 70 **17.** 95 **19.** 120 **21.** 90; 86; 84; 100 **25.** 3:1

WRITTEN EXERCISES, PAGES 323-325

1. 10 **3.** $\sqrt{21}$ **5.** 6 **7.** 6 **9.** 5 **13.** 12 or 4 **15.** 6 **17.** 9 **19.** 4 **21. a.** $5\sqrt{5}$; **b.** $\sqrt{29}$ **23.** 2

CHAPTER REVIEW, PAGES 329-330

1. chord; secant **2.** radius **3.** diam. **4.** inscribed in **5.** $\perp$ **6.** 10 **7.** $13\sqrt{2}$ **8.** 42 **9.** 110 **10. a.** 180 **b.** $\cong$ **11.** 60 **12.** 120 **13.** $<$ **14.** 13 **15.** 75 **16.** 50; 50 **17.** 220; 140 **18.** 50 **19.** 66 **20.** 30 **21.** 40 **22.** 12 **23.** 9 **24.** $\sqrt{55}$

MIXED REVIEW, PAGE 332

1. $3 < y < 2x + 3$ **3.** 1. $\overline{MN}$ is the median of trap. $WXYZ$ (Given) 2. $\overline{ZY} \parallel \overline{MN} \parallel \overline{WX}$ (Th. 4–16(1)) 3. N is midpt. of $\overline{YX}$ (Def. median of trap.) 4. $\overline{MN}$ bis. $\overline{WY}$ (Th. 4–8 Cor.) **5.** 7, $7\sqrt{3}$ **7.** 160; 20
9. $2\sqrt{34}$, 8 **11.** 1. $\angle 1 \cong \angle 2$; $\angle 2 \cong \angle 3$ (Given) 2. $\overline{AB} \parallel \overline{DC}$ (Th. 2–5) 3. $\overline{AD} \parallel \overline{BC}$ (Post. 11) 4. $ABCD$ is ▱ (Def. ▱) 5. $\overline{AB} \cong \overline{DC}$ (Th. 4–1) **13. a.** inside **b.** on **c.** on **15.** $t = 20$ **17.** $RS = 14$; $ST = 7$

ALGEBRA REVIEW, PAGE 333

1. 1.69 **3.** $\frac{19}{3}$ **5.** $18\sqrt{2}$ **7.** 2826 **9.** 42 **11.** $-\frac{1}{2}$ **13.** 54 **15.** $15\sqrt{2}$ **17.** 96
19. cd **21.** πrl **23.** πd^2 **25.** $\frac{c - by}{a}$ **27.** $\frac{S + 360}{180}$ **29.** $\pm\sqrt{xy}$ **31.** $\frac{2A}{b}$

Chapter 8

WRITTEN EXERCISES, PAGES 338–339

9.

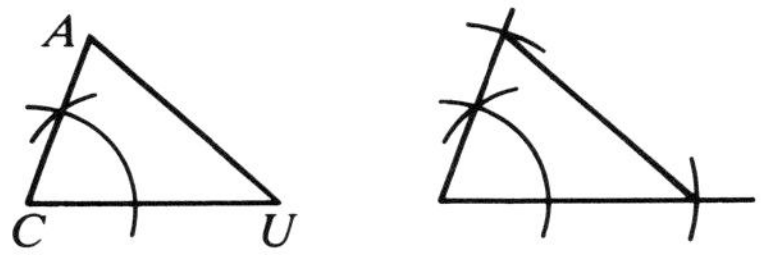

13.

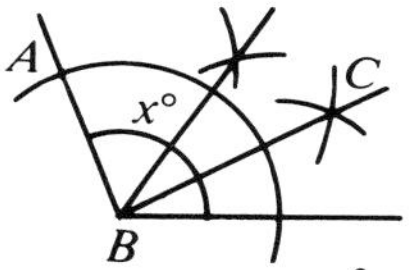

$m\angle ABC = \frac{3}{4}x$

15. a.

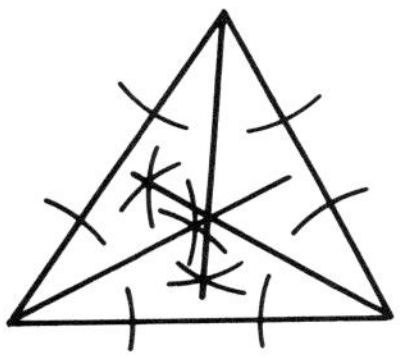

c. The pts. of int. of the ∠ bis. are equidistant from the sides of the △.

19. Methods will vary; e.g.,

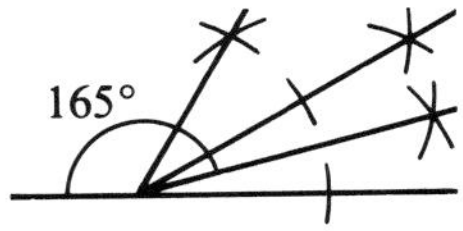

WRITTEN EXERCISES, PAGES 342–343

7. Const. 6 **11.** Methods will vary; e.g.,

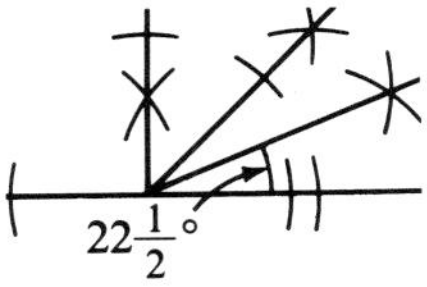

15. a.

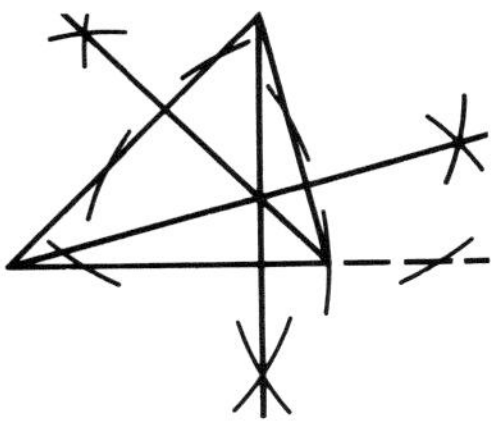

b. Yes; yes **c.**

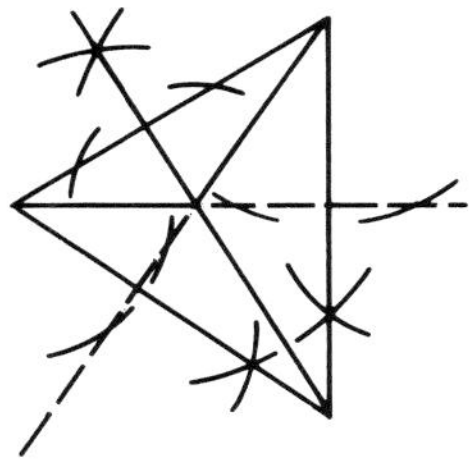

19. $p = 2a$, so $s = \frac{1}{4}(2a) = \frac{1}{2}a$. Methods will vary; e.g., Draw a line and construct $\overline{AB}$ such that $AB = a$. Const. the $\perp$ bis. of $\overline{AB}$, int. $\overline{AB}$ at M. Const. a $\perp$ to $\overline{AB}$ at A. Const. $\overline{AD}$ and $\overline{MC}$ both $\cong$ to $\overline{AM}$. Draw $\overline{DC}$.

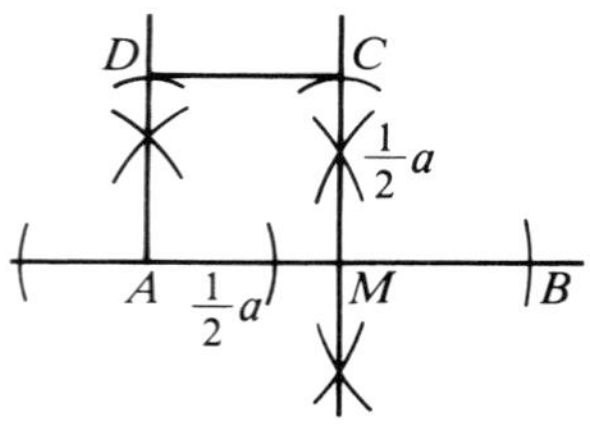

WRITTEN EXERCISES, PAGES 346–347

1. a. any acute △ **b.** any obt. △ **c.** any rt. △ **3.** 2; 4 **5.** 3.8; 5.7 **9.** Let X, Y, and Z be collinear. **11.** $PU = \frac{1}{2}AP$; $2x = \frac{1}{2}x^2$, $x = 4$ **13.** $CP = \frac{2}{3}CW$; $z^2 - 15 = \frac{2}{3}(2z^2 - 5z - 12)$; $3z^2 - 45 = 4z^2 - 10z - 24$; $z^2 - 10z + 21 = 0$; $(z - 7)(z - 3) = 0$; $z = 7$ or $z = 3$; if $z = 3$, $CP = -6$ so reject $z = 3$; $z = 7$; $CW = 51$; $PW = \frac{1}{3}(51) = 17$. **15. a.** pts. in the interior of $\angle XPY$ **b.** pts. in the interior of the $\angle$ vert. to $\angle XPY$

WRITTEN EXERCISES, PAGES 352–353

7. Const. 11 **9.** Draw ⊙ with radius r. Choose pt. A on ⊙O and with ctr. A and radius r, mark off $\widehat{AB}$. With ctr. B and radius r, mark off $\widehat{BC}$. Continue to divide the ⊙ into 6 $\cong$ arcs. Draw chords $\overline{AC}$, $\overline{EC}$ and $\overline{AE}$.

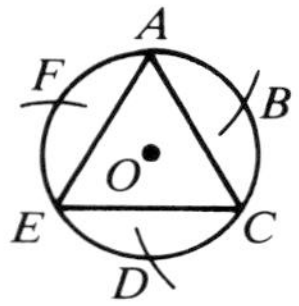

11. Draw diam. $\overline{AC}$. Const. the $\perp$ bis. of $\overline{AC}$, int. ⊙O at B and D. Const. tangents to ⊙O at A, B, C, and D, int. at $EFGH$ as shown.

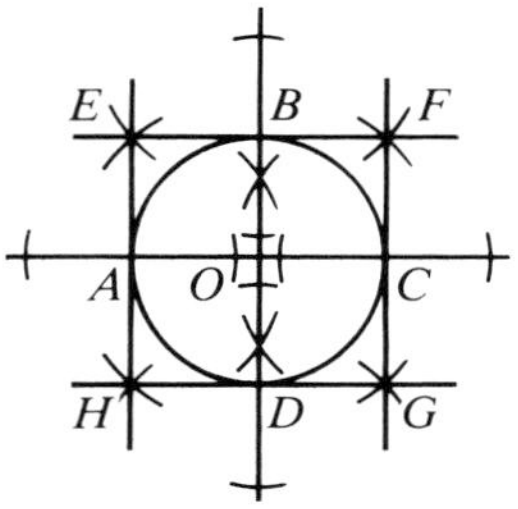

Ex. 11

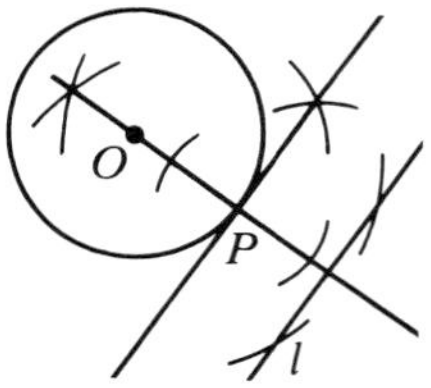

Ex. 15

15. Const. a $\perp$ to l through O, int. ⊙O at P. Const. a tangent to ⊙O at P.

WRITTEN EXERCISES, PAGE 356

3. a. Use Const. 12 to divide $\overline{AB}$ into 5 $\cong$ parts, $\overline{AW}$, $\overline{WX}$, $\overline{XY}$, $\overline{YZ}$, and $\overline{ZB}$ **b.** No **c.** $AX:XB = 2:3$ **9.** $\frac{z}{w} = \frac{y}{x}$ or $\frac{z}{y} = \frac{w}{x}$; use Const. 13 **11.** Use Const. 14 to const. $\overline{XY}$ with length $\sqrt{yp}$; use Const. 12 to divide $\overline{XY}$ into 3 $\cong$ parts with length x. **13.** Use Const. 1 to const. 2 segs., the prod. of whose lengths is $6yz$, for example, $2y$ and $3z$ or $6y$ and z; use Const. 14. **15.** Draw a line and const. $\overline{AB}$ so that $AB = 3$ and $\overline{BC}$ so that $BC = 5$. Use Const. 14.

WRITTEN EXERCISES, PAGES 360–362

1. the $\perp$ bis. of $\overline{AB}$ **3.** a circle with ctr. O and $r = 2$ cm **5.** the seg. joining midpoints of $\overline{BC}$ and $\overline{DA}$ **7.** diag. $\overline{BD}$ **9.** a plane that is $\parallel$ to both given planes halfway between them **11.** a sphere with ctr. E and $r = 3$ cm **13. a.** Bisect $\angle HEX$ **b.** Bisect the angles formed by j and k (two lines). **15.** Construct the ⊙ (excluding pts. A and B) with ctr. the midpt. of $\overline{AB}$ and $r = \frac{1}{2}AB$.

17. Construct a ⊙ concentric to the given ⊙ with radius half as great. (See diagram at right.)

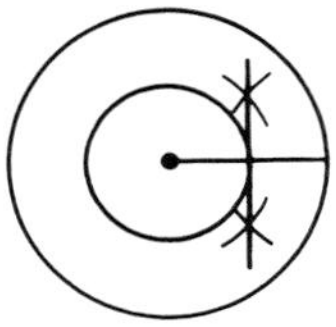

19. A line ⊥ to the plane of the square at the int. of the diag.

WRITTEN EXERCISES, PAGES 364–366

3. **a.** a ⊙ with ctr. D, $r = 1$ cm **b.** a ⊙ with ctr. E, $r = 2$ cm

c.

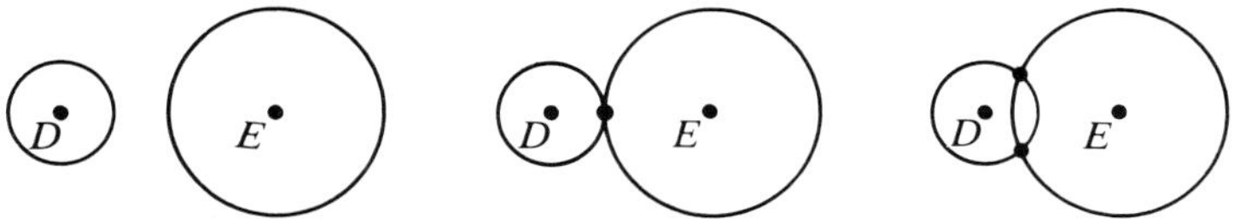

$DE > 3$ cm $DE = 3$ cm $DE < 3$ cm

d. The locus is 0, 1, or 2 pts., depending on the int. of ⊙D and ⊙E.

5. The locus is int. of ⊙P, with $r = 3$ cm, and l. (2 pts.)

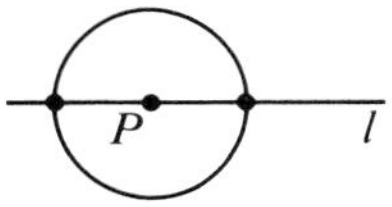

7. The locus is the int. of ⊙A and ⊙B, each with $r = 2$ cm. (2 pts.)

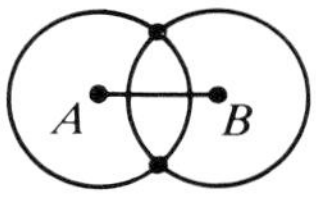

9. The locus is the int. of ⊙A, with $r = 2$ cm, and the bis. of ∠A. (1 pt.)

11.

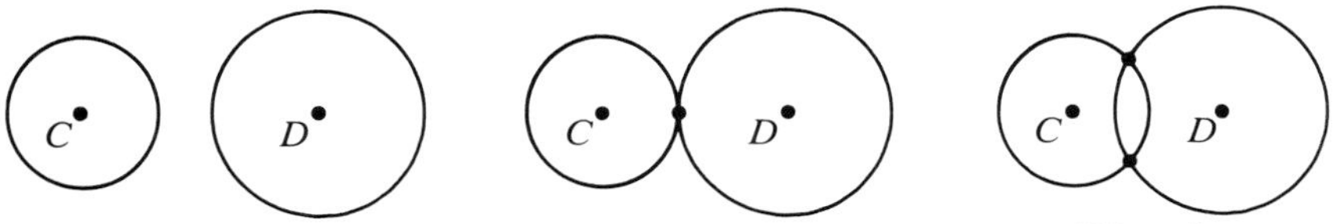

$CD > 5$ cm $CD = 5$ cm $CD < 5$ cm

13. Figures may vary. Example:

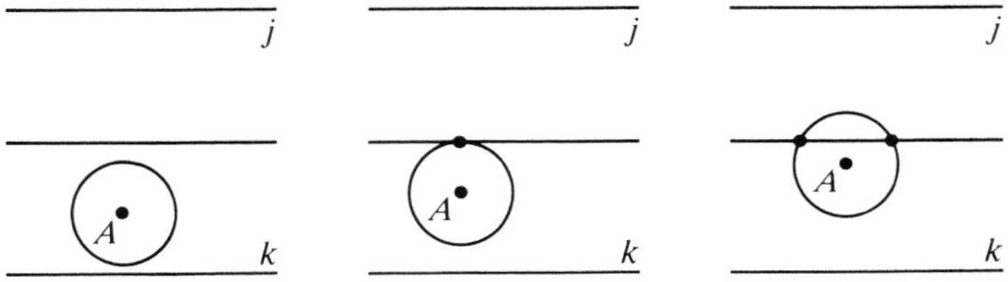

15. no pts., 1 pt., or a ⊙ **19. a.** a plane ⊥ $\overline{RS}$ that bis. $\overline{RS}$ **b.** a plane ⊥ $\overline{RT}$ that bis. $\overline{RT}$ **c.** line; line **d.** a plane ⊥ $\overline{RW}$ that bis. $\overline{RW}$ **e.** point; point

WRITTEN EXERCISES, PAGES 368–369

1. The locus is j and k.

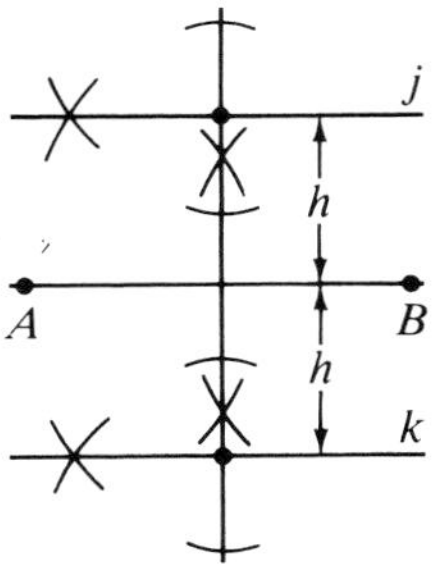

3. Const. the bis. of $\angle XYZ$. Const. a line $\parallel$ to, and a units from, $\overleftrightarrow{YZ}$. $\odot W$ is the required $\odot$.

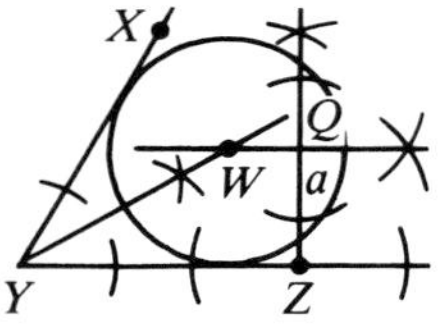

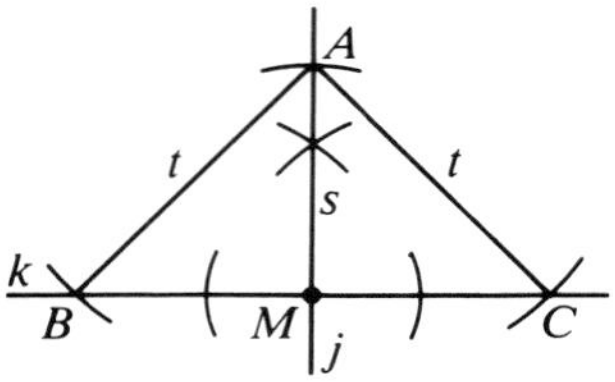

7. Construct line $j \perp k$. Mark off $MA = s$, then construct $\overline{AB}$ and $\overline{AC}$.

11. Const. $\angle A$. Const. a line $\parallel$ to and s units from $\overrightarrow{AX}$ in order to locate pt. C. Const. $\overline{BC} \perp \overline{AC}$

Ex. 11

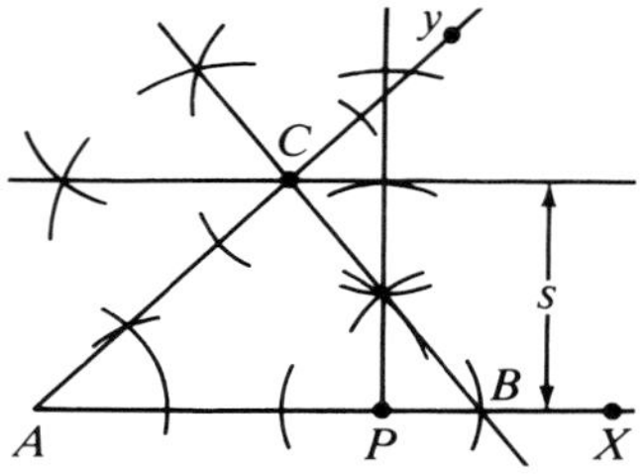

Ex. 13

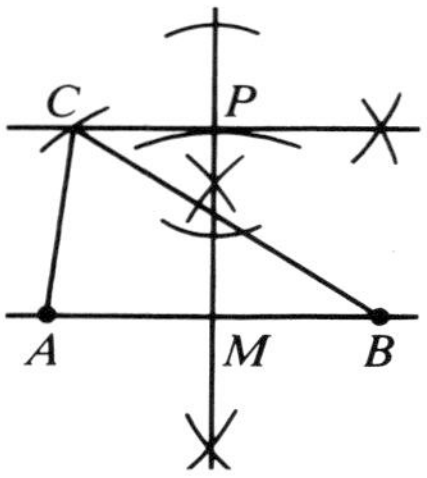

13. Const. $\overline{AB}$ such that $AB = t$. Const. the $\perp$ bis. of $\overline{AB}$ intersecting $\overline{AB}$ at M. Const. a line that is $\parallel$ to, and r units from, $\overleftrightarrow{AB}$. Let C be the intersection of this line and $\odot M$ with radius s.

EXTRA, PAGES 370–371

3. Some of the points are the same: L and R, M and S, N and T. **5.** 1. $NM = \frac{1}{2}AB = XY$; $\overline{NM} \parallel \overline{AB} \parallel \overline{XY}$ (Th. 4–17) 2. $XYMN$ is a ▱ (Th. 4–5) 3. $\overline{CR} \perp \overline{AB}$ (Def. alt.) 4. $\overline{CR} \perp \overline{XY}$ (Th. 2–4). 5. $\overline{NX} \parallel \overline{CR}$ (Th. 4–17) 6. $\overline{NX} \perp \overline{XY}$ (Th. 2–4) 7. $XYMN$ is a rect. (Th. 4–13).

CHAPTER REVIEW, PAGES 372–373

1. Const. 1 **2.** Const. 2 **3.** Const. 3 **4.** Const. 4 **5.** Const. 5 **6.** Const. 6 **7.** Const. 7 **8.** $\perp$ bisectors of the sides **9.** bisectors of the ∠s **10.** 18 **11.** 1:2 **12.** Const. 8 **13.** Const. 9 **14.** Const. 11 **15.** Const. 10 **16.** $t = \sqrt{bc}$; $\frac{b}{t} = \frac{t}{c}$ or $\frac{c}{t} = \frac{t}{b}$; Const. 14 **17.** $\frac{a}{b} = \frac{c}{t}$ or $\frac{a}{c} = \frac{b}{t}$; Const. 13 **18.** Const. 1, Const. 12 **19.** a line $\parallel$ to both given lines located halfway between them **20.** the $\perp$ bisecting plane of $\overline{AB}$ **21.** a plane $\parallel$ to both given planes located halfway between them. **22.** a line $\perp$ to the plane of $\triangle HJK$ at the pt. where the $\perp$ bis. of the sides int. **23.** the int. of the $\perp$ bis. of $\overline{PQ}$ and the $\odot$ with ctr. P and $r = 8$ cm (2 pts.) **24.** a $\odot$ with ctr. R and $r = 8$ cm that lies in a plane $\perp$ l **25.** The locus is 2 circles, a $\odot$ and a pt., a $\odot$, 1 pt., or no pts., depending on the int. of (1) a sphere with crt. Z and $r = 2$ m and (2) 2 planes $\parallel$ to, and 1 m from, the given plane.

26. Const. a rt. $\angle$, $\angle C$. Bis. $\angle C$ and mark off $\overline{CX}$ such that $CX = \frac{1}{2}a$. Const. a $\perp$ $\overline{CX}$ at X, int. the sides of $\angle C$ at A and B. (By const., the meas. of alt. to the hyp. of rt. $\triangle ABC$ is $\frac{1}{2}a$. $\triangle AXC \cong \triangle BXC$ by ASA, so $AC = BC$.)

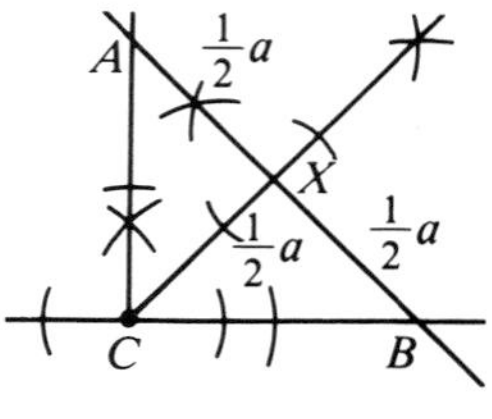

27. Const. $\overline{RS}$ such that $RS = a$, and then bis. $\overline{RS}$. Using R as ctr. and radius c, swing an arc. Using M as ctr. and radius b, swing another arc. Locate T at the int. of these arcs. Draw $\overline{RT}$ and $\overline{ST}$.

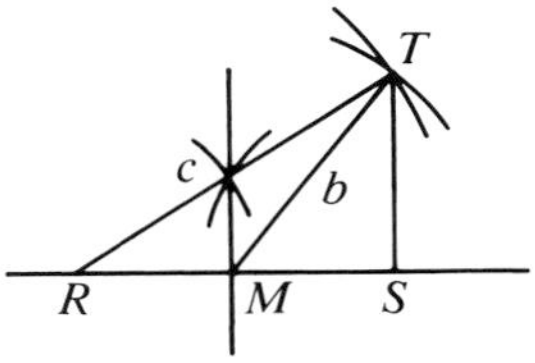

PREPARING FOR COLLEGE ENTRANCE EXAMS, PAGE 375

1. B **2.** C **3.** E **4.** A **5.** B **6.** C **7.** A **8.** C **9.** E

CUMULATIVE REVIEW: CHAPTERS 1-8, PAGES 376-377

1. S **3.** S, A **5.** N **7.** A **9.** S **11.** A **13.** S **15.** A **17.** N **19.** A **21.** S **23.** 9 **25.** 39 **27.** 5 possibilities: 2 circles, 1 circle and a point, a circle, a point, no points **29. a.** 3.6 **b.** $\frac{30}{7}$ **31.** 18 **33.** Plan for proof: $m\widehat{PQ} = 2 \cdot m\angle 1$; $m\angle POQ = m\widehat{PQ}$; $\angle POQ$ is rt. $\angle$; $\triangle OPQ$ is isos. rt. $\triangle$ **35.** 8 **37.** $4\sqrt{6}$ **39.** Plan for const.: Bis. $\overline{XY}$ (Const. 4); draw line through midpt. M; const. $\overline{AM} \cong \overline{XM}$ and $\overline{BM} \cong \overline{XM}$ (Const. 1); draw rect. $AXBY$

Chapter 9

WRITTEN EXERCISES, PAGES 382-384

1. 60 cm² **3.** 5 cm **5.** 24 **7.** $2x^2 - 6x$ **9.** 36 cm², 26 cm **11.** 5 cm, 80 cm² **13.** $a^2 - 9$, $4a$ **15.** $x - 3$, $4x - 6$ **17.** 130 **19.** 48 **21.** $24\frac{1}{2}$ **23.** $40\,xy$ **25.** \$408 **27.** 36 **29.** $36\sqrt{3}$ **31.** 12 cm by 36 cm **33.** $8\sqrt{3}$, 8 **35. a.** $100 - 2x$; $A = x(100 - 2x)$ **c.** 25 m by 50 m

WRITTEN EXERCISES, PAGES 387-390

1. 28 cm² **3.** 12 **5.** 6 **7.** $6y$ **9.** 48 **11.** $18\sqrt{3}$ **13.** 44 **15.** 10 cm **17. a.** area $\triangle ABC = 40$; area $\triangle ABM = 20$ **b.** $BC = 2\,BM$; area $\triangle ABC = \frac{1}{2}BC \cdot h = BM \cdot h$; area $\triangle ABM = \frac{1}{2}BM \cdot h = \frac{1}{2} \cdot$ area $\triangle ABC$ **19.** 18 **21.** 20 **23.** $8\sqrt{3}$ **25.** $48\sqrt{3}$ **27.** 240 **29.** Diag. of rhombus are $\perp$; diag. $\overline{AC}$ forms $2 \cong$ ⚠. Thus, area of $ABCD = 2 \cdot$ area $\triangle ABC = 2\left(\frac{1}{2} \cdot d_1 \cdot \frac{1}{2} d_2\right) = \frac{1}{2} d_1 d_2$ **31. a.** 100 **b.** $100\sqrt{2}$ **c.** $100\sqrt{3}$ **d.** 200 **e.** $100\sqrt{3}$ **f.** $100\sqrt{2} \approx 140$, $100\sqrt{3} \approx 170$ **33.** 504 cm², 936 cm² **37. b.** $x + y + z$ has constant value = alt. of given $\triangle$.

WRITTEN EXERCISES, PAGES 392-393

1. 70 **3.** 6 **5.** 5 **7.** 1 **9.** 10; 5; $3\frac{3}{4}$ **11.** 180 **13.** 64 **15.** $\frac{27\sqrt{3}}{4}$ **17.** 84 **19. a.** 15 **b.** 74 **21.** 80 **23.** $36\sqrt{3}$, $72\sqrt{3}$, $108\sqrt{3}$ **27.** 204 **29.** 122

WRITTEN EXERCISES, PAGES 398-399

1. 8; 256 **3.** $\frac{7\sqrt{2}}{2}$; $\frac{7}{2}$ **5.** 3; $18\sqrt{3}$; $27\sqrt{3}$ **7.** $\frac{4\sqrt{3}}{3}$; $\frac{2\sqrt{3}}{3}$; $4\sqrt{3}$ **9.** $2\sqrt{3}$; 24; $24\sqrt{3}$ **11.** $4\sqrt{3}$; $24\sqrt{3}$; $72\sqrt{3}$ **13.** $36\sqrt{3}$ **15.** $216\sqrt{3}$ **17.** 0.5; 5.19; 1.30 **19.** 0.865; 6; 2.595 **21. a.** $m\angle AOX = \frac{1}{2}m\angle AOB = \frac{1}{2}\left(\frac{360}{10}\right) = 18$ **b.** $AX \approx 0.3090$, $OX \approx 0.9511$ **c.** 6.18 **d.** 0.2939 **e.** 2.939 **23. a.** $m\angle AOX = \frac{1}{2}m\angle AOB = \frac{1}{2}\left(\frac{360}{n}\right) = \frac{180}{n}$ **b.** $\sin\,(m\angle AOX)^\circ = \sin\left(\frac{180}{n}\right)^\circ = \frac{AX}{1}$; $AX = \sin\left(\frac{180}{n}\right)^\circ$ **c.** $\cos\,(m\angle AOX)^\circ = \cos\left(\frac{180}{n}\right)^\circ = \frac{OX}{1}$; $OX = \cos\left(\frac{180}{n}\right)^\circ$ **d.** $p = n \cdot AB = 2n \cdot AX = 2n \cdot \sin\left(\frac{180}{n}\right)^\circ$ **e.** $A = n\left(\frac{1}{2}AB \cdot OX\right) = n \cdot AX \cdot OX = n \cdot \sin\left(\frac{180}{n}\right)^\circ \cdot \cos\left(\frac{180}{n}\right)^\circ$

WRITTEN EXERCISES, PAGES 404–406

1. 14π, 49π **3.** 5π, $\frac{25}{4}\pi$ **5.** 10, 100π **7.** 5, 10π **9.** 264; 5544 **11.** 8; $5\frac{1}{11}$ **13.** 62.8; 314.0 **15.** 1.6; 0.2 **17.** 15-inch **19.** Area I $= \frac{9\pi}{2}$, Area II $= 8\pi$, Area III $= \frac{25\pi}{2}$. Area I + Area II = Area III **21.** $4r^2 + \pi r^2$ **23.** $2\pi r^2$ **25.** 2200 cm (22 m) **27.** 9π; 18π **29.** 27π; 36π **31. a.** r **b.** $A = \pi r^2$ **33.** $A = \frac{C^2}{4\pi}$ **35.** Radius is hyp. of rt. △ of which radii of O and P are legs.

WRITTEN EXERCISES, PAGES 408–409

1. 5π, 25π **3.** 2π, 12π **5.** 2π, 3π **7.** 5π, $\frac{25}{2}\pi$ **9.** $\frac{4\pi}{3}$, 4π **11.** 15π, $\frac{375\pi}{2}$ **13.** 6 **15.** $4\pi - 8$ **17.** $\frac{25\pi}{8}$ **19.** $18\pi - 18\sqrt{3}$ **21.** $100\pi - 192$ **23.** 15.2 **25.** $4\pi - 8$ **27.** $36\sqrt{3} - 18\pi$ **29.** $40\sqrt{3} + \frac{110\pi}{3}$

WRITTEN EXERCISES, PAGES 412–414

1. 1:4; 1:16 **3.** 6:7; 36:49 **5.** 3:13; 9:169 **7.** 3:8; 3:8 **9.** 49:121 **11.** 1:2; 1:4 **13.** $x^2:y^2$ **15.** Answers may vary. $\triangle ABG \sim \triangle CEG$, 9:25; $\triangle AFG \sim \triangle CBG$, 9:25; $\triangle EFD \sim \triangle EBC$, 4:25; $\triangle ABF \sim \triangle DEF$, 9:4 **17. a.** 1:4 **b.** 1:2 **c.** 1:2 **d.** 1:1 **19.** 16:65 **21.** 1:3

CHAPTER REVIEW, PAGE 418

1. 64 **2.** $8\sqrt{5}$ **3.** 18 cm² **4.** 240 **5.** 9 **6.** $24 + 8\sqrt{3}$; $32\sqrt{3}$ **7.** 7 **8.** 24 **9.** $30 + 4\sqrt{2}$; 52 **10.** 36 m² **11.** $9\sqrt{3}$ **12.** $6\sqrt{3}$ cm² **13.** 188.4; 2826 **14.** 22 cm **15.** $8\pi\sqrt{2}$ **16.** 18π **17.** $24\pi + 9\sqrt{3}$ **18.** $\frac{169\pi}{4} - 30$ **19. a.** 9:16 **b.** 3:4 **20.** 1:4 **21.** 75

MIXED REVIEW, PAGE 420

1. 1. $\overline{AB} \perp \overline{BC}$; $\overline{DC} \perp \overline{BC}$ (Given) 2. $m\angle ABC = 90 = m\angle DCB$ (Def. ⊥ lines, rt. ∠) 3. $\overline{AC} \cong \overline{BD}$ (Given) 4. $\overline{BC} \cong \overline{BC}$ (Reflex.) 5. $\triangle ABC \cong \triangle DCB$ (HL Th.) 6. $\angle 1 \cong \angle 2$ (Corr. parts. ≅ ▵) 7. $\overline{CE} \cong \overline{BE}$ (Th. 3–2) 8. $\triangle BCE$ is isos. (Def. isos. △) **3. a.** $\overline{BE}$ **b.** $\overline{EV}$ **5.** $4\sqrt{3}$ **7.** 61 **9.** 1. $\overline{EF} \cong \overline{HG}$; $\overline{EF} \parallel \overline{HG}$ (Given) 2. $EFGH$ is ▱ (Th. 4–5) 3. $\overline{EH} \parallel \overline{FG}$ (Def. ▱) 4. $\angle EHF \cong \angle GFH$ (If lines ∥ alt. int. ∠s ≅) **11. a.** $\frac{40}{3}\pi$ **b.** 80π **13.** 1440 **15.** sphere P with $r = 4$ cm and its interior **17.** 32°

ALGEBRA REVIEW, PAGE 421

1. 9π **3.** 75π **5.** $\frac{400}{9}\pi$ **7.** $\frac{49}{8}\pi$ **9.** 36π **11.** 32π **13.** $5\pi\sqrt{2}$ **15.** 12π **17.** 1256 **19.** $\frac{792}{7}$ **21.** 11 **23.** $\frac{A}{2\pi r}$ **25.** 30

Chapter 10

WRITTEN EXERCISES, PAGES 426–428

1. 40; 88; 48 **3.** 3; 54; 90 **5.** 6; 168; 108 **7.** 54; 27 **9.** 10; 600 **11.** 5; 125 **13.** 4; 8 **15.** 240; $240 + 32\sqrt{3}$; $160\sqrt{3}$ **17.** 252; 372; 420 **19.** 576; $576 + 192\sqrt{3}$; $1152\sqrt{3}$ **21.** 2700 cm³ **23.** 1.8 kg **25.** 19 kg **27.** $50x^3$; $120x^2$ **29.** 3 cm × 6 cm × 9 cm **31. a.** 192; 288 **b.** $216 + 16\sqrt{13}$ **33.** 64 cm³ **35.** apoth. of base $= \frac{s\sqrt{3}}{2}$, p of base $= 6s$; $B = \frac{1}{2} \cdot \frac{s\sqrt{3}}{2} \cdot 6s = \frac{3s^2\sqrt{3}}{2}$; $V = \frac{3s^2\sqrt{3}}{2} \cdot h = \frac{3}{2}\sqrt{3}s^2h$

WRITTEN EXERCISES, PAGES 432–434

1. 27 **3.** 192 **5.** 6; $\sqrt{34}$ **7.** 25; $\sqrt{674}$ **9.** 3; $\sqrt{41}$ **11.** 60; 96; 48 **13.** 480; 736; $\frac{256\sqrt{161}}{3}$ **15. a.** $VX = 15$ cm, $VY = 13$ cm **b.** 384 cm²; pyr. is not reg. **17.** Vol. of pyr. $= \frac{1}{6}$ of vol. of rect. solid **19.** $h = 8$, $l = \sqrt{73}$ **21. a.** $OM = 3$, $OA = 6$, $BC = 6\sqrt{3}$ **b.** L.A. $= 45\sqrt{3}$, $V \doteq 36\sqrt{3}$

23. L.A. = 144, $V = 24\sqrt{39}$ **25.** $\frac{e^3\sqrt{2}}{12}$ **27.** $V_1:V_2 = h_1:h_2$; diag. of base $= 10\sqrt{2}$; $\tan 40° = \frac{h_1}{5\sqrt{2}}$, $\tan 80° = \frac{h_2}{5\sqrt{2}}$; $h_1:h_2 = 5\sqrt{2}\tan 40° : 5\sqrt{2}\tan 80° = \tan 40° : \tan 80°$

WRITTEN EXERCISES, PAGES 439–441

1. 40π, 72π, 80π **3.** 24π, 56π, 48π **5.** $r = 4$ **7.** 48π **9.** 5; 20π; 36π; 16π **11.** 5; 156π; 300π; 240π **13.** 12; 9; 324π; 432π **15. a.** $\frac{1}{4}$ **b.** $\frac{1}{4}$ **c.** $\frac{1}{8}$ **17. b.** can: 62.5π cm³; bottle: 48π cm³ **19.** 2200π cm³ **21.** 9; 27 **23.** $r = 3$ **25. a.** 15π; 12π **b.** 20π; 16π **c.** No **27.** 16π **29.** cone: $l = \sqrt{6^2 + 8^2} = 10$; L.A. $= \pi rl = (6\pi) \cdot 10 = 60\pi$; pyr.: $l = \sqrt{10^2 - 3^2} = \sqrt{91}$; L.A. $= \frac{1}{2}pl = \frac{1}{2} \cdot 36 \cdot \sqrt{91} = 18\sqrt{91}$ **31.** $\frac{16\pi\sqrt{2}}{3}$

WRITTEN EXERCISES, PAGES 446–448

1. 36π; 36π **3.** π; $\frac{\pi}{6}$ **5.** 4; $\frac{256\pi}{3}$ **7.** 8π; $\frac{8\sqrt{2}\pi}{3}$ **9.** 4; 8 **11.** 21π cm² **13.** V of hemi. $= 4 \cdot V$ of sphere **15.** 6 **17.** 8 **19.** 1:2 **21.** Yes **23.** A of sphere $= 4\pi r^2$; L.A. of cyl. $= 2\pi rh = 2\pi r(2r) = 4\pi r^2$ **25. a.** $\frac{1324\pi}{3}$ cm³ **b.** $V \approx 400\pi$ cm³ **c.** thin **27.** $h = 2x$; $r^2 = 10^2 - x^2 = 100 - x^2$; $V = \pi r^2 h = \pi(100 - x^2)(2x) = 2\pi x(100 - x^2)$ **b.** $\frac{4000\sqrt{3}\pi}{9}$ **29.** 144π

WRITTEN EXERCISES, PAGES 455–457

1. Yes **3. a.** 3:4 **b.** 3:4 **c.** 9:16 **d.** 27:64 **5. a.** 4:1 **b.** 16:1 **c.** 64:1 **7.** 9,953,280 in.³ (or 213 yd³) **9. a.** 2:3 **b.** 2:3 **c.** 4:9 **11.** large one **13.** 48 kg **15.** 108 **17. a.** 9:16 **b.** 9:16 **c.** 9:7 **d.** 27:64 **e.** 27:37 **19.** upper, 54 cm³; lower, 196 cm³ **21.** $V_1:V_2 = \frac{4}{3}\pi a^3 : \frac{4}{3}\pi b^3 = a^3 : b^3$ **23.** $r_1:r_2 = l_1:l_2$; $A_1:A_2 = \pi r_1 l_1 : \pi r_2 l_2 = {r_1}^2 : {r_2}^2$ **25.** $B_1:B_2 = {e_1}^2:{e_2}^2$; $h_1:h_2 = e_1:e_2$; $V_1:V_2 = B_1h_1:B_2h_2 = {e_1}^2e_1:{e_2}^2e_2 = {e_1}^3:{e_2}^3$ **27.** $6\sqrt[3]{4}$

EXTRA, PAGE 461

1. 22 **3.** $\frac{56\pi}{3}$ **5.** 8 cm²

CHAPTER REVIEW, PAGE 463

1. lateral edge **2.** 672 **3.** 236; 240 **4.** 558 **5.** $\frac{160\sqrt{3}}{3}$ **6.** 4; 60 **7.** 900; 1020; 17 **8.** 8; 2400 **9.** 24π; 56π **10.** 60π; 96π; 96π **11.** $2\sqrt{10}$ **12.** doubles **13.** 616 **14.** 288π ft³ **15.** 20 **16.** $\frac{5324\pi}{3}$ cm³ **17.** 1:3 **18.** 1:9 **19.** 1:26 **20.** 64:27

PREPARING FOR COLLEGE ENTRANCE EXAMS, PAGE 465

1. A **2.** C **3.** C **4.** C **5.** B **6.** C **7.** B

CUMULATIVE REVIEW: CHAPTERS 1–10, PAGES 466–467

1. F **3.** T **5.** F **7.** F **9.** T **11.** T **13.** F **15.** T **17.** T **19. a.** 46 **b.** 1:4 **21.** 1. $\overline{WZ} \perp \overline{ZY}$; $\overline{WX} \perp \overline{XY}$ (Given) 2. $m\angle Z = 90 = m\angle X$ (Def. ⊥ lines, rt. ∠) 3. $\overline{WX} \cong \overline{YZ}$ (Given) 4. $\overline{WY} \cong \overline{WY}$ (Reflex.) 5. $\triangle WXY \cong \triangle YZW$ (HL Th.) 6. $\angle XYW \cong \angle ZWY$ (Corr. parts ≅ ⚠ are ≅) 7. $\overline{WZ} \parallel \overline{XY}$ (Th. 2–5) **23.** 360; 400 **25.** 129 **27.** 125π cm³ **29.** 1. $ABCD$ is ▱, $\overline{AC} \perp \overline{BD}$ at P (Given) 2. $\overline{AP} \cong \overline{PC}$ (Th. 4–3) 3. $\overline{BD}$ is ⊥ bis. of $\overline{AC}$ (Def. ⊥ bis.) 4. $\overline{AB} \cong \overline{BC}$ (Th. 3–5) 5. $\overline{DC} \cong \overline{AB}$; $\overline{BC} \cong \overline{AD}$ (Th. 4–1) 6. $\overline{DC} \cong \overline{AB} \cong \overline{BC} \cong \overline{AD}$ (Trans. Prop.) 7. $ABCD$ is a rhombus (Def. rhombus) **31.** Use Const. 14 with y as the length of the alt. and z as the length of one seg. of the hyp. **33.** 13 **35. a.** $12 + 3\pi$ **b.** $9\pi - 18$

Chapter 11

WRITTEN EXERCISES, PAGES 471–472

1. 7 **3.** 5 **5.** 4 **7.** $\sqrt{10}$ **9.** $4\sqrt{2}$ **11.** $2\sqrt{5}$ **13.** $\sqrt{130}$ **15.** 10 **17.** $\sqrt{(a-c)^2+(b-d)^2}$ **19.** (10, 0), (8, 6), (6, 8), (0, 10), (−6, 8), (−8, 6), (−10, 0), (−8, −6), (−6, −8), (0, −10), (6, −8), (8, −6) **21.** $AM = 3\sqrt{5} = AY$ **23.** $AB = 2\sqrt{2}$; $BC = 6$; $AC = 2\sqrt{5}$ **25.** $\frac{JA}{RF} = \frac{6}{3} = \frac{2}{1}$; $\frac{AN}{FK} = \frac{2\sqrt{5}}{\sqrt{5}} = \frac{2}{1}$; $\frac{JN}{RK} = \frac{4\sqrt{2}}{2\sqrt{2}} = \frac{2}{1}$ **27.** 39 **29.** −11, −1, 2, 5, 15 **31.** (3, 2)

WRITTEN EXERCISES, PAGES 477–478

1. (4, 3); 9 **3.** (−3, 0); 7 **5.** $(j, -14)$; $\sqrt{17}$ **7.** $x^2 + y^2 = 4$ **9.** $(x-2)^2 + (y+1)^2 = 1$ **11.** $(x+4)^2 + (y+7)^2 = g^2$ **15.** $(x-5)^2 + (y-5)^2 = 25$ **17.** $x^2 + (y-6)^2 = 100$ **19.** $x^2 + (y-2)^2 = 4$ **21.** −2, 6 **23.** The ⊙ with ctr. (0, 3) and $r = 3$, excluding points (0, 0) and (0, 6). **25.** (−2, 4); 6

WRITTEN EXERCISES, PAGES 480–481

1. (3, 3) **3.** (4, 5) **5.** (1.9, 0.4) **7.** $B(4, 10)$ **9.** $B(-13, 2b-2)$ **11.** 8 **13. a.** $OK = 6 = SE$; $OS = 2\sqrt{5} = KE$ **b.** ▱ **c.** (4, 2) **d.** (4, 2) **e.** The diag. of a ▱ bis. each other. **15. a.** $\left(-\frac{5}{2}, \frac{7}{2}\right)$; $\left(-\frac{5}{2}, \frac{7}{2}\right)$ **b.** ▱; If the diag. of a quad. bis. each other, then the quad. is a ▱. **17.** $T(4, 5)$

WRITTEN EXERCISES, PAGES 485–486

1. 1 **3.** −1 **5.** −1 **7.** 0 **9.** $-\frac{a}{b}$ **11.** $\frac{s-v}{r-t}$ **13.** 9 **15.** $m(r-p)+q$ **17.** Answers will vary; e.g., **a.** (0, −1), (−1, −3), (2, 3) **b.** (0, −7), (1, −10), (−1, −4) **19.** 1, −1; $(1)\cdot(-1) = -1$ **21.** $\frac{6}{5}$, $-\frac{5}{6}$; $\left(\frac{6}{5}\right)\cdot\left(-\frac{5}{6}\right) = -1$ **23.** No. Slope of $\overline{DE} = \frac{7}{3}$ = slope of $\overline{FG}$; slope of $\overline{EF}$ must be $\frac{7}{3}$ for all 4 pts. to be collinear, but slope of $\overline{EF} = \frac{1}{2}$. **25.** $2\sqrt{3} + 1$

WRITTEN EXERCISES, PAGES 488–490

1. a. $\frac{1}{2}$ **b.** $\frac{1}{2}$ **c.** −2 **3.** $\frac{7}{2}$; 0; $\frac{7}{2}$; 0 **5. a.** −3; −3 **b.** Slope of $\overline{PN}$ = slope of $\overline{LM}$ **c.** $\frac{1}{3}$; $-\frac{1}{7}$ **d.** Slope of $\overline{MN}$ ≠ slope of $\overline{LP}$ **e.** trap. **7.** Slope of $\overline{AC} = -\frac{5}{2}$, slope of $\overline{AB} = \frac{3}{7}$, slope of $\overline{BC} = \frac{8}{5}$; slope of alt. to $\overline{AC} = \frac{2}{5}$, slope of alt. to $\overline{AB} = -\frac{7}{3}$, slope of alt. to $\overline{BC} = -\frac{5}{8}$ **9.** Slope of $\overline{DE} = 0$, slope of $\overline{EF} = -3$, slope of $\overline{DF} = 1$; slope of med. to $\overline{DE} = 3$, slope of med. to $\overline{EF} = \frac{3}{7}$, slope of med. to $\overline{DF} = -\frac{3}{5}$ **11.** −1 **13.** trap.; $\overline{HI} \parallel \overline{JK}$ (= slopes) and $\overline{HK} \nparallel \overline{IJ}$ (≠ slopes) **15.** rect.; $\overline{HI} \parallel \overline{JK}$ (= slopes), $\overline{HK} \parallel \overline{IJ}$ (= slopes) and $\overline{HI} \perp \overline{IJ}$ (prod. of slopes is −1) **17.** $\frac{3}{4}$ **19. a.** T **b.** T **c.** T **21.** $T(0, 0)$, $A(16, 8)$, and $Y(14, 6)$

WRITTEN EXERCISES, PAGES 494–496

1. line passes through pts. (0, 2) and (6, 0) **3.** line passes through pts. (3, 0) and (0, 2) **5.** line passes through pts. (0, 5) and (−15, 0) **7.** $y - 3 = \frac{3}{4}(x-2)$ **9.** $y - 7 = 4x$ Answers will vary in Exs. 11–13; e.g.: **11.** $y - 2 = \frac{1}{4}(x-1)$ **13.** $y + 2 = -\frac{2}{5}(x-7)$ **15.** $y = \frac{5}{7}x - 9$ **17.** (1, 2) **19.** (0, 4) **21.** $5x - 2y = 13$ **23.** Answers will vary; e.g., $2x + y = 8$ **25.** $y = 1$ **27.** $y = 8$ **29.** $x = 7$

Answers will vary in Exs. 31–37; e.g.: **31.** $2x + 5y = 13$ **33.** $x + 2y = -1$ **35.** $-4x + 3y = 12$ **37. a.** $x + 5y = 18$; $x - 4y = 0$; $2x + y = 18$ **b.** (8, 2) is pt. of int. of $\overleftrightarrow{EA}$ and $\overleftrightarrow{OB}$. **c.** $\overleftrightarrow{FC}$ passes through (8, 2) since $2(8) + 2 = 18$. **d.** $FP = 2\sqrt{5}$, $FC = 3\sqrt{5}$; $\frac{2}{3}FC = \frac{2}{3} \cdot 3\sqrt{5} = FP$

WRITTEN EXERCISES, PAGES 500–501

1. $(0, b)$, $(a, 0)$ **3.** $(-f, 2f)$, $(f, 2f)$ **5.** $(h + m, n)$ **7.** $\left(\frac{s}{2}, \frac{s\sqrt{3}}{2}\right)$ **9.** $C(\sqrt{a^2 - b^2}, b)$, $B(a + \sqrt{a^2 - b^2}, b)$

WRITTEN EXERCISES, PAGES 503–504

1. Plan for proof: Use the distance formula twice to show that each distance is $\sqrt{a^2 + b^2}$.

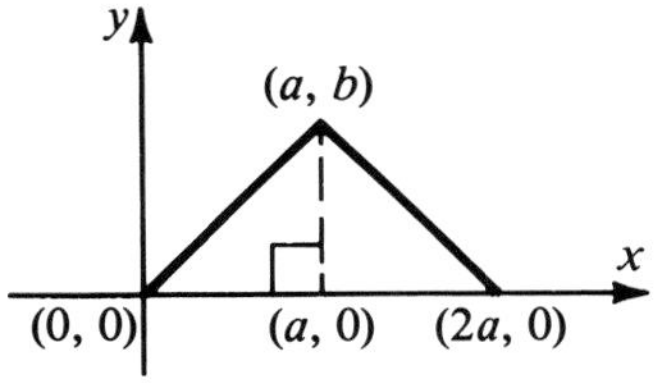

Ex. 1

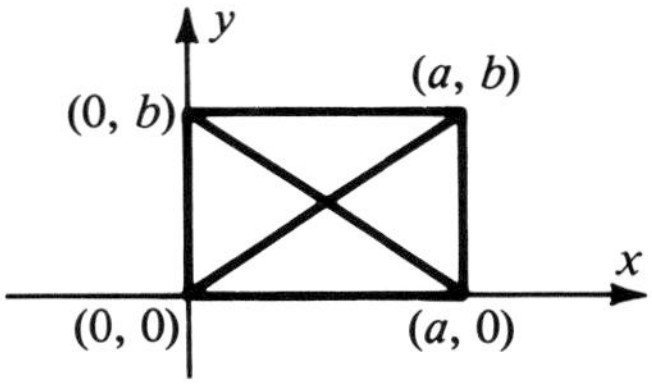

Ex. 3

3. Plan for proof: Use the distance formula twice to show that the length of each diagonal is $\sqrt{a^2 + b^2}$.

CHAPTER REVIEW, PAGES 505–506

1.

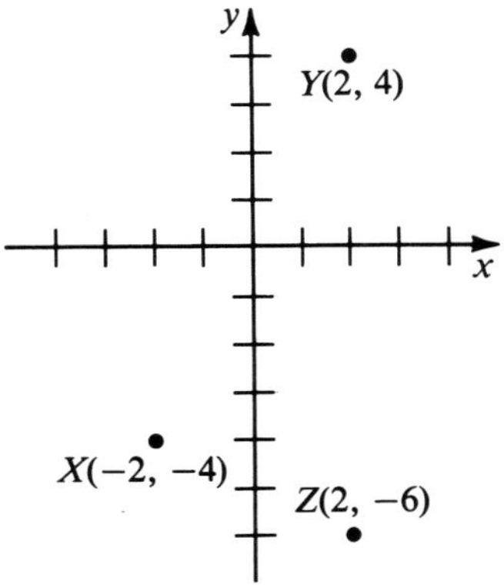

2. $4\sqrt{5}$, 10, $2\sqrt{5}$ **3.** $(4\sqrt{5})^2 + (2\sqrt{5})^2 = 10^2$ **4.** $(-3, 0)$; 10 **5.** $(5, -1)$; 7 **6.** $(x - 2)^2 + (y + 3)^2 = 9$ **7.** $(x + 6)^2 + (y + 1)^2 = 25$ **8.** $\left(4, -\frac{3}{2}\right)$ **9.** $(-1, 0)$ **10.** $(0, b)$ **11.** $(-11, 11)$ **12.** $-\frac{1}{4}$ **13.** $(0, -19)$ **14.** Answers will vary; e.g., (2, 8), (−1, −7), (1, 3) **15.** 0 **16.** slope of $\overline{TQ} = \frac{3}{4}$ = slope of $\overline{SR}$, $\overline{TQ} \parallel \overline{SR}$; slope of $\overline{ST} = -1$, slope of $\overline{QR} = 0$, $\overline{ST} \nparallel \overline{QR}$ **17.** $\frac{3}{4}$, $-\frac{4}{3}$ **18.** (3, 1)

19.

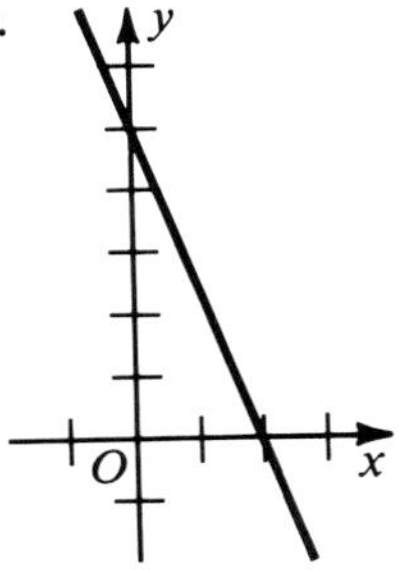

Answers will vary in Exs. 20–21; e.g.: **20.** $y - 2 = -2(x - 5)$ **21.** $x + 3y = -11$ **22.** $\frac{2}{3}$; -2 **23.** (3, 1) **24.** $C(0, e)$, $A(d, 0)$ **25.** $(a, a\sqrt{3})$ or $(a, -a\sqrt{3})$ **26.** Plan for proof: Using $G(h, h)$, $E(2h, 0)$, and $F(2h, h)$ show that the slope of $\overline{OG} \cdot$ slope of $\overline{GE} = -1$. **27.** Plan for proof: Let the coordinates be $E(2k, 0)$, $F(2k, h)$, and $G(k, h)$. Use dist. formula to show that $OG = GE = \sqrt{k^2 + h^2}$. **28.** Plan for proof: Let the coordinates of the vertices be (0, 0), (2b, 2c), (a + 2b, 2c), and (a, 0). Midpts. of two opp. sides are (b, c) and (a + b, c). Show that the slopes of the seg. and the other two sides are =.

MIXED REVIEW, PAGE 508

1. a. $\frac{AE}{CE} = \frac{BE}{DE} = \frac{2}{3}$; $\angle AEB \cong \angle CED$ (vert. ∠s $\cong$); $\triangle AEB \sim \triangle CED$ (SAS Sim. Th.); $\angle B \cong \angle D$ (Def. sim. polygons) **b.** $x = 18$ **c.** 4:9 **3.** rt. **5.** $x = 6\sqrt{5}$; $y = 8$ **7.** 1. In $\triangle ABC$, $\overline{CD}$ bis. $\angle C$; $\overline{CD} \perp \overline{AB}$ (Given) 2. $\angle ACD \cong \angle BCD$ (Def. $\angle$ bis.) 3. $m\angle ADC = 90 = m\angle BDC$ (Def. $\perp$ lines, rt. $\angle$) 4. $\overline{CD} \cong \overline{CD}$ (Reflex.) 5. $\triangle ACD \cong \triangle BCD$ (ASA) 6. $\overline{AC} \cong \overline{BC}$ (Corr. parts $\cong$ △s) 7. $\triangle ABC$ is isos. (Def. isos. $\triangle$)
9. $6\frac{2}{3}$ **11.** 89° **15. a.** $\frac{1}{3}$ **b.** $\frac{1}{3}$ **c.** $2\sqrt{2}$ **d.** $\frac{2\sqrt{2}}{3}$

ALGEBRA REVIEW, PAGE 509

1. -216 **3.** 32 **5.** $\frac{1}{16}$ **7.** r^5 **9.** 1 **11.** $\frac{1}{x^3}$ **13.** 81 **15.** 1 **17.** 8 **19.** $6y^6$ **21.** 1
23. 7 **25.** $\frac{1}{x^4}$ **27.** $16x^6y^4$ **29.** $-32s^7t^3$ **31.** $\frac{27}{8}$

Chapter 12

WRITTEN EXERCISES, PAGES 513-515

1. a. $A'(4, 2)$, $B'(8, 4)$, $C'(6, -2)$ **b.** isometry **c.** (8, 8) **3. a.** $A'(0, 12)$, $B'(12, 18)$, $C'(6, 0)$ **b.** not an isometry **c.** (4, 2) **5. a.** $A'(12, 4)$, $B'(8, 6)$, $C'(10, 0)$ **b.** isometry **c.** (0, 6) **7. a.** Yes **b.** Yes **c.** No **9.** $\overline{DB}$ and $\overline{AC}$ int. at X; P on $\overline{DC}$ maps to pt. where $\overrightarrow{PX}$ int. $\overline{AB}$. Does not preserve distance. **11.** $A'(5, 3)$, $B'(7, 3)$, $C'(9, 5)$, $D'(7, 5)$; $ABCD$ has area 4, $p = 8$; $A'B'C'D'$ has area 4, $p = 4 + 4\sqrt{2}$ **13.** Yes **15.** No

WRITTEN EXERCISES, PAGES 518-520

1. 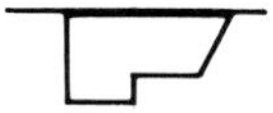 **3.** 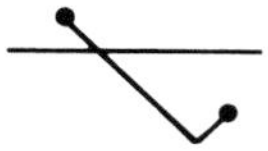 **5.**  **7. a.** $(2, -4)$ **b.** $(-2, 4)$

9. a. (0, 2) **b.** $(0, -2)$ **11. a.** $(-3, 2)$ **b.** $(3, -2)$ **13.** (4, 2) **15.** $(-2, 0)$ **17.** $(-2, -3)$ **19.** Since $P = P'$, prove $PQ = PQ'$. 1. j bis. $\overline{QQ'}$ (Def. reflection) 2. Let j int. $\overline{QQ'}$ at X, then $\overline{QX} \cong \overline{Q'X}$ and $\angle QXP \cong \angle Q'XP$ (Def. $\perp$ bis.; all rt. ∠s $\cong$) 3. $\overline{PX} \cong \overline{PX}$ (Reflex.) 4. $\triangle QXP \cong \triangle Q'XP$ (SAS) 5. $PQ = PQ'$ (Corr. parts $\cong$ △s) **23.** t is $\perp$ bis. of $\overline{BB'}$ **25.** Path from ball to hole hits walls first at X, then at Y. By Th. 12-1, $YH = YH'$, $XH' = XH''$. Dist. traveled $= BX + XY + YH = BX + XY + YH' = BX + XH' = BX + XH'' = BH''$. **27.** Aim for image of hole under reflection in two walls, as in Ex. 25. **29.** Yes
31. $y = -x - 5$ **33. a.** (6, 3) **b.** $(10, -2)$ **c.** (13, 1) **d.** $(10 - x, y)$ **35.** $x = 0$ **37.** $x = 2$
39. $y = x$ **41.** $y = -\frac{5}{3}x + 6$

WRITTEN EXERCISES, PAGES 524-525

1. a. $A'(6, 6)$, $B'(8, 10)$, $C'(11, 5)$ **c.** Yes, Yes **3.** (8, 4) **5.** $(-4, -3)$ **7.** $A'(1, 4)$, $B'(4, 6)$, $C'(5, 10)$; $A''(-1, 4)$, $B''(-4, 6)$, $C''(-5, 10)$ **9.** $(-x, y + 4)$ **11.** To translate $\odot A \parallel$ to $\overline{CD}$ through dist. CD toward B, const. line t through $A \parallel$ to $\overline{CD}$ and take A' on t so that $AA' = CD$. Draw $\odot A' \cong \odot A$, int. $\odot B$ in two pts., call one Y. To locate X, the preimage of Y on $\odot A$, const. line s through $Y \parallel$ to $\overline{CD}$; X is int. of s and $\odot A$ with $XY = CD$. **13.** $(x - 4, y + 9)$ **15.** Midpts. lie on reflection line. Proof: Let X be glide image of A. In $\triangle AXA'$, refl. line is $\parallel \overline{AX}$ and bis. $\overline{XA'}$, so it bis. $\overline{AA'}$ (Th. 5-3).

WRITTEN EXERCISES, PAGES 527-529

1. $\mathcal{R}_{O,440}$ **3.** $\mathcal{R}_{A,90}$ **5.** $\mathcal{R}_{O,180}$ **7.** C **9.** D **11.** E **13.** D **15.** reflection **17.** translation
19. half-turn **21.** translation **23.** (3), (8) **25.** 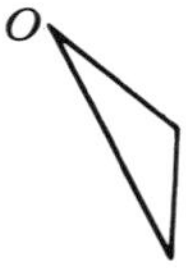

27. (4, 2) **29. b.** $A'(-3, 0)$, $B'(-1, 4)$ **c.** slopes $-\frac{1}{2}$ and 2; lines ⊥ **d.** Th. 12–3 **31.** Label the intersection of l and l' point X, and label an angle between l and l' adjacent to $\angle FXF'$ as $\angle 1$. Then $x + 90 + 90 + m\angle FXF' = 360$, so $x = 180 - m\angle FXF'$ (sum of meas. of ∡ of quad. = 360; Subtr. Prop. =). Also $m\angle 1 = 180 - m\angle FXF'$, so $x = m\angle 1$ (Angle Add. Post.; Subst.). **33. a.** $\mathcal{R}_{C,90}$ **b.** A rotation is an isometry. **c.** $\overline{AD}$ is image of $\overline{EB}$ under 90° rotation. **35.** Rotate l through 90° about A, intersecting k at X. Let Z be preimage of X on l. Const. ⊥s to $\overline{AX}$ at X and $\overline{AZ}$ at Z, int. at Y.

WRITTEN EXERCISES, PAGES 532–533

1. $A'(12, 0)$, $B'(8, 4)$, $C'(4, -4)$ **3.** $A'(3, 0)$, $B'(2, 1)$, $C'(1, -1)$ **5.** $A'(-12, 0)$, $B'(-8, -4)$, $C'(-4, 4)$ **7.** $A'(6, 0)$, $B'(7, -1)$, $C'(8, 1)$ **9.** 4, expansion **11.** $\frac{1}{3}$, contraction **13.** 4, expansion **15.** $P'(-3, 3)$, $Q'(0, -3)$, $R'(12, 0)$, $S'(6, 6)$ **17.** $P'(-6, 0)$, $Q'(-6, -8)$, $R'(-12, -12)$, $S'(-10, 2)$ **19. a.** 3:1 **b.** 9:1 **21. a.** $\frac{2}{3}$ **b.** $\frac{1}{2}$ **c.** $D_{G,-\frac{1}{2}}$ **d.** N; P

WRITTEN EXERCISES, PAGES 538–540

1. F' is to the right of j, F'' is to the right of k. **3.** F' is to the right of k, F'' is to the left of F. **9. a.** Q **b.** S **c.** M **d.** Q **11.** Q **13.** (1, 2) **15.** $(-5, -1)$ **17.** $(-1, 2)$ **19.** $(-1, 1)$ **21.** *c.* By Th. 12–6, each prod. translates P through twice the dist. between reflection lines. These dist. are =, and each prod. uses right line of pair for first reflection, so images coincide. **23. a.** The seg. that joins the midpts. of two sides of a △ has a length = $\frac{1}{2}$ the length of the third side. (Th. 4–17(2)) **b.** Yes **c.** Yes; translation **25. a.** one; infinitely many (all pts. on y-axis); one; none **b.** (3, 0)

WRITTEN EXERCISES, PAGES 542–543

1. $\frac{1}{4}$ **3.** $\frac{3}{2}$ **5.** C **7.** A **9.** C **11.** A **13.** C **15.** I **17.** H_O **19.** $(x + 6, y + 8)$ **21.** $S^{-1}:(x, y) \rightarrow (x + 3, y + 1)$ **23.** $S^{-1}:(x, y) \rightarrow (4x, 4y)$ **25.** $S^{-1}(x, y) \rightarrow (y, x)$ **27.** $\mathcal{R}_{O,72}$ **29. a.** They are inverses.

WRITTEN EXERCISES, PAGES 546–547

1. a. 5 **b.** No **c.** $\mathcal{R}_{O,0} = I$, $\mathcal{R}_{O,72}$, $\mathcal{R}_{O,144}$, $\mathcal{R}_{O,216}$, $\mathcal{R}_{O,288}$ **3. a.** 4 **b.** Yes **c.** $\mathcal{R}_{O,0} = I$, $\mathcal{R}_{O,90}$, $\mathcal{R}_{O,180} = H_O$, $\mathcal{R}_{O,270}$ **5.** A, B, C, D, E, K, M, T, U, V, W, Y **7.** H, I, N, O, S, X, Z

9.
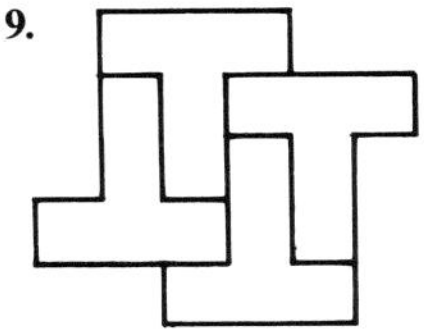

11.
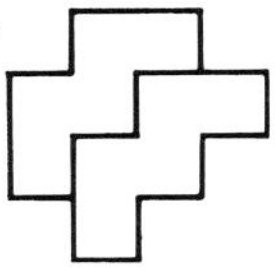

13.

15.
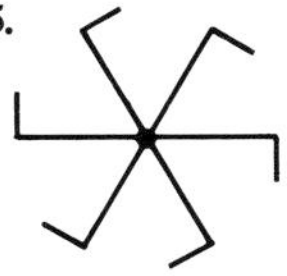

17.
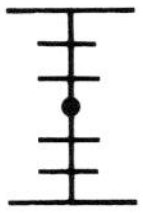

19. a.

b.
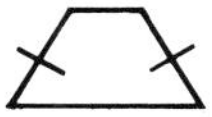

c. not poss. **21. a.** a reg. octagon **b.**

c.
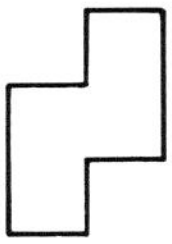

23. a. at the midpts. of the sides **b.** trans. sym.
27. $(50 \cdot n)°$ rot. sym. for all integers n, which is equivalent to $(10 \cdot n)°$ rot. sym. for all integers n

EXTRA, PAGES 549–550

1. Let t be the ⊥ bis. of the base.

∘	I	R_t
I	I	R_t
R_t	R_t	I

3.

∘	I	$\mathcal{R}_{O,120}$	$\mathcal{R}_{O,240}$
I	I	$\mathcal{R}_{O,120}$	$\mathcal{R}_{O,240}$
$\mathcal{R}_{O,120}$	$\mathcal{R}_{O,120}$	$\mathcal{R}_{O,240}$	I
$\mathcal{R}_{O,240}$	$\mathcal{R}_{O,240}$	I	$\mathcal{R}_{O,120}$

7.

$\circ$	I	$\mathcal{R}_{O,120}$	$\mathcal{R}_{O,240}$	R_j	R_k	R_l
I	I	$\mathcal{R}_{O,120}$	$\mathcal{R}_{O,240}$	R_j	R_k	R_l
$\mathcal{R}_{O,120}$	$\mathcal{R}_{O,120}$	$\mathcal{R}_{O,240}$	I	R_l	R_j	R_k
$\mathcal{R}_{O,240}$	$\mathcal{R}_{O,240}$	I	$\mathcal{R}_{O,120}$	R_k	R_l	R_j
R_j	R_j	R_k	R_l	I	$\mathcal{R}_{O,120}$	$\mathcal{R}_{O,240}$
R_k	R_k	R_l	R_j	$\mathcal{R}_{O,240}$	I	$\mathcal{R}_{O,120}$
R_l	R_l	R_j	R_k	$\mathcal{R}_{O,120}$	$\mathcal{R}_{O,240}$	I

9. a. No **b.** Yes **c.** I and $\mathcal{R}_{O,180}$

CHAPTER REVIEW, PAGES 551–552

1. $\cong$ **2.** (6, 1) **3.** $\left(\frac{3}{2}, 5\right)$ **4.** No

5.

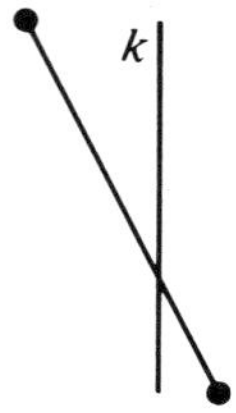

6.

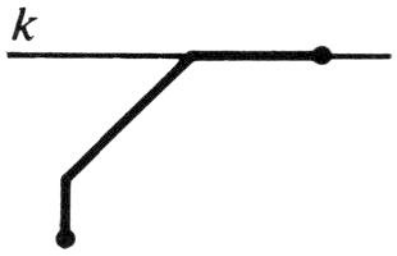

7.

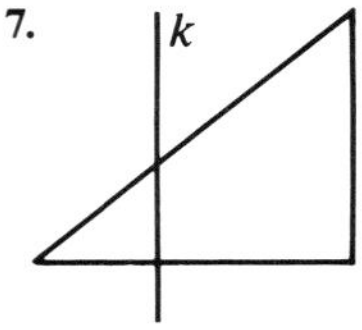

8. (7, 5)

9. (5, 0) **10. a.** (2, −4) **b.** $(x + 2, y - 4)$ **c.** (1, 6) **11.** (12, 2) **12. a.** $A'(-2, 3)$, $B'(-1, -1)$, $C'(3, 1)$ **b.** $A'(-3, -2)$, $B'(1, -1)$, $C'(-1, 3)$ **13.** $\mathcal{R}_{O,500}$ and $\mathcal{R}_{O,-220}$ **14.** (0, 4) **15.** (0, −1) **16.** (2, 4) **17.** (−3, −1) **18.** (3, −1) **19.** (1, 3) **20.** $(x + 1, y - 6)$ **21.** $O, \frac{1}{5}$ **22.** I **23.** −75 **24.** No **25.** Yes **26.** Yes **27.** reg. pentagon

PREPARING FOR COLLEGE ENTRANCE EXAMS, PAGE 554

1. B **2.** A **3.** E **4.** C **5.** C **6.** E **7.** A **8.** E **9.** B **10.** D

CUMULATIVE REVIEW: CHAPTERS 1–12, PAGES 555–559

True-False Exercises **1.** F **3.** T **5.** F **7.** T **9.** T **11.** T **13.** F **15.** F **17.** F **19.** F **21.** F **23.** T **25.** T **Multiple-Choice Exercises** **1.** b **3.** a **5.** c **7.** d **9.** a **11.** d **13.** b **15.** b **Completion Exercises** **1.** $-\frac{3}{2}$ **3.** vertical **5.** (−6, 1) **7.** rt. **9.** $n - 3$ **11.** 33 **13.** 512 cm^3 **15.** SAS, ASA, and AAS **17.** 6 **19.** $6\sqrt{3}$ **21.** $22\frac{1}{2}$ **23.** $10\sqrt{3}$ **25.** $\frac{15}{17}$ **27.** $27\sqrt{7}$ **29.** (−5, 2) **Always-Sometimes-Never Exercises** **1.** N **3.** S **5.** S **7.** S **9.** N **11.** S **13.** N **15.** N **17.** N **19.** S **21.** A **23.** S **Construction Exercises** **3.** Const. $\angle BAC$ with $m\angle BAC = 30$ by const. a 60° $\angle$ and bisecting it; mark off $AD = x$ on $\overrightarrow{AB}$ and $AE = x$ on $\overrightarrow{AC}$; draw $\overline{DE}$. **5.** Const. $l \perp m$ at A; on m mark off $AB = y$; from B, locate D on l such that $BD = x$; const. $n \perp l$ at D; on n mark off $DC = y$; draw $\overline{BC}$. **7.** Const. $l \perp m$ at D; locate A on m such that $AD = x$; locate B and C on l such that $DB = y = DC$; draw $\overline{AB}$ and $\overline{AC}$. **9.** Plan for Const.: Use Const. 12 to divide $\overline{AB}$ into 10 = seg.; mark off a length consisting of 3 = seg. and a length consisting of 2 = seg.; use these lengths for consecutive sides of a ▱. **Proof Exercises** **1.** 1. $\overline{OP} \cong \overline{OQ}$; $\overline{OS} \cong \overline{OR}$ (Given) 2. $\angle POS \cong \angle QOR$ (Vert. $\angle$s are $\cong$) 3. $\triangle POS \cong \triangle QOR$ (SAS Post.) 4. $\overline{PS} \cong \overline{QR}$ (Corr. parts $\cong$ $\triangle$s are $\cong$) **3.** 1. $\overline{PR} \perp \overline{QS}$ (Given) 2. $m\angle POS = 90 = m\angle QOR$ (Def. $\perp$ lines, rt. $\angle$) 3. $\overline{PS} \cong \overline{QR}$; $\overline{OS} \cong \overline{OR}$ (Given) 4. $\triangle POS \cong \triangle QOR$ (HL Th.) 5. $\angle PSO \cong \angle QRO$ (Corr. parts $\cong$ $\triangle$s are $\cong$) **5.** 1. $ABCD$ is a rect. (Given) 2. X is midpt. of $\overline{AC}$ and $\overline{BD}$ (The diag. of a ▱ bis. each other.) 3. $AC = 2AX = 2CX$; $BD = 2BX = 2DX$ (Midpt. Th.) 4. $AC = BD$ (The diag. of a rect. are $\cong$.) 5. $2AX = 2BX = 2CX = 2DX$ (Subst. Prop.) 6. $AX = BX = CX = DX$ (Div. Prop. =)

Logic

EXERCISES, PAGES 577–578

1. I like the city and you like the country. **3.** You do not like the country. **5.** I like the city or you do not like the country. **7.** I do not like the city or you do not like the country. **9.** It is not true that "I like the city or you like the country." **11.** $p \vee q$ **13.** $\sim(p \vee q)$ **15.** $\sim(p \wedge q)$ **17.** Yes

19.

p	q	$\sim q$	$p \vee \sim q$
T	T	F	T
T	F	T	T
F	T	F	F
F	F	T	T

25.

p	q	r	$q \vee r$	$p \wedge (q \vee r)$
T	T	T	T	T
T	T	F	T	T
T	F	T	T	T
T	F	F	F	F
F	T	T	T	F
F	T	F	T	F
F	F	T	T	F
F	F	F	F	F

EXERCISES, PAGES 579–580

1. If you like to paint, then you are an artist. **3.** If you are not an artist, then you do not draw landscapes. **5.** If you like to paint and you are an artist, then you draw landscapes. **7.** If you draw landscapes or you are an artist, then you like to paint. **9.** $b \rightarrow w$ **11.** $(\sim b \vee \sim w) \rightarrow s$ **13.** $\sim(b \rightarrow s)$ **15. a.** yes; no **b.** yes; yes

17.

p	q	$p \rightarrow q$	$\sim(p \rightarrow q)$
T	T	T	F
T	F	F	T
F	T	T	F
F	F	T	F

19. $\sim(p \rightarrow q)$ and $p \wedge \sim q$

EXERCISES, PAGES 581–582

1. 1. Given; 2. Step 1, Simplification; 3. Given; 4. Steps 2, 3, Modus Ponens **5.** $a \wedge b$ (Given); a (Step 1 and Simplification); $a \rightarrow \sim c$ (Given); $\sim c$ (Steps 2, 3, Modus Ponens); $c \vee d$ (Given); d (Steps 4, 5, Disj. Syllogism) **7.** Given: $j \rightarrow d$; $d \rightarrow c$; $j \wedge w$. Prove: c. **9.** Given: $t \vee r$; $t \rightarrow p$; $r \rightarrow s$; $\sim p$. Prove: s.

EXERCISES, PAGE 583

1. 1. Given; 2. Step 1, Double Neg. 3. Given 4. Steps 2, 3, Modus Tollens **5.** $p \vee (\sim q)$ (Given); $(\sim q) \vee p$ (Step 1, Comm. Rule); q (Given); $\sim(\sim q)$ (Step 3, Double Neg.); p (Steps 2, 4, Disj. Syllogism) **9.** Given: $s \rightarrow w$; $\sim s \rightarrow p$; $\sim p$. Prove: w.

EXERCISES, PAGE 585

1. $p \wedge r$ **3.** $s \wedge (t \vee p)$ **5.** $(t \vee s) \wedge (\sim t \vee s)$ **7.** Electricity passes through $p \vee \sim p$ but never through $p \wedge \sim p$.

Projections; Dihedral Angles

EXERCISES, PAGES 586–587

1. yes **3.** yes **5.** no **7.** yes **9.** yes **15.** 2 $\parallel$ lines, a line, 2 pts. **17.** an equilateral $\triangle$, an isos. $\triangle$, a scalene $\triangle$, a line seg. **19.** a ▱, a rhombus, a rectangle, a square, a line seg.

EXERCISES, PAGE 588

5. Planes X and Y are $\perp$ if a plane $\angle$ of the dihedral $\angle$ formed by X and Y is a rt. $\angle$. **7.** If 2 $\parallel$ planes are cut by a third plane, alt. int. dihedral $\angle$s are $\cong$, and same-side dihedral $\angle$s are supp.; if a plane is $\perp$ to one of 2 $\parallel$ planes, it is $\perp$ to the other, also. **9. b.** 55 **11. b.** 55

Technical Drawing

EXERCISES, PAGE 605

1.

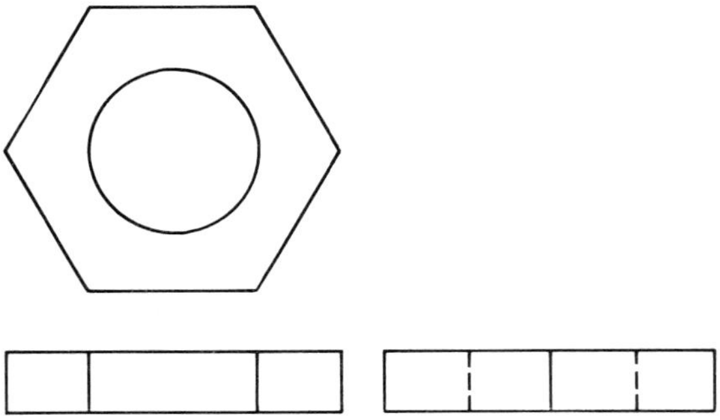

3.

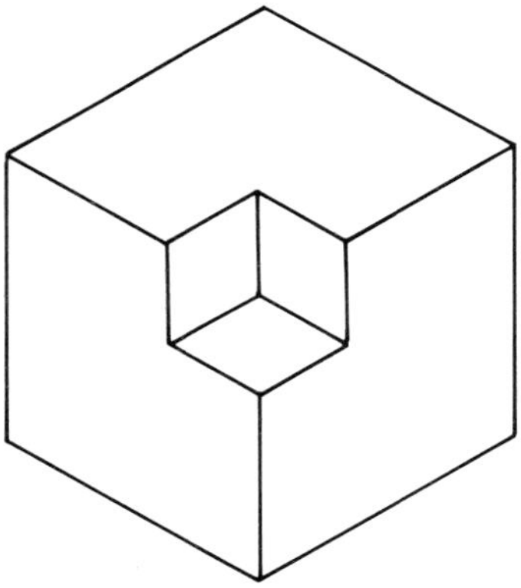